KV-042-058

WITHDRAWN
FROM
UNIVERSITIES
AT
MEDWAY
LIBRARY

DRILL HALL LIBRARY
MEDWAY

PHA
1500
ST37

Human Emerging and Re-emerging Infections

4731039

Human Emerging and Re-emerging Infections
Bacterial and Mycotic Infections

Volume II

Edited by

Sunit Kumar Singh

Laboratory of Human Molecular Virology and Immunology
Molecular Biology Unit, Faculty of Medicine, Institute of Medical Sciences
Banaras Hindu University, Varanasi, India

WILEY Blackwell

Copyright © 2016 by John Wiley & Sons, Inc. All rights reserved

Published by John Wiley & Sons, Inc., Hoboken, New Jersey
Published simultaneously in Canada

No part of this publication may be reproduced, stored in a retrieval system, or transmitted in any form or by any means, electronic, mechanical, photocopying, recording, scanning, or otherwise, except as permitted under Section 107 or 108 of the 1976 United States Copyright Act, without either the prior written permission of the Publisher, or authorization through payment of the appropriate per-copy fee to the Copyright Clearance Center, Inc., 222 Rosewood Drive, Danvers, MA 01923, (978) 750-8400, fax (978) 750-4470, or on the web at www.copyright.com. Requests to the Publisher for permission should be addressed to the Permissions Department, John Wiley & Sons, Inc., 111 River Street, Hoboken, NJ 07030, (201) 748-6011, fax (201) 748-6008, or online at http://www.wiley.com/go/permission.

Limit of Liability/Disclaimer of Warranty: While the publisher and author have used their best efforts in preparing this book, they make no representations or warranties with respect to the accuracy or completeness of the contents of this book and specifically disclaim any implied warranties of merchantability or fitness for a particular purpose. No warranty may be created or extended by sales representatives or written sales materials. The advice and strategies contained herein may not be suitable for your situation. You should consult with a professional where appropriate. Neither the publisher nor author shall be liable for any loss of profit or any other commercial damages, including but not limited to special, incidental, consequential, or other damages.

For general information on our other products and services or for technical support, please contact our Customer Care Department within the United States at (800) 762-2974, outside the United States at (317) 572-3993 or fax (317) 572-4002.

Wiley also publishes its books in a variety of electronic formats. Some content that appears in print may not be available in electronic formats. For more information about Wiley products, visit our web site at www.wiley.com.

Library of Congress Cataloging-in-Publication Data:

Human emerging and re-emerging infections / edited by Sunit K. Singh.
　　p. ; cm.
　Includes index.
　ISBN 978-1-118-64471-3 (cloth)
　I. Singh, Sunit K., editor.
　[DNLM: 1. Communicable Diseases, Emerging.　WA 110]
　RA643
　616.9–dc23

　　　　　　　　　　　　　　　　　　　2015028631

Printed in Singapore by C.O.S. Printers Pte Ltd

oBook ISBN: 9781118644843
ePDF ISBN: 9781118644829
ePub ISBN: 9781118644645

10　9　8　7　6　5　4　3　2　1

UNIVERSITIES AT MEDWAY
1 9 MAY 2016
DRILL HALL LIBRARY

Dedicated to my parents

Contents

FOR USE IN THE LIBRARY ONLY

List of Contributors

Shaw M. Akula
Department of Microbiology and Immunology, Brody School of Medicine at East Carolina University, Greenville, NC, USA

Zeal T. Akula
Department of Microbiology and Immunology, Brody School of Medicine at East Carolina University, Greenville, NC, USA

Fatih Anfasa
Department of Viroscience, Erasmus Medical Center, Rotterdam, The Netherlands
Department of Internal Medicine, Faculty of Medicine, Universitas Indonesia, Jakarta, Indonesia

Juana Angel
Instituto de Genética Humana, Facultad de Medicina, Pontificia Universidad Javeriana, Bogotá, Colombia

Olympia Apostolopoulou
Department of Critical Care Medicine, Medical School, University of Athens, Chaidari-Athens, Greece

George Arabatzis
Mycology Research Laboratory, Department of Microbiology, Medical School, National and Kapodistrian University, Athens, Greece

Alicia I. Arechavala
Mycology Unit, Francisco J. Muniz Hospital, Buenos Aires City, Argentine Republic

Ricardo Ataíde
Burnet Institute, Center for Biomedical Research, Melbourne, Victoria Australia

Lucilla Baldassarri
Department of Infectious, Parasitic and Immune-Mediated Diseases, Istituto Superiore di Sanità, Rome, Italy

Monique Barel
Université Paris Descartes, Sorbonne Paris Cité, Bâtiment Leriche, Paris, France
INSERM, U1151, Unité de Pathogénie des Infections Systémiques, Paris, France

Alfonso Barreto
Departamento de Microbiología, Facultad de Ciencias, Pontificia Universidad Javeriana, Bogotá, Colombia

Daniela Basso
Department of Medicine-DIMED, University of Padova, Padova, Italy

Sadia Benamrouz
Biologie et Diversité des Pathogènes Eucaryotes Emergents (BDEEP), Centre d'Infection et d'Immunité de Lille (CIIL), Institut Pasteur de Lille, INSERM U1019, CNRS UMR 8402, Université de Lille, France
Ecologie et Biodiversité, Faculté Libre des Sciences et Technologies de Lille, Université Catholique de Lille, France

Marlene Benchimol
Universidade do Grande Rio, UNIGRANRIO, Rio de Janeiro, Brazil
Centro Nacional de Biologia Estrutural e Bioimagem (CENABIO), Universidade Federal do Rio de Janeiro, Rio de Janeiro, Brazil
Instituto Nacional de Metrologia, Qualidade e Tecnologia – Inmetro, Duque de Caxias, Rio de Janeiro, Brazil

Mark Eric Benbow
Department of Entomology and Department of Osteopathic Medical Specialties, Michigan State University, MI, USA

Carlos Fernández Benítez
Unidad de Gestión Clínica Centro de Salud de Laviana, Asturias, Spain

Alberto Berardi
Neonatal Intensive Care Unit, Polyclinic University Hospital, Modena, Italy

Philippe Boeuf
Burnet Institute, Center for Biomedical Research, Melbourne, Victoria Australia

José Antonio Boga
Servicio de Microbiología, Hospital Universitario Central de Asturias, Oviedo, Spain

Michael S. Bronze
Department of Internal Medicine, University of Oklahoma Health Sciences Center, Oklahoma City, OK, USA

Andreas Burkovski
Department Biologie, Friedrich-Alexander-Universität Erlangen-Nürnberg, Erlangen, Germany

Gabriela Certad
Biologie et Diversité des Pathogènes Eucaryotes
Emergents (BDEEP), Centre d'Infection et d'Immunité
de Lille (CIIL), Institut Pasteur de Lille, INSERM U1019,
CNRS UMR 8402, Université de Lille, France

Alain Charbit
Université Paris Descartes, Sorbonne Paris Cité,
Bâtiment Leriche, Paris, France
INSERM, U1151, Unité de Pathogénie des Infections
Systémiques, Paris, France

Kaw Bing Chua
Temasek Life Sciences Laboratory, National University
of Singapore, Singapore

Jennifer Cnops
Laboratory for Cellular and Molecular Immunology,
Vrije Universiteit Brussel, Brussels, Belgium
Department of Structural Biology, VIB, Brussels,
Belgium

Alexandra Correia
Instituto de Biologia Molecular e Celular, Universidade
do Porto, Porto, Portugal
Instituto de Ciências Biomédicas de Abel Salazar,
Universidade do Porto, Porto, Portugal

Christina M. Coyle
Departments of Medicine, Jacobi Medical Center and
the Montefiore Medical Center, Albert Einstein College
of Medicine, Bronx, NY, USA

Roberta Creti
Department of Infectious, Parasitic and
Immune-Mediated Diseases, Istituto Superiore di
Sanità, Rome, Italy

Filipe Dantas-Torres
Aggeu Magalhães Research Centre, Oswaldo Cruz
Foundation, Recife, Pernambuco, Brazil

Frank R. DeLeo
Laboratory of Human Bacterial Pathogenesis, Rocky
Mountain Laboratories, National Institute of Allergy
and Infectious Diseases, National Institutes of Health,
Hamilton, MT, USA

Wanderley de Souza
Centro Nacional de Biologia Estrutural e Bioimagem
(CENABIO), Universidade Federal do Rio de Janeiro,
Rio de Janeiro, Brazil
Instituto Nacional de Metrologia, Qualidade e
Tecnologia – Inmetro, Duque de Caxias, Rio de Janeiro,
Brazil
Laboratório de Ultraestrutura Celular Hertha Meyer,
Universidade Federal do Rio de Janeiro, Brazil

Elizabeth S. Didier
Division of Microbiology, Tulane National Primate
Research Center, Covington, LA, USA
Department of Tropical Medicine, School of Public
Health and Tropical Medicine, Tulane University, New
Orleans, LA, USA

George Dimopoulos
Department of Critical Care Medicine, Medical School,
University of Athens, Chaidari-Athens, Greece

Esteban Domingo
Centro de Biología Molecular "Severo Ochoa"
(CSIC-UAM), Consejo Superior de Investigaciones
Científicas (CSIC), Campus de Cantoblanco, Madrid,
Spain

John Doorbar
Department of Pathology, University of Cambridge,
Cambridge, United Kingdom

Laurent Dortet
INSERM U914 "Emerging Resistance to Antibiotics", Le
Kremlin-Bicêtre, Paris, France

Douglas A. Drevets
Department of Internal Medicine, University of
Oklahoma Health Sciences Center, Oklahoma City, OK,
USA
Department of Veterans Affairs Medical Center,
Oklahoma City, OK, USA

Taylor Eddens
Richard King Mellon Foundation Institute for Pediatric
Research, Children's Hospital of Pittsburgh of UPMC,
Pittsburgh, PA, USA

Peter Q. Eichacker
Critical Care Medicine Department, Clinical Center,
National Institutes of Health, Bethesda, MD, USA

Brenda L. Fredericksen
Maryland Pathogen Research Institute, University of
Maryland, College Park, MD, USA
Department of Cell Biology and Molecular Genetics,
University of Maryland College Park, College Park,
MD, USA

Chi-Ling Fu
Department of Urology, Stanford University School of
Medicine, Stanford, CA, USA

Joaquim Gascón
Global Health Institute (ISGlobal), Hospital
Clínic-Universitat de Barcelona, Barcelona, Spain

Thomas W. Geisbert
Department of Microbiology and Immunology, The
University of Texas Medical Branch, Galveston, TX, USA

Giovanni Gherardi
Centro Integrato di Ricerche, Laboratory of
Microbiology, University Campus Biomedico, Rome,
Italy

Namraj Goire
Microbiology Department, PathWest Laboratory
Medicine WA, Queen Elizabeth II Medical Centre,
Nedlands, Western Australia, Australia
Queensland Paediatric Infectious Diseases Laboratory,
Queensland Children's Medical Research Institute,
Royal Children's Hospital, Brisbane, Queensland,
Australia

Thaddeus G. Golos
Department of Comparative Biosciences, Wisconsin National Primate Research Center, Madison, WI, USA

Christopher R. Gourley
School of Molecular Biosciences, College of Veterinary Medicine, Washington State University, Pullman, WA, USA

Nathalie Grall
Laboratoire de Bactériologie, Hôpital Bichat-Claude Bernard, APHP, Paris, France
EA 3964 - Université Paris-Diderot, Paris, France

Luis Otero Guerra
Servicio de Microbiología, Hospital de Cabueñes, Gijón, Asturias, Spain

Belinda Hall
Department of Microbial and Cellular Sciences and School of Biosciences and Medicine, University of Surrey, Guildford, UK

John J. Halperin
Department of Neurosciences, Overlook Medical Center, Summit, NJ, USA
Sidney Kimmel Medical College of Thomas Jefferson University, Philadelphia, PA, USA

Olfat Hammam
Department of Urology, Stanford University School of Medicine, Stanford, CA, USA

Alistair Harrison
The Center for Microbial Pathogenesis, The Research Institute at Nationwide Children's Hospital, Columbus, OH, USA

Elizabeth L. Hartland
Department of Microbiology and Immunology, University of Melbourne, Victoria, Australia

Caitlin Hicks
Johns Hopkins Hospital, Department of Surgery, Baltimore, MD, USA

Michael Hsieh
Division of Urology, Children's National Health System, Washington, DC, USA
Departments of Urology and Pediatrics, The George Washington University, Washington, DC, USA
The Biomedical Research Institute, Rockville, MD, USA

Katherine L. Hussmann
Department of Cell Biology and Molecular Genetics, University of Maryland College Park, College Park, MD, USA

Tetsuro Ikegami
Department of Pathology, The University of Texas Medical Branch at Galveston, Galveston, TX, USA

J. Igor Iruretagoyena
Department of Obstetrics and Gynecology, University of Wisconsin-Madison, Madison, WI, USA

Cong Jin
National Institute for Viral Disease Control and Prevention, Chinese Center for Disease Control and Prevention, Beijing, China

Heather Williamson Jordan
Department of Biological Sciences, Mississippi State University, MS, USA

Aniket Kaloti
Richard King Mellon Foundation Institute for Pediatric Research, Children's Hospital of Pittsburgh of UPMC, Pittsburgh, PA, USA

Sophia Kathariou
Department of Food, Bioprocessing and Nutrition Sciences, North Carolina State University, Raleigh, NC, USA

Kevin R. Kazacos
Department of Comparative Pathobiology, Purdue University College of Veterinary Medicine, West Lafayette, IN, USA

Jay K. Kolls
Richard King Mellon Foundation Institute for Pediatric Research, Children's Hospital of Pittsburgh of UPMC, Pittsburgh, PA, USA

Michael E. Konkel
School of Molecular Biosciences, College of Veterinary Medicine, Washington State University, Pullman, WA, USA

Stefan Kunz
Institute of Microbiology, University Hospital Center and University of Lausanne, Lausanne, Switzerland

Monica M. Lahra
Neisseria Reference Laboratory and WHO Collaborating Centre for STD, Microbiology Department, South Eastern Area Laboratory Services, Prince of Wales Hospital, Sydney, New South Wales, Australia

Thien-Linh Le
Department of Urology, Stanford University School of Medicine, Stanford, CA, USA

Dexin Li
National Institute for Viral Disease Control and Prevention, Chinese Center for Disease Control and Prevention, Beijing, China

Caroline Lin Lin Chua
Taylor's University Lakeside Campus, School of Biosciences, Subang Jaya, Selangor, Malaysia

David S. Lindsay
Center for Molecular Medicine and Infectious Diseases,
Department of Biological Sciences and Pathobiology,
Virginia-Maryland Regional College of Veterinary
Medicine, Virginia Tech, Blacksburg, VA, USA

Camille Locht
Univ. Lille, U1019 - UMR 8204 - CIIL - Centre
d'Infection et d'Immunité de Lille, F-59000 Lille, France
CNRS, UMR 8204, F-59000 Lille, France
Inserm, U1019, F-59000 Lille, France
CHU Lille, F-59000 Lille, France
Institut Pasteur de Lille, F-59000 Lille, France

Thea Lu
Laboratory of Human Bacterial Pathogenesis, Rocky
Mountain Laboratories, National Institute of Allergy
and Infectious Diseases, National Institutes of Health,
Hamilton, MT, USA

Igor S. Lukashevich
Department of Pharmacology and Toxicology, School of
Medicine, Center for Predictive Medicine, NIH Regional
Bio-containment Laboratory, University of Louisville,
Louisville, USA

Fabiana S. Machado
Department of Biochemistry and Immunology,
Institute of Biological Sciences, Faculty of Medicine,
Federal University of Minas Gerais, Belo Horizonte,
Brazil

Stefan Magez
Laboratory for Cellular and Molecular Immunology,
Vrije Universiteit Brussel, Brussels, Belgium
Department of Structural Biology, VIB, Brussels,
Belgium

Hélène Marquis
Department of Microbiology and Immunology, Cornell
University, Ithaca, NY, USA

Anandi Martin
Laboratory of Microbiology, Department of
Biochemistry and Microbiology, Ghent University,
Ghent, Belgium

Verónica Martín
Centro de Investigación en Sanidad Animal
(CISA-INIA), Instituto Nacional de Investigación
Agraria y Alimentaria, Valdeolmos, Madrid, Spain

Byron E. E. Martina
Department of Viroscience, Erasmus Medical Center,
Rotterdam, The Netherlands
Artemis One Health Research Foundation, Utrecht,
The Netherlands

Yuri C. Martins
Departments of Pathology, Jacobi Medical Center and
the Montefiore Medical Center, Albert Einstein College
of Medicine, Bronx, NY, USA

Kevin M. Mason
The Center for Microbial Pathogenesis, The Research
Institute at Nationwide Children's Hospital,
Department of Pediatrics, Columbus, OH, USA
The Center for Microbial Interface Biology, The Ohio
State University, Columbus, OH, USA

Dimitrios K. Matthaiou
Department of Critical Care Medicine, Medical School,
University of Athens, Chaidari-Athens, Greece

Jean-Louis Mege
Unité de Recherche sur les Maladies Infectieuses
Tropicales et Emergentes, Aix-Marseille Université,
Centre National de la Recherche Scientifique Unité
Mixte de Recherche 7278, Institut National de la Santé et
de la Recherche Scientifique Unité 1095, Marseille,
France

Martha C. Mesa
Departamento de Microbiología, Facultad de Ciencias,
Pontificia Universidad Javeriana, Bogotá, Colombia

Nathalie Mielcarek
Univ. Lille, U1019 - UMR 8204 - CIIL - Centre
d'Infection et d'Immunité de Lille, F-59000 Lille, France
CNRS, UMR 8204, F-59000 Lille, France
Inserm, U1019, F-59000 Lille, France
CHU Lille, F-59000 Lille, France
Institut Pasteur de Lille, F-59000 Lille, France

Chad E. Mire
Department of Microbiology and Immunology, The
University of Texas Medical Branch, Galveston, TX,
USA

Lydia Mosi
University of Ghana, Legon, Ghana
West African Center for Cell Biology of Infectious
Pathogens, University of Ghana, Legon

Nikolaos Moussas
Department of Critical Care Medicine, Medical School,
University of Athens, Chaidari-Athens, Greece

Leonard Nainggolan
Department of Internal Medicine, Faculty of Medicine,
Universitas Indonesia, Jakarta, Indonesia

Carlos Fernando Narváez
Facultad de Salud, Programa de Medicina, Universidad
Surcolombiana, Neiva, Colombia

Gayathri Natarajan
Department of Microbiology, Ohio State University,
Columbus, OH, USA

Ricardo Negroni
Mycology Unit, Francisco J. Muniz Hospital, Buenos
Aires City, Argentine Republic

Patrice Nordmann
INSERM U914 "Emerging Resistance to Antibiotics", Le Kremlin-Bicêtre, Paris, France
Medical and Molecular Microbiology Unit, Department of Medicine, Faculty of Science, University of Fribourg, Fribourg, Switzerland

Steve Oghumu
Department of Environmental Health Sciences, College of Public Health, Ohio State University, Columbus, OH, USA
Department of Pathology, Ohio State University Medical Center, Columbus, OH, USA

Marwan Osman
Biologie et Diversité des Pathogènes Eucaryotes Emergents (BDEEP), Centre d'Infection et d'Immunité de Lille (CIIL), Institut Pasteur de Lille, INSERM U1019, CNRS UMR 8402, Université de Lille, France
Centre AZM pour la Recherche en Biotechnologie et ses Applications, Laboratoire Microbiologie, Santé et Environnement, Université Libanaise, Tripoli, Lebanon

Domenico Otranto
Department of Veterinary Medicine, University of Bari, Valenzano (Bari), Italy

Célia Pais
Department of Biology, Centre of Molecular and Environmental Biology (CBMA), University of Minho, Braga, Portugal

Juan Carlos Palomino
Laboratory of Microbiology, Department of Biochemistry and Microbiology, Ghent University, Ghent, Belgium

Akue Jean Paul
International Center for Medical Research of Franceville (CIRMF), Franceville, Gabon

Sabine Pellett
Department of Bacteriology, University of Wisconsin-Madison, Madison, WI, USA

Michela Pelloso
Department of Medicine-DIMED, University of Padova, Padova, Italy

Hong-Juan Peng
Department of Pathogen Biology, School of Public Health and Tropical Medicine, Southern Medical University, Guangdong Province, P.R. China

Antonio Pereira-Neves
Fiocruz Pernambuco, Centro de Pesquisas Aggeu Magalhães, Departamento de Microbiologia, Laboratório de Biologia Celular de Patógenos, Recife, Brazil
Centro Nacional de Biologia Estrutural e Bioimagem (CENABIO), Universidade Federal do Rio de Janeiro, Rio de Janeiro, Brazil

Inaia Phoenix
Department of Pathology, The University of Texas Medical Branch at Galveston, Galveston, TX, USA

María-Jesús Pinazo
Global Health Institute (ISGlobal), Hospital Clínic-Universitat de Barcelona, Barcelona, Spain

Mario Plebani
Department of Medicine-DIMED, University of Padova, Padova, Italy

Alexander Plyusnin
Department of Virology, Haartman Institute, University of Helsinki, Helsinki, Finland

Laurent Poirel
INSERM U914 "Emerging Resistance to Antibiotics", Le Kremlin-Bicêtre, Paris, France
Medical and Molecular Microbiology Unit, Department of Medicine, Faculty of Science, University of Fribourg, Fribourg, Switzerland

Garyphalia Poulakou
Department of Internal Medicine, Medical School, University of Athens, Chaidari-Athens, Greece

Kenneth E. Remy
Critical Care Medicine Department, Clinical Center, National Institutes of Health, Bethesda, MD, USA

Sophie Roberts
Royal Liverpool Hospital, Liverpool, United Kingdom

Luz-Stella Rodríguez
Instituto de Genética Humana, Facultad de Medicina, Pontificia Universidad Javeriana, Bogotá, Colombia

Tais Berelli Saito
Department of Pathology, University of Texas Medical Branch, Galveston, TX, USA

Paula Sampaio
Department of Biology, Centre of Molecular and Environmental Biology (CBMA), University of Minho, Braga, Portugal

Abhay R. Satoskar
Department of Pathology, Ohio State University Medical Center, Columbus, OH, USA
Department of Microbiology, Ohio State University, Columbus, OH, USA

Benjamin A. Satterfield
Department of Microbiology and Immunology, The University of Texas Medical Branch, Galveston, TX, USA

Noemí Sevilla
Centro de Investigación en Sanidad Animal (CISA-INIA), Instituto Nacional de Investigación Agraria y Alimentaria, Valdeolmos, Madrid, Spain

Henry Shikani
Division of Parasitology and Tropical Medicine, Department of Pathology, Albert Einstein College of Medicine, Bronx, NY, USA

Rachel Simmonds
Department of Microbial and Cellular Sciences and School of Biosciences and Medicine, University of Surrey, Guildford, UK

Kumara Singaravelu
Departments of Medicine, Jacobi Medical Center and the Montefiore Medical Center, Albert Einstein College of Medicine, Bronx, NY, USA

Sunit Kumar Singh
Laboratory of Human Molecular Virology and Immunology, Molecular Biology Unit, Faculty of Medicine, Institute of Medical Sciences, Banaras Hindu University, Varanasi, India

Tarja Sironen
Department of Virology, Haartman Institute, University of Helsinki, Helsinki, Finland

Susanne H. Sokolow
Department of Biology, Stanford University, Stanford, CA, USA

David J. Speers
Microbiology Department, PathWest Laboratory Medicine WA, Queen Elizabeth II Medical Centre, Nedlands, Western Australia, Australia

Jonathan Fernández Suárez
Servicio de Microbiología, Hospital Universitario Central de Asturias, Oviedo, Spain

Anshul V. Subramanya
Department of Microbiology and Immunology, Brody School of Medicine at East Carolina University, Greenville, NC, USA

Chong Tin Tan
Department of Medicine, University Malaya Medical Centre, Kuala Lumpur, Malaysia

Feng Tan
Department of Parasitology, Wenzhou Medical University, Zhejiang province, P.R. China

Herbert B. Tanowitz
Department of Medicine and Pathology, Albert Einstein College of Medicine, Bronx, NY, USA

Wiwit Tantibhedhyabgkul
Unité de Recherche sur les Maladies Infectieuses Tropicales et Emergentes, Aix-Marseille Université, Centre National de la Recherche Scientifique Unité Mixte de Recherche 7278, Institut National de la Santé et de la Recherche Scientifique Unité 1095, Marseille, France
Department of Immunology, Faculty of Medicine Siriraj Hospital, Mahidol University, Bangkok, Thailand

Alberto Tessari
Department of Medicine-DIMED, University of Padova, Padova, Italy

Faustino Torrico
School of Medicine, San Simón University (UMSS), Cochabamba, Bolivia

Alexandra J. Umbers
Department of Medicine, University of Melbourne, Parkville, Victoria, Australia

Fernando Vázquez
Servicio de Microbiología, Hospital Universitario Central de Asturias, Oviedo, Spain
Departamento Biología Funcional, Área de Microbiología, Facultad de Medicina, Oviedo, Spain

Rianna Vandergaast
Department of Cell Biology and Molecular Genetics, University of Maryland College Park, College Park, MD, USA

Olli Vapalahti
Department of Virology, Haartman Institute, University of Helsinki, Helsinki, Finland

Aristea Velegraki
Mycology Research Laboratory, Department of Microbiology, Medical School, National and Kapodistrian University, Athens, Greece

Manuel Vilanova
Instituto de Biologia Molecular e Celular, Universidade do Porto, Porto, Portugal
Instituto de Ciências Biomédicas de Abel Salazar, Universidade do Porto, Porto, Portugal

Adam J. Vogrin
Department of Microbiology and Immunology, University of Melbourne, Victoria, Australia

David H. Walker
Department of Pathology, University of Texas Medical Branch, Galveston, TX, USA

Lia R. Walker
Department of Microbiology and Immunology, Brody School of Medicine at East Carolina University, Greenville, NC, USA

Louis M. Weiss
Division of Infectious Diseases, Department of Medicine, Albert Einstein College of Medicine, Bronx, NY, USA

David M. Whiley
Queensland Paediatric Infectious Diseases Laboratory, Queensland Children's Medical Research Institute, Royal Children's Hospital, Brisbane, Queensland, Australia

E.D. Williamson
Defence Science and Technology Laboratory, Porton Down, Salisbury, United Kingdom

Kum Thong Wong
Department of Pathology, University Malaya Medical Centre, Kuala Lumpur, Malaysia

Pablo Yagupsky
Clinical Microbiology Laboratory, Soroka University Medical Center, Ben-Gurion University of the Negev, Beer-Sheva, Israel

Carlo-Federico Zambon
Department of Medicine-DIMED, University of Padova, Padova, Italy

Preface

Infectious diseases are a global problem and represent a continuous and increasing threat to human health and welfare. Due to deforestation, migration of populations, travel and trade, and changes in agricultural practices, the infectious diseases outbreaks are taking place frequently. Although the rate of morbidity and mortality due to infectious diseases have decreased over the past decade, the worldwide impact of infectious diseases remains substantial.

Advances in the development of drugs and use of vaccines to prevent various infections have eased the burden of infectious diseases. The evolution of pathogens with resistance to antibacterial and antiviral agents continues to challenge us for the better understanding of the mechanisms of drug resistance and to devise new ways to circumvent the problem.

We have witnessed the re-emergence of malaria and tuberculosis, and the emergence of new viral infections leading to the viral hemorrhagic fevers and respiratory complications. The root causes of the emergence and re-emergence of infectious diseases include environmental changes due to urbanization and deforestation, rapid population growth, and migration of populations. Estimation of the infectious disease outbreaks must include factors such as population susceptibility, infective dose, incubation period, modes of transmission, routes of transmission, mortality rate, effectiveness of treatment interventions and population movement. Communicable infectious diseases have added another layer of complexity in the form of transmission parameters such as periodicity of infections and secondary attack rates. There is need of new disease-outbreak models to understand the periodicity and patterns of outbreaks of infectious diseases. The capacity of all nations to recognize, prevent, and respond to the threat of emerging and re-emerging infectious diseases is the critical foundation for an effective global response. Biomedical scientists and clinicians should join their hands for collaborative research in order to understand the molecular mechanisms of pathogenesis and for developing new diagnostic and therapeutic tools.

There is a need to enhance global investment in both developed and developing countries to improve their research capacities, diagnosis, and response to the emerging and re-emerging infections. Better coordination aligned to health system needs and translational research capabilities are required to meet the challenges posed by emerging and re-emerging infections. Strong communication and surveillance strategies are required among national and international agencies for enhanced implementation of different disease control programs.

Significant progress has been made over the last several years in dissecting out the molecular biology and pathogenesis of the many emerging and re-emerging pathogens. This book includes most common emerging and re-emerging infections caused by bacteria, virus, fungi, and parasites. The book has been published in two volumes. The volume one includes viral and parasitic infections, whereas volume two includes bacterial and fungal infections.

This book is primarily aimed to virologists, clinicians, biomedical researchers, health-care workers, microbiologists, immunologists, and students of medicine or biomedical sciences wishing to gain rapid overview. This book will serve as a useful resource in the field of infectious diseases.

A comprehensive book such as this is clearly beyond the capacity of an individual's efforts. The large panel of internationally renowned infectious disease experts have contributed their chapters, whose detailed knowledge in diverse areas of infectious diseases have greatly enriched this book.

Sunit Kumar Singh

Acknowledgments

I acknowledge the support provided by Virologists, Microbiologists, Immunologists and Infectious Diseases experts, whose willingness to share their work and expertise has made this extensive overview on *Human Emerging and Re-emerging Infections* possible. My appreciations extend to my family and parents for their understanding and support during the preparation of this book. Above all I thank my wife Seema, daughter Eshita and son Shaurya, who supported and encouraged me despite all the time it took me away from them. It was a long and difficult journey for them. I am also thankful to Ms. Stephanie Dollan and Ms. Mindy Okura-Marszycki of Wiley-Blackwell for their help and professional support.

About the Editor

Dr. Sunit Kumar Singh completed his bachelor's degree at GB Pant University of Agriculture and Technology, Pantnagar, India, and his master's degree at the CIFE, Mumbai, India. After receiving his master's degree, Dr. Singh joined the Department of Pediatric Rheumatology, Immunology, and Infectious Diseases, Children's Hospital, University of Würzburg, Würzburg, Germany, as a biologist. He completed his PhD degree at the University of Würzburg in the area of molecular infection biology.

Dr. Singh has completed his postdoctoral trainings at the Department of Internal Medicine, Yale University, School of Medicine, New Haven, CT, USA, and the Department of Neurology, University of California Davis Medical Center, Sacramento, CA, USA, in the areas of vector-borne infectious diseases and neuroinflammation, respectively. He has also worked as visiting scientist at the Department of Pathology, Albert Einstein College of Medicine, NY, USA; Department of Microbiology, College of Veterinary Medicine, Chonbuk National University, Republic of Korea; Department of Arbovirology, Institute of Parasitology, Ceske Budejovice, Czech Republic; and Department of Genetics and Laboratory Medicine, University of Geneva, Geneva,

Switzerland. He has extensive experience in the area of virology and immunology. Dr. Singh served as a scientist and led a research group in the area of molecular neurovirology and inflammation biology at the prestigious CSIR–Centre for Cellular and Molecular Biology (CCMB), Hyderabad, India. Presently, he is working as Associate Professor (Molecular Immunology) and leading a research group in the area of human molecular virology and immunology, in the Molecular Biology Unit, Faculty of Medicine, Institute of Medical Sciences (IMS), Banaras Hindu University (BHU), Varanasi, India. His main areas of research interest are host–pathogen interaction in hemorrhagic fever viral infections, molecular neurovirology and immunology. There are several awards to his credit, including the Skinner Memorial Award, Travel Grant Award, NIH-Fogarty Fellowship, and Young Scientist Award. Dr. Singh has published many research papers in the areas of neurovirology and inflammation biology in various peer-reviewed journals. He has edited several books such as *Neuroviral Infections*, *Viral Hemorrhagic Fevers*, *Human Respiratory Viral Infections* and *Viral Infections and Global Change*, etc. Dr. Singh is associated with several international journals of repute as associate editor and editorial board member.

Chapter 27

Pathogenesis of *Haemophilus influenzae* in Humans

Alistair Harrison[1] and Kevin M. Mason[2,3]

[1]The Center for Microbial Pathogenesis, The Research Institute at Nationwide Children's Hospital, Columbus, OH, USA

[2]The Center for Microbial Pathogenesis, The Research Institute at Nationwide Children's Hospital, Department of Pediatrics, Columbus, OH, USA

[3]The Center for Microbial Interface Biology, The Ohio State University, Columbus, OH, USA

27.1 Introduction

27.1.1 Pfeiffer's bacillus and the influenza pandemic of 1918–1919

Haemophilus influenzae, a small gram-negative cocobacillus in the Pasteurellaceae family, was first identified by Richard Pfeiffer. In 1892, Pfeiffer, then working under Robert Koch, identified a bacillus presumed to be the causative agent of the influenza pandemic of 1889–1890. Initially named *Bacillus influenzae*, or more colloquially Pfeiffer's bacillus, the role of this bacterium in the etiology of influenza was not disputed until the great influenza pandemic of 1918–1919 (Fildes, 1956). Despite the scale of the 1918–1919 pandemic and the exhaustive efforts of the medical community to identify the agent of the disease, Pfeiffer's bacillus was rarely isolated from patients (Tognotti, 2003). Subsequently, the etiological agent of the influenza pandemic was identified as a virus that could be separated from bacteria by filtration (Olitsky and Gates, 1921).

Bacillus influenzae was renamed *Haemophilus influenzae* in 1920 (Winslow et al., 1920) and the critical breakthrough in correlating the presence of *H. influenzae* with disease followed in 1933. Smith and colleagues infected ferrets with bacteriologically sterile filtrates of throat washings from influenza patients and found the animals became sick. Moreover, the disease could then be transmitted between animals. These data again suggested that the etiological agent of influenza was viral (Smith et al., 1933). Smith and colleagues subsequently showed that the addition of *H. influenzae* to viral filtrates, or *H. influenzae* (suis) to swine influenza virus, did not result in appreciable differences in the symptoms ferrets demonstrated (Smith et al., 1933). These observations led to a startling but correct supposition. Lewis and Shope showed that *H. influenzae* (suis) could be isolated from pigs suffering from swine influenza. *H. influenzae* (suis) alone, however, was incapable of generating influenza when pigs were experimentally infected, whereas bacteriologically sterile filtrates purified from extracts of lung and bronchial lymph nodes of pigs with influenza produced only a mild experimental infection in pigs. However, when *H. influenzae* (suis) was added to the filtrate and the mixture used to infect pigs, the pigs demonstrated classic symptoms of swine influenza (Lewis and Shope, 1931; Shope, 1931). Due to the perceived similarities between *H. influenzae* and *H. influenzae* (suis) as well as the pathologies of human and swine influenza, both Shope and Smith suggested that human influenza was caused by a combination of a virus and *H. influenzae* (Shope, 1931; Smith et al., 1933). The polymicrobial nature of virus and *H. influenzae* in disease had been established.

27.1.2 *Haemophilus influenzae* type b (Hib) and meningitis

The role of *H. influenzae* in disease was suggested prior to the work of Smith and Shope. An initial report from 1892 suggested that *H. influenzae* contributed to inflammation of the meninges. A study at the Rockefeller Institute in New York early in the 20th century identified *H. influenzae* in the cerebrospinal fluids of patients suffering from meningitis (Wollstein, 1911). The role of *H. influenzae* in meningitis was not truly codified until Margaret Pittman determined that clinical isolates of *H. influenzae* could be divided

---

Human Emerging and Re-emerging Infections: Bacterial & Mycotic Infections, Volume II, First Edition. Edited by Sunit K. Singh.
© 2016 John Wiley & Sons, Inc. Published 2016 by John Wiley & Sons, Inc.

into strains with or without capsule. Moreover, the encapsulated strains could be further divided into two immunologically distinct types, type a and type b, with all of the type b strains initially isolated from the spinal fluid of patients with meningitis (Pittman, 1931). Four additional serotypes were subsequently identified; types c, d, e, and f, but the role for each serotype in disease was limited. Early work showed that the type b strains were more virulent than other strains tested and were almost exclusively isolated form meningitis patients (Chandler et al., 1937; Pittman, 1933). Moreover, it was suggested that increased virulence of Hib was due to the possession of the type b capsule that increases the ability of a bacterium to avoid the bactericidal effects of complement (Sutton et al., 1982). A series of capsular transformants, in which a capsule-deficient strain of *H. influenzae* was engineered to produce each of the capsule serotypes, showed that virulence was capsule-dependent, with type b capsule–carrying transformants being the most virulent (Zwahlen et al., 1989).

Until the latter part of the 20th century, the epidemiology of *Haemophilus* disease focused on *H. influenzae* type b (Hib). This approach was understandable due to the high number of cases of life-threatening invasive disease observed in young children that were caused by Hib. In the United States and Canada, Hib was primarily the cause of meningitis (up to 60% of Hib disease), with epiglottitis being of secondary importance (up to 30% of Hib disease). Less frequent illnesses were arthritis, facial cellulitis, and bacteremia (Wenger, 1998). In two- and three-year population-based studies in the late 1970s and the early 1980s, between 67 and 131 cases/100,000 cases of invasive Hib disease were identified in US children under the age of 5 years, with disease associated with Hib peaking in children between 6 and 8 months of age. Moreover, approximately two-thirds of these cases were meningitis (Broome, 1987; Wenger et al., 1990). Morbidity due to *H. influenzae*–induced meningitis included significant neurological sequelae (Sell et al., 1972). Critically, in a large multistate survey carried out by Wenger and colleagues, 3% of patients with Hib-derived meningitis died (Wenger et al., 1990). The problems associated with Hib disease were further exacerbated with the increased number of disease isolates that exhibited antibiotic resistance. For example, two large studies of bacterial meningitis showed that approximately 30% of Hib strains isolated were ampicillin-resistant (Doern et al., 1988; Wenger et al., 1990). As an alternative to antibiotic treatment, there were proposals to modify societal or behavioral risk factors to attempt to minimize Hib diseases (reviewed in Wenger, 1998). However, these approaches were deemed unworkable. Due to the scourge of Hib disease, the rise in the number of antibiotic-resistant Hib strains isolated, and the limited number of behavior modifiers available to curb Hib disease, a great amount of effort was put into the development of an effective Hib vaccine.

27.2 The advent of the Hib vaccine

In the early to mid-20th century, it was recognized that sera from animals immunized with Hib was protective against subsequent Hib infection (Alexander et al., 1944; Pittman, 1933). This protection was also abrogated if the anti-carbohydrate antibodies were depleted by absorption with carbohydrate prepared from Hib (Alexander et al., 1944). Additional observations indicated that infants are protected from Hib by maternally acquired IgG that decline soon after birth, thus leaving infants susceptible to Hib disease (Fothergill and Wright, 1933; Schneerson et al., 1971). The capsule of Hib contains a polymer of polyribosylribitol phosphate (PRP) (Crisel et al., 1975), critical for the Hib-specific immune response. Anderson et al. (1972) showed sera isolated from subjects immunized with purified PRP to be bactericidal toward Hib strains. It was therefore thought that early immunization with PRP would be an efficacious way to protect infants from Hib disease. Inoculation of human subjects with purified PRP did indeed produce high bactericidal titers that were Hib-specific. However, an effective titer of anti-PRP antibody was age-dependent, low in children under 1 year of age (Anderson et al., 1972, 1977). Further, anti-PRP antibodies generated against Hib were elicited in animals. Early experiments demonstrated that PRP immunization of rabbits produced high antibody titers, although PRP preparations were contaminated with small amounts of protein and LPS. Both PRP and the contaminating proteins were necessary for a strong immune response (Anderson and Insel, 1981; Anderson and Smith, 1977). Immunization of rabbits with purified PRP resulted in clearance of Hib, via complement and antibody-enhanced phagocytosis and subsequent clearance by the spleen and liver (Weller et al., 1978). A PRP-based vaccine was thus generated and used in extensive clinical trials, the data of which were worrisome. The protection elicited by the PRP vaccine was age-dependent with little protection generated in children under the age of 18 months. The PRP vaccine could not generate a strong antibody response at the age when children were most susceptible to Hib infections (Peltola et al., 1977; Robbins et al., 1973; Smith et al., 1973). This lack of protection afforded by PRP vaccines at the age when Hib disease was most prevalent, coupled with its inability to be subsequently boosted and to reduce nasopharyngeal carriage of Hib, resulted in alternative approaches (Makela et al., 1977; Takala et al., 1991).

Four PRP conjugate vaccines were incrementally introduced. The first anti-Hib conjugate vaccine contained a diphtheria toxoid conjugate, subsequently followed by vaccines containing mutant diphtheria toxin conjugate, meningococcal outer membrane protein conjugate, and tetanus toxoid conjugate. An early

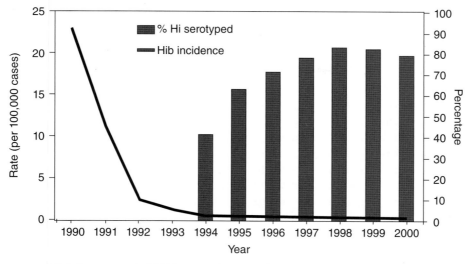

Fig. 27-1 Incidence rate of *H. influenzae* type b (Hib) invasive disease and percentage of *H. influenzae* (Hi) isolates serotyped among children aged <5 years in the United States from 1990 to 2000. Data obtained from CDC (2002).

retrospective survey carried out in Helsinki, Finland, investigated the effect of vaccination with PRP conjugated with either diphtheria or tetanus toxoid given between 3 and 6 months of age and then with a boosting dose between 14 and 18 months. At the conclusion of the study, no cases of meningitis due to Hib could be identified (Peltola et al., 1992). Large studies in the United Kingdom, the United States, France, Chile, and Bangladesh, among others, noted a significant decrease in Hib meningitis after the introduction of conjugated Hib vaccines (Black et al., 1991; Booy et al., 1994; Lagos et al., 1996; Murphy et al., 1993a; Reinert et al., 1993; Sultana et al., 2013). As well as impacting Hib disease, vaccination with conjugate PRP vaccines reduced nasopharyngeal carriage of Hib (Adegbola et al., 1998; Mohle-Boetani et al., 1993; Murphy et al., 1993b; Takala et al., 1993). Reduced carriage in a partially vaccinated population helps prevent the spread of Hib disease due to a reduction in the rate of Hib transmission. Overall estimates of the effectiveness of the Hib vaccine in 2000 predicted that, in the developed world, 38,000 cases of Hib disease, 21,000 of which were meningitis, were prevented by vaccination (Peltola, 2000). Despite these great advances in reducing both the mortality and morbidity of Hib disease, a large number of children still succumb to Hib disease. Estimates of Hib disease in the post-Hib vaccine era vary. In 2000, it was estimated that there were approximately 8 million cases of Hib disease worldwide, resulting in the deaths of 371,000 children under 5 years of age. These incidences of Hib disease occurred predominantly in Africa and Asia, where the use of conjugate Hib vaccines was not widespread (Watt et al., 2009). Estimates in these developing countries whereby use of the Hib vaccine is limited found a mortality rate of approximately 30%, 100,000 of which were from Hib meningitis. Moreover, in those that survived Hib meningitis, 30% of patients exhibited significant sequelae, which again represents 100,000 people (Peltola, 2000).

27.2.1 *Haemophilus influenzae* disease in the post-Hib vaccine era

After introduction of the conjugate Hib vaccine, incidences of invasive Hib disease decreased significantly. Staggeringly, in the United States, the number of invasive cases of Hib disease decreased by 99% following the introduction of the vaccine, an achievement which had a major impact on both the mortality and morbidity due to Hib, as well as on the financial costs associated with Hib disease (CDC, 2002; Zhou et al., 2002). This decrease in Hib prevalence has contributed to shifts in the age of susceptible populations, as well as the identification of *H. influenzae* strains responsible for non-Hib disease (see Figure 27-1). A study of diseases caused by *H. influenzae* in the United States between 1989 and 2008 found an increase in isolation of serotypes a, e, and f, with the greatest increase in serotype f in adults 18 years of age and older. Despite the increase in isolation of non-Hib strains, it was apparent that there was no significant increase in non-Hib disease (MacNeil et al., 2011). Additional studies found the rates of invasive *H. influenzae* disease caused by serotypes e and f did increase in both children and adults, although the number of cases were relatively low (Resman et al., 2011; Urwin et al., 1996; Waggoner-Fountain et al., 1995).

The role of serotype a (Hia) in disease in the post-Hib era was demonstrated by several large-scale studies. Studies in the United States, Canada, the North American Arctic, and others found an increase in the proportion of Hia strains isolated from invasive disease patients (Bender et al., 2010; Bruce et al., 2008; McConnell et al., 2007; Millar et al., 2005). In the US and Arctic studies, there was also an age bias in the patient populations whereby Hia was isolated primarily from young children (Bender et al., 2010; Bruce et al., 2008; Sill et al., 2007). Moxon and colleagues showed that strains carrying the type a capsule were more virulent than the other four non-type b serotypes (Zwahlen et al., 1989). Moreover, both the type a and the type b capsule contain the five-carbon sugar ring, ribitol (Branefors-Helander et al., 1977; Crisel et al., 1975). The increase in the number of Hia strains that caused invasive disease was not due to capsule switching as revealed by multilocus sequence typing (Sill et al., 2007).

Since the introduction of the Hib vaccine, there have been significant increases in the rate of infections caused by typeable strains in older populations. For example, ~25% of the typeable strains isolated in a population 65 years of age or older were Hib (Dworkin et al., 2007). In a further study, patients over the age of 80 years were predominantly susceptible to invasive Hib disease. These data suggested that immunocompromised patients were most at risk from invasive Hib disease (Bisgard et al., 1998; Campos et al., 2004). Moreover, older populations with higher incidences of immunodeficiencies also have decreased rates of vaccination against Hib (Nix et al., 2012).

Introduction of the Hib vaccines has subsequently led to a more prominent role for nontypeable strains of *H. influenzae* (NTHi) in invasive disease in all age groups studied (MacNeil et al., 2011). For example, the majority of invasive disease in Canadian children was due to non-Hib strains, of which NTHi had a predominant role (McConnell et al., 2007; Shuel et al., 2011; Sill et al., 2007; Tsang et al., 2007). Both biotyping and multilocus sequence typing of invasive NTHi strains showed great strain diversity, a small subset of strains was not predominant in invasive disease (Sill et al., 2007). An increase in invasive *H. influenzae* disease due to NTHi in adults has also been demonstrated (Bender et al., 2010; Bruce et al., 2008; Dworkin et al., 2007). Studies in the United States, Spain, Sweden, and England showed a significant increase in NTHi-based invasive disease in the population older than 65 years of age (Campos et al., 2004; Dworkin et al., 2007; Resman et al., 2011; Rubach et al., 2011; Slack et al., 1998).

27.2.2 The emergence of diseases caused by nontypeable *H. influenzae* strains

The near eradication of Hib in populations in which comprehensive vaccination has occurred has shifted the burden of *H. influenzae* disease onto NTHi (Dajani et al., 1979; Klein, 1997). As outlined in Section 27.2.1, NTHi has a role in invasive disease, particularly in older populations. However, NTHi is also prevalent in respiratory tract illnesses, mediating diseases of the upper and lower human airway, diseases such as otitis media (OM), acute sinusitis, conjunctivitis, community-acquired pneumonia, and exacerbations in patients with chronic obstructive pulmonary disease (COPD) or cystic fibrosis (Das et al., 2013; Kilpi et al., 2001; Klein, 1997; Murphy, 2003; Roman et al., 2004; Sethi and Murphy, 2001; St Geme, 2000). The socioeconomic impact of managing NTHi-induced diseases in both children and adults is extremely

significant. Direct and indirect costs of diagnosing and managing acute OM (AOM) exceeds $2 billion annually in the United States alone, with estimates of all OM cases in the United States being in excess of $5 billion (Ahmed et al., 2013; Kaplan et al., 1997; Klein, 2000). NTHi is the pathogen in approximately one-third of cases of AOM and has a significant role in OM with effusion (OME) (Del Beccaro et al., 1992; Kilpi et al., 2001; Spinola et al., 1986). In the developed world, 80% of children under the age of 3 years have one incidence of AOM, whereas 40% of children under the age of 7 years have six or more incidences of AOM. Globally, it is estimated that there are 709 million cases of AOM per year (Monasta et al., 2012). OM is the most common reason for a doctor's visit in school-age children, and morbidity associated with OM is significant with long-term hearing loss being one consequence of untreated OME, particularly in poorer countries (Auinger et al., 2003; Monasta et al., 2012). Further, clinical management of this highly prevalent disease has relied heavily on antibiotic therapies (Stool et al., 1994). Likely the result of the large quantity of antibiotics used to treat disease(s) due to NTHi (Finkelstein et al., 2003), there has been an emergence of antibiotic-resistant strains (Fuste et al., 1996; Nazir et al., 2004; Watanabe et al., 2004a). Development of therapies to treat NTHi-mediated diseases necessitates a better understanding of NTHi pathogenic virulence determinants and host microenvironmental cues that stage a dynamic interplay in bacteria–host interactions.

27.3 The NTHi pan-genome

H. influenzae, strain Rd, a rough derivative of a type d serotype that lost the ability to synthesize capsule (Wilcox and Smith, 1975), was the first free-living organism to have its genome sequenced (Fleischmann et al., 1995). Strain Rd was chosen, as its 1.8 Mb genome was representative of the sizes of other bacterial genomes but had a G + C content similar to that of the human genome. Also, a physical clone map did not exist, so it would be a true *de novo* assembly. The sequencing and assembly of the strain Rd genome was an important achievement in demonstrating experimental approaches that would be critical in the sequencing and assembly of subsequent bacterial genomes. Sequencing the strain Rd genome was also very important in understanding the basic biology of *H. influenzae*. However, strain Rd is an avirulent, highly passaged laboratory strain with little clinical relevance. It was thus important to sequence a clinical isolate of *H. influenzae*. Thus, a second *H. influenzae* strain sequenced was the NTHi strain, 86–028NP, a clinical isolate from a child with OM (Harrison et al., 2005). When the two genomes were compared, the genome of strain 86–028NP was found to be ~100 kb larger than that of strain Rd, containing 280 additional open reading frames (ORFs) that are absent in strain Rd. A number of these

ORFs are encoded in a genomic island (ICEHin1056) in strain 86–028NP, common with Hib strains, but absent in strain Rd (Harrison et al., 2005). This ICEHin1056 subfamily of islands is important, as a common feature of these islands is the possession of antibiotic-resistant genes which, coupled with mobility of the island, could have important implications for the spread of antibiotic resistance between *H. influenzae* strains (Juhas et al., 2007). Of note, strain 86–028NP also possesses two high molecular weight (HMW) adhesin gene clusters and *lav*, which encodes an autotransporter protein, also shown to play a role in adherence. These genes are all absent in strain Rd and may have important roles in virulence (Harrison et al., 2005). After the genome of strain 86–028NP was published, the genomes of two additional NTHi strains were completed. R2846 is an NTHi strain isolated from the middle ear fluid of a child with AOM, whereas R2866 (Int1), although nonencapsulated, is an invasive strain that was isolated from a child with meningitis. Finally, the genome sequence of a Hib strain (10810) was completed in 2010.

The completion of these NTHi genomes and the partial completion of many other NTHi genomes led to the development of the NTHi pan-genome. The pan-genome hypothesis suggests that a species possesses more genes than any one single member of the species alone. Every member of the species has a core set of genes, as well as an additional set of contingency genes that varies between members of the species (Medini et al., 2005). These genes are freely shared between members of the species via horizontal gene transfer and it is hypothesized that no two members of the species have an identical complement of genes (Hogg et al., 2007; Shen et al., 2005). Analyses of the *H. influenzae* pan-genome showed that the core set of genes was conserved absolutely with *H. influenzae*, no genes of foreign origin could be identified (Hogg et al., 2007). Conversely, the contingency genes contained genes whose origin was clearly foreign. However, the number of foreign genes identified was limited by the small number of cohabitating bacteria in the human respiratory tract that could share DNA. In addition, uptake of DNA by *H. influenzae* is dependent on uptake signal sequences. Thus, *H. influenzae* is more likely to take up DNA from other *H. influenzae* cells (Hogg et al., 2007).

To be able to share DNA, NTHi uses DNA restriction to recognize NTHi DNA and generally exclude foreign DNA. Type II restriction enzymes were first identified and characterized from *H. influenzae* strain Rd. Smith et al. noted that extracts from *H. influenzae* strain Rd contained endonuclease activity that was active against non–*H. influenzae* DNA but would not cleave *H. influenzae* DNA (Kelly and Smith, 1970). The cleavage site for this endonuclease (endonuclease R, or HindII) was subsequently identified (Smith and Wilcox, 1970). The mechanism that allows restriction enzymes to selectively cleave foreign DNA was also elucidated in *H. influenzae* with the observations

that a small percentage of the adenine bases in *H. influenzae* strain Rd DNA were methylated in the 6-amino position. Subsequent analyses identified the methylated sequence to correspond to the recognition site of HindII. These observations provided a mechanism of DNA restriction in bacteria: a bacterium protects its own DNA through methylation of the recognition sites for its own restriction enzymes while leaving foreign DNA susceptible to cleavage (Roy and Smith, 1973a, 1973b).

Through the process of transformation and protection by restriction, exchange of genes within the pan-genome between *H. influenzae* cells will therefore drive inter-strain variation and produce great strain diversity. This strain diversity will have important implications for behavior of *H. influenzae* as a commensal and as an opportunistic pathogen (Power et al., 2012).

27.4 NTHi disease virulence determinants

27.4.1 Biofilm-mediated disease

Biofilms are a complex community of bacteria, anchored to a surface and surrounded by matrix material such as bacteria-derived exopolysaccharides and DNA of bacterial and host origin. Biofilms contribute tremendously to pathogenesis and, importantly, to the prolonged, recurrent and difficult-to-treat nature of these diseases. However, bacterial and host elements that dictate the complexity and fitness of NTHi biofilms and their impact on NTHi persistence and disease severity are not completely understood. Therefore, the impact of biofilms on disease sequelae is a prerequisite to the identification of novel therapeutic approaches to combat this highly recalcitrant disease.

As observed with many other chronic, recurring infections, NTHi-mediated OM results from the development of robust organized biofilms in the middle ear (Bakaletz, 2007; Ehrlich et al., 2002; Hall-Stoodley et al., 2006; Post et al., 2007). NTHi has the propensity to build and reside within an organized biofilm, as demonstrated in animal models of experimental NTHi-induced OM and in clinical samples obtained from disease states (Ehrlich et al., 2002; Hall-Stoodley et al., 2006; Jurcisek and Bakaletz, 2007) (see Figure 27-2). Biofilm formation was prevalent in nearly all patients with a history of chronic OM undergoing tympanostomy tube placement compared with those without a history of OM (Hall-Stoodley et al., 2006). Nistico and colleagues (2011) provided further evidence that *H. influenzae* resided in biofilms on adenoids from chronic OM patients compared with controls. Das and colleagues (2013) have utilized production of proteins from NTHi biofilms as biomarkers to identify sinusitis infection in an animal model of NTHi-induced sinusitis.

The NTHi biofilm matrix consists of host components, particularly polymorphonuclear cells, DNA,

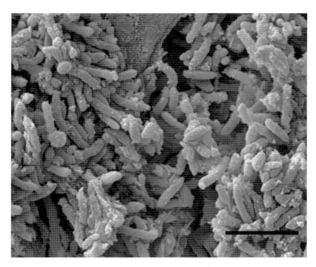

Fig. 27-2 NTHi biofilm formation on the surface of chinchilla nasopharyngeal cells. NTHi strain 86–028NP was inoculated onto the surface of chinchilla nasopharyngeal cells, incubated for 24 h, washed to remove planktonic cells, fixed, and prepared for scanning electron microscopy. Bar = 30 μm.

and neutrophil extracellular traps (NETs). Juneau and colleagues (2011) demonstrated that NTHi endotoxin initiates NET formation and that NTHi entrapped within NET structures were resistant to both extracellular killing and phagocytic killing, promoting formation of viable multicellular biofilm communities. These data suggest that NETs contribute to NTHi persistence. NTHi express type IV pili proteins that associate with extracellular DNA (Jurcisek and Bakaletz, 2007). Loss of these proteins alters biofilm architecture and stability and decreases biofilm formation *in vitro* (Carruthers et al., 2012; Jurcisek et al., 2007). NTHi strain 86–028NP expresses integration host factor (IHF). IHF belongs to the DNABII family of nucleic acid–binding proteins and has been shown to associate with extracellular DNA within the NTHi biofilm (Goodman et al., 2011). Antibody directed against IHF destabilizes the NTHi biofilm promoting clearance *in vitro* and in an animal model of OM (Brandstetter et al., 2013; Goodman et al., 2011). In addition, the expression of high-molecular-weight adhesins, adhesion and penetration protein, and outer membrane P5 and P6 proteins within biofilms make these attractive therapeutic targets (Gallaher et al., 2006; Webster et al., 2006). Sharpe et al. (2011) examined production of NTHi outer membrane vesicles, particularly as a result of innate immune assault. The authors suggest that vesicles may play a significant role in NTHi biofilm architecture, thereby protecting NTHi from host defense mechanisms.

In addition to the protective nature of the biofilm from innate mechanisms of host clearance, lipooligosaccharide (LOS) modifications provide a survival benefit. Modification of LOS with sialic acid promotes NTHi biofilm production *in vivo* (Jurcisek et al., 2005). Loss of sialylation decreased biofilm stability and in some cases promoted serum sensitivity that affected viability (Bouchet et al., 2003; Figueira

et al., 2007). NTHi phosphorylcholine (PCho) is added to some LOS forms in a phase-variable manner, providing enhanced persistence. Hong and colleagues (2007) found that PCho promotes NTHi persistence by reducing the host inflammatory response and by promoting formation of stable biofilm communities. Soluble mediators, secreted by NTHi, also modulate LOS composition, affecting biofilm maturation and persistence (Armbruster et al., 2009, 2011; Unal et al., 2010).

Eradication of biofilms remains an attractive option to quell NTHi-mediated diseases. Slinger and colleagues (2006) examined the ability of eight antibiotics singly and in combination to eradicate biofilms and demonstrated that biofilm cultures were more resistant to antibiotics than planktonic cells. Slinger and Chan showed that antibiotic combinations containing rifampin and ciprofloxacin were most effective against biofilms, whereas Wang and colleagues (2009) demonstrated an inhibition of biofilm synthesis using ciprofloxacin and azithromycin at concentrations higher than twofold the minimal inhibitory concentration. From these studies, it is clear that conventional susceptibility testing of antibiotics on planktonic NTHi is limited in predictive capacity to clear NTHi when grown as a biofilm community.

In addition to combination antibiotic therapies, therapeutic immunization to resolve OM biofilms remains an attractive option. Novotny and colleagues (2011) recently demonstrated efficacy for transcutaneous immunization to prevent and resolve experimental NTHi-induced OM. This immunization approach targeted the outer membrane protein P5 adhesin and the type IV pili of NTHi and resulted in resolution of NTHi biofilms from the middle ears of chinchillas. Recently, Novotny and colleagues (2013) described a mechanism of migration of dermal dendritic cells to the nasal-associated lymphoid tissue, expansion of host T cells, and subsequent production of specific antibody that facilitates NTHi clearance.

Microenvironmental changes associated with disease progression have been shown to influence bacterial physiological responses including increased antibiotic resistance, virulence factor expression, as well as induction of biofilm formation (Banin et al., 2005; Moreau-Marquis et al., 2009; Reid et al., 2009; Wakeman and Skaar, 2012). Lack of available nutrients in the host may significantly influence the NTHi pathogenic lifestyle. The development of structured biofilms in other pathogens (e.g., *Pseudomonas aeruginosa*) is conditional upon availability of essential nutrients (Banin et al., 2005; Moreau-Marquis et al., 2009; Wu and Outten, 2009). In particular, nutritional immunity influences microbial growth and biofilm formation (Ong et al., 2006). Thereby, microbes adapt to fluctuations in host environments with specific mechanisms to obtain essential nutrients. Vogel and colleagues (2012) have shown that deficiency in the import of heme–iron leads to the formation of NTHi biofilms with a lace-like architecture, suggesting a role for heme–iron availability in modulation of biofilm structure. Recently, Szelestey and colleagues (2013) demonstrated that transient heme–iron restriction has dramatic effects on biofilm development and influences NTHi persistence and disease progression. Transient heme–iron restriction also promoted NTHi invasion and intracellular growth leading to the development of intracellular bacterial communities that may contribute to the chronic nature of disease (Szelestey et al., 2013). These studies begin to address how host immune pressure and nutrient availability influence pathogenic behaviors that impact disease severity.

27.5 *Haemophilus influenzae*–host responses to disease

27.5.1 Oxidative stress

H. influenzae can grow aerobically and so experience metabolism-derived oxidative stress. Within the host, *H. influenzae* also encounters oxidative stress derived from host cells and from copathogens (Craig et al., 2001; Naylor et al., 2007; Pericone et al., 2000). To survive, *H. influenzae* has therefore evolved multiple overlapping defenses against oxidative stress (Harrison et al., 2012).

Orthologs of the peroxide-sensitive transcriptional regulator OxyR have been characterized in both Hib and NTHi strains (Harrison et al., 2007; Maciver and Hansen, 1996). The type b and the NTHi *oxyR* mutants were more sensitive to the effects of H_2O_2, as compared with their respective parental strains but the *H. influenzae* type b *oxyR* mutant exhibited similar levels of catalase activity as the parent strain. Conversely, the NTHi *oxyR* mutant had reduced catalase activity in the NTHi *oxyR* mutant (Maciver and Hansen, 1996). The NTHi strain 86–028NP *oxyR* mutant was used to define the OxyR regulon. As expected, H_2O_2-induced OxyR-regulated genes included those with roles in mitigating the effects of oxidative stress, as well as those genes that encode proteins that aid in the repair of proteins and DNA (Harrison et al., 2007).

A second regulator, the two-component Arc system of *H. influenzae* strain Rd is functionally homologous to that of *Escherichia coli*, in which gene expression is regulated due to the redox condition of the growth medium (Georgellis et al., 2001). In *H. influenzae* strain Rd, expression of genes in both the respiratory chain and the tricarboxylic acid cycle was repressed by the Arc system, whereas expression of genes involved in polyamine metabolism and iron sequestration increased. Repressed expression of the former groups of genes modulates respiratory activity and so mitigates the production of reactive oxygen species as the bacteria transition from anaerobiosis to an aerobic lifestyle. In the latter situation, ArcA promoted the expression of a Dps-like protein (Wong et al., 2007), which protects *E. coli* from DNA damage mediated by iron and H_2O_2 (Chiancone and Ceci, 2010). The Arc system also mediates serum resistance

in NTHi, through increased expression of Lic2B, a glycosyltransferase that adds galactose to the outer core of the LOS which protects from deposition of iC3b, a cleavage product of complement component factor C3b, on the bacterial surface (Wong et al., 2011). Lic2B is important for NTHi survival in a mouse model of bacteremia and for serum survival during anaerobic growth. This observation suggested changes in host environments signaled by alterations of redox conditions were sensed by the Arc system, which then coordinately regulated expression of genes critical for immune evasion. Possibly, the Arc system is critical when *H. influenzae* residing within low-oxygen environments, for example, in venous blood, or mucosal biofilms is subject to increasing oxidative stress (Wong et al., 2011).

H. influenzae encodes a single superoxide dismutase (SOD). *H. influenzae* lacks either the gene, *sodC*, encoding the periplasmic [Cu, Zn]-SOD, or the ability to synthesize functional [Cu, Zn]-SOD (Kroll et al., 1991). *sodC* is absent in all NTHi strains investigated (Norskov-Lauritsen et al., 2009), whereas all currently available NTHi genome sequences lack an identifiable ortholog of *sodC*. *H. influenzae* does however encode a functional ortholog of an [Mn]-SOD ortholog, found in Hib, *H. influenzae* strain Rd KW20, NTHi strain 86–028NP, and all *Haemophilus*-type strains tested (Cattoir et al., 2006; Kroll et al., 1993). To date, orthologs of [Fe]-SOD or [Ni]-SOD have not been found in any *Haemophilus* strain.

Classically, the catalases HPI and HPII are encoded by *katG* and *katE*. In *E. coli* and *Salmonella typhimurium*, *katE* is expressed constitutively during exponential growth with increased expression during stationary phase. Conversely, *katG* is normally expressed at low levels and is induced, via OxyR, upon peroxide stress (Christman et al., 1985; Loewen et al., 1985; Vijayakumar et al., 2004). *H. influenzae* only has an ortholog of HPII, encoded by *hktE*. Unlike in *E. coli*, *hktE* expression is maximal during exponential growth and reduced during stationary phase (Bishai et al., 1994). When an *hktE* mutant of *H. influenzae* strain Rd was cultured anaerobically in minimal medium, then shifted to an aerobic growth condition, there was a rise in H_2O_2 in the medium and the mutant exhibited a growth defect (Vergauwen et al., 2003). Expression of *hktE* is also inducible with H_2O_2 (Bishai et al., 1994; Harrison et al., 2007). However, as *hktE* has no appreciable role in an animal model of Hib virulence (Bishai et al., 1994), catalase appears to be the enzyme used primarily to scavenge high concentrations of endogenously generated H_2O_2 (Vergauwen et al., 2003, 2006).

Glutaredoxins can reduce disulfide bonds, utilizing the reducing power of glutathione (Dalle-Donne et al., 2009; Fernandez de Henestrosa et al., 1997). *H. influenzae* imports glutathione from the environment via the ABC-like dipeptide transporter DppBCDF (Vergauwen et al., 2010). The imported glutathione protects NTHi against oxidative stress induced by H_2O_2 or *tert*-butyl hydroperoxide (tBuOOH) (Pauwels et al., 2003, 2004; Vergauwen et al., 2003). A catalase mutant of *H. influenzae* was not impaired in growth in rich medium, colonized infant rats at parental levels and caused similar levels of bacteremia (Bishai et al., 1994). These data suggested that *H. influenzae* must have an additional method of scavenging peroxide, with glutathione being the electron donor for this peroxidase (Vergauwen et al., 2003). This led to the identification of the peroxidase PgdX. A *pgdX* mutant generated in *H. influenzae* strain Rd was more susceptible to tBuOOH and cumene hydroperoxide (CHP), when compared with the parent strain (Pauwels et al., 2003, 2004; Vergauwen et al., 2001, 2006). A *pgdX* mutant of *H. influenzae* strain Rd however showed increased resistance to H_2O_2, loss of PgdX was compensated by an increase in catalase expression (Pauwels et al., 2004). PgdX is optimized to scavenge smaller concentrations of H_2O_2 than catalase. In the absence of PgdX, low levels of H_2O_2 cannot be efficiently scavenged by catalase and OxyR activity is induced. Catalase activity is therefore upregulated. As catalase and PgdX scavenge metabolically generated H_2O_2, but PgdX is protective during stationary phase, the lower efficiency of PgdX in scavenging H_2O_2 suggests that stationary phase cultures of *H. influenzae* are more susceptible to oxidative stress (Pauwels et al., 2004; Vergauwen et al., 2003, 2006).

The alkyl hydroperoxidase AhpC was identified in *Salmonella* spp. and degrades the organic hydroperoxides tBuOOH and CHP as well as H_2O_2 (Jacobson et al., 1989; Niimura et al., 1995; Seaver and Imlay, 2001a). AhpC primarily scavenges low levels of endogenous H_2O_2 in *E. coli*, whereas catalase scavenges high levels of H_2O_2 (Seaver and Imlay, 2001a, 2001b). Scavenging reactive oxygen species leads to oxidation of AhpC. AhpC is therefore reduced with electrons from NADPH, transferred via the flavoprotein AhpF (Jacobson et al., 1989). *H. influenzae* orthologs of *ahpC* (known as *tsaA*) have been identified in NTHi strains, 1885MEE and 86–028NP, but expression of *tsaA* in NTHi strain 86–028NP was not inducible by H_2O_2 (Harrison et al., 2005, 2007; Munson et al., 2004).

Dps binds iron, which is then oxidized to Fe^{3+} by H_2O_2, which prevents hydroxyl radical generation via the Fenton reaction (Chiancone and Ceci, 2010; Park et al., 2005). Dps also binds DNA, which both protects the DNA from damage and slows initiation of chromosome replication to allow time for any DNA damage to be repaired (Chodavarapu et al., 2008; Grant et al., 1998; Martinez and Kolter, 1997). Expression of *dps* in NTHi strain 86–028NP is induced by the presence of H_2O_2, via OxyR, whereas expression of *dps* in *H. influenzae* strain Rd is also induced by ArcA (Harrison et al., 2007; Wong et al., 2007). Thus, a *dps* mutant of *H. influenzae* strain Rd is more sensitive to H_2O_2, as compared with the parent strain, when grown anaerobically (Wong et al., 2007).

27.5.2 Iron and virulence

Haemophilus, a strictly obligate human commensal, lacks the ability to synthesize the protoporphyrin ring required for iron sequestration and heme production (Panek and O'Brian, 2002; White and Granick, 1963) and therefore must acquire both heme and iron. Intracellular heme is incorporated into cytochromes for respiration, catalase for enzymatic neutralization of oxygen radicals, and is essential for sequestration of free iron to minimize oxygen radical production (Stojiljkovic and Perkins-Balding, 2002). As *H. influenzae* causes diseases of the upper and lower respiratory tract, mechanisms to acquire heme and iron are essential for virulence. Heme and free iron are sequestered in the human host by binding to transport or storage proteins. Free hemoglobin is rapidly bound in the host by the serum protein haptoglobin and the hemoglobin–haptoglobin complex, and free heme is bound by hemopexin. In addition, most iron is bound intracellularly by heme or ferritin or complexed to iron-containing proteins such as transferrin and lactoferrin (see Figure 27-3).

Haemophilus strains do not produce siderophores as a means to scavenge iron. *Haemophilus* species produce many proteins implicated in acquisition of iron and heme which includes ferric iron ABC transport systems, heme and heme–hemopexin binding proteins, transferrin-binding proteins, hemoglobin-haptoglobin binding proteins, and proteins involved in heme binding at the outer membrane and transport into the cell (Harrison et al., 2005). The outer membrane proteins HxuC and Hup are essential for heme acquisition (Cope et al., 1994; Morton et al., 2004b). The periplasmic heme-binding lipoprotein (HbpA) binds heme at low affinity and was shown to mediate heme utilization to support growth (Hanson et al., 1992; Morton et al., 2005). Recently, HbpA was shown to bind glutathione at higher affinity, implicating a role for this protein in oxidative stress (Vergauwen et al., 2010). Hemoglobin utilization is mediated by binding a family of hemoglobin-binding proteins (Hgp) on the outer membrane surface (Maciver et al., 1996; Morton et al., 2004a; Ren et al., 1998). Although playing a role in utilization of hemoglobin, deletion of all Hgp proteins does not abolish hemoglobin utilization suggesting a role for additional proteins in hemoglobin uptake (Morton et al., 1999). In addition to bacterial surface binding and transport of heme across the outer membrane, mechanisms to acquire heme across the inner membrane have been demonstrated. The *dppBCDF* gene cluster constitutes part of the periplasmic heme acquisition system of *H. influenzae*, likely dependent upon the delivery of heme to the DppBC inner membrane permease via the independently expressed heme-bound periplasmic protein HbpA (Morton et al., 2009). However, since heme utilization was only partially impaired, the authors indicated that additional periplasmic systems must be available to transport heme. The inner

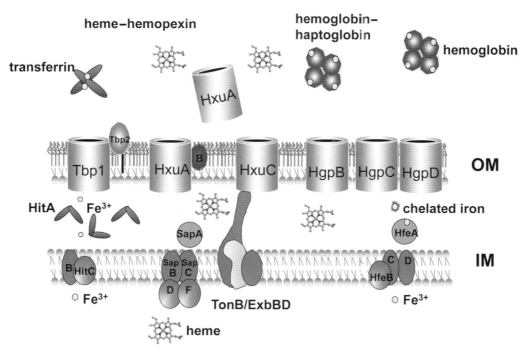

Fig. 27-3 Mechanisms of heme and iron uptake in *H. influenzae*. Tbp, transferrin-binding protein; Hxu, heme-binding outer membrane complex; Hgp, hemoglobin-binding outer membrane proteins; Hit and Hfe complexes, inner membrane iron transport systems; Sap, inner membrane ABC transporter.

membrane ABC transporter, Sap, has been implicated in heme–iron acquisition through transport into the bacterial cytoplasm (Mason et al., 2011). NTHi contains a periplasmic protein, SapA, which binds heme and through transport dependent upon the permease components, SapBC, acquires heme–iron (Mason et al., 2005, 2006, 2011; Shelton et al., 2011; Vogel et al., 2012). Loss of genes that encode the Sap transporter impose a heme–iron-starved phenotype on NTHi (Mason et al., 2011; Vogel et al., 2012). Loss of Sap transporter function results in clearance of NTHi from the chinchilla middle ear and nasopharynx (Mason et al., 2005), implicating a critical role for heme–iron acquisition in NTHi survival at these host sites. Further, NTHi upregulates iron uptake systems in the middle ear and NTHi deficient in iron acquisition are less fit to survive in this environment (Harrison et al., 2013; Mason et al., 2005, 2006).

Regulation of iron uptake and utilization is essential for NTHi survival in an animal model of human OM (Harrison et al., 2013). Recent work globally defined the Fur regulon in NTHi through microarray and proteomic analysis (Harrison et al., 2013). In addition to regulation of known iron utilization proteins, Fur was shown to regulate expression of multiple genes that encode proteins annotated as having hypothetical functions.

27.5.3 Innate immune resistance mechanisms and NTHi survival

The ability of opportunistic pathogens to survive in hostile host environments has prompted studies on bacterial response mechanisms to components of innate immunity. Toll-like receptors (TLRs) are important mediators of the host innate immune response. TLR9 recognizes CpG DNA motifs in bacterial DNA. TLR-9 genes were upregulated during OM, in middle ear mucosal cells and infiltrating leukocytes. Leichtle and colleagues (2011) demonstrated that TLR-9 depletion significantly prolonged the inflammatory response induced by NTHi in the middle ear and delayed bacterial clearance.

Innate immune molecules such as antimicrobial peptides are a key component of the host innate immune defense of the middle ear. NTHi induces β-defensin-2 production via TLR-2 signaling in human middle ear epithelial cells. The authors previously demonstrated that IL-1α induces β-defensin-2 gene expression. Bacterial stimulation and IL-1α act synergistically to induce hBD-2 (Lee et al., 2008). As a first line of innate defense, antimicrobial peptides serve to limit bacterial colonization of mucosal surfaces (Laube et al., 2006; Lehrer, 2004; Nizet et al., 2001; Salzman et al., 2003; Selsted and Ouellette, 2005). Bacteria therefore adapt to resist peptide lethality by remodeling the bacterial outer membrane surface to dampen charge and alter hydrophobicity (Brogden, 2005; Ernst et al., 2001; Kawasaki, 2006; Peschel, 2002). Bacteria also export antimicrobial peptides via multiple transferable resistance (MTR)–mediated efflux pumps (Shafer et al., 1998), secrete exoproteases to degrade antimicrobial peptides (Sieprawska-Lupa et al., 2004), secrete bacterial molecules that suppress the host innate defense (Attia et al., 2010; Islam et al., 2001), and release proteins that adsorb extracellular antimicrobial peptides (Frick et al., 2003). In NTHi, outer membrane remodeling provides a first line of defense against cationic peptides. Lysenko and colleagues (2000) demonstrated that the presence of phosphoryl-choline alters outer membrane hydrophobicity conferring resistance to the cathelicidin LL-37. Additionally, HtrB is required for hexa-acylation of NTHi lipid A, so NTHi mutants which lack HtrB are unable to fully acylate their lipid A and so are susceptible to AMP-mediated killing (Starner et al., 2002). Examination of *Haemophilus* genomes show the lack of genes with predicted roles in resistance mechanisms such as peptide efflux or exoprotease activity. Yet, further work has identified an ABC transporter system, the Sap transporter, as a mechanism to bind and transport host-derived antimicrobial peptides away from the inner membrane target to the cytoplasm where they are degraded (Shelton et al., 2011). SapA, the periplasmic substrate-binding protein of the Sap transporter, binds antimicrobial peptides (Mason et al., 2006). A Sap-deficient NTHi strain was therefore attenuated for infection of the chinchilla middle ear, primarily due to increased susceptibility of NTHi to host antimicrobial peptides (Shelton et al., 2011).

Bacterial IgA proteases are important virulence factors in respiratory tract infections. An enzyme which cleaves immunoglobulin A1 (IgA1) has been identified in *H. influenzae* (Kilian et al., 1979; Male, 1979). Since IgA1 is primarily associated with mucosal surfaces, IgA protease activity confers a survival advantage during colonization. Fernaays and colleagues characterized a second IgA protease (IgAB) in NTHi that predominately associates with strains adapted to infection in COPD (Fernaays et al., 2006; Murphy et al., 2011). Recently, Zhang and colleagues (2012) demonstrated the *igaB* gene to be encoded on a genetic island common to strains recovered from adults with COPD. Similar screens identified genetic islands more prevalent in NTHi isolates from children with OM than in those obtained from throats of healthy children (Bergman and Akerley, 2003; Sandstedt et al., 2010). Linkage of strain genetic signatures to disease etiology may ultimately provide information to specifically target disease-causing strains or provide biomarkers for disease.

27.5.4 Cytokine stimulation and inflammation

NTHi stimulates a prolonged immune response that often leads to inflammatory responses that have deleterious effects on host mucosal cells (reviewed in Wang et al., 2012). Through release of inflammatory mediators and recruitment of leukocytes to the site of infection, typically the lungs (COPD) or the middle ear (OM), host cell damage may occur in the absence of tight regulation of the immune response. Proinflammatory mediators such as IL-8, TNF-α, and IL-1β were upregulated in human airway epithelial cells upon synergistic activity of NTHi and TNF-α (Watanabe et al., 2004b). A similar synergistic effect was observed in middle ear cells exposed to NTHi

and IL-1α, resulting in upregulation of antimicrobial peptides (Moon et al., 2006). Importantly, upregulation of inflammation by NTHi can increase susceptibility to colonization of other microorganisms. Loria and colleagues recently showed that NTHi infection upregulates the NLRP3 inflammasome which triggers IL-1β release in a caspase-1-dependent manner (Rotta Detto Loria et al., 2013). The transforming growth factor (TGF)-β pathway is regulated in the middle ear cleft during bacterial OM. Lee and colleagues (2011) showed there were higher amounts of TGF-β in *Haemophilus*-infected ears compared with pneumococcus-infected ears. Coincident with these increased levels of TGF-β was the increased incidence of granulation tissue in NTHi-induced disease. Tight control of inflammation then is critical for epithelial cell homeostasis. Lim and colleagues (2007, 2008) have shown that expression of the tumor suppressor deubiquitinase, CYLD, acts as a negative regulator of inflammation by downregulation of NF-κB. Further, Kyo and colleagues (2012) reported that a mucin present on the apical surface of mucosal epithelia suppresses NTHi-induced TLR signaling. These data suggest additional immunosuppressive effects that control bacteria-induced inflammation.

Inner ear infection, affecting language development and motor coordination, can be induced by proinflammatory responses to NTHi in the middle ear. Moon and colleagues hypothesized that NTHi PAMPs trigger host spiral ligament fibrocytes (SLF) to release inflammatory cytokines that attract effector cells and cause inner ear dysfunction. The secretion of monocyte chemotactic protein 1 (MCP-1) by SLF cells is dependent upon NTHi binding of TLR-2, with subsequent NF-κB activation and binding of NF-κB to the enhancer region of MCP-1 (Moon et al., 2007). Further studies demonstrated that MCP-1 induction contributes to inner ear inflammation via CCR2-mediated recruitment of monocytes. Inhibition of MCP-1 or CCR2 did not abrogate inner ear inflammation since targeting these molecules resulted in compensatory upregulation of alternative genes such as MCP-2/CCL8 (Woo et al., 2010). Further work by Oh and colleagues (2012) provided a mechanistic basis for polymorphonuclear cell influx demonstrating that NTHi stimulates release of CXCL2 from SLF cells resulting in the activation of c-Jun and consequent influx of polymorphonuclear cells into the cochlea of the middle ear.

27.6 *Haemophilus influenzae* and polymicrobial disease

OM is a polymicrobial disease, often exacerbated by viral infection followed by bacterial infection of the middle ear. Although most studies to date focus on a single pathogen coincident with OM, recent work has monitored the ability of copathogens to infect, form biofilms, and cause disease in the middle ear. Weimer and colleagues monitored the ability of

Streptococcus pneumoniae and *H. influenzae* to colonize, form biofilms, and persist in a chinchilla model of OM. Coinfection reduced the incidence of systemic disease by *S. pneumoniae* compared with those infected with pneumococcus alone. Importantly, the presence of *H. influenzae* increased biofilm formation in the middle ear, and significantly increased the ability of pneumococcus to form a biofilm. Further, a greater proportion of the translucent pneumococcal colony type that is primarily known to associate with mucosal surfaces as biofilms, yet with a reduced incidence of invasive disease, was observed. These data suggest that systemic pneumococcal infection may result not only from pneumococcal infection but may be influenced by bacterial populations in which pneumococcus predominate (Weimer et al., 2010).

NTHi also provides passive protection for pneumococcus in the chinchilla model of OM. In a further study, Weimer and colleagues utilized a β-lactamase-deficient strain of NTHi to monitor the bactericidal effect of amoxicillin when delivered to the middle ear containing NTHi or pneumococcus alone, or as mixed biofilms of both strains. The β-lactamase-deficient NTHi strain was susceptible to antibiotic treatment, indicated by complete eradication of both planktonic bacteria in middle ear effusions as well as the bacteria in the biofilm. Biofilms formed by NTHi deficient in β-lactamase production were not protected from clearance. However, mixed biofilms generated *in vitro* were protected from clearance (both NTHi and pneumococcus), as long as NTHi were expressing β-lactamase. In stark contrast, biofilms generated *in vivo* were protected from clearance (both NTHi and pneumococcus) even if NTHi was not expressing β-lactamase. These data suggest two mechanisms of NTHi protection of pneumococcus from amoxicillin killing *in vivo*; first via β-lactamase production *in vivo* and the second via formation of biofilm communities (Weimer et al., 2011). Further, pneumococcus is protective toward NTHi in *in vivo* biofilms (Weimer et al., 2011).

The polymicrobial nature of biofilms present during OM disease has also been studied using models of *H. influenzae* and *Moraxella catarrhalis* coinfections. Armbruster and colleagues recently demonstrated that polymicrobial infection impacts biofilm development and antibiotic resistance during OM disease. *H. influenzae* promotes *M. catarrhalis* persistence via *Haemophilus*-induced quorum signals that increased the resistance of *M. catarrhalis* in biofilms (Armbruster et al., 2010). A better understanding of *H. influenzae* interactions with other bacteria, and the polymicrobial influence on disease progression, will be necessary to fully appreciate disease etiology in the host.

References

Adegbola, R.A., Mulholland, E.K., Secka, O., Jaffar, S., and Greenwood, B.M. 1998. Vaccination with a *Haemophilus influenzae* type b conjugate vaccine reduces oropharyngeal carriage of *H. influenzae* type b among Gambian children. *J. Infect. Dis.*, 177, 1758–1761.

Ahmed, S., Shapiro, N.L., and Bhattacharyya, N. 2013. Incremental health care utilization and costs for acute otitis media in children. *Laryngoscope*, 124, 301–305.

Alexander, H.E., Heidelberger, M., and Leidy, G. 1944. The protective or curative element in type b *H. influenzae* rabbit serum. *Yale J. Biol. Med.*, 16, 425–434.

Anderson, P., and Insel, R.A. 1981. A polysaccharide-protein complex from *Haemophilus influenzae* type b. I. Activity in weanling rabbits and human T lymphocytes. *J. Infect. Dis.*, 144, 509–520.

Anderson, P., Johnston Jr., R.B., and Smith, D.H. 1972. Human serum activities against *Hemophilus influenzae*, type b. *J. Clin. Invest.*, 51, 31–38.

Anderson, P., and Smith, D.H. 1977. Immunogenicity in weanling rabbits of a polyribophosphate complex from *Haemophilus influenzae* type b. *J. Infect. Dis.*, 136(Suppl), S63–S70.

Anderson, P., Smith, D.H., Ingram, D.L., Wilkins, J., Wehrle, P.F., and Howie, V.M. 1977. Antibody of polyribophate of *Haemophilus influenzae* type b in infants and children: Effect of immunization with polyribophosphate. *J. Infect. Dis.*, 136(Suppl), S57–S62.

Armbruster, C.E., Hong, W., Pang, B., Weimer, K.E., Juneau, R.A., Turner, J., and Swords, W.E. 2010. Indirect pathogenicity of *Haemophilus influenzae* and *Moraxella catarrhalis* in polymicrobial otitis media occurs via interspecies quorum signaling. *MBio*, 1.

Armbruster, C.E., Hong, W.Z., Pang, B., Dew, K.E., Juneau, R.A., Byrd, M.S., Love, C.F., Kock, N.D., and Swords, W.E. 2009. LuxS promotes biofilm maturation and persistence of nontypeable *Haemophilus influenzae in vivo* via modulation of lipooligosaccharides on the bacterial surface. *Infect. Immun.*, 77, 4081–4091.

Armbruster, C.E., Pang, B., Murrah, K., Juneau, R.A., Perez, A.C., Weimer, K.E., and Swords, W.E. 2011. RbsB (NTHI_0632) mediates quorum signal uptake in nontypeable *Haemophilus influenzae* strain 86–028NP. *Mol. Microbiol.*, 82, 836–850.

Attia, A.S., Benson, M.A., Stauff, D.L., Torres, V.J., and Skaar, E.P. 2010. Membrane damage elicits an immunomodulatory program in *Staphylococcus aureus*. *PLoS Pathog.*, 6, e1000802.

Auinger, P., Lanphear, B.P., Kalkwarf, H.J., and Mansour, M.E. 2003. Trends in otitis media among children in the United States. *Pediatrics*, 112, 514–520.

Bakaletz, L.O. 2007. Bacterial biofilms in otitis media: Evidence and relevance. *Pediatr. Infect. Dis. J.*, 26, S17–S19.

Banin, E., Vasil, M.L., and Greenberg, E.P. 2005. Iron and *Pseudomonas aeruginosa* biofilm formation. *Proc. Natl. Acad. Sci. USA*, 102, 11076–11081.

Bender, J.M., Cox, C.M., Mottice, S., She, R.C., Korgenski, K., Daly, J.A., and Pavia, A.T. 2010. Invasive *Haemophilus influenzae* disease in Utah children: An 11-year population-based study in the era of conjugate vaccine. *Clin. Infect. Dis.*, 50, e41–e46.

Bergman, N.H., and Akerley, B.J. 2003. Position-based scanning for comparative genomics and identification of genetic islands in *Haemophilus influenzae* type b. *Infect. Immun.*, 71, 1098–1108.

Bisgard, K.M., Kao, A., Leake, J., Strebel, P.M., Perkins, B.A., and Wharton, M. 1998. *Haemophilus influenzae* invasive disease in the United States, 1994–1995: Near disappearance of a vaccine-preventable childhood disease. *Emerg. Infect. Dis.*, 4, 229–237.

Bishai, W.R., Howard, N.S., Winkelstein, J.A., and Smith, H.O. 1994. Characterization and virulence analysis of catalase mutants of *Haemophilus influenzae. Infect. Immun.*, 62, 4855–4860.

Black, S.B., Shinefield, H.R., Fireman, B., Hiatt, R., Polen, M., and Vittinghoff, E. 1991. Efficacy in infancy of oligosaccharide conjugate *Haemophilus influenzae* type b (HbOC) vaccine in a United States population of 61,080 children. The Northern California Kaiser Permanente Vaccine Study Center Pediatrics Group. *Pediatr. Infect. Dis. J.*, 10, 97–104.

Booy, R., Hodgson, S., Carpenter, L., Mayon-White, R.T., Slack, M.P., Macfarlane, J.A., Haworth, E.A., Kiddle, M., Shribman, S., Roberts, J.S., and Moxon, E.R. 1994. Efficacy of *Haemophilus influenzae* type b conjugate vaccine PRP-T. *Lancet*, 344, 362–366.

Bouchet, V., Hood, D.W., Li, J., Brisson, J.R., Randle, G.A., Martin, A., Li, Z., Goldstein, R., Schweda, E.K., Pelton, S.I., Richards, J.C., and Moxon, E.R. 2003. Host-derived sialic acid is incorporated into *Haemophilus influenzae* lipopolysaccharide and is a major virulence factor in experimental otitis media. *Proc. Natl. Acad. Sci. USA*, 100, 8898–8903.

Brandstetter, K.A., Jurcisek, J.A., Goodman, S.D., Bakaletz, L.O., and Das, S. 2013. Antibodies directed against integration host factor mediate biofilm clearance from Nasopore. *Laryngoscope*, 123, 2626–2632.

Branefors-Helander, P., Erbing, C., Kenne, L., and Linberg, R. 1977. The structure of the capsular antigen from *Haemophilus influenzae* type a. *Carbohydr. Res.*, 56, 117–122.

Brogden, K.A. 2005. Antimicrobial peptides: Pore formers or metabolic inhibitors in bacteria? *Nat. Rev. Microbiol.*, 3, 238–250.

Broome, C.V. 1987. Epidemiology of *Haemophilus influenzae* type b infections in the United States. *Pediatr. Infect. Dis. J.*, 6, 779–782.

Bruce, M.G., Deeks, S.L., Zulz, T., Navarro, C., Palacios, C., Case, C., Hemsley, C., Hennessy, T., Corriveau, A., Larke, B., Sobel, I., Lovgren, M., Debyle, C., Tsang, R., and Parkinson, A.J. 2008. Epidemiology of *Haemophilus influenzae* serotype a, North American Arctic, 2000–2005. *Emerg. Infect. Dis.*, 14, 48–55.

Campos, J., Hernando, M., Roman, F., Perez-Vazquez, M., Aracil, B., Oteo, J., Lazaro, E., and de Abajo, F. 2004. Analysis of invasive *Haemophilus influenzae* infections after extensive vaccination against *H. influenzae* type b. *J. Clin. Microbiol.*, 42, 524–529.

Carruthers, M.D., Tracy, E.N., Dickson, A.C., Ganser, K.B., Munson Jr., R.S., and Bakaletz, L.O. 2012. Biological roles of nontypeable *Haemophilus influenzae* type IV pili encoded by the *pil* and *com* operons. *J. Bacteriol.*, 194, 1927–1933.

Cattoir, V., Lemenand, O., Avril, J.L., and Gaillot, O. 2006. The *sodA* gene as a target for phylogenetic dissection of the genus *Haemophilus* and accurate identification of human clinical isolates. *Int. J. Med. Microbiol.*, 296, 531–540.

Centers for Disease Control and Prevention (CDC). 2002. Progress towards elimination of *Haemophilus influenzae* type b invasive disease among infants and children, United States, 1998–2000. *Morb. Mortal. Wkly. Rep.*, 51, 234–237.

Chandler, C.A., Fothergill, L.D., and Dingle, J.H. 1937. Studies on *Haemophilus influenzae*: II. A comparative study of the virulence of smooth, rough, and respiratory strains of *Haemophilus influenzae* as determined by infection of mice with mucin suspensions of the organism. *J. Exp. Med.*, 66, 789–799.

Chiancone, E., and Ceci, P. 2010. The multifaceted capacity of Dps proteins to combat bacterial stress conditions: Detoxification of iron and hydrogen peroxide and DNA binding. *Biochim. Biophys. Acta*, 1800, 798–805.

Chodavarapu, S., Gomez, R., Vicente, M., and Kaguni, J.M. 2008. *Escherichia coli* Dps interacts with DnaA protein to impede initiation: A model of adaptive mutation. *Mol. Microbiol.*, 67, 1331–1346.

Christman, M.F., Morgan, R.W., Jacobson, F.S., and Ames, B.N. 1985. Positive control of a regulon for defenses against oxidative stress and some heat-shock proteins in *Salmonella typhimurium. Cell*, 41, 753–762.

Cope, L.D., Thomas, S.E., Latimer, J.L., Slaughter, C.A., Muller-Eberhard, U., and Hansen, E.J. 1994. The 100 kDa haem: Haemopexin-binding protein of *Haemophilus influenzae*: Structure and localization. *Mol. Microbiol.*, 13, 863–873.

Craig, J.E., Cliffe, A., Garnett, K., and High, N.J. 2001. Survival of nontypeable *Haemophilus influenzae* in macrophages. *FEMS Microbiol. Lett.*, 203, 55–61.

Crisel, R.M., Baker, R.S., and Dorman, D.E. 1975. Capsular polymer of *Haemophilus influenzae*, type b. I. Structural characterization of the capsular polymer of strain Eagan. *J. Biol. Chem.*, 250, 4926–4930.

Dajani, A.S., Asmar, B.I., and Thirumoorthi, M.C. 1979. Systemic *Haemophilus influenzae* disease: An overview. *J. Pediatr.*, 94, 355–364.

Dalle-Donne, I., Rossi, R., Colombo, G., Giustarini, D., and Milzani, A. 2009. Protein S-glutathionylation: A regulatory device from bacteria to humans. *Trends Biochem. Sci.*, 34, 85–96.

Das, S., Rosas, L.E., Jurcisek, J.A., Novotny, L.A., Green, K.B., and Bakaletz, L.O. 2013. Improving patient care via development of a protein-based diagnostic test for microbe-specific detection of chronic rhinosinusitis. *Laryngoscope*, 124, 608–615.

Del Beccaro, M.A., Mendelman, P.M., Inglis, A.F., Richardson, M.A., Duncan, N.O., Clausen, C.R., and Stull, T.L. 1992. Bacteriology of acute otitis media: A new perspective. *J. Pediatr.*, 120, 81–84.

Doern, G.V., Jorgensen, J.H., Thornsberry, C., Preston, D.A., Tubert, T., Redding, J.S., and Maher, L.A. 1988. National collaborative study of the prevalence of antimicrobial resistance among clinical isolates of *Haemophilus influenzae. Antimicrob. Agents Chemother.*, 32, 180–185.

Dworkin, M.S., Park, L., and Borchardt, S.M. 2007. The changing epidemiology of invasive *Haemophilus influenzae* disease, especially in persons > or = 65 years old. *Clin. Infect. Dis.*, 44, 810–816.

Ehrlich, G.D., Veeh, R., Wang, X., Costerton, J.W., Hayes, J.D., Hu, F.Z., Daigle, B.J., Ehrlich, M.D., and Post, J.C. 2002. Mucosal biofilm formation on middle-ear mucosa in the chinchilla model of otitis media. *JAMA*, 287, 1710–1715.

Ernst, R.K., Guina, T., and Miller, S.I. 2001. *Salmonella typhimurium* outer membrane remodeling: Role in resistance to host innate immunity. *Microbes. Infect.*, 3, 1327–1334.

Fernaays, M.M., Lesse, A.J., Cai, X., and Murphy, T.F. 2006. Characterization of igaB, a second immunoglobulin A1 protease gene in nontypeable *Haemophilus influenzae. Infect. Immun.*, 74, 5860–5870.

Fernandez de Henestrosa, A.R., Badiola, I., Saco, M., Perez de Rozas, A.M., Campoy, S., and Barbe, J. 1997. Importance of the *galE* gene on the virulence of *Pasteurella multocida. FEMS Microbiol. Lett.*, 154, 311–316.

Figueira, M.A., Ram, S., Goldstein, R., Hood, D.W., Moxon, E.R., and Pelton, S.I. 2007. Role of complement in defense of the middle ear revealed by restoring the virulence of nontypeable *Haemophilus influenzae* siaB mutants. *Infect. Immun.*, 75, 325–333.

Fildes, P. 1956. Richard Friedrich Johannes Pfeiffer. 1858–1945. *Biogr. Mem. Fellows R. Soc.*, 2, 237–247.

Finkelstein, J.A., Stille, C., Nordin, J., Davis, R., Raebel, M.A., Roblin, D., Go, A.S., Smith, D., Johnson, C.C., Kleinman, K., Chan, K.A., and Platt, R. 2003. Reduction in antibiotic use among US children, 1996–2000. *Pediatrics*, 112, 620–627.

Fleischmann, R.D., Adams, M.D., White, O., Clayton, R.A., Kirkness, E.F., Kerlavage, A.R., Bult, C.J., Tomb, J.F., Dougherty, B.A., Merrick, J.M., Mckenney, K., Sutton, G., FitzHugh, W., Fields, C., Gocyne, J.D., Scott, J., Shirley, R., Liu, L.I., Glodek, A., Kelley, J.M., Weidman, J.F., Phillips, C.A., Spriggs, T., Hedblom, E., and Cotton, M.D. 1995. Whole-genome random sequencing and assembly of *Haemophilus influenzae* Rd. *Science*, 269, 496–512.

Fothergill, L.D., and Wright, J. 1933. Influenzal meningitis. The relation of age incidence to the bactericidal power of blood against the causal organism. *J. Immunol.*, 24, 273–284.

Frick, I.M., Akesson, P., Rasmussen, M., Schmidtchen, A., and Bjorck, L. 2003. SIC, a secreted protein of *Streptococcus pyogenes* that inactivates antibacterial peptides. *J. Biol. Chem.*, 278, 16561–16566.

Fuste, M.C., Pineda, M.A., Palomar, J., Vinas, M., and Loren, J.G. 1996. Clonality of multidrug-resistant nontypeable strains of *Haemophilus influenzae*. *J. Clin. Microbiol.*, 34, 2760–2765.

Gallaher, T.K., Wu, S., Webster, P., and Aguilera, R. 2006. Identification of biofilm proteins in non-typeable *Haemophilus Influenzae*. *BMC Microbiol.*, 6, 65.

Georgellis, D., Kwon, O., Lin, E.C., Wong, S.M., and Akerley, B.J. 2001. Redox signal transduction by the ArcB sensor kinase of *Haemophilus influenzae* lacking the PAS domain. *J. Bacteriol.*, 183, 7206–7212.

Goodman, S.D., Obergfell, K.P., Jurcisek, J.A., Novotny, L.A., Downey, J.S., Ayala, E.A., Tjokro, N., Li, B., Justice, S.S., and Bakaletz, L.O. 2011. Biofilms can be dispersed by focusing the immune system on a common family of bacterial nucleoid-associated proteins. *Mucosal Immunol.*, 4, 625–637.

Grant, R.A., Filman, D.J., Finkel, S.E., Kolter, R., and Hogle, J.M. 1998. The crystal structure of Dps, a ferritin homolog that binds and protects DNA. *Nat. Struct. Biol.*, 5, 294–303.

Hall-Stoodley, L., Hu, F.Z., Gieseke, A., Nistico, L., Nguyen, D., Hayes, J., Forbes, M., Greenberg, D.P., Dice, B., Burrows, A., Wackym, P.A., Stoodley, P., Post, J.C., Ehrlich, G.D., and Kerschner, J.E. 2006. Direct detection of bacterial biofilms on the middle-ear mucosa of children with chronic otitis media. *JAMA*, 296, 202–211.

Hanson, M.S., Slaughter, C., and Hansen, E.J. 1992. The *hbpA* gene of *Haemophilus influenzae* type b encodes a heme-binding lipoprotein conserved among heme-dependent *Haemophilus* species. *Infect. Immun.*, 60, 2257–2266.

Harrison, A., Bakaletz, L.O., and Munson Jr., R.S. 2012. *Haemophilus influenzae* and oxidative stress. *Front. Cell. Infect. Microbiol.*, 2, 40.

Harrison, A., Dyer, D.W., Gillaspy, A., Ray, W.C., Mungur, R., Carson, M.B., Zhong, H., Gipson, J., Gipson, M., Johnson, L.S., Lewis, L., Bakaletz, L.O., and Munson Jr., R.S. 2005. Genomic sequence of an otitis media isolate of nontypeable *Haemophilus influenzae*: Comparative study with *H. influenzae* serotype d, strain KW20. *J. Bacteriol.*, 187, 4627–4636.

Harrison, A., Ray, W.C., Baker, B.D., Armbruster, D.W., Bakaletz, L.O., and Munson Jr., R.S. 2007. The OxyR regulon in nontypeable *Haemophilus influenzae*. *J. Bacteriol.*, 189, 1004–1012.

Harrison, A., Santana, E.A., Szelestey, B.R., Newsom, D.E., White, P., and Mason, K.M. 2013. Ferric uptake regulator and its role in the pathogenesis of nontypeable *Haemophilus influenzae*. *Infect. Immun.*, 81, 1221–1233.

Hogg, J.S., Hu, F.Z., Janto, B., Boissy, R., Hayes, J., Keefe, R., Post, J.C., and Ehrlich, G.D. 2007. Characterization and modeling of the *Haemophilus influenzae* core and supragenomes based on the complete genomic sequences of Rd and 12 clinical nontypeable strains. *Genome Biol.*, 8, R103.

Hong, W., Mason, K., Jurcisek, J., Novotny, L., Bakaletz, L.O., and Swords, W.E. 2007. Phosphorylcholine decreases early inflammation and promotes the establishment of stable biofilm communities of nontypeable *Haemophilus influenzae* strain 86–028NP in a chinchilla model of otitis media. *Infect. Immun.*, 75, 958–965.

Islam, D., Bandholtz, L., Nilsson, J., Wigzell, H., Christensson, B., Agerberth, B., and Gudmundsson, G. 2001. Downregulation of bactericidal peptides in enteric infections: A novel immune escape mechanism with bacterial DNA as a potential regulator. *Nat. Med.*, 7, 180–185.

Jacobson, F.S., Morgan, R.W., Christman, M.F., and Ames, B.N. 1989. An alkyl hydroperoxide reductase from *Salmonella typhimurium* involved in the defense of DNA against oxidative damage. Purification and properties. *J. Biol. Chem.*, 264, 1488–1496.

Juhas, M., Power, P.M., Harding, R.M., Ferguson, D.J., Dimopoulou, I.D., Elamin, A.R., Mohd-Zain, Z., Hood, D.W., Adegbola, R., Erwin, A., Smith, A., Munson, R.S., Harrison, A., Mansfield, L., Bentley, S., and Crook, D.W. 2007. Sequence and functional analyses of *Haemophilus* spp. genomic islands. *Genome Biol.*, 8, R237.

Juneau, R.A., Pang, B., Weimer, K.E., Armbruster, C.E., and Swords, W.E. 2011. Nontypeable *Haemophilus influenzae* initiates formation of neutrophil extracellular traps. *Infect. Immun.*, 79, 431–438.

Jurcisek, J., Greiner, L., Watanabe, H., Zaleski, A., Apicella, M.A., and Bakaletz, L.O. 2005. Role of sialic acid and complex carbohydrate biosynthesis in biofilm formation by nontypeable *Haemophilus influenzae* in the chinchilla middle ear. *Infect. Immun.*, 73, 3210–3218.

Jurcisek, J.A., and Bakaletz, L.O. 2007. Biofilms formed by nontypeable *Haemophilus influenzae* in vivo contain both double-stranded DNA and type IV pilin protein. *J. Bacteriol.*, 189, 3868–3875.

Jurcisek, J.A., Bookwalter, J.E., Baker, B.D., Fernandez, S., Novotny, L.A., Munson Jr., R.S., and Bakaletz, L.O. 2007. The PilA protein of non-typeable *Haemophilus influenzae* plays a role in biofilm formation, adherence to epithelial cells and colonization of the mammalian upper respiratory tract. *Mol. Microbiol.*, 65, 1288–1299.

Kaplan, B., Wandstrat, T.L., and Cunningham, J.R. 1997. Overall cost in the treatment of otitis media. *Pediatr. Infect. Dis. J.*, 16, S9–S11.

Kawasaki, K. 2006. [Outer membrane remodeling of *Salmonella typhimurium* and host innate immunity]. *Yakugaku Zasshi*, 126, 1227–1234.

Kelly Jr., T.J., and Smith, H.O. 1970. A restriction enzyme from *Hemophilus influenzae*. II. *J. Mol. Biol.*, 51, 393–409.

Kilian, M., Mestecky, J., and Schrohenloher, R.E. 1979. Pathogenic species of the genus *Haemophilus* and *Streptococcus pneumoniae* produce immunoglobulin A1 protease. *Infect. Immun.*, 26, 143–149.

Kilpi, T., Herva, E., Kaijalainen, T., Syrjanen, R., and Takala, A.K. 2001. Bacteriology of acute otitis media in a cohort of Finnish children followed for the first two years of life. *Pediatr. Infect. Dis. J.*, 20, 654–662.

Klein, J.O. 1997. Role of nontypeable *Haemophilus influenzae* in pediatric respiratory tract infections. *Pediatr. Infect. Dis. J.*, 16, S5–S8.

Klein, J.O. 2000. The burden of otitis media. *Vaccine*, 19(Suppl 1), S2–S8.

Kroll, J.S., Langford, P.R., and Loynds, B.M. 1991. Copper-zinc superoxide dismutase of *Haemophilus influenzae* and *H. parainfluenzae*. *J. Bacteriol.*, 173, 7449–7457.

Kroll, J.S., Langford, P.R., Saah, J.R., and Loynds, B.M. 1993. Molecular and genetic characterization of superoxide dismutase in *Haemophilus influenzae* type b. *Mol. Microbiol.*, 10, 839–848.

Kyo, Y., Kato, K., Park, Y.S., Gajghate, S., Umehara, T., Lillehoj, E.P., Suzaki, H., and Kim, K.C. 2012. Antiinflammatory role of MUC1 mucin during infection with nontypeable *Haemophilus influenzae*. *Am. J. Respir. Cell. Mol. Biol.*, 46, 149–156.

Lagos, R., Horwitz, I., Toro, J., San Martin, O., Abrego, P., Bustamante, C., Wasserman, S.S., Levine, O.S., and Levine, M.M. 1996. Large scale, postlicensure, selective vaccination of Chilean infants with PRP-T conjugate vaccine: Practicality and effectiveness in preventing invasive *Haemophilus influenzae* type b infections. *Pediatr. Infect. Dis. J.*, 15, 216–222.

Laube, D.M., Yim, S., Ryan, L.K., Kisich, K.O., and Diamond, G. 2006. Antimicrobial peptides in the airway. *Curr. Top. Microbiol. Immunol.*, 306, 153–182.

Lee, H.Y., Takeshita, T., Shimada, J., Akopyan, A., Woo, J.I., Pan, H., Moon, S.K., Andalibi, A., Park, R.K., Kang, S.H., Kang, S.S., Gellibolian, R., and Lim, D.J. 2008. Induction of beta defensin 2 by NTHi requires TLR2 mediated MyD88 and IRAK-TRAF6-p38MAPK signaling pathway in human middle ear epithelial cells. *BMC Infect. Dis.*, 8, 87.

Lee, Y.W., Chung, Y., Juhn, S.K., Kim, Y., and Lin, J. 2011. Activation of the transforming growth factor beta pathway in bacterial otitis media. *Ann. Otol. Rhinol. Laryngol.*, 120, 204–213.

Lehrer, R.I. 2004. Primate defensins. *Nat. Rev. Microbiol.*, 2, 727–738.

Leichtle, A., Lai, Y., Wollenberg, B., Wasserman, S.I., and Ryan, A.F. 2011. Innate signaling in otitis media: Pathogenesis and recovery. *Curr. Allergy Asthma Rep.*, 11, 78–84.

Lewis, P.A., and Shope, R.E. 1931. Swine influenza: II. A hemophilic bacillus from the respiratory tract of infected swine. *J. Exp. Med.*, 54, 361–371.

Lim, J.H., Ha, U.H., Woo, C.H., Xu, H., and Li, J.D. 2008. CYLD is a crucial negative regulator of innate immune response in *Escherichia coli* pneumonia. *Cell. Microbiol.*, 10, 2247–2256.

Lim, J.H., Jono, H., Koga, T., Woo, C.H., Ishinaga, H., Bourne, P., Xu, H., Ha, U.H., Xu, H., and Li, J.D. 2007. Tumor suppressor CYLD acts as a negative regulator for non-typeable *Haemophilus influenza*-induced inflammation in the middle ear and lung of mice. *PLoS One*, 2, e1032.

Loewen, P.C., Switala, J., and Triggs-Raine, B.L. 1985. Catalases HPI and HPII in *Escherichia coli* are induced independently. *Arch. Biochem. Biophys.*, 243, 144–149.

Lysenko, E.S., Gould, J., Bals, R., Wilson, J.M., and Weiser, J.N. 2000. Bacterial phosphorylcholine decreases susceptibility to the antimicrobial peptide LL-37/hCAP18 expressed in the upper respiratory tract. *Infect. Immun.*, 68, 1664–1671.

Maciver, I., and Hansen, E.J. 1996. Lack of expression of the global regulator OxyR in *Haemophilus influenzae* has a profound effect on growth phenotype. *Infect. Immun.*, 64, 4618–4629.

Maciver, I., Latimer, J.L., Liem, H.H., Muller-Eberhard, U., Hrkal, Z., and Hansen, E.J. 1996. Identification of an outer membrane protein involved in utilization of hemoglobin–haptoglobin complexes by nontypeable *Haemophilus influenzae*. *Infect. Immun.*, 64, 3703–3712.

MacNeil, J.R., Cohn, A.C., Farley, M., Mair, R., Baumbach, J., Bennett, N., Gershman, K., Harrison, L.H., Lynfield, R., Petit, S., Reingold, A., Schaffner, W., Thomas, A., Coronado, F., Zell, E.R., Mayer, L.W., Clark, T.A., and Messonnier, N.E. 2011. Current epidemiology and trends in invasive *Haemophilus influenzae* disease–United States, 1989–2008. *Clin. Infect. Dis.*, 53, 1230–1236.

Makela, P.H., Peltola, H., Kayhty, H., Jousimies, H., Pettay, O., Ruoslahti, E., Sivonen, A., and Renkonen, O.V. 1977. Polysaccharide vaccines of group A *Neisseria meningtitidis* and *Haemophilus influenzae* type b: A field trial in Finland. *J. Infect. Dis.*, 136(Suppl), S43–S50.

Male, C.J. 1979. Immunoglobulin A1 protease production by *Haemophilus influenzae* and *Streptococcus pneumoniae*. *Infect. Immun.*, 26, 254–261.

Martinez, A., and Kolter, R. 1997. Protection of DNA during oxidative stress by the nonspecific DNA-binding protein Dps. *J. Bacteriol.*, 179, 5188–5194.

Mason, K.M., Bruggeman, M.E., Munson, R.S., and Bakaletz, L.O. 2006. The non-typeable *Haemophilus influenzae* Sap transporter provides a mechanism of antimicrobial peptide resistance and SapD-dependent potassium acquisition. *Mol. Microbiol.*, 62, 1357–1372.

Mason, K.M., Munson Jr., R.S., and Bakaletz, L.O. 2005. A mutation in the *sap* operon attenuates survival of nontypeable *Haemophilus influenzae* in a chinchilla model of otitis media. *Infect. Immun.*, 73, 599–608.

Mason, K.M., Raffel, F.K., Ray, W.C., and Bakaletz, L.O. 2011. Heme utilization by nontypeable *Haemophilus influenzae* is essential and dependent on Sap transporter function. *J. Bacteriol.*, 193, 2527–2535.

McConnell, A., Tan, B., Scheifele, D., Halperin, S., Vaudry, W., Law, B., and Embree, J. 2007. Invasive infections caused by *Haemophilus influenzae* serotypes in twelve Canadian IMPACT centers, 1996–2001. *Pediatr. Infect. Dis. J.*, 26, 1025–1031.

Medini, D., Donati, C., Tettelin, H., Masignani, V., and Rappuoli, R. 2005. The microbial pan-genome. *Curr. Opin. Genet. Dev.*, 15, 589–594.

Millar, E.V., O'Brien, K.L., Watt, J.P., Lingappa, J., Pallipamu, R., Rosenstein, N., Hu, D., Reid, R., and Santosham, M. 2005. Epidemiology of invasive *Haemophilus influenzae* type A disease among Navajo and White Mountain Apache children, 1988–2003. *Clin. Infect. Dis.*, 40, 823–830.

Mohle-Boetani, J.C., Ajello, G., Breneman, E., Deaver, K.A., Harvey, C., Plikaytis, B.D., Farley, M.M., Stephens, D.S., and Wenger, J.D. 1993. Carriage of *Haemophilus influenzae* type b in children after widespread vaccination with conjugate *Haemophilus influenzae* type b vaccines. *Pediatr. Infect. Dis. J.*, 12, 589–593.

Monasta, L., Ronfani, L., Marchetti, F., Montico, M., Vecchi Brumatti, L., Bavcar, A., Grasso, D., Barbiero, C., and Tamburlini, G. 2012. Burden of disease caused by otitis media: Systematic review and global estimates. *PLoS One*, 7, e36226.

Moon, S.K., Lee, H.Y., Pan, H., Takeshita, T., Park, R., Cha, K., Andalibi, A., and Lim, D.J. 2006. Synergistic effect of interleukin 1 alpha on nontypeable *Haemophilus influenzae*-induced up-regulation of human beta-defensin 2 in middle ear epithelial cells. *BMC Infect. Dis.*, 6, 12.

Moon, S.K., Woo, J.I., Lee, H.Y., Park, R., Shimada, J., Pan, H., Gellibolian, R., and Lim, D.J. 2007. Toll-like receptor 2-dependent NF-kappaB activation is involved in nontypeable *Haemophilus influenzae*-induced monocyte chemotactic protein 1 up-regulation in the spiral ligament fibrocytes of the inner ear. *Infect. Immun.*, 75, 3361–3372.

Moreau-Marquis, S., O'Toole, G.A., and Stanton, B.A. 2009. Tobramycin and FDA-approved iron chelators eliminate *Pseudomonas aeruginosa* biofilms on cystic fibrosis cells. *Am. J. Respir. Cell. Mol. Biol.*, 41, 305–313.

Morton, D.J., Bakaletz, L.O., Jurcisek, J.A., VanWagoner, T.M., Seale, T.W., Whitby, P.W., and Stull, T.L. 2004a. Reduced severity of middle ear infection caused by nontypeable *Haemophilus influenzae* lacking the hemoglobin/hemoglobin-haptoglobin binding proteins (Hgp) in a chinchilla model of otitis media. *Microb. Pathog.*, 36, 25–33.

Morton, D.J., Madore, L.L., Smith, A., Vanwagoner, T.M., Seale, T.W., Whitby, P.W., and Stull, T.L. 2005. The heme-binding lipoprotein (HbpA) of *Haemophilus influenzae*: Role in heme utilization. *FEMS Microbiol. Lett.*, 253, 193–199.

Morton, D.J., Seale, T.W., Vanwagoner, T.M., Whitby, P.W., and Stull, T.L. 2009. The dppBCDF gene cluster of *Haemophilus influenzae*: Role in heme utilization. *BMC Res. Notes*, 2, 166.

Morton, D.J., Smith, A., Ren, Z., Madore, L.L., VanWagoner, T.M., Seale, T.W., Whitby, P.W., and Stull, T.L. 2004b. Identification of a haem-utilization protein (Hup) in *Haemophilus influenzae*. *Microbiology*, 150, 3923–3933.

Morton, D.J., Whitby, P.W., Jin, H., Ren, Z., and Stull, T.L. 1999. Effect of multiple mutations in the hemoglobin- and hemoglobin-haptoglobin-binding proteins, HgpA, HgpB, and HgpC, of *Haemophilus influenzae* type b. *Infect. Immun.*, 67, 2729–2739.

Munson Jr., R.S., Harrison, A., Gillaspy, A., Ray, W.C., Carson, M., Armbruster, D., Gipson, J., Gipson, M., Johnson, L., Lewis, L., Dyer, D.W., and Bakaletz, L.O. 2004. Partial analysis of the genomes of two nontypeable *Haemophilus influenzae* otitis media isolates. *Infect. Immun.*, 72, 3002–3010.

Murphy, T.F. 2003. Respiratory infections caused by non-typeable *Haemophilus influenzae*. *Curr. Opin. Infect. Dis.*, 16, 129–134.

Murphy, T.F., Lesse, A.J., Kirkham, C., Zhong, H., Sethi, S., and Munson Jr., R.S. 2011. A clonal group of nontypeable *Haemophilus influenzae* with two IgA proteases is adapted to infection in chronic obstructive pulmonary disease. *PLoS One*, 6, e25923.

Murphy, T.V., Pastor, P., Medley, F., Osterholm, M.T., and Granoff, D.M. 1993a. Decreased *Haemophilus* colonization in children vaccinated with *Haemophilus influenzae* type b conjugate vaccine. *J. Pediatr.*, 122, 517–523.

Murphy, T.V., White, K.E., Pastor, P., Gabriel, L., Medley, F., Granoff, D.M., and Osterholm, M.T. 1993b. Declining incidence of *Haemophilus influenzae* type b disease since introduction of vaccination. *JAMA*, 269, 246–248.

Naylor, E.J., Bakstad, D., Biffen, M., Thong, B., Calverley, P., Scott, S., Hart, C.A., Moots, R.J., and Edwards, S.W. 2007. *Haemophilus influenzae* induces neutrophil necrosis: A role in chronic obstructive pulmonary disease? *Am. J. Respir. Cell Mol. Biol.*, 37, 135–143.

Nazir, J., Urban, C., Mariano, N., Burns, J., Tommasulo, B., Rosenberg, C., Segal-Maurer, S., and Rahal, J.J. 2004. Quinolone-resistant *Haemophilus influenzae* in a long-term care facility: Clinical and molecular epidemiology. *Clin. Infect. Dis.*, 38, 1564–1569.

Niimura, Y., Poole, L.B., and Massey, V. 1995. *Amphibacillus xylanus* NADH oxidase and *Salmonella typhimurium* alkyl-hydroperoxide reductase flavoprotein components show extremely high scavenging activity for both alkyl hydroperoxide and hydrogen peroxide in the presence of *S. typhimurium* alkyl-hydroperoxide reductase 22-kDa protein component. *J. Biol. Chem.*, 270, 25645–25650.

Nistico, L., Kreft, R., Gieseke, A., Coticchia, J.M., Burrows, A., Khampang, P., Liu, Y., Kerschner, J.E., Post, J.C., Lonergan, S., Sampath, R., Hu, F.Z., Ehrlich, G.D., Stoodley, P., and Hall-Stoodley, L. 2011. Adenoid reservoir for pathogenic biofilm bacteria. *J. Clin. Microbiol.*, 49, 1411–1420.

Nix, E.B., Hawdon, N., Gravelle, S., Biman, B., Brigden, M., Malik, S., McCready, W., Ferroni, G., and Ulanova, M. 2012. Risk of invasive *Haemophilus influenzae* type b (Hib) disease in adults with secondary immunodeficiency in the post-Hib vaccine era. *Clin. Vaccine Immunol.*, 19, 766–771.

Nizet, V., Ohtake, T., Lauth, X., Trowbridge, J., Rudisill, J., Dorschner, R.A., Pestonjamasp, V., Piraino, J., Huttner, K., and Gallo, R.L. 2001. Innate antimicrobial peptide protects the skin from invasive bacterial infection. *Nature*, 414, 454–457.

Norskov-Lauritsen, N., Overballe, M.D., and Kilian, M. 2009. Delineation of the species *Haemophilus influenzae* by phenotype, multilocus sequence phylogeny, and detection of marker genes. *J. Bacteriol.*, 191, 822–831.

Novotny, L.A., Clements, J.D., and Bakaletz, L.O. 2011. Transcutaneous immunization as preventative and therapeutic regimens to protect against experimental otitis media due to nontypeable *Haemophilus influenzae*. *Mucosal Immunol.*, 4, 456–467.

Novotny, L.A., Clements, J.D., and Bakaletz, L.O. 2013. Kinetic analysis and evaluation of the mechanisms involved in the resolution of experimental nontypeable *Haemophilus influenzae*-induced otitis media after transcutaneous immunization. *Vaccine*, 31, 3417–3426.

Oh, S., Woo, J.I., Lim, D.J., and Moon, S.K. 2012. ERK2-dependent activation of c-Jun is required for nontypeable *Haemophilus influenzae*-induced CXCL2 upregulation in inner ear fibrocytes. *J. Immunol.*, 188, 3496–3505.

Olitsky, P.K., and Gates, F.L. 1921. Experimental studies of the nasopharyngeal secretions from influenza patients: II. Filterability and resistance to glycerol. *J. Exp. Med.*, 33, 361–372.

Ong, S.T., Ho, J.Z., Ho, B., and Ding, J.L. 2006. Iron-withholding strategy in innate immunity. *Immunobiology*, 211, 295–314.

Panek, H., and O'Brian, M.R. 2002. A whole genome view of prokaryotic haem biosynthesis. *Microbiology*, 148, 2273–2282.

Park, S., You, X., and Imlay, J.A. 2005. Substantial DNA damage from submicromolar intracellular hydrogen peroxide detected in Hpx- mutants of *Escherichia coli*. *Proc. Natl. Acad. Sci. USA*, 102, 9317–9322.

Pauwels, F., Vergauwen, B., and Van Beeumen, J.J. 2004. Physiological characterization of *Haemophilus influenzae* Rd deficient in its glutathione-dependent peroxidase PGdx. *J. Biol. Chem.*, 279, 12163–12170.

Pauwels, F., Vergauwen, B., Vanrobaeys, F., Devreese, B., and Van Beeumen, J.J. 2003. Purification and characterization of a chimeric enzyme from *Haemophilus influenzae* Rd that exhibits glutathione-dependent peroxidase activity. *J. Biol. Chem.*, 278, 16658–16666.

Peltola, H. 2000. Worldwide *Haemophilus influenzae* type b disease at the beginning of the 21st century: Global analysis of the disease burden 25 years after the use of the polysaccharide vaccine and a decade after the advent of conjugates. *Clin. Microbiol. Rev.*, 13, 302–317.

Peltola, H., Kayhty, H., Sivonen, A., and Makela, H. 1977. *Haemophilus influenzae* type b capsular polysaccharide vaccine in children: A double-blind field study of 100,000 vaccinees 3 months to 5 years of age in Finland. *Pediatrics*, 60, 730–737.

Peltola, H., Kilpi, T., and Anttila, M. 1992. Rapid disappearance of *Haemophilus influenzae* type b meningitis after routine childhood immunisation with conjugate vaccines. *Lancet*, 340, 592–594.

Pericone, C.D., Overweg, K., Hermans, P.W., and Weiser, J.N. 2000. Inhibitory and bactericidal effects of hydrogen peroxide production by *Streptococcus pneumoniae* on other inhabitants of the upper respiratory tract. *Infect. Immun.*, 68, 3990–3997.

Peschel, A. 2002. How do bacteria resist human antimicrobial peptides? *Trends Microbiol.*, 10, 179–186.

Pittman, M. 1931. Variation and type specificity in the bacterial species *Hemophilus influenzae*. *J. Exp. Med.*, 53, 471–492.

Pittman, M. 1933. The action of type-specific *Hemophilus influenzae* antiserum. *J. Exp. Med.*, 58, 683–706.

Post, J.C., Hiller, N.L., Nistico, L., Stoodley, P., and Ehrlich, G.D. 2007. The role of biofilms in otolaryngologic infections: Update 2007. *Curr. Opin. Otolaryngol. Head Neck Surg.*, 15, 347–351.

Power, P.M., Bentley, S.D., Parkhill, J., Moxon, E.R., and Hood, D.W. 2012. Investigations into genome diversity of *Haemophilus influenzae* using whole genome sequencing of clinical isolates and laboratory transformants. *BMC Microbiol.*, 12, 273.

Reid, D.W., O'May, C., Roddam, L.F., and Lamont, I.L. 2009. Chelated iron as an anti-*Pseudomonas aeruginosa* biofilm therapeutic strategy. *J. Appl. Microbiol.*, 106, 1058–1058.

Reinert, P., Liwartowski, A., Dabernat, H., Guyot, C., Boucher, J., and Carrere, C. 1993. Epidemiology of *Haemophilus influenzae* type b disease in France. *Vaccine*, 11(Suppl 1), S38–S42.

Ren, Z., Jin, H., Morton, D.J., and Stull, T.L. 1998. hgpB, a gene encoding a second *Haemophilus influenzae* hemoglobin- and hemoglobin-haptoglobin-binding protein. *Infect. Immun.*, 66, 4733–4741.

Resman, F., Ristovski, M., Ahl, J., Forsgren, A., Gilsdorf, J.R., Jasir, A., Kaijser, B., Kronvall, G., and Riesbeck, K. 2011. Invasive disease caused by *Haemophilus influenzae* in Sweden 1997–2009; evidence of increasing incidence and clinical burden of non-type b strains. *Clin. Microbiol. Infect.*, 17, 1638–1645.

Robbins, J.B., Parke Jr., J.C., Schneerson, R., and Whisnant, J.K. 1973. Quantitative measurement of "natural" and immunization-induced *Haemophilus influenzae* type b capsular polysaccharide antibodies. *Pediatr. Res.*, 7, 103–110.

Roman, F., Canton, R., Perez-Vazquez, M., Baquero, F., and Campos, J. 2004. Dynamics of long-term colonization of respiratory tract by *Haemophilus influenzae* in cystic fibrosis patients shows a marked increase in hypermutable strains. *J. Clin. Microbiol.*, 42, 1450–1459.

Rotta Detto Loria, J., Rohmann, K., Droemann, D., Kujath, P., Rupp, J., Goldmann, T., and Dalhoff, K. 2013. *Haemophilus influenzae* infection upregulates the NLRP3 inflammasome and leads to caspase-1-dependent secretion of interleukin-1beta – A possible pathway of exacerbations in COPD. *PLoS One*, 8, e66818.

Roy, P.H., and Smith, H.O. 1973a. DNA methylases of *Hemophilus influenzae* Rd. I. Purification and properties. *J. Mol. Biol.*, 81, 427–444.

Roy, P.H., and Smith, H.O. 1973b. DNA methylases of *Hemophilus influenzae* Rd. II. Partial recognition site base sequences. *J. Mol. Biol.*, 81, 445–459.

Rubach, M.P., Bender, J.M., Mottice, S., Hanson, K., Weng, H.Y., Korgenski, K., Daly, J.A., and Pavia, A.T. 2011. Increasing incidence of invasive *Haemophilus influenzae* disease in adults, Utah, USA. *Emerg. Infect. Dis.*, 17, 1645–1650.

Salzman, N.H., Ghosh, D., Huttner, K.M., Paterson, Y., and Bevins, C.L. 2003. Protection against enteric salmonellosis in transgenic mice expressing a human intestinal defensin. *Nature*, 422, 522–526.

Sandstedt, S.A., Marrs, C.F., Patel, M., Hirasawa, H., Zhang, L., Davis, G.S., and Gilsdorf, J.R. 2010. Prevalence of *Haemophilus influenzae* type b genetic islands among clinical and commensal *H. influenzae* and *H. haemolyticus* isolates. *J. Clin. Microbiol.*, 48, 2565–2568.

Schneerson, R., Rodrigues, L.P., Parke Jr., J.C., and Robbins, J.B. 1971. Immunity to disease caused by *Hemophilus influenzae* type b. II. Specificity and some biologic characteristics of "natural," infection-acquired, and immunization-induced antibodies to the capsular polysaccharide of *Hemophilus influenzae* type b. *J. Immunol.*, 107, 1081–1089.

Seaver, L.C., and Imlay, J.A. 2001a. Alkyl hydroperoxide reductase is the primary scavenger of endogenous hydrogen peroxide in *Escherichia coli*. *J. Bacteriol.*, 183, 7173–7181.

Seaver, L.C., and Imlay, J.A. 2001b. Hydrogen peroxide fluxes and compartmentalization inside growing *Escherichia coli*. *J. Bacteriol.*, 183, 7182–7189.

Sell, S.H., Merrill, R.E., Doyne, E.O., and Zimsky Jr., E.P. 1972. Long-term sequelae of *Hemophilus influenzae* meningitis. *Pediatrics*, 49, 206–211.

Selsted, M.E., and Ouellette, A.J. 2005. Mammalian defensins in the antimicrobial immune response. *Nat. Immunol.*, 6, 551–557.

Sethi, S., and Murphy, T.F. 2001. Bacterial infection in chronic obstructive pulmonary disease in 2000: A state-of-the-art review. *Clin. Microbiol. Rev.*, 14, 336–363.

Shafer, W.M., Qu, X., Waring, A.J., and Lehrer, R.I. 1998. Modulation of *Neisseria gonorrhoeae* susceptibility to vertebrate antibacterial peptides due to a member of the resistance/nodulation/division efflux pump family. *Proc. Natl. Acad. Sci. USA*, 95, 1829–1833.

Sharpe, S.W., Kuehn, M.J., and Mason, K.M. 2011. Elicitation of epithelial-derived immune effectors by outer membrane vesicles of nontypeable *Haemophilus influenzae*. *Infect. Immun.*, 79, 4361–4369.

Shelton, C.L., Raffel, F.K., Beatty, W.L., Johnson, S.M., and Mason, K.M. 2011. Sap transporter mediated import and subsequent degradation of antimicrobial peptides in *Haemophilus*. *PLoS Pathog.*, 7, e1002360.

Shen, K., Antalis, P., Gladitz, J., Sayeed, S., Ahmed, A., Yu, S., Hayes, J., Johnson, S., Dice, B., Dopico, R., Keefe, R., Janto, B., Chong, W., Goodwin, J., Wadowsky, R.M., Erdos, G., Post, J.C., Ehrlich, G.D., and Hu, F.Z. 2005. Identification, distribution, and expression of novel genes in 10 clinical isolates of nontypeable *Haemophilus influenzae*. *Infect. Immun.*, 73, 3479–3491.

Shope, R.E. 1931. Swine influenza: III. Filtration experiments and etiology. *J. Exp. Med.*, 54, 373–385.

Shuel, M., Hoang, L., Law, D.K., and Tsang, R. 2011. Invasive *Haemophilus influenzae* in British Columbia: Non-Hib and non-typeable strains causing disease in children and adults. *Int. J. Infect. Dis.*, 15, e167–e173.

Sieprawska-Lupa, M., Mydel, P., Krawczyk, K., Wojcik, K., Puklo, M., Lupa, B., Suder, P., Silberring, J., Reed, M., Pohl, J., Shafer, W., McAleese, F., Foster, T., Travis, J., and Potempa, J. 2004. Degradation of human antimicrobial peptide LL-37 by *Staphylococcus aureus*-derived proteinases. *Antimicrob. Agents Chemother*, 48, 4673–4679.

Sill, M.L., Law, D.K., Zhou, J., Skinner, S., Wylie, J., and Tsang, R.S. 2007. Population genetics and antibiotic susceptibility of invasive *Haemophilus influenzae* in Manitoba, Canada, from 2000 to 2006. *FEMS Immunol. Med. Microbiol.*, 51, 270–276.

Slack, M.P., Azzopardi, H.J., Hargreaves, R.M., and Ramsay, M.E. 1998. Enhanced surveillance of invasive *Haemophilus influenzae* disease in England, 1990 to 1996: Impact of conjugate vaccines. *Pediatr. Infect. Dis. J.*, 17, S204–S207.

Slinger, R., Chan, F., Ferris, W., Yeung, S.W., St Denis, M., Gaboury, I., and Aaron, S.D. 2006. Multiple combination antibiotic susceptibility testing of nontypeable *Haemophilus influenzae* biofilms. *Diagn. Microbiol. Infect. Dis.*, 56, 247–253.

Smith, D.H., Peter, G., Ingram, D.L., Harding, A.L., and Anderson, P. 1973. Responses of children immunized with the capsular polysaccharide of *Hemophilus influenzae*, type b. *Pediatrics*, 52, 637–644.

Smith, H.O., and Wilcox, K.W. 1970. A restriction enzyme from *Hemophilus influenzae*. I. Purification and general properties. *J. Mol. Biol.*, 51, 379–391.

Smith, W., Andrewes, C.H., and Laidlaw, P.P. 1933. A virus obtained from influenza patients. *Lancet*, 222, 59–112.

Spinola, S.M., Peacock, J., Denny, F.W., Smith, D.L., and Cannon, J.G. 1986. Epidemiology of colonization by nontypable *Haemophilus influenzae* in children: A longitudinal study. *J. Infect. Dis.*, 154, 100–109.

St Geme 3rd, J.W. 2000. The pathogenesis of nontypable *Haemophilus influenzae* otitis media. *Vaccine*, 19(Suppl 1), S41–S50.

Starner, T.D., Swords, W.E., Apicella, M.A., and McCray Jr., P.B. 2002. Susceptibility of nontypeable *Haemophilus influenzae* to human beta-defensins is influenced by lipooligosaccharide acylation. *Infect. Immun.*, 70, 5287–5289.

Stojiljkovic, I., and Perkins-Balding, D. 2002. Processing of heme and heme-containing proteins by bacteria. *DNA Cell. Biol.*, 21, 281–295.

Stool, S.E., Berg, A.O., Berman, S., Carney, C.J., Cooley, J.R., Culpepper, L., Eavey, R.D., Feagans, L.V., Finitzo, T., Friedman, E., Goertz, J.A., Goldstein, A.J., Grundfast, K.M., Long, D.G., Macconi, L.L., Melton, L.B., Roberts, J.E., Sherrod, J.L., and Sisk, J.E. 1994. Managing otitis-media with effusion in young-children. *Pediatrics*, 94, 766–773.

Sultana, N.K., Saha, S.K., Al-Emran, H.M., Modak, J.K., Sharker, M.A., El-Arifeen, S., Cohen, A.L., Baqui, A.H., and Luby, S.P. 2013. Impact of introduction of the *Haemophilus influenzae* Type b conjugate vaccine into childhood immunization on meningitis in Bangladeshi infants. *J. Pediatr.*, 163, S73–S78.

Sutton, A., Schneerson, R., Kendall-Morris, S., and Robbins, J.B. 1982. Differential complement resistance mediates virulence of *Haemophilus influenzae* type b. *Infect. Immun.*, 35, 95–104.

Szelestey, B.R., Heimlich, D.R., Raffel, F.K., Justice, S.S., and Mason, K.M. 2013. *Haemophilus* responses to nutritional immunity: Epigenetic and morphological contribution to biofilm architecture, invasion, persistence and disease severity. *PLoS Pathog.*, 9, e1003709.

Takala, A.K., Eskola, J., Leinonen, M., Kayhty, H., Nissinen, A., Pekkanen, E., and Makela, P.H. 1991. Reduction of oropharyngeal carriage of *Haemophilus influenzae* type b (Hib) in children immunized with an Hib conjugate vaccine. *J. Infect. Dis.*, 164, 982–986.

Takala, A.K., Santosham, M., Almeido-Hill, J., Wolff, M., Newcomer, W., Reid, R., Kayhty, H., Esko, E., and Makela, P.H. 1993. Vaccination with *Haemophilus influenzae* type b meningococcal protein conjugate vaccine reduces oropharyngeal carriage of *Haemophilus influenzae* type b among American Indian children. *Pediatr. Infect. Dis. J.*, 12, 593–599.

Tognotti, E. 2003. Scientific triumphalism and learning from facts: Bacteriology and the "Spanish flu" challenge of 1918. *Soc. Hist. Med.*, 16, 97–110.

Tsang, R.S., Sill, M.L., Skinner, S.J., Law, D.K., Zhou, J., and Wylie, J. 2007. Characterization of invasive *Haemophilus influenzae* disease in Manitoba, Canada, 2000–2006: Invasive disease due to non-type b strains. *Clin. Infect. Dis.*, 44, 1611–1614.

Unal, C.M., Schaar, V., and Riesbeck, K. 2010. Bacterial outer membrane vesicles in disease and preventive medicine. *Semin. Immunopathol.*, 33, 395–408.

Urwin, G., Krohn, J.A., Deaver-Robinson, K., Wenger, J.D., and Farley, M.M. 1996. Invasive disease due to *Haemophilus influenzae* serotype f: Clinical and epidemiologic characteristics in the *H. influenzae* serotype b vaccine era. The *Haemophilus influenzae* Study Group. *Clin. Infect. Dis.*, 22, 1069–1076.

Vergauwen, B., Elegheert, J., Dansercoer, A., Devreese, B., and Savvides, S.N. 2010. Glutathione import in *Haemophilus influenzae* Rd is primed by the periplasmic heme-binding protein HbpA. *Proc. Natl. Acad. Sci. USA*, 107, 13270–13275.

Vergauwen, B., Herbert, M., and Van Beeumen, J.J. 2006. Hydrogen peroxide scavenging is not a virulence determinant in the pathogenesis of *Haemophilus influenzae* type b strain Eagan. *BMC Microbiol.*, 6, 3.

Vergauwen, B., Pauwels, F., Jacquemotte, F., Meyer, T.E., Cusanovich, M.A., Bartsch, R.G., and Van Beeumen, J.J. 2001. Characterization of glutathione amide reductase from *Chromatium gracile*. Identification of a novel thiol peroxidase (Prx/Grx) fueled by glutathione amide redox cycling. *J. Biol. Chem.*, 276, 20890–20897.

Vergauwen, B., Pauwels, F., and Van Beeumen, J.J. 2003. Glutathione and catalase provide overlapping defenses for protection against respiration-generated hydrogen peroxide in *Haemophilus influenzae*. *J. Bacteriol.*, 185, 5555–5562.

Vijayakumar, S.R., Kirchhof, M.G., Patten, C.L., and Schellhorn, H.E. 2004. RpoS-regulated genes of *Escherichia coli* identified by random *lacZ* fusion mutagenesis. *J. Bacteriol.*, 186, 8499–8507.

Vogel, A.R., Szelestey, B.R. Raffel, F.K., Sharpe, S.W., Gearinger, R.L., Justice, S.S., and Mason, K.M. 2012. SapF-mediated heme-iron utilization enhances persistence and coordinates biofilm architecture of *Haemophilus*. *Front. Cell. Infect. Microbiol.*, 2, 42.

Waggoner-Fountain, L.A., Hendley, J.O., Cody, E.J., Perriello, V.A., and Donowitz, L.G. 1995. The emergence of *Haemophilus influenzae* types e and f as significant pathogens. *Clin. Infect. Dis.*, 21, 1322–1324.

Wakeman, C.A., and Skaar, E.P. 2012. Metalloregulation of Gram-positive pathogen physiology. *Curr. Opin. Microbiol.*, 15, 169–174.

Wang, D., Wang, Y., and Liu, Y.N. 2009. Activity of ciprofloxacin and azithromycin on biofilms produced in vitro by *Haemophilus influenzae*. *Chin. Med. J.*, 122, 1305–1310.

Wang, W.Y., Lim, J.H., and Li, J.D. 2012. Synergistic and feedback signaling mechanisms in the regulation of inflammation in respiratory infections. *Cell. Mol. Immunol.*, 9, 131–135.

Watanabe, H., Hoshino, K., Sugita, R., Asoh, N., Watanabe, K., Oishi, K., and Nagatake, T. 2004a. Possible high rate of transmission of nontypeable *Haemophilus influenzae*, including beta-lactamase-negative ampicillin-resistant strains, between children and their parents. *J. Clin. Microbiol.*, 42, 362–365.

Watanabe, T., Jono, H., Han, J., Lim, D.J., and Li, J.D. 2004b. Synergistic activation of NF-kappaB by nontypeable *Haemophilus influenzae* and tumor necrosis factor alpha. *Proc. Natl. Acad. Sci. USA*, 101, 3563–3568.

Watt, J.P., Wolfson, L.J., O'Brien, K.L., Henkle, E., Deloria-Knoll, M., McCall, N., Lee, E., Levine, O.S., Hajjeh, R., Mulholland, K., and Cherian, T. 2009. Burden of disease caused by *Haemophilus influenzae* type b in children younger than 5 years: Global estimates. *Lancet*, 374, 903–911.

Webster, P., Wu, S., Gomez, G., Apicella, M., Plaut, A.G., and St Geme 3rd, J.W. 2006. Distribution of bacterial proteins in biofilms formed by non-typeable *Haemophilus influenzae*. *J. Histochem. Cytochem.*, 54, 829–842.

Weimer, K.E., Armbruster, C.E., Juneau, R.A., Hong, W., Pang, B., and Swords, W.E. 2010. Coinfection with *Haemophilus influenzae* promotes pneumococcal biofilm formation during experimental otitis media and impedes the progression of pneumococcal disease. *J. Infect. Dis.*, 202, 1068–1075.

Weimer, K.E., Juneau, R.A., Murrah, K.A., Pang, B., Armbruster, C.E., Richardson, S.H., and Swords, W.E. 2011. Divergent mechanisms for passive pneumococcal resistance to beta-lactam antibiotics in the presence of *Haemophilus influenzae*. *J. Infect. Dis.*, 203, 549–555.

Weller, P.F., Smith, A.L., Smith, D.H., and Anderson, P. 1978. Role of immunity in the clearance of bacteremia due to *Haemophilus influenzae*. *J. Infect. Dis.*, 138, 427–436.

Wenger, J.D. 1998. Epidemiology of *Haemophilus influenzae* type b disease and impact of *Haemophilus influenzae* type b conjugate vaccines in the United States and Canada. *Pediatr. Infect. Dis. J.*, 17, S132–S136.

Wenger, J.D., Hightower, A.W., Facklam, R.R., Gaventa, S., and Broome, C.V. 1990. Bacterial meningitis in the United States, 1986: Report of a multistate surveillance study. *J. Infect. Dis.*, 162, 1316–1323.

White, D.C., and Granick, S. 1963. Hemin biosynthesis in *Haemophilus*. *J. Bacteriol.*, 85, 842–850.

Wilcox, K.W., and Smith, H.O. 1975. Isolation and characterization of mutants of *Haemophilus influenzae* deficient in an adenosine 5′-triphosphate-dependent deoxyribonuclease activity. *J. Bacteriol.*, 122, 443–453.

Winslow, C.E., Broadhurst, J., Buchanan, R.E., Krumwiede, C., Rogers, L.A., and Smith, G.H. 1920. The families and genera of the bacteria: Final report of the Committee of the Society of American Bacteriologists on characterization and classification of bacterial types. *J. Bacteriol.*, 5, 191–229.

Wollstein, M. 1911. Serum treatment of influenzal meningitis. *J. Exp. Med.*, 14, 73–82.

Wong, S.M., Alugupalli, K.R., Ram, S., and Akerley, B.J. 2007. The ArcA regulon and oxidative stress resistance in *Haemophilus influenzae*. *Mol. Microbiol.*, 64, 1375–1390.

Wong, S.M., St Michael, F., Cox, A., Ram, S., and Akerley, B.J. 2011. ArcA-regulated glycosyltransferase Lic2B promotes complement evasion and pathogenesis of nontypeable *Haemophilus influenzae*. *Infect. Immun.*, 79, 1971–1983.

Woo, J.I., Pan, H., Oh, S., Lim, D.J., and Moon, S.K. 2010. Spiral ligament fibrocyte-derived MCP-1/CCL2 contributes to inner ear inflammation secondary to nontypeable *H. influenzae*-induced otitis media. *BMC Infect. Dis.*, 10, 314.

Wu, Y., and Outten, F.W. 2009. IscR controls iron-dependent biofilm formation in *Escherichia coli* by regulating type I fimbria expression. *J. Bacteriol.*, 191, 1248–1257.

Zhang, L., Xie, J., Patel, M., Bakhtyar, A., Ehrlich, G.D., Ahmed, A., Earl, J., Marrs, C.F., Clemans, D., Murphy, T.F., and Gilsdorf, J.R. 2012. Nontypeable *Haemophilus influenzae* genetic islands associated with chronic pulmonary infection. *PLoS One*, 7, e44730.

Zhou, F., Bisgard. K.M., Yusuf, H.R., Deuson, R.R., Bath, S.K., and Murphy, T.V. 2002. Impact of universal *Haemophilus influenzae* type b vaccination starting at 2 months of age in the United States: An economic analysis. *Pediatrics*, 110, 653–661.

Zwahlen, A., Kroll, J.S., Rubin, L.G., and Moxon, E.R. 1989. The molecular basis of pathogenicity in *Haemophilus influenzae*: Comparative virulence of genetically-related capsular transformants and correlation with changes at the capsulation locus cap. *Microb. Pathog.*, 7, 225–235.

Chapter 28

Campylobacter jejuni: Molecular Mechanisms and Animal Models of Colonization and Disease

Christopher R. Gourley and Michael E. Konkel

School of Molecular Biosciences, College of Veterinary Medicine, Washington State University, Pullman, WA, USA

28.1 Introduction

Campylobacter jejuni is a major gastrointestinal pathogen throughout the world. *Campylobacter* infects and sickens millions of people per year and causes significant mortality in the children of developing nations. Human infection results from ingestion of *C. jejuni*–contaminated material and causes diarrheal disease. Infection with some *C. jejuni* strains increases the risk of developing the flaccid neuronal paralysis Guillain–Barré syndrome (GBS). Recent advances in genetic sequencing tools and mutant generation, coupled with investigation using *in vitro* models, has greatly advanced our knowledge of the molecular mechanisms of *Campylobacter* colonization and disease. Besides *in vitro* models, *in vivo* models of *C. jejuni* colonization and disease are required for our increased understanding. Birds are commensal hosts for *C. jejuni*, and chickens function as informative *in vivo* models for understanding avian colonization by *Campylobacter*. While many other types of animals become transiently colonized with *C. jejuni*, additional treatment of those animals is often necessary before they will exhibit human-like disease symptoms. Animals currently used to model *C. jejuni*–mediated disease include mice, rabbits, ferrets, piglets, and monkeys. In this chapter, we discuss general characteristics of *C. jejuni*, describe colonization and virulence factors, outline a model for *C. jejuni* invasion of host cells, and close with a discussion of *C. jejuni* colonization and disease models focusing on the different *in vivo* hosts currently utilized.

28.2 Pathogen

C. jejuni is a major cause of human gastroenteritis across the globe. First identified in 1972, specific isolation from humans and animals was not achieved until 1979 (Butzler and Skirrow, 1979; Dekeyser et al., 1972). *C. jejuni* endemically colonizes commercial chicken flocks. Human exposure to infected or cross-contaminated food and water is an ongoing problem even in the developed world.

C. jejuni is a Gram-negative, curved rod exhibiting motility through polar flagella. These microaerophiles favor growth in low oxygen (3–5%) environments, including animal digestive tracts. A single circular chromosome of ~1.64 million base pairs (~31% G or C) and ~1700 open reading frames comprises the *C. jejuni* genome (Parkhill et al., 2000). Additional genetic elements, including plasmids and bacteriophages, are present in some isolates (Fouts et al., 2005). Pervasive genetic variation between strains is due to hypervariable regions within the genome and promiscuous uptake and exchange of genetic material through natural competence involving the Type II secretion system (Avrain et al., 2004; de Boer et al., 2002; Parkhill et al., 2000; Wilson et al., 2003). Regions of hypervariability often encode genes of external structural components, such as lipooligosaccharide (LOS), capsule, and the flagellum (Duong and Konkel, 2009; Parkhill et al., 2000). The surface exposure of these molecules may enable recognition by host innate immune cells, however, the genetic variability demonstrated by *C. jejuni* isolates may also partly explain variability in host colonization and disease severity (Huizinga et al., 2012; Maue et al., 2013). Similarities between *C. jejuni* LOS and human neuronal gangliosides are believed to result in a host immune response that can mistakenly target nerves and lead to GBS (Fujimoto and Amako, 1990; Komagamine and Yuki, 2006; Yuki and Odaka, 2005). GBS occurs at a frequency of approximately 0.1% in *C. jejuni*–infected individuals, but as many as 40% of individuals with GBS were first infected with *Campylobacter* (Allos, 1997; Buzby et al., 1997). The cost associated with the care and treatment of individuals with acute *C. jejuni* infections and GBS in the United States is estimated to be $1.2 billion per year (Frenzen, 2008).

28.3 Clinical presentation and treatment

C. jejuni and other *Campylobacter* species can be found throughout the environment and in many different animal hosts. Humans become infected with *C. jejuni* by eating contaminated meat, drinking contaminated water or milk, or consuming food cross-contaminated during preparation. The clinical symptoms for *C. jejuni*–mediated disease include mild to severe bloody diarrhea with leukocytes, fever, nausea, and abdominal cramps. Symptoms generally abate within 5 days and are confined to the individual (Allos, 2001; Blaser, 1990). Diagnosis is made with direct or enriched culturing of stool samples or rectal swabs on selective media containing blood under microaerobic conditions at 42°C. *C. jejuni*–infected individuals may be treated with erythromycin or ciprofloxacin, but this is rarely required (Allos, 2001; Blaser, 1990). Proper palliative care includes fluid replacement.

28.4 Epidemiology

The incidence of *C. jejuni*–mediated enteritis in the United States is sporadic, but can be cluster-associated when groups of people are infected from the same source. According to the Centers for Disease Control and Prevention, *Campylobacter* species are a significant cause of foodborne illness in the United States with over 845,000 cases per year and an estimated 76 deaths (Scallan et al., 2011). In 2012, the incidence of *C. jejuni*–mediated disease was 14.3 cases/100,000 persons in the United States, the greatest incidence since 2000 (http://www.cdc.gov/foodnet /data/trends/tables/2012/table2a-b.html#table-2b). In children under 5 years of age, the incidence of disease is estimated to be over a hundred times greater in developing countries than in developed countries (Coker et al., 2002). Infants and individuals (15–29

years) are the most often reported to be infected, and groups with compromised immune systems may be at increased risk of infection (Allos and Blaser, 2009; Tauxe, 1992).

28.5 Prevention and control

Controlling the spread of *C. jejuni* to humans depends on mitigating the presence of *C. jejuni* in food sources, such as poultry, and preventing the direct consumption of already contaminated food. Poultry are the most prominent reservoir for *C. jejuni*. Most commercially reared poultry are colonized by *C. jejuni* at 2–3 weeks of age, and some estimate close to 90% of domestic chicken carcasses are contaminated with *C. jejuni* at the time of sale (Doyle and Jones, 1992; Stern, 1992). Decreasing the number and bacterial load of infected carcasses may lead to positive downstream effects, including decreased infections from both direct meat consumption and cross-contamination (FAO/WHO, 2009). Food prepared by cooking to proper temperature and handled in a hygienic environment, free from sources of cross-contamination, carries little direct risk of infection. This is possible through enforced regulation and education in the developed world but more difficult in underdeveloped regions of the world where infrastructure is lacking. While antibiotic treatment of poultry may have initially seemed a promising strategy, increased resistance by *Campylobacter* has been reported to multiple classes of antibiotics, including the cephalosporins and fluoroquinolones (Ruiz-Palacios, 2007). The Food and Drug Administration banned the use of enrofloxacin (Baytril) in poultry in 2005 due to an increase in the number of ciprofloxacin-resistant isolates of *C. jejuni* (http://www.fda.gov/NewsEvents/ Newsroom/PressAnnouncements/2005/ucm108467 .htm). New strategies, including targeting transmission within commercial flocks, vaccines, probiotics, and other therapeutics, should be explored.

28.6 Colonization and virulence factors

C. jejuni synthesize unique molecules and employ specific strategies for adherence to cells, invasion of cells, colonization in the commensal avian host, and pathogenesis. Discussed in sections 28.6 through 28.8 are a selection of these colonization and virulence factors.

28.6.1 Capsule and lipooligosaccharide

As in many bacteria, *C. jejuni* produces varied surface-exposed sugar molecules, including capsular polysaccharides (CPSs) and LOS, for survival and virulence functions.

Although once thought to be high-molecular-weight lipopolysaccharides, it is now clear from genetic analysis, electron microscopic characterization, and biological analysis that many *C. jejuni* isolates produce CPSs (Bacon et al., 2001; Karlyshev et al., 1999, 2000, 2001; Karlyshev and Wren, 2001). A capsule is produced by the organization of CPSs into a membrane-linked, extracellular layer surrounding the bacterial cell (Roberts, 1996; Whitfield and Valvano, 1993). *C. jejuni* capsular mutants have shown decreased adherence and invasion *in vitro* (INT 407 cells), decreased *in vivo* fitness effects (in chickens (slightly decreased cecal and colon colonization), in ferrets (decreased diarrhea)), and sensitivity to hyperosmotic conditions (in culture) (Bachtiar et al., 2007; Bacon et al., 2001; Cameron et al., 2012). Emerging evidence supports involvement of *C. jejuni* capsular expression in protection from innate and acquired immune responses (e.g., complement-mediated lysis) (Keo et al., 2011; Maue et al., 2013). Continued characterization of differences between capsule expressing and non-expressing (and mutated) isolates will more completely define CPS function.

C. jejuni also produces externally exposed LOS with strain-dependent structural variation (Gilbert et al., 2002; Parker et al., 2008; Semchenko et al., 2010). Similar to the flagellar and CPS genes, many LOS genes are encoded in areas with homopolymeric DNA tracts that can induce phase variation through slipped-strand mispairing (Gilbert et al., 2002; Guerry et al., 2002; Karlyshev et al., 2002, 2005; Linton et al., 2000). As mentioned earlier, similarity in structure between *C. jejuni* LOS and human neuronal gangliosides can lead to the autoimmune disorder GBS (Fujimoto and Amako, 1990; Komagamine and Yuki, 2006; Yuki and Odaka, 2005). Ongoing research has revealed that the *C. jejuni* LOS is involved in immune cell recognition (Toll-like receptor 4 (TLR4) and sialoadhesin), bacteriophage resistance, and resistance to cationic antimicrobials (defensins) (Cullen et al., 2013; de Zoete et al., 2010; Heikema et al., 2010; Huizinga et al., 2012; Keo et al., 2011; Rathinam et al., 2009; Stephenson et al., 2013).

28.6.2 Flagella and flagellar-mediated protein secretion

C. jejuni flagellar and flagellar-related genes encoding structural and accessory components are involved in motility, chemotaxis, adherence, biofilm formation, and protein secretion (Type III secretion). *C. jejuni* motility is potentiated by unipolar or bipolar flagella. Motility has long been considered an essential component of *C. jejuni* virulence given the need for *C. jejuni* to move through the intestinal mucus in order to adhere to and invade intestinal epithelial cells (Lee et al., 1986; Morooka et al., 1985; Newell et al., 1985; Szymanski et al., 1995; Yao et al., 1994). Flagellar gene expression and assembly is tightly controlled by three separate sigma factors (σ^{28}, σ^{54}, σ^{70}) and is dependent on two-component signaling through FlgS/R (Wosten et al., 2004). Interestingly, expression of the *C. jejuni* flagellin genes is controlled by two sigma factors (Nuijten et al., 1990). FlaA and FlaB comprise the

flagellar filament. FlaA is the most abundant component and required for full-length assembly of the filament (Guerry et al., 1991). Deletion of *flaA*, but not *flaB*, appreciably decreases *C. jejuni* invasion of INT 407 cells (Wassenaar et al., 1991). A FlgS mutant is deficient in chicken colonization and FlgS/R signal activation is associated with production of the flagellar export apparatus (Hendrixson, 2008; Joslin and Hendrixson, 2009). CheY, a putative chemotaxis system member found in many bacteria, drives *C. jejuni* flagellar rotation. Mutant *cheY* strains display altered adherence and invasion of INT 407 cells, decreased colonization in mice, and decreased ability to cause diarrhea in ferrets (Yao et al., 1997). Proper regulation of flagellar genes is essential to *C. jejuni* colonization and invasion.

The ability of *C. jejuni* to form microcolonies and biofilms on the surface of intestinal epithelial cells is partially dependent on the flagellum (Haddock et al., 2010). These formations are enabled by the glycosylation-mediated (see Section 28.7) autoagglutination of flagellar proteins (Guerry et al., 2006; Kalmokoff et al., 2006; Thibault et al., 2001). Additionally, *C. jejuni* involvement in and production of biofilms may promote survival and provide an environmental reservoir that leads to host colonization (Kalmokoff et al., 2006; Teh et al., 2010; Trachoo et al., 2002).

Flagellar movement in combination with adhesins, including CapA and the fibronectin-binding proteins CadF and FlpA, ensures *C. jejuni*–host cell contact leading to invasion (Fauchere et al., 1989; Jin et al., 2001; Konkel et al., 1997, 2010). It has been shown that *C. jejuni* secrete proteins from the flagellar Type III secretion system (T3SS). The flagellum is the only T3SS in *C. jejuni* (Desvaux et al., 2006; Konkel et al., 2004; Minamino and Namba, 2004; Neal-McKinney et al., 2010). This complex is made up of the basal body, hook, and hook-associated proteins, and delivers nonflagellar *C. jejuni* proteins to host cells (Konkel et al., 2004; Neal-McKinney and Konkel, 2012). *Campylobacter* invasion antigens (Cia proteins) are synthesized in response to host cues and secreted upon *C. jejuni*–host cell contact (Malik-Kale et al., 2008; Rivera-Amill and Konkel, 1999, 2001). Currently, it has been shown that CiaB is necessary for full virulence in a piglet model of infection, CiaC is necessary for maximal cellular invasion and causes membrane ruffling by initiating cytoskeletal rearrangement of host cells, and CiaI aids in *C. jejuni* intracellular survival (Buelow et al., 2011; Christensen et al., 2009; Konkel et al., 2001). The characterization of additional purported flagellar secreted proteins, including FspA1, FspA2, and FedABCD, continues (Barrero-Tobon and Hendrixson, 2012; Poly et al., 2007a).

Researchers have only begun to define the importance of the *C. jejuni* flagellum. However, it is clear that *C. jejuni* depends on flagella for motility, host cell invasion, and protein secretion.

28.7 Glycosylation

C. jejuni contains two separate glycosylation systems for modifying proteins. First, O-linked glycosylation modifies *C. jejuni* flagellar proteins with the addition of pseudaminic acid and is necessary for proper flagella assembly. Mutations in this system result in attenuated motility, adherence and invasion of INT 407 cells, and virulence (ferrets) (Goon et al., 2003; Guerry et al., 2006; Szymanski et al., 1999). Additionally, *C. jejuni* demonstrates N-linked glycosylation through the function of proteins encoded in *pgl* genes (Szymanski et al., 1999; Wacker et al., 2002). N-linked glycosylation modifies asparagine residues with glycans, such as heptasaccharides. A diverse selection of more than 60 *Campylobacter* proteins, many predicted as periplasmic, are modified in this manner (Nothaft et al., 2012; Young et al., 2002). A specific motif is required, but in itself is not fully predictive, for protein N-linked glycosylation (Kowarik et al., 2006; Nita-Lazar et al., 2005). Glycosylation modifies protein function while contributing to protein complexity. N-linked glycosylation mutants have displayed decreased adherence and invasion of INT 407 cells, reduced colonization (in both mice and chickens), less natural competence, and it also alters immune recognition of proteins (Hendrixson and DiRita, 2004;

Kakuda and DiRita, 2006; Karlyshev et al., 2004; Kelly et al., 2006; Larsen et al., 2004; Linton et al., 2002; Nothaft et al., 2012; Szymanski et al., 2002). The continued characterization of *C. jejuni* glycosylation will better define this additional layer of protein function and regulation.

28.8 Model of human cell invasion

C. jejuni utilize bacterial colonization and virulence factors to manipulate host cell components during the invasion of host cells (Figure 28-1). *C. jejuni* moves through the intestinal mucus toward the intestinal epithelium using unipolar or bipolar flagella. Bacterial factors, some of which are secreted, induce a host inflammatory response that disrupts the integrity of the host epithelial cell barrier (Figure 28-1, Steps 1 and 2) (Allos, 2001; Baqar et al., 2001; Boehm et al., 2012; Fleckenstein and Kopecko, 2001; Konkel and Cieplak, 1992; Konkel et al., 1993; Larson et al., 2008; Wassenaar and Blaser, 1999). Barrier disruption is proposed to allow *C. jejuni* to translocate across the epithelium and gain access to the basal cell surface (Figure 28-1, Step 3) (Konkel et al., 1992c). *C. jejuni* invasion is enabled by binding to host cells and secretion and delivery of bacterial effector proteins (Figure 28-1, Step 4). The

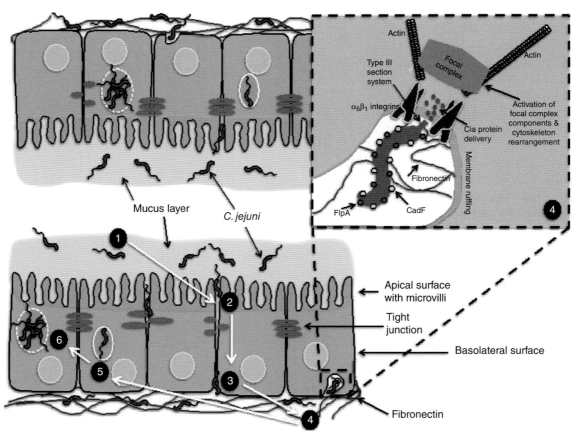

Fig. 28-1 *Campylobacter* pathogenesis and invasion. Step 1. *C. jejuni* penetrates intestinal mucus using flagellar motility. Step 2. *C. jejuni* synthesizes and secretes novel invasion proteins (Cia proteins). *C. jejuni*–secreted proteins and host inflammatory responses lead to tight junction barrier disruption. Step 3. *C. jejuni* translocates across the epithelial barrier. Step 4. *C. jejuni* binds fibronectin and delivers the Cia proteins to the host cell cytosol initiating focal complex–mediated invasion and cytoskeleton rearrangement of host cells. Step 5. *C. jejuni* resides within a host cell vacuole. Step 6. *C. jejuni* propagation and vacuole escape eventually results in host cell death.

C. jejuni adhesins (e.g., CadF and FlpA) set the stage for cell invasion by binding to the extracellular matrix component fibronectin (Eucker and Konkel, 2012; Fauchere et al., 1989; Flanagan et al., 2009; Jin et al., 2001; Konkel et al., 1997, 2010; Monteville and Konkel, 2002). Effector proteins (e.g., Cia proteins), secreted through the flagellum, trigger activation of the host cell focal complex (Barrero-Tobon and Hendrixson, 2012; Buelow et al., 2011; Christensen et al., 2009; Konkel et al., 1999, 2001, 2004; Malik-Kale et al., 2008; Rivera-Amill et al., 2001; Rivera-Amill and Konkel, 1999). The focal complex is a set of proteins that serves as a nexus for signaling, adhesion, and host cell movement. Focal complex proteins, including β_1-integrin, focal adhesion kinase (FAK), and paxillin, have been implicated in *C. jejuni* host cell invasion (Eucker and Konkel, 2012; Krause-Gruszczynska et al., 2011; Monteville et al., 2003). Fibronectin binding is proposed to lead to β_1-integrin signaling through FAK to paxillin (Eucker and Konkel, 2012; Krause-Gruszczynska et al., 2011). The formation of this signaling complex results in further cell signaling and ultimately membrane ruffling, cytoskeleton rearrangement, and internalization of *C. jejuni* (Eucker and Konkel, 2012; Monteville et al., 2003). Invaded cells eventually die after *C. jejuni* propagates and escapes the host cell vacuole (Figure 28-1, Steps 5 and 6) (Buelow et al., 2011; Konkel et al., 1992b).

28.8.1 Cytolethal distending toxin

Some isolates of *C. jejuni* produce cytolethal distending toxin (CDT). *C. jejuni* CDT comprises three separate proteins (CdtA, B, and C) that form an active holotoxin complex (Lara-Tejero and Galan, 2000). Homologs of this toxin are found in many other bacteria. CDT stalls cell division in the G2/M phase and causes cells to exhibit a distended morphology. CdtA and CdtC bind host cells, and CdtB functions as a DNase (Lara-Tejero and Galan, 2001; Lee et al., 2003). CDT proteins stimulate secretion of interleukin-8 (IL-8) (Zheng et al., 2008). IL-8 is a proinflammatory cytokine, and IL-8 secretion is a hallmark of bacterial infection. CDT can cause apoptosis in monocytes and is involved in NF-κB–deficient mouse colonization (Fox et al., 2004; Hickey et al., 2005). The impact of CDT on human disease has thus far proven to be minimal, as isolates lacking CDT remain pathogenic (Mortensen et al., 2011). Study of *Campylobacter* CDT function in colonization and disease continues.

28.9 *In vitro* modeling

C. jejuni–mediated disease in humans is proposed to be dependent on the ability of the bacterium to adhere to and invade intestinal epithelial cells (Konkel et al., 2001; O'Croinin and Backert, 2012). *In vitro* modeling has been essential for dissection of *C. jejuni* adherence and invasion mechanisms. Intestinal epithelial cells

are polarized, having distinct apical and basal surface characteristics. Replicating consistent cell polarization under *in vitro* culture conditions can be challenging. Now we highlight representative examples of selected human cells used to study *C. jejuni* adherence and invasion. Additional discussion of *in vitro* models can be found elsewhere (reviewed in Konkel et al., 2008). We will also briefly describe the general assay used to determine *C. jejuni* adherence to and invasion of host cells.

28.9.1 Nonpolarized cells

INT 407 cells were derived from the intestinal cells of a 2-month-old human embryo. INT 407 cells are now considered to be genetically indistinguishable from HeLa cells likely due to culture contamination (Masters, 2002). HeLa cells were derived from adult cervical cancer cells. INT 407, HeLa, and other nonpolarized cells not discussed here, provide great utility because of their robust and hearty nature in culture and proven ability for modeling cell signaling pathways and *C. jejuni* adherence, invasion, and trafficking (Eucker and Konkel, 2012; O'Croinin and Backert, 2012).

28.9.2 Polarized epithelial cells

Polarized cells provide an *in vitro* model for the study of *C. jejuni* adherence and invasion, as well as translocation (Konkel et al., 1992c). Caco-2 cells, derived from heterogeneous human epithelial colorectal adenocarcinoma cells, are a commonly used human cell line. Caco-2 cells produce polarized monolayers, similar to columnar intestinal epithelium with brush borders, tight junctions, and intestinal epithelia–like protein production under proper culture conditions. These polarized cells have distinct apical and basal cell surfaces. *C. jejuni* translocation across polarized Caco-2 monolayers can serve as an informative *in vitro* model for *C. jejuni* invasion of the intestinal epithelium. However, compared with INT 407 and HeLa cells, Caco-2 cells require more specialized culturing techniques to produce the desired polarized state.

Availability of both polarized epithelial cells and nonpolarized cells provide researchers options when determining the most appropriate *in vitro* model to study a particular aspect of *C. jejuni*–host cell interaction.

28.10 Adherence and invasion assay

The study of *C. jejuni* colonization and disease utilizes an ever-increasing number of experimental techniques. Still, *in vitro* adherence and invasion assays are standardly used to determine *C. jejuni* association with host cells and characterize isolates and mutants (Konkel et al., 1992a). Briefly, host cells are grown to the desired density in culture dishes or plates. To determine total host cell–associated bacteria, *C. jejuni* are added to the medium overlaying the host cells. Some protocols promote bacteria–host cell contact by centrifugation. After a set incubation time, the inoculum is removed and host cells are washed. Host cells are then lysed with a low concentration of detergent (e.g., Triton X-100) that does not lyse the bacteria. Counting *C. jejuni* colonies, grown from plated serial dilutions of lysed host cells, provides total cell-associated bacteria. To determine the number of *C. jejuni* internalized in host cells, the above protocol is modified where a bactericidal concentration of gentamicin is added to host cells after the post-inoculation washes. After a set incubation time, the gentamicin-containing medium is removed and cells are washed and lysed as before. The lysates containing bacteria are serially diluted and plated on growth media for enumeration of viable internalized bacteria.

The use of human cell lines and *C. jejuni* adherence and invasion assays provide quantifiable and repeatable data applicable for the dissection of *C. jejuni*–mediated disease mechanisms. Additional *in vivo* studies are necessary for a more complete understanding of *C. jejuni* colonization and disease mechanisms, along with host response to infection.

28.11 *In vivo* modeling

28.11.1 *In vivo* models of colonization and disease

In vivo models of both *Campylobacter* colonization and disease are essential to better understand *Campylobacter* virulence factors and pathogenesis relating to human infection. These models allow for controlled study of symptoms, evaluation of host immune responses, determination of bacterial virulence factors, comparison to *in vitro* model results, and evaluation of new therapeutics or vaccines. In Table 28-1, we list a selection of *in vivo* animal studies organized by host animal. Birds are commensal hosts and thought to be the primary environmental reservoir for transmission of *Campylobacter* to other animals, including humans. The chicken is a very informative and the most widely used avian colonization model. With varying utility and success, many animals, including chickens, mice, ferrets, rabbits, pigs, rats, guinea pigs, hamsters, monkeys, and dogs, have been experimentally infected with *Campylobacter* (Aguero-Rosenfeld et al., 1990; Newell, 2001; Plummer et al., 2012; Prescott et al., 1981; Sung et al., 2013). The ideal *in vivo* disease model would demonstrate enteritis symptoms and immune response similar to humans, be cost effective, and provide consistent results across lab groups. Modeling human-like *C. jejuni*–mediated disease in animals has proven extremely challenging. Although many different types of animals become transiently colonized with *Campylobacter*, modification or chemical treatment of that animal is often necessary before it will exhibit human-like clinical symptoms. Work continues to determine the most practical and informative *in vivo* disease models for virulence factor characterization.

Discussed in sections 28.11 through 28.14 are selected *in vivo* models used to study *C. jejuni* colonization and disease with an emphasis on colonization and virulence factor determination.

28.11.2 Avian colonization model

28.11.2.1 Chickens

Most human infections are traced to an avian source (Mughini Gras et al., 2012; Wilson et al., 2008). Although *Campylobacter* colonization of the human gut causes clinical symptoms, including bloody diarrhea and fever, *Campylobacter* colonization of the avian gut is generally considered to be asymptomatic. Chickens provide an excellent and relevant model for colonization by *Campylobacter*. The colonization of commercially grown, market-age chickens approaches 100% (Jacobs-Reitsma, 1997; Jacobs-Reitsma et al., 1995). Therefore, a better functional understanding of the colonization dynamics between birds and bacteria is needed. Strategies designed to mitigate avian bacterial load may greatly decrease the possibility for downstream exposure of naïve birds and consumers.

Newly hatched chicks can be immediately challenged with *Campylobacter* and colonized in research settings, but commercially grown chicks generally do not become colonized until 2 or 3 weeks post-hatch (Sahin et al., 2002; Shanker et al., 1988; Stern et al., 1988; Young et al., 1999). This finding is most likely attributable to infectious dose and maternal antibodies. Birds in a research setting usually receive a high dose (in the millions of colony-forming units; CFU) of bacteria to ensure colonization. Commercially grown birds are likely exposed to fewer bacteria and have maternal antibodies that have been shown to, at least partially, shield them from colonization during weeks 1–2 of life (Cawthraw and Newell, 2010; Sahin et al., 2001, 2003; Shoaf-Sweeney et al., 2008).

Once colonized, *Campylobacter* resides in the chicken cecum at levels of 10^6–10^8 CFU/g. Fewer *C. jejuni* are found throughout the intestinal tract and may reach other organs (Achen et al., 1998; Beery et al., 1988; Meade et al., 2009; Sahin et al., 2002; Young et al., 1999). *C. jejuni* primarily resides in chicken intestinal mucus and invades fewer epithelial cells than during human infection (Beery et al., 1988; Meinersmann et al., 1991; Sahin et al., 2002; Van Deun et al., 2008). Colonized birds can remain colonized for life (Jacobs-Reitsma et al., 1995; Sahin et al., 2002).

Chickens have been and continue to be utilized as models to access the colonization potential (non-colonizers, transient colonizers, and stable colonizers) of *Campylobacter* strains (Hanel et al., 2009; Korolik et al., 1998; Meinersmann et al., 1991; Stern et al., 1988; Wilson et al., 2010). Recently reviewed by

Table 28-1 Selected *C. jejuni* animal studies

Host animal	Administration	Study subject	Reference
Chickens (*Gallus gallus domesticus*)			
White leghorn	Oral	Colonization	Beery et al. (1988)
White leghorn	Oral/cloacal	Colonization	Shanker et al. (1988)
Chicken	Oral	Colonization	Stern et al. (1988)
Chicken	Direct	*Ex vivo* cecal tissue colonization	Meinersmann et al. (1991)
Chicken	Oral	Colonization of FlaA and FlaB mutants	Wassenaar et al. (1993)
Broiler	Environment	Epidemiology	Jacobs-Reitsma et al. (1995)
Broiler	Environment	Epidemiology	Jacobs-Reitsma (1997)
Broiler	Oral	Colonization and shedding	Achen et al. (1998)
Chicken	Oral	Colonization	Korolik et al. (1998)
Hyline W-36®	Oral/cloacal	Colonization	Young et al. (1999)
Broiler	Environment	Epidemiology	Evans and Sayers (2000)
Broiler	Environment	Epidemiology and maternal antibodies	Sahin et al. (2001)
Hyline W-36®	Oral	Colonization and protection by CadF, DnaJ, PldA, and CiaB mutants	Ziprin et al. (2001)
Broiler	Oral	Colonization and maternal antibodies	Sahin et al. (2003)
Leghorn	Oral	Colonization, flagella mutants	Carrillo et al. (2004)
White leghorn strain Δ	Oral	Colonization, identification of 22 mutants	Hendrixson and DiRita (2004)
Light Sussex	Oral	Colonization, colonization of Maf5, FlaA, KpsM, and PglH mutants	Jones et al. (2004)
Ross 308	Oral	Colonization of FlgR mutant	Wosten et al. (2004)
Leghorn	Oral	Colonization of FilA, RpoN, and FlgK mutants	Fernando et al. (2007)
Broiler	Oral	Maternal antibody characterization	Shoaf-Sweeney et al. (2008)
White leghorn	Oral	Colonization	Van Deun et al. (2008)
Chicken	Oral	Colonization of CadF, CapA, JlpA, FlpA, Peb1A, and Cj1349c mutants	Flanagan et al. (2009)
White leghorn	Oral	Colonization	Hanel et al. (2009)
Light Sussex	Oral	Colonization of glycosylation island mutants	Howard et al. (2009)
Ross 308	Oral	Colonization and immune response	Meade et al. (2009)
Ross, white leghorn	Oral	Colonization and maternal antibodies	Cawthraw and Newell (2010)
Ross 308	Oral	Colonization	Wilson et al. (2010)
Mice (*Mus musculus*)			
Crl: (CFW) (SW)BR	Oral	Colonization and antibiotic treatment	Field et al. (1984)
BALB/c, CD1	Oral	Colonization and pathogenesis	Lee et al. (1986)
BALB/c nu/nu and nu/+	Oral	Colonization and pathogenesis	Yrios and Balish (1986c)
BALB/c nu/nu and nu/+	Oral	Colonization and pathogenesis	Yrios and Balish (1986a)
BALB/c nu/nu and nu/+	Oral	Immune response and protection	Yrios and Balish (1986b)
C3H	Oral	Colonization and pathogenesis	Youssef et al. (1987)
Swiss white	Oral	Colonization and pathogenesis	Jesudason et al. (1989)
BALB/c ByJ	Oral	Colonization and immune response	Baqar et al. (1993)
BALB/c ByJ	Oral	Vaccine	Rollwagen et al. (1993)
BALB/c, C3H/HeJ, CBA/CAJ, C58/J	Intranasal	Pathogenesis and immune response	Baqar et al. (1996)
BALB/c, CB-17-SCID-Beige	Oral	Colonization and pathogenesis	Hodgson et al. (1998)
BALB/c, C57BL/6, DBA/2	Intraperitoneal injection	Pathogenesis and pathology	Vuckovic et al. (1998)
BALB/c	Oral	Vaccine	Lee et al. (1999)
CB-17-SCID-Beige	Oral	Pathogenesis of CdtB mutant	Purdy et al. (2000)
C57BL/129, NF-κB-deficient	Oral	Pathogenesis and immune response to CdtB mutant	Fox et al. (2004)
C3H, SCID	Oral	Pathogenesis of DccR and DccS mutants	MacKichan et al. (2004)
C3H, SCID	Oral	Colonization and pathogenesis	Chang and Miller (2006)
Unspecified	Oral	Vaccine	Prokhorova et al. (2006)
C57BL/6, IL-10$^{-/-}$	Oral	Pathogenesis and immune response	Mansfield et al. (2007)
BALB/c	Intranasal	Vaccine	Baqar et al. (2008)
BALB/c	Oral	Vaccine	Du et al. (2008)
C57BL/6 IL-10$^{+/+}$, IL-10$^{-/-}$, C3Bir IL-10$^{-/-}$, C3H/HeJ IL-10$^{+/+}$, NOD IL-10$^{+/+}$, IL-10$^{+/-}$, and IL-10$^{-/-}$	Oral	Pathogenesis and immune response	Mansfield et al. (2008)

(continued)

Table 28-1 (*Continued*)

Host animal	Administration	Study subject	Reference
C57BL/6 IL-10$^{-/-}$	Oral	Colonization and pathogenesis	Bell et al. (2009)
BALB/c	Intranasal	Vaccine	Monteiro et al. (2009)
BALB/c	Oral	Vaccine	Islam et al. (2010)
C57BL/10ScSn, C57BL/6 TLR9$^{-/-}$, TRIF$^{-/-}$, MYD88$^{-/-}$	Oral	Pathogenesis and immune response FdhD and Cj0952c mutant characterization	Bereswill et al. (2011)
C57BL/6 IL-10$^{-/-}$	Oral	Colonization and pathogenesis	Jerome et al. (2011)
C57BL/10 IL-10$^{-/-}$	Oral	Pathogenesis and immune response	Haag et al. (2012a)
C57BL/6, IL-10$^{-/-}$	Oral	Colonization, pathogenesis, and immune response	Haag et al. (2012b)
C57BL/6 IL-10$^{-/-}$	Oral	Colonization and pathogenesis	Kim et al. (2012)
BALB/c	Oral	Vaccine	Albert et al. (2013)
C57BL/6 IL-10$^{-/-}$	Oral	Colonization and pathogenesis	Bell et al. (2013)
129S6/SvEv IL-10$^{-/-}$	Oral	Pathogenesis of FlpA mutant	Larson et al. (2013)
Rabbits			
New Zealand White	RITARD	Pathogenesis and immune response	Caldwell et al. (1983)
Rabbit	RITARD	Colonization and symptoms	Pang et al. (1987)
New Zealand White	RITARD	Pathogenesis and immune response	Burr et al. (1988)
Rabbit	RITARD	Antigenic variation	Logan et al. (1989)
Rabbit	RITARD	Update of model	Davis and Banks (1991)
New Zealand White	RITARD	Pathogenesis and immune response	Pavlovskis et al. (1991)
New Zealand White	RILT	Protein expression during infection	Panigrahi et al. (1992)
New Zealand White	RILT	Colonization and pathogenesis	Everest et al. (1993)
New Zealand White	Oral	Vaccine	Rollwagen et al. (1993)
New Zealand White	RITARD	Vaccine and RecA mutants	Guerry et al. (1994)
New Zealand White	Inoculation	Ganglioside mimicry	Caporale et al. (2006)
Ferrets (*Mustela putorius furo*)			
Ferret	Oral	Pathogenesis and immune response	Fox et al. (1987)
Ferret	Oral	Colonization and pathogenesis	Bell and Manning (1990a)
Ferret	Inoculation/Oral	Reproductive complications	Bell and Manning (1990b)
Ferret	Oral	Pathogenesis of PspA mutant	Dolg et al. (1996)
Ferret	Oral	Pathogenesis of CheY mutant	Yao et al. (1997)
Ferret	Oral	Plasmid-based virulence	Bacon et al. (2000)
Ferret	Oral	Pathogenesis of KpsM mutant	Bacon et al. (2001)
Ferret	Oral	Vaccine	Burr et al. (2005)
Ferret	Oral	Pathogenesis of Cj0977 mutant	Goon et al. (2006)
Ferret	Oral	Glycosylation, pathogenesis PseAm mutant	Guerry et al. (2006)
Ferret	Oral	Pathogenesis and immune response	Poly et al. (2007b)
Ferret	Oral	Pathogenesis and immune response	Nemelka et al. (2009)
Pigs (*Sus scrofa domesticus*)			
Pig	Oral	Colonization and pathogenesis	Boosinger and Powe (1988)
Large White × Landrace	Oral	Colonization and pathogenesis	Vitovec et al. (1989)
Pig	Oral	Colonization and pathogenesis	Babakhani et al. (1993)
Pig	Oral	Pathogenesis of CiaB mutant	Konkel et al. (2001)
Yorkshire	Oral	Colonization and pathogenesis	Law et al. (2009)
Pig	Oral	Pathogenesis of Dps mutant	Theoret et al. (2011)
Monkeys			
Macaca mulatta	Oral	Colonization and pathogenesis	Fitzgeorge et al. (1981)
Macaca nemestrina	Oral	Pathogenesis and immune response	Russell et al. (1989)
Macaca mulatta	Oral	Vaccine	Baqar et al. (1995)
Macaca mulatta	Oral	Pathogenesis and immune response	Islam et al. (2006)
Aotus nancymae	Oral	Pathogenesis and immune response	Jones et al. (2006)
Aotus nancymae	Oral	Vaccine	Monteiro et al. (2009)
Human volunteers (*Homo sapiens*)			
Human	Oral	Colonization and pathogenesis	Black et al. (1988)
Human	Oral	Vaccine	Scott (1997)
Human	Oral	Vaccine	Prendergast et al. (2004)
Human	Oral	Development of antibiotic resistance	Lindow et al. (2010)
Human	Oral	Pathogenesis and immune response	Tribble et al. (2010)

RITARD, removable intestinal tie adult rabbit diarrhea model; RILT, rabbit intestinal loop test.

Hermans et al. (2011), numerous genes involved with drug and bile resistance, chemotaxis, motility, surface structure production, two-component signaling, invasion, adhesion, iron acquisition, and stress response (temperature, oxidative, and nitrosative) have been identified as important for *Campylobacter* colonization of chickens. For example, the fibronectin-binding proteins CadF and FlpA have been identified as adhesins involved in *C. jejuni* binding to the epithelium (Flanagan et al., 2009; Konkel et al., 1997). Furthermore, mutations in genes encoding secreted proteins (CiaB and FedsABCD), phospholipase PldA, the chaperone DnaJ, and numerous regulators or structure components of flagella, including FlaA, FilA, FlgK, FlgR, Maf5, and RpoN, have been shown to negatively affect chicken colonization (Fernando et al., 2007; Hendrixson and DiRita, 2004; Hermans et al., 2011; Jones et al., 2004; Wassenaar et al., 1993; Wosten et al., 2004; Ziprin et al., 2001). Continued characterization of *Campylobacter* wild-type strains and mutants will provide increased insight into the mechanisms and dynamics of the *Campylobacter*–avian interaction. This work should lead to development and testing of vaccines, new therapeutics, and strategies to decrease *Campylobacter* colonization of commercial flocks.

28.12 Disease models

28.12.1 Murine disease models

Infection models using mice have several distinct advantages. Mice are comparatively inexpensive to house, have a rapid propagation time, and the number of biological tools available for mouse study are much greater than for other model animals. There are also a plethora of different mouse strains, including various parental (wild-type) strains and genetically modified strains that researchers can choose. With all these advantages, work has logically focused on the development of a mouse model that mimics human-like *C. jejuni*–mediated disease. Numerous mouse backgrounds (wild-type and genetically modified strains), infection methods, and experimental treatments have been used to attempt to create a mouse model that demonstrates human-like *C. jejuni*–mediated disease. Although progress has been made, no single mouse model is agreed to be the standard for modeling *C. jejuni*–mediated disease. Over the years, several obstacles have become apparent. The mouse gut, specifically normal microbiota residing in the gut, has been shown to be a major barrier preventing human-like infection (Bereswill et al., 2011; Chang and Miller, 2006; Haag et al., 2012b). Even between identical mouse strains, differences in gut microbiota may alter experimental results (Chang and Miller, 2006). Many potential vaccines and therapeutic candidates have been tested in mice, but the mouse strains utilized, methods of administration, and quantifications of protection are far from being standardized (Albert et al., 2013; Baqar et al., 1993, 2008; Du

et al., 2008; Islam et al., 2010; Lee et al., 1999; Monteiro et al., 2009; Prokhorova et al., 2006; Rollwagen et al., 1993). Discussed in section 28.12 are selected strategies used to model *C. jejuni*–mediated disease in mice.

28.12.1.1 Intranasal or intraperitoneal infection

Oral challenge most closely mimics the natural route of *C. jejuni* infection. Some researchers have explored other routes of *Campylobacter* infection well outside the accepted norm for natural infection. For example, intranasal inoculation or intraperitoneal injection of mice with high doses of *Campylobacter* can induce bacteremia, systemic disease, and death (intranasal) (Baqar et al., 1996; Vuckovic et al., 1998). The utility of intranasal inoculation is of questionable value for understanding enteritis, but has been used to test immune response related to vaccine efficacy (Baqar et al., 2008; Monteiro et al., 2009). The continued use of these infection routes will likely be limited because they do not produce symptoms that mimic human-like *C. jejuni*–mediated disease.

28.12.1.2 Germ-free, gnotobiotic, and mice with altered intestinal microbiota

Oral infection of germ-free, gnotobiotic, and mice with altered intestinal microbiota has gained popularity for the study of *C. jejuni* infection. Germ-free mice contain no intestinal microbiota, whereas gnotobiotic mice contain a defined population of intestinal microbiota. Germ-free mice are considered gnotobiotic in that they lack all microbiota. It must also be understood that gnotobiotic mice of identical strains from different sources may not contain the same microbiota. Antibiotic or other chemical treatment can also be used to alter mouse intestinal microbiota.

Germ-free mice have been shown to allow robust intestinal colonization and translocation of *C. jejuni* to organs beyond the intestinal tract (Jesudason et al., 1989; Lee et al., 1986; Youssef et al., 1987). However, germ-free mice have altered immune function that confounds analysis of natural immune responses and signaling during infection (Savidge et al., 1991; Shroff and Cebra, 1995; Szeri et al., 1976). Generation and maintenance of germ-free mice and gnotobiotic mice also requires special isolator housing and sterile feeding procedures. Germ-free athymic mice infected with *C. jejuni* demonstrate intestinal colonization, diarrhea, and splenomegaly (Yrios and Balish, 1986a, 1986b, 1986c). More recently, other genetically modified germ-free or gnotobiotic mouse strains have been used to model *C. jejuni* infection, and those models will be discussed in Section 28.12.1.3.

It has been known for some time that antibiotic treatment of wild-type mice alters the composition of their intestinal microbiota and makes them more susceptible to colonization by pathogens (Que and

Hentges, 1985; van der Waaij et al., 1971). Pretreatment of mice with tobramycin was shown to increase *C. jejuni* colonization, whereas pretreatment with streptomycin does not increase colonization (Field et al., 1984; Jesudason et al., 1989). This indicates that not all antibiotics will enable enhanced *C. jejuni* colonization of mice. Mice from a colony with so-called limited flora have also been shown to be more susceptible to *C. jejuni* colonization (Chang and Miller, 2006). Recently, the administration of five different antibiotics over the course of 6 weeks was used to clear culturable bacteria and increase *C. jejuni* colonization (Bereswill et al., 2011). The authors then antibiotically treated another group of mice and recolonized them with human or mouse flora, from fecal samples, and challenged with *C. jejuni*. Mice with human flora were colonized at a greater level and for a longer duration with *C. jejuni* than mice recolonized with mouse flora (Bereswill et al., 2011). The antibiotically treated mice and the antibiotically treated mice recolonized with human flora also allowed translocation of *C. jejuni* to mesenteric lymph nodes during infection (Bereswill et al., 2011). Increased numbers and recruitment of T-lymphocytes, B-lymphocytes, T-regulatory cells, and neutrophils, and increased secretion of pro-inflammatory cytokines including tumor necrosis factor-α (TNF-α), interleukin-6 (IL-6), and monocyte chemoattractant protein-1 (MCP-1) was noted for mice recolonized with human flora and subsequently infected with *C. jejuni* (Bereswill et al., 2011). The results of this study are quite interesting, but it remains to be determined if the technical and logistical procedure used to produce the treated and reconstituted mice can be replicated by others.

The historical and emerging data clearly support the importance of the mouse intestinal microbiome for resistance to *C. jejuni* colonization and disease. Mice without intestinal microbiota or with limited or altered intestinal microbiota are more susceptible to *C. jejuni* colonization. More consistent and informative mouse models of *C. jejuni* colonization and disease are likely to result from careful consideration of resident microbiota.

28.12.1.3 Genetically modified mice

Genetically modified mice, sometimes lacking or having altered intestinal microbiota, are emerging as models for *C. jejuni*–mediated disease (Bereswill et al., 2011; Fox et al., 2004; Haag et al., 2012a, 2012b; MacKichan et al., 2004; Mansfield et al., 2007; Purdy et al., 2000). The genetically modified mice used to study *C. jejuni* generally carry a mutation(s) that decreases or alters their immune function making them susceptible to infection and disease. For example, the adaptive immunity of severe combined immunodeficient (SCID) mice is compromised by a mutation that results in poor maturation of T- and B-lymphocytes. SCID mice are readily colonized

by *C. jejuni* and can demonstrate diarrhea, intestinal inflammation, lesions, and bacterial translocation beyond the intestine (Hodgson et al., 1998; Purdy et al., 2000). Colonization and *C. jejuni*–mediated disease in SCID mice seems to be enhanced by limitation of intestinal flora (Chang and Miller, 2006; MacKichan et al., 2004). Characterization of *C. jejuni* CDT toxin and signal transduction mutants has been conducted in SCID mice (MacKichan et al., 2004; Purdy et al., 2000). An alternative to SCID mice is interleukin-10 (IL-10)–deficient mice. IL-10 acts as an anti-inflammatory cytokine preventing an overactive inflammatory response (reviewed in Moore et al., 2001). However, some mice lacking IL-10 develop enterocolitis even without the introduction of pathogens (Kuhn et al., 1993). IL-10-deficient mice are colonized by *C. jejuni* and exhibit diarrhea, intestinal inflammation, lesions, bacterial translocation beyond the intestine, and an antibody response that is repeatable but is not identical across different parental mouse strain backgrounds (Mansfield et al., 2007, 2008). This model has been able to identify differences in the colonization and virulence of *C. jejuni* strains (Bell et al., 2009; Wilson et al., 2010). Ongoing research with IL-10-deficient mice is working to correlate variable gene expression of *Campylobacter* strains during infection with differences in colonization and virulence (Bell et al., 2013; Jerome et al., 2011; Kim et al., 2012). IL-10-deficient mice, cleared of microbiota by multi-antibiotic treatment, demonstrate even more acute disease including severe inflammation, rectal bleeding, ulcerative colitis, and death after challenge with *C. jejuni* (Haag et al., 2012a). These mice also had increased numbers of apoptotic cells, innate immune cells, effector cells, and increased secretion of TNF-α, interferon-γ (IFN-γ), and MCP-1 from colon tissue (Haag et al., 2012a). Analysis of immune response and function in genetically modified mice can be complicated; however, many valuable insights can be gained as long as the immunological context is considered.

Slow but continued development of a standard mouse model of *Campylobacter* enteritis is ongoing, and in the interim, characterization of *Campylobacter* wild-type strains and mutants continues using the available models. In Table 28-2, we outline a proposed scoring system that would consistently evaluate *C. jejuni*–mediated disease in any mouse model. Improvements in genetic characterization and increased *Campylobacter* mutant generation reinforce the need for a consistent and informative small animal model for *Campylobacter* infection.

28.13 Other small animal disease models

28.13.1 Rabbits

Rabbits do not demonstrate acute *C. jejuni*–mediated disease without modification. Nevertheless, *in vivo* models utilizing rabbits have played an important role in studying *Campylobacter* virulence. Rabbits

Table 28-2 Evaluation of *C. jejuni*–mediated disease in mice

Analysis	Increasing severity of disease		
	Score 1	Score 2	Score 3
Clinical disease	Hunched posture Ruffled coat	Score 1 plus diarrhea or loose stool	Score 2 plus dehydration, weight loss, blood in stool
Intestinal colonization	Recover fewer bacteria than challenge	Recover more bacteria than challenge	Recover significantly more bacteria than challenge
Invasion	[a]None to minimal detected	Detectable	Score 2 plus (translocation to MLN, spleen, etc.)
Intestinal gross pathology	Some fluid accumulation	Moderate fluid accumulation	Score 2 plus hyperemia
Histopathology	**Score 1**	**Score 2**	**Score 3**
Lumen	Mucus	Score 1 plus neutrophils, exudate, and sloughed epithelium	Score 2 plus increased immune cell recruitment (neutrophils and monocytes)
Epithelial layer	Goblet cell hyperplasia	Goblet cell depletion	Score 2 plus abnormal crypts, necrosis/ulcers, and monocytes
Lamina propria	Neutrophils	Score 1 plus monocytes and submucosal extension	Score 2 plus diffuse cell distribution and increased submucosal extension
Submucosa	[a]None	Moderate inflammation presence of moncytes and neutrophils	Score 2 plus severe inflammation with edema and fibrosis
[b]Cytokines and chemokines	Increasing production of cytokines and chemokines (e.g., MIP-2, TNF-α, IL-6, IFN-γ, MCP-1)		

[a]None means not detected or no symptoms.

[b]Concentrations of cytokines and chemokines are expected to vary depending on assay conditions and methods of measurement. Relative differences between infected and uninfected animals may still demonstrate trends.

MLN, mesenteric lymph node; MIP-2, macrophage inflammatory protein 2 (also known as chemokine (C-X-C motif) ligand 2 (CXCL2)); TNF-α, tumor necrosis factor-α; IL-6, interleukin-6; IFN-γ, interferon-γ; MCP-1, monocyte chemoattractant protein-1 (also known as chemokine (C-C motif) ligand 2 (CCL2)). Examples of specific cytokine and chemokine levels in *C. jejuni* infected IL-10$^{-/-}$ deficient mice can be found in Haag et al. (2012a).

were used to demonstrate the mechanisms underlying GBS-related paralysis induced by the ganglioside mimicry of *Campylobacter* LOS (Caporale et al., 2006; Wang et al., 2000). Separately, two surgical models have been used to induce human-like acute *C. jejuni*–mediated enteritis in rabbits. The removable intestinal tie adult rabbit diarrhea (RITARD) model temporarily blocks part of the rabbit intestine and prevents normal peristalsis (Caldwell et al., 1983). This blockage is thought to increase contact time of *C. jejuni* with the intestinal surface. Bloody mucoid diarrhea, lesions, bacteremia, and sometimes death may result if the challenged animals are maintained a few days after surgery (Burr et al., 1988; Caldwell et al., 1983; Davis and Banks, 1991). The RITARD model has been utilized to study the immune response (IgG, IgA), antigenic variation, the role of flagella, *in vivo* pathogenesis of cytotoxin-producing strains, and characteristics of RecA mutants (Burr et al., 1988; Guerry et al., 1994; Logan et al., 1989; Pang et al., 1987; Pavlovskis et al., 1991; Rollwagen et al., 1993). The rabbit ileal loop test (RILT) is conducted by sequestering several ileal loops, inoculating them with *Campylobacter,* and later harvesting the fluid produced from inflammation for analysis of the acute immune response and differential protein production (Everest et al., 1993; Panigrahi et al., 1992). These surgical methods require

great technical skill and are not high in throughput. Although the symptoms demonstrated are similar to those during human *C. jejuni*–mediated disease, the level of manipulation required to achieve the desired results is significant. Thus, these models are not well suited for routine screening of mutants.

28.13.2 Ferrets

Adult and juvenile (kits) domestic ferrets have been utilized to study *Campylobacter* infection. Both adults and juveniles can be readily colonized with *C. jejuni* by oral gavage, resulting in bloody, mucoid diarrhea with leukocytes and bacteremia (Bell and Manning, 1990a, 1991; Fox et al., 1987). Ferrets mount an aggressive immune response (IgG, IgA) to colonization, with previous exposure seeming to provide some protection from subsequent infection (Bell and Manning, 1990a; Nemelka et al., 2009). *Campylobacter* mutants involved in diverse processes, such as glycosylation (PspA), chemotaxis/flagella (CheY), protein secretion (VirB11), capsule production (KpsM), and σ^{28} regulated (Cj0977), have all shown decreased virulence in ferrets versus a *C. jejuni* wild-type strain (Bacon et al., 2000, 2001; Dolg et al., 1996; Goon et al., 2006; Guerry et al., 2006; Yao et al., 1997). In addition to defining strain virulence and assessing host immune responses, ferrets have been used to assess *Campylobacter* whole cell vaccines (Burr et al., 2005; Nemelka et al., 2009; Poly et al., 2007b). The ferret model has several challenges. Disease determination related to diarrhea can be subjective, specifically when considering that ferrets may suffer colitis without introduction of *Campylobacter*. Additionally, the availability and maintenance of pathogen-free ferrets is expensive and there are limited reagents available for immunological study.

28.14 Large animal disease models

28.14.1 Piglets

Although older pigs may become transiently colonized by *Campylobacter*, very young or gnotobiotic neonatal piglets are used to model human-like *C. jejuni*–mediated disease. Newborn or very young piglets prevented from nursing and orally challenged with *C. jejuni* exhibit bloody, mucoid diarrhea, intestinal damage including lesions, and neutrophil recruitment (Babakhani et al., 1993; Boosinger and Powe, 1988; Vitovec et al., 1989). Currently this model has been utilized to characterize the virulence of *C. jejuni* wild-type strains and mutants that separately lack virulence factors, including CiaB (an invasion-related secreted protein) and Dps (iron-binding protein) (Konkel et al., 2001; Law et al., 2009; Theoret et al., 2011). As with ferrets, gnotobiotic piglets display disease similar to humans. Unfortunately, this model suffers from logistical challenges, including having to coordinate experiments to coincide with the availability of newborn piglets. The expense and maintenance of a breeding program also hinders this model from being routinely used to test *Campylobacter* mutants.

28.14.2 Primates

Several types of monkeys, including *Macaca nemestrina*, *M. fascicularis*, and *M. mulatta*, and *Aotus nancymaae* have been used to study *C. jejuni*–mediated disease. *Campylobacter* challenged monkeys show clinical symptoms very similar to humans, produce an antibody response, and have shown promise as prehuman trial vaccine subjects (Baqar et al., 1995; Islam et al., 2006; Jones et al., 2006; Monteiro et al., 2009; Russell et al., 1989). Work with monkeys is challenging and expensive, requiring special training and facilities. Although probably the closest model organisms to humans, *Campylobacter*-related nonhuman primate experiments are likely to remain limited.

28.14.3 Human volunteer studies

Human volunteer studies have been used to determine strain infectivity, minimum disease-causing dosage, disease symptoms, host immune response, study development of antibiotic resistance, and evaluate vaccine candidates (Black et al., 1988; Lindow et al., 2010; Prendergast et al., 2004; Scott, 1997; Tribble et al., 2010). These are technically challenging experiments and often involve relatively small subject numbers. Special care must be taken to mitigate possible experimental complications, including bacteremia. *Campylobacter* strains that do not exhibit glycoside mimicry may also be chosen to limit the possibility of *Campylobacter*-related GBS in experimental subjects (Lindow et al., 2010). Volunteer human experiments will never be a high-throughput system for studying *Campylobacter* enteritis, but for certain situations, such as vaccine testing, these studies provide unparalleled utility.

Post hoc clinical studies of individuals presenting with acute or chronic disease offer fewer hurdles, but the information gained is largely descriptive because of the inability to completely design experiments. Results of observational studies must also be evaluated with the knowledge that naïve and epidemically exposed populations may manifest different clinical symptoms (Janssen et al., 2008). Previously exposed adult populations in underdeveloped parts of the world less often demonstrate acute disease features of high fever and severe diarrhea, whereas newly infected populations of adults and children are more likely to demonstrate acute disease (Janssen et al., 2008). Still, these studies can provide useful insight into disease spread, related complications (e.g., GBS), and disease prevalence.

28.15 Context and future perspectives

Campylobacters infect and sicken millions of people per year and cause a great deal of mortality in the youth of developing nations. The large commensal avian reservoir ensures pathogen survival and enables zoonotic transmission. Strategies of control focusing on hygiene can effectively mitigate transmission risks, but with widespread residence in birds and other animals, the potential for transmission persists. Understanding the bacteria–avian interaction dynamic will be key to developing novel therapeutic, vaccination, and other colonization-limiting strategies for birds.

With the advent of enhanced genetic sequencing tools and mutant creation coupled to investigation using *in vitro* infection and colonization models, our knowledge of the molecular mechanisms of *Campylobacter* colonization and infection has never been greater. Somewhat lagging has been the development of a standardized animal model for *C. jejuni*–mediated disease. Many disease models have been proposed, but none have achieved the standard status that chickens occupy for modeling colonization. Some success and utility has been shown with the primate, ferret, pig, and rabbit models; however, each of these models suffers from specific hurdles that prevent widespread usage. Monkeys best mimic human disease and immune response, but the expense and special facilities used to house them limit usage. Ferrets and piglets can demonstrate disease symptoms similar to humans but also require special facilities and protocol considerations, discouraging their broad usage. The surgical modification of rabbits provides a model that enables disease symptoms similar to humans, but this surgical modification requires specific technical skills, limiting widespread usage of this model. Mice and the facilities to house and care for them are readily available within the research community. Treatment or genetic modification of mice is necessary to make them consistently susceptible to *C. jejuni*–mediated disease. IL-10-deficient mice have emerged as the most frequently utilized mouse model of *C. jejuni*–mediated disease. These mice lack IL-10-mediated immune suppression and demonstrate an exaggerated immune response to infection. IL-10-deficient mice reproducibly demonstrate *C. jejuni*–mediated disease symptoms and are readily available to the research community. Of all the disease models we have discussed, IL-10-deficient mice currently provide the most consistent choice for

high-throughput characterization of *C. jejuni* strains and mutants. Regardless of the animal model chosen, the contribution of *Campylobacter* virulence factors in colonization and disease requires study using *in vivo* models.

Acknowledgments

We appreciate the careful reading and constructive critique of this chapter by Dr. Tyson Eucker, Dr. Jason Neal-McKinney, Nicholas Negretti, Mark S. Nissen, Dr. Jason L. O'Loughlin, Dr. Derrick Samuelson, and Dr. Jennifer Stone. This work is supported from funds awarded to Dr. Michael E. Konkel from the National Institutes of Health (NIH, Award Number R56 AI088518–01A1), and from the United States Department of Agriculture, National Institute of Food and Agriculture (Award Number 2011-67015-30772). The content is solely the responsibility of the authors and does not necessarily represent the official views of the NIH or USDA.

References

Achen, M., Morishita, T.Y., and Ley, E.C. 1998. Shedding and colonization of *Campylobacter jejuni* in broilers from day-of-hatch to slaughter age. *Avian Dis.*, 42(4), 732–737.

Aguero-Rosenfeld, M.E., Yang, X.H., and Nachamkin, I. 1990. Infection of adult Syrian hamsters with flagellar variants of *Campylobacter jejuni*. *Infect. Immun.*, 58(7), 2214–2219.

Albert, M.J., Mustafa, A.S., Islam, A., and Haridas, S. 2013. Oral immunization with cholera toxin provides protection against *Campylobacter jejuni* in an adult mouse intestinal colonization model. *MBio*, 4(3), e00246–e00213.

Allos, B.M. 1997. Association between *Campylobacter* infection and Guillain–Barre syndrome. *J. Infect. Dis.*, 176(Suppl 2), S125–S128.

Allos, B.M. 2001. *Campylobacter jejuni* infections: Update on emerging issues and trends. *Clin. Infect. Dis.*, 32(8), 1201–1206.

Allos, B.M., and Blaser, M.J. 2009. *Campylobacter jejuni* and related species, In: Mandell, G.L., Douglas, R.G., and Bennett, J.E., editors. *Principles and Practice of Infectious Diseases*, 7th ed. New York: Churchill Livingstone. pp. 2793–2802.

Avrain, L., Vernozy-Rozand, C., and Kempf, I. 2004. Evidence for natural horizontal transfer of *tetO* gene between *Campylobacter jejuni* strains in chickens. *J. Appl. Microbiol.*, 97(1), 134–140.

Babakhani, F.K., Bradley, G.A., and Joens, L.A. 1993. Newborn piglet model for campylobacteriosis. *Infect. Immun.*, 61(8), 3466–3475.

Bachtiar, B.M., Coloe, P.J., and Fry, B.N. 2007. Knockout mutagenesis of the *kpsE* gene of *Campylobacter jejuni* 81116 and its involvement in bacterium–host interactions. *FEMS Immunol. Med. Microbiol.*, 49(1), 149–154.

Bacon, D.J., Alm, R.A., Burr, D.H., Hu, L., Kopecko, D.J., Ewing, C.P., Trust, T.J., and Guerry, P., 2000. Involvement of a plasmid in virulence of *Campylobacter jejuni* 81–176. *Infect. Immun.*, 68(8), 4384–4390.

Bacon, D.J., Szymanski, C.M., Burr, D.H., Silver, R.P., Alm, R.A., and Guerry, P. 2001. A phase-variable capsule is involved in virulence of *Campylobacter jejuni* 81–176. *Mol. Microbiol.*, 40(3), 769–777.

Baqar, S., Applebee, L.A., Gilliland Jr., T.C., Lee, L.H., Porter, C.K., and Guerry, P. 2008. Immunogenicity and protective efficacy of recombinant *Campylobacter jejuni* flagellum-secreted proteins in mice. *Infect. Immun.*, 76(7), 3170–3175.

Baqar, S., Bourgeois, A.L., Applebee, L.A., Mourad, A.S., Kleinosky, M.T., Mohran, Z., and Murphy, J.R. 1996. Murine intranasal challenge model for the study of *Campylobacter* pathogenesis and immunity. *Infect. Immun.*, 64(12), 4933–4939.

Baqar, S., Bourgeois, A.L., Schultheiss, P.J., Walker, R.I., Rollins, D.M., Haberberger, R.L., and Pavlovskis, O.R. 1995. Safety and immunogenicity of a prototype oral whole-cell killed *Campylobacter* vaccine administered with a mucosal adjuvant in non-human primates. *Vaccine*, 13(1), 22–28.

Baqar, S., Pacheco, N.D., and Rollwagen, F.M. 1993. Modulation of mucosal immunity against *Campylobacter jejuni* by orally administered cytokines. *Antimicrob. Agents Chemother.*, 37(12), 2688–2692.

Baqar, S., Rice, B., Lee, L., Bourgeois, A.L., El Din, A.N., Tribble, D.R., Heresi, G.P., Mourad, A.S., and Murphy, J.R. 2001. *Campylobacter jejuni* enteritis. *Clin. Infect. Dis.*, 33(6), 901–905.

Barrero-Tobon, A.M., and Hendrixson, D.R. 2012. Identification and analysis of flagellar coexpressed determinants (Feds) of *Campylobacter jejuni* involved in colonization. *Mol. Microbiol.*, 84(2), 352–369.

Beery, J.T., Hugdahl, M.B., and Doyle, M.P. 1988. Colonization of gastrointestinal tracts of chicks by *Campylobacter jejuni*. *Appl. Environ. Microbiol.*, 54(10), 2365–2370.

Bell, J.A., Jerome, J.P., Plovanich-Jones, A.E., Smith, E.J., Gettings, J.R., Kim, H.Y., Landgraf, J.R., Lefebure, T., Kopper, J.J., Rathinam, V.A., St Charles, J.L., Buffa, B.A., Brooks, A.P., Poe, S.A., Eaton, K.A., Stanhope, M.J., and Mansfield, L.S. 2013. Outcome of infection of C57BL/6 IL-10(−/−) mice with *Campylobacter jejuni* strains is correlated with genome content of open reading frames up- and down-regulated *in vivo*. *Microb. Pathog.*, 54, 1–19.

Bell, J.A., and Manning, D.D. 1990a. A domestic ferret model of immunity to *Campylobacter jejuni*-induced enteric disease. *Infect. Immun.*, 58(6), 1848–1852.

Bell, J.A., and Manning, D.D. 1990b. Reproductive failure in mink and ferrets after intravenous or oral inoculation of *Campylobacter jejuni*. *Can. J. Vet. Res.*, 54(4), 432–437.

Bell, J.A., and Manning, D.D. 1991. Evaluation of *Campylobacter jejuni* colonization of the domestic ferret intestine as a model of proliferative colitis. *Am. J. Vet. Res.*, 52(6), 826–832.

Bell, J.A., St Charles, J.L., Murphy, A.J., Rathinam, V.A., Plovanich-Jones, A.E., Stanley, E.L., Wolf, J.E., Gettings, J.R., Whittam, T.S., and Mansfield, L.S. 2009. Multiple factors interact to produce responses resembling spectrum of human disease in *Campylobacter jejuni* infected C57BL/6 IL-10−/− mice. *BMC Microbiol.*, 9, 57.

Bereswill, S., Fischer, A., Plickert, R., Haag, L.M., Otto, B., Kuhl, A.A., Dasti, J.I., Zautner, A.E., Munoz, M., Loddenkemper, C., Gross, U., Gobel, U.B., and Heimesaat, M.M. 2011. Novel murine infection models provide deep insights into the "menage a trois" of *Campylobacter jejuni*, microbiota and host innate immunity. *PLoS One*, 6(6), e20953.

Black, R.E., Levine, M.M., Clements, M.L., Hughes, T.P., and Blaser, M.J. 1988. Experimental *Campylobacter jejuni* infection in humans. *J. Infect. Dis.*, 157(3), 472–479.

Blaser, M. 1990. *Campylobacter* species. In: Mandell, G.L., Dolin, R., Bennett, J.E., editors. *Principles and Practice of Infectious Diseases*, 3rd ed. New York: Churchill Livingstone. pp. 1649–1658.

Boehm, M., Hoy, B., Rohde, M., Tegtmeyer, N., Baek, K.T., Oyarzabal, O.A., Brondsted, L., Wessler, S., and Backert, S. 2012. Rapid paracellular transmigration of *Campylobacter jejuni* across polarized epithelial cells without affecting TER: Role of proteolytic-active HtrA cleaving E-cadherin but not fibronectin. *Gut Pathog.*, 4(1), 3.

Boosinger, T.R., and Powe, T.A. 1988. *Campylobacter jejuni* infections in gnotobiotic pigs. *Am. J. Vet. Res.*, 49(4), 456–458.

Buelow, D.R., Christensen, J.E., Neal-McKinney, J.M., and Konkel, M.E. 2011. *Campylobacter jejuni* survival within human epithelial cells is enhanced by the secreted protein CiaI. *Mol. Microbiol.*, 80(5), 1296–1312.

Burr, D.H., Caldwell, M.B., Bourgeois, A.L., Morgan, H.R., Wistar Jr., R., and Walker, R.I. 1988. Mucosal and systemic immunity to *Campylobacter jejuni* in rabbits after gastric inoculation. *Infect. Immun.*, 56(1), 99–105.

Burr, D.H., Rollins, D., Lee, L.H., Pattarini, D.L., Walz, S.S., Tian, J.H., Pace, J.L., Bourgeois, A.L., and Walker, R.I. 2005. Prevention of disease in ferrets fed an inactivated whole cell *Campylobacter jejuni* vaccine. *Vaccine*, 23(34), 4315–4321.

Butzler, J.P., and Skirrow, M.B. 1979. *Campylobacter* enteritis. *Clin. Gastroenterol.*, 8(3), 737–765.

Buzby, J.C., Allos, B.M., and Roberts, T. 1997. The economic burden of *Campylobacter*-associated Guillain−Barre syndrome. *J. Infect. Dis.*, 176(Suppl 2), S192–S197.

Caldwell, M.B., Walker, R.I., Stewart, S.D., and Rogers, J.E. 1983. Simple adult rabbit model for *Campylobacter jejuni* enteritis. *Infect. Immun.*, 42(3), 1176–1182.

Cameron, A., Frirdich, E., Huynh, S., Parker, C.T., and Gaynor, E.C. 2012. Hyperosmotic stress response of *Campylobacter jejuni*. *J. Bacteriol.*, 194(22), 6116–6130.

Caporale, C.M., Capasso, M., Luciani, M., Prencipe, V., Creati, B., Gandolfi, P., De Angelis, M.V., Di Muzio, A., Caporale, V., and Uncini, A. 2006. Experimental axonopathy induced by immunization with *Campylobacter jejuni* lipopolysaccharide from a patient with Guillain−Barre syndrome. *J. Neuroimmunol.*, 174(1−2), 12–20.

Carrillo, C.D., Taboada, E., Nash, J.H., Lanthier, P., Kelly, J., Lau, P.C., Verhulp, R., Mykytczuk, O., Sy, J., Findlay, W.A., Amoako, K., Gomis, S., Willson, P., Austin, J.W., Potter, A., Babiuk, L., Allan, B., and Szymanski, C.M. 2004. Genome-wide expression analyses of *Campylobacter jejuni* NCTC11168 reveals coordinate regulation of motility and virulence by flhA. *J. Biol. Chem.*, 279(19), 20327–20338.

Cawthraw, S.A., and Newell, D.G. 2010. Investigation of the presence and protective effects of maternal antibodies against *Campylobacter jejuni* in chickens. *Avian Dis.*, 54(1), 86–93.

Chang, C., and Miller, J.F. 2006. *Campylobacter jejuni* colonization of mice with limited enteric flora. *Infect. Immun.*, 74 (9), 5261–5271.

Christensen, J.E., Pacheco, S.A., and Konkel, M.E. 2009. Identification of a *Campylobacter jejuni*-secreted protein required

for maximal invasion of host cells. *Mol. Microbiol.*, 73(4), 650–662.

Coker, A.O., Isokpehi, R.D., Thomas, B.N., Amisu, K.O., and Obi, C.L. 2002. Human campylobacteriosis in developing countries. *Emerg. Infect. Dis.*, 8(3), 237–244.

Cullen, T.W., O'Brien, J.P., Hendrixson, D.R., Giles, D.K., Hobb, R.I., Thompson, S.A., Brodbelt, J.S., and Trent, M.S. 2013. EptC of *Campylobacter jejuni* mediates phenotypes involved in host interactions and virulence. *Infect. Immun.*, 81(2), 430–440.

Davis, J.A., and Banks, R.E. 1991. Modification to the RITARD surgical model. *J. Invest. Surg.*, 4(4), 499–504.

de Boer, P., Wagenaar, J.A., Achterberg, R.P., van Putten, J.P., Schouls, L.M., and Duim, B. 2002. Generation of *Campylobacter jejuni* genetic diversity *in vivo*. *Mol. Microbiol.*, 44(2), 351–359.

de Zoete, M.R., Keestra, A.M., Roszczenko, P., and van Putten, J.P. 2010. Activation of human and chicken toll-like receptors by *Campylobacter* spp. *Infect. Immun.*, 78(3), 1229–1238.

Dekeyser, P., Gossuin-Detrain, M., Butzler, J.P., and Sternon, J. 1972. Acute enteritis due to related vibrio: First positive stool cultures. *J. Infect. Dis.*, 125(4), 390–392.

Desvaux, M., Hebraud, M., Henderson, I.R., and Pallen, M.J. 2006. Type III secretion: What's in a name? *Trends Microbiol.*, 14(4), 157–160.

Dolg, P., Yao, R., Burr, D.H., Guerry, P., and Trust, T.J. 1996. An environmentally regulated pilus-like appendage involved in *Campylobacter* pathogenesis. *Mol. Microbiol.*, 20(4), 885–894.

Doyle, M.P., and Jones, M.P. 1992. *Food-Borne Transmission and Antibiotic Resistance of Campylobacter jejuni. Campylobacter jejuni. Current Status and Future Trends*. Washington, DC: American Society for Microbiology. pp. 45–48.

Du, L.F., Li, Z.J., Tang, X.Y., Huang, J.Q., and Sun, W.B. 2008. Immunogenicity and immunoprotection of recombinant PEB1 in *Campylobacter jejuni*-infected mice. *World J. Gastroenterol.*, 14(40), 6244–6248.

Duong, T., and Konkel, M.E. 2009. Comparative studies of *Campylobacter jejuni* genomic diversity reveal the importance of core and dispensable genes in the biology of this enigmatic food-borne pathogen. *Curr. Opin. Biotechnol.*, 20(2), 158–165.

Eucker, T.P., and Konkel, M.E. 2012. The cooperative action of bacterial fibronectin-binding proteins and secreted proteins promote maximal *Campylobacter jejuni* invasion of host cells by stimulating membrane ruffling. *Cell. Microbiol.*, 14(2), 226–238.

Evans, S.J., and Sayers, A.R. 2000. A longitudinal study of *Campylobacter* infection of broiler flocks in Great Britain. *Prev. Vet. Med.*, 46(3), 209–223.

Everest, P.H., Goossens, H., Sibbons, P., Lloyd, D.R., Knutton, S., Leece, R., Ketley, J.M., and Williams, P.H. 1993. Pathological changes in the rabbit ileal loop model caused by *Campylobacter jejuni* from human colitis. *J. Med. Microbiol.*, 38(5), 316–321.

FAO/WHO. 2009. *Risk Assessment of Campylobacter spp. in Broiler Chickens: Interpretative Summary, Microbiological Risk Assessment Series*. Geneva: Food and Agriculture Organization of the United Nations/World Health Organization. p. 35.

Fauchere, J.L., Kervella, M., Rosenau, A., Mohanna, K., and Veron, M. 1989. Adhesion to HeLa cells of *Campylobacter jejuni* and *C. coli* outer membrane components. *Res. Microbiol.*, 140(6), 379–392.

Fernando, U., Biswas, D., Allan, B., Willson, P., and Potter, A.A. 2007. Influence of *Campylobacter jejuni fliA, rpoN* and *flgK* genes on colonization of the chicken gut. *Int. J. Food Microbiol.*, 118(2), 194–200.

Field, L.H., Underwood, J.L., and Berry, L.J. 1984. The role of gut flora and animal passage in the colonisation of adult mice with *Campylobacter jejuni*. *J. Med. Microbiol.*, 17(1), 59–66.

Fitzgeorge, R.B., Baskerville, A., and Lander, K.P. 1981. Experimental infection of Rhesus monkeys with a human strain of *Campylobacter jejuni*. *J. Hyg.*, 86(3), 343–351.

Flanagan, R.C., Neal-McKinney, J.M., Dhillon, A.S., Miller, W.G., and Konkel, M.E. 2009. Examination of *Campylobacter jejuni* putative adhesins leads to the identification of a new protein, designated FlpA, required for chicken colonization. *Infect. Immun.*, 77(6), 2399–2407.

Fleckenstein, J.M., and Kopecko, D.J. 2001. Breaching the mucosal barrier by stealth: An emerging pathogenic mechanism for enteroadherent bacterial pathogens. *J. Clin. Invest.*, 107(1), 27–30.

Fouts, D.E., Mongodin, E.F., Mandrell, R.E., Miller, W.G., Rasko, D.A., Ravel, J., Brinkac, L.M., DeBoy, R.T., Parker, C.T., Daugherty, S.C., Dodson, R.J., Durkin, A.S., Madupu, R., Sullivan, S.A., Shetty, J.U., Ayodeji, M.A., Shvartsbeyn, A., Schatz, M.C., Badger, J.H., Fraser, C.M., and Nelson, K.E. 2005. Major structural differences and novel potential virulence mechanisms from the genomes of multiple *Campylobacter* species. *PLoS Biol.*, 3(1), e15.

Fox, J.G., Ackerman, J.I., Taylor, N., Claps, M., and Murphy, J.C. 1987. *Campylobacter jejuni* infection in the ferret: An animal model of human campylobacteriosis. *Am. J. Vet. Res.*, 48(1), 85–90.

Fox, J.G., Rogers, A.B., Whary, M.T., Ge, Z., Taylor, N.S., Xu, S., Horwitz, B.H., and Erdman, S.E. 2004. Gastroenteritis in NF-kappaB-deficient mice is produced with wild-type *Campylobacter jejuni* but not with *C. jejuni* lacking cytolethal distending toxin despite persistent colonization with both strains. *Infect. Immun.*, 72(2), 1116–1125.

Frenzen, P.D. 2008. Economic cost of Guillain–Barré syndrome in the United States. *Neurology*, 71(1), 21–27.

Fujimoto, S., and Amako, K. 1990. Guillain–Barre syndrome and *Campylobacter jejuni* infection. *Lancet*, 335(8701), 1350.

Gilbert, M., Karwaski, M.F., Bernatchez, S., Young, N.M., Taboada, E., Michniewicz, J., Cunningham, A.M., and Wakarchuk, W.W. 2002. The genetic bases for the variation in the lipo-oligosaccharide of the mucosal pathogen, *Campylobacter jejuni*. Biosynthesis of sialylated ganglioside mimics in the core oligosaccharide. *J. Biol. Chem.*, 277(1), 327–337.

Goon, S., Ewing, C.P., Lorenzo, M., Pattarini, D., Majam, G., and Guerry, P. 2006. A sigma28-regulated nonflagella gene contributes to virulence of *Campylobacter jejuni* 81–176. *Infect. Immun.*, 74(1), 769–772.

Goon, S., Kelly, J.F., Logan, S.M., Ewing, C.P., and Guerry, P. 2003. Pseudaminic acid, the major modification on *Campylobacter* flagellin, is synthesized via the Cj1293 gene. *Mol. Microbiol.*, 50(2), 659–671.

Guerry, P., Alm, R.A., Power, M.E., Logan, S.M., and Trust, T.J. 1991. Role of two flagellin genes in *Campylobacter* motility. *J. Bacteriol.*, 173(15), 4757–4764.

Guerry, P., Ewing, C.P., Schirm, M., Lorenzo, M., Kelly, J., Pattarini, D., Majam, G., Thibault, P., and Logan, S. 2006. Changes in flagellin glycosylation affect *Campylobacter* autoagglutination and virulence. *Mol. Microbiol.*, 60(2), 299–311.

Guerry, P., Pope, P.M., Burr, D.H., Leifer, J., Joseph, S.W., and Bourgeois, A.L. 1994. Development and characterization of recA mutants of *Campylobacter jejuni* for inclusion in attenuated vaccines. *Infect. Immun.*, 62(2), 426–432.

Guerry, P., Szymanski, C.M., Prendergast, M.M., Hickey, T.E., Ewing, C.P., Pattarini, D.L., and Moran, A.P. 2002. Phase variation of *Campylobacter jejuni* 81–176 lipooligosaccharide affects ganglioside mimicry and invasiveness in vitro. *Infect. Immun.*, 70(2), 787–793.

Haag, L.M., Fischer, A., Otto, B., Plickert, R., Kuhl, A.A., Gobel, U.B., Bereswill, S., and Heimesaat, M.M. 2012a. *Campylobacter jejuni* induces acute enterocolitis in gnotobiotic IL-10−/− mice via Toll-like-receptor-2 and -4 signaling. *PLoS One*, 7(7), e40761.

Haag, L.M., Fischer, A., Otto, B., Plickert, R., Kuhl, A.A., Gobel, U.B., Bereswill, S., and Heimesaat, M.M. 2012b. Intestinal microbiota shifts towards elevated commensal *Escherichia coli* loads abrogate colonization resistance against *Campylobacter jejuni* in mice. *PLoS One*, 7(5), e35988.

Haddock, G., Mullin, M., MacCallum, A., Sherry, A., Tetley, L., Watson, E., Dagleish, M., Smith, D.G., and Everest, P. 2010.

Campylobacter jejuni 81–176 forms distinct microcolonies on in vitro-infected human small intestinal tissue prior to biofilm formation. *Microbiology*, 156(Pt 10), 3079–3084.

Hanel, I., Borrmann, E., Muller, J., Muller, W., Pauly, B., Liebler-Tenorio, E.M., and Schulze, F. 2009. Genomic and phenotypic changes of *Campylobacter jejuni* strains after passage of the chicken gut. *Vet. Microbiol.*, 136(1–2), 121–129.

Heikema, A.P., Bergman, M.P., Richards, H., Crocker, P.R., Gilbert, M., Samsom, J.N., van Wamel, W.J., Endtz, H.P., and van Belkum, A. 2010. Characterization of the specific interaction between sialoadhesin and sialylated *Campylobacter jejuni* lipooligosaccharides. *Infect. Immun.*, 78(7), 3237–3246.

Hendrixson, D.R. 2008. Restoration of flagellar biosynthesis by varied mutational events in *Campylobacter jejuni*. *Mol. Microbiol.*, 70(2), 519–536.

Hendrixson, D.R., and DiRita, V.J. 2004. Identification of *Campylobacter jejuni* genes involved in commensal colonization of the chick gastrointestinal tract. *Mol. Microbiol.*, 52(2), 471–484.

Hermans, D., Van Deun, K., Martel, A., Van Immerseel, F., Messens, W., Heyndrickx, M., Haesebrouck, F., and Pasmans, F. 2011. Colonization factors of *Campylobacter jejuni* in the chicken gut. *Vet. Res.*, 42(1), 82.

Hickey, T.E., Majam, G., and Guerry, P. 2005. Intracellular survival of *Campylobacter jejuni* in human monocytic cells and induction of apoptotic death by cytolethal distending toxin. *Infect. Immun.* 73(8), 5194–5197.

Hodgson, A.E., McBride, B.W., Hudson, M.J., Hall, G., and Leach, S.A. 1998. Experimental *Campylobacter* infection and diarrhoea in immunodeficient mice. *J. Med. Microbiol.* 47(9), 799–809.

Howard, S.L., Jagannathan, A., Soo, E.C., Hui, J.P., Aubry, A.J., Ahmed, I., Karlyshev, A., Kelly, J.F., Jones, M.A., Stevens, M.P., Logan, S.M., and Wren, B.W. 2009. *Campylobacter jejuni* glycosylation island important in cell charge, legion-aminic acid biosynthesis, and colonization of chickens. *Infect. Immun.*, 77(6), 2544–2556.

Huizinga, R., Easton, A.S., Donachie, A.M., Guthrie, J., van Rijs, W., Heikema, A., Boon, L., Samsom, J.N., Jacobs, B.C., Willison, H.J., and Goodyear, C.S. 2012. Sialylation of *Campylobacter jejuni* lipo-oligosaccharides: Impact on phagocytosis and cytokine production in mice. *PLoS One*, 7(3), e34416.

Islam, A., Raghupathy, R., and Albert, M.J. 2010. Recombinant PorA, the major outer membrane protein of *Campylobacter jejuni*, provides heterologous protection in an adult mouse intestinal colonization model. *Clin. Vaccine Immunol.*, 17(11), 1666–1671.

Islam, D., Lewis, M.D., Srijan, A., Bodhidatta, L., Aksomboon, A., Gettayacamin, M., Baqar, S., Scott, D., and Mason, C.J. 2006. Establishment of a non-human primate *Campylobacter* disease model for the pre-clinical evaluation of *Campylobacter* vaccine formulations. *Vaccine*, 24(18), 3762–3771.

Jacobs-Reitsma, W.F. 1997. Aspects of epidemiology of *Campylobacter* in poultry. *Vet. Q.*, 19(3), 113–117.

Jacobs-Reitsma, W.F., van de Giessen, A.W., Bolder, N.M., and Mulder, R.W. 1995. Epidemiology of *Campylobacter* spp. at two Dutch broiler farms. *Epidemiol. Infect.*, 114(3), 413–421.

Janssen, R., Krogfelt, K.A., Cawthraw, S.A., van Pelt, W., Wagenaar, J.A., and Owen, R.J. 2008. Host-pathogen interactions in *Campylobacter* infections: The host perspective. *Clin. Microbiol. Rev.*, 21(3), 505–518.

Jerome, J.P., Bell, J.A., Plovanich-Jones, A.E., Barrick, J.E., Brown, C.T., and Mansfield, L.S. 2011. Standing genetic variation in contingency loci drives the rapid adaptation of *Campylobacter jejuni* to a novel host. *PLoS One*, 6(1), e16399.

Jesudason, M.V., Hentges, D.J., and Pongpech, P. 1989. Colonization of mice by *Campylobacter jejuni*. *Infect. Immun.*, 57(8), 2279–2282.

Jin, S., Joe, A., Lynett, J., Hani, E.K., Sherman, P., and Chan, V.L. 2001. JlpA, a novel surface-exposed lipoprotein specific to *Campylobacter jejuni*, mediates adherence to host epithelial cells. *Mol. Microbiol.*, 39(5), 1225–1236.

Jones, F.R., Baqar, S., Gozalo, A., Nunez, G., Espinoza, N., Reyes, S.M., Salazar, M., Meza, R., Porter, C.K., and Walz, S.E. 2006. New World monkey *Aotus nancymae* as a model for *Campylobacter jejuni* infection and immunity. *Infect. Immun.*, 74(1), 790–793.

Jones, M.A., Marston, K.L., Woodall, C.A., Maskell, D.J., Linton, D., Karlyshev, A.V., Dorrell, N., Wren, B.W., and Barrow, P.A. 2004. Adaptation of *Campylobacter jejuni* NCTC11168 to high-level colonization of the avian gastrointestinal tract. *Infect. Immun.*, 72(7), 3769–3776.

Joslin, S.N., and Hendrixson, D.R. 2009. Activation of the *Campylobacter jejuni* FlgSR two-component system is linked to the flagellar export apparatus. *J. Bacteriol.*, 191(8), 2656–2667.

Kakuda, T., and DiRita, V.J. 2006. Cj1496c encodes a *Campylobacter jejuni* glycoprotein that influences invasion of human epithelial cells and colonization of the chick gastrointestinal tract. *Infect. Immun.*, 74(8), 4715–4723.

Kalmokoff, M., Lanthier, P., Tremblay, T.L., Foss, M., Lau, P.C., Sanders, G., Austin, J., Kelly, J., and Szymanski, C.M. 2006. Proteomic analysis of *Campylobacter jejuni* 11168 biofilms reveals a role for the motility complex in biofilm formation. *J. Bacteriol.*, 188(12), 4312–4320.

Karlyshev, A.V., Everest, P., Linton, D., Cawthraw, S., Newell, D.G., and Wren, B.W. 2004. The *Campylobacter jejuni* general glycosylation system is important for attachment to human epithelial cells and in the colonization of chicks. *Microbiology*, 150(Pt 6), 1957–1964.

Karlyshev, A.V., Henderson, J., Ketley, J.M., and Wren, B.W. 1999. Procedure for the investigation of bacterial genomes: Random shot-gun cloning, sample sequencing and mutagenesis of *Campylobacter jejuni*. *Biotechniques*, 26(1), 50–52, 54, 56.

Karlyshev, A.V., Ketley, J.M., and Wren, B.W. 2005. The *Campylobacter jejuni* glycome. *FEMS Microbiol. Rev.*, 29(2), 377–390.

Karlyshev, A.V., Linton, D., Gregson, N.A., Lastovica, A.J., and Wren, B.W. 2000. Genetic and biochemical evidence of a *Campylobacter jejuni* capsular polysaccharide that accounts for Penner serotype specificity. *Mol. Microbiol.*, 35(3), 529–541.

Karlyshev, A.V., Linton, D., Gregson, N.A., and Wren, B.W. 2002. A novel paralogous gene family involved in phase-variable flagella-mediated motility in *Campylobacter jejuni*. *Microbiology*, 148(Pt 2), 473–480.

Karlyshev, A.V., McCrossan, M.V., and Wren, B.W. 2001. Demonstration of polysaccharide capsule in *Campylobacter jejuni* using electron microscopy. *Infect. Immun.*, 69(9), 5921–5924.

Karlyshev, A.V., and Wren, B.W. 2001. Detection and initial characterization of novel capsular polysaccharide among diverse *Campylobacter jejuni* strains using alcian blue dye. *J. Clin. Microbiol.*, 39(1), 279–284.

Kelly, J., Jarrell, H., Millar, L., Tessier, L., Fiori, L.M., Lau, P.C., Allan, B., and Szymanski, C.M. 2006. Biosynthesis of the N-linked glycan in *Campylobacter jejuni* and addition onto protein through block transfer. *J. Bacteriol.*, 188(7), 2427–2434.

Keo, T., Collins, J., Kunwar, P., Blaser, M.J., and Iovine, N.M. 2011. Campylobacter capsule and lipooligosaccharide confer resistance to serum and cationic antimicrobials. *Virulence*, 2(1), 30–40.

Kim, J.S., Artymovich, K.A., Hall, D.F., Smith, E.J., Fulton, R., Bell, J., Dybas, L., Mansfield, L.S., Tempelman, R., Wilson, D.L., and Linz, J.E. 2012. Passage of *Campylobacter jejuni* through the chicken reservoir or mice promotes phase variation in contingency genes Cj0045 and Cj0170 that strongly associates with colonization and disease in a mouse model. *Microbiology*, 158(Pt 5), 1304–1316.

Komagamine, T., and Yuki, N. 2006. Ganglioside mimicry as a cause of Guillain–Barre syndrome. *Curr. Drug Targets: CNS Neurol. Disord.*, 5(4), 391–400.

Konkel, M.E., and Cieplak Jr., W. 1992. Altered synthetic response of *Campylobacter jejuni* to cocultivation with human epithelial cells is associated with enhanced internalization. *Infect. Immun.*, 60(11), 4945–4949.

Konkel, M.E., Corwin, M.D., Joens, L.A., and Cieplak, W. 1992a. Factors that influence the interaction of *Campylobacter jejuni* with cultured mammalian cells. *J. Med. Microbiol.*, 37(1), 30–37.

Konkel, M.E., Garvis, S.G., Tipton, S.L., Anderson Jr., D.E., and Cieplak Jr., W. 1997. Identification and molecular cloning of a gene encoding a fibronectin-binding protein (CadF) from *Campylobacter jejuni*. *Mol. Microbiol.*, 24(5), 953–963.

Konkel, M.E., Hayes, S.F., Joens, L.A., and Cieplak Jr., W. 1992b. Characteristics of the internalization and intracellular survival of *Campylobacter jejuni* in human epithelial cell cultures. *Microb. Pathog.*, 13(5), 357–370.

Konkel, M.E., Kim, B.J., Rivera-Amill, V., and Garvis, S.G. 1999. Bacterial secreted proteins are required for the internaliztion of *Campylobacter jejuni* into cultured mammalian cells. *Mol. Microbiol.*, 32(4), 691–701.

Konkel, M.E., Klena, J.D., Rivera-Amill, V., Monteville, M.R., Biswas, D., Raphael, B., and Mickelson, J. 2004. Secretion of virulence proteins from *Campylobacter jejuni* is dependent on a functional flagellar export apparatus. *J. Bacteriol.*, 186(11), 3296–3303.

Konkel, M.E., Larson, C.L., and Flanagan, R.C. 2010. *Campylobacter jejuni* FlpA binds fibronectin and is required for maximal host cell adherence. *J. Bacteriol.*, 192(1), 68–76.

Konkel, M.E., Mead, D.J., and Cieplak Jr., W. 1993. Kinetic and antigenic characterization of altered protein synthesis by *Campylobacter jejuni* during cultivation with human epithelial cells. *J. Infect. Dis.*, 168(4), 948–954.

Konkel, M.E., Mead, D.J., Hayes, S.F., and Cieplak Jr., W. 1992c. Translocation of *Campylobacter jejuni* across human polarized epithelial cell monolayer cultures. *J. Infect. Dis.*, 166(2), 308–315.

Konkel, M.E., Monteville, M.R., Klena, J.D., and Joens, L.A. 2008. *In Vitro and in Vivo Models Used to Study Campylobacter jejuni Virulence Properties, Microbial Food Safety in Animal Agriculture.* Blackwell Publishing. pp. 195–210.

Konkel, M.E., Monteville, M.R., Rivera-Amill, V., and Joens, L.A. 2001. The pathogenesis of *Campylobacter jejuni*-mediated enteritis. *Curr. Issues Intest. Microbiol.*, 2(2), 55–71.

Korolik, V., Alderton, M.R., Smith, S.C., Chang, J., and Coloe, P.J. 1998. Isolation and molecular analysis of colonising and non-colonising strains of *Campylobacter jejuni* and *Campylobacter coli* following experimental infection of young chickens. *Vet. Microbiol.*, 60(2–4), 239–249.

Kowarik, M., Young, N.M., Numao, S., Schulz, B.L., Hug, I., Callewaert, N., Mills, D.C., Watson, D.C., Hernandez, M., Kelly, J.F., Wacker, M., and Aebi, M. 2006. Definition of the bacterial N-glycosylation site consensus sequence. *EMBO J.*, 25(9), 1957–1966.

Krause-Gruszczynska, M., Boehm, M., Rohde, M., Tegtmeyer, N., Takahashi, S., Buday, L., Oyarzabal, O.A., and Backert, S. 2011. The signaling pathway of *Campylobacter jejuni*-induced Cdc42 activation: Role of fibronectin, integrin beta1, tyrosine kinases and guanine exchange factor Vav2. *Cell Commun. Signal.*, 9, 32.

Kuhn, R., Lohler, J., Rennick, D., Rajewsky, K., and Muller, W. 1993. Interleukin-10-deficient mice develop chronic enterocolitis. *Cell*, 75(2), 263–274.

Lara-Tejero, M., and Galan, J.E. 2000. A bacterial toxin that controls cell cycle progression as a deoxyribonuclease I-like protein. *Science*, 290(5490), 354–357.

Lara-Tejero, M., and Galan, J.E. 2001. CdtA, CdtB, and CdtC form a tripartite complex that is required for cytolethal distending toxin activity. *Infect. Immun.*, 69(7), 4358–4365.

Larsen, J.C., Szymanski, C., and Guerry, P. 2004. N-linked protein glycosylation is required for full competence in *Campylobacter jejuni* 81–176. *J. Bacteriol.*, 186(19), 6508–6514.

Larson, C.L., Samuelson, D.R., Eucker, T.P., O'Loughlin, J.L., and Konkel, M.E. 2013. The fibronectin-binding motif within FlpA facilitates Campylobacter jejuni adherence to host cell and activation of host cell signaling. *Emerg. Microbes Infect.*, 2, e6.

Larson, C.L., Shah, D.H., Dhillon, A.S., Call, D.R., Ahn, S., Haldorson, G.J., Davitt, C., and Konkel, M.E. 2008. *Campylobacter jejuni* invade chicken LMH cells inefficiently and stimulate differential expression of the chicken CXCLi1 and CXCLi2 cytokines. *Microbiology*, 154(Pt 12), 3835–3847.

Law, B.F., Adriance, S.M., and Joens, L.A. 2009. Comparison of *in vitro* virulence factors of *Campylobacter jejuni* to *in vivo* lesion production. *Foodborne Pathog. Dis.*, 6(3), 377–385.

Lee, A., O'Rourke, J.L., Barrington, P.J., and Trust, T.J. 1986. Mucus colonization as a determinant of pathogenicity in intestinal infection by *Campylobacter jejuni*: A mouse cecal model. *Infect. Immun.*, 51(2), 536–546.

Lee, L.H., Burg 3rd, E., Baqar, S., Bourgeois, A.L., Burr, D.H., Ewing, C.P., Trust, T.J., and Guerry, P. 1999. Evaluation of a truncated recombinant flagellin subunit vaccine against *Campylobacter jejuni*. *Infect. Immun.*, 67(11), 5799–5805.

Lee, R.B., Hassane, D.C., Cottle, D.L., and Pickett, C.L. 2003. Interactions of *Campylobacter jejuni* cytolethal distending toxin subunits CdtA and CdtC with HeLa cells. *Infect. Immun.*, 71(9), 4883–4890.

Lindow, J.C., Poly, F., Tribble, D.R., Guerry, P., Carmolli, M.P., Baqar, S., Porter, C.K., Pierce, K.K., Darsley, M.J., Sadigh, K.S., Dill, E.A., Campylobacter Study, T., and Kirkpatrick, B.D. 2010. Caught in the act: *In vivo* development of macrolide resistance to *Campylobacter jejuni* infection. *J. Clin. Microbiol.*, 48(8), 3012–3015.

Linton, D., Allan, E., Karlyshev, A.V., Cronshaw, A.D., and Wren, B.W. 2002. Identification of N-acetylgalactosamine-containing glycoproteins PEB3 and CgpA in *Campylobacter jejuni*. *Mol. Microbiol.*, 43(2), 497–508.

Linton, D., Gilbert, M., Hitchen, P.G., Dell, A., Morris, H.R., Wakarchuk, W.W., Gregson, N.A., and Wren, B.W. 2000. Phase variation of a beta-1,3 galactosyltransferase involved in generation of the ganglioside GM1-like lipo-oligosaccharide of *Campylobacter jejuni*. *Mol. Microbiol.*, 37(3), 501–514.

Logan, S.M., Guerry, P., Rollins, D.M., Burr, D.H., and Trust, T.J. 1989. *In vivo* antigenic variation of *Campylobacter* flagellin. *Infect. Immun.*, 57(8), 2583–2585.

MacKichan, J.K., Gaynor, E.C., Chang, C., Cawthraw, S., Newell, D.G., Miller, J.F., and Falkow, S. 2004. The *Campylobacter jejuni dccRS* two-component system is required for optimal *in vivo* colonization but is dispensable for in vitro growth. *Mol. Microbiol.*, 54(5), 1269–1286.

Malik-Kale, P., Parker, C.T., and Konkel, M.E. 2008. Culture of *Campylobacter jejuni* with sodium deoxycholate induces virulence gene expression. *J. Bacteriol.*, 190(7), 2286–2297.

Mansfield, L.S., Bell, J.A., Wilson, D.L., Murphy, A.J., Elsheikha, H.M., Rathinam, V.A., Fierro, B.R., Linz, J.E., and Young, V.B. 2007. C57BL/6 and congenic interleukin-10-deficient mice can serve as models of *Campylobacter jejuni* colonization and enteritis. *Infect. Immun.*, 75(3), 1099–1115.

Mansfield, L.S., Patterson, J.S., Fierro, B.R., Murphy, A.J., Rathinam, V.A., Kopper, J.J., Barbu, N.I., Onifade, T.J., and Bell, J.A. 2008. Genetic background alters host-pathogen interactions with *Campylobacter jejuni* and influences disease phenotype. *Microb. Pathog.*, 45(4), 241–257.

Masters, J.R. 2002. HeLa cells 50 years on: The good, the bad and the ugly. *Nat. Rev. Cancer*, 2(4), 315–319.

Maue, A.C., Mohawk, K.L., Giles, D.K., Poly, F., Ewing, C.P., Jiao, Y., Lee, G., Ma, Z., Monteiro, M.A., Hill, C.L., Ferderber, J.S., Porter, C.K., Trent, M.S., and Guerry, P. 2013. The polysaccharide capsule of *Campylobacter jejuni* modulates the host immune response. *Infect. Immun.* 81(3), 665–672.

Meade, K.G., Narciandi, F., Cahalane, S., Reiman, C., Allan, B., and O'Farrelly, C. 2009. Comparative *in vivo* infection models yield insights on early host immune response to *Campylobacter* in chickens. *Immunogenetics*, 61(2), 101–110.

Meinersmann, R.J., Rigsby, W.E., Stern, N.J., Kelley, L.C., Hill, J.E., and Doyle, M.P. 1991. Comparative study of colonizing and noncolonizing *Campylobacter jejuni*. *Am. J. Vet. Res.*, 52(9), 1518–1522.

Minamino, T., and Namba, K. 2004. Self-assembly and type III protein export of the bacterial flagellum. *J. Mol. Microbiol. Biotechnol.*, 7(1–2), 5–17.

Monteiro, M.A., Baqar, S., Hall, E.R., Chen, Y.H., Porter, C.K., Bentzel, D.E., Applebee, L., and Guerry, P. 2009. Capsule polysaccharide conjugate vaccine against diarrheal disease caused by *Campylobacter jejuni*. *Infect. Immun.*, 77(3), 1128–1136.

Monteville, M.R., and Konkel, M.E. 2002. Fibronectin-facilitated invasion of T84 eukaryotic cells by *Campylobacter jejuni* occurs preferentially at the basolateral cell surface. *Infect. Immun.*, 70(12), 6665–6671.

Monteville, M.R., Yoon, J.E., and Konkel, M.E. 2003. Maximal adherence and invasion of INT 407 cells by *Campylobacter jejuni* requires the CadF outer-membrane protein and microfilament reorganization. *Microbiology*, 149(Pt 1), 153–165.

Moore, K.W., de Waal Malefyt, R., Coffman, R.L., and O'Garra, A. 2001. Interleukin-10 and the interleukin-10 receptor. *Annu. Rev. Immunol.*, 19, 683–765.

Morooka, T., Umeda, A., and Amako, K. 1985. Motility as an intestinal colonization factor for *Campylobacter jejuni*. *J. Gen. Microbiol.*, 131(8), 1973–1980.

Mortensen, N.P., Schiellerup, P., Boisen, N., Klein, B.M., Locht, H., Abuoun, M., Newell, D., and Krogfelt, K.A. 2011. The role of *Campylobacter jejuni* cytolethal distending toxin in gastroenteritis: Toxin detection, antibody production, and clinical outcome. *Acta Pathol. Microbiol. Immunol. Scand.*, 119(9), 626–634.

Mughini Gras, L., Smid, J.H., Wagenaar, J.A., de Boer, A.G., Havelaar, A.H., Friesema, I.H., French, N.P., Busani, L., and van Pelt, W. 2012. Risk factors for campylobacteriosis of chicken, ruminant, and environmental origin: A combined case-control and source attribution analysis. *PLoS One*, 7(8), e42599.

Neal-McKinney, J.M., Christensen, J.E., and Konkel, M.E. 2010. Amino-terminal residues dictate the export efficiency of the *Campylobacter jejuni* filament proteins via the flagellum. *Mol. Microbiol.*, 76(4), 918–931.

Neal-McKinney, J.M., and Konkel, M.E. 2012. The *Campylobacter jejuni* CiaC virulence protein is secreted from the flagellum and delivered to the cytosol of host cells. *Front. Cell. Infect. Microbiol.*, 2, 31.

Nemelka, K.W., Brown, A.W., Wallace, S.M., Jones, E., Asher, L.V., Pattarini, D., Applebee, L., Gilliland Jr., T.C., Guerry, P., and Baqar, S. 2009. Immune response to and histopathology of *Campylobacter jejuni* infection in ferrets (*Mustela putorius furo*). *Comp. Med.*, 59(4), 363–371.

Newell, D.G. 2001. Animal models of *Campylobacter jejuni* colonization and disease and the lessons to be learned from similar *Helicobacter pylori* models. *Symp. Ser. Soc. Appl. Microbiol.*, (30), 57S–67S.

Newell, D.G., McBride, H., and Dolby, J.M. 1985. Investigations on the role of flagella in the colonization of infant mice with *Campylobacter jejuni* and attachment of *Campylobacter jejuni* to human epithelial cell lines. *J. Hyg.*, 95(2), 217–227.

Nita-Lazar, M., Wacker, M., Schegg, B., Amber, S., and Aebi, M. 2005. The N-X-S/T consensus sequence is required but not sufficient for bacterial N-linked protein glycosylation. *Glycobiology*, 15(4), 361–367.

Nothaft, H., Scott, N.E., Vinogradov, E., Liu, X., Hu, R., Beadle, B., Fodor, C., Miller, W.G., Li, J., Cordwell, S.J., and Szymanski, C.M. 2012. Diversity in the protein N-glycosylation pathways within the *Campylobacter* genus. *Mol. Cell. Proteomics*, 11(11), 1203–1219.

Nuijten, P.J., van Asten, F.J., Gaastra, W., and van der Zeijst, B.A. 1990. Structural and functional analysis of two *Campylobacter jejuni* flagellin genes. *J. Biol. Chem.*, 265(29), 17798–17804.

O'Croinin, T., and Backert, S. 2012. Host epithelial cell invasion by *Campylobacter jejuni*: Trigger or zipper mechanism? *Front. Cell. Infect. Microbiol.*, 2, 25.

Pang, T., Wong, P.Y., Puthucheary, S.D., Sihotang, K., and Chang, W.K. 1987. In-vitro and in-vivo studies of a cytotoxin from *Campylobacter jejuni*. *J. Med. Microbiol.*, 23(3), 193–198.

Panigrahi, P., Losonsky, G., DeTolla, L.J., and Morris Jr., J.G. 1992. Human immune response to *Campylobacter jejuni* proteins expressed *in vivo*. *Infect. Immun.*, 60(11), 4938–4944.

Parker, C.T., Gilbert, M., Yuki, N., Endtz, H.P., and Mandrell, R.E. 2008. Characterization of lipooligosaccharide-biosynthetic loci of *Campylobacter jejuni* reveals new lipooligosaccharide classes: Evidence of mosaic organizations. *J. Bacteriol.*, 190(16), 5681–5689.

Parkhill, J., Wren, B.W., Mungall, K., Ketley, J.M., Churcher, C., Basham, D., Chillingworth, T., Davies, R.M., Feltwell, T., Holroyd, S., Jagels, K., Karlyshev, A.V., Moule, S., Pallen, M.J., Penn, C.W., Quail, M.A., Rajandream, M.A., Rutherford, K.M., van Vliet, A.H., Whitehead, S., and Barrell, B.G. 2000. The genome sequence of the food-borne pathogen *Campylobacter jejuni* reveals hypervariable sequences. *Nature*, 403(6770), 665–668.

Pavlovskis, O.R., Rollins, D.M., Haberberger Jr., R.L., Green, A.E., Habash, L., Strocko, S., and Walker, R.I. 1991. Significance of flagella in colonization resistance of rabbits immunized with *Campylobacter* spp. *Infect. Immun.*, 59(7), 2259–2264.

Plummer, P., Sahin, O., Burrough, E., Sippy, R., Mou, K., Rabenold, J., Yaeger, M., and Zhang, Q. 2012. Critical role of LuxS in the virulence of *Campylobacter jejuni* in a guinea pig model of abortion. *Infect. Immun.*, 80(2), 585–593.

Poly, F., Ewing, C., Goon, S., Hickey, T.E., Rockabrand, D., Majam, G., Lee, L., Phan, J., Savarino, N.J., and Guerry, P. 2007a. Heterogeneity of a *Campylobacter jejuni* protein that is secreted through the flagellar filament. *Infect. Immun.*, 75(8), 3859–3867.

Poly, F., Read, T., Tribble, D.R., Baqar, S., Lorenzo, M., and Guerry, P. 2007b. Genome sequence of a clinical isolate of *Campylobacter jejuni* from Thailand. *Infect. Immun.*, 75(7), 3425–3433.

Prendergast, M.M., Tribble, D.R., Baqar, S., Scott, D.A., Ferris, J.A., Walker, R.I., and Moran, A.P. 2004. *In vivo* phase variation and serologic response to lipooligosaccharide of *Campylobacter jejuni* in experimental human infection. *Infect. Immun.*, 72(2), 916–922.

Prescott, J.F., Barker, I.K., Manninen, K.I., and Miniats, O.P. 1981. *Campylobacter jejuni* colitis in gnotobiotic dogs. *Can. J. Comp. Med.*, 45(4), 377–383.

Prokhorova, T.A., Nielsen, P.N., Petersen, J., Kofoed, T., Crawford, J.S., Morsczeck, C., Boysen, A., and Schrotz-King, P. 2006. Novel surface polypeptides of *Campylobacter jejuni* as traveller's diarrhoea vaccine candidates discovered by proteomics. *Vaccine*, 24(40–41), 6446–6455.

Purdy, D., Buswell, C.M., Hodgson, A.E., McAlpine, K., Henderson, I., and Leach, S.A. 2000. Characterisation of cytolethal distending toxin (CDT) mutants of *Campylobacter jejuni*. *J. Med. Microbiol.*, 49(5), 473–479.

Que, J.U., and Hentges, D.J. 1985. Effect of streptomycin administration on colonization resistance to *Salmonella typhimurium* in mice. *Infect. Immun.*, 48(1), 169–174.

Rathinam, V.A., Appledorn, D.M., Hoag, K.A., Amalfitano, A., and Mansfield, L.S. 2009. *Campylobacter jejuni*-induced activation of dendritic cells involves cooperative signaling through Toll-like receptor 4 (TLR4)-MyD88 and TLR4-TRIF axes. *Infect. Immun.*, 77(6), 2499–2507.

Rivera-Amill, V., Kim, B.J., Seshu, J., and Konkel, M.E. 2001. Secretion of the virulence-associated *Campylobacter* invasion antigens from *Campylobacter jejuni* requires a stimulatory signal. *J. Infect. Dis.*, 183(11), 1607–1616.

Rivera-Amill, V., and Konkel, M.E. 1999. Secretion of *Campylobacter jejuni* Cia proteins is contact dependent. *Adv. Exp. Med. Biol.*, 473, 225–229.

Roberts, I.S. 1996. The biochemistry and genetics of capsular polysaccharide production in bacteria. *Annu. Rev. Microbiol.*, 50, 285–315.

Rollwagen, F.M., Pacheco, N.D., Clements, J.D., Pavlovskis, O., Rollins, D.M., and Walker, R.I. 1993. Killed *Campylobacter* elicits immune response and protection when administered with an oral adjuvant. *Vaccine*, 11(13), 1316–1320.

Ruiz-Palacios, G.M. 2007. The health burden of *Campylobacter* infection and the impact of antimicrobial resistance: Playing chicken. *Clin. Infect. Dis.*, 44(5), 701–703.

Russell, R.G., Blaser, M.J., Sarmiento, J.I., and Fox, J. 1989. Experimental *Campylobacter jejuni* infection in *Macaca nemestrina*. *Infect. Immun.*, 57(5), 1438–1444.

Sahin, O., Luo, N., Huang, S., and Zhang, Q. 2003. Effect of *Campylobacter*-specific maternal antibodies on *Campylobacter jejuni* colonization in young chickens. *Appl. Environ. Microbiol.*, 69(9), 5372–5379.

Sahin, O., Morishita, T.Y., and Zhang, Q. 2002. *Campylobacter* colonization in poultry: Sources of infection and modes of transmission. *Anim. Health Res. Rev.*, 3(2), 95–105.

Sahin, O., Zhang, Q., Meitzler, J.C., Harr, B.S., Morishita, T.Y., and Mohan, R. 2001. Prevalence, antigenic specificity, and bactericidal activity of poultry anti-*Campylobacter* maternal antibodies. *Appl. Environ. Microbiol.*, 67(9), 3951–3957.

Savidge, T.C., Smith, M.W., James, P.S., and Aldred, P. 1991. *Salmonella*-induced M-cell formation in germ-free mouse Peyer's patch tissue. *Am. J. Pathol.*, 139(1), 177–184.

Scallan, E., Hoekstra, R.M., Angulo, F.J., Tauxe, R.V., Widdowson, M.A., Roy, S.L., Jones, J.L., and Griffin, P.M. 2011. Foodborne illness acquired in the United States–Major pathogens. *Emerg. Infect. Dis.*, 17(1), 7–15.

Scott, D.A. 1997. Vaccines against *Campylobacter jejuni*. *J. Infect. Dis.* 176(Suppl 2), S183–S188.

Semchenko, E.A., Day, C.J., Wilson, J.C., Grice, I.D., Moran, A.P., and Korolik, V. 2010. Temperature-dependent phenotypic variation of *Campylobacter jejuni* lipooligosaccharides. *BMC Microbiol.*, 10, 305.

Shanker, S., Lee, A., and Sorrell, T.C. 1988. Experimental colonization of broiler chicks with *Campylobacter jejuni*. *Epidemiol. Infect.*, 100(1), 27–34.

Shoaf-Sweeney, K.D., Larson, C.L., Tang, X., and Konkel, M.E. 2008. Identification of *Campylobacter jejuni* proteins recognized by maternal antibodies of chickens. *Appl. Environ. Microbiol.*, 74(22), 6867–6875.

Shroff, K.E., and Cebra, J.J. 1995. Development of mucosal humoral immune responses in germ-free (GF) mice. *Adv. Exp. Med. Biol.*, 371A, 441–446.

Stephenson, H.N., John, C.M., Naz, N., Gundogdu, O., Dorrell, N., Wren, B.W., Jarvis, G.A., and Bajaj-Elliott, M. 2013. *Campylobacter jejuni* lipooligosaccharide sialylation, phosphorylation and amide/ester linkage modifications fine-tune human Toll-like receptor 4 activation. *J. Biol. Chem.*, 288(27), 19661–19672.

Stern, N.J. 1992. Reservoirs for *Campylobacter jejuni* and Approaches for Intervention in Poultry, *Campylobacter jejuni*: Current Status and Future Trends. Washington, DC: American Society for Microbiology. pp. 49–60.

Stern, N.J., Bailey, J.S., Blankenship, L.C., Cox, N.A., and McHan, F. 1988. Colonization characteristics of *Campylobacter jejuni* in chick ceca. *Avian Dis.*, 32(2), 330–334.

Sung, J., Morales, W., Kim, G., Pokkunuri, V., Weitsman, S., Rooks, E., Marsh, Z., Barlow, G.M., Chang, C., and Pimentel, M. 2013. Effect of repeated *Campylobacter jejuni* infection on gut flora and mucosal defense in a rat model of post infectious functional and microbial bowel changes. *Neurogastroenterol. Motil.*, 25(6), 529–537.

Szeri, I., Anderlik, P., Banos, Z., and Radnai, B. 1976. Decreased cellular immune response of germ-free mice. *Acta Microbiol. Acad. Sci. Hung.*, 23(3), 231–234.

Szymanski, C.M., Burr, D.H., and Guerry, P. 2002. *Campylobacter* protein glycosylation affects host cell interactions. *Infect. Immun.*, 70(4), 2242–2244.

Szymanski, C.M., King, M., Haardt, M., and Armstrong, G.D. 1995. *Campylobacter jejuni* motility and invasion of Caco-2 cells. *Infect. Immun.*, 63(11), 4295–4300.

Szymanski, C.M., Yao, R., Ewing, C.P., Trust, T.J., and Guerry, P. 1999. Evidence for a system of general protein glycosylation in *Campylobacter jejuni*. *Mol. Microbiol.*, 32(5), 1022–1030.

Tauxe, R.V. 1992. Epidemiology of Campylobacter jejuni infections in the United States and other industrial nations. In: Nachamkin, I., Blaser, M.J., and Tompkins, L.S., editors. *Campylobacter jejuni: Current and Future Trends*. Washington, DC: American Society for Microbiology. pp. 9–12.

Teh, K.H., Flint, S., and French, N. 2010. Biofilm formation by *Campylobacter jejuni* in controlled mixed-microbial populations. *Int. J. Food Microbiol.*, 143(3), 118–124.

Theoret, J.R., Cooper, K.K., Glock, R.D., and Joens, L.A. 2011. A *Campylobacter jejuni* Dps homolog has a role in intracellular survival and in the development of campylobacteriosis in neonate piglets. *Foodborne Pathog. Dis.*, 8(12), 1263–1268.

Thibault, P., Logan, S.M., Kelly, J.F., Brisson, J.R., Ewing, C.P., Trust, T.J., and Guerry, P. 2001. Identification of the carbohydrate moieties and glycosylation motifs in *Campylobacter jejuni* flagellin. *J. Biol. Chem.*, 276(37), 34862–34870.

Trachoo, N., Frank, J.F., and Stern, N.J. 2002. Survival of *Campylobacter jejuni* in biofilms isolated from chicken houses. *J. Food Prot.*, 65(7), 1110–1116.

Tribble, D.R., Baqar, S., Scott, D.A., Oplinger, M.L., Trespalacios, F., Rollins, D., Walker, R.I., Clements, J.D., Walz, S., Gibbs, P., Burg 3rd, E.F., Moran, A.P., Applebee, L., and Bourgeois, A.L. 2010. Assessment of the duration of protection in *Campylobacter jejuni* experimental infection in humans. *Infect. Immun.*, 78(4), 1750–1759.

van der Waaij, D., Berghuis-de Vries, J.M., and Lekkerkerk, L.-V. 1971. Colonization resistance of the digestive tract in conventional and antibiotic-treated mice. *J. Hyg.*, 69(3), 405–411.

Van Deun, K., Pasmans, F., Ducatelle, R., Flahou, B., Vissenberg, K., Martel, A., Van den Broeck, W., Van Immerseel, F., and Haesebrouck, F. 2008. Colonization strategy of *Campylobacter jejuni* results in persistent infection of the chicken gut. *Vet. Microbiol.*, 130(3–4), 285–297.

Vitovec, J., Koudela, B., Sterba, J., Tomancova, I., Matyas, Z., and Vladik, P. 1989. The gnotobiotic piglet as a model for the pathogenesis of *Campylobacter jejuni* infection. *Int. J. Med. Microbiol.*, 271(1), 91–103.

Vuckovic, D., Abram, M., and Doric, M. 1998. Primary *Campylobacter jejuni* infection in different mice strains. *Microb. Pathog.*, 24(4), 263–268.

Wacker, M., Linton, D., Hitchen, P.G., Nita-Lazar, M., Haslam, S.M., North, S.J., Panico, M., Morris, H.R., Dell, A., Wren, B.W., and Aebi, M. 2002. N-linked glycosylation in *Campylobacter jejuni* and its functional transfer into *E. coli. Science*, 298(5599), 1790–1793.

Wang, Q., Wang, H., and Zhang, H. 2000. Experimental study on an animal model of axonal form Guillain–Barre syndrome. *Zhonghua Yi Xue Za Zhi*, 80(12), 947–949.

Wassenaar, T.M., and Blaser, M.J. 1999. Pathophysiology of *Campylobacter jejuni* infections of humans. *Microbes Infect.*, 1(12), 1023–1033.

Wassenaar, T.M., Bleumink-Pluym, N.M., and van der Zeijst, B.A. 1991. Inactivation of *Campylobacter jejuni* flagellin genes by homologous recombination demonstrates that flaA but not flaB is required for invasion. *EMBO J.*, 10(8), 2055–2061.

Wassenaar, T.M., van der Zeijst, B.A., Ayling, R., and Newell, D.G. 1993. Colonization of chicks by motility mutants of *Campylobacter jejuni* demonstrates the importance of flagellin A expression. *J. Gen. Microbiol.*, 139(Pt 6), 1171–1175.

Whitfield, C., and Valvano, M.A. 1993. Biosynthesis and expression of cell-surface polysaccharides in Gram-negative bacteria. *Adv. Microb. Physiol.*, 35, 135–246.

Wilson, D.J., Gabriel, E., Leatherbarrow, A.J., Cheesbrough, J., Gee, S., Bolton, E., Fox, A., Fearnhead, P., Hart, C.A., and Diggle, P.J. 2008. Tracing the source of campylobacteriosis. *PLoS Genet.*, 4(9), e1000203.

Wilson, D.L., Bell, J.A., Young, V.B., Wilder, S.R., Mansfield, L.S., and Linz, J.E. 2003. Variation of the natural transformation frequency of *Campylobacter jejuni* in liquid shake culture. *Microbiology*, 149(Pt 12), 3603–3615.

Wilson, D.L., Rathinam, V.A., Qi, W., Wick, L.M., Landgraf, J., Bell, J.A., Plovanich-Jones, A., Parrish, J., Finley, R.L., Mansfield, L.S., and Linz, J.E. 2010. Genetic diversity in *Campylobacter jejuni* is associated with differential colonization of broiler chickens and C57BL/6J IL10-deficient mice. *Microbiology*, 156(Pt 7), 2046–2057.

Wosten, M.M., Wagenaar, J.A., and van Putten, J.P. 2004. The FlgS/FlgR two-component signal transduction system regulates the fla regulon in *Campylobacter jejuni. J. Biol. Chem.*, 279(16), 16214–16222.

Yao, R., Burr, D.H., Doig, P., Trust, T.J., Niu, H., and Guerry, P. 1994. Isolation of motile and non-motile insertional mutants of *Campylobacter jejuni*: The role of motility in adherence and invasion of eukaryotic cells. *Mol. Microbiol.*, 14(5), 883–893.

Yao, R., Burr, D.H., and Guerry, P. 1997. CheY-mediated modulation of *Campylobacter jejuni* virulence. *Mol. Microbiol.*, 23(5), 1021–1031.

Young, C.R., Ziprin, R.L., Hume, M.E., and Stanker, L.H. 1999. Dose response and organ invasion of day-of-hatch Leghorn chicks by different isolates of *Campylobacter jejuni. Avian Dis.*, 43(4), 763–767.

Young, N.M., Brisson, J.R., Kelly, J., Watson, D.C., Tessier, L., Lanthier, P.H., Jarrell, H.C., Cadotte, N., St Michael, F., Aberg, E., and Szymanski, C.M. 2002. Structure of the N-linked glycan present on multiple glycoproteins in the Gram-negative bacterium, *Campylobacter jejuni. J. Biol. Chem.*, 277(45), 42530–42539.

Youssef, M., Corthier, G., Goossens, H., Tancrede, C., Henry-Amar, M., and Andremont, A. 1987. Comparative translocation of enteropathogenic *Campylobacter* spp. and *Escherichia coli* from the intestinal tract of gnotobiotic mice. *Infect. Immun.*, 55(4), 1019–1021.

Yrios, J.W., and Balish, E. 1986a. Colonization and infection of athymic and euthymic germfree mice by *Campylobacter jejuni* and *Campylobacter fetus* subsp. *Fetus. Infect. Immun.*, 53(2), 378–383.

Yrios, J.W., and Balish, E. 1986b. Immune response of athymic and euthymic germfree mice to *Campylobacter* spp. *Infect. Immun.*, 54(2), 339–346.

Yrios, J.W., and Balish, E. 1986c. Pathogenesis of *Campylobacter* spp. in athymic and euthymic germfree mice. *Infect. Immun.*, 53(2), 384–392.

Yuki, N., and Odaka, M. 2005. Ganglioside mimicry as a cause of Guillain–Barre syndrome. *Curr. Opin. Neurol.*, 18(5), 557–561.

Zheng, J., Meng, J., Zhao, S., Singh, R., and Song, W. 2008. *Campylobacter*-induced interleukin-8 secretion in polarized human intestinal epithelial cells requires *Campylobacter*-secreted cytolethal distending toxin- and Toll-like receptor-mediated activation of NF-kappaB. *Infect. Immun.*, 76(10), 4498–4508.

Ziprin, R.L., Young, C.R., Byrd, J.A., Stanker, L.H., Hume, M.E., Gray, S.A., Kim, B.J., and Konkel, M.E. 2001. Role of *Campylobacter jejuni* potential virulence genes in cecal colonization. *Avian Dis.*, 45(3), 549–557.

Chapter 29

Pathogenesis of *Francisella tularensis* in Humans

Monique Barel[1,2], Nathalie Grall[3,4], and Alain Charbit[1,2]

[1]Université Paris Descartes, Sorbonne Paris Cité, Bâtiment Leriche, Paris, France

[2]INSERM, U1151, Unité de Pathogénie des Infections Systémiques, Paris, France

[3]Laboratoire de Bactériologie, Hôpital Bichat-Claude Bernard, APHP, Paris, France

[4]EA 3964 - Université Paris-Diderot, Paris, France

29.1 Introduction

Tularemia is an endemo-epidemic zoonosis in the Northern Hemisphere. It is a rare and poorly understood disease whose manifestations are highly polymorphic, ranging from ulcerative ganglion with benign forms to fatal pulmonary forms. The causative agent, *Francisella tularensis* is a small Gram-negative bacillus, aerobic, nonspore-forming and nonmotile. This intracellular pathogen is one of the most infectious for humans. It is considered a potential Class A agent in bioterrorism by the Centers for Disease Control (CDC) in the United States, as well as pathogen responsible for smallpox, plague, anthrax, botulism, or hemorrhagic fevers (Dennis *et al.*, 2001). Since the events of September 11, 2001 in the United States, this disease has taken on greater importance. In this context, it appears necessary to better understand the mechanisms of the pathogenesis of this disease in order to develop appropriate preventive and curative treatments.

29.1.1 Taxonomy

At the taxonomic level, the position of *F. tularensis* is complex and has changed over time. This bacterium has undergone successive different names: *Bacterium tularense, Pasteurella tularensis, Francisella tularense,* and finally *Francisella tularensis*, a tribute to Sir Edward Francis. However, the study of 16S rRNA indicates that the genus *Francisella* belongs to the group of γ-Proteobacteria (Keim et al., 2007).

This type contains two *Francisella* species: *F. tularensis* and *F. philomiragia*. *F. philomiragia* is an opportunistic pathogen, which is rarely responsible for disease in humans. Four subspecies (ssp. or biovars) of *F. tularensis* are currently listed: *F. tularensis* ssp. *tularensis* (type A), *F. tularensis* ssp. *holarctica* (type B), *F. tularensis* ssp. *mediasiatica*, and *F. tularensis* ssp. *novicida* (McLendon et al., 2006). The four subspecies differ in their virulence and geographical origin, but are very close to a phylogenetic perspective. *F. tularensis* type A is the most virulent subspecies, with a mortality rate of 30–60% in severe forms (septic tularemia and respiratory tularemia) without antibiotic treatment. *F. tularensis* type B is also responsible for severe diseases in humans, but rarely lethal, with a mortality rate of about 1% in the absence of treatment. The other two subspecies are less well known, but do not cause severe disease in humans (Oyston et al., 2004). The live vaccine strain (LVS) strain of *F. tularensis* is derived from the type B strain. The Soviets used live attenuated strains of *F. tularensis* ssp. *holarctica* as vaccines in the 1940s and 1950s. One of these strains was provided to the United States in 1956, which selected a blue variant on agar-containing peptones and cysteine. The strain of *F. tularensis* LVS drifted off this blue variant after complex manipulations, including lyophilization and repeated passages in mice (Wayne Conlan and Oyston, 2007). *F. tularensis* LVS is no longer virulent for humans, but has retained its virulence in mice. Intraperitoneal injection of *F. tularensis* LVS in mice causes fulminate fatal disease that appears similar to human tularemia at the histopathological level.

29.1.2 Bacteriological characters

F. tularensis is a small Gram-negative bacillus (0.2–1.7 μm in length by 0.2–0.7 μm in diameter), strictly aerobic, nonspore-forming and nonmotile. In addition to the Gram staining, *F. tularensis* can be more easily identified by May Grunewald Giemsa (MGG). Virulent forms of *F. tularensis* are surrounded with a capsule of 0.02–0.04 μm thickness. Its loss does not affect its viability but is accompanied by a loss in virulence. The lipids in the capsule and wall are unusually in high proportion for a Gram-negative (50% and 70%, respectively) and the nature of the fatty acids is particular to the genus *Francisella* (fatty acid with long carbon chains (C18–C26) saturated or monounsaturated and with two types of hydroxy fatty acids, C16: 0 3OH and C18: 0 3OH) (Jantzen et al., 1979). Bacteria are resistant to cold, drying, and freezing. Thus, below 0°C, it can persist for several months in water, mud, or in infected animal carcasses. Above 10°C, persistence does not exceed a few days. It is destroyed by heating for 30 min at 56°C and by common disinfectants and antiseptics. At the biochemical level, *F. tularensis* is urease negative, weakly catalase positive, oxidase negative, indole negative, and does not hydrolyze gelatin, nor does it reduce nitrates. It acidifies slowly and lowly glucose without gas production. The *tularensis* subspecies, unlike the subspecies *holarctica*, has a citrulline–ureidase activity. Subspecies *holarctica* has three biotypes: biotype I (sensitive to erythromycin, glucose + maltose + glycerol −), biotype II (resistant to erythromycin, glucose + maltose + glycerol −), and biotype III (sensitive to erythromycin, glucose + maltose + + glycerol). Culture of *F. tularensis* is slow (2–5 days), difficult, and cannot be performed on the usual media. It is therefore carried out on enriched media, at the optimum temperature of 37°C, aerobically, and sometimes in CO_2-enriched atmosphere (5–7%). Existing media are Francis medium (1% glucose, 0.1% cysteine, and 10% defibrinated rabbit blood), McCoy and Chapin medium (based on egg yolk), Kudo medium (based on egg yolk and rabbit serum), or blood agar supplemented with Polyvitex. The latter is used the most, and small pearly gray colonies are obtained in 2–5 days.

29.1.3 Molecular biological aspects including details of genome

The genome of the *F. tularensis* ssp. *tularensis* Schu S4 strain was sequenced and analyzed in 2005 (Larsson et al., 2005). The circular double-stranded DNA genome is relatively small, measuring only 1,892,819 base pairs and has an overall low GC% (32%). *In silico* analysis of the genome has highlighted a high proportion of incomplete pathways, explaining the tedious nature of the nutritional requirements of the bacteria. These observations suggest an evolution by reducing the coding capacity of the genome, which could lead

ultimately *F. tularensis* to become an obligate intracellular pathogen. Sequence of *F. tularensis* ssp. *holarctica* OSU18 was completed in 2006 (Petrosino et al., 2006), *F. novicida* U112 in 2007 (Rohmer et al., 2007), *F. tularensis* ssp. *mediasiatica* FSC147 in 2009 (Larsson et al., 2009), and *F. philomiragia* in 2012 (Zeytun et al., 2012). The number of protein genes ranges from 1754 to 1604 for ssp. *holartica* (LVS strain) or ssp. *tularensis* (Schu 4 strain), respectively, and the number of RNA genes ranges from 53 to 48 for LVS or Schu 4, respectively. A series of *Francisella* lineages have been recently identified, which are not associated with human disease. In particular, strains closely related to *F. philomiragia* have been identified as etiological agents of Francisellosis in fish and designated as *Francisella noatunensis*. With the increasing facility and rapidity to sequence bacterial genomes, it is likely that novel species and subspecies will be discovered in the coming years, considerably expanding the boundaries of the genus *Francisella* (Sjodin et al., 2012 and references therein).

29.2 Epidemiology

29.2.1 Incidence and geographic distribution

Tularemia is found in forested areas of the Northern Hemisphere, between the 30th and 71st parallel (North America, Continental Europe, Russia, China, Japan). *F. tularensis* ssp. *tularensis* (type A) is found mainly in North America, including the United States, *F. tularensis* ssp. *holarctica* (type B) is present mainly in Europe but also in the Northern United States and Canada; *F. tularensis* ssp. *mediasiatica* is found in Central Asia, and *F. tularensis* ssp. *novicida* is present in North America and Australia. Due to benign cases going unnoticed on the one hand, and the difficulties of bacteriological diagnosis on the other, the number of cases is probably underestimated. In addition, many countries do not register cases of tularemia, and the exact amount of the impact in the world remains unknown. In North America, the United States is particularly concerned with tularemia, although the incidence declined steadily since 1950, from more than 1000 cases per year to 100–200 cases per year currently. This decrease in the incidence of tularemia was also observed in the same way in other countries. The disease is present in all states outside of Hawaii, but the Southern states, Arkansas, Missouri, and Massachusetts, are the most affected. In Europe, the Northern states (Sweden, Norway, Finland) and the Eastern European countries are most concerned. However, the disease has never been reported in Britain. There is also a seasonal distribution of cases of tularemia, with a winter peak from November to February, corresponding to the hunting season, and a summer peak from June to August, corresponding to the period of activity of ticks and mosquitoes (Scandinavia and the United States mainly). Tularemia occurs most often in sporadic form but epidemics have also been observed. An outbreak occurred in

Kosovo in 1999, with 327 cases serologically confirmed. The infection was transmitted by ingestion of water or food contaminated with sick rodents, and favored by poor sanitary conditions in the period of post-war (Reintjes et al., 2002). In 2000, an outbreak of 270 confirmed cases was described in Sweden. The risk factors found were mosquito bites, work on farm, and possession of cat (Eliasson et al., 2002). In United States, an outbreak of pneumonic tularemia was declared in 2000 on Martha's Vineyard Island. Fifteen cases were identified, with 11 cases with pneumonia and 1 death. This outbreak was linked to the inhalation of dust aerosols caused by lawnmowers (Feldman et al., 2001; Matyas et al., 2007). Outbreaks also occurred in Spain in 2007 and in northern Norway in 2011.

29.2.2 Reservoirs and vector

The natural reservoir of *F. tularensis* is extremely wide and varied, and the natural cycle of the bacteria involves the environment, animals, and biting arthropods (Figure 29-1).

F. tularensis was isolated from more than 250 species of mammals, birds, amphibians, and invertebrates. According to the species, the susceptibility to the disease varies, and it is possible to distinguish three groups. Group 1 consists of the most susceptible animals, such as rodents and lagomorphs. These animals develop septicemia form of the disease and die within 1–2 weeks. They represent an important reservoir of the disease (wild rabbits, beavers, and muskrats in the United States; rats, rabbits, and rodents in Europe). Group 2 consists of other species

of rodents, birds, sheep, cattle, and dogs. These animals are susceptible, but the fatality rate is low. Finally, group 3 consists of carnivores. These animals are very receptive, and rarely develop clinical disease, unless there is a very high infective dose. They play only a minor role in the cycle of tularemia. Blood-sucking arthropods (fleas, lice, bugs, horseflies, mosquitoes, and ticks) play a role of vector and reservoir of the disease. Indeed, the trans-ovarian transmission of *F. tularensis* has been shown in the tick. The numerous species of ticks provide an inexhaustible reservoir of the disease, such as *Ixodes ricinus*, *Dermacentor reticularis*, *Dermacentor variabilis*, *Dermacentor andersoni*, or *Amblyomma americanum*. The bacterium is present in the feces, and especially in the salivary glands of the tick, allowing rapid injection at the bite. The environment is contaminated with feces, mainly from rodents. The bacterial survival then depends on the outside temperature. Below 0°C, it can survive for several months in the bodies of infected animals, mud, water, earth. However, above 10°C, survival is only a few days. The persistence of the bacteria in the aquatic environment could also be related to its survival in aquatic protozoa such as amoebae (Greub and Raoult, 2004).

29.2.3 Population at risk

Exposed activities are among other risk factors: laboratory staff, veterinarians, foresters, gamekeepers, farmers, shepherds, butchers, and quarterers. Some leisure activities are also a risk factor. Hunters are particularly vulnerable, by handling contaminated game or animals that died from tularemia. The outdoor activities like gardening, hiking, or camping promote arthropod bites and therefore contamination by *F. tularensis*. Finally, adverse socioeconomic conditions may lead to contamination of food and water by the development of the rodent population and the lack of health facilities. Thus, an outbreak occurred in Kosovo in 1999–2000, in the post-war period (Reintjes et al., 2002).

29.3 Clinical features

29.3.1 Modes of transmission

Modes of contamination and entry of the bacteria are multiple. The most common penetration pathway is the cutaneous way. The bacteria then enters through intact skin, when there is direct contact with infected live or dead animal, after handling contaminated objects, or through skin tissue gap from scratch from an infected animal, or bite from blood-sucking arthropods (mainly ticks and mosquitoes). Conjunctive mucous membranes and throat are also routes of contamination, in case of contaminated liquid projection after eye rubbing with contaminated hands, or immersion in fresh water contaminated by feces of infected animals. Contamination from the

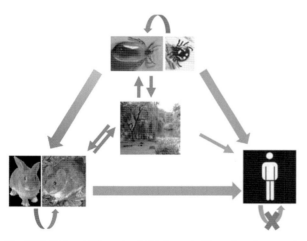

Fig. 29-1 Natural life cycle of *F. tularensis*. The natural reservoir of *F. tularensis* is extremely wide and varied, and the natural cycle of the bacteria involves the environment, animals, and biting arthropods. More than 250 species are susceptible to the bacteria. Three groups are distinguished. Group 1 is the most susceptible with rodents and lagomorphs. Group 2 (birds, sheep, cattle, and dogs) has a lower fatality rate. Group 3 rarely develops clinical disease. Blood-sucking arthropods (fleas, lice, bugs, horseflies, mosquitoes, and ticks) play a role of vector and reservoir of the disease. The environment is contaminated with feces, mainly from rodents. There is no interhuman transmission, and it is not necessary to isolate patients with tularemia.

gastrointestinal tract occurs after consumption of contaminated water or undercooked meat from an infected animal. This mode of transmission is the most common cause of sporadic or familial cases, but outbreaks have also been described, such as occurred in the Czech Republic in 1978 (131 cases), following the consumption of apple juice craft made with apples contaminated by rodents (Cerny, 2001). Pulmonary contamination by inhalation aerosol is caused by the manipulation of animals or contaminated products (fur, wool, litter, seeds), rural work or gardening (lawn mowing, clearing). The epidemic, which occurred in Spain in 2007, with 507 cases is an example of an outbreak of pneumonic tularemia associated with the inhalation of infected aerosols during farm work. This route of infection, which can cause a severe form of pneumonic tularemia, makes this bacterium a class A bioterrorism agent. However, there is no inter-human transmission, and it is not necessary to isolate patients with tularemia. The infective dose depends on the route of infection: 5–10 bacteria by airway are enough while 10^6–10^8 bacteria in the gastrointestinal tract are necessary (Figure 29-2).

29.3.2 Incubation phase and invasion

The incubation period is generally 3–5 days, with a range from 1 to 15 days. Then the disease begins with abrupt fever, which can be undulating, chills, headache, fatigue, and anorexia. The presence of nausea and/or vomiting, diarrhea, myalgia, or rash is also possible. After the invasion phase, six main clinical forms can be distinguished. They are extremely varied, both by location and gravity, and depend on the entrance door of the bacteria.

29.3.2.1 Ulcerative form node

The ulcero-glandular form is the most common and accounts for approximately 80% of cases. A lesion type papule appears at the site of inoculation of the bacteria, and then progresses to ulceration. This painful and oozing ulceration is generally unique, but can be multiple. From this injury, the bacteria gain the node draining the territory affected by the lymphatic system and multiply. Then lymphadenopathy appears which progresses to abscess, fistula, and sclerosis.

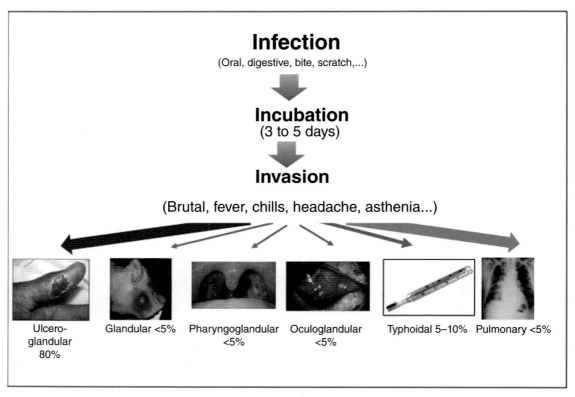

Fig. 29-2 Clinical features. Infection occurs after multiple modes of contamination and entry. The most common penetration pathway is the cutaneous way, through a skin tissue gap (after bite or scratch) or through intact skin. Conjunctive mucous membranes and throat are also routes of contamination, as the gastrointestinal tract after consumption of contaminated water or undercooked meat from an infected animal. Pulmonary contamination by inhalation aerosol is caused by the manipulation of animals or contaminated products (fur, wool, litter, seeds), rural work, or gardening (lawn mowing, clearing). The incubation period is generally 3–5 days. The disease begins with abrupt fever, with chills, headache, fatigue, anorexia and possible nausea and/or vomiting, diarrhea, myalgia, or rash. After the invasion phase, six main clinical forms can be distinguished. The ulceroglandular form is the most common and accounts for approximately 80% of cases. Then lymphadenopathy appears. The oropharyngeal form is quite rare, as the oculoganglionar form. The typhoid form is a serious septicemia, which represents 5–10% of cases. The pulmonar form is most often secondary to inhalation of infected aerosols, but can also complicate other clinical forms of tularemia after hematogenous spread of bacteria. Very few bacteria are needed to develop these severe clinical forms. This is rare, but probably underdiagnosed because of the lack of specific signs and the difficulty of diagnosis. This respiratory disease may be life-threatening because of the development of acute respiratory distress syndrome.

These nodes are single or multiple, painful, and surrounded by a periadenitis. In case of contact with infected animal or equipment, the lesion is usually in the upper limbs, with an axillary lymphadenopathy. In case of tick bite, the lesion is more likely located in the lower limbs or trunk. Injury to the head and neck rather evokes a mosquito bite.

29.3.2.2 Pharyngeal lymph form

The oropharyngeal form is quite rare and is most often due to contamination by oral ingestion of undercooked meat, water, or raw contaminated vegetables. It presents as a febrile pharyngitis, erythematous, or ulcerative pulpy, with the appearance of submandibular lymph and/or carotid-jugular nodes. Sometimes, it is complicated with abdominal pain (mesenteric lymph nodes), nausea, vomiting, or diarrhea.

29.3.2.3 Oculoganglionar form

The oculoganglionar form represents less than 5% of cases. It follows conjunctive inoculation through contaminated fingers or more rarely by projection of contaminated material, and comes in the form of a usually painful unilateral conjunctivitis, with pus. It is accompanied with lymphadenopathy satellites, and severe infectious syndrome. Ocular forms may be however complicated by corneal ulcers, dacryocystitis, or dacryoadenitis.

29.3.2.4 Node form

The nodal form is similar to the ulcerative node form, but with no lesions. Tularemia is one of the etiologies of adenitis and can be confused with cat scratch disease.

29.3.2.5 Pulmonary form

The pulmonary form is most often secondary to inhalation of infected aerosols, but can also complicate other clinical forms of tularemia after hematogenous spread of bacteria (10–15% of ulcerative nodes tularemia, 30–70% of typhoid tularemia). Very few bacteria are needed to develop these severe clinical forms: 5–10 bacteria of *F. tularensis* ssp. *tularensis*, 100–1000 bacteria of *F. tularensis* ssp. *holarctica*. This is rare, but probably underdiagnosed because of the lack of specific signs and the difficulty of diagnosis. Clinically, fever is high and may be accompanied by shortness of breath, dry cough, or blood stained sputum and chest pain. Pulmonary auscultation is low, and chest radiography may show unilateral or bilateral infiltrates, hilar lymphadenopathy, and/or pleural effusions. This respiratory disease may be life-threatening because of the development of acute respiratory distress syndrome.

29.3.2.6 Typhoid form

This serious septicemia form represents 5–10% of cases. This is a febrile pure form, without skin ulceration or lymphadenopathy. It is associated with a state of prostration, weight loss, sometimes with abdominal pain and pulmonary form in 30–80% of cases.

29.3.2.7 Other clinical forms

Other subclinical or asymptomatic forms are regularly demonstrated in epidemic areas and in epidemiological surveys. Rashes, with kind of papules, pustules, nodosum erythema, urticarial lesions, can be observed in 10–20% of cases. They are called tularemides. Other atypical forms have also been described: meningitis and meningoencephalitis, endocarditis, pericarditis, pleurisy, peritonitis, or hepatitis.

29.3.2.8 Prognosis

Without treatment, the disease progresses over 3–5 weeks, more rarely to several months. The recovery is long, a few weeks to several months, with persistent fatigue, and sometimes bouts of intermittent fever. Treatment should be very early, so that the disease does not take a slow and crippling pace. The mortality rate in the absence of treatment varies significantly depending on the clinical presentation and the virulence of the causative agent. Indeed, mortality without treatment is about 1% for infections by *F. tularensis* ssp. *holarctica*, whereas it can reach 30–60% of infections with *F. tularensis* ssp. tularensis. Poor prognostic factors include presence of a weakened state, with typhoid or neuro-brain forms, a delay in diagnosis, inappropriate antibiotic treatment, or rhabdomyolysis.

29.4 Pathogenesis and immunity

29.4.1 Host–pathogen interaction

F. tularensis is a facultative intracellular bacteria, which is able to infect numerous cell types such as dendritic cells, neutrophils, hepatocytes, alveolar macrophages, endothelial cells, and fibroblasts. However, it replicates *in vivo* mainly in macrophages (Santic et al., 2006). Two well-differentiated mechanisms of entry exist for invasive bacteria: the 'zipper' or the 'trigger' mechanisms. *Listeria monocytogenes* invade host cells by the zipper mechanism. *Salmonella* and *Shigella* use the trigger mechanism. The two entry pathways share a requirement for actin polymerization (Veiga and Cossart, 2006). Unlike these traditional mechanisms, *F. tularensis* is captured by macrophages whose membrane forms large loops or pseudopodia (Clemens et al., 2005) (Figure 29-3).

This mechanism is also dependent on actin polymerization. Efficient uptake of *F. tularensis* by human macrophages is dependent on the presence of serum

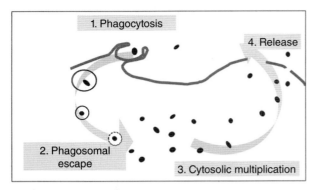

Fig. 29-3 *F. tularensis* is a facultative intracellular bacteria. After efficient uptake of *F. tularensis* by human macrophages through large loops of membrane and presence of serum in the medium, *F. tularensis* is taken up into spacious phagosomes. About 30–60 min after the onset of infection, the bacteria gradually escape from the phagosome by an unknown mechanism to multiply in the cytosol. Release occurs after apoptosis or pyroptosis.

whether in the culture medium or on preopsonized bacteria (Clemens and Horwitz, 2007). Adherence and uptake of the bacteria increase with increasing concentration of serum in the medium. When heat-inactivated, human serum does not support efficient uptake of *F. tularensis* by human macrophages (Barel et al., 2008). Apart from the complement receptor CR3 (CD11b/CD18) (Ben Nasr et al., 2006; Clemens et al., 2005), different phagocytic receptors have been identified by several laboratories and are involved in *Francisella* uptake: the mannose receptor (MR) (Schulert and Allen, 2006), the scavenger receptor A (SRA) (Pierini, 2006), FcγRs (Balagopal et al., 2006), and surface-exposed nucleolin (Barel et al., 2008).

29.4.2 Mechanisms and route of infection

After engulfment by phagocytic cells, *F. tularensis* is taken up into spacious phagosomes, which sequentially acquire markers of early and late endosomes, such as EEA-1 and LAMP (1 and 2). The phagosome or *F. tularensis*–containing phagosome (FCP) is processed until the late endosome stage, but the phagosome–lysosome fusion is inhibited. The progressive shrinking of the phagosome that was observed by electron microscopy (Clemens et al., 2009) helps to fit to the bacterium size. The importance and the mechanisms by which this volume restriction occurs are still unknown. While the importance of phagosomal acidification has been shown to be an absolute prerequisite for *L. monocytogenes* escape from the phagosomal compartment (Kayal and Charbit, 2006), some controversy still exists on the importance of this process for *F. tularensis* subspecies' abilities to access the host cytosol (Bonquist et al., 2008; Chong et al., 2008; Clemens et al., 2004, 2009; Cremer et al., 2009; Santic et al., 2008). Indeed, there is a transient acidification of the phagosome within the first 15–30 min, when bacteria are unopsonized (Santic et al., 2008, 2010). This transient acidification has been reported to be essential for the subsequent escape

of *F. tularensis* into the cytosol of the macrophage. Unopsonized *F. tularensis* also escapes the phagosome rapidly and replicates robustly in the cytosol (Jones et al., 2012). In contrast, there is only a modest acidification of only 20–30% of the phagosomes when bacteria are opsonized (Clemens et al., 2009) and this modest acidification seems not to be required for escape. Furthermore, opsonized bacteria seem to escape the phagosome with delayed kinetics and replicate only modestly in the host cell cytosol. Therefore, the route of uptake appears to have a profound impact on the outcome of infection (Celli and Zahrt, 2013). The oxidative burst is a process generally triggered in the phagosome. Intracellular pathogens like *F. tularensis* have developed different strategies to resist the reactive oxygen species generated. *Francisella* acid phosphatases regulate the generation of the oxidative burst in human phagocytes (Mohapatra et al., 2010). *F. tularensis* also inhibits the respiratory burst in polynuclear neutrophils by preventing the assembly of the NADPH oxidase by a still unknown mechanism (McCaffrey et al., 2010). About 30–60 min after the onset of infection, the bacteria gradually escape from the phagosome by an unknown mechanism to multiply in the cytosol (Oyston, 2008; Price et al., 2011). In the cytosol, the bacteria can feed themselves and multiply to a high number, after up to 7 or 9 cycles of replication. Genes encoding either metabolic pathways or predicted membrane proteins have been identified in genetic screens of *F. tularensis*. An important role for metabolic adaptation and nutrient acquisition system in intracellular survival of *Francisella* is therefore supported (Meibom and Charbit, 2010a). In response to deprivation of nutrients by intracellular pathogens, the host cell adapts its metabolism by manipulating its cytosolic content (Barel et al., 2012). In the cytosol, the bacteria also inhibit the production of proinflammatory cytokines by macrophages, including IL-1, TNF, IL-6, and IL-12 (Bublitz et al., 2010; Telepnev et al., 2005). In macrophages from murine bone marrow, the bacteria could also re-enter the path of endocytosis by autophagy after intracytoplasmic multiplication, through the creation of a multiple membrane structure, "*Francisella*-containing vacuole" (FCV) (Checroun et al., 2006). The function of these organelles remains poorly understood. *F. tularensis* causes the death of macrophages by apoptosis through the intrinsic pathway (Allen and McCaffrey, 2007; Lai and Sjostedt, 2003; Schwartz et al., 2013) or by pyroptosis via the inflammasome pathway (Broz and Monack, 2011). Finally, the bacterium is also able to delay the activation of the inflammasome and thus cell death and stimulation of the immune system (Weiss et al., 2007).

29.4.3 Molecular basis of virulence in *F. tularensis*

The major determinants of virulence known are the lipopolysaccharide (LPS), which show little

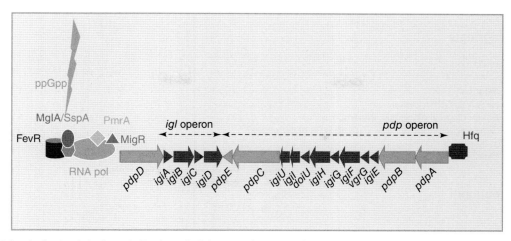

Fig. 29-4 Molecular basis of virulence in *F. tularensis*. Schematic depiction of *Francisella* pathogenicity island (FPI) and its regulation. The proteins MglA and SspA heterodimerize and associate with RNA polymerase (RNAP). The regulatory activity of the MglA–SspA–RNAP complex is dependent on the production of the alarmone ppGpp, which promotes interaction of the master regulator FevR (also called PigR in *F novicida*) with the MglA–SspA–RNAP complex. Upregulation of the FPI also depends on the association of PmrA and MigR to the complex. The RNA chaperon Hfq is the only protein known to negatively regulate the expression of one part of the FPI. The *igl* genes are in dark grey the *pdp* genes in light grey. The *arrows* indicate the orientation and approximate sizes of the FPI-encoded genes.

inflammatory effect, a 33 kb pathogenicity island (designated Francisella pathogenicity island or FPI) containing genes essential for intracellular survival, and a series of more than 200 genes scattered along the *Francisella* chromosome, participating to various extents to pathogenesis. As for many other pathogenic bacteria, the total number of genes connected to virulence, which correspond to ≥10% of the total genome, include notably virulence genes *per se*, regulatory genes, stress-related genes, as well as numerous metabolic genes that concur to bacterial intracellular adaptation. The focus of this paragraph is not to describe in detail the contribution of these numerous factors (see, for recent reviews on the topic, Celli and Zahrt, 2013; Meibom and Charbit, 2010b; Pechous et al., 2009). We will only briefly describe the FPI and its complex regulation (Figure 29-4).

Remarkably, the FPI is duplicated in the subspecies *tularensis, holarctica,* and *mediasiatica*, while it is present in a single copy in environmental strains *F. philomiragia* and *F. tularensis* ssp. *novicida* (Larsson et al., 2009). The FPI consists of two operons: *pdpD-iglABCD* operon, and the *pdp*A operon containing 11 genes (Barker et al., 2007). Except for *pdpD*, all these genes are required for intracellular survival and virulence. *In silico* analysis of the pathogenicity island showed homology of some genes (*pdpB, dotU, hcp clpV, vrgG, iglA, iglB*, and *iglD*) with genes known to be involved in a system of type VI secretion genes in other pathogenic bacteria such as *Pseudomonas aeruginosa* or *Vibrio cholerae*. The contribution of each of these genes to pathogenesis has been extensively studied in the past few years (see, e.g., Barker et al., 2009; Bröms et al., 2012; de Bruin et al., 2011; Lindgren et al., 2013). However, in spite of all the efforts devoted, the exact nature and mechanism of action of the hypothetical FPI-encoded Type VI secretion system is still unclear. Expression of the FPI is essentially under the control of a series of positive regulator proteins (MglA and SspA, which associate with subunit of the core structure of RNA polymerase, RNAP). This interaction depends on the master regulator FevR and requires the production of the alarmone ppGpp (mediated by the pair of proteins RelA/SpoT) (Dai et al., 2011). Upregulation of the FPI also depends on the action of at least two other regulators PmrA and MigR, which modulate the production of ppGpp (Faron et al., 2013). The RNA chaperone Hfq is the only protein known to negatively regulate the expression of half of the FPI genes (Meibom et al., 2009).

29.5 Diagnosis

Tularemia is a disease with a difficult diagnosis, especially as the incidence is low, the disease is unknown with highly polymorphic clinical forms. The presumptive diagnosis is clinical, but the definitive diagnosis is biological (Figure 29-5).

29.5.1 Clinical diagnosis

In opposition to Brucellosis, Lyme disease, or Leptospirosis, tularemia is wrongly not considered as pathology related to rural environment in general. For the majority of clinicians and microbiologists, it remains a disease particularly related to hares and hunters. For quick and effective treatment, diagnosis of tularemia should be considered, not only in front of a classic ulcerative form node, but also with all influenza-like illness in the context of contact with an animal or in an endemic area. The notions about risk (laboratory staff, veterinarians, hunters, farmers), area at risk (wetland ponds, lakes, marshes, forests), or blood-sucking arthropod bites should also be suggested for the diagnosis.

Diagnosis

• Clinical presumption

• Biological diagnosis:
        NFS, VS, CRP: normal
        Bacteriological diagnosis: uncertain
        Serological diagnosis: +++ (70% to 90% of the cases)

• Molecular biology: PCR RNA 16S, qRT-PCR (*fop*, *tul4*, 23 kDa, ISFtu2...)

Treatment

• No naturally acquired resistance described

• Treatment based on clinical experience

> Usually for adults:
> - Ciprofloxacine (per os: 1 g/day) or
> - Ofloxacine (per os: 800 mg/day) or
> - Levofloxacine (per os: 500 mg/day) or
> - Doxycycline (per os: 200 mg/day)

Fig. 29-5 Diagnosis and treatment. While diagnosis is difficult, due to low incidence and highly polymorphic clinical forms, biological diagnosis may be obtained after nonspecific or specific laboratory examinations. Molecular biology tests may be necessary but are restricted to specialized laboratories. Effective antibiotic treatment should be started as soon as possible.

29.5.2 Biological diagnosis

Due to the risk of contamination of laboratory personnel, handling of infected samples, cultures, and strains of *Francisella* must be performed in a biological safety laboratory type P2 for *F. tularensis* ssp. *holarctica*, and P3 for *F. tularensis* ssp. *tularensis*. It is therefore necessary to announce laboratory personnel in case of clinical suspicion of tularemia.

29.5.2.1 Nonspecific

Blood count, erythrocyte sedimentation rate, and CRP are usually normal. In severe cases, neutrophil leukocytosis can be observed. CPK and liver enzymes may be increased.

29.5.2.2 Specific

The samples to be taken will vary depending on the clinical presentation: swabbing the skin secretions at inoculation point, punctures of lymphadenopathy (the earliest is the best as cultures become negative at day 10th), or swab exudates in the conjunctive or throat forms. Other samples are rarely performed, such as sputum, bronchoalveolar wash (BAL), or cerebrospinal fluid (CSF). In case of clinical sepsis, blood cultures will be taken.

29.5.2.2.1 Direct examination

Gram staining, although difficult, can show a small Gram-negative bacillus. MGG staining type may help identify more easily the strain.

It is also possible to detect *F. tularensis* in various samples by direct immunofluorescence, but few laboratories have labeled antibodies.

29.5.2.2.2 Culture

Culture should be performed aerobically at 37°C on enriched media, particularly with cysteine, such as blood agar supplemented with Polyvitex. The growth of the bacteria is slow, and small pearly colonies are obtained in 2–5 days. It should be noted that the bottles of blood cultures should be kept for at least 21 days. After culture, the identification of the organism is based on the bacteriological characters (non-motile bacillus small Gram-negative nonspore-forming, oxidase negative, weak positive catalase), and especially on the rapid agglutination on slide with *Francisella* anti-serum and on molecular biology techniques. The inoculation of laboratory animals is dangerous and should not be used.

29.5.2.2.3 Serology

Serology is positive during the second week of illness, reaching a peak in 1–2 months. The antibodies may persist for more than 10 years. An increase in antibody titer of at least four times on two successive analyses to 3 weeks apart is necessary for diagnosis. The specificity of serology is poor because of cross-reactions with *Brucella* spp., *Yersinia* spp., or *Proteus vulgaris* OX19. Specific *F. tularensis* anti-protein 17 kDa and anti-protein 43 kDa antibodies are evidenced by Western blot analysis. But these tests are restricted to more specialized laboratories.

29.5.2.2.4 Method of PCR

Gene amplification by polymerase chain reaction (PCR) can be performed on clinical specimens or after culture. The amplification and sequencing of the 16S rRNA gene allows differentiation of *F. tularensis* ssp. *tularensis* from *F. tularensis* ssp. *holarctica*, whose sequences differ by only one base. Detecting *F. tularensis* can also be performed by quantitative PCR, using primers and probes specific for the genes encoding either lipoprotein membrane Fop, tul4, 23 kDa protein, or ISFtu2.

29.5.2.2.5 Method of RFLP

The restriction fragment length polymorphism (RFLP) technique is based on both nature of ISFtu1 and ISFtu2 sequences and their insertion, used to characterize subspecies of *F. tularensis* (Thomas et al., 2003).

29.6 Treatment

Effective antibiotic treatment should be started as soon as possible. Antibiogram is carried out on Mueller–Hinton blood-enriched Polyvitex incubated aerobically at 37°C for 48 h. *F. tularensis* is sensitive *in vitro* to aminoglycosides (streptomycin, gentamicin, kanamycin), tetracyclines, macrolides (erythromycin, except strains of *F. tularensis* ssp. *holarctica* biotype II), synergistin (pristinamycin), chloramphenicol, furans, and fluoroquinolones (ofloxacin, ciprofloxacin, levofloxacin). Natural resistance is observed toward all penicillins, cephalosporins, of first and second generation, and co-trimoxazole. Synergy between amoxicillin or ticarcillin and clavulanic acid was observed. The third-generation cephalosporins, active *in vitro*, are not effective *in vivo*. The *in vitro* susceptibility is not always correlated with therapeutic success, but there is no acquired resistance currently described. Historically, the treatment was based on aminoglycosides (streptomycin or gentamicin). These antibiotics are administered only by parenteral route, have toxicity on kidney and on cochleovestibular system, and are not indicated during pregnancy. In severe cases, the combination with an aminoglycoside is possible. Along with antibiotics, local treatment of ulcers by wet dressings or surgical drainage of lymphadenopathy is sometimes necessary.

29.7 Prevention

As reservoir of *F. tularensis* is very large, it is difficult to design a highly effective prevention. However, different measures are used to limit the risks.

29.7.1 Prophylactic measures

Prevention is largely based on the fight against animal reservoir and vectors. Any suspicion of tularemia in a live or dead animal, or mortality in lagomorphs and rodents, must be reported to veterinary services. Farms and agricultural areas must be protected from the intrusion of wild rodents. The importation of rodents and lagomorphs, live or dead, wild or domestic, is subject to approval of the Departments of Veterinary Services. An epidemic was declared in the United States in 2002 among prairie dogs sold in pet stores, with secondary human cases, showing the need for control of trade in animals (Avashia et al.,

2004). In humans, it is necessary to inform at-risk populations, such as hunters, veterinarians, and laboratory personnel, on the risks of contamination, symptoms, and means of protection. Consumption of lagomorphs in outbreak periods should also be prohibited. Finally, only trained personnel should handle infected tissues and samples, and only in a microbiological safety laboratory type P2 for *F. tularensis* ssp. *holarctica* and P3 for *F. tularensis* ssp. *tularensis*.

29.7.2 Individual measures

On an individual basis, basic hygiene measures, especially hand washing, should be respected. Handling dead wild animals requires wearing gloves. Game, either mammal or bird, should be consumed only when cooked. It should also be remembered that even the plants and the water may be contaminants, hence the need for washing plants if they are eaten raw and for not drinking untreated water in suspicious wild game area. Protection against tick bites or mosquitoes may be achieved with protective clothing (long sleeves, pants), and the use of repellents such as diethyl toluamide (DEET). However, bites can occur through clothing and repellents do not guarantee complete protection. Once a tick is found on the body, it is necessary to extract it carefully and as early as possible. In case of strong suspicion of contamination (epidemiologically linked to a confirmed contact with a dead hare in an endemic area), antibiotic prophylaxis should be started. The molecules used and dosages are identical to cure, but duration of treatment is 14 days, regardless of the chosen molecule.

29.7.3 Vaccination

There is currently no effective vaccine against tularemia. Three types of vaccines against tularemia were studied: the killed vaccines, the vaccine antigen fraction, and live attenuated vaccines. The first vaccines to be developed were killed vaccines in United States in the 1930s and 1940s (Foshay, 1950). These vaccines were never able to provide adequate protection, probably due to insufficient induction of cell-mediated immunity, needed for host defense against an intracellular bacterium development as *F. tularensis*. Many highly immunogenic antigens have been identified as LPS, FOPA (Fulop et al., 1995), or Tul4 (Golovliov et al., 1995). Only LPS is capable of inducing a protective immune response. Immunization with LPS induces protection against infection with *F. tularensis* ssp. *holarctica*, but is much less effective against infections by *F. tularensis* ssp. *tularensis* (Conlan et al., 2002; Fulop et al., 2001). Therefore, today, no effective vaccine, containing antigen fractions, exists, like killed vaccines, probably due to the lack of induction of cell-mediated immunity. Several live attenuated strains of *F. tularensis* ssp. *holarctica* have been tested and used as a vaccine in the 1940s in the USSR. They stimulate both humoral and cellular immune response and are therefore likely to have a real protective power. Between 1946 and 1960, 60 million people were successfully vaccinated in the USSR. The strain of *F. tularensis* LVS has been developed as a vaccine in the United States in the 1950s and 1960s from one of these strains originated from the USSR. The efficacy of this vaccine was then proven in mice, guinea pigs, rabbits, monkeys, and humans (Tulis et al., 1969). This vaccine has been used in the personnel at risk in the US army, and led to a significant reduction in the number of cases of pneumonic tularemia, the incidence decreasing from 5.7 to 0.27 cases per 1000 people at risk (Burke, 1977). The number of cases with ulcerative node tularemia did not change, but the symptoms were less severe in vaccinated individuals. Despite its effectiveness, the vaccine poses many problems: the mechanism of attenuation is unknown, with a significant risk of reversion; the variable immunogenicity and safety of the vaccine have not been evaluated in immunocompromised individuals. It is therefore no more commercialized. Research is currently underway to develop new attenuated effective and safe live vaccines, especially with metabolic mutants (mutations inhibiting the biosynthesis of purines or the synthesis of aromatic amino acids).

29.8 Tularemia and bioterrorism

Given its virulence, low infectious dose, pathogenicity, and easy spread by aerosol, *F. tularensis* has long been considered as being able to be used in acts of terrorism or biological warfare. Indeed, inhalation from May to October of bacteria subspecies tularensis is sufficient to cause a human disease whose mortality can reach 30–60% without treatment. Studies on its use by the military were done by Japanese, Soviet,

United States, and perhaps by other countries (Dennis et al., 2001). It seems to have been used in 1942 by Soviet Union at the Battle of Stalingrad, where 30,000 German soldiers and 100,000 people and Soviet troops were contaminated. It should be noted that a surge in cases of tularemia is observed in many wars, probably due to poor hygiene. An outbreak has occurred in Kosovo in 1999 (Reintjes et al., 2002), without the use of germ as bacteriological weapon ever been mentioned. After the war, the military study of *F. tularensis* has continued in the United States until the early 1970s, and in the 1990s in the USSR according to Ken Alibek, a scientist and senior responsible in Soviet weapons program. The Soviets have developed strains of *F. tularensis* resistant to antibiotics and vaccines. After the use of sarin gas in Tokyo in 1995, the attacks of September 11, 2001 in New York, and the mailing of anthrax spores in the United States, the world became aware of the potential use of infectious agents, including *F. tularensis* during bioterrorism. *F. tularensis* is now considered a potential bioterrorism class A agent by the CDC. It could be used to contaminate food or water supply systems, but its use as aerosol would be more realistic. According to a report by the World Health Organization in 1969, 50 kg of *F. tularensis* ssp. *tularensis* spread over an urban area of 5 million would imply 250,000 cases of tularemia, with pulmonary and septicemia forms, and 19,000 deaths. Based on this report, the CDC recently estimated the economic cost of a bioterrorist attack by aerosol of *F. tularensis* ssp. *tularensis* to 5.4 billion dollars per 100,000 people exposed.

In conclusion, as *F. tularensis* is capable to adapt to various and extreme environments, while missing numerous metabolic pathways, the knowledge of the exact mechanisms, which permit the different responses of the bacteria, is therefore a challenge for the scientists.

References

Allen, L.-A.H., and McCaffrey, R.L. 2007. To activate or not to activate: Distinct strategies used by *Helicobacter pylori* and *Francisella tularensis* to modulate the NADPH oxidase and survive in human neutrophils. *Immunol. Rev.*, 219, 103–117.

Avashia, S., Petersen, J., Lindley, C., Schriefer, M., Gage, K., and Cetron, M. 2004. First reported prairie dog–to-human tularemia transmission, Texas, 2002. *Emerg. Infect. Dis.*, 10, 483–486.

Balagopal, A., MacFarlane, A.S., Mohapatra, N., Soni, S., Gunn, J.S., and Schlesinger, L.S. 2006. Characterization of the receptor-ligand pathways important for entry and survival of *Francisella tularensis* in human macrophages. *Infect. Immun.*, 74, 5114–5125.

Barel, M., Hovanessian, A., Meibom, K., Briand, J.-P., Dupuis, M., and Charbit, A. 2008. A novel receptor-ligand pathway for entry of *Francisella tularensis* in monocyte-like THP-1 cells: Interaction between surface nucleolin and bacterial elongation factor Tu. *BMC Microbiol.*, 8, 145–153.

Barel, M., Meibom, K., Dubail, I., Botella, J., and Charbit, A. 2012. *Francisella tularensis* regulates the expression of the amino acid transporter SLC1A5 in infected THP-1 human monocytes. *Cell. Microbiol.*, 14, 1769–1783.

Barker, J.R., and Klose, K.E. 2007. Molecular and genetic basis of pathogenesis in *Francisella tularensis*. *Ann. NY Acad. Sci.*, 1105, 138–159.

Barker, J.R., Chong, A., Wehrly, T.D., Yu, J.-J., Rodriguez, S.A., Liu, J., et al. 2009. The *Francisella tularensis* pathogenicity island encodes a secretion system that is required for phagosome escape and virulence. *Mol. Microbiol.*, 74, 1459–1470.

Ben Nasr, A., Haithcoat, J., Masterson, J.E., Gunn, J.S., Eaves-Pyles, T., and Klimpel, G.R. 2006. Critical role for serum opsonins and complement receptors CR3 (CD11b/CD18) and CR4 (CD11c/CD18) in phagocytosis of *Francisella tularensis* by human dendritic cells (DC): Uptake of *Francisella* leads to activation of immature DC and intracellular survival of the bacteria. *J. Leukoc. Biol.*, 80, 774–786.

Bonquist, L., Lindgren, H., Golovliov, I., Guina, T., and Sjostedt, A. 2008. MglA and Igl proteins contribute to the modulation of *Francisella tularensis* live vaccine strain-containing phagosomes in murine macrophages. *Infect. Immun.*, 76, 3502–3510.

Bröms, J.E., Meyer, L., Lavander, M., Larsson, P.R., and Sjostedt, A. 2012. DotU and VgrG, core components of type VI secretion systems, are essential for *Francisella* LVS pathogenicity. *PLoS One*, 7, e34639.

Broz, P., and Monack, D.M. 2011. Molecular mechanisms of inflammasome activation during microbial infections. *Immunol. Rev.*, 243, 174–190.

Bublitz, D.C., Noah, C.E., Benach, J.L., and Furie, M.B. 2010. *Francisella tularensis* suppresses the proinflammatory response of endothelial cells via the endothelial protein C receptor. *J. Immunol.*, 185, 1124–1131.

Burke, D.S. 1977. Immunization against tularemia: Analysis of the effectiveness of Live *Francisella tularensis* vaccine in prevention of laboratory-acquired tularemia. *J. Infect. Dis.*, 135, 55–60.

Celli, J., and Zahrt, T.C. 2013. Mechanisms of *Francisella tularensis* intracellular pathogenesis. *Cold Spring Harb. Perspect. Med.*, 3, a010314.

Cerny, Z. 2001. Changes of the epidemiology and the clinical picture of tularemia in Southern Moravia (the Czech Republic) during the period 1936–1999. *Eur. J. Epidemiol.*, 17, 637–642.

Checroun, C., Wehrly, T.D., Fischer, E.R., Hayes, S.F., and Celli, J. 2006. Autophagy-mediated reentry of *Francisella tularensis* into the endocytic compartment after cytoplasmic replication. *Proc. Natl. Acad. Sci. USA*, 103, 14578–14583.

Chong, A., Wehrly, T.D., Nair, V., Fischer, E.R., Barker, J.R., Klose, K.E., and Celli, J. 2008. The early phagosomal stage of *Francisella tularensis* determines optimal phagosomal escape and *Francisella* pathogenicity island protein expression. *Infect. Immun.*, 76, 5488–5499.

Clemens, D.L., and Horwitz, M.A. 2007. Uptake and intracellular fate of *Francisella tularensis* in human macrophages. *Ann. N.Y. Acad. Sci.*, 1105, 160–186.

Clemens, D.L., Lee, B.Y., and Horwitz, M.A. 2004. Virulent and avirulent strains of *Francisella tularensis* prevent acidification and maturation of their phagosomes and escape into the cytoplasm in human macrophages *Infect. Immun.*, 72, 3204–3217.

Clemens, D.L., Lee, B.Y., and Horwitz, M.A. 2005. *Francisella tularensis* enters macrophages via a novel process involving pseudopod loops. *Infect. Immun.*, 73, 5892–5902.

Clemens, D.L., Lee, B.-Y., and Horwitz, M.A. 2009. *Francisella tularensis* phagosomal escape does not require acidification of the phagosome. *Infect. Immun.*, 77, 1757–1773.

Conlan, J.W., KuoLee, R., Shen, H., and Webb, A. 2002. Different host defences are required to protect mice from primary systemic vs pulmonary infection with the facultative intracellular bacterial pathogen, *Francisella tularensis* LVS. *Microb. Pathog.*, 32, 127–134.

Cremer, T.J., Amer, A.O., Tridandapani, S., and Butchar, J.P. 2009. *Francisella tularensis* regulates autophagy-related host cell signaling pathways. *Autophagy*, 5, 125–128.

Dai, S., Mohapatra, N., Schlesinger, L., and Gunn, J. 2011. Regulation of *Francisella tularensis* virulence. *Front. Microbiol.*, 1, 1–10.

de Bruin, O.M., Duplantis, B.N., Ludu, J.S., Hare, R.F., Nix, E.B., Schmerk, C.L., et al. 2011. The biochemical properties of the *Francisella* pathogenicity island (FPI)-encoded proteins IglA, IglB, IglC, PdpB and DotU suggest roles in type VI secretion. *Microbiology*, 157, 3483–3491.

Dennis, D.T., Inglesby, T.V., Henderson, D.A., Bartlett, J.G., Ascher, M.S., Eitzen, E., et al. 2001. Tularemia as a biological weapon: Medical and public health management. *JAMA*, 285, 2763–2773.

Eliasson, H., Lindbäck, J., Nuorti, P., Arneborn, M., Giesecke, J., and Tegnell, A. 2002. The 2000 tularemia outbreak: A case-control study of risk factors in disease-endemic and emergent areas, Sweden. *Emerg. Infect. Dis.*, 8, 956–960.

Faron, M., Fletcher, J.D., Rasmussen, J.A., Long, M.E., Allen, L.-A.H., and Jones, B.D. 2013. The *Francisella tularensis* migR, trmE, and cphA genes contribute to FPI gene regulation and intracellular growth by modulation of the stress alarmone ppGpp. *Infect. Immun.*, 81, 2800–2811.

Feldman, K.A., Enscore, R.E., Lathrop, S.L., Matyas, B.T., McGuill, M., Schriefer, M.E., et al. 2001. An outbreak of primary pneumonic tularemia on Martha's Vineyard. *N. Engl. J. Med.*, 345, 1601–1606.

Foshay, L. 1950. Tularemia. *Annu. Rev. Microbiol.*, 4, 313–330.

Fulop, M., Manchee, R., and Titball, R. 1995. Role of lipopolysaccharide and a major outer membrane protein from *Francisella tularensis* in the induction of immunity against tularemia. *Vaccine*, 13, 1220–1225.

Fulop, M., Mastroeni, P., Green, M., and Titball, R.W. 2001. Role of antibody to lipopolysaccharide in protection against low- and high-virulence strains of *Francisella tularensis*. *Vaccine*, 19, 4465–4472.

Golovliov, I., Ericsson, M., Akerblom, L., Sandström, G., Tärnvik, A., and Sjöstedt, A. 1995. Adjuvanticity of ISCOMs incorporating a T cell-reactive lipoprotein of the facultative intracellular pathogen *Francisella tularensis*. *Vaccine*, 13, 261–267.

Greub, G., and Raoult, D. 2004. Microorganisms resistant to free-living amoebae. *Clin. Microbiol. Rev.*, 17, 413–433.

Jantzen, E., Berdal, B.P., and Omland, T. 1979. Cellular fatty acid composition of *Francisella tularensis*. *J. Clin. Microbiol.*, 10, 928–930.

Jones, C.L., Napier, B.A., Sampson, T.R., Llewellyn, A.C., Schroeder, M.R., and Weiss, D.S. 2012. Subversion of host recognition and defense systems by *Francisella* spp. *Microbiol. Mol. Biol. Rev.*, 76, 383–404.

Kayal, S., and Charbit, A. 2006. Listeriolysin O: A key protein of *Listeria monocytogenes* with multiple functions. *FEMS Microbiol. Rev.*, 30, 514–529.

Keim, P., Johansson, A., and Wagner, D.M. 2007. Molecular epidemiology, evolution, and ecology of *Francisella*. *Ann. N.Y. Acad. Sci.*, 1105, 30–66.

Lai, X.H., and Sjostedt, A. 2003. Delineation of the molecular mechanisms of *Francisella tularensis*-induced apoptosis in murine macrophages. *Infect. Immun.*, 71, 4642–4646.

Larsson, P., Oyston, P.C., Chain, P., Chu, M.C., Duffield, M., Fuxelius, H.H., et al. 2005. The complete genome sequence of *Francisella tularensis*, the causative agent of tularemia. *Nat. Genet.*, 37, 153–159.

Larsson, P.r., Elfsmark, D., Svensson, K., Wikström, P., Forsman, M., Brettin, T., et al. 2009. Molecular evolutionary consequences of niche restriction in *Francisella tularensis*, a facultative intracellular pathogen. *PLoS Pathog.*, 5, e1000472.

Lindgren, M., Eneslt, K., Broms, J.E., and Sjostedt, A. 2013. Importance of *PdpC, IglC, IglI*, and *IglG* for modulation of a host cell death pathway induced by *Francisella tularensis*. *Infect. Immun.*, 81, 2076–2084.

Matyas, B.T., Nieder, H.S., and Telford, S.R. 2007. Pneumonic tularemia on Martha's Vineyard. *Ann. N.Y. Acad. Sci.*, 1105, 351–377.

McCaffrey, R.L., Schwartz, J.T., Lindemann, S.R., Moreland, J.G., Buchan, B.W., Jones, B.D., and Allen, L.-A.H. 2010. Multiple mechanisms of NADPH oxidase inhibition by type A and type B *Francisella tularensis*. *J. Leukoc. Biol.*, 88, 791–805.

McLendon, M.K., Apicella, M.A., and Allen, L.-A.H. 2006. *Francisella tularensis*: Taxonomy, genetics, and immunopathogenesis of a potential agent of biowarfare. *Ann. Rev. Microbiol.*, 60, 167–185.

Meibom, K.L., and Charbit, A. 2010a. *Francisella tularensis* metabolism and its relation to virulence. *Front. Microbiol.*, 1, 140.

Meibom, K.L., and Charbit, A. 2010b. The unraveling panoply of *Francisella tularensis* virulence attributes. *Curr. Opin. Microbiol.*, 13, 11–17.

Meibom, K.L., Forslund, A.-L., Kuoppa, K., Alkhuder, K., Dubail, I., Dupuis, M., et al. 2009. Hfq, a novel pleiotropic regulator of virulence-associated genes in *Francisella tularensis*. *Infect. Immun.*, 77, 1866–1880.

Mohapatra, N.P., Soni, S., Rajaram, M.V.S., Dang, P.M.-C., Reilly, T.J., El-Benna, J., et al. 2010. *Francisella* acid phosphatases inactivate the NADPH oxidase in human phagocytes. *J. Immunol.*, 184, 5141–5150.

Oyston, P.C., Sjostedt, A., and Titball, R.W. 2004. Tularaemia: Bioterrorism defence renews interest in *Francisella tularensis*. *Nat. Rev. Microbiol.*, 2, 967–978.

Oyston, P.C.F. 2008. *Francisella tularensis*: Unravelling the secrets of an intracellular pathogen. *J. Med. Microbiol.*, 57, 921–930.

Pechous, R.D., McCarthy, T.R., and Zahrt, T.C. 2009. Working toward the Future: Insights into *Francisella tularensis* pathogenesis and vaccine development. *Microbiol. Mol. Biol. Rev.*, 73, 684–711.

Petrosino, J.F., Xiang, Q., Karpathy, S.E., Jiang, H., Yerrapragada, S., Liu, Y., et al. 2006. Chromosome rearrangement and diversification of *Francisella tularensis* revealed by the type B (OSU18) genome sequence. *J. Bacteriol.*, 188, 6977–6985.

Pierini, L.M. 2006. Uptake of serum-opsonized *Francisella tularensis* by macrophages can be mediated by class A scavenger receptors. *Cell. Microbiol.*, 8, 1361–1370.

Price, C.T.D., Al-Quadan, T., Santic, M., Rosenshine, I., and Abu Kwaik, Y. 2011. Host proteasomal degradation generates amino acids essential for intracellular bacterial growth. *Science*, 334, 1553–1557.

Reintjes, R., Dedushaj, I., Gjini, A., Jorgensen, T., Cotter, B., Lieftucht, A., et al. 2002. Tularemia outbreak investigation in Kosovo: Case control and environmental studies. *Emerg. Infect. Dis.*, 8, 69–73.

Rohmer, L., Fong, C., Abmayr, S., Wasnick, M., Larson Freeman, T., Radey, M., et al. 2007. Comparison of *Francisella tularensis* genomes reveals evolutionary events associated with the emergence of human pathogenic strains. *Genome Biol.*, 8, R102.

Santic, M., Al Khodor, S., and Abu Kwaik, Y. 2010. Cell biology and molecular ecology of *Francisella tularensis*. *Cell. Microbiol.*, 12, 129–139.

Santic, M., Asare, R., Skrobonja, I., Jones, S., and Abu Kwaik, Y. 2008. Acquisition of the vacuolar ATPase proton pump and phagosome acidification are essential for escape of *Francisella tularensis* into the macrophage cytosol. *Infect. Immun.*, 76, 2671–2677.

Santic, M., Molmeret, M., Klose, K.E., and Abu Kwaik, Y. 2006. *Francisella tularensis* travels a novel, twisted road within macrophages. *Trends Microbiol.*, 14, 37–44.

Schulert, G.S., and Allen, L.-A.H. 2006. Differential infection of mononuclear phagocytes by *Francisella tularensis*: Role of the macrophage mannose receptor. *J. Leukoc. Biol.*, 80, 563–571.

Schwartz, J.T., Bandyopadhyay, S., Kobayashi, S.D., McCracken, J., Whitney, A.R., DeLeo, F.R., and Allen, L.A.H. 2013. *Francisella tularensis* alters human neutrophil gene expression: Insights into the molecular basis of delayed neutrophil apoptosis. *J. Inn. Immun.*, 5, 124–136.

Sjodin, A., Svensson, K., Ohrman, C., Ahlinder, J., Lindgren, P., Duodu, S., et al. 2012. Genome characterisation of the genus *Francisella* reveals insight into similar evolutionary paths in pathogens of mammals and fish. *BMC Genomics*, 13, 268.

Telepnev, M., Golovliov, I., and Sjostedt, A. 2005. *Francisella tularensis* LVS initially activates but subsequently down-regulates intracellular signaling and cytokine secretion in mouse monocytic and human peripheral blood mononuclear cells. *Microb. Pathog.*, 38, 239–247.

Thomas, R., Johansson, A., Neeson, B., Isherwood, K., Sjö√∂stedt, A., Ellis, J., and Titball, R.W. 2003. Discrimination of human pathogenic subspecies of *Francisella tularensis* by using restriction fragment length polymorphism. *J. Clin. Microbiol.*, 41, 50–57.

Tulis, J., Eigelsbach, H., and Hornick, R. 1969. Oral vaccination against tularemia in the monkeys. *Proc. Soc. Exp. Biol. Med.*, 132, 893–897.

Veiga, E., and Cossart, P. 2006. The role of clathrin-dependent endocytosis in bacterial internalization. *Trends Cell. Biol.*, 16, 499–504.

Wayne Conlan, J., and Oyston, P.C.F. 2007. Vaccines against *Francisella tularensis*. *Ann. N.Y. Acad. Sci.*, 1105, 325–350.

Weiss, D.S., Henry, T., and Monack, D.M. 2007. *Francisella tularensis*: Activation of the inflammasome. *Ann. N.Y. Acad. Sci.*, 1105, 219–237.

Zeytun, A., Malfatti, S.A., Vergez, L.M., Shin, M., Garcia, E., and Chain, P.S.G. 2012. Complete genome sequence of *Francisella philomiragia* ATCC 25017. *J. Bacteriol.*, 194, 3266.

Chapter 30

Pathogenesis of *Yersinia pestis* in Humans

E.D. Williamson

Defence Science and Technology Laboratory, Porton Down, Salisbury, United Kingdom

30.1 Epidemiology of plague

Plague is caused by the Gram-negative bacterium *Yersinia pestis*. The bacterium is thought to have originated in or near China (Morelli et al., 2010) as early as 600 BC and its dissemination along the Silk routes from Asia to other continents is thought to have started in approximately 540 AD. The bacterium was carried by infected rats along with supplies of food and clothing to reach Europe in the early 14th century. As it was disseminated at different rates to near and further-flung countries, country-specific lineages developed, leading to the classification of *Y. pestis* into three biovars: Antigua, Medievalis, and Orientalis. Strains of *Y. pestis* in the world today are predominantly of the Orientalis biovar.

Y. pestis was disseminated from Asia via central Africa to China and reached the European continent in the early 14th century. A third pandemic is thought to have arisen in China in 1855, spreading plague to Hong Kong in 1894 and then via maritime routes to Madagascar in 1898 and to the west coast of N. America thereafter. Infected rats arriving on board cargo ships from Hong Kong disembarked and effectively spread the bacterium among the wild rodent population in the San Francisco area. This led to endemic disease which still remains in the southwestern regions of the United States until this day. The climate in this area of the United States is propitious for the bacterium, with the seasons changing from mild winter conditions to extremely hot summers. There have been reports of a "rat fall" (Gage, 1999) in the summer months with potentially infected rats being found dead under decking or in basements. The removal of these carcasses and their flea burden is then potentially hazardous for homeowners and those living in close proximity to these sources of infection and provides a food source for wild and domesticated animals.

Although this illustrates well the intransigent nature of endemic disease, annual cases of plague in the southwestern United States account for only about 0.5% of the world's total annual number of cases reported to the World Health Organization (WHO, 2005). Plague, although an ancient disease, is still present in several other endemic areas of the world today including China, India, Madagascar, and Africa and can cause outbreaks from time to time, most often associated with major climate change or geological disturbance (e.g., earthquake) which disturbs the wild rodent population and may bring humans into closer proximity than usual with them, particularly under the conditions of limited water and poor sanitation that prevail after such disasters (Perry and Fetherston, 1997).

Approximately 3000 cases of plague globally are reported to the World Health Organization each year (WHO, 2005). When the statistics are analyzed further, it appears that the reporting of plague cases from the African continent is increasing, whilst remaining steady in Asian countries (Alvarez and Cardineau, 2010). An outbreak at a diamond mine was reported in the Democratic Republic of the Congo, which caused the death of 53 miners from pneumonic plague (Bertherat et al., 2005).

On many counts, therefore, the causative bacterium *Y. pestis* is notorious. In the course of three pandemics, *Y. pestis* has caused the deaths of millions of people, decimating the European population in the 13th–14th centuries. Additionally, *Y. pestis* is still present in the modern world and still causing outbreaks of disease. Having originated as a member of the enterobacteriaceae, *Y. pestis* has evolved away

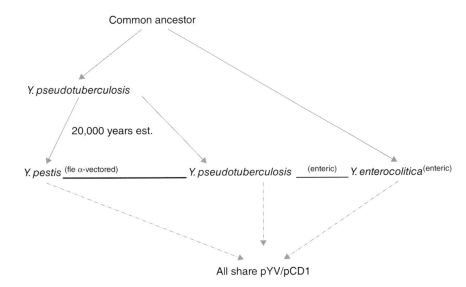

Fig. 30-1 Loss and gain of virulence factors during the evolution of *Y. pestis* from an enteropathogenic to flea -vectored lifestyle. In the process, *Y. pestis* has gained 2 new plasmids (pMT1 and pPCP); lost adhesins (*yadA*), invasins (*inv*), and suppressed LPS genes (*LpxL*); retained the haem storage locus (*hms*) and inactivated O-antigen genes, allowing plasminogen activator protein (PLA) to function fully.

from the enteropathogenic phenotype over an estimated 5000–15,000 years (Achtmann et al., 1999) to become a flea-vectored pathogen with enhanced virulence (Parkhill et al., 2001).

30.2 Virulence factors in *Y. pestis*

Y. pestis has adapted to this new lifestyle with the loss of invasin genes, the retention of the key adhesin Ail, which confers resistance to complement and modulates phagocytes in infected lymph nodes and the anti-phagocytic frimbrial adhesin (ph6 antigen, Lindler and Tall, 1993) and the inactivation of other mechanisms (e.g., haem storage and O-antigen genes on the bacterial surface), which in turn allows the unrestrained activity of the plasminogen activator (PLA) protein to degrade complement and plasminogen and promote adhesion (Cowan et al., 2000; Galvan et al., 2008; Lahteenmaki et al., 2001; Sodeinde et al., 1992) (Figure 30-1). However, in addition to these genome-encoded factors, it is the possession by *Y. pestis* of three plasmids which encode significant additional virulence properties which arguably confers maximum survival advantages. These plasmids govern

the production of the pesticin–coagulase complex (pPCP–pPST) involved in the transmission and mobilization of bacteria in the mammalian host, the secretion and assembly of the anti-phagocytic F1 antigen on the bacterial surface (pMT1/pFra), and the type three secretion system (TTSS) (pCD1/pYV) to translocate immunomodulatory and cytotoxic proteins into mammalian cells (Figure 30-2). Of these plasmids, the pMT1/pFra plasmid is largest at 90 kDa and governs the secretion of murine toxin, a phospholipase D which aids the persistence of *Y. pestis* bacteria in the flea gut (Hinnebusch et al., 2002). Additionally, this plasmid encodes the fraction 1 (F1) antigen, which is secreted and assembles as a large polymer of more than 200 kDa on the capsule surrounding the bacterium (Tito et al., 2001). F1 antigen is therefore a dominant structure on the bacterial surface and is strongly antiphagocytic toward host cells, thus preserving the bacteria *in vivo*. The second largest plasmid is the pCD1/pYV plasmid which encodes for the TTSS. The latter is a system common to a number of Gram-negative bacteria and which allows them to produce a physical injectisome when they make close contact with host cells (Cornelis, 2006). The injectisome is a

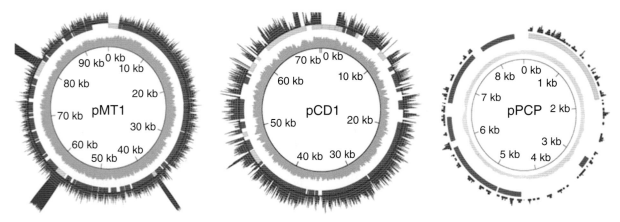

Fig. 30-2 Plasmid-encoded virulence factors in *Y. pestis*. *Y. pestis* has acquired plasmids which determine the production of murine toxin and fraction 1 antigen (pMT1), the TTSS, and yersinia outer proteins, including the V antigen (pCD1) and the pesticin–coagulase complex (pPCP).

hollow structure, composed of Yersinia secretory factor F (YscF), which forms a bridge between the bacterium and the host cell to allow the delivery of anti-host factors into the host cell. For *Y. pestis*, these anti-host factors include the yersinia outer proteins Yop B and Yop D which are able to form pores in the host cell to allow delivery of other Yops and cytotoxic factors (Goure et al., 2005; Lawton et al., 2002; Matson and Nilles, 2001; Nilles et al., 1998). In *Y. pestis*, the construction of the injectisome and the secretion of Yops is closely regulated intracellularly by the interaction of the V antigen (also known in this context as the low calcium response V antigen or LcrV) with the LcrG and other proteins. As well as acting intracellularly, the V antigen is also transported through the injectisome to its tip (Mueller et al., 2005), where it is thought to have a pivotal role in making contact with host cells and where it has been hypothesized to adopt a pentameric structure (Broz et al., 2007). The host cell sensing and translocation process is tightly regulated (Kopaskie et al., 2013). It is proposed that the needle cap LcrV forms a tether between YscF and YopB-YopD to constitute the injectisome and that it is the insertion of LcrV-capped needles from a calcium-rich environment into host cells which may trigger the low-calcium signal for effector translocation (Ligtenberg et al., 2013). Indeed, a secretion-competent, translocation-defective mutant strain of Yersinia, which was able to secrete Yops and sense host cell membranes, but unable to form complexes between YscF, LcrV, and Yop and unable to insert Yop B into host cell membranes, indicated that the LcrV tip complexed with YscF is essential for translocon formation and Yop B insertion (Harmon et al., 2013). Further, discrete sites within LcrV have been attributed with its functioning as a secretion substrate, needle cap, and injectisome assembly factor (Ligtenberg et al., 2013). Despite the array of virulence factors produced by *Y. pestis*, which are either genome and plasmid-encoded, only a selected few of these have been shown to be protective immunogens. It is clear from the above that both YscF and V antigen are exposed to the human immune system early in infection, as constituents of the injectisome, which is perhaps why both are protective immunogens (Anderson et al., 1996; Matson et al., 2001; Wang et al., 2008; Williamson et al., 1995). Additionally, a Yop with adhesive properties, YadC, and an ABC transporter protein, OppA, have been shown to be protective immunogens (Forman et al., 2008; Tanabe et al., 2006). Predominantly, however, the F1 and V antigens are protective immunogens individually and which are additive as such in their effect (reviewed in Williamson and Oyston, 2013). Additionally, the inhibition of PLA expression prolonged the survival of C57BL/6 mice infected with *Y. pestis* (Lathem et al., 2007), and PLA was recognized in convalescent human sera (Easterbrook et al., 1995) but immunization with a DNA construct expressing PLA did not protect Balb/c mice (Wang et al., 2008).

Y. pestis is therefore a successful pathogen, although man is not the principal host species and has been described as an "accidental host" (Perry and Fetherston, 1997). *Y. pestis* is a zoonotic infection, transmitted to man by the bite of an infected flea, where that flea has fed on an infected wild animal. In endemic areas, *Y. pestis* is maintained in the wild rodent population, principally in the wild rat population, but also and depending on territory, the wild populations of ground squirrels, prairie dogs, rabbits, etc. From these populations of wild animals, *Y. pestis* can be transmitted to domestic animals, particularly to cats by the bite of an infected flea (Gasper et al., 1993). More recently, there have been a number of reports of the transmission of *Y. pestis* to humans from infected domestic cats by aerosolized bacteria (Doll et al., 1994; Gage et al., 2000).

The human disease of plague is potentially lethal and needs to be detected and diagnosed to instigate prompt treatment before individuals become symptomatic. Plague occurs in several forms: bubonic, septicemic, and pneumonic, but the most common form by far in endemic regions is the bubonic form. An infected flea can deliver more than a million bacteria in a single bite and this causes the development of excessively swollen draining lymph nodes (buboes), most commonly in the axillae or groins of the affected host. If left untreated, the mortality rate in bubonic plague cases is in the range of 40–60%. Without treatment, bubonic plague can develop into a secondary pneumonic plague which could be lethal and also could be transmitted to a naive individual in aerosolized droplets, caused by coughing. However, the person-to-person transmission of pneumonic plague is a rare event in the modern world. The third classical presentation of plague is the septicaemic form, which can develop from the bubonic syndrome but is typically not characterized by the development of buboes. In all of these presentations, however, infection needs to be detected and treated promptly to ensure survival.

30.3 Clinical manifestations

The human syndrome of plague infection is characterized by influenza-like symptoms with raised temperature, headache, and general malaise. In the few well-documented cases of human plague, infection has initially been mistaken for influenza, thus delaying the early seeking of medical advice and treatment. Clinical isolates of *Y. pestis* have been collected from two separate and well-documented cases. The first involved a researcher who contracted a laboratory-acquired infection, most likely by aerosol and after suffering influenza-like symptoms over a weekend, died a few days later (Anon, 1962). The second, more recent case, involved a vet who contracted the disease when treating an infected domestic cat. Unaware of his infection, the vet traveled from Colorado to Arizona, where he became symptomatic and died of plague (Doll et al., 1994). Although personal tragedies, these well-documented cases provided much information about the human syndrome of plague and from both cases, clinical isolates have been used subsequently to support the development of vaccines and therapies for the disease (Heath et al., 1998; Williamson et al., 1995). Both these clinical isolates of *Y. pestis* belong to the biovar Orientalis. The clinical isolate from the Colorado case has been fully genome-sequenced and as *Y. pestis* Co92, has become the type strain for the *Y. pestis* biovar Orientalis (Parkhill et al., 2001).

The idiosyncrasies of the human syndrome of plague are illustrated in a more recent fatal case of plague infection, recorded as a laboratory-acquired infection (Ritger et al., 2011). In this case, a researcher was working with a pigmentation-deficient, attenuated strain of *Y. pestis* KIM. This strain is a routinely used laboratory strain, thought to be highly attenuated through loss of its iron-acquisition ability which is required for growth *in vivo*. However, the researcher was diabetic and also had an undiagnosed hereditary hemochromatosis condition, of which he was unaware. During his work, he was exposed to this strain of *Y. pestis* which became virulent for him due to his excessive iron burden compensating for the iron deficiency in the mutated strain. This case highlights the virulence of *Y. pestis* and its ability to exploit vulnerabilities in potential hosts.

If diagnosed in time, however, plague can be treated and this was illustrated by the survival of a couple who had become infected with plague at home in Santa Fe in 2002 and subsequently traveled to New York where they became ill, were hospitalized, treated, and survived (Hartocollis, 2004).

30.4 Prospects for vaccination and therapy

A number of killed whole cell vaccines (KWCV) have been available for plague, since the first (Haffkine's vaccine), used in a large vaccination campaign in Bombay in 1897 (Demeurre, 2012). Since then, a number of KWCV have been fielded in the United States (e.g., the Army vaccine in 1946; the plague vaccine USP manufactured by the Cutter laboratories until 1994 and thereafter by the Greer Laboratories until 1999). A very similar vaccine, comprising heat-killed *Y. pestis*, was produced by the Commonwealth Serum Laboratories in Australia until 2005, when production was discontinued. Although epidemiological and experimental evidence suggests that such KWCV formulations were effective against bubonic plague, they appeared to offer little protection against pneumonic plague, attributable at least in part to lack of the virulence (V) antigen (Jones et al., 2003; Williamson and Oyston, 2012). The latter is a pivotal component of the TTSS and has been shown to be a protective immunogen in its own right (Anderson et al., 1996).

Currently, recombinant vaccines comprising a combination of the F1 and V antigens either as a genetic fusion or as individual co-administered proteins are being developed as the next generation of vaccines for plague and have the potential to be highly efficacious against all forms of plague (including pneumonic) and ultimately to provide a defined, licensable product (Hart et al., 2012; Williamson and Oyston, 2013).

A range of antibiotics can be used to treat a suspected plague infection, although treatment started before the individual becomes symptomatic has far greater potential for success. Thus, patients may be started on streptomycin intramuscularly and followed up with tetracycline orally, or given ciprofloxacin twice daily by either the i.v. or oral routes, whereas chloramphenicol would be the antibiotic of choice for plague meningitis (Williamson and Oyston, 2012). Advances in the identification of new targets in *Y. pestis* for antibiotic therapy, to defeat the threat of antibiotic resistance, have recently been reviewed (Williamson and Oyston, 2012).

30.5 Correlates of protection and surrogate markers

The prediction of human efficacy for a new candidate vaccine for a disease such as plague poses some difficult questions. It is neither ethical nor practical to deliberately expose human volunteers to the causative organisms and although natural outbreaks of plague occur in endemic foci around the world, outbreaks in these locations are unpredictable in occurrence and scale and so cannot be guaranteed to provide a statistically valid population on which to test a vaccine for impact. However, there is now a range of animal models of plague infection, which together authentically replicate the human infection and these can be immunized and challenged, so that immune correlates of protection can be derived (reviewed in Oyston and Williamson, 2012).

30.6 Surrogate markers of protection

Having used the animal models to derive immune correlates of protective efficacy against plague, it is possible to identify surrogate markers of protection for use in man in place of the endpoint of survival in the animal model. For current recombinant vaccines in development which incorporate the F1 and V proteins, serological surrogate markers have been proposed based around the ability of the vaccinee's serum to inhibit cytotoxicity *in vitro* and to passively transfer immunity *in vivo* (Williamson and Oyston, 2012). These assays rely predominantly on the development of neutralizing antibodies for the cytotoxic effect of the V antigen when expressed from Yersinia species or on the opsonophagocytosis of bacteria coated with antibody to the surface-exposed F1 antigen.

The assay of cellular responses in clinical trial volunteers receiving the rF1+rV vaccine failed to detect any significant changes in the display of activation markers by immune effector cells post-immunization (Williamson et al., 2005). This negative result was attributed either to the fact that immunization did not induce marked changes in circulating leucocytes immediately after dosing or that those leucocytes transiently lost activation markers (Williamson et al., 2005). However, there is evidence from nonclinical studies which directly indicates the importance of specific cell-mediated immunity in the development of protective efficacy to plague (Levy et al., 2011; Parent et al., 2005; Philipovskiy and Smiley, 2007; Smiley, 2007; Williamson et al., 2007; Williamson and Oyston, 2012). The development of a specific CMI response is predicted to be an essential component of human recovery from infection with *Y. pestis*.

30.7 B- and T-cell epitope-mapping of the protective F1 and V antigens

At least one neutralizing B-cell epitope exists on F1 antigen and at least two neutralizing B-cell epitopes have been identified on the V antigen (Hill et al., 2009). These functional B-cell epitopes have been identified by the derivation of neutralizing monoclonal antibodies, which have also become important tools in the analysis of animal and human serological responses to the F1 and V antigens.

The mapping of T-cell epitopes in F1 has been facilitated by the availability of HLA transgenic mice carrying full-length genomic constructs for HLA-DRA1*0101 and HLA-DRB1*0101 and crossed for more than six generations to C57BL/6 Ab-null mice, and thus lacking expression of endogenous mouse MHC class II molecules (Altmann et al., 1995; Musson et al., 2010). Using these mice, an HLA-DR1-restricted T-cell epitope was identified in the C-terminus of F1 antigen (141–160) (Musson et al., 2010).

For V antigen, protective H-2k-restricted T-cell epitopes have been identified at sequence positions 71–86, 101–166, and 166–181 (Parent et al., 2005). Subsequently, immunodominant T-cell epitopes in V antigen were identified for $H-2^d$ and $H-2^k$ mice in regions which overlap the sequences previously identified: 102–121, 162–181, and additionally in sequence position 212–231 (Shim et al., 2006) with a subsequent study involving deletion mutants, indicating that the sequence 218–234 either encompasses B- and/or T-cell epitopes essential for protection, or exerts a structural influence essential for protection (Vernazza et al., 2009). Rao et al. have combined peptides comprising some of these sequences from the V antigen into a multiple antigen peptide, which when used to immunize mice, induced Th1 and Th17 responses (Shreewastav et al., 2012). The significance of these efforts is that the identification of such epitopes may provide cellular targets against which to assay the

Table 30-1 Immunogenicity and efficacy of the circularly permutated F1 (cpF1) compared with F1 and a C-terminal truncate of F1 (ΔC–F1)

Group	Strain	Survival	IgG titer to F1
cpF1	Balb/c	1/6	1:83
cpF1	C57Bl6	6/6	1:1967
rF1	Balb/c	6/6	1:15,333
rF1	C57Bl6	6/6	1:15,333
ΔC-F1	Balb/c	0/6	–
ΔC-F1	C57Bl6	1/6	1:75
Alhydrogel	Balb/c	0/6	–
Alhydrogel	C57Bl6	0/6	–

Taken from Chalton et al. (2006). Copyright © 2006, American Society for Microbiology.

recall response of vaccinees to confirm the development of an appropriate and potentially protective CMI response.

One approach to confirming the unique requirement for an epitope was demonstrated using the $H-2^d$-restricted T-cell epitope previously identified in the N-terminal sequence (7–20) of F1 antigen (Musson et al., 2010). A stable, circularly permutated and monomeric form of F1 antigen (cpF1) was prepared by excising the N-terminal sequence, and after inserting a linker, joining it to the C-terminal sequence of the protein (Chalton et al., 2006). This had the effect of disrupting the $H-2^d$-restricted epitope so that the cpF1 was no longer a protective immunogen in Balb/c mice, although it was still immunogenic and protective in C57Bl6 mice, since $H-2^k$ epitopes were unaffected (Table 30-1). A similar exercise can be carried out in HLA-transgenic mice to confirm the unique requirement for individual epitopes. Immunization of HLA-transgenic mice with peptides encompassing putative individual B- or T-cell epitopes could be used to confirm the requirement of individual epitopes, or combinations of epitopes, in protection.

While most publications concerning epitope mapping have applied to the key protective proteins F1 and V, Rao's group has gone further in preparing B–T cell epitope fusion constructs not only for these proteins but also for the YscF protein; in mice such constructs induced immunological readouts which indicate that they have the potential to induce protective immune responses (Ali et al., 2013; Shreewastav et al., 2012).

30.8 Prospects for licensure

Successful licensure of next-generation vaccines for plague will depend on the successful use of the FDA's Animal Rule (US FDA HHS, 2002), the satisfactory demonstration of immune correlates of protection across the animal species used as efficacy models, and the subsequent identification of suitable surrogate markers of efficacy with which to monitor the responses of clinical trial volunteers. To date several medical countermeasures have been licensed on the basis of the Animal Rule: levofloxacin for inhalational plague (US FDA, April 2012, Ciprofloxacin for inhalational plague and anthrax) and Raxibacumab and Anthim (US FDA 2012 and 2015) for anthrax intoxication. The licensure of a next-generation vaccine for plague by this means would mean that there is a prospect of protecting future generations from the risk of contracting bubonic or pneumonic plague from environmental exposure to *Y. pestis*.

References

Achtman, M., Zurth, K., Morelli, C., Torrea, G., Guiyoule, A., and Carniel, E. 1999. *Yersinia pestis*, the cause of plague, is a recently emerged clone of *Yersinia pseudotuberculosis*. *Proc. Natl. Acad. Sci. USA*, 96, 14043–14048.

Ali, R., Kumar, S., and Rao, D.N. 2013. B and T cell mapping and study of the humoral and cell mediated immune response to B-T constructs of YscF antigen of *Yersinia pestis*. *Comp. Immunol. Infect. Dis.*, 36(4), 365–378.

Altmann, D.M., Douek, D.C., Frater, A.J., Hetherington, C.M., Inoko, H., and Elliott J.I. 1995. The T-cell response of HLA-DR transgenic mice to human myelin basic protein and other antigens in the presence and absence of human CD4. *J. Exp. Med.*, 181, 867–875.

Alvarez, M.L., and Cardineau, G.A. 2010. Prevention of bubonic and pneumonic plague using plant-derived vaccines. *Biotechnol. Adv.*, 28, 184–196.

Anderson, G.W., Leary, S.E.C., Williamson, E.D., Titball, R.W., Welkos, S.L., Worsham, P.L., and Friedlander, A.M. 1996. Recombinant V antigen protects mice against pneumonic and bubonic plague caused by F1-capsule-positive and -negative strains of *Yersinia pestis*. *Infect. Immun.*, 64, 4580–4585.

Anon. 1962. Death of a porton scientist. *Lancet*, 2, 463.

Bertherat, E., Lamine, KM., Formenty, P., Thuier, P., Mondonge, V., Mitifu, A., and Rahalison, L. 2005. Major pulmonary plague outbreak in a mining camp in the Democratic Republic of Congo: Brutal awakening of an old scourge. *Med. Trop. (Mars)*, 65, 511–514.

Broz, P., Mueller, C.A., Muller, S.A., Philippsen, A., Sorg, I., Engel, A., and Cornelis, G.R. 2007. Function and molecular architecture of the *Yersinia* injectisome tip complex. *Mol. Microbiol.*, 65, 1311–1320.

Chalton, D.A., Musson, J.A., Flick-Smith, H.C., Walker, N.J., McGregor, A., Williamson, E.D., Miller, J., Robinson, J.H., and Lakey, J.H. 2006. A functional monomeric plague vaccine created by circular permutation. *Infect. Immun.*, 74, 6624–6631.

Cornelis, G.R. 2006. The type III secretion injectisome. *Nat. Rev. Microbiol.*, 4, 811–825.

Cowan, C., Jones, H.A., Kaya, Y.H., Perry, R.D., and Straley, S.D. 2000. Invasion of epithelial cells by *Yersinia pestis*: Evidence for a *Yersinia pestis*-specific invasin. *Infect. Immun.*, 68, 4523–4530.

Demeurre, C.E. 2012. Live vaccines against plague and pseudotuberculosis. In: Carniel, E., and Hinnebusch, B.J., editors. *Yersinia Systems Biology and Control*. UK: Caister Academic Press. pp. 143–68.

Doll, J.M., Zeitz, P.S., Ettestad, P., Bucholtz, A.L., Davis, T., and Gage, K. 1994. Cat-transmitted fatal pneumonic plague in a person who traveled from Colorado to Arizona. *Am. J. Trop. Med. Hyg.*, 51, 109–114.

Easterbrook, T.J., Reddin, K., Robinson, A., and Modi, N. 1995. Studies on the immunogenicity of the Pla protein from *Yersinia pestis*. In: Ravagnan, G., and Chiesa, C., editors. *Yersiniosis: Present and Future*. Basel, Switzerland: Karger. pp. 214–215.

Forman, S., Wulff, C.R., Myers-Morales, T., Cowan, C., Perry, R.D., and Straley, S.C. 2008. *yadBC* of *Yersinia pestis*, a new virulence determinant for bubonic plague. *Infect. Immun.*, 76, 578–587.

Gage, K.L. 1999. Plague surveillance. *Plague Manual. Epidemiology, distribution, surveillance and control.* WHO/CDS/CSR/EDC.99.2.

Gage, K.L., Dennis, D.T., Orloski, K.A., Ettestad, P., Brown, T.L., Reynolds, P.J., Pape, W.J., Fritz, C.L., Carter, L.G., and Stein, J.D. 2000. Cases of cat-associated human plague in the Western US, 1977–1998. *Clin. Infect. Dis.*, 30, 893–900.

Galvan, E.M., Lasaro, M.A.S., and Schifferli, D.M. 2008. Capsular antigen fraction 1 and PLA modulate the susceptibility of *Yersinia pestis* to pulmonary antimicrobial peptides such as cathelicidin. *Infect. Immun.*, 76, 1456–1464.

Gasper, P.W., Barnes, A.M., Quan, T.J., Benziger, I.P., Carter, L.G., Beard, M.L., Maupin, G.O. 1993. Plague (*Yersinia pestis*) in cats: Description of experimentally induced disease. *J. Med. Entomol.*, 30, 20–26.

Goure, J., Broz, P., Attree, O., Cornelis, G.R., and Attree, I. 2005. Protective anti-V antibodies inhibit *Pseudomonas* and *Yersinia* translocon assembly within host membranes. *J. Infect. Dis.*, 192, 218–225.

Harmon, D.E., Murphy, J.L., Davis, A.J., and Mecsas, J. 2013. A mutant with aberrant extracellular LcrV-YscF interactions fails to form pores and translocate Yop effector proteins but retains the ability to trigger Yop secretion in response to host cell contact. *J. Bacteriol.*, 195, 2244–2254.

Hart, M.K., Saviolakis, G.A., Welkos, S.L., and House, R.V. 2012. Advanced development of the rF1V and rBV A/B vaccines: Progress and challenges. *Adv. Prev. Med.*, 2012, 731604.

Hartocollis, A. Plague Survivors Are Back in Hospital, to Say Thanks. *New York Times*, February 11, 2004.

Heath, D.G., Anderson, G.W., Mauro, J.M., Welkos, S.L., Andrews, G.P., Adamovicz, J., and Friedlander, A.M. 1998. Protection against experimental bubonic and pneumonic plague by a recombinant capsular F1–V antigen fusion protein vaccine. *Vaccine*, 16, 1131–1137.

Hill, J., Leary, S.E.C., Smither, S., Hewer, J., Best, A., Weeks, S., Pettersson, J., Forsberg, A., Lingard, B., Brown, K.A., Lipka, A., Williamson, E.D., and Titball, R.W. 2009. N255 is a key residue for recognition by a monoclonal antibody which protects against *Yersinia pestis* infection. *Vaccine*, 27, 7073–7079.

Hinnebusch, B.J., Rudolph, A.E., Cherepanov, P., Dixon, J.E., Schwan, T.G., and Forsberg, A. 2002. Role of *Yersinia* murine toxin in survival of *Yersinia pestis* in the midgut of the flea vector. *Science*, 296, 733–735.

Jones, S.M., Griffin, K.F., Hodgson, I., and Williamson, E.D. 2003. Protective efficacy of a fully recombinant plague vaccine in the guinea pig. *Vaccine*, 21, 3912–3918.

Kopaskie, K.S., Ligtenburg, K.G., and Schneewind, O. 2013. Translational regulation of *Yersinia enterocolitica* mRNA encoding type III secretion substrate. *J. Biol. Chem.*, 288, 35478–35488.

Lahteenmaki, K., Kukkonen, M., and Korhonen, T.K. 2001. The Pla surface protease/adhesin of *Yersinia pestis* mediates bacterial invasion into human endothelial cells. *FEBS Lett.*, 504, 69–72.

Lathem, W.W., Price, P.A., Miller, V.L., and Goldman, W.E. 2007. A plasminogen-activating protease specifically controls the development of primary pneumonic plague. *Science*, 315, 509–513.

Lawton, D.G., Longstaff, C., Wallace, B.A., Hill, J., Leary, S.E.C., Titball, R.W., and Brown, K.A. 2002. Interactions of the type III secretion pathway proteins LcrV and LcrG from *Yersinia pestis* are mediated by coiled-coil domains. *J. Biol. Chem.*, 277, 38714–38722.

Levy, Y., Flashner, Y., Tidhar, A., Zauberman, A., Aftalion, M., Lazar, S., Gur, D., Shafferman, A., and Mamroud, E. 2011. T cells play an essential role in anti-F1 mediated rapid protection against bubonic plague. *Vaccine*, 29, 6866–6873.

Ligtenberg, K.G., Miller, N.C., Mitchell, A., Plano, G.V., and Schneewind, O. 2013. LcrV mutants that abolish *Yersinia* type III injectisome function. *J. Bacteriol.*, 195, 777–787.

Lindler, L.E., and Tall, B.D. 1993. *Yersinia pestis* pH-6 antigen forms fimbriae and is induced by intracellular association with macrophages. *Mol. Microbiol.* 8, 311–324.

Matson, J.S., and Nilles, M.L. 2001. LcrG–LcrV interaction is required for control of yops secretion in *Yersinia pestis. J. Bacteriol.*, 183, 5082–5091.

Morelli, G., Song, Y., Mazzoni, C.J., Eppinger, M., Roumagnac, P., Wagner, D.M., Feldkamp, M., Kusecek, B., Vogler, A.J., Li, Y., Cui, Y., Thomson, N.R., Jombart, T., Leblois, T.R., Lichtner, R.P., Rahalison, L., Petersen, J.M., Balloux, F., Keim, P., Wirth, T., Ravel, J., Yang, R., Carniel, E., and Achtman, M. 2010. *Yersinia pestis* genome sequencing identifies patterns of global phylogenetic diversity. *Nat. Genet.*, 42(12), 1140–1143.

Mueller, C.A., Broz, P., Muller, S.A., Ringler, P., Erne-Brand, F., Sorg, I., Kuhn, M., Engel, A., and Cornelis, G.R. 2005. The V-antigen of *Yersinia* forms a distinct structure at the tip of injectisome needles. *Science*, 310, 674–676.

Musson, J.A., Ingram, R., Durand, G., Ascough, S., Waters, E.L., Hartley, M.G., Robson, T., Maillere, B., Williamson, E.D., Sriskandan, S., Altmann, D., and Robinson, J.H. 2010. The repertoire of DR1-restricted CD4 T cell epitopes of the capsular Caf1 antigen of *Y. pestis* in human leucocyte antigen transgenic mice. *Infect. Immun.*, 78, 4356–4362.

Nilles, M.L., Fields, K.A., and Straley, S.C. 1998. The V antigen of *Yersinia pestis* regulates Yop vectorial targeting as well as Yop secretion through effects on Yopb and Lcrg. *J. Bacteriol.*, 180, 3410–3420.

Oyston, P.C.F., and Williamson, E.D. 2012. Modern advances against plague. In: Sariaslani, S., and Gadd, G.M., editors. *Advances in Applied Microbiology.* Elsevier Inc.: Academic Press. pp. 209–241.

Parent, M.A., Berggren, K.N., Kummer, L.W., Wilhelm, L.B., Szaba, F.M., Mullarky, I.K., and Smiley, S.T. 2005. Cell-mediated protection against pulmonary *Yersinia pestis* infection. *Infect. Immun.*, 73, 7304–7310.

Parkhill, J., Wren, B.W., Thomson, N.R., Titball, R.W., Holden, M.T.G., Prentice, M.B., Sebaihia, M., James, K.D., Churcher, C., Mungall, K.L., et al. 2001. Genome sequence of *Yersinia pestis*, the causative agent of plague. *Nature*, 413, 523–527.

Perry, R.D., and Fetherston, J.D. 1997. *Yersinia pestis*–Etiologic agent of plague. *Clin. Microbiol. Rev.*, 10, 35–66.

Philipovskiy, A.V., and Smiley, S.T. 2007. Vaccination with live *Yersinia pestis* primes CD4 an CD8 T cells that synergistically protect against lethal pulmonary *Yersinia pestis* infection. *Infect. Immun.*, 75, 878–885.

Ritger, K., Black, S., Weaver, K., Jones, J., Gerber, S., Conover, C., Soyemi, K., Metzger, K., King, B., Mead, P., Molins, C., Schriefer, M., Shieh, W.-J., Zaki, S., and Medina-Marino, A. 2011. Fatal laboratory-acquired infection with an attenuated *Yersinia pestis* strain. *Morb. Mortal. Wkly. Rep.*, 60, 201–205.

Shim, H.K., Musson, J.A., Harper, H.M., McNeil, H.V., Walker, N., Flick-Smith, H.C., Von Delvig, A., Williamson, E.D., and Robinson, J.H. 2006. Mechanisms of MHC class II-restricted processing and presentation of the V antigen of *Yersinia pestis. Immunology*, 119(3), 385–392.

Shreewastav, R.K., Ali, R., Babu, J., and Rao, D.N. 2012. Cell-mediated immune response to epitopic MAP (multiple antigen peptide) construct of LcrV antigen of *Yersinia pestis* in murine model. *Cell Immunol.*, 278, 55–62.

Smiley, S.T. 2007. Cell-mediated defense against *Yersinia pestis* infection. *Adv. Exp. Med. Biol.*, 603, 376–386.

Sodeinde, O.A., Subrahmanyam, Y.V., Stark, K., Quan, T., Bao, Y., and Goguen, J.D. 1992. A surface protease and the invasive character of plague. *Science*, 258, 1004–1007.

Tanabe, M., Atkins, H.S., Harland, D.N., Elvin, S.J., Stagg, A.J., Mirza, O., Titball, R.W., Byrne, B., and Brown, K.A. 2006. The ABC transporter protein OppA provides protection against experimental *Yersinia pestis* infection. *Infect. Immun.*, 74, 3687–3691.

Tito, M.A., Miller, J., Griffin, K.F., Williamson, E.D., Titball, R.W., and Robinson, C.V. 2001. Macromolecular organisation of the *Yersinia pestis* capsular F1-antigen: Insights from ToF mass spectrometry. *Protein Sci.*, 10, 2408–2413.

US FDA HHS. 2002. New drug and biological drug products; evidence needed to demonstrate effectiveness of new drugs when human efficacy studies are not ethical or feasible. Final rule. *Fed. Regist.*, 67, 37988–37998.

US FDA. April 2012: http://www.fda.gov/newsevents/newsroom/pressannouncements/ucm302220.htm

US FDA. December 2012: http://www.raps.org/focusonline/news/news-article-view/article/2649/in-first-for-animal-rule-pathway-fda-approves-gsks-raxibacumab.aspx

Vernazza, C., Lingard, B., Flick-Smith, H.C., Baillie, L.W., Hill, J., and Atkins, H.S. 2009. Small protective fragments of the *Yersinia pestis* V antigen. *Vaccine*, 27, 2775–2780.

Wang, S.X., Joshi, S., Mboudjeka, I., Liu, F.J., Ling, F.T., Goguen, J.D., and Lu, S. 2008. Relative immunogenicity and protection potential of candidate *Yersinia pestis* antigens against lethal mucosal plague challenge in Balb/C mice. *Vaccine*, 26, 1664–1674.

WHO. 2005. Plague. Democratic Republic of the Congo. *Wkly. Epidemiol. Rec.*, 80, 138–140.

Williamson, E.D., Eley, S.M., Griffin, K.F., Green, M., Russell, P., Leary, S.E.C., Oyston, P.C.F., Easterbrook, T., Reddin, K.M., Robinson, A., and Titball, R.W. 1995. A new improved sub-unit vaccine for plague: the basis of protection. *FEMS Immunol. Med. Microbiol.*, 12(3–4), 223–230.

Williamson, E.D., Flick-Smith, H.C., LeButt, C., Rowland, C.A., Jones, S.M., Waters, E.L., Gwyther, R.J., Miller, J., Packer, P.J., and Irving, M. 2005. Human immune response to a plague vaccine comprising recombinant F1 and V antigens. *Infect. Immun.*, 73, 3598–3608.

Williamson, E.D., Flick Smith, H.C., Waters, E.L., Miller, J., Hodgson, I., Smith, S., LeButt, C.S., and Hill, J. 2007. Immunogenicity of the rF1+rV vaccine with the identification of potential immune correlates of protection. *Microb. Pathog.*, 42(1), 12–22.

Williamson, E.D., and Oyston, P.C.F. July 2012. Acellular vaccines against plague. In: Carniel, E., and Hinnebusch, J.B., editors. *Yersinia, Systems Biology and Control*, Chapter 9. UK: Caister Academic Press. pp. 123–142.

Williamson, E.D., and Oyston, P.C.F. 2013. Protecting against plague: Towards a next generation vaccine. *Clin. Exp. Immunol.*, 172(1), 1–8.

Chapter 31

Pathogenesis of *Legionella pneumophila* in Humans

Adam J. Vogrin and Elizabeth L. Hartland

Department of Microbiology and Immunology, University of Melbourne, Victoria, Australia

31.1 Introduction

In July 1976, a convention of the American Legion was held at the Bellevue-Stratford Hotel in Philadelphia, USA. In the following days, a large number of individuals present at the convention and other city visitors in the vicinity of the hotel presented with a severe form of pneumonia. In total, there were 182 cases reported with 29 deaths and the illness was subsequently termed Legionnaires' disease (Fraser et al., 1977). The cause of the epidemic remained unknown for some months, inciting great public concern. Much speculation was made about the cause, which ranged from nickel carbonyl poisoning to a chemical warfare experiment gone wrong. Eventually, in December 1976, Joseph McDade determined that the causative agent of the outbreak was a previously unknown bacterial species, which was subsequently named *Legionella pneumophila* (McDade et al., 1977).

L. pneumophila has since been identified as the aetiological agent of unexplained outbreaks of pneumonia prior to 1976. The earliest documented outbreak occurred in 1957 in Austin, Minnesota, USA, where 78 townspeople were hospitalized with acute respiratory disease and two people died. The source of the epidemic was not identified until 1979 when a study of survivors showed that they had elevated levels of antibodies to *L. pneumophila* compared to matched controls (Osterholm et al., 1983). This investigation along with clinical and epidemiological observations led to *L. pneumophila* being attributed to the outbreak.

L. pneumophila is a Gram-negative rod-shaped bacterium belonging to the gamma-subgroup of proteobacteria. The genus *Legionella* of the family Legionellaceae contains over 50 species belonging to over 70 serogroups, most of which are harmless to humans (Fields et al., 2002b). Some species, however, are known to cause disease, with *L. pneumophila* being the most common infectious agent of Legionnaires' disease worldwide. *Legionella* species also cause the relatively less severe flu-like illness, Pontiac fever. Collectively, the diseases caused by *Legionella* are termed legionellosis.

Legionella species are ubiquitously found in freshwater environments, including man-made water supply systems, where they are able to parasitize various protozoan species. The ability of *Legionella* to multiply intracellularly within amoebae was first described in 1980 (Rowbotham, 1980). This discovery led to a novel concept that bacteria able to parasitize protozoa may use the same mechanisms to infect humans.

Technologies that use water at higher than ambient temperature and cause water aerosolization, such as large air conditioning systems, have contributed to the emergence of *Legionella* as a human pathogen. The inhalation of *Legionella*-contaminated water droplets enables the bacterium to enter the human lung, where it infects and replicates within alveolar macrophages (Figure 31-1). Outbreaks of Legionnaires' disease almost exclusively originate from contaminated man-made water supply systems and have caused worldwide public health concern since the discovery of the pathogen.

31.2 Epidemiology

31.2.1 Incidence and risk factors

Annual reported legionellosis cases in the United States have increased 217%, from 1110 cases in the year 2000 to 3522 in 2009 (CDC, 2011). This may be due to the fact that the reporting of legionellosis cases has been improving with many laboratories now routinely using a *Legionella* urinary antigen test in the

Human Emerging and Re-emerging Infections: Bacterial & Mycotic Infections, Volume II, First Edition. Edited by Sunit K. Singh.
© 2016 John Wiley & Sons, Inc. Published 2016 by John Wiley & Sons, Inc.

(a)

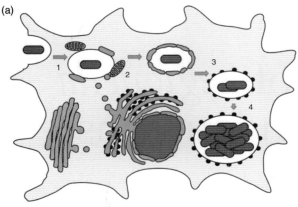

(b)

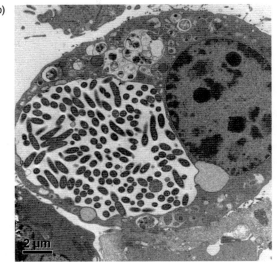

Fig. 31-1 Intracellular replication of *Legionella pneumophila*. (a) Formation of the *L. pneumophila*-containing vacuole (LCV). (1) After internalization, the LCV avoids endosomal fusion, transiently associates with mitochondria, and intercepts endoplasmic reticulum-derived early secretory vesicles prior to their transport to the Golgi (2) The ER-derived vesicles fuse with the LCV membrane. (3) The LCV associates with ribosomes and contains resident ER markers such as calnexin and Sec61. (4) The mature replicative vacuole is formed and *L. pneumophila* enters the replicative phase. After multiple rounds of replication, *L. pneumophila* then lyses the host cell, allowing bacteria to infect nearby cells. (b) Transmission electron micrograph of a macrophage infected with *L. pneumophila* for 24 hours containing a large LCV. Courtesy of Dr. Ralf Schuelein and Ms. Vicki Bennett-Wood.

diagnosis of pneumonia cases, rather than an increase in the incidence of the bacterium itself. In Europe, the number of reported cases has been steady over the years 2005–2008 (Joseph and Ricketts, 2010). However, as legionellosis is a largely under-diagnosed disease, primarily due to symptomatic similarities to other respiratory illnesses, it is likely that the actual number of legionellosis cases is much higher in all countries. There are usually more cases of legionellosis reported in the summer and early autumn, but the disease can occur at any time around a year.

While the mortality rate of Legionnaires' disease varies greatly from 5% to 30%, during the period 1980–1998, the average case-fatality rate of Legionnaires' disease in the United States decreased from 34% to 12% (Benin et al., 2002). Early recognition

of Legionnaires disease from increased testing and rapid diagnosis has likely contributed to this decrease in mortality. Increased awareness of *Legionella* outbreaks and more timely and appropriate responses help resolve infection by early treatment, thereby reducing mortality.

People at high risk of infection are the elderly, current or former smokers, people with a prior respiratory illness, and immune-compromised individuals. Elderly people are at greatest risk because of the reduced capacity of their immune system to fight infection and their compromised pulmonary function. Many reported legionellosis cases are hospital acquired, due to the prevalence of high-risk patients such as those receiving immunosuppressive treatments for various conditions, including cancer.

31.2.2 Transmission

Mostly, *Legionella* species are found naturally in freshwater environments but also in damp soil. In Australia, New Zealand, and Thailand, many cases of legionellosis are attributed to *L. longbeachae*, which is primarily found in soil (Amodeo et al., 2010; Li et al., 2002; Phares et al., 2007). Consequently, compost and potting mixes now carry a mandatory warning describing the risk of infection from the inhalation of dust. *Legionella* is commonly isolated from man-made aquatic environments where it is able to survive chlorination (Kuchta et al., 1983) and the bacteria are often found in biofilms (Declerck, 2010).

Legionella exploit free-living protozoa as hosts for replication. These natural hosts provide nutrients and a protective niche for the bacteria that allows for abundant intracellular replication. Some have the ability to form protective cysts and this is especially beneficial to *Legionella* as it allows the bacteria to survive high temperatures, disinfection procedures, and drying (Marciano-Cabral and Cabral, 2003). *Legionella* species grow best in warm water, and hence infection is associated with water held at higher than ambient temperature, including in spas, showers, and air conditioning cooling towers of large buildings (Breiman et al., 1990; Fallon and Rowbotham, 1990; Ferre et al., 2009). The water aerosols generated by these systems leads to transmission of *Legionella* to humans after inhalation of contaminated water droplets. However, the bacteria do not spread from person to person.

The close biological association between *Legionella* and protozoa has pre-adapted the bacteria for human infection due to similarities in function between protozoa and human macrophages. *L. pneumophila*, in particular, is able to infect a wide range of protozoan species and it is widely accepted that *Legionella* has accumulated a diverse range of virulence factors in order to survive and replicate within distinctly different protozoa (Fields, 1996; Rowbotham, 1980; Valster et al., 2010). However, our knowledge of exactly what is required for bacterial survival within protozoan host cells and the mechanisms involved in supporting *Legionella* replication is still developing. The inhalation of protozoa containing *Legionella* may also be an effective vehicle for transmission to humans, as well as playing a crucial role in the continuing presence of *Legionella* in aquatic systems as a host for replication (Rowbotham, 1986).

Legionellosis is a disease that has emerged in the last half of the 20th century; this is thought to be primarily due to human alteration of the environment. Left in their natural environment, *Legionella* would be a very rare cause of human infection, as natural freshwater environments have not been implicated as significant reservoirs of outbreaks of legionellosis.

31.3 Genome organization of *Legionella pneumophila*

The genomes of several clinical isolates of *L. pneumophila* have been sequenced. By comparison, they are well conserved being 3.3–3.5 Mbp in length and possessing a GC content of approximately 38%. The genomes contain 2900–3200 protein-coding genes with an average length of 980–1080 bp and a coding density of 87–88% (Cazalet et al., 2004; Chien et al., 2004; Gomez-Valero et al., 2011; Schroeder et al., 2010).

A DNA array of 217 *L. pneumophila* strains has shown a high degree of conservation amongst genes known to be associated with virulence, suggesting a strong selection pressure for their maintenance (Cazalet et al., 2008). No overall genomic profile distinguishes clinical and environmental samples or strains of different serogroups. Interestingly, the genes responsible for the core and O side-chain synthesis of lipopolysaccharide (LPS) in serogroup 1 strains have a significant level of genetic variation which suggests the likelihood of horizontal transfer of the LPS cluster (Cazalet et al., 2008). Indeed horizontal gene transfer of mobile genetic elements is the main source of genetic diversity amongst *L. pneumophila* strains. In addition, plasmid excision and integration constitutes another source of genome plasticity in *L. pneumophila*. The gene cluster encoding a Lvh Type IVa secretion system (T4aSS) can be present on a plasmid or in an integrated form in the genome (Cazalet et al., 2004; Chien et al., 2004) and two mobile elements carrying a T4ASS have also been reported in *L. pneumophila* strain Corby (Glockner et al., 2008). A 100-kb region comprising several genes encoding efflux transporters for heavy metals and toxins has also been identified. Fringing this region are tRNA, and phage-related and transposase genes, indicating a possible acquisition via horizontal transfer (Chien et al., 2004). This region may be responsible for protecting *Legionella* from toxins present in plumbing and man-made water supply systems. *Legionella* possesses more genes encoding efflux transporters for heavy metals and toxins relative to many other bacterial species, possibly as these are required for survival within protozoan hosts that accumulate heavy metals from the environment (Fernandez-Leborans and Herrero, 2000).

Legionella possesses a Type IVb secretion system (T4bSS) termed the Dot/Icm (Defective in Organelle Trafficking/Intracellular Multiplication) that injects a large number of bacterial virulence "effector" proteins into the host cell. The vast majority of *dot/icm* genes are conserved amongst different strains and are present in the same chromosomal location (Gomez-Valero et al., 2011; Morozova et al., 2004). More than 200 Dot/Icm effectors are conserved in all *L. pneumophila* strains with 95–100% nucleotide similarity (Gomez-Valero et al., 2011). Despite an apparent high level of functional redundancy amongst these proteins, their conservation highlights their communal importance to the *L. pneumophila* life cycle and broad host range. In addition to the Dot/Icm T4bSS, the *L. pneumophila* genome encodes several other T4SSs and conjugative elements that likely contribute to genome plasticity as T4SSs are involved in DNA uptake and transfer as well as the spread of conjugative plasmids and protein translocation (Backert and Meyer, 2006).

31.4 Evolution

The *L. pneumophila* genome encodes many homologues of eukaryotic proteins or proteins with motifs found primarily in eukaryotes, and as a prokaryote it possesses the widest variety of these homologues (Cazalet et al., 2004; Chien et al., 2004). These include a number F-box and U box proteins, two CD39 ecto-nucleoside triphosphate diphosphohydrolases, a sphingosine-1-phosphate lyase, a Sec7 domain protein, a mitochondrial carrier protein, and a Soluble N-ethylmaleimide-sensitive factor Attachment protein REceptor (SNARE) protein (Cazalet et al., 2004; Hubber and Roy, 2010; Xu and Luo, 2013). The eukaryotic-like proteins are likely to have been obtained through the close association of *Legionella* with its protozoan host and presumably interfere with host cell processes by mimicking the function of eukaryotic proteins. An analysis of several *L. pneumophila* genomes found that more than 50% of genes encoding eukaryotic-like proteins are conserved amongst strains with a high level (89–100%) of nucleotide identity, suggesting a high selection pressure for their maintenance (Gomez-Valero et al., 2011).

The majority of Legionnaires' disease cases worldwide are caused by *L. pneumophila* serogroup 1. A multi-genome analysis of six *L. pneumophila* serogroup 1 strains has shown that the strains not only possess a highly conserved core genome of housekeeping genes but also many eukaryotic-like genes, *dot/icm* genes, and secreted effectors (Gomez-Valero et al., 2011). The core genome of the analyzed strains included 2434 genes, approximately 80% of the total number of predicted genes in each genome. The gene order in the strains was also highly conserved, except for a 260 kb inversion in the Lens strain.

Frequent horizontal gene transfer and recombination events have contributed to a diverse accessory genome. It has been suggested that the numerous secretion systems of *L. pneumophila* facilitated the distribution of large chromosomal fragments of over 200 kb via conjugal transfer. In addition, *L. pneumophila* is naturally competent and possesses functional recombination machinery required for the integration of foreign DNA into the *Legionella* genome (Sexton and Vogel, 2004; Stone and Kwaik, 1999). This important evolutionary ability is permissive for *L. pneumophila*'s dynamic accessory genome, allowing for frequent horizontal transfer and recombination events. Analysis of nucleotide polymorphisms amongst the six strains identified numerous large fragments from different origins including eukaryotes, other prokaryotes, as well as different strains and species of *Legionella*. The diverse accessory genome of about 300 genes comprises mobile genetic elements, genomic islands, and many genes of unknown function, including Dot/Icm effector proteins (Gomez-Valero et al., 2011). The dynamic genome of *L. pneumophila* reflects its capacity to survive in a diverse range of environments and hosts.

31.5 Pathogenesis

L. pneumophila is an intracellular pathogen that replicates within a host eukaryotic cell, utilizing it as a source of nutrients. The pathogen possesses a remarkable ability to manipulate various host cell processes to form a protective, membrane-bound vacuole that supports intracellular replication known as the *Legionella*-containing vacuole (LCV).

After internalization by phagocytes, ingested bacteria are normally destroyed via a degradative process known as the endocytic or lysosomal pathway. The LCV avoids the endocytic pathway, and hence destruction in lysosomes by intercepting vesicles that are trafficking in the secretory pathway of the host cell (Roy et al., 1998) (Figure 31-1). The recruitment and fusion of ER exit vesicles to the early LCV results in a vacuole positive for Rab1 and Sec22b. The LCV membrane then matures into a membrane that resembles the endoplasmic reticulum (ER) (Kagan and Roy, 2002). The cytoplasmic face of the LCV membrane becomes lined with ribosomes, and resident ER markers are present, including calnexin, BiP, and Sec61 (Shin and Roy, 2008) (Figure 31-1). Once the mature LCV is formed, *L. pneumophila* enters a replicative growth phase and after multiple rounds of replication will exit the host cell, usually via lysis, allowing for infection of nearby cells where the infection process will begin anew (Shin and Roy, 2008).

L. pneumophila alternates between two growth states during infection, namely a non-motile, thin-walled replicative form and a motile, thick-walled infectious form (Garduno et al., 2002; Rowbotham, 1986). Bacteria differentiate into the replicative state after entering a host cell and establishing the LCV. In this state, bacteria are avirulent and non-flagellated (Molofsky and Swanson, 2004). After numerous rounds of replication, nutrients become limited, triggering the switch to the infectious form. The bacterial two-component gene regulators, LetA and LetS (*Legionella* transmission activator and sensor, respectively), govern this differentiation through activation of two small regulatory RNAs, RsmY and RsmZ (Rasis and Segal, 2009; Sahr et al., 2009). In this state, bacteria are highly virulent and flagellated, allowing for release and transmission to a new host (Edwards et al., 2010; Hammer et al., 2002; Molofsky and Swanson, 2004).

31.5.1 The Dot/Icm type IV secretion system and protein translocation

Legionella possesses the Dot/Icm T4bSS that is absolutely required for intracellular replication within protozoa and human macrophages (Andrews et al., 1998; Segal and Shuman, 1999). The Dot/Icm system is ancestrally related to DNA conjugation systems but transfers around 300 bacterial effector proteins into the host cell, rather than conjugating plasmid DNA into a bacterial recipient. The Dot/Icm apparatus comprises around 27 proteins, including several bacterial inner and outer membrane components and presumably proteins that span LCV membrane (Qiu and Luo, 2013). The translocation of effector proteins requires recognition of a C-terminal secretion signal that varies in its amino acid composition amongst different effectors (Huang et al., 2011; Kubori et al., 2008; Nagai et al., 2005). Rather than the amino acid sequence itself, it is the physicochemical properties of the C-terminal 35 amino acids that are most important for recognition by the Dot/Icm system (Lifshitz et al., 2013). Effectors lacking a strong translocation signal require chaperones for their secretion (Lifshitz et al., 2013). The cytosolic Dot/Icm components, IcmS and IcmW, form a putative chaperone complex that is necessary for the translocation of a subset of effector proteins into the host cell (Cambronne and Roy, 2007). *icmS* and *icmW* knockout mutants still recruit early secretory vesicles and replicate to some extent, but the mutant vacuoles eventually fuse with lysosomes (Coers et al., 2000). This suggests that effector proteins chaperoned by IcmS/IcmW are necessary for avoiding lysosomal fusion.

31.5.2 Dot/Icm effector proteins and LCV trafficking

Given the essential contribution of the Dot/Icm system to *L. pneumophila* pathogenesis, there is much current investigation into the cellular processes that support intracellular survival and replication of the pathogen and the role of Dot/Icm effectors in modulating these pathways. The translocated Dot/Icm effector proteins contribute to formation of the LCV by interfering with multiple cellular operations (Albert-Weissenberger et al., 2007; Hubber and Roy, 2010; Xu and Luo, 2013). In fact, *L. pneumophila* appears to influence almost every aspect of host cell physiology, including vesicle trafficking, membrane fusion, gene transcription, protein translation, and cell survival. Dissecting the function of individual effectors has proved difficult as many effectors are found in families of paralogues and even unrelated effectors can target the same cellular process. Nevertheless, research over the past few years has revealed the function of many newly identified *L. pneumophila* effectors. Some of the best characterized effectors have novel enzymatic activities that introduce post-translational modifications and act in concert to exert an exquisite control over the function of their substrate (Mukherjee et al., 2011; Rolando and Buchrieser, 2012). Interestingly, many Dot/Icm effectors also share amino acid sequence similarity with eukaryotic proteins or are predicted to carry a motif or domain that is found predominantly in eukaryotic proteins.

31.5.3 Manipulation of host cell trafficking pathways

The LCV intercepts ER-derived early secretory vesicles, which involves the recruitment of host factors Arf1 (ADP Ribosylation Factor-1) and Rab1 (a member of the RAS oncogene family), both of which are regulators of ER–Golgi vesicle traffic (Kagan and Roy, 2002). These host factors are found on wild-type but not on *dot/icm*-mutant LCVs (Kagan and Roy, 2002), indicating that Dot/Icm-translocated effectors are responsible for their recruitment.

Arf1 is a host GTPase involved in the regulation of vesicle trafficking between the ER and the Golgi. Arf proteins are activated by Arf-specific guanine nucleotide exchange factors (GEFs) containing Sec7 protein domains. The Sec7 domain is required for the exchange of GDP for GTP on Arf proteins, switching them to an active state (Jackson and Casanova, 2000). The Dot/Icm-translocated effector RalF contains a Sec7 domain that is homologous to mammalian Sec7 domains (Amor et al., 2005). RalF functions as a guanine exchange factor that localizes to the LCV where it is able to recruit and activate Arf1 (Donaldson and Jackson, 2000; Nagai et al., 2002). A structural C-terminal capping domain regulates the activity and localization of RalF following translocation which confers GEF function in the secretory pathway of the infected cell (Alix et al., 2012).

The host GTPase, Rab1, is involved in the recruitment of proteins that facilitate the transport, adherence, and fusion of vesicles (Zerial and McBride, 2001). DrrA (or SidM) is a Dot/Icm-translocated effector that recruits Rab1 to the LCV via the displacement of GDP association inhibitors that maintain inactive Rab1 in the cytosol (Machner and Isberg, 2006). Inactivation of Rab1 is mediated by the Dot/Icm effector LepB, which acts as a GTPase-activating protein (GAP). LepB binds activated Rab1-GTP and hydrolyzes GTP to GDP, thereby converting Rab1 into an inactive form (Ingmundson et al., 2007).

Aside from its GEF activity, DrrA contains a second region that is responsible for the activation of Rab1, once it has been recruited to the LCV, via its GEF activity (Machner and Isberg, 2007). The GEF activity of DrrA is complemented by an N-terminal domain that mediates AMPylation of tyrosine 77 in the switch II region of Rab1 (Muller et al., 2010). This posttranslational modification blocks access for GAPs, including LepB, and leads to constitutive activation of Rab1.

Rab1 activity is exquisitely regulated by another *L. pneumophila* effector, SidD, which deAMPylates Rab1 leading to the release of Rab1 from the LCV (Neunuebel et al., 2011; Tan and Luo, 2011). In addition, the FIC domain protein, AnkX, modifies Ser76 in the switch II region of Rab1 with phosphocholine, which is removed by the Dot/Icm effector Lem3 (Mukherjee et al., 2011; Tan et al., 2011). Phosphocholination of Rab1 does not appear to affect binding by DrrA, suggesting that the modifications may shape Rab1 interactions with distinct sets of proteins. Despite these seemingly important functions, deletion mutants of these effector genes are able to replicate at normal levels within macrophages and protozoa (Machner and Isberg, 2006; Nagai et al., 2002), indicating the likely existence of functional redundancy amongst translocated effectors targeting Rab1.

Various other Dot/Icm-translocated effectors have also been linked to an interaction with ER-derived vesicles and the formation of the replicative vacuole. LidA is required for efficient formation of the replicative vacuole (Derre and Isberg, 2005) and enhances the activity of DrrA by promoting the tethering of ER-derived vesicles to the LCV via Rab1 association (Machner and Isberg, 2006). LidA has been shown to bind multiple Rab GTPases (Chen and Machner, 2013; Cheng et al., 2012; Schoebel et al., 2011). LidA may be recruited to the LCV via its association with Rab GTPases, where it could play a role in the tethering of ER vesicles to the LCV. Altogether, these findings indicate a comprehensive capacity of *Legionella* to control the activity of Rab1 at several levels, in dictating whether it is present in an active or inactive form on the LCV.

The fusion of ER-derived vesicles to the LCV is believed to require the attachment of a SNARE protein on the target membrane to a SNARE on the vesicle membrane (Kagan et al., 2004). Sec22b is a SNARE located on ER-derived vesicles that is recruited to the LCV in a Dot/Icm-dependant manner, contributing to the formation of the replicative vacuole (Kagan et al., 2004). Vesicle fusion may be achieved via the plasma membrane-localized SNARE proteins, syntaxin 2, syntaxin 3, syntaxin 4, and SNAP23, which localize to the LCV where they interact with Sec22b (Arasaki and Roy, 2010). It is also possible that a *Legionella* effector mimics SNARE function, facilitating the fusion of ER-derived vesicles to the LCV. For example, the effectors YlfA (LegC7) and YlfB (LegC2) possess a protein domain similar to the IncA protein family. IncA is a bacterial SNARE-like molecule involved in the fusion of vesicles in *Chlamydia*-infected cells (Delevoye et al., 2004; 2008). YlfA/LegC7 associates with ER membranes and may target ER-derived vesicles (de Felipe et al., 2008).

Another host GTPase implicated in LCV biogenesis is Sar1, which regulates the formation of coat protein II (COPII)-coated vesicles derived from the ER (Sato and Nakano, 2007). Sar1 plays several roles in the cell including COPII coat recruitment and cargo sorting. A siRNA knockdown of Sar1 and a dominant interfering Sar1 variant have been used to show that impeding Sar1 function interferes with the intracellular replication of *L. pneumophila* (Dorer et al., 2006; Kagan and Roy, 2002). The LCV in cells expressing Sar1H79G, a GTP-restricted Sar1 variant, does not associate with ER-derived vesicles (Robinson and Roy, 2006), suggesting that the vesicles destined to interact with the LCV are generated by a Sar1-dependant process. Further study is required to determine what bacterial factors are involved in this process. Collectively, these examples demonstrate several complex strategies employed by *L. pneumophila* to subvert the trafficking of ER-derived vesicles for the benefit of phagosome maturation.

31.5.4 Phagosomal pH regulation

A phagosome undergoing endocytic maturation will become acidified by the vacuolar ATPase (vATPase) proton pump acquired at the late endosomal stage (Forgac, 2007). A low luminal pH is important for lysosome maturation as well as for the activity of lysosomal enzymes. In order to avoid lysosomal degradation, many intracellular pathogens regulate pH in the phagosomal lumen (Huynh and Grinstein, 2007; Ohkuma and Poole, 1978). *L. pneumophila* and *L. longbeachae* are able to maintain a phagosome of neutral pH. However, reports differ on the association of vATPase with the LCV during infection. One report has stated that the majority of *L. pneumophila* and *L. longbeachae* containing vacuoles do not co-localize with the vATPase protein pump (Asare and Abu Kwaik, 2007). However, proteomic analysis of LCVs purified from *Dictyostelium* and macrophages by immunomagnetic separation showed that several components of the vATPase were associated with the LCV, despite neutral pH within the vacuole and the absence of other late endosomal features on the LCV (Shevchuk et al., 2009). This suggested that although the vATPase may be present on the LCV, it is not actively lowering pH. Later, it was shown that the Dot/Icm-translocated effector SidK specifically targets host vATPase via an interaction with VatA, the component of vATPase that is responsible for hydrolyzing ATP (Xu et al., 2010). Binding of SidK to VatA results in the inhibition of ATP hydrolysis and proton translocation into the LCV. The same study also demonstrated that microinjection of bone marrow-derived macrophages with SidK impaired their ability to digest non-pathogenic *Escherichia coli* (Xu et al., 2010). Thus, SidK contributes to the protection of internalized *L. pneumophila* by subverting the function of host vATPase, thereby blocking the acidification of the LCV.

31.5.5 Manipulation of host cell ubiquitination and autophagy

Ubiquitination of proteins is a post-translational modification that regulates numerous host cell processes by altering the activity of proteins or directing proteins for degradation by the cell proteasome. The importance of this function to *Legionella* replication is illustrated by two early findings that polyubiquitinated proteins decorate the LCV shortly after infection and proteosome inhibitors limit *Legionella* intracellular replication (Dorer et al., 2006). During infection, *L. pneumophila* translocates a number of effectors with presumed or proven ability to interfere with or exploit the host cell ubiquitination machinery (Hubber et al., 2013). For example, the *L. pneumophila* genome encodes functional mimics of eukaryotic E3 ubiquitin ligases that act in concert with components of the host ubiquitination machinery to target both host and bacterial proteins for polyubiquitination. These include F-box- and U-box-containing proteins and proteins containing ankyrin repeat domains, a motif that has been implicated in numerous eukaryotic protein–protein interactions (Hubber et al., 2013; Mosavi et al., 2002). The identification of substrates for these E3 ligases is key to understanding their function. For example, Clk1 is a substrate of the U-box protein, LubX, and Clk family inhibitors limit intracellular growth, although the molecular role of Clk kinases in *Legionella* replication is not known (Kubori et al., 2008). Another idea based on a severe replication defect of a mutant lacking the F-box protein, AnkB (LegAU13), is that protein ubiquitination and degradation promotes the supply of amino acids for *L. pneumophila* replication in the LCV (Price et al., 2011). However, this finding may be strain-specific, as *ankB* mutants in other strains show little to no defect in intracellular replication.

Host proteins are not the only targets of *Legionella* U-box and F-box proteins, and a mechanism has been suggested whereby Dot/Icm effector activity is regulated by targeted ubiquitination and degradation by the host cell proteasome. For example, the effector SidH is also a target of LubX, which has led to the description of LubX as a "metaeffector," namely an effector that controls the activity of another subset of effectors (Kubori et al., 2010).

Another much studied aspect of *Legionella*–host cell interactions is the association of the LCV with the autophagy pathway. Eukaryotic cells may use autophagy to sequester cytosolic organelles, pathogens, and pathogen-modified vacuoles into a membrane-bound compartment termed the autophagosome (Joshi and Swanson, 2011). Autophagosomes are recognizable by a double membrane and various membrane markers such as Atg5,

Atg7, Atg9, and LC3/Atg8. Autophagy is linked to the ubiquitin-conjugation system and is a means for the cell to dispose of non-functional organelles and to recycle components by fusing the autophagosome with lysosomes to promote degradation of their cargo machinery (Hubber et al., 2013; Joshi and Swanson, 2011). Autophagosomes also provide a defense mechanism against intracellular pathogens (Swanson, 2006).

Autophagy appears to limit *L. pneumophila* replication as mutants of the amoeba, *Dictyostelium discoideum*, that lack Atg9 are more permissive for bacterial replication (Tung et al., 2010), and depletion of Atg5 by siRNA treatment in macrophages promotes *L. pneumophila* replication (Matsuda et al., 2009). Likewise, the induction of autophagy limits *L. pneumophila* replication and this depends on a functional Dot/Icm secretion system (Matsuda et al., 2009). The nascent LCV becomes positive for autophagy markers 2–4 hours after infection, but fusion with lysosomes is inhibited (Amer and Swanson, 2005; Joshi and Swanson, 2011). A key step in autophagosome development is the cleavage of cytosolic LC3, and its conjugation to phosphatidylethanolamine by Atg3 and Atg7. Recently, the Dot/Icm effector, RavZ, was described as a cysteine protease that cleaves the amide bond between the carboxyl-terminal glycine residue and an adjacent aromatic residue in LC3 (Choy et al., 2012). This results in a form of LC3 that can no longer be conjugated by Atg3 and Atg7 (Choy et al., 2012). Hence, *L. pneumophila* irreversibly modifies LC3 to inhibit autophagosome development and fusion with lysosomes.

31.5.6 Lipid metabolism

Phosphoinositide metabolism plays a role in membrane dynamics, actin remodelling, and cell signalling (Di Paolo and De Camilli, 2006). Phosphoinositides will also anchor target proteins to specific membranes, contributing to the identity of subcellular compartments (Yeung et al., 2006). *L. pneumophila* exploits host phosphoinositides to localize secreted effector proteins to the cytoplasmic face of the LCV (Weber et al., 2009). There is much evidence that phosphatidylinositol-4 phosphate (PtdIns(4)P), normally produced by the host PtdIns-4-kinase IIIβ, is present on the LCV membrane. Antibodies that bind PtdIns(4)P and a tagged PtdIns(4)P-binding protein, GST-FAPP1, both bind to phagosomes isolated from *L. pneumophila*-infected *D. discoideum* and RAW 264.7 cells (Weber et al., 2009).

The Dot/Icm-translocated effector SidC and its paralogue SdcA anchor to the LCV via PtdIns(4)P (Ragaz et al., 2008). *sidC-sdcA* deletion mutants are less efficient at recruiting ER-derived vesicles and establishing the LCV. Beads coated with SidC, or its 70-kDa N-terminal fragment, recruit ER vesicles in *Dictyostelium* and macrophage lysates, suggesting that the effector plays an important role in formation of the ER-derived vacuole. SidC harbours a 20-kDa PtdIns(4)P-binding domain near the C-terminus for anchoring to the cytoplasmic face of the LCV and recruits ER vesicles to the LCV via a 70-kDa N-terminal fragment (Ragaz et al., 2008).

Other *L. pneumophila* effector proteins also exploit host phosphoinositides. SidM directly binds PtdIns(4)P, LpnE binds PtdIns(3)P, and LidA binds PtdIns(3)P as well as PtdIns(4)P (Brombacher et al., 2009; Weber et al., 2009). These examples demonstrate the ability of *L. pneumophila* to exploit host cell phosphatidylinositol lipids, using them as a means to anchor various effector proteins to the membrane of the LCV. *L. pneumophila* undoubtedly controls the composition and timing of PtdIns flux on the LCV membrane. For example, PtdIns(4)P is generated on the LCV by the effector, SidF, which is a phosphatidylinositol polyphosphate 3-phosphatase that specifically hydrolyzes the D3 phosphate of PtdIns(3,4)P(2) and PtdIns(3,4,5)P(3) (Hsu et al., 2012). The PI phosphatase activity of SidF is necessary for anchoring PI(4)P-binding effectors to the LCV.

Sphingolipid metabolism is involved in several physiological functions, including proliferation, inflammation, cell survival, and apoptosis (Bandhuvula and Saba, 2007). Recent evidence has suggested that *L. pneumophila* may modulate sphingolipid metabolism, resulting in the manipulation of one or more of these processes (Degtyar et al., 2009). The Dot/Icm secreted effector LegS2 is homologous to the conserved eukaryotic enzyme, sphingosine-1-phosphatase lyase (SPL). SPL is an enzyme that is involved in the metabolism of sphingolipids and it is possible that *Legionella* manipulates sphingolipid metabolism to acquire a degradation product for its virulence. *L. pneumophila* translocates LegS2 into the host cytosol where it then localizes to the mitochondria (Degtyar et al., 2009). This is in contrast to eukaryotic SPL, which localizes to the ER. Exactly what function LegS2 performs at the mitochondria is yet to be determined. *L. pneumophila* encodes several other proteins putatively involved in sphingolipid metabolism, including proteins highly similar to sphingomyelinase and sphingosine kinase (Gomez-Valero et al., 2009). While it is not completely understand exactly why *L. pneumophila* modulates sphingolipid metabolism, it serves as another example of the manipulation of a host process by *Legionella*.

31.5.7 Protein synthesis and the stress response

eEF1A and eEF1Bγ are two eukaryotic elongation factors that are involved in polypeptide chain elongation. The Dot/Icm-secreted effector SidI targets both of these elongation factors to inhibit host protein synthesis (Shen et al., 2009). Interaction with eEF1A simultaneously induces a stress response in host cells and it has been shown that eEF1A is required for the activation of heat shock factor 1 (HSF1), a major stress response regulatory protein (Shamovsky et al., 2006). *L. pneumophila* infection activates HSF1 in host cells but infection with a *sidI* deletion mutant leads to a reduction in the level of HSF1 activation (Shen et al., 2009). Various other host stress response genes are also induced by *L. pneumophila* infection of both amoebae and mammalian cells (Farbrother et al., 2006; Losick and Isberg, 2006).

A family of effector proteins, Lgt1, Lgt2, and Lgt3, functions as glucosyltransferases that are structurally similar to the large clostridial toxins and use UDP-glucose as a substrate (Belyi et al., 2013). The *L. pneumophila* effector protein Lgt1 glycosylates serine 53 in the GTP-binding domain of eEF1A, resulting in the inhibition of protein synthesis and induction of cell death (Belyi et al., 2006; Tzivelekidis et al., 2011). Lgt1 is also able to modify the heat shock protein 70 subfamily B suppressor Hbs1, adding to the induction of a stress response in host cells (Belyi et al., 2009). However, the toxicity induced by Lgt1 in yeast and mammalian cells seems to be due to its effect on eEF1A rather than Hbs1 (Belyi et al., 2012). Like many Dot/Icm effector mutants, deletion of all three *lgt* genes had no impact on intracellular replication (Ivanov and Roy, 2009), and hence the contribution of these effectors to *L. pneumophila* infection is still unclear. It is possible that the induction of a stress response may make the host cell environment more favourable for *bacterial* replication; however, the induction of cell death would seem to be counterproductive.

31.5.8 Cell death and cytotoxicity

The manipulation of host signalling in order to avoid untimely host cell apoptosis is important in preventing the premature termination of *L. pneumophila* replication. Microarray analysis of *L. pneumophila*-infected cells has shown that genes involved in NF-κB signalling and genes with anti-apoptotic function are transcriptionally upregulated (Abu-Zant et al., 2007; Losick and Isberg, 2006). The transcription factor NF-κB is involved in the promotion of host cell survival and *L. pneumophila*-infected cells have been shown to activate NF-κB signalling in a Dot/Icm-dependant manner (Abu-Zant et al., 2007; Losick and Isberg, 2006). This indicates that a Dot/Icm-translocated effector may be responsible for the manipulation of NF-κB signalling, thus influencing host cell survival.

One effector putatively involved in preventing apoptosis was SdhA, a paralogue of SidH (Laguna et al., 2006). Macrophages infected with an *L. pneumophila sdhA* deletion mutant displayed increased nuclear degradation, membrane permeability, mitochondrial disruption, and caspase activation (Laguna et al., 2006), suggesting a role for SdhA in the prevention of host cell death. Mutants lacking *sdhA* display severe intracellular growth defects due to the rapid induction of host cell death and are attenuated *in vivo* in A/J mice as well as in *Galleria mellonella* (Harding et al., 2013; Laguna et al., 2006). More recently,

SdhA was found to be critical for maintaining the integrity of the LCV (Creasey and Isberg, 2012). *sdhA* mutants become cytosolic as the LCV membrane is degraded. How SdhA maintains the LCV membrane is not known, but cell death results from the recognition of cytosolic *Legionella* DNA by the AIM2 inflammasome (Ge et al., 2012). Another *Legionella* effector, SidF, has been reported to inhibit apoptosis by interfering with the function of BNIP3 and Bcl-rambo, two pro-apoptotic members of the Bcl2 protein family (Banga et al., 2007). How SidF mediates this function is unknown, particularly given its activity as a phosphatidylinositol polyphosphate 3-phosphatase (Hsu et al., 2012).

The activation of host NF-κB by *L. pneumophila* may also contribute to the inhibition of apoptosis (Losick and Isberg, 2006). NF-κB is a transcriptional regulator that contributes to inflammation and host cell survival and activates several anti-apoptotic genes. NF-κB is initially activated upon recognition of *Legionella* flagellin, but activation is also sustained for several hours after infection in a flagellin-independent manner (Bartfeld et al., 2009). LegK1 is a substrate of the Dot/Icm T4SS that has demonstrated an ability to activate NF-κB by mimicking host IKK. LegK1 directly phosphorylates the IκB family of NF-κB inhibitor proteins resulting in their ubiquitination and degradation by the cell proteasome (Ge et al., 2009). Other Dot/Icm effectors have also been implicated in NF-κB activation. SdbA and LubX were both found to contribute to the sustained activation of NF-κB in A549 epithelial cells (Bartfeld et al., 2009). Hence, it is evident that *L. pneumophila* manipulates multiple host pathways associated with apoptosis using an array of secreted effector proteins.

31.6 Immunity

L. pneumophila is an accidental pathogen of humans and to date human-to-human transmission has not been observed. Hence, people remain an evolutionary dead end for the bacteria and there is no selective pressure from the mammalian immune system. Much of our knowledge of the immune response to *Legionella* infection has arisen from studies using mice, which have been extrapolated to humans. These have been extensively reviewed elsewhere (Brown et al., 2013; Luo, 2012; Newton et al., 2010), so here we will restrict discussion to what is known about human infection.

In healthy individuals, *Legionella* infections are usually effectively cleared by the immune system with few symptoms (Palusinska-Szysz and Janczarek, 2010). A robust inflammatory response followed by cell-mediated immunity is the primary mechanism of host defence against *Legionella*. Healthy individuals also generate anti-*Legionella* antibodies in their serum; however, their contribution to clearance is not clear (Rudbeck et al., 2009).

After inhalation, *Legionella* is phagocytosed by alveolar macrophages present in the lungs. *Legionella* poses a challenge to the human immune system as the bacteria have the ability to survive and replicate within the very immune cells designed to destroy bacteria. Immune cells such as macrophages, natural killer (NK) cells, and immature dendritic cells are primarily involved in the initial activation of the innate immune system triggered by bacterial surface antigens. Toll-like receptors (TLRs) present on immune cells recognize components of bacteria such as flagellum, LPS, and peptidoglycan. The stimulation of TLRs leads to the production of proinflammatory cytokines and the expression of co-stimulatory molecules, recruiting lymphocytes to the site of infection and activating immune cells. Analysis of patient genotypes after a Legionnaires' disease outbreak showed that a common TLR5 polymorphism, which introduces a premature stop codon (TLR5392STOP), is associated with a small increased risk of Legionnaire's disease (Hawn et al., 2003). The dominant TLR5392STOP polymorphism likely increases an individual's risk of Legionnaires' disease, as TLR5 is no longer able to mediate flagellin signalling in lung epithelial cells, impairing the production of proinflammatory cytokines. Other innate immune signalling pathways associated with genetic susceptibility or resistance to Legionnaires' disease are TLR6 and TLR4. While the TLR6 polymorphism, 359T>C, is associated with an elevated risk of Legionnaires' disease which is further enhanced by smoking (Misch et al., 2013), certain TLR4 alleles are associated with protection (Hawn et al., 2005).

Patients diagnosed with Legionnaires' disease have increased serum levels of interferon γ and IL-12, indicative of a Th1-type response (Tateda et al., 1998). *In vitro*, *L. pneumophila* infection assays of human monocytes and macrophages show that a Th1-type cytokine response ultimately inhibits *L. pneumophila* replication (Bhardwaj et al., 1986; Matsiota-Bernard et al., 1993; Nash et al., 1984).

T cells are also believed to be important for clearance of bacteria during infection given that depletion of CD4 and CD8 T cells results in a decreased survival rate for mice infected with *L. pneumophila* compared to non-treated mice (Susa et al., 1998). However, in humans, the contribution of T cells is not completely established as people with low CD4 T cells counts, such as those infected with HIV, are not necessarily more susceptible to *Legionella* infection. In general, immune compromised people, such as those receiving immunosuppressive therapies, including anti-cancer chemotherapies, are at greater risk of serious infection (Ginevra et al., 2009).

There is no vaccine currently available for the prevention of legionellosis. However, several studies have demonstrated protective immunity in animal models using live avirulent bacteria, *Legionella* major secretory proteins, membrane fragments, and flagellin (Blander and Horwitz, 1991a; Blander and Horwitz, 1991b, 1993; Blander et al., 1989; Ricci et al., 2005).

31.7 Clinical features

The incubation period for Legionnaires' disease is from 2 to 14 days. Patients with Legionnaires' disease commonly develop muscle aches, fever, chills, and a cough. As the illness progresses, pneumonia and severe respiratory distress develop. *Legionella* may disseminate to other organs in the body via the blood stream and lymphatic system and so non-pulmonary symptoms may arise. This leads to some patients experiencing headache, diarrhoea, tiredness, confusion, and loss of appetite (Cunha, 2010; Ginevra et al., 2009; Jespersen et al., 2010). Pontiac fever, on the other hand, is an acute self-limiting illness that causes flu-like symptoms (fever, chills, and malaise) without pneumonia. Incubation time is 24–28 hours.

The majority of patients presenting with legionellosis are elderly males who may be smokers, immuno-suppressed, and/or patients with a pre-existing illness. The major cause of death in patients with Legionnaires' disease is respiratory failure. Many patients display abnormal chest X-rays, although an absence does not exclude *Legionella* infection (Jespersen et al., 2010). A comparison of pneumonia patients with Legionnaires' disease compared to non-Legionnaires' disease found that patients with Legionnaires' disease displayed an increased prevalence of central nervous system symptoms (headache, confusion, and drowsiness) and diarrhoea in comparison to the non-Legionnaires' disease patients at the time of admission to hospital (Hugosson et al., 2007). Legionnaires' disease patients also displayed a higher fever and raised C-reactive protein levels. Hyponatraemia and elevated liver enzymes were also more frequent at the time of admission.

31.8 Diagnosis

Legionnaires' disease is difficult to diagnose without further testing, as symptoms are often identical to many other forms of pneumonia. Elevated erythrocyte sedimentation rates of over 90 mm/h have been shown to distinguish Legionnaires' disease from viral pneumonias (Cunha et al., 2010). Various abnormalities may be produced such as haematuria, hypophosphataemia, thrombocytopaenia, hyponatraemia, and abnormal liver function tests. The absence of these symptoms does not exclude *Legionella* infection and so the use of laboratory-based diagnostic methods is important for early detection of legionellosis.

The *Legionella* urinary antigen test is a simple and rapid test now used by many diagnostic laboratories to detect antigens of *L. pneumophila* serogroup 1 (the most common causative agent of Legionnaires' disease worldwide) in urine. This test detects the presence of the *L. pneumophila* serogroup 1 antigen, present in urine during infection, using an enzyme immunoassay. It is highly specific for *L. pneumophila* serogroup 1 and so where possible is coupled with bacterial culture from respiratory secretions or pleural fluid to give a more definitive diagnosis of legionellosis. Although the urine antigen test will identify the most common cause of Legionnaires' disease, it is sometimes used in conjunction with other diagnostic methods, particularly in countries where non-*L. pneumophila* species are prevalent. For example, *L. longbeachae* is a common cause of legionellosis in Australia, Thailand, and New Zealand, and this organism cannot be identified by the urine antigen test (Phares et al., 2007; Yu et al., 2002).

In vitro bacteriological culture of *L. pneumophila* requires specialized media, as the bacteria are fastidious in their growth requirements. Bacteria are cultured on buffered charcoal yeast extract (BCYE) media (Feeley et al., 1979) in the presence and absence of L-cysteine to determine whether they belong to the genus *Legionella*. L-cysteine is needed for the growth of *Legionella* species and provides a mechanism to distinguish *Legionella* species from other bacterial genera in the same sample. *Legionella* species produce characteristic branched-chain fatty acids in their cell wall, giving them a distinct morphology helping to distinguish them from similar bacteria. Although the culture is specific for *Legionella*, sensitivity varies likely due to varying levels of bacteria present in patient samples. Culture of *Legionella* takes at least 3–5 days to obtain positive results (Jarraud et al., 2013).

Other diagnostic methods include serology, latex agglutination assays, direct immunofluorescence assays, and various polymerase chain reaction (PCR) detection methods. Some rely on previous isolation of the strain by culture whereas other may be applied directly to a patient sample. Serology may be used to detect the presence of *Legionella* antibodies in patient serum. While being highly specific, sensitivity varies and it is not often used in clinical settings as it takes several weeks to obtain the results due to the amount of time needed for seroconversion (Jarraud et al., 2013).

Direct fluorescence assays involve the staining of patient samples with fluorescent antibodies specific for individual *Legionella* strains. Samples are analyzed using fluorescence microscopy to detect the presence of *Legionella*. This assay may be completed in less than 4 hours and is highly specific. Again, sensitivity varies likely due to the number of bacteria present in the patient sample but also due to the skill of the laboratory staff performing the test as it is quite technically demanding (Murdoch, 2003; Sethi et al., 2007; She et al., 2007).

PCR is also used for rapid detection of *Legionella* DNA in patient samples and can be a highly effective diagnostic tool. PCR amplifies target regions of DNA and can be used to detect specific genes in samples. As *Legionella*-specific genes are targeted, this test is highly specific with varying levels of sensitivity depending on the sample used (i.e., lower respiratory tract secretions are generally more reliable than serum or urine samples) (Benitez and Winchell, 2013; Diederen et al., 2008; Zhou et al., 2011). Quantitative real-time polymerase chain reaction (qPCR) has also been developed for rapid identification of *Legionella* in patient samples. One system targeting the macrophage infectivity potentiator (*mip*) gene of *L. pneumophila* demonstrated a degree of specificity. The test was able to positively identify all *L. pneumophila* subspecies and 16 serogroups tested (39/39 strains) with no cross-reaction from non-*pneumophila* strains (0/69 strains) and non-*Legionella* strains (0/58 strains). Culture-positive patient samples were tested using this system and were also all found to be qPCR positive (81/81). Culture-negative samples were also tested and 47/80 samples were found to be positive, indicating qPCR to be the more sensitive method (Mentasti et al., 2012). With the wealth of genomic information now available, it is possible to design rapid PCR tests to identify *Legionella* species and distinguish one species and even serogroup from another (Yong et al., 2010). Diagnostic testing of clinical samples for *Legionella* has been extensively outlined in a recent methods handbook (Jarraud et al., 2013).

31.9 Treatment

Pontiac fever is a mild self-limited illness that is not treated with antibiotics. Supportive therapy may be used to treat symptoms if required. Complete recovery will usually occur within 1 week (Glick et al., 1978; Remen et al., 2011). Legionnaires' disease can generally be treated successfully with antibiotics, and healthy individuals usually recover from infection without complication, although recovery time varies and may take several weeks. β-Lactam antibiotics such as penicillin are ineffective as they are unable to penetrate macrophages efficiently and most *Legionella* isolates produce β-lactamases (Marre et al., 1982). Instead, Levofloxacin (or other fluoroquinolones) or azithromycin are commonly used in the treatment of Legionnaires' disease (Arora et al., 2012; Cunha, 2010). Levofloxacin is a broad-spectrum antibiotic that inhibits the function of prokaryotic DNA gyrase and topoisomerase IV, preventing cell division. Azithromycin inhibits bacterial protein synthesis by binding the 50s ribosomal subunit, thereby inhibiting the translation of mRNA into peptides. Azithromycin was found to be well tolerated and was effective in treating patients with community-acquired Legionnaires' disease. The overall cure rate after 10–14 days of therapy was 95% and after 4–6 weeks of therapy was 96% (Plouffe et al., 2003).

Combined antibiotic therapy can be used in severe unresponsive cases. Although there are few comprehensive studies looking at combined therapy, it has been observed that combined therapy reduced the Intensive Care Unit (ICU) mortality rates in patients with severe community-acquired pneumonia caused by *L. pneumophila* in comparison to monotherapy (Rello et al., 2013). Care must be taken when administering combined therapy as there is a risk of additional toxicity and drug interactions.

Antibiotic therapy should be administered intravenously for three to five days or until clinical stability is reached and then may be substituted with oral antibiotic therapy. Longer therapy duration of up to three weeks may be required for immunosuppressed individuals and patients with advanced disease (Arora et al., 2012; Cunha, 2010). Prognosis of Legionnaires' disease depends largely on the patient's cardiopulmonary and immune function as well as the initial number of bacteria the patient was infected with and also the early administration of effective antibiotic therapy. Fatalities most often occur in individuals with a compromised immune system, prior respiratory conditions, or in patients presenting with advanced disease prior to treatment. However, if cardiopulmonary function is good, early treatment of Legionnaires' disease, even in compromised individuals, delivers a good prognosis.

31.10 Prevention and control

There is no current commercially available vaccine for *Legionella* infection, so control of the bacteria in the environment is extremely important for preventing disease. The key to the prevention of legionellosis is in the proper maintenance of water supply systems at risk. *Legionella* becomes a problem when the bacteria grow to high numbers. This can be avoided with suitable maintenance, including routine cleaning and disinfection procedures as well as appropriate testing to ensure effective microbiological control (Kozak et al., 2013). Testing for the presence of *Legionella* has led to the identification of previously undetected potential sources of infection. The current gold standard for testing is bacteriological culture, which can take up to 10 days to deliver the result. This lengthy period is a problem in outbreak investigations where there is an urgent need for a more rapid test for environmental samples. Molecular-based testing such as PCR is ineffective, as the test cannot reliably distinguish live from dead bacteria (Keer and Birch, 2003). The detection of live bacteria prompts intervention to prevent any (or additional) cases of legionellosis. In most countries, legionellosis is a notifiable disease to the relevant national health authority. The surveillance and reporting of clinical legionellosis cases is extremely important for quick identification and control of outbreaks.

After detection, *Legionella* may be eradicated from aquatic systems via hyperchlorination treatment and superheating water above 70°C. Several other eradication strategies exist, including other chemical treatments and the continuous copper–silver ionization of water supply systems (Arora et al., 2012; Carson and Mumford, 2010; Dupuy et al., 2011; Lin et al., 2011; Marchesi et al., 2012). Due to the exploitation by *Legionella* of environmental amoeba species as replicative and protective hosts, amoebicidal agents may also be considered in the control of legionellosis (Dupuy et al., 2011). Devices that continuously distribute chlorine dioxide into water supply systems have been used to control microbial contamination. However, the use of a monochloramine alternative has proved more effective. A recent study found that, over a one-year period, a reduction in the number of *Legionella*-contaminated sites decreased from 97.0% to 13.3% with the use of monochloramine in comparison to a reduction of 100% to 56.7–60.8% with the use of chlorine dioxide devices (Marchesi et al., 2012).

To prevent the risk of human exposure to bacteria in water supply systems, conditions that produce aerosols and promote bacterial growth are minimized in modern water systems. These controls may be achieved through the use of adiabatic cooling systems, dry cooling plants, point of use heaters, and routine disinfection protocols. Bacterial growth may also be reduced by avoiding temperatures between 20°C and 45°C, designing pipework to prevent water stagnation, and minimizing organic contaminants (Carson and Mumford, 2010; Fields et al., 2002a).

Testing of water systems for the presence of *Legionella* is generally carried out at regular intervals to verify whether the maintenance protocols implemented are working effectively. If chemical disinfectants are used, concentrations are usually tested at multiple points in the system to evaluate the effectiveness. It is clear that properly maintained systems reduce the risks of Legionnaires' disease (Dupuy et al., 2011; Fields et al., 2002a; Flannery et al., 2006). Ideally, maintenance protocols for water supply are designed in conjunction with infection control practitioners to ensure that the appropriate disinfection methods are implemented.

31.11 Conclusion

Legionnaires' disease remains a significant public health concern, as it is impossible to eradicate the bacteria from the environment. Even with adequate monitoring and surveillance, sporadic cases and unexpected outbreaks still occur. Rapid diagnosis and early intervention with antimicrobial treatment is critical to achieving the best outcome for patients, and the design of water supply systems, particularly in hospitals, needs to take *Legionella* infection risk into account. Independently of its importance as a human pathogen, *Legionella* has increased our understanding of many new and important aspects of cell biology and immunity, including mechanisms of vesicle trafficking and membrane fusion as well as inflammasome activation. Despite these recent advances in our knowledge of *L. pneumophila* host–pathogen interactions, many aspects of *L. pneumophila* intracellular survival and replication are still not well defined. Individually, Dot/Icm effector proteins are not essential for intracellular growth, with single effector mutants still replicating at equivalent levels to wild-type bacteria. This functional redundancy likely indicates that additional unidentified effector proteins are important and act in a complementary fashion to other effector proteins. It is clear that there is still a great deal to learn about the intracellular survival and replication mechanisms of *Legionella* that leads to Legionnaires' disease.

Acknowledgements

We would like to thank Dr. Ralf Schuelein and Ms. Vicki Bennett-Wood for providing us with the illustration and electron micrograph in Figure 31-1. This work was supported by an Australian Research Council (ARC) Future Fellowship and ARC Discovery grant awarded to E.L.H. A.J.V. was supported by University of Melbourne MDHS Faculty Research (Trust) Scholarship.

References

Abu-Zant, A., Jones, S., Asare, R., Suttles, J., Price, C., Graham, J., and Kwaik, Y.A. 2007. Anti-apoptotic signalling by the Dot/Icm secretion system of *L. pneumophila*. *Cell. Microbiol.*, 9, 246–264.

Albert-Weissenberger, C., Cazalet, C., and Buchrieser, C. 2007. *Legionella pneumophila* – a human pathogen that co-evolved with fresh water protozoa. *Cell. Mol. Life Sci.*, 64, 432–448.

Alix, E., Chesnel, L., Bowzard, B.J., Tucker, A.M., Delprato, A., Cherfils, J., Wood, D.O., Kahn, R.A., and Roy, C.R. 2012. The capping domain in RalF regulates effector functions. *PLoS Pathog.*, 8, e1003012.

Amer, A.O., and Swanson, M.S. 2005. Autophagy is an immediate macrophage response to *Legionella pneumophila*. *Cell. Microbiol.*, 7, 765–778.

Amodeo, M.R., Murdoch, D.R., and Pithie, A.D. 2010. Legionnaires' disease caused by *Legionella longbeachae* and *Legionella pneumophila*: Comparison of clinical features, host-related risk factors, and outcomes. *Clin. Microbiol. Infect.*, 16, 1405–1407.

Amor, J.C., Swails, J., Zhu, X., Roy, C.R., Nagai, H., Ingmundson, A., Cheng, X., and Kahn, R.A. 2005. The structure of RalF, an ADP-ribosylation factor guanine nucleotide exchange factor from *Legionella pneumophila*, reveals the presence of a cap over the active site. *J. Biol. Chem.*, 280, 1392–1400.

Andrews, H.L., Vogel, J.P., and Isberg, R.R. 1998. Identification of linked *Legionella pneumophila* genes essential for intracellular growth and evasion of the endocytic pathway. *Infect. Immun.*, 66, 950–958.

Arasaki, K., and Roy C.R. 2010. *Legionella pneumophila* promotes functional interactions between plasma membrane syntaxins and Sec22b. *Traffic*, 11, 587–600.

Arora, B., Kaur, K.P., and Sethi, B. 2012. Review article Legionellosis: An update. *J. Clin. Diagn. Res.*, 6, 1331–1336.

Asare, R., and Abu Kwaik, Y. 2007. Early trafficking and intracellular replication of *Legionella longbeachaea* within an ER-derived late endosome-like phagosome. *Cell. Microbiol.*, 9, 1571–1587.

Backert, S., and Meyer, T.F. 2006. Type IV secretion systems and their effectors in bacterial pathogenesis. *Curr. Opin. Microbiol.*, 9, 207–217.

Bandhuvula, P., and Saba, J.D. 2007. Sphingosine-1-phosphate lyase in immunity and cancer: Silencing the siren. *Trends Mol. Med.*, 13, 210–217.

Banga, S., Gao, P., Shen, X., Fiscus, V., Zong, W.X., Chen, L., and Luo, Z.Q. 2007. *Legionella pneumophila* inhibits macrophage apoptosis by targeting pro-death members of the Bcl2 protein family. *Proc. Natl. Acad. Sci. USA*, 104, 5121–5126.

Bartfeld, S., Engels, C., Bauer, B., Aurass, P., Flieger, A., Bruggemann, H., and Meyer, T.F. 2009. Temporal resolution of two-tracked NF-kappaB activation by *Legionella pneumophila*. *Cell. Microbiol.*, 11, 1638–1651.

Belyi, Y., Jank, T., and Aktories, K. 2013. Cytotoxic glucosyltransferases of *Legionella pneumophila*. *Curr. Top. Microbiol. Immunol.*, 376, 211–226.

Belyi, Y., Niggeweg, R., Opitz, B., Vogelsgesang, M., Hippenstiel, S., Wilm, M., and Aktories, K. 2006. *Legionella pneumophila* glucosyltransferase inhibits host elongation factor 1A. *Proc. Natl. Acad. Sci. USA*, 103, 16953–16958.

Belyi, Y., Stahl, M., Sovkova, I., Kaden, P., Luy, B., and Aktories, K. 2009. Region of elongation factor 1A1 involved in substrate recognition by *Legionella pneumophila* glucosyltransferase Lgt1: Identification of Lgt1 as a retaining glucosyltransferase. *J. Biol. Chem.*, 284, 20167–20174.

Belyi, Y., Tartakovskaya, D., Tais, A., Fitzke, E., Tzivelekidis, T., Jank, T., Rospert, S., and Aktories, K. 2012. Elongation factor 1A is the target of growth inhibition in yeast caused by *Legionella pneumophila* glucosyltransferase Lgt1. *J. Biol. Chem.*, 287, 26029–26037.

Benin, A.L., Benson, R.F., and Besser, R.E. 2002. Trends in Legionnaires disease, 1980–1998: Declining mortality and new patterns of diagnosis. *Clin. Infect. Dis.*, 35, 1039–1046.

Benitez, A.J., and Winchell, J.M. 2013. Clinical application of a multiplex real-time PCR assay for simultaneous detection of *Legionella* species, *Legionella pneumophila*, and *Legionella pneumophila* serogroup 1. *J. Clin. Microbiol.*, 51, 348–351.

Bhardwaj, N., Nash, T.W., and Horwitz, M.A. 1986. Interferon-gamma-activated human monocytes inhibit the intracellular multiplication of *Legionella pneumophila*. *J. Immunol.*, 137, 2662–2669.

Blander, S.J., and Horwitz, M.A. 1991a. Vaccination with *Legionella pneumophila* membranes induces cell-mediated and protective immunity in a guinea pig model of Legionnaires' disease. Protective immunity independent of the major secretory protein of *Legionella pneumophila*. *J. Clin. Invest.*, 87, 1054–1059.

Blander, S.J., and Horwitz, M.A. 1991b. Vaccination with the major secretory protein of *Legionella* induces humoral and cell-mediated immune responses and protective immunity across different serogroups of *Legionella pneumophila* and different species of *Legionella*. *J. Immunol.*, 147, 285–291.

Blander, S.J., and Horwitz, M.A. 1993. Major cytoplasmic membrane protein of *Legionella pneumophila*, a genus common antigen and member of the hsp 60 family of heat shock proteins, induces protective immunity in a guinea pig model of Legionnaires' disease. *J. Clin. Invest.*, 91, 717–723.

Blander, S.J., Breiman, R.F., and Horwitz, M.A. 1989. A live avirulent mutant *Legionella pneumophila* vaccine induces protective immunity against lethal aerosol challenge. *J. Clin. Invest.*, 83, 810–815.

Breiman, R.F., Fields, B.S., Sanden, G.N., Volmer, L., Meier, A., and Spika, J.S. 1990. Association of shower use with Legionnaires' disease. Possible role of amoebae. *J. Am. Med. Asocc.*, 263, 2924–2926.

Brombacher, E., Urwyler, S., Ragaz, C., Weber, S.S., Kami, K., Overduin, M., and Hilbi, H. 2009. Rab1 guanine nucleotide exchange factor SidM is a major phosphatidylinositol 4-phosphate-binding effector protein of *Legionella pneumophila*. *J. Biol. Chem.*, 284, 4846–4856.

Brown, A.S., van Driel, I.R., and Hartland, E.L. 2013. Mouse models of Legionnaires' disease. *Curr. Top. Microbiol. Immunol.*, 376, 271–291. DOI: 10.1007/82_2013_349

Cambronne, E.D., and Roy, C.R. 2007. The *Legionella pneumophila* IcmSW complex interacts with multiple Dot/Icm effectors to facilitate type IV translocation. *PLoS Pathog.*, 3, e188.

Carson, P., and Mumford, C. 2010. Legionnaires' disease: Causation, prevention and control. *Loss Prevent. Bull.*, 216, 20–28.

Cazalet, C., Jarraud, S., Ghavi-Helm, Y., Kunst, F., Glaser, P., Etienne, J., and Buchrieser, C. 2008. Multigenome analysis identifies a worldwide distributed epidemic *Legionella pneumophila* clone that emerged within a highly diverse species. *Genome Res.*, 18, 431–441.

Cazalet, C., Rusniok, C., Bruggemann, H., Zidane, N., Magnier, A., Ma, L., Tichit, M., Jarraud, S., Bouchier, C., Vandenesch, F., Kunst, F., Etienne, J., Glaser, P., and Buchrieser, C. 2004. Evidence in the *Legionella pneumophila* genome for exploitation of host cell functions and high genome plasticity. *Nat. Genet.*, 36, 1165–1173.

CDC. 2011. Legionellosis – United States, 2000–2009. *MMWR Morb. Mortal. Wkly. Rep.*, 60, 1083–1086.

Chen, Y., and Machner, M.P. 2013. Targeting of the small GTPase Rab6A' by the *Legionella pneumophila* effector LidA. *Infect. Immun.*, 81, 2226–2235.

Cheng, W., Yin, K., Lu, D., Li, B., Zhu, D., Chen, Y., Zhang, H., Xu, S., Chai, J., and Gu, L. 2012. Structural insights into a unique *Legionella pneumophila* effector LidA recognizing both GDP and GTP bound Rab1 in their active state. *PLoS Pathog.*, 8, e1002528.

Chien, M., Morozova, I., Shi, S., Sheng, H., Chen, J., Gomez, S.M., Asamani, G., Hill, K., Nuara, J., Feder, M., Rineer, J., Greenberg, J.J., Steshenko, V., Park, S.H., Zhao, B., Teplitskaya, E., Edwards, J.R., Pampou, S., Georghiou, A., Chou, I.C., Iannuccilli, W., Ulz, M.E., Kim, D.H., Geringer-Sameth, A., Goldsberry, C., Morozov, P., Fischer, S.G., Segal, G., Qu, X., Rzhetsky, A., Zhang, P., Cayanis, E., De Jong, P.J., Ju, J., Kalachikov, S., Shuman, H.A., and Russo, J.J. 2004. The genomic sequence of the accidental pathogen *Legionella pneumophila*. *Science*, 305, 1966–1968.

Choy, A., Dancourt, J., Mugo, B., O'Connor, T.J., Isberg, R.R., Melia, T.J., and Roy, C.R. 2012. The *Legionella* effector RavZ inhibits host autophagy through irreversible Atg8 deconjugation. *Science*, 338, 1072–1076.

Coers, J., Kagan, J.C., Matthews, M., Nagai, H., Zuckman, D.M., and Roy, C.R. 2000. Identification of Icm protein complexes that play distinct roles in the biogenesis of an organelle permissive for *Legionella pneumophila* intracellular growth. *Mol. Microbiol.*, 38, 719–736.

Creasey, E.A., and Isberg, R.R. 2012. The protein SdhA maintains the integrity of the *Legionella*-containing vacuole. *Proc. Natl. Acad. Sci. USA*, 109, 3481–3486.

Cunha, B.A. 2010. Legionnaires' disease: Clinical differentiation from typical and other atypical pneumonias. *Infect. Dis. Clin. North. Am.*, 24, 73–105.

Cunha, B.A., Strollo, S., and Schoch, P. 2010. Extremely elevated erythrocyte sedimentation rates (ESRs) in Legionnaires' disease. *Eur. J. Clin. Microbiol. Infect. Dis.*, 29, 1567–1569.

Declerck, P. 2010. Biofilms: The environmental playground of *Legionella pneumophila*. *Environ. Microbiol.*, 12, 557–566.

de Felipe, K.S., Glover, R.T., Charpentier, X., Anderson, O.R., Reyes, M., Pericone, C.D., and Shuman, H.A. 2008. *Legionella* eukaryotic-like type IV substrates interfere with organelle trafficking. *PLoS Pathog.*, 4, e1000117.

Degtyar, E., Zusman, T., Ehrlich, M., and Segal, G. 2009. A *Legionella* effector acquired from protozoa is involved in sphingolipids metabolism and is targeted to the host cell mitochondria. *Cell. Microbiol.*, 11, 1219–1235.

Delevoye, C., Nilges, M., Dautry-Varsat, A., and Subtil, A. 2004. Conservation of the biochemical properties of IncA from *Chlamydia trachomatis* and *Chlamydia caviae*: Oligomerization of IncA mediates interaction between facing membranes. *J. Biol. Chem.*, 279, 46896–46906.

Delevoye, C., Nilges, M., Dehoux, P., Paumet, F., Perrinet, S., Dautry-Varsat, A., and Subtil, A. 2008. SNARE protein mimicry by an intracellular bacterium. *PLoS Pathog.*, 4, e1000022.

Derre, I., and Isberg, R.R. 2005. LidA, a translocated substrate of the *Legionella pneumophila* type IV secretion system, interferes with the early secretory pathway. *Infect. Immun.*, 73, 4370–4380.

Diederen, B.M., Kluytmans, J.A., Vandenbroucke-Grauls, C.M., and Peeters, M.F. 2008. Utility of real-time PCR for diagnosis of Legionnaires' disease in routine clinical practice. *J. Clin. Microbiol.*, 46, 671–677.

Di Paolo, G., and De Camilli, P. 2006. Phosphoinositides in cell regulation and membrane dynamics. *Nature*, 443, 651–657.

Donaldson, J.G., and Jackson, C.L. 2000. Regulators and effectors of the ARF GTPases. *Curr. Opin. Cell. Biol.*, 12, 475–482.

Dorer, M.S., Kirton, D., Bader, J.S., and Isberg, R.R. 2006. RNA interference analysis of *Legionella* in Drosophila cells: Exploitation of early secretory apparatus dynamics. *PLoS Pathog.*, 2, e34.

Dupuy, M., Mazoua, S., Berne, F., Bodet, C., Garrec, N., Herbelin, P., Menard-Szczebara, F., Oberti, S., Rodier, M.H., Soreau, S., Wallet, F., and Hechard, Y. 2011. Efficiency of water disinfectants against *Legionella pneumophila* and *Acanthamoeba*. *Water Res.*, 45, 1087–1094.

Edwards, R.L., Jules, M., Sahr, T., Buchrieser, C., and Swanson, M.S. 2010. The *Legionella pneumophila* LetA/LetS two-component system exhibits rheostat-like behavior. *Infect. Immun.*, 78, 2571–2583.

Fallon, R.J., and Rowbotham, T.J. 1990. Microbiological investigations into an outbreak of Pontiac fever due to *Legionella micdadei* associated with use of a whirlpool. *J. Clin. Pathol.*, 43, 479–483.

Farbrother, P., Wagner, C., Na, J., Tunggal, B., Morio, T., Urushihara, H., Tanaka, Y., Schleicher, M., Steinert, M., and Eichinger, L. 2006. *Dictyostelium* transcriptional host cell response upon infection with *Legionella*. *Cell. Microbiol.*, 8, 438–456.

Feeley, J.C., Gibson, R.J., Gorman, G.W., Langford, N.C., Rasheed, J.K., Mackel, D.C., and Baine, W.B. 1979. Charcoal-yeast extract agar: Primary isolation medium for *Legionella pneumophila*. *J. Clin. Microbiol.*, 10, 437–441.

Fernandez-Leborans, G., and Herrero, Y.O. 2000. Toxicity and bioaccumulation of lead and cadmium in marine protozoan communities. *Ecotoxicol. Environ. Saf.*, 47, 266–276.

Ferre, M.R., Arias, C., Oliva, J.M., Pedrol, A., Garcia, M., Pellicer, T., Roura, P., and Dominguez, A. 2009. A community outbreak of Legionnaires' disease associated with a cooling tower in Vic and Gurb, Catalonia (Spain) in 2005. *Eur. J. Clin. Microbiol. Infect. Dis.*, 28, 153–159.

Fields, B.S. 1996. The molecular ecology of legionellae. *Trends Microbiol.*, 4, 286–290.

Fields, B.S., Benson, R.F., and Besser, R.E. 2002a. *Legionella* and Legionnaires' disease: 25 years of investigation. *Clin. Microbiol. Rev.*, 15, 506–526.

Fields, B.S., Benson, R.F., and Besser, R.E. 2002b. *Legionella* and Legionnaires' disease: 25 years of investigation. *Clin. Microbiol. Rev.*, 15, 506–526.

Flannery, B., Gelling, L.B., Vugia, D.J., Weintraub, J.M., Salerno, J.J., Conroy, M.J., Stevens, V.A., Rose, C.E., Moore, M.R., Fields, B.S., and Besser, R.E. 2006. Reducing *Legionella* colonization in water systems with monochloramine. *Emerg. Infect. Dis.*, 12, 588–596.

Forgac, M. 2007. Vacuolar ATPases: Rotary proton pumps in physiology and pathophysiology. *Nat. Rev. Mol. Cell. Biol.*, 8, 917–929.

Fraser, D.W., Tsai, T.R., Orenstein, W., Parkin, W.E., Beecham, H.J., Sharrar, R.G., Harris, J., Mallison, G.F., Martin, S.M., McDade, J.E., Shepard, C.C., and Brachman, P.S. 1977. Legionnaires' disease: Description of an epidemic of pneumonia. *N. Engl. J. Med.*, 297, 1189–1197.

Garduno, R.A., Garduno, E., Hiltz, M., and Hoffman, P.S. 2002. Intracellular growth of *Legionella pneumophila* gives rise to a differentiated form dissimilar to stationary-phase forms. *Infect. Immun.*, 70, 6273–6283.

Ge, J., Gong, Y.N., Xu, Y., and Shao, F. 2012. Preventing bacterial DNA release and absent in melanoma 2 inflammasome activation by a *Legionella* effector functioning in membrane trafficking. *Proc. Natl. Acad. Sci. USA*, 109, 6193–6198.

Ge, J., Xu, H., Li, T., Zhou, Y., Zhang, Z., Li, S., Liu, L., and Shao, F. 2009. A *Legionella* type IV effector activates the NF-kappaB pathway by phosphorylating the IkappaB family of inhibitors. *Proc. Natl. Acad. Sci. USA*, 106, 13725–13730.

Ginevra, C., Duclos, A., Vanhems, P., Campese, C., Forey, F., Lina, G., Che, D., Etienne, J., and Jarraud, S. 2009. Host-related risk factors and clinical features of community-acquired Legionnaires' disease due to the Paris and Lorraine endemic strains, 1998–2007, France. *Clin. Infect. Dis.*, 49, 184–191.

Glick, T.H., Gregg, M.B., Berman, B., Mallison, G., Rhodes Jr., W.W., and Kassanoff, I. 1978. Pontiac fever. An epidemic of unknown etiology in a health department: I. Clinical and epidemiologic aspects. *Am. J. Epidemiol.*, 107, 149–160.

Glockner, G., Albert-Weissenberger, C., Weinmann, E., Jacobi, S., Schunder, E., Steinert, M., Hacker, J., and Heuner, K. 2008. Identification and characterization of a new conjugation/type IVA secretion system (trb/tra) of *Legionella pneumophila* Corby localized on two mobile genomic islands. *Int. J. Med. Microbiol.*, 298, 411–428.

Gomez-Valero, L., Rusniok, C., and Buchrieser, C. 2009. *Legionella pneumophila*: Population genetics, phylogeny and genomics. *Infect. Genet. Evol.*, 9, 727–739.

Gomez-Valero, L., Rusniok, C., Jarraud, S., Vacherie, B., Rouy, Z., Barbe, V., Medigue, C., Etienne, J., and Buchrieser, C. 2011. Extensive recombination events and horizontal gene transfer shaped the *Legionella pneumophila* genomes. *BMC Genomics*, 12, 536.

Hammer, B.K., Tateda, E.S., and Swanson, M.S. 2002. A two-component regulator induces the transmission phenotype of stationary-phase *Legionella pneumophila*. *Mol. Microbiol.*, 44, 107–118.

Harding, C.R., Stoneham, C.A., Schuelein, R., Newton, H., Oates, C.V., Hartland, E.L., Schroeder, G.N., and Frankel, G. 2013. The Dot/Icm effector SdhA is necessary for virulence of *Legionella pneumophila* in *Galleria mellonella* and A/J mice. *Infect. Immun.*, 81, 2598–2605.

Hawn, T.R., Verbon, A., Janer, M., Zhao, L.P., Beutler, B., and Aderem, A. 2005. Toll-like receptor 4 polymorphisms are associated with resistance to Legionnaires' disease. *Proc. Natl. Acad. Sci. USA*, 102, 2487–2489.

Hawn, T.R., Verbon, A., Lettinga, K.D., Zhao, L.P., Li, S.S., Laws, R.J., Skerrett, S.J., Beutler, B., Schroeder, L., Nachman, A., Ozinsky, A., Smith, K.D., and Aderem, A. 2003. A common dominant TLR5 stop codon polymorphism abolishes flagellin signaling and is associated with susceptibility to Legionnaires' disease. *J. Exp. Med.*, 198, 1563–1572.

Hsu, F., Zhu, W., Brennan, L., Tao, L., Luo, Z.Q., and Mao, Y. 2012. Structural basis for substrate recognition by a unique *Legionella* phosphoinositide phosphatase. *Proc. Natl. Acad. Sci. USA*, 109, 13567–13572.

Huang, L., Boyd, D., Amyot, W.M., Hempstead, A.D., Luo, Z.Q., O'Connor, T.J., Chen, C., Machner, M., Montminy, T., and Isberg, R.R. 2011. The E block motif is associated with *Legionella pneumophila* translocated substrates. *Cell. Microbiol.*, 13, 227–245.

Hubber, A., Kubori, T., and Nagai, H. 2013. Modulation of the ubiquitination machinery by *Legionella*. *Curr. Top. Microbiol. Immunol.*, 376, 227–247.

Hubber, A., and Roy, C.R. 2010. Modulation of host cell function by *Legionella pneumophila* type IV effectors. *Annu. Rev. Cell. Dev. Biol.*, 26, 261–283.

Hugosson, A., Hjorth, M., Bernander, S., Claesson, B.E., Johansson, A., Larsson, H., Nolskog, P., Pap, J., Svensson, N., and Ulleryd, P. 2007. A community outbreak of Legionnaires' disease from an industrial cooling tower: Assessment of clinical features and diagnostic procedures. *Scand. J. Infect. Dis.*, 39, 217–224.

Huynh, K.K., and Grinstein, S. 2007. Regulation of vacuolar pH and its modulation by some microbial species. *Microbiol. Mol. Biol. Rev.*, 71, 452–462.

Ingmundson, A., Delprato, A., Lambright, D.G., and Roy, C.R. 2007. *Legionella pneumophila* proteins that regulate Rab1 membrane cycling. *Nature*, 450, 365–369.

Ivanov, S.S., and Roy, C.R. 2009. Modulation of ubiquitin dynamics and suppression of DALIS formation by the *Legionella pneumophila* Dot/Icm system. *Cell. Microbiol.*, 11, 261–278.

Jackson, C.L., and Casanova, J.E. 2000. Turning on ARF: The Sec7 family of guanine-nucleotide-exchange factors. *Trends Cell. Biol.*, 10, 60–67.

Jarraud, S., Descours, G., Ginevra, C., Lina, G., and Etienne, J. 2013. Identification of *Legionella* in clinical samples. *Methods Mol. Biol.*, 954, 27–56.

Jespersen, S., Sogaard, O.S., Schonheyder, H.C., Fine, M.J., and Ostergaard, L. 2010. Clinical features and predictors of mortality in admitted patients with community- and hospital-acquired legionellosis: A Danish historical cohort study. *BMC Infect. Dis.*, 10, 124.

Joseph, C.A., and Ricketts, K.D. 2010. Legionnaires disease in Europe 2007–2008. *Euro. Surveill.*, 15, 19493.

Joshi, A.D., and Swanson, M.S. 2011. Secrets of a successful pathogen: *Legionella* resistance to progression along the autophagic pathway. *Front. Microbiol.*, 2, 138.

Kagan, J.C., and Roy, C.R. 2002. *Legionella* phagosomes intercept vesicular traffic from endoplasmic reticulum exit sites. *Nat. Cell. Biol.*, 4, 945–954.

Kagan, J.C., Stein, M.P., Pypaert, M., and Roy, C.R. 2004. *Legionella* subvert the functions of Rab1 and Sec22b to create a replicative organelle. *J. Exp. Med.*, 199, 1201–1211.

Keer, J.T., and Birch, L. 2003. Molecular methods for the assessment of bacterial viability. *J. Microbiol. Methods*, 53, 175–183.

Kozak, N.A., Lucas, C.E., and Winchell, J.M. 2013. Identification of *Legionella* in the environment. *Methods Mol. Biol.*, 954, 3–25.

Kubori, T., Hyakutake, A., and Nagai, H. 2008. *Legionella* translocates an E3 ubiquitin ligase that has multiple U-boxes with distinct functions. *Mol. Microbiol.*, 67, 1307–1319.

Kubori, T., Shinzawa, N., Kanuka, H., and Nagai, H. 2010. *Legionella* metaeffector exploits host proteasome to temporally regulate cognate effector. *PLoS Pathog.*, 6, e1001216.

Kuchta, J.M., States, S.J., McNamara, A.M., Wadowsky, R.M., and Yee, R.B. 1983. Susceptibility of *Legionella pneumophila* to chlorine in tap water. *Appl. Environ. Microbiol.*, 46, 1134–1139.

Laguna, R.K., Creasey, E.A., Li, Z., Valtz, N., and Isberg, R.R. 2006. A *Legionella pneumophila*-translocated substrate that is required for growth within macrophages and protection from host cell death. *Proc. Natl. Acad. Sci. USA*, 103, 18745–18750.

Li, J.S., O'Brien, E.D., and Guest, C. 2002. A review of national legionellosis surveillance in Australia, 1991 to 2000. *Commun. Dis. Intell. Q. Rep.*, 26, 461–468.

Lifshitz, Z., Burstein, D., Peeri, M., Zusman, T., Schwartz, K., Shuman, H.A., Pupko, T., and Segal, G. 2013. Computational modeling and experimental validation of the *Legionella* and *Coxiella* virulence-related type-IVB secretion signal. *Proc. Natl. Acad. Sci. USA*, 110, E707–E715.

Lin, Y.E., Stout, J.E., and Yu, V.L. 2011. Prevention of hospital-acquired legionellosis. *Curr. Opin. Infect. Dis.*, 24, 350–356.

Losick, V.P., and Isberg, R.R. 2006. NF-kappaB translocation prevents host cell death after low-dose challenge by *Legionella pneumophila*. *J. Exp. Med.*, 203, 2177–2189.

Luo, Z.Q. 2012. *Legionella* secreted effectors and innate immune responses. *Cell. Microbiol.*, 14, 19–27.

Machner, M.P., and Isberg, R.R. 2006. Targeting of host Rab GTPase function by the intravacuolar pathogen *Legionella pneumophila*. *Dev. Cell.*, 11, 47–56.

Machner, M.P., and Isberg, R.R. 2007. A bifunctional bacterial protein links GDI displacement to Rab1 activation. *Science*, 318, 974–977.

Marchesi, I., Cencetti, S., Marchegiano, P., Frezza, G., Borella, P., and Bargellini, A. 2012. Control of *Legionella* contamination in a hospital water distribution system by monochloramine. *Am. J. Infect. Control*, 40, 279–281.

Marciano-Cabral, F., and Cabral, G. 2003. *Acanthamoeba* spp. as agents of disease in humans. *Clin. Microbiol. Rev.*, 16, 273–307.

Marre, R., Medeiros, A.A., and Pasculle, A.W. 1982. Characterization of the beta-lactamases of six species of *Legionella*. *J. Bacteriol.*, 151, 216–221.

Matsiota-Bernard, P., Lefebre, C., Sedqui, M., Cornillet, P., and Guenounou, M. 1993. Involvement of tumor necrosis factor alpha in intracellular multiplication of *Legionella pneumophila* in human monocytes. *Infect. Immun.*, 61, 4980–4983.

Matsuda, F., Fujii, J., and Yoshida, S. 2009. Autophagy induced by 2-deoxy-D-glucose suppresses intracellular multiplication of *Legionella pneumophila* in A/J mouse macrophages. *Autophagy*, 5, 484–493.

McDade, J.E., Shepard, C.C., Fraser, D.W., Tsai, T.R., Redus, M.A., and Dowdle, W.R. 1977. Legionnaires' disease: Isolation of a bacterium and demonstration of its role in other respiratory disease. *N. Engl. J. Med.*, 297, 1197–1203.

Mentasti, M., Fry, N.K., Afshar, B., Palepou-Foxley, C., Naik, F.C., and Harrison, T.G. 2012. Application of *Legionella pneumophila*-specific quantitative real-time PCR combined with direct amplification and sequence-based typing in the diagnosis and epidemiological investigation of Legionnaires' disease. *Eur. J. Clin. Microbiol. Infect. Dis.*, 31, 2017–2028.

Misch, E.A., Verbon, A., Prins, J.M., Skerrett, S.J., and Hawn, T.R. 2013. A TLR6 polymorphism is associated with increased risk of Legionnaires' disease. *Genes Immun.*, 14, 420–426.

Molofsky, A.B., and Swanson, M.S. 2004. Differentiate to thrive: Lessons from the *Legionella pneumophila* life cycle. *Mol. Microbiol.*, 53, 29–40.

Morozova, I., Qu, X., Shi, S., Asamani, G., Greenberg, J.E., Shuman, H.A., and Russo, J.J. 2004. Comparative sequence analysis of the icm/dot genes in *Legionella*. *Plasmid*, 51, 127–147.

Mosavi, L.K., Minor Jr., D.L., and Peng, Z.Y. 2002. Consensus-derived structural determinants of the ankyrin repeat motif. *Proc. Natl. Acad. Sci. USA*, 99, 16029–16034.

Mukherjee, S., Liu, X., Arasaki, K., McDonough, J., Galan, J.E., and Roy, C.R. 2011. Modulation of Rab GTPase function by a protein phosphocholine transferase. *Nature*, 477, 103–106.

Muller, M.P., Peters, H., Blumer, J., Blankenfeldt, W., Goody, R.S., and Itzen, A. 2010. The *Legionella* effector protein DrrA AMPylates the membrane traffic regulator Rab1b. *Science*, 329, 946–949.

Murdoch, D.R. 2003. Diagnosis of *Legionella* infection. *Clin. Infect. Dis.*, 36, 64–69.

Nagai, H., Cambronne, E.D., Kagan, J.C., Amor, J.C., Kahn, R.A., and Roy, C.R. 2005. A C-terminal translocation signal required for Dot/Icm-dependent delivery of the *Legionella* RalF protein to host cells. *Proc. Natl. Acad. Sci. USA*, 102, 826–831.

Nagai, H., Kagan, J.C., Zhu, X., Kahn, R.A., and Roy, C.R. 2002. A bacterial guanine nucleotide exchange factor activates ARF on *Legionella* phagosomes. *Science*, 295, 679–682.

Nash, T.W., Libby, D.M., and Horwitz, M.A. 1984. Interaction between the Legionnaires' disease bacterium (*Legionella pneumophila*) and human alveolar macrophages. Influence of antibody, lymphokines, and hydrocortisone. *J. Clin. Invest.*, 74, 771–782.

Neunuebel, M.R., Chen, Y., Gaspar, A.H., Backlund Jr., P.S., Yergey, A., and Machner, M.P. 2011. De-AMPylation of the small GTPase Rab1 by the pathogen *Legionella pneumophila*. *Science*, 333, 453–456.

Newton, H.J., Ang, D.K., van Driel, I.R., and Hartland, E.L. 2010. Molecular pathogenesis of infections caused by *Legionella pneumophila*. *Clin. Microbiol. Rev.*, 23, 274–298.

Ohkuma, S., and Poole, B. 1978. Fluorescence probe measurement of the intralysosomal pH in living cells and the perturbation of pH by various agents. *Proc. Natl. Acad. Sci. USA*, 75, 3327–3331.

Osterholm, M.T., Chin, T.D., Osborne, D.O., Dull, H.B., Dean, A.G., Fraser, D.W., Hayes, P.S., and Hall, W.N. 1983. A 1957 outbreak of Legionnaires' disease associated with a meat packing plant. *Am. J. Epidemiol.*, 117, 60–67.

Palusinska-Szysz, M., and Janczarek, M. 2010. Innate immunity to *Legionella* and toll-like receptors – Review. *Folia Microbiol. (Praha)*, 55, 508–514.

Phares, C.R., Wangroongsarb, P., Chantra, S., Paveenkitiporn, W., Tondella, M.L., Benson, R.F., Thacker, W.L., Fields, B.S., Moore, M.R., Fischer, J., Dowell, S.F., and Olsen, S.J. 2007. Epidemiology of severe pneumonia caused by *Legionella longbeachae*, *Mycoplasma pneumoniae*, and *Chlamydia pneumoniae*: 1-year, population-based surveillance for severe pneumonia in Thailand. *Clin. Infect. Dis.*, 45, e147–155.

Plouffe, J.F., Breiman, R.F., Fields, B.S., Herbert, M., Inverso, J., Knirsch, C., Kolokathis, A., Marrie, T.J., Nicolle, L., and Schwartz, D.B. 2003. Azithromycin in the treatment of *Legionella pneumonia* requiring hospitalization. *Clin. Infect. Dis.*, 37, 1475–1480.

Price, C.T., Al-Quadan, T., Santic, M., Rosenshine, I., and Abu Kwaik, Y. 2011. Host proteasomal degradation generates amino acids essential for intracellular bacterial growth. *Science*, 334, 1553–1557.

Qiu, J., and Luo, Z.Q. 2013. Effector translocation by the *Legionella* Dot/Icm type IV secretion system. *Curr. Top. Microbiol. Immunol.*, 376, 103–115.

Ragaz, C., Pietsch, H., Urwyler, S., Tiaden, A., Weber, S.S., and Hilbi, H. 2008. The *Legionella pneumophila* phosphatidylinositol-4 phosphate-binding type IV substrate SidC recruits endoplasmic reticulum vesicles to a replication-permissive vacuole. *Cell. Microbiol.*, 10, 2416–2433.

Rasis, M., and Segal, G. 2009. The LetA-RsmYZ-CsrA regulatory cascade, together with RpoS and PmrA, post-transcriptionally regulates stationary phase activation of *Legionella pneumophila* Icm/Dot effectors. *Mol. Microbiol.*, 72, 995–1010.

Rello, J., Gattarello, S., Souto, J., Sole-Violan, J., Valles, J., Peredo, R., Zaragoza, R., Vidaur, L., Parra, A., and Roig, J. 2013. Community-acquired *Legionella* pneumonia in the intensive care unit: Impact on survival of combined antibiotic therapy. *Med. Intensiva*, 37, 320–326.

Remen, T., Mathieu, L., Hautemaniere, A., Deloge-Abarkan, M., Hartemann, P., and Zmirou-Navier, D. 2011. Pontiac fever among retirement home nurses associated with airborne *Legionella*. *J. Hosp. Infect.*, 78, 269–273.

Ricci, M.L., Torosantucci, A., Scaturro, M., Chiani, P., Baldassarri, L., and Pastoris, M.C. 2005. Induction of protective immunity by *Legionella pneumophila* flagellum in an A/J mouse model. *Vaccine*, 23, 4811–4820.

Robinson, C.G., and Roy, C.R. 2006. Attachment and fusion of endoplasmic reticulum with vacuoles containing *Legionella pneumophila*. *Cell. Microbiol.*, 8, 793–805.

Rolando, M., and Buchrieser, C. 2012. Post-translational modifications of host proteins by *Legionella pneumophila*: A sophisticated survival strategy. *Future Microbiol.*, 7, 369–381.

Rowbotham, T.J. 1980. Preliminary report on the pathogenicity of *Legionella pneumophila* for freshwater and soil amoebae. *J. Clin. Pathol.*, 33, 1179–1183.

Rowbotham, T.J. 1986. Current views on the relationships between amoebae, legionellae and man. *Isr. J. Med. Sci.*, 22, 678–689.

Roy, C.R., Berger, K.H., and Isberg, R.R. 1998. *Legionella pneumophila* DotA protein is required for early phagosome trafficking decisions that occur within minutes of bacterial uptake. *Mol. Microbiol.*, 28, 663–674.

Rudbeck, M., Molbak, K., and Uldum, S.A. 2009. Dynamics of *Legionella* antibody levels during 1 year in a healthy population. *Epidemiol. Infect.*, 137, 1013–1018.

Sahr, T., Bruggemann, H., Jules, M., Lomma, M., Albert-Weissenberger, C., Cazalet, C., and Buchrieser, C. 2009. Two small ncRNAs jointly govern virulence and transmission in *Legionella pneumophila*. *Mol. Microbiol.*, 72, 741–762.

Sato, K., and Nakano, A. 2007. Mechanisms of COPII vesicle formation and protein sorting. *FEBS Lett.*, 581, 2076–2082.

Schoebel, S., Cichy, A.L., Goody, R.S., and Itzen, A. 2011. Protein LidA from *Legionella* is a Rab GTPase supereffector. *Proc. Natl. Acad. Sci. USA*, 108, 17945–17950.

Schroeder, G.N., Petty, N.K., Mousnier, A., Harding, C.R., Vogrin, A.J., Wee, B., Fry, N.K., Harrison, T.G., Newton, H.J., Thomson, N.R., Beatson, S.A., Dougan, G., Hartland, E.L., and Frankel, G. 2010. *Legionella pneumophila* strain 130b possesses a unique combination of type IV secretion systems and novel Dot/Icm secretion system effector proteins. *J. Bacteriol.*, 192, 6001–6016.

Schuelein, R., Ang, D., van Driel, I.R., and Hartland, E.L. 2011. Immune control of *Legionella* infection: An *in vivo* perspective. *Front. Microbiol.*, 2, 126.

Segal, G., and Shuman, H.A. 1999. *Legionella pneumophila* utilizes the same genes to multiply within *Acanthamoeba castellanii* and human macrophages. *Infect. Immun.*, 67, 2117–2124.

Sethi, S., Gore, M.T., and Sethi, K.K. 2007. Increased sensitivity of a direct fluorescent antibody test for *Legionella pneumophila* in bronchoalveolar lavage samples by immunomagnetic separation based on BioMags. *J. Microbiol. Methods*, 70, 328–335.

Sexton, J.A., and Vogel, J.P. 2004. Regulation of hypercompetence in *Legionella pneumophila*. *J. Bacteriol.*, 186, 3814–3825.

Shamovsky, I., Ivannikov, M., Kandel, E.S., Gershon, D., and Nudler, E. 2006. RNA-mediated response to heat shock in mammalian cells. *Nature*, 440, 556–560.

She, R.C., Billetdeaux, E., Phansalkar, A.R., and Petti, C.A. 2007. Limited applicability of direct fluorescent-antibody testing for *Bordetella* sp. and *Legionella* sp. specimens for the clinical microbiology laboratory. *J. Clin. Microbiol.*, 45, 2212–2214.

Shen, X., Banga, S., Liu, Y., Xu, L., Gao, P., Shamovsky, I., Nudler, E., and Luo, Z.Q. 2009. Targeting eEF1A by a *Legionella pneumophila* effector leads to inhibition of protein synthesis and induction of host stress response. *Cell. Microbiol.*, 11, 911–926.

Shevchuk, O., Batzilla, C., Hagele, S., Kusch, H., Engelmann, S., Hecker, M., Haas, A., Heuner, K., Glockner, G., and Steinert, M. 2009. Proteomic analysis of *Legionella*-containing phagosomes isolated from *Dictyostelium*. *Int. J. Med. Microbiol.*, 299, 489–508.

Shin, S., and Roy, C.R. 2008. Host cell processes that influence the intracellular survival of *Legionella pneumophila*. *Cell. Microbiol.*, 10, 1209–1220.

Stone, B.J., and Kwaik, Y.A. 1999. Natural competence for DNA transformation by *Legionella pneumophila* and its association with expression of type IV pili. *J. Bacteriol.*, 181, 1395–1402.

Susa, M., Ticac, B., Rukavina, T., Doric, M., and Marre, R. 1998. *Legionella pneumophila* infection in intratracheally inoculated T cell-depleted or -nondepleted A/J mice. *J. Immunol.*, 160, 316–321.

Swanson, M.S. 2006. Autophagy: Eating for good health. *J. Immunol.*, 177, 4945–4951.

Tan, Y., Arnold, R.J., and Luo, Z.Q. 2011. *Legionella pneumophila* regulates the small GTPase Rab1 activity by reversible phosphorylcholination. *Proc. Natl. Acad. Sci. USA*, 108, 21212–21217.

Tan, Y., and Luo, Z.Q. 2011. *Legionella pneumophila* SidD is a deAMPylase that modifies Rab1. *Nature*, 475, 506–509.

Tateda, K., Matsumoto, T., Ishii, Y., Furuya, N., Ohno, A., Miyazaki, S., and Yamaguchi, K. 1998. Serum cytokines in patients with *Legionella pneumonia*: Relative predominance of Th1-type cytokines. *Clin. Diagn. Lab. Immunol.*, 5, 401–403.

Tung, S.M., Unal, C., Ley, A., Pena, C., Tunggal, B., Noegel, A.A., Krut, O., Steinert, M., and Eichinger, L. 2010. Loss of *Dictyostelium* ATG9 results in a pleiotropic phenotype affecting growth, development, phagocytosis and clearance and replication of *Legionella pneumophila*. *Cell. Microbiol.*, 12, 765–780.

Tzivelekidis, T., Jank, T., Pohl, C., Schlosser, A., Rospert, S., Knudsen, C.R., Rodnina, M.V., Belyi, Y., and Aktories, K. 2011. Aminoacyl-tRNA-charged eukaryotic elongation factor 1A is the bona fide substrate for *Legionella pneumophila* effector glucosyltransferases. *PLoS One*, 6, e29525.

Valster, R.M., Wullings, B.A., and van der Kooij, D. 2010. Detection of protozoan hosts for *Legionella pneumophila* in engineered water systems by using a biofilm batch test. *Appl. Environ. Microbiol.*, 76, 7144–7153.

Weber, S.S., Ragaz, C., and Hilbi, H. 2009. The inositol polyphosphate 5-phosphatase OCRL1 restricts intracellular growth of *Legionella*, localizes to the replicative vacuole and binds to the bacterial effector LpnE. *Cell. Microbiol.*, 11, 442–460.

Xu, L., and Luo, Z.Q. 2013. Cell biology of infection by *Legionella pneumophila*. *Microbes Infect.*, 15, 157–167.

Xu, L., Shen, X., Bryan, A., Banga, S., Swanson, M.S., and Luo, Z.Q. 2010. Inhibition of host vacuolar H+-ATPase activity by a *Legionella pneumophila* effector. *PLoS Pathog.*, 6, e1000822.

Yeung, T., Ozdamar, B., Paroutis, P., and Grinstein, S. 2006. Lipid metabolism and dynamics during phagocytosis. *Curr. Opin. Cell. Biol.*, 18, 429–437.

Yong, S.F., Tan, S.H., Wee, J., Tee, J.J., Sansom, F.M., Newton, H.J., and Hartland, E.L. 2010. Molecular detection of *Legionella*: Moving on from mip. *Front. Microbiol.*, 1, 123.

Yu, V.L., Plouffe, J.F., Pastoris, M.C., Stout, J.E., Schousboe, M., Widmer, A., Summersgill, J., File, T., Heath, C.M., Paterson, D.L., and Chereshsky, A. 2002. Distribution of *Legionella* species and serogroups isolated by culture in patients with sporadic community-acquired legionellosis: An international collaborative survey. *J. Infect. Dis.*, 186, 127–128.

Zerial, M., and McBride, H. 2001. Rab proteins as membrane organizers. *Nat. Rev. Mol. Cell. Biol.*, 2, 107–117.

Zhou, G., Cao, B., Dou, Y., Liu, Y., Feng, L., and Wang, L. 2011. PCR methods for the rapid detection and identification of four pathogenic *Legionella* spp. and two *Legionella pneumophila* subspecies based on the gene amplification of gyrB. *Appl. Microbiol. Biotechnol.*, 91, 777–787.

Chapter 32

Pathogenesis of *Kingella kingae* Infections

Pablo Yagupsky

Clinical Microbiology Laboratory, Soroka University Medical Center, Ben-Gurion University of the Negev, Beer-Sheva, Israel

32.1 Introduction

In the 1960s, Elizabeth O. King from the US Centers for Disease Control in Atlanta isolated a novel Gram-negative coccobacillus from human respiratory secretions, blood, and bone and joint exudates (Henriksen and Bøvre, 1968, 1976). The bacterium, initially allocated to the genus *Moraxella* and named *Moraxella kingii* in honor of King's pioneering work, was later transferred to a separate genus and rebaptized *Kingella kingae* (Henriksen and Bøvre, 1976).

For many years, the organism was considered a rare cause of human disease, exceptionally isolated from patients with skeletal system infections and endocarditis (Yagupsky, 2004). In the late 1980s, the serendipitous discovery that inoculation of synovial fluid samples aspirated from young children with culture-negative arthritis into blood culture vials (BCVs) improves detection of the organism resulted in the recognition of *K. kingae* as an important invasive pediatric pathogen (Yagupsky et al., 1992a, 1995a). More recently, development of sensitive nucleic acid amplification (NAA) assays firmly established the role of *K. kingae* as the leading etiology of bone and joint infections below 4 years of age (Chometon et al., 2007; Ilharreborde et al., 2009; Verdier et al., 2005). In addition, the organism causes bacteremia with no focal infection, lower respiratory infections, endocarditis and, less frequently, meningitis, peritonitis, cellulitis, and a variety of ocular and soft tissue infections (Yagupsky, 2004).

32.2 Bacteriology

K. kingae is a facultative anaerobic member of the Neisseriaceae family that appears as pairs or short chains of plump bacilli with tapered ends (Figure 32-1). The organism is β-hemolytic, exhibits weak oxidase activity, negative catalase, urease, and indole tests, and produces acid from glucose and maltose but not from other sugars. *K. kingae* hydrolyses indoxyl phosphate and L-prolyl-β-naphthylamide, exhibits positive alkaline and acid phosphatase reactions, and its identification can be confirmed by commercial systems such as the API NH kit (bioMérieux, Marcy-l'Étoile, France), the VITEK 2 instrument (bioMérieux, Marcy-l'Étoile, France) (Valenza et al., 2007), matrix-assisted laser desorption ionization time-of-flight mass spectrometry (MALDI-TOF) (Couturier et al., 2011), and *16S rDNA* gene sequencing (Clarridge et al., 2004).

K. kingae grows on routine trypticase-soy agar with added hemoglobin (blood-agar medium) and chocolate-agar, producing marked pitting of the agar surface, and fails to grow on MacConkey medium. Its growth is accelerated by a CO_2-enriched atmosphere, but only a small fraction of strains are truly capnophilic (Yagupsky, 2004).

A study in which *K. kingae* isolates from Israel, France, Russia, Norway, and the United States were studied by multilocus sequence typing (MLST) of six housekeeping genes demonstrated that *K. kingae* organisms display genomic variability and, whereas some clones have spread widely, others have a restricted geographic distribution (Basmaci et al., 2012b). Comparison of more than 40 genomes of asymptomatically carried K. kingae and invasive organisms from diverse geographic origins revealed that the genome size of the species ranges between 1,990,794 bp and 2,096,758 bp, comprises 1981–2300 protein-encoding genes and between 43 and 52 RNA genes, and has a GC content between 46.8% and 46.9%

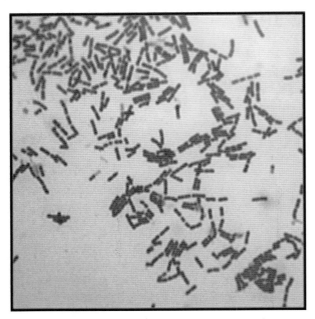

Fig. 32-1 Typical Gram's stain of *K. kingae* organisms depicting short Gram-negative coccobacilli with tapered ends arranged in pairs or short chains.

(Fournier et al., 2012; Kaplan et al., 2012; Raouli et al., unpublished data).

32.2.1 *Kingella kingae* colonization site

Asymptomatic colonization of the upper respiratory tract is the common strategy shared by many bacterial pathogens such as *Streptococcus pneumoniae*, *Neisseria meningitidis*, and *Haemophilus influenzae* to establish a foothold on the body's mucosal surfaces, persist, and spread from person to person. The colonized epithelial surfaces are also the portal of entry of bacteria to the bloodstream, an event that may be followed by dissemination of circulating organisms to remote normally sterile body sites, such as the skeletal system and the central nervous system.

Based on anecdotal isolation of *K. kingae* from patients with pneumonia and children with respiratory symptoms (Yagupsky, 2004), and the observation that other members of the bacterial Neisseriaceae family are upper respiratory tract commensals, it was presumed that the organism could be also part of the residing respiratory microbiota. In a pioneering study, biweekly nasopharyngeal and oropharyngeal cultures were obtained from young attendees to a day care center facility in southern Israel over an 11-month period. A total of 109 of 624 (17.5%) oropharyngeal specimens grew *K. kingae*, but the organism was not recovered from any of the nasopharyngeal cultures (Yagupsky et al., 1995b). These results have been confirmed in a recent study in which 4472 oropharyngeal and nasopharyngeal specimens were prospectively obtained from a cohort of 716 healthy children. A total of 388 (8.7%) oropharyngeal cultures, but only a single nasopharyngeal culture, grew *K. kingae*, indicating that the organism occupies a narrow niche in the upper respiratory tract (Amit et al., in press).

32.2.2 Detection of *K. kingae* in the respiratory tract

32.2.2.1 Culture detection

Because of the high density of the resident bacterial flora and the relative slow growth of *K. kingae*, detection of the organism in pharyngeal cultures is difficult. A selective medium consisting of blood-agar medium with added 2 mg/mL of vancomycin (BAV medium) has been developed to improve identification of the organism from respiratory cultures (Yagupsky et al., 1995c). The rationale behind this formulation is to facilitate recognition of β-hemolytic *K. kingae* organisms by reducing the presence of competitive Gram-positive bacteria. In a blinded evaluation, the BAV medium detected 43 of 44 (97.7%) pharyngeal cultures positive for *K. kingae*, whereas only 10 of 44 (22.7%) cultures were identified on routine blood-agar plates ($P < 0.001$). The original BAV and similar selective media (Basmaci et al., 2012a) have been successfully employed in many epidemiological studies aimed to investigate the carriage of the organism by the healthy children's population.

32.2.2.2 Detection by NAA

In recent years, novel molecular detection assays have enabled diagnosis of *K. kingae* infections in patients in which cultures of synovial fluid on routine and BAV media failed to reveal the presence of the organism (Cherkaoui et al., 2009). It was natural that this sensitive approach was consequently adopted to study the colonization of the respiratory tract by *K. kingae* and its connection to invasive disease. The RTX toxin, a putative virulence factor, is produced by all *K. kingae* strains examined so far and, therefore, the encoding RTX locus genes appear as pertinent targets for molecular detection of both invasive infections and mucosal colonization by the organism (Basmaci et al., 2012a; Cherkaoui et al., 2009; Ceroni et al., 2010a; Lehours et al., 2011). The NAA assays developed to amplify conserved segments of the RTX toxin encoding genes *rtxA* and/or *rtxB* are able to detect as few as 30 colony forming units (c.f.u.) of the organism, exhibiting a higher sensitivity than the PCR tests that target the *16S rDNA* (Cherkaoui et al., 2009; Lehours et al., 2011) or the *cpn60* genes (Lehours et al., 2011), are highly specific, can be applied to a variety of clinical specimens, and allow detection of strains exhibiting *RTX* locus polymorphisms (Basmaci et al., 2012a; Cherkaoui et al., 2009; Ceroni et al., 2010a, 2010b; Lehours et al., 2011).

Using a real-time PCR assay that targets the *rtxA* and *rtxB* genes, Ceroni et al. (2012) detected *K. kingae* DNA sequences in the pharynx of 8.1% of 431 young asymptomatic Swiss children and in all 27 patients with culture-positive and/or NAA-proven invasive disease. In a second study by the same group comprising 123 patients aged 6–48 months with skeletal system complaints, in all 30 children with

K. kingae-proven osteoarthritis (diagnosed either by culture and/or NAA), the oropharyngeal specimen was also positive, while eight pharyngeal samples derived from 84 patients with microbiologically unconfirmed joint or bone infections or with skeletal infections caused by other bacteria were also positive for *K. kingae* DNA, reflecting the background respiratory carriage of the organism in the young pediatric population (Ceroni et al., 2013a). These results imply that failure to detect *K. kingae*-specific genomic sequences in a pharyngeal specimen may practically exclude the organism as the etiology of an osteoarticular infection. However, because *K. kingae* is a frequent resident of the respiratory mucosa of children in the relevant age group, the predictive value of a positive pharyngeal NAA assay is limited.

In a refined study, Basmaci et al. grew *K. kingae* in the pharyngeal cultures of 8 of 12 NAA-positive/culture-negative synovial fluid specimens of young children with arthritis (Cherkaoui et al., 2009). They succeeded in extracting and sequencing the *rtxA* gene amplicons from six PCR-positive synovial fluid samples and compared them with those of the pharyngeal isolates. The six paired pharyngeal and synovial fluid amplicons were found to contain identical sequences, establishing a firm link between *K. kingae* organisms colonizing the pharynx and those invading the skeletal tissues.

More recently, the real-time PCR assay targeting the toxin-encoding *rtxB* gene was employed to assess the bacterial load of asymptomatic pharyngeal carriers and children in whom the diagnosis of *K. kingae* arthritis or osteomyelitis was confirmed by a positive PCR test performed in either blood, synovial fluid, or bone exudates (Ceroni et al., 2013c). The number of amplification cycles required to obtain a positive result was used as a surrogate for colonization density. Contrarily to what has been demonstrated in other respiratory pathogens such as pneumococcus or *H. influenzae*, no differences between healthy carriers and sick children were found, suggesting that factors other than bacterial density are more important for the development of an invasive *K. kingae* infection (Ceroni et al., 2013c). Using the same approach, it was shown that the bacterial density was remarkably stable in colonized children between the ages of 8 months and 4 years, despite substantial differences in the risk to develop an invasive *K. kingae* infection during this wide age interval (Ceroni et al., 2013b).

The sensitivity of cultures and NAA assays for detecting *K. kingae* colonization has been compared in a single study in which the yield of both approaches was assessed during the investigation of a large outbreak of invasive infections in a French day care center facility (Bidet et al., 2013). Overall, 12 of 18 pharyngeal specimens were positive by real-time PCR compared to 6 of 18 positive by culture on modified BAV medium ($P < 0.01$), suggesting that NAA assays are more sensitive than cultures to determine the true carriage rate. However, when multiple strains are found to be circulating in the population, the culture approach has the advantage of enabling a comprehensive genotypic comparison of identified isolates. In addition, when evaluating the efficacy of prophylactic antibiotics administration for eradicating *K. kingae* from colonized children, cultures have the advantage of detecting living bacteria, whereas the viability of *K. kingae* organisms in NAA-positive/culture-negative specimens is questionable (Bidet et al., 2013).

32.3 Mechanisms of colonization and virulence

32.3.1 Pili

To colonize the human mucosae, bacteria have first to adhere to epithelial surfaces and avoid been washed out. In a series of elegant studies, St. Geme et al. have shown that *K. kingae* expresses type IV pili, and that these surfaced-exposed fibers are essential for the attachment of the organism to respiratory and synovial cells (Kehl-Fie et al., 2008). The investigators disclosed a chromosomal gene cluster homologous to that found in other Gram-negative organisms consisting of a *pilA1* gene that encodes the major pilin subunit, and *pilA2* and *FimB* genes of unknown functions that are dispensable for adherence and expression of pili (Kehl-Fie et al., 2008). As it is the case for other surface-exposed virulence factors, the PilA subunit exhibits significant strain-to-strain variation in sequence and antibody reactivity, suggesting that it is subjected to selective pressure by the immune system (Kehl-Fie et al., 2009). The expression of pili in *K. kingae* appears to be finely regulated by three genes (σ^{54}, *PilS*, and *PilR*) (Kehl-Fie et al., 2009), and the majority of colonizing strains and those isolated from individuals with bacteremia express piliation, whereas those derived from patients with bone, joint, or endocardial infections are non-piliated (Kehl-Fie et al., 2010). This observation brings to mind that piliation offers a selective advantage in the colonization process and at the early mucosal and bloodstream stages of the infection, but is detrimental to the bacterium for invading deep body tissues. Additional work by the same research group identified two other genes named *pilC1* and *pilC2* in physically separated chromosomal locations that encode homologs of the *Neisseria* PilC proteins that are essential for adherence and piliation (Kehl-Fie et al., 2008; Porsch et al., 2013). *K. kingae* PilC1 and PilC2 proteins have a low level of homology to each other, contain calcium-binding sites, and are dispensable for pilus assembly (Kehl-Fie et al., 2008; Porsch et al., 2013). While the PilC1 site is necessary for twitching motility and adherence, the PilC2 site has only a minor influence on twitching motility and no influence on adherence (Porsch et al., 2013). Additional research indicated that a trimeric autotransporter protein called Knh is crucial for a firm adherence of *K. kingae* to the epithelium (Porsch et al., 2012). However, Knh is covered by the bacterial carbohydrate capsule, rendering it inaccessible for attachment to the host's cell. The adherence process is initiated by attachment of the long type IV pili to specific membrane receptor on the epithelium of the host cells. This is followed by a strong retraction of the pili fibers, which pulls the bacterium into close contact with the host cell membrane, physically displacing the capsule aside and unmasking the Knh element that can, then, anchor to the respiratory host cell surface (Porsch et al., 2012).

32.3.2 Capsule

Synthesis of a polysaccharide capsule is a convergent evolutionary strategy of many important human pathogens that inhabit the respiratory tract. These lipid-anchored surface-exposed carbohydrate structures confer encapsulated organisms antiphagocytic properties, providing a key protection against the host's immune response and, thus, enabling survival on the mucosal surfaces and, incidentally, in the bloodstream and deep body tissues. Polysaccharide capsules show chemical and antigenic heterogeneity within members of the same species to evade the immune response, and this diversity is the basis for serotyping of pneumococci, *H. influenzae*, and *N. meningitidis* strains. It has been recently demonstrated that *K. kingae* organisms are also coated with a polysaccharide capsule (Porsch et al., 2012). Searching of the draft genome of the organism detected a locus that exhibited homology to the ABC-type capsule export operon *ctrABCD* and unlinked genes with homology to the *ctrE/lipA* and *ctrF/lipB* genes of the taxonomically closely related *N. meningitidis* species (Adeolu and Gupta, 2013; Porsch et al., 2012). Wild-type *K. kingae* organisms stained with cationic ferritin exhibited an electron-dense rim on the bacterial surface when examined by thin-section Transmission Electron Microscopy (TEM), consisting with an anionic bacterial capsule. When the *ctrA* gene was insertionally inactivated, non-mucoid colonies were visualized and no capsule could be demonstrated (Porsch et al., 2012). Preliminary results indicate that all *K. kingae* isolates produce a polysaccharide capsule, disregarding their clinical source (asymptomatic carriage or a variety of invasive diseases) (St. Geme JW, personal communication). The composition of the capsule shows chemical heterogeneity suggesting that there are multiple *K. kingae* capsular types, similar to the case with many other bacterial pathogens (Bendaoud et al., 2011). These important findings may cast light to the peculiar epidemiological curves of *K. kingae* carriage and disease. Immunity to polysaccharides in humans matures between the ages of 2 and 4 years, explaining the increased susceptibility of young children to mucosal colonization and invasive diseases caused by encapsulated organisms in general, and *K. kingae* in particular, whereas chemical and,

thus, antigenic variability may enable immune evasion, enabling replacement of colonizing organisms by new strains harboring a different capsular type.

32.3.3 Biofilm production

Biofilm formation is the predominant mode of growth for bacteria in most environments and particularly on the human body surfaces. Colonizing bacteria build up biofilms containing large quantities of tightly packed organisms encased in a polymeric matrix and attached to the epithelium. Life in such enclosed and crowded conditions protects bacteria from the immune system, desiccation, and antimicrobial drugs. Consequently, biofilms play an important role in mucosal colonization and pathogenesis of persisting infections, such as chronic osteomyelitis or lung disease in cystic fibrosis patients (Bendaoud et al., 2011; Kaplan, 2010). The sequence of biofilm establishment, growth, and architectural remodeling is a precisely regulated and highly dynamic process. Periodic inhibition of biofilm formation is crucial in the life cycle of many pathogens because it makes possible the release of trapped bacterial cells, and subsequent dispersion and colonization of new body niches and hosts. In has been shown that two linear polysaccharides of *K. kingae* show potent anti-biofilm activity (Bendaoud et al., 2011). It is speculated that these compounds could play a role in the regulation of biofilm formation by colonizing *K. kingae*, enabling release and dissemination of the organism by intimate person-to-person contact, and/or inhibiting biofilm production by other bacterial species competing for the same niche on the pharyngeal mucosa.

32.3.4 RTX toxin

In a study carried out by Kehl-Fie and St. Geme, it was demonstrated that the *K. kingae* genome comprises a genetic locus encoding a RTX (repeat in toxin) system (Kehl-Fie et al., 2007). The locus consists of five genes designated *rtxA*, *rtxB*, *rtxC*, *rtxD*, and *tolC*, is flanked by insertion elements, and has a reduced G+C content compared to the whole genome, suggesting that it has been horizontally transferred from a donor species (Kehl-Fie et al., 2007). Sequencing of the locus genes revealed that *rtxA*, *rtxB*, and *rtxC* genes encode proteins that share >70% identity with their homologs in *Moraxella bovis*, whereas *rtxD* and *tolC* genes encode proteins that share homology with their *N. meningitidis* counterparts (Kehl-Fie et al., 2007). The *K. kingae* RTX toxin exhibits a wide range of cytotoxic activities, especially in macrophage-like and synovial cells, and to a lesser degree in respiratory epithelial cells, and appears to be secreted in the extracellular environment as a soluble protein, as well as a component of outer membrane vesicles (OMVs) that are internalized by host's cells, indicating that the bacterium may utilize different mechanisms for toxin delivery (Maldonado et al., 2011). It is probable that this bacterial constituent is universally conserved among members of the species because it improves colonization fitness by disrupting the pharyngeal epithelium, perhaps with an enhanced effect in the setting of a viral coinfection. Incidentally, the RTX toxin may also enable survival of *K. kingae* in the bloodstream and invasion of skeletal system and cardiac and other body tissues and, therefore, has a disease-promoting effect (Kehl-Fie et al., 2007).

32.3.5 Outer membrane vesicles

Gram-negative bacteria interact with the surroundings by releasing toxins and other proteins that exert distal effects without the need to expending energy in moving themselves. In addition, secreted material may reach sections of the environment that are inaccessible to the whole bacterium. Frequently, secretion of these bacterial products to the extracellular space implies formation of small (20–250 nm) spherical structures named OMVs. These OMVs are produced by small portions of the outer membrane enclosing the periplasm and entrapped proteins that bulge away from the cell, pinch off, and are, then, released (Kulp and Kuehn, 2010). Maldonado et al. described blebbing OMV in clinical isolates of *K. kingae* containing several major proteins, including the RTX toxin and the PilC2 pilus adhesin (Maldonado et al., 2011) (Figure 32-2). These OMVs are hemolytic and leukotoxic, and are internalized by human osteoblasts and synoviocytes. Upon internalization, the cells produce increased levels of human-granulocyte-macrophage colony-stimulating factor (GM-CSF) and interleukin 6 (IL-6), suggesting that these cytokines may be involved in the inflammatory response of skeletal system tissues in the course of invasive *K. kingae* infections (Maldonado et al., 2011).

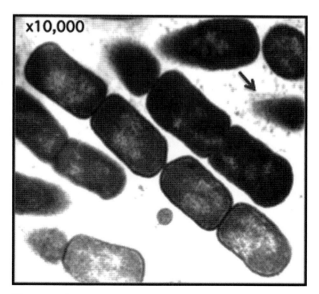

Fig. 32-2 Electron microscope picture of *K. kingae* organisms showing outer membrane vesicles (arrow). *Source:* Amit et al. (in press).

32.4 Antibiotic resistance

Because of the selective pressure exerted by the mounting use and abuse of antimicrobial drugs, many bacterial species have acquired antibiotic resistance as a means to survive in the human host. Young children are repeatedly exposed to antibiotics, and colonizing respiratory species such as pneumococci, *Staphylococcus aureus*, *Moraxella catarrhalis*, and *H. influenzae* currently exhibit alarming rates of resistance to β-lactam drugs and other antimicrobials, causing difficult-to-treat infections.

K. kingae has been traditionally considered exquisitely susceptible to antibiotics, and β-lactamase production has been rarely described (Banerjee et al., in press; Birgisson et al., 1997; Sordillo et al., 1993). When β-lactamase production and genomic clonality were studied in a large collection of *K. kingae* isolates from Israeli patients with a variety of clinical infections and asymptomatic pharyngeal carriers, the enzyme was detected in only 2 of 190 (1.1%) invasive isolates but in 68 of 428 (15.9%) randomly chosen carriage organisms ($P < 0.001$) (Yagupsky et al., 2013). The enzyme, a TEM β-lactamase, is located in a plasmid, its effect is inhibited by clavulanate, and its production in Israeli isolates was found to be limited to four distinct pulsed-field electrophoresis (PFGE) clones which are common among carriage strains but rare among invasive ones (Yagupsky et al., 2013). This association suggests that antibiotic resistance may confer a biological advantage to *K. kingae* organisms residing on the pharynx of young children, coinciding with the age of increased susceptibility to pharyngeal colonization and enhanced antimicrobial drugs consumption (Rossignoli et al., 2007; Yagupsky, 2004; Yagupsky et al., 1995a). On the other hand, strains expressing β-lactamase appear to be less capable of invading the bloodstream and deep inside the host's tissues, suggesting that improved colonization ability does not necessarily imply enhanced virulence. The prevalence of β-lactamase production appears to show wide geographic differences, and the trait is widespread among invasive isolates from Minneapolis (USA) and Reykjavik (Iceland). Strikingly, the strain responsible for β-lactamase production in Minnesota is identical by PFGE and MLST to the one isolated in Iceland and exhibited the same *rtx* and *por* gene alleles (Banerjee et al., in press; Bonacorsi et al., personal communication). The genotypic identity of *K. kingae* isolates derived from these relatively isolated sites located one ocean apart suggests a transatlantic crossing, although the direction (from Iceland to the United States or the way around) and the time of the event remain entirely speculative. It is possible that this phenomenon could be related to the large human migration from Iceland to North America, and particularly to Minnesota, during the last decades of the 19th century or, alternatively, may reflect more recent population contacts between the two locations.

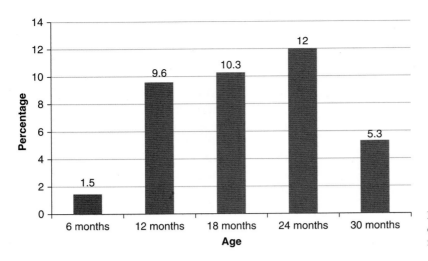

Fig. 32-3 Prevalence of pharyngeal *K. kingae* colonization among children aged 0–36 months (Amit et al., in press).

32.5 Prevalence of *K. kingae* carriage

Because bacterial colonization of the respiratory mucosal surfaces is established at the interface between humans and their surroundings, the forces that shape the acquisition, composition, and elimination of residing organisms reflect a complex array of host, microbial, environmental, and socioeconomic determinants. Factors such as sampling season, population's age and health status, living quarters crowding, number of young children at home, day care attendance, antibiotic consumption, and smoke exposure may profoundly influence the results of prevalence studies (Garcia-Rodriguez and Fresnadillo-Martinez, 2002). Monitoring the presence of individual components of the pharyngeal flora is also strongly dependent on technical and methodological aspects such as sampling site, specimen collection technique, quality of swabs, transport time to the laboratory, use of selective media, and number of bacterial colonies examined (Garcia-Rodriguez and Fresnadillo-Martinez, 2002). Not surprisingly, results of epidemiological investigations on *K. kingae* colonization have found discrepancies in the prevalence rate, although the overall picture indicates that the pharyngeal carriage is strongly age dependent (Amit et al., in press). In an early study carried out in southern Israel, the organism was not detected in infants younger than six months attending a Well-Baby Care Clinic (Yagupsky et al., 1995b). The prevalence of *K. kingae* was 10.0% in the 6 months to 4 years group, and decreased to 6.0% in older children (Yagupsky et al., 1995b). In a later study, oropharyngeal specimens submitted to a clinical microbiology laboratory for isolation of *Streptococcus pyogenes* were also plated onto a BAV medium (Yagupsky et al., 2001). *K. kingae* prevalence decreased with increasing age: the organism was detected in 22 of 694 (3.2%) samples obtained from children younger than 4 years, in 10 of 679 (1.5%) of those derived from patients aged 4–17 years, and in 5 of 671 (0.8%) cultures from adults (*P* for trend <0.001) (Yagupsky et al., 2001). In a longitudinal study in which the younger age group was targeted, a cohort of 716 healthy children living in southern Israel was repeatedly sampled between the ages of 2 months and 30 months (Amit et al., in press). *K. kingae* was not isolated before the age of 6 months, the colonization rate was low at 6 months, increased in 12-month-old children, remained relatively stable between 12 and 24 months of age, and decreased significantly at 30 months (*P* < 0.001) (Amit et al., in press) (Figure 32-3). These figures are comparable to the prevalence rate found in a Swiss population of children aged less than 4 years studied by real-time PCR (Ceroni et al., 2012).

In a recent study aimed to identify the risk factors for *K. kingae* colonization, age 6–29 months was strongly and independently associated with carriage in the multivariate analysis (Amit et al., 2013). This age-dependent carriage rate overlaps remarkably with the epidemiological curve of invasive *K. kingae* disease found in a study comprising 291 previously healthy Israeli children with culture-proven infections (Figure 32-4) (Dubnov-Raz et al., 2010).

It has been demonstrated that most invasive *K. kingae* infections are diagnosed between July and December (Yagupsky, 2004). In a study aimed to investigate the temporal pattern of *K. kingae* carriage (Yagupsky et al., 2002), throat cultures were obtained between February and May to represent the time of the year when only a small fraction of invasive *K. kingae* infections are diagnosed, and from October to December, coinciding with the peak attack rate of disease. Overall, 21 of 1020 (2.1%) specimens cultured between February and May and 16 of 1024 (1.6%) of those studied from October to December grew the organism (*P* > 0.4). The lack of seasonal *K. kingae* carriage was confirmed in a recent investigation of potential risk factors for colonization in children that showed no significant association between month of the year and colonization rate (Amit et al., 2013). It appears, then, that the striking temporal distribution of invasive *K. kingae* disease cannot be explained solely on the basis of the characteristics of the pharyngeal carriage of the organism, implying interplay of mucosal colonization with still unidentified cofactors, perhaps seasonal viral respiratory infections.

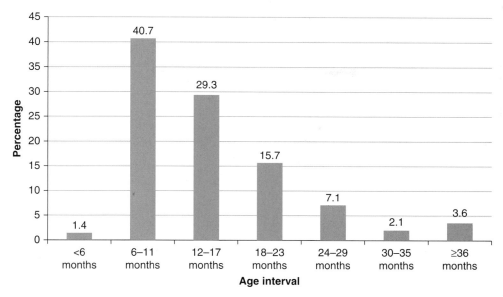

Fig. 32-4 Age distribution of 140 children with invasive *K. kingae* infections detected in a single medical center in Southern Israel between 1988 and 2013.

32.6 Colonization dynamics and turnover of strains

The dynamics of carriage of respiratory bacteria can be studied in longitudinal surveys in which the population is repeatedly sampled over a prolonged period of time and isolates are analyzed by highly discriminative methods. In the aforementioned prospective study, pharyngeal swabs were obtained from attendees to a day care facility in southern Israel over an 11-month period, and the *K. kingae* isolates detected were genotyped (Slonim et al., 1998). Using a combination of PFGE with three different restriction enzymes, immunoblotting with rabbit immune serum, and ribotyping with two restriction enzymes, and employing stringent criteria for defining genotypic identity, it was convincingly demonstrated that two distinct strains, characterized by specific typing profiles, represented 28.0% and 46.0% of the isolates typed and, once they took roots in the day care center, disseminated effectively, displacing older strains (Slonim et al., 1998).

In a large cohort study in which healthy children were periodically cultured between 2 and 30 months of age, the potential turnover of *K. kingae* strains over time was assessed by PFGE of isolates in the subset of colonized children in which the organism was recovered in >1 occasion (Amit et al., in press). Overall, 283 of 716 (39.5%) children carried *K. kingae* at least once, of whom 64 were colonized twice, 13 had three positive visits, and three children had four positive visits. Comparison of isolates showed that genotypic concordance was lost over time, and sequential carriage of as many as four different clones was detected. Carriage of *K. kingae* appears, then, as a dynamic process with frequent turnovers of colonizing strains that, after been carried continuously or intermittently for weeks or months, are replaced by newly acquired organisms, similar to observations made with other respiratory pathogens. Because in these two studies only a single *K. kingae* colony from each positive pharyngeal culture was typed, unrecognized colonization by multiple clones and persistence of "old" strains at a low level, rather than complete replacement, cannot be definitely excluded. It should be pointed out, however, that re-isolation of a clone that was previously carried and lost was infrequent (Amit et al., in press; Slonim et al., 1998). This pattern, instead of a random temporal distribution of PFGE genotypes, suggests clearance or at least quantitative reduction in the colonizing density of individual strains. It is possible that prolonged carriage induces strain-specific immunity that facilitates elimination of the carried organism but does not prevent acquisition of an antigenically different strain. This possibility is supported by the demonstration of strain-to-strain variability of *K. kingae* outer membrane proteins (OMPs) (Slonim et al., 2003), the *PilA*1 gene encoding the major pilus subunit (Kehl-Fie et al., 2010), the RTX toxin (Basmaci et al., 2012b; Lehours et al., 2011), and the capsule (St Geme JW, personal communication), suggesting that immunogenic surface-exposed bacterial components involved in pharyngeal colonization undergo antigenic variation to evade the host's immune response.

32.7 Immunity to *K. kingae* colonization

The high rates of *K. kingae* carriage and invasive infections observed in the first 3 years of life followed by a sharp fall in older children, the scarcity of cases among adults, and the occurrence of disease in immunocompromised patients suggest that an acquired immune response is necessary to protect from both mucosal colonization and disease (Yagupsky, 2004).

Although the specific functions of *K. kingae* OMPs have not been investigated yet, studies conducted with other respiratory pathogens have shown that surface-exposed OMPs are frequent targets for antibodies that exert protection against mucosal and invasive infections. Testing sera from children with bacteremia and osteoarticular infections by an ELISA assay using *K. kingae* OMPs as antigen showed a statistically significant increase in IgG in the convalescent period, indicating an immunological response (Slonim et al., 2003). When levels of IgG antibodies to the infecting isolate and heterologous organisms were measured, increased levels were found against the homologous strain, suggesting a more effective immune response (Yagupsky and Slonim, 2005). Serum IgA antibodies tended to decrease between the acute and convalescent phases of the disease, indicating that the immune response of this antibody type is transient and short living (Yagupsky and Slonim, 2005).

In a study in which the dynamics of IgG and IgA antibody levels to OMPs were investigated, serum samples were obtained from children at different ages with no history of invasive *K. kingae* infection (Slonim et al., 2003). Because IgA does not cross the placenta, serum IgA levels were compared to those of IgG type to determine whether antibodies detected in young children represent exposure to the organism or vertically acquired immunity. The mean IgG level was high at 2 months, gradually decreased thereafter reaching the lowest level at 6–7 months of age, remained low until the age of 18 months, and increased in 24-month-old children. Serum IgA levels were lowest at 2 months and gradually increased between the ages of 4 and 7 months. Further increment of both antibody types was observed in children older than 24 months (Slonim et al., 2003). Antibody levels of healthy *K. kingae* carriers were in the upper range of those observed in the general population of children of the same age (Slonim et al., 2003).

Low prevalence of respiratory carriage and invasive infection rates in the first 6 months of life, coupled with undetectable serum IgA and high IgG levels, indicate protection conferred by maternally derived immunity. High colonization rate and increased attack rate of invasive infection in 6–24-month-old children coincide with the age of lowest antibody levels. Low carriage rate, decreasing incidence of disease, and increasing antibody levels found in older children probably represent mounting immunological maturity and accumulative experience with *K. kingae* antigens through respiratory colonization, and more rarely, after invasive infection (Slonim et al., 2003).

When OMPs of different isolates were reacted with convalescent sera, OMPs of 108, 43, and 41 kDa were recognized by most convalescent sera, whereas recognition of other proteins varied between strains and depended on the serum employed (Yagupsky and Slonim, 2005). These results suggest that some immunogenic *K. kingae* OMPs are common to all strains and have highly conserved epitopes, while the presence of other OMPs is variable and/or the

exposed epitopes are polymorphic, resulting in substantial differences in the affinity of antibodies and, probably, in the efficacy of the immune response. Development of immunity to a carried strain probably contributes to its eradication from the mucosal surface but does not prevent acquisition and re-colonization of the respiratory tract by strains that do not share the same immune determinants (Yagupsky and Slonim, 2005; Yagupsky et al., 2013).

32.8 Transmission

In the last few decades, development of molecular typing techniques made it possible for the first time discrimination between genetically distinct strains of colonizing organisms, enabling deeper understanding of the geographic and temporal transmission of individual lineages in the population. The dissemination of *K. kingae* in the open community was studied in two ethnic groups living side by side in southern Israel (Yagupsky et al., 2009b). Organisms recovered from oropharyngeal cultures obtained from healthy young Jewish and Bedouin children during a 12-month period were typed by PFGE and compared. Isolates from Bedouin children usually differed from those derived from Jews, confirming the relative social isolation of the two populations (Yagupsky et al., 2009b). Typing of isolates also show significant spatial clustering of clones in the Bedouin towns' neighborhoods and households, indicating person-to-person transmission between family members and playmates and confirming the importance of close mingling in the spread of *K. kingae*. However, no significant geographic clustering of *K. kingae* clones was detected in the Jewish city quarters. Because the traditionally nomadic Bedouins have recently settled in towns but have kept the traditional tribal divisions, geographic clustering can be considered a surrogate for family ties and social intercourse. Bedouin youngsters do not usually attend day care facilities and most of their social interaction takes place within their extended families and clans, explaining the similarity of *K. kingae* organisms isolated in households located within a short radius. On the contrary, Jewish children live in westernized urban conditions, attend day care centers since early age, and have multiple social contacts outside their immediate physical environment. Under these circumstances, Jewish children living throughout the different city neighborhoods are connected by numerous and complex social networks, conspicuous and occult as well, through which respiratory bacteria may circulate and *K. kingae* and other organisms may be acquired. The multiplicity of potential sources of transmission in the Jewish community could, then, blur the connection between place of residence and distribution of *K. kingae* clones (Yagupsky et al., 2009b).

In a recent study by Kampouroglu et al., the potential intrafamilial transmission of the organism was studied by NAA in a Swiss population. The research found that 55% of children with PCR-documented invasive *K. kingae* disease and 40% of asymptomatic carriers have siblings with a positive pharyngeal PCR assay, suggesting dissemination of the organism by close contact (Kampouroglou et al., 2013).

32.9 From colonization to clinical disease

The link between colonization of the pharyngeal epithelium by *K. kingae* organisms and the development of an invasive infection was convincingly demonstrated in three bacteremic children in whom the organism was also recovered from an upper respiratory culture. Typing of paired blood and pharyngeal isolates by PFGE demonstrated genomic identity (Yagupsky et al., 2009a). This finding was later supported by the results of two studies employing NAA technology. In a study in which 27 young children were diagnosed with PCR-proven *K. kingae* osteoarthritis, the assay performed on the pharyngeal swab was positive for *rtxB* gene DNA in all cases (Ceroni et al., 2013a). In a second and more definitive study, young children with arthritis and culture-negative/PCR-positive synovial fluid samples had *K. kingae* cultured from the pharynx. The *rtxA* gene sequences of the synovial exudate amplicons and those of the pharyngeal isolate were undistinguishable, suggesting that the oropharynx is likely the site from which virulent *K. kingae* organisms enter the bloodstream and disseminate, causing focal skeletal system and endocardial infections (Basmaci et al., 2012a).

Employing an animal model, Basmaci et al. (2012b) demonstrated wide strain-to-strain differences in terms of invasive capability. When 5-day-old albino Sprague Dawley rats were inoculated intraperitoneally in the left inferior quadrant with 10^7 *K. kingae* c.f.u., the ATCC strain 2330 (a respiratory isolate) proved to be non-virulent, whereas two other strains, derived from patients with invasive infections, showed significant differences in terms of animals' survival rate (Basmaci et al., 2012b).

In a study conducted to determine the association of *K. kingae* genotypes with specific clinical syndromes and the temporal and geographic distribution of invasive clones, a collection of 181 invasive strains, isolated between 1991 and 2012 from Israeli patients with bacteremia, skeletal system infections, or endocarditis, were typed by PFGE (Amit et al., 2012b). A total of 32 different *K. kingae* clones were identified, of which five (B, H, K, N, and P) caused 72.9% of all invasive infections, and were recovered during the 21-year period from different regions of the country. Clone K was significantly associated with bacteremia, clone N with skeletal system infections, and clone P with bacterial endocarditis. Other clones, which are frequently isolated from healthy carriers, were seldom, if ever, detected in patients with invasive diseases (Amit et al., 2012b; Yagupsky et al., 2009b). This discrepancy suggests a trade-off between transmissibility and virulence because, by invading deep organs, bacteria lose access to human body surfaces and, therefore, cannot propagate any further. Moreover, sick individuals are isolated from other members of the community, treated with antimicrobial drugs, and may even succumb to the disease, interrupting the chain of person-to-person transmission and resulting in extinction of the pathogen. Conversely, clone N, which was a common etiology of invasive *K. kingae* disease in Israel in the 1990s, is almost never found as a pharyngeal colonizer in healthy children (Yagupsky et al., 2009b), was negatively associated with bacteremia, but showed significant affinity for bone and joint infections (Amit et al., 2012b). This observation indicates that strains that show remarkable tissue invasiveness may be rapidly cleared from the respiratory tract and the bloodstream, implying that persistence in these niches may require a different biological specialization. Clone K organisms, on the other hand, appear to exhibit an optimal balance between transmissibility and invasiveness. This PFGE clone was the predominant strain detected as early as 1993 in a southern Israel day care center, lasting in the pharynx of colonized attendees for up to 4 months (Slonim et al., 1998), and ranked second among strains carried by healthy Jewish children in a 2006–2007 study (Yagupsky et al., 2009b). Clone K organisms, which represented 41.7% of all invasive strains isolated in southern Israel over the last two decades, were responsible for the excess of *K. kingae* morbidity observed in the Jewish population of the region (Amit et al., 2012a). Interestingly, all clone K strains harbor a 33 bp duplication or triplication in a fragment of the *rtx*A gene that is exceptionally found in other clones (Basmaci et al., 2012a). Whether this repetition in a gene encoding for a virulence factor contributes to the epidemic success of the clone remains speculative.

In two recent reports, *K. kingae* organisms belonging to sequence type (ST)-25, an MLST clone so far detected only in France, was found to be strongly associated with childhood osteomyelitis causing four bone infections among five affected day care attendees and one sporadic case (Bidet et al., 2013), and a ST-6 strain caused a cluster of three cases of osteomyelitis in an Israeli day care center (Yagupsky et al., 2006). These observations imply that carriage of particular strains entails increased risk for clinical disease and invasion of specific body tissues.

32.10 Viral infections as facilitators of *K. kingae* invasion

Unbiased observations of children with invasive *K. kingae* infections at the time of presentation to the Emergency Department and before culture results were known revealed high prevalence of symptoms consistent with an intercurrent viral upper respiratory infection, evidence of oral varicella blisters, or concomitant buccal aphthous ulcers (Yagupsky, 2004). In a study aimed to assess the effect of acyclovir administration on shortening symptoms and viral shedding in children with culture-proven herpetic gingivostomatitis, 29 affected patients underwent bacteriological blood cultures. Four (13.8%) of the cultured patients were bacteremic, and *K. kingae* was isolated in all cases (Amir and Yagupsky, 1998). Because the peak incidence of primary *Herpes simplex* infections and many respiratory viruses coincides with the age of *K. kingae* carriage, it seems plausible that viral-induced damage to the mucosal layer facilitates penetration of colonizing organisms, invasion of the bloodstream, and dissemination to distant sites for which individual strains exhibit particular tropism (Yagupsky, 2004; Yagupsky et al., 2002, 2009a). In a recent study, Basmaci et al. (2013) confirmed the link between viral diseases and *K. kingae* morbidity by detecting rhinoviruses in two young children with concomitant PCR-proven *K. kingae* infections of bone and soft tissues.

32.11 Epidemiology of invasive *K. kingae* infections

Between 1988 and 2013, a total of 141 patients with culture-confirmed invasive *K. kingae* disease were identified at the Soroka University Medical Center (SUMC) in southern Israel in which the BCV method has been routinely employed for more than two decades. The age-related incidence of invasive *K. kingae* infections was markedly skewed to the left (Figure 32-4): 140 of the 141 patients were children, of whom 139 were younger than 4 years, and only two were younger than 6 months. Based on these data, an annual incidence of 9.4/100,000 children younger than 4 years residents of the region has been calculated. (Dubnov-Raz et al., 2010). However, because even when the BCV technique is used, detection of *K. kingae* remains suboptimal, this figure can only be considered a minimal estimate.

Most of the old medical literature on *K. kingae* infections consists of reports of single cases, or a small series of patients in whom unusual clinical manifestations were probably over-represented. Based on the large experience accumulated at the SUMC over the years, a more accurate picture can be drawn. Among the 141 patients, 78 (55.3%) presented with skeletal system infections, 56 (39.7%) with bacteremia with no apparent focus, 5 (3.5%) had bacteremic lower respiratory tract infections, and 2 (1.4%) patients, including the only adult in the series, presented with endocarditis. Other rare manifestations of *K. kingae* disease that have been described in other reports but were not detected in this series include meningitis (Van Erps et al., 1992), ocular infections (Carden et al., 1991), peritonitis (Bofinger et al., 2007), and pericarditis (Matta et al., 2007).

Invasive diseases caused by *K. kingae* occur throughout the year peaking during the months of November and December and exhibit a nadir during February, March, and April (Dubnov-Raz et al., 2008, 2010).

32.12 Day care center attendance, a risk factor for *K. kingae* colonization and disease

During the last decades, a growing proportion of children attend day care centers outside the home. This trend has substantial public health significance because the risk of person-to-person transmission and occurrence of mucosal and invasive infection is significantly increased among day care center attendees (Nafstad et al., 1999; Robinson, 2001). Respiratory pathogens spread from child-to-child by direct contact or through fomite transmission among youngsters with poor hygienic habits sharing toys coated with respiratory secretions or saliva (Robinson, 2001). Because of age stratification, day care classes comprise children of approximately the same age and, therefore, with similar degrees of immunological immaturity and susceptibility to infectious agents. Under these circumstances, introduction of a virulent bacterium in a crowded day care facility attended by immunologically naïve children may result in rapid propagation of the organism and cause outbreaks of disease.

In a prospective study conducted among 1277 children younger than 5 years referred to a Pediatric Emergency Department in southern Israel, day care attendance was strongly associated with *K. kingae* carriage after controlling for other variables (odds ratio: 9.66 [95% confidence intervals: 2.99–31.15], $P < 0.001$) (Amit et al., 2013). In an 11-month longitudinal study, 35 of 48 (72.9%) day care center attendees carried *K. kingae* in the pharynx at least once and, on average, 27.5% of the children harbored the organism at any given time, a much larger proportion than that detected in the general pediatric population (Yagupsky et al., 1995b). Molecular typing of the colonized attendees isolates showed genotypic similarities, demonstrating person-to-person transmission of the organism (Slonim et al., 1998). In an ongoing study, temporal clustering of PFGE clones was observed among *K. kingae* isolates of otherwise unrelated classmates attending 1st, 2nd, and 3rd elementary school grades, indicating that person-to-person transmission of *K. kingae* continues well beyond preschool age (Yagupsky P, unpublished data).

In recent years, clusters of invasive *K. kingae* infections including the entire spectrum of the disease (septic arthritis, osteomyelitis, spondylodiscitis, endocarditis, and meningitis) have been reported from French (Bidet et al., 2013), American (Kiang et al., 2005; Seña, et al., 2010), and Israeli day care facilities (Yagupsky et al., 2006; Miron, D., personal communication). The epidemiological investigation of these events revealed that the colonization rate among attendees to the classrooms where morbidity was detected was unusually high (up to 45% in a US cluster, as determined by culture (Kiang et al., 2005), and 85% in the French outbreak, as demonstrated by a sensitive NAA method (Bidet et al., 2013), and all pharyngeal isolates detected in the classrooms where disease occurred were genotypically identical and indistinguishable from the patients' clinical isolates, demonstrating that the invasive outbreak strains were both successful respiratory colonizers and highly virulent organisms.

32.13 Clinical presentation of *K. kingae* infections

Most young children with invasive *K. kingae* disease have been otherwise healthy. In contrast, affected children older than 4 years and adults often have underlying chronic diseases, immunosuppressing conditions, malignancy, or cardiac valve pathology (Dubnov-Raz et al., 2010; Yagupsky, 2004). The clinical presentation of patients with invasive *K. kingae* infections other than endocarditis is usually mild, requiring a high index of suspicion. A normal body temperature or low-grade fever, and minimal or no constitutional symptoms are commonly observed (Dubnov-Raz et al., 2008, 2010; Yagupsky, 2004). The blood leukocyte count, C-reactive protein levels (CRP), and the erythrocyte sedimentation rate are frequently normal, and the yield of blood cultures in patients with focal disease is low (Dubnov-Raz et al., 2010; Yagupsky, 2004). With the exception of patients with endocarditis who develop embolic phenomena, a single focus of infection is usually detected (Sarda et al., 1998).

32.13.1 Skeletal system infections

Sixty-two of the 78 (79.5%) children with skeletal system involvement diagnosed in southern Israel since 1987 had septic arthritis, six had osteomyelitis, one had both, two children had tenosynovitis, one patient had dactylitis, and five children presented with *K. kingae* bacteremia and transient skeletal system symptoms, suggesting an abortive infection.

K. kingae arthritis generally involves large joints such as the knee, hip, ankle, or shoulder (Dubnov-Raz et al., 2010; Yagupsky, 2004). However, the small metacarpophalangeal, sternoclavicular, and tarsal joints, which are rarely invaded by other bacteria, are over-represented in *K. kingae* disease (Yagupsky, 2004). The leukocyte count of the synovial fluid shows <50,000 WBC/mm^3 in one quarter of patients and the Gram-stain is rarely positive (Dubnov-Raz et al., 2010; Yagupsky, 2004).

Although *K. kingae* osteomyelitis usually invades the long bones, involvement of the calcaneus, thalus, sternum, or clavicle may occur (Dubnov-Raz et al., 2010; Luhmann and Luhmann, 1999; Lundy and Kehl, 1998; Mallet et al., 2014; Yagupsky, 2004). Onset of

K. kingae osteomyelitis is generally more insidious than that observed in septic arthritis, and most children are diagnosed after more than one week of evolution (Dubnov-Raz et al., 2010). Formation of large intraosseous abscesses involving the epiphyses and/or metaphyses and frequent fistulizing through the contiguous joint have been recently detected by MRI studies in 9 of 40 (22.5%) children with *K. kingae* bone infections (Mallet et al., 2014).

K. kingae is currently responsible for more than a quarter of all cases of hematogenous spondylodiscitis in children younger than 5 years (Ceroni et al., 2010b; Fuursted et al., 2008; Garron et al., 2002). It is presumed that the organism penetrates into the rich network of blood vessels that traverse the cartilaginous vertebral endplates and enter the annulus in young children during a bacteremic episode. The disease usually involves the lumbar discs, and patients present with limping, low back pain, refusal to sit or walk, or neurological symptoms. X-ray or MRI studies demonstrate narrowing of the intervertebral space (Chanal et al., 1987; Garron et al., 2002; Yagupsky, 2004).

Although bone and joint infections are not considered self-limited diseases, transient involvement of the skeletal system during an episode of *K. kingae* bacteremia may occur (Baticle et al., 2008; Lebel et al., 2006; Yagupsky and Press, 2004). Children with this condition present with limping or refusal to walk or bear weight but without objective signs of osteoarthritis. In the few published cases, by the time blood cultures became positive, patients were already afebrile and the skeletal complaints had resolved without antimicrobial therapy, suggesting an abortive clinical course. It is noticeable that a few children were not prescribed antibiotics after the positive culture results were known and made an uneventful recovery (Lebel et al., 2006). Despite this favorable experience, caution is recommended, and adequate antibiotics should be probably administered to all patients in which *K. kingae* is recovered from a normally sterile or body fluid.

32.13.2 Bacteremia

K. kingae bacteremia without evidence of endocarditis or other focal infection is the second most frequent presentation of pediatric *K. kingae* disease (Yagupsky, 2004). A maculopapular rash resembling disseminated meningococcal or gonococcal infection has been reported in a few patients (Shanson and Gazzard, 1984). Mild to moderate fever is usually measured and the mean WBC count is less than 15,000/mm^3 (Dubnov-Raz et al., 2010; Yagupsky and Dagan, 1994). Therefore, the current guidelines for managing young febrile children with no apparent focus, which rely on the height of body temperature and presence of leukocytosis for obtaining blood cultures [80], are not sensitive enough for detecting *K. kingae* bacteremia.

32.13.3 Endocarditis

K. kingae stands for the "K" of the HACEK acronym that denotes a group of oral and upper respiratory organisms that is collectively responsible for up to 6% of cases of bacterial endocarditis (Chambers et al., 2013). In contrast to other *K. kingae* infections, endocarditis has been mostly diagnosed in older children and adults, whereas occurrence of the disease below the age of 4 years is exceptional (Yagupsky, 2004). Native and prosthetic valve involvement has been reported with similar frequency. The left side of the heart, and the mitral valve in particular, is usually involved. Cardiac malformations and rheumatic fever are common predisposing factors, but many patients had no antecedent valvular disease (Berkun et al., 2004; Marom et al., 2013; Rotstein et al., 2010; Seña et al., 2010; Wells et al., 2001; Yagupsky, 2004).

32.14 Diagnosis of invasive *K. kingae* infections

32.14.1 Culture-detection of *K. kingae*

The recovery of *K. kingae* from clinical specimens seeded onto routine solid culture media is suboptimal (Yagupsky, 2004; Yagupsky et al., 1992a). The yield of cultures can be significantly improved by inoculating clinical specimens into aerobic BCV from a variety of blood culture systems such as BACTEC (Becton Dickinson, Cockeysville, MD), BacT/Alert (Organon Teknika Corporation, Durham, NC), and Hémoline DUO (bioMérieux, Marcy l'Étoile, France) or by seeding an Isolator 1.5 Microbial Tube (Wampole Laboratories, Cranbury, NJ) (Yagupsky, 2004). When BCV detected as positive by the automated blood culture instrument are subcultured onto blood-agar or chocolate-agar plates, *K. kingae* grows readily, indicating that routine solid media are able to support the organism. This observation suggests that skeletal system exudates exert an inhibitory effect on the bacterium and that dilution of purulent material in a large volume of broth decreases the concentration of inhibitory factors, improving the recovery of this fastidious organism (Yagupsky, 2004; Yagupsky et al., 1992a). In studies conducted in Israel and France in which BCV were routinely inoculated with synovial fluid aspirates from young children with arthritis, *K. kingae* was isolated in almost half of the patients with culture-proven disease (Moumile et al., 2005; Yagupsky et al., 1995a).

32.14.2 Detection of *K. kingae* by NAA assays

Because of the risk for long-term orthopedic sequelae, prompt bacteriological confirmation of the diagnosis of pediatric bone and joint infections and early initiation of appropriate antibiotic therapy are critical to prevent disability. However, even when the BCV method is used, a substantial fraction infections remain bacteriologically unconfirmed and, in any case, complete identification of the causative agent and antibiotic susceptibility testing require a minimum of 2–3 days.

In recent years, the use of conventional and real-time PCR has enabled identification of the etiologic agents of septic arthritis and osteomyelitis within 24 hours (Fenollar et al., 2008). This approach involves extracting DNA from clinical samples, incubation of the DNA with broad range oligonucleotide primers that anneal to constant regions of the *16S rDNA* gene, and amplification of the intervening sequence, which varies according to the bacterial species (Moumile et al., 2005). The resulting amplification products are either sequenced and compared with sequences deposited in the Genbank database, or hybridized with organism-specific probes. In addition, PCR assays that amplify *K. kingae*-specific targets such as *cpn60* or the RTX toxin genes have been developed and have been associated with high reliability (Ceroni et al., 2010a, 2010b; Chometon et al., 2007; Cherkaoui et al., 2009; Ferroni et al., 2013; Ilhaerreborde et al., 2009; Lehours et al., 2011).

Chometon et al. (2007) sequentially analyzed bone and joint samples using BCVs, conventional PCR with universal primers, and real-time PCR with primers derived from the *K. kingae cpn60* gene and achieved a detection rate that increased from 29% to 41% to 45% of microbiologically proven cases, making *K. kingae* the most common cause of septic arthritis and osteomyelitis in their series and accounting for ~80% of cases in children younger than 3 years. In a study by Ilhaerreborde et al. (2009) that included 89 children with suspected septic arthritis, joint fluid was cultured using BCV and examined by real-time PCR reaction employing highly specific primers that amplify a 169-bp fragment of the *K. kingae cpn60* gene. The diagnosis of septic arthritis was confirmed by culture in 36 (40%) specimens, including 7 that grew *K. kingae*. Real-time PCR identified another 24 cases of *K. kingae* among the 53 culture-negative patients. Of note, the NAA assay detected *K. kingae* in all seven samples that recovered the bacterium and was negative in all other children in whom other microorganisms were identified. In total, *K. kingae* was present in 31 (52%) of the 60 bacteriologically documented cases (Ilhaerreborde et al., 2009).

In a large study conducted in a French pediatric orthopedic unit, bone and joint exudates from children with suspected skeletal system infections were inoculated into BacT/Alert BCV and subjected to amplification of the *K. kingae cpn60* gene. *K. kingae* was detected by BCV and/or NAA tests in 35 of 46 (76.1%) children aged <4 years with septic arthritis and in 9 of 17 (52.9%) children of the same age with osteomyelitis or concurrent septic arthritis and osteomyelitis (Ferroni et al., 2013). Of notice, in 39 of the 44 patients, *K. kingae* was diagnosed by NAA only and the organism was not detected in any child older than 4 years. These results clearly demonstrate that the culture recovery of *K. kingae* from skeletal system exudates by culture remains unsatisfactory, even when specimens are inoculated into BCV. Use of NAA methods substantially improve detection of the organism, reduces the fraction of culture-negative infections, and convincingly demonstrates that *K. kingae* is the leading bacterial etiology of suppurative arthritis and osteomyelitis in children aged 6–48 months.

32.15 Treatment and prognosis

K. kingae is usually highly susceptible to penicillins and cephalosporins, and β-lactamase production has been rarely reported in invasive isolates, although its prevalence shows wide geographic differences (Banerjee et al., in press; Yagupsky, 2004; Yagupsky et al., 2013). With rare exceptions, the organism is also susceptible to aminoglycosides, macrolides, trimethoprim-sulfamethoxazole, fluoroquinolones, tetracycline, and chloramphenicol (Jensen et al., 1994; Kugler et al., 1999; Yagupsky, 2004; Yagupsky et al., 2001). *K. kingae* exhibits decreased susceptibility to oxacillin (MIC_{50}: 3 mcg/mL and MIC_{90}: 6 mcg/mL), 40% of isolates are resistant to clindamycin, and all strains are highly resistant to glycopeptide antibiotics, a serious concern in areas where community-associated methicillin-resistant *S. aureus* is prevalent and clindamycin or vancomycin are empirically administered to patients with skeletal system infections (Pääkkönen and Peltola, 2013; Saphyakhajon et al., 2008; Yagupsky, 2012).

Because of the lack of specific guidelines for treating *K. kingae* disease, patients have been administered a variety of antibiotic regimens according to protocols developed for infections caused by traditional pathogens. The empirical therapy for skeletal infections in young children usually consists of intravenous administration of a combination of a penicillinase-stable β-lactam drug such as oxacillin and a broad-spectrum second-generation (cefuroxime) or third-generation cephalosporin (ceftriaxone), pending on culture results (Pääkkönen and Peltola, 2013; Saphyakhajon et al., 2008). The initial antibiotic regimen is frequently changed to ampicillin or cefuroxime once *K. kingae* is identified, and production of β-lactamase is excluded. Clinical response and a variety of acute-phase reactants, and especially serum CRP levels falling below 2 mg/dL, are used to guide switching to oral antibiotics and determine duration of therapy (Pääkkönen and Peltola, 2013). Traditionally, patients with *K. kingae* arthritis have been administered 2–3 weeks of antibiotics. Recent evidence, however, suggests that a total 10-day course of sequential parenteral and oral antibiotic drugs may suffice for uncomplicated articular disease caused by usual pathogens, but the experience with this novel approach in the management of *K. kingae* is still limited (Pääkkönen and Peltola, 2013). Although some children with *K. kingae* septic arthritis have been managed with repeat joint aspirations and lavage (Lebel et al., 2006), most patients promptly respond to conservative treatment with appropriate antibiotics and do not require invasive surgical procedures (Yagupsky, 2004).

Children with *K. kingae* osteomyelitis have been treated from 3 to 6 weeks. Despite the frequent diagnostic delay and severe radiological picture, no chronicity or residual orthopedic disabilities have been reported (Luhmann and Luhmann, 1999; Lundy and Kehl, 1998; Yagupsky, 2004).

Pediatric patients with *K. kingae* spondylodiscitis have been treated from 3 to 12 weeks (Yagupsky, 2004). The disease usually runs a benign clinical course living no permanent neurologic sequelae (Garron et al., 2002; Yagupsky, 2004).

Children with bacteremia have been generally given intravenous β-lactam antibiotics followed by oral therapy once the clinical condition improved and an endocardial focus has been ruled out. Total duration of therapy has ranged from 1 to 2 weeks and the clinical course is benign (Yagupsky, 2004; Yagupsky and Dagan, 1994).

Patients with *K. kingae* endocarditis are usually administered intravenous β-lactam antibiotics alone or in combination with an aminoglycoside for 4–7 weeks (Yagupsky, 2004). Despite the benign course observed in other *K. kingae* infections and the remarkable susceptibility of the bacterium to antibiotics, life-threatening complications such as cardiac failure, septic shock, mitral valve rupture or perforation, paravalvular abscesses, mycotic aneurisms, pulmonary infarcts, meningitis, cerebrovascular accidents, and other peripheral embolic phenomena are common (Yagupsky, 2004). Early surgical interventions have been necessary for life-threatening complications unresponsive to medical therapy and the overall mortality rate is unusually high (around 16%) (Rotstein et al., 2010; Yagupsky, 2004). This clinical aggressiveness appears to be related, in part, to the unfamiliarity of many laboratories with the identification of *K. kingae*, which is probably dismissed as a blood culture contaminant, a problem often encountered with other members of the HACEK group. Because of the potential severity of *K. kingae* endocarditis, routine echocardiographic evaluation of all patients from whom the organism is isolated from a normally sterile site is strongly recommended (Yagupsky, 2004).

32.16 Prevention

As it is the case with other pathogens of respiratory origin, the population of asymptomatic carriers at any given time is huge compared to that of diseased individuals. Despite a background carriage rate as high as 5–10%, the incidence of invasive *K. kingae* infections is low (Dubnov-Raz et al., 2010), and the calculated annual risk of young Swiss carriers to develop osteomyelitis or septic arthritis was found to be <1% (Ceroni et al., 2012). Therefore, in the absence of clinical disease, there is no indication to eradicate the organism from the colonized mucosal surfaces. However, the risk of acquisition of *K. kingae* and progression to a severe and even life-threatening infection appears to be greatly increased among youngsters in day care. When data of the five clusters of invasive disease detected in day care centers are pooled, a total of 16 of 90 (17.8%) classmates developed a documented or presumptive *K. kingae* infection, including fatal endocarditis, within a 1-month period (Bidet et al., 2013; Kiang et al., 2005; Miron, personal communication; Seña et al., 2010; Yagupsky et al., 2006). This unusual attack rate, coupled with the finding that a large proportion of attendees to the classes where the index cases occurred carried the infecting strain, indicated that the organism combined unusual colonization ability, remarkable transmissibility, and high virulence. Under these circumstances, administration of prophylactic antibiotics aimed to eradicate colonization in contacts and prevent further cases of disease was attempted. Either 10 mg/kg (Bidet et al., 2013) or 20 mg/kg of rifampin twice daily for 2 days, alone (Kiang et al., 2005) or in combination with amoxicillin (80 mg/kg per day) in two divided doses for 2 days (Seña et al., 2010) or 4 days (Yagupsky et al., 2006), was used. The effectiveness of these regimens, however, was limited and ranged between 47% (Bidet et al., 2013) and 80% (Yagupsky et al., 2006), indicating that eradication of *K. kingae* from colonized mucosal surfaces is difficult to achieve.

References

Adeolu, M., and Gupta, R.S. 2013. Phylogenomics and molecular signatures for the order *Nesisseriales*: Proposal for division of the order *Neissseriales* into the emended family *Neisseriaceae* and *Chromobacteriaceae* fam. nov. *Antonie van Leeuwenhoek* 104, 1–24.

Amir, J., and Yagupsky, P. 1998. Invasive *Kingella kingae* infection associated with stomatitis in children. *Pediatr. Infect. Dis. J.*, 17, 757–758.

Amit, U., Dagan, R., Porat, N., Trefler, R., and Yagupsky, P. 2012a. Epidemiology of invasive *Kingella kingae* infections in two distinct pediatric populations cohabiting in one geographic area. *Pediatr. Infect. Dis J.*, 31, 415–417.

Amit, U., Porat, N., Basmaci, R., Bidet, P., Bonacorsi, S., Dagan, R., and Yagupsky, P. 2012b. Genotyping of invasive *Kingella kingae* isolates reveals predominant clones and association with specific clinical syndromes. *Clin. Infect. Dis.*, 55, 1074–1079.

Amit, U., Dagan, R., and Yagupsky, P. 2013. Prevalence of pharyngeal carriage of *Kingella kingae* in young children and risk factors for colonization. *Pediatr. Infect. Dis. J.*, 32, 191–193.

Amit, U., Flaishmakher, S., Dagan, R., Porat, N., and Yagupsky, P. in press. Age-dependent carriage of *Kingella kingae* in young children and turnover of colonizing strains. *J. Pediatr. Infect. Dis. Soc.*

Banerjee, A., Kaplan, J.B., Soherwardy, A., Nudell, Y., MacKenzie, G.A., Johnson, S., and Balashova, N.V. in press. Characterization of a TEM-1 β-lactamase producing *Kingella kingae* clinical isolates. *Antimicrob. Agents Chemother.*

Baraff, L.J., Bass, J.W., Fleisher, G.R., Klein, J.O., McCracken Jr. G.H., Powell, K.R., and Schriger, D.L. 1993. Practice guideline for the management of infants and children 0 to 36 months of age with fever without source. Agency for Health Policy and Research. *Ann. Emerg. Med.*, 22, 1198–1210.

Basmaci, R., Ilharreborde, B., Bidet, P., Doit, C., Lorrot, M., Mazda, K., Bingen, E., and Bonacorsi, S. 2012a. Isolation of *Kingella kingae* in the oropharynx during *K. kingae* arthritis on children. *Clin. Microbiol. Infect.*, 18, e134–e136.

Basmaci, R., Yagupsky, P., Ilharreborde, B., Guyot, K., Porat, N., Chomton, M., Thiberge, J.M., Mazda, K., Bingen, E., Bonacorsi, S., and Bidet, P. 2012b. Multilocus sequence typing and *rtxA* toxin gene sequencing analysis of *Kingella kingae* isolates demonstrates genetic diversity and international clones. *PLoS One*, 7, e38078.

Basmaci, R., Ilharreborde, B., Doit, C., Presedo, A., Lorrot, M., Alison, M., Mazda, K., Bidet, P., and Bonacorsi, S. 2013. Two atypical cases of *Kingella kingae* invasive infection with concomitant human rhinovirus infection. *J. Clin. Microbiol.*, 51, 3137–3139.

Baticle, E., Courtivron, B., Baty, G., Holstein, A., Morange, V., Mereghetti, L., Goudeau, A., and Lanotte, P. 2008. Pediatric osteoarticular infections caused by *Kingella kingae* from 1995 to 2006 at CHRU de Tours. *Ann. Biol. Clin.*, 66, 454–458.

Bendaoud, M., Vinogradov, E., Balashova, N.V., Kadouri, D.E., Kachlany, S.C., and Kaplan, J.B. 2011. Broad-spectrum biofilm inhibition by *Kingella kingae* exopolysaccharide. *J. Bacteriol.*, 193, 3879–3886.

Berkun, Y., Brand, A., Klar, A., Halperin, E., and Hurvitz, H. 2004. *Kingella kingae* endocarditis and sepsis in an infant. *Eur. J. Pediatr.*, 163, 687–688.

Bidet, P., Collin, E., Basmaci, R., Courroux, C., Prisse, V., Dufour, V., Bingen, E., Grimprel, E., and Bonacorsi, S. 2013. Investigation of an outbreak of osteoarticular infections caused by *Kingella kingae* in a childcare center using molecular techniques. *Pediatr. Infect. Dis. J.*, 32, 558–560.

Birgisson, H., Steingrimsson, O., and Gudnason, T. 1997. *Kingella kingae* infections in paediatric patients: 5 cases of septic arthritis, osteomyelitis and bacteraemia. *Scand. J. Infect. Dis.*, 29, 495–499.

Bofinger, J.J., Fekete, T., and Samuel, R. 2007. Bacterial peritonitis caused by *Kingella kingae*. *J. Clin. Microbiol.*, 45, 3118–3120.

Carden, S.M., Colville, D.J., Gonis, G., and Gilbert, G.L. 1991. *Kingella kingae* endophtalmitis in an infant. *Aust. N. Z. J. Ophthalmol.*, 19, 217–220.

Ceroni, D., Cherkaoui, A., Ferey, S., Kaelin, A., and Schrenzel, J. 2010a. *Kingella kingae* osteoarticular infections in young children: Clinical features and contribution of a new specific real-time PCR assay to the diagnosis. *J. Pediatr. Orthop.*, 30, 301–304.

Ceroni, D., Cherkaoui, A., Kaelin, A., and Schrenzel, J. 2010b. *Kingella kingae* spondylodiscitis in young children: Toward a new approach for bacteriological investigations? A preliminary report. *J. Child. Orthop.*, 4, 173–175.

Ceroni, D., Dubois-Ferriere, V., Anderson, R., Combescure, C., Lamah, L., Cherkaoui, A., and Schrenzel, J. 2012. Small risk of osteoarticular infections in children with asymptomatic carriage of *Kingella kingae*. *Pediatr. Infect. Dis. J.*, 31, 983–985.

Ceroni, D., Dubois-Ferriere, V., Cherkaoui, A., Gesuele, R., Combescure, C., Lamah, L., Manzano, S., Hibbs, J., and Schrenzel, J. 2013a. Detection of *Kingella kingae* osteoarticular infections in children by oropharyngeal swab PCR. *Pediatrics*, 131, e230–e235.

Ceroni, D., Dubois-Ferriere, V., Kherad, O., Llana, R.A., Kherad, O., Lascombes, P., Renzi, G., Manzano, S., Cherkaoui, A., and Schrenzel, J. 2013b. Oropharyngeal colonization density of *Kingella kingae*. *Pediatr. Infect. Dis. J.*, 32, 803–804.

Ceroni, D., Llana, R.A., Kherad, O., Dubois-Ferriere, V., Las-combes, P., Renzi, G., Manzano, S., Cherkaoui, A., and Schrenzel, J. 2013c. Comparing the oropharyngeal colonization density of *Kingella kingae* between asymptomatic carriers and children with invasive osteoarticular infections. *Pediatr. Infect. Dis. J.*, 32, 412–414.

Chambers, S.T., Murdoch, D., Morris, A., Holland, D., Pappas, P., Almela, M., Fernández-Hidalgo, N., Almirante, B., Bouza, E., Forno, D., del Rio, A., Hannan, M.M., Harkness, J., Kanafani, Z.A., Lalani, T., Lang, S., Raymond, N., Read, K., Vinogradova, T., Woods, C.W., Wray, D., Corey, G.R., Chu, V.H., and International Collaboration on Endocarditis Prospective Cohort Study Investigators. 2013. HACEK infective endocarditis: Characteristics and outcomes from a large, multi-national cohort. *PLoS One*, 8, e63181, 1–9.

Chanal, C., Tiget, F., Chapuis, P., Campagne, D., Jan, M., and Sirot, J. 1987. Spondylitis and osteomyelitis caused by *Kingella kingae* in children. *J. Clin. Microbiol.*, 25, 2407–2409.

Cherkaoui, A., Ceroni, D., Emonet, S., Lefevre, Y., and Schrenzel, J. 2009. Molecular diagnosis of *Kingella kingae* osteoarticular infections by specific real-time PCR assay. *J. Med. Microbiol.*, 58, 65–68.

Chometon, S., Benito, Y., Chaker, M., Boisset, S., Ploton, C., Bérard, J., Vandenesch, F., and Freydière, A.M. 2007. Specific real-time polymerase chain reaction places *Kingella kingae* as the most common cause of osteoarticular infections in young children. *Pediatr. Infect. Dis. J.*, 26, 377–381.

Clarridge, J.E. 2004. Impact of 16S rDNA sequence analysis for identification of bacteria on clinical microbiology and infectious diseases. *Clin. Microbiol. Rev.*, 17, 840–862.

Couturier, M.R., Mehinovic, E., Croft, A.C., and Fisher, M.A. 2011. Identification of HACEK clinical isolates by matrix-assisted laser desorption ionization-time of flight mass spectrometry. *J. Clin. Microbiol.*, 49, 1104–1106.

Dubnov-Raz, G., Scheuerman, O., Chodick, G., Finkelstein, Y., Samra, Z., and Garty, B.Z. 2008. Invasive *Kingella kingae* infections in children; clinical and laboratory characteristics. *Pediatrics*, 122, 1305–1309.

Dubnov-Raz, G., Ephros, M., Garty, B.Z., Schlesinger, Y., Maayan-Metzger, A., Hasson, J., Kassis, I., Schwartz-Harari, O., and Yagupsky, P. 2010. Invasive pediatric *Kingella kingae* infections: A nationwide collaborative study. *Pediatr. Infect. Dis. J.*, 29, 639–643.

Fenollar, F., Lévy, P.Y., and Raoult, D. 2008. Usefulness of broad-range PCR for the diagnosis of osteoarticular infections. *Curr. Opin. Rheumatol.*, 20, 463–470.

Ferroni, A., Al Khouri, H., Dana, C., Quesne, G., Berche, P., Glorion, C., and Pejin, Z. 2013. Prospective survey of acute osteoarticular infections in a French paediatric orthopedic surgery unit. *Clin. Microbiol. Infect.*, 9, 822–828.

Fournier, P.E., Rouli, L., El Karkouri, K., Nguyen, T.T., Yagupsky, P., and Raoult, D. 2012. Genomic comparison of *Kingella kingae* strains. *J. Bacteriol.*, 195, 5972.

Fuursted, K., Arpi, M., Lindblad, B.E., and Pedersen, L.N. 2008. Broad-range PCR as a supplement to culture for detection of bacterial pathogens in patients with a clinically diagnosed spinal infection. *Scand. J. Infect. Dis.*, 40, 772–777.

Garcia-Rodriguez, J.A., and Fresnadillo Martinez, M.J. 2002. Dynamics of nasopharyngeal colonization by potential respiratory pathogens. *J. Antimicrob. Chemother.*, 50(suppl. S2), 59–73.

Garron, E., Viehweger, E., Launay, F., Guillaume, J.M., Jouve, J.L., and Bollini, G. 2002. Nontuberculous spondylodiscitis in children. *J. Pediatr. Orthop.*, 22, 321–328.

Henriksen, S.D., and Bøvre, K. 1968. *Moraxella kingii* sp. nov. A haemolytic saccharolytic species of the genus *Moraxella*. *J. Gen. Microbiol.*, 51, 377–385.

Henriksen, S.D., and Bøvre, K. 1976. Transfer of *Moraxella kingii* Henriksen and Bøvre to the genus *Kingella* gen. nov. in the family *Neisseriaceae*. *J. Syst. Bacteriol.*, 6, 447–450.

Ilhaerreborde, B., Bidet, P., Lorrot, M., Even, J., Mariani-Kurkdjian, P., Ligouri, S., Vitoux, C., Lefevre, Y., Doit, C., Fitoussi, F., Pennecot, G., Bingen, E., Mazda, K., and Bonacorsi, S. 2009. A new real-time PCR-based method for *Kingella kingae* DNA detection: Application to a prospective series of 89 children with acute arthritis. *J. Clin. Microbiol.*, 47, 1837–1841.

Jensen, K.T., Schonheyder, H., and Thomsen, V.F. 1994. In-vitro activity of β-lactam and other antimicrobial agents against *Kingella kingae*. *J. Antimicrob. Chemother.*, 33, 635–640.

Kampouroglou, G., Dubois-Ferriere, V., Anderson de la Llana, R., Renzi, G., Manzano, S., Cherkaoui, A., Schrenzel, J., and Ceroni, D. 2013. Intrafamilial oropharyngeal transmission of *Kingella kingae*; a prospective study. *Pediatr. Infect. Dis. J.*, 33, 410–411.

Kaplan, J.B. 2010. Biofilm dispersal: Mechanisms, clinical implications, potential therapeutic uses. *J. Dent. Res.*, 89, 205–218.

Kaplan, J.B., Lo, G., Xie, G., Johnson, S.L., Chain, P.S.G., Donnelly, R., Kachlany, S.C., and Balashova, N.V. 2012. Genome sequence of *Kingella kingae* septic arthritis isolate PYKK081. *J. Bacteriol.*, 194, 3017.

Kehl-Fie, T.E., and St Geme 3rd., J.W. 2007. Identification and characterization of an RTX toxin in the emerging pathogen *Kingella kingae*. *J. Bacteriol.*, 189, 430–436.

Kehl-Fie, T.E., Miller, S.E., and St Geme 3rd., J.W. 2008. *Kingella kingae* expresses type IV pili that mediate adherence to respiratory epithelial and synovial cells. *J. Bacteriol.*, 190, 7157–7163.

Kehl-Fie, T.E., Porsch, E.A., Miller, S.E., and St. Geme 3rd., J.W. 2009. Expression of *Kingella kingae* type IV pili is regulated by σ^{45}, PilS, and PilR. *J. Bacteriol.*, 191, 4976–4986.

Kehl-Fie, T.E., Porsch, E.A., Yagupsky, P., Grass, E.A., Obert, C., Benjamin Jr., D.K., and St Geme 3rd, J.W. 2010. Examination of type IV pilus expression and pilus-associated phenotypes in *Kingella kingae* clinical isolates. *Infect. Immun.*, 78, 1692–1699.

Kiang, K.M., Ogunmodede, F., Juni, B.A., Boxrud, D.J., Glennen, A., Bartkus, J.M., Cebelinski, E.A., Harriman, K., Koop, S., Faville, R., Danila, R., and Lynfield, R. 2005. Outbreak of osteomyelitis/septic arthritis caused by *Kingella kingae* among child care center attendees. *Pediatrics*, 116, e206–e213.

Kugler K.C., Biedenbach, D.J., and Jones, R.N. 1999. Determination of the antimicrobial activity of 29 clinical important compounds tested against fastidious HACEK group organisms. *Diagn. Microbiol. Infect. Dis.*, 34, 73–76.

Kulp, A., and Kuehn, M.J. 2010. Biological functions and biogenesis of secreted outer membrane vesicles. *Annu. Rev. Microbiol.*, 64, 163–184.

Lebel, E., Rudensky, B., Karasik, M., Itzchaki, M., and Schlesinger, Y. 2006. *Kingella kingae* infections in children. *J. Pediatr. Orthop. B*, 15, 289–292.

Lehours, P., Freydière, A.M., Richer, O., Burucoa, C., Boisset, S., Lanotte, F., Prère, M.P., Ferroni, A., Lafuente, C., Vandenesch, F., Mégraud, F., and Ménard, A. 2011. The *rtx*A toxin gene of *Kingella kingae*: A pertinent target for molecular diagnosis of osteoarticular infections. *J. Clin. Microbiol.*, 49, 1245–1250.

Luhmann, J.D., and Luhmann, S.J. 1999. Etiology of septic arthritis in children: An update for the 1990s. *Pediatr. Emerg. Care*, 15, 40–42.

Lundy, D.W., and Kehl, D.K. 1998. Increasing prevalence of *Kingella kingae* in osteo-articular infections in young children. *J. Pediatr. Orthop.*, 18, 262–267.

Maldonado, R., Wei, R., Kachlani, S.C., Kazi, M., and Balashova, N.V. 2011. Cytotoxic effects of *Kingella kingae* outer membrane vesicles on human cells. *Microb. Pathog.*, 51, 22–30.

Mallet, C., Ceroni, D., Litzelmann, E., Dubois-Ferriere V., Lorrot, M., Bonacorsi, S., Mazda, K., and Ilharreborde, B. 2014. Unusually severe cases of *Kingella kingae* osteoarticular infections in children. *Pediatr. Infect. Dis. J.*, 33, 1–4.

Marom, D., Ashkenazi, S., Samra, Z., and Birk, E. 2013. Infective endocarditis in previously healthy children with structurally normal hearts. *Pediatr. Cardiol.*, 34, 1415–1421.

Matta, M., Wermert, D., Podglajen, I., Sanchez, O., Buu-Hoï, A., Gutmann, L., Meyer G., and Mainardi, J.L. 2007. Molecular diagnosis of *Kingella kingae* pericarditis by amplification and sequencing of the 16S rRNA gene. *J. Clin. Microbiol.*, 45, 3133–3134.

Moumile, K., Merckx, J., Glorion, C., Pouliquen, J.C., Berche, P., and Ferroni, A. 2005. Bacterial aetiology of acute osteoarticular infections in children. *Acta Paediatr.*, 94, 419–422.

Nafstad, P., Hagen, J.A., Oie, L., Magnus, P., and Jaakola, J.K. 1999. Day care and respiratory health. *Pediatrics*, 103, 753–758.

Pääkkönen, M., and Peltola, H. 2013. Treatment of acute septic arthritis. *Pediatr. Infect. Dis.*, 32, 684–685.

Porsch, E.A., Kehl-Fie, T.E., and St. Geme 3rd, J.W. 2012. Modulation of *Kingella kingae* adherence to human epithelial cells by type IV pili, capsule, and a novel trimeric autotransporter. *mBio*, 3, e00372-12.

Porsch, E.A., Johnson, M.D.L., Broadnax, A.D., Garrett, C.K., Redinbo, M.R., and St. Geme 3rd, J.W. 2013. Calcium binding properties of the *Kingella kingae* PilC1 and PilC2 proteins have differential effect on type IV pilus-mediated adherence and twitching motility. *J. Bacteriol.*, 195, 886–895.

Robinson, J. 2001. Infectious diseases in schools and child care facilities. *Pediatr. Rev.*, 22, 39–45.

Rossignoli, A., Clavenna, A., and Bonati, M. 2007. Antibiotic prescription and prevalence rate in the outpatient paediatric population: Analysis of surveys published during 2000–2005. *Eur. J. Clin. Pharmacol.*, 63, 1099–1106.

Rotstein, A., Konstantinov, I.E., and Penny, D.J. 2010. *Kingella*-infective endocarditis resulting in a perforated aortic root abscess and fistulous connection between the sinus of Valsalva and the left atrium in a child. *Cardiol. Young*, 20, 332–333.

Saphyakhajon, P., Joshi, A.Y., Huskins, W.C., Henry, N.K., and Boyce, T.G. 2008. Empiric antibiotic therapy for acute osteoarticular infections with suspected methicillin-resistant *Staphylococcus aureus* or *Kingella*. *Pediatr. Infect. Dis. J.*, 27, 765–767.

Sarda, H., Ghazali, D., Thibault, M., Leturdu, F., Adams, C., and Le Loc'h, H. 1998. Infection multifocale invasive à *Kingella kingae*. *Arch. Pediatr.*, 5, 159–162.

Seña, A.C., Seed, P., Nicholson, B., Joyce, M., and Cunningham, C.K. 2010. *Kingella kingae* endocarditis and a cluster investigation among daycare attendees. *Pediatr. Infect. Dis. J.*, 29, 86–88.

Shanson, D.C., and Gazzard, B.G. 1984. *Kingella kingae* septicaemia with a clinical presentation resembling disseminated gonococcal infection. *Br. Med. J.*, 289, 730–731.

Slonim. A., Walker, E.S., Mishori, E., Porat, N., Dagan, R., and Yagupsky, P. 1998. Person-to-person transmission of *Kingella kingae* among day care center attendees. *J. Infect. Dis.*, 178, 1843–1846.

Slonim, A., Steiner, M., and Yagupsky, P. 2003. Immune response to invasive *Kingella kingae* infections, age-related incidence of disease, and levels of antibody to outer-membrane proteins. *Clin. Infect. Dis.*, 37, 521–527.

Sordillo, E.M., Rendel, M., Sood, R., Belinfanti, J., Murray, O., and Brook, D. 1993. Septicemia due to β-lactamase- positive *Kingella kingae*. *Clin. Infect. Dis.* 17, 818–819.

Valenza, G., Ruoff, C., Vogel, U., Frosch, M., and Abele-Horn, M. 2007. Microbiological evaluation of the new VITEK 2 *Neisseria–Haemophilus* identification card. *J. Clin. Microbiol.*, 45, 3493–3497.

Van Erps, J., Schmedding, E., Naessens, A., and Keymeulen, B. 1992. *Kingella kingae*, a rare cause bacterial meningitis. *Clin. Neurol. Neurosurg.*, 94, 173–175.

Verdier, I., Gayet-Ageron, A., Ploton, C., Taylor, P., Benito, Y., Freydière, A.M., Chotel, F., Bérard, J., Vanhems, P., and Vandenesch, F. 2005. Contribution of a broad range polymerase chain reaction to the diagnosis of osteoarticular infections caused by *Kingella kingae*: Description of twenty-four recent pediatric diagnoses. *Pediatr. Infect. Dis. J.*, 24, 692–696.

Wells, L., Rutter, N., and Donald, F. 2001. *Kingella kingae* endocarditis in a sixteen-month-old-child. *Pediatr. Infect. Dis. J.*, 20, 454–455.

Yagupsky, P., and Dagan, R. 1994. *Kingella kingae* bacteremia in children. *Pediatr. Infect. Dis J.*, 13, 1148–1149.

Yagupsky, P., Dagan, R., Howard, C.W., Einhorn, M., Kassis, I., and Simu, A. 1992a. High prevalence of *Kingella kingae* in joint fluid from children with septic arthritis revealed by the BACTEC blood culture system. *J. Clin. Microbiol.*, 30, 1278–1281.

Yagupsky, P., Bar-Ziv, Y., Howard, C.B., and Dagan, R. 1995a. Epidemiology, etiology, and clinical features of septic arthritis in children younger than 24 months. *Arch. Pediatr. Adolesc. Med.*, 149, 537–540.

Yagupsky, P., Dagan, R., Prajgrod, F., and Merires, M. 1995b. Respiratory carriage of *Kingella kingae* among healthy children. *Pediatr. Infect. Dis. J.*, 14, 673–678.

Yagupsky, P., Merires, M., Bahar, J., and Dagan, R. 1995c. Evaluation of a novel vancomycin-containing medium for primary isolation of *Kingella kingae* from upper respiratory tract specimens. *J. Clin. Microbiol.*, 31, 426–427.

Yagupsky, P., Katz, O., and Peled, N. 2001. Antibiotic susceptibility of *Kingella kingae* isolates from respiratory carriers and patients with invasive infections. *J. Antimicrob. Chemother.*, 47, 191–193.

Yagupsky, P., Peled, N., and Katz, O. 2002. Epidemiological features of invasive *Kingella kingae* infections and respiratory carriage of the organism. *J. Clin. Microbiol.*, 40, 4180–4184.

Yagupsky, P. 2004. *Kingella kingae*: From medical rarity to an emerging paediatric pathogen. *Lancet Infect. Dis.*, 4, 32–41.

Yagupsky, P. 2012. Antibiotic susceptibility of *Kingella kingae* isolates from children with skeletal system infections. *Pediatr. Infect. Dis. J.*, 31, 212.

Yagupsky, P., and Press, J. 2004. Unsuspected *Kingella kingae* infections in afebrile children with mild skeletal symptoms: The importance of blood cultures. *Eur. J. Pediatr.* 163, 563–564.

Yagupsky, P., and Slonim, A. 2005. Characterization and immunogenicity of *Kingella kingae* outer-membrane proteins. *FEMS Immunol. Med. Microbiol.*, 43, 45–50.

Yagupsky, P., Erlich, Y., Ariela, S., Trefler, R., and Porat, N. 2006. Outbreak of *Kingella kingae* skeletal system infections in children in daycare. *Pediatr. Infect. Dis. J.*, 25, 526–532.

Yagupsky, P., Porat, N., and Pinco, E. 2009a. Pharyngeal colonization by *Kingella kingae* in children with invasive disease. *Pediatr. Infect. Dis. J.*, 28, 155–157.

Yagupsky, P., Weiss-Salz, I., Fluss, R., Freedman, L., Peled, N., Trefler R., Porat N., and Dagan, R. 2009b. Dissemination of *Kingella kingae* in the community and long-term persistence of invasive clones. *Pediatr. Infect. Dis. J.*, 28, 707–710.

Yagupsky, P., Slonim, A., Amit, U., Porat, N., and Dagan, R. 2013. β-Lactamase production by *Kingella kingae* in Israel is clonal and common in carriage organisms but rare among invasive strains. *Eur. J. Clin. Microbiol. Infect. Dis.*, 32, 1049–1053.

Chapter 33

Pathogenesis of *Helicobacter pylori* in Humans

Carlo-Federico Zambon, Daniela Basso, Michela Pelloso, Alberto Tessari, and Mario Plebani

Department of Medicine-DIMED, University of Padova, Padova, Italy

33.1 Introduction

Helicobacter pylori, an S-shaped or curved microaerophilic, flagellated, Gram-negative ε-proteobacterium, chronically colonizes the human gastric mucosa, a highly hostile niche for bacterial survival and growth. The Australian researchers, Marshall and Warren (1984) were the first to isolate the bacterium in 1982 by culture of biopsy samples of human gastric mucosa, for which they were awarded the Nobel prize in 2005. Initially, the bacterium was assigned to the genus *Campylobacter*, the new genus *Helicobacter* being defined a few years later.

Currently, *H. pylori* infection is the most prevalent known chronic condition in humans worldwide, but the bacterium originally colonized humans in the African continent 88,000–116,000 years ago, presumably being acquired via a single host jump from a still unknown non-human host. When our ancestors left Africa some 60,000 years ago they carried the infection with them and, since then, *H. pylori* and humans have undergone a co-evolution (Moodley et al., 2012).

At present, there are seven genetically distinct *H. pylori* populations, hpEurope, hpSahul, hpEastAsia, hpAsia2, hpNEAfrica, hpAfrica1, and hpAfrica2 (Moodley et al., 2012), their geographic distribution clearly reflecting past human migration and ethnic distribution (Linz et al., 2007). This phenomenon depends, on the one hand, on the "environmental isolation" of *H. pylori*, which has no non-human reservoir and, on the other, on its high genetic fluidity leading to intra-species differentiation. The distinct *H. pylori* populations are therefore the fruit of thousands of years of microevolution, during which time genetic variants are continuously produced and selected depending on the fitness advantage conferred on bacterial strains populating the varying physiology of the gastric niche in human hosts. *H. pylori* displays an extraordinary genetic heterogeneity compared to other bacterial species, as proven by the reported genetic diversity between bacteria isolated from non-related patients and from the same patients at different time points during the long course of chronic infection (Suerbaum and Josenhans, 2007). Moreover, one patient may concomitantly host multiple, genetically distinct, strains (Suerbaum and Josenhans, 2007). Interestingly, the observed allelic diversity of *H. pylori* is due to an extraordinary rate of homologous recombination after natural transformation between co-infecting strains rather than intra-strain mutation acquisition (Suerbaum and Josenhans, 2007; Yahara et al., 2012). Genomic heterogeneity is an additional feature of the genetic diversity of *H. pylori*. The recent availability of next-generation sequencing platforms has rapidly increased the number of complete genomes of *H. pylori* strains, isolated from patients both with different gastro-duodenal pathologies and living in distinct geographic areas. The average size of the *H. pylori* genome is 1.62 Mb, but it ranges from 1.51 to 1.71 Mb and the G+C content is around 39% (Alm et al., 1999; Tomb et al., 1997). Almost 90% of the genome is occupied by coding sequences; however, the numbers of Open Reading Frames (ORF) and encoded proteins vary accordingly to genome size (from 1429 to 1749 and from 1382 to 1707, respectively). Only about 1200 genes are considered to constitute the core genome, the remaining ones being strain-specific sequences (Ahmed et al., 2013). The genome size and the number of coding sequences of the different *H. pylori*

Human Emerging and Re-emerging Infections: Bacterial & Mycotic Infections, Volume II, First Edition. Edited by Sunit K. Singh.
© 2016 John Wiley & Sons, Inc. Published 2016 by John Wiley & Sons, Inc.

populations appear to differ, being larger in hpEurope than in hpEastAsia or hpAfrica1 (Dong et al., 2012).

A further aspect of *H. pylori* genomic variability is the presence (in varying percentages of the strains) of the so-called cag Pathogenicity Island (cag PAI): a chromosome segment of >30 kb containing approximately 28 genes encoding for a type IV secretion system (T4SS). As for other genetic features of the bacterium, the prevalence of cag PAI differs among distinct *H. pylori* population, in particular, between hpEurope and hpEastAsia being approximately 60% and 100%, respectively.

Regarding transcriptional control, *H. pylori* appears to bear fewer genes encoding for transcription regulation proteins than other bacterial species, such as *Escherichia coli* (Alm et al., 1999; Tomb et al., 1997) and the small bacterial genome is organized into an apparently simple operon structure with 337 operons containing almost 88% of genes (Sharma et al., 2010). However, transcriptional start sites (TSS) analysis has shown additional levels of complexity in the *H. pylori* transcriptome due to (1) uncoupling of polycistrons by internal TSS and (2) genome-wide antisense transcription. Moreover, almost 60 small noncoding RNAs (sRNAs) have been discovered and found to be potentially relevant in regulating messenger RNAs (mRNAs) expression (Sharma et al., 2010). Gene expression modulation, set by *H. pylori* in response to external stimuli (e.g., pH fluctuations, nutrient depletion) or host immune response, plays a fundamental role in allowing the bacterium to initially colonize human gastric mucosa and establish chronic infection.

During the early phases of infection of a new host, once the bacterium arrives in the human stomach, it must quickly abandon the lumen to escape its extremely low pH and avoid being discharged in the intestine. Therefore, the bacterium crosses the gastric mucous layer and migrates to the gastric epithelial surface, where pH is only slightly acidic. This migration is directed by chemotactic signals due to pH, urea, and bicarbonate gradients. To penetrate the mucous layer, the bacterium makes a corkscrew movement, thanks to its helical shape and flagella. Although it is not an acidophilic microorganism, *H. pylori* has developed a resistance to acidic pH by means of its cytosolic urease enzyme, which is multimeric being composed of six UreA subunits (26.5 kDa) and six UreB subunits (60.3 kDa), and which hydrolyzes urea into carbon dioxide (CO_2) and ammonia (NH_3). The ammonia produced buffers the pH of the cytosol, periplasm, and of the close proximity of the bacterial wall, thereby aiding the bacterium to survive and swim through mucous by solubilizing it (Celli et al., 2009). Urea is imported by the bacterium through a proton-gated pump working only at acidic pH, thus avoiding any over-alkalization of the bacterium. The relevance of the urease activity for *H. pylori* survival is clarified by the facts that the knockout mutant for the enzyme cannot colonize any host and that the

protein makes up 10% of the total amount of protein of the bacterium. On reaching the gastric mucosa, *H. pylori* adheres to the epithelial cells by means of a complex pattern of bacterial outer membrane proteins (adhesins) that can recognize and bind human carbohydrate antigens exposed on surface glycoproteins. Two relevant and well-studied adhesins, BabA and SabA, recognize Lewis-B and sialyl Lewis antigens, respectively, as ligands (Basso et al., 2010). The pattern of expression of adhesins is under flexible genetic control allowing the bacterium to avoid shedding into intestine and to adapt to changes in gastric mucosa during chronic infection.

The contact between gastric epithelium and *H. pylori* triggers a host immune response that invariably leads to gastritis and other perturbations in gastric physiology, such as acid secretion. From this moment, bacterial and human effectors are engaged in continuous "crosstalk" between the host and the microorganism, resembling a lifelong war of position with no benefit for the *H. pylori* host. Although *H. pylori* may originally have conferred an advantage on our human ancestors, bacterial infections of the gastric mucosa are now clearly associated with the risk of developing severe diseases, such as peptic ulcer, MALToma, and gastric adenocarcinoma (Montecucco and Rappuoli, 2001). The link between *H. pylori* infection and gastric cancer prompted the World Health Organization and International Agency for Research on Cancer consensus group to classify the bacterium as type I carcinogen in 1994 (IARC, 1994). More recently, *H. pylori* infection has been associated with extra-gastric diseases such as unexplained iron deficiency anemia and idiopathic thrombocytopenic purpura (Arnold et al., 2009; George, 2009; Muhsen and Cohen, 2008; Pellicano et al., 2009; Qu et al., 2010). Other extragastric manifestations of *H. pylori* infection, including cardiovascular and neurological disorders, have been investigated, but the scientific evidence reported is as yet inconclusive (Banić et al., 2012; Malfertheiner et al., 2012b; Roubaud Baudron et al., 2013; Selgrad et al., 2012; Venerito et al., 2013).

33.2 Epidemiology

H. pylori infection, which affects more than 50% of the human population, is usually acquired during childhood, lasting a lifetime if not treated.

The prevalence of infection varies greatly worldwide, being higher in the developing countries, where it is estimated at 70%, than in developed countries, where it affects about 30–40% of population with some exceptions, like eastern European countries which tend to report a higher prevalence. Overall, however, a decreasing trend in the prevalence of infection has been reported throughout the world. The incidence of new infections in developing countries is 3–10% while in developed countries it appears to be 0.5% (Rosenberg, 2010).

In the **Asian pacific region**, prevalence rates are high, several studies on Asian children and adolescents reporting percentages ranging from 20 to 84. In the adult population, the percentage of infected subjects is nearly 90 in Bangladesh, India, Siberia, and Taiwan and 55–70 in Japan. The population in Australia has a prevalence of 20%, while rates vary in New Zealand, depending on subjects' origin, the rate being 49% in Pacific Islands citizens, about 25% in Asians and Maoris, and 14% in European.

In the **Middle East**, the prevalence rate for infected adults is 90% in Egypt and at least 80% in Libya, Saudi Arabia, and Turkey.

Studies from **Africa** have also reported a high overall prevalence rate (66.1–91.3% in adults and 44.3% in children), while data from **South America** show high prevalence in the adult population of Chile and Brazil (78% and 66%, respectively), but lower rates in children (24.3% and 40.7%, respectively).

In **Northern America**, the prevalence is nearly 30% (Canada and the United States), but it is much higher (79%) in the subpopulation of poor Americans (mainly blacks) in the United States; the higher rate in the latter is probably linked to their African ancestry.

The prevalence of *H. pylori* infection in **Western European** adult populations ranges from 30 to 50% while in **Eastern Europe** it is about 70%. The lowest prevalence has been found in children in the Netherlands (1.2%) and adults in Sweden (11%), the highest in the Albanian population (70.7%) (Goh et al., 2011; Hunt et al., 2011).

Differences between different countries, and within them, for incidence and prevalence are attributable to variations in the characteristics of *H. pylori* strains and to the bacterium's interaction with the host, but are also linked to socioeconomic conditions and environmental factors, which contribute to determining differences in disease expression. Of the various risk factors explored, several have been demonstrated to significantly affect the outcome of *H. pylori* infection:

- **Low socioeconomic status** – the most clearly established risk factor for infection acquisition, especially during childhood (transmission, which is mainly fecal or oral, is favored by poor hygienic conditions, household overcrowding, limited education, difficult access to sanitized water, and poor diet);
- **Age** – the incidence of infection, more prevalent in the older age categories, increases proportionate to age, although in developing countries infection is more prevalent in younger age categories, most children having infection by the time they are 10 years of age;
- **Gender** – in adults worldwide, *H. pylori* infection is reportedly more prevalent in males, but no such gender difference is reported in children (de Martel and Parsonnet, 2006).

- The presence of **infected siblings** and different **ethnicities** (especially African ancestry in some groups of US citizens) have also been reported as risk factors for infection (IARC, 2012).

33.2.1 Transmission

It is believed that the **person-to-person route** is the most important transmission convector in developed countries and, although the exact mechanism is still unknown, close contact with other humans, family members, is a strong risk factor for the transmission of infection.

Infection may be transmitted via the oral–oral (*H. pylori* can be detected in the saliva of infected subjects) or fecal–oral route. Also **waterborne** and **iatrogenic transmission** may contribute to the spread of infection (IARC, 2012).

33.3 Clinical features

H. pylori, which is etiologically associated with several major gastro-duodenal diseases related to each other in a continuum in terms of the infection's natural history, affects more than 50% of population worldwide and invariably causes gastritis. However, few infected subjects have symptoms and even fewer develop relevant clinical complications. Most symptomatic patients complain of "gastritis episodes," with unspecific symptoms like bloating, belching, a burning sensation, epigastric pain with nausea, vomiting, and/or a vague sense of abdominal discomfort. These symptoms, which are usually categorized and reported by patients under the heading "dyspepsia," often cease spontaneously, and can then recur, although in some cases they persist and progress.

Only a small proportion of infected patients develop *H. pylori*-related major gastro-duodenal diseases (i.e., gastric or duodenal ulcer, gastric adenocarcinoma, MALT lymphoma), due to the complex interrelationships between the genetic characteristics of the *H. pylori* infecting strain and environmental factors, the patient's lifestyle, and immunological response against infection.

A number of studies have also demonstrated that *H. pylori* infection is often associated with several extra-gastrointestinal diseases, but its real role in the physiopathology and development of these conditions is still uncertain and warrants further studies (Suzuki et al., 2011; Tan and Goh, 2012; Venerito et al., 2013).

The gastro-duodenal diseases related to *H. pylori* infection are as follows:

1. Dyspepsia
2. Acute infection
3. Peptic ulcer disease (duodenal and gastric ulcers)
4. Non atrophic chronic gastritis (active)
5. Chronic atrophic gastritis
6. Gastric adenocarcinoma (intestinal or diffuse)
7. MALT lymphoma

H. pylori seems to be negatively associated to Gastro-Oesophageal Reflux Disease (GORD) and its sequelae (Barret's esophagus and esophageal adeno-carcinoma). However, it has also been reported that *H. pylori* eradication therapy neither causes nor exacerbates these diseases (Malfertheiner et al., 2012b).

33.3.1 Dyspepsia

The clinical definition of dyspepsia has changed over time: traditionally, the term referred to a series of unspecific symptoms localized at the superior abdominal region. Based on the Roma III criteria, the term dyspepsia actually refers to a clinical condition characterized by the presence of one or more clearly defined symptoms: postprandial fullness, early satiation, epigastric pain, and epigastric burning.

Dyspepsia can be divided into two main categories: organic and functional. Possible causes of organic dyspepsia include peptic ulcer, GORD, gastric or esophageal cancer, pancreatic or hepatobiliary diseases, and food intolerance. Functional dyspepsia, which can only be diagnosed on completion of a diagnostic work-up (which must include an endoscopic examination), is defined as a condition in which upper abdominal symptoms occur in the absence of organic, systemic, or metabolic condition(s) that is (are) likely to explain them (Drossman, 2006; Tack et al., 2006). Some patients with functional dyspepsia have *H. pylori* infection requiring eradication therapy, proven to be of benefit in improving functional symptoms (Mazzoleni et al., 2011). In agreement with this approach, the Asian and the Maastricht IV/Florence Consensus reports consider *H. pylori*-positive dyspepsia as an independent disease and suggests that the "test and treat" strategy should be used before making a diagnosis of functional dyspepsia (Malfertheiner et al., 2012b; Miwa et al., 2012).

33.3.2 Peptic ulcer disease (PUD)

Peptic ulcer disease comprises lesions developing in the acid-peptic gastro-duodenal environment. Ulcers can be identified by endoscopic examination, which reveals a loss of mucosal integrity with a diameter measuring at least 5 mm that is often covered with fibrin.

H. pylori infection and therapy with non-steroidal anti-inflammatory drugs (NSAIDs) or with low-dose acetylsalicylic acid (ASA) are independent risk factors for the development of peptic ulcer disease with consequent bleeding. A small proportion of cases are related to Zollinger–Ellison disease. Ninety-five percent of duodenal and almost 70% of gastric ulcers are etiologically related to *H. pylori* infection, but only 5–10% of *H. pylori*-infected patients develop a peptic ulcer disease during their lifetime (Farinha and Gascoyne, 2005; Ford et al., 2006; Malfertheiner et al., 2005). In recent years, thanks to the increased *H. pylori*

eradication rate, the incidence of peptic ulcer disease is rapidly shrinking in the western world.

Numerous factors contribute to mucosal defense against the aggression of acid gastric juice, and ulceration is the result of the imbalance between aggression and defense mechanisms, the most important probably being the presence of the mucosal phospholipid surfactant barrier. *H. pylori* strains (e.g., with CagA and VacA expression) causing peptic ulcer disease are highly virulent, have a strong adhesive capacity, and are able to produce large amounts of toxic enzymes; they can lower mucus hydrophobicity by the conversion of lecithin into lysolecithin due to the action of phospholipase A and C and, by inducing inflammation, they can damage the tight junction of the luminal epithelial membrane and impair its impermeability, thereby allowing acid to damage the epithelium (Malfertheiner et al., 2009; Tytgat, 2011).

Host polymorphisms of genes involved in pathogen recognition and in the inflammatory response might predispose patients to, or protect them from, *H. pylori*-associated peptic ulcer disease: HLA DQA 1301, *NOD1* polymorphisms, and *IL-8* 251 A/T polymorphisms increase, while *IL-1β* and *TLR4*-gene polymorphisms seem to have a protective role.

Patients with duodenal ulcer usually present more severe density of infection and inflammation in the distal antral region with a subsequent increase in basal and stimulated gastric acid secretion. It has been suggested that this non-atrophic antral gastritis is more likely to develop in patients with a high constitutive acid secretory capacity, which promotes antral colonization and the development of antral predominant gastritis (Malfertheiner et al., 2009). Normally, in the antrum when the pH is low, D cells are stimulated to secrete somatostatin with a subsequent paracrine inhibitory control on adjacent gastrin-releasing G cells and a reduction in gastric acid output from parietal cells. *H. pylori* infection impairs this negative feedback regulation of gastrin release through different mechanisms: (1) by producing large amounts of alkaline ammonia in the antrum, it reduces acidity; (2) antral inflammation impairs the inhibitory neuronal control with interruption of the antral–fundic neuronal connections that normally control acid secretion; (3) the inflammatory cytokines IL-1β, TNF-α, and IL-8, largely produced in the inflamed antrum, significantly impact on the somatostatin–gastrin–acid axis; (4) this bacterium is also capable of causing a reduction in antral D cells. Overall, *H. pylori* leads to hyperplasia of both enterochromaffin-like and acid-secreting parietal cells and therefore to excessive acid secretion; the resultant increase in acid secretion leads to an acid overload in the duodenum and to the subsequent development of focal gastric metaplasia, a prerequisite for *H. pylori* colonization and for acid damage and ulceration of the epithelium. Eradication at this stage solves hypergastrinemia (after a maximum of 1 month), acid hypersecretion (after about 6 months), and eliminates the risk of ulcer, but it does not affect duodenal gastric metaplasia (Malfertheiner et al., 2009, 2011).

The typical location of gastric ulcer in *H. pylori*-infected patients is the angulus, but the condition can potentially occur in any part of the stomach. A higher grade of inflammation, often found in the antrum–corpus junction area (the transitional zone) with the involvement of acid-secreting mucosa, leads to gastric acid secretion impairment.

The main symptom of uncomplicated peptic ulcer is epigastric pain, which is reported by 80–90% of patients; it may also be accompanied by dyspepsia. Epigastric pain in PUD is usually rhythmic and recurrent, periods of exacerbation alternating with periods of well-being, occurring during the night in 60% of patients with gastric ulcer and 30% with duodenal ulcer. Changes in pain characteristics should raise the suspicion of complicated disease. On clinical examination, the only detectable sign of uncomplicated ulcer may be epigastric pain at deep palpation.

In patients with duodenal ulcer, epigastric pain typically occurs during the fasting state and is usually relieved by food or antacid drugs intake, while patients with gastric ulcer usually tend to have postprandial pain that is exacerbated by food intake.

A small percentage of patients (1–2%) with peptic ulcer are asymptomatic, while in others, the development of a complication may lead to the first sign of the disease. The most frequent peptic disease complication is gastrointestinal bleeding, which occurs in about 10% of patients, 20% of these patients requiring major medical treatment. Bleeding may be acute and massive, with an overall mortality of 6–10% or more in the presence of co-morbidities or it may be chronic, causing iron deficiency anemia.

Perforation is a complication in about 5% of patients with peptic ulcer (especially duodenal ulcers and gastric ulcers localized on the anterior wall of the stomach); typically, it presents with sudden and acute generalized abdominal pain due to the onset of a chemical peritonitis, which can evolve to bacterial peritonitis with hypotension, sepsis, and shock.

Clinical abdominal examination usually reveals signs of peritonitis (abdominal tenderness and abdominal guarding) with abdominal pain, which may be exacerbated by coughing, hip flexion, or eliciting the Blumberg sign with rebound tenderness, but the most useful sign is obtained at abdominal X-ray, showing a pneumoperitoneum with gas under the infradiaphragmatic region.

Ulcers on the posterior walls can perforate the stomach, and penetrate the pancreas, liver, or biliary tract, the accompanying changes in epigastric pain becoming more persistent and irradiating to the back. Obstruction, which occurs in about 2% of patients, is characterized by early satiety and discomfort after eating a small amount of food, with nausea and vomiting a few hours after eating, and weight loss. The presence of gastric or extra-gastric malignancies must be ruled out in patients with these signs and symptoms.

It is now widely agreed that patients with *H. pylori*-associated peptic ulcer disease need eradication therapy to prevent recurrent bleeding and complications. *H. pylori* eradication is also recommended for reducing the risk of peptic ulcer disease in NSAIDs and low-dose ASA users (Gisbert et al., 2004; Malfertheiner et al., 2012b).

33.3.3 From gastritis to gastric cancer

Gastritis, the common feature of all *H. pylori* infections, is defined as histologically confirmed inflammation of the gastric mucosa. Gastritis is at the wide bottom of a pyramid and cancer at the narrow top. The cascade of events moving from gastritis, passing through precancerous lesions, and culminating in gastric cancer was described by Correa et al. in 1970s (Correa et al., 1975), but after Marshall's discovery of the pathogenic role of *H. pylori* in the establishment of gastritis, this pattern has been updated and the infection is recognized as the most important risk factor for the development of both intestinal-type and diffuse-type gastric adenocarcinoma (Figure 33-1).

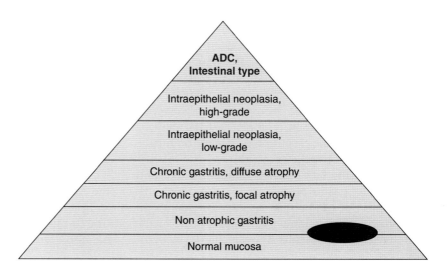

Fig. 33-1 The pyramid of events from gastritis to gastric adenocarcinoma (intestinal type).

Although the bacterium preferentially colonizes the antrum, it could potentially infect any part of the stomach. Early acute gastritis is characterized by the presence of marked neutrophilic infiltrates in the mucous neck region and lamina propria, also with the possible formation of small pit abscesses. The presence of lymphoid follicles with germinal centers is another typical finding in biopsies from *H. pylori*-infected patients.

Acute gastritis, which occurs around 72 hours after effective contact with the bacterium, is usually asymptomatic but sometimes patients experience a transient clinical manifestation, characterized by episodes of epigastric pain or a general epigastric discomfort with or without nausea and vomiting. This early phase of infection is associated with increased gastric acid secretion and, endoscopically, with severe acute inflammation of the gastric mucosa. After a few days, injury to the parietal cells making them unable to secrete acid causes the onset of achlorhydria and the symptoms diminish. This damage is probably due to the action of *H. pylori* itself, but it could also be a secondary effect of inflammation, characterized by elevated levels of IL-1β that can strongly inhibit acid secretion. The recovery of a normal acid secretion may take months (Marshall and Windsor, 2005; Marshall et al., 1985; Mitchell et al., 1992; Morris and Nicholson, 1987; Sobala et al., 1991).

If not treated with appropriate therapy, the infection persists and leads to a chronic form of gastritis defined as "active" because of the persistence of a neutrophilic infiltration of the mucosa in addition to the appearance of the mononuclear (lymphocytes and plasma) cells typical of chronic inflammation. Chronic gastritis may remain asymptomatic throughout the patient's lifetime, although a relatively large percentage (at least 30%) of subjects experience dyspepsia. *H. pylori*-associated chronic gastritis is not a steady-state condition, but must be considered a continuously evolving phenomenon. As stated above, *H. pylori* colonization usually starts in the antrum and initially affects the corpus mucosa only to a limited extent. Subsequently, *H. pylori* can migrate from the antrum to the antrum–corpus junction and then to the corpus mucosa. Despite the inflammatory infiltrate, the chronically infected gastric mucosa may maintain or lose its morphological and functional characteristics because appropriate glands become replaced by inappropriate intestinal-type glands, this phenomenon being known as atrophy. Initial focal atrophy can progress into diffuse atrophy. Therefore, based on its localization, chronic gastritis may be antrum predominant, corpus predominant, or diffuse, and may be non-atrophic or atrophic.

In view of the fact that the different extent and distribution of gastric atrophy is associated with different grades of cancer risk, an international group of gastroenterologists and pathologists (OLGA, the Operative Link for Gastritis Assessment) developed a system to report gastritis in terms of stage, using the OLGA staging system (Rugge et al., 2005; Rugge et al., 2008).

Corpus atrophy affects gastric secretion by destroying the chief and parietal cells, thereby impairing the secretion of pepsinogen I, hydrochloric acid, and the intrinsic factor. These pathological changes in gastric secretion can lead to megaloblastic anemia with vitamin B12 deficiency. Moreover, high intragastric pH due to impaired gastric acid secretion and persistent inflammation deplete the ascorbic acid normally present in the gastric juice, compromising iron absorption and entailing a risk of iron deficiency anemia.

Anemia is often the only clinical sign of atrophic gastritis: patients with chronic atrophic gastritis do not usually complain of any specific or significant symptoms, and only 20–30% of them have dyspepsia.

H. pylori eradication, followed by the gradual healing of acute and chronic inflammation, prevents the development and spread of atrophic precancerous lesions, but appears less effective in reversing mucosal metaplastic changes (Malfertheiner et al., 2012b; Shiotani et al., 2013).

Intestinal metaplasia may further progress into advanced precancerous lesions, originally grouped under the general term of "epithelial dysplasia" and now subdivided into low-grade and high-grade intraepithelial neoplasia. These lesions can develop into invasive cancer (Rugge et al., 2013).

33.3.4 Gastric adenocarcinoma

According to WHO Globocan, gastric cancer is the fourth most common cancer worldwide, and the second leading cause of death from cancer (9.7% of the total mortality). Although one million new cases were expected worldwide in 2008, the incidence is steadily declining (WHO and IARC, 2012).

At least 70% of cases occur in developing countries but more than 50% of the total number of cases occur in Eastern Asia, mainly China, which also has the highest mortality rate (28.1/100,000 in men and 13/100,000 in women), the lowest being in North America (2.8 and 1.5/100,000 in men and women, respectively). Mortality rates are high also in Central and Eastern Europe (including the Baltic States Estonia, Latvia, and Lithuania), and in Central and South America.

Countries with a high incidence rate are Japan, Korea, China, Eastern Europe, and the Andean parts of Latin America while incidences are lower in Africa, Oceania, North America, and Brazil (Piazuelo et al., 2010).

A low rate of gastric cancer has been described in the African continent despite the high prevalence of *H. pylori* infections, usually acquired at an early age (the so-called African Enigma), and a similar pattern has also been found in other regions in Colombia and

Costa Rica. In addition to differences in the virulence of *H. pylori* and host susceptibility, factors considered to contribute to the variations are diet (excessive salt intake, insufficient fresh vegetables, and fruit intake) and infection from intestinal parasites that are able to modulate the host immune response toward the Th2 pattern, in opposition to the Th1 pattern linked to the onset and development of gastric cancer (Piazuelo et al., 2010).

Gastric tumors are classified anatomically as proximal (or "cardia") and distal (or "non-cardia"), the latter type being commonly associated with *H. pylori* infection, which is thought to be responsible for more than 60% of total gastric cancer cases and from 75 to 90% of non-cardia cancer cases (de Martel et al., 2012). Histologically, the two main types recognized are intestinal and diffuse, *H. pylori* being the strongest known risk factor for both.

In 1994, the World Health Organization and International Agency for Research on Cancers defined *H. pylori* as *carcinogenic to humans* (group 1) (IARC, 1994) and it is estimated that approximately 1% of infected patients will develop gastric cancer after several decades of infection. Childhood infection incurs an extremely high risk of gastric cancer because it is frequently asymptomatic and can persist unrecognized for decades (Blaser et al., 1995).

Since the 1970s, when first described by Correa, the pathway leading to intestinal adenocarcinoma has been updated.

According to this pattern of progression, and as described above, intestinal-type adenocarcinoma is typically preceded by a prolonged and severe infection that eventually leads to multifocal atrophic gastritis with intestinal metaplasia, bacterial virulence, and environmental- and host-related factors all being involved in determining the cancer outcome (Fox and Wang, 2007).

Patients with early stage gastric cancer are asymptomatic or have aspecific symptoms due to the presence of associated lesions (for example, chronic gastritis). The high mortality rate from gastric cancer is primarily due to late-stage diagnosis. Two-thirds of gastric cancer cases in the United States are diagnosed when the tumor is already invading the gastric wall (the muscularis propria). The most frequent symptoms reported by patients at the time of cancer diagnosis are weight loss and persistent abdominal pain, sometimes accompanied by vomiting, melena, or other signs of bleeding (hematemesis or anemia) and dysphagia. In the advanced stages of disease, clinical examination may indicate an abdominal mass and/or lymph nodal localization, the most typical being the Troisier lymph node, which usually presents as a left-sided supraclavicular tumefaction. Neoplasia should also be suspected in the presence of signs of secondary lesions, for example, jaundice (hepatic metastasis), ascites (peritoneal involvement), or dyspnea (lung or pleural localizations).

According to the Asian Consensus Report, alarm symptoms indicating that an endoscopic examination is required in a patient with dyspepsia are: unplanned weight loss, progressive dysphagia, recurrent or persistent vomiting, evidence of gastrointestinal bleeding, anemia, fever, and also a family history of gastric cancer.

New onset dyspepsia in a patient over 40 years of age in a country with a high prevalence of upper gastrointestinal malignancy, or in a patient aged over 45 or 50 years in a population with an intermediate or low prevalence respectively, should also be further investigated (Miwa et al., 2012).

33.3.5 Mucosa-associated lymphoid tissue (MALT) lymphoma

Low-grade MALT lymphoma, the most common gastrointestinal lymphoma, accounts for approximately 50% of cases of gastrointestinal non-Hodgkin's lymphoma. It is typically an indolent and slowly progressive disease and most cases (nearly 100% according to IARC data) are linked to *H. pylori* infection (de Martel et al., 2012). The low-grade forms have an unspecific clinical presentation characterized by symptoms such as dyspepsia, vomiting, and eventually bleeding, while patients with high-grade disease have a palpable abdominal masses and/or weight loss.

Histologically, MALT lymphoma is a small B-cell lymphocytic lymphoma in which the mucosa is diffusely infiltrated by lymphocytes that also create lympho-epithelial lesions, destroying the normal mucosal architecture. Endoscopically, the mucosa appears irregular, and often presents multiple ulcerations with healing.

In (1993), Wotherspoon et al. clearly described the association between MALT lymphoma and *H. pylori*, and also IARC in (2012) states that *H. pylori* causes low-grade MALT gastric lymphoma in humans in the upgrade of its monograph. Findings reported in intervention studies demonstrate that treatment of *H. pylori* infection in the early stages is strongly associated with remission in 60–80% of cases when translocation t(11,18) is absent, whereas high-grade disease lymphocytes become more undifferentiated and no longer respond to eradication therapy, with alternative treatment being required (Malfertheiner et al., 2012b). MALT lymphoma can ultimately metastatize to lymphatic tissue outside the stomach, for example, the tonsils or salivary glands, but there is some evidence in the literature that, also in these localizations, the disease may regress after eradication therapy (Marshall and Windsor, 2005).

33.3.6 Extra-gastric manifestation of *Helicobacter pylori* infection

H. pylori has been associated with a wide spectrum of non-digestive system diseases (Table 33-1), and has

Table 33-1 *H. pylori* and extra-gastrointestinal diseases (Malfertheiner et al., 2012a, 2012b; Tan and Goh, 2012; Venerito et al., 2013)

Disease	Unequivocal causative associations	Demonstrated effect of treatment
Idiopathic thrombocytopenic purpura	Yes	Yes
Iron deficiency anemia	Yes	Yes
Vitamin B12 deficiency	Yes	Yes
Stroke	No	Inconclusive
Alzheimer's disease	No	Inconclusive
Idiopathic Parkinson's disease	No	Inconclusive
Ischemic heart disease	No	Inconclusive
Asthma	No	Inverse – Inconclusive
Obesity and related illness	No	Inverse – Inconclusive
Diabetes mellitus	No	Inconclusive
Extra-gastric malignancies	No	Inconclusive

been demonstrated to play a role in the pathogenesis of idiopathic thrombocytopenic purpura (ITP) and unexplained iron deficiency anemia. The Maastricht IV Consensus Report recommended that a diagnostic workup was necessary for these two diseases and that the "test and treat strategies" should be used (Malfertheiner et al., 2012b).

33.3.6.1 Idiopathic thrombocytopenic purpura

The prevalence of *H. pylori* infection in ITP patients is higher than might be expected, and it has been well demonstrated that eradication therapy in these cases improves the platelet response rate. However, as yet there is no evidence that *H. pylori*-infected patients have a more aggressive form of ITP (Malfertheiner et al., 2012b; Tan and Goh, 2012).

At least four different possible relationships between *H. pylori* infection and the onset and persistence of ITP have been proposed:

1. the anti *H. pylori* antibody appears to cross-react with platelet glycoprotein antigens, and its eradication leads to the disappearance of auto-antibodies in most cases (Tan and Goh, 2012);
2. *cagA*-positive strains are usually associated with a higher degree of inflammation and can shift the Th1/Th2 immunologic balance in favor of the Th1 response, which is also involved in the development of ITP and it may therefore be important to create a favorable microenvironment for both the onset and persistence of ITP itself. These patients, who have high titers of anti-CagA antibodies, seem

to respond better to eradication therapy (Stasi et al., 2009);
3. a molecular mimicry that has been observed between some platelets peptides and VacA (Scandellari et al., 2009) may lead to the production of cross-reacting antibodies;
4. in the presence of platelet antibodies, the lipopolysaccharide of Gram-negative bacteria is able to enhance the Fc-dependent platelet phagocytosis (Asahi et al., 2008; Semple et al., 2007).

33.3.6.2 Iron deficiency anemia

Iron deficiency anemia might be consequent to achlorhydria and low ascorbic acid levels, both described in long-standing *H. pylori* infections. A recent meta-analysis demonstrated the increase in hemoglobin level after eradication in patients with *H. pylori* infection and unexplained iron deficiency anemia, with normalization of iron absorption tests (Muhsen and Cohen, 2008).

The Maastricht IV/Florence consensus and Second Asian Pacific Consensus Guidelines recommend the identification and the treatment of *H. pylori* infection in patients with unexplained iron deficiency anemia (Fock et al., 2009; Malfertheiner et al., 2012b).

33.4 Pathogenesis and immunity

In general, *H. pylori* is not spontaneously cleared from the infected stomach, and can survive for decades in this ecologic niche, causing gastric damage and disease. All *H. pylori*-infected individuals have mild to severe gastric mucosal inflammation but, in a subset of subjects, it can also cause peptic ulcer, gastric adenocarcinoma, or MALToma (Montecucco and Rappuoli, 2001). The relationship between *H. pylori* infection and these severe diseases has been clearly demonstrated. Ninety-five percent of duodenal and 70% of gastric ulcers are associated with *H. pylori* (Farinha and Gascoyne, 2005; Ford et al., 2006; Malfertheiner et al., 2005), while from 75 to 90% of noncardia gastric cancers and 75 to 100% of gastric diffuse large B-cell lymphomas are attributable to the infection (de Martel et al., 2012). However, these severe diseases are present in only 10–20% of infected patients, those affected by malignant disease being a very small minority (1–2%) (Cover and Peek, 2013). Moreover, the incidence of gastric adenocarcinoma differs greatly worldwide, being highest in areas of East Asia, Latin America, and Caribbean and East Europe, and is partly independent of the prevalence of *H. pylori* infection (Soerjomataram et al., 2012).

The development of gastric cancer is a complex multistep process involving different pathophysiological mechanisms related to host immune response, chronic gastric inflammation, DNA damage, direct effect of *H. pylori* virulence factors on cell growth/apoptosis, and gastric overgrowth/imbalance of bacteria other than *H. pylori*. Many factors have been associated so far with gastric cancer development and they often interplay in favoring this severe *H. pylori*-associated clinical outcome acting at multiple level:

33.4.1 The duration of the infection

Infections contracted in childhood are at higher risk of severe diseases than infections contracted in adult life (Blaser et al., 1995). Asymptomatic infections are at higher adenocarcinoma risk, since they may be diagnosed only occasionally thus persisting for long periods in the host stomach.

33.4.2 Host-related factors

The risk of severe *H. pylori*-associated diseases, specifically gastric adenocarcinoma, is related both to the hosts' habits and their genetic makeup.

A diet rich in salt has been associated with an increased risk of gastric cancer (Cover and Peek, 2013) and this association may, at least in part, explain the high incidence of gastric cancer in countries such as Japan and Colombia. On the contrary, diets rich in fruits and vegetables reduce the risk of gastric adenocarcinoma (Jenab et al., 2006; Zhang et al., 2013b). There is also a significant correlation between gastric cancer and tobacco consumption, which has been demonstrated to act as an independent risk factor (Ladeiras-Lopes et al., 2008), but also to work synergistically with *H. pylori* infection in determining gastric cancer risk (Brenner et al., 2002). On the other hand, genetic inter-patient variability is directly connected to severe *H. pylori*-associated gastric diseases. The host genotype modulates the risks by modifying the physiological response and the individual's susceptibility to *H. pylori* infection (Mayerle et al., 2013; Persson et al., 2011).

33.4.3 Bacterial-related factors

H. pylori strains are genetically different in terms of virulence, with several genetically determined virulence factors having been described (Yamaoka, 2010). The most relevant factors in gastric bacterial colonization and the risk of severe *H. pylori*-associated diseases are the urease enzyme, the cag PAI, and the vacuolating cytotoxin A.

33.4.3.1 Urease

All fresh *H. pylori* isolates produce large quantities of the enzyme urease (urea amidohydrolase, EC 3.5.1.5), which makes up about 10% of the total bacterial proteins. This enzyme is an essential colonization factor of *H. pylori*: isogenic urease-negative mutants are unable to colonize the stomach of their specific target host (Montecucco and Rappuoli, 2001). *H. pylori* urease, an Ni^{2+}-containing enzyme with a molecular mass of approximately 540 kDa, hydrolyzes urea into NH_3 and CO_2. *H. pylori* urease, which has two functional subunits, UreA and UreB, is a dodecamer made up of six UreA (26.5 kDa) and six UreB (60.3 kDa) subunits, arranged in a double ring with a diameter of 13 nm (Marais et al., 1999; Mobley et al., 1995). The synthesis of active urease requires the expression of not only the two structural genes, *ureA* and *ureB*, but of another five accessory genes, *ureE*, *ureF*, *ureG*, *ureH*, and *ureI*. UreE, F, G, and H are required for the incorporation of nickel ions into the apo-enzyme, and their removal leads to loss of urease activity (Marais et al., 1999). *ureI* encodes a six-transmembrane segment inner membrane protein, not essential for active urease biosynthesis. *H. pylori* urease activity, an important *H. pylori* virulence factor, allows this non-acidophilic bacterium to survive in an acidic gastric environment. *H. pylori* can survive for several hours at pH 1, but only in the presence of physiological concentrations of urea (about 3 mM in the human gut and 1 mM in gastric juice), thus confirming the role of urease in acidic tolerance. Urease is found mainly in the bacterial cytoplasm, adhering to the outer surface (Dunn and Phadnis, 1998), has a neutral pH optimum activity, and is inactive at pH below 4.5, a value higher than that normally found in the gastric environment. The cooperation between the different urease proteins is essential in *H. pylori* tolerance to acid. A pH below 6.0 activates UreI, which rapidly transports urea from outside to inside the bacterial cell. Intracellular urea is then hydrolyzed into NH_3 and CO_2 by intra-cytoplasmic urease. By forming NH_4^+ ions, NH_3 buffers acid in both the cytosol and in the periplasm, where it is rapidly diffused. At pH values ranging from 2.5 to 6.0 and a urea concentration of 1 mM, the periplasmic pH is about 6.2, a value sufficient to allow the growth and survival of *H. pylori* (Voland et al., 2003). When the external pH is higher than 6.0, UreI is inactive and this prevents high urease activity at a neutral pH, thus limiting toxic intra-bacterial alkalinization. Ammonia generated by urea allows *H. pylori* to survive in the gastric acid environment, but is toxic for the host cells, causing swelling of intracellular acidic compartments, alterations in vesicular membrane transport, reduced protein synthesis and ATP production, and cell-cycle arrest (Montecucco and Rappuoli, 2001). Ammonia can also form carcinogenetic compounds by reacting with intermediates released through the myeloperoxidase activity of the polymorphonuclear cells infiltrating the inflamed gastric mucosa (Montecucco and Rappuoli, 2001). Finally, urease is involved in the recruitment of neutrophils and monocytes in the inflamed mucosa, and in triggering the production of pro-inflammatory cytokines (Tanahashi et al., 2000).

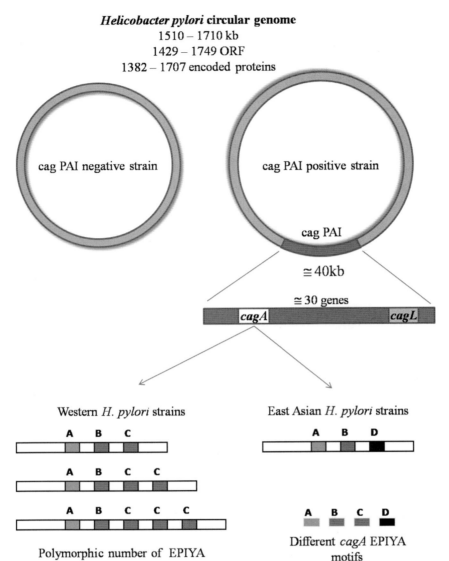

Fig. 33-2 Diagram representing *Helicobacter pylori* genome of strains with and without the cag Pathogenicity Island (cag PAI). Virulence factors *cagA* and *cagL* encoded by cag PAI are also shown.

33.4.3.2 cag Pathogenicity Island (cag PAI)

One of the main virulence determinants of *H. pylori*, cag PAI is a ≅40-kb DNA fragment (Figure 33-2) present in the genome of a fraction of bacterial strains, probably horizontally acquired from a different bacterial species in the phylogenesis of *H. pylori* (Censini et al., 1996). The cag PAI is present in almost all East Asian *H. pylori* strains, but in a smaller percentage (60–70%) of Western strains (Noto and Peek, 2012). Patients with cag PAI-positive strain infection are at a higher risk of developing severe gastritis, atrophy, dysplasia, and non-cardia gastric cancer than patients with cag PAI-negative strain infection (Basso et al., 1998; Montecucco and Rappuoli, 2001; Palli et al., 2007; Zambon et al., 2003).

The cag PAI genes of *H. pylori* encode a functional type IV secretion system (TFSS) that translocates the bacterial oncoprotein CagA into the cytosol of gastric epithelial cells.

CagA is a 120–140-kDA protein encoded by the cag PAI that, once inoculated into the host epithelial cells, alters several signaling pathways, thereby promoting a series of epithelial responses with a carcinogenic potential (Noto and Peek, 2012; Salama et al., 2013; Tegtmeyer et al., 2011; Yamaoka, 2010): (1) stimulation of cellular proliferation; (2) cytoskeletal rearrangements morphologically characterized by elongation and spreading of cultured cells (hummingbird phenotype); (3) enhanced apoptosis; (4) disruption of the apical–junctional complexes, thereby altering the barrier function and cellular polarity. In the long term, these events may contribute to the development of dysplasia, and disturb the balance between cell proliferation/apoptosis, thus making cells more prone to the damaging effects on DNA of carcinogens, and ultimately favoring carcinogenesis.

Moreover, *H. pylori* strains bearing functional cag PAI induce a massive release of several pro-inflammatory cytokines from gastric mucosa, including IL-1β, IL-6, and IL-8 and a set of genes involved in TNF-α regulation (Cox et al., 2001; Israel and Peek, 2001).

Some of the host cell signaling perturbations exerted by CagA require its phosphorylation. Once injected in the host cell, CagA is phosphorylated by Src and Abl family kinases in the C-terminal region on tyrosine phosphorylation motifs, containing the Glu–Pro–Ile–Tyr–Ala (EPIYA) amino acid sequence (Hatakeyama, 2004). CagA is characterized by structural diversity in the EPIYA-repeat segment, which results in two major forms: Western and East Asian CagA (Figure 33-2). Four types of EPIYA segment have been described, and referred to as A, B, C, and D. Most Western strains have an ABC pattern and East Asian strains an ABD pattern (Hatakeyama, 2004). C segments are often duplicated, and Western strains with ABCC and ABCCC patterns are relatively common. As the EPIYA-C segment number increases, the magnitude of CagA phosphorylation increases and CagA-induced cellular effects become more marked, thus leading to an enhanced risk of gastric cancer (Basso et al., 2008; Hatakeyama, 2004). However, the EPIYA-D motif, possessed by virtually all East Asian strains, elicits the release of even higher levels of IL-8 and more severe gastric mucosa inflammation (Argent et al., 2008; Hatakeyama, 2004). It has, moreover, been suggested that this phenomenon may partly explain the high incidence of *H. pylori*–associated gastric cancer reported in East Asian countries.

In addition to the CagA protein, the TFSS is also used by *H. pylori* to inject peptidoglycan (PGN) into the host cells. PGN induces an innate immune pathway mediated by the intra-cytoplasmic pathogen pattern recognition molecule NOD1 (nucleotide binding oligomerization domain 1). In fact, once inside the cell, PGN is targeted by NOD1, activating its signaling pathway resulting in the production of a series of pro-inflammatory cytokines as well as type I interferons (Viala et al., 2004; Watanabe et al., 2010).

A body of recent scientific evidence shows that, in addition to the above-described CagA- or PGN-dependent effects, *H. pylori* T4SS exerts a series of host cell perturbations (Tegtmeyer et al., 2011). Among the proteins involved in these processes, CagL appears to play a relevant role; this suggests that CagL is also a virulence factor for *H. pylori*. It has been shown *in vitro* that this protein interacts both with α5β1 and αvβ1 integrins at the host cell surface. By binding α5β1 integrin, CagL promotes T4SS pilus elongation, CagA translocation into host cells, and activates SRC kinase involved in the subsequent CagA phosphorylation, thus acting as a cofactor for CagA activity. Moreover, CagL–α5β1 integrin interaction directly results in the induction of the release of the pro-inflammatory cytokine IL-8 and the repression of the expression of the (H$^+$+K$^+$) ATPase by host cells. The perturbation of gastric acid secretion is further exploited by CagL through the interaction with αvβ1 integrin, leading to the induction of gastrin expression (Salama et al., 2013). The relevance of CagL as virulence factor is further supported by disease association studies linking gastric cancer risk to specific amino acid CagL polymorphisms in Indian and Taiwanese patients (Shukla et al., 2013; Yeh et al., 2011).

33.4.3.3 Vacuolating cytotoxin A (VacA)

H. pylori strains produce and release in the surrounding medium an active cytotoxin causing cellular vacuolization *in vitro* and *in vivo*, named vacuolating cytotoxin A (VacA).

This toxin, encoded by the *vacA* gene as a pro-toxin of 140 kDa made of the signal peptide, a p33 fragment, a p55 fragment, and a p40 carboxy-terminal domain, is exported, partially retained on the bacterial cell surface and also released, after proteolytic cleavage as a mature toxin composed of the p33 and p55 domains.

After binding to the host epithelial cell surface, VacA oligomerizes and is inserted in the plasma membrane at the lipid raft level via hydrophobic protein–lipid interactions. This insertion leads to the formation of membrane permeability pores, which allow the release of anions (bicarbonate, chloride, and urea) from the cell cytosol to the surface epithelial wall (Papini et al., 2001). The VacA toxin channel may be internalized by cells and localized in the late endosomal compartment, where it enhances the vacuolar ATPase proton pumping activity. In the presence of weak bases, particularly ammonia generated by *H. pylori* urease, osmotically active NH4$^+$ ions accumulate in the endosomes, causing water influx and vesicle swelling, the essential steps in vacuole formation (Cover and Peek, 2013). VacA also targets other cytoplasmic organelles, including endoplasmic reticulum, Golgi, and mitochondria. Overall, at the gastroepithelial level, VacA alters cellular polarity, promotes cellular death (via both apoptosis and necrosis), and autophagy (Boquet and Ricci, 2012; Cover and Blanke, 2005). Furthermore, this *H. pylori* toxin enhances epithelial permeability, which may facilitate the diffusion of nutrients from the gastric mucosa to the mucus layer, thus favoring *H. pylori* survival. At the gastric mucosal level, besides gastric epithelial cells, VacA also targets T and B lymphocytes; by inhibiting their stimulation-induced proliferation, it exerts an immune-modulatory action relevant for the establishment and continuation of chronic *H. pylori* infection (Cover and Blanke, 2005; Gebert et al., 2003; Torres et al., 2007).

All *H. pylori* strains bear a *vacA* gene, and almost all secrete a VacA product. The vacuolating toxin activity of the strains, however, differs significantly, and this may depend on differences in levels of *vacA* transcription and in sequence variations in *vacA* alleles. The *vacA* gene presents sequence polymorphisms localized in the 5′ end (s = signal region), in the mid region of the gene (m = middle region), and between the s and m region (i = intermediate region). Two main s, m, and i alleles have been identified, and named as s1, s2, m1, m2, i1, and i2, respectively (Figure 33-3). Almost all the possible combinations have been

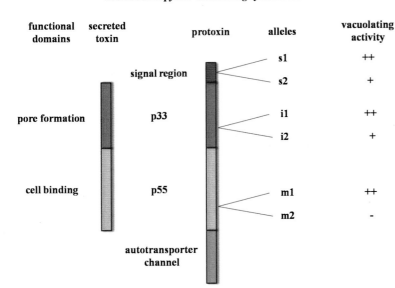

Fig. 33-3 The vacuolating cytotoxin A (VacA). Protoxin and toxin structures and their domains are shown. VacA polymorphisms and their effect on vacuolating activity are also represented.

described, although the most prevalent are s1/i1/m1 and s2/i2/m2 (Basso et al., 2008). *H. pylori* strains bearing the s1, m1, and i1 alleles have a higher cytotoxic activity, and infections sustained by them have been associated with an increased risk of duodenal and gastric ulcer, atrophic gastritis, preneoplastic lesions (intestinal metaplasia), and gastric cancer (Atherton et al., 1997; Basso et al., 1998; Kim and Blanke, 2012; Zambon et al., 2003).

A strong genetic association between s1, i1, and m1 *vacA* alleles and *cagA* has been described, although the *vacA* gene is located outside cag PAI and at a considerable distance in the *H. pylori* genome. Molecular evolutionary studies suggest that the *vacA* gene was acquired by *H. pylori* before the cag PAI. Moreover, findings reported in several recent studies show that VacA and CagA functionally interact in several of their activities, sometimes acting synergistically, but mainly in an antagonistic way (Boquet and Ricci, 2012; Kim and Blanke, 2012; Salama et al., 2013). All these observations suggest that a strong *H. pylori* VacA activity must be somehow counterbalanced by CagA action in order to avoid excessive gastric mucosal damage. *H. pylori* must establish a dynamic fine-tuning of its virulence to ensure chronic long-term colonization of the gastric niche (Kim and Blanke, 2012).

33.4.4 Innate and adaptive immune response to *Helicobacter pylori* infection

The hallmark of *H. pylori* infection is the presence of gastric mucosal inflammation characterized by the co-existence of chronic (lymphocytes and monocytes) and active (neutrophils) inflammatory infiltrates (Wilson and Crabtree, 2007). The type of inflammation and its degree depend on complex interactions between bacterial virulence factors and the host immune response, which is unable to resolve the infection, favoring its persistence, and potential evolution into cancer. In this section, we describe the present knowledge and understanding of the main immune cell types, their role in the maintenance of chronic inflammation and bacterial tolerance, and in piloting carcinogenesis.

Immune cells in the lamina propria of the gastric mucosa interplay with bacteria, the epithelial cell layer, and stromal cells, both directly, by cell-to-cell contact, and indirectly, by soluble mediators. Among the large bulk of immune cells involved in *H. pylori*-associated gastritis, neutrophils, macrophages, monocytes, lymphocyte subsets, and dendritic cells (DCs) appear to play a major role. *H. pylori* recruits neutrophils in the gastric mucosa through the secretion of the virulence factor HP-NAP and by inducing IL-8 release, mainly from gastric epithelial cells (Handa et al., 2010; Peek et al., 2010; Wilson and Crabtree, 2007). This induction is dependent on urease, commonly released by all *H. pylori* strains, and on OipA, a protein whose production is strictly associated with the cag PAI. The consequent recruitment of neutrophils, always in the infected gastric mucosa, is much more pronounced when the infection is caused by more virulent strains. Neutrophils are deputed to phagocytosis and the killing of bacteria by oxygen-dependent and/or oxygen-independent mechanisms. Reactive oxygen (ROS) and nitrogen (RNS) species are produced in large amounts within the neutrophil phagosomes, in which they kill invaginated pathogens. This innate immune response to invading pathogens, however, is not efficient enough to kill *H. pylori*, which not only promotes the migration of neutrophils to the gastric mucosa but can also manipulate phagocytosis and the subsequent oxidative burst. Among the mechanisms impeding *H. pylori* phagocytosis are: low gastric pH, which obviates opsonization; mucins present in the gastric microenvironment which prevent antibody binding; urease produced by *H. pylori*, which prevents deposition of complement C3; the unique lipid composition of the *H. pylori* outer membrane, which allows a reduced uptake of the bacterium. Moreover, phagocytosis is prevented by cag PAI, which confers further growth advantages on more virulent strains. By impeding the accumulation of ROS within the phagosomes and favoring their release in the extracellular space, *H. pylori* can survive phagocytosis by neutrophils. It is believed that these ROS in excess are a major cause of gastric mucosal damage, and that their effects on DNA are primarily involved in gastric carcinogenesis.

In the presence of *H. pylori* infection, *macrophages and monocytes*, which coordinate the innate immune response to pathogens, produce factors such as IL-12 that stimulate Th1 cells; they also play a part in the amplification of the inflammatory response by producing IL-1β, TNFα, and IL-6 (Wilson and Crabtree, 2007). *H. pylori* targets macrophages, thereby inducing the expression of inducible nitric oxide synthase (iNOS), an enzyme that catalyzes the production of nitric oxide (NO) using L-arginine as substrate (Handa et al., 2010; Wilson and Crabtree, 2007). NO kills *H. pylori in vitro*, but not *in vivo*, thanks to intriguing biochemical pathways: (1) in the gastric mucosa, NO rapidly reacts with superoxide ($O_2^{\cdot-}$) to produce peroxynitrite ($ONOO^-$), which is highly toxic. By using CO_2 produced by urease catalysis, *H. pylori* converts peroxynitrite into non-toxic, short-lived $ONOOCO_2^-$ and quench $ONOO^-$ by the synthesis of alkyl-hydroperoxide reductase; (2) *H. pylori* produces arginase, which converts L-arginine into urea, thus causing a depletion of L-arginine for NO production and an increased availability of urea, which is then utilized by urease to synthesize ammonia, which in turn is required to neutralize the gastric luminal HCl (Handa et al., 2010). The bacterium also has the ability to evade phagocytosis by macrophages, using strategies similar to those described above for evading neutrophils phagocytosis. More virulent cag PAI *H. pylori* strains producing VacA can survive within the macrophages in large structures called megasomes deriving from the fusion of phagosomes (Wilson and Crabtree, 2007).

A frequent finding at microscopy, in the presence of *H. pylori*-infected gastric mucosa, is infiltrating lymphocytes. Sometimes organized into follicles, these infiltrating lymphocytes are made up of B and T cells, although the latter are dominant in *H. pylori* infection (Peek et al., 2010; Suarez et al., 2006; Wilson and Crabtree, 2007). While CD8+ T cells play a minor role in *H. pylori* immunity, CD4+ Th1 and Th17 play a major role in protective immunity against *H. pylori*, but their action is counteracted by DCs and Treg cells (Algood et al., 2007; Hitzler et al., 2011; Rolig et al., 2011). The phenotypic characterization of T lymphocyte subsets infiltrating *H. pylori*-infected gastric mucosa is based on specific surface markers, genetic signature, and cytokines production, as shown in Table 33-2.

Th17 cells are CD4+ T cells characterized by RorγT transcription factor expression and IL-17A production, which triggers stromal and epithelial cells into releasing pro-inflammatory cytokines and chemokines, such as IL-8, TNFα, IL-1β, IL-6, CXCL1, CXCL2, CCL2, and CCL7. An increase in Th17 has been found in *H. pylori*-infected human gastric mucosa and animal models of *H. pylori* infection (Otani et al., 2009; Sugimoto et al., 2009), and also in human gastric cancer specimens (Pinchuk et al., 2013). The expansion of this T cell subtype, which depends on IL-1β, IL-6, and TGF-β1, was shown to be induced by *H. pylori*-primed stromal fibroblasts and myofibroblasts (Pinchuk et al., 2013) and to be dependent on *H. pylori* chemotaxis-induced apoptosis, chemotaxis being the bacterial ability to move toward beneficial environment signals and away from the harmful ones (Rolig et al., 2011). It has been observed that the *H. pylori*-specific Th17 response persists in the gastric mucosa of patients with a history of *H. pylori* infection and that this is dependent on IL-1β, thus indicating that this cytokine and the T cell subset may play a role in causing a persistent

Table 33-2 Types of T lymphocytes and their properties

T cell types	Surface markers	Genetic signature	Functional properties	Cytokines production	Cytokines inducers
CD8+			Killing of unwanted target cells		
	CD3+CD45+CD8+				
CD4+					
	CD3+CD45+CD4+				
CD4+ Th1			Promotion of inflammation by cytokine production	IL-2, IFN-γ, TNF-β	IL-12
	CD3+CD45+CD4+				
CD4+ Th2			Helping B lymphocytes proliferate, produce antibody, and become memory cells	IL-4, IL-5, IL-6, IL-9, IL-10, IL-13	IL-4
	CD3+CD45+CD4+				
CD4+ Th17		Related orphan receptor-γT (RorγT)	Promotion of inflammation by cytokine production	IL-17A, IL-21, IL-22	IL-1β, IL-6, TGF-β
	CD3+CD45+CD4+				
CD4+ Treg	CD3+CD45+ CD4+CD25+	Forkhead box P3 (FoxP3)	Regulating immunosuppressive responses	IL-10, TGF-β	TGF-β

Th, T helper; Treg, T regulatory.

inflammation and gastric cancer risk despite eradication of the bacterium (Serelli-Lee et al., 2012). FoxP3 expression and the CD25 surface marker characterize Treg cells. Th17 and Treg cells, which are developmentally related, co-exist in a delicate balance that can determine the outcome of a bacterial infection (Littman and Rudensky, 2010). Treg cells also accumulate and proliferate in the infected gastric mucosa upon the exposure of $CD4^+$ naive cells to TGF-β, which is produced by gastric epithelial cells that have been exposed to the bacterium (Beswick et al., 2011; Jang, 2010; Lundgren et al., 2005; Mitchell et al., 2012). *H. pylori* alters the Th17/Treg balance, leading to a Treg-biased response: irrespective of CagA or VacA, *H. pylori* has little inhibitory effect on TGF-β, and little stimulatory effects on IL-6 and IL-23 expression by DCs, this favoring a Treg over a Th17 response (Kao et al., 2010). Moreover, by inducing IL-1β release from DCs, *H. pylori* significantly stimulates the proliferation of Treg cells, which inhibit activated T cells through direct contact, by competition for antigen-presenting cells (APC) binding and/or by secreted cytokines such as IL-10 and TGF-β (Mitchell et al., 2012; Sakaguchi et al., 2010); in doing so, they may favor *H. pylori* evasion of immune-mediated clearance, thus promoting prolonged infection and carcinogenesis (Jang, 2010). However, although Treg cells are mainly recognized as inhibitors of the immune response to pathogens, their expansion in infected gastric mucosa might be interpreted from another viewpoint: it has been suggested that, rather than being promoted by Treg, the persistent infection promotes and expands the Treg population in the attempt to limit the damage caused by a prolonged and excessive inflammatory response. This hypothesis is supported by the findings of Mitchell et al. (2012), who demonstrated that IL-1β, which appears to play a central role in *H. pylori* biology, triggers Treg proliferation yet diminishes Treg suppressive function.

The imbalance between Th17 and regulatory Treg cells has recently been recognized as one of the major immune components linking inflammation and cancer in *H. pylori* infection. DCs, the main regulators of the Th17/Treg balance, are potent antigen presenting cells that are functionally targeted by *H. pylori*. By binding to the dendritic cell receptor SIGN, the *H. pylori* fucose-containing glycans, which can be switched on and off through phase variation expression, induce IL-10 expression, not IL-12 or IL-6 expression, thus tailoring Th1 polarization (Gringhuis et al., 2009), and the vacuolating cytotoxin VacA has been shown to suppress DCs maturation, which is necessary for adaptive immune response, thus contributing to establishing a persistent infection in the gastric mucosa (Kim et al., 2011). *H. pylori* also induces impaired phagocytosis and enhanced secretion of IL-1β, IL-6, IL-10, IL-23, and IL-12 by DCs, and enhances the ability of DCs to induce Tregs, thus significantly contributing to a tolerogenic ground, which inhibits protective immunity upon vaccination (Fehlings et al., 2012; Hitzler et al., 2011) Figure 33-4 depicts part of stromal cell interplay in *H. pylori*-infected gastric mucosa.

The interactions of *H. pylori* with the host immune system cells are not an all-or-nothing phenomenon, since the type of inflammatory response and its degree depend on several bacterial- and host-related factors. Bacterial factors mainly include *H. pylori* virulence determinants, but it has recently been suggested that the commensal flora may also play a role, while host-related factors are mainly represented by host genetics, which might enhance or inhibit the gastric mucosal production of inflammatory mediators, cytokines in particular, in response to *H. pylori* infection.

33.4.4.1 Commensal flora

The commensal bacterial flora have been implicated in aspects of immune system regulation and development, and an altered microbiota has been reported to both ameliorate and worsen symptoms, especially those of inflammatory bowel diseases (Atarashi et al., 2011; Ivanov et al., 2009; Lathrop et al., 2011; Round and Mazmanian, 2009). The human stomach is known to have its own microbiota and, reportedly, gastric microbial communities are both unaffected and affected by *H. pylori* infection (Bik et al., 2006; Maldonado-Contreras et al., 2011). Studies on humans, however, have not allowed scientists to verify whether specific microbial communities create a favorable ground for *H. pylori* infection and disease evolution. On comparing uninfected with chronically *H. pylori*-infected stomach samples, any difference in microbiota communities does not necessarily indicate their role as *H. pylori* susceptibility factors, but they might rather be the consequence of *H. pylori* infection: on determining a reduced acid secretion, this infection may favor gastric colonization by the commensal bacterial flora. However, studies on well-established *H. pylori* mouse models with *H. pylori*-infected INS-GAS (gastrin overexpression which leads to gastric atrophy and subsequent gastrointestinal intraepithelial neoplasia by 20 months, which is accelerated by *H. pylori*) showed that gastric atrophy, metaplasia, and dysplasia were more severe and gastrointestinal intraepithelial neoplasia (GIN) more frequent in animals hosting commensal flora than in germ-free animals (Lofgren et al., 2011). Subsequently, in another murine model, Rolig et al. (2013) confirmed that inflammation is less severe after *H. pylori* infection in germ-free animals, with a reduced expansion of $CD4^+$ Th1 cells, but they also demonstrated that C57BL/6N mice from different vendors mount different inflammatory responses to *H. pylori* infection. The cause of this difference was attributed to differences in commensal flora, which might modulate $CD4^+$ T cell recruitment to the gastric compartment, and a potential role was also suggested for *Clostridia* species. Several *Lactobacillus* strains were shown to be protective in inflammatory bowel diseases (Round and Mazmanian, 2009), and *L. acidophilus* was found

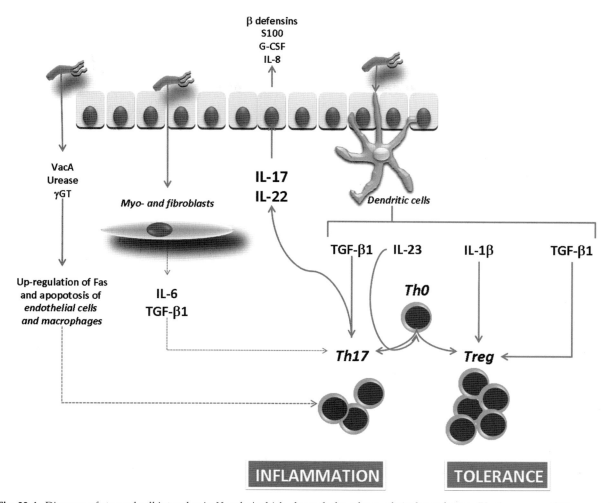

Fig. 33-4 Diagram of stromal cell interplay in *H. pylori* which, through the release of virulence factors (VacA, urease, γGT), may act directly or indirectly on dendritic cells, myofibroblasts, fibroblasts, endothelial cells, and macrophages. By releasing different cytokines, stromal cells induce the production of Th17 and Treg from resident naive Th0 cells. Persistent inflammation is favored by Th17, immune tolerance is favored by Treg cell expansion, and the prevalence of Treg cells versus Th17 cells favors *H. pylori* tolerance despite inflammation.

to reduce IL-8 expression and inactivate the NF-κB inflammatory pathway in gastric epithelial cells exposed to *H. pylori* (Yang et al., 2012), further supporting the claim that multiple interactions taking place between *H. pylori* and other bacterial species might modulate the overall inflammatory response to *H. pylori* infection. These interactions might be a potential target for manipulation aiming to prevent disease and improve outcome.

33.4.4.2 Host genetics

H. pylori-associated inflammation is characterized by an accumulation in the gastric mucosa of pro- and anti-inflammatory cytokines, the imbalance of which may concur in promoting different *H. pylori* infection outcomes (i.e., chronic non-atrophic gastritis, chronic atrophic gastritis, gastric ulcer, duodenal ulcer, gastric adenocarcinoma, and MALT lymphoma). Increased mucosal levels of pro-inflammatory cytokines, including IL-1β, IL-4, IL-6, IL-8, and TNF-α, are invariably associated with *H. pylori* infection, but increased levels of the anti-inflammatory IL-10 and of the Th1 IFN-γ are also present (Chiba et al., 2008). Pro-inflammatory

cytokines are believed to play a key role in linking inflammation and cancer, through the activation of NF-κB, mitogen-activated protein kinase (MAPK), and c-Jun N-terminal kinase (JNK) cascades (Chiba et al., 2012). The degree of cytokines synthesis in response to stimuli depends on the stimulus type, but it might also depend on polymorphisms of their respective genes or of genes encoding cytokine antagonists. Three di-allelic polymorphisms in the *IL-1B* gene (−511C/T, −31C/T, and +3954C/T) reported to significantly affect IL-1β production are significantly associated with the risk of gastric cancer, not duodenal ulcer, in the presence of *H. pylori* infection (Camargo et al., 2006; Hwang et al., 2002; Peleteiro et al., 2010; Witkin et al., 2002; Zambon et al., 2002; Zhang et al., 2012a, 2012b). In particular, the two *IL-1B*-511T and *IL-1B*-31C alleles, which are functional SNPs in near-complete linkage disequilibrium affecting the promoter activity of the *IL-1B* gene, enhance gastric cancer risk in Caucasians, but not in Asians (Camargo et al., 2006; Chen et al., 2006). However, the most relevant genetic polymorphisms involving the IL-1β pathway and determining an increased gastric cancer risk are those of the *IL-1RN* gene, which encodes the IL-1β-specific receptor antagonist IL-1RA (Camargo et al., 2006; Peleteiro et al., 2010; Persson et al., 2011; Zhang et al., 2012b). *IL-1B* and *IL-1RN* genes are clustered with the *IL-1A* gene on chromosome 2q14; the second intron of the *IL-1RN* gene contains a variable number of tandem repeats (VNTR) with an 86-base pair nucleotide sequence as its repeating element. The *IL-1RN* alleles may be classified on the basis of the exact number of VNTR repeats as follows: allele 1, four repeats; allele 2, two repeats; allele 3, five repeats; allele 4, three repeats; allele 5, six repeats. Alternatively, *IL-1RN* alleles may be divided into two categories: long (L, including alleles 1, 3, 4, and 5) and short (S, allele 2). Allele 2, which has high circulating IL-1Ra and even higher IL-1β levels (Witkin et al., 2002), significantly enhances the risk of gastric cancer, and also gastric precancerous lesions (Camargo et al., 2006; Peleteiro et al., 2010; Persson et al., 2011; Zhang et al., 2012b). *TNF-A* cytokine gene polymorphisms are also involved in *H. pylori*-associated gastric carcinogenesis. Of the several promoter region SNPs studied, those most extensively evaluated are *TNF-A*-308G/A and *TNF-A*-857C/T: the rare alleles A and T, respectively, appear to slightly increase the risk of gastric cancer, but not that of duodenal ulcer (Persson et al., 2011; Zhang et al., 2008, 2013a). Pro-inflammatory cytokine IL-8, always induced in *H. pylori*-infected gastric mucosa, is a potent chemoattractant and activator of neutrophils and has been shown to be the most upregulated gene in the gastric cancer cell line AGS following the exposure to *H. pylori* (Eftang et al., 2012). The A/T SNP at position −251 of the *IL-8* gene, located on chromosome 4q12–21, could affect *IL-8* gene transcription and secretion (Ohyauchi et al., 2005; Taguchi et al., 2005). The *IL-8* -251A allele tends to be associated with increased IL-8 production, but also with an increased risk of gastric cancer and peptic ulcer (Xue et al., 2012a; Yin et al., 2013).

Among the cytokine gene polymorphisms investigated to identify contributory causes of disease development, those of the anti-inflammatory IL-10 have been widely studied. IL-10 inhibits Th1 lymphocytes and stimulates Th2 and B lymphocytes, thus downregulating the pro-inflammatory response. Among the three genetic polymorphisms of the promoter (−1082 A/G, −819 C/T, −592 A/C), *IL-10* −592 A is significantly associated with a reduced gastric cancer risk (Persson et al., 2011; Xue et al., 2012b). The reported polymorphisms of cytokines genes have been the focus of more intense investigation, also in meta-analyses. However, genetic polymorphisms of other cytokines, such as *IL-12*, *IL-4*, and *IFN-G*, have been evaluated and reported to have a different effects on the risk of *H. pylori*-associated diseases (Canedo et al., 2008; El-Omar et al., 2003; García-González et al., 2005; Hou et al., 2007; Kato et al., 2006; Navaglia et al., 2005; Pessi et al., 2005; Queiroz et al., 2009; Seno et al., 2007; Thye et al., 2003; Yang et al., 2013; Zambon et al., 2008).

33.4.4.3 Cytokines and gastric physiology

The underlying mechanism linking IL-1β/IL-1RN, TNF-α, and IL-8 genetics with gastric cancer risk may be related to the effects of these cytokines on signaling and physiology of gastric mucosa. IL-1β, which inhibits gastric acid secretion 100-fold and 6000-fold more potently than proton pump inhibitors and H_2 antagonists, respectively, inhibits acid-secreting cells directly and indirectly by causing apoptosis of enterochromaffin-like cells (ECLs) through the induction of bax and iNOS and causes the reduction in gastrin expression via the activation of NF-κBand SMAD7 (Beales and Calam, 1998; Calam et al., 1997; Chakravorty et al., 2009; Datta et al., 2011; Wilson and Crabtree, 2007). TNF-α is less potent than IL-1β in inhibiting acid secretion, probably because it has opposite effects, reducing histamine by inducing ECLs apoptosis, but simultaneously stimulating G cells to produce gastrin (Calam et al., 1997). IL-8, unlike IL-1β and TNF-α, stimulates acid secretion by inducing gastrin release (Figure 33-5).

The inhibition of parietal cell function due to the effects of IL-1β and TNF-α is one of the mechanisms linking *H. pylori* infection and gastric carcinogenesis. In fact, persistently reduced gastric acid secretion creates a favorable ground for gastric colonization by different bacterial species, which might enable the nitrosation of a number of naturally occurring guanidines and ʟ-arginine food-containing polypeptides, thus producing mutagenic compounds. Moreover, gastric acid is implicated in the conversion of nitrate into carcinogenic N-nitroso compounds and in determining iron deficiency anemia (Nagini, 2012; Queiroz et al., 2013). IL-1β might favor gastric carcinogenesis also because it perturbs intracellular signaling by promoting NF-κB and MAPK signaling cascades, impairs host immune surveillance, recruits PMN cells, and induces ROS production, all of which are considered part of the process of malignant transformation (Dinarello, 2006). Like IL-1β, TNF-α is a well-known key player in cancer development and progression (Bauer et al., 2012; Lippitz, 2013).

33.5 Diagnosis

There are two basic tests for diagnosing *H. pylori* infection: one type is invasive (requiring endoscopy and biopsy) and the other is non-invasive. The tests are based on the direct identification of the bacterium (culture, histology), or indirect, being based on the identification of bacterial antigens (stool testing), bacterial nucleic acids (molecular testing), antibody production (serology), or the identification of *H. pylori*-derived enzyme activities (rapid urease test, ^{13}C urea breath test (UBT)). The choice of the appropriate diagnostic test(s) should be in line with epidemiological, clinical, availability, and cost–benefit issues, as summarized in Table 33-3. The non-invasive UBT and monoclonal stool antigen testing are recommended

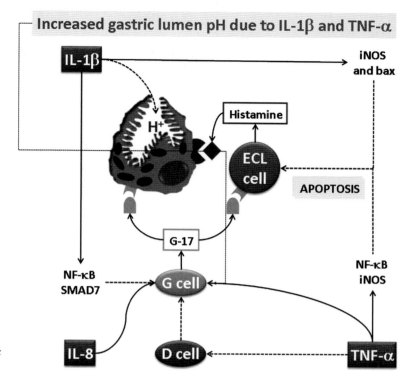

Fig. 33-5 Impact of IL-1β and TNF-α on gastric physiology.

when the test-and-treat strategy is appropriate. On the basis of the statements provided by the Maastricht IV/Florence consensus report (Malfertheiner et al., 2012b), a test-and-treat strategy is warranted where *H. pylori* prevalence is >20% and the risk of gastric cancer is low; in many countries, this means dyspeptic patients below a locally determined age cut-off (depending on the local incidence of gastric cancer for age groups) and without "alarm" symptoms or signs which are associated with an increased gastric cancer risk. Among adults, the prevalence of *H. pylori* infection is lower than 20% in Australia, while it is higher than 80% in Africa (Ethiopia, Nigeria, Egypt, Libya), in South America (Brazil, Chile), in Asia (Bangladesh, India, Siberia), and in the Middle East (Saudi Arabia, Turkey). In North America, *H. pylori* prevalence is about 30%, comparable to that in Western Europe (30–50%), but not that in Eastern Europe (70%). The overall prevalence of *H. pylori* infection among children and young adults is above 20% in developing countries, but lower than 20% in the majority of developed countries (Hunt et al., 2011). For this reason, in developed countries, the use of a test-and-treat strategy for younger patients presenting with dyspepsia is declining in favor of the immediate use of proton pump inhibitors (PPI) (Hunt et al., 2011; Koletzko et al., 2011). However, it must be borne in mind that in low-incidence countries, the prevalence of *H. pylori* among immigrants mirrors that of the country of origin, and these patients should be managed accordingly. For patients aged 50 years and older and for younger subjects with alarm symptoms, endoscopy should be performed to rule out gastrointestinal malignancy, and histology testing for *H. pylori* infection, if no malignancy is found, is the most appropriate and logical

approach, supported, if necessary, by the rapid urease test which, if positive, is sufficient to warrant initiation of treatment (Malfertheiner et al., 2007). Culture with an antibiogram of gastric biopsies should be deemed appropriate whenever attempts to eradicate *H. pylori* have repeatedly failed and resistance to antibiotic is suspected. The most appropriate test for confirming eradication is UBT; however, if it is unavailable, stool antigen testing can be used. Serological tests are recommended to assess *H. pylori* in patients with a bleeding ulcer and conditions (e.g., mucosal atrophy and MALT lymphoma) known to be associated with a low bacterial density (Malfertheiner et al., 2007). They should also be considered a reliable diagnostic tool in high-prevalence countries and in subjects who have not previously undergone eradication protocols or in those undergoing PPI treatment. Anti-*H. pylori* antibodies are part of the GastroPanel test, a serologically based diagnostic algorithm that also includes pepsinogens and gastrin-17 and allows the detection of the presence of *H. pylori*-associated and non-*H. pylori*-associated non-atrophic and atrophic gastritis. This test, which is suitable for screening subjects at risk of gastric cancer in countries with a high incidence of this cancer type (Fock et al., 2008), is also indicated in low-incidence areas to detect gastric atrophy. In patients with partial gastrectomy, the urease-based tests (UBT and rapid urease test) are not recommended due to their low sensitivity, histology being much more accurate in these cases (Tian et al., 2012). Except for serology, PPI and antibiotics should be discontinued two weeks before checking for *H. pylori* infection with invasive and non-invasive diagnostic tests, because they can give false negative results (Malfertheiner et al., 2007).

Table 33-3 Appropriate diagnostic test(s) on the basis of epidemiological and clinical evidences as well as on availability and cost–benefit issues

	Resident country or country of origin if recent immigrant	Appropriate first-line test	Appropriate second-line test	Comments
Dyspeptic children or young adults	Developing countries (*H. pylori* prevalence > 20%)	^{13}C-UBT or stool antigen testing	Serology	
	Developed countries (*H. pylori* prevalence < 20%)	*H. pylori* infection diagnosis is not recommended		*H. pylori* infection diagnosis by first-line tests may be considered in children with first-degree relatives with gastric cancer or with refractory iron deficiency anemia
Dyspeptic adults less than 50 years without alarm symptoms	Living or recently coming from countries with *H. pylori* prevalence > 20%	^{13}C-UBT or stool antigen testing	Serology	
	High gastric cancer risk countries	^{13}C-UBT or stool antigen testing	GastroPanel	GastroPanel offers the opportunity to verify whether the pre-cancerous atrophic gastritis has developed. Gastric cancer is rare in the absence of chronic active gastritis and the extent and severity of gastritis together with atrophy and intestinal metaplasia is positively associated with cancer
	Living or recently coming from countries with *H. pylori* prevalence < 20%	*H. pylori* infection diagnosis is not recommended		
Dyspeptic adults over 50 years or younger with alarm symptoms	Any country	EGDS and histology	Rapid urease test	EGDS is required to verify the presence of gastric cancer. A positive rapid urease test is sufficient to initiate treatment
Patients with bleeding ulcer	Any country	Serology		
Patients suspected of having low bacterial load (e.g., mucosal atrophy, MALT lymphoma)	Any country	GastroPanel	EGDS and histology	EGDS and histology are mandatory when GastroPanel indicates the presence of atrophy and when the MALT lymphoma is suspected
Patients with partial gastrectomy	Any country	EGDS and histology	Serology	^{13}C UBT is not recommended because the sensitivity is lower than 80%. The volume of the stomach is reduced and the gastric anatomy is altered causing a less acidic environment, hostile to the colonization of *H. pylori*. A more rapid urea transit through the residual stomach associated with a reduced *H. pylori* load create a bias for UBT
Post-eradication control of dyspeptic patients	Any country	^{13}C UBT	Stool antigen testing	Testing should be performed not earlier than 4 weeks from the end of therapy, but a longer period is advisable (6–8 weeks)
Post-eradication control of gastric ulcer or gastric MALT lymphoma	Any country	EGDS and histology		
Patients who require third-line treatment for failure of first- and second-line treatments	Any country	EGDS and *H. pylori* culture and antibiotic susceptibility testing		Third-line treatment should be based on resistance testing

33.5.1 ^{13}C-urea breath test (UBT)

The sensitivity and specificity of this non-invasive diagnostic test exceed 95% (Bytzer et al., 2011; Malfertheiner et al., 2007). UBT allows the identification of *H. pylori* in the gastric lumen by detecting the activity of one of the main bacterial products: the enzyme urease, produced in large amounts by *H. pylori* that catalyzes the hydrolysis of urea into CO_2 and NH_3. CO_2, rapidly absorbed by the stomach, is eliminated in the breath. The principle underlying UBT is based on the assumption that *H. pylori* is the most common urease-containing gastric pathogen able to hydrolyze orally administered labeled urea to produce isotopically labeled CO_2, which can be detected in breath samples (Gisbert and Pajares, 2004). Urea can be labeled with two different isotopes, ^{13}C (non-radioactive) or ^{14}C (radioactive). While ^{13}C-urea is innocuous and can be used repeatedly in the same patient and can also be used safely in children or pregnant women, ^{14}C-urea is not authorized by most Health Authorities because it involves radiation. After overnight fasting, a basal breath sample should be collected before ^{13}C-urea ingestion. The most widely used UBT is based on the administration of 75 mg of ^{13}C-urea in a citric acid solution, although a high accuracy has been reported also for lower (50 mg) and higher (100 mg) urea doses and for other nutrient meals, such as orange juice or apple juice (Gisbert and Pajares, 2004). A second breath sample is obtained 30 minutes after urea ingestion, although excellent accuracy may be obtained when samples are collected as early as 10–15 minutes after urea ingestion. In the basal and post-load breath samples, the ^{12}C and ^{13}C isotopes of CO_2 are measured by a mass spectrometer and the results are usually expressed as parts per thousand (‰). An algebraic difference between a post-load and basal ^{13}C/^{12}C ratio above 5‰ is recommended for establishing the presence of *H. pylori* infection.

33.5.2 Serology

Serology, based on the measurement in sera of IgG antibodies against *H. pylori* antigens (while IgM antibody detection has no role in this infection), has a lower accuracy (80–84%) than UBT, with a sensitivity of 90–97% and a specificity of 50–96% (Bytzer et al., 2011; Leal et al., 2008; Malfertheiner et al., 2007); it has some advantages over the non-invasive diagnostic methods since its accuracy is not influenced by: (1) on-going PPI or antibiotic treatment; (2) bleeding; (3) previous gastric resection; (4) low bacterial load. One of the major limitations of this approach is its inability to distinguish between active and previous infections, especially in patients who have undergone recent eradication: a significant drop in antibody titers post-therapy takes 4–6 months or more and levels may remain high for years (Rautelin et al., 2003). Although the detection of specific *H. pylori* antibodies in urine and saliva has been proposed, their sensitivity

is too low and this approach is not recommended for diagnostic or screening purposes (Malfertheiner et al., 2007).

33.5.3 GastroPanel

To obtain insight in the severity and extension of gastric mucosa inflammation, it has been suggested that the so-called "GastroPanel" should be requested. This biochemical integrated test, which can confirm a clinical suspicion of gastritis, is based on the measurement in serum by ELISAs of pepsinogen A and C (PGA and PGC), Gastrin-17 (G-17), and anti *H. pylori* antibodies, all of which are entered in a classification algorithm analysis (Agréus et al., 2012); it allows the identification of the presence of gastritis, and its type: either chronic atrophic or chronic non-atrophic gastritis, thus providing a useful tool for detecting subjects at higher risk of gastric cancer, since chronic atrophic gastritis is a well-known pre-cancerous condition, and *H. pylori*-associated non-atrophic gastritis is also a gastric cancer risk condition (Watanabe et al., 2012). GastroPanel is characterized by a very high negative predictive value, in the region of 90%, in both adults and children, and also by a high positive predictive value, ranging from 62 to 97%, in distinguishing between normal and diseased gastric mucosa (Guariso et al., 2009; Iijima et al., 2009; Storskrubb et al., 2008). When this test is used to detect the presence of chronic atrophic gastritis and therefore for cancer screening, data interpretation should be made while bearing in mind the epidemiology of gastric cancer and chronic atrophic gastritis: extremely low PGA values ($\leq$25 μg/L) are indices of severe corpus atrophic gastritis worldwide in subjects of any age. During the course of chronic atrophic gastritis, mucosal atrophy advances from the side of the pyloric gland towards the oral side, and PGA levels and PGA/PGC ratios progressively decrease as mucosal atrophy advances (Mukoubayashi et al., 2007). To detect mild/moderate forms of atrophic corpus gastritis, stringent criteria should be applied in geographic areas with a high incidence, while they may be less stringent in areas with a low incidence. In Japan, where the incidence of severe chronic atrophic gastritis is high (17–29%), PGA values lower than 60 μg/L and a PGA/PGC ratio <4 are indices of gastric cancer, and have a sensitivity of 71% and a specificity of 69.2% (Shikata et al., 2012), whereas in Europe, where severe CAG is less frequent (5–10%) (Palli et al., 2007; Telaranta-Keerie et al., 2010; Weck et al., 2007), a PGA/PGC ratio of about 5 is recommended for diagnosing this condition (Broutet et al., 2003).

33.5.4 Stool antigen testing

This non-invasive tool allowing the detection of *H. pylori* antigens in stool is characterized by an elevated sensitivity (93–96%) and specificity (96–97%) for both the diagnosis and confirmation of eradication after therapy when monoclonal antibodies are used (Gisbert et al., 2006). It is therefore recommended, especially as its diagnostic accuracy is equivalent to that of UBT if the test used is a validated laboratory-based ELISA monoclonal test (Malfertheiner et al., 2012b).

33.5.5 Rapid urease test

This test, which allows the detection of *H. pylori* urease in biopsy specimens within 1 hour, has shown a high sensitivity (96%) and specificity (around 100%) (Calvet et al., 2010). A positive finding is considered sufficient to initiate treatment (Malfertheiner et al., 2012b).

33.5.6 Histology

Histology is highly reliable for the diagnosis of *H. pylori* infection, with a sensitivity of 93% and specificity of 85–90% (Tian et al., 2012). As it is invasive, it is recommended in specific clinical settings (i.e., in patients with alarm symptoms, elderly dyspepsia patients from countries with a high gastric cancer incidence). At histology, the bacterium is usually detectable (with Giemsa staining modified for *H. pylori*) within the mucous gel layer covering the gastric mucosa. *H. pylori* may be difficult to identify (even with special stains) in cases of extensive intestinal metaplasia, or during PPI therapy; in such cases, the *H. pylori* infection is suggested by the presence of both mononuclear and neutrophilic (active) inflammation. After a successful eradication therapy, the neutrophils quickly disappear; persistent neutrophils and/or mononuclear infiltrate indicate an unsuccessful treatment. In routine diagnostic practice, any semiquantitative score of the bacterium's density has no clinically significant implications and a distinction between *H. pylori* negative versus positive status is considered adequate (Rugge et al., 2011). A careful examination of multiple biopsy specimens (two antral and two corpus) has a higher probability of establishing the correct *H. pylori* status especially in cases of patchy colonization of the gastric mucosa (Dixon et al., 1996). However, when histology is performed, the aim should be not only to identify *H. pylori* infection, but mainly to detect cancer, *H. pylori*-associated gastritis, and precancerous lesions, and this is why standardized biopsy protocols are recommended (the most widely used is the Sydney System protocol), and mucosal samples are taken from the oxyntic, antral, and incisura angularis areas; whenever this approach is planned, a careful consideration should be made of the need to take additional specimens from any focal lesions (Rugge et al., 2011).

33.5.7 Culture

The culture of the bacterium in selective media identifies active infections with excellent specificity but less sensitivity than histology or rapid urease test.

It is the standardized method allowing the analysis of antibiotic sensitivities of the infecting strain. Maastricht IV/Florence Consensus Report recommends *H. pylori* culture and standard susceptibility testing to antimicrobial agents before considering clarithromycin-containing triple therapy in regions or population of high clarithromycin resistance (Cla^R). If endoscopy is performed, culture and antibiotics susceptibility testing should be considered in all regions before second-line treatment, while they should be performed in all cases when a second-line treatment fails. If culture is not possible, molecular tests can be used to detect *H. pylori* and clarithromycin and/or fluoroquinolone resistances (Malfertheiner et al., 2012a).

33.5.8 Molecular methods

The identification of *H. pylori* genes by means of PCR and its variant, real-time PCR, may be a suitable approach for diagnosing and assessing clarithromycin resistance. For diagnostic purposes, DNA sequences shared by all *H. pylori* strains should be PCR amplified; these include *ureA* and 16S. The most suitable starting templates are bacterial colonies and gastric biopsies. Saliva, feces, and other body fluids might also be used, but in these cases *H. pylori* detection often fails due to a low DNA content and DNA degradation phenomena. Molecular testing can be used to detect clarithromycin and/or fluoroquinolone resistance in gastric biopsies if culture and standard susceptibility testing is not feasible (Malfertheiner et al., 2012b).

Clarithromycin resistance in *H. pylori* patients is due to point mutations in the gene encoding the 23S rRNA; three major mutations have been described: A2146C, A2146G, and A2147G. The resistance of *H. pylori* to quinolones is due to point mutations in the so-called quinolone resistance-determining region of the *gyrA* gene coding for the A subunit of the DNA gyrase, mainly at codons 87 and 91 (Cambau et al., 2009; Oleastro et al., 2003; Zambon et al., 2007).

33.6 *Helicobacter pylori* eradication therapy

The triple approach, comprising PPI-clarithromycin and amoxicillin or metronidazole, and proposed in 1997 at the first Maastricht conference for the treatment of *H. pylori* infection, has become universal, being recommended by all the consensus conferences held worldwide. However, the most recent available data show that this combination has lost some efficacy, often curing a maximum of 70% of patients, which is less than the original target of 80%, and far below the percentage that should be expected for patients with an infectious disease. Of the several explanations given for the decrease in efficacy of the standard triple therapy, including poor compliance, high gastric acidity, high bacterial load, and the type of strains, the most relevant is the increase resistance of *H. pylori* to

Table 33-4 Treatment options for *H. pylori* eradication

Regimen	Duration (days)	Drugs (daily dosage)	Eradication rate	Comments
Triple therapy with amoxicillin	7–14	• PPI (standard dose, bid) • Amoxicillin (1 g, bid) • Clarithromycin (500 mg, bid)	90% to 70–80% Steadily decline in treatment efficacy	First-line therapy in areas with a low rate of resistance to clarithromycin High-dose PPIs (twice a day) or lengthening the duration of treatment from 7 to 10–14 days improve eradication of 6–12% and 5%, respectively
Triple therapy with metronidazole	7–14	• PPI (standard dose, bid) • Clarithromycin (500 mg, bid) • Metronidazole (400 mg, bid)	90% to 70–80% Steadily decline in treatment efficacy	First-line therapy in areas with a low rate of resistance to clarithromycin To be preferred in patients with penicillin allergy
Sequential therapy	10	• PPI (standard dose, bid) • Amoxicillin (1 g, bid) for 5 days; then for another 5 days: • PPI (standard dose, bid) • Clarithromycin (500 mg, bid) • Metronidazole (500 mg, bid)	83–86%	First-line therapy in areas with a high rate of resistance to clarithromycin if Bismuth-based quadruple therapy is not available
Concomitant therapy	7–10	• PPI (standard dose, bid) • Amoxicillin (1 g, bid) • Clarithromycin (500 mg, bid) • Metronidazole (500 mg, bid)	75–87%	First-line therapy in areas with a high rate of resistance to clarithromycin if Bismuth-based quadruple therapy is not available. More simple regimen than the sequential therapy
Bismuth-based quadruple therapy	10–14	• PPI (standard dose, bid) • Bismuth (120 mg, qid) • Metronidazole (250 mg, qid) • Tetracycline (500 mg, qid)	79–90%	First-line therapy in areas with a high rate of resistance to clarithromycin. This empirical approach is based on the observations that no resistance to bismuth salts has been described, resistance to tetracycline is rarely found, and resistance to metronidazole, common *in vitro*, can be overcome by increasing the duration of treatment
Bismuth-containing rescue therapy	10–14	• PPI (standard dose, bid) • Bismuth (120 mg, qid) • Metronidazole (500 mg, tid) • Tetracycline (500 mg, qid)	76%	Second-line therapy. Well tolerated, good alternative for patients who failed with a first-line therapy Clarithromycin is abandoned based on the assumption that the first-line treatment might have selected clarithromycin-resistant strains
Hybrid therapy	14	• PPI (standard dose, bid) • Amoxicillin (1 g, bid) for 7 days; then for another 7 days: • PPI (standard dose bid) • Amoxicillin (1 g, bid) • Clarithromycin (500 mg, bid) • Metronidazole (500 mg, bid)	82–90%	Second-line therapy
Levofloxacin-based triple therapy	10	• PPI (standard dose, bid) • Amoxicillin (1 g, bid) • Levofloxacin (500 mg, qid)	81–87%	Second-line therapy. Good alternative for patients who failed with standard treatment. Lower incidence in side effects and better eradication rate than the levofloxacin-based quadruple therapy Levofloxacin should be used with care, because the rapid acquisition of resistance may compromise future efficacy, and in patients who had previously received fluoroquinolones, resistance might have been already developed
Levofloxacin-based quadruple therapy	10	• PPI (standard dose, bid) • Amoxicillin (500 mg, qid) • Bismuth (120 mg, qid) • Levofloxacin (500 mg, qid)	84%	Second-line therapy

(continued)

Table 33-4 (*Continued*)

Regimen	Duration (days)	Drugs (daily dosage)	Eradication rate	Comments
Quadruple cultured-guided therapy	10	• PPI (standard dose, bid) • Bismuth (120 mg, qid) • Two antibiotics selected by antimicrobial sensitivity tests	90%	Third-line therapy
Rifabutin-based triple therapy	14	• PPI (standard dose, bid) • Amoxicillin (1 g, bid) • Rifabutin (150 mg, bid)	79%	Third-line therapy. Serious myelotoxicity and ocular adverse events have been reported with rifabutin therapy. Risk of development of more resistant strains to *Mycobacterium tuberculosis* and *Mycobacterium avium*
Furazolidone-based quadruple therapy	7	• PPI (standard dose, bid) • Furazolidone (200 mg, bid) • Tetracycline (1 g, bid) • Tripotassium dicitratobismuthate (240 mg, bid)	90%	Third-line therapy

PPI, proton pump inhibitor; bid, twice a day; tid, three times a day; qid, four times a day.

Standard dosages for PPIs are: lansoprazole 30 mg, omeprazole 20 mg, pantoprazole 40 mg, rabeprazole 20 mg, esomeprazole 40 mg.

PPIs should be administered 20–30 minutes before meals.

Antibiotics should be administered during meals.

clarithromycin, which has now reached a high prevalence (>20%) in most countries in Central, Western, and Southern Europe, while remaining low (<10%) in Northern European countries.

While no new drug has been developed for this condition, a number of studies have been carried out in recent years using different combinations of known antibiotics. First-, second-, and third-line treatment options, which take into consideration the prevalence of clarithromycin resistance, are shown in Table 33-4 (Chuah et al., 2011; Gatta et al., 2009, 2013; Luther et al., 2010; Malfertheiner et al., 2012b).

References

Agréus, L., Kuipers, E.J., Kupcinskas, L., Malfertheiner, P., Di Mario, F., Leja, M., Mahachai, V., Yaron, N., van Oijen, M., Perez Perez, G., Rugge, M., Ronkainen, J., Salaspuro, M., Sipponen, P., Sugano, K., and Sung, J. 2012. Rationale in diagnosis and screening of atrophic gastritis with stomach-specific plasma biomarkers. *Scand. J. Gastroenterol.*, 47, 136–147.

Ahmed, N., Loke, M.F., Kumar, N., and Vadivelu, J., 2013. *Helicobacter pylori* in 2013: Multiplying genomes, emerging insights. *Helicobacter*, 18(suppl. 1), 1–4.

Algood, H.M., Gallo-Romero, J., Wilson, K.T., Peek Jr., R.M., and Cover, T.L. 2007. Host response to *Helicobacter pylori* infection before initiation of the adaptive immune response. *FEMS Immunol. Med. Microbiol.*, 51, 577–586.

Alm, R.A., Ling, L.S., Moir, D.T., King, B.L., Brown, E.D., Doig, P.C., Smith, D.R., Noonan, B., Guild, B.C., deJonge, B.L., Carmel, G., Tummino, P.J., Caruso, A., Uria-Nickelsen, M., Mills, D.M., Ives, C., Gibson, R., Merberg, D., Mills, S.D., Jiang, Q., Taylor, D.E., Vovis, G.F., and Trust, T.J. 1999. Genomic-sequence comparison of two unrelated isolates of the human gastric pathogen *Helicobacter pylori*. *Nature*, 397, 176–180.

Argent, R.H., Hale, J.L., El-Omar, E.M., and Atherton, J.C. 2008. Differences in *Helicobacter pylori* CagA tyrosine phosphorylation motif patterns between western and East Asian strains, and influences on interleukin-8 secretion. *J. Med. Microbiol.*, 57, 1062–1067.

Arnold, D.M., Bernotas, A., Nazi, I., Stasi, R., Kuwana, M., Liu, Y., Kelton, J.G., and Crowther, M.A. 2009. Platelet count response to *H. pylori* treatment in patients with immune thrombocytopenic purpura with and without *H. pylori* infection: A systematic review. *Haematologica*, 94, 850–856.

Asahi, A., Nishimoto, T., Okazaki, Y., Suzuki, H., Masaoka, T., Kawakami, Y., Ikeda, Y., and Kuwana, M. 2008. *Helicobacter pylori* eradication shifts monocyte Fcgamma receptor balance toward inhibitory FcgammaRIIB in immune thrombocytopenic purpura patients. *J. Clin. Invest.*, 118, 2939–2949.

Atarashi, K., Tanoue, T., Shima, T., Imaoka, A., Kuwahara, T., Momose, Y., Cheng, G., Yamasaki, S., Saito, T., Ohba, Y., Taniguchi, T., Takeda, K., Hori, S., Ivanov, I.I., Umesaki, Y., Itoh, K., and Honda, K., 2011. Induction of colonic regulatory T cells by indigenous *Clostridium* species. *Science*, 331, 337–341.

Atherton, J.C., Peek, R.M., Tham, K.T., Cover, T.L., and Blaser, M.J. 1997. Clinical and pathological importance of heterogeneity in vacA, the vacuolating cytotoxin gene of *Helicobacter pylori*. *Gastroenterology*, 112, 92–99.

Banić, M., Franceschi, F., Babić, Z., and Gasbarrini, A. 2012. Extragastric manifestations of *Helicobacter pylori* infection. *Helicobacter*, 17(suppl. 1), 49–55.

Basso, D., Navaglia, F., Brigato, L., Piva, M.G., Toma, A., Greco, E., Di Mario, F., Galeotti, F., Roveroni, G., Corsini, A., and Plebani, M. 1998. Analysis of *Helicobacter pylori* vacA and cagA genotypes and serum antibody profile in benign and malignant gastroduodenal diseases. *Gut*, 43, 182–186.

Basso, D., Plebani, M., and Kusters, J.G. 2010. Pathogenesis of *Helicobacter pylori* infection. *Helicobacter*, 15(suppl. 1), 14–20.

Basso, D., Zambon, C-F., Letley, D.P., Stranges, A., Marchet, A., Rhead, J.L., Schiavon, S., Guariso, G., Ceroti, M., Nitti, D., Rugge, M., Plebani, M., and Atherton, J.C. 2008. Clinical relevance of *Helicobacter pylori* cagA and vacA gene polymorphisms. *Gastroenterology*, 135, 91–99.

Bauer, J., Namineni, S., Reisinger, F., Zöller, J., Yuan, D., and Heikenwälder, M. 2012. Lymphotoxin, NF-κB, and cancer: The dark side of cytokines. *Dig. Dis.*, 30, 453–468.

Beales, I.L., and Calam, J. 1998. Interleukin 1 beta and tumour necrosis factor alpha inhibit acid secretion in cultured rabbit parietal cells by multiple pathways. *Gut*, 42, 227–234.

Beswick, E.J., Pinchuk, I.V., Earley, R.B., Schmitt, D.A., and Reyes, V.E. 2011. Role of gastric epithelial cell-derived transforming growth factor beta in reduced CD4+ T cell proliferation and development of regulatory T cells during *Helicobacter pylori* infection. *Infect. Immun.*, 79, 2737–2745.

Bik, E.M., Eckburg, P.B., Gill, S.R., Nelson, K.E., Purdom, E.A., Francois, F., Perez-Perez, G., Blaser, M.J., and Relman, D.A. 2006. Molecular analysis of the bacterial microbiota in the human stomach. *Proc. Natl. Acad. Sci. USA*, 103, 732–737.

Blaser, M.J., Chyou, P.H., and Nomura, A., 1995. Age at establishment of *Helicobacter pylori* infection and gastric adenocarcinoma, gastric ulcer, and duodenal ulcer risk. *Cancer Res.*, 55, 562–565.

Boquet, P., and Ricci, V. 2012. Intoxication strategy of *Helicobacter pylori* VacA toxin. *Trends Microbiol.*, 20, 165–174.

Brenner, H., Arndt, V., Bode, G., Stegmaier, C., Ziegler, H., and Stümer, T. 2002. Risk of gastric cancer among smokers infected with *Helicobacter pylori*. *Int. J. Cancer*, 98, 446–449.

Broutet, N., Plebani, M., Sakarovitch, C., Sipponen, P., Mégraud, F., and Eurohepygast Study Group. 2003. Pepsinogen A, pepsinogen C, and gastrin as markers of atrophic chronic gastritis in European dyspeptics. *Br. J. Cancer*, 88, 1239–1247.

Bytzer, P., Dahlerup, J.F., Eriksen, J.R., Jarbøl, D.E., Rosenstock, S., and Wildt, S., Danish Society for Gastroenterology. 2011. Diagnosis and treatment of *Helicobacter pylori* infection. *Dan. Med. Bull.*, 58, C4271.

Calam, J., Gibbons, A., Healey, Z.V., Bliss, P., and Arebi, N. 1997. How does *Helicobacter pylori* cause mucosal damage? Its effect on acid and gastrin physiology. *Gastroenterology*, 113(suppl. 6), S43–S49.

Calvet, X., Lehours, P., Lario, S., and Mégraud, F. 2010. Diagnosis of *Helicobacter pylori* infection. *Helicobacter*, 15(suppl. 1), 7–13.

Camargo, M.C., Mera, R., Correa, P., Peek Jr., R.M., Fontham, E.T., Goodman, K.J., Piazuelo, M.B., Sicinschi, L., Zabaleta, J., and Schneider, B.G. 2006. Interleukin-1beta and interleukin-1 receptor antagonist gene polymorphisms and gastric cancer: A meta-analysis. *Cancer Epidemiol. Biomarkers Prev.*, 15, 1674–1687.

Cambau, E., Allerheiligen, V., Coulon, C., Corbel, C., Lascols, C., Deforges, L., Soussy, C.J., Delchier, J.C., and Megraud, F. 2009. Evaluation of a new test, genotype HelicoDR, for molecular detection of antibiotic resistance in *Helicobacter pylori*. *J. Clin. Microbiol.*, 47, 3600–3607.

Canedo, P., Corso, G., Pereira, F., Lunet, N., Suriano, G., Figueiredo, C., Pedrazzani, C., Moreira, H., Barros, H., Carneiro, F., Seruca, R., Roviello, F., and Machado, J.C. 2008. The interferon gamma receptor 1 (IFNGR1) −56C/T gene polymorphism is associated with increased risk of early gastric carcinoma. *Gut*, 57, 1504–1508.

Celli, J.P., Turner, B.S., Afdhal, N.H., Keates, S., Ghiran, I., Kelly, C.P., Ewoldt, R.H., McKinley, G.H., So, P., Erramilli, S., and Bansil, R. 2009. *Helicobacter pylori* moves through mucus by reducing mucin viscoelasticity. *Proc. Natl. Acad. Sci. USA*, 106, 14321–14326.

Censini, S., Lange, C., Xiang, Z., Crabtree, J.E., Ghiara, P., Borodovsky, M., Rappuoli, R., and Covacci, A. 1996. cag, a Pathogenicity Island of *Helicobacter pylori*, encodes type I-specific and disease-associated virulence factors. *Proc. Natl. Acad. Sci. USA*, 93, 14648–14653.

Chakravorty, M., Datta, De D., Choudhury, A., and Roychoudhury, S. 2009. IL1B promoter polymorphism regulates the expression of gastric acid stimulating hormone gastrin. *Int. J. Biochem. Cell. Biol.*, 41, 1502–1510.

Chen, H., Wilkins, L.M., Aziz, N., Cannings, C., Wyllie, D.H., Bingle, C., Rogus, J., Beck, J.D., Offenbacher, S., Cork, M.J., Rafie-Kolpin, M., Hsieh, C.M., Kornman, K.S., and Duff, G.W. 2006. Single nucleotide polymorphisms in the human interleukin-1B gene affect transcription according to haplotype context. *Hum. Mol. Genet.*, 15, 519–529.

Chiba, T., Marusawa, H., Seno, H., and Watanabe, N. 2008. Mechanism for gastric cancer development by *Helicobacter pylori* infection. *J. Gastroenterol. Hepatol.*, 23, 1175–1181.

Chiba, T., Marusawa, H., and Ushijima, T. 2012. Inflammation-associated cancer development in digestive organs: Mechanisms and roles for genetic and epigenetic modulation. *Gastroenterology*, 143, 550–563.

Chuah, S.K., Tsay, F.W., Hsu, P.I., and Wu, D.C. 2011. A new look at anti-*Helicobacter pylori* therapy. *World J. Gastroenterol.*, 17, 3971–3975.

Correa, P., Haenszel, W., Cuello, C., Tannenbaum, S., and Archer, M. 1975. A model for gastric cancer epidemiology. *Lancet*, 2, 58–60.

Cover, T.L., and Blanke, S.R. 2005. *Helicobacter pylori* VacA, a paradigm for toxin multifunctionality. *Nat. Rev. Microbiol.*, 3, 320–332.

Cover, T.L., and Peek Jr., R.M. 2013. Diet, microbial virulence and *Helicobacter pylori*-induced gastric cancer. *Gut. Microbes*, 3, 4–6.

Cox, J.M., Clayton, C.L., Tomita, T., Wallace, D.M., Robinson, P.A., and Crabtree, J.E. 2001. cDNA array analysis of cag Pathogenicity Island-associated *Helicobacter pylori* epithelial cell response genes. *Infect. Immun.*, 69, 6970–6980.

Datta, De D., Bhattacharjya, S., Maitra, M., Datta, A., Choudhury, A., Dhali, G.K., and Roychoudhury, S. 2011. IL1B induced Smad 7 negatively regulates gastrin expression. *PLoS. One*, 6, e14775.

de Martel, C., and Parsonnet, J. 2006. *Helicobacter pylori* infection and gender: A meta-analysis of population-based prevalence surveys. *Dig. Dis. Sci.*, 51, 2292–2301.

de Martel, C., Ferlay, J., Franceschi, S., Vignat, J., Bray, F., Forman, D., and Plummer, M., 2012. Global burden of cancers attributable to infections in 2008: A review and synthetic analysis. *Lancet Oncol.*, 13, 607–615.

Dinarello, C.A. 2006. The paradox of pro-inflammatory cytokines in cancer. *Cancer Metastasis Rev.*, 25, 307–313.

Dixon, M.F., Genta, R.M., Yardley, J.H., and Correa, P. 1996. Classification and grading of gastritis. The updated Sydney System. International workshop on the histopathology of gastritis, Houston 1994. *Am. J. Surg. Pathol.*, 20, 1161–1181.

Dong, Q.J., Zhan, S.H., Wang, L.L., Xin, Y.N., Jiang, M., and Xuan, S.Y. 2012. Relatedness of *Helicobacter pylori* populations to gastric carcinogenesis. *World J. Gastroenterol.*, 18, 6571–6576.

Drossman, D.A. 2006. The functional gastrointestinal disorders and the Rome III process. *Gastroenterology*, 130, 1377–1390.

Dunn, B.E., and Phadnis, S.H. 1998. Structure, function and localization of *Helicobacter pylori* urease. *Yale J. Biol. Med.*, 71, 63–73.

Eftang, L.L., Esbensen, Y., Tannæs, T.M., Bukholm, I.R., and Bukholm, G. 2012. Interleukin-8 is the single most up-regulated gene in whole genome profiling of H. pylori exposed gastric epithelial cells. *BMC Microbiol.*, 12, 9.

El-Omar, E.M., Rabkin, C.S., Gammon, M.D., Vaughan, T.L., Risch, H.A., Schoenberg, J.B., Stanford, J.L., Mayne, S.T., Goedert, J., Blot, W.J., Fraumeni Jr., J.F., and Chow, W.H. 2003. Increased risk of noncardia gastric cancer associated with proinflammatory cytokine gene polymorphisms. *Gastroenterology*, 124, 1193–1201.

Farinha, P., and Gascoyne, R.D. 2005. *Helicobacter pylori* and MALT lymphoma. *Gastroenterology*, 128, 1579–1605.

Fehlings, M., Drobbe, L., Moos, V., Renner Viveros, P., Hagen, J., Beigier-Bompadre, M., Pang, E., Belogolova, E., Churin, Y., Schneider, T., Meyer, T.F., Aebischer, T., and Ignatius, R. 2012. Comparative analysis of the interaction of *Helicobacter pylori* with human dendritic cells, macrophages, and monocytes. *Infect. Immun.*, 80, 2724–2734.

Fock, K.M., Talley, N., Moayyedi, P., Hunt, R., Azuma, T., Sugano, K., Xiao, S.D., Lam, S.K., Goh, K.L., Chiba, T., Uemura, N., Kim, J.G., Kim, N., Ang, T.L., Mahachai, V., Mitchell, H., Rani, A.A., Liou, J.M., Vilaichone, R.K., Sollano, J., and Asia-Pacific Gastric Cancer Consensus Conference. 2008. Asia-Pacific consensus guidelines on gastric cancer prevention. *J. Gastroenterol. Hepatol.*, 23, 351–365.

Fock, K.M., Katelaris, P., Sugano, K., Ang, T.L., Hunt, R., Talley, N.J., Lam, S.K., Xiao, S.D., Tan, H.J., Wu, C.Y., Jung, H.C., Hoang, B.H., Kachintorn, U., Goh, K.L., Chiba, T., Rani, A.A.,and Second Asia-Pacific Conference. 2009. Second Asia-Pacific consensus guidelines for *Helicobacter pylori* infection. *J. Gastroenterol. Hepatol.*, 24, 1587–1600.

Ford, A.C., Delaney, B.C., Forman, D., and Moayyedi, P. 2006. Eradication therapy for peptic ulcer disease in *Helicobacter pylori* positive patients. *Cochrane Database Syst. Rev.*, 2, CD003840.

Fox, J.G., and Wang, T.C. 2007. Inflammation, atrophy, and gastric cancer. *J. Clin. Invest.*, 117, 60–69.

García-González, M.A., Lanas, A., Wu, J., Benito, R., Santolaria, S., Crusius, B., and Peña, S. 2005. Lack of association of IL-12 p40 gene polymorphism with peptic ulcer disease. *Hum. Immunol.*, 6, 72–76.

Gatta, L., Vakil, N., Leandro, G., Di Mario, F., and Vaira, D. 2009. Sequential therapy or triple therapy for *Helicobacter pylori* infection: Systematic review and meta-analysis of randomized controlled trials in adults and children. *Am. J. Gastroenterol.*, 104, 3069–3079.

Gatta, L., Vakil, N., Vaira, D., and Scarpignato, C. 2013. Global eradication rates for *Helicobacter pylori* infection: Systematic review and meta-analysis of sequential therapy. *Brit. Med J.*, 347, f4587.

Gebert, B., Fischer, W., Weiss, E., Hoffmann, R., and Haas, R. 2003. *Helicobacter pylori* vacuolatingcytotoxin inhibits T lymphocyte activation. *Science*, 301, 1099–1102.

George, J.N. 2009. Definition, diagnosis and treatment of immune thrombocytopenic purpura. *Haematologica*, 94, 759–762.

Gisbert, J.P., and Pajares, J.M. 2004. Review article: 13C-urea breath test in the diagnosis of *Helicobacter pylori* infection – a critical review. *Aliment. Pharmacol. Ther.*, 20, 1001–1017.

Gisbert, J.P., de la Morena, F., and Abraira, V. 2006. Accuracy of monoclonal stool antigen test for the diagnosis of *H. pylori* infection: A systematic review and meta-analysis. *Am. J. Gastroenterol.*, 101, 1921–1930.

Gisbert, J.P., Khorrami, S., Carballo, F., Calvet, X., Gené, E., and Dominguez-Munoz, E. 2004. *Helicobacter pylori* eradication therapy vs. antisecretory non-eradication therapy (with or without long-term maintenance antisecretory therapy) for the prevention of recurrent bleeding from peptic ulcer (Review). *Cochrane Database Syst. Rev.*, CD004062.

Goh, K.L., Chan, W.K., Shiota, S., and Yamaoka, Y. 2011. Epidemiology of *Helicobacter pylori* infection and public health implications. *Helicobacter*, 16(suppl. 1), 1–9.

Gringhuis, S.I., den Dunnen, J., Litjens, M., van der Vlist, M., and Geijtenbeek, T.B., 2009. Carbohydrate-specific signaling through the DC-SIGN signalosome tailors immunity to *Mycobacterium tuberculosis*, HIV-1 and *Helicobacter pylori*. *Nat. Immunol.*, 10, 1081–1088.

Guariso, G., Basso, D., Bortoluzzi, C.F., Meneghel, A., Schiavon, S., Fogar, P., Farina, M., Navaglia, F., Greco, E., Mescoli, C., Zambon, C.-F., and Plebani, M., 2009. GastroPanel: Evaluation of the usefulness in the diagnosis of gastro-duodenal mucosal alterations in children. *Clin. Chim. Acta*, 402, 54–60.

Handa, O., Naito, Y., and Yoshikawa, T. 2010. *Helicobacter pylori*:

A ROS-inducing bacterial species in the stomach. *Inflamm. Res.*, 59, 997–1003.

Hatakeyama, M. 2004. Oncogenic mechanisms of the *Helicobacter pylori* CagA protein. *Nat. Rev. Cancer*, 4, 688–694.

Hitzler, I., Oertli, M., Becher, B., Agger, E.M., and Müller, A., 2011. Dendritic cells prevent rather than promote immunity conferred by a helicobacter vaccine using a mycobacterial adjuvant. *Gastroenterology*, 141, 186–196.

Hou, L., El-Omar, E.M., Chen, J., Grillo, P., Rabkin, C.S., Baccarelli, A., Yeager, M., Chanock, S.J., Zatonski, W., Sobin, L.H., Lissowska, J., Fraumeni Jr., J.F., and Chow, W.H. 2007. Polymorphisms in Th1-type cell-mediated response genes and risk of gastric cancer. *Carcinogenesis*, 28, 118–123.

Hunt, R.H., Xiao, S.D., Megraud, F., Leon-Barua, R., Bazzoli, F., van der Merwe, S., Vaz Coelho, L.G., Fock, M., Fedail, S., Cohen, H., Malfertheiner, P., Vakil, N., Hamid, S., Goh, K.L., Wong, B.C., Krabshuis, J., Le Mair, A., and World Gastroenterology Organization. 2011. *Helicobacter pylori* in developing countries. World Gastroenterology Organization global guideline. *J. Gastrointestin. Liver Dis.*, 20, 299–304.

Hwang, I.R., Kodama, T., Kikuchi, S., Sakai, K., Peterson, L.E., Graham, D.Y., and Yamaoka, Y. 2002. Effect of interleukin 1 polymorphisms on gastric mucosal interleukin 1beta production in *Helicobacter pylori* infection. *Gastroenterology*, 123, 1793–1803.

IARC Working Group on the Evaluation of Carcinogenic Risks to Humans. 1994. Schistosomes, liver flukes and *Helicobacter pylori*. *IARC Monogr. Eval. Carcinog. Risks Hum.*, 61, 1–241.

IARC Working Group on the Evaluation of Carcinogenic Risks to Humans. 2012. A review of human carcinogens: Biological agents, *Helicobacter pylori*. *IARC Monogr. Eval. Carcinog. Risks Hum.*, 100B, 385–435.

Iijima, K., Abe, Y., Kikuchi, R., Koike, T., Ohara, S., Sipponen, P., and Shimosegawa, T. 2009. Serum biomarker tests are useful in delineating between patients with gastric atrophy and normal, healthy stomach. *World J. Gastroenterol.*, 15, 853–859.

Israel, D.A., and Peek, R.M. 2001. Review article: Pathogenesis of *Helicobacter pylori*-induced gastric inflammation. *Aliment. Pharmacol. Ther.*, 15, 1271–1290.

Ivanov, I.I., Atarashi, K., Manel, N., Brodie, E.L., Shima, T., Karaoz, U., Wei, D., Goldfarb, K.C., Santee, C.A., Lynch, S.V., Tanoue, T., Imaoka, A., Itoh, K., Takeda, K., Umesaki, Y., Honda, K., and Littman, D.R. 2009. Induction of intestinal Th17 cells by segmented filamentous bacteria. *Cell*, 139, 485–498.

Jang, T.J. 2010. The number of Foxp3-positive regulatory T cells is increased in *Helicobacter pylori* gastritis and gastric cancer. *Pathol. Res. Pract.*, 206, 34–38.

Jenab, M., Riboli, E., Ferrari, P., Friesen, M., Sabate, J., Norat, T., Slimani, N., Tjønneland, A., Olsen, A., Overvad, K., Boutron-Ruault, M.C., Clavel-Chapelon, F., Boeing, H., Schulz, M., Linseisen, J., Nagel, G., Trichopoulou, A., Naska, A., Oikonomou, E., Berrino, F., Panico, S., Palli, D., Sacerdote, C., Tumino, R., Peeters, P.H., Numans, M.E., Bueno-de-Mesquita, H.B., Büchner, F.L., Lund, E., Pera, G., Chirlaque, M.D., Sánchez, M.J., Arriola, L., Barricarte, A., Quirós, J.R., Johansson, I., Johansson, A., Berglund, G., Bingham, S., Khaw, K.T., Allen, N., Key, T., Carneiro, F., Save, V., Del Giudice, G., Plebani, M., Kaaks, R., and Gonzalez, C.A. 2006. Plasma and dietary carotenoid, retinol and tocopherol levels and the risk of gastric adenocarcinomas in the European prospective investigation into cancer and nutrition. *Br. J. Cancer*, 95, 406–415.

Kao, J.Y., Zhang, M., Miller, M.J., Mills, J.C., Wang, B., Liu, M., Eaton, K.A., Zou, W., Berndt, B.E., Cole, T.S., Takeuchi, T., Owyang, S.Y., and Luther, J. 2010. *Helicobacter pylori* immune escapeis mediated by dendritic cell-induced Treg skewing and Th17 suppression in mice. *Gastroenterology*, 138, 1046–1054.

Kato, I., Canzian, F., Franceschi, S., Plummer, M., van Doorn, L.J., Lu, Y., Gioia-Patricola, L., Vivas, J., Lopez, G., Severson, R.K.,

Schwartz, A.G., and Muñoz, N. 2006. Genetic polymorphisms in anti-inflammatory cytokine signaling and the prevalence of gastric precancerous lesions in Venezuela. *Cancer Causes Control*, 17, 1183–1191.

Kim, I.J., and Blanke, S.R., 2012. Remodeling the host environment: Modulation of the gastric epithelium by the *Helicobacter pylori* vacuolating toxin (VacA). *Front. Cell. Infect. Microbiol.*, 2, 37.

Kim, J.M., Kim, J.S., Yoo, D.Y., Ko, S.H., Kim, N., Kim, H., and Kim, Y.J. 2011. Stimulation of dendritic cells with *Helicobacter pylori* vacuolating cytotoxin negatively regulates their maturation via the restoration of E2F1. *Clin. Exp. Immunol.*, 166, 34–45.

Koletzko, S., Jones, N.L., Goodman, K.J., Gold, B., Rowland, M., Cadranel, S., Chong, S., Colletti, R.B., Casswall, T., Elitsur, Y., Guarner, J., Kalach, N., Madrazo, A., Megraud, F., Oderda, G., and *H pylori* Working Groups of ESPGHAN and NASPGHAN. 2011. Evidence-based guidelines from ESPGHAN and NASPGHAN for *Helicobacter pylori* infection in children. *J. Pediatr. Gastroenterol. Nutr.*, 53, 230–243.

Ladeiras-Lopes, R., Pereira, A.K., Nogueira, A., Pinheiro-Torres, T., Pinto, I., Santos-Pereira, R., and Lunet, N. 2008. Smoking and gastric cancer: Systematic review and meta-analysis of cohort studies. *Cancer Causes Control*, 19, 689–701.

Lathrop, S.K., Bloom, S.M., Rao, S.M., Nutsch, K., Lio, C.W., Santacruz, N., Peterson, D.A., Stappenbeck, T.S., and Hsieh, C.S. 2011. Peripheral education of the immune system by colonic-commensal microbiota. *Nature*, 478, 250–254.

Leal, Y.A., Flores, L.L., García-Cortés, L.B., Cedillo-Rivera, R., and Torres, J. 2008. Antibody-based detection tests for the diagnosis of *Helicobacter pylori* infection in children: A meta-analysis. *PLoS One*, 3, e3751.

Linz, B., Balloux, F., Moodley, Y., Manica, A., Liu, H., Roumagnac, P., Falush, D., Stamer, C., Prugnolle, F., van der Merwe, S.W., Yamaoka, Y., Graham, D.Y., Perez-Trallero, E., Wadstrom, T., Suerbaum, S., and Achtman, M. 2007. An African origin for the intimate association between humans and *Helicobacter pylori*. *Nature*, 445, 915–918.

Lippitz, B.E. 2013. Cytokine patterns in patients with cancer: A systematic review. *Lancet Oncol.*, 14, e218–e228.

Littman, D.R., and Rudensky, A.Y. 2010. Th17 and regulatory T cells in mediating and restraining inflammation. *Cell*, 140, 845–858.

Lofgren, J.L., Whary, M.T., Ge, Z., Muthupalani, S., Taylor, N.S., Mobley, M., Potter, A., Varro, A., Eibach, D., Suerbaum, S., Wang, T.C., and Fox, J.G. 2011. Lack of commensal flora in *Helicobacter pylori*-infected INS-GAS mice reduces gastritis and delays intraepithelial neoplasia. *Gastroenterology*, 140, 210–220.

Lundgren, A., Strömberg, E., Sjöling, A., Lindholm, C., Enarsson, K., Edebo, A., Johnsson, E., Suri-Payer, E., Larsson, P., Rudin, A., Svennerholm, A.M., and Lundin, B.S. 2005. Mucosal FOXP3-expressing CD4+ CD25 high regulatory T cells in *Helicobacter pylori*-infected patients. *Infect. Immun.*, 73, 523–531.

Luther, J., Higgins, P.D., Schoenfeld, P.S., Moayyedi, P., Vakil, N., and Chey, W.D. 2010. Empiric quadruple vs. triple therapy for primary treatment of *Helicobacter pylori* infection: Systematic review and meta-analysis of efficacy and tolerability. *Am. J. Gastroenterol.*, 105, 65–73.

Maldonado-Contreras, A., Goldfarb, K.C., Godoy-Vitorino, F., Karaoz, U., Contreras, M., Blaser, M.J., Brodie, E.L., and Dominguez-Bello, M.G. 2011. Structure of the human gastric bacterial community in relation to *Helicobacter pylori* status. *ISME J.*, 5, 574–579.

Malfertheiner, P. 2011. The intriguing relationship of *Helicobacter pylori* infection and acid secretion in peptic ulcer disease and gastric cancer. *Dig. Dis.*, 29, 459–464.

Malfertheiner, P., Megraud, F., and O'Morain, C., 2005. The Maastricht 3 Consensus Report: Guidelines for the management of *Helicobacter pylori* infection. *Eur. Gastroenterol. Rev.*, 2005, 59–62.

Malfertheiner, P., Megraud, F., O'Morain, C., Bazzoli, F., El-Omar, E., Graham, D., Hunt, R., Rokkas, T., Vakil, N., and Kuipers, E.J., 2007. Current concepts in the management of *Helicobacter pylori* infection: The Maastricht III Consensus Report. *Gut*, 56, 772–781.

Malfertheiner, P., Chan, F.K., and McColl, K.E. 2009. Peptic ulcer disease. *Lancet*, 374, 1449–1461.

Malfertheiner, P., Selgrad, M., and Bornschein, J. 2012a. *Helicobacter pylori*: Clinical management. *Curr. Opin. Gastroenterol.*, 28, 608–614.

Malfertheiner, P., Megraud, F., O'Morain, C.A., Atherton, J., Axon, A.T., Bazzoli, F., Gensini, G.F., Gisbert, J.P., Graham, D.Y., Rokkas, T., El-Omar, E.M., Kuipers, E.J., and European Helicobacter Study Group. 2012b. Management of *Helicobacter pylori* infection – the Maastricht IV/Florence Consensus Report. *Gut*, 61, 646–664.

Marais, A., Mendz, G.L., Hazell, S.L., and Mégraud, F., 1999. Metabolism and genetics of *Helicobacter pylori*: The genome era. *Microbiol. Mol. Biol. Rev.*, 63, 642–674.

Marshall, B.J., and Warren, J.R., 1984. Unidentified curved bacilli in the stomach of patients with gastritis and peptic ulceration. *Lancet*, 1, 1311–1315.

Marshall, B.J., and Windsor, H.M. 2005. The relation of *Helicobacter pylori* to gastric adenocarcinoma and lymphoma: Pathophysiology, epidemiology, screening. clinical presentation, treatment, and prevention. *Med. Clin. North Am.*, 89, 313–344.

Marshall, B.J., Armstrong, J.A., McGechie, D.B., and Glancy, R.J. 1985. Attempt to fulfil Koch's postulates for pyloric Campylobacter. *Med. J. Aust.*, 142, 436–439.

Mayerle, J., den Hoed, C.M., Schurmann, C., Stolk, L., Homuth, G., Peters, M.J., Capelle, L.G., Zimmermann, K., Rivadeneira, F., Gruska, S., Völzke, H., de Vries, A.C., Völker, U., Teumer, A., van Meurs, J.B., Steinmetz, I., Nauck, M., Ernst, F., Weiss, F.U., Hofman, A., Zenker, M., Kroemer, H.K., Prokisch, H., Uitterlinden, A.G., Lerch, M.M., and Kuipers, E.J. 2013. Identification of genetic loci associated with *Helicobacter pylori* serologic status. *J. Am. Med. Assoc.*, 309, 1912–1920.

Mazzoleni, L.E., Sander, G.B., Francesconi, C.F., Mazzoleni, F., Uchoa, D.M., De Bona, L.R., Milbradt, T.C., Von Reisswitz, P.S., Berwanger, O., Bressel, M., Edelweiss, M.I., Marini, S.S., Molina, C.G., Folador, L., Lunkes, R.P., Heck, R., Birkhan, O.A., Spindler, B.M., Katz, N., Colombo Bda, S., Guerrieri, P.P., Renck, L.B., Grando, E., Hocevar de Moura, B., Dahmer, F.D., Rauber, J., and Prolla, J.C. 2011. *Helicobacter pylori* eradication in functional dyspepsia: HEROES trial. *Arch. Intern. Med.*, 171, 1929–1936.

Mitchell, J.D., Mitchell, H.M., and Tobias, V. 1992. Acute *Helicobacter pylori* infection in an infant, associated with gastric ulceration and serological evidence of intra-familial transmission. *Am. J. Gastroenterol.*, 87, 382–386.

Mitchell, P.J., Afzali, B., Fazekasova, H., Chen, D., Ali, N., Powell, N., Lord, G.M., Lechler, R.I., and Lombardi, G. 2012. *Helicobacter pylori* induces in-vivo expansion of human regulatory T cells through stimulating interleukin-1β production by dendritic cells. *Clin. Exp. Immunol.*, 170, 300–309.

Miwa, H., Ghoshal, U.C., Fock, K.M., Gonlachanvit, S., Gwee, K.A., Ang, T.L., Chang, F.Y., Hongo, M., Hou, X., Kachintorn, U., Ke, M., Lai, K.H., Lee, K.J., Lu, C.L., Mahadeva, S., Miura, S., Park, H., Rhee, P.L., Sugano, K., Vilaichone, R.K., Wong, B.C., and Bak, Y.T., 2012. Asian consensus report on functional dyspepsia. *J. Gastroenterol. Hepatol.*, 27, 626–641.

Mobley, H.L.T., Island, M.D., and Hausinger, R.P. 1995. Molecular biology of microbial ureases. *Microbiol. Rev.*, 59, 451–480.

Montecucco, C., and Rappuoli, R., 2001. Living dangerously: How *Helicobacter pylori* survives in the human stomach. *Nat. Rev. Mol. Cell. Biol.*, 2, 457–466.

Moodley, Y., Linz, B., Bond, R.P., Nieuwoudt, M., Soodyall, H., Schlebusch, C.M., Bernhöft, S., Hale, J., Suerbaum, S., Mugisha, L., van der Merwe, S.W., and Achtman, M. 2012.

Age of the association between *Helicobacter pylori* and man. *PLoS Pathog.*, 8, 1002693.

Morris, A., and Nicholson, G. 1987. Ingestion of *Campylobacter pyloridis* causes gastritis and raised fasting gastric pH. *Am. J. Gastroenterol.*, 82, 192–199.

Muhsen, K., and Cohen, D. 2008. *Helicobacter pylori* infection and iron stores: A systematic review and meta-analysis. *Helicobacter*, 13, 323–340.

Mukoubayashi, C., Yanaoka, K., Ohata, H., Arii, K., Tamai, H., Oka, M., and Ichinose, M. 2007. Serum pepsinogen and gastric cancer screening. *Intern. Med.*, 46, 261–266.

Nagini, S. 2012. Carcinoma of the stomach: A review of epidemiology, pathogenesis, molecular genetics and chemoprevention. *World J. Gastrointest Oncol.*, 4, 156–169.

Navaglia, F., Basso, D., Zambon, C-F., Ponzano, E., Caenazzo, L., Gallo, N., Falda, A., Belluco, C., Fogar, P., Greco, E., Di Mario, F., Rugge, M., and Plebani, M. 2005. Interleukin 12 gene polymorphisms enhance gastric cancer risk in *H. pylori* infected individuals. *J. Med. Genet.*, 42, 503–510.

Noto, J.M. and Peek Jr., R.M. 2012. The *Helicobacter pylori* cag pathogenicity island. *Methods Mol. Biol.*, 921, 41–50.

Ohyauchi, M., Imatani, A., Yonechi, M., Asano, N., Miura, A., Iijima, K., Koike, T., Sekine, H., Ohara, S., and Shimosegawa, T., 2005. The polymorphism interleukin 8–251 A/T influences the susceptibility of *Helicobacter pylori* related gastric diseases in the Japanese population. *Gut*, 54, 330–335.

Oleastro, M., Ménard, A., Santos, A., Lamouliatte, H., Monteiro, L., Barthélémy, P., and Mégraud, F. 2003. Real-time PCR assay for rapid and accurate detection of point mutations conferring resistance to clarithromycin in *Helicobacter pylori*. *J. Clin. Microbiol.*, 41, 397–402.

Otani, K., Watanabe, T., Tanigawa, T., Okazaki, H., Yamagami, H., Watanabe, K., Tominaga, K., Fujiwara, Y., Oshitani, N., and Arakawa, T. 2009. Anti-inflammatory effects of IL-17A on *Helicobacter pylori*-induced gastritis. *Biochem. Biophys. Res. Commun.*, 382, 252–258.

Palli, D., Masala, G., Del Giudice, G., Plebani, M., Basso, D., Berti, D., Numans, M.E., Ceroti, M., Peeters, P.H., Bueno de Mesquita, H.B., Buchner, F.L., Clavel-Chapelon, F., Boutron-Ruault, M.C., Krogh, V., Saieva, C., Vineis, P., Panico, S., Tumino, R., Nyrén, O., Simán, H., Berglund, G., Hallmans, G., Sanchez, M.J., Larrãnaga, N., Barricarte, A., Navarro, C., Quiros, J.R., Key, T., Allen, N., Bingham, S., Khaw, K.T., Boeing, H., Weikert, C., Linseisen, J., Nagel, G., Overvad, K., Thomsen, R.W., Tjonneland, A., Olsen, A., Trichoupoulou, A., Trichopoulos, D., Arvaniti, A., Pera, G., Kaaks, R., Jenab, M., Ferrari, P., Nesi, G., Carneiro, F., Riboli, E., and Gonzalez, C.A. 2007. CagA+ *Helicobacter pylori* infection and gastric cancer risk in the EPIC-EURGAST study. *Int. J. Cancer*, 120, 859–867.

Papini, E., Zoratti, M., and Cover, T.L. 2001. In search of the *Helicobacter pylori* VacA mechanism of action. *Toxicon*, 39, 1757–1767.

Peek Jr., R.M., Fiske, C., and Wilson, K.T. 2010. Role of innate immunity in *Helicobacter pylori*-induced gastric malignancy. *Physiol. Rev.*, 90, 831–858.

Peleteiro, B., Lunet, N., Carrilho, C., Durães, C., Machado, J.C., La Vecchia, C., and Barros, H. 2010. Association between cytokine gene polymorphisms and gastric precancerous lesions: Systematic review and meta-analysis. *Cancer Epidemiol. Biomarkers Prev.*, 19, 762–776.

Pellicano, R., Franceschi, F., Saracco, G., Fagoonee, S., Roccarina, D., and Gasbarrini, A. 2009. Helicobacters and extragastric diseases. *Helicobacter*, 14(suppl. 1), 58–68.

Persson, C., Canedo, P., Machado, J.C., El-Omar, E.M., and Forman, D. 2011. Polymorphisms in inflammatory response genes and their association with gastric cancer: A HuGE systematic review and meta-analyses. *Am. J. Epidemiol.*, 173, 259–270.

Pessi, T., Virta, M., Adjers, K., Karjalainen, J., Rautelin, H., Kosunen, T.U., and Hurme, M. 2005. Genetic and environmental factors in the immunopathogenesis of atopy: Interaction of *Helicobacter pylori* infection and IL4 genetics. *Int. Arch. Allergy. Immunol.*, 137, 282–288.

Piazuelo, M.B., Epplein, M., and Correa, P. 2010. Gastric cancer: An infectious disease. *Infect. Dis. Clin. North Am.*, 24, 853–869.

Pinchuk, I.V., Morris, K.T., Nofchissey, R.A., Earley, R.B., Wu, J.Y., Ma, T.Y., and Beswick, E.J. 2013. Stromal cells induce Th17 during *Helicobacter pylori* infection and in the gastric tumor microenvironment. *PLoS. One*, 8, e53798.

Qu, X.H., Huang, X.L., Xiong, P., Zhu, C.Y., Huang, Y.L., Lu, L.G., Sun, X., Rong, L., Zhong, L., Sun, D.Y., Lin, H., Cai, M.C., Chen, Z.W., Hu, B., Wu, L.M., Jiang, Y.B., and Yan, W.L. 2010. Does *Helicobacter pylori* infection play a role in iron deficiency anemia? A meta-analysis. *World J. Gastroenterol.*, 16, 886–896.

Queiroz, D.M., Saraiva, I.E., Rocha, G.A., Rocha, A.M., Gomes, L.I., Melo, F.F., and Bittencourt, P.F. 2009. IL2–330G polymorphic allele is associated with decreased risk of *Helicobacter pylori* infection in adulthood. *Microbes Infect.*, 11, 980–987.

Queiroz, D.M., Rocha, A.M., Melo, F.F., Rocha, G.A., Teixeira, K.N., Carvalho, S.D., Bittencourt, P.F., Castro, L.P., and Crabtree, J.E. 2013. Increased gastric IL-1β concentration and iron deficiency parameters in *H. pylori* infected children. *PLoS One*, 8, e57420.

Rautelin, H., Lehours, P., and Mégraud, F. 2003. Diagnosis of *Helicobacter pylori* infection. *Helicobacter*, 8(suppl. 1), 13–20.

Rolig, A.S., Carter, J.E., and Ottemann, K.M. 2011. Bacterial chemotaxis modulates host cell apoptosis to establish a T-helper cell, type 17 (Th17)-dominant immune response in *Helicobacter pylori* infection. *Proc. Natl. Acad. Sci. USA*, 108, 19749–19754.

Rolig, A.S., Cech, C., Ahler, E., Carter, J.E., and Ottemann, K.M., 2013. The degree of Helicobacter pylori-triggered inflammation is manipulated by preinfection host microbiota. *Infect. Immun.*, 81, 1382–1389.

Rosenberg, J.J. 2010. *Helicobacter pylori*. *Pediatr. Rev.*, 31, 85–86.

Roubaud Baudron, C., Franceschi, F., Salles, N., and Gasbarrini, A. 2013. Extragastric diseases and *Helicobacter pylori*. *Helicobacter*, 18(suppl. 1), 44–51.

Round, J.L., and Mazmanian, S.K. 2009. The gut microbiota shapes intestinal immune responses during health and disease. *Nat. Rev. Immunol.*, 9, 313–323.

Rugge, M., Genta, R.M., and OLGA Group. 2005. Staging gastritis: An international proposal. *Gastroenterology*, 129, 1807–1808.

Rugge, M., Correa, P., Di Mario, F., El-Omar, E., Fiocca, R., Geboes, K., Genta, R.M., Graham, D.Y., Hattori, T., Malfertheiner, P., Nakajima, S., Sipponen, P., Sung, J., Weinstein, W., and Vieth, M., 2008. OLGA staging for gastritis: A tutorial. *Dig. Liver Dis.*, 40, 650–658.

Rugge, M., Pennelli, G., Pilozzi, E., Fassan, M., Ingravallo, G., Russo, V.M., Di Mario, F., Gruppo Italiano Patologi Apparato Digerente (GIPAD), and Società Italiana di Anatomia Patologica e Citopatologia Diagnostica/International Academy of Pathology, Italian division (SIAPEC/IAP). 2011. Gastritis: The histology report. *Dig. Liver Dis.*, 43(suppl. 4), S373–S384.

Rugge, M., Capelle, L.G., Cappellesso, R., Nitti, D., and Kuipers, E.J., 2013. Precancerous lesions in the stomach: From biology to clinical patient management. *Best Pract. Res. Clin. Gastroenterol.*, 27, 205–223.

Sakaguchi, S., Miyara, M., Costantino, C.M., and Hafler, D.A. 2010. FOXP3+ regulatory T cells in the human immune system. *Nat. Rev. Immunol.*, 10, 490–500.

Salama, N.R., Hartung, M.L., and Müller, A. 2013. Life in the human stomach: Persistence strategies of the bacterial pathogen *Helicobacter pylori*. *Nat. Rev. Microbiol.*, 11, 385–399.

Scandellari, R., Allemand, E., Vettore, S., Plebani, M., Randi, M.L., and Fabris, F., 2009. Platelet response to *Helicobacter pylori* eradication therapy in adult chronic idiopathic thrombocytopenic purpura seems to be related to the presence of anticytotoxin-associated gene A antibodies. *Blood Coagul. Fibrinolysis*, 20, 108–113.

Selgrad, M., Bornschein, J., Rokkas, T., and Malfertheiner, P. 2012. *Helicobacter pylori*: Gastric cancer and extragastric intestinal malignancies. *Helicobacter*, 17(suppl. 1), 30–35.

Semple, J.W., Aslam, R., Kim, M., Speck, E.R., and Freedman, J. 2007. Platelet-bound lipopolysaccharide enhances Fc receptor-mediated phagocytosis of IgG-opsonized platelets. *Blood*, 109, 4803–4805.

Seno, H., Satoh, K., Tsuji, S., Shiratsuchi, T., Harada, Y., Hamajima, N., Sugano, K., Kawano, S., and Chiba, T. 2007. Novel interleukin-4 and interleukin-1 receptor antagonist gene variations associated with non-cardia gastric cancer in Japan: Comprehensive analysis of 207 polymorphisms of 11 cytokine genes. *J. Gastroenterol. Hepatol.*, 22, 729–737.

Serelli-Lee, V., Ling, K.L., Ho, C., Yeong, L.H., Lim, G.K., Ho, B., and Wong, S.B. 2012. Persistent *Helicobacter pylori* specific Th17 responses in patients with past *H. pylori* infection are associated with elevated gastric mucosal IL-1β. *PLoS One*, 7, e39199.

Sharma, C.M., Hoffmann, S., Darfeuille, F., Reignier, J., Findeiss, S., Sittka, A., Chabas, S., Reiche, K., Hackermüller, J., Reinhardt, R., Stadler, P.F., and Vogel, J. 2010. The primary transcriptome of the major human pathogen *Helicobacter pylori*. *Nature*, 464, 250–255.

Shikata, K., Ninomiya, T., Yonemoto, K., Ikeda, F., Hata, J., Doi, Y., Fukuhara, M., Matsumoto, T., Iida, M., Kitazono, T., and Kiyohara, Y. 2012. Optimal cutoff value of the serum pepsinogen level for prediction of gastric cancer incidence: The Hisayama Study. *Scand. J. Gastroenterol.*, 47, 669–675.

Shiotani, A., Cen, P., and Graham, D.Y. 2013. Eradication of gastric cancer is now both possible and practical. *Semin. Cancer Biol.*, 23, 492–501. DOI: 10.1016/j.semcancer.2013.07.004.

Shukla, S.K., Prasad, K.N., Tripathi, A., Jaiswal, V., Khatoon, J., Ghsohal, U.C., Krishnani, N., and Husain, N. 2013. *Helicobacter pylori* cagL amino acid polymorphisms and its association with gastroduodenal diseases. *Gastric. Cancer*, 16, 435–439.

Sobala, G.M., Crabtree, J.E., Dixon, M.F., Schorah, C.J., Taylor, J.D., Rathbone, B.J., Heatley, R.V., and Axon, A.T. 1991. Acute *Helicobacter pylori* infection: Clinical features, local and systemic immune response, gastric mucosal histology, and gastric juice ascorbic acid concentrations. *Gut*, 32, 1415–1418.

Soerjomataram, I., Lortet-Tieulent, J., Parkin, D.M., Ferlay, J., Mathers, C., Forman, D., and Bray, F. 2012. Global burden of cancer in 2008: A systematic analysis of disability-adjusted life-years in 12 world regions. *Lancet*, 380, 1840–1850.

Stasi, R., Sarpatwari, A., Segal, J.B., Osborn, J., Evangelista, M.L., Cooper, N., Provan, D., Newland, A., Amadori, S., and Bussel, J.B. 2009. Effects of eradication of *Helicobacter pylori* infection in patients with immune thrombocytopenic purpura: A systematic review. *Blood*, 113, 1231–1240.

Storskrubb, T., Aro, P., Ronkainen, J., Sipponen, P., Nyhlin, H., Talley, N.J., Engstrand, L., Stolte, M., Vieth, M., Walker, M., and Agréus, L. 2008. Serum biomarkers provide an accurate method for diagnosis of atrophic gastritis in a general population: The Kalixanda study. *Scand. J. Gastroenterol.*, 43, 1448–1455.

Suarez, G., Reyes, V.E., and Beswick, E.J. 2006. Immune response to *H. pylori*. *World J. Gastroenterol.*, 12, 5593–5598.

Suerbaum, S., and Josenhans, C. 2007. *Helicobacter pylori* evolution and phenotypic diversification in a changing host. *Nat. Rev. Microbiol.*, 5, 441–452.

Sugimoto, M., Ohno, T., Graham, D.Y., and Yamaoka, Y., 2009. Gastric mucosal interleukin-17 and -18 mRNA expression in *Helicobacter pylori*-induced Mongolian gerbils. *Cancer Sci.*, 100, 2152–2159.

Suzuki, H., Franceschi, F., Nishizawa, T., and Gasbarrini, A. 2011. Extragastric manifestations of *Helicobacter pylori* infection. *Helicobacter*, 16(suppl. 1), 65–69.

Tack, J., Talley, N.J., Camilleri, M., Holtmann, G., Hu, P., Malagelada, J.R., and Stanghellini, V. 2006. Functional gastroduodenal disorders. *Gastroenterology*, 130, 1466–1479.

Taguchi, A., Ohmiya, N., Shirai, K., Mabuchi, N., Itoh, A., Hirooka, Y., Niwa, Y., and Goto, H., 2005. Interleukin-8 promoter polymorphism increases the risk of atrophic gastritis and gastric cancer in Japan. *Cancer Epidemiol. Biomarkers Prev.*, 14, 2487–2493.

Tan, H.J., and Goh, K.L. 2012. Extragastrointestinal manifestations of *Helicobacter pylori* infection: Facts or myth? A critical review. *J. Dig. Dis.*, 13, 342–349.

Tanahashi, T., Kita, M., Kodama, T., Yamaoka, Y., Sawai, N., Ohno, T., Mitsufuji, S., Wei, Y.P., Kashima, K., and Imanishi, J. 2000. Cytokine expression and production by purified *Helicobacter pylori* urease in human gastric epithelial cells. *Infect. Immun.*, 68, 664–671.

Tegtmeyer, N., Wessler, S., and Backert, S. 2011. Role of the cag-Pathogenicity Island encoded type IV secretion system in *Helicobacter pylori* pathogenesis. *FEBS J.*, 278, 1190–1202.

Telaranta-Keerie, A., Kara, R., Paloheimo, L., Härkönen, M., and Sipponen, P., 2010. Prevalence of undiagnosed advanced atrophic corpus gastritis in Finland: An observational study among 4,256 volunteers without specific complaints. *Scand. J. Gastroenterol.*, 45, 1036–1041.

Thye, T., Burchard, G.D., Nilius, M., Müller-Myhsok, B., and Horstmann, R.D. 2003. Genomewide linkage analysis identifies polymorphism in the human interferon-gamma receptor affecting *Helicobacter pylori* infection. *Am. J. Hum. Genet.*, 72, 448–453.

Tian, X.Y., Zhu, H., Zhao, J., She, Q., and Zhang, G.X. 2012. Diagnostic performance of urea breath test, rapid urea test, and histology for *Helicobacter pylori* infection in patients with partial gastrectomy: A meta-analysis. *J. Clin. Gastroenterol.*, 46, 285–292.

Tomb, J.F., White, O., Kerlavage, A.R., Clayton, R.A., Sutton, G.G., Fleischmann, R.D., Ketchum, K.A., Klenk, H.P., Gill, S., Dougherty, B.A., Nelson, K., Quackenbush, J., Zhou, L., Kirkness, E.F., Peterson, S., Loftus, B., Richardson, D., Dodson, R., Khalak, H.G., Glodek, A., McKenney, K., Fitzegerald, L.M., Lee, N., Adams, M.D., Hickey, E.K., Berg, D.E., Gocayne, J.D., Utterback, T.R., Peterson, J.D., Kelley, J.M., Cotton, M.D., Weidman, J.M., Fujii, C., Bowman, C., Watthey, L., Wallin, E., Hayes, W.S., Borodovsky, M., Karp, P.D., Smith, H.O., Fraser, C.M., and Venter, J.C., 1997. The complete genome sequence of the gastric pathogen *Helicobacter pylori*. *Nature*, 388, 539–547.

Torres, V.J., VanCompernolle, S.E., Sundrud, M.S., Unutmaz, D., and Cover, T.L. 2007. *Helicobacter pylori* vacuolating cytotoxin inhibits activation-induced proliferation of human T and B lymphocyte subsets. *J. Immunol.*, 179, 5433–5440.

Tytgat, G.N., 2011. Etiopathogenetic principles and peptic ulcer disease classification. *Dig. Dis.*, 29, 454–458.

Venerito, M., Selgrad, M., and Malfertheiner, P. 2013. *Helicobacter pylori*: Gastric cancer and extragastric malignancies – clinical aspects. *Helicobacter*, 18(suppl. 1), 39–43.

Viala, J., Chaput, C., Boneca, I.G., Cardona, A., Girardin, S.E., Moran, A.P., Athman, R., Mémet, S., Huerre, M.R., Coyle, A.J., DiStefano, P.S., Sansonetti, P.J., Labigne, A., Bertin, J., Philpott, D.J., and Ferrero, R.L., 2004. Nod1 responds to peptidoglycan delivered by the *Helicobacter pylori* cag Pathogenicity Island. *Nat. Immunol.*, 5, 1166–1174.

Voland, P., Weeks, D.L., Marcus, E.A., Prinz, C., Sachs, G., and Scott, D., 2003. Interactions among the seven *Helicobacter pylori* proteins encoded by the urease gene cluster. *Am. J. Physiol. Gastrointest. Liver Physiol.*, 284, G96–G106.

Watanabe, T., Asano, N., Kitani, A., Fuss, I.J., Chiba, T., and Strober, W. 2010. NOD1-mediated mucosal host defense against *Helicobacter pylori*. *Int. J. Inflam.*, 2010, 476482.

Watanabe, M., Kato, J., Inoue, I., Yoshimura, N., Yoshida, T., Mukoubayashi, C., Deguchi, H., Enomoto, S., Ueda, K., Maekita, T., Iguchi, M., Tamai, H., Utsunomiya, H., Yamamichi, N., Fujishiro, M., Iwane, M., Tekeshita, T., Mohara, O., Ushijima, T., and Ichinose, M., 2012. Development of gastric cancer in nonatrophic stomach with highly active inflammation identified by serum levels of pepsinogen and *Helicobacter pylori* antibody together with endoscopic rugal hyperplastic gastritis. *Int. J. Cancer*, 131, 2632–2642.

Weck, M.N., Stegmaier, C., Rothenbacher, D., and Brenner, H., 2007. Epidemiology of chronic atrophic gastritis: Population-based study among 9444 older adults from Germany. *Aliment. Pharmacol. Ther.*, 26, 879–887.

WHO, and IARC. 2012. Globocan 2008: Estimated cancer incidence, mortality, prevalence and disability-adjusted life years (DALYs) worldwide in 2008. http://globocan.iarc.fr/v3.0.

Wilson, K.T., and Crabtree, J.E., 2007. Immunology of *Helicobacter pylori*: Insights into the failure of the immune response and perspectives on vaccine studies. *Gastroenterology*, 133, 288–308.

Witkin, S.S., Gerber, S., and Ledger, W.J., 2002. Influence of interleukin-1 receptor antagonist gene polymorphism on disease. *Clin. Infect. Dis.*, 34, 204–209.

Wotherspoon, A.C., Doglioni, C., Diss, T.C., Pan, L., Moschini, A., de Boni, M., and Isaacson, P.G. 1993. Regression of primary low-grade B-cell gastric lymphoma of mucosa-associated lymphoid tissue type after eradication of *Helicobacter pylori*. *Lancet*, 342, 575–577.

Xue, H., Liu, J., Lin, B., Wang, Z., Sun, J., and Huang, G., 2012a. A meta-analysis of interleukin-8 −251 promoter polymorphism associated with gastric cancer risk. *PLoS One*, 7, e28083.

Xue, H., Wang, Y.C., Lin, B., An, J., Chen, L., Chen, J., and Fang, J.Y., 2012b. A meta-analysis of interleukin-10 −592 promoter polymorphism associated with gastric cancer risk. *PLoS. One*, 7, e39868.

Yahara, K., Kawai, M., Furuta, Y., Takahashi, N., Handa, N., Tsuru, T., Oshima, K., Yoshida, M., Azuma, T., Hattori, M., Uchiyama, I., and Kobayashi, I., 2012. Genome-wide survey of mutual homologous recombination in a highly sexual bacterial species. *Genome Biol. Evol.*, 4, 628–640.

Yamaoka, Y. 2010. Mechanisms of disease: *Helicobacter pylori* virulence factors. *Nat. Rev. Gastroenterol. Hepatol.*, 7, 629–641.

Yang, Y.J., Chuang, C.C., Yang, H.B., Lu, C.C., and Sheu, B.S., 2012. *Lactobacillus acidophilus* ameliorates *H. pylori*-induced gastric inflammation by inactivating the Smad 7 and NFκB pathways. *BMC Microbiol.*, 12, 38.

Yang, C.A., Scheibenbogen, C., Bauer, S., Kleinle, C., Wex, T., Bornschein, J., Malfertheiner, P., Hellmig, S., Schumann, R.R., and Hamann, L., 2013. A frequent Toll-like receptor 1 gene polymorphism affects NK- and T-cell IFN-γ production and is associated with *Helicobacter pylori*-induced gastric disease. *Helicobacter*, 18, 13–21.

Yeh, Y.C., Chang, W.L., Yang, H.B., Cheng, H.C., Wu, J.J., and Sheu, B.S., 2011. *H. pylori* cagL amino acid sequence polymorphism Y58E59 induces a corpus shift of gastric integrin α5β1

related with gastric carcinogenesis. *Mol. Carcinog.*, 50, 751–759.

Yin, Y.W., Hu, A.M., Sun, Q.Q., Zhang, B.B., Wang, Q., Liu, H.L., Zeng, Y.H., Xu, R.J., Zhang, S.J., and Shi, L.B., 2013. Association between interleukin-8 gene −251 T/A polymorphism and the risk of peptic ulcer disease: A meta-analysis. *Hum. Immunol.*, 174, 125–130.

Zambon, C.-F., Basso, D., Navaglia, F., Germano, G., Gallo, N., Milazzo, M., Greco, E., Fogar, P., Mazza, S., Di Mario, F., Basso, G., Rugge, M., and Plebani, M., 2002. *Helicobacter pylori* virulence genes and host IL-1RN and IL-1beta genes interplay in favouring the development of peptic ulcer and intestinal metaplasia. *Cytokine*, 18, 242–251.

Zambon, C.-F., Navaglia, F., Basso, D., Rugge, M., and Plebani, M. 2003. *Helicobacter pylori* babA2, cagA, and s1 vacA genes work synergistically in causing intestinal metaplasia. *J. Clin. Pathol.*, 56, 287–291.

Zambon, C.-F., Fasolo, M., Basso, D., D'Odorico, A., Stranges, A., Navaglia, F., Fogar, P., Greco, E., Schiavon, S., Padoan, A., Fadi, E., Sturniolo, G.C., Plebani, M., and Pedrazzoli, S., 2007. Clarithromycin resistance, tumor necrosis factor alpha gene polymorphism and mucosal inflammation affect *H. pylori* eradication success. *J. Gastrointest. Surg.*, 11, 1506–1514.

Zambon, C.-F., Basso, D., Marchet, A., Fasolo, M., Stranges, A., Schiavon, S., Navaglia, F., Greco, E., Fogar, P., Falda, A., D'Odorico, A., Rugge, M., Nitti, D., and Plebani, M., 2008. IL-4 −588C>T polymorphism and IL-4 receptor alpha [Ex5+14A>G; Ex11+828A>G] haplotype concur in selecting *H. pylori* cagA subtype infections. *Clin. Chim. Acta*, 389, 139–145.

Zhang, J., Dou, C., Song, Y., Ji, C., Gu, S., Xie, Y., and Mao, Y. 2008. Polymorphisms of tumornecrosis factor-alpha are associated with increased susceptibility to gastriccancer: A meta-analysis. *J. Hum. Genet.*, 53, 479–489.

Zhang, B.B., Wang, J., Bian, D.L., and Chen, X.Y. 2012a. No association between IL-1β −31 C/T polymorphism and the risk of duodenal ulcer: A meta-analysis of 3793 subjects. *Hum. Immunol.*, 73, 1200–1206.

Zhang, Y., Liu, C., Peng, H., Zhang, J., and Feng, Q. 2012b. IL1 receptor antagonist gene IL1-RN variable number of tandem repeats polymorphism and cancer risk: A literature review and meta-analysis. *PLoS One*, 7, e46017.

Zhang, B.B., Liu, X.Z., Sun, J., Yin, Y.W., and Sun, Q.Q., 2013a. Association between TNF α gene polymorphisms and the risk of duodenal ulcer: A meta-analysis. *PLoS One*, 8, e57167.

Zhang, Z., Xu, G., Ma, M., Yang, J., and Liu, X. 2013b. Dietary fiber intake reduces risk for gastric cancer: A meta-analysis. *Gastroenterology*, 145, 113–120.

Chapter 34

Pathogenesis of *Chlamydia trachomatis* in Humans

Luis Otero Guerra[1], José Antonio Boga[2], Jonathan Fernández Suárez[2], Carlos Fernández Benítez[3], and Fernando Vázquez[2,4]

[1]Servicio de Microbiología, Hospital de Cabueñes, Gijón, Asturias, Spain

[2]Servicio de Microbiología, Hospital Universitario Central de Asturias, Oviedo, Spain

[3]Unidad de Gestión Clínica Centro de Salud de Laviana, Asturias, Spain

[4]Departamento Biología Funcional, Área de Microbiología, Facultad de Medicina, Oviedo, Spain

34.1 Introduction

The Chlamydiaceae family includes, as the main human pathogenic species, *Chlamydia trachomatis*, *Chlamydia pneumoniae*, and *Chlamydia psittaci*. They compose a very divergent division within the evolutionary tree of the eubacteria, and, together with species in the mycoplasma group, are the bacteria with the smallest genomes. Chlamydiae are metabolically deficient in their ability to synthesize amino acids and ATP. They are energy parasites and thus require an exogenous source of energy, which they obtain from host cells.

C. trachomatis infection is the most commonly reported sexually transmitted infection in the United States and Europe, and also produces trachoma, a contagious eye infection that is the principal infectious cause of blindness worldwide, generally in poor populations in developing nations. In women, genital chlamydial infections commonly result in asymptomatic cervicitis and urethritis, and reinfections or untreated cases progress to pelvic inflammatory disease (PID) and sequelae such as ectopic pregnancy, infertility, and chronic pelvic pain. Chlamydial infection in men is the leading cause of non-gonococcal urethritis (NGU). In both men and women, chlamydial infection is usually asymptomatic and, as with other inflammatory sexually transmitted infections, chlamydial infection facilitates the transmission of HIV infection among both men and women in both the HIV carrier and recipient. Chlamydiae are susceptible to antibiotics, particularly tetracyclines and macrolides, but because the chlamydial cell wall is different from that of many bacteria, beta-lactam antibiotics such as penicillin lack bactericidal activity *in vitro*.

In this chapter, we review the epidemiology, pathogenesis, clinical features, diagnosis, treatment, and prevention of this important infectious disease.

34.2 Pathogen

C. trachomatis belongs to the order Chlamydiales, family Chlamydiaceae, and genus *Chlamydia*. It is an obligate intracellular bacterium with a structure of inner and outer membranes similar to that of other Gram-negative bacteria, but lacking or almost lacking peptidoglycans (despite having genes for their synthesis). *C. trachomatis* organisms possess DNA, RNA, and prokaryotic ribosomes. They can synthesize protein, nucleic acids, and lipids, but require amino acids that cannot be synthesized and must be supplied by the host cell; they also require ATP (Solomon et al., 2004).

One of the most important characteristics of this microorganism is that it has a biphasic life cycle, the first form being the elemental body (EB), which is infectious, but can neither replicate nor divide. It must penetrate the eukaryotic cell to continue its life cycle as the second form, the non-infectious reticular body (RB).

34.2.1 Genome of *C. trachomatis*

From the genetic point of view, the analysis of environmental *Chlamydia* genomes reveals that, about 700 million years ago, there was a common ancestor from

Human Emerging and Re-emerging Infections: Bacterial & Mycotic Infections, Volume II, First Edition. Edited by Surit K. Singh.
© 2016 John Wiley & Sons, Inc. Published 2016 by John Wiley & Sons, Inc.

which evolved species adapted to eukaryotic cells. It had virulence factors found in modern *Chlamydia*, including a type III secretion system (Horn et al., 2004). The phylogenetic mosaic of chlamydial genes, including a large number of genes with phylogenetic origins from eukaryotes, implies a complex evolution for adaptation to obligate intracellular parasitism (Stephens et al., 1998).

C. trachomatis contains a single 1,043,000 bp chromosome (Stephens et al., 1998), with a G+C content of approximately 40%. This small genome encodes 893 chromosomal and 8 plasmid open reading frames (ORFs) (Belland et al., 2003). The first complete *C. trachomatis* genome sequence was published in 1998 (Stephens et al., 1998) but there are now more than 100 completed sequences for *C. trachomatis* strains, archived and available on online databases (NCBI genome database, Sanger Institute, European Nucleotide Archive at EMBL). All sequenced genomes demonstrate similar size and nucleotide sequence similarity (>99% identical) (Abdelsamed et al., 2013). The high degree of genomic conservation implies that few genetic features are involved in phenotypic dissimilarities (Nunes et al., 2013). The *C. trachomatis* genome has a plasticity zone (PZ), which is the site of the most extensive variation in sequence and gene content between chlamydial genomes. The variation in size of the *C. trachomatis* PZ is largely due to differential deletion of the cytotoxin gene(s) (Carlson et al., 2004).

In addition to the chromosome, *C. trachomatis* commonly possesses, as an extra-chromosomal genetic element, the 7.4-kb plasmid pCT, which was first isolated from a *C. trachomatis* L2 strain in 1986 (Palmer and Falkow, 1986). Their studies identified pCT DNA in laboratory strains of all *C. trachomatis* serovars. The plasmid is highly conserved (less than 1% variation in nucleotide sequence). This suggests that the plasmid might be essential because it may contribute to the regulation of chlamydial chromosomal gene expression, and a direct impact of the plasmid gene product on virulence is also possible (Byrne, 2010). However, several *C. trachomatis* strains lacking the plasmid have been isolated.

34.3 Structure of *C. trachomatis*

The EB of *C. trachomatis* is spherical (or, rarely, pear-shaped) and 0.2–0.3 μm in diameter. They lack a peptidoglycan layer, and they are perfectly matched to the extracellular medium (an outer membrane with a large amount of protein rich in sulfur-containing amino acids, such as cysteine). It has an eccentrically located nucleoid containing condensed DNA and cytoplasm with 70S ribosomes. The cell envelope is double layered. The cell wall can itself be resolved into two layers: an inner layer composed of hexagonally arrayed structures, and a granular outer layer containing the outer membrane (Solomon et al.,

2004). The inner layer therefore delimits the periplasmic space.

The RB is of 0.5–1.0 μm in diameter within a membrane-bound vacuole known as an inclusion. They are osmotically fragile, but protected by their intracellular localization. They contain diffuse, fibrillar DNA plus a high concentration of ribosomes. The cell envelope appears less complex than that of the EB, lacking the hexagonally packed structures of the EB periplasmic space.

34.3.1 Proteins and antigens of *C. trachomatis*

34.3.1.1 Lipopolysaccharide (LPS)

This is a thermostable antigen common to all members of the family *Chlamydiaceae*, and as an endotoxin it has low activity. In the absence of LPS, *C. trachomatis* is unable to make the transition from RB to EB, which is the infective form (Nguyen et al., 2011).

34.3.1.2 Outer membrane proteins

34.3.1.2.1 MOMP (major outer membrane protein), encoded by the ompA gene
It is the major outer membrane protein (MOMP), representing 60% of the dry weight of the membrane. It is exposed on the surface of the membrane and therefore is highly immunogenic.

MOMP presents two types of structures:

- Trimeric in EBs, acting as an adhesin, providing non-specific interactions and enabling penetration of the EB into the eukaryotic cell.
- Monomeric in RBs, acting as a porin, facilitating permeability of nutrients and ATP.

The MOMP protein topology is as follows. It contains five conserved transmembrane domains (CDI-V) and four variable domains (VDI-IV) on the outer exposed surface. The variability of epitopes located in this protein seems to respond to factors related to immune pressure.

34.3.1.2.2 PMPs: polymorphic membrane proteins
The PMPs are a family of nine membrane proteins (PmpA-I) (Abdelsamed et al., 2013), with a high degree of diversity between them (>50%), which have an important role in the biology of *C. trachomatis*. They have a transport function and act as adhesins. Phylogenetic reconstructions of PMPs allow *C. trachomatis* to be grouped in different pathotypes, for which reason it has been suggested that these proteins might modulate tissue tropism.

34.3.1.2.3 COMC: complex of outer membrane proteins rich in cysteine residues
The EB is surrounded by this protein complex, which contributes to its rigidity and osmotic stability. It participates in the early stages of the process of anchoring

the EB to the eukaryotic cell as another adhesin. Up to 28 proteins are part of this complex (including PmpB, PmpC, PmpE, PmpF, PmpG, PmpH, OprB, and PorB), but the two most prevalent proteins are OmcA and OmcB. OmcB (also called Omp2 or envB) is the most abundant in the complex and is highly immunogenic. It appears to be involved in the transition from RB to EB (Belland et al., 2003).

34.3.1.3 Cellular process proteins: Hsp proteins

These are proteins with a structure that has been highly conserved during evolution. They are classified into different families according to their molecular weight. They have two principal functions: they act as chaperones, and they are induced in response to stress.

Hsp60 or "GroEL-like" appears during persistent chronic infections.
Hsp70 or "Dnak-like," is located in the cytoplasm and in the outer membrane of the EB.

34.3.1.4 Type III secretion system

The type III secretion system (TTSS) consists of >40 ORFs that release effector proteins directly into the cytosol or in the inclusion membrane. They facilitate the interaction between the host and the bacterial pathogen. Some of the most important proteins are the Inc proteins (IncA-G), related to the transition from EB to RB; TarP, related to invasion; CrpA, which seems to activate the CD8+ cells; and the recently identified CPAF (*chlamydial protease-like activity factor*), which interferes with the ability of the host to respond to the chlamydial infection.

34.3.1.5 Cytotoxin

A putative chlamydial cytotoxin has been described, with amino acid sequence similarities to the clostridial toxin B protein. Significantly, cytotoxin gene variations are useful for defining *C. trachomatis* disease phenotypes. Genital biovars contain a single gene with a large central deletion, while lymphogranuloma venereum (LGV) strains lack the toxin gene. In addition, cytotoxin gene polymorphisms have been useful in distinguishing ocular and genital *C. trachomatis* isolates (Byrne, 2010).

Depending on the different serological responses of the variable domains of MOMP, *C. trachomatis* is classified into different serovars. *C. trachomatis* serovars can infect and survive in diverse host niches corresponding to different tissue tropism, causing a wide spectrum of diseases in humans (Abdelsamed et al., 2013), but, on the other hand, MOMP serovars fail to show relationships with virulence (Byrne, 2010).

The different serovars are divided into three biovars or pathotypes (based on the PMPs), each responsible for a different infectious process:

a) trachoma (serovars A, B, Ba, and C)
b) lymphogranuloma venereum, LGV (serovars L1 and L2, and the genovariants L2a, L2b, L2c, L2d, L2e, L2f, L2g, and L3).
c) oculo-genital noninvasive disease (serovars D, Da, E, F, G, H, I, Ia, and J, and the genovariants Ja and K).

These biovars or pathotypes have a different geographical distribution and affect different population groups.

34.4 Pathogenesis and immunity

The *Chlamydiae* share a unique biphasic developmental cycle. The EB is the infectious form, infects eukaryotic host cells, and can survive for only a limited period of time outside the host cell. This particle changes to a metabolically active and dividing form, the RB.

The growth cycle is divided into three phases.

1. EB attachment and penetration into susceptible host cells, involving specific receptor sites.
2. Intracellular change to RB, multiplication by binary fission, and reorganization back to EB.
3. EB release from the host cells.

In the first phase, the infectious EBs attach to the host cell. Bacterial factors such as glycosaminoglycan (GAG) (Menozzi et al., 2002), MOMP (Su et al., 1996), OmcB (Fadel and Eley, 2007), and PmpD (Wehrl et al., 2004) have been proposed as adhesins or ligands for receptor interactions. Although several molecules, including heparan sulfate, mannose receptor, mannose 6-phosphate receptor, and estrogen receptor have been proposed as host receptors (Campbell and Kuo, 2006), recently it has been reported that the host membrane protein disulfide isomerase (PDI), which is a component of the estrogen receptor alpha (ERalpha) complex, targets cross-linked EB outer membrane proteins and associates with *C. trachomatis* attached to human endometrial epithelial cells. This supports its role in EB entry (Abromaitis and Stephens, 2009).

Although multiple redundant strategies probably exist to ensure chlamydial entry, and the route is dependent on the chlamydia species or features of the host cell type being invaded, recent research describes a process of electrostatic binding using the cystic fibrosis transmembrane conductance regulator (CFTR) membrane protein to enter cells by endocytosis that does not depend on the microtubule system (Ajonuma et al., 2010). The unifying feature of EB entry, regardless of cell type or species, is the Rac1-dependent actin remodeling at attachment sites (Carabeo et al., 2004).

In the second growth cycle phase, the EB becomes surrounded by host cell membrane to form the inclusion body and *C. trachomatis* proteins are inserted into the inclusion body membrane. The inclusion body fuses with host cell vesicles containing nutrients necessary for replication (Cocchiaro and Valdivia, 2009). Halfway through the 48–72-hour replication cycle, some RBs make the transition to infectious EBs. Despite the accumulation of 500–1000 infectious EBs in the inclusion, the host cell function undergoes minimal disruption. Within the host cell, the EB remains within a membrane-lined phagosome that does not fuse with the host cell lysosome. Although no host cell markers are found in the inclusion membrane, a functional interaction with the Golgi apparatus is suggested by the traffic of lipid markers to the inclusion. C. trachomatis appears to elude the host endocytic pathway and inhabits a non-acidic vacuole that is dissociated from late endosomes and lysosomes. Inside the host cells, C. trachomatis circumvents endogenous stress mechanisms, prevents lysosomal fusion, inhibits apoptosis, and evades intracellular innate immunity. Thus, CPAF, an RB protease secreted into the cytoplasm, degrades cytoskeletal elements and processes intermediate-filament proteins to stabilize the inclusion bodies and minimize exposure to intracellular innate immune mechanisms.

The third growth cycle phase begins approximately 48 hours after infection. Release may occur by cytolysis or by a process of exocytosis or extrusion of the whole inclusion, which leaves the host cell intact. At developmental cycle completion, EBs must exit the cell to initiate subsequent rounds of infection. Egress can occur via two discrete mechanisms. Cell lysis involves the sequential disruption of inclusion and cellular membranes by cysteine proteases and the host cell is destroyed. Alternatively, the inclusion can remain membrane-bound and be pushed out, or "extruded," from the host cell. This process is dependent on actin-polymerization and myosin, and the host cell is often left intact (Hybiske and Stephens, 2007). This strategy is very successful and enables the organism to cause essentially silent, chronic infection. C. trachomatis may also enter a persistent state after treatment with certain cytokines such as interferon-γ, treatment with antibiotics, or with restriction of certain nutrients.

Although the pathological consequences of C. trachomatis genital infection are well established, the mechanism(s) that result in Chlamydia-induced tissue damage are not fully understood. Abundant in vitro data suggest that the inflammatory response to C. trachomatis is initiated and sustained by actively infected non-immune host epithelial cells. Nevertheless, the presence of neutrophils in endometrial surface epithelium and in gland lumens, dense subepithelial stromal lymphocytic infiltration, stromal plasma cells (PCs), and germinal centers containing transformed lymphocytes in C. trachomatis-infected human female genital tract tissue samples hinder the determination of specific factors responsible for the disease sequelae (Kiviat et al., 1990).

Animal models of chlamydial infection have been used to examine the nature and timing of the inflammatory response that occurs in the female genital tract after in vivo infection. Mouse and guinea pig models show that the response to primary chlamydial infection occurs within 1–2 days of infection and is initiated and sustained by epithelial cells, which are the primary targets of chlamydial infection, and act as first responders, initiating and propagating immune responses. The fact that in vitro infection of cervical and colonic epithelial cells with C. trachomatis induced the secretion of an array of proinflammatory cytokines and chemokines that have chemoattractant and proinflammatory functions supports the concept of the epithelial cell as an important and early component of the host response to chlamydial infection (Rasmussen et al., 1997). Moreover, in contrast to invasion by other bacteria that induce a rapid but transient proinflammatory cytokine response at entry, chlamydial invasion alone was not sufficient to elicit a response. Instead, intracellular chlamydial growth was required and the epithelial cytokine response was sustained throughout the chlamydial developmental cycle.

Chlamydial infection results in the production of cytokines and chemokines, which induce increased expression of endothelial adhesion molecules that aid in the attraction of immune cells. Thus, infection of fallopian tube cell cultures results in epithelial cell release of IL-1 and cellular damage, independent of inflammatory cell influx. The addition of IL-1 receptor antagonist to the cultures completely eliminates tissue destruction induced by infection, indicating a direct role for this cytokine in pathogenesis (Hvid et al., 2007). In vitro infection of fallopian tube epithelium also results in the production of tumor necrosis factor (TNF) and increased expression of adhesion molecules on oviduct endothelial cells (Kelly et al., 2001). This milieu could easily lead to activation and recruitment of first innate and, later, adaptive immune cells to effect resolution of the infection. However, subsets of these responses may also induce collateral damage to genital tract tissue. Defining the specific responses that promote tissue damage and differentiating them from those that lead to benign resolution of infection is an important ongoing research goal.

Although the presence of several chemokines and cytokines has been found in chlamydial infection, as has been described above, a recent report using C. trachomatis-infected polA2EN cells, an established polarized, immortalized, endocervical epithelial cell model, indicates that C. trachomatis infection results in a modest increase in pro-inflammatory cytokine response IL1α protein levels, which does not occur until 72 hours after infection. Furthermore, protein levels of the pro-inflammatory cytokines and chemokines IL6, TNFα, and CXCL8 were not significantly different between C. trachomatis-infected polA2EN cells and mock infected cells at any time during the chlamydial developmental cycle up to

120 hours post-infection (Buckner et al., 2013). These results sugges t that *C. trachomatis* can use evasion strategies to circumvent a robust pro-inflammatory cytokine and chemokine response. These evasion strategies, together with the inherent immune repertoire of endocervical epithelial cells, may aid *C. trachomatis* in establishing, and possibly sustaining, an intracellular niche in the microenvironments of the endocervix *in vivo*.

In any case, the production of the above-mentioned cytokines and chemokines induces the recruitment of neutrophils, NK cells, and a mixture of CD4, CD8, B cells, and PCs, which produce antibodies able to inactivate extracellular EBs, and, IFN-γ, which inhibits intracellular *C. trachomatis* replication. Antibody responses require CD4 T lymphocyte "help." B lymphocytes and antibodies recognize "native-protein" epitopes which are not necessarily linear, whereas T lymphocytes recognize short linear peptides in the context of self MHC-I for CD8 T cells and self MHC-II for CD4 T cells. CD4 T cells can be activated as pro-inflammatory Th17 and Th1 cells or as Th2 cells. Both Th1 and Th2 cells help B cells to make antibody but the immunoglobulin subclasses produced differ. As different immunoglobulin subclasses vary in biological activity, this may be relevant to pathology following infection. Antibodies and T lymphocyte responses to several *C. trachomatis* proteins have been described. Thus, antibodies to any of the 50 inclusion body (Inc) proteins encoded by *C. trachomatis*, 22 of which are in inclusion membranes, are found in sera from infected women (Li et al., 2008) and both CD4 and CD8 T cells respond to Inc epitopes (Gupta et al., 2009a). Although, not all *C. trachomatis* proteins generate antibodies in every individual, anti-EB antibodies were correlated with fallopian tube damage in sub-fertile women (El Hakim et al., 2010). Infected women have IgG antibodies to MOMP, IncB, and IncC; and IgG2 (a Th1-dependent subclass) is the prevalent subclass of anti-Inc antibodies. Antibody titers to IncB and IncC correlate with cervicitis and PID, being highest in cervicitis (Gupta et al., 2009a). CD4 T-cell responses to IncB and IncC in infertile women generated a Th2 cytokine pattern, whereas fertile women showed a Th1 pattern (Gupta et al., 2009b). In guinea pigs, antibody protects against reinfection and serum IgG is more important than local IgA. In mice, both Th1 cells and antibody contribute directly to protection. MOMP is immunodominant, but MOMP immunization protects only if the protein is non-reduced, emphasizing the importance of native-protein B cell epitopes. Antibiotic therapy clears experimental infection but prevents generation of immunity (Rank and Whittum-Hudson, 2010).

Autoimmune mechanisms could be of interest in chlamydial pathogenesis. *C. trachomatis* heat shock proteins (cHSPs) are analogous to human HSPs, and antibodies to cHSP60 are associated with autoimmunity (Campanella et al., 2009). *C. trachomatis* infection is associated with antibodies to sperm in semen and IgG antibodies to cHSP60 are associated with reproductive failure. Sperm have several surface HSPs and cross-reaction with cHSPs may initiate autoimmune responses to sperm (Naaby-Hansen and Herr, 2010). IgG responses to cHSP60 do not correlate with pathology in women (Huston et al., 2010) but women with tubal infertility have a high prevalence of antibodies to cHSP60 (Rodgers et al., 2010).

After infection has resolved, antigens may persist for some time, inducing continued neutrophil inflammation and ongoing release of tissue-damaging molecules in the host. The release of proteases, clotting factors, and tissue growth factors from infected host cells and infiltrating inflammatory cells leads to tissue damage and eventual scarring – the hallmark of chlamydial-induced disease. There is no distinction between damage induced by professional innate immune cells (neutrophils and monocytes) and adaptive lymphocyte populations and both cell populations contribute to pathogenesis. Thus, inflammation may kill bacteria but chronic inflammation can cause tissue damage. Unfortunately, the adaptive response induced by infection is not effective in preventing reinfection and the host's oviducts remain vulnerable to repeated inflammatory insult. Defining the specific responses that promote tissue damage and differentiating them from those that lead to benign resolution of infection is an important ongoing research goal.

34.5 Epidemiology

C. trachomatis is one of the most common sexually transmitted bacterial organisms in the world (Malhotra et al., 2013), causing significant morbidity and socioeconomic burden. In the United States, in 2009, 1,244,180 cases were reported to the Centers for Disease Control and Prevention (CDC), with an incidence of 409.2 per 100,000 population (CDC, 2010a). However, it must be remembered that many infections are not detected due to the frequently asymptomatic nature of the disease, incomplete screening coverage, and underreporting. During 2001–2002, a sample of 14,322 young people (18–26 years), representative of young US adults, presented an overall prevalence of *C. trachomatis* of 4.19%. Women (4.7%) were more likely to be infected than men (3.6%) (Miller et al., 2004).

The number of new cases is growing every year, especially since 2008 (de Barbeyrac, 2013). This increase may be related to higher rates of transmission, but also to the widespread implementation of more sensitive diagnostic techniques, especially that of molecular diagnosis, which has led to an estimated improvement in sensitivity of 10–30%.

Each pathotype has its own clinical and epidemiological characteristics. Non-invasive ocular–genital infection is the leading cause of STI in the world (much more prevalent than the other presentations of *C. trachomatis* trachoma and LGV):

a) Trachoma, which manifests as a chronic follicular conjunctivitis, is endemic in more than 50 countries, especially in rural areas of sub-Saharan Africa, the Middle East, Asia, and South America. Its transmission is related to poor hygienic conditions, either within the same family or between individuals whose interaction is very close. It occurs via direct contact with fomites, contact with eyes, or indirectly through flies (flies collect the mucopurulent secretions from infected eyes with their feet and can transmit the infection to others even at long distances). In endemic areas, children and women (but fewer of the latter) are mainly affected. According to the WHO, there are 1.3 million people in the world who are blind due to trachoma and it remains the leading cause of preventable blindness in developing countries. It is estimated that 15% of all cases of blindness in the world are due to trachoma (Solomon et al., 2004).

b) LGV is a sexually transmitted infection that invades the lymphoid tissues. In the pre-antibiotic era, it was endemic in Europe and the USA (Nieuwenhuis et al., 2004), but now it is considered endemic to Africa, India, South America, and the Caribbean. Until the beginning of the 21st century (2003), cases detected in the West were considered to have been imported from these regions. Transmission usually occurs through sexual contact. In the United States, the prevalence is estimated at 0.1 cases/100,000 people, although there is a higher proportion in active homosexuals and people returning from abroad. Since 2003 a new presentation of the disease has spread among men who have sex with men (MSM), numerous outbreaks having been detected in Europe, North America, and Australia, caused mainly by the new L2b genovariant (Martin-Iguacel et al., 2010; Nieuwenhuis et al., 2004; Stark et al., 2007).

c) The oculo-genital disease has the following non-invasive clinical presentations: genital infection in adults and conjunctivitis/pneumonia in newborns. It is estimated that in the United States, there are from 2 to 3 million cases/year (declared and undeclared cases).

Genital infection mainly affects women aged 15–24 years (75% of cases) (de Barbeyrac, 2013), with 50–85% of asymptomatic cases. It is directly related to the onset of sexual activity. Those who have a higher risk of becoming infected are sex workers, individuals with a new sexual partner, individuals with promiscuous behavior, and persons interned in prisons or correctional institutions (Joesoef et al., 2009).

As we have already mentioned, prevalence tends to be higher in females (Mylonas, 2012), but not all studies have found this to be the case (Dielissen et al., 2013; Fernández-Benítez et al., 2013). The prevalence ranges between 1.1% and 10.6% in women and between 0.1% and 12.1% in men (the difference in each case being mainly due to geographical factors).

Many cases are detected during routine screening, and women undergo screening much more frequently than men. This indicates that this imbalance towards women might not be real.

The other clinical presentations, conjunctivitis/pneumonia, are associated with children under 2 years infected in the birth canal of the mother carrying D-K serovars of *C. trachomatis*.

34.6 Clinical features

C. trachomatis serovars A, B, Ba, and C cause trachoma. The clinical manifestations of trachoma are divided into two phases: acute, characterized by active follicular keratoconjunctivitis, and late-stage manifestations characterized by a cicatricial tarsoconjunctival disease with trichiasis, entropium, and subsequent loss of vision (Solomon et al., 2004).

The acute phase is most commonly seen in children. The incubation period is 5–12 days, and conjunctivitis is characterized by mild self-limited follicular tarsal conjunctivitis (upper eyelid), sometimes with mucopurulent secretion but predominantly asymptomatic. This initial stage may resolve spontaneously, although superinfection is possible. Without treatment, the disease progresses to the chronic phase.

The late-stage manifestations of trachoma are seen in adults. The severity, duration, and reincidence of acute phase in childhood predict the progression of the tarsoconjunctival scarring in adulthood. Important conjunctival inflammation leads to eyelid scarring, retraction, and subsequent entropion (inward rolling) and trichiasis (ingrown eyelashes). When eyelashes touch the cornea, it leads to abrasion, edema, ulceration, scarring, corneal pannus, and, without treatment, blindness.

The clinical signs are used to categorize the severity of the disease in the Simplified WHO Trachoma Grading System (Thylefors et al., 1987), which is used for decisions regarding individual and population interventions.

The *C. trachomatis* serovars D through K, including the serovars Da and Ia and the genovariant Ja, are associated with genital tract disease, typically non-gonoccocal urethritis in men and cervicitis in women, although the majority of affected persons are asymptomatic and therefore remain undetected (Peipert 2003) and provide an ongoing reservoir for infection. Other clinical syndromes in men include acute epididymitis, proctitis, conjunctivitis, reactive arthritis, and Reiter's syndrome. Male infertility, chronic prostatitis, and urethral strictures are possible results of infection. In women, cervicitis is the commonest presentation and other clinical manifestations include urethritis, bartholinitis, uppergenital tract infection (endometritis, salpingitis, and PID), perihepatitis, proctitis, conjunctivitis, and reactive arthritis. In pregnant women, it is responsible for premature births and newborn infections: conjunctivitis and pneumonia (Table 34-1).

Table 34-1 *C. trachomatis* serovars D through K. Clinical manifestations and sequelae

Men	Women	New born
Urethritis	Cervicitis	Premature births
Epididymitis	Urethritis	Low birth weight
Proctitis	Bartholinitis	Neonatal death
Conjunctivitis	Endometritis	Conjunctivitis
Reactive arthritis	Salpingitis	Pneumonia
Reiter's syndrome	Pelvic inflammatory disease	
Chronic prostatitis	Perihepatitis	
Urethral strictures	Proctitis	
Infertility	Conjunctivitis	
	Reactive arthritis	
	Infertility	
	Chronic pelvic pain	
	Pregnant women	
	Ectopic pregnancy	
	Premature rupture of membranes	
	Preterm labor	
	Postpartum endometritis	

C. trachomatis is the most common cause of NGU in men. It accounts for 25–60% (usually 30–40%) of NGU in men and 4–35% (usually 15–25%) of cases of coinfection with gonococcal urethritis (Martin and Bowie, 1999). The incubation period after exposure is 5–21 days, which is longer than is the case in gonococcal urethritis. The main sign is mild-to-moderate mucoid or watery urethral discharge, and in certain cases meatitis is also observed. Symptoms included dysuria and itching (Varela et al., 2003). Asymptomatic cases are frequent and range among series from 40% to 96% (Cecil et al., 2001; Detels et al., 2001).

Epididymitis is another clinical syndrome caused by *C. trachomatis*, especially among sexually active men <35 years of age. Asymptomatic urethritis frequently accompanies sexually transmitted epididymitis and patients typically have unilateral testicular pain and tenderness. Oligospermia is observed in acute phase but is not a cause of infertility.

Chlamydial proctitis occurs almost exclusively in MSM who have had receptive anal intercourse, and the clinical presentation depends on the infecting chlamydial serovar. L1, L2, and L3 serovars cause LGV (discussed in detail separately). Non-LGV serovars cause infection of the rectum that is usually asymptomatic or minimally symptomatic, and this is the reason for recommendations for routine screening in MSM, particularly in HIV-infected patients. Symptoms include anal rectal pain, discharge, tenesmus, and constipation.

C. trachomatis may be an etiology of chronic prostatitis, although this area remains highly speculative.

Approximately 1% of men with urethritis develop reactive arthritis, and approximately one-third of these patients have the complete manifestations

formerly referred to as Reiter syndrome (arthritis, uveitis, and urethritis). *C. trachomatis* appears to be the most common inciting pathogen.

The clinical significance and transmissibility of *C. trachomatis* detected at oropharyngeal sites is unclear (Bernstein et al., 2009). Oral sex has become more common among young persons, usually carried out without barrier protection. It is known that other sexual infections such as gonorrhoea, syphilis, and herpes can infect the pharynx, which plays an important role in the transmission route. However, there is limited knowledge regarding the prevalence of pharyngeal infection and the importance of oral sex as a route of chlamydial transmission. Based on nucleic acid amplification technology, *C. tracomatis* were found in the pharynx of MSM (1.5–4.8%) and in heterosexual men (3–7%) and women (7–12%) who had had recent unprotected active oral sex and confirmed or suspected genital *C. trachomatis* infection (Bernstein et al., 2009; Karlsson et al., 2011; Peters et al., 2011; Wikström et al., 2010). *C. trachomatis* serovars L1–L3 (LGV) were also found in the pharynx of MSM (Haar et al., 2013). Infected patients could be asymptomatic or only with pharyngeal discomfort.

Cervicitis caused by *C. trachomatis* in women is the equivalent of the NGU in men and is the most common chlamydial syndrome. Approximately 50–85% of women with *C. trachomatis* infection have neither signs nor symptoms. When symptoms do occur, they are often mild: vaginal discharge, intermenstrual vaginal bleeding, and post-coital bleeding. Signs are mucopurulent cervical discharge, cervical friability, and cervical edema. Associated urethritis can also occur and patients complain of the typical symptoms of a urinary tract infection such as frequency and dysuria accompanied by pyuria, but urine culture is negative.

In 45–50% of women with untreated *C. trachomatis* cervicitis, the infection persists one year after the detection. This can facilitate disseminating *C. trachomatis* into the endometrium, causing endometritis, into the fallopian tubes causing salpingitis, or into the peritoneum causing peritonitis. Peritoneal diffusion can develop into perihepatitis (Fitz–Hugh–Curtis syndrome) characterized by inflammation of the liver capsule and adjacent peritoneal surfaces. Infections of the female upper genital tract are collectively called pelvic inflammatory disease (PID) and occur in 10–20% untreated cases of cervicitis. The long-term consequences of PID include chronic pelvic pain, ectopic pregnancy, prematurity, and infertility. It has been estimated that two-thirds of all cases of infertility due to tubal factor and one-third of all cases of ectopic pregnancy may be due to undiagnosed *C. trachomatis* infection. There is evidence that screening women who are at risk for *C. trachomatis* infection can prevent reproductive sequelae by reducing the incidence of PID.

Genital *C. trachomatis* infection in pregnant women can increase the risk for premature rupture of the membranes and low birth weight. Infants born vaginally to infected mothers with genital disease are at risk for infection. Perinatal infection initially affects the mucous membranes of the eyes, causing inclusion conjunctivitis in 20–45% of cases. Clinical manifestations of conjunctivitis arise 5–12 days after birth as swelling of the conjunctiva with clear discharge which becomes mucopurulent, and marked swelling of the eyelids with red and thickened conjunctivae. Treatment of conjunctivitis usually results in healing without complications. However, untreated infection may persist for months and cause corneal and conjunctival scarring.

Pneumonia appears in 10–20% of cases probably spread from the nasopharynx (70% of children infected have positive cultures at this location). Infants present with hypoxemia and eosinophilia, and the chest radiograph shows an interstitial bilateral infiltrate (Attenburrow and Barker, 1985). Serology shows increased titers of IgM.

LGV is a sexually transmitted infection caused by serovars L1, L2, and L3 of *C. trachomatis*. LGV produces a systemic infection (the bacteria may be recovered from the blood or CSF) and invasion of lymphoid nodes (causing lymphadenitis). Clinically, three phases can be differentiated:

- The first phase (3–12 days after sexual contact) is characterized by a lesion or injury ulcerative in the genital mucosa or adjacent skin at the site of inoculation. This initial injury may be also intraurethral, cervical or rectal causing urethritis, cervicitis or proctitis, respectively. Usually, this stage is asymptomatic or with very few symptoms, and lesions spontaneously heal within a few days.
- The second stage occurs 2–4 weeks later and is characterized by local direct extension of the infection to regional lymph nodes, causing lymphadenopathy, erythema, and pain (buboes), usually unilateral. In contrast to the urogenital infections due to *C. trachomatis* (serovars D through K) that are often mild, LGV can cause severe inflammation and invasive infection, often with systemic symptoms (Van der Bij et al., 2006). Buboes may break spontaneously, draining pus. Systemic manifestations during this stage are fever, headache, and myalgia.
- The third stage involves severe complications such as fibrosis, strictures in the anogenital tract, genital elephantiasis, frozen pelvis, and infertility.

34.7 Diagnosis and treatment

The diagnosis of the chlamydial infections is made by different diagnostic methods. Basically, there are four types of diagnostic assays: nucleic acid amplification techniques (NAATs), cell culture, enzyme immunoassays (EIA), and direct fluorescence assays.

Only NAATs can be recommended (level I, grade A) (Watson et al., 2002), although a perfect gold standard has not been defined (Lanjouw et al., 2010).

NAATs have the best turnaround times, and although sensitivity and specificity do vary slightly between different NAATs, other factors like cost, hands-on time, combined testing for other agents, and degree of automation play an important role in choosing a specific NAAT (Levett et al., 2008).

One problem that existed in the past was the appearance of a *C. trachomatis* variant (the Swedish variant) with a 377 bp deletion in the target area for some *C. trachomatis* NAAT tests. Logically, this meant that they were unable to detect this particular variant. Nonetheless, the latest versions of the NAATs are capable of detecting the Swedish variant (Lanjouw et al., 2010).

Since it has been suggested that hormonal levels influence detection of *C. trachomatis* by NAATs, the optimal period for taking vaginal swabs would be four weeks after the last menstrual bleeding (level III).

Another concern (competitive inhibition) is raised by the use of duplex or multiplex assays detecting more than one target. If one of the targets is present in excess, other targets may be reported as falsely negative (Gaydos et al., 2003). In these cases, the use of monoplex assays is needed to achieve the desired sensitivity (level II).

For specimens with a high bacterial load, all types of confirmatory testing will be positive and, therefore, confirmatory testing is unnecessary and expensive. Confirmatory testing of specimens with a low bacterial load does not solve the issue of true positivity and is therefore not recommended (level II, grade B) (Schachter et al., 2006).

Except in pregnant women, a test-of-cure is not advised, unless therapeutic compliance is in question, symptoms persist, or reinfection is suspected. Since a previous *C. trachomatis* infection is a risk factor for future STIs, a follow-up visit after three months can be considered (level II) (CDC, 2010b).

In LGV, it is recommended that all MSM with a positive NAAT from rectal and urethral/urine samples who report rectal symptoms and/or who are in contact with someone with LGV should send a sample to check for LGV. The sensitivity of the NAATs is 95%. Other diagnostic methods for LGV are as follows (Lanjouw et al., 2010):

a) Culture (sensitivity 30% in referral centers).
b) Chlamydia serology (four types of techniques): complement fixation (CF) test, single L-type immunofluorescence test, microimmunofluorescence test (micro-IF), and anti-MOMP IgA assay. In general, a fourfold rise in antibody or single-point titers of >1/6436 and >128 for the micro-IF test has been considered positive. The anti-MOMP IgA is the most useful assay for rectal LGV infection but sensitivity and specificity reached only 75% in asymptomatic MSM with rectal *C. trachomatis*.
c) Frei skin test (low sensitivity and specificity, not produced commercially).

d) Rectal polymorphonuclear leukocytes (PMNLs) from rectal swabs is predictive of LGV proctitis, especially in HIV-positive MSM, with levels of >10–20 PMNLs per high-power field.

LGV diagnostic techniques have certain limitations: for example, cultures are not frequently carried out, NAAT from rectal samples is not currently approved by the FDA, CF and microIF do not distinguish between the serotypes, and serology cannot necessarily distinguish past from current LGV infection, which might prove restrictive given the high number of recurrent LGV infections now seen in MSM.

In order to distinguish LGV from non-LGV serovars, there exist two techniques. Restriction fragment length polymorphism (RFLP) analysis of *C. trachomatis*-positive specimens is now used to distinguish LGV-associated serovars from oculogenital serovars. Sequencing, which is becoming widely available, is the method now recommended by the UK health protection agency (HPA) for genotyping, and various assays have been developed for this purpose. These techniques have been applied with great success in anorectal specimens collected from patients with proctitis during the recent LGV outbreaks in Western Europe (White et al., 2013).

In trachoma, the diagnostic gold standard in endemic areas is clinical, and alternative tests include IF cytology, PCR, Giemsa cytology, and culture. In neonatal conjunctivitis and pneumonia in infants, the gold standard to be used is culture (sensitivity and specificity 100%) but now it is the NAAT, although there are insufficient data in newborns for this test to replace culture as the gold standard. Alternative methods include antigen detection method, direct fluorescent antibody staining, and serum antibody with the micro-immunofluorescence method.

Another area of interest is the type of specimen on which the diagnostic test is carried out. The first choice in men is first-void urine (first 10 mL) (sensitivity 85–95%, the concordance is highest for symptomatic men) (level I, grade A); early-morning urine does not seem to be more sensitive than urine sampled at the time of visit. Thus, male urines can be collected at the time of the visit (level II). The patient should not have voided in the two hours prior to specimen collection.

The first choice in women is a (self-collected) vaginal swab because the sensitivity of testing first-void urine is slightly lower than that for males: 80–90% (level I, grade A) (Lanjouw et al., 2010) and in some cases a vaginal swab is better than endocervical swabs (sensitivity: 86%, specificity: 99.6%).

Other specimens are (Lanjouw et al., 2010):

a) Pharyngeal and conjunctival specimens. NAATs have now been adequately validated for these specimens (level II).
b) Rectal specimens. Specificity is less than 95% and confirmation by another assay might be appropriate (level II). In MSM, positive rectal specimens should be genotyped for LGV according to local guidelines. If available, it is recommended in MSM with symptomatic proctitis (level II, grade B).
c) Up to 10% of semen specimens might contain inhibitors for NAATs, and since a good correlation exists between first-void urine positivity and semen positivity, testing of semen specimens is not recommended (level II, grade B).
d) Pooling of urine specimens cannot be recommended (level II, grade B).
e) Pap-smear cannot be recommended for specific screening programs, nor for diagnostic purposes (level II).

Serology has limitations; in general, only invasive disease will lead to antibody levels that are useful for diagnostic purposes. Only MOMP-derived synthetic peptide-based EIAs show no cross-reactions but the duration of antibody-positivity is not known and for this reason they are of no value in the diagnosis of uncomplicated cervicitis and urethritis, and of limited value in the diagnosis of ascending infections or for infertility workup. In the case of LGV, high titers (IgG and/or IgA) can be diagnostic; and in neonatal pneumonia, the IgM can be diagnostic. Especially when direct detection by NAAT is not possible or not reliable, antibody testing may be helpful in the diagnosis of invasive disease, such as LGV involving the lymph nodes and neonatal pneumonia (level I, grade A) (Lanjouw et al., 2010; Nieuwenhuis et al., 2004).

34.8 Treatment

Tetracyclines, fluoroquinolones, and macrolides are used for antibiotic treatment of chlamydial infections. Chlamydial resistance to recommended antimicrobial agents appears to be rare, and its evaluation is complicated by the lack of standardized tests, as well as by the fact that *in vitro* resistance does not correlate with clinical outcome.

The treatment of chlamydial infections is shown in Table 34-2, following the CDC guidelines (CDC, 2010b).

34.9 Prevention and control

As part of its global strategy to combat STIs 2006–2015 (WHO, 2007), the WHO recommends investment in the prevention and control of STIs, including HIV, since they have long been recognized as constituting a serious public health problem. If the present social, demographic, and migratory trends are maintained, the population exposed to STIs will continue increasing dramatically. The burden of morbidity is particularly heavy in the less developed parts of the world, but in the industrialized countries, an increase is also predicted due to the prevalence of chronic viral infections, changes in patterns of sexual behavior, and a rise in the frequency of international travel. The socioeconomic costs of STIs and of their complications

Table 34-2 Treatment of *C. trachomatis* infections

Syndrome	Recommended regime	Alternative regime	Comment and special considerations
Nongonococcal urethritis and cervicitis	Azithromycin 1 g orally in a single dose, or Doxycycline 100 mg orally twice a day for 7 days	Erythromycin base 500 mg orally four times a day for 7 days or Erythromycin ethylsuccinate 800 mg four times a day for 7 days or Levofloxacin 500 mg one time a day orally for 7 days or Ofloxacin 300 mg orally twice a day for 7 days	HIV persons should receive the same regime as those who are HIV negative
LGV	Doxycycline 100 mg orally twice a day for 21 days	Erythromycin base 500 mg orally four times a day for 21 days Probably effective: Azithromycin 1g orally weekly for 3 weeks or Fluoroquinolones	Buboes might require aspiration or incision and drainage Follow-up: Until signs and symptoms have resolved Sex partners should be examined, tested, and treated with azithromycin 1 g orally twice a day for 7 days Pregnancy and lactating women should be treated with Erythromycin HIV persons should receive the same regime as those who are HIV negative
Ophthalmia neonatorum, pneumonia in infants	Erythromycin base or Erythromycin ethylsuccinate 50 mg/kg/ day orally divided into four doses daily for 14 days	Azithromycin might be effective	Infants treated with Erythromycin should be followed for signs and symptoms of hypertrophic pyloric stenosis
Sexual abuse in children	Erythromycin base or Erythromycin ethylsuccinate 50 mg/kg/day orally divided into four doses daily for 14 days	Azithromycin 1 g orally in a single dose	
Epididymitis	Doxycycline 100 mg orally twice a day for 10 days		
Proctitis	Doxycycline 100 mg orally twice a day for 7 days		
Trachoma	Depends on the stage of the disease; in early stages, Tetracycline eye ointment and oral Azithromycin single dose	Treatment of later stages of trachoma may require surgery	The World Health Organization (WHO) guidelines recommend giving antibiotics to an entire community when more than 10% of children have been affected by trachoma

and sequels are considerable and are even greater if the role of these diseases as cofactors in the transmission of HIV is taken into account.

Individuals with an STI may also have HIV. The prevention and treatment of STIs reduce the risk of transmission of HIV by sexual contact (Buchacz et al., 2004; Cohen et al., 1997; WHO, 2000). There is sufficient evidence that condoms, if used correctly and systematically, afford effective protection against the transmission of HIV to men and women. What is more, they reduce the risk of infection by other STIs (Holmes et al., 2004).

Objective 6A of the Sixth Objective of the Millennium Goals (UN, 2010) asks countries to detain the growth in HIV/SIDA propagation and to begin to reduce it by 2015. Thus, the prevention of STIs is one of the attainable and cost-effective interventions that will contribute to the achievement of Objective 6A. For all these reasons, in countries with high prevalence rates among the general population and in high-risk groups, both these population groups must be offered a combination of measures to provide strategies for safer sexual relations. Among these measures should be included the promotion of the adequate use of condoms and their distribution. In those cases where infection is concentrated in the high-risk groups, these must be given priority but without prejudicing the educational and prevention services aimed at the population as a whole.

Owing to the risk of sequelae after *C. trachomatis* infection in women, in many developed countries, screening programs have been proposed, with the ultimate objective of avoiding serious long-term consequences, principally infertility, and also to reduce the risk of transmission. The U.S. preventive services task force (USPSTF) recommends the detection of genital infection by *C. trachomatis* in all sexually active, non-pregnant women of 24 years or less and also in older, non-pregnant women at high risk of infection. The USPSTF considers the following to be factors of increased risk: a history of chlamydial or other sexually transmitted infection, new or multiple sexual partners, inconsistent condom use, and exchanging sex for money or drugs. They also recommend screening for all pregnant women who are at increased risk. But for men, the USPSTF concludes that the current evidence is insufficient to recommend screening (U.S. Preventive Services Task Force, 2007).

Treatment for *C. trachomatis* should be carried out immediately in all patients with infection, since delays have been associated with complications (Geisler et al., 2008).

Patients must be told to communicate to partners with whom they have had sexual relations during the 60 days previous to the appearance of symptoms or to being diagnosed with *C. trachomatis* that they should see their doctor for evaluation, analysis, and treatment. Although the defined intervals of exposure for the identification of sexual partners considered to be at risk are based on a limited evaluation, the most recent partners should be evaluated and treated, even if the last sexual contact was more than 60 days before the appearance of symptoms or diagnosis (CDC, 2010b).

For the capture of sexual partners, the index patient is asked to attend together with his/her partner "*bring your own partner*" (BYOP). However, in the case of heterosexual partners, if there is a possibility of the partner not attending for evaluation and thus not receiving treatment, it is worth considering giving the index patient antibiotics for the other partner. This is known as "*expedited partner therapy*" (EPT) and implies the treatment of sexual partners of people infected with *C. trachomatis* without conducting a medical evaluation (CDC, 2011). Patients must inform their sexual partners about their *C. trachomatis* infection and provide them with written information explaining the importance of proper medical evaluation if any symptom suggesting possible complications appear (testicular pain in men or pelvic pain in women). They should also be made aware that they must avoid having sexual relations for seven days after a single-dose treatment or until the end of a seven-day treatment regime

and until all their sexual partners have been treated (CDC, 2006, 2010b, 2011). EPT is not recommended in the case of MSM due to the high risk of their sexual partners having other coexisting undiagnosed STIs, especially HIV infection. Kretzschmar carried out a study to determine which clinical intervention would be most effective in diminishing the prevalence of *C. trachomatis*. They compared an increase in screening coverage percentages with the effects of a similar increase in the percentage of treatment of sexual partners. The principal conclusion that they drew was that an effort made to effectively treat the partners of infected patients can contribute towards the control of *C. trachomatis* infection at least as much as an increase in the screening coverage. In general, it can be said that the benefits of a reduction in prevalence obtained by extending screening coverage are relatively small in view of the considerable effort required to achieve this goal (Kretzschmar et al., 2012). Furthermore, an increase in the treatment of sexual partners reduces reinfection rates and this in turn improves the effectiveness of screening (Heijne et al., 2011).

Since *C. trachomatis* infections do not confer long-term immunity, people who have been infected and treated may become reinfected. In fact, a high rate of reinfection, in both men and women, has been observed during the months following treatment for *C. trachomatis*. During a follow-up period of one year after treatment, a reinfection rate as high as 20% was found in young women (Hosenfeld et al., 2009). The majority of reinfections are caused by the failure of sexual partners to undergo treatment, by a new infection of the partner, by the patient initiating a new relationship with an infected partner, or by failure of the treatment (Batteiger et al., 2010). With each reinfection, the likelihood of developing serious complications becomes greater (Haggerty et al., 2010). The repetition of the *C. trachomatis* test in recently infected men and women must be a priority and it should be carried out approximately 6 months after treatment, irrespective of whether the patient believes his/her partner was treated (Hosenfeld et al., 2009). If it is not possible to repeat the test after three months, it can be done in any health center during the 12 months after the initial treatment (CDC, 2010b).

34.10 Conclusion

In this chapter, we have reviewed the main aspects of *C. trachomatis* infection, including pathogenesis, epidemiology, diagnosis, treatment, and prevention. Our knowledge of this infection has improved in recent years, but there remain important aspects that our research efforts should focus on. For example, the mechanisms that result in chlamydia-induced tissue damage are not fully understood and differentiating them from those that lead to benign resolution of infection is an important research goal. As in other infections such as gonorrhea and syphilis, it is important to determine the importance that pharyngeal infection has in transmission. It is necessary to improve diagnostic assays and use non-invasive types of specimen, seeking greater sensitivity and specificity and specimens that are more acceptable to the patient and which will therefore improve the acceptance of screening programs. Concerning screening, efforts should be made to determine the benefits of programs that target men in order to decrease infection among women, to find the most appropriate intervals for screening in non-pregnant women, and to measure the risks associated with screening programs.

A deeper understanding of these aspects, together with advances in the areas of diagnosis and testing, will lead to new progress in the fight against this globally important infection.

Acknowledgment

The authors wish to thank Mr. Nicholas Airey for English language corrections.

References

Abdelsamed, H., Peters, J., and Byrne, G.I. 2013. Genetic variation in *Chlamydia trachomatis* and their hosts: Impact on disease severity and tissue tropism. *Future Microbiol.*, 8, 1129–1146.

Abromaitis, S., and Stephens, R.S. 2009. Attachment and entry of *Chlamydia* have distinct requirements for host protein disulfide isomerase. *PLoS Pathog.*, 5, e1000357.

Ajonuma, L.C., Fok, K.L., Ho, L.S., Chan, P.K., Chow, P.H., Tsang, L.L., Wong, C.H., Chen, J., Li, S., Rowlands, D.K., Chung, Y.W., and Chan, H.C. 2010. CFTR is required for cellular entry and internalization of *Chlamydia trachomatis*. *Cell. Biol. Int.*, 34, 593–600.

Attenburrow, A.A., and Barker, C.M. 1985. Chlamydial pneumonia in the low birthweight neonate. *Arch. Dis. Child*, 60, 1169.

de Barbeyrac, B. 2013. Current aspects of *Chlamydia trachomatis* infection. *Presse Med.*, 42(4 Pt 1), 440–445.

Batteiger, B.E., Tu, W., Ofner, S., Van Der Pol, B., Stothard, D.R., Orr, D.P., Katz, B.P., and Fortenberry, J.D. 2010. Repeated *Chlamydia trachomatis* genital infections in adolescent women. *J. Infect. Dis.*, 201, 42–51.

Belland, R.J., Zhong, G., Crane, D.D., Hogan, D., Sturdevant, D., Sharma, J., Beatty, W.L., and Caldwell, H.D. 2003. Genomic transcriptional profiling of the developmental cycle of *Chlamydia trachomatis*. *Proc. Natl. Acad. Sci. USA*, 100, 8478–8483.

Bernstein, K.T., Stephens, S.C., Barry, P.M., Kohn, R., Philip, S.S., Liska, S., and Klausner, J.D. 2009. *Chlamydia trachomatis* and *Neisseria gonorrhoeae* transmission from the oropharynx to the urethra among men who have sex with men. *Clin. Infect. Dis.*, 49, 1793–1797.

Buchacz, K., Patel, P., Taylor, M., Kerndt, P.R., Byers, R.H., Holmberg, S.D., and Klausner, J.D. 2004. Syphilis increases HIV viral load and decreases CD4 cell counts in HIV-infected patients with new syphilis infections. *AIDS*, 18, 2075–2079.

Buckner, L.R., Lewis, M.E., Greene, S.J., Foster, T.P., and Quayle, A.J. 2013. *C. trachomatis* infection results in a modest pro-inflammatory cytokine response and a decrease in T cell chemokine secretion in human polarized endocervical epithelial cells. *Cytokine*, 63, 151–165.

Byrne, G.I. 2010. *Chlamydia trachomatis* strains and virulence: Rethinking links to infection prevalence and disease severity. *J. Infect. Dis.*, 201(suppl. 2), S126–S133.

Campanella, C., Gammazza A.M., Mularoni, L., Cappello, F., Zummo, G., and Di, F.V. 2009. A comparative analysis of the products of GROEL-1 gene from *Chlamydia trachomatis* serovar D and the HSP60 var1 transcript from *Homo sapiens* suggests a possible autoimmune response. *Int. J. Immunogenet*, 36, 73–78.

Campbell, L.A., and Kuo, C.C. 2006. Interactions of *Chlamydia* with the host cells that mediate attachment and uptake. In: Bavoil, P.M., and Wyrick, P.B., editors. *Chlamydia: Genomics and Pathogenesis*. Wymondham, UK: Horizon Bioscience. pp. 505–522.

Carabeo, R.A., Grieshaber, S.S., Hasenkrug, A., Dooley, C., and Hackstadt, T. 2004. Requirement for the Rac GTPase in *Chlamydia trachomatis* invasion of non-phagocytic cells. *Traffic*, 5, 418–425.

Carlson, J.H., Hughes, S., Hogan, D., Cieplak, G., Sturdevant, D.E., McClarty, G., Caldwell, H.D., and Belland, R.J. 2004. Polymorphisms in the *Chlamydia trachomatis* cytotoxin locus associated with ocular and genital isolates. *Infect. Immun.*, 72, 7063–7072.

Cecil, J.A., Howell, M.R., Tawes, J.J., Gaydos, J.C., McKee Jr., K.T., Quinn, T.C., and Gaydos, C.A. 2001. Features of *Chlamydia trachomatis* and *Neisseria gonorrhoeae* infection in male Army recruits. *J. Infect. Dis.*, 184, 1216–1219.

Centers for Disease Control and Prevention. 2006. *Expedited Partner Therapy in the Management of Sexually Transmitted Diseases*. Atlanta, GA: US Department of Health and Human Services. Available at: http://www.cdc.gov/std/treatment/eptfinalreport2006.pdf. Accessed 3 November 2013.

Centers for Disease Control and Prevention. 2010a. *Sexually Transmitted Disease Surveillance 2009*. Atlanta, GA: U.S. Department of Health and Human Services.

Centers for Disease Control and Prevention. 2010b. Sexually transmitted diseases treatment guidelines, 2010. *MMWR Morb. Mortal. Wkly. Rep.*, 59(RR-12), 1–116.

Centers for Disease Control and Prevention (CDC). 2011. CDC grand rounds: *Chlamydia* prevention: Challenges and strategies for reducing disease burden and sequelae. *MMWR Morb. Mortal. Wkly. Rep.*, 60, 370–373.

Cocchiaro, J.L., and Valdivia, R.H. 2009. New insights into *Chlamydia* intracellular survival mechanisms. *Cell. Microbiol.*, 11, 1571–1578.

Cohen, M.S., Hoffman, I.F., Royce, R.A., Kazembe, P., Dyer, J.R., Daly, C.C., Zimba, D., Vernazza, P.L., Maida, M., Fiscus, S.A., and Eron Jr., J.J. 1997. Reduction of concentration of HIV-1 in semen after treatment of urethritis: Implications for prevention of sexual transmission of HIV-1. AIDSCAP Malawi Research Group. *Lancet*, 349, 1868–1873.

Detels, R., Green, A.M., Klausner, J.D., Katzenstein, D., Gaydos, C., Handsfield, H., Pequegnat, W., Mayer, K., Hartwell, T.D., and Quinn, T.C. 2001. The incidence and correlates of symptomatic and asymptomatic *Chlamydia trachomatis* and *Neisseria gonorrhoeae* infections in selected populations in five countries. *Sex Transm. Dis.*, 38, 503–509.

Dielissen, P.W., Teunissen, D.A., and Lagro-Janssen, T.L. 2013. *Chlamydia* prevalence in the general population: Is there a sex difference? A systematic review. *BMC Infect. Dis.*, 13, 534.

El Hakim, E.A., Gordon, U.D., and Akande, V.A. 2010. The relationship between serum *Chlamydia* antibody levels and severity of disease in infertile women with tubal damage. *Arch. Gynecol. Obstet.*, 281, 727–733.

Fadel, S., and Eley, A. 2007. *Chlamydia trachomatis* OmcB protein is a surface-exposed glycosaminoglycan-dependent adhesin. *J. Med. Microbiol.*, 56, 15–22.

Fernández-Benítez, C., Mejuto-López, P., Otero-Guerra, L., Margolles-Martins, M.J., Suárez-Leiva, P., Vazquez, F., and Chlamydial Primary Care Group. 2013. Prevalence of genital *Chlamydia trachomatis* infection among young men and women in Spain. *BMC Infect. Dis.*, 13, 388.

Gaydos, C.A., Quinn, T.C., Willis, D., Weissfeld, A., Hook, E.W., Martin, D.H., Ferrero, D.V., and Schachter, J. 2003. Performance of the APTIMA Combo 2 assay for detection of *Chlamydia trachomatis* and *Neisseria gonorrhoeae* in female urine and endocervical swab specimens. *J. Clin. Microbiol.*, 41, 304–309.

Geisler, W.M., Wang, C., Morrison, S.G., Black, C.M., Bandea, C.I., and Hook 3rd., E.W. 2008. The natural history of untreated *Chlamydia trachomatis* infection in the interval between screening and returning for treatment. *Sex Transm. Dis.*, 35, 119–123.

Gupta, R., Salhan, S., and Mittal, A. 2009a. Seroprevalence of antibodies against *Chlamydia trachomatis* inclusion membrane proteins B and C in infected symptomatic women. *J. Infect. Dev. Ctries*, 3, 191–198.

Gupta, R., Vardhan, H., Srivastava, P., Salhan, S., and Mittal, A. 2009b. Modulation of cytokines and transcription factors (T-Bet and GATA3) in CD4 enriched cervical cells of *Chlamydia trachomatis* infected fertile and infertile women upon stimulation with chlamydial inclusion membrane proteins B and C. *Reprod. Biol. Endocrinol.*, 7, 84.

Haar, K., Dudareva-Vizule, S., Wisplinghoff, H., Wisplinghoff, F., Sailer, A., Jansen, K., Henrich, B., and Marcus, U. 2013. Lymphogranuloma venereum in men screened for pharyngeal and rectal infection, Germany. *Emerg. Infect. Dis.*, 19, 488–492. doi:10.3201/eid1903.121028.

Haggerty, C.L., Gottlieb, S.L., Taylor, B.D., Low, N., Xu, F., and Ness, R.B. 2010. Risk of sequelae after *Chlamydia trachomatis* genital infection in women. *J. Infect. Dis.*, 201(suppl. 2), S134–S155.

Heijne, J.C., Althaus, C.L., Herzog, S.A., Kretzschmar, M., and Low, N. 2011. The role of reinfection and partner notification in the efficacy of *Chlamydia* screening programs. *J. Infect. Dis.*, 203, 372–377.

Holmes, K.K., Levine, R., and Weaver, M. 2004. Effectiveness of condoms in preventing sexually transmitted infections. *Bull. World Health Organ.*, 82, 454–461.

Horn, M., Collingro, A., Schmitz-Esser, S., Beier, C.L., Purkhold, U., Fartmann, B., Brandt, P., Nyakatura, G.J., Droege, M., Frishman, D., Rattei, T., Mewes, H.W., and Wagner, M. 2004. Illuminating the evolutionary history of *Chlamydiae*. *Science*, 304, 728–730.

Hosenfeld, C.B., Workowski, K.A., Berman, S., Zaidi, A., Dyson, J., Mosure, D., Bolan, G., and Bauer, H.M. 2009. Repeat infection with *Chlamydia* and gonorrhea among females: A systematic review of the literature. *Sex Transm. Dis.*, 36, 478–489.

Huston, W.M., Armitage, C.W., Lawrence, A., Gloeckl, S., Bell, S.J., Debattista, J., Allan, J.A., Timms, P., and Queensland Clinical Chlamydia Research Network. 2010. HtrA, RseP, and Tsp proteins do not elicit a pathology-related serum IgG response during sexually transmitted infection with *Chlamydia trachomatis*. *J. Reprod. Immunol.*, 85, 168–171.

Hvid, M., Baczynska, A., Deleuran, B., Fedder, J., Knudsen, H.J., Christiansen, G., and Birkelund, S. 2007. Interleukin-1 is the initiator of fallopian tube destruction during *Chlamydia trachomatis* infection. *Cell. Microbiol.*, 9, 2795–2803.

Hybiske, K., and Stephens, R.S. 2007. Mechanisms of host cell exit by the intracellular bacterium Chlamydia. *Proc. Natl. Acad. Sci. USA*, 104, 11430–11435.

Joesoef, M.R., Weinstock, H.S., Kent, C.K., Chow, J.M., Boudov, M.R., Parvez, F.M., Cox, T., Lincoln, T., Miller, J.L., Sternberg, M., and Corrections STD Prevalence Monitoring Group. 2009. Sex and age correlates of *Chlamydia* prevalence in adolescents and adults entering correctional facilities, 2005:

Implications for screening policy. *Sex Transm. Dis.*, 36(suppl. 2), S67–S71.

Karlsson, A., Österlund, A., and Forssén, A. 2011. Pharyngeal *Chlamydia trachomatis* is not uncommon any more. *Scand. J. Infect. Dis.*, 43, 344–348.

Kelly, K.A., Natarajan, S., Ruther, P., Wisse, A., Chang, M.H., and Ault, K.A. 2001. *Chlamydia trachomatis* infection induces mucosal addressin cell adhesion molecule-1 and vascular cell adhesion molecule-1, providing an immunologic link between the fallopian tube and other mucosal tissues. *J. Infect. Dis.*, 184, 885–891.

Kiviat, N.B., Wolner-Hanssen, P., Eschenbach, D.A., Wasserheit, J.N., Paavonen, J.A., Bell, T.A., Critchlow, C.W., Stamm, W.E., Moore, D.E., and Holmes, K.K. 1990. Endometrial histopathology in patients with culture-proved upper genital tract infection and laparoscopically diagnosed acute salpingitis. *Am. J. Surg. Pathol.*, 14, 167–175.

Kretzschmar, M., Satterwhite, C., Leichliter, J., and Berman, S. 2012. Effects of screening and partner notification on *Chlamydia* positivity in the United States: A modeling study. *Sex Transm. Dis.*, 39, 325–331.

Lanjouw, E., Ossewaarde, J.M., Stary, A., Boag, F., and van der Meijden, W.I. 2010. 2010 European guideline for the management of *Chlamydia trachomatis* infections. *Int. J. STD AIDS*, 21, 729–737.

Levett, P.N., Brandt, K., Olenius, K., Brown, C., Montgomery, K., and Horsman, G.B. 2008. Evaluation of three automated nucleic acid amplification systems for detection of *Chlamydia trachomatis* and *Neisseria gonorrhoeae* in first-void urine specimens. *J. Clin. Microbiol.*, 46, 2109–2111.

Li, Z., Chen, C., Chen, D., Wu, Y., Zhong, Y., and Zhong, G. 2008. Characterization of fifty putative inclusion membrane proteins encoded in the *Chlamydia trachomatis* genome. *Infect. Immun.*, 76, 2746–2757.

Malhotra, M., Sood, S., Mukherjee, A., Muralidhar, S., and Bala, M. 2013. Genital *Chlamydia trachomatis*: An update. *Indian J. Med. Res.*, 138, 303–316.

Martin, D.H., and Bowie, W.R. 1999. Urethritis in males. In: Holmes, K.K., Sparling, P.F., and Mardh, P-A., editors. *Sexually Transmitted Diseases*, 3rd ed. New York: McGraw-Hill. pp. 833–845.

Martin-Iguacel, R., Llibre, J.M., Nielsen, H., Heras, E., Matas, L., Lugo, R., Clotet, B., and Sirera, G. 2010. Lymphogranuloma venereum proctocolitis: A silent endemic disease in men who have sex with men in industrialized countries. *Eur. J. Clin. Microbiol. Infect. Dis.*, 29, 917–925.

Menozzi, F.D., Pethe, K., Bifani, P., Soncin, F., Brennan, M.J., and Locht, C. 2002. Enhanced bacterial virulence through exploitation of host glycosaminoglycans. *Mol. Microbiol.*, 43, 1379–1386.

Miller, W.C., Ford, C.A., Morris, M., Handcock, M.S., Schmitz, J.L., Hobbs, M.M., Cohen, M.S., Harris, K.M., and Udry, J.R. 2004. Prevalence of chlamydial and gonococcal infections among young adults in the United States. *J. Am. Med. Assoc.*, 291, 2229–2236.

Mylonas, I. 2012. Female genital *Chlamydia trachomatis* infection: Where are we heading? *Arch. Gynecol. Obstet.*, 285, 1271–1285.

Naaby-Hansen, S., and Herr, J.C. 2010. Heat shock proteins on the human sperm surface. *J. Reprod. Immunol.*, 84, 32–40.

Nguyen, B.D., Cunningham, D., Liang, X., Chen, X., Toone, E.J., Raetz, C.R., Zhou, P., and Valdivia, E.H. 2011. Lipooligosaccharide is required for the generation of infectious elementary bodies in *Chlamydia trachomatis*. *Proc. Natl. Acad. Sci. USA*, 108, 10284–10289.

Nieuwenhuis, R.F., Ossewaarde, J.M., Götz, H.M., Dees, J., Thio, H.B., Thomeer, M.G., den Hollander, J.C., Neumann, M.H., and van der Meijden, W.I. 2004. Resurgence of lymphogranuloma venereum in Western Europe: An outbreak of *Chlamydia trachomatis* serovar l2 proctitis in The Netherlands among men who have sex with men. *Clin. Infect. Dis.*, 39, 996–1003.

Nunes, A., Borrego, M.J., and Gomes, J.P. 2013. Genomic features beyond *Chlamydia trachomatis* phenotypes: What do we think we know? *Infect. Genet. Evol.*, 16, 392–400.

Palmer, L., and Falkow, S. 1986. A common plasmid of *Chlamydia trachomatis*. *Plasmid*, 16, 52–62.

Peipert, J.F. 2003. Clinical practice. Genital chlamydial infections. *N. Engl. J. Med.*, 349, 2424–2430.

Peters, R.P., Verweij, S.P., Nijsten, N., Ouburg, S., Mutsaers, J., Jansen, C.L., van Leeuwen, A.P., and Morré, S.A. 2011. Evaluation of sexual history-based screening of anatomic sites for *Chlamydia trachomatis* and *Neisseria gonorrhoeae* infection in men having sex with men in routine practice. *BMC Infect. Dis.*, 11, 203.

Rank, R.G., and Whittum-Hudson, J.A. 2010. Protective immunity to chlamydial genital infection: Evidence from animal studies. *J. Infect. Dis.*, 201(suppl. 2), S168–S177.

Rasmussen, S.J., Eckmann, L., Quayle, A.J., Shen, L., Zhang, Y.X., Anderson, D.J., Fierer, J., Stephens, R.S., and Kagnoff, M.F. 1997. Secretion of proinflammatory cytokines by epithelial cells in response to *Chlamydia* infection suggests a central role for epithelial cells in chlamydial pathogenesis. *J. Clin. Invest.*, 99, 77–87.

Rodgers, A.K., Wang, J., Zhang, Y., Holden, A., Berryhill, B., Budrys, N.M., Schenken, R.S., and Zhong, G. 2010. Association of tubal factor infertility with elevated antibodies to *Chlamydia trachomatis* caseinolytic protease P. *Am. J. Obstet. Gynecol.*, 203, 494.e7–494.e14.

Schachter, J., Chow, J.M., Howard, H., Bolan, G., and Moncada, J. 2006. Detection of *Chlamydia trachomatis* by nucleic acid amplification testing: Our evaluation suggests that CDC-recommended approaches for confirmatory testing are ill-advised. *J. Clin. Microbiol.*, 44, 2512–2517.

Solomon, A.W., Peeling, R.W., Foster, A., and Mabey, D.C. 2004. Diagnosis and assessment of trachoma. *Clin. Microbiol. Rev.*, 17, 982–1011.

Stark, D., van Hal, S., Hillman, R., Harkness, J., and Marriott, D. 2007. Lymphogranuloma venereum in Australia: Anorectal *Chlamydia trachomatis* Serovar L2b in men who have sex with men. *J. Clin. Microbiol.*, 45, 1029–1031.

Stephens, R.S., Kalman, S., Lammel, C., Fan, J., Marathe, R., Aravind, L., Mitchell, W., Olinger, L., Tatusov, R.L., Zhao, Q., Koonin, E.V., and Davis, R.W. 1998. Genome sequence of an obligate intracellular pathogen of humans: *Chlamydia trachomatis*. *Science*, 282, 754–759.

Su, H., Raymond, L., Rockey, D.D., Fischer, E., Hackstadt, T., and Caldwell, H.D. 1996. A recombinant Chlamydia trachomatis major outer membrane protein binds to heparan sulfate receptors on epithelial cells. *Proc. Natl. Acad. Sci. USA*, 93, 11143–11148.

Thylefors, B., Dawson, C.R., Jones, B.R., West, S.K., and Taylor, H.R. 1987. A simple system for the assessment of trachoma and its complications. *Bull. World Health Organ.*, 65, 477–483.

UN. 2010. The Millennium Development Goals Report. United Nations. 2010 Report. Available at: http://www.un.org/millenniumgoals/pdf/MDG%20Report%202010%20En%20r15%20-low%20res%2020100615%20-.pdf. Accessed 17 November 2013.

US Preventive Services Task Force. 2007. Screening for chlamydial infection: U.S. Preventive Services Task Force recommendation statement. *Ann. Intern. Med.*, 147, 128–134.

Van der Bij, A.K., Spaargaren, J., Morré, S.A., Fennema, H.S., Mindel, A., Coutinho, R.A., and de Vries, H.J. 2006. Diagnostic and clinical implications of anorectal lymphogranuloma venereum in men who have sex with men: A retrospective case–control study. *Clin. Infect. Dis.*, 42, 186–194.

Varela, J.A., Otero, L., García, M.J., Palacio, V., Carreño, F., Cuesta, M., Sánchez, C., and Vázquez, F. 2003. Trends in the prevalence of pathogens causing urethritis in Asturias, Spain, 1989–2000. *Sex Transm. Dis.*, 30, 280–283.

Watson, E.J., Templeton, A., Russell, I., Paavonen, J., Mardh, P.A., Stary, A., and Pederson, B.S. 2002. The accuracy and efficacy of screening tests for *Chlamydia trachomatis*: A systematic review. *J. Med. Microbiol.*, 51, 1021–1031.

Wehrl, W., Brinkmann, V., Jungblut, P.R., Meyer, T.F., and Szczepek, A.J. 2004. From the inside out-processing of the Chlamydial autotransporter PmpD and its role in bacterial adhesion and activation of human host cells. *Mol. Microbiol.*, 51, 319–334.

White, J., O'Farrell, N., and Daniels, D. 2013. 2013 UK National Guideline for the management of lymphogranuloma venereum. *Int. J. STD AIDS*, 24, 593–601.

Wikström, A., Rotzén-Ostlund, M., and Marions, L. 2010. Occurrence of pharyngeal *Chlamydia trachomatis* is uncommon in patients with a suspected or confirmed genital infection. *Acta Obstet. Gynecol. Scand.*, 89, 78–81.

World Health Organization. 2000. Consultation on STD interventions for preventing HIV: What is the evidence? Geneva, Joint United Nations Programme on HIV/AIDS (UNAIDS) and World Health Organization (WHO), 2000. Available at: http://www.who.int/hiv/pub/sti/who_hsi_2000_02.pdf. Accessed 10 November 2013

World Health Organization (WHO). 2007. Global strategy for the prevention and control of sexually transmitted infections: 2006–2015. Available at: http://whqlibdoc.who.int/publications/2007/9789241563475_eng.pdf. Accessed 15 October 2013.

Chapter 35

Pertussis or Whooping Cough

Camille Locht[1,2,3,4,5] and Nathalie Mielcarek[1,2,3,4,5]

[1]Univ. Lille, U1019 - UMR 8204 - CIIL - Centre d'Infection et d'Immunité de Lille, F-59000 Lille, France

[2]CNRS, UMR 8204, F-59000 Lille, France

[3]Inserm, U1019, F-59000 Lille, France

[4]CHU Lille, F-59000 Lille, France

[5]Institut Pasteur de Lille, F-59000 Lille, France

35.1 Introduction

Pertussis or whooping cough is a severe and highly contagious respiratory tract disease that can be fatal, especially during the first months of life. The main etiological agent of this disease is the Gram-negative bacterium *Bordetella pertussis*. However, whooping cough-like diseases can also be caused by the relatively closely related *B. parapertussis*, although *B. parapertussis*-induced disease is usually milder than that induced by *B. pertussis* (Heininger et al., 1994). In recent years, other, more distantly related members of the *Bordetella* genus, such as *B. holmesii*, have also been associated with pertussis (Mazengia et al., 2000). There is increasing evidence that the incidences of *B. parapertussis* (Cherry and Seaton, 2012) and *B. holmesii* (Njamkepo et al., 2011) infections are largely underestimated. While *B. pertussis* is a strictly upper respiratory pathogen, *B. parapertussis* (Wallihan et al., 2013) and *B. holmesii* (Shepard et al., 2004) can cause bacteremia, albeit rarely and mostly in immunocompromised subjects. Other members of the *Bordetella* genus, including *B. bronchiseptica* (Mattoo and Cherry, 2005), *B. hinzii* (Funke et al., 1996), and *B. petrii* (Le Coustumier et al., 2011), can also cause respiratory, as well as disseminated infections in humans. Still other *Bordetella* species, such as *B. trematum* (Vandamme et al., 1996) and *B. ansorpii* (Ko et al., 2005) have occasionally been isolated from wounds or epidermal cysts. However, these infections usually occur in patients with other underlying diseases.

Whooping cough is a relatively recent disease. It was first described in the second half of the 15th century (Cherry and Heininger, 2004), although descriptions of symptoms that resemble pertussis date back to Hippocrates around 400 BC (Versteegh, 2005). *B. pertussis* was first isolated in the beginning of the 20th century, and its role as the causative agent of whooping cough was conclusively established in 1906 (Bordet and Gengou, 1906).

Up to the first half of the 20th century, the calculated average pertussis attack rate was close to 1% of the population, with epidemic peaks occurring approximately every 3 years (Cherry, 1999). By far, most of the cases occurred in children less than 5 years of age, and only a very small proportion of cases were reported in persons who were older than 15 years. Moreover, more than 95% of all pertussis-linked deaths occurred in less than 5-year-old children. The introduction of mass vaccination programs in the 1950s and 1960s has dramatically changed the epidemiology of the disease and has in some countries reduced the incidence to nearly negligible levels, with a concomitant decline of infant deaths due to pertussis. Nevertheless, the disease has not been totally controlled, and in no country has the disease been eliminated despite several decades of mass vaccination. In fact, since the 1980s, a constant increase in incidence has been reported in countries with high immunization coverage, especially in adolescents and adults, as well as in pre-vaccination infants (Crowcroft and Pebody, 2006), making pertussis today the most poorly controlled vaccine-preventable childhood disease in the world.

35.2 Clinical manifestations

The symptoms of pertussis depend on age and on the immune status of the infected individual. Typical disease progression can be separated into three stages: catarrhal, paroxysmal, and convalescent. Symptoms during the first stage, which lasts usually for 1–2 weeks, are very similar to those of a common cold, with a runny nose and occasional non-productive cough. In the majority of cases, pertussis is only suspected and diagnosed during the following, paroxysmal stage when the most typical and severe clinical manifestations occur. These include intense bouts of coughing followed by an inspiratory "whooping" sound (Rutledge and Keen, 2012), vomiting, and, sometimes, cyanosis in young children. This stage can be particularly exhausting for the infected person, especially following an episode of paroxysmal cough. Up to 15 episodes of coughing bouts can be experienced each day and this can go on from 2 to 10 weeks. Very young infected infants not yet vaccinated usually do not cough because their chest wall musculature is not sufficient to create a whoop. They are, nevertheless, particularly prone to develop serious and sometimes life-threatening clinical complications, such as pneumonia, otitis media, apnea, hypotension, dehydration, weight loss due to excessive vomiting, and brain damage due to oxygen deprivation (Heininger et al., 1997; Smith and Vyas, 2000). More than 90% of infants younger than 2 months who are suspected to have pertussis are therefore hospitalized to prevent or treat complications (Long and Edwards, 2003). The final stage of pertussis is a long convalescence lasting up to 1 year for a full recovery. During that stage, the clinical symptoms gradually decrease in severity and frequency.

35.3 *Bordetella* virulence

The clinical manifestations of whooping cough result from the colonization of the human respiratory tract by *B. pertussis*. Successful infection of the host by *B. pertussis* requires the production of a large number of virulence factors, including adhesins and toxins (for recent review, see Mielcarek and Locht, 2013). The bacteria are transmitted through inhalation of droplets produced during the cough of an infected individual. Initial adherence to ciliated epithelial cells of the nasopharynx is promoted by several adhesins, among which filamentous hemagglutinin (FHA), the fimbriae, and pertactin are well studied.

FHA is encoded by the *fhaB* gene as a large, 370-kDa precursor, which is processed to yield an approximately 230-kDa mature FHA protein after the removal of a large C-terminal fragment by a specific protease of the subtilisin family, called SphB1 (Coutte et al., 2001; Domenighini et al., 1990). This monomeric protein is both surface-associated and secreted via a two-partner secretion (TPS) pathway involving a transporter located in the outer membrane and called FhaC (Delattre et al., 2011; Hodak et al., 2006). Binding of FHA to host cell-surface receptors is mediated by several binding domains present on the molecule, such as an Arg-Gly-Asp (RGD) domain interacting with the integrin CR3 ($\alpha_M\beta_2$, CD11bCD18) present at the surface of macrophages (Relman et al., 1990), a carbohydrate-binding site, and a heparin-binding domain involved in the interaction with sulfated carbohydrates present at the surface of

respiratory epithelial cells and sulfated proteoglycans contained in the extracellular matrix (Hannah et al., 1994; Menozzi et al., 1991). Binding of FHA requires cholesterol-containing lipid rafts in the host cell membrane (Lamberti et al., 2009).

In addition to FHA, *B. pertussis* also expresses fimbriae, initially called agglutinogens. Two closely related but serologically distinct fimbriae can be produced and are defined by either Fim2 or Fim3 major subunits associated to a common minor subunit called FimD (Blom et al., 1983). The major and the minor subunits bind to sulfated sugars (Geuijen et al., 1996, 1997, 1998), whereas FimD was also shown to bind to very late antigen-5 (VLA-5) integrin on monocytes (Hazenbos et al., 1995).

The role of pertactin in the pathogenesis of *B. pertussis* is still unclear. Pertactin belongs to the autotransporter family and is non-covalently attached at the surface of the bacteria. This protein contains an RGD domain potentially involved in adherence of *B. pertussis* to monocytes (Leininger et al., 1991). On the other hand, a natural pertactin-deficient clinical isolate was shown to present higher invasion ability of human dendritic cells compared to the reference strain (Stefanelli et al., 2009). Pertactin does not seem to play a significant role in bacterial colonization of the host respiratory tract (Khelef et al., 1994; Nicholson et al., 2009; Roberts et al., 1991). Using a pertactin-deficient *B. bronchiseptica* strain, Inatsuka et al. (2010) showed that pertactin plays a role in survival of the bacteria in the host respiratory tract by overcoming neutrophil-mediated clearance, rather than contributing to initial bacterial adherence to host cells.

In addition to adhesins, *B. pertussis* toxins are also key players in the early host–pathogen interaction (Hewlett and Donato, 2007). Among the toxins expressing cytotoxic and immunomodulatory activities, three of them were shown to be critical for the establishment of a niche in the host respiratory tract: pertussis toxin (PTX), adenylate cyclase toxin (ACT), and tracheal cytotoxin (TCT).

PTX is a complex, multimeric protein exclusively produced by *B. pertussis* and composed of five different subunits, named S1–S5, according to their decreasing molecular weights (for review, see Locht et al., 2011). It is a member of the ADP-ribosylating proteins with a typical A–B structure (Tamura et al., 1982). The majority of the biological and toxic activities are carried by the A protomer corresponding to the S1 subunit, which catalyzes ADP-ribosylation of the α subunit of trimeric G proteins in target host cells (Katada and Ui, 1982). PTX assembly and secretion relies on a complex mechanism. The subunits are first translocated into the periplasm, assembled, and then secreted through the bacterial outer membrane via a specific machinery composed of the Ptl proteins, which belong to a type IV secretion system (T4SS) family (Farizo et al., 2000; Llosa et al., 2009). Once secreted, PTX is able to interact with virtually all host cells through its B oligomer and disturbs their metabolic functions. The S2 and S3 subunits contained in the B oligomer possess at least two carbohydrate-binding sites (Saukkonen et al., 1992; Stein et al., 1994) but they are not essential for bacterial adherence, since *B. pertussis* mutants deficient in PTX production bind as efficiently to epithelial cell lines (van den Berg et al., 1999) and colonize as efficiently the mouse respiratory tract (Alonso et al., 2001) as their isogenic PTX-producing parental strains. However, immunomodulatory activities of PTX, such as neutrophil chemotaxis, may impact on the early steps of bacterial colonization (Carbonetti et al., 2003). PTX is also able to modulate adaptive immune responses. It promotes the maturation of dendritic cells, resulting in the production of pro-inflammatory cytokines followed by an enhanced Th1-type immune response (Hou et al., 2003). PTX has been used as an adjuvant able to enhance both Th1- and Th2-type CD4$^+$ T cells (Ryan et al., 1998) and to promote Th-17 differentiation through IL-6 induction (Chen et al., 2007).

ACT is another key toxin for *B. pertussis* to establish infection (Ladant and Ullmann, 1999). This 177-kDa monomeric protein belongs to the RTX (Repeat in ToXin) family (Ladant et al., 1986) and is secreted through a specific type I secretion system (Glaser et al., 1988). The C-terminal domain of the toxin is responsible for the binding to various target cells, including macrophages, neutrophils, dendritic, natural killer cells, and lymphocytes (Guermonprez et al., 2001; Paccani et al., 2011). The N-terminal region, which expresses the catalytic activity of the toxin, is cleaved off within the target cells by calcium-activated cellular calpain, and this enzymatically active domain then localizes inside the nucleus of the target cells and with mitochondria (Uribe et al., 2013). The catalytic domain is activated by calmodulin and induces a rapid increase of the intracellular cAMP concentration. As a consequence, the macrophages display reduced proinflammatory and bactericidal functions, such as phagocytic and oxidative activities, chemotaxis, and apoptosis (Gueirard et al., 1997). Consistently, ACT-deficient *B. pertussis* is cleared faster than an ACT-producing strain from the lungs of infected mice (Carbonetti et al., 2005; Khelef et al., 1992). In addition, it has been shown that ACT interacts with and may enhance FHA adherence properties (Perez Vidakovics et al., 2006; Zaretzky et al., 2002). More recently, Henderson et al. (2012) showed that both ACT and FHA are involved in the suppression of IL17-mediated inflammation as a strategy to facilitate the persistence of the bacteria in the respiratory tract of the infected host.

The destruction of ciliated respiratory epithelial cells in the trachea is a key step in the infection process, since it prevents mechanical clearance of *B. pertussis* early after infection. In addition, this toxic effect might well be involved in the cough syndrome of pertussis, since the destruction of ciliated cells results in the accumulation of mucus. Using hamster tracheal rings, Cookson et al. (1989) demonstrated that purified TCT induces the loss of ciliary activity and ciliated cell extrusion. TCT is a muramyl peptide breakdown product of peptidoglycan constitutive of the *B. pertussis* cell wall and released during bacterial growth (Rosenthal et al., 1987). TCT and endotoxins act synergistically to trigger inflammatory responses through the induction of nitric oxide and interleukin-1 production by respiratory epithelial cells (Flak et al., 2000; Heiss et al., 1993a, 1993b).

35.4 Pertussis as a re-emergent disease

35.4.1 Pertussis in the pre-vaccine era

Before the widespread use of pertussis vaccines, the number of reported pertussis cases was as high as 200,000–250,000 every year in a country like the United States (CDC, 1980), resulting in an annual incidence of more than 150 cases per 100,000 population. However, it may be possible that only around 20% of cases were reported which would bring the actual attack rate up to almost 900/100,000 population (Cherry, 1999). Pertussis occurred typically in epidemic peaks of 2–5 years, with an average of every 3 years (Fine and Clarkson, 1982).

Death occurred in approximately 4–5% of the pertussis cases, and in 1900, the infant mortality rate from pertussis in the United States was 4.5 deaths/1000. In the first part of the 20th century, whooping cough ranked as the leading cause of death from communicable disease among children younger than 14 years in the United States (Black, 1997). Strikingly, the attack rate in females was higher than that in males, unlike most other childhood infectious diseases, which show a greater prevalence in boys. This gender difference increased with age.

Globally, pertussis disease was responsible for more deaths during the first year of life than were measles, diphtheria, or scarlet fever. In contrast to these latter diseases, to which newborn babies had a high degree of resistance, newborns were susceptible to whooping cough from the day of birth, and 40% of all pertussis deaths occurred in babies under 5 months of age. Above 14 years, pertussis-linked death was rare, and only 3% of all pertussis cases occurred at ages above 14 years (Cherry, 1999).

Adult pertussis was occasionally seen, but was rare, suggesting that childhood infection did protect against second attacks. Early studies have shown that only a very small fraction (less than 0.5%) of pertussis cases were second attacks (Cherry, 1999). However, second attacks may have been more common in the elderly, suggesting that infection-induced immunity is not lifelong, although they were usually less severe than the first attacks.

35.4.2 Decline of pertussis through mass vaccination and change in epidemiology

Following a large-scale vaccination during the 1950s, a dramatic reduction by more than 90% of incidence and mortality from pertussis was observed in most of the industrialized countries (WHO, 2010). In some countries, such as the United States, the incidence of reported pertussis cases fell by more than 99% in the 1970s, compared to the pre-vaccine era, and reached a nadir, down to 0.47 cases/100,000 population in 1979 (Black, 1997). However, even before the introduction of pertussis vaccines, death rates from pertussis steadily declined by 70% from the early 1900s to the late 1930s (Mortimer and Jones, 1979). Nevertheless, there was an accelerated decline in mortality from the 1940s onward, which is most likely explained by the vaccine effectiveness.

Curiously and unlike other diseases that have been controlled by vaccination, pertussis epidemics continued to occur regularly every 3–4 years, with an unchanged inter-epidemic periodicity compared to the pre-vaccination era (Fine and Clarkson, 1982). This suggests that vaccination against pertussis is more effective in protecting against disease than in controlling the circulation of B. pertussis, as effectiveness of vaccine programs in reducing pathogen transmission is expected to result in an increase in inter-epidemic periods. However, these initial observations are in contrast with a more recent study, showing an increase in periodicity of pertussis outbreaks by 1.27 years after vaccination (Broutin et al., 2010), pointing to reduced disease transmission following the implementation of immunization programs, due to strong herd immunity effects.

Mass vaccination changed the epidemiology of the disease, and adolescent and adult pertussis became increasingly more common. Nevertheless, most pertussis cases still occurred in less than 1-year-old infants. In the years 1982 and 1983, 53.1% of the reported pertussis cases in the United States occurred in that age group, whereas in Germany and Italy, where vaccination was limited at that time, the less than 1-year-old infants accounted for fewer than 15% of total cases (Black, 1997).

However, underreporting of pertussis is frequent. Clinical diagnosis is complicated by the wide heterogeneity of the disease expression and its modification by immunization or prior B. pertussis infection. Therefore, it may remain unnoticed by many physicians. The classic form of the disease in the pre-vaccine era was described as the occurrence of paroxysmal cough, posttussive vomiting, inspiratory whoop, and a duration of cough lasting from more than 28 days up to 3 months. However, in the era of pertussis vaccination, the disease is often atypical or mild, of shorter duration, and not necessarily associated with whoop (Cherry et al., 2005). The atypical forms of pertussis are now more commonly seen in adolescents and adults who had experienced pertussis or had been vaccinated during childhood. More than 10% of adolescents with cough that lasts for at least 1–2 weeks may actually be infected bt B. pertussis. Proper diagnosis of pertussis should therefore not only rely on clinical criteria, but also on laboratory studies.

Serological investigations have been particularly useful to establish a link between prolonged atypical cough and B. pertussis infection, and sero-epidemiological studies based on the very B. pertussis-specific PTX serology have indicated that B. pertussis infection is more common in adolescents and adults than assumed solely by clinical diagnosis. Interestingly, infection in adolescents and adults is more frequent in high-coverage countries than in low-coverage countries, where infection is more frequent in children than in adolescents and adults (Pebody et al., 2005).

Clinical evaluations together with laboratory confirmation may lead to case definitions that can be adapted to specific groups of patients, such as different age cohorts, and a case definition algorithm has been proposed taking into account the various specificities among different cohorts (Cherry et al., 2012).

35.4.3 The re-emergence of pertussis

Despite continued vaccination in most of the industrialized countries, pertussis rates have started to rise again in the United States in the 1980s, and this increase was apparent in all age groups, although infants between 1 and 2 months were at the highest risk for pertussis (Tan et al., 2005). In other countries, such as Canada, Australia, Argentina, Israel, Finland, Norway, and the Netherlands (Mooi and de Greeff, 2007), a similar increase in pertussis incidence has been observed after a period of apparent happy honey moon.

In the vaccine era, pertussis is increasingly reported in older children, adolescents, and adults. However, since in these age groups pertussis is often atypical, it may frequently remain unrecognized. A serological study from the United States showed that more than 20% of adults with prolonged cough of more than 2 weeks may have pertussis (Wright et al., 1995). The incidence also continues to increase among the young infants, in whom essentially all mortality occurs, and there has been a considerable shift in morbidity towards infants less than 6 months of age (Elliott et al., 2004). In the United States, the case fatality rate for pertussis was 0.2% in the period 1997–2000, whereas in developing countries, it can reach 4% in infants less than 1 year of age, and 1% in infants between 1 and 4 years (Crowcroft and Pebody, 2006). In France, pertussis is currently the primary infectious cause of mortality in newborns and infants younger than 2 months, and 13 pertussis deaths were reported in 2000 (Floret et al., 2001). Clearly, pertussis vaccination did not reduce infection and transmission rate to the same extend as has been seen for other vaccine-preventable childhood diseases.

The World Health Organization estimated that, in 2008, about 16 million pertussis cases occurred worldwide, of which 95% were in the developing countries, and about 195,000 children died from the disease (Black et al., 2010). The global vaccination coverage with three doses of pertussis vaccine has reached 83% in 2011, thanks to the Expanded Programme on Immunization by the World Health Organization (WHO, 2012). However, in Africa and South-East Asia, the countries counting for 95% of the global pertussis cases, vaccination coverage reached only 71% and 75%, respectively, and in some countries vaccination coverage was less than 70%, probably explaining why these countries currently pay the highest toll.

In very recent years, outbreaks of pertussis with concomitant pertussis deaths have been spectacular in several parts of the world where pertussis vaccination is widely used. In 2010, California experienced a large outbreak with over 9000 cases, the highest number in 60 years, resulting in 10 deaths and 809 hospitalizations. The incidence rate was 23.4 per 100,000 inhabitants (Winter et al., 2012). In the first half of 2012, the state of Washington experienced an epidemic with a total of more than 2500 cases reported between January and June (CDC, 2012). The highest incidence was observed in infants less than 1 year and children aged 10–14 years, and roughly 40% of hospitalized children were less than 2 months old. Over the entire US population, approximately 50,000 cases of pertussis were reported in 2012, the highest number since 1955, with 18 deaths (Allen, 2013). Strong epidemic peaks started to appear in 2004 (Cherry, 2012). Other countries reported similar outbreaks. From the third quarter of 2011 into 2012, the number of laboratory-confirmed pertussis cases in England and Wales was the highest for more than 12 years. Most of the pertussis cases were reported in the adolescent and adult ages, but also found extending into infants less than 3 months of age, resulting in 13 deaths in these infants (Van Hoek et al., 2013).

35.4.3.1 Reasons for the re-emergence of pertussis

The reasons for the re-emergence of pertussis are still a matter of debate and may include increased awareness, improved diagnostics, a change from first-generation to second-generation vaccines, changes in vaccine quality, waning of immunity, and *B. pertussis* strain adaptation (Cherry, 2012). Although increased awareness and better diagnostic tools may account in part for the apparent increased incidence in atypical pertussis, especially in adolescents and adults, it is more difficult to imagine that they would be the main reason for the increase in complications and deaths associated with pertussis in unimmunized infants.

A change in *B. pertussis* population structure has first been observed in the Netherlands and proposed to be a consequence of the introduction of pertussis vaccines in the 1950s (Mooi et al., 2001). Changes in genes coding for surface or secreted proteins, especially pertactin and PTX, were observed between strains circulating before and after vaccine introduction in the Netherlands. Both proteins are present in most pertussis vaccines. Polymorphisms in PTX were very limited, whereas polymorphisms in pertactin were somewhat more substantial. Little variation was seen in other proteins studied. Differences between vaccine strains and circulating strains have also been detected in other countries, such as Finland, Australia, and the United States (Mooi et al., 2013). However, except for the Netherlands, these variants did not appear to be associated with changes in vaccine efficacy in other countries.

More recent polymorphisms were discovered in the promoter region of the PTX gene, and one of which, designated *ptxP3*, led to increased production of PTX, and hence perhaps to increased virulence of *B. pertussis* (King et al., 2013; Mooi et al., 2008). Although hospitalization and death rates were higher in the Netherlands in a period where *ptxP3* strains dominated, compared to earlier periods, globally, the case fatality rate has remained constant over recent years, in spite of the increased rates in pertussis deaths in young infants (Crowcroft and Pebody, 2006). These observations suggest that the modern *B. pertussis* strains are not necessarily more virulent than the ancient strains. Whether such nucleotide polymorphisms have, in fact, a global impact on vaccine escape remains thus still to be confirmed, especially for escape from the first-generation, whole-cell pertussis vaccines, which contain thousands of different antigens, many of which may perhaps contribute to vaccine-induced protection.

However, as whole-cell vaccines are being replaced by the new-generation, acellular vaccines (see below), polymorphisms of genes encoding the antigens included in these acellular vaccines may perhaps be more likely to contribute to vaccine effectiveness loss. It is striking to see that very recently in countries where acellular pertussis vaccines are massively used, clinical strains lacking the genes coding for pertactin or PTX, both included in most acellular pertussis vaccines, have emerged. A PTX-deficient strain was isolated from an infant with suspected pertussis, but who did not display hyperlymphocytosis (Bouchez et al., 2009). While strains lacking PTX are less virulent than PTX-positive strains, the lack of pertactin does not appear to affect *B. pertussis* virulence in humans. No major clinical differences were observed between <6 month-old infants infected with pertactin-deficient strains compared to age-matched infants infected with pertactin-positive strains (Bodilis and Guiso, 2013). Pertactin-deficient isolates were first discovered in Japan in 1997 (Otsuka et al., 2012). Now they have been isolated in France (Bouchez et al., 2009), Italy (Stefanelli et al., 2009), Finland (Barkoff et al., 2012), and the United States (Queenan et al., 2013), in addition to Japan. The emergence of pertactin-deficient isolates is not anecdotal, as in countries such as Japan and France, the proportion of such isolates has been steadily increasing since 1997 to reach 13% of all *B. pertussis* isolates in France in 2011 (Hegerle et al., 2012) and 27% in Japan

(Otsuka et al., 2012). A study on 12 isolates from children hospitalized in Philadelphia during 2011 and 2012 revealed that 11 of them did not produce pertactin, although several of them are likely to be clustered (Queenan et al., 2013).

Finally, waning immunity after vaccination and the change from whole-cell to acellular vaccines are also likely contributors to the current pertussis re-emergence, especially the increase in incidence in older children and young teenagers, as discussed in Section 35.6 Shortcomings of current pertussis vaccines.

35.5 Pertussis vaccines

35.5.1 Whole-cell vaccines

Shortly after the first description of *B. pertussis* as the main etiological agent of whooping cough, attempts were undertaken by Bordet and Gengou to develop pertussis vaccines. Killed whole-cell vaccines were first tested in children in France, Tunisia, and Denmark in the second decade of the 20th century (Howson et al., 1991). In the 1930s, the whole-cell vaccines were further refined by Kendrick and Eldering and combined with diphtheria and tetanus toxoids since 1942 to produce the combined diphtheria–pertussis–tetanus (DPT) vaccines, which has been recommended by the American Academy of Pediatrics since 1947. This resulted in a massive use of DPT vaccines since the 1950s in the United States and in the 1960s in many other countries, which, in turn, resulted in a spectacular decline of pertussis morbidity and mortality.

However, adverse events, including serious adverse events and death subsequent to the use of DPT vaccines including a whole-cell pertussis component, have been reported as early as the 1933 (Madsen, 1933; Sauer, 1933). It was in the 1970s and early 1980s, when the pertussis incidence had reached a nadir that the question about the safety of whole-cell pertussis vaccines was widely publicized in the lay press, leading to a different meaning of the DPT acronym as "Dissatisfied Parents Together" (Coulter and Fischer, 1985). As a consequence, vaccination coverage dropped tremendously in several countries (British Medical Journal, 1981) and resulted in a complete vaccination stop in some countries such as Japan and Sweden. Immediately, pertussis epidemics increased in these countries, thereby, in fact, illustrating the effectiveness of high coverage with whole-cell pertussis vaccines (Storsaeter et al., 2007). The controversy about the alleged, rare severe adverse events of whole-cell pertussis vaccines, including encephalopathy and death, has never been fully settled since (Baker, 2003). More frequent, milder adverse reactions, such as fever, local pain, redness, and swelling, associated with the whole-cell pertussis vaccines (Cody et al., 1981), were probably due to the lipopolysaccharides contained in these vaccines.

35.5.2 Acellular vaccines

These allegations, together with the variability of efficacy between whole-cell vaccines produced by different manufacturers, have led to the development of new-generation, acellular vaccines, composed of purified protective *B. pertussis* antigens. The first acellular pertussis vaccines were developed in Japan in 1970 and have shown efficacy after three doses beginning at 2 years of age (Sato et al., 1984). These first acellular pertussis vaccines were composed essentially of chemically detoxified PTX and FHA. A clearly improved safety profile and comparable efficacy to that of whole-cell vaccines (Zhang et al., 2011) have resulted in the progressive replacement of whole-cell pertussis DPT vaccines by acellular DTaP vaccines in most industrialized countries. In Europe, all countries, except for Poland, now use DTaP vaccines for the primary vaccination series (http://www.euvac.net).

However, the available acellular vaccines vary in their antigen content and composition (for review, see Locht and Mielcarek, 2012). All contain at least detoxified PTX, most contain FHA in addition, and many further contain pertactin and sometimes fimbriae. The amounts of the different antigens may vary between different vaccines.

As for the whole-cell vaccines, the primary vaccination series usually comprises three immunizations given at one- or two-months intervals, starting at two or three months of age and often followed by additional administrations in the second year of life. However, in spite of the improved safety profile of DTaP compared to DPT vaccines, repeated administrations of DTaP vaccines may nevertheless sometimes induce local reactions, including whole limb swelling (Rennels, 2003). Therefore, reduced antigen content vaccines (Tdap) were developed and are now recommended for use as booster vaccines for adolescents and adults (CDC, 2011).

35.6 Shortcomings of current pertussis vaccines

Although the initial efficacy trials with DTaP vaccines suggested that they were as efficacious as good-quality DPT vaccines (Zhang et al., 2011), more recent evidence indicates that DTaP vaccines are less effective in controlling pertussis than whole-cell vaccines. In particular, the duration of vaccine-induced immunity appears to be less long after immunization with DTaP, as compared to DPT vaccines. During a recent pertussis outbreak in California in 2010–2011, it became apparent that 11–17-year-old teenagers who had received four doses of DTaP vaccines were six times as likely to be diagnosed with pertussis than those who had received DTP vaccines (Klein et al., 2013; Witt et al., 2013). Teenagers who had received mixed DTP/DTaP vaccines had an intermediate likelihood for being diagnosed with pertussis, and the decreasing number of DTP vaccine doses was associated with increased pertussis risk. These observations confirmed earlier fears based on a parallel study during the same outbreak providing evidence that the effectiveness of DTaP vaccines given during the preschool-age was markedly lower than that expected in the 8–12-year-old group (Witt et al., 2012). Efficacy after five doses of DTaP vaccine dropped from 98% in the first 12 months after vaccination to 70% 5 years later (Misegades et al., 2012). The increase in odds of acquiring pertussis has been estimated to be 42% each year after the fifth dose of DTaP vaccine (Klein et al., 2012).

Similar conclusions on rapid waning of DTaP vaccine-induced immunity have been reached by investigating pertussis outbreaks in other states, such as Minnesota and Oregon, where risk ratios increased by fourfold to fivefold within 6 years after the five-dose series of solely DTaP vaccine administration (Tartof et al., 2013). Again, the reported rates of pertussis starting at 10 years of age were significantly lower among children who had started with a DTP vaccine than among those who had started with a DTaP vaccine, and this difference remained even after Tdap boosters (Liko et al., 2013).

The difference in waning immunity after DTaP compared to DTP vaccines is not unique to the United States. Australia also experienced recent outbreaks with the highest incidence rates in 6–11-year-old children. Here again, children who had received three doses of DTaP vaccine during the primary vaccination course had higher rates of pertussis than those who had received three doses of DTP vaccine (Sheridan et al., 2012). Among those who had received mixed courses, the incidence rates were highest for children who had received the DTaP vaccine as the first dose, whereas children who had received the DTP vaccine as the initial dose had rates that were between those of pure DTaP and pure DTP recipients. This difference persisted for more than a decade. Thus, priming with DTP vaccines substantially diminishes the risk of subsequent pertussis as opposed to vaccination with DTaP vaccine only, indicating that the initial vaccine received may be the most important factor for sustained immunity. Receiving DTP vaccines as part of the primary series of pertussis vaccines also enhanced the effectiveness of the booster doses with Tdap vaccines in the United States (Witt et al., 2013).

Rapid waning of DTaP vaccine-induced protection had previously also been noticed in Sweden, where incidence rates increased from 8/100,000 person-years 1 year after three doses of DTaP vaccine administrations to 48/100,000 person-years 5–6 years later (Gustafsson et al., 2006). The relative risk of pertussis was again higher for DTaP vaccine recipients, compared to DTP vaccine recipients. In Senegal, it has been observed early on that DTaP vaccine-induced immunity starts to wane 18 months after the last dose, and the DTaP vaccine afforded significantly less protection than the DTP vaccine comparator (Simondon et al., 1997). However, in this study, the DTaP vaccine was a two-component vaccine containing only PTX and FHA as *B. pertussis* antigens, whereas most of the current DTaP vaccines also contain pertactin, and head-to-head efficacy trials have shown that the efficacy of DTaP vaccines increases with the number of pertussis components (Olin et al., 1997). Nevertheless, even when compared to five-component DTaP vaccines, an efficacious whole-cell vaccine appears to be more protective in long-term follow-up studies (Gustafsson et al., 2006).

The reasons for these differences between whole-cell and acellular vaccines are not entirely clear. The differences may be a consequence of the type of immune reactions generated by these vaccines (Vickers et al., 2006). While DTaP vaccines typically induce Th2-type responses that may even affect the response to non-related antigens (Mascart et al., 2007), DTP vaccines, similar to natural infection, essentially induce Th1-type responses (Mascart et al., 2003), and several lines of evidence suggest that Th1-type T cell responses, in addition to antibody responses, contribute to protection against pertussis (Higgs et al., 2012).

35.7 Attempted solutions to the pertussis problem in teenagers

As a potential solution to the problem of the increasing pertussis incidence in young teenagers in the absence of a more effective and durable pertussis vaccine, repeated administrations of booster doses have been proposed. However, although acellular vaccines have extensively been documented to be safer than whole-cell vaccines (Zhang et al., 2011), local reactions still occur after repeated administrations of DTaP vaccines (Rennels 2003). First reported in 1997 (Schmitt et al., 1997), the mechanisms leading to these local reactions are still not fully elucidated. However, there have been suggestions that they could be related to the Th2-type responses induced by the DTaP vaccines (Rowe et al., 2005).

Tdap vaccines with a reduced antigen content have therefore been recommended for booster doses in several countries, as they have been shown to be less reactogenic than the full antigen-content DTaP vaccines (Halperin et al., 2011a). In the United States, an adolescent booster dose with Tdap was recommended at the age of 11–12 years since 2006 (Broder et al., 2006). However, in view of the recent studies, this may not be the most adequate vaccination schedule to protect the young teenagers, and boosting earlier than at the age of 11 may be necessary. This may pose a major logistic challenge, as there is usually no established routine health care visit for adolescent vaccination before 11–12 years (Misegades et al., 2012), which in turn may reduce coverage. Furthermore, the effectiveness and duration of protection of Tdap have not yet been well established. A recent study concluded that Tdap vaccination was only moderately effective in preventing pertussis in adolescents and adults who had received DTaP vaccines in the primary vaccination series (Baxter et al., 2013). This contrasts with earlier reports showing high effectiveness of Tdap boosting for adolescents and adults (Wei et al., 2010). However, in the earlier studies, only a small number of pertussis cases were reported, and all individuals had received DTP vaccines as children. Predicted estimates on the longevity of protection suggest that the duration of protection from adolescent pertussis boosters would be at best in the range of 6–10 years (Lavine et al., 2012). Boosting of <11-year olds would therefore shift the susceptibility age to older teenagers and young adults, including adults of child-bearing age, thereby increasing the source of infection to newborns and infants, unless more frequent, decennial boosting is recommended (Zepp et al., 2011). Finally, extensive injection site reactions still occur following Tdap boosting in children, albeit with lower frequencies than following DTaP vaccination (Quinn et al., 2011). It is not yet known whether repeated injections of Tdap will result in increased local reactions. These issues will be important to address in order to establish a cost–benefit ratio of generalized repeated Tdap administrations.

35.8 Attempted solutions to the pertussis problem in infants

Although adolescent and adult pertussis can sometimes be severe, and the burden of adult pertussis is not negligible (Rothstein and Edwards, 2005; McGuinnes et al., 2013), the most severe cases occur generally in the newborn population, where the case fatality rate ranges between 0.2% and 4% (Blangiardi and Ferrera et al., 2009). More than 90% of all pertussis-linked deaths in countries with high vaccination coverage occur in infants less than 4 months of age (Broder et al., 2006). The primary role of pertussis vaccination strategies should therefore be to protect this most vulnerable age group. As in most countries pertussis vaccination starts at 2 months and requires three to four doses for the completion of the primary course, the most vulnerable age group is thus not adequately protected by the current vaccination schedules.

35.8.1 Earlier and at-birth vaccination

A slight acceleration of pertussis vaccination by advancing the first dose from 8 to 6 weeks of age reduced the average severe disease notification rates by roughly 10% in Australia (Foxwell et al., 2011). Earlier vaccination, possibly at birth, has been proposed, and stand-alone acellular pertussis vaccines given at birth were shown to be safe and able to induce anti-pertussis antibodies in a fraction of vaccinated children (Belloni et al., 2003; Wood et al., 2010). However, antibody responses were only measured after two or three months. Importantly, neonatal pertussis vaccination with acellular vaccines did not lead

to the feared immune tolerance to the pertussis antigens, as had been previously shown after neonatal vaccination with whole-cell vaccines. However, it elicited bystander interference on immune responses to other co-administrated antigens within the combination vaccine given in the subsequent routine primary vaccination schedule (Knuf et al., 2010).

In contrast to a stand-alone acellular pertussis vaccine, a combination DTaP vaccine given at birth was associated with lower responses to the diphtheria and pertussis antigens after the routine DTaP vaccination, when compared to children who did not receive a DTaP vaccine injection at birth (Halasa et al., 2008). Although antibody production appeared, thus, to be accelerated when a stand-alone acellular pertussis vaccine was given at birth, the effect of this acceleration on earlier protection against pertussis has not been determined yet. When T cell responses were measured, neonatal immunization with acellular pertussis vaccines was found to elicit a strong Th2 bias with the production of high levels of IL-5 and IL-13 (White et al., 2010), which has the potential to antagonize the development of Th1-type immune responses, thereby having a negative effect on the longevity of protection and potentially the safety of repeated acellular vaccine injections (Rowe et al., 2005).

Given the uncertainties about effectiveness and potential safety issues of neonatal vaccination with the current DTaP pertussis vaccines and given the fact that stand-alone acellular pertussis vaccines are not licenced for routine use, it is difficult to recommend at-birth vaccination with the current vaccines. In addition, analyses performed in the Netherlands have suggested that the cost-effectiveness of at-birth immunization as an addition to the current Dutch national immunization program is highly unfavorable (Westra et al., 2010).

35.8.2 Postpartum and cocoon vaccination

Another strategy to control pertussis within the most vulnerable age group is based on the observation that mothers are the most frequent source of pertussis among young infants (Izureita et al., 1996). Hence, postpartum women can be regarded as one of the most important target groups to protect newborns against pertussis. Consequently, attempts have been undertaken to implement Tdap immunization to postpartum women before hospital discharge (Healy et al., 2009). In addition to boosting maternal protection, thereby decreasing the likelihood of transmission of pertussis to the newborn, this strategy may also lead to the induction of antibodies that may be transferred to the newborn through breastfeeding, which in turn may provide passive protection before active immunization through the current vaccination schedules.

However, even via an anamnestic response, antibodies to the pertussis antigens present in Tdap are not detectable until 5–7 days after vaccination and reach a peak only by approximately two weeks (Halperin et al., 2011b), which may not be sufficiently rapid for effective herd protection of the infants within the first weeks of life. A study carried out on 514 infants with pertussis concluded that immunization of postpartum mothers with Tdap did not reduce pertussis disease significantly in infants less than 6 months of age (Castagnini et al., 2012), suggesting that significant impact on infant pertussis requires a cocoon strategy that is more complete than by only vaccination of the mothers and should be extended to other household contacts.

Initial computer simulation studies have suggested that a cocoon strategy including all household contacts would result in a 70% reduction in incidence of pertussis in infants up to 3 months of age (Van Rie and Hethcote, 2004). However, more recent calculations show that the number needed to vaccinate in order to prevent severe infant pertussis through cocoon immunization would be such that in the context of a low pertussis incidence this strategy would be too resource intensive and rather inefficient (Skowronski et al., 2012). By combining the data from natural infection with a contact network structure, a mathematical model indicated that regular boosters up to once every 5 years starting at age 11 would reduce the global pertussis burden by only approximately 15%, even with a logistically unfeasible vaccination coverage of 75% (Rohani et al., 2010). This reduction would even be more modest for protection against infant pertussis.

35.8.3 Pregnancy vaccination

Postpartum cocoon immunization was compared for its effectiveness in reducing infant pertussis to pregnancy vaccination, which was found to be substantially more cost-effective (Terranella et al., 2013). The U.S. Advisory Committee on Immunization Practices now recommends pregnancy vaccination as a preferred alternative to postpartum vaccination for preventing infant pertussis. The committee recommends a dose of Tdap should be administered during each pregnancy, irrespective of the prior history of Tdap immunization, and that the optimal timing for Tdap administration is between 27 and 36 weeks of gestation (American College of Obstetricians and Gynecologists, 2013).

The concept of maternal vaccination during pregnancy to protect the infants against pertussis is not new and was already tempted in the 1930s (Lichty et al., 1938). A significant proportion (75%) of infants born to mothers vaccinated with a whole-cell vaccine preparation during the latter part of pregnancy had opsonic antibodies in their sera at levels approaching those of their mothers. Interestingly, a similar proportion was observed in infant sera from mothers who were not vaccinated during pregnancy, but had experienced pertussis in their childhood, and the proportion reached 100% in infants born from mothers with pertussis during childhood and vaccination during pregnancy. Although it is not known whether this directly translates into better protection against pertussis disease in the newborns, this study shows that vaccination during pregnancy leads to the transfer of maternal anti-pertussis antibodies to the offspring in a substantial proportion of cases.

However, in this study, the mothers were immunized three times with high doses of the whole-cell vaccine. This resulted in common local reactions, sometimes very painful and lasting for several days (Mooi and de Greeff, 2007). These and other adverse events observed after vaccination during pregnancy have rendered this strategy inacceptable for large-scale implementation. Liability issues also hamper the application of maternal vaccination (Paradiso, 2002).

Nevertheless, maternal immunization offers a number of advantages. In addition to theoretically offering protection to the mother and the newborn child from birth until completion of the primary immunization series, it has the logistic advantage of easy access to the target group by the frequent visits of health care centers during pregnancy. With the advent of the more recent less-reactogenic acellular vaccines, especially of the reduced-dose Tdap vaccines, maternal immunization has gained impetus during the last 2 years. Initial studies showed that the administration of Tdap during the second trimester of pregnancy was safe and resulted in higher antibody levels to PTX, FHA, pertactin, and fimbriae in the sera of the newborns, compared to the sera of newborns from unvaccinated mothers (Gall et al., 2011). A comprehensive review of the U.S. Vaccine Adverse Event Reporting System on adverse events in women who had received Tdap during pregnancy confirmed the safety of Tdap administration during pregnancy (Zheteyeva et al., 2012). The most frequent adverse event reported was spontaneous abortion, but it was not more frequent than its occurrence in all pregnancies.

Due to the relatively short half-life of maternally acquired anti-PTX IgG, the timing between vaccination and delivery appears to be critical (Healy et al., 2013). Immunization during the third trimester is optimal for maximal transfer to the fetus. Immunization before conception or early during pregnancy is unlikely to lead to sufficient amounts of placental pertussis-specific antibody transport to protect the newborns through 2 and 3 months of age. This implies also that Tdap vaccination would have to be repeated at each pregnancy, which in turn may again raise safety issues of repeated Tdap vaccination. Moreover, it is difficult to infer that transfer of maternal antibodies will necessarily result in significant protection of the newborns, since serological correlates of protection are not established for pertussis. However, data suggest that infants who acquire pertussis in the first 3–6 months of life tend to have lower pertussis-specific antibody levels at birth than matched controls (Heininger et al., 2013).

One concern for immunization during pregnancy is the potential interference of maternal antibodies on the vaccine-induced immune response in the infants. A recent study shows that infants born from mothers vaccinated with Tdap during pregnancy had a slightly lower antibody response to pertussis antigens after the primary vaccination series with DTaP (Hardy-Fairbanks et al., 2013). However, this may not be clinically relevant and disappeared prior to and after the booster doses. There was no effect on the infant responses to antigens not included in the Tdap.

A simulation based on a cohort model reflecting U.S. 2009 births and the DTP schedule suggests that Tdap vaccination during pregnancy would reduce the annual infant pertussis incidence by roughly 1/3, which was predicted to be more effective than postpartum vaccination (Terranella et al., 2013). The cost of quality-adjusted life-years saved was also substantially less for pregnancy vaccination than postpartum vaccination. However, cost-effectiveness of vaccination strategies may vary tremendously depending on the incidence of the disease (Kowalzik et al., 2007). Nevertheless, even though a reduction by 1/3 of infant pertussis is significant, it is not sufficient to bring the disease under control, and better vaccines are therefore desperately needed.

35.9 Conclusions

Pertussis is a severe respiratory illness that can lead to death, especially in the most vulnerable age group of newborns, although the burden of adolescent and adult pertussis is far from negligible. The introduction of first-generation, whole-cell pertussis vaccines have tremendously reduced the disease incidence,

but have failed to completely bring pertussis under control. During the last 20 years, these whole-cell vaccines have now gradually been replaced by acellular pertussis vaccines, with an improved safety profile. However, recent evidence suggests that these new-generation vaccines are even less effective in global pertussis control, and spectacular outbreaks have recently been reported in regions with high immunization coverage. Several approaches, including neonatal vaccination, cocooning, and maternal immunization are being tempted, however, with limited success. Unless substantially improved vaccines become available soon, pertussis is thus likely here to stay for more generations, and eradication of the disease is out of reach. Over the last decades, there has been little research on improving pertussis vaccines. Only in recent years have new vaccine candidates been investigated, the most advanced of which is a live attenuated vaccine currently in clinical development (for review, see Locht and Mielcarek, 2012). However, it will still take many years before such vaccines become widely available.

After this chapter has been accepted for publication, Amirthalingam et al. (2014) reported a 90% effectiveness of maternal vaccination between 28 and 38 weeks of pregnancy against clinical pertussis in children younger than 2 months during the U.K. outbreak of 2011/2012. The study also points to the importance of timing, as vaccination effectiveness dropped substantially when the vaccine was administered less than 7 days before delivery. Despite these promising first results, acceptance rates of vaccination during pregnancy, even in countries where it is recommended, are still low and remain a challenge (Laenen et al., 2015), and the effect of maternal vaccination on the primary vaccine responses of the infants is still insufficiently well known.

References

Allen, A. 2013. The pertussis paradox. *Science*, 314, 454–456.

Alonso, S., Pethe, K., Mielcarek, N., Raze, D., and Locht, C. 2001. Role of ADP-ribosyltransferase activity of pertussis toxin in toxin–adhesin redundancy with filamentous hemagglutinin during *Bordetella pertussis* infection. *Infect. Immun.*, 69, 6038–6043.

American College of Obstetricians and Gynecologists. Committee Opinion No. 566. 2013. Update on immunization and pregnancy: Tetanus, diphtheria, and pertussis vaccination. *Obstet. Gynecol.*, 121, 1411–1414.

Amirthalingam, G., Andrews, N., Campbell, H., Ribeiro, S., Kara, E., Donegan, K., Fry, N.K., Miller, E., and Ramsay, M. 2014. Effectiveness of maternal pertussis vaccination in England: an observational study. *Lancet*, 384, 1521–1528.

Baker, JP. 2003. The pertussis vaccine controversy in Great Britain, 1974–1986. *Vaccine*, 21, 4003–4010.

Barkoff, A.M., Mertsola, J., Guillot, S., Guiso, N., Berbers, G., and He, Q. 2012. Appearance of *Bordetella pertussis* strains not expressing the vaccine antigen pertactin in Finland. *Clin. Vaccine Immunol.*, 19, 1703–1704.

Baxter, R., Barlett, J., Rowhani-Rahbar, A., Fireman, B., and Klein, NP. 2013. Effectiveness of pertussis vaccines for adolescents and adults: Case–control study. *Brit. Med. J.*, 347, f4249.

Belloni, C., De Silvestri, A., Tinelli, C., Avanzini, M.A., Marconi, M., Strano, F., Rondini, C., and Chirico, G. 2003. Immunogenicity of a three-component acellular pertussis vaccine administered at birth. *Pediatrics*, 111, 1042–1045.

Black, S. 1997. Epidemiology of pertussis. *Pediatr. Infect. Dis. J.*, 16(suppl.), S85–S89.

Black, R.E., Cousens, S., Johnson, H.L., Lawn, J.E., Rudan, I., Bassani, D.G., Jha, P., Campbell, H., Walker, C.F., Cibulskis, R., Eisele, T., Liu, L., Mathers, C., and Child Health Epidemiology Reference Group of WHO and UNICEF. 2010. Global, regional, and national causes of child mortality in 2008: A systematic analysis. *Lancet*, 375, 1969–1987.

Blangiardi, F., and Ferrera, G. 2009. Reducing the risk of pertussis in newborn infants. *J. Prev. Med. Hyg.*, 50, 206–216.

Blom, J., Hansen, G.A., and Poulsen, FM. 1983. Morphology of cells and hemagglutinogens of *Bordetella* species: Resolution of substructural units in fimbriae of *Bordetella pertussis*. *Infect. Immun.*, 42, 308–317.

Bodilis, H., and Guiso, N. 2013. Virulence of pertactin-negative *Bordetella pertussis* isolated from infants, France. *Emerg. Infect. Dis.*, 19, 471–474.

Bordet, J., and Gengou, O. 1906. Le microbe de la coqueluche. *Ann. Inst. Pasteur (Paris)*, 20, 731–741.

Bouchez, V., Brun, D., Cantinelli, T., Dore, G., Njamkepo, E., and Guiso, N. 2009. First report and detailed characterization of *B. pertussis* isolates not expressing pertussis toxin or pertactin. *Vaccine*, 27, 6034–6041.

British Medical Journal. 1981. *Pertussis vaccine* (editorial). *Brit. Med. J.*, 282, 1563–1564.

Broder, K.R., Cortese, M.M., Iskander, J.K., and Advisory Committee on Immunization Practices. 2006. Preventing tetanus, diphtheria, and pertussis among adolescents: Use of tetanus toxoid, reduced diphtheria toxoid and acellular pertussis vaccine recommendations of the advisory committee on immunization practices. *MMWR Recomm. Rep.*, 55, 1–34.

Broutin, H., Vigoud, C., Grenfell, B.T., Miller, M.A., and Rohani, P. 2010. Impact of vaccination and birth rate on the epidemiology of pertussis: A comprehensive study in 64 countries. *Proc. R. Soc. B.*, 277, 3239–3245.

Carbonetti, N.H., Artamonova, G.V., Mays, R.M., and Worthington, Z.E. 2003. Pertussis toxin plays an early role in respiratory tract colonization by *Bordetella pertussis*. *Infect. Immun.*, 71, 6358–6366.

Carbonetti, N.H., Artamonova, G.V., Andreasen, C., and Bushar, N. 2005. Pertussis toxin and adenylate cyclase toxin provide a one-two punch for establishment of *Bordetella pertussis* infection of the respiratory tract. *Infect. Immun.*, 73, 2698–2703.

Castagnini, L.A., Healy, C.M., Rench, M.A., Wooton, S.H., Munoz, F.M., and Baker, C.L. 2012. Impact of maternal postpartum tetanus and diphtheria toxoids and acellular pertussis immunization on infant pertussis infection. *Clin. Infect. Dis.*, 54, 78–84.

CDC. 1980. Reported morbidity and mortality rates in the United States. *Morb. Mortal. Wkly. Rep.*, 28, 1–120.

CDC. 2011. Recommended adult immunization schedule – United States, 2011. *Morb. Mortal. Wkly. Rep.*, 60, 1–4.

CDC. 2012. Pertussis epidemic – Washington. *Morb. Mortal. Wkly. Rep.*, 61, 517–522.

Chen, X., Howard, O.M., and Oppenheim, J.J. 2007. Pertussis toxin by inducing IL-6 promotes the generation of IL-17-producing CD4 cells. *J. Immunol.*, 178, 6123–6129.

Cherry, J.D. 1999. Pertussis in the preantibiotic and prevaccine era, with emphasis on adult pertussis. *Clin. Infect. Dis.*, 28(suppl. 2), S107–S111.

Cherry, J.D. 2012. Epidemic pertussis in 2012. The resurgence of a vaccine-preventable disease. *N. Engl. J. Med.*, 367, 785–787.

Cherry, J.D., and Heininger, U. 2004. Pertussis and other *Bordetella* infections. In: Feigin, R.D., Cherry, J.D., Demmler, G.J., and Kaplan, S.L., editors. *Textbook of Pediatric Infectious Diseases*. Philadelphia, PA: Saunders. pp. 1588–1608.

Cherry, J.D., and Seaton, B.L. 2012. Patterns of *Bordetella parapertussis* respiratory illnesses: 2008–2010. *Clin. Infect. Dis.*, 54, 534–537.

Cherry, J.D., Grimprel, E., Guiso, N., Heininger, U., and Mertsola, J. 2005. Defining pertussis epidemiology, clinical, microbiologic and serologic perspectives. *Pediatr. Infect. Dis. J.*, 24, S25–S34.

Cherry, J.D., Tan, T., Wirsing von König, C.H., Forsyth, K.D., Thisyakorn, U., Greenberg, D., Johnson, D., Marchant, C., and Plotkin, S. 2012. Clinical definitions of pertussis: Summary of a global pertussis initiative roundtable meeting, February 2011. *Clin. Infect. Dis.*, 54, 1756–1764.

Cody, C.L., Baraff, L.J., Cherry, J.D., Marcy, S.M., and Manclark, C.R. 1981. Nature and rates of adverse reactions associated with DTP and DT immunizations in infants and children. *Pediatrics*, 68, 650–660.

Cookson, B.T., Cho, H.L., Herwaldt, L.A., and Goldman, W.E. 1989. Biological activities and chemical composition of purified tracheal cytotoxin of *Bordetella pertussis*. *Infect. Immun.*, 57, 2223–2229.

Coulter, H.L., and Fischer, B.L. 1985. *DPT: A Shot in the Dark*. San Diego, CA: Harcourt Brace Jovanovich.

Coutte, L., Antoine, R., Drobecq, H., Locht, C., and Jacob-Dubuisson, F. 2001. Subtilisin-like autotransporter serves as maturation protease in a bacterial secretion pathway. *EMBO J.*, 20, 5040–5048.

Crowcroft, N.S., and Pebody, R.G. 2006. Recent developments in pertussis. *Lancet*, 367, 1926–1936.

Delattre, A.S., Saint, N., Clantin, B., Willery, E., Lippens, G., Locht, C., Villeret, V., and Jacob-Dubuisson, F. 2011. Substrate recognition by the POTRA domains of TpsB transporter FhaC. *Mol. Microbiol.* 81, 99–112.

Domenighini, M., Relman, D., Capiau, C., Falkow, S., Prugnola, A., Scarlato, V., and Rappuoli, R. 1990. Genetic characterization of *Bordetella pertussis* filamentous haemagglutinin: A protein processed from an unusually large precursor. *Mol. Microbiol.*, 4, 787–800.

Elliott, E., McIntyre, P., Ridley, G., Morris, A., Massie, J., McEniery, J., and Knight, G. 2004. National study of infants hospitalized with pertussis in the acellular vaccine era. *Pediatr. Infect. Dis. J.*, 23, 246–252.

Farizo, K.M., Huang, T., and Burns, D.L. 2000. Importance of holotoxin assembly in Ptl-mediated secretion of pertussis toxin from *Bordetella pertussis*. *Infect. Immun.*, 68, 4049–4054.

Fine, P.E.M., and Clarkson, J.A. 1982. Recurrence of whooping cough: Possible implications for assessment of vaccine efficacy. *Lancet*, 319, 666–669.

Flak, T.A., Heiss, L.N., Engle, J.T., and Goldman, W.E. 2000. Synergistic epithelial responses to endotoxin and a naturally occurring muramyl peptide. *Infect. Immun.*, 68, 1235–1242.

Floret, D., Groupe de Pathologie Infectieuse Pédiatrique, and Groupe Francophone de Réanimation et d'urgence Pédiatrique. 2001. Les décès par infection bactérienne communautaire. Enquête dans les services de réanimation pédiatrique Français. *Arch. Pediatr.*, 8, 705–711.

Foxwell, A.R., McIntyre, P., Quinn, H., Roper, K., and Clements, M.S. 2011. Severe pertussis in infants: Estimated impact of first vaccine dose at 6 versus 8 weeks in Australia. *Pediatr. Infect. Dis. J.*, 30, 161–163.

Funke, G., Hess, T., von Graevenitz, A., and Vandamme, P. 1996. Characteristics of *Bordetella hinzii* strains isolated from a cystic fibrosis patient over a 3-year period. *J. Clin. Microbiol.*, 34, 966–969.

Gall, S.A., Myers, J., and Pichichero, M. 2011. Maternal immunization with tetanus–diphtheria–pertussis vaccine: Effect on maternal and neonatal serum antibodies. *Am. J. Obstet. Gynecol.*, 204, 334.e1–5.

Geuijen, C.A., Willems, R.J., and Mooi, F.R. 1996. The major fimbrial subunit of *Bordetella pertussis* binds to sulfated sugars. *Infect. Immun.*, 64, 2657–2665.

Geuijen, C.A., Willems, R.J., Bongaerts, M., Top, J., Gielen, H., and Mooi, F.R. 1997. Role of the *Bordetella pertussis* minor fimbrial subunit, FimD, in colonization of the mouse respiratory tract. *Infect. Immun.*, 65, 4222–4228.

Geuijen, C.A., Willems, R.J., Hoogerhout, P., Puijk, W.C., Meloen, R.H., and Mooi, F.R. 1998. Identification and characterization of heparin binding regions of the Fim2 subunit of *Bordetella pertussis*. *Infect. Immun.*, 66, 2256–2263.

Glaser, P., Sakamoto, H., Ballalou, J., Ullmann, A., and Danchin, A. 1988. Secretion of cyclolysin, the calmodulin-sensitive adenylate cyclase-haemolysin bifunctional protein of Bordetella pertussis. *EMBO J.*, 7, 3997–4004.

Gueirard, P., Druilhe, A., Pretolani, M., and Guiso, N. 1997. Role of adenylate cyclase-hemolysin in alveolar macrophage apoptosis during *Bordetella pertussis* infection *in vivo*. *Infect. Immun.*, 66, 1718–1725.

Guermonprez, P., Khelef, N., Blouin, E., Rieu, P., Ricciardi-Castagnoli, P., Guiso, N., Ladant, D., and Leclerc, C. 2001. The adenylate cyclase toxin of *Bordetella pertussis* binds to target cells via the alpha(M)beta(2) integrin (CD11b/CD18). *J. Exp. Med.*, 193, 1035–1044.

Gustafsson, L., Hessel, L., Storsaeter, J., and Olin, P. 2006. Long-term follow-up of Swedish children vaccinated with acellular pertussis vaccines at 3, 5, and 12 months of age indicates the need for a booster dose at 5 to 7 years of age. *Pediatrics*, 118, 978–984.

Halasa, N.B., O'Shea, A., Shi, J.R., LaFleur, B.J., and Edwards, K.M. 2008. Poor immune responses to a birth dose of diphtheria, tetanus, and acellular pertussis vaccine. *J. Pediatr.*, 153, 327–332.

Halperin, S.A., McNeil, S., Langely, J., Blatter, M., Dionne, M., Embree, J., Johnson, R., Latiolais, T., Meekison, W., Noya, F., Senders, S., Zickler, P., and Johnson, D.R. 2011a. Tolerability and antibody response in adolescents and adults revaccinated with tetanus toxoid, reduced diphtheria toxoid, and acellular pertussis vaccine adsorbed (Tdap) 4–5 years after a previous dose. *Vaccine*, 29, 8459–8465.

Halperin, B.A., Morris, A., MacKinnon-Cameron, D., Mutch, J., Langley, J.M., McNeil, S.A., MacDougall, S., and Halperin, S.A. 2011b. Kinetics of the antibody response to tetanus-diphtheria-acellular pertussis vaccine in women of childbearing age and postpartum women. *Clin. Infect. Dis.*, 53, 885–892.

Hannah, J.H., Menozzi, F.D., Renauld, G., Locht, C., and Brennan, M.J. 1994. Sulfated glycoconjugate receptors for the *Bordetella pertussis* adhesin filamentous hemagglutinin (FHA) and mapping of the heparin-binding domain on FHA. *Infect. Immun.*, 62, 5010–5019.

Hardy-Fairbanks, A.J., Pan, S.J., Decker, M.D., Johnson, D.R., Greenberg, D.P., Kirkland, K.B., Talbot, E.A., and Bernstein, H.H. 2013. Immune responses in infants whose mothers received Tdap vaccine during pregnancy. *Pediatr. Infect. Dis. J.*, 32, 1257–1260.

Hazenbos, W.L., Geuijen, C.A., van den Berg, B.M., Mooi, F.R., and van Furth, R. 1995. *Bordetella pertussis* fimbriae bind to human monocytes via the minor fimbrial subunit FimD. *J. Infect. Dis.*, 171, 924–929.

Healy, C.M., Rench, M.A., Castagnini, L.A., and Baker, C.J. 2009. Pertussis immunization in a high-risk postpartum population. *Vaccine*, 27, 5599–2602.

Healy, C.M., Renc, M.A., and Baker, C.J. 2013. Importance of timing of maternal combined tetanus, diphtheria, and acellular pertussis (Tdap) immunization and protection of young infants. *Clin. Infect. Dis.*, 56, 539–544.

Hegerle, N., Paris, A.S., Brun, D., Dore, G., Njamkepo, E., Guillot, S., and Guiso, N. 2012. Evolution of French *Bordetella pertussis* and *Bordella parapertussis* isolates: Increase of Bordetellae not expressing pertactin. *Clin. Microbiol. Infect.*, 18, E340–E346.

Heininger, U., Stehr, K., Schmitt-Grohé, S., Lorenz, C., Rost, R., Christenson, P.D., Uberall, M., and Cherry, J.D. 1994. Clinical characteristics of illness caused by *Bordetella parapertussis* compared with illness caused by *Bordetella pertussis*. *Pediatr. Infect. Dis. J.*, 13, 306–309.

Heininger, U., Klich, K., Stehr, K., and Cherry, J.D. 1997. Clinical findings in *Bordetella pertussis* infections: Results of a prospective multicenter surveillance study. *Pediatrics*, 100, e10.

Heininger, U., Riffelmann, M., Bär, G., Rudin, C., and Wirsing von König, W.H. 2013. The protective role of maternally derived antibodies against *Bordetella pertussis* in young infants. *Pediatr. Infect. Dis. J.*, 32, 695–698.

Heiss, L.N., Flak, T.A., Lancaster Jr., J.R., McDaniel, M.L., and Goldman, W.E. 1993a. Nitric oxide mediates *Bordetella pertussis* tracheal cytotoxin damage to the respiratory epithelium. *Infect. Agents Dis.*, 2, 173–177.

Heiss, L.N., Moser, S.A., Unanue, E.R., and Goldman, W.E. 1993b. Interleukin-1 is linked to the respiratory epithelial cytopathology of pertussis. *Infect. Immun.*, 61, 3123–3128.

Henderson, M.W., Inatsuka, C.S., Sheets, A.J., Williams, C.L., Benaron, D.J., Donato, G.M., Gray, M.C., Hewlett, E.L., and Cotter, P.A. 2012. Contribution of *Bordetella* filamentous hemagglutinin and adenylatecyclase toxin to suppression and evasion of interleukin-17-mediated inflammation. *Infect. Immun.*, 80, 2061–2075.

Hewlett, E.L., and Donato, G.M. 2007. *Bordetella* toxins. In: Locht, C., editor. *Bordetella Molecular Microbiology*. Norfolk, UK: Horizon Bioscience. pp. 97–118.

Higgs, R., Higgins, S.C., Ross, P.J., and Mills, K.H. 2012. Immunity to the respiratory pathogen *Bordetella pertussis*. *Mucosal Immunol.*, 5, 485–500.

Hodak, H., Clantin, B., Willery, E., Villeret, V., Locht, C., and Jacob-Dubuisson, F. 2006. Secretion signal of the filamentous haemagglutinin, a model two-partner secretion substrate. *Mol. Microbiol.*, 61, 368–382.

Hou, W., Wu, Y., Sun, S., Shi, M., Sun, Y., Yang, C., Pei, G., Gu, Y., Zhong, C., and Sun, B. 2003. Pertussis toxin enhances Th1 responses by stimulation of dendritic cells. *J. Immunol.*, 170, 1728–1736.

Howson, C.P., Howe, C.J., and Fineberg, H.V. 1991. *Adverse Effects of Pertussis and Rubella Vaccines*. Washington, DC: National Academy Press.

Inatsuka, C.S., Xu, Q., Vujkovic-Cvijin, I., Wong, S., Stibitz, S., Miller, J.F., and Cotter, P.A. 2010. Pertactin is required for *Bordetella* species to resist neutrophil-mediated clearance. *Infect. Immun.*, 78, 2901–2909.

Izureita, H.S., Kenyon, T.A., Strebel, P.M., Baughman, A.L., Shulman, S.T., and Wharton, M. 1996. Risk factors for pertussis in young infants during an outbreak in Chicago in 1993. *Clin. Infect. Dis.*, 22, 503–507.

Katada, T., and Ui, M. 1982. ADP ribosylation of the specific membrane protein of C6 cells by islet-activating protein associated with modification of adenylate cyclase activity. *J. Biol. Chem.*, 257, 7210–7216.

Khelef, N., Bachelet, C.M., Vargaftig, B.B., and Guiso, N. 1994. Characterization of murine lung inflammation after infection with parental *Bordetella pertussis* and mutants deficient in adhesins or toxins. *Infect. Immun.*, 62, 2893–2900.

Khelef, N., Sakamoto, H., and Guiso, N. 1992. Both adenylate cyclase and hemolytic activities are required by *Bordetella pertussis* to initiate infection. *Microb. Pathog.*, 12, 227–235.

King, A.J., van der Lee, S., Mohangoo, A., van Gent, M., van der Ark, A., and van de Waterbeemd, B. 2013. Genome-wide gene expression analysis of *Bordetella pertussis* isolates associated with a resurgence in pertussis: Elucidation of factors involved in the increased fitness of epidemic strains. *PLoS One*, 8, e66150.

Klein, N.P., Barlett, J., Rowhani-Rahbar, A., Fireman, B., and Baxter, R. 2012. Waning protection after fifth dose of acellular pertussis vaccine in children. *N. Engl. J. Med.*, 367, 1012–1019.

Klein, N.P., Bartlett, J., Fireman, B., Rowhani-Rahbar, A., and Baxter, R. 2013. Comparative effectiveness of acellular versus whole-cell pertussis vaccines in teenagers. *Pediatrics*, 131, e1716–e1722.

Knuf, M., Schmitt, H.J., Jacquet, J.M., Collard, A., Kieninger, D., Meyer, C.U., Siegrist, C.A., and Zepp, F. 2010. Booster vaccination after neonatal priming with acellular pertussis vaccine. *J. Pediatr.*, 156, 675–678.

Ko, K.S., Peck, K.R., Oh, W.S., Lee, N.Y., Lee, J.H., and Song, J.H. 2005. New species of *Bordetella, Bordetella ansorpii* sp. nov., isolated from the purulent exudate of an epidermal cyst. *J. Clin. Microbiol.*, 43, 2516–2519.

Kowalzik, F., Barbosa, A.P., Fernandes, V.R., Carvalho, P.R., Avila-Aguero, M.L., Goh, D.Y., Goh, A., de Miguel, J.G., Moraga, F., Roca, J., Campins, M., Huang, M., Quian, J., Riley, N., Beck, D., and Verstraeten, T. 2007. Prospective multinational study of pertussis infection in hospitalized infants and their household contacts. *Pediatr. Infect. Dis. J.*, 26, 238–242.

Ladant, D., and Ullmann, A. 1999. *Bordetella pertussis* adenylate cyclase: A toxin with multiple talents. *Trends Microbiol.*, 7, 172–176.

Ladant, D., Brezin, C., Alonso, J.M., Crenon, I., and Guiso, N. 1986. *Bordetella pertussis* adenylate cyclase. Purification, characterization, and radioimmunoassay. *J. Biol. Chem.*, 261, 16264–16269.

Laenen, J., Roelants, M., Devlieger, R., and Vandermeulen, C. 2015. Influenza and pertussis vaccination coverage in pregnant women. *Vaccine*, 33, 2125–2131.

Lamberti, Y., Alvarez Hayes, J., Perez Vidakovics, M.L., and Rodriguez, M.E. 2009. Cholesterol-dependent attachment of human respiratory cells by *Bordetella pertussis*. *FEMS Immunol. Med. Microbiol.*, 56, 143–150.

Lavine, J.S., Bjornstad, O.N., de Blasio, B.F., and Storsaeter, J. 2012. Short-lived immunity against pertussis, age-specific routes of transmission, and utility of teenage booster vaccine. *Vaccine*, 30, 544–551.

Le Coustumier, A., Njamkepo, E., Cattoir, V., Guillot, S., and Guiso, N. 2011. *Bordetella petrii* infection with long-lasting presistance in human. *Emerg. Infect. Dis.*, 17, 612–618.

Leininger, E., Roberts, M., Kenimer, J.G., Charles, I.G., Fairweather, N., Novotny, P., and Brennan, M.J. 1991. Pertactin, an Arg-Gly-Asp-containing Bordetella pertussis surface protein that promotes adherence to mammalian cells. *Proc. Natl. Acad. Sci. USA*, 88, 345–349.

Lichty Jr., J.A., Slavin, B., and Bradford, W.L. 1938. An attempt to increase resistance to pertussis in newborn infants by immunizing their mothers during pregnancy. *J. Clin. Invest.*, 17, 613–621.

Liko, J., Robison, S.G., and Cieslak, P.R. 2013. Priming with whole-cell versus acellular pertussis vaccine. *N. Engl. J. Med.*, 368, 581–582.

Llosa, M., Roy, C., and Dehio, C. 2009. Bacterial type IV secretion systems in human disease. *Mol. Microbiol.*, 73, 141–151.

Locht, C., and Mielcarek, N. 2012. New pertussis vaccination approaches: En route to protect newborns? *FEMS Immunol. Med. Microbiol.*, 66, 121–133.

Locht, C., Coutte, L., and Mielcarek, N. 2011. The ins and outs of pertussis toxin. *FEBS J.*, 278, 4668–4682.

Long, S.S., and Edwards, K.M. 2003. *Bordetella pertussis* and other species. In: Long, S., Pickering, L., and Prober, C., editors. *Principles and Practice of Pediatric Infectious Diseases*, 2nd ed. Philadelphia, PA: Churchill Livingstone. pp. 880–888.

Madsen, T. 1933. Vaccination against whooping cough. *J. Am. Med. Assoc.*, 101, 187–188.

Mascart, F., Verscheure, V., Malfroot, A., Hainaut, M., Pierard, D., Temmerman, S., Peltier, A., Debrie, A.S., Levy, J., Del Giudice, G., and Locht, C. 2003. *Bordetella pertussis* infection in 2-months-old infants promotes type 1 T cell responses. *J. Immunol.*, 170, 1504–1509.

Mascart, F., Hainaut, M., Peltier, A., Verscheure, V., Levy, J., and Locht, C. 2007. Modulation of the infant immune responses by the first pertussis vaccine administrations. *Vaccine*, 25, 391–398.

Mattoo, S., and Cherry, J.D. 2005. Molecular pathogenesis, epidemiology and clinical manifestations of respiratory infections due to *Bordetella pertussis* and other *Bordetella* subspecies. *Clin. Microbiol. Rev.*, 18, 326–382.

Mazengia, E., Silva, E.A., Peppe, J.A., Timpen, R., and George, H. 2000. Recovery of *Bordetella holmesii* from patients with pertussis-like symptoms: Use of pulse-field gel electrophoresis to characterize circulating strains. *J. Clin. Microbiol.*, 38, 2330–2333.

McGuinnes, C.B., Hill, J., Fonseca, E., Hess, G., Hitchcock, W., and Krischnarajah, G. 2013. The disease burden of pertussis in adults 50 years old and older in the United States: A retrospective study. *BMC Infect. Dis.*, 13, 32.

Menozzi, F.D., Gantiez, C., and Locht, C. 1991. Interaction of the *Bordetella pertussis* filamentous hemagglutinin with heparin. *FEMS Microbiol. Lett.*, 62, 59–64.

Mielcarek, N., and Locht, C. 2013. Whooping cough. In: Rosenberg, E., DeLong, E.F., Lory, S., Stackebrandt, E., and Thompson, F., editors. *The Prokaryotes*, 4th ed., volume 5: *Human Microbiology*. Springer Verlag, Berlin, Heidelberg. pp. 291–307.

Misegades, L.K., Winter, K., Harriman, K., Talarico, J., Messonnier, N.E., Clark, T.A., and Martin, S.W. 2012. Association of childhood pertussis with receipt of 5 doses of pertussis vaccine by time since last vaccine dose, California, 2010. *J. Am. Med. Assoc.*, 308, 2126–2132.

Mooi, F.R., and de Greeff, S.C. 2007. The case for maternal vaccination against pertussis. *Lancet Infect. Dis.*, 7, 614–624.

Mooi, F.R., van Loo, I.H.M., and King, A.J. 2001. Adaptation of *Bordetella pertussis* to vaccination: A cause for its reemergence? *Emerg. Infect. Dis.*, 7, 526–528.

Mooi, F.R., van Loo, I.H., van Gent, M., He, Q., Bart, M.J., Heuvelman, K.J., de Greeff, S.C., Diavatopoulos, D., Teunis, P., Nagelkerke, N., and Mertsola, J. 2008. *Bordetella pertussis* strains with increased toxin production associated with pertussis resurgence. *Emerg. Infect. Dis.*, 15, 1206–1213.

Mooi, F.R., van der Maas, N.A., and De Melker, H.A. 2013. Pertussis resurgence: Waning immunity and pathogen adaptation – two sides of the same coin. *Epidemiol. Infect.*, 13, 1–10.

Mortimer Jr., E.A., and Jones, P.K. 1979. An evaluation of pertussis vaccine. *Rev. Infect. Dis.*, 1, 927–934.

Nicholson, T.L., Brockmeier, S.L., and Loving, C.L. 2009. Contribution of *Bordetella bronchiseptica* filamentous hemagglutinin and pertactin to respiratory disease in swine. *Infect. Immun.*, 77, 2136–2146.

Njamkepo, E., Bonacorsi, S., Debruyne, M., Gibaud, G.A., Guillot, S., and Guiso, N. 2011. Significant finding of *Bordetella holmesii* DNA in nasopharyngeal samples from French patients with suspected pertussis. *J. Clin. Microbiol.*, 49, 4347–4348.

Olin, P., Rasmussen, F., Gustafsson, L., Hallander, H.O., and Heijbel, H. 1997. Randomised controlled trial of two-component, three-component, and five-component acellular pertussis vaccines compared with whole-cell pertussis vaccine. Ad hoc group for the study of pertussis vaccines. *Lancet*, 350, 1569–1577.

Otsuka, N., Han, H.J., Toyoizumi-Ajisaka, H., Nakamura, Y., Arakawa, Y., Shibayama, K., and Kamachi, K. 2012. Prevalence and genetic characterization of pertactin-deficient *Bordetella pertussis* in Japan. *PLoS One*, 7, e31985.

Paccani, S.R., Finetti, F., Davi, M., Patrussi, L., D'Elios, M.M., Ladant, D., and Baldari, C.T. 2011. The *Bordetella pertussis* adenylate cyclase toxin binds to T cells via LFA-1 and induces its disengagement from the immune synapse. *J. Exp. Med.*, 208, 1317–1330.

Paradiso, P.R. 2002. Maternal immunization: The influence of liability issues on vaccine development. *Vaccine*, 20(suppl. 1), S73–S74.

Pebody, R.G., Gay, N.J., Giammanco, A., Baron, S., Schellekens, J., Tischer, A., Olander, R.M., Andrews, N.J., Edmunds, W.J., Lecoeur, H., Lévy-Bruhl, D., Maple, P.A., de Melker, H., Nardone, A., Rota, M.C., Salmaso, S., Conyn-van Spaendonck, M.A., Swidsinski, S., and Miller, E. 2005. The seroepidemiology of *Bordetella pertussis* infection in Western Europe. *Epidemiol. Infect.*, 133, 159–171.

Perez Vidakovics, M.L., Lamberti, Y., van der Pol, W.L., Yantorno, O., and Rodriguez, M.E. 2006. Adenylate cyclase influences filamentous haemagglutinin-mediated attachment of *Bordetella pertussis* to epithelial alveolar cells. *FEMS Immunol. Med. Microbiol.*, 48, 140–147.

Queenan, A.M., Cassiday, P.K., and Evangelista, A. 2013. Pertactin-negative variants of *Bordetella pertussis* in the United States. *N. Engl. J. Med.*, 368, 583–584.

Quinn, P., Gold, M., Royle, J., Buttery, J., Richmond, P., McIntyre, P., Wood, N., Lee, S.S., and Marshall, H. 2011. Recurrence of extensive injection site reactions following DTPa or dTpa vaccine in children 4–6 years old. *Vaccine*, 29, 4230–4237.

Relman, D., Tuomanen, E., Falkow, S., Golenbock, D.T., Saukkonen, K., and Wright, S.D. 1990. Recognition of a bacterial adhesion by an integrin: Macrophage CR3 (alpha M beta 2,

CD11b/CD18) binds filamentous hemagglutinin of *Bordetella pertussis*. *Cell*, 61, 1375–1382.

Rennels, M.B. 2003. Extensive swelling reactions occurring after booster doses of diphtheria–tetanus–acellular pertussis vaccines. *Semin. Pediatr. Infect. Dis.*, 14, 196–198.

Roberts, M., Fairweather, N.F., Leininger, E., Pickard, D., Hewlett, E.L., Robinson, A., Hayward, C., Dougan, G., and Charles, I.G. 1991. Construction and characterization of *Bordetella pertussis* mutants lacking the *vir*-regulated P.69 outer membrane protein. *Mol. Microbiol.*, 5, 1393–1404.

Rohani, P., Zhong, X., and King, A.A. 2010. Contact network structure explains the changing epidemiology of pertussis. *Science*, 330, 982–985.

Rosenthal, R.S., Nogami, W., Cookson, B.T., Goldman, W.E., and Folkening, W.J. 1987. Major fragment of soluble peptidoglycan released from growing *Bordetella pertussis* is tracheal cytotoxin. *Infect. Immun.*, 55, 2117–2120.

Rothstein, E., and Edwards, K. 2005. Health burden of pertussis in adolescents and adults. *Pediatr. Infect. Dis. J.*, 24, S44–S47.

Rowe, J., Yerkovich, S.T., Richmond, P., Suriyaarachchi, D., Fischer, E., Feddema, L., Loh, R., Sly, P.D., and Holt, PG. 2005. Th2-associated local reactions to the acellular diphtheria–tetanus–pertussis vaccine in 4- to 6-year-old children. *Infect. Immun.*, 73, 8130–8135.

Rutledge, R.K., and Keen, E.C. 2012. Images in clinical medicine. Whooping cough in an adult. *N. Engl. J. Med.*, 366, e39.

Ryan. M., McCarthy, L., Rappuoli, R., Mahon, B.P., and Mills, K.H. 1998. Pertussis toxin potentiates Th1 and Th2 responses to co-injected antigen: Adjuvant action is associated with enhanced regulatory cytokine production and expression of the co-stimulatory molecules B7–1, B7–2 and CD28. *Int. Immunol.*, 10, 651–662.

Sato, Y., Kimura, M., and Fukumi, H. 1984. Development of pertussis component vaccine in Japan. *Lancet*, 1, 122–126.

Sauer, L. 1933. Immunization with *Bacillus* pertussis vaccine. *J. Am. Med. Assoc.*, 101, 1449–1451.

Saukkonen, K., Burnette, W.N., Mar, V.L., Masure, H.R., and Tuomanen, E.I. 1992. Pertussis toxin has eukaryotic-like carbohydrate recognition domains. *Proc. Natl. Acad. Sci. USA*, 89, 118–122.

Schmitt, H.J., Beutel, K., Schuind, A., Knuf, M., Wagner, S., Müschenborn, S., Bogaerts, H., Bock, H.L., and Clemens, R. 1997. Reactogenicity and immunogenicity of a booster dose of a combined diphtheria, tetanus, and tricomponent acellular pertussis vaccine at fourteen to twenty-eight months of age. *J. Pediatr.*, 130, 616–623.

Shepard, C.W., Daneshvar, M.I., Kaiser, R.M., Ashford, D.A., Lonsway, D., Patel, B., Morey, R.E., Jordan, J.G., Weyant, R.S., and Fischer, M. 2004. *Bordetella holmesii* bacteremia: A newly recognized clinical entity among asplenic patients. *Clin. Infect. Dis.*, 38, 799–804.

Sheridan, S.L., Ware, R.S., Grimwood, M.B., and Lambert, S.B. 2012. Number and order of whole cell pertussis vaccine in infancy and disease protection. *J. Am. Med. Assoc.*, 308, 454–456.

Simondon, F., Preziosi, M.P., Yam, A., Kane, C.T., Chabirand, L., Iteman, I., Sanden, G., Mboup, S., Hoffenbach, A., Knudsen, K., Guiso, N., Wassilak, S., and Cadoz, M. 1997. A randomized double-blind trial comparing a two-component acellular and a whole-cell pertussis vaccine in Senegal. *Vaccine*, 15, 1606–1612.

Skowronski, D.M., Janjua, N.Z., Tsafack, E.P., Ouakki, M., Hoang, L., and De Serres, G. 2012. The number needed to vaccinate to prevent infant pertussis hospitalization and death through parent cocoon immunization. *Clin. Infect. Dis.*, 54, 318–327.

Smith, C., and Vyas, H. 2000. Early infantile pertussis; increasingly prevalent and potentially fatal. *Eur. J. Pediatr.*, 159, 898–900.

Stefanelli, P., Fazio, C., Fedele, G., Spensieri, F., Ausiello, C., and Mastrantonio, P. 2009. A natural pertactin deficient strain of *Bordetella pertussis* shows improved entry in human monocyte-derived dendritic cells. *New Microbiol.*, 32, 159–166.

Stein, P.E., Boodhoo, A., Armstrong, G.D., Heerze, L.D., Cockle, S.A., Klein, M.H., and Read, R.J. 1994. Structure of a pertussis toxin-sugar complex as a model for receptor binding. *Nat. Struct. Biol.*, 1, 591–596.

Storsaeter, J., Wolter, J., and Locht, C. 2007. Pertussis vaccines. In: Locht, C., editor. *Bordetella Molecular Microbiology*. Horizon Press. Norfolk, U.K. pp. 245–288.

Tamura, M., Nogimori, K., Murai, S., Yajima, M., Ito, K., Katada, T., Ui, M., and Ishii, S. 1982. Subunit structure of islet-activating protein, pertussis toxin, in conformity with the A–B model. *Biochemistry*, 21, 5516–5522.

Tan, T., Trindade, E., and Skowronski, D. 2005. Epidemiology of pertussis. *Pediatr. Infect. Dis. J.*, 24, S10–S18.

Tartof, S.Y., Lewis, M., Kenyon, C., White, K., Osborn, A., Liko, J., Zell, E., Martin, S., Messonnier, N.E., Clark, T.A., and Skoff, T.H. 2013. Waning immunity to pertussis following 5 doses of DTaP. *Pediatrics*, 131, e1047–e1052.

Terranella, A., Asay, G.R., Messonnier, M.L., Clark, T.A., and Liang, J.L. 2013. Pregnancy dose tdap and postpartum cocooning to prevent infant pertussis: A decision analysis. *Pediatrics*, 131, e1748–1756.

Uribe, K.B., Etxebarria, A., Martín, C., and Ostolaza, H. 2013. Calpain-mediated processing of adenylate cyclase toxin generates a cytosolic soluble catalytically active N-terminal domain. *PLoS One*, 8, e67648.

Van den Berg, B.M., Beekhuizen, H., Willems, R.J., Mooi, F.R., and van Furth, R. 1999. Role of *Bordetella pertussis* virulence factors in adherence to epithelial cell lines derived from the human respiratory tract. *Infect. Immun.*, 67, 1056–1062.

Vandamme, P., Heyndrickx, M., Vancanneyt, M., Hoste, B., De Vos, P., Falsen, E., Kesters, K., and Hinz, K.H. 1996. *Bordetella trematum* sp. nov., isolated from wounds and ear infections in humans, and reassessment of *Alcaligenes denitrificans* Rüger and Tan 1983. *Int. J. Syst. Bacteriol.*, 46, 849–858.

Van Hoek, A.J., Campbell, H., Amirthalingam, G., Andrews, N., and Miller, E. 2013. The number of deaths among infants under one year of age in England with pertussis: Results of a capture/recapture analysis for the period 2001–2011. *Euro Surveill.*, 18, pii=20414.

Van Rie, A., and Hethcote, H.W. 2004. Adolescent and adult pertussis vaccination: Computer simulations of five new strategies. *Vaccine*, 22, 3154–3165.

Versteegh, F.G.A. 2005. Pertussis: New insights in diagnosis, incidence and clinical manifestations. PhD Thesis, Free University of Amsterdam.

Vickers, D., Ross, A.G., Mainar-Jaime, R.C., Neudorf, C., and Shah, S. 2006. Whole-cell and acellular pertussis vaccination programs and rates of pertussis among infants and young children. *Can. Med. Assoc. J.*, 175, 1213–1217.

Wallihan, R., Selvarangan, R., Marcon, M., Koranyi, K., Spicer, K., and Jackson, M.A. 2013. *Bordetella parapertussis* bacteremia: Two case reports. *Pediatr. Infect. Dis. J.*, 32, 796–798.

Wei, S.C., Tatti, K., Cushing, K., Rosen, J., Brown, K., Cassiday, P., Clark, T., Olans, R., Pawloski, L., Martin, M., Tondella, M.L., and Martin, S.W. 2010. Effectiveness of adolescent and adult tetanus, reduced-dose diphtheria, and acellular pertussis vaccine against pertussis. *Clin. Infect. Dis.*, 51, 315–321.

Westra, T.A., de Vries, R., Tamminga, J., Sauboin, C.J., and Postma, M.J. 2010. Cost-effectiveness analysis of various pertussis vaccination strategies primarily aimed at protecting infants in the Netherlands. *Clin. Ther.*, 32, 1479–1495.

White, O.J., Rowe, J., Richmond, P., Marshall, H., McIntyre, P., Wood, N., and Holt, P.G. 2010. Th2-polarisation of cellular immune memory to neonatal pertussis vaccination. *Vaccine*, 28, 2648–2652.

WHO. 2010. Pertussis vaccines. WHO position paper. *Wkly. Epidemiol. Rec.*, 40, 385–400.

WHO. 2012. Global routine vaccination coverage, 2011. *Wkly. Epidemiol. Rec.*, 44, 432–435.

Winter, K., Harriman, K., Zipprich, J., Schechter, R., Talarico, J., Watt, J., and Chavez, G. 2012. California pertussis epidemic, 2010. *J. Pediatr.*, 161, 1091–1096.

Witt, M.A., Arias, L., Katz, P.H., Truong, E.T., and Witt, D.J. 2013. Reduced risk of pertussis among persons ever vaccinated with whole cell pertussis vaccine compared to recipients of acellular pertussis vaccines in a large US cohort. *Clin. Infect. Dis.*, 56, 1248–1254.

Witt, M.A., Katz, P.H., and Witt, D.J. 2012. Unexpectedly limited durability of immunity following acellular pertussis vaccination in preadolescents in a North American outbreak. *Clin. Infect. Dis.*, 54, 1730–1735.

Wood, N., McIntyre, P., Marshall, H., and Roberton, D. 2010. Acellular pertussis vaccine at birth and one month induces antibody responses by two months of age. *Pediatr. Infect. Dis. J.*, 29, 209–215.

Wright, S.W., Edwards, K.M., Decker, M.D., and Zeldin, M.H. 1995. Pertussis infection in adults with persistent cough. *J. Am. Med. Assoc.*, 273, 1044–1046.

Zaretzky, F.R., Gray, M.C., and Hewlett, E.L. 2002. Mechanism of association of adenylate cyclase toxin with the surface of *Bordetella pertussis*: A role for toxin-filamentous haemagglutinin interaction. *Mol. Microbiol.*, 45, 1589–1598.

Zepp, F., Heininger, U., Mertsola, J., Bernatowska, E., Guiso, N., Roord, J., Tozzi, A.E., and Van Damme, P. 2011. Rationale for pertussis booster vaccination throughout life in Europe. *Lancet Infect. Dis.*, 11, 557–570.

Zhang, L., Prietsch, S.O., Axelsson, I., and Halperin, S.A. 2011. Acellular vaccines for preventing whooping cough in children. *Cochrane Database Syst. Rev.*, 1, CD001478.

Zheteyeva, Y.A., Moro, P.L., Tepper, N.K., Rasmussen, S.A., Barash, F.E., Revzina, N.V., Kissin, D., Lewis, P.W., Yue, X., Haber, P., Tokars, J.I., Vellozzi, C., and Broder, K.R. 2012. Adverse event reports after tetanus toxoid, reduced diphtheria toxoid, and acellular pertussis vaccines in pregnant women. *Am. J. Obstet. Gynecol.*, 207, 59.e1–e7.

Chapter 36

Pathogenesis of Human Ehrlichioses

Tais Berelli Saito and David H. Walker

Department of Pathology, University of Texas Medical Branch, Galveston, TX, USA

36.1 Introduction

The first human infection with a member of the genus *Ehrlichia* was recognized in 1986 (Fishbein et al., 1987; Maeda et al., 1987). By 1990 ehrlichial infection was recognized as an important tick-transmitted human disease (Dumler and Bakken, 1995). Subsequently, other ehrlichial species have been discovered to infect humans (Buller et al., 1999; Pritt et al., 2011). Infection with *Ehrlichia chaffeensis* has a potential fatal outcome if timely appropriate treatment is not administered (Dumler and Bakken, 1995).

Ehrlichioses are emergent vector-borne infectious diseases with a wide range of vertebrate hosts and worldwide distribution. Humans have been reported to be infected by five *Ehrlichia* species: *E. chaffeensis*, *E. ewingii*, *E. canis*, *E. ruminantium*, and *E. muris*-like agent (EMLA) (Allsopp et al., 2005; Buller et al., 1999; Perez et al., 1996, 2006; Pritt et al., 2011). Life-threatening disease occurs in humans infected with *E. chaffeensis*, the agent of the human monocytic ehrlichiosis (HME). *E. ewingii*, a pathogen that infects dogs, was first recognized as a human pathogen in 1998 (Buller et al., 1999). Only a few human cases of ehrlichiosis in Venezuela have been reported to be caused by *E. canis* (Perez et al., 2006). *E. ruminantium* DNA was detected in fatal human cases of suspected ehrlichiosis in South Africa as well as in less severe disease associated with an *E. ruminantium*-like agent (Panola Mountain agent) (Allsopp et al., 2005; Reeves et al., 2008). EMLA was recently described as a human pathogen in Wisconsin and Minnesota (Pritt et al., 2011).

36.2 Pathogen

Ehrlichioses are infections caused by alpha-1 proteobacteria of the order Rickettsiales, family Anaplasmataceae, which multiply within membrane-bound vacuoles in the cytoplasm of eukaryotic cells (Rikihisa, 1991). The family Anaplasmataceae comprises five genera: *Anaplasma*, *Ehrlichia*, *Wolbachia*, *Neorickettsia*, and *Neoehrlichia* (Dumler et al., 2001).

36.2.1 Genus and species

Ehrlichiae have a cell wall that lacks lipopolysaccharide (LPS) and peptidoglycan but contains tandem repeat proteins (TRPs), including TRP47 and TRP32, and a family of 28-kDa proteins. *E chaffeensis* contains 22 proteins of ~28 kDa with three hydrophilic, surface-exposed hypervariable domains, a surface-exposed 75–145 kDa protein with two to five tandem repeat units and a 200-kDa ankyrin-containing protein (McBride et al., 2003; Yu et al., 1997, 2000).

The complete genomes of *E. chaffeensis*, *E. canis*, and *E. ruminantium* have been determined (Collins et al., 2005; Dunning Hotopp et al., 2006; Mavromatis et al., 2006). All *Ehrlichiae* have a circular chromosome with a genome size of $1.2–1.5 \times 10^6$ bp (Dunning Hotopp et al., 2006). The majority of the genes are conserved in all genomes. More than 400 genes are classified in ortholog clusters associated with housekeeping gene functions. The *E. chaffeensis* genome has 312 open reading frames that encode hypothetical or uncharacterized proteins (Dunning Hotopp et al., 2006). A large part of the evolutionarily reduced genome of the obligately intracellular alpha-proteobacteria is necessary for vital activities of the pathogen, such as genes for biosynthesis of nucleotides and vitamins and protein synthesis. The abundance of major surface proteins in *Ehrlichiae* appears to be related to bacterial adaptation to different hosts (Dunning Hotopp et al., 2006).

The limited genetic variability of *E. chaffeensis* based on the 16S rRNA gene in isolates from the United States is 99.4–99.6% identity (Kawahara et al., 2009; Rar and Golovljova, 2011; Wen et al., 2003). The groESL operon sequences reveal the difference between isolates, demonstrating 95% identity between American and Japan isolates (Kawahara et al., 2009). The nucleotide differences are located in a predicted cell envelope protein (Miura and Rikihisa, 2007). Comparison among *Ehrlichia* species using the 16S rRNA gene sequences demonstrates 97% ± 7% similarity for *E. canis*, *E. chaffeensis*, *E. ewingii*, *E. muris*, and *E. ruminantium*, which were used for the reclassification of the species in the genus *Ehrlichia* in 2001 (Anderson et al., 1991; Dame et al., 1992; Dumler et al., 2001; Roux and Raoult, 1995; Shibata et al., 2000; Wen et al., 1995a, b).

36.2.2 Target cells

Each *Ehrlichia* species infects leukocytes or endothelial cells. *E. canis*, *E. chaffeensis*, and *E. muris* are usually detected in macrophages and monocytes *in vivo* and can be propagated *in vitro* in macrophage cell lines (Barnewall and Rikihisa, 1994; Dawson et al., 1991a, 1991b; Heimer et al., 1998). *E. ewingii* is detected in neutrophils, and it has yet to be propagated in cell culture (Ewing et al., 1971). *E. ruminantium* is detected in endothelial cells, neutrophils, or macrophages *in vivo*, and can be propagated in endothelial or macrophage cell lines *in vitro* (Bezuidenhout et al., 1985; Logan et al., 1987; Prozesky and Du Plessis, 1987; Sahu, 1986).

36.2.3 Life cycle

Ehrlichiae are small gram-negative bacteria with variable often pleomorphic shapes and forms, varying from coccoid to ellipsoidal. *Ehrlichiae* are obligately intracellular bacteria with a diameter of 0.5–1.5 μm and reside within cytoplasmic vacuoles, either singly and more often in compact inclusions (morulae) in mature or immature hematopoietic cells, especially mononuclear phagocytes such as monocytes and macrophages. For some ehrlichial species, neutrophils are the target cells. *Ehrlichiae* are present in peripheral blood and in tissues, usually mononuclear phagocyte-rich organs (spleen, liver, bone marrow, lymph node) of mammalian hosts (Dumler and Bakken, 1995; Rikihisa, 1991; Walker et al., 2004). *Ehrlichiae* present two ultrastructural forms: a larger coccoid or an elongated reticulate cell (RC) with ribosomes and DNA strands dispersed in the protoplasm and a smaller coccoid dense-core cell (DC) with the ribosomes and DNA condensed in the center of the cell (Figures 36-1a and

36-1b). The RC is 0.4–0.6 × 0.7–1.9 μm, and the DC is 0.4–0.6 μm in diameter. Both forms have a cell wall with a smooth inner membrane and wavy outer membrane. The DC is the infectious form and expresses the 75–145 kDa (usually designated as 120 kDa) tandem-repeat-containing protein (Popov et al., 1995, 2000). The RC undergoes vegetative growth, and the DC form is a stationary stage (Popov et al., 1998; Rar and Golovljova, 2011).

Developmental forms of *Ehrlichiae* have been observed in ixodid ticks (Figure 36-1c), and are considered part of the bacterial life cycle in the vector (Popov et al., 2007; Rikihisa, 1991). The persistence of *Ehrlichiae* in their principal vertebrate host is important for the maintenance of the transmission cycle as the critical pathogen reservoir. *Ehrlichiae* are transmitted transstadially in the vector tick, but not transovarially, which emphasizes the importance of the vertebrate reservoir. During vector tick feeding on infected animals, the bacterium is acquired, and it is maintained through ecdysis in the ixodid tick, prior to potential transmission to another vertebrate host during next developmental stage of the vector (Paddock and Childs, 2003; Rar and Golovljova, 2011; Rar et al., 2011). The dynamics of the ehrlichial life cycle in the vector tick have not been characterized.

Ehrlichial pathogens infect mammals of a broad range of species, inducing a range of severity of clinical manifestations. *E. chaffeensis* causes symptomatic infections in humans and persistent and subclinical infections in deer (Dawson et al., 1996a,b; Dawson and Ewing, 1992; Ewing et al., 1995; Fishbein et al., 1994; Lockhart et al., 1997). *E. ewingii* causes disease in canids and humans (Buller et al., 1999; Ewing et al., 1971). Each of these species is known to be transmitted and maintained in a tick vector, including *Amblyomma variegatum* for *E. ruminantium* (Bezuidenhout, 1987), *A. americanum* (Figure 36-2a) for *E. chaffeensis*, and *E. ewingii* (Anziani et al., 1990; Ewing et al., 1995) and *Ixodes scapularis* (Figure 36-2b) for EMLA (Dumler et al., 2001; Pritt et al., 2011).

36.3 Epidemiology

Two-thirds of cases of human ehrlichiosis in the United States occur between May and August (Figure 36-3a) during the season of tick activity (Dumler and Bakken, 1995; Dumler et al., 2007). In North America the principal vector of HME is *A. americanum* (lone star tick); however, other species of ticks can carry *E. chaffeensis*, for example, *I. pacificus* and *D. variabilis*. *E. chaffeensis* DNA has also been detected in ticks in South Korea (*Hemaphysalis longicornis* and *I. persulcatus*), Cameroon (*Rhipicephalus sanguineus*), China (*A. testudinarium* and *H. yeni*), and Argentina (*A. parvum*) (Cao et al., 2000; Kim et al., 2003; Kramer et al., 1999; Lee et al., 2005; Ndip et al., 2010; Tomassone et al., 2008). Cases of HME have also been reported outside of the range of *A. americanum* ticks in the United States, suggesting alternative vectors or other

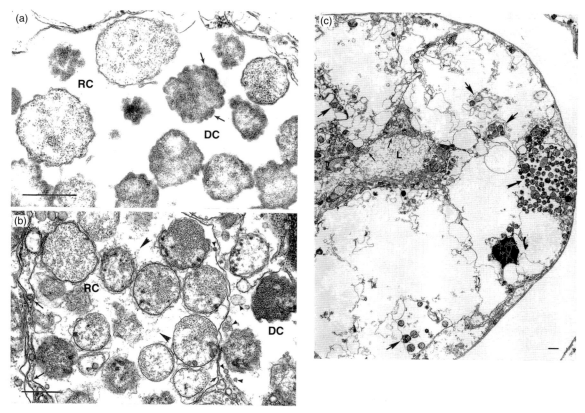

Fig. 36-1 *E. muris* ultrastructure of RCs and DCs in tick tissues. Bars = 0.5 μm. (a) Morula containing the developmental forms of *Ehrlichia*: RCs and DCs; arrows show uneven outlines of the DC surface. (b) Morula in the salivary gland of the tick; thin arrows indicate morular membranes. Large arrowheads show intramorular fibrils, and small arrowheads indicate vesicles inside a morula. (c) Salivary gland alveolus of tick heavily infected with *Ehrlichiae*. Arrowheads indicate morulae. The thick arrow shows a morula. L = lumen. Thin arrows indicate microvilli of epithelial cells protruding into the lumen. Bar = 1 μm. Reproduced with permission from Popov et al. (2007).

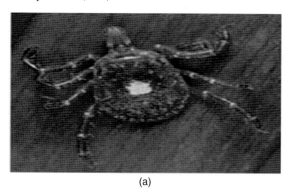

(a)

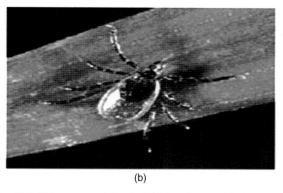

(b)

Fig. 36-2 Tick vectors of the ehrlichial pathogen. (a) *A. americanum* adult female showing the characteristic white spot in the back ("lone star" tick). (b) *I. scapularis* adult female. *Source:* Centers for Disease Control and Prevention (CDC); National Center for Emerging and Zoonotic Infectious Diseases (NCEZID); Division of Vector-Borne Diseases (DVBD).

antigenically related *Ehrlichiae* (Figure 36-3b). Cases have also been reported in Europe and Africa, and recently *E. chaffeensis* DNA was detected in patients in Venezuela and Cameroon (Dumler and Bakken, 1995; Martinez et al., 2008; Ndip et al., 2009). *E. chaffeensis* has been confirmed by an isolation of the agent only in the United States (Dawson et al., 1991a).

The white-tailed deer is considered the principal host of *E. chaffeensis* in the United States; however, natural infection with *E. chaffeensis* has also been documented by isolation of the bacteria from a goat in an endemic area in the United States, and *E. chaffeensis* DNA has been detected in naturally infected coyotes in the United States, sika deer in South Korea and Japan, and rodents in Korea and China (Dugan et al., 2000; Kawahara et al., 2009; Kim et al., 2006; Kocan et al., 2000; Lee et al., 2009; Wen et al., 2003). Other mammals have been experimentally infected with *E. chaffeensis*, but their ability to serve as a reservoir host has not been confirmed by transmission studies (Dugan et al., 2000). Calves develop clinical signs in a dose-dependent manner, and red foxes (*Vulpes vulpes*) do not show clinical manifestations of infection (Davidson et al., 1999; delos Santos et al., 2007). Experimentally infected dogs develop thrombocytopenia without clinical signs; however, in rare cases, epistaxis, lymphadenopathy, lethargy, hematuria, and generalized purpura have been observed in

(a)

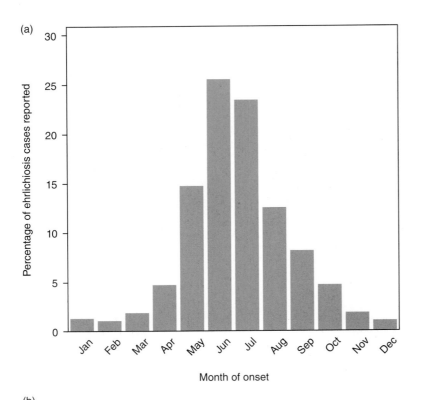

Month of onset

(b)

Ixodes scapularis distribution

Ixodes pacificus distribution

Amblyomma americanum distribution

overlapping distribution (*I. scapularis* and *A. americanum*)

Fig. 36-3 Epidemiologic data of ehrlichial infection in the United States. (a) Percentage of cases reported between 1994 and 2010 by month of onset demonstrates the seasonality of cases. There are cases reported in each month of the year; however, most are reported between May and August. (b) Approximate distribution of vector tick species for human ehrlichiosis. (c) Incidence of HME cases by state in 2010 per million persons. (d) Number of human cases of ehrlichiosis reported to CDC annually between 1994 and 2010. (e) Fatality rate of human ehrlichiosis reported annually to CDC between 2000 and 2010. NN = Not notifiable.

naturally infected dogs in the United States and South Korea (Breitschwerdt et al., 1998; Yu et al., 2008; Zhang et al., 2003b). Dogs can remain infected with *E. chaffeensis* for 2–4 months, suggesting that dogs may serve as a natural host for *E. chaffeensis* (Zhang et al., 2003b).

Most cases of infection by *E. chaffeensis* are reported in regions corresponding to the presence of the vector lone star tick and reservoir white-tailed deer across the south-central, southeastern, and mid-Atlantic areas (Dumler et al., 2007; Paddock and Childs, 2003)

and appear to be spreading northward with *A. americanum*. White-tailed deer become persistently infected subclinically and serve as a source of *Ehrlichiae* for uninfected ticks (Ewing et al., 1995).

E. ewingii is transmitted by *A. americanum* ticks, and it is a more prevalent species in nature than *E. chaffeensis* in some regions (Dumler et al., 2007).

Human infections with *E. chaffeensis* are reported in small numbers in many states in the United States (Figure 36-3c), suggesting that HME is a rare disease;

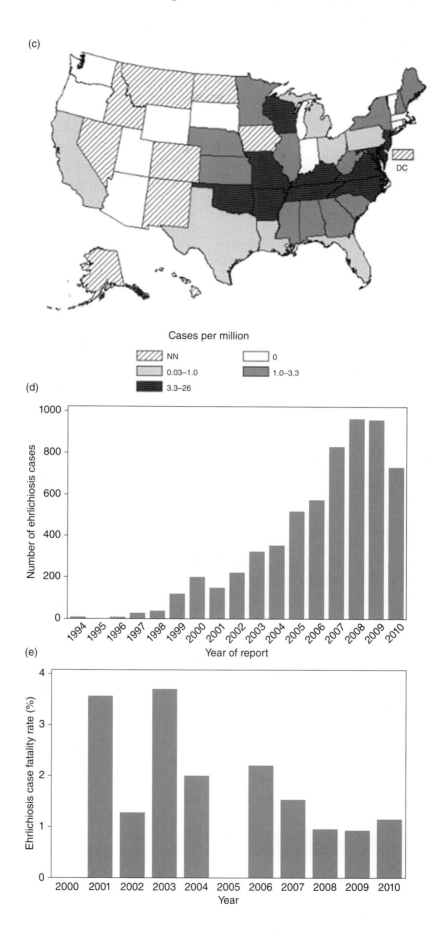

Fig. 36-3 (Continued)

however, active prospective studies suggest that the incidence is substantially higher owing to misdiagnosis, underreporting, and patients treated empirically without laboratory evaluation for ehrlichial infection (Figures 36-3d and 36-3e) (Dumler et al., 2007; Walker et al., 2004).

36.4 Clinical features

The initial clinical presentation of HME consists of non-specific flu-like symptoms such as fever, headache, myalgia, and malaise, and occurs around 9 days after tick exposure. Other multisystem symptoms that manifest in 10%–40% of the cases include nausea, vomiting, diarrhea, abdominal tenderness, regional lymphadenopathy, cough, pharyngitis, arthralgia, rash, stiff neck, photophobia, dizziness, and confusion (Dumler et al., 2007; Fishbein et al., 1994; Olano et al., 2003a). Rash is observed in 30%–40% of the cases as petechiae, macules, and diffuse erythema (Table 36-1). Central nervous system (CNS) involvement occurs in 20% of the cases as meningitis or meningoencephalitis, in some cases associated with coma (Paddock and Childs, 2003). Infections are described slightly more often in males than females, at the median age of 50 years (Demma et al., 2005). The laboratory manifestations of HME are leukopenia, lymphocytopenia followed by γ/δ T lymphocytosis, thrombocytopenia, and elevated serum transaminase concentrations (Dumler et al., 2007; Walker et al., 2004).

Patients with fever, leukopenia and/or thrombocytopenia and increase in hepatic transaminases in serum and potential tick exposure should have ehrlichiosis in the differential diagnoses (Dumler et al., 2007).

Patients infected with *E. chaffeensis* develop a moderate to severe disease with a 41–63% requiring hospitalization (Dahlgren et al., 2011; McQuiston et al., 1999). Immunocompetent patients that control the infection develop granulomas, and in lethal infections in immunocompetent patients inflammation is present in tissues with a paucity of bacteria (Dumler et al., 1993a). In contrast, immunocompromised patients' tissues contain massive ehrlichial burdens. Severe clinical manifestations of HME are toxic or

septic shock-like syndrome, CNS involvement, and respiratory distress syndrome, which lead to fatality in approximately 2% of infections, especially in immunocompromised patients (Demma et al., 2005; Dumler, 2005; Fordham et al., 1998; Paddock and Childs, 2003; Paddock et al., 2001; Patel and Byrd, 1999; Stone et al., 2004).

36.5 Pathogenesis and immunity (description of host–pathogen interaction, details of mechanisms, and routes of infections)

36.5.1 Initial infection

Ehrlichiosis, a vector-borne disease, is transmitted to a vertebrate host by a feeding tick. The events taking place within the vector as well as the initial events at the initial host site of infection are not well elucidated.

Ehrlichiae enter the host cell by endocytosis via glycophosphoinositol-anchored receptors, surviving and multiplying in endosomes (Rikihisa, 2006).

The 120-kDa TRP in the outer membrane of the DC wall serves as an adhesin that mediates internalization of *Ehrlichiae* (Popov et al., 1998, 2000). Selectin proteins (E- and L-selectin) are suggested to act as co-receptors in ehrlichial attachment and entry (Figure 36-4) (Zhang et al., 2003a).

Ehrlichiae enter into host cells by receptor-mediated endocytosis, dependent on transglutaminase activity. Bacterial proteins, such as caveolae and glycosylphosphatidylinositol-anchored proteins, with protein cross-linking by transglutaminase induce cell signaling events that enable entry into the host cell, associated with tyrosine phosphorylation, phospholipase C (PLC)-γ2 activation, and inositol 1,4,5-trisphosphate production, which increases host cytosolic free calcium, an essential step for *E. chaffeensis* entry into the host cell (Lin and Rikihisa, 2003b; Lin et al., 2002; Mott et al., 1999).

36.5.2 Interaction with host cell

E. chaffeensis replicates in the early endosomes, which are characterized by the presence of cytoplasmic small GTPase, Rab5, vacuolar-type (H+) ATPase, and early endosomal antigen 1 (EEA1). These endosomes are slightly acidic and, importantly, do not fuse with lysosomes (Mott et al., 1999). This mechanism of evasion of host cell defense favors the accumulation of a transferrin receptor (TfR) in the ehrlichial inclusions, which is not present at the time of bacterial internalization, but progressively accumulates until eventually all cytoplasmic TfR is directed to the inclusions (Barnewall et al., 1997). Also, *E. chaffeensis* infection induces activation of the iron responsive protein 1, by binding iron-responsive elements in the 3'-untranslated region of the TfR mRNA. As a consequence, the degradation of the mRNA of TfR is inhibited, leading to upregulation of TfR mRNA in the host

Table 36-1 Symptoms and signs of human ehrlichiosis infection by *E. chaffeensis* (HME)

Symptoms/signs	HME (%)
Rash	26–40
CNS involvement	20
Fever	96
Headache	72
Myalgia	68
Malaise	77
Nausea, vomiting, diarrhea	25–57
Cough	28

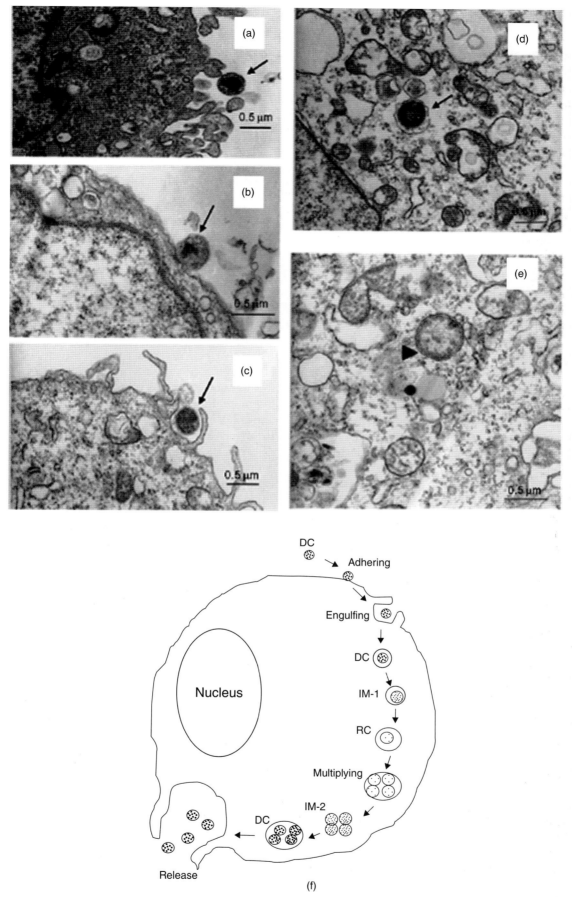

Fig. 36-4 *E. chaffeensis* interaction with vertebrate cells. (a) At 0 hour post-inoculation, a DC is attached to a DH82 cell and surrounded by microvilli. (b) A DC closely adherent to the membrane of the host cell at 1 hour post-inoculation. (c) A DC being engulfed by the host cell at 1 hour post-inoculation. (d) A DC is in a vacuole at 1 hour post-inoculation. (e) An *Ehrlichia* in a vacuole presumably transforming from a DC to an RC (intermediate phase) (arrowhead). Dense-cored *Ehrlichiae* are indicated by arrows. (f) Illustration of the developmental cycle of *E. chaffeensis*. A DC attaches to the host cells, and enters into the host through phagocytosis. In the phagosome, the DC transforms into the intermediate phase (IM)-1 and subsequently into an RC. The RC multiplies by binary fission for 48 hours and transforms into an intermediate phase (IM)-2 cell and eventually matures into a DC at 72 hours, which is released by lysis of the host cell. Reproduced with permission from Zhang et al. (2007).

cell, and induction of more TfR protein and TfR endosomes. *Ehrlichiae* depend on the host cell as a source of nutrients, which need to be directed to the endosomal inclusions. One of the most critical required nutrients for ehrlichial survival is iron. *E. chaffeensis* is able to accumulate iron-transferrin (Tf) in the early endosome, which may be the source of iron, which is suggested to be correlated with the increase of TfR endosomes (Barnewall et al., 1997, 1999). The mechanism by how IFN-γ induces ehrlichial death is associated with depletion of host cytoplasmic iron from Tf. As another defense mechanism, *Ehrlichiae* block Jak-Stat signal transduction in target cells (especially monocytes), which decreases IFN-γ production (Barnewall and Rikihisa, 1994; Lee and Rikihisa, 1998).

In vitro studies using cell culture of *E. chaffeensis*-infected human monocyte cells (THP1 cells) demonstrate that the bacteria produce proteins that bind host protein and modify gene expression to favor intracellular survival and replication as an obligately intracellular pathogen (Lin and Rikihisa, 2003b; Lin et al., 2002). *E. chaffeensis* downregulates molecules that are important in membrane trafficking, such as SNAP 23 (synaptosomal-associated protein, 23 kDa), Rab5A (a member of the RAS oncogene family), STX16 (syntaxin 16), and Th1 cytokines (IL-12 and IL-18), suggesting mechanisms of avoiding maturation and fusion of its endosome with lysosomes, as a protective evasion from the host cell defenses. Also *E. chaffeensis* upregulates cell cyclins and apoptosis inhibitors, allowing survival of its host cell for a longer time (Mott et al., 1999; Rikihisa, 2003, 2006; Zhang et al., 2004).

36.5.3 Distribution/dissemination

The study of ehrlichial dissemination in naturally infected hosts is limited, and the animal models described do not mimic the natural route of infection by a tick bite. Ehrlichial infection by the intraperitoneal route in a murine model has been evaluated by *in situ* hybridization, to demonstrate the distribution of infected cells in tissues during infection. The initial detection of infection is found in Kupffer cells and endothelial cells in the liver on day 6 post-infection (p.i.). Infected mononuclear cells are present in hepatic sinusoids and central hepatic veins, and sinuses of the spleen and lymph nodes. Endothelial cells and macrophages in the bone marrow are also infected on day 6 p.i. After the progression of the infection, further distribution of the bacteria is identified. On day 9 p.i. infected cells are widely distributed in the liver, in endothelial cells of capillaries of Peyer's patches and follicles of intestine, and in veins of perifollicular areas of the spleen. Infected mononuclear cells in lymphatic sinuses of lymph nodes and endothelial cells in the bone marrow are also identified at this time in the course of infection. Lung cells, such as alveolar macrophages, and monocytes and endothelial cells, are infected on day 9 p.i. in this animal model. Other

studies have demonstrated bacterial distribution in the brain (Brown and Skowronek, 1990; Jardine et al., 1995; Okada et al., 2003). Animal models of fatal and persistent ehrlichial infection have also shown an earlier distribution of the bacterium in spleen, liver, and lung at day 3–4 p.i. (Olano et al., 2004; Sotomayor et al., 2001). Although some information is available about ehrlichial distribution, how the dissemination occurs, especially from the site of infection by tick transmission, is not known.

36.5.4 Pathogenesis of disease/immune response (immunopathogenesis)

The immunopathogenesis of ehrlichiosis is not well understood, especially the mechanisms involved in disease development, and the interaction of the *Ehrlichiae* and host immune system. In human cases the levels of bacteria in blood and tissues of immunocompetent patients are very low, which could suggest that the disease manifestations are mediated by the immune system (Sehdev and Dumler, 2003). Furthermore, the cytokine and chemokine responses to ehrlichial infection are dependent on non-classical pathogen-derived molecular patterns (PAMPs), since *Ehrlichiae* do not contain LPS and peptidoglycan, or other classical PAMPs in the cell wall (Dunning Hotopp et al., 2006; Lin and Rikihisa, 2003a). *In vivo* studies of infection of *E. chaffeensis* in TLR4-function deficient mice demonstrated a prolonged course of infection; however, the role of pattern-recognition receptors in the development of ehrlichiosis and in the pathogenesis of the disease is not understood (Telford and Dawson, 1996). *In vitro* studies using a human cell line infected with *E. chaffeensis* indicated an increase in expression of cytokines, such as IL-1β, IL-8, and IL-10, as early as 2 and 24 hours p.i. (Lee and Rikihisa, 1996). The erhlichial infection of bone marrow-derived macrophages demonstrated the induction of cytokine and chemokine production through the MyD88 pathway, which indicates a role of this pathway in pathogen recognition by innate immune cells; however, the induction of these factors is independent of TRIF signaling. Interestingly neither IL-1R nor IL-18R, which have MyD88 as a signaling adaptor, is required for the induction of cytokines and a chemokine in response to infection. This observation is not consistent in all ehrlichial species. In *E. muris* infection, natural killer T cells secrete IFN-γ by a CD1d-dependent, but not MyD88-dependent, process stimulated by endogenous glycolipid as the PAMP (Mattner et al., 2005).

Ehrlichia infection induces production of proinflammatory cytokines and chemokines; however, the majority of the TLRs (TLR1, 2, 3, 4, 5, 6, 7, 9, 11, 12, or 13) are not responsible for mediating this process (Dunning Hotopp et al., 2006; Miura et al., 2011). An emergent concept of TLR-independent activation of innate immune cells by *Ehrlichiae* has been supported by induction of different cytokines independent of the TLRs, which does not even require bacterial infection (Koh et al., 2010). Other forms of induction of downstream signaling pathways have been suggested in ehrlichial infections. The TRPs induce chemokine responses, such as IL-8, MCP-1, and MIP-1β. Other TRPs do not induce cytokines; however, they can interact with some cytoplasmic proteins that may activate the ERK pathway and NF-κB directly, via the MyD88 pathway (Wakeel et al., 2009). Studies have demonstrated the critical role of NF-κB, as well as the ERK pathway, in induction of cytokines and chemokines in ehrlichial infection (Lee and Rikihisa, 1996; Miura et al., 2011).

Upon interaction with new host cells, *E. chaffeensis* genes encoding the type IV secretion (T4S) apparatus are upregulated. Bacterial T4S systems transport macromolecules across the membrane in an ATP-dependent manner and are increasingly recognized as a virulence factor delivery mechanism that allows pathogens to modulate eukaryotic cell functions for their own benefit. The delivered macromolecules are referred to as T4S substrates, or effectors, because they affect and alter basic host cellular processes, resulting in disease development (Rikihisa et al., 2009).

From the point of view of pathogenesis, the most important changes in the host cell caused by *E. chaffeensis* infection are downregulation of the innate immune system and a differentially regulated cell cycle. The most striking feature of *E. chaffeensis* infection is repression of host cell cytokines that modulate innate and adaptive immunity to intracellular bacteria.

E. chaffeensis avoids stimulation of IL-12 production and represses IL-15 and IL-18 production. These cytokines play fundamental roles in stimulating NK cells and T helper 1 cells to produce gamma interferon (IFN-γ), which then activates macrophages to kill phagocytosed bacteria.

IL-12 and IL-15 also activate NK cells and cytotoxic T lymphocytes to kill cells infected with intracellular bacteria. Thus, repression of IL-12, IL-15, and IL-18 may play a role in *E. chaffeensis* evasion of host innate and adaptive immunity. *E. chaffeensis* induces the production of apoptosis inhibitors such as NF-κB, BCL2A1, BIRC3, IER3, and MCL1. In the early stage of infection (7 hours), *E. chaffeensis* represses the BCL2 antagonists BIK and BNIP3L, which induce apoptosis by inactivating BCL2 proteins. The expression of BCL2 proteins and their antagonists returns to normal levels gradually in the late stages of infection. However, hematopoietic cell kinase is induced during the late stages of infection (Zhang et al., 2004).

36.6 Immunity

Little information is available about the immune response of humans infected with ehrlichial pathogens. Observed tissue injury and abundance of inflammatory cells in organs of certain patients with a low bacterial load suggests that immune responses may contribute to immune-mediated pathology (Rikihisa, 1999). Severe ehrlichial diseases occur in immunocompromised patients, such as HIV-infected individuals or those receiving immunosuppressive therapy, apparent with a higher incidence than in the general population, indicating the importance of host immune responses in resistance to ehrlichial infections (Lawrence et al., 2009; Paddock et al., 2001).

Human patients with acute infection with *E. chaffeensis* expand activated T cells expressing the T-cell receptor (TCR) gamma/delta, CD45RO, and HLA-DR antigens (Caldwell et al., 1995); however, these cells (primed and activated) are removed from the body by programmed or apoptotic cell death (Caldwell et al., 1996). The significance of this response in resistance to infection is not clear, and similar responses have been observed in patients recovering from infections caused by other intracellular pathogens (Caldwell et al., 1995).

A recent study has described changes in soluble mediators in two human patients infected with *E. chaffeensis*, one with mild disease and another that succumbed to fatal infection. The evaluation of lethal HME revealed substantially increased levels of IL-1α, IL-1β, TNF-α, and IL6 (pro-inflammatory cytokines) compared to non-lethal infection and uninfected controls. Also the anti-inflammatory cytokines (IL-10 and IL-13) were increased in lethal infection compared to non-lethal infection and uninfected controls. The concentration of Th1 cytokines (IL-2 and IFN-γ) was lower in a lethal infection and was associated with a slight increase in the levels of Th2 cytokines (IL-4 and IL-9) and the Th17 cytokine (IL-17) compared to non-lethal HME and healthy controls. Neutrophil chemoattractant and neutrophil growth factor (IL-8 and G-CSF) were increased in lethal disease, while the concentration of GM-CSF was lower than that in non-lethal infection. MCP-1α, MIP-1α, and MIP-1β (chemoattractants for macrophages/monocytes and NK cells) were highest in lethal compared to non-lethal infection and controls. Serum concentrations of T-cell chemokine (RANTES) and IFN-γ dependent chemokine (IP-10) were lower in lethal compared to non-lethal infection, but higher than in uninfected controls (Ismail et al., 2012).

Patients with HME frequently present with leukopenia, neutropenia, and lymphocytopenia followed by γ/δ T lymphocytosis, and thrombocytopenia. The cause of these blood count changes was suggested to be due to peripheral sequestration, consumption, or destruction of circulating blood cells in the absence of vasculitis (Dumler et al., 1993b). The decreased number of leukocytes has been suspected to be due to apoptosis. Patients with milder infection expressed a higher level of FAS mRNA compared to the patient with fatal ehrlichial infection and healthy controls; however, the patient with fatal *Ehrlichia* infection expressed a higher level of TNFR mRNA than that expressed in leukocytes from the patient with milder *Ehrlichia* infection or healthy controls, suggesting that signaling via TNFR, but not Fas/FasL signaling, may be relevant to observed low levels of lymphocytes and weak Th1 response (Ismail et al., 2012).

Taken together these findings reveal that the presence of uncontrolled elevation of inflammatory cytokines in a patient with fatal HME could contribute to prolonged activation of *Ehrlichia* target cells, which could possibly contribute to tissue injury and multi-organ failure (Ismail et al., 2012).

Severe ehrlichial disease occurs in immunocompromised patients, such as HIV-infected individuals or those receiving immunosuppressive therapy. *E. ewingii* has a higher incidence in immunocompromised persons than in the general population, indicating the importance of host immune responses in resistance to ehrlichial infections.

The evaluation of pathogenic mechanisms and the immune response in ehrlichial infection requires development of appropriate animal models. *E. chaffeensis* infection in mice is transient and subclinical (Winslow et al., 1998). Severely immunodeficient (SCID) mice that lack both T and B cells develop persistent infection with *E. chaffeensis* and also develop splenomegaly, lymphadenopathy, and foci of liver inflammation and necrosis, suggesting the importance of T cells and B cells in resistance to ehrlichial infections (Winslow et al., 1998). Major histocompatibility class II (MHC II) gene knockout mice and Toll-like receptor (TLR) 4 gene deficient mice develop persistent ehrlichial infection, demonstrating a critical role for CD4 T cells and TLR4-dependent activation of innate immunity (Ganta et al., 2002, 2004). However, genomic analysis of *E. chaffeensis* reveals a lack of genes for synthesis of LPS, a known ligand for TLR4 (Lin and Rikihisa, 2003a).

Immunity and immunopathogenic mechanisms in ehrlichial infections have been demonstrated in murine models, using very closely related *Ehrlichia* species. *E. muris* and *I. ovatus Ehrlichiae* have been used extensively in elucidating host responses to *Ehrlichiae* (Olano et al., 2004; Sotomayor et al., 2001). *E. muris* causes persistent infection in immunocompetent AKR and C57BL/6 mice, similar to persistent *E. chaffeensis* infection in white-tailed deer, the natural host (Olano et al., 2004). A decrease in ehrlichial loads is associated with an increase in antibody titers; however, low ehrlichial loads and inflammatory lesions persist for a month or longer (Feng and Walker, 2004; Kawahara et al., 1996). A high portion (81%) of MHC Class I gene knockout on C3H background mice infected with *E. muris* died compared to none of the wild-type mice, demonstrating the importance of CD8 T lymphocytes in protective immunity (Feng and Walker, 2004). Furthermore, mice depleted of both CD4 and CD8 T lymphocytes had a higher fatality rate (80%) than mice depleted of only CD4 T cells (44%), revealing the additive effect of CD4 and CD8 T cells in immunity to *Ehrlichiae*. Also, a combination of cytokines is important for protection, as

demonstrated by higher fatality (75%) after depletion of both interferon (IFN)-γ and tumor-necrosis factor (TNF)-α combined than in single cytokine depletions (Feng and Walker, 2004).

I. ovatus Ehrlichia (IOE) strain HF565 infection has been described as a lethal model for ehrlichial disease, since it mimics the clinical manifestations and histopathological changes observed in severe or lethal cases of *E. chaffeensis* infection in humans (Sotomayor et al., 2001). Infected mice develop leukopenia, thrombocytopenia, and elevation in serum concentrations of hepatic transaminases. Infection with IOE causes severe tissue damage to multiple organ systems associated with low bacterial levels, similar to what is observed in immunocompetent HME patients, who develop multisystem organ failure without an overwhelming infection (Ismail et al., 2004; Stone et al., 2004). Histopathological changes in infected mice are interstitial pneumonitis, hepatocellular necrosis and apoptosis, erythrophagocytosis in the liver, and myeloid hyperplasia of bone marrow (Sotomayor et al., 2001). Associated weight loss and hypoglycemia suggest shock-like syndrome as the cause of death (Fichtenbaum et al., 1993; Ismail et al., 2004). This severe toxic shock-like disease associated with IOE infection is associated with overproduction of TNF-α and IL-10, a high frequency of *Ehrlichia*-specific TNF-α-producing CD8(+) T cells, a low frequency of *Ehrlichia*-specific CD4(+) T cells, and low levels of IL-12 in the spleen (Ismail et al., 2004). TNF receptor gene knockout mice are more resistant than wild-type animals to liver injury caused by IOE infection (Ismail et al., 2006).

Protective immunity to ehrlichial infections requires CD4 T-cell production of gamma interferon (IFN-γ), and is dependent on IL-12p40-dependent cytokines (IFN-γ and TNF-α). Further studies have demonstrated the role of B cells and antibodies in immunity against ehrlichial infection (Bitsaktsis et al., 2004, 2007). Effector and/or memory phenotype CD8 T cells are responsible for a lethal outcome in mice that have recovered from a sublethal infection when rechallenged with an ordinarily sublethal dose of IOE; moreover, cytokine production by CD8 T cells and/or other host cells is suggested to induce chemokine-dependent disease (Bitsaktsis and Winslow, 2006). Taken together, fatal ehrlichiosis caused by IOE is characterized by CD8 T-cell-mediated tissue damage, overproduction of TNF-α, and interleukin (IL)-10, and a deficient CD4 Th1 cell response (Stevenson et al., 2010).

Natural killer T (NKT) cells, a type of innate-like lymphocytes, are stimulated by persistent *E. muris* infection. NKT cells decrease in number initially after a lethal IOE challenge, followed by a late expansion and migration of activated NK cells to the liver, which coincides with the development of hepatic injury. These changes in NK and NKT cells are correlated with an increase in NK cell cytotoxic activity, higher expression of granzyme B, increased production of IFN-γ and TNF-α, increased hepatic population of dendritic cells and CD8 T cells, decreased splenic CD4 T cells, increased serum concentrations of IL-12p40, IL-18, RANTES, and monocyte chemotactic protein-1, and increased production of IL-18 by hepathic mononuclear cells (Stevenson et al., 2008, 2010).

Neutrophils may play a role in pathogenic changes in the immune response. Depletion of neutrophils from IOE-infected mice improves bacterial elimination, decreases immune-mediated pathology, and prolongs survival in mice infected with a lethal dose of IOE. Furthermore, neutrophil depletion is associated with decreased numbers of TNF-α-producing CD8 T cells, which mediate immunopathology and toxic shock in fatal murine ehrlichiosis (Yang et al., 2013). Interleukin (IL)-18 has been proposed to play a pathogenic role in *Ehrlichia*-induced toxic shock, as demonstrated by its role in hepatic leukocyte recruitment, apoptosis and bacterial loads and decreased survival (Ghose et al., 2011).

B cells and antibodies play an important role in immunity to ehrlichiosis. Mice deficient in B cells or FcγRI are susceptible to a low dose of IOE, suggesting an important role for B cells and antibodies in resistance to primary infection (Yager et al., 2005). Passive transfer of immune sera protects SCID mice against *E. muris* or *E. chaffeensis* infection (Feng and Walker, 2004; Winslow et al., 2000). Monoclonal IgG antibodies of different isotypes directed against an epitope within the amino terminus of the first hypervariable region of P28–19 (OMP-1g) provide protection against *E. chaffeensis* infection in SCID mice, and IgG2c antibodies are more effective than IgG2b or IgG3 antibodies and exhibit high binding affinity (Li et al., 2002; Li et al., 2001). Passive transfer of polyclonal antibodies predominantly containing IgG1 isotypes directed against the tandem repeats of *E. chaffeensis*, TRP120, TRP47, and TRP32, significantly reduces bacterial burdens following infection of SCID and immunocompetent C57BL/6 mice (Kuriakose et al., 2012).

Recent studies, using the *E. muris* infection model, demonstrated that IgM antibodies are sufficient for short- and long-term protection against ehrlichial infection. Splenic plasmablasts are responsible for the generation of IgM produced during acute infection (Racine et al., 2008, 2011). These IgM antibodies are polyreactive, which has not been described previously for infection-induced IgM. The polyreactive IgM binds pathogen antigens, but also binds a number of unrelated self and foreign antigens and is produced during human ehrlichiosis (Jones et al., 2012).

36.7 Diagnosis

A tentative clinical diagnosis can be made based on clinical manifestations and clinical laboratory findings associated with the history of exposure to ticks or a tick-endemic region during the tick season. It is important to initiate treatment with an effective antibiotic as early as possible to avoid the progressive evolution of the infection, which can lead to a fatal outcome (Dumler et al., 2007).

36.7.1 Differential diagnosis (Chapman et al., 2006; Olano et al., 2003b)

Fever, headache, myalgia, and malaise
- Viral syndromes
- Rocky Mountain spotted fever
- Upper respiratory illness, including influenza
- Urinary tract infection
- Sepsis

Tick bite history (plus above symptoms)
- Rocky Mountain spotted fever
- *Rickettsia parkeri* infection
- Relapsing fever
- Tularemia
- Lyme borreliosis
- Colorado tick fever
- Babesiosis

Rash (plus symptoms above)
- Rocky Mountain spotted fever
- *Rickettsia parkeri* infection
- Meningococcemia
- Toxic shock syndrome
- Murine typhus
- Q fever
- Typhoid fever
- Leptospirosis
- Hepatitis
- Enteroviral infection
- Endocarditis
- Kawasaki disease
- Collagen-vascular diseases
- Immune thrombocytopenic purpura

36.7.2 Clinical laboratory abnormalities

Blood counts are commonly used in suspected cases as a presumptive method for supporting a diagnosis of ehrlichial infection. Pancytopenia occurs early in the course of infection, and anemia can be found in 50% of the cases after 2 weeks of illness. Leukopenia, especially due to lymphopenia, is present in 60%–70% of patients during the first week of the disease. Thrombocytopenia, a consistent finding in HME, is present in 70%–90% of cases. Increased levels of hepatic transaminases are detected in 90% of patients, however, at only mildly or moderately elevated levels. Other findings are increased serum concentrations of alkaline phosphatase and bilirubin, hyponatremia, and organ-specific abnormalities depending on the involvement (Chapman et al., 2006; Dumler, 2005; Olano et al., 2003a, 2003b; Paddock and Childs, 2003).

36.7.3 Cytology

A peripheral blood smear, bone marrow, or cerebrospinal fluid (CSF) can be examined for the presence of ehrlichial morulae in mononuclear cells by Wright, Diff-Quik, or Giemsa-Romanowsky stains (Lin and Rikihisa, 2003a; Paddock et al., 1997). These intracellular inclusions representing vacuoles containing microcolonies of *Ehrlichiae* are basophilic, stain blue, and can be of different sizes and numbers in a single cell. Although it is a rapid method, the examination of the blood smear has very low sensitivity (less than 10%) to detect *Ehrlichiae*-infected patients, except those who are immunocompromised (Standaert et al., 2000; Tan et al., 2001).

36.7.4 Polymerase chain reaction (PCR)

Ehrlichial DNA can be detected in blood samples early in the infection, making PCR a sensitive approach for early diagnosis. The sensitivity varies between 60% and 85% for *E. chaffeensis* detection (Olano et al., 2003b; Standaert et al., 2000). In *E. ewingii* the sensitivity of molecular diagnosis is unknown; however, PCR is the definitive method to establish the diagnosis of ehrlichiosis caused by *E. ewingii*, since the bacteria have not been isolated in cell culture, and a specific serologic method is not available (Buller et al., 1999; Paddock et al., 2001). *Ehrlichiae* have been detected in CSF by PCR; however, the sensitivity is significantly lower than in blood likely due to a lower quantity of infected cells in the CSF (Dawson et al., 1991a; Standaert et al., 2000; Tan et al., 2001). Several target genes have been used for the diagnosis of ehrlichiosis by PCR, including 16S rRNA, *groESL* heat shock operon, genus-specific disulfide bond formation protein (*dsb*), 120-kDa and TRP32 protein genes (specific for *E. chaffeensis*), and 28-kDa outer membrane protein genes (Sumner et al., 1997, 1999). Samples for diagnosis should be collected before or at the initiation of antibiotic therapy to increase diagnostic sensitivity because of the efficiency of treatment in controlling the levels of bacteria in blood. Diagnostic tests using PCR are particularly valuable in the early stage of infection when immunity reflected by the presence of antibodies has yet to develop (Anderson et al., 1991; Standaert et al., 2000).

36.7.5 Isolation in cell culture

E. chaffeensis can be cultured from patient blood or CSF (Standaert et al., 2000). However, *E. ewingii* has not been successfully cultivated *in vitro* (Dumler et al., 2007). This method is not available in most clinical laboratories. Several cell lines can be used to isolate *E. chaffeensis*, which can be detected initially in cultures by Diff-Quik staining of morulae. Isolation is the gold standard for diagnosis of ehrlichial infection; however, it is not a rapid method, and the sensitivity is very low compared to PCR. The same concepts

concerning the time of sample collection for PCR apply for the isolation of *Ehrlichiae* (Dawson et al., 1991a; Dumler, 2005; Dumler and Bakken, 1995).

36.7.6 Serodiagnosis

Indirect immunofluorescence assay (IFA) is considered the "gold standard" method and is the most common test for the diagnosis of HME with a sensitivity of 88%–90% for paired acute and convalescent sera (Dawson et al., 1990; Dumler et al., 2007). However, IFA has several important limitations, as the IgG and IgM antibodies are usually absent during the early stage of infection. Also, antibodies against one species of *Ehrlichia* frequently cross-react with other species, and even in some cases with bacteria of the genus *Anaplasma*. Paired serology is preferred for the confirmation of active infection. The diagnosis of ehrlichial disease is confirmed if the serum sample tested for IFA has a titer of 256 or higher, seroconversion from negative to positive antibody status (minimum titer, 64), or fourfold increase in antibody titer (Chapman et al., 2006; Childs et al., 1999).

36.7.7 Immunohistochemistry (IHC)

Biopsy samples including bone marrow biopsies or autopsy tissues can be used to demonstrate ehrlichial antigens by immunohistochemistry. This method has good sensitivity for the diagnosis of fatal cases of ehrlichial infection. The presence of the pathogen has been observed in the spleen, liver, lung, lymph nodes, and bone marrow (Dawson et al., 2001; Olano et al., 2003a; Walker and Dumler, 1997).

36.8 Treatment

Empirical treatment should be given on the basis of clinical suspicion of the diagnosis of ehrlichiosis and should not await laboratory confirmation. Drugs of the tetracycline group are used to treat ehrlichial infections. The drug of choice is doxycycline. Studies of antimicrobial susceptibility *in vitro* have demonstrated high sensitivity of all *Ehrlichia* species to doxycycline. Doxycycline also has few side effects, provides an easy therapeutic regime, and has better tolerance *in vivo* than other tetracyclines (Brouqui and Raoult, 1992; Dumler et al., 2007; Walker et al., 2004). The dosage for adult patients is 100 mg every 12 hours for at least 3 days after remission of fever, or up to 14 days for the total course. In children, although tetracycline can cause dental discoloration, several courses of doxycycline do not. Thus, doxycycline is the drug of choice to treat ehrlichial infection in children at a dose of 2.2 mg/kg orally every 12 hours (Dumler et al., 2007). In cases in which treatment is initiated early in the course of infection, clinical signs and symptoms resolve in 24–48 hours after the initiation of antibiotic therapy (Dumler et al., 2007). Relapse of symptoms after complete treatment has not been documented; however, some patients may experience headache,

Table 36-2 Drugs studied for the treatment of HME

Drug	Comments
Doxycycline	Minimum 3–5 days after defervescence
	Adults: 100 mg 12/12 hours PO or IV
	Children: 2.2 mg/kg 12/12 hours PO
Rifampin	Poorly studied
Chloramphenicol	Use controversial
Fluoroquinolones	Use controversial

malaise, and weakness for weeks. Drug resistance to doxycycline has not been identified. The use of tetracyclines is generally contraindicated for pregnant women due to the risks associated with malformation of teeth and bones in the fetus and hepatotoxicity and pancreatitis in the mother. However, tetracycline has been used successfully to treat HME in pregnant women (Smith Sehdev et al., 2002), and the use of tetracyclines might be warranted during pregnancy in life-threatening situations where the clinical suspicion of tick transmitted diseases is high. Some studies of chloramphenicol and fluoroquinolones discourage their use to treat HME; however, rifampin could be tried in a patient in whom doxycycline is contraindicated (Dumler et al., 2007) (Table 36-2).

36.9 Prevention and control

Ehrlichioses as tick-borne diseases can be controlled by avoidance of these arthropods in endemic areas, by use of repellents, protective clothing and careful search of the body, and immediate removal of ticks.

Prophylactic antibiotic therapy for ehrlichial infection following a tick bite is not recommended due to the low chance of infection after a tick bite (Wolf et al., 2000)

36.10 Future directions

Ehrlichioses are emergent infectious diseases with a worldwide distribution and with large range of animal hosts. These vector-borne diseases are implicated in life-threatening infections in humans and important economic losses in livestock animals. Several aspects of the infection and pathogenesis have been characterized, but very little is known about the natural disease, especially related to the establishment of the infection and the complete pathogenesis of the disease. Few studies have indicated some modification induced by the pathogen during the passage through the tick vector, but there is much still to be understood. Identification of detailed cell entry pathways in ehrlichial infection, vector–host–pathogen interaction, pathogen evasion and immune mechanisms in ehrlichial diseases, and preventive strategies have been the focus of research in ehrlichial infections. The understanding of the aspects that involve disease establishment is essential for the identification of long-term disease prevention and/or efficient treatment.

References

Allsopp, M.T., Louw, M., and Meyer, E.C. 2005. *Ehrlichia ruminantium*: an emerging human pathogen? *Ann. NY Acad. Sci.*, 1063, 358–360.

Anderson, B.E., Dawson, J.E., Jones, D.C., and Wilson, K.H. 1991. *Ehrlichia chaffeensis*, a new species associated with human ehrlichiosis. *J. Clin. Microbiol.*, 29, 2838–2842.

Anziani, O.S., Ewing, S.A., and Barker, R.W. 1990. Experimental transmission of a granulocytic form of the tribe Ehrlichieae by Dermacentor variabilis and Amblyomma americanum to dogs. *Am. J. Vet. Res.*, 51, 929–931.

Barnewall, R.E., and Rikihisa, Y. 1994. Abrogation of gamma interferon-induced inhibition of *Ehrlichia chaffeensis* infection in human monocytes with iron-transferrin. *Infect. Immun.*, 62, 4804–4810.

Barnewall, R.E., Rikihisa, Y., and Lee, E.H. 1997. *Ehrlichia chaffeensis* inclusions are early endosomes which selectively accumulate transferrin receptor. *Infect. Immun.*, 65, 1455–1461.

Barnewall, R.E., Ohashi, N., and Rikihisa, Y. 1999. *Ehrlichia chaffeensis* and *E. sennetsu*, but not the human granulocytic ehrlichiosis agent, colocalize with transferrin receptor and up-regulate transferrin receptor mRNA by activating iron-responsive protein 1. *Infect. Immun.*, 67, 2258–2265.

Bezuidenhout, J.D. 1987. Natural transmission of heartwater. *Onderstepoort. J. Vet. Res.*, 54, 349–351.

Bezuidenhout, J.D., Paterson, C.L., and Barnard, B.J. 1985. In vitro cultivation of *Cowdria ruminantium*. *Onderstepoort. J. Vet. Res.*, 52, 113–120.

Bitsaktsis, C., and Winslow, G. 2006. Fatal recall responses mediated by CD8 T cells during intracellular bacterial challenge infection. *J. Immunol.*, 177, 4644–4651.

Bitsaktsis, C., Huntington, J., and Winslow, G. 2004. Production of IFN-gamma by CD4 T cells is essential for resolving ehrlichia infection. *J. Immunol.*, 172, 6894–6901.

Bitsaktsis, C., Nandi, B., Racine, R., MacNamara, K.C., and Winslow, G. 2007. T-cell-independent humoral immunity is sufficient for protection against fatal intracellular ehrlichia infection. *Infect. Immun.*, 75, 4933–4941.

Breitschwerdt, E.B., Hegarty, B.C., and Hancock, S.I. 1998. Sequential evaluation of dogs naturally infected with *Ehrlichia canis, Ehrlichia chaffeensis, Ehrlichia equi, Ehrlichia ewingii*, or *Bartonella vinsonii*. *J. Clin. Microbiol.*, 36, 2645–2651.

Brouqui, P., and Raoult, D. 1992. In vitro antibiotic susceptibility of the newly recognized agent of ehrlichiosis in humans, *Ehrlichia chaffeensis*. *Antimicrob. Agents Chemother.*, 36, 2799–2803.

Brown, C.C., and Skowronek, A.J. 1990. Histologic and immunochemical study of the pathogenesis of heartwater (*Cowdria ruminantium* infection) in goats and mice. *Am. J. Vet. Res.*, 51, 1476–1480.

Buller, R.S., Arens, M., Hmiel, S.P., Paddock, C.D., Sumner, J.W., Rikhisa, Y., Unver, A., Gaudreault-Keener, M., Manian, F.A., Liddell, A.M., Schmulewitz, N., and Storch, G.A. 1999. *Ehrlichia ewingii*, a newly recognized agent of human ehrlichiosis. *N. Engl. J. Med.*, 341, 148–155.

Caldwell, C.W., Everett, E.D., McDonald, G., Yesus, Y.W., and Roland, W.E. 1995. Lymphocytosis of gamma/delta T cells in human ehrlichiosis. *Am. J. Clin. Pathol.*, 103, 761–766.

Caldwell, C.W., Everett, E.D., McDonald, G., Yesus, Y.W., Roland, W.E., and Huang, H.M. 1996. Apoptosis of gamma/delta T cells in human ehrlichiosis. *Am. J. Clin. Pathol.*, 105, 640–646.

Cao, W.C., Gao, Y.M., Zhang, P.H., Zhang, X.T., Dai, Q.H., Dumler, J.S., Fang, L.Q., and Yang, H. 2000. Identification of *Ehrlichia chaffeensis* by nested PCR in ticks from Southern China. *J. Clin. Microbiol.*, 38, 2778–2780.

Chapman, A.S., Bakken, J.S., Folk, S.M., Paddock, C.D., Bloch, K.C., Krusell, A., Sexton, D.J., Buckingham, S.C., Marshall, G.S., Storch, G.A., Dasch, G.A., McQuiston, J.H., Swerdlow, D.L., Dumler, S.J., Nicholson, W.L., Walker, D.H., Eremeeva, M.E., and Ohl, C.A. 2006. Diagnosis and management of tickborne rickettsial diseases: Rocky Mountain spotted fever, ehrlichioses, and anaplasmosis – United States: A practical guide for physicians and other health-care and public health professionals. *MMWR Recomm. Rep.*, 55, 1–27.

Childs, J.E., Sumner, J.W., Nicholson, W.L., Massung, R.F., Standaert, S.M., and Paddock, C.D. 1999. Outcome of diagnostic tests using samples from patients with culture-proven human monocytic ehrlichiosis: Implications for surveillance. *J. Clin. Microbiol.*, 37, 2997–3000.

Collins, N.E., Liebenberg, J., de Villiers, E.P., Brayton, K.A., Louw, E., Pretorius, A., Faber, F.E., van Heerden, H., Josemans, A., van Kleef, M., Steyn, H.C., van Strijp, M.F., Zweygarth, E., Jongejan, F., Maillard, J.C., Berthier, D., Botha, M., Joubert, F., Corton, C.H., Thomson, N.R., Allsopp, M.T., and Allsopp, B.A. 2005. The genome of the heartwater agent *Ehrlichia ruminantium* contains multiple tandem repeats of actively variable copy number. *Proc. Natl. Acad. Sci. USA*, 102, 838–843.

Dahlgren, F.S., Mandel, E.J., Krebs, J.W., Massung, R.F., and McQuiston, J.H. 2011. Increasing incidence of *Ehrlichia chaffeensis* and *Anaplasma phagocytophilum* in the United States, 2000–2007. *Am. J. Trop. Med. Hyg.*, 85, 124–131.

Dame, J.B., Mahan, S.M., and Yowell, C.A. 1992. Phylogenetic relationship of Cowdria ruminantium, agent of heartwater, to Anaplasma marginale and other members of the order Rickettsiales determined on the basis of 16S rRNA sequence. *Int. J. Syst. Bacteriol.*, 42, 270–274.

Davidson, W.R., Lockhart, J.M., Stallknecht, D.E., and Howerth, E.W. 1999. Susceptibility of red and gray foxes to infection by *Ehrlichia chaffeensis*. *J. Wildl. Dis.*, 35, 696–702.

Dawson, J.E. 1996. Human ehrlichiosis in the United States. *Curr. Clin. Top. Infect. Dis.*, 16 164–171.

Dawson, J.E., Anderson, B.E., Fishbein, D.B., Sanchez, J.L., Goldsmith, C.S., Wilson, K.H., and Duntley, C.W. 1991a. Isolation and characterization of an *Ehrlichia* sp. from a patient diagnosed with human ehrlichiosis. *J. Clin. Microbiol.*, 29, 2741–2745.

Dawson, J.E., and Ewing, S.A. 1992. Susceptibility of dogs to infection with *Ehrlichia chaffeensis*, causative agent of human ehrlichiosis. *Am. J. Vet. Res.*, 53, 1322–1327.

Dawson, J.E., Fishbein, D.B., Eng, T.R., Redus, M.A., and Green, N.R. 1990. Diagnosis of human ehrlichiosis with the indirect fluorescent antibody test: Kinetics and specificity. *J. Infect. Dis.*, 162, 91–95.

Dawson, J.E., Paddock, C.D., Warner, C.K., Greer, P.W., Bartlett, J.H., Ewing, S.A., Munderloh, U.G., and Zaki, S.R. 2001. Tissue diagnosis of *Ehrlichia chaffeensis* in patients with fatal ehrlichiosis by use of immunohistochemistry, *in situ* hybridization, and polymerase chain reaction. *Am. J. Trop. Med. Hyg.*, 65, 603–609.

Dawson, J.E., Rikihisa, Y., Ewing, S.A., and Fishbein, D.B. 1991b. Serologic diagnosis of human ehrlichiosis using two *Ehrlichia canis* isolates. *J. Infect. Dis.*, 163, 564–567.

Dawson, J.E., Warner, C.K., Baker, V., Ewing, S.A., Stallknecht, D.E., Davidson, W.R., Kocan, A.A., Lockhart, J.M., and Olson, J.G. 1996. Ehrlichia-like 16S rDNA sequence from wild white-tailed deer (Odocoileus virginianus). *J. Parasitol.*, 82, 52–58.

delos Santos, J.R., Boughan, K., Bremer, W.G., Rizzo, B., Schaefer, J.J., Rikihisa, Y., Needham, G.R., Capitini, L.A., Anderson,

D.E., Oglesbee, M., Ewing, S.A., and Stich, R.W. 2007. Experimental infection of dairy calves with *Ehrlichia chaffeensis*. *J. Med. Microbiol.*, 56, 1660–1668.

Demma, L.J., Holman, R.C., McQuiston, J.H., Krebs, J.W., and Swerdlow, D.L. 2005. Epidemiology of human ehrlichiosis and anaplasmosis in the United States, 2001–2002. *Am. J. Trop. Med. Hyg.*, 73, 400–409.

Dugan, V.G., Little, S.E., Stallknecht, D.E., and Beall, A.D. 2000. Natural infection of domestic goats with *Ehrlichia chaffeensis*. *J. Clin. Microbiol.*, 38, 448–449.

Dumler, J.S. 2005. *Anaplasma* and *Ehrlichia* infection. *Ann. N.Y. Acad. Sci.*, 1063, 361–373.

Dumler, J.S., and Bakken, J.S. 1995. Ehrlichial diseases of humans: Emerging tick-borne infections. *Clin. Infect. Dis.*, 20, 1102–1110.

Dumler, J.S., Dawson, J.E., and Walker, D.H. 1993a. Human ehrlichiosis: Hematopathology and immunohistologic detection of *Ehrlichia chaffeensis*. *Hum. Pathol.*, 24, 391–396.

Dumler, J.S., Sutker, W.L., and Walker, D.H. 1993b. Persistent infection with Ehrlichia chaffeensis. *Clin. Infect. Dis.*, 17, 903–905.

Dumler, J.S., Barbet, A.F., Bekker, C.P., Dasch, G.A., Palmer, G.H., Ray, S.C., Rikihisa, Y., and Rurangirwa, F.R. 2001. Reorganization of genera in the families Rickettsiaceae and Anaplasmataceae in the order Rickettsiales: Unification of some species of *Ehrlichia* with *Anaplasma*, *Cowdria* with *Ehrlichia* and *Ehrlichia* with *Neorickettsia*, descriptions of six new species combinations and designation of *Ehrlichia equi* and 'HGE agent' as subjective synonyms of *Ehrlichia phagocytophila*. *Int. J. Syst. Evol. Microbiol.*, 51, 2145–2165.

Dumler, J.S., Madigan, J.E., Pusterla, N., and Bakken, J.S. 2007. Ehrlichioses in humans: Epidemiology, clinical presentation, diagnosis, and treatment. *Clin. Infect. Dis.*, 45(suppl. 1), S45–S51.

Dunning Hotopp, J.C., Lin, M., Madupu, R., Crabtree, J., Angiuoli, S.V., Eisen, J.A., Seshadri, R., Ren, Q., Wu, M., Utterback, T.R., Smith, S., Lewis, M., Khouri, H., Zhang, C., Niu, H., Lin, Q., Ohashi, N., Zhi, N., Nelson, W., Brinkac, L.M., Dodson, R.J., Rosovitz, M.J., Sundaram, J., Daugherty, S.C., Davidsen, T., Durkin, A.S., Gwinn, M., Haft, D.H., Selengut, J.D., Sullivan, S.A., Zafar, N., Zhou, L., Benahmed, F., Forberger, H., Halpin, R., Mulligan, S., Robinson, J., White, O., Rikihisa, Y., and Tettelin, H. 2006. Comparative genomics of emerging human ehrlichiosis agents. *PLoS Genet.*, 2, e21.

Ewing, S.A., Roberson, W.R., Buckner, R.G., and Hayat, C.S. 1971. A new strain of *Ehrlichia canis*. *J. Am. Vet. Med. Assoc.*, 159, 1771–1774.

Ewing, S.A., Dawson, J.E., Kocan, A.A., Barker, R.W., Warner, C.K., Panciera, R.J., Fox, J.C., Kocan, K.M., and Blouin, E.F. 1995. Experimental transmission of *Ehrlichia chaffeensis* (Rickettsiales: Ehrlichieae) among white-tailed deer by *Amblyomma americanum* (Acari: Ixodidae). *J. Med. Entomol.*, 32, 368–374.

Feng, H.M., and Walker, D.H. 2004. Mechanisms of immunity to *Ehrlichia muris*: A model of monocytotropic ehrlichiosis. *Infect. Immun.*, 72, 966–971.

Fichtenbaum, C.J., Peterson, L.R., and Weil, G.J. 1993. Ehrlichiosis presenting as a life-threatening illness with features of the toxic shock syndrome. *Am. J. Med.*, 95, 351–357.

Fishbein, D.B., Sawyer, L.A., Holland, C.J., Hayes, E.B., Okoroanyanwu, W., Williams, D., Sikes, K., Ristic, M., and McDade, J.E. 1987. Unexplained febrile illnesses after exposure to ticks. Infection with an *Ehrlichia? JAMA*, 257, 3100–3104.

Fishbein, D.B., Dawson, J.E., and Robinson, L.E. 1994. Human ehrlichiosis in the United States, 1985 to 1990. *Ann. Intern. Med.*, 120, 736–743.

Fordham, L.A., Chung, C.J., Specter, B.B., Merten, D.F., and Ingram, D.L. 1998. Ehrlichiosis: Findings on chest radiographs in three pediatric patients. *AJR Am. J. Roentgenol.*, 171, 1421–1424.

Ganta, R.R., Cheng, C., Wilkerson, M.J., and Chapes, S.K. 2004. Delayed clearance of *Ehrlichia chaffeensis* infection in CD4+ T-cell knockout mice. *Infect. Immun.*, 72, 159–167.

Ganta, R.R., Wilkerson, M.J., Cheng, C., Rokey, A.M., and Chapes, S.K. 2002. Persistent *Ehrlichia chaffeensis* infection occurs in the absence of functional major histocompatibility complex class II genes. *Infect. Immun.*, 70, 380–388.

Ghose, P., Ali, A.Q., Fang, R., Forbes, D., Ballard, B., and Ismail, N. 2011. The interaction between IL-18 and IL-18 receptor limits the magnitude of protective immunity and enhances pathogenic responses following infection with intracellular bacteria. *J. Immunol.*, 187, 1333–1346.

Heimer, R., Tisdale, D., and Dawson, J.E. 1998. A single tissue culture system for the propagation of the agents of the human ehrlichioses. *Am. J. Trop. Med. Hyg.*, 58, 812–815.

Ismail, N., Soong, L., McBride, J.W., Valbuena, G., Olano, J.P., Feng, H.M., and Walker, D.H. 2004. Overproduction of TNF-alpha by CD8+ type 1 cells and down-regulation of IFN-gamma production by CD4+ Th1 cells contribute to toxic shock-like syndrome in an animal model of fatal monocytotropic ehrlichiosis. *J. Immunol.*, 172, 1786–1800.

Ismail, N., Stevenson, H.L., and Walker, D.H. 2006. Role of tumor necrosis factor alpha (TNF-alpha) and interleukin-10 in the pathogenesis of severe murine monocytotropic ehrlichiosis: Increased resistance of TNF receptor p55- and p75-deficient mice to fatal ehrlichial infection. *Infect. Immun.*, 74, 1846–1856.

Ismail, N., Walker, D.H., Ghose, P., and Tang, Y.W. 2012. Immune mediators of protective and pathogenic immune responses in patients with mild and fatal human monocytotropic ehrlichiosis. *BMC Immunol.*, 13, 13–26.

Jardine, J.E., Vogel, S.W., van Kleef, M., and van der Lugt, J.J. 1995. Immunohistochemical identification of *Cowdria ruminantium* in formalin-fixed tissue sections from mice, sheep, cattle and goats. *Onderstepoort. J. Vet. Res.*, 62, 277–280.

Jones, D.D., Deiulio, G.A., and Winslow, G.M. 2012. Antigen-driven induction of polyreactive IgM during intracellular bacterial infection. *J. Immunol.*, 189, 1440–1447.

Kawahara, M., Suto, C., Shibata, S., Futohashi, M., and Rikihisa, Y. 1996. Impaired antigen specific responses and enhanced polyclonal stimulation in mice infected with *Ehrlichia muris*. *Microbiol. Immunol.*, 40, 575–581.

Kawahara, M., Tajima, T., Torii, H., Yabutani, M., Ishii, J., Harasawa, M., Isogai, E., and Rikihisa, Y. 2009. *Ehrlichia chaffeensis* infection of sika deer, Japan. *Emerg. Infect. Dis.*, 15, 1991–1993.

Kim, C.M., Kim, M.S., Park, M.S., Park, J.H., and Chae, J.S. 2003. Identification of *Ehrlichia chaffeensis*, *Anaplasma phagocytophilum*, and *A. bovis* in *Haemaphysalis longicornis* and *Ixodes persulcatus* ticks from Korea. *Vector Borne Zoonotic Dis.*, 3, 17–26.

Kim, C.M., Yi, Y.H., Yu, D.H., Lee, M.J., Cho, M.R., Desai, A.R., Shringi, S., Klein, T.A., Kim, H.C., Song, J.W., Baek, L.J., Chong, S.T., O'Guinn, M.L., Lee, J.S., Lee, I.Y., Park, J.H., Foley, J., and Chae, J.S. 2006. Tick-borne rickettsial pathogens in ticks and small mammals in Korea. *Appl. Environ. Microbiol.*, 72, 5766–5776.

Kocan, A.A., Levesque, G.C., Whitworth, L.C., Murphy, G.L., Ewing, S.A., and Barker, R.W. 2000. Naturally occurring *Ehrlichia chaffeensis* infection in coyotes from Oklahoma. *Emerg. Infect. Dis.*, 6, 477–480.

Koh, Y.S., Koo, J.E., Biswas, A., and Kobayashi, K.S. 2010. MyD88-dependent signaling contributes to host defense against ehrlichial infection. *PLoS One*, 5, e11758.

Kramer, V.L., Randolph, M.P., Hui, L.T., Irwin, W.E., Gutierrez, A.G., and Vugia, D.J. 1999. Detection of the agents of human ehrlichioses in ixodid ticks from California. *Am. J. Trop. Med. Hyg.*, 60, 62–65.

Kuriakose, J.A., Zhang, X., Luo, T., and McBride, J.W. 2012. Molecular basis of antibody mediated immunity against *Ehrlichia chaffeensis* involves species-specific linear epitopes in tandem repeat proteins. *Microbes. Infect.*, 14, 1054–1063.

Lawrence, K.L., Morrell, M.R., Storch, G.A., Hachem, R.R., and Trulock, E.P. 2009. Clinical outcomes of solid organ transplant recipients with ehrlichiosis. *Transpl. Infect. Dis.*, 11, 203–210.

Lee, E.H., and Rikihisa, Y. 1996. Absence of tumor necrosis factor alpha, interleukin-6 (IL-6), and granulocyte-macrophage colony-stimulating factor expression but presence of IL-1beta, IL-8, and IL-10 expression in human monocytes exposed to viable or killed *Ehrlichia chaffeensis*. *Infect. Immun.*, 64, 4211–4219.

Lee, E.H., and Rikihisa, Y. 1998. Protein kinase A-mediated inhibition of gamma interferon-induced tyrosine phosphorylation of Janus kinases and latent cytoplasmic transcription factors in human monocytes by *Ehrlichia chaffeensis*. *Infect. Immun.*, 66, 2514–2520.

Lee, S.O., Na, D.K., Kim, C.M., Li, Y.H., Cho, Y.H., Park, J.H., Lee, J.H., Eo, S.K., Klein, T.A., and Chae, J.S. 2005. Identification and prevalence of *Ehrlichia chaffeensis* infection in *Haemaphysalis longicornis* ticks from Korea by PCR, sequencing and phylogenetic analysis based on 16S rRNA gene. *J. Vet. Sci.*, 6, 151–155.

Lee, M., Yu, D., Yoon, J., Li, Y., Lee, J., and Park, J. 2009. Natural co-infection of *Ehrlichia chaffeensis* and *Anaplasma bovis* in a deer in South Korea. *J. Vet. Med. Sci.*, 71, 101–103.

Li, J.S., Chu, F., Reilly, A., and Winslow, G.M. 2002. Antibodies highly effective in SCID mice during infection by the intracellular bacterium *Ehrlichia chaffeensis* are of picomolar affinity and exhibit preferential epitope and isotype utilization. *J. Immunol.*, 169, 1419–1425.

Li, J.S., Yager, E., Reilly, M., Freeman, C., Reddy, G.R., Reilly, A.A., Chu, F.K., and Winslow, G.M. 2001. Outer membrane protein-specific monoclonal antibodies protect SCID mice from fatal infection by the obligate intracellular bacterial pathogen *Ehrlichia chaffeensis*. *J. Immunol.*, 166, 1855–1862.

Lin, M., and Rikihisa, Y. 2003a. *Ehrlichia chaffeensis* and *Anaplasma phagocytophilum* lack genes for lipid A biosynthesis and incorporate cholesterol for their survival. *Infect. Immun.*, 71, 5324–5331.

Lin, M., and Rikihisa, Y. 2003b. Obligatory intracellular parasitism by *Ehrlichia chaffeensis* and *Anaplasma phagocytophilum* involves caveolae and glycosylphosphatidylinositol-anchored proteins. *Cell. Microbiol.*, 5, 809–820.

Lin, M., Zhu, M.X., and Rikihisa, Y. 2002. Rapid activation of protein tyrosine kinase and phospholipase C-gamma2 and increase in cytosolic free calcium are required by *Ehrlichia chaffeensis* for internalization and growth in THP-1 cells. *Infect. Immun.*, 70, 889–898.

Lockhart, J.M., Davidson, W.R., Stallknecht, D.E., Dawson, J.E., and Howerth, E.W. 1997. Isolation of *Ehrlichia chaffeensis* from wild white-tailed deer (*Odocoileus virginianus*) confirms their role as natural reservoir hosts. *J. Clin. Microbiol.*, 35, 1681–1686.

Logan, L.L., Whyard, T.C., Quintero, J.C., and Mebus, C.A. 1987. The development of *Cowdria ruminantium* in neutrophils. *Onderstepoort. J. Vet. Res.*, 54, 197–204.

Maeda, K., Markowitz, N., Hawley, R.C., Ristic, M., Cox, D., and McDade, J.E. 1987. Human infection with *Ehrlichia canis*, a leukocytic rickettsia. *N. Engl. J. Med.*, 316, 853–856.

Martinez, M.C., Gutierrez, C.N., Monger, F., Ruiz, J., Watts, A., Mijares, V.M., Rojas, M.G., and Triana-Alonso, F.J. 2008. *Ehrlichia chaffeensis* in child, Venezuela. *Emerg. Infect. Dis.*, 14, 519–520.

Mattner, J., Debord, K.L., Ismail, N., Goff, R.D., Cantu 3rd, C., Zhou, D., Saint-Mezard, P., Wang, V., Gao, Y., Yin, N., Hoebe, K., Schneewind, O., Walker, D., Beutler, B., Teyton, L., Savage, P.B., and Bendelac, A. 2005. Exogenous and endogenous glycolipid antigens activate NKT cells during microbial infections. *Nature*, 434, 525–529.

Mavromatis, K., Doyle, C.K., Lykidis, A., Ivanova, N., Francino, M.P., Chain, P., Shin, M., Malfatti, S., Larimer, F., Copeland, A., Detter, J.C., Land, M., Richardson, P.M., Yu, X.J., Walker, D.H., McBride, J.W., and Kyrpides, N.C. 2006. The genome of the obligately intracellular bacterium *Ehrlichia canis* reveals themes of complex membrane structure and immune evasion strategies. *J. Bacteriol.*, 188, 4015–4023.

McBride, J.W., Comer, J.E., and Walker, D.H. 2003. Novel immunoreactive glycoprotein orthologs of *Ehrlichia* spp. *Ann. N.Y. Acad. Sci.*, 990, 678–684.

McQuiston, J.H., Paddock, C.D., Holman, R.C., and Childs, J.E. 1999. The human ehrlichioses in the United States. *Emerg. Infect. Dis.*, 5, 635–642.

Miura, K., and Rikihisa, Y. 2007. Virulence potential of *Ehrlichia chaffeensis* strains of distinct genome sequences. *Infect. Immun.*, 75, 3604–3613.

Miura, K., Matsuo, J., Rahman, M.A., Kumagai, Y., Li, X., and Rikihisa, Y. 2011. *Ehrlichia chaffeensis* induces monocyte inflammatory responses through MyD88, ERK, and NF-kappaB but not through TRIF, interleukin-1 receptor 1 (IL-1R1)/IL-18R1, or toll-like receptors. *Infect. Immun.*, 79, 4947–4956.

Mott, J., Barnewall, R.E., and Rikihisa, Y. 1999. Human granulocytic ehrlichiosis agent and *Ehrlichia chaffeensis* reside in different cytoplasmic compartments in HL-60 cells. *Infect. Immun.*, 67, 1368–1378.

Ndip, L.M., Labruna, M., Ndip, R.N., Walker, D.H., and McBride, J.W. 2009. Molecular and clinical evidence of *Ehrlichia chaffeensis* infection in Cameroonian patients with undifferentiated febrile illness. *Ann. Trop. Med. Parasitol.*, 103, 719–725.

Ndip, L.M., Ndip, R.N., Esemu, S.N., Walker, D.H., and McBride, J.W. 2010. Predominance of *Ehrlichia chaffeensis* in *Rhipicephalus sanguineus* ticks from kennel-confined dogs in Limbe, Cameroon. *Exp. Appl. Acarol.*, 50, 163–168.

Okada, H., Usuda, H., Tajima, T., Kawahara, M., Yoshino, T., and Rikihisa, Y. 2003. Distribution of ehrlichiae in tissues as determined by *in-situ* hybridization. *J. Comp. Pathol.*, 128, 182–187.

Olano, J.P., Hogrefe, W., Seaton, B., and Walker, D.H. 2003a. Clinical manifestations, epidemiology, and laboratory diagnosis of human monocytotropic ehrlichiosis in a commercial laboratory setting. *Clin. Diagn. Lab. Immunol.*, 10, 891–896.

Olano, J.P., Masters, E., Hogrefe, W., and Walker, D.H. 2003b. Human monocytotropic ehrlichiosis, Missouri. *Emerg. Infect. Dis.*, 9, 1579–1586.

Olano, J.P., Wen, G., Feng, H.M., McBride, J.W., and Walker, D.H. 2004. Histologic, serologic, and molecular analysis of persistent ehrlichiosis in a murine model. *Am. J. Pathol.*, 165, 997–1006.

Paddock, C.D., and Childs, J.E. 2003. *Ehrlichia chaffeensis*: A prototypical emerging pathogen. *Clin. Microbiol. Rev.*, 16, 37–64.

Paddock, C.D., Sumner, J.W., Shore, G.M., Bartley, D.C., Elie, R.C., McQuade, J.G., Martin, C.R., Goldsmith, C.S., and Childs, J.E. 1997. Isolation and characterization of *Ehrlichia chaffeensis* strains from patients with fatal ehrlichiosis. *J. Clin. Microbiol.*, 35, 2496–2502.

Paddock, C.D., Folk, S.M., Shore, G.M., Machado, L.J., Huycke, M.M., Slater, L.N., Liddell, A.M., Buller, R.S., Storch, G.A., Monson, T.P., Rimland, D., Sumner, J.W., Singleton, J., Bloch, K.C., Tang, Y.W., Standaert, S.M., and Childs, J.E. 2001. Infections with *Ehrlichia chaffeensis* and *Ehrlichia ewingii* in persons coinfected with human immunodeficiency virus. *Clin. Infect. Dis.*, 33, 1586–1594.

Patel, R.G., and Byrd, M.A. 1999. Near fatal acute respiratory distress syndrome in a patient with human ehrlichiosis. *South. Med. J.*, 92, 333–335.

Perez, M., Rikihisa, Y., and Wen, B. 1996. *Ehrlichia canis*-like agent isolated from a man in Venezuela: Antigenic and genetic characterization. *J. Clin. Microbiol.*, 34, 2133–2139.

Perez, M., Bodor, M., Zhang, C., Xiong, Q., and Rikihisa, Y. 2006. Human infection with *Ehrlichia canis* accompanied by clinical signs in Venezuela. *Ann. N.Y. Acad. Sci.*, 1078, 110–117.

Popov, V.L., Chen, S.M., Feng, H.M., and Walker, D.H. 1995. Ultrastructural variation of cultured *Ehrlichia chaffeensis*. *J. Med. Microbiol.*, 43, 411–421.

Popov, V.L., Han, V.C., Chen, S.M., Dumler, J.S., Feng, H.M., Andreadis, T.G., Tesh, R.B., and Walker, D.H. 1998. Ultrastructural differentiation of the genogroups in the genus *Ehrlichia*. *J. Med. Microbiol.*, 47, 235–251.

Popov, V.L., Yu, X., and Walker, D.H. 2000. The 120 kDa outer membrane protein of *Ehrlichia chaffeensis*: Preferential expression on dense-core cells and gene expression in Escherichia coli associated with attachment and entry. *Microb. Pathog.*, 28, 71–80.

Popov, V.L., Korenberg, E.I., Nefedova, V.V., Han, V.C., Wen, J.W., Kovalevskii, Y.V., Gorelova, N.B., and Walker, D.H. 2007. Ultrastructural evidence of the ehrlichial developmental cycle in naturally infected *Ixodes persulcatus* ticks in the course of coinfection with *Rickettsia*, *Borrelia*, and a flavivirus. *Vector Borne Zoonotic Dis.*, 7, 699–716.

Pritt, B.S., Sloan, L.M., Johnson, D.K., Munderloh, U.G., Paskewitz, S.M., McElroy, K.M., McFadden, J.D., Binnicker, M.J., Neitzel, D.F., Liu, G., Nicholson, W.L., Nelson, C.M., Franson, J.J., Martin, S.A., Cunningham, S.A., Steward, C.R., Bogumill, K., Bjorgaard, M.E., Davis, J.P., McQuiston, J.H., Warshauer, D.M., Wilhelm, M.P., Patel, R., Trivedi, V.A., and Eremeeva, M.E. 2011. Emergence of a new pathogenic *Ehrlichia* species, Wisconsin and Minnesota, 2009. *N. Engl. J. Med.*, 365, 422–429.

Prozesky, L., and Du Plessis, J.L. 1987. Heartwater. The development and life cycle of *Cowdria ruminantium* in the vertebrate host, ticks and cultured endothelial cells. *Onderstepoort. J. Vet. Res.*, 54, 193–196.

Racine, R., Chatterjee, M., and Winslow, G.M. 2008. CD11c expression identifies a population of extrafollicular antigen-specific splenic plasmablasts responsible for CD4 T-independent antibody responses during intracellular bacterial infection. *J. Immunol.*, 181, 1375–1385.

Racine, R., McLaughlin, M., Jones, D.D., Wittmer, S.T., MacNamara, K.C., Woodland, D.L., and Winslow, G.M. 2011. IgM production by bone marrow plasmablasts contributes to long-term protection against intracellular bacterial infection. *J. Immunol.*, 186, 1011–1021.

Rar, V., and Golovljova, I. 2011. *Anaplasma*, *Ehrlichia*, and "Candidatus Neoehrlichia" bacteria: Pathogenicity, biodiversity, and molecular genetic characteristics, a review. *Infect. Genet. Evol.*, 11, 1842–1861.

Rar, V.A., Epikhina, T.I., Livanova, N.N., Panov, V.V., Doroshenko, E.K., Pukhovskaia, N.M., Vysochina, N.P., and Ivanov, L.I. 2011. Study of the heterogeneity of 16s rRNA gene and groESL operone in the DNA samples of *Anaplasma phagocytophilum*, *Ehrlichia muris*, and "Candidatus Neoehrlichia mikurensis" determined in the *Ixodes persulcatus* ticks in the area of Urals, Siberia, and far east of Russia. *Mol. Gen. Mikrobiol. Virusol.*, 2, 17–23.

Reeves, W.K., Loftis, A.D., Nicholson, W.L., and Czarkowski, A.G. 2008. The first report of human illness associated with the Panola Mountain *Ehrlichia* species: A case report. *J. Med. Case Rep.*, 2, 139.

Rikihisa, Y. 1991. The tribe *Ehrlichieae* and ehrlichial diseases. *Clin. Microbiol. Rev.*, 4, 286–308.

Rikihisa, Y. 1999. Clinical and biological aspects of infection caused by *Ehrlichia chaffeensis*. *Microbes. Infect.*, 1, 367–376.

Rikihisa, Y. 2003. Mechanisms to create a safe haven by members of the family Anaplasmataceae. *Ann. N.Y. Acad. Sci.*, 990, 548–555.

Rikihisa, Y. 2006. *Ehrlichia* subversion of host innate responses. *Curr. Opin. Microbiol.*, 9, 95–101.

Rikihisa, Y., Lin, M., Niu, H., and Cheng, Z. 2009. Type IV secretion system of *Anaplasma phagocytophilum* and *Ehrlichia chaffeensis*. *Ann. N.Y. Acad. Sci.*, 1166, 106–111.

Roux, V., and Raoult, D. 1995. Phylogenetic analysis of the genus Rickettsia by 16S rDNA sequencing. *Res. Microbiol.*, 146, 385–396.

Sahu, S.P. 1986. Fluorescent antibody technique to detect *Cowdria ruminantium* in *in vitro*-cultured macrophages and buffy coats from cattle, sheep, and goats. *Am. J. Vet. Res.*, 47, 1253–1257.

Sehdev, A.E., and Dumler, J.S. 2003. Hepatic pathology in human monocytic ehrlichiosis. *Ehrlichia chaffeensis* infection. *Am. J. Clin. Pathol.*, 119, 859–865.

Shibata, S., Kawahara, M., Rikihisa, Y., Fujita, H., Watanabe, Y., Suto, C., and Ito, T. 2000. New *Ehrlichia* species closely related to *Ehrlichia chaffeensis* isolated from Ixodes ovatus ticks in Japan. *J. Clin. Microbiol.*, 38, 1331–1338.

Smith Sehdev, A.E., Sehdev, P.S., Jacobs, R., and Dumler, J.S. 2002. Human monocytic ehrlichiosis presenting as acute appendicitis during pregnancy. *Clin. Infect. Dis.*, 35, e99–e102.

Sotomayor, E.A., Popov, V.L., Feng, H.M., Walker, D.H., and Olano, J.P. 2001. Animal model of fatal human monocytotropic ehrlichiosis. *Am. J. Pathol.*, 158, 757–769.

Standaert, S.M., Yu, T., Scott, M.A., Childs, J.E., Paddock, C.D., Nicholson, W.L., Singleton Jr., J., and Blaser, M.J. 2000. Primary isolation of *Ehrlichia chaffeensis* from patients with febrile illnesses: Clinical and molecular characteristics. *J. Infect. Dis.*, 181, 1082–1088.

Stevenson, H.L., Crossley, E.C., Thirumalapura, N., Walker, D.H., and Ismail, N. 2008. Regulatory roles of CD1d-restricted NKT cells in the induction of toxic shock-like syndrome in an animal model of fatal ehrlichiosis. *Infect. Immun.*, 76, 1434–1444.

Stevenson, H.L., Estes, M.D., Thirumalapura, N.R., Walker, D.H., and Ismail, N. 2010. Natural killer cells promote tissue injury and systemic inflammatory responses during fatal *Ehrlichia*-induced toxic shock-like syndrome. *Am. J. Pathol.*, 177, 766–776.

Stone, J.H., Dierberg, K., Aram, G., and Dumler, J.S. 2004. Human monocytic ehrlichiosis. *JAMA*, 292, 2263–2270.

Sumner, J.W., Nicholson, W.L., and Massung, R.F. 1997. PCR amplification and comparison of nucleotide sequences from the *groESL* heat shock operon of *Ehrlichia* species. *J. Clin. Microbiol.*, 35, 2087–2092.

Sumner, J.W., Childs, J.E., and Paddock, C.D. 1999. Molecular cloning and characterization of the *Ehrlichia chaffeensis* variable-length PCR target: An antigen-expressing gene that exhibits interstrain variation. *J. Clin. Microbiol.*, 37, 1447–1453.

Tan, H.P., Dumler, J.S., Maley, W.R., Klein, A.S., Burdick, J.F., Fred Poordad, F., Thuluvath, P.J., and Markowitz, J.S. 2001. Human monocytic ehrlichiosis: An emerging pathogen in transplantation. *Transplantation*, 71, 1678–1680.

Telford 3rd, S.R., and Dawson, J.E. 1996. Persistent infection of C3H/HeJ mice by *Ehrlichia chaffeensis*. *Vet. Microbiol.*, 52, 103–112.

Tomassone, L., Nunez, P., Gurtler, R.E., Ceballos, L.A., Orozco, M.M., Kitron, U.D., and Farber, M. 2008. Molecular detection of *Ehrlichia chaffeensis* in *Amblyomma parvum* ticks, Argentina. *Emerg. Infect. Dis.*, 14, 1953–1955.

Wakeel, A., Kuriakose, J.A., and McBride, J.W. 2009. An *Ehrlichia chaffeensis* tandem repeat protein interacts with multiple host targets involved in cell signaling, transcriptional regulation, and vesicle trafficking. *Infect. Immun.*, 77, 1734–1745.

Walker, D.H., and Dumler, J.S. 1997. Human monocytic and granulocytic ehrlichioses. Discovery and diagnosis of emerging tick-borne infections and the critical role of the pathologist. *Arch. Pathol. Lab. Med.*, 121, 785–791.

Walker, D.H., Ismail, N., Olano, J.P., McBride, J.W., Yu, X.J., and Feng, H.M. 2004. *Ehrlichia chaffeensis*: A prevalent, life-threatening, emerging pathogen. *Trans. Am. Clin. Climatol. Assoc.*, 115, 375–382; discussion 382–374.

Wen, B., Cao, W., and Pan, H. 2003. *Ehrlichiae* and ehrlichial diseases in china. *Ann. N.Y. Acad. Sci.*, 990, 45–53.

Wen, B., Rikihisa, Y., Fuerst, P.A., and Chaichanasiriwithaya, W. 1995a. Diversity of 16S rRNA genes of new Ehrlichia strains isolated from horses with clinical signs of Potomac horse fever. *Int. J. Syst. Bacteriol.*, 45, 315–318.

Wen, B., Rikihisa, Y., Mott, J., Fuerst, P.A., Kawahara, M., and Suto, C. 1995b. *Ehrlichia muris sp.* nov., identified on the basis of 16S rRNA base sequences and serological, morphological, and biological characteristics. *Int. J. Syst. Bacteriol.*, 45, 250–254.

Winslow, G.M., Yager, E., Shilo, K., Collins, D.N., and Chu, F.K. 1998. Infection of the laboratory mouse with the intracellular pathogen *Ehrlichia chaffeensis*. *Infect. Immun.*, 66, 3892–3899.

Winslow, G.M., Yager, E., Shilo, K., Volk, E., Reilly, A., and Chu, F.K. 2000. Antibody-mediated elimination of the obligate intracellular bacterial pathogen *Ehrlichia chaffeensis* during active infection. *Infect. Immun.*, 68, 2187–2195.

Wolf, L., McPherson, T., Harrison, B., Engber, B., Anderson, A., and Whitt, P. 2000. Prevalence of *Ehrlichia ewingii* in *Amblyomma americanum* in North Carolina. *J. Clin. Microbiol.*, 38, 2795.

Yang, Q., Ghose, P., and Ismail, N. 2013. Neutrophils mediate immunopathology and negatively regulate protective immune responses during fatal bacterial infection-induced toxic shock. *Infect. Immun.*, 81, 1751–1763.

Yu, X.J., Crocquet-Valdes, P., and Walker, D.H. 1997. Cloning and sequencing of the gene for a 120-kDa immunodominant protein of *Ehrlichia chaffeensis*. *Gene*, 184, 149–154.

Yu, X., McBride, J.W., Zhang, X., and Walker, D.H. 2000. Characterization of the complete transcriptionally active *Ehrlichia chaffeensis* 28 kDa outer membrane protein multigene family. *Gene*, 248, 59–68.

Yu, D.H., Li, Y.H., Yoon, J.S., Lee, J.H., Lee, M.J., Yu, I.J., Chae, J.S., and Park, J.H. 2008. *Ehrlichia chaffeensis* infection in dogs in South Korea. *Vector Borne Zoonotic. Dis.*, 8, 355–358.

Zhang, J.Z., McBride, J.W., and Yu, X.J. 2003a. L-selectin and E-selectin expressed on monocytes mediating *Ehrlichia chaffeensis* attachment onto host cells. *FEMS Microbiol. Lett.*, 227, 303–309.

Zhang, X.F., Zhang, J.Z., Long, S.W., Ruble, R.P., and Yu, X.J. 2003b. Experimental *Ehrlichia chaffeensis* infection in beagles. *J. Med. Microbiol.*, 52, 1021–1026.

Zhang, J.Z., Sinha, M., Luxon, B.A., and Yu, X.J. 2004. Survival strategy of obligately intracellular *Ehrlichia chaffeensis*: Novel modulation of immune response and host cell cycles. *Infect. Immun.*, 72, 498–507.

Chapter 37

Neisseria gonorrhoeae: The Pathogen, Diagnosis, and Antimicrobial Resistance

Namraj Goire[1,2], David J. Speers[1], Monica M. Lahra[3], and David M. Whiley[2]

[1]Microbiology Department, PathWest Laboratory Medicine WA, Queen Elizabeth II Medical Centre, Nedlands, Western Australia, Australia

[2]Queensland Paediatric Infectious Diseases Laboratory, Queensland Children's Medical Research Institute, Royal Children's Hospital, Brisbane, Queensland, Australia

[3]Neisseria Reference Laboratory and WHO Collaborating Centre for STD, Microbiology Department, South Eastern Area Laboratory Services, Prince of Wales Hospital, Sydney, New South Wales, Australia

37.1 Introduction

Gonorrhoea remains the second most common bacterial sexually transmitted disease in the world today. *Neisseria gonorrhoeae*, the etiological agent of gonorrhoea, has proved itself to be a challenging bacterium in terms of antimicrobial therapy. As an obligate human pathogen, this bacterium is known for its ability to survive the host's immune responses via phase and antigenic variations and to utilize genetic transformation to acquire genes associated with antibiotic resistance. Over time the gonococcus has acquired resistance to the penicillins, tetracyclines, macrolides, quinolones, and more recently, the cephalosporins. In the current context, there is increasing resistance, increasing disease rates, and a known association of gonorrhoea with increased spread of HIV. Gonococcal disease is a significant public health issue.

37.1.1 The organism *Neisseria gonorrhoeae*

Neisseria gonorrhoeae is a Gram-negative bacterium belonging to the genus *Neisseria* of the family *Neisseriaceae* of class *beta-Protobacteria*. Most members of the genus *Neisseria* are Gram-negative cocci, found as commensal flora inhabiting the surfaces of mucous membranes of warm-blooded hosts (Versalovic et al., 2011). Of the 19 members in this genus, only 2 species are significant human pathogens, *N. gonorrhoeae* and *N. meningitidis*, whilst the remainder have

negligible infection rates. Some non-pathogenic strains of the meningococcus may exist as an asymptomatic colonizer of humans, with 10–30% of the adolescent population acting as carriers (Virji, 2009). Hence, unlike the meningococcus, regardless of where in the human body it is cultured or detected, the gonococcus is always considered a pathogen, and its presence regarded as an infection requiring treatment (CDC, 2008).

N. gonorrhoeae is a non-motile and non-spore-forming bacterium usually identified as Gram-negative intracellular diplococci within polymorphonuclear neutrophils (PMNs) on microscopy of clinical specimens (Holder, 2008). The bacterium is primarily isolated from the mucosal surfaces of humans, including the urethra, cervix, and oropharynx. Occasionally, the organism is also isolated from a variety of other sites including joint fluids, the conjunctiva, and the blood stream (Versalovic et al., 2011).

37.1.2 Historical perspective: gonorrhoea through the ages

Gonorrhoea is one of the oldest known human diseases, and gonorrhoea-like infections have been recorded in ancient texts dating back to 5 BC (Kline et al., 2004; Lewis, 2010; Tobiason and Seifert, 2006). The ancient Greeks believed that mucopurulent discharge associated with gonococcal infections in males was a loss of semen, and christened the disease "gonorrhoea," which translates as "the loss of human seed." The colloquial name for the disease "the clap" is thought to be derived from the French *les clapiers* (literally hutch), referring to the small huts in which prostitutes lived and worked in Medieval France (Lee, 2012). For centuries the disease was considered an early symptom of the venereal disease syphilis and the possible link between an infectious agent and gonorrhoea was not considered until the germ theory of disease gained acceptance in late 19th century (Richmond, 1954). It was only with the discovery of the bacterium in 1879, and the demonstration of fulfilment of Koch's postulates in 1882 by German bacteriologist Albert Neisser, that the identity of the causative organism was finally established. Subsequently, with the introduction of Gram's staining method in 1884, the *in vitro* culture of the gonococcus by Leistikow and Bumm in the 1880s, and the development of fermentation tests for the differentiation of the different *Neisseria* species by Elser and Huntoon in 1909, the scope for the microbiological study of the gonococcus was widened (Harkness, 1948).

37.2 Cell structure

N. gonorrhoeae is 0.6–1.0 μm in diameter, usually seen in pairs with adjacent flattened sides and with a typical Gram-negative cell wall structure (Kraus and Glassman, 1974; Sparling and Handsfield, 2000; Versalovic et al., 2011). The cytoplasm contains numerous granules, ribosomes, and a nucleus with deoxyribonucleic acid (DNA) threads surrounded by a cytoplasmic membrane to which mesosomes are attached. The rigid peptidoglycan layer is located external to the cytoplasmic membrane and is surrounded by a highly complex, triple-layered, outer membrane composed of lipopolysaccharides, phospholipids, and proteins. This acts as the interface between the pathogen and its environment and contains a host of molecules that play a vital role in the pathobiological characteristics of the organism (Wolf-Watz et al., 1975). These molecules determine interactions with host phagocyte cells, and antibacterial agents targeting the bacterium. The wide antigenic variation of the surface components together with their extensive virulence and antibacterial properties has made them popular targets for epidemiological, serological, and pathological studies (Murphy and Cannon, 1988).

37.2.1 Surface components of the outer membrane

The outer membrane harbours the gonococcal endotoxin lipooligopolysaccharide (LOS), numerous fimbriae such as the *N*-methylphenylalanine pili, and a variety of other proteins embedded in a phospholipid layer. The most prominent of these surface proteins is the porin B (PorB) protein, which may exist as one of two major PorB protein types: IA (PIA) and PIB. The porins act as hydrophilic channels for the essential solutes to pass through the hydrophobic outer membrane, which otherwise repels polar macromolecules (Sparling and Handsfield, 2000; Murphy and Cannon, 1988). Also present on the surface are the opacity-associated (Opa) proteins, which are encoded by up to 12 copies of the *Opa* gene (Hauck and Meyer, 2003). In addition to contributing to virulence, the Opa proteins are responsible for the opaque appearance of the bacterial colonies growing on agar.

The outer membrane hosts an array of efflux pumps, which together with the porins control the movement of molecules across the membrane. The gonococcus has been shown to possess four types of efflux pumps – *mtr* (multiple transferable resistance), *far* (fatty acid resistance), *mac* (macrolide specific), and *nor* (norfloxacin resistance). These toxic waste removal pumps also play an important role in reducing the antibacterial susceptibility of the gonococcus (Folster et al., 2009; Rouquette-Loughlin et al., 2003, 2005; Shafer and Folster 2006; Veal et al., 2002). Other outer membrane proteins include two transferrin receptors (encoded by the genes *Tbpb1 and Tbpb2*) and a lactoferrin receptor (*Lbp*), which are induced in iron-deficient environments (Lee and Bryan, 1989).

37.3 Genetics

The circular gonococcal genome is estimated to be around 2219 kilobases (kb) with a G + C content of 50% (Bihlmater et al., 1991). The bacterium possesses more than one copy of the genome and hence is polyploid (Tobiason and Seifert, 2006). Genetic insertions highlight the readily transformable nature of the gonococcus (Dillard and Seifert, 2001). Recognized insertions, including genetic islands that may contribute to antimicrobial resistance (AMR) and virulence, are mostly attributed to genetic exchange with other *Neisseria* species such as *N. meningitidis*.

Most gonococci possess a 24.5 mega-Dalton (mD) conjugative plasmid that can facilitate the transfer of other, usually non-transferable, plasmids without mobilizing the chromosomal genes. A 2.6 mD cryptic plasmid is also present, although its exact function is yet to be elucidated. The wide extent of chromosomal, plasmid, and naked DNA gene acquisition from other gonococci and unrelated bacteria demonstrates the ability of the gonococcus to readily undergo genetic transformation. Thus, the gonococcus is truly a panmictic organism. This attribute has serious implications for effective antimicrobial therapy for gonorrhoea, enabling acquisition and incorporation of various resistance genes from other related commensal species (Cannon and Sparling, 1984; Lewis, 2010; Weinstock and Workowski, 2009).

37.4 Pathology

The *N. gonorrhoeae* organism differs from other members of its family by being an obligate human pathogen. Moreover, several other *Neisseria* species are found only in humans, leading to the suggestion by Dillard et al. that these organisms most likely evolved *within* humans from a common ancestor (Dillard and Seifert, 2001). It has been further suggested that gonococci evolved from a meningococcal strain that colonized the human genitourinary tract, explaining the significant genetic similarity between the two species (Feavers and Maiden, 1998). Recently, human genomic elements have been discovered in the gonococcus, demonstrating the extraordinary ability of the bacterium to undergo horizontal genetic transfer, even with its mammalian host (Anderson and Seifert, 2011). This "familiarity" with the human body over several eons has seemingly bestowed the gonococcus with the ability to adapt readily to the human environment, and resist the innate immune systems of human mucosal surfaces. The success of the gonococcus as a human pathogen continues to this day despite the arrival of the antimicrobial era a number of decades ago.

37.5 Virulence factors and pathogenesis

The gonococcus possesses several virulence determinants that aid in invading host cells and propagating infection. The first stage of infection, involving adherence and invasion, is mediated by the surface components. The bacterium initially attaches to the cuboidal or columnar epithelial cells by means of its fimbriae or type 4 pili, which exhibit great affinity for most human cells. Following this, the gonococcus uses the outer membrane protein Opa (also known as protein II) to closely bind and then invade the epithelial cells (Merz and So, 2000; Mosleh et al., 1997). Gonococci are also capable of forming an inter-organism bridge by extending the Opa protein to bind the LOS of the neighbouring bacterium, thus forming links to produce a microcolony, similar to biofilms formed by other pathogens (Higashi et al., 2007).

Once established, gonococci employ several mechanisms to avoid the host cellular and humoral immunity. Protein IA (Por A), protein IB (Por B), protein III (Rmp), and the gonococcal capsule play a vital role in avoiding phagocytic activity (Baron, 1996). The gonococcal endotoxin, LOS, is responsible for most of the disease symptoms associated with gonorrhoea, and also facilitates release of enzymes such as proteases and phospholipases, which play an important role in pathogenesis (Paz et al., 1995; Patrone and Stein, 2007). Whilst triggering the intense inflammatory response that results in ciliary loss and subsequent death of epithelial cells, LOS activates complements and thus recruits PMNs. The gonococcus is capable of manipulating the recruited PMN and survive inside them by withstanding their oxidative and non-oxidative components. (Criss and Seifert, 2008). The resultant lysis of PMN generates the purulent discharge typical of a gonococcal infection (Holder, 2008; Sparling and Handsfield, 2000).

37.6 Clinical manifestations

N. gonorrhoeae can infect the mucous membranes of the urethra, endocervix, and less commonly the anorectum, pharynx, and conjunctiva. Complications of infection include pelvic inflammatory disease, urethral strictures, abscesses in the urogenital tract, prostatitis and epididymo-orchitis, and disseminated infection (Lewis, 2010; Tapsall, 2006). Untreated gonococcal infection in pregnancy is associated with stillbirth and preterm delivery in up to 35%, and perinatal death in up to 10% of cases. Gonococcal ocular infection (ophthalmia neonatorum) occurs in the neonates of up to 50% mothers with untreated gonorrhoea, which can result in blindness (WHO, 2011a). Asymptomatic gonococcal infection occurs in women, men who have sex with men (MSM), and transgender people (WHO, 2011b). In the presence of gonorrhoea, HIV co-transmission rates are reported to be amplified by a factor of 3–5 (WHO, 2011a).

Pharyngeal gonococcal infection is usually asymptomatic and mostly self-limiting, but it may present as clinical pharyngitis. Isolation and identification of gonococci from the pharynx can be challenging due to the plethora of related bacterial commensal flora at this site. Pharyngeal infection may contribute towards the persistence of the pathogen' and to its increasing AMR (Barry and Klausner, 2009; Weinstock and Workowski, 2009).

In females, about 20–40% of untreated gonorrhoea cases result in pelvic inflammatory disease (PID), with potential complications including tuboovarian abscess, infertility, and ectopic pregnancy. Acute PID is often polymicrobic as co-infection with other sexually transmitted pathogens such as *Chlamydia trachomatis* and *Mycoplasma genitalium* is common. About 1% of mucosal gonorrhoea infections develop into disseminated infection, commonly presenting with arthritis–dermatitis syndrome. In rare cases, further complications such as meningitis, septic shock, or endocarditis may occur (Sparling and Handsfield, 2000; Tapsall, 2006). Gonorrhoea is usually a sexually transmitted disease with the exception of vertical transmission during delivery causing neonatal conjunctival infection, and, rarely, horizontal transmission within communities in settings of poor hygiene (Goodyear-Smith, 2007). Females may be asymptomatically infected for up to 12 months, thus contributing greatly to disease transmission (Jakopanec et al., 2009). The risk of female-to-male transmission is approximately 20% per sexual contact, while the male-to-female transmission rate is estimated to be around 60% (Holmes et al., 1970). The high prevalence among the MSM population indicates that gonorrhoea is efficiently transmitted via anal and oropharyngeal intercourse, although definitive studies are lacking in this field (Holder, 2008).

37.7 Epidemiology

After *C. trachomatis*, gonorrhoea is the most common bacterial sexually transmitted infection worldwide. In 2008, the World Health Organization (WHO) estimated that globally there are 106 million new cases of *Neisseria gonorrhoeae* infection amongst men and women 15–49 years old, annually, with a trend towards increased infection rates over the last decade (WHO, 2011c). The highest incidence of infection occurs in developing and under-developed countries, where reliable disease-surveillance systems are not always in place, hampering effective monitoring and likely underestimating the prevalence of the disease. In the developed world, the United States reports much higher rates of infection (100.8 per 100,000 (CDC, 2011b) than the European Union region (9.7 per 100,000 (ECDC, 2011)) with Australian infection rates high and rising: 60.0 per 100, 000 (NNDSS, 2012). Prevalence of gonorrhoea is influenced by the ethnic and economic backgrounds and sexual habits of the population (WHO, 2007a). In some population groups, there is a comparatively higher incidence than the background rate of disease, as has been reported in the United States in African-Americans and MSM (CDC, 2011b; Jones et al., 2000).

37.8 Diagnosis

Clinical diagnosis of gonorrhoea, based on signs and symptoms, and with consideration of disease demographics, and behavioural risk factors is common practice in parts of the world where adequate diagnostic laboratory infrastructures are lacking. The *WHO Training Modules for Syndromic Management of Sexually Transmitted Infections 2007* recommends physicians commence gonococcal therapy for men presenting with mucopurulent urethral discharge immediately. As women are often asymptomatic, a comprehensive assessment of risk factors and infection rates is recommended (WHO, 2007b). Overall, empirical diagnosis of gonorrhoea can be difficult and therefore laboratory-based diagnosis, where available, is important for the management and control of gonorrhoea. Several different microbiological, immunological, and nucleic acid amplification tests (NAAT) are routinely used for this purpose.

37.8.1 Microbiological techniques

Isolation of *N. gonorrhoeae* from the patient specimen remains the gold standard for diagnosis of gonorrhoea. Depending on the symptoms and the patient's sexual practices, the appropriate specimens include urethral, endocervical, and rectal swabs. Where oral intercourse has been reported, obtaining a pharyngeal swab is recommended, particularly if symptoms of pharyngitis are present. In symptomatic males, gonococcal urethritis can be diagnosed by the presence of Gram-negative diplococci within or in close association with PMNL on a smear prepared from urethral exudate. A properly performed Gram's stain has a sensitivity of 90–95% in such instances, and a specificity of 50–70% depending on the adequacy of the specimen, site of collection, underlying symptoms, and the patient population (Orellana et al., 2012; Versalovic et al., 2011).

Diagnosis of *N. gonorrhoeae* by culture requires selective media, and its presence is confirmed by specific biochemical and physiological tests. Thayer–Martin agar, its variation Martin–Lewis agar, and New York City agar are the most commonly used selective culture media for isolation of *N. gonorrhoeae*. Ideal conditions for the growth of gonococcus in media include incubation at 35–37°C in a 5% CO_2 rich atmosphere (Bonin et al., 1984; Versalovic et al., 2011). The most important tests in the differentiation of *N. gonorrhoea* from other *Neisseria* species (which are all Gram-negative diplococcic and oxidase test positive) are a positive superoxol test, the ability to utilize glucose and reduce nitrites, specific enzymatic detection, and fastidious growth requirements including fatty acid intolerance (Versalovic et al., 2011; WHO, 2013).

Antimicrobial susceptibility testing can be performed on cultured isolates to detect and monitor the resistance of antimicrobial therapy. Antimicrobial susceptibility testing is of increasing clinical importance in the current context of escalating gonococcal AMR, and increasing use of non-culture-based diagnostic test methods that cannot currently provide this information. The WHO and the CDC recommend repeat culture and antimicrobial susceptibility testing on any suspected case of treatment failure, along with a careful interpretation of NAAT-based results from non-genital sites (CDC, 2011a; Whiley et al., 2006; WHO, 2012).

37.8.2 Non-culture diagnostic techniques

Non-culture-based diagnosis of gonorrhoea is made by direct detection methods, mostly NAAT, but also enzyme-linked immunosorbant assays (EIA) directed against gonococcal antigens. Although the commercially available EIA tests are less time-consuming than traditional culture methods and less expensive than NAAT, they have reduced accuracy with lower sensitivity and specificity, limiting their use in routine laboratory diagnosis (Herring et al., 2006; Papasian et al., 1984).

37.8.3 The nucleic acid detection tests

NAAT detect a unique region in the gonococcal genome. Two different molecular technologies are currently used to detect gonococcal deoxyribose nucleic acid (DNA): DNA hybridization and DNA amplification.

For DNA hybridization, an oligonucleotide probe is constructed and labelled with a chemiluminescence marker. The presence of gonococci in the clinical specimen is demonstrated by annealing of this probe to a complementary sequence present on the gonococcal genome, removal of unhybridized probe, and subsequent detection of emitted light via the marker. Although commercial DNA hybridization assays targeting gonococci are available, they have been shown to have lower sensitivity compared to traditional bacterial culture methods and nucleic acid amplification methods. For this reason, DNA hybridization technology has not gained wide acceptance for diagnosing gonorrhoea (Koumans et al., 1998).

DNA amplification based technologies target and amplify a unique region of the genome, or its ribosomal RNA, with subsequent detection of the amplified product. Several commercial tests are available for *N. gonorrhoeae* that use either strand displacement amplification (SDA), transcription mediated amplification (TMA), or polymerase chain reaction (PCR) amplification methods (Gaydos, 2005; Whiley et al., 2006). In addition, a number of in-house PCR-based assays have been described and are widely used for diagnosing gonorrhoea in the developed world.

The popularity of nucleic acid detection tests for diagnosing gonorrhoea can be attributed to the higher sensitivity and faster turn-around time involved when compared to culture and other non-culture-based methods. In comparison to *N. gonorrhoeae* NAAT, the sensitivity of gonococcal culture ranges from 85% to 95% for acute infections but may be as low as 50% for females with chronic infections. Also, most NAAT do not require the invasive collection associated with specimens such as urethral swabs, nor do they require viable organisms, providing flexible collection and transportation options, especially for patients in remote locations. These attributes of sensitivity, convenience, and robustness make NAAT excellent screening tests for gonorrhoea, and hence play an invaluable role in disease control and eradication (Tapsall, 2001; Whiley et al., 2006).

37.8.3.1 Challenges associated with NAAT

37.8.3.1.1 General issues
NAAT for gonorrhoea come with their own set of challenges that have become evident through their increased use. A major impediment to instituting NAAT remains the costs involved, including infrastructure and reagents, and the level of training of laboratory staff required for successfully running the tests. In addition, when performing NAAT, specimen handling and the testing procedure must be meticulous, as this testing is prone to contamination by mishandling.

The three main steps involved in NAAT, nucleic acid capture or extraction, target amplification, and amplicon detection, harbour the potential for introduced errors, including operational mistakes, reagent failure, and carry-over contamination between specimens. Clinical specimens may also contain interfering

chemicals such as beta-human gonadotropin, crystals, haemoglobin, and nitrites that are antagonistic towards one or more of the reagents used, thereby inhibiting the overall reaction and reducing the sensitivity. This is especially true for urine, a common specimen for gonococcal NAAT that contains several human metabolic by-products capable of inhibiting the amplification reaction (Mahony et al., 1998; Whiley et al., 2006).

37.8.3.1.2 Sequence-related issues
37.8.3.1.2.1 Sequence variation
NAAT rely on the recognition of a very small genetic sequence, making it vulnerable to the genetic variability of *N. gonorrhoeae* and the presence of the target region in a related organism, which may lead to false-negative and false-positive results, respectively. The best example of sequence variation in the gonococcus affecting NAAT-based diagnosis is the use of the cryptic plasmid protein B (*cppB*) gene target. The *cppB* gene was a popular target for *N. gonorrhoeae* based assays until discrepancies in detection were revealed between assays using this target and NAAT using genome targets. Isolates that lack the *cppB* gene have since been reported in various regions, leading to recommendations that the *cppB* gene not be used for NAAT-based detection (Bruisten et al., 2004; Smith et al., 2005). Moreover, the *cppB* problem has also shown that different patient populations may harbour different gonococcal strains as the predominant subtype. In Australia, Tabrizi et al. (2011) found the *cppB* gene to be highly conserved in South Australian and Victorian gonococcal populations, whereas Lum et al. (2005) showed that there was a high prevalence of *cppB* gene lacking auxotypes in the Northern Territory. More recently, variant *N. gonorrhoeae* strains have been described in Australia and elsewhere that harbour mismatches in the target region of the *porA* pseudogene, also a popular target for gonococcal NAAT (Whiley et al., 2011). This has fueled the belief that NAAT should target multiple conserved genes of the gonococcus (Goire et al., 2008; Whiley et al., 2008).

37.8.3.1.2.2 Cross reaction with other species
Human genitourinary and pharyngeal endogenous flora includes a host of *Neisseria* species that have a high degree of sequence homology with the gonococcus. In addition, gonococci are highly competent in genomic exchange within the *Neisseria* genus throughout their life cycle (Kline et al., 2004). This results in commensal *Neisseria* species acquiring gonococcal genomic sequences, thereby potentially reducing the specificity of NAAT (Linz et al., 2000). In 1999, Farrell (Farrell, 1999) described cross reactions between the gonococcus and at least two different commensal *Neisseria* species with a *cppB*-based assay. More recently, Tabrizi et al. (2011) evaluated six commercial *N. gonorrhoeae* NAAT against clinical gonococcal and commensal *Neisseria* and other related species, demonstrating cross reactions to varying degrees

amongst the assays. Similarly, cross reaction with other species remains an issue in the diagnosis of gonorrhoea by NAAT. However, the extent of cross reaction depends on the type of specimen, gene targeted by the assay, type of commensal species present in the specimen, and the patient population being tested (Tabrizi et al., 2011).

37.8.3.1.2.3 *Antimicrobial susceptibility, test-of-cure, and NAAT*

A major limitation of NAAT is the inability to determine the AMR of the infecting organism. As only nucleic acid is required for NAAT diagnosis, viability of organisms is not paramount. However, the lack of a viable organism precludes antimicrobial susceptibility tests. With global increases in gonococcal AMR, surveillance to detect resistance is increasingly important (Dillon, 2011; Unemo and Dillon, 2011; WHO, 2012). Antibiotic resistance data from culture need to be available in a routine diagnostic setting to inform treatment guidelines and to monitor the effects of interventions. Further, effective AMR surveillance is key to understanding both the magnitude and trends of gonococcal resistance (WHO, 2012). It is increasingly important that antimicrobial susceptibility testing and surveillance be maintained in the face of the growing popularity of NAAT.

Interpretation of the gonococcal test-of-cure may be compromised by NAAT, as DNA from non-viable organisms may be detected for days after effective treatment with the potential for a false assumption of failed therapy and further antibiotic use, whereas cultures are negative after effective antimicrobial therapy (Bachmann et al., 2002).

37.9 Treatment and antimicrobial resistance

Over the last few decades there has been a dramatic change in the recommended antimicrobials for the empiric treatment of gonorrhoea. From being a readily treatable disease with a single oral dose of an antibacterial, gonorrhoea has evolved into a therapeutically challenging disease with diminishing treatment options. The WHO recommends treatment with drugs with a cure rate of 95% and above in a given population for the treatment of gonorrhoea (WHO, 2007b). However, there are very few drugs available to treat gonorrhoea that now fit that criteria.

The most reliably effective antibacterials currently available are third-generation cephalosporins, spectinomycin and azithromycin (Chisholm et al., 2009; WHO, 2012; Zenilman et al., 1987). However, the gonococcus has exhibited resistance in some cases to spectinomycin; in addition, there have been several reports of high-level resistance to azithromycin (Easmon et al., 1984; Katz et al., 2012). Critically, the emergence of resistance to the oral third-generation cephalosporin agents, and continuing increases in

minimum inhibitory concentrations (MIC) to the injectable third-generation cephalosporin ceftriaxone raised considerable alarm. The recent emergence and spread of cases of ceftriaxone-resistant gonococci (H041 and F89) has raised fears that a pan-resistant strain may emerge rapidly and spread in the future (Barry and Klausner, 2009; Ohnishi et al., 2011a; Whiley et al., 2012).

37.9.1 *N. gonorrhoeae* and antibacterial agents

One of the oldest known human diseases, gonorrhoea has been treated with a wide range of therapies over the course of history. Before the discovery of gonococcal pathogenesis, treatments were directed at the symptoms of the disease, confined mostly to flushing procedures and exposing the genitalia to chemicals with putative antiseptic properties (Fekete, 1993; Lewis, 2010).

The introduction of antibiotics had a profound effect on the treatment and control of gonorrhoea. However, the recommended gonococcal antibacterials have changed several times over the last few decades with the emergence of resistance (Table 37-1).

37.9.1.1 Sulphonamides

Sulphonamides were the first antibacterial agents used specifically for the treatment of gonorrhoea, and they gained wide acceptance within a short span of time (Blake, 1940; Deakin et al., 1941). At the time of their introduction, sulphonamides had almost 100% efficacy in treating the disease, leading to widespread use. However, within a decade of introduction, the first clinical treatment failures with sulphonamides were reported. Subsequently, it was shown that the gonococcus is able to alter the dihidropteroate synthase enzyme target of sulphonamides by acquisition of a mutated form of the encoding *folP* gene, thereby becoming sulphonamide resistant (Sköld, 2001). The gonococcus is believed to have acquired this altered gene by horizontal genetic transfer from other commensal *Neisseria* species. It is now widely accepted that the regular exposure to antibacterials of a large number of genetically similar *Neisseria* species on the human body eventually gave rise to mutant *N. gonorrhoeae* strains that are resistant to these drugs (Aminov and Mackie, 2007; Weinstock and Workowski, 2009). By the late 1940s, sulphonamides had fallen into disuse due to the high rates of treatment failure and the discovery that penicillin was very effective against the gonococcus.

37.9.1.2 Penicillin

Penicillin was the first of the beta-lactam antibiotics that target the penicillin-binding proteins (PBP),

Table 37-1 Major antibacterials used against gonorrhoea, timeline of resistance development, and current CDC treatment recommendations

Antibacterial agents	Approximate year of introduction	Mechanism of action	Approximate year when the emergence of resistance first detected	Mechanism of resistance	Current CDC recommendations regarding use (CDC, 2011a)
Sulfonamides	1935 (Deakin et al., 1941)	Disruption of folic acid pathway by inhibition of dihydropteroate synthase (Sköld, 2001)	1944 (Sköld, 2001)	Chromosomal – mutations in the *folP* gene (Sköld, 2001)	Not recommended
Penicillin	Clinical use in the 1940s (Lewis, 2010)	Inhibition of cell wall synthesis by binding to PBP, especially PBP2, (Dougherty, 1985)	Increased MIC circa 1950, treatment failure circa 1960, fully resistant strains circa 1970 (Barry and Klausner, 2009)	(a) Chromosomal – mutations resulting in altered PBP, increased efflux pump activity, reduced permeability of porins (Lee et al., 2010). (b) Plasmid mediated – via acquisition of TEM-1 carrying beta-lactamase plasmid (Percival et al., 1976; CDC, 1976)	Not recommended
Spectinomycin	First introduced in 1960 for clinical investigation (Ribeiro and Mathieson, 1975)	Inhibition of protein synthesis by binding to 30S ribosomal subunit (Galimand et al., 2000)	1973 (Ashford et al., 1981)	Chromosomal – alteration of ribosomal targets by mutation in 16s rRNA genes (Galimand et al., 2000)	Not available in the United States
Tetracyclines	1954 (oral), 1955 (intra-muscular) (Marmell and Prigot, 1955).	Inhibition of attachment of aminoacyl-tRNA, affecting peptide chain elongation (Chopra and Roberts, 2001)	Circa 1970 (chromosomally mediated), 1984 (plasmid mediated) (Chopra and Roberts, 2001)	Chromosomally and plasmid mediated. Alteration of target by the ribosomal protection proteins and *tetM* plasmid (Hu et al., 2005; Morse et al., 1986)	Single-dose treatment of ceftriaxone (250 mg) or cefixime (400 mg) along with azithromycin (1 g) or doxycycline 100 mg daily for 7 days
Fluoroquinolones	1980 (Dan, 2004)	Inhibition of *gyrA* and *parC* enzymes, thereby preventing DNA replication (Dan, 2004)	1982 (Martin et al., 2005)	Chromosomal – single or multiple mutations in *gyrA* and *parC* genes (Belland et al., 1994)	Not recommended
Azithromycin	1994 (Handsfield et al., 1994)	Inhibition of protein synthesis by binding to the 23S ribosome (Ng et al., 2002; Roberts et al., 1999)	2001 – reduced susceptibility (Palmer et al., 2008), 2004 – high-level resistance (Dillon et al., 2001)	Chromosomal – alteration of target by point mutations or due to inactivating methylase enzymes (Ng et al., 2002; Roberts et al., 1999)	Single-dose treatment of ceftriaxone (250 mg) or cefixime (400 mg) with azithromycin (1 g) or doxycycline 100 mg daily for 7 days
Cephalosporins (broad spectrum)	Circa 1976 (extended spectrum) (Barry and Klausner, 2009)	Inhibition of cell wall synthesis by binding to PBP (Barry and Klausner, 2009)	First reported treatment failure in Japan in 2001 to oral third-generation cephalosporin, followed by reports of raised MIC, then emergence of H041 and F89 ceftriaxone-resistant strains in 2009 and 2011, respectively (Akasaka et al., 2001; Ohnishi et al., 2011b; Unemo et al., 2012)	Chromosomal mutations resulting in altered PBP (mosaics and point mutations), increased efflux pump activity, reduced permeability of porins (Unemo et al., 2012)	Single-dose treatment of ceftriaxone (250 mg) or cefixime (400 mg) with azithromycin (1 g) or doxycycline 100 mg daily for 7 days

which are instrumental in building of the bacterial cell wall (Barbour, 1981; Dougherty, 1985). Penicillin was an effective treatment for gonorrhoea until the mid-1950s when reports of treatment failure with conventionally used doses appeared. In most instances, increasing the dose of penicillin led to treatment success. Subsequent *in vitro* studies showed that there had been a gradual increase in the MIC of penicillin, explaining the increased dosages required to achieve cure *in vivo* (Lewis, 2010). By the early 1970s the gonococcus had acquired a plasmid harbouring a TEM-1 penicillinase gene, and penicillinase-producing gonococcal strains, which did not respond to higher dose therapy, were subsequently reported from different parts of the world (Shafer and Folster, 2006; Siegel et al., 1978).

However, there are also a small number of isolated populations around the world that owing to factors ranging from limited use of antimicrobials to geographical and epidemiological isolation have endemic gonococcal strains that remain largely susceptible to most antibiotic therapies. One such example is the remote indigenous communities of the north-western region of Australia, where rates of penicillin resistance have consistently remained well below the national average (AGSP, 2012) and where penicillin is still used for first-line treatment.

37.9.1.3 Spectinomycin

Spectinomycin, an injectable aminocyclitol compound similar to the aminoglycoside class, was routinely used since its inception in the 1960s for patients intolerant of penicillins. Spectinomycin inhibits protein synthesis in susceptible bacteria by binding to 30S ribosomal subunits; however, by the 1980s low numbers of spectinomycin-resistant gonococci due to single-step chromosomal mutations leading to altered binding sites emerged (Ashford et al., 1981).

37.9.1.4 Tetracyclines

Tetracyclines, a group of oral antibiotics developed in the 1950s that inhibit peptide chain elongation in susceptible bacteria, were previously prescribed for gonorrhoea and are currently in use to treat syphilis and chlamydia. Tetracycline resistance in gonococci developed in the 1970s, mediated either chromosomally via a point mutation in the *rpsJ* gene resulting in an altered target site or via the *tetM* plasmid (Hu et al., 2005). Interestingly, the gonococcus is known to have derived the *tetM* determinant responsible for high-level tetracycline resistance from streptococci, further highlighting its genetic receptiveness (Morse et al., 1986). With the emergence of penicillinase-producing, tetracycline-resistant gonococci, the scope of drugs available to treat gonorrhoea further narrowed.

37.9.1.5 Azithromycin

After azithromycin was shown to be active against *N. gonorrhoeae* in the 1990s, it was used as a treatment of choice for gonorrhoea in some South American countries (Dillon et al., 2001). In other parts of the world, until recently, it was only prescribed in conjunction with other anti-gonococcal drugs for the management of dual infection with *N. gonorrhoea* and *C. trachomatis*. Azithromycin, like other antimicrobials of the macrolide–lincosamide–streptogramin group, inhibits protein synthesis by binding to the active site of the 23S rRNA component of the 50S bacterial ribosome. The first cases of reduced gonococcal susceptibility were reported from Latin America, followed by the description of high-level resistance in Scotland in 2004 (Galarza et al., 2009; Palmer et al., 2008). The gonococcus becomes resistant to azithromycin by alteration of the 23S rRNA target, either by mutation or via enzymatic modification. Two rRNA mutations, a C2611T (cytosine to thymine substitution at position 2611) and an A2059G (adenine to guanine substitution at position 2059) together with rRNA overexpression due to an *mtrR* promoter gene adenine deletion, have been linked to azithromycin resistance (Chisholm et al., 2010b).

37.9.1.6 Fluoroquinolones

Fluoroquinolones were first used to treat gonorrhoea in the early 1980s and were initially very successful in controlling gonococcal infections around the world. Their mechanism of action involves inhibition of DNA replication by binding to topoisomerases and thus inhibiting efficient replication. The bacterium becomes resistant by acquiring mutations in the *gyrA* and *parC* genes, which code for the topoisomerases DNA gyrase and topoisomerase IV, respectively (Belland et al., 1994). Remarkably, within a few years of their introduction, the first fluoroquinolone-resistant strains were isolated, which was followed by a rapid global spread, relegating this newly discovered class of drug to the list of antimicrobials largely ineffective for gonorrhoea (Martin et al., 2005).

37.9.1.7 Cephalosporins

The cephalosporin antibiotics were discovered in the 1970s, and are a class of beta-lactam antibiotics effective against penicillin-resistant gonococcal strains The cephalosporin antibiotics are derived from the fermentation products of the fungus *Acremonium chrysogenum* (formerly *Cephalosporium acremonium*). The basic structure of all cephalosporins is a 7-aminocephalosporanic acid nucleus, which consists of a beta-lactam ring, common to all beta-lactam antibiotics including penicillin, fused to a dihydrothiazine ring. On the basis of their range of activity against micro-organisms, the cephalosporins have been divided into "generations," with the first- and the second-generation antibiotics more active against Gram-positive and Gram-negative organisms, respectively (Barry and Klausner, 2009; Sparling and Handsfield, 2000; O'Callaghan, 1979). The later "third" and "fourth" generations of cephalosporins exhibit true broad-spectrum bactericidal activity against a wide range of bacteria, including the gonococcus. The "second-generation" cefuroxime and the "third-generation" ceftriaxone, cefixime, and ceftibuten are the four most commonly used cephalosporins for the treatment of gonorrhoea, although the efficacy of cefuroxime for gonorrhoea remains questionable (Ison et al., 2004).

Ceftriaxone and cefixime are currently the mainstay of gonococcal antibiotic regimens in many parts of the world due to the widespread resistance to many other classes of antibiotics. The first case of treatment failure associated with cefixime was reported in Japan in 2002, followed by similar case reports from other parts of the world and subsequent more widespread detection of resistance effectively leaving ceftriaxone as the only proven reliable antibacterial agent in clinical practice (Barry and Klausner, 2009; Ito et al., 2004; Tanaka et al., 2002; Unemo et al., 2011).

Compared to the other "third-generation" or extended spectrum cephalosporins (ESC), ceftriaxone has a longer serum half-life as its chemical structure facilitates extensive protein binding (Fekete, 1993; Ison et al., 2004; Sparling and Handsfield, 2000; Rajan et al., 1982). In addition, ceftriaxone has been proven clinically as an anti-gonococcal agent and has emerged as the recommended first-line treatment for gonorrhoea by the health authorities of the United Kingdom, United States, and the WHO (Barry and Klausner, 2009; Bignell and FitzGerald, 2011; CDC, 2011a; Eichmann et al., 1981; Khan et al., 1982). However, with the recent isolation of strains resistant to ceftriaxone and several other classes of anti-gonococcal drugs (the H041 and F89 strains) in Japan and France, respectively, there are concerns that gonorrhoea may eventually become untreatable in certain circumstances (Ohnishi et al., 2011a, 2011a; Unemo et al., 2012).

Gonococcal cephalosporin resistance is currently the most significant threat to the management of gonnorhoea. Since their introduction in the 1970s, the ESCs have been successfully used as anti-gonococcal agents around the world in populations with high resistance rates to other drugs (CDC, 2011a; Kirkcaldy et al., 2011; Lewis, 2010). Oral ESCs such as cefixime and cefdinir were initially selected over the injectable ESCs such as ceftriaxone due to the comparative ease of use and administration. However, the first case of oral cephalosporin treatment failure to cefdinir was reported from Japan in 2001. Following this report there has been a steady increase in the number of Japanese isolates with reduced susceptibility to oral cephalosporins, including cefixime, with current resistance rates reported to be at around 30% (Akasaka et al., 2001; Ohnishi et al., 2010a). Since then,

several cases of treatment failure with cefixime have been reported from Europe, leaving the parenteral form of ESC as the treatment of choice (Unemo et al., 2011; Yokoi et al., 2007).

Not unexpectedly, fully ceftriaxone-resistant strains have now been reported from Japan and France, preceded by reports of a gradual rise in the MIC of ceftriaxone from around the world. The two most widely used antimicrobial susceptibility reporting standards, the Clinical Laboratory Standards Institute (CLSI) and the European Committee on Antimicrobial Susceptibility Testing (EUCAST), define a raised MIC to ceftriaxone as levels greater than 0.5 μg/mL and greater than 0.125 μg/mL (considered resistant above this breakpoint), respectively (EUCAST, 2012; Kirkcaldy et al., 2011). Reflecting the worldwide trend, the proportion of Australian isolates with reduced susceptibility to ceftriaxone (defined as an MIC of 0.06–0.125 μg/mL) has increased substantially, from 0.6% in 2006 to 0.8% in 2007 to 1.1% in 2008 to 2.0% in 2009 to 4.2% in 2012 (AGSP, 2013). In 2009, Tapsall et al. (Tapsall et al., 2009a) reported two Australian cases of ceftriaxone treatment failure, although this was considered a reflection of difficulty in treating pharyngeal infections rather than an outright failure of ceftriaxone therapy in uncomplicated gonorrhoea.

All cephalosporin resistance mechanisms of gonococci are yet to be fully understood (Ohnishi et al., 2011a). The genetic markers associated with resistance or reduced susceptibility to cephalosporins described so far are chromosomally mediated as plasmid-mediated cephalosporin resistance is yet to be observed. It is now understood that multiple gene mutations are required for high-level cephalosporin resistance as each mutation (*penA*, *mtrR* promoter, *penB*) in isolation are incapable of producing high-level resistance (Liao et al., 2011; Lindberg et al., 2007; Zhao et al., 2009). High-level resistance to cephalosporins is therefore generated by a combination of altered PBP targets with reduced cephalosporin entry and enhanced removal from the bacterial cell wall periplasmic space.

The *penA* gene, which encodes PBP2, may undergo point mutation to produce either a single amino acid substitution or may acquire one of a set of mosaic *penA* sequences via transformation with commensal *Neisseria* species, resulting in decreased binding of beta-lactams (Ameyama et al., 2002; Ito et al., 2005; Lee et al., 2010; Ohnishi et al., 2011a; Unemo et al., 2012Whiley et al., 2007a, 2007b). The *mtrR* gene codes for the MtrCD-E efflux pump where an adenine deletion (A-deletion) in the 13 base-pair inverted region of the *mtrR* promoter gene results in over-expression of the MtrC-D-E efflux pump. The resultant selective expulsion of hydrophobic agents such as cephalosporins contributes to reduced susceptibility to these agents (Tanaka et al., 2006; Warner et al., 2008). A mutation of the *penB* gene, which encodes the PorB porin, confers reduced permeability to hydrophobic agents and tetracyclines and has also been suggested to affect activity of the Mtr efflux pumps (Olesky et al., 2006; Zhao et al., 2009). The pilus gene *pilQ* was previously implicated but is no longer considered a relevant marker associated with cephalosporin resistance in the gonococcus (Whiley et al., 2010a; Zhao et al., 2005).

The *penA* gene is of particular interest in gonococcal resistance to "third-generation" cephalosporins as all isolates reported with reduced susceptibility harboured at least one form of *penA* alteration, as well as the first two reported strains with high-level ceftriaxone resistance, F89 and H041, from France and Japan, respectively (Unemo et al., 2012). The mosaic *penA* gene, so named due to the presence of a host of sequences that have been acquired from commensal *Neisseria* and related species, highlights both the genetic versatility of the gonococcus and the role played by commensal species in acting as a reservoir for resistance determinants. At the same time, the role of non-mosaic *penA* point mutations in mediating ceftriaxone resistance is now also understood to be important (Whiley et al., 2010b).

In 2002, Ameyama et al. first described gonococcal strains exhibiting reduced susceptibility to cefixime that harboured a unique *penA* sequence that resembled a mosaic-like structure formed from the combination of *penA* genes from *Neisseria perflava* (*N. sicca*), *N. cinerea*, *N. flavescens*, and *N. meningitidis* (Ameyama et al., 2002). This mosaic *penA* gene has since been implicated in conferring resistance to cefixime in several other studies and clones containing this mosaic gene are now known to have successfully spread internationally (Ohnishi et al., 2010b; Unemo et al., 2010).

The F89 strain isolated in France harboured a mosaic *penA* gene with a point mutation at nucleotide position 501, resulting in an alanine to proline amino acid (AA) substitution in PBP2, rather than alanine to valine as previously suggested (Unemo et al., 2012). Similarly, the H041 strain isolated from Japan also harboured a mosaic PBP2 sequence, although with four additional AA substitutions, three of which never been described in *N. gonorrhoeae* or related *Neisseria* species before. In addition, the investigators suggest that as yet unidentified, marker (marker "X") may also have an important role to play in ceftriaxone resistance in H041 (Ohnishi et al., 2011a). Tomberg et al. have suggested that replacement of alanine at AA position 501 by bulkier amino acids results in interference with the R side chain of ceftriaxone, thereby leading to reduced binding to PBP2 (Tomberg et al., 2010). It is therefore evident that the role of *penA* mosaic sequences and point mutations, especially those that result in AA changes at position 501, is very important in ceftriaxone resistance.

Although penicillinase-producing *N. gonorrhoeae* (PPNG) (with a plasmid-mediated TEM-1 gene) are widely distributed, cephalosporinase production has not yet been identified.

37.10 The depleting antibiotic armamentarium

Acquisition of genetic markers of AMR often leads to a compromise in the replicative fitness of the recipient bacterial strain due to the resultant physiological changes. Therefore, once the antibiotic pressure has been relieved, susceptible wild-type strains can often outcompete the less-fit resistant strains, a phenomenon widely observed in different bacterial species. However, mutant bacteria may develop compensatory mutations restoring their fitness (Andersson, 2006) negating any selective advantage for the susceptible genotypes. Worryingly, gonococcal strains harbouring the topoisomerase mutation for ciprofloxacin resistance and the efflux pump Mtr-CDE mutation responsible for multiple antibiotic resistances demonstrate enhanced fitness when compared to the wild-type strains (Kunz et al., 2012; Warner et al., 2008). This evolutionary advantage means that the AMR machinery is retained even in the absence of antimicrobial pressure, leading to an accumulation of AMR mechanisms in the gonococcal population. Hence, monotherapy with successive anti-gonococcal agents has, over time, significantly contributed to the selection of the mutant strains in circulation around the world today. This further means that, once resistant strains have appeared, successful reintroduction of antibacterial agents such as fluoroquinolones after a period of disuse would not likely be possible.

Whilst the antibiotic armamentarium against gonorrhoea is getting depleted, there is a serious lack of new promising antibiotics in the pipeline. This has renewed interest in two injectable antibiotics with long serum half-lives that would potentially allow single dosing for gonorrhoea, the carbapenem ertapenem and the aminoglycoside gentamicin. Gentamicin has been successfully used as first-line syndromic management as part of the dual treatment strategy for 18 years in Malawi (Chisholm et al., 2010a; Livermore et al., 2004; Ross and Lewis, 2012), but it is unclear if this success could be transferred to developed countries with their more mobile populations and higher antimicrobial use.

Seven decades after the introduction of antibiotic therapy, gonorrhoea remains a significant sexually transmitted infection and a continued and escalating public health concern. Rising AMR in *N. gonorrhoeae* can be primarily attributed to the exceptional evolutionary adaptability of the gonococcus, and the unrestricted and inappropriate use of antibiotics in some settings (WHO, 2012). The unwarranted use of antimicrobials serves to increase the AMR in commensal flora, which can potentially be transferred to pathogenic organisms such as gonococci.

Syndromic management of gonorrhoea is a necessary practice in many parts of the world that lack the infrastructure for laboratory diagnosis, and in many circumstances it is in these settings that antimicrobial use is least regulated. It is from these regions with a limited capacity for culture and unregulated use of antibiotics that AMR data are often lacking. Ease of travel facilitates the rapid and widespread propagation of resistant infections. In developed settings, the increasing use of NAAT platforms restricts culture-based antibiotic resistance surveillance and epidemiological typing of gonococcus. It is therefore imperative that appropriate plans for monitoring the development and spread of resistant strains be urgently put in place to help prevent the emergence and spread of pandemic, multi-resistant gonorrhoea (Tapsall et al., 2009b; Kirkcaldy et al., 2011; Ohnishi et al., 2011a; Whiley et al., 2012; WHO, 2012).

The onus therefore is on developing public health infrastructure, and maintaining and enhancing vigilant surveillance systems globally, to facilitate prompt detection of emerging new resistant strains and to help contain their spread (Tapsall, 2006, 2009). The capacity of directed NAAT to identify resistance markers is evident, and whilst this cannot replace culture-based surveillance, it can enhance it, and this should be exploited (Goire et al., 2011, 2012, 2013). As the spectre of pandemic spread of untreatable *N. gonorrhoeae* looms large, it is imperative that effective measures, including antibiotic control and robust AMR surveillance strategies, be put into place internationally.

References

AGSP (Australian Gonococcal Surveillance Programme) 2012. Australian Gonococcal Surveillance Programme annual report, 2011. *Commun. Dis. Intell.*, 36, E166–E173.

AGSP (Australian Gonococcal Surveillance Programme) 2013. Australian Gonococcal Surveillance Programme annual report, 2012. *Commun. Dis. Intell.*, 37, E233–E239.

Akasaka, S., Muratani, T., Yamada, Y., Inatomi, H., Takahashi, K., and Matsumoto, T. 2001. Emergence of cephem- and aztreonam-high-resistant *Neisseria gonorrhoeae* that does not produce beta-lactamase. *J. Infect. Dis. Chemother.*, 7, 49–50.

Ameyama, S., Onodera, S., Takahata, M., Minami, S., Maki, N., Endo, K., Goto, H., Suzuki, H., and Oishi, Y. 2002. Mosaic-like structure of penicillin-binding protein 2 gene (penA) in clinical isolates of *Neisseria gonorrhoeae* with reduced susceptibility to cefixime. *Antimicrob. Agents Chemother.*, 46, 3744–3749.

Aminov, R.I., and Mackie, R.I. 2007. Evolution and ecology of antibiotic resistance genes. *FEMS Microbiol. Lett.*, 271, 147–161.

Anderson, M.T., and Seifert, H.S. 2011. *Neisseria gonorrhoeae* and humans perform an evolutionary LINE dance. *Mob. Genet. Elements*, 1, 85–87.

Andersson, D.I. 2006. The biological cost of mutational antibiotic resistance: Any practical conclusions? *Curr. Opin. Microbiol.*, 9, 461–465.

Ashford, W., Adams, H.U., Johnson, S., Thornsberry, C., Potts, D., English, J.C., Biddle, J., and Jaffe, H. 1981. Spectinomycin-resistant penicillinase-producing *Neisseria gonorrhoeae*. *Lancet*, 318, 1035–1037.

Bachmann, L.H., Desmond, R.A., Stephens, J., Hughes, A., and Hook III, E.W. 2002. Duration of persistence of gonococcal DNA detected by ligase chain reaction in men and women following recommended therapy for uncomplicated gonorrhea. *J. Clin. Microbiol.*, 40, 3596–3601.

Barbour, A.G. 1981. Properties of penicillin-binding proteins in *Neisseria gonorrhoeae*. *Antimicrob. Agents Chemother.*, 19, 316–322.

Baron, S., editor. 1996. *Medical Microbiology*, 4th ed. Galveston, TX: Univesity of Texas Medical Branch at Galveston.

Barry, P.M., and Klausner, J.D. 2009. The use of cephalosporins for *gonorrhea*: The impending problem of resistance. *Expert Opin. Pharmacother.*, 10, 555–577.

Belland, R.J., Morrison, S.G., Ison, C., and Huang, W.M. 1994. *Neisseria gonorrhoeae* acquires mutations in analogous regions of gyrA and parC in fluoroquinolone-resistant isolates. *Mol. Microbiol.*, 14, 371–380.

Bignell, C., and Fitzgerald, M. 2011. UK national guideline for the management of *gonorrhoea* in adults, 2011. *Int. J. STD AIDS*, 22, 541–547.

Bihlmater, A., Römling, U., Meyer, T.F., Tümmler, B., and Gibbs, C.P. 1991. Physical and genetic map of the *Neisseria gonorrhoeae* strain MS11-N198 chromosome. *Mol. Microbiol.*, 5, 2529–2539.

Blake, F.G. 1940. Chemotherapy with the sulfonamide derivatives: General principles. *Bull. N. Y. Acad. Med.*, 6, 22–33.

Bonin, P., Tanino, T.T., and Handsfield, H.H. 1984. Isolation of *Neisseria gonorrhoeae* on selective and nonselective media in a sexually transmitted disease clinic. *J. Clin. Microbiol.*, 19, 218–220.

Bruisten, S.M., Noordhoek, G.T., van den Brule, A.J., Duim, B., Boel, C.H., El-Faouzi, K., du Maine, R., Mulder, S., Luijt, D., and Schirm, J. 2004. Multicenter validation of the cppB gene as a PCR target for detection of *Neisseria gonorrhoeae*. *J. Clin. Microbiol.*, 42, 4332–4334.

Cannon, J.G., and Sparling, P.F. 1984. The genetics of the gonococcus. *Annu. Rev. Microbiol.*, 38, 111–33.

CDC (Centers for Disease Control and Prevention) 1976. Penicillinase-producing *Neisseria gonorrhoeae*. *MMWR Morb. Mortal. Weekly Rep.*, 33, 408–410.

CDC (Centers for Disease Control and Prevention) 2008. *Gonorrhea Laboratory Information Identification of N. gonorrhoeae and Related Species: Pathogenicity of Neisseria and Related Species of Human Origin* [Online]. Centers for Disease Control and Prevention. Available from: http://www.cdc.gov/std/gonorrhea/lab/pathogenicity.htm. Accessed September 5, 2013.

CDC (Centers for Disease Control and Prevention) 2011a. *2010 Sexually Transmitted Diseases Surveillance* [Online]. Centers for Disease Control and Prevention. Available from: http://www.cdc.gov/std/stats10/gonorrhea.htm. Accessed March 20, 2012.

CDC (Centers for Disease Control and Prevention) 2011b. *Sexually Transmitted Infections Treatment Guidelines 2010 – Gonococcal Infections* [Online]. Centers for Disease Control and Prevention. Available from: http://www.cdc.gov/std/treatment/2010/gonococcal-infections.htm. Accessed March 21, 2012.

Chisholm, S.A., Neal, T.J., Alawattegama, A.B., Birley, H.D.L., Howe, R.A., and Ison, C.A. 2009. Emergence of high-level azithromycin resistance in *Neisseria gonorrhoeae* in England and Wales. *J. Antimicrob. Chemother.*, 64, 353–358.

Chisholm, S.A., Dave, J., and Ison, C.A. 2010a. High-level azithromycin resistance occurs in *Neisseria gonorrhoeae* as a result of a single point mutation in the 23S rRNA genes. *Antimicrob. Agents Chemother.*, 54, 3812–3816.

Chisholm, S.A., Mouton, J.W., Lewis, D.A., Nichols, T., Ison, C.A., and Livermore, D.M. 2010b. Cephalosporin MIC creep among gonococci: Time for a pharmacodynamic rethink? *J. Antimicrob. Chemother.*, 65, 2141–2148.

Chopra, I., and Roberts, M. 2001. Tetracycline antibiotics: mode of action, applications, molecular biology, and epidemiology of bacterial resistance. *Microbiol. Mol. Biol. Rev.*, 65(2), 232–260.

Criss, A.K., and Seifert, H.S. 2008. *Neisseria gonorrhoeae* suppresses the oxidative burst of human polymorphonuclear leukocytes. *Cell. Microbiol.*, 10, 2257–2270.

Dan, M. 2004. The use of fluoroquinolones in *gonorrhoea*: The increasing problem of resistance. *Expert Opin. Pharmacother.*, 5, 829–854.

Deakin, R., Wortman, M., and Laforce, R. 1941. Sulfonamide therapy in male gonorrhea. *Am. J. Public Health Nations Health*, 31, 682–686.

Dillard, J.P., and Seifert, H.S. 2001. A variable genetic island specific for *Neisseria gonorrhoeae* is involved in providing DNA for natural transformation and is found more often in disseminated infection isolates. *Mol. Microbiol.*, 41, 263–277.

Dillon, J.-a.R. 2011. Sustainable antimicrobial surveillance programs essential for controlling *Neisseria gonorrhoeae* superbug. *Sex. Transm. Dis.*, 38, 899–901. doi:10.1097/OLQ .0b013e318232459b

Dillon, J.-a.R., Rubabaza, J.-P.A., Benzaken, A.S., Sardinha, J.C.G., Li, H., Bandera, M.G.C., and Filho, E.D.S.F. 2001. Reduced susceptibility to azithromycin and high percentages of penicillin and tetracycline resistance in *Neisseria gonorrhoeae* isolates from Manaus, Brazil, 1998. *Sex. Transm. Dis.*, 28, 521–526.

Dougherty, T.J. 1985. Involvement of a change in penicillin target and peptidoglycan structure in low-level resistance to beta-lactam antibiotics in *Neisseria gonorrhoeae*. *Antimicrob. Agents Chemother.*, 28, 90–95.

Easmon, C.S., Forster, G.E., Walker, G.D., Forster, G.E., Walker, G.D., Ison, C.A., Harris, J.R., and Munday, P.E. 1984. Spectinomycin as initial treatment for *gonorrhoea*. *Br. Med. J.*, 289, 1032–1034.

ECDC (European Centre for Disease Prevention and Control) 2011. *Annual Epidemiological Report 2011 – Reporting on 2009 Surveillance Data and 2010 Epidemic Intelligence Data* [Online]. European Centre for Disease Prevention and Control. Available from: http://ecdc.europa.eu/en/publications/Publications/1111_SUR_Annual_Epidemiological_Report_on_Communicable_Diseases_in_Europe.pdf. Accessed March 20, 2012.

Eichmann, A., Weidmann, G., and Havas, L. 1981. One-dose treatment of acute uncomplicated *gonorrhoea* of male patients with ceftriaxone Ro 13–9904, a new parenteral cephalosporin. *Chemotherapy*, 27, 62–69.

EUCAST (European Committee on Antimicrobial Susceptibility Testing) 2012. *European Committee on Antimicrobial Susceptibility Testing, Clinical Breakpoints 2012* [Online]. European Society of Clinical Microbiology and Infectious Diseases. Available from: http://www.eucast.org/clinical_breakpoints/. Accessed April 1, 2012.

Farrell, D.J. 1999. Evaluation of AMPLICOR *Neisseria gonorrhoeae* PCR using cppB Nested PCR and 16S rRNA PCR. *J. Clin. Microbiol.*, 37, 386–390.

Feavers, I.M., and Maiden, M.C.J. 1998. A gonococcal porA pseudogene: Implications for understanding the evolution and pathogenicity of *Neisseria gonorrhoeae*. *Mol. Microbiol.*, 30, 647–656.

Fekete, T. 1993. Antimicrobial susceptibility testing of *Neisseria gonorrhoeae* and implications for epidemiology and therapy. *Clin. Microbiol. Rev.*, 6, 22–33.

Folster, J.P., Johnson, P.J., Jackson, L., Dhulipali, V., Dyer, D.W., and Shafer, W.M. 2009. MtrR modulates rpoH expression and levels of antimicrobial resistance in *Neisseria gonorrhoeae*. *J. Bacteriol.*, 191, 287–297.

Galarza, P.G., Alcalá, B., Salcedo, C., Canigia, L.F., Buscemi, L., Pagano, I., Oviedo, C., and Vazquez, J.A. 2009. Emergence of high level azithromycin-resistant Neisseria gonorrhoeae strain isolated in Argentina. *Sex. Transm. Dis.*, 36, 787–788.

Galimand, M., Gerbaud, G., and Courvalin, P. 2000. Spectinomycin resistance in *Neisseria* spp. due to mutations in 16S rRNA. *Antimicrob. Agents Chemother.*, 44, 1365–1366.

Gaydos, C.A. 2005. Nucleic acid amplification tests for gonorrhea and chlamydia: Practice and applications. *Infect. Dis. Clin. North Am.*, 19, 367–386.

Goire, N., Nissen, M.D., Lecornec, G.M., Sloots, T.P., and Whiley, D.M. 2008. A duplex *Neisseria gonorrhoeae* real-time polymerase chain reaction assay targeting the gonococcal porA pseudogene and multicopy opa genes. *Diagn. Microbiol. Infect. Dis.*, 61, 6–12.

Goire, N., Freeman, K., Tapsall, J.W., Lambert, S.B., Nissen, M.D., Sloots, T.P., and Whiley, D.M. 2011. Enhancing gonococcal antimicrobial resistance surveillance: A real-time PCR assay for detection of penicillinase-producing *Neisseria gonorrhoeae* by use of non-cultured clinical samples. *J. Clin. Microbiol.*, 2, 513–518.

Goire, N., Ohnishi, M., Limnios, A.E., Lahra, M.M., Lambert, S.B., Nimmo, G.R., Nissen, M.D., Sloots, T.P., and Whiley, D.M. 2012. Enhanced gonococcal antimicrobial surveillance in the era of ceftriaxone resistance: A real-time PCR assay for direct detection of the *Neisseria gonorrhoeae* H041 strain. *J. Antimicrob. Chemother.*, 67, 902–905.

Goire, N., Lahra, M.M., Ohnishi, M., Hogan, T., Limnios, A.E., Nissen, M.D., Sloots, T.P., and Whiley, D.M. 2013. Polymerase chain reaction-based screening for the ceftriaxone-resistant *Neisseria gonorrhoeae* F89 strain. *EuroSurveillance*, 18. Available from: http://www.eurosurveillance.org/ViewArticle.aspx?ArticleId=20444

Goodyear-Smith, F. 2007. What is the evidence for non-sexual transmission of gonorrhoea in children after the neonatal period? A systematic review. *J. Forensic Leg. Med.*, 148, 489–502.

Handsfield, H.H., Dalu, Z.A., Martin, D.H., Douglas, J.M., Mccarty, J.M., and Schlossberg, D. (Azithromycin Gonorrhea Study Group). 1994. Multicenter trial of single-dose azithromycin vs. ceftriaxone in the treatment of uncomplicated gonorrhea. *Sex. Transm. Dis.*, 21, 107–111.

Harkness, A. 1948. The pathology of gonorrhoea. *Br. J. Vener. Dis.*, 24, 137–147.

Hauck, C.R., and Meyer, T.F. 2003. 'Small' talk: Opa proteins as mediators of *Neisseria*–host-cell communication. *Curr. Opin. Microbiol.*, 6, 43–49.

Herring, A., Ballard, R., Mabey, D., and Peeling, R.W. 2006. Evaluation of rapid diagnostic tests: Chlamydia and gonorrhoea. *Nat. Rev. Microbiol.*, 4, S41–S48.

Higashi, D.L., Lee, S.W., Snyder, A., Weyand, N.J., Bakke, A., and So, M. 2007. Dynamics of *Neisseria gonorrhoeae* attachment: Microcolony development, cortical plaque formation, and cytoprotection. *Infect. Immun.*, 75, 4743–4753.

Holder, N.A. 2008. Gonococcal infections. *Pediatr. Rev.*, 29, 228–234.

Holmes, K.K., Johnson, D.W., and Trostle, T.J. 1970. An estimate of the risk of men acquiring gonorrhea by sexual contact with females. *Am. J. Epidemiol.*, 91, 170–174.

Hu, M., Nandi, S., Davies, C., and Nicholas, R.A. 2005. High-level chromosomally mediated tetracycline resistance in *Neisseria gonorrhoeae* results from a point mutation in the rpsJ gene encoding ribosomal protein S10 in combination with the mtrR and penB resistance determinants. *Antimicrob. Agents Chemother.*, 49, 4327–4334.

Ison, C.A., Mouton, J.W., Jones, K., Fenton, K.A., and Livermore, D.M. 2004. Which cephalosporin for gonorrhoea? *Sex. Transm. Infect.*, 80, 386–388.

Ito, M., Yasuda, M., Yokoi, S., Ito, S.-I., Takahashi, Y., Ishihara, S., Maeda, S.-I., and Deguchi, T. 2004. Remarkable increase in Central Japan in 2001–2002 of *Neisseria gonorrhoeae* isolates with decreased susceptibility to penicillin, tetracycline, oral cephalosporins, and fluoroquinolones. *Antimicrob. Agents Chemother.*, 48, 3185–3187.

Ito, M., Deguchi, T., Mizutani, K.-S., Yasuda, M., Yokoi, S., Ito, S.-I., Takahashi, Y., Ishihara, S., Kawamura, Y., and Ezaki, T. 2005. Emergence and spread of *Neisseria gonorrhoeae* clinical isolates harboring mosaic-like structure of penicillin-binding protein 2 in central Japan. *Antimicrob. Agents Chemother.*, 49, 137–143.

Jakopanec, I., Borgen, K., and Aavitsland, P. 2009. The epidemiology of gonorrhoea in Norway, 1993–2007: Past victories, future challenges. *BMC Infect. Dis.*, 9, 33.

Jones, C.A., Knaup, R.C., Hayes, M., and Stoner, B.P. 2000. Urine screening for gonococcal and chlamydial infections at community-based organizations in a high-morbidity area. *Sex. Transm. Dis.*, 27, 146–151.

Katz, A.R., Komeya, A.Y., Soge, O.O. Katz, A.R., Komeya, A.Y., Soge, O.O., Kiaha, M.I., Lee, M.V., Wasserman, G.M., Maningas, E.V., Whelen, A.C., Kirkcaldy, R.D., Shapiro, S.J., Bolan, G.A., and Holmes, K.K. 2012. *Neisseria gonorrhoeae* with high-level resistance to azithromycin: Case report of the first isolate identified in the United States. *Clin. Infect. Dis.*, 54, 841–843.

Khan, M.Y., Siddiqui, Y., and Gruninger, R.P. 1982. Comparative in-vitro activity of selected new beta-lactam antimicrobials against *Neisseria gonorrhoeae*. *Br. J. Vener. Dis.*, 58, 228–230.

Kirkcaldy, R., Ballard, R., and Dowell, D. 2011. Gonococcal resistance: Are cephalosporins next? *Curr. Infect. Dis. Rep.*, 13, 196–204.

Kline, K.A., Sechman, E.V., Skaar, E.P., and Seifert, H.S. 2004. Recombination, repair and replication in the pathogenic *Neisseriae*: The 3 R's of molecular genetics of two human-specific bacterial pathogens. *Mol. Microbiol.*, 51, 297–297.

Koumans, E.H., Johnson, R.E., Knapp, J.S., and St. Louis, M.E. 1998. Laboratory testing for *Neisseria gonorrhoeae* by recently introduced nonculture tests: A performance review with clinical and public health considerations. *Clin. Infect. Dis.*, 27, 1171–1180.

Kraus, S., and Glassman, L. 1974. Scanning electron microscope study of *Neisseria gonorrhoeae*. *Appl. Microbiol.*, 27, 584–592.

Kunz, A.N., Begum, A.A., Wu, H., D'ambrozio, J.A., Robinson, J.M., Shafer, W.M., Bash, M.C., and Jerse, A.E. 2012. Impact of fluoroquinolone resistance mutations on gonococcal fitness and in vivo selection for compensatory mutations. *J. Infect. Dis.*, 205, 1821–1829.

Lee, K.C. 2012. The clap heard around the world. *Arch. Dermatol.*, 148, 223.

Lee, B.C., and Bryan, L.E. 1989. Identification and comparative analysis of the lactoferrin and transferrin receptors among clinical isolates of gonococci. *J. Med. Microbiol.*, 28, 199–204.

Lee, S.-G., Lee, H., Jeong, S.H., Yong, D., Chung, G.T., Lee, Y.S., Chong, Y., and Lee, K. 2010. Various penA mutations together with mtrR, porB and ponA mutations in Neisseria gonorrhoeae isolates with reduced susceptibility to cefixime or ceftriaxone. *J. Antimicrob. Chemother.*, 65, 669–675.

Lewis, D.A. 2010. The Gonococcus fights back: Is this time a knock out? *Sex. Transm. Infect.*, 86, 415–421.

Liao, M., Gu, W.-M., Yang, Y., and Dillon, J.-a.R. 2011. Analysis of mutations in multiple loci of *Neisseria gonorrhoeae* isolates reveals effects of PIB, PBP2 and MtrR on reduced susceptibility to ceftriaxone. *J. Antimicrob. Chemother.*, 66, 1016–1023.

Lindberg, R., Fredlund, H., Nicholas, R., and Unemo, M. 2007. *Neisseria gonorrhoeae* isolates with reduced susceptibility to cefixime and ceftriaxone: Association with genetic polymorphisms in penA, mtrR, porB1b, and ponA. *Antimicrob. Agents Chemother.*, 51, 2117–2122.

Linz, B., Schenker, M., Zhu, P., and Achtman, M. 2000. Frequent interspecific genetic exchange between commensal Neisseriae and *Neisseria meningitidis*. *Mol. Microbiol.*, 36, 1049–1058.

Livermore, D.M., Alexander, S., Marsden, B., James, D., Warner, M., Rudd, E., and Fenton, K. 2004. Activity of ertapenem against *Neisseria gonorrhoeae*. *J. Antimicrob. Chemother.*, 54, 280–281.

Lum, G., Freeman, K., Nguyen, N.L., Limnios, E.A., Tabrizi, S.N., Carter, I., Chambers, I.W., Whiley, D.M., Sloots, T.P., Garland, S.M., and Tapsall, J.W. 2005. A cluster of culture positive gonococcal infections but with false negative cppB gene based PCR. *Sex. Transm. Infect.*, 81, 400–402.

Mahony, J., Chong, S., Jang, D., Luinstra, K., Faught, M., Dalby, D., Sellors, J., and Chernesky, M. 1998. Urine specimens from pregnant and nonpregnant women inhibitory to amplification of chlamydia trachomatis nucleic acid by PCR, ligase chain reaction, and transcription-mediated amplification: Identification of urinary substances associated with inhibition and removal of inhibitory activity. *J. Clin. Microbiol.*, 36, 3122–3126.

Marmell, M., and Prigot, A. 1955. Intramuscular tetracycline (achromycin) in the treatment of acute gonorrhoea in the male*. *Br. J. Vener. Dis.* 31, 188.

Martin, I.M.C., Ison, C.A., Aanensen, D.M., Fenton, K.A., and Spratt, B.G. 2005. Changing epidemiologic profile of quinolone-resistant *Neisseria gonorrhoeae* in London. *J. Infect. Dis.*, 192, 1191–1195.

Merz, A.J., and So, M. 2000. Interactions of pathogenic *Neisseriae* with epithelial cell membranes. *Annu. Rev. Cell Dev. Biol.*, 16, 423–457.

Morse, S.A., Johnson, S.R., Biddle, J.W., and Roberts, M.C. 1986. High-level tetracycline resistance in *Neisseria gonorrhoeae* is result of acquisition of streptococcal tetM determinant. *Antimicrob. Agents Chemother.*, 30, 664–670.

Mosleh, I.M., Boxberger, H.J., Sessler, M.J., and Meyer, T.F. 1997. Experimental infection of native human ureteral tissue with *Neisseria gonorrhoeae*: Adhesion, invasion, intracellular fate, exocytosis, and passage through a stratified epithelium. *Infect. Immun.*, 65, 3391–3398.

Murphy, G.L., and Cannon, J.G. 1988. Genetics of surface protein variation in *Neisseria gonorrhoeae*. *BioEssays*, 9, 7–11.

Nakayama, S.-I., Tribuddharat, C., Prombhul, S., Shimuta, K., Srifuengfung, S., Unemo, M., and Ohnishi, M. 2012. Molecular analyses of TEM genes and their corresponding penicillinase-producing *Neisseria gonorrhoeae* isolates in Bangkok, Thailand. *Antimicrob. Agents Chemother.*, 56, 916–920.

Ng, L.-K., Martin, I., Liu, G., and Bryden, L. 2002. Mutation in 23S rRNA Associated with Macrolide Resistance in Neisseria gonorrhoeae. *Antimicrob. Agents Chemother.*, 46, 3020–3025.

NNDSS 2012. *National Notifiable DIseases Surveillance System, Australian Government Department of Health and Ageing* – Notification rates 2012. Available from: http://www9.health.gov.au/cda/source/rpt_4.cfm. Accessed September 9, 2013.

O'callaghan, C.H. 1979. Description and classification of the newer cephalosporins and their relationships with the established compounds. *J. Antimicrob. Chemother.*, 5, 635–671.

Ohnishi, M., Ono, E., Shimuta, K., Watanabe, H., and Okamura, N. 2010a. Identification of TEM-135 β-lactamase in penicillinase-producing *Neisseria gonorrhoeae* strains in Japan. *Antimicrob. Agents Chemother.*, 54, 3021–3023.

Ohnishi, M., Watanabe, Y., Ono, E., Takahashi, C., Oya, H., Kuroki, T., Shimuta, K., Okazaki, N., Nakayama, S.-I., and Watanabe, H. 2010b. Spread of a chromosomal cefixime-resistant penA gene among different *Neisseria gonorrhoeae* lineages. *Antimicrob. Agents Chemother.*, 54, 1060–1067.

Ohnishi, M., Saika, T., Hoshina, S., Iwasaku, K., Nakayama, S., Watanabe, H., and Kitawaki, J. 2011a. Ceftriaxone-resistant *Neisseria gonorrhoeae*, Japan. *Emerg. Infect. Dis.*, 17, 148–149.

Ohnishi, M., Golparian, D., Shimuta, K., Saika, T., Hoshina, S., Iwasaku, K., Nakayama, S.-I., Kitawaki, J., and Unemo, M. 2011b. Is *Neisseria gonorrhoeae* initiating a future era of untreatable gonorrhea? Detailed characterization of the first strain with high-level resistance to ceftriaxone. *Antimicrob. Agents Chemother.*, 55, 3538–3545.

Olesky, M., Zhao, S., Rosenberg, R.L., and Nicholas, R.A. 2006. Porin-mediated antibiotic resistance in *Neisseria gonorrhoeae*:

Ion, solute, and antibiotic permeation through PIB proteins with penB mutations. *J. Bacteriol.*, 188, 2300–2308.

Orellana, M.A., Gómez-Lus, M.L., and Lora, D. 2012. Sensitivity of Gram stain in the diagnosis of urethritis in men. *Sex. Transm. Infect.*, 88, 284–287.

Palmer, H.M., Young, H., Winter, A., and Dave, J. 2008. Emergence and spread of azithromycin-resistant *Neisseria gonorrhoeae* in Scotland. *J. Antimicrob. Chemother.*, 62, 490–494.

Papasian, C.J., Bartholomew, W.R., and Amsterdam, D. 1984. Modified enzyme immunoassay for detecting *Neisseria gonorrhoeae* antigens. *J. Clin. Microbiol.*, 20, 641–643.

Patrone, J., and Stein, D. 2007. Effect of gonococcal lipooligosaccharide variation on human monocytic cytokine profile. *BMC Microbiol.*, 7, 7.

Paz, H.D.L., Cooke, S.J., and Heckels, J.E. 1995. Effect of sialylation of lipopolysaccharide of *Neisseria gonorrhoeae* on recognition and complement-mediated killing by monoclonal antibodies directed against different outer-membrane antigens. *Microbiology*, 141, 913–920.

Percival, A., Rowlands, J., Corkill, J.E., Alergant C.D.,Arya, O.P., Rees, E., and Annels, E.H. 1976. Penicillinase-producing gonococci in Liverpool. *Lancet*, 2, 1379–1382.

Rajan, V.S., Sng, E.H., Thirumoorthy, T., and Goh, C.L. 1982. Ceftriaxone in the treatment of ordinary and penicillinase-producing strains of *Neisseria gonorrhoeae*. *Br. J. Vener. Dis.*, 58, 314–316.

Ribeiro, J., and Mathieson, S.I. 1975. Spectinomycin in gonorrhoea. *Br. J. Vener. Dis.*, 51, 272–273.

Richmond, P. 1954. American attitude towards germ theory of disease (1860–1880). *J. Hist. Med.*, 9, 58–84.

Roberts, M.C., Chung, W.O., Roe, D., Xia, M., Marquez, C., Borthagaray, G., Whittington, W.L., and Holmes, K.K. 1999. Erythromycin-resistant *Neisseria gonorrhoeae* and oral commensal *Neisseria* spp. carry known rRNA methylase genes. *Antimicrob. Agents Chemother.*, 43, 1367–1372.

Ross, J.D.C., and Lewis, D.A. 2012. Cephalosporin resistant *Neisseria gonorrhoeae*: Time to consider gentamicin? *Sex. Transm. Infect.*, 88, 6–8.

Rouquette-Loughlin, C., Dunham, S.A., Kuhn, M., Balthazar, J.T., and Shafer, W.M. 2003. The NorM efflux pump of *Neisseria gonorrhoeae* and *Neisseria meningitidis* recognizes antimicrobial cationic compounds. *J. Bacteriol.*, 185, 1101–1106.

Rouquette-Loughlin, C.E., Balthazar, J.T., and Shafer, W.M. 2005. *Characterization of the MacA-MacB efflux system in Neisseria gonorrhoeae*. Oxford University Press.

Shafer, W.M., and Folster, J.P. 2006. Towards an understanding of chromosomally mediated penicillin resistance in *Neisseria gonorrhoeae*: Evidence for a porin–efflux pump collaboration. *J. Bacteriol.*, 188, 2297–2299.

Siegel, M.S., Thornsberry, C., Biddle, J.W., O'mara, P.R., Perine, P.L., and Wiesner, P.J. 1978. Penicillinase-producing *Neisseria gonorrhoeae*: Results of surveillance in the United States. *J. Infect. Dis.*, 137, 170–175.

Sköld, O. 2001. Resistance to trimethoprim and sulfonamides. *Vet. Res.*, 32, 261–273.

Smith, D.W., Tapsall, J.W., and Lum, G. 2005. Guidelines for the use and interpretation of nucleic acid detection tests for *Neisseria gonorrhoeae* in Australia: A position paper on behalf of the Public Health Laboratory Network. *Commun. Dis. Intell.*, 29, 358–365.

Sparling, P.F., and Handsfield, H.H. 2000. In: Mandell, G.L., Bennett, J.E. and Dolin R., editors. *Principles and Practice of Infectious Disease* (5 ed., pp. 2242–2258). United States of America: Churchill Livingstone.

Tabrizi, S.N., Unemo, M., Limnios, A.E., Hogan, T.R., Hjelmevoll, S.-O., Garland, S.M., and Tapsall, J. 2011. Evaluation of six commercial nucleic acid amplification tests for detection of *Neisseria gonorrhoeae* and other *Neisseria* species. *J. Clin. Microbiol.*, 49, 3610–3615.

Tanaka, M., Nakayama, H., Tunoe, H., Egashira, T., Kanayama, A., Saika, T., Kobayashi, I., and Naito, S. 2002. A remarkable reduction in the susceptibility of *Neisseria gonorrhoeae* isolates to cephems and the selection of antibiotic regimens for the single-dose treatment of gonococcal infection in Japan. *J. Infect. Chemother.*, 8, 81–86.

Tanaka, M., Nakayama, H., Huruya, K., Konomi, I., Irie, S., Kanayama, A., Saika, T., and Kobayashi, I. 2006. Analysis of mutations within multiple genes associated with resistance in a clinical isolate of *Neisseria gonorrhoeae* with reduced ceftriaxone susceptibility that shows a multidrug-resistant phenotype. *Int. J. Antimicrob. Agents*, 27, 20–26.

Tapsall, J.W. 2001. Antimicrobial resistance in *Neisseria gonorrhoeae*. In: (ed.). *Department of Communicable Disease Surveillance and Response*, WHO.

Tapsall, J. 2006. Antibiotic resistance in *Neisseria gonorrhoeae* is diminishing available treatment options for gonorrhea: Some possible remedies. *Expert Rev. Anti. Infect. Ther.*, 4, 619–628.

Tapsall, J.W. 2009. *Neisseria gonorrhoeae* and emerging resistance to extended spectrum cephalosporins. *Curr. Opin. Infect. Dis.*, 22, 87–91. doi:10.1097/QCO.0b013e328320a836.

Tapsall, J., Read, P., Carmody, C., Bourne, C., Ray, S., Limnios, A., Sloots, T., and Whiley, D. 2009a. Two cases of failed ceftriaxone treatment in pharyngeal gonorrhea verified by molecular microbiological methods. *J. Med. Microbiol.*, 58, 683–687.

Tapsall, J.W., Ndowa, F., Lewis, D.A., and Unemo, M. 2009a. Meeting the public health challenge of multidrug- and extensively drug-resistant *Neisseria gonorrhoeae*. *Expert Rev. Anti. Infect. Ther.*, 7, 821–834.

Tobiason, D.M., and Seifert, H.S. 2006. The obligate human pathogen, *Neisseria gonorrhoeae*, is polyploid. *PLoS Biol.*, 4, e185.

Tomberg, J., Unemo, M., Davies, C., and Nicholas, R.A. 2010. Molecular and structural analysis of mosaic variants of penicillin-binding protein 2 conferring decreased susceptibility to expanded-spectrum cephalosporins in *Neisseria gonorrhoeae*: Role of epistatic mutations. *Biochemistry*, 49, 8062–8070.

Unemo, M., and Dillon, J.-a.R. 2011. Review and international recommendation of methods for typing *Neisseria gonorrhoeae* isolates and their implications for improved knowledge of gonococcal epidemiology, treatment, and biology. *Clin. Microbiol. Rev.*, 24, 447–458.

Unemo, M., Golparian, D., Syversen, G., Vestrheim, D., and Moi, H. 2010. Two cases of verified clinical failures using internationally recommended first-line cefixime for gonorrhoea treatment, Norway, 2010. *Euro Surveillance* [Online], 15. Available from: http://www.eurosurveillance.eu/images/dynamic/EE/V15N47/art19721.pdf

Unemo, M., Golparian, D., Stary, A., and Eigentler, A. 2011. First *Neisseria gonorrhoeae* strain with resistance to cefixime causing gonorrhoea treatment failure in Austria, 2011. *Euro Surveillance* [Online], 16. Available from: http://www.eurosurveillance.eu/images/dynamic/EE/V16N43/art19998.pdf

Unemo, M., Golparian, D., Nicholas, R., Ohnishi, M., Gallay, A., and Sednaoui, P. 2012. High-level cefixime- and ceftriaxone-resistant *Neisseria gonorrhoeae* in France: Novel penA mosaic allele in a successful international clone causes treatment failure. *Antimicrob. Agents Chemother.*, 56, 1273–1280.

Veal, W.L., Nicholas, R.A., and Shafer, W.M. 2002. Overexpression of the MtrC-MtrD-MtrE efflux pump due to an mtrR mutation is required for chromosomally mediated penicillin resistance in *Neisseria gonorrhoeae*. *J. Bacteriol.*, 184, 5619–5624.

Versalovic, J., Carroll, K., Funke, G., Jorgensen, J., Landry, M., and Warncock, D., editors. 2011. *Manual of Clinical Microbiology*. Washington, DC: American Society for Microbiology.

Virji, M. 2009. Pathogenic *Neisseriae*: Surface modulation, pathogenesis and infection control. *Nat. Rev. Micro.*, 7, 274–286.

Warner, D.M., Shafer, W.M., and Jerse, A.E. 2008. Clinically relevant mutations that cause derepression of the *Neisseria gonorrhoeae* MtrC–MtrD–MtrE efflux pump system confer different levels of antimicrobial resistance and in vivo fitness. *Mol. Microbiol.*, 70, 462–478.

Weinstock, H., and Workowski, K.A. 2009. Pharyngeal gonorrhea: An important reservoir of infection? *Clin. Infect. Dis.*, 49, 1798–1800.

Whiley, D.M., Tapsall, J.W., and Sloots, T.P. 2006. Nucleic acid amplification testing for *Neisseria gonorrhoeae*: An ongoing challenge. *J. Mol. Diagn.*, 8, 3–15.

Whiley, D., Bates, J., Limnios, A., Nissen, M.D., Tapsall, J., and Sloots, T.P. 2007a. Use of a novel screening PCR indicates presence of *Neisseria gonorrhoeae* isolates with a mosaic penA gene sequence in Australia. *Pathology*, 39, 445–446. doi:10.1080/00313020701444515.

Whiley, D.M., Limnios, E.A., Ray, S., Sloots, T.P., and Tapsall, J.W. 2007b. Diversity of penA alterations and subtypes in *Neisseria gonorrhoeae* strains from Sydney, Australia, that are less susceptible to ceftriaxone. *Antimicrob. Agents Chemother.*, 51, 3111–3116.

Whiley, D.M., Lambert, S.B., Bialasiewicz, S., Goire, N., Nissen, M.D., and Sloots, T.P. 2008. False-negative results in nucleic acid amplification tests – Do we need to routinely use two genetic targets in all assays to overcome problems caused by sequence variation? *Crit. Rev. Microbiol.*, 34, 71–76.

Whiley, D.M., Jacobsson, S., Tapsall, J.W., Nissen, M.D., Sloots, T.P., and Unemo, M. 2010a. Alterations of the pilQ gene in *Neisseria gonorrhoeae* are unlikely contributors to decreased susceptibility to ceftriaxone and cefixime in clinical gonococcal strains. *J. Antimicrob. Chemother.*, 65, 2543–2547.

Whiley, D.M., Goire, N., Lambert, S.B., Ray, S., Limnios, E.A., Nissen, M.D., Sloots, T.P., and Tapsall, J.W. 2010b. Reduced susceptibility to ceftriaxone in *Neisseria gonorrhoeae* is associated with mutations G542S, P551S and P551L in the gonococcal penicillin-binding protein 2. *J. Antimicrob. Chemother.*, 65, 1615–1618.

Whiley, D. M., Limnios, E., Moon, N., Gehrig, N., Goire, N., Hogan, T., Lam, A., Jacob, K., Lambert, S., Nissen, M., and Sloots, T.P. 2011. False-negative results using *Neisseria gonorrhoeae* porA pseudogene PCR – A clinical gonococcal isolate with an *N. meningitidis* porA sequence, Australia, March 2011. *Euro Surveillance* [Online], 16. Available from: http://www.eurosurveillance.org/ViewArticle.aspx?ArticleId=19874

Whiley, D.M., Goire, N., Lahra, M.M., Donovan, B., Limnios, A.E., Nissen, M.D., and Sloots, T.P. 2012. The ticking time bomb: Escalating antibiotic resistance in *Neisseria gonorrhoeae* is a public health disaster in waiting. *J. Antimicrob. Chemother.*, 67, 2059–2061.

WHO (World Health Organization). 2007a. *Global Strategy for the Prevention and Control of Sexually Transmitted Infections: 2006–2015: Breaking the Chain of Transmission.* World Health Organization. Available from: http://whqlibdoc.who .int/publications/2007/9789241563475_eng.pdf. Accessed September 6, 2013.

WHO (World Health Organization). 2007b. *Training Modules for the Syndromic Management of Sexually Transmitted Infections,* 2nd ed. World Health Organization. Available from: http://www.who.int/reproductivehealth/publications/rtis /9789241593407/en/index.html. Accessed September 9, 2013.

WHO (World Health Organization). 2011a. *Sexually Transmitted Infections, Fact Sheet No. 110.* World Health Organization. Available from: http://www.who.int/mediacentre/ factsheets/fs110/en/. Accessed March 20, 2012.

WHO (World Health Organization). 2011b. *Prevention and Treatment of HIV and Other Sexually Transmitted Infections Among Men Who Have Sex With Men and Transgender People. Recommendations for a Public Health Approach.* Geneva: WHO Document Production Services. Available from: http://apps.who .int/iris/bitstream/10665/44619/1/9789241501750_eng.pdf. Accessed September 6, 2013.

WHO (World Health Organization). 2011c. *Global HIV/AIDS Response Epidemic Update and Health Sector Progress Towards Universal Access Report 2011.* World Health Organization. Available from: http://whqlibdoc.who.int/publications/ 2011/9789241502986_eng.pdf. Accessed September 9, 2013.

WHO (World Health Organization). 2012. *Global Action Plan to Control the Spread and Impact of Antimicrobial Resistance in Neisseria gonorrhoeae.* WHO Department of Reproductive Health and Research. Available from: http://www.who.int/ reproductivehealth/publications/rtis/9789241503501/en/in dex.html. Accessed September 9, 2013.

WHO (World Health Organization). 2013. *Laboratory Diagnosis of Sexually Transmitted Infections Including Humanimmunodeficiency Virus.* Available from: http://apps.who.int/iris/bits tream/10665/85343/1/9789241505840_eng.pdf. Accessed September 9, 2013.

Wolf-Watz, H., Elmros, T., Normark, S., and Bloom, G.D. 1975. Cell envelope of *Neisseria gonorrhoeae*: Outer membrane and peptidoglycan composition of penicillin-sensitive and – resistant strains. *Infect. Immun.*, 11, 1332–1341.

Yokoi, S., Deguchi, T., Ozawa, T., Yasuda, M., Ito, S.-I., Kubota, Y., Tamaki, M., and Maeda, S.-I. 2007. *Threat to Cefixime Treatment for Gonorrhea.* Centers for Disease Control and Prevention.

Zenilman, J.M., Nims, L.J., Menegus, M.A., Nolte, F., and Knapp, J.S. 1987. Spectinomycin-resistant gonococcal infections in the United States, 1985–1986. *J. Infect. Dis.*, 156, 1002–1004.

Zhao, S., Tobiason, D.M., Hu, M., Seifert, H.S., and Nicholas, R.A. 2005. The penC mutation conferring antibiotic resistance in *Neisseria gonorrhoeae* arises from a mutation in the PilQ secretin that interferes with multimer stability. *Mol. Microbiol.*, 57, 1238–1251.

Zhao, S., Duncan, M., Tomberg, J., Davies, C., Unemo, M., and Nicholas, R.A. 2009. Genetics of chromosomally mediated intermediate resistance to ceftriaxone and cefixime in *Neisseria gonorrhoeae. Antimicrob. Agents Chemother.*, 53, 3744–3751.

Chapter 38

Pathogenesis of *Corynebacterium diphtheriae* and *Corynebacterium ulcerans*

Andreas Burkovski

Department Biologie, Friedrich-Alexander-Universität Erlangen-Nürnberg, Erlangen, Germany

38.1 Introduction

Diphtheria, the "strangling angel of children," has been known for more than 2500 years and already mentioned in writings by Hippocrates and in the Talmud (for recent reviews, see Burkovski, 2013a, 2013b). The disease was most likely disseminated from the Middle East to Europe, where numerous waves of epidemics occurred, often in cycles that include gaps of 100 years or more (English, 1985). From Europe, it reached all other continents, for example, North America, where George Washington became a victim of this disease in 1799. Although often connected to wars, outbreaks of diphtheria were unpredictable from ancient times to the 20th century. Interestingly and in contrast to many other infections, diphtheria became much more prevalent at the end of the 19th century, most likely favored by poverty, and poor housing and living conditions in the quickly developing and expanding cities during industrialization. In fact, at the end of the 19th century and start of the 20th century, diphtheria developed into a major cause of infant mortality (English, 1985; Osler, 1892). Some areas lost 5% of their population to the disease, with the average mortality rate reaching 10% (English, 1985). This situation improved dramatically based on the development of antitoxin treatment, vaccination, and antibiotics, and in the midst of the 20th century diphtheria was almost eradicated in industrialized countries. However, despite this tremendous success story, diphtheria was never completely defeated. Now, several thousand cases are reported to the World Health Organization every year and the historic breakdown of the Union of the Socialist Soviet Republics (USSR) in 1989 led to a massive re-emergence of diphtheria in Russia and Ukraine as well as later in Belarus, the Baltic States, and other newly independent states of the former Soviet Union.

38.2 Epidemiology of diphtheria

Diphtheria is typically spread from person to person by close physical contact. Bacteria are usually distributed by respiratory droplets produced by coughing or sneezing, and secretions and contaminated material might also be sources of infection. Establishment of the disease after infection takes 2–5 days. Untreated, patients remain infectious for 2–4 weeks, while application of antibiotics usually renders patients non-infectious within 1 day (Burkovski, 2013a; Murphy, 1996).

Before the advent of vaccination, diphtheria was a typical disease of infants. This indicates a wide distribution of the pathogen within the population, leading to early contact and infection. Once the infection was overcome, a high antibody titer was guaranteed due to steady contact with the pathogen. Later, introduction of immunization programs reduced the number of cases dramatically. The last major epidemic occurred in the 1940s due to World War II. Thereafter, only local and relatively small epidemics were observed with an all-time low incidence of only 623 cases reported by the World Health Organization in 1980. This situation changed dramatically, with the breakdown of the former USSR. In 1990, a large-scale outbreak started with the Russian federation and Ukraine as centers of this epidemic (Eskola et al., 1998; Galazka et al., 1995; Popovic et al., 2000). The outbreak spread quickly to neighboring countries and diphtheria infections were observed in Azerbaijan, Belarus, Estonia, Finland, Kazakhstan, Latvia, Lithuania, Poland, Tajikistan, Turkey, and Uzbekistan (Galazka et al., 1995). Between 1990 and 1998 more than 157,000 cases and 5000 deaths were reported (Dittmann et al., 2000; Markina et al., 2000; Vitek and Wharton, 1998). Interestingly, during this epidemic more adolescents and adults than children were infected, which was especially pronounced in the Baltic States, Belarus, Russia, and Ukraine. In these countries, diphtheria cases among people 15 years and older ranged from two-thirds to four-fifths of total cases (Dittmann et al., 2000; Hardy et al., 1996). By mass immunization – especially of adults – starting in 1993, the epidemic was finally stopped. Today, diphtheria is again uncommon in developed countries and the worldwide incidence is restricted to a few thousand cases annually.

38.3 Responsible pathogens

Corynebacterium diphtheriae, the classical etiological agent of respiratory diphtheria, is the type species of the genus *Corynebacterium* (Barksdale, 1970; Lehmann and Neumann, 1896). This genus belongs to the class of *Actinobacteria* (high G+C Gram-positive bacteria) and comprises a collection of morphologically similar, irregular- or club-shaped, non-sporulating, noncapsulated (mico-)aerobic microorganisms (Ventura et al., 2007; Zhi et al., 2009). Today, about 88 *Corynebacterium* species are described with about 51 having more or less medical importance (Bernard, 2012).

Based on colony morphology and physiological differences, *C. diphtheriae* isolates are traditionally grouped into four biovars (or biotypes), that is, *C. diphtheriae* biovar *mitis, gravis, intermedius,* and *belfanti* (Funke et al., 1997a; Goodfellow et al., 2012). Biovar classification is easy for *intermedius*, which is lipophilic and non-haemolytic, and for *belfanti*, which is the only biovar not able to reduce nitrate or use glycogen and starch as carbon sources, while a clear identification of and discrimination between *mitis* and *intermedius* is difficult and often not reliable (Funke et al., 1997a; Goodfellow et al., 2012; Sangal et al., 2013). Sequencing approaches leading to genome information of a number of strains (Cerdeno-Tarraga et al., 2003; Sangal et al., 2012a; 2012b; Trost et al., 2012) provided a possible explanation for this observation. Bioinformatic analyses revealed that the genome of *C. diphtheriae* is highly variable (Trost et al., 2012; Trost and Tauch, 2013) and consequently biovar differentiation does not correlate reliably with gene content (Sangal et al., 2013).

During the last years, the frequency of human diphtheria-like infections associated with *C. ulcerans* has appeared to be increasing in different countries (Bonmarin et al., 2009; Hatanaka et al., 2003; Hogg et al., 2009; Komiya et al., 2010; Wagner et al., 2001, 2010, 2011). This species, which was recognized before as a commensal of different animal species (Dixon, 2010; Schuhegger et al., 2008; Sykes et al., 2010) is

closely related to *C. diphtheriae* and may also produce diphtheria toxin, the major virulence factor involved in the pathogenesis of diphtheria infections.

38.4 Pathogenesis and immunity

C. diphtheriae is typically spread from person to person by close physical contact. Besides patients with respiratory or cutaneous diphtheria, asymptomatic carriers and animals may serve as a reservoir of the pathogen. Isolations of *C. diphtheriae* strains were reported, for example, from cows (Corboz et al., 1996), domestic cats (Hall et al., 2010), and horses (Henricson et al., 2000; Leggett et al., 2010). The existence of animal reservoirs is even more often observed, when other toxigenic *Corynebacterium* species are taken into consideration (Bonmarin et al., 2009). Respiratory diphtheria-like illnesses caused by toxigenic *C. ulcerans* strains are increasingly reported from various industrialized countries (De Zoysa et al., 2005; Hatanaka et al., 2003; Tiwari et al., 2008; Wagner et al., 2001) and have become even more common than *C. diphtheriae* infections in the United Kingdom (Wagner et al., 2010). *C. ulcerans* has been detected as a commensal in domestic and wild animals (Dixon, 2010; Schuhegger et al., 2008; Sykes et al., 2010) that may serve as a reservoir for zoonotic infections. Infections with toxigenic *C. ulcerans* usually occur in adults, who have consumed raw milk or have close contact with domestic animals (Bostock et al., 1984; Hart, 1984; Wagner et al., 2010), while person-to-person transmission has not been reported.

The main factor, responsible for the severe neurological and cardiac complications and the high mortality rate of diphtheria, is the diphtheria toxin. This protein is synthesized depending on the iron concentration in the environment. When iron becomes limiting, as it is the case in the human host, the bacteria start to synthesize the toxin, which is then secreted into the extracellular medium as a single polypeptide chain (Holmes, 2000; Pappenheimer, 1977; Schmitt, 2013). In this form, the toxin is inactive, and can be absorbed into the circulatory system and disseminated to distant parts of the body. When binding to its receptor, the uncleaved precursor of the heparin-binding EGF-like growth factor (Naglich et al., 1992), it can enter the cell by endocytosis. Once inside the endosome, the inactive diphtheria toxin, consisting of an A and B chain, is cleaved, and the A chain is released into the cytoplasm, where it is activated by further cleavage into the active toxin and deactivates its cellular target elongation factor 2 (EF-2) by ADP-ribosylation (Falnes and Sandvig, 2000; Lord et al., 1999; Pappenheimer, 1977; Varol et al., 2013). Infection of *C. diphtheriae* or *C. ulcerans* by a lysogenic phage is a crucial step for diphtheria pathogenesis, since the toxin is encoded on a temperate bacteriophage (for a recent review of corynebacteriophages, see Sangal and Hoskisson, 2013).

The interaction of *C. diphtheriae* (and *C. ulcerans*) with its host is, however, not restricted to toxin secretion. Molecular mechanisms influencing colonization and pathogenicity of toxigenic corynebacteria have been recently summarized (Ott and Burkovski, 2013). Best investigated adhesion factors of *C. diphtheriae* might be the pili structures (for review, see Readon-Robinson and Ton-That, 2013). These determine – at least partially – the adhesion rate and host cell preference (Mandlik et al., 2007). Interestingly, the pili repertoire is highly variable in different *C. diphtheriae* isolates (Ott et al., 2010a; Trost et al., 2012; Trost and Tauch, 2013). Besides pili formation the non-fimbrial protein 67–72p (DIP0733) might be an important adhesin. *C. diphtheriae* is able to bind erythrocytes due to this protein, a process called hemagglutination, which facilitates *C. diphtheriae* to disseminate throughout the body via the blood stream. Furthermore, the 67–72p protein seems to be involved in the internalization process and might contribute to apoptosis and necrosis of host cells (Sabbadini et al., 2012). Agglutination of human erythrocytes is also effectively induced by neuraminidase-secreting strains. However, in the case of the *C. diphtheriae* sialidase NanH, it is still unclear whether it is involved in sialic acids decoration and pathogenesis or in metabolism (Kim et al., 2013). Besides proteins as adhesion factors, *C. diphtheriae* exhibits a 10-kDa lipoglycan on its surface, the CdiLAM, which binds to epithelial cells (Burkovski, 2013c; Moreira et al., 2008). Additionally, *C. diphtheriae* is able to bind fibrinogen and convert it to fibrin, resulting in a fibrin layer on the bacterial surface that may be an efficient method to avoid phagocytosis. Furthermore, this process is involved in pseudomembrane formation (Sabbadini et al., 2010).

C. diphtheriae was generally considered as an extracellular colonizer of the upper respiratory tract; however, failure of penicillin treatment in cases of severe pharyngitis and tonsillitis could only partially be explained by penicillin tolerance (von Hunolstein et al., 2002), and the occurrence of diphtheria among vaccinated adults in Rio de Janeiro as well as an increasing number of cases of endocarditis indicated the possibility of deeper colonization of the human host (Mattos-Guaraldi et al., 2000, 2001). In fact, *C. diphtheriae* is able to invade non-phagocytic cells in a strain-specific manner (Bertuccini et al., 2004; Hirata et al., 2002; Ott et al., 2010b, 2013). The invasion process seems to involve actin polymerization along with a phosphotyrosine signaling event in cultured respiratory cells (Mattos-Guaraldi et al., 2002). Furthermore, infection of epithelial cells leads to inflammatory host response, especially the activation of the NFκB signal transduction pathway (Ott et al., 2013). NFκB induction is strictly dependent on intracellular bacteria, since bacterial adhesion is not sufficient to activate inflammatory pathways.

C. diphtheriae is not only able to invade and persist at least for some time in epithelial cells, but also bacteria taken up by human macrophages via non-opsonic phagocytosis are able to persist inside the macrophages and may induce apoptosis and necrosis (dos Santos et al., 2010).

Data on natural immunity against *C. diphtheriae* are scarce. Already Loeffler isolated the bacteria from swabs of symptom-free school children in Berlin and also today the bacterium may be harbored asymptomatically by 3–5% of healthy individuals in endemic areas (Long, 2000). The reasons for lacking symptoms development are unclear and deserve further investigation.

38.5 Clinical features of diphtheria

Symptoms of an infection with *C. diphtheriae* – or with *C. ulcerans* – develop over a period of 1–7 days with an average incubation period of 2–4 days. Manifestations are either local – with diphtheria of the upper respiratory tract and skin – or systemic due to dissemination of the toxin or the pathogen to different organs and tissues. As a result of the distribution of the diphtheria toxin by the circulatory system, life-threatening systemic complications such as loss of motor functions and congestive heart failure may develop (Burkovski, 2013a, 2013b; George, 2005; Murphy, 1996.

The classical manifestation is diphtheria of the respiratory tract. In more than 90% of patients, the primary foci of the infection are tonsils or pharynx, and the nose and larynx are the next common sites. Unusual sites of infections are buccal mucosa, upper and lower lips, hard and soft palates, and tongue (Hadfield et al., 2000; Nikolaeva et al., 1995). Diphtheria begins with sore throat and low fever, which is rarely higher than 39 °C. Malaise, dysphasia, and headache are not pronounced symptoms. Tissues infected show inflammatory symptoms and 2–5 days after infection, pseudomembrane formation starts. Later, secretion of the diphtheria toxin results in inflammation of deeper lying capillaries and consequently the release of fibrin. Together with the fibrin protein, components of the destroyed epithelial cells and bacteria lead to an extension of the pseudomembrane, which appears grayish or yellowish-white at the beginning of infection. Later, damage of the capillaries due to inflammatory processes causes bleeding and due to decaying erythrocytes, the color of the pseudomembrane may turn into a dirty reddish-brown and the breath of the patients becomes foul-smelling. As a result of the progressive pseudomembrane formation reaching the larynx, a croupy cough develops and hoarseness is observed. Later, trachea, the main bronchi, and also the smaller bronchi may be covered by firmly or loosely attached pseudomembranes, leading to severe hypoxia due to airway obstruction, cyanosis, and suffocation. In severe cases, the air passages might be completely blocked (Burkovski, 2013a, 2013b; Demirci and Abuhammour, 2013; George, 2005; Kole and Roy, 2013).

In parallel to pseudomembrane formation, the tissue of throat and neck becomes edematous and lymphadenopathy develops. In advanced cases, strong swelling of the neck tissue is observed, designated as bull neck diphtheria. This is induced by soft tissue edema and swelling of lymph nodes. Affected lymph nodes might be enlarged, appear blackish-red, and be hemorrhagic (Burkovski, 2013a, 2013b; Demirci and Abuhammour, 2013; Hadfield et al., 2000; Kole and Roy, 2013).

Besides the local effects of diphtheria, absorption and dissemination of diphtheria toxin by the circulatory system may lead to neurological and cardiac symptoms, which are directly proportional to the severity of local infections. About three-fourths of patients with severe diphtheria infection develop neurologic complications. First symptoms of neuropathy are paralysis of the soft palate and posterior pharyngeal wall; later, oculomotor paralysis and dysfunction of facial and pharyngeal nerves may be observed. In later stages, the nerves of trunk, neck, arms, and hands might be affected leading to paralysis. In between 10% and 25% of cases, the myocardium is damaged by the action of diphtheria toxin, and cardiac symptoms often prove fatal weeks after acute infection during convalescence or even thereafter (Burkovski, 2013a, 2013b; Demirci and Abuhammour, 2013; George, 2013; Hadfield et al., 2000; MacGregor, 1995; Perles et al., 2000).

In tropical areas, nasopharyngeal diphtheria infections might be outnumbered by diphtheria of the skin (Burkovski, 2013a, 2013b; Höfler, 1991). Common sites for cutaneous diphtheria are lower legs, feet, and hands due to the infection of insect bites or other small wounds by the pathogenic corynebacteria (Burkovski, 2013a; Connell et al., 2005; Hamour et al., 1995). The infection starts with a fluid-filled pustule, which breaks down quickly and progresses as a punched-out ulcer. During the first weeks the site of infection is covered by a hard, grayish pseudomembrane, which falls off later, leaving a hemorrhagic base with a surrounding of edematous grayish-white, pink, or purple-colored tissue (Burkovski, 2013a, 2013b; George, 2013; Hadfield et al., 2000; Höfler, 1991). The infection is often chronic and healing takes several weeks to months (George, 2013; Höfler, 1991). The long-lasting ulcers may favor dissemination of cutaneous diphtheria and possibly explain high infection rates of skin lesions with *C. diphtheriae* in some African and Asian countries (Bezjak and Farsey, 1970a, 1970b; Höfler, 1991; Liebow et al., 1946; Livingood et al., 1946). Additionally, due to the long infection time cutaneous diphtheria might play a role as a reservoir for the nasopharyngeal form.

38.6 Diagnosis and treatment

Until Loeffler isolated and identified *C. diphtheriae* as the etiological agent of diphtheria (Loeffler, 1884), diagnosis was solely based on the classical symptoms of low fever, sore throat, hoarse voice, bull neck edema, and pseudomembrane formation. Also the foul smell evoked by decaying epithelial and blood

cells as well as neurological disorders such as paralysis of the soft palate, oculomotor paralysis, and dysfunction of facial, pharyngeal and laryngeal nerves was indicative for the disease (Burkovski, 2013a, 2013b; Hadfield et al., 2000). Diagnosis was improved by the introduction of selective tellurite-containing agar. Later testing of nitrate reduction, and pyrazinamidase, cystinase, and urease activity was applied to validate diagnosis (Berger et al., 2013). These tests were supplemented by biochemical differentiation methods such as the API Coryne system, the Anaerobe, and *Corynebacterium* card for Vitek2 or others (Adderson et al., 2008; Freney et al., 1991; Funke et al., 1997b; Gavin et al., 1992; Hudspeth et al., 1998; Rennie et al., 2008; Soto et al., 1994). The diagnostic toolbox was further improved by toxin detection using immunological toxicity testing by immune-diffusion or immune-precipitation assays (Elek, 1948, 1949; Ouchterlony, 1948) and molecular biological methods, including the detection of the 16S rRNA and *tox* genes (Adderson et al., 2008; Aravena-Roman et al., 1995; Bolt et al., 2010; De Zoysa et al., 2008; Khamis et al., 2004, 2005; Mothershed et al., 2002; Pallen et al., 1994; Schuhegger et al., 2008; Sing et al., 2011; Tang et al., 2000).

Today, the challenge is not only to simply identify the pathogenic species, but also to distinguish between several species and subspecies and explore the clonal relatedness of clinical isolates and their transmission routes (Berger et al., 2013). This task is best performed by a combination of microbiology techniques, standard biochemical tests, and molecular methods for species identification and toxigenicity testing. A workflow for the rapid testing of *Corynebacterium* species has been presented recently (Berger et al., 2013): Starting from a suspicious colony of Gram-positive, coryneform, catalase-positive bacteria, standard biochemical tests as described above are performed, while in parallel DNA is extracted and real-time PCR approaches are used to analyze the presence or absence of a *tox* gene. In the case of a positive result of the PCR, Elek tests are carried out to analyze toxigenicity.

The recent introduction of matrix-assisted laser desorption/ionization time-of-flight mass spectrometry (MALDI-TOF MS) into laboratory diagnosis will allow – at least specialized laboratories – a fast, reliable, and specific identification of *C. diphtheriae* and related species from growing colonies. Quality-controlled reference spectra of over 3200 medically relevant microorganisms, including more than 150 *Corynebacterium* subspecies, were already available in 2012 (Alatoom et al., 2012) and a continuous expansion of reference databases can be expected. MALDI-TOF MS for *Corynebacterium* differentiation was already evaluated in four studies (Alatoom et al., 2012; Farfour et al., 2012a; Konrad et al., 2010; Theel et al., 2012) and this technique will be most likely the standard approach in future, due to its speed, reliability, and low running costs.

Two targets are important for a successful treatment of diphtheria, the secreted extremely potent diphtheria toxin, which has an estimated lethal dose for humans of ≤ 0.1 µg/kg (Pappenheimer, 1984), and the bacterial pathogen itself.

To combat the detrimental effect of bacterial toxins, in 1890, von Behring and Kitasato already suggested the application of antitoxin (von Behring and Kitosato, 1890; for an overview, see Grundbacher, 1992). Later, von Behring was also the first who used antitoxin from a horse to successfully treat diphtheria. The immunoglobulin preparation neutralizes the toxin, prevents its detrimental action on protein synthesis, and reduces mortality dramatically. However, only freely circulating toxin is bound by the antibodies directed against the protein, while receptor-bound or internalized diphtheria toxin stays active. Consequently, an early administration of antitoxin is crucial and should be started in suspicious cases even without final identification of the pathogen and characterization of its toxigenicity (Wagner et al., 2009; Zasada, 2013). In fact, diphtheria antitoxin administration was found to be inactive after the second days of neurological diphtheria symptoms (Logina and Donaghy, 1999).

In parallel, elimination of bacteria is important to prevent both continuation of diphtheria toxin secretion and transmission of the pathogen to other, healthy individuals. Antimicrobial drugs of choice in this respect are penicillin and erythromycin (Zasada, 2013). In cases of cutaneous diphtheria, additionally, local application of bacitracin or gentian violet is recommended (Höfler, 1991). Resistance of *C. diphtheriae* strains to penicillin is rarely found, while erythromycin resistance is observed more frequently (Coyle et al., 1979; Farfour et al., 2012bGladin et al., 1999; Kneen et al., 1998; Patey et al., 1995; Pereira et al., 2008). Multiresistant *C. diphtheriae* strains were isolated only very rarely with a concentration in Brazil (Pereira et al., 2008), but were also reported from Russia (Gladin et al., 1999), France (Patey et al., 1995), Vietnam (Kneen et al., 1998), and Canada (Mina et al., 2011). In general, antibiotic resistance seems to be connected to drug prescription and use in different countries and not to *C. diphtheriae* biotype or toxigenicity (Engler et al., 2001; Pereira et al., 2008; Zasada, 2013). High prevalence of tetracycline resistance was, for example, observed in Indonesia (Rockhill et al., 1982), Western Europe (Funke et al., 1999), and Brasil (Pereira et al., 2008), while from India an extremely high rate of trimethoprim-sulphamethoxazole resistant strains was reported (Sharma et al., 2007) and reduced susceptibility to cefoxamine was found in Italy (von Hunolstein et al., 2003) and Poland (Zasada et al., 2010).

38.7 Prevention and control

Diphtheria management and control are based on three columns: vaccination, surveillance, and eradication of the pathogen in infected patients. Mass vaccination is obviously the most effective approach to prevent diphtheria. At the beginning of the 20th century, the work of scientists in North America and Europe, including Ehrlich, Fraenkel, Park, Ramon, von Behring, and others, laid the basis for vaccination. Today, the diphtheria toxoid vaccine, which is in most formulations combined with tetanus and pertussis antigens, is one of the oldest and safest vaccines known (for an excellent review of diphtheria vaccine development, see Malito and Rappouli, 2013). Immunization of children, which is effective in approximately 97% of cases, is typically achieved with four doses of the diphtheria toxoid vaccine within the first two years. With increasing levels of antitoxin immunity, the frequency of isolation of toxigenic *C. diphtheriae* strains decreased and the annual incidence of diphtheria dropped to 0.1–0.2 per million (Höfler, 1991; Kwantes, 1984; Murphy, 1996; Vitek, 2006). Since antibody titers wane over time, a large percentage of adults may have antitoxin levels below the protective level and are at risk of an infection. Therefore, booster immunization of adults is recommended, at least for persons traveling to regions with high rates of endemic diphtheria (Murphy, 1996; von Hunolstein et al., 2000).

The second cornerstone of diphtheria management is an ongoing monitoring and surveillance (for recent review, see Wagner et al., 2013), since (i) non-toxigenic strains, which may be lyogenized and become toxigenic, circulate in population (e.g., Edwards et al., 2011; Reacher et al., 2000), (ii) single cases and small outbreaks of respiratory and cutaneous diphtheria are constantly observed worldwide (e.g., Farfour et al., 2012b; Frelund et al., 2011; Lowe et al., 2011; Sears et al., 2012), and (iii) most important, diphtheria has already demonstrated its tremendous potential to re-emerge in epidemic proportions with the breakdown of the former Soviet Union. To make surveillance across countries, regions, and continents efficient, similar use of case definitions and methods in laboratory diagnosis prove extremely helpful and consequently international networks have been established to coordinate diphtheria surveillance. For example, the European diphtheria surveillance network DIP-NET brings together 46 partner countries with the aim to establish a joint surveillance policy and practice (Neal and Efstratiou, 2007).

The third approach of diphtheria prevention and control is an appropriate and fast antibiotic treatment to eliminate the focus of infection and prevent transmission (George, 2005; Zasada, 2013).

38.8 Population dynamics and changes in pathology of *C. diphtheriae* and *C. ulcerans*

The introduction of diphtheria toxoid vaccine led to a dramatic reduction of diphtheria cases. Interestingly, also the number of isolated toxigenic

C. diphtheriae strains decreased, suggesting an effect of the vaccine not only against the toxin, but also, at least partially, against the pathogen. In contrast to this development, non-toxigenic strains were increasingly isolated in western developed countries (Gilbert, 1997; Hadfield et al., 2000; Wagner et al., 2011; Zuber et al., 1992). These non-toxigenic *C. diphtheriae* strains have been persisting over years in different, often poor populations (Lowe et al., 2011; Romney et al., 2006; Shashikala et al., 2011), where they have been associated especially with skin infections. Also an increase in systemic infections by non-toxigenic *C. diphtheriae* strains has been observed.

A shift in *C. diphtheriae* population was also observed after the outbreak by a unique clonal group of *C. diphtheriae* in Russia in 1990 (Popovic et al., 2000). A rising heterogenicity of circulating strains after the epidemic, emergence of new toxigenic variants, and persistence of invasive non-toxigenic strains were also observed (Mokrousov, 2009).

An explanation for the observed rapid changes in *C. diphtheriae* populations and quick adaptation to changing environments was provided by recent genome sequencing projects. Compared to the USSR outbreak strain, which was sequenced in 2003 (Cerdeno-Tarraga et al., 2003), recently determined genome sequences showed significant differences in gene content and genome island distribution (Sangal et al., 2012a, 2012b; 2013; Trost et al., 2012). Based on these data, it was concluded that *C. diphtheriae* is an organism with an open pan-genome (Trost et al., 2012; Trost and Tauch, 2013), which is characterized by a high degree of genetic variability.

As mentioned above, a shift from *C. diphtheriae* to *C. ulcerans* as the etiological agent of diphtheria has been observed at least for some western countries. However, the bacterium is increasingly isolated worldwide and *C. ulcerans* diphtheria of the respiratory tract and skin is considered as an emerging zoonosis (Dias et al., 2011; Mattos-Guaraldi et al., 2013; Wagner et al., 2010).

Recently, a hypothesis was presented that infection with toxigenic *C. diphtheriae* strains is connected to the development of Alzheimer's disease (Merril, 2013). Sporadic Alzheimer's disease is associated with aging and results in dementia. By the age of 65 years about 1 in 10 individuals is affected by dementia, and this ratio is doubling about every 5 years and by the age of 85 years 1 in 3 individuals develops symptoms of dementia. Although in the case of Alzheimer's disease an unequivocal diagnosis is only possible post-mortem, it can be concluded that Alzheimer's disease is widespread in population and it was estimated that the worldwide prevalence was between 11 and 60 million individuals in 2006 and will reach between 47 and 221 million cases by 2050 (Brookmeyer et al., 2007). Due to the widespread epidemiology and the global impact of this disease on society, extensive research was carried out and a number of risk factors discussed (Lindsay et al., 2002), including poor education,

variations in apolipoprotein E and the amyloid-β protein precursor as well as infections with herpes simplex virus I (Wozniak and Itzhaki, 2010) and *Chlamydia pneumoniae* (Shima et al., 2010). Several lines of evidence support the recent hypothesis that also *C. diphtheriae* might play a role as an additional agent of Alzheimer's disease (Merril, 2013).

The loss of smell is an early symptom of Alzheimer's disease (Serby et al., 1991). When the disease progresses from the entorhinal cortex to hippocampus and neocortical areas, olfactory dysfunctions also become more severe (Murphy et al., 1990). It was proposed that these symptoms might be the result of toxigenic *C. diphtheriae* colonizing the nasopharyngeal cavity (Merril, 2013). The secretion of diphtheria toxin by these bacteria might then lead to a sporadic or chronic exposure of nasopharyngeal tissue to diphtheria toxin and consequent degradation of neurons of the olfactory bulb and central nervous system. Since these cannot be replaced by new cells, the slow but progressive degeneration process characteristic of Alzheimer's disease is evoked. The idea of diphtheria toxin involvement is supported by the fact that protein synthesis in cortical regions from cases of Alzheimer's disease is impaired (Landstrom et al., 1989). In fact, a positive effect of diphtheria vaccination with respect to Alzheimer's disease development was observed (Tyas et al., 2001; Verreault et al., 2001). While toxoid vaccination results in protection against diphtheria toxin, the decline of the immune system in elderly persons might result in a renewed sensitivity to the toxin and consequently might favor Alzheimer's disease development. Vice versa, booster vaccination directed against the toxin might have a protective effect. However, although attractive, the hypothesis of an association of sporadic Alzheimer's disease with *C. diphtheriae* deserves further investigation.

38.9 Conclusion

Although almost eradicated in western industrialized countries by mass vaccination, diphtheria is a continuous threat, requiring constant surveillance, mass immunization of children, as well as booster immunization of adults to prevent re-emergence. Obviously, *C. diphtheriae* strains are circulating in population and are persisting in the environment. Besides groups in populations that refuse vaccination, for example, for religious reasons, or that lack access to medical treatment, numerous reports of isolations from pet and wild animals suggest that these might be an important reservoir for *C. diphtheriae*. This is also true for *C. ulcerans*, where only zoonotic infections reported and no man-to-man transmission.

Since overcrowding, poor diet and living conditions, as well as a low standard of health protection are risk factors associated with diphtheria infection and transmission (Harnisch et al., 1989; Lowe et al., 2011; Romney et al., 2006), diphtheria may re-emerge in the future in different parts of the world, especially in connection to major population shifts and failure to immunize children and adults (Byard, 2013). Recent examples of outbreaks have been from India (John, 2008) and Nigeria (Sadoh and Oladokun, 2012), where the immunization rate and frequency per individual are poor. Unfortunately, many countries do not hold diphtheria antitoxin stocks nowadays, due to the relative low incidence of cases (Wagner et al., 2009). This should be of public concern, since it further increases the individual risk of a fatal diphtheria out-come, which remains with more than 10% very high also in recent outbreaks (World Health Organization, 2006).

Today, diphtheria is in principle easy to treat with antitoxin and antibiotics. An occurrence of antibiotics-resistant strains, as increasingly observed for other bacterial pathogens, would change this situation and might contribute to diphtheria re-emergence. The identification of new putative targets for antimicrobial drugs (Astegno et al., 2013) is a positive development in this respect.

References

Adderson, E.E., Boudreaux, J.W., Cummings, J.R., Pounds, S., Wilson, D.A., Procop, G.W., and Hayden, R.T. 2008. Identification of clinical coryneform bacterial isolates: Comparison of biochemical methods and sequence analysis of 16S rRNA and *rpoB* genes. *J. Clin. Microbiol.*, 46, 921–927.

Alatoom, A.A., Cazanave, C.J., Cunningham, S.A., Ihde, S.M., and Patel, R. 2012. Identification of non-*Corynebacterium diphtheriae* by use of matrix-assisted laser desorption ionization-time of flight mass spectrometry. *J. Clin. Microbiol.*, 50, 160–163.

Aravena-Roman, M., Bowman, R., and O'Neill, G. 1995. Polymerase chain reaction for the detection of toxigenic *Corynebacterium diphtheriae*. *Pathology*, 27, 71–73.

Astegno, A., Giorgetti, A., Allegrini, A., Cellini, B., and Dominici, P. 2013. Characterization of C-S lyase from *C. diphtheriae*: A possible target for new antimicrobial drugs. *BioMed. Res. Int.*, 2013, Article ID 701536. doi:10.1155/2013/701536

Barksdale, L. 1970. *Corynebacterium diphtheriae* and its relatives. *Bacteriol. Rev.*, 34, 378–422.

Berger, A., Hogardt, M., Konrad, R., and Sing, A. 2013. Detection methods for laboratory diagnosis of diphtheria. In: Burkovski, A., editor. *Corynebacterium Diphtheriae and Related Toxigenic Species*. Dordrecht: Springer. pp. 171–206.

Bernard, K. 2012. The genus *Corynebacterium* and other medically relevant coryneform-like bacteria. *J. Clin. Microbiol.*, 50, 3152–3158.

Bertuccini, L., Baldassarri, L., and von Hunolstein, C. 2004. Internalization of non-toxigenic *Corynebacterium diphtheriae* by cultured human respiratory epithelial cells. *Microb. Pathog.*, 37, 111–108.

Bezjak, V., and Farsey, S.J. 1970a. *Corynebacterium diphtheriae* carriership in Ugandan children. *J. Trop. Pediatr.*, 16, 12–16.

Bezjak, V., and Farsey, S.J. 1970b. *Corynebacterium diphtheriae* in skin lesions in Ugandan children. *Bull. World Health Organ.*, 43, 643–650.

Bolt, F., Cassiday, P., Tondella, M.L., De Zoysa, A., Efstratiou, A., Sing, A., Zasada, A., Bernard, K., Guiso, N., Badell, E., Rosso, M.L., Baldwin, A., and Dowson, C. 2010. Multilocus sequence typing identifies evidence for recombination and two distinct lineages of *Corynebacterium diphtheriae*. *J. Clin. Microbiol.*, 48, 4177–4185.

Bonmarin, I., Guiso, N., Le Fleche-Mateos, A., Patey, O., Patrick, A.D.G., and Levy-Bruhl, D. 2009. Diphtheria: A zoonotic disease in France? *Vaccine*, 27, 4196–4200.

Bostock, A.D., Gilbert, F.R., Lewis, D., and Smith, D.C. 1984. *Corynebacterium ulcerans* infection associated with untreated milk. *J. Infect.*, 9, 286–288.

Brookmeyer, R., Johnson, E., Ziegler-Graham, K., and Arrighi, H.M. 2007. Forecasting the global burden of Alzheimer's disease. *Alzheimers Dement.*, 3, 186–191.

Burkovski, A. 2013a. Diphtheria. In: Rosenberg, E., DeLong, E.F., Thompson, F., Lory, S., and Stackebrandt, E., editors. *The Prokaryotes, 4th ed., Vol. 5, Human Microbiology*. New York: Springer. pp. 237–246.

Burkovski, A. 2013b. Diphtheria and its etiological agents. In: Burkovski, A., editor. *Corynebacterium diphtheriae and Related Toxigenic Species*. Dordrecht: Springer. pp. 1–14.

Burkovski, A. 2013c. Cell envelope of corynebacteria: Structure and influence on pathogenicity. *ISRN Microbiol.*, Article ID 935736. doi:10.1155/2013/935736

Byard, R.W. 2013. Diphtheria – 'The strangling angel' of children. *J. Forensic Leg. Med.*, 20, 65–68.

Cerdeno-Tarraga, A.M., Efstratiou, A., Dover, L.G., Holden, M.T., Pallen, M., Bentley, S.D., Besra, G.S., Churcher, C., James, K.D., De Zoysa, A., Chillingworth, T., Cronin, A., Dowd, L., Feltwell, T., Hamlin, N., Holroyd, S., Jagels, K., Moule, S., Quail, M.A., Rabbinowitsch, E., Rutherford, K.M., Thomson, N.R., Unwin, L., Whitehead, S., Barrell, B.G., and Parkhill, J. 2003. The complete genome sequence and analysis of *Corynebacterium diphtheriae* NCTC13129. *Nucleic Acids Res.*, 31, 6516–6523.

Connell, T.G., Rele, M., Daley, A.J., and Curtis, N. 2005. Skin ulcers in a returned traveler. *Lancet*, 365, 726.

Corboz, L., Thoma, R., Braun, U., and Zbinden, R. 1996. Isolation of *Corynebacterium diphtheriae* subsp. *belfanti* from a cow with chronic active dermatitis [article in German]. *Schweiz. Arch. Tierheilkd.*, 138, 596–599.

Coyle, M.B., Minshew, B.H., Bland, J.A., and Hsu, P.C. 1979. Erythromycin and clindamycin resistance in *Corynebacterium diphtheriae* from skin lesions. *Antimicrob. Agents Chemother.*, 16, 525–527.

De Zoysa, A., Hawkey, P.M., Engler, K., George, R., Mann, G., Reilly, W., Taylor, D., and Efstratiou, A. 2005. Characterization of toxigenic *Corynebacterium ulcerans* strains isolated from humans and domestic cats in the United Kingdom. *J. Clin. Microbiol.*, 43, 4377–4381.

De Zoysa, A., Hawkey, P., Charlett, A., and Efstratiou, A. 2008. Comparison of four molecular typing methods for characterization of *Corynebacterium diphtheriae* and determination of transcontinental spread of *C. diphtheriae* based on BstEII rRNA gene profiles. *J. Clin. Microbiol.*, 46, 3626–3635.

Demirci C.S., and Abuhammour, W. 2013. Pediatric diphtheria. Available from: http://emedicine.medscape.com/article /963334.

Dias, A.A., Santos, L.S., Sabbadini, P.S., Santos, C.S., Silva Jr., F.C., Napoleão, F., Nagao, P.E., Villas-Bôas, M.H., Hirata Jr., R., and Guaraldi, A.L. 2011. *Corynebacterium ulcerans* diphtheria: An emerging zoonosis in Brazil and worldwide. *Rev. Saude Publica*, 45, 1176–1191.

Dittmann, S., Wharton, M., Vitek, C., Ciotti, M., Galazka, A., Guichard, S., Hardy, I., Kartoglu, U., Koyama, S., Kreysler, J., Martin, B., Mercer, D., Ronne, T., Roure, C., Steinglass, R., Strebel, P., Sutter, R., and Trostle, M. 2000. Successful control of epidemic diphtheria in the states of the Former Union of Soviet Socialist Republics: Lessons learned. *J. Infect. Dis.*, 181(Suppl. 1), S10–S22.

Dixon, B. 2010. Sick as a dog. *Lancet*, 10, 73.

dos Santos, C.S., dos Santos, L.S., de Souza, M.C., dos Santos Dourado, F., de Souza de Oliveira Dias, A.A., Sabbadini, P.S., Pereira, G.A., Cabral, M.C., Hirata Jr., R., and Mattos-Guaraldi, A.L. 2010. Non-opsonic phagocytosis of homologous non-toxigenic and toxigenic *Corynebacterium diphtheriae* strains by human U-937 macrophages. *Microbiol. Immunol.*, 54, 1–10.

Edwards, B., Hunt, A.C., and Hoskisson, P.A. 2011. Recent cases of non-toxigenic *Corynebacterium diphtheriae* in Scotland: Justification for continued surveillance. *J. Med. Microbiol.*, 60, 561–562.

Elek, S.D. 1948. The recognition of toxigenic bacterial strains *in vitro*. *Br. Med. J.*, 1, 493–496.

Elek, S.D. 1949. The plate virulence test for diphtheria. *J. Clin. Pathol.*, 2, 250–258.

Engler, K.H., Warner, M., and George, R.C. 2001. *In vitro* activity of ketolides HMR 3004 and HMR 3647 and seven other antimicrobial agents against *Corynebacterium diphtheriae*. *J. Antimicrob. Chemother.*, 47, 27–31.

English, P.C. 1985. Diphtheria and theories of infectious disease: Centennial appreciation of the critical role of diphtheria in the history of medicine. *Pediatrics*, 76, 1–9.

Eskola, J., Lumio, J., and Vuopio-Varkila, J. 1998. Resurgent diphtheria – Are we safe? *Br. Med. Bull.*, 54, 635–645.

Falnes, P.O., and Sandvig, K. 2000. Penetration of protein toxins into cells. *Curr. Opin. Cell. Biol.*, 12, 407–413.

Farfour, E., Leto, J., Barritault, M., Barberis, C., Meyer, J., Dauphin, B., Le Guern, A.S., Leflèche, A., Badell, E., Guiso, N., Leclercq, A., Le Monnier, A., Lecuit, M., Rodriguez-Nava, V., Bergeron, E., Raymond, J., Vimont, S., Bille, E., Carbonnelle, E., Guet-Revillet, H., Lécuyer, H., Beretti, J.L., Vay, C., Berche, P., Ferroni, A., Nassif, X., and Join-Lambert, O. 2012a. Evaluation of the Andromas matrix-assisted laser desorption ionization-time of flight mass spectrometry system for identification of aerobically growing Gram-positive bacilli. *J. Clin. Microbiol.*, 50, 2702–2707.

Farfour, E., Badell, E., Zasada, A., Hotzel, H., Tomaso, H., Guillot, S., and Guiso, N. 2012b. Characterization and comparison of invasive *Corynebacterium diphtheriae* isolates from France and Poland. *J. Clin. Microbiol.*, 50, 173–175.

Frelund, H., Noren, T., Lepp, T., Morfeldt, E., and Henriques Normark, B. 2011. A case of diphtheria in Sweden, October 2011. *Eurosurveillance*, 16, 20038.

Freney, J., Duperron, M.T., Courtier, C., Hansen, W., Allard, F., Boeufgras, J.M., Monget, D., and Fleurette, J. 1991. Evaluation of API Coryne in comparison with conventional methods for identifying coryneform bacteria. *J. Clin. Microbiol.*, 29, 38–41.

Funke, G., von Graevenitz, A., Clarridge 3rd, J.E., and Bernard, K.A. 1997a. Clinical microbiology of Coryneform bacteria. *Clin. Microbiol. Rev.*, 10, 125–159.

Funke, G., Renaud, F.N., Freney, J., and Riegel, P. 1997b. Multicenter evaluation of the updated and extended API (RAPID) Coryne database 2.0. *J. Clin. Microbiol.*, 35, 3122–3126.

Funke, G., Altwegg, M., Frommelt, L., and von Graevenitz, A. 1999. Emergence of related nontoxigenic *Corynebacterium diphtheriae* biotype *mitis* strains in Western Europe. *Emerg. Infect. Dis.*, 5, 477–480.

Galazka, A.M., Robertson, S.E., and Oblapenko, G.P. 1995. Resurgence of diphtheria. *Eur. J. Epidemiol.*, 11, 95–105.

Gavin, S.E., Leonard, R.B., Briselden, A.M., and Coyle, M.B. 1992. Evaluation of the rapid CORYNE identification system for *Corynebacterium* species and other coryneforms. *J. Clin. Microbiol.*, 30, 1692–1695.

George, R.C. 2005. Diphtheria. *Medicine*, 33, 31–33.

Gilbert, L. 1997. Infections with *Corynebacterium diphtheriae* – Changing epidemiology and clinical manifestations. Report of the third international meeting of the European Laboratory Working Group on Diphtheria (ELWGD), Institute Pasteur, Paris 7–8 June 1996. *Commun. Dis. Intell.*, 21, 161–164.

Gladin, D.P., Kozlova, N.S., Zaitseva, T.K., Cherednichenko, A.S., and Khval, S.A. 1999. Sensitivity of *Corynebacterium diphtheriae* isolated in Saint-Petersburg to antibacterial drugs [in Russian]. *Antibiot. Khimioter.*, 44:17–21.

Goodfellow, M., Kaempfer, P., Busse, H.-J., Trujillo, M.E., Suzuki, K.-I., Ludwig, W., and Whitman, W.B. 2012. *The actinobacteria, Part A.* In: Whitman, W.B., editor. *Bergey's Manual of Systematic Bacteriology*, 2nd ed., Vol. 5. London: Springer. pp. 1034.

Grundbacher, F.J. 1992. Behring's discovery of diphtheria and tetanus antitoxins. *Immunol. Today* 13, 188–190.

Hadfield, T.L., McEvoy, P., Polotsky, Y., Tzinserling, V.A., and Yakovlev, A.A. 2000. The pathology of diphtheria. *J. Infect. Dis.*, 181 (Suppl. 1), S116–S120.

Hall, A.J., Cassiday, P.K., Bernard, K.A., Bolt, F., Steigerwalt, A.G., Bixler, D., Pawloski, L.C., Whitney, A.M., Iwaki, M., Baldwin, A., Dowson, C.G., Komiya, T., Takahashi, M., Henrikson, H.P., and Tondella, M.L. 2010. Novel *Corynebacterium diphtheriae* in domestic cats. *Emerg. Infect. Dis.*, 16, 688–691.

Hamour, A.A., Efstratiou, A., Neill, R., and Dunbar, E.M. 1995. Epidemiology and molecular characterization of toxigenic *Corynebacterium diphtheriae var mitis* from a case of cutaneous diphtheria in Manchester. *J. Infect.*, 31, 153–157.

Hardy, I.R., Dittmann, S., and Sutter, R.W. 1996. Current situation and control strategies for resurgence of diphtheria in newly independent states of the former Soviet Union. *Lancet*, 347, 1739–1744.

Harnisch, J.P., Tronca, E., Nolan, C.M., Turck, M., and Holmes, K.K. 1989. Diphtheria among alcoholic urban adults. A decade of experience in Seattle. *Ann. Intern. Med.*, 111, 71–82.

Hart, R.J. 1984. *Corynebacterium ulcerans* in humans and cattle in North Devon. *J. Hyg. (Lond.)*, 92, 161–164.

Hatanaka, A., Tsunoda, A., Okamoto, M., Ooe, K., Nakamura, A., Miyakoshi, M., Komiya, T., and Takahashi, M. 2003. *Corynebacterium ulcerans* in Japan. *Emerg. Infect. Dis.*, 9, 752–753.

Henricson, B., Segarra, M., Garvin, J., Burns, J., Jenkins, S., Kim, C., Popovic, T., Golaz, A., and Akey, A. 2000. Toxigenic *Corynebacterium diphtheriae* associated with an equine wound infection. *J. Vet. Diagn. Invest.*, 12, 253–257.

Hirata, R., Napoleao, F., Monteiro-Leal, L.H., Andrade, A.F., Nagao, P.E., Formiga, L.C., Fonseca, L.S., and Mattos-Guaraldi, A.L. 2002. Intracellular viability of toxigenic *Corynebacterium diphtheriae* strains in HEp-2 cells. *FEMS Microbiol. Lett.*, 215, 115–119.

Höfler, W. 1991. Cutaneous diphtheria. *Int. J. Dermatol.*, 30, 845–847.

Hogg, R.A., Wessels, J., Hart, J., Efstratiou, A., De Zoysa, A., Mann, G., Allen, T., and Pritchard, G.C. 2009. Possible zoonotic transmission of toxigenic *Corynebacterium ulcerans* from companion animals in a human case of fatal diphtheria. *Vet. Rec.*, 165, 691–692.

Holmes, R.K. 2000. Biology and molecular epidemiology of diphtheria toxin and the *tox* gene. *J. Infect. Dis.*, 181 (Suppl. 1), S156–S167.

Hudspeth, M.K., Hunt Gerardo, S., Citron, D.M., and Goldstein, E.J. 1998. Evaluation of the RapID CB Plus system for identification of *Corynebacterium* species and other gram-positive rods. *J. Clin. Microbiol.*, 36, 543–547.

Janda, M.W. 1999. The corynebacteria revisited: New species, identification kits, and antimicrobial susceptibility testing. *Clin. Microbiol. Newslett.*, 21, 175–182.

John, T.J. 2008. Resurgence of diphtheria in India in the 21st century. *Indian J. Med. Res.*, 128, 669–670.

Khamis, A., Raoult, D., and La Scola, B. 2004. *rpoB* gene sequencing for identification of *Corynebacterium* species. *J. Clin. Microbiol.*, 42, 3925–3931.

Khamis, A., Raoult, D., and La Scola, B. 2005. Comparison between *rpoB* and 16S rRNA gene sequencing for molecular identification of 168 clinical isolates of *Corynebacterium*. *J. Clin. Microbiol.*, 43, 1934–1936.

Kim, S., Oh, D.-B., and Kwon, O. 2013. Sialidases of corynebacteria and their biotechnological applications. In: Burkovski,

A., editor. *Corynebacterium diphtheriae and Related Toxigenic Species*. Dordrecht: Springer. pp. 247–262.

Kneen, R., Pham, N.G., Solomon, T., Tran, T.M., Nguyen, T.T., Tran, B.L., Wain, J., Day, N.P., Tran, T.H., Parry, C.M., and White, N.J. 1998. Penicillin vs. erythromycin in the treatment of diphtheria. *Clin. Infect. Dis.*, 27, 845–850.

Kole, A.K., and Roy, R. 2013. Respiratory diphtheria. *N. Engl. J. Med.*, 369, 16.

Komiya, T., Seto, Y., de Zoysa, A., Iwaki, M., Hatanaka, A., Tsunoda, A., Arakawa, Y., Kozaki, S., and Takahashi, M. 2010. Two Japanese *Corynebacterium ulcerans* isolates from the same hospital: Ribotype, toxigenicity, and serum antitoxin titre. *J. Med. Microbiol.*, 59, 1497–1504.

Konrad, R., Berger, A., Huber, I., Boschert, V., Hörmansdorfer, S., Busch, U., Hogardt, M., Schubert, S., and Sing, A. 2010. Matrix-assisted laser desorption/ionisation time-of-flight (MALDI-TOF) mass spectrometry as a tool for rapid diagnosis of potentially toxigenic *Corynebacterium* species in the laboratory management of diphtheria-associated bacteria. *Eurosurveillance*, 15, pii 19699.

Kwantes, W. 1984. Diphtheria in Europe. *J. Hyg. (Cambridge)*, 93, 433–437.

Landstrom, N.S., Anderson, J.P., Lindroos, H.G., Winblad, B., and Wallace, W.C. 1989. Alzheimer's disease-associated reduction of polysomal mRNA translation. *Mol. Brain Res.*, 5, 259–269.

Leggett, B.A., De Zoysa, A., Abbott, Y.E., Leonard, N., Markey, B., and Efstratiou, A. 2010. Toxigenic *Corynebacterium diphtheriae* isolated from a wound in a horse. *Vet. Rec.*, 166, 656–657.

Lehmann, K.B., and Neumann, R. 1896. *Atlas und Grundriss der Bakteriologie und Lehrbuch der speziellen bakteriologischen Diagnostik*. Munich: Lehmann.

Liebow, A.A., MacLean, P.D., and Bumstead, J.M. 1946. Tropical ulcers and cutaneous diphtheria. *Arch. Intern. Med.*, 78, 255–295.

Lindsay, J., Laurin, D., Verreault, R., Hebert, R., Helliwell, B., Hill, G.B., and McDowell, I. 2002. Risk factors for Alzheimer's disease: A prospective analysis from the Canadian Study of Health and Aging. *Am. J. Epidemiol.*, 156, 445–453.

Livingood, C.S., Perry, D., and Forrester, J.S. 1946. Cutaneous diphtheria: Report of 140 cases. *J. Invest. Dermatol.*, 7, 341–364.

Loeffler, F. 1884. Untersuchungen über die Bedeutung der Mikroorganismen für die Entstehung der Diphtherie beim Menschen, bei der Taube und beim Kalbe. *Mitteilungen an dem Kaiserlichen Gesundheitsamte*, 2, 421–499.

Logina, I., and Donaghy, M. 1999. Diphtheritic polyneuropathy: A clinical study and comparison with Guillain-Barre syndrome. *J. Neurol. Neurosurg. Psychiatry*, 67, 433–438.

Long, S.S. 2000. Diphtheria (*Corynebacterium diphtheriae*). In: Behrman, R.E., Kliegman, R.M., and Jensen, H.B., editors. *Nelson Textbook of Pediatrics*. Philadelphia, PA: N.B. Saunders. pp. 817–820.

Lord, J.M., Smith, D.C., and Roberts, L.M. 1999. Toxin entry: How bacterial proteins get into mammalian cells. *Cell. Microbiol.*, 1, 85–91.

Lowe, C.F., Bernard, K.A, and Romney, M.G. 2011. Cutaneous diphtheria in the urban poor population of Vancouver, British Columbia: A 10-year review. *J. Clin. Microbiol.*, 49, 2664–2666.

MacGregor, R.R. 1995. *Corynebacterium diphtheriae*. In: Mandell, G.L., Douglas, J.E., and Dolin, R., editors. *Principles and Practice of Infectious Diseases*, 4th ed., New York: Churchill Livingston. pp. 1866–1869.

Malito, E., and Rappouli, R. 2013. History of diphtheria vaccine development. In: Burkovski, A., editor. *Corynebacterium diphtheriae and Related Toxigenic Species*. Dordrecht: Springer. pp. 225–238.

Mandlik, A., Swierczynski, A., Das, A., and Ton-That, H. 2007. *Corynebacterium diphtheriae* employs specific minor pilins to target human pharyngeal epithelial cells. *Mol. Microbiol.*, 64, 111–124.

Markina, S.S., Maksimova, N.M., Vitek, C.R., Bogatyreva, E.Y., and Monisov, A.A. 2000. Diphtheria in the Russian Federation in the 1990s. *J. Infect. Dis.*, 181(Suppl. 1), S27–S34.

Mattos-Guaraldi, A.L., Duarte Formiga, L.C., and Pereira, G.A. 2000. Cell surface components and adhesion in *Corynebacterium diphtheriae*. *Microbes Infect.*, 2, 1507–1512.

Mattos-Guaraldi, A.L., Formiga, L.C., Camello, T.C., Pereira, G.A., Hirata Jr., R., and Halpern, M. 2001. *Corynebacterium diphtheriae* threats in cancer patients. *Rev. Argent. Microbiol.*, 33, 96–100.

Mattos-Guaraldi, A.L., Formiga, L.C.D., Andrade, A.F.B., and Hirata, R.J. 2002. Patterns of adherence to HEp-2 cells and ability to induce actin polymerization by toxigenic *Corynebacterium diphtheriae* strains. Seventh International Meeting of the European Laboratory Working Group on Diphtheria – ELWGD, Vienna.

Mattos-Guaraldi, A.L., Hirata Jr., R., and Azevedo, V.A. 2013. *Corynebacterium diphtheriae*, *Corynebacterium ulcerans* and *Corynebacterium pseudotuberculosis* – General aspects. In: Burkovski, A., editor. *Corynebacterium diphtheriae and Related Toxigenic Species*. Dordrecht: Springer. pp. 15–38.

Merril, C.R. 2013. Is sporadic Alzheimer's disease associated with diphtheria toxin? *J. Alzheimer Dis.*, 34, 595–600.

Mina, N.V., Burdz, T., Wiebe, D., Rai, J.S., Rahim, T., Shing, F., Hoang, L., and Bernard, K. 2011. Canada's first case of a multidrug-resistant *Corynebacterium diphtheriae* strain, isolated from a skin abscess. *J. Clin. Microbiol.*, 49, 4003–4005.

Mokrousov, I. 2009. *Corynebacterium diphtheriae*: Genome diversity, population structure and genotyping perspectives. *Infect. Genet. Evol.*, 9, 1–15.

Moreira, L.O., Mattos-Guaraldi, A.L., and Andrade, A.F. 2008. Novel lipoarabinomannan-like lipoglycan (CdiLAM) contributes to the adherence of *Corynebacterium diphtheriae* to epithelial cells. *Arch. Microbiol.*, 190, 521–530.

Mothershed, E.A., Cassiday, P.K., Pierson, K., Mayer, L.W., and Popovic, T. 2002. Development of a real-time fluorescence PCR assay for rapid detection of the diphtheria toxin gene. *J. Clin. Microbiol.*, 40, 4713–4719.

Murphy, J.R. 1996. *Corynebacterium diphtheriae*. In: Baron, S., editor. *Medical Microbiology*, 4th ed. Galveston, TX: University of Texas Medical Branch at Galveston. pp. 99–100.

Murphy, C., Gilmore, M.M., Seery, C.S., Salmon, D.P., and Lasker, B.R. 1990. Olfactory thresholds are associated with degree of dementia in Alzheimer's disease. *Neurobiol. Aging*, 11, 465–469.

Naglich, J.G., Metherall, J.E., Russell, D.W., and Eidels, L. 1992. Expression cloning of a diphtheria toxin receptor: Identity with a heparin-binding EGF-like growth factor precursor. *Cell*, 69, 1051–1061.

Neal, S., and Efstratiou, A. 2007. DIPNET – Establishment of a dedicated surveillance network for diphtheria in Europe. *Eurosurveillance*, 12, E9–E10.

Nikolaeva, I.N., Astaf'eva, N.V., Barer, G.M., Parkhomenko, I.G., Iushchuck, N.D., Petina, G.K., and Vorob'ev, A.S. 1995. Diphtheria of the oral mucosa. *Stomatologiia*, 74, 26–28.

Osler, W. 1892. *Principles and Practice of Medicine*. New York: D. Appleton.

Ott, L., and Burkovski, A. 2013. Toxigenic corynebacteria: Adhesion, invasion and host response. In: Burkovski, A., editor. *Corynebacterium diphtheriae and Related Toxigenic Species*. Dordrecht: Springer. pp. 143–170.

Ott, L., Höller, M., Rheinlaender, J., Schäffer, T.E., Hensel, M., and Burkovski, A. 2010a. Strain-specific differences in pili formation and the interaction of *Corynebacterium diphtheriae* with host cells. *BMC Microbiol.*, 10, 257.

Ott, L., Höller, M., Gerlach, R.G., Hensel, M., Rheinlaender, J., Schäffer, T.E., and Burkovski, A. 2010b. *Corynebacterium diphtheriae* invasion-associated protein (DIP1281) is involved in cell surface organization, adhesion and internalization in epithelial cells. *BMC Microbiol.* 10, 2.

Ott, L., Scholz, B., Höller, M., Hasselt, K., Ensser, A., and Burkovski, A. 2013. Induction of the NFκ-B signal transduction pathway in response to *Corynebacterium diphtheriae* infection. *Microbiology*, 159, 126–135.

Ouchterlony, O. 1948. In vitro method for testing the toxin-producing capacity of diphtheria bacteria. *Acta Pathol. Microbiol. Scand.*, 25, 186–191.

Pallen, M.J., Hay, A.J., Puckey, L.H., and Efstratiou, A. 1994. Polymerase chain reaction for screening clinical isolates of corynebacteria for the production of diphtheria toxin. *J. Clin. Pathol.*, 47, 353–356.

Pappenheimer Jr., A.M. 1977. Diphtheria toxin. *Annu. Rev. Biochem.*, 46, 69–94.

Pappenheimer Jr., A.M. 1984. The diphtheria bacillus and its toxin: A model system. *J. Hyg. (Lond.)*, 93, 397–404.

Patey, O., Bimet, F., Emond, J.P., Estrangin, E., Riegel, P.H., Halioua, B., Dellion, S., and Kiredjian, M. 1995. Antibiotic susceptibilities of 38 non-toxigenic strains of *Corynebacterium diphtheriae*. *J. Antimicrob. Chemother.*, 36, 1108–1110.

Pereira, G.A., Pimeta, F.P., Santos, F.R., Damasco, P.V., Hirata Jr., R., and Mattos-Guaraldi, A.L. 2008. Antimicrobial resistance among Brazilian *Corynebacterium diphtheriae* strains. *Mem. Inst. Oswaldo Cruz*, 103, 507–510.

Perles, Z., Nir, A., Cohen, E., Bashary, A., and Engelhard, D. 2000. Atriventricular block in a toxic child: Do not forget diphtheria. *Pediatr. Cardiol.*, 21, 282–283.

Popovic, T., Mazurova, I.K., Efstratiou, A., Vuopio-Varkila, J., Reeves, M.W., De Zoysa, A., Glushkevich, T., and Grimont, P. 2000. Molecular epidemiology of diphtheria. *J. Infect. Dis.*, 181 (Suppl. 1), S168–S177.

Reacher, M., Ramsay, M., White, J., De Zoysa, A., Efstratiou, A., Mann, G., Mackay, A., and George, R.C. 2000. Nontoxigenic *Corynebacterium diphtheriae*: An emerging pathogen in England and Wales? *Emerg. Infect. Dis.*, 6, 640–645.

Readon-Robinson, M.E., and Ton-That, H. 2013. Assembly and function of *Corynebacterium diphtheriae* pili. In: Burkovski, A., editor. *Corynebacterium diphtheriae and Related Toxigenic Species*. Dordrecht: Springer. pp. 123–142.

Rennie, R.P., Brosnikoff, C., Turnbull, L., Reller, L.B., Mirrett, S., Janda, W., Ristow, K., and Krilcich, A. 2008. Multicenter evaluation of the Vitek 2 anaerobe and *Corynebacterium* identification card. *J. Clin. Microbiol.*, 46, 2646–2651.

Rockhill, R.C., Hadiputranto, S.H., Siregar, S.P., and Muslihun, B. 1982. Tetracycline resistance of *Corynebacterium diphtheriae* isolated from diphtheria patients in Jakarta, Indonesia. *Antimicrob. Agents Chemother.*, 21, 842–843.

Romney, M.G., Roscoe, D.L., Bernard, K., Lai, S., Efstratiou, A., and Clarke, A.M. 2006. Emergence of an invasive clone of nontoxigenic *Corynebacterium diphtheriae* in the urban poor population of Vancouver, Canada. *J. Clin. Microbiol.*, 44, 1625–1629.

Sabbadini, P.S., Genovez, M.R., Silva, C.F., Adelino, T.L., Santos, C.S., Pereira, G.A., Nagao, P.E., Dias, A.A., Mattos-Guaraldi, A.L., and Hirata Jr., R. 2010. Fibrinogen binds to nontoxigenic and toxigenic *Corynebacterium diphtheriae* strains. *Mem. Inst. Oswaldo Cruz*, 105, 706–711.

Sabbadini, P.S., Assis, M.C., Trost, E., Gomes, D.L., Moreira, L.O., Dos Santos, C.S., Pereira, G.A., Nagao, P.E., Azevedo, V.A., Hirata Jr., R., Dos Santos, A.L., Tauch, A., and Mattos-Guaraldi, A.L. 2012. *Corynebacterium diphtheriae* 67–72p hemagglutinin, characterized as the protein DIP0733, contributes to invasion and induction of apoptosis in HEp-2 cells. *Microb. Pathog.*, 52, 165–176.

Sadoh, E.E., and Oladokun, R.E. 2012. Re-emergence of diphtheria and pertussis: Implications for Nigeria. *Vaccine*, 30, 7221–7228.

Sangal, V., and Hoskisson, P.A. 2013. Corynephages: Infections of infectors. In: Burkovski, A., editor. *Corynebacterium diphtheriae and Related Toxigenic Species*. Dordrecht: Springer. pp. 67–82.

Sangal, V., Tucker, N.P., Burkovski, A., and Hoskisson, P.A. 2012a. The draft genome sequence of *Corynebacterium diphtheriae mitis* NCTC 3529 reveals significant diversity between the primary disease causing biovars. *J. Bacteriol.*, 194, 3269.

Sangal, V., Tucker, N.P., Burkovski, A., and Hoskisson, P.A. 2012b. The genome of *Corynebacterium diphtheriae* biovar *intermedius* NCTC 5011. *J. Bacteriol.*, 194, 4738.

Sangal, V., Burkovski, A., Hunt, A.C., Edwards, B., Blom, J., and Hoskisson, P.A. 20134. A lack of genetic basis of biovar differentiation in clinically important *Corynebacterium diphtheriae* from whole genome sequencing. *Infect. Genet. Evol.*, in press21, 54–57.

Schmitt, M.P. 2013. Iron acquisition and iron-dependent gene regulation in *Corynebacterium diphtheriae*. In: Burkovski, A., editor. *Corynebacterium diphtheriae and Related Toxigenic Species*. Dordrecht: Springer. pp. 95–122.

Schuhegger, R., Lindermayer, M., Kugler, R., Heesemann, J., Busch, U., and Sing, A. 2008. Detection of toxigenic *Corynebacterium diphtheriae* and *Corynebacterium ulcerans* strains by a novel real-time PCR. *J. Clin. Microbiol.*, 46, 2822–2823.

Sears, A., McLean, M., Hingston, D., Eddie, B., Short, P., and Jones, M. 2012. Cases of cutaneous diphtheria in New Zealand: Implications for surveillance and management. *N. Z. Med. J.*, 125, 64–71.

Serby, M., Larson, P., and Kalstein, D.T. 1991. The nature and course of olfactory deficits in Alzheimer's disease. *Am. J. Psychiatry*, 148, 357–360.

Sharma, N.C., Banavaliker, J.N., Ranjan, R., and Kumar, R. 2007. Bacteriological and epidemiological characteristics of diphtheria cases in and around Delhi – Aretrospective study. *Indian J. Med. Res.*, 126, 545–552.

Shashikala, P., Reddy, P.V., Prashanth, K., Kanungo, R., Devi, S., Anitha, P., Rajarajeshweri, N., and Cherian, T.M. 2011. Persistence of nontoxigenic *Corynebacterium diphtheriae* biotype *gravis* strains in Pondicherry, Southern India. *J. Clin. Microbiol.*, 49, 763–764.

Shima, K., Kuhlenbäumer, G., and Rupp, J. 2010. *Chlamydia pneumoniae* infection and Alzheimer's disease: A connection to remember? *Med. Microbiol. Immunol. (Berlin)*, 199, 283–289.

Sing, A., Berger, A., Schneider-Brachert, W., Holzmann, T., and Reischl, U. 2011. Rapid detection and molecular differentiation of toxigenic *Corynebacterium diphtheriae* and *Corynebacterium ulcerans* strains by LightCycler PCR. *J. Clin. Microbiol.*, 49, 2485–2489.

Soto, A., Zapardiel, J., and Soriano, F. 1994. Evaluation of API Coryne system for identifying coryneform bacteria. *J. Clin. Pathol.*, 47, 756–975.

Sykes, J.E., Mapes, S., Lindsay, L.L., Samitz, E., and Byrne, B.A. 2010. *Corynebacterium ulcerans* bronchopneumonia in a dog. *J. Vet. Int. Med.*, 24, 973–976.

Tang, Y.W., von Graevenitz, A., Waddington, M.G., Hopkins, M.K., Smith, D.H., Li, H., Kolbert, C.P., Montgomery, S.O., and Persing, D.H. 2000. Identification of coryneform bacterial isolates by ribosomal DNA sequence analysis. *J. Clin. Microbiol.*, 38, 1676–1678.

Theel, E.S., Schmitt, B.H., Hall, L., Cunningham, S.A., Walchak, R.C., Patel, R., and Wengenack, N.L. 2012. Formic acid-based direct, on-plate testing of yeast and *Corynebacterium* species by Bruker Biotyper matrix-assisted laser desorption ionization-time of flight mass spectrometry. *J. Clin. Microbiol.*, 50, 3093–3095.

Tiwari, T.S.P., Golaz, A., Yu, D.T., Ehresmann, K.R., Jones, T.F., Hill, H.E., Cassiday, P.K., Pawloski, L.C., Moran, J.S., Popovic, T., and Wharton, M. 2008. Investigations of 2 cases of diphtheria-like illness due to toxigenic *Corynebacterium ulcerans*. *Clin. Infect. Dis.*, 46, 395–401.

Trost, E., and Tauch, A. 2013. Comparative genomics and pathogenicity islands of *Corynebacterium diphtheriae*, *Corynebacterium ulcerans* and *Corynebacterium pseudotuberculosis*. In: Burkovski, A., editor. *Corynebacterium diphtheriae and Related Toxigenic Species*. Dordrecht: Springer. pp. 143–170.

Trost, E., Blom, J., Soares, S. de C., Huang, I.H., Al-Dilaimi, A., Schröder, J., Jaenicke, S., Dorella, F.A., Rocha, F.S., Miyoshi, A., Azevedo, V., Schneider, M.P., Silva, A., Camello, T.C., Sabbadini, P.S., Santos, C.S., Santos, L.S., Hirata Jr., R., Mattos-Guaraldi, A.L., Efstratiou, A., Schmitt, M.P., Ton-That, H., and Tauch, A. 2012. Pangenomic study of *Corynebacterium diphtheriae* that provides insights into the genomic diversity of pathogenic isolates from cases of classical diphtheria, endocarditis, and pneumonia. *J. Bacteriol.*, 194, 3199–3215.

Tyas, S.L., Manfreda, J., Strainc, C.A., and Montgomery, P.R. 2001. Risk factors for Alzheimer's disease: A population-based, longitudinal study in Manitoba, Canada. *Int. J. Epidemiol.*, 30, 590–597.

Varol, B., Özerman, E., and Bektas, M. 2013. Toxin structure, delivery and action. In: Burkovski, A., editor. *Corynebacterium diphtheriae and Related Toxigenic Species*. Dordrecht: Springer. pp. 83–94.

Ventura, M., Canchaya, C., Tauch, A., Chandra, G., Fitzgerald, G.F., Chater, K.F., and van Sinderen, D. 2007. Genomics of actinobacteria: Tracing the evolutionary history of an ancient phylum. *Microbiol. Mol. Biol. Rev.*, 71, 495–548.

Verreault, R., Laurin, D., Lindsay, J., and De Serres, G. 2001. Past exposure to vaccines and subsequent risk of Alzheimer's disease. *Can. Med. Assoc. J.*, 165, 1495–1498.

Vitek, C.R. 2006. Diphtheria. *Curr. Top. Microbiol. Immunol.*, 34, 71–94.

Vitek, C.R., and Wharton, M. 1998. Diphtheria in the former Soviet Union: Reemergence of a pandemic disease. *Emerg. Infect. Dis.*, 4, 539–550.

von Behring, E.A., and Kitosato, S. 1890. Über das Zustandekommen der Diphtherie-Immunität und der Tetanus-Immunität bei Thieren. *Dtsch. Med. Wochenschr.*, 16, 113–114.

von Hunolstein, C., Rota, M.C., Alfarone, G., Ricci, M.L., and Salmaso, S., and the Italian Serology Working Group. 2000. Diphtheria antibody levels in the Italian population. *Eur. J. Clin. Microbiol. Infect. Dis.*, 19, 433–437.

von Hunolstein, C., Scopetti, F., Efstratiou, A., and Engler, K. 2002. Penicillin tolerance amongst non-toxigenic *Corynebacterium diphtheriae* isolated from cases of pharyngitis. *J. Antimicrob. Chemother.*, 50, 125–128.

von Hunolstein, C., Alfarone, G., Scopetti, F., Pataracchia, M., La Valle, R., Franchi, F., Pacciani, L., Manera, A., Giammanco, A., Farinelli, S., Engler, K., De Zoysa, A., and Efstratiou, A. 2003. Molecular epidemiology and characteristics of *Corynebacterium diphtheriae* and *Corynebacterium ulcerans* strains isolated in Italy during the 1990s. *J. Med. Microbiol.*, 52, 181–188.

Wagner, J., Ignatius, R., Voss, S., Höpfner, V., Ehlers, S., Funke, G., Weber, U., and Hahn, H. 2001. Infection of the skin caused by *Corynebacterium ulcerans* and mimicking classical cutaneous diphtheria. *Clin. Infect. Dis.*, 33, 1598–1600.

Wagner, K.S., Stickings, P., White, J.M., Neal, S., Crowcroft, N.S., Sesardic, D., and Efstratiou, A. 2009. A review of the international issues surrounding the availability and demand for diphtheria antitoxin for therapeutic use. *Vaccine*, 28, 14–20.

Wagner, K.S., White, J.M., Crowcroft, N.S., De Martin, S., Mann, G., and Efstratiou, A. 2010. Diphtheria in the United Kingdom, 1986–2008: The increasing role of *Corynebacterium ulcerans*. *Epidemiol. Infect.*, 138, 1519–1530.

Wagner, K.S., White, J.M., Neal, S., Crowcroft, N.S., Kupreviciene, N., Paberza, R., Lucenko, I., Joks, U., Akbas, E., Alexandrou-Athanassoulis, H., Detcheva, A., Vuopio, J., von Hunolstein, C., Murphy, P.G., Andrews, N., and Efstratiou, A. 2011. Screening for *Corynebacterium diphtheriae* and *Corynebacterium ulcerans* in patients with upper respiratory tract infections 2007–2008: A multicentre European study. *Clin. Microbiol. Infect.*, 17, 519–525.

Wagner, K.S., Zakikhany, K., White, J.M., Amirthalingam, G., Crowcroft, N.S., and Efstratiou, A. 2013. Diphtheria surveillance. In: Burkovski, A., editor. *Corynebacterium diphtheriae and Related Toxigenic Species*. Dordrecht: Springer. pp. 207–224.

World Health Organization. 2006. Diphtheria vaccine: WHO position paper. *Wkly. Epidemiol. Rec.*, 81, 24–32.

Wozniak, M.A., and Itzhaki, R.F. 2010. Antiviral agents in Alzheimer's disease: Hope for the future? *Ther. Adv. Neurol. Disord.*, 3, 141–152.

Zasada, A.A. 2013. Antimicrobial susceptibility and treatment. In: Burkovski, A., editor. *Corynebacterium diphtheriae and Related Toxigenic Species*. Dordrecht: Springer. pp. 239–248.

Zasada, A.A., Baczewska-Rej, M., and Wardak, S. 2010. An increase in non-toxigenic *Corynebacterium diphtheriae* infections in Poland – Molecular epidemiology and antimicrobial susceptibility of strains isolated from past outbreaks and those currently circulation in Poland. *Int. J. Infect. Dis.*, 14, e907–e912.

Zhi, X.Y., Li, W.J., and Stackebrandt, E. 2009. An update of the structure and 16S rRNA gene sequence-based definition of higher ranks of the class Actinobacteria, with the proposal of two new suborders and four new families and emended descriptions of the existing higher taxa. *Int. J. Syst. Evol. Microbiol.*, 59, 589–608.

Zuber, P.L., Gruner, E., Altwegg, M., and von Graevenitz, A. 1992. Invasive infection with non-toxigenic *Corynebacterium diphtheriae* among drug users. *Lancet*, 339, 1359.

Chapter 39

Pathogenesis of *Staphylococcus aureus* in Humans

Thea Lu and Frank R. DeLeo

Laboratory of Human Bacterial Pathogenesis, Rocky Mountain Laboratories, National Institute of Allergy and Infectious Diseases, National Institutes of Health, Hamilton, MT, USA

39.1 Introduction

Staphylococcus aureus is a leading cause of human infections worldwide (Diekema et al., 2001). The pathogen is also a commensal organism and approximately 30% of non-institutionalized individuals are colonized asymptomatically by *S. aureus* in the anterior nares (Gorwitz et al., 2008). Nasal carriage is associated with infection in susceptible individuals (von Eiff et al., 2001). The pathogen can cause peripheral to serious infections in almost all tissues, especially in immunocompromised people (Lowy, 1998). Disease can manifest as skin and soft-tissue infections, pneumonia, surgical-site infections, bloodstream infections, endocarditis, septic shock, and many others (CDC, 2003; Fey et al., 2003; Herold et al., 1998; Lowy, 1998). As well as causing a myriad of infections, *S. aureus* is well-known for its ability to acquire resistance to antibiotics (Chambers and DeLeo, 2009; Noble et al., 1992). Methicillin-resistant *S. aureus* (MRSA) strains resistant to most clinically applicable antibiotics pose a significant health risk, as is typical for MRSA strains that are endemic in hospitals (Levy and Marshall, 2004). Additionally, MRSA bacteremia is associated with significant increases in length and cost of hospitalization (Cosgrove et al., 2005). In the past 60 years, several waves of antibiotic-resistant *S. aureus* have caused serious epidemics and pandemics (Chambers and DeLeo, 2009; DeLeo and Chambers, 2009). More recently, there has been emergence of community-associated MRSA (CA-MRSA), which by definition occurs in otherwise healthy individuals. This attribute suggests that CA-MRSA strains are endowed with enhanced virulence and/or transmissibility (DeLeo et al., 2010; Herold et al., 1998). In the subsequent sections, we describe the key characteristics of molecular biology and genome, epidemiology, and pathogenesis important in understanding the emergence and success of the spread of *S. aureus*.

39.2 Molecular biology and genome characteristics

Multilocus sequence typing (MLST), which is based on the single nucleotide variations in seven housekeeping genes, has been widely used for typing of *S. aureus* isolates (Chambers and DeLeo, 2009; Enright et al., 2000). MLST indexes nucleotide variation that accumulates slowly and is thus useful for tracking bacterial evolution over an extended period of time. The MLST profile is used to group *S. aureus* into clonal complexes, which are defined as having nucleotide identity in five or more of the housekeeping genes. HA-MRSA lineages can be grouped in five major clonal complexes (CCs): CC5, CC8, CC22, CC30, and CC45 (Chambers and DeLeo, 2009; Enright et al., 2002). The predominant CA-MRSA strains circulating belong to CC1, CC8, CC30, CC59, and CC80 lineages (David and Daum, 2010; DeLeo et al., 2010; Witte, 2009). Despite the known strain variance by MLST and other typing methods, more than 95% of *S. aureus* strains share about 75% of the genes, which are known collectively as the core genome (Grumann et al., 2014). The *S. aureus* core genome contains housekeeping genes, genes required for growth, surface-associated genes (including microbial surface components), regulator genes, and several conserved virulence determinants, totaling approximately 2.3 Mbp. By comparison, the accessory genome – that derived from mobile genetic elements (MGEs) – encodes exotoxins, superantigens, antimicrobial-resistant genes, and other factors that promote infection and inhibit the innate immune response (Lindsay and Holden, 2004). Although MLST has been highly useful for understanding *S. aureus* evolution and strain diversity, the method indexes variation at small number of nucleotides. Therefore, comparison of *S. aureus* genomes – especially a comparison of all nucleotide changes in the core genome – is now widely used for analysis of strain evolution, genetic diversity, and tracking outbreaks (Adem et al., 2005; DeLeo et al., 2011; Espadinha et al., 2013; Kennedy et al., 2008; Miller et al., 2007; Wertheim et al., 2005).

39.2.1 The core genome

39.2.1.1 Secreted toxins

A number of virulence determinants encoded by the core genome have been reported to mediate staphylococcal pathogenicity. Such molecules include exotoxins and regulator genes, and alpha-hemolysin (Hla) and alpha-type phenol-soluble modulins (PSMα) are key exotoxins largely conserved among *S. aureus* strains. Hla is known to cause lysis of erythrocytes, platelets, monocytes, and endothelial cells (Bhakdi and Tranum Jensen, 1991; Song et al., 1996). Findings in animal infection models provide strong support to the idea that Hla contributes to *S. aureus* pathogenesis (Bartlett et al., 2008; Kennedy et al., 2010; Kernodle et al., 1997; Kobayashi et al., 2011; Powers et al., 2012; Rauch et al., 2012; Wardenburg and Schneewind, 2008). PSMα peptides are secreted by amphipathic alpha helical molecules that are ~20 amino acids in length. These detergent-like peptides have cytolytic activity toward host cells such as erythrocytes and neutrophils (Cheung et al., 2012; Wang et al., 2007). Importantly, they contribute significantly to virulence in mouse and rabbit infection models (Kobayashi et al., 2011; Wang et al., 2007). Although genes encoding PSMα peptides are present in virtually all *S. aureus* isolates and strains, the molecules are expressed at high levels in prominent CA-MRSA strains (Clarke, 2010; Li et al., 2009). Thus, it is likely that they have contributed to the success of CA-MRSA strains such as USA300.

39.2.1.2 Two-component gene regulatory systems

The genome of *S. aureus* encodes 16 two-component gene regulatory systems (TCSs), including several known to contribute to antibiotic resistance and virulence (Fournier and Hooper, 2000; Kuroda et al., 2003; Nygaard et al., 2010). *S. aureus* accessory gene regulator (Agr) is a well-characterized global gene regulator that encodes two divergent transcriptional units driven by the promoters P2 and P3 (Novick, 2003a). The P2 promoter, activated by an autoinduction mechanism, drives the transcription of *agrDBCA*, which encodes the cell density sensing signal transduction system. The autoinduction peptide (AIP) is encoded by *agrD*, and AIP is secreted into the culture supernatant by the transmembrane protein AgrB (Zhang et al., 2002). AIP then binds to the signal receptor, AgrC, which in turn activates the response regulator AgrA (Ji et al., 1995; Lina et al., 1998; Lyon et al., 2000). Activated AgrA upregulates P2 and P3. P3 then drives production of RNAIII, which has been shown to regulate the expression of downstream virulence factors (Nastaly et al., 2010; Novick, 2003a). For example, the *agr* system plays a critical role in the virulence of the CA-MRSA strain USA300 by upregulating numerous putative and proven virulence factors, including toxins, exoenzymes, fibrinogen-binding proteins, and methicillin-resistant genes (Chatterjee et al., 2011; Cheung et al., 2011; Genestier et al., 2005; Montgomery et al., 2010; Tsuji et al., 2007; Wirtz et al., 2009). Whole genome sequencing of CC30 isolates also revealed the importance of *agr* in the evolution of virulence in healthcare settings (DeLeo et al., 2011; McAdam et al., 2012). Additionally, the role of *agr* in mediating virulence *in vivo* has been shown in several staphylococcal genetic backgrounds and animal models of endocarditis, musculoskeletal infection, subcutaneous abscess, and pneumonia (Blevins et al., 2003; Cheung et al., 1994; Kobayashi et al., 2011; Montgomery et al., 2010; Wright et al., 2005).

The *saeRS* TCS is also known to play an important role in virulence and is involved in *S. aureus* adherence, hemolytic activity, and production of multiple exotoxins, including Hla. In studies with CA-MRSA strains USA400 and USA300, *saeRS* regulates expression of numerous molecules that likely contribute to the success of these microbes as human pathogens (Voyich et al., 2009). For example, *saeRS* influences expression of exotoxins and adhesins in USA400, while the corresponding TCS in USA300 has been shown to regulate methicillin-resistant genes and other extracellular virulence genes (Nygaard et al., 2010; Voyich et al., 2009).

Other TCSs have functions outside of virulence *per se*. For example, the *vraRS* TCS has been implicated in antibiotic resistance. This regulon can be induced by exposure of bacterium to antibiotics that target the cell wall. Notably, a gene knockout strain lacking *vraRS* has increased susceptibility to cell wall synthesis inhibitors (Kuroda et al., 2003; Yin et al., 2006).

39.2.1.3 The accessory genome

The accessory genome of *S. aureus* plays an important role in the evolution of virulence, particularly in the development of antibiotic resistance. However, the acquisition of MGEs has also enabled *S. aureus* to obtain additional virulence factors which may also allow for changes in virulence in emerging strains (Cameron et al., 2011). The high genetic and phenotypic plasticity of the pathogen, which is important for its ability to adapt to diverse and hostile environments, is in part due to the relatively large accessory genome that comprises up to 25% of the staphylococcal chromosome (Feng et al., 2008; Lindsay, 2010; Malachowa and DeLeo, 2010; Suzuki et al., 2012). *S. aureus* most likely acquires new genetic material through transduction via bacteriophages, since conjugation and DNA transformation are assumed to occur rarely (Lindsay and Holden, 2006; Morikawa et al., 2012). However, virulence determinants are not necessarily associated with a particular phage and may be exchanged by horizontal gene transfer and recombination (Brüssow et al., 2004; Feng et al., 2008; Malachowa and DeLeo, 2010). Hence, staphylococcal toxin genes differ in their horizontal genetic mobility and distribution among *S. aureus* strains (Grumann et al., 2014). Studies have demonstrated that infection conditions favor strains that possess atypical genomic integration and phage mobility over colonizing strains in healthy carriers (Goerke et al., 2009, 2006; Goerke and Wolz, 2004). There is also segregation of strain genotypes based on geographic location and/or demographics. For example, SCC*mec*IV and Panton–Valentine leukocidin, which are encoded on MGEs, were common among CA-MRSA strains from three continents, but early on were rarely – if ever – found in HA-MRSA isolates (Vandenesch et al., 2003). It is also noteworthy that these strains do not share a common genetic lineage.

The *S. aureus* genome contains 1–4 prophages (Lindsay, 2010). These prophages often encode molecules that facilitate interaction with the host, including virulence determinants such as staphylokinase (*sak*), chemotaxis inhibitory protein of *S. aureus* CHIPS (*chp*), and staphylococcal inhibitor of complement SCIN (*scn*) (Brüssow et al., 2004; Coleman et al., 1989; Feng et al., 2008; Lindsay, 2010; van Wamel et al., 2006). Other prophages encode superantigen toxins (*sea, seg, sek, sek2, sep, seq*), Panton–Valentine leukocidin (*lukS-PV* and *lukF-PV*), and exfoliative toxin A (*eta*) (Feng et al., 2008; Lindsay, 2010; Malachowa and DeLeo, 2010). Phages often encode a single virulence factor, although phiSa3 phages, such as phiN315, carry as many as five virulence factor genes that form an immune escape complex (IEC) (Coleman et al., 1989; Goerke et al., 2006; Malachowa and DeLeo, 2010; van Wamel et al., 2006). The phage-associated gene *sasX*, involved with the acquisition of tetracycline and methicillin resistance, was identified in ST239 strains (Li et al., 2012). This gene is important in immune evasion because it mediates staphylococcal aggregation and adhesion and has been demonstrated to play a role in virulence in mouse skin and lung infection models (Li et al., 2012).

SCC*mec* (staphylococcal cassette chromosome *mec*) is an MGE that has been shown to modulate transmission, virulence, and multidrug resistance in MRSA. It is integrated into *orfX*, an *S. aureus* gene of unknown function, and consists of three elements: (1) the *mec* complex including the *mecA* gene encoding a unique penicillin-binding protein that effectively catalyzes cell wall synthesis despite the presence of anti-staphylococcal penicillins, cephalosporins, and carbapenems; (2) the cassette chromosome recombinase (*ccr*) genes; and (3) joining regions which may carry additional antibiotic-resistant genes (Chambers and DeLeo, 2009; Ito et al., 2003; Malachowa and DeLeo, 2010). SCC*mec* is classified based upon the *ccr* genes and the genetic composition of the *mec* complex with SCC*mec* subtypes differentiated by the joining regions (Ito et al., 2001; IWC-SCC, 2009; Katayama et al., 2000; Okuma et al., 2002). To date, at least 11 types of SCC*mec* allotypes have been identified (Hao et al., 2012; Ito et al., 2014; Turlej et al., 2011). HA-MRSA strains, especially those isolated before 1990, contain SCC*mec*I-SCC*mec*III, while SCC*mec*IV, and to some extent SCC*mec*V, predominate in CA-MRSA isolates (Chambers and DeLeo, 2009). Several CA-MRSA lineages similarly contain SCC*mec*IV and PVL-containing prophages, suggesting acquisition of these MGEs from a single source (Chua et al., 2011b). The conditions that select for methicillin resistance may help drive the selection of virulence determinants associated with SCC*mec*. These factors include the arginine catabolic mobile element (ACME), which contains the gene clusters *arc* and *opp-3* postulated to be involved in virulence (Diep et al., 2006, 2008). Recently, Thurlow et al. (2013) demonstrated that ACME promotes survival of USA300 in acidic conditions that mimic human skin, and it interferes with host polyamines that contribute to the host healing process. CA-MRSA strains containing SCC*mec*IV grow faster than nosocomial strains carrying other SCC*mec* allotypes and have growth rates similar to methicillin-susceptible *S. aureus* (MSSA) isolates (Diep et al., 2008; Okuma et al., 2002). This lack of cost in fitness to the organism as well as its wide distribution among *S. aureus* isolates are likely major contributors to the success of SCC*mec*IV-containing organisms in the community (Chambers and DeLeo, 2009).

S. aureus pathogenicity islands (SaPIs) comprise another group of MGEs that insert into different sites in the *S. aureus* chromosome. They are highly mobile with the assistance of helper phages and are known to insert into six different sites (Lindsay et al., 1998; Novick et al., 2010). The core structure of SaPIs is highly conserved and may contain virulence factors such as toxic shock syndrome toxin-1 TSST-1 (*tst*) and staphylococcal enterotoxins (Novick, 2003b). A more recent analysis revealed the existence of seven novel SaPIs that can each harbor multiple superantigens, including *seb, selk*, and *selq* (Sato'o et al., 2013).

Some MRSA strains also harbor integrative and conjugative elements (ICE) (Schijffelen et al., 2010; Smyth and Robinson, 2009). ICEs typically contain conjugation components that may play a role in the conjugative transfer of antibiotic-resistant genes between strains unable to undergo transformation and transduction (Smyth and Robinson, 2009). One particular ICE, ICE6013, integrates into multiple loci and encodes predicted secreted and surface proteins, suggesting that ICEs may also contribute to phenotypic variation in *S. aureus* (Smyth and Robinson, 2009).

39.3 History and epidemiology

S. aureus was identified and named by Sir Alexander Ogston in the latter part of the 19th century (Newsom, 2008; Smith, 1982). It is perhaps fitting that Ogston, a Scottish surgeon, first cultured the organism from the pus of unopened human abscesses, since abscesses are a hallmark of *S. aureus* infections. It is notable that the mortality associated with *S. aureus* infections was quite high prior to the antibiotic era (e.g., Skinner and Keefer 1941, reported ~82% mortality for *S. aureus* infections that had associated bacteremia). Thus, it is not surprising that Alexander Fleming's discovery of penicillin in 1928 was one of the most important scientific discoveries in history. By 1945, the United States had distributed 650 billion units of penicillin (Grossman, 2008). The introduction of penicillin and other antibiotics coincided with the decline of mortality due to infectious diseases in the United States and elsewhere in the industrialized world (Armstrong et al., 1999; Cohen, 2000). Although penicillin was highly effective for treatment of *S. aureus* infections in the 1940s, the organism soon developed resistance to penicillin, a phenomenon conferred by genes encoding β-lactamase (Murray and Moellering, 1978). In the 1950s, penicillin was no longer universally effective for treatment of *S. aureus* infections and new antibacterial agents such as methicillin were introduced to address this escalating problem.

39.4 The rise of MRSA

S. aureus rapidly acquired resistance to methicillin – first described in 1961 – 2 years after methicillin was first used clinically (Jevons, 1961). MRSA has since spread worldwide. Highly successful strains emerged in the 1970s and this phenomenon led to numerous outbreaks in hospitals. By the late 1990s to early 2000s, the annual rate of nosocomial staphylococcal infections in the United States was 7.1–11.0% (Noskin et al., 2007). In 2005, the rate of invasive MRSA infections in the United States was estimated at 31.8/100,000 individuals, with a mortality rate of 6.3/100,000 individuals (Klevens et al., 2008). This relatively high level of *S. aureus*–associated mortality – approximately 19,000 deaths per year – exceeded that of HIV infections in the United States (Boucher and Corey, 2008). In 2011, the incidence and mortality rates of overall *S. aureus* infections decreased to 25.82/100,000 and 3.62/100,000 individuals, respectively (Dantes et al., 2013).

MRSA epidemics have typically been associated with specific staphylococcal clones rather than a diverse number of strains (Chambers and DeLeo, 2009). Successive epidemic waves occurred in the mid-1950s with the penicillin-resistant phage type 80/81 strains, the 1960s with the archaic MRSA strains, the 1980s with the toxic shock syndrome clones, the 1980s to the present with HA-MRSA, the 1990s until present with CA-MRSA, and the early 2000s with LA-MRSA. An early MLST analysis of a global collection of 912 MRSA and MSSA isolates indicated that there are 11 major MRSA clones from 5 groups of related genotypes (Enright et al., 2002).

39.4.1 CA-MRSA

The first descriptions of CA-MRSA originated from Western Australia in the early 1990s (Udo et al., 1993). One unexpected finding was that these individuals lived in remote rural areas and had no recent contact with healthcare facilities. Soon afterward, CA-MRSA was reported in North America (Herold et al., 1998) and then globally (Chua et al., 2011a; David and Daum, 2010; DeLeo et al., 2010) (Table 39-1). The earliest CA-MRSA infections were not caused by multidrug-resistant strains, and infections could be treated readily with non-β-lactam antibiotics (Herold et al., 1998). However, over time some of these strains acquired increased resistance to multiple antimicrobial agents, a characteristic associated with the movement of these strains into the healthcare setting (McDougal et al., 2010).

CA-MRSA isolates are comprised of at least 20 distinct genetic lineages, but 5 lineages, ST-1-IV (WA-1, USA400), ST8-IV (USA300), ST30-IV (South West Pacific clone), ST59-V (Taiwan clone), and ST80-IV (European clone), make up the majority of these cases (David and Daum, 2010; DeLeo et al., 2010; Witte, 2009). ST8 and ST30 are considered pandemic (DeLeo et al., 2010; Monecke et al., 2011), and ST8 (USA300) is currently the most prominent cause of community-associated bacterial infections in the United States.

39.4.2 USA300 – an epidemic CA-MRSA clone

The first reported USA300 infections occurred among football players in Pennsylvania and prisoners in Los Angeles and Mississippi (Tenover and Goering, 2009). CA-MRSA transmission results from direct contact with the organism, typically by skin-to-skin contact with a colonized or infected individual. In some cases, fomites contaminated with CA-MRSA may have a role in transmission. The Centers for Disease Control and Prevention in Atlanta, GA, USA, have delineated five factors that facilitate MRSA transmission: crowding, frequent skin-to-skin contact, compromised skin integrity, contaminated items and surfaces, and lack of cleanliness (CDC, 2013). These factors are present in the diverse populations that have increased numbers of CA-MRSA infections (DeLeo et al., 2010). The vast majority of USA300 infections (75–95%) are those affecting skin and soft-tissues (Kazakova et al., 2005; McDougal et al., 2003; Pan et al., 2003, 2005; Tenover et al., 2006). However, USA300 is replacing the traditional HA-MRSA strains as the primary cause of

Table 39-1 Pandemic and epidemic MRSA lineages

CC	ST	SCC*mec*	Common names	Locations	Epidemiology
1	1	IV	CMRSA-7, MW2, USA400, WA-1, WA MRSA-1, WA MRSA-45	Africa, Asia, Australia, Europe, Middle East, North America, Oceania	CA
	573	V	WA MRSA-10	Australia	CA
	772	V	Bengal Bay Clone, WA MRSA-60	Australia, Asia, Europe, Middle East	CA
5	5	I	Chilean Clone, Cordobes Clone, EMRSA-3, Geraldine Clone, Southern Germany Clone, WA MRSA-18, WA MRSA-21	Africa, Australia, Caribbean, Europe, Middle East, South America	HA, CA
		II	CMRSA-2, EMRSA-3, Irish AR07.3, Irish AR07.4, Irish AR11, New York–Japan Clone, Rhine-Hesse Epidemic Strain, USA100	Asia, Australia, Caribbean, Europe, Middle East, North America, South America	HA, CA
		IV	Marseille Cystic Fibrosis Clone, Pediatric Clone, USA800, USA900, WA MRSA-3, WA MRSA-74	Australia, Caribbean, Europe, Middle East, North America, South America	HA, CA
		V	WA MRSA-11, WA MRSA-14, WA MRSA-34, WA MRSA-35, WA MRSA-90, WA MRSA-108, WA MRSA-109	Africa, Asia, Australia, Europe	CA
		VI	HDE288, Pediatric Clone	Australia, Europe, North America	HA, CA
		VII	JCSC6082	Europe	
	73	IV	WA MRSA-50, WA MRSA-65	Australia	CA
	105	II	USA100	Europe, North America	HA
	225	II	USA100	Asia, Europe, North America	HA
	228	I	Italian Clone, South German Epidemic Strain	Europe	HA
	231	II	USA100	North America	
	526	IV	WA MRSA-39	Australia	
	575	IV	WA MRSA-25	Australia	CA
	835	I	WA MRSA-40, WA MRSA-46, WA MRSA-48	Australia	
	835	IV	WA MRSA-48	Australia	CA
6	6	IV	WA MRSA-51	Asia, Australia, Middle East	CA
8	8	II	Irish-1, Irish AR05, Irish AR13, Irish AR14, Irish New03	Europe, North America	HA
		IV	CMRSA-5, CMRSA-10, EMRSA-2, EMRSA-6, EMRSA-12, EMRSA-13, EMRSA-14, Irish AR43, Irish-02, Lyon Clone, Spanish PFGE type A, USA300, USA500, WA MRSA-5, WA MRSA-6, WA MRSA-12, WA MRSA-31, WA MRSA-62	Africa, Asia, Australia, Caribbean, Europe, Middle East, North America, Oceania, South America	HA, CA
		V	WA MRSA-53	Africa, Australia, Europe	CA
		VIII	CMRSA-9, WA MRSA-16	Australia, Europe, North America	HA
	72	IV	USA700, WA MRSA-44	Asia, Australia, Caribbean, Europe, North America, South America	HA, CA
	239	III	AUS-2, AUS-3, Brazilian Clone, CMRSA-6, Czech Clone, EMRSA-1, EMRSA-4, EMRSA-7, EMRSA-9, EMRSA-11, Hungarian Clone, Irish Phenotype III, Irish AR01, Irish AR09, Irish AR15, Irish AR23, Portuguese Clone, Vienna Clone	Africa, Asia, Australia, Caribbean, Europe, Middle East, North America, South America	HA, CA
	247	I	EMRSA-5, EMRSA-8, EMRSA-17, Iberian Clone, Irish AR22, Irish New02, North German Epidemic Strain, Roman Clone, Spanish PFGE type E1	Australia, Europe, North America, South America	HA
	250	I	Archaic, Early or Ancestral MRSA, Irish AR02, Irish Phenotype I and II	Australia, Europe	HA
	254	Irregular	EMRSA-10, Hannover Epidemic Strain	Europe	HA
	576	IV	EMRSA-14, WA MRSA-5, WA MRSA-6, WA MRSA-31	Australia, Europe, North America	
9	834	IV	WA MRSA-13	Asia, Australia	CA
12	12	Irregular	WA MRSA-59	Australia, Europe, North America	

(continued)

Table 39-1 (*Continued*)

CC	ST	SCCmec	Common names	Locations	Epidemiology
15	15	I	USA900	Australia, Europe	HA
		V	USA900	Middle East	CA
22	22	IV	Barnim Epidemic Strain, EMRSA-15, Irish AR06, Spanish PFGE type E13, CMRSA-8	Africa, Asia, Australia, Europe, Middle East, North America	HA, CA
30	30	IV	Mexican Clone, OSPC, Southwest Pacific Clone, UR6, USA1100, WA1, West Samoan Phage Pattern (WSPP) Clone	Asia, Australia, Europe, Middle East, North America, South America	CA
	36	II	CMRSA-4, EMRSA-16, Irish AR7.0, Irish AR07.2, Spanish PFGE type E12, USA200	Africa, Australia, Europe, North America, South America	HA
	36	IV	EMRSA-16, USA200	Australia, Europe, North America	HA
	39	II	CMRSA-4, EMRSA-16, Irish AR7.0, Irish AR07.2, Spanish PFGE type E12, USA200	Europe, North America	HA
45	45	II	CMRSA-1, USA600	Europe, North America	HA
		IV	Berlin Clone, Berlin Epidemic Strain, USA600, WA MRSA-23, WA MRSA-75	Asia, Australia, Europe, Middle East, North America	HA, CA
		V	Vic CA-MRSA, WA MRSA-4, WA MRSA-84	Asia, Australia, Europe	CA
	1970	V	WA MRSA-106	Australia	CA
59	59	Composite	WA MRSA-15	Asia, Australia	HA, CA
		IV	USA1000, WA MRSA-15, WA MRSA-55, WA MRSA-56, WA MRSA-73	Asia, Australia, Europe, North America	HA, CA
		V	Taiwan Clone, WA MRSA-9	Asia, Australia, Caribbean, Europe, North America	CA
	87	IV	WA MRSA-24	Australia, North America	
	952	V	Taiwan Clone	Asia, Australia, Europe	
75	75	IV	WA MRSA-8 and -79	Australia	CA
	883	IV	WA MRSA-47	Australia	CA
	1304	IV	WA MRSA-72	Australia	CA
80	80	IV	European CA-MRSA Clone, EUST80	Africa, Australia, Europe, Middle East	CA
	583	IV	WA MRSA-17	Australia	CA
	728	IV	WA MRSA-30	Australia	CA
88	78	IV	WA MRSA-2	Australia, Europe	CA
	88	IV	WA-MRSA-2	Africa, Asia, Australia, Europe, Middle East	HA, CA
93	93	IV	Queensland Clone	Australia, Europe	CA
97	97	IV	WA MRSA-54	Australia, Europe, Middle East	LA, CA
	953	IV	WA MRSA-54	Australia	CA
121	121	V	USA1200	Asia, Australia, Europe, Middle East	CA
	577	V	WA MRSA-22	Asia, Australia	CA
152	152	V	WA MRSA-89	Europe	CA
188	188	IV	WA MRSA-38, -78	Australia	HA, CA
361	361	VII	WA MRSA-28	Australia	
	672	IV	ST672-MRSA-IV, WA MRSA-29	Australia, Europe	
398	398	V	S0385, LA-MRSA	Asia, Europe, North America, South America	LA, CA

Data were gathered from Adler et al. (2012), Ateba Ngoa et al. (2012), Brauner et al. (2013), Cañas-Pedrosa et al. (2012), Chambers and DeLeo (2009), Chroboczek et al. (2013), Chuang and Huang (2013), Coombs et al. (2013), David et al. (2013), El-Mahdy et al. (2014), Espadinha et al. (2013), Fankhauser et al. (2013), Fernandez et al. (2013), García et al. (2012), González-Domínguez et al. (2012), Hall et al. (2009), He et al. (2013), Ho et al. (2012), Hudson et al. (2013), Jansen van Rensburg et al. (2012), Kawaguchiya et al. (2013), Khalil et al. (2012), Mediavilla et al. (2012), Medina et al. (2013), Menegotto et al. (2012), Monecke et al. (2011), Nichol et al. (2013), Nimmo et al. (2013), Portillo et al. (2013), Rodríguez-Noriega et al. (2010), Senn et al. (2013), Sonnevend et al. (2012), Stefani et al. (2012), Sung et al. (2012), Tavares et al. (2013), Teixeira et al. (2012), Velasco et al. (2012), Vergison et al. (2012), Williamson et al. (2013), Wu et al. (2013), Xiao et al. (2013). CC, clonal complex; ST, sequence type; SCCmec, staphylococcal cassette chromosome *mec*; HA, hospital acquired; CA, community acquired; LA, livestock associated.

bloodstream infections in nosocomial settings (Tenover et al., 2012). In Canada, USA300 first emerged in the western provinces, but since 2004 has spread in a manner similar to that in the United States, and the strain has become the predominant CA-MRSA strain (Christianson et al., 2007; Laupland et al., 2008; Simor et al., 2010). Although USA300 isolates have been recovered from patients in Latin America, and notably in Columbia, these isolates are variants of the US and Canadian epidemic USA300 clone known as USA300-LV (Nimmo, 2012). The strain was found in rural scavenging pigs in Peru, a finding that provides support to the idea that these animals could serve as a possible reservoir (Arriola et al., 2011). While the prevalence of

USA300 is low in countries outside of the Americas, the success of USA300 as a human pathogen makes it a major cause of global concern.

The first European report of USA300 occurred in Denmark in 2000 and was soon followed by a number of cases spanning 15 countries (Nimmo, 2012). In contrast to North America, USA300 is far less prevalent in Europe, with the proportion of isolates ranging from 0.8% to 3.5% in various countries (Nimmo, 2012). There is no clear explanation for the limited success of USA300 in Europe. One possibility is that the niche filled by USA300 in North America is occupied by other unrelated strains in Europe. Indeed, the predominant CA-MRSA strains circulating in Europe are genetically distinct from those found in other parts of the world (Tristan et al., 2007). The European CA-MRSA clone, known as ST80-IV, was first identified in 2003 in Greece – the country with the highest incidence of European MRSA infections (Chua et al., 2011a; Otter and French, 2010). ST80 appears to have originated from North Africa or the Middle East, as it is the predominant clone in these regions, and the first European reports of this clone came from North African immigrants (Goering et al., 2009). USA300 has been recently imported to the Middle East, as the only cases reported in the region involved US military personnel and one child in Israel (Glikman et al., 2010; Huang et al., 2011; Murray et al., 2010). USA300 first appeared in the Western Pacific in 2004–2005 (China) and then in 2007 (South Korea and Japan) (Nimmo, 2012). The first reports of USA300 in Australia were recorded in 2003 (Monecke et al., 2009). However, the most common isolates recovered from patients in Asia and Australia are ST59-V and ST30-IV (Monecke et al., 2011).

39.4.3 CA-MRSA in domestic pets and livestock

S. aureus is a commensal microbe and pathogen of a number of animal species aside from humans. MRSA was first reported in animals in 1972, when it was isolated from bovine milk (Devriese et al., 1972). Since that time, MRSA has been recovered from colonized or infected cattle, cats, chickens, dogs, horses, goats, pigs, sheep and others (Buyukcangaz et al., 2013; Chu et al., 2012; Eriksson et al., 2013; Faires et al., 2010; Foster, 2012; Gharsa et al., 2012; Persoons et al., 2009; Richter et al., 2012; Rutland et al., 2009; van de Giessen et al., 2009; Vitale et al., 2006; Weese et al., 2005). Importantly, MRSA can be readily transmitted from domestic pets and livestock to humans. For example, Weese et al. (2006b) found that there is animal to human and human to animal transmission of MRSA between humans and cats or dogs. The same group also reported an outbreak of MRSA skin infections among veterinary personnel that was linked to a colonized and ultimately infected foal (Weese et al., 2006a). More recently, an ST398 strain known as livestock-associated MRSA

(LA-MRSA) emerged as a significant cause of acute infections among pig farmers in Europe (Voss et al., 2005). LA-MRSA typically causes superficial skin and soft-tissue infections and there is little or no transmission of this strain among humans. By comparison, an ST398 MSSA clone emerged recently as a cause of community-associated infections in northern Manhattan, NY (Bhat et al., 2009; Uhlemann et al., 2013). These infections were not linked to previous contact with livestock and this clone transmitted readily among humans (Bhat et al., 2009; Uhlemann et al., 2012). Comparison of the genomes of LA-MRSA ST398 and ST398 MSSA revealed key differences between these clones that likely causes tropism, including the presence of genes encoding functional adhesion molecules and human-specific immune evasion molecules in the human-adapted clone (Price et al., 2012; Uhlemann et al., 2012). In Asia, the predominant LA-MRSA strain appears to be ST9, which has been found in pigs (Cui et al., 2009; Guardabassi et al., 2009; Wagenaar et al., 2009). The main risk for infection is related to contact with pigs and veal calves (Graveland et al., 2011; Van Den Broek et al., 2009). While LA-MRSA is a poor persistent colonizer in humans, current practices in farm hygiene and antibiotic use may ultimately contribute to an adaptation to new hosts and acquisition of additional virulence determinants (Graveland et al., 2011).

39.4.4 Community-associated MRSA strains in the healthcare setting

The known ability of *S. aureus* strains to transfer from healthcare settings to the community or from the community to the hospital (less frequent), has led to naming infections caused by such strains as "hospital-acquired community onset" and "community-acquired hospital onset" (Klevens et al., 2006; Otter and French, 2012; Scanvic et al., 2001). This nomenclature is based on clinical features of the pathogen rather than molecular characteristics, which is far too imprecise for identification and tracking of outbreak strains. Therefore, genotype-based definitions of *S. aureus*, such as those from molecular typing methods, are preferred (Otter and French, 2012). Pulsed-field gel electrophoresis (PFGE) generates a DNA fingerprint for a bacterial isolate and has been traditionally used to identify strains, particularly at a local level (Struelens et al., 1992). The US Centers for Disease Control and Prevention uses PFGE to define MRSA strains (Tenover et al., 2006). Typing by this method allows high discrimination of the variant PFGE patterns, which serve as markers for epidemiological investigations (Enright and Spratt, 1999). The sequence-based techniques of MLST (Section 39.2 and Table 39-1) and SCC*mec* typing, however, are most often used to describe global *S. aureus* epidemiology (Chua et al., 2014). Single-locus DNA sequencing, such as *spa* typing, is also used for reliable and accurate typing of

MRSA (Harmsen et al., 2003). Along with software algorithms developed for ST grouping, *spa* typing is a valuable epidemiological tool used to analyze local or widespread *S. aureus* outbreaks (Mellmann et al., 2007; Strommenger et al., 2008). Whole genome sequencing has become widely used to characterize clinical isolates, track outbreaks, and understand the evolution of specific *S. aureus* lineages (Harris et al., 2013; Köser et al., 2012; McCarthy et al., 2012; Price et al., 2013). Molecular typing methods and genome sequence data clearly indicate that CA-MRSA did not originate in the healthcare setting (Baba et al., 2002; Diep et al., 2006; Naimi et al., 2003). Thus, the selective pressures that have driven the emergence of MRSA in healthcare settings are different from those that have facilitated the emergence and success of CA-MRSA. A number of theories have been put forth for the emergence of CA-MRSA, including the overuse of fluoroquinolones, exploitation of a vacant ecological niche, and/or acquisition of PVL and SCC*mec*IV (Boyle-Vavra and Daum, 2006; Chambers and DeLeo, 2009; David and Daum, 2010; DeLeo et al., 2010). In any case, CA-MRSA has been increasingly implicated in nosocomial infections (Otter and French, 2011). Mathematical models predict that CA-MRSA may ultimately displace traditional HA-MRSA strains in healthcare settings (D'Agata et al., 2009; Skov and Jensen, 2009).

39.5 Clinical features

One in three people are permanently – and asymptomatically – colonized with *S. aureus* in the anterior nares (Graham et al., 2006; Kluytmans et al., 1997). The bacterium can also colonize the axillae, vagina, pharynx, and damaged skin surfaces (Cursino et al., 2012; Mermel et al., 2011; Miko et al., 2012). In general, individuals colonized with *S. aureus* have a higher risk of developing infections compared with non-carriers, although the risk of serious, life-threatening infections is reduced (Kluytmans et al., 1997). *S. aureus* colonization is also an important factor in the spread of infection, and isolates from repeat infections are typically caused by carriage strains (Huang et al., 2008). Infections occur when a break in the skin or mucosal barrier allows the pathogen to enter the tissues or bloodstream (Lowy, 1998). Risk factors for infection, particularly for HA-MRSA strains, include hospitalization or surgery, presence of foreign objects such as intravenous catheters and indwelling percutaneous medical devices, dialysis, and residence in a long-term care facility (Brumfitt and Hamilton-Miller, 1989; Lowy, 1998). By comparison, CA-MRSA infections occur in otherwise healthy individuals who lack predisposing risk factors (DeLeo et al., 2010).

S. aureus can infiltrate and establish disease in virtually any tissue (Gordon and Lowy, 2008; Lowy, 1998). The majority of *S. aureus* infections manifest as skin and soft-tissue, bone, respiratory, and endovascular infections. Bloodstream infections are more commonly associated with the healthcare setting, whereas skin and soft-tissue infections are primarily community-associated infections and thus caused by community-associated MSSA or MRSA strains (Naimi et al., 2003). Respiratory tract and urinary tract infections are less likely associated with CA-MRSA strains (Cooke and Brown, 2010; Naimi et al., 2003).

39.5.1 Bacteremia and sepsis

S. aureus–associated bacteremia is correlated with relatively high morbidity and mortality. The incidence of *S. aureus* bacteremia is 20–50 cases per 100,000 persons per year, and 10–30% of these cases end in death (van Hal et al., 2012). Age is the strongest and most consistent predictor of mortality caused by *S. aureus* bacteremia (van Hal et al., 2012). Interestingly, MRSA has only added to the baseline of MSSA infection rates for nosocomial staphylococcal bacteremia (Wyllie et al., 2006). Compared with MSSA, infection with MRSA was associated with an increase in the risk of mortality (Saiman et al., 2003; Whitby et al., 2001). While the factors underlying the increase in death associated with MRSA have yet to be satisfactorily elucidated, several explanations have been suggested including increased pathogen virulence, decreased efficacy of administered therapy, delay in applying the appropriate antibiotics, and differences in host health status between those with MRSA and MSSA infections (Boucher et al., 2010). For instance, a study evaluating the effects of delayed treatment on HA-MRSA–associated bacteremia determined that the delay of therapy was associated with an almost twofold increase in mortality (Lodise et al., 2003).

S. aureus is a common cause of sepsis (Bone, 1994). Some infections may progress to sepsis, and typical symptoms include fever, hypotension, tachycardia, and tachypnea (Lowy, 1998). Those at risk for developing staphylococcal sepsis include the elderly, the immunocompromised, and patients undergoing chemotherapy and surgery (Balk, 2000; Lowy, 1998). Severe sepsis can lead to multi-organ failure, disseminated intravascular coagulation, lactic acidosis, and death (Bone, 1994; Mandell et al., 2010).

39.5.2 Endocarditis

The overall incidence of endocarditis in the United States has remained stable over time, ranging from 5 to 7.9 cases per 100,000 persons per year from 1970 to 2006 (Correa de Sa et al., 2010). The epidemiology, however, has changed within the last four decades. Skin organisms such as *S. aureus* have been reported to cause more cases of infective endocarditis than oral streptococci (Habib et al., 2009; Que and Moreillon, 2011). While overall mortality remained stable, *S. aureus*–associated infective endocarditis increased from 37.6% in 1998 to 49.3% in 2009 (Bor et al.,

2013). Individuals at risk for *S. aureus*–associated endocarditis include patients undergoing hemodialysis, patients with a history of hospitalization, intravenous drug users, elderly patients, and patients with prosthetic valves, degenerative valvular disease and rheumatic heart disease (Lowy, 1998; Murdoch et al., 2009). Initial diagnosis can be difficult, as symptoms can be limited to fever and malaise. However, staphylococcal endocarditis frequently involves vascular phenomena related to the cardiac valves with a rapid onset and high fever (Chambers et al., 1983; Pierce et al., 2012).

39.5.3 Metastatic infections

S. aureus has the capability to cause metastatic infections, including those affecting bones, joints, kidneys, and lungs (Chambers et al., 1983; Libman and Arbeit, 1984; Musher et al., 1994). The suppurative collections at these locations may serve as potential sites for recurrent infections (Musher et al., 1994). The ability of *S. aureus* to colonize tissues and persist after initial infection may be due to its capability to invade host cells and/or survive following internalization. For example, *S. aureus* has been shown to survive, at least for a short period of time, within neutrophils, macrophages, osteoblasts, keratinocytes, and endothelial cells (Hudson et al., 1995; Kintarak et al., 2004; Kubica et al., 2008; Menzies and Kourteva, 1998; Voyich et al., 2005).

39.5.4 Osteomyelitis

Osteomyelitis, which can involve a single localized infection of bone or a more widespread infection including marrow, cortex, periosteum and the surrounding soft-tissue, is commonly caused by *S. aureus* (Lew and Waldvogel, 2004). Presentation of symptoms range from the exposure of fractured bone in open wounds, indolent draining fistula, and local swelling, to bone pain with no skin lesion (Lew and Waldvogel, 2004). Depending on the location and population studied, 2.8–43% of invasive CA-MRSA infections originate in bone and joints (Vardakas et al., 2013). These studies indicated that most of the patients with CA-MRSA bone and joint infections are <2 years of age (Vardakas et al., 2013). *S. aureus* has been shown to survive in osteoblasts during culture *in vitro* for an extended period of time (Hudson et al., 1995; Webb et al., 2007). Antibiotics are far less effective against intracellular *S. aureus* (Ellington et al., 2006; Wilson et al., 1982), and it has been suggested that an intracellular reservoir may play a role in chronic and recurrent infections (Hamza et al., 2013). *In vitro*, osteoblasts containing internalized *S. aureus* produce cytokines that are known to play a significant role in the inflammatory response (Bost et al., 1999). Whether *S. aureus* invades non-phagocytic cells during human infection *in vivo* has not been shown conclusively and remains an open question.

39.5.5 Toxic shock syndrome

Toxic shock syndrome (TSS) is an uncommon complication of staphylococcal infection, particularly of toxin-producing strains. Enterotoxins, such as TSS toxin-1 (TSST-1), act as superantigens that precipitate an overreaction of the immune response and production of cytokines (Silversides et al., 2010). The disease is characterized by rapid onset of fever over the course of several hours, often in young previously healthy individuals, with symptoms identical to septic shock, including multi-organ dysfunction, desquamation, and hypotension (Lowy, 1998; Silversides et al., 2010). TSS was first identified in *S. aureus* in 1978, with cases peaking in the 1980s at an incidence between 6 and 12 cases per 100,000 persons per year (Hajjeh et al., 1999; Lowy, 1998; Todd et al., 1978). TSS frequently develops from the site of colonization rather than infection (Chesney et al., 1981). The majority of TSS cases are menstrual related, and notably, the increased incidence of TSS in the 1980s was linked to the use of super absorbent tampons. However, TSS can be associated with postpartum infections, intrauterine devices, surgical wounds, soft-tissue infections, infected burns, nasal packs, and pneumonia (Strausbaugh, 1993).

39.5.6 Asthma and rhinitis

S. aureus may also trigger or enhance respiratory system and/or skin allergies. Superantigen-producing strains of *S. aureus* have been associated with atopic dermatitis (Bunikowski et al., 1999; Zollner et al., 2000). Several other studies implicated superantigens as a possible factor in the development of chronic airway inflammation (Bachert et al., 2010; Barnes, 2009; Gevaert et al., 2005; Zhang et al., 2011). A meta-analysis of clinical studies link *S. aureus* enterotoxins and atopic diseases (Pastacaldi et al., 2011). Nine out of 10 studies analyzed showed that patients with asthma or allergic rhinitis were more likely than controls to have been exposed to *S. aureus* and/or its enterotoxins (Pastacaldi et al., 2011).

39.6 Clinical manifestations of CA-MRSA infections

CA-MRSA strains cause the same types of infections as MSSA strains. The majority of CA-MRSA infections (75–90% of cases) affect skin and soft-tissues and manifest as abscesses or cellulitis with purulent drainage (Fridkin et al., 2005; Moran et al., 2006). Severe invasive syndromes, including osteomyelitis (Arnold et al., 2006; Buck et al., 2005), pneumonia and necrotizing pneumonia (Buck et al., 2005; Francis et al., 2005; Fridkin et al., 2005; Gonzalez et al., 2005; Hageman et al., 2006; Herold et al., 1998; Naimi et al., 2003), necrotizing fasciitis (Miller et al., 2005), purpura fulminans (Adem et al., 2005; Kravitz et al., 2005), and pyomyositis and myositis (Pannaraj et al., 2006), have also been associated with CA-MRSA infections. However, such infections are relatively infrequent.

39.7 Pathogenesis and immunity

All *S. aureus* strains have the capacity to produce molecules that are directed to promote evasion of innate and adaptive immunity, including those that inhibit the function of phagocytic leukocytes, serum complement, and antimicrobial peptides. In addition, *S. aureus* produces numerous adhesion molecules, cytolytic toxins, and superantigens, which are involved in colonization and/or contribute to disease (DeLeo et al., 2009; Foster, 2005) (Table 39-2).

39.7.1 Host defense

Neutrophils are the most prominent cellular component of innate immunity and are essential for host defense against *S. aureus* infections (Rigby and DeLeo, 2012). Nearly 60% of the leukocytes that mature within bone marrow are granulocyte precursors (Athens et al., 1961; Bainton et al., 1971), and correspondingly, granulocytes comprise the majority of leukocytes in blood. Neutrophils are among the first leukocytes to migrate to the site of infection, phagocytose microbes, and in turn destroy them with reactive oxygen species (ROS) and antimicrobial granule proteins. During infection, host cells such as monocytes, macrophages, mast cells, fibroblasts, keratinocytes, endothelial cells, and epithelial cells produce numerous inflammatory molecules that recruit neutrophils. These molecules are produced and/or secreted in part following ligation of host cell pattern recognition receptors by *S. aureus* molecules. For example, Toll-like receptor 2 (TLR2) recognizes teichoic acids such as LTA present in the staphylococcal cell wall (Ginsburg, 2002; Hajjar et al., 2001), Nod-like receptors recognize the muramyl dipeptide motif in bacterial peptidoglycan (Girardin et al., 2003), and TNF-α receptor 1 (TNFR1) has been shown to interact with protein A, which triggers the release of cytokines (Perfetto et al., 2003; Tufano et al., 1991). Staphylococcal surface components as well as secreted toxins are recognized by host cells and elicit IL-8 production (Krakauer, 1998; Soell et al., 1995; Standiford et al., 1994; Yao et al., 1996). CD4+ T-cells also produce cytokines that recruit neutrophils to the site of infection in response to *S. aureus* capsular polysaccharide (McLoughlin et al., 2006; Tzianabos et al., 2001). Host-derived factors, such as these chemotactic molecules, and bacterial factors, such as LTA, prime neutrophils for enhanced function (Colotta et al., 1992; Kobayashi et al., 2005).

Following phagocytosis, neutrophils employ oxygen-dependent and oxygen-independent processes to destroy microbes. Activated neutrophils generate microbicidal ROS, which are produced by a multisubunit NADPH-dependent oxidase that assembles at the phagosome membrane (Quinn

et al., 2006). Although superoxide, the initial product of NADPH oxidase, is weakly microbicidal, secondarily derived ROS, including hypochlorous acid, hydroxyl radical, chloramines, and singlet oxygen, have microbicidal activity toward microbes *in vitro* (Klebanoff, 1968, 2005; Marcinkiewicz, 1997; Rosen and Klebanoff, 1977). Although most ingested microorganisms are killed by neutrophils, *S. aureus* is relatively resistant to the effects of phagosomal ROS and antimicrobial peptides (Palazzolo Ballance et al., 2008). Indeed, it is well documented that 15–50% of ingested *S. aureus* survive and ultimately cause neutrophil lysis (Gresham et al., 2000; Kobayashi et al., 2010; Rogers and Tompsett, 1952; Voyich et al., 2005). This phenomenon likely contributes to the success of *S. aureus* as a human pathogen.

39.7.2 *S. aureus* molecules involved in colonization

Wall teichoic acid (WTA) has been suggested to play a major role in the adherence of the *S. aureus* to host tissues, while later stages of colonization are mediated by sortase-anchored microbial surface components that recognize adhesive matrix molecules (MSCRAMMs) (Burian et al., 2010a; Weidenmaier et al., 2008). *S. aureus* clumping factor B (ClfB) is expressed during nasal colonization and adheres to host cytokeratin (Burian et al., 2010b; Haim et al., 2010; O'Brien et al., 2002b; Wertheim et al., 2008). Iron-regulated surface determinant A (IsdA) is also expressed during nasal colonization and promotes adhesion to human desquamated epithelial cells (Burian et al., 2010b; Clarke et al., 2009, 2006). Several other *S. aureus* proteins, including SdrC, SdrD, SdrH, SasD, and SasG, have been shown to interact with host factors and serum constituents such as fibrinogen (Corrigan et al., 2009; Heilmann, 2011; Muthukrishnan et al., 2011; Roche et al., 2003). The type of MSCRAMMs and adhesins expressed and expression levels of these factors vary among *S. aureus* strains (McCarthy and Lindsay, 2010; Muthukrishnan et al., 2011). This variation may explain why some strains are more likely to be isolated from *S. aureus* carriers. *S. aureus* adhesion to host cells is also varied among individuals (Aly et al., 1977; Weidenmaier et al., 2008), adding further complexity to the factors that underlie colonization.

39.7.3 Inhibition of chemotactic signaling

More than 60% of *S. aureus* clinical strains produce chemotaxis inhibitory protein of staphylococci (CHIPS), which *in vitro* blocks neutrophil interaction with the formyl peptide receptor (FPR) and C5a receptor (C5aR) (de Haas et al., 2004). Similarly, neutrophil migration can be blocked by *S. aureus* extracellular adherence protein (Eap) (also called the major histocompatibility class II analog protein, Map). Eap binds to intercellular adhesion molecule-1 (ICAM-1)

Table 39-2 *S. aureus* virulence factors

Gene	Virulence factor	Function
Adhesins		
aaa	Autolysin/adhesin	Binds to fibrinogen, fibronectin, vitronectin
atl	Autolysin	Binds to fibrinogen, fibronectin, vitronectin, endothelial cells
bap	Biofilm-associated protein	Intercellular adhesion; prevents binding to fibrinogen, fibronectin, and host tissue and internalization
bbp	Bone sialoprotein-binding protein	Adhesin for bone sialoprotein, binds fibrinogen
clfA	Clumping factor A	Binds to fibrinogen, platelet aggregation
clfB	Clumping factor B	Binds to fibrinogen, cytokeratin 10, desquamated nasal epithelia cells; platelet aggregation
cna	Collagen-binding adhesin	Binds to collagen (type I and IV)
coa	Coagulase	Binds and activates prothrombin; promotes conversion of fibrinogen to fibrin
eap/map	Extracellular adherence protein/MHC-II analog protein	Binds to fibrinogen, fibronectin, vitronectin, collagen, ICAM-1, eukaryotic cell surfaces, staphylococcal cells; promotes uptake of *S. aureus* by eukaryotic cells; impairs angiogenesis and wound healing; stimulates production of TNFα and IL-6; involved in biofilm formation
ebh	Emp homolog	Binds to extracellular matrix of host cells
ebpS	Elastin-binding protein	Binds to elastin and tropoelastin
efb	Extracellular fibrinogen-binding protein	Binds to fibrinogen; inhibits complement
emp	Extracellular matrix protein-binding protein	Binds extracellular matrix of host cells; involved in biofilm formation
fnbA, fnbB	Fibronectin binding protein A and B	Binds to fibrinogen, fibronectin, elastin; intercellular adhesion
icaADB/C	Intercellular adhesion gene cluster	Produces polysaccharide intercellular adhesin (PIA) which is involved in aggregation and biofilm formation
isdA	Iron-regulated surface determinant A	Binds to fibrinogen, fibronectin, fetuin, hemoglobin, transferrin, hemin, desquamated nasal epithelial cells
isdB	Iron-regulated surface determinant B	Binds to hemoglobin, hemin, platelet integrin GPIIb/IIIa
isdC	Iron-regulated surface determinant C	Binds to hemin
isdH	Iron-regulated surface determinant H	Binds to haptoglobin, haptoglobin–hemoglobin complex
pls	Plasmin-sensitive protein	Promotes binding to nasal epithelial cells; intercellular adhesion; prevents binding to IgG and Fn and internalization; binds lipids of host cells
sasABCDFGHK (sraP)	*S. aureus* surface protein A (serine-rich surface protein)	Binding to platelets, extracellular matrix
sbi	Second immunoglobulin-binding protein	Binds Fc domain of immunoglobulin; binds complement protein C3 and promotes C3–C3b conversion
sdrC	Serine-aspartate repeat protein C	Binding to nasal epithelial cells and β-neurexin
sdrD	Serine-aspartate repeat protein D	Binding to nasal epithelial cells
sdrE	Serine-aspartate repeat protein E	Platelet aggregation
sdrH	Serine-aspartate repeat protein H	Binds to extracellular matrix
spa	*S. aureus* protein A	Binding to IgG, IgM, von Willebrand factor, platelet receptor gC1qR; bacterial cell aggregation
vwp	Von Willebrand factor binding protein (vWbp)	Binds and activates prothrombin; binds fibrinogen and vW factor
Evasion of host defense		
apsSRX	Antimicrobial peptide sensor	Binds and impairs antimicrobial peptides
aur	Aureolysin	LL-37 cleavage
cap	Capsular polysaccharide (CPS)	Alter C3 (CPS5 and 8) or C3b (CPS1) deposition
catA, ahpC	Catalase, alkyl hydroxide reductase	Inactivate hydrogen peroxide; pivotal for nasal colonization
chp	Chemotaxis inhibitory protein (CHIPS)	Blocks C5a receptor and formyl peptide receptors
crtOPQMN	Carotenoid biosynthesis pathway	Produces staphyloxanthin to protect against ROS
dltABCD	D-alanyl-lipoteichoic acid cell wall modification pathway	Inserts D-alanine into teichoic acids
ecb	Extracellular complement-binding protein	Inhibits convertase
efb	Extracellular fibrinogen-binding protein	Binds fibrinogen; inhibits C3 and C5 convertases
flr	FPR-like 1 inhibitory protein (FLIPr)	Binds formyl peptide receptor
graRS	Glycopeptide resistance associated (regulatory system)	Impairs phagocytosis and LL-37
ltaS	Lipoteichoic acid (LTA)	Binds to TLR2; LTA can elicit secretion of cytokines and chemoattractants through TLR2

(continued)

Table 39-2 (*Continued*)

Gene	Virulence factor	Function
mprF	Multiple peptide resistance factor F	Inserts lysine into teichoic acids
oatA	O-acetyltransferase	Peptidoglycan O-acetylation
sak	Staphylokinase	Plasminogen activator; α-defensin binding; cleaves complement factors
scn	Staphylococcal complement inhibitor	Inhibits convertase
sdrE	*S. aureus* surface protein E	Binds complement regulator factor H
spa	*S. aureus* protein A	Binds Fc domain of immunoglobulin, von Willebrand factor and TNFR-1; binds complement protein C3 and promotes C3–C3b conversion
ssl10	Staphylococcal superantigen-like 10	Binds to chemokine receptors
ssl11	Staphylococcal superantigen-like 11	Targets neutrophils
ssl5	Staphylococcal superantigen-like 5	Specific binding to P-selectin glycoprotein ligand-1 blocking PMN rolling
ssl7	Staphylococcal superantigen-like 7	Binds to the Fc region of IgA and block recognition by neutrophils
tar	Teichoic acid ribitol	Biosynthesis pathway produces wall teichoic acid (WTA) which can bind to TLR2
trx, trxR	Thioredoxin, thioredoxin reductase	Inactivates ROS
Toxins		
eta, etb, etd	Exfoliative toxins A, B, and D	Exotoxins with superantigen activity; glutamate-specific serine proteases that digest desmoglein 1
fmt	Formyltransferase	Produces formyl peptides which are ligands to formyl peptide receptor
hla	Hemolysin-α (α toxin)	Cytolytic pore-forming toxin
hlb	Hemolysin-β (β toxin)	Sphingomyelinase with cytolytic activity
hld	Hemolysin-δ (δ toxin)	Cytolytic toxin; binds neutrophils and monocytes
hlg	Hemolysin-γ (γ toxin)	Bicomponent pore-forming toxin
lukDE/M	Leukocidins D, E, and M	Bicomponent pore-forming leukotoxin that kills PMNs
lukGH (lukAB)	Leukocidins GH (AB)	Bicomponent pore-forming leukotoxin that kills PMNs
lukSF-PV	Panton–Valentine leukocidin (PVL)	Bicomponent pore-forming leukotoxin that kills PMNs
psm	Phenol-soluble modulins	Pore-forming toxins
sea-see, seg-sej, ser-set	*S. aureus* enterotoxins	Gastroenteric toxicity; immunomodulation via superantigen activity; binds to class II MHC
selK-selQ, selU-selX	*S. aureus* enterotoxin-like proteins	Immunomodulation via superantigen activity
ssl10	Staphylococcal superantigen-like 10	Binds to chemokine receptors
ssl11	Staphylococcal superantigen-like 11	Targets neutrophils
ssl5	Staphylococcal superantigen-like 5	Specific binding to P-selectin glycoprotein ligand-1 blocking PMN rolling
ssl7	Staphylococcal superantigen-like 7	Binds to the Fc region of IgA and block recognition by neutrophils
tst	Toxic shock syndrome toxin-1 (TSST-1)	Endothelial toxicity (direct and cytokine-mediated); superantigen activity
Other enzymes and proteins		
–	Fatty acid–modifying enzyme (FAME)	Fatty acids modification
ahpC	Alkyl hydroxide reductase	Residual catalase activity; with CatA, required for survival, persistence, and nasal colonization
arc, opp-3	Arginine deiminase system and oligopeptide permease system (arginine catabolic mobile element, ACME-I/-II/-III)	May aid in colonization
catA	Catalase	Inactivates free hydrogen peroxide; with AhpC, required for survival, persistence, and nasal colonization
eno	Enolase	Catalyzes phosphorglycerate to phosphoenolpyruvate; binds to laminin
geh	Glycerol ester hydrolase	Triacylglycerol degradation
lip	Lipase	Triacylglycerol degradation
mntABC/H	Manganese transporters	Minimizes host-imposed nutrient metal starvation
plc	PtdIns-phospholipase C	Phosphatidylinositol-specific lipase activity
scpA	Staphopain A	Cysteine protease
sspA	*S. aureus* serine protease (V8 protease)	Serine protease
sspB	Staphopain B	Cysteine protease

Data were gathered from Arciola et al. (2011), Athanasopoulos et al. (2006), Barbu et al. (2010), Bera et al. (2005), Bestebroer et al. (2007), Chavakis et al. (2002, 2007), Clarke and Foster (2006), Cooney et al. (1993), Corrigan et al. (2009), Cucarella et al. (2002), Downer et al. (2002), Dryla et al. (2003), Edwards et al. (2012), Foster (2005), Gordon and Lowy (2008), Haggar et al. (2004), Haim et al. (2010), Harraghy et al. (2003), Heilmann (2011), Heilmann et al. (2005), Hirschhausen et al. (2010), Huesca et al. (2002), Hussain et al. (2009), Johnson et al. (2008), Jönsson et al. (1991), Kehl-Fie et al. (2013), Koch et al. (2012), Kraus et al. (2008), Li et al. (2007a, 2007b), Mazmanian et al. (2003), McAdow et al. (2012), McDevitt et al. (1997), Merino et al. (2009), Miajlovic et al. (2010), Nguyen et al. (2000), Ní Eidhin et al. (1998), O'Brien et al. (2002a, 2002b), Patti et al. (1995), Peacock et al. (2002), Pinchuk et al. (2010), Roche et al. (2003, 2004), Sakamoto et al. (2008), Scriba et al. (2008), Sharp et al. (2012), Siboo et al. (2008), Sieprawska-Lupa et al. (2004), Signäs et al. (1989), Smith et al. (2011), Speziale et al. (2009), Thompson et al. (2010), Thurlow et al. (2012), Tung et al. (2000), Urushibara et al. (2012), Vazquez et al. (2011), Ventura et al. (2010), Vergara-Irigaray et al. (2009), Walenkamp et al. (2009), Wang et al. (2012a), Wann et al. (2000), Zecconi and Scali (2013).

on the surface of endothelial cells and thereby inhibits the interaction of these cells with neutrophils (Chavakis et al., 2002). *S. aureus* also evades the immune response by producing the staphylococcal complement inhibitor (SCIN) proteins. SCIN targets the C3 convertases of the classical and alternative pathway and traps them in a catalytically inactive state (Ricklin et al., 2009; Rooijakkers et al., 2009). Although *S. aureus* produces multiple molecules that inhibit chemotaxis *in vitro*, sites of *S. aureus* infection are typically replete with neutrophils, which brings into question the contribution of these molecules to the pathogenesis of infection.

39.7.4 Production of cytolytic toxins

S. aureus has the capacity to produce several toxins that form pores in the plasma membrane, which in turn can cause cytolysis *in vitro*. Hla or α-toxin is one of the most studied *S. aureus* cytolytic toxins. Historically, it has been linked to human deaths and can target a wide range of cell types. The *hla* gene is regulated by at least three global regulatory systems including *agr* (Xiong et al., 2006). Hla is secreted as a monomer, but forms a heptameric pore in the host cell membrane, which results in cytolysis (Montoya and Gouaux, 2003). The binding of Hla to host cells requires the expression of the host receptor ADAM10 (a disintegrin and metalloprotease 10) (Wilke and Wardenburg, 2010). The Hla–ADAM10 complex interferes with intracellular signaling events leading to the disruption of focal adhesions (Inoshima et al., 2011). The resulting degradation of the epithelial barrier facilitates *S. aureus* invasion. Hla is known to play a prominent role in animal models of skin infection and pneumonia (Bubeck Wardenburg et al., 2007; Inoshima et al., 2011; Powers et al., 2012; Wardenburg et al., 2007).

S. aureus also produces several two-component pore-forming leukotoxins/leukocidins. These molecules are comprised of two polypeptides, named LukS (slow) and LukF (fast) due to their chromatographic elution properties (Noda et al., 1980). These toxin subunits assemble at the plasma membrane to form an octomeric pore consisting of repeat LukS/F subunits (Miles et al., 2002; Yamashita et al., 2011). Gamma-hemolysin (γ-hemolysin or Hlg) is the only two-component leukocidin that is encoded by three genes (*hlgA*, *hlgB*, *hlgC*), which can yield two possible combinations of the LukS component (HlgA or HlgC) with a single LukF component (HlgB) (Cooney et al., 1993). Hlg targets erythrocytes, monocytes, PMNs, and lymphocytes and lyses these cells *in vitro* (Cooney et al., 1993; Ferreras et al., 1998; Potrich et al., 2009; Szmigielski et al., 1998). Hlg appears to play a role in the development of septic arthritis and weight loss in animal infection models (Malachowa et al., 2011; Nilsson et al., 1999; Supersac et al., 1998). The other two-component leukocidins, PVL (consisting of LukS-PV and LukF-PV) (Panton and Valentine,

1932), LukDE (Gravet et al., 1998; Morinaga et al., 2003), and LukGH/LukAB (Ventura et al., 2010), are well-known to cause cytolysis *in vitro*, but the exact role of these molecules during host interaction *in vivo* remains unknown. *In vitro*, these leukotoxins cause lysis of neutrophils, monocytes, and macrophages (Dumont et al., 2011; Foster, 2005; Malachowa et al., 2012; Meyer et al., 2009; Ventura et al., 2010). PVL has been closely scrutinized, because the genes encoding this molecule are present in many strains that cause CA-MRSA infections (Diep et al., 2010; Lo and Wang, 2011; Otto, 2013; Vandenesch et al., 2003). PVL is linked to rare necrotizing *S. aureus* pneumonia, which also likely involves antecedent respiratory virus infection (Gillet et al., 2002). On the other hand, more recent evidence *in vitro* and in animal infection models suggests that PVL can prime the innate immune response for enhanced clearance of *S. aureus*, findings consistent with the outcome of large multicenter clinical trials (Bae et al., 2009; Graves et al., 2012; Malachowa et al., 2012). Most *S. aureus* strains encode LukDE and LukGH (also known as LukAB), and LukDE contributes to virulence in mouse infection models (Alonzo et al., 2012; Dumont et al., 2011). CCR5, expressed on immune cells and also a target for the human immunodeficiency virus, has been identified as a binding target of LukDE (Alonzo et al., 2013). The exact function of LukGH/AB during *S. aureus* infection remains unknown. Data from animal infection models are at variance, depending on the model and animal species, and there is evidence that LukGH and PVL function cooperatively to prime host responses to infection (Brown et al., 2009; Bubeck Wardenburg et al., 2007; Crémieux et al., 2009; Kobayashi et al., 2011; Labandeira-Rey et al., 2007; Li et al., 2010; Lipinska et al., 2011; Voyich et al., 2006; Wardenburg and Schneewind, 2008).

Delta-hemolysin (Hld) and the phenol-soluble modulins (PSMs) are amphipathic cytolytic peptides encoded within the core genome of most *S. aureus* lineages. PSMs are divided into two subfamilies based on length, the PSMα peptides, including Hld, PSMα1–4, and PSM-mec, which range from 20–26 amino acids, and the PSMβ peptides, including PSMβ1 and PSMβ2, which are 43–44 amino acids long (Queck et al., 2009; Wang et al., 2007). Hld is a well-known cytolytic peptide of *S. aureus*, and it was purified more than 60 years ago (Wiseman, 1975). The hemolytic activity of Hld has been attributed to the formation of membrane pores and/or its ability to act as a detergent at high concentration (Verdon et al., 2009). PSMα peptides have cytolytic activity toward white and red blood cells and the mechanism is similar to that of Hld (Cheung et al., 2010; Wang et al., 2007). PSMs are regulated by *agr*, and the relative high production of PSMs is associated with the most successful CA-MRSA lineages (Queck et al., 2008; Rautenberg et al., 2011; Wang et al., 2007). Indeed, CA-MRSA strains in general produce greater quantities of PSMs compared with typical HA-MRSA strains (Li et al., 2009; Wang et al., 2007). PSMα peptides, in particular, were shown to play a role in the virulence of the murine skin infection model (Wang et al., 2007). These peptides induce a proinflammatory response and promote chemotaxis and activation of neutrophils, which is mediated by binding to the human formyl peptide receptor 2 (FPR2) (Cheung et al., 2010; Kretschmer et al., 2010; Queck et al., 2009; Wang et al., 2007). PSMs also have demonstrated antimicrobial activity, which may endow S. aureus with a fitness advantage over other microbes (Joo et al., 2011). Moreover, PSMs are implicated in the formation of biofilms, aggregates of bacteria embedded in an extracellular matrix, which contribute to the difficulty of treating infections (Periasamy et al., 2012). PSMβ peptides are hydrophobic and surfactant properties likely promote biofilm maturation and pathogen dissemination in biofilm-associated infections (Wang et al., 2011).

39.7.5 Inhibition of opsonophagocytosis?

S. aureus produces molecules, surface bound or freely secreted, that have the potential to inhibit opsonophagocytosis. For example, surface-associated proteins such as SpA interfere with the deposition of antibodies and complement in vitro (Atkins et al., 2008; Schifferli and Peters, 1983). In addition, S. aureus capsule polysaccharides can render the bacterium resistant to phagocytosis in vitro (Baddour et al., 1992; Luong and Lee, 2002; Nilsson et al., 1997; O'Riordan and Lee, 2004; Thakker et al., 1998). Many strains of S. aureus produce the polysaccharide intercellular adhesion (PIA), a positively charged molecule that promotes biofilm formation by binding negatively charged surfaces of the bacterial cells (Cramton et al., 1999; Cue et al., 2012; Vuong et al., 2004). This polysaccharide can partially protect S. aureus from phagocytosis and killing by neutrophils in vitro (Ulrich et al., 2007).

Protein A (SpA) is anchored to the bacterial cell wall envelope and possesses five immunoglobulin-binding domains that associate with the Fc region of IgG and the Fab domains of IgG and IgM (Cary et al., 1999; Sjödahl, 1977; Ton-That et al., 1999). The binding of SpA to IgG blocks the interaction of Ab with the Fc receptor of neutrophils, thereby blocking phagocytosis (Forsgren and Quie, 1974). S. aureus strains lacking SpA are phagocytosed more efficiently by neutrophils in vitro and have decreased virulence in animal infection models (Palmqvist et al., 2002; Patel et al., 1987).

S. aureus clumping factor A and clumping factor B (ClfA and ClfB) are fibrinogen-binding MSCRAMMs that inhibit deposition of serum opsonins in vitro. ClfA has been shown to prevent phagocytosis of S. aureus by macrophages and neutrophils (Foster, 2005; Palmqvist et al., 2004). ClfA and ClfB presumably protect the pathogen by coating the cell surface with fibrinogen, which in turn inhibits deposition of

opsonins. Consistent with these proposed functions, ClfA contributes to virulence in a S. aureus infection model in mice (Josefsson et al., 2001).

The S. aureus complement inhibitor (SCIN) family of proteins, consisting of SCIN-A, SCIN-B, and SCIN-C, inactivates complement in vitro and have capacity to block phagocytosis (Rooijakkers et al., 2007). SCIN proteins actively inhibit the classical and alternative pathways by binding to the C3 convertases, C4b2b and C3bBb, and ultimately preventing the formation of C3b, an important opsonin (Jongerius et al., 2007; Rooijakkers et al., 2005a). Extracellular fibrinogen-binding protein (Efb) also blocks C3 deposition on the bacterial cell surface (Jongerius et al., 2007). The C-terminal region of Efb binds directly to the complement factors C3, C3b, and C3d (Hammel et al., 2007b; Lee et al., 2004a, 2004b). A homologous protein, Ehp, binds to complexes containing C3d, ultimately inhibiting opsonization (Hammel et al., 2007a). Both Efb and Ehp influence virulence in mouse models of S. aureus infection (Jongerius et al., 2007; Mamo et al., 1994; Palma et al., 1996). S. aureus can also inactivate C3b and IgG after they are bound to the surface of opsonized bacteria. Staphylokinase, a plasminogen activator protein secreted by S. aureus, binds to host plasminogen molecules that adhere to the bacterial surface. As a result, the activated serine protease of plasmin cleaves surface-associated C3b and IgG, thus reducing the likelihood of phagocytosis (Rooijakkers et al., 2005b).

Although S. aureus produces multiple molecules that have the potential to inhibit the function of serum complement and inhibit opsonization, in reality S. aureus is ingested readily by human phagocytic leukocytes in vitro and in vivo during infection (Kobayashi et al., 2010; Vandenbroucke Grauls et al., 1984; Verhoef et al., 1977; Voyich et al., 2005). Thus, the relative contribution of these molecules to human infections remains unknown.

39.7.6 Resistance to host bactericidal processes

S. aureus utilizes several strategies to avoid killing by the innate immune system. For example, the pathogen can modify the cell surface to avoid contact with antimicrobial peptides. These antimicrobial peptides can be sensed by the pathogen via the Aps sensor/regulator system, which induces resistance mechanisms in S. aureus (Li et al., 2007a, 2007b). The staphylococcal dlt operon (D-alanyl-lipoteichoic acid), which is regulated by the two-component regulatory system GraRS, is involved in a cell wall modification pathway that transfers D-alanine into teichoic acids (Kraus et al., 2008; Peschel et al., 1999). The subsequent neutralization of the negative charge of the bacterial cell surface renders the pathogen less sensitive to the bactericidal effects of cationic antimicrobial peptides (Peschel et al., 1999). S. aureus lacking these D-alanine modifications are more susceptible to killing by neutrophils and exhibit attenuated virulence in mice (Collins et al., 2002). The cell surface can also be modified by the protein MprF (multiple peptide resistance factor), which adds an L-lysine to phosphatidylglycerol and thus reduces its affinity to cationic peptides (Peschel et al., 2001; Staubitz et al., 2004). MprF may also contribute to the evasion of killing by neutrophils (Kristian et al., 2003). Staphylokinase binds defensins and as a result, neutralizes bactericidal activity (Bokarewa et al., 2006). Aureolysin, an extracellular metalloprotease, cleaves the cathelicidin LL-37, an antimicrobial peptide present in the specific granules of human neutrophils (Sieprawska-Lupa et al., 2004). *S. aureus* is also remarkable for its resistance to lysosome, a well-known bactericidal enzyme that normally hydrolyzes 1,4-beta-linkages in peptidoglycan. *S. aureus* membrane-bound *O*-acetyltransferase (encoded by the *oatA*) modifies *N*-acetylmuramic acid on lysozyme, thereby rendering the enzyme inactive (Bera et al., 2005).

NADPH oxidase–derived ROS produced within the phagosome combined with antimicrobial components of neutrophils granules are typically sufficient for killing of ingested microbes. Consistent with this idea, patients that have chronic granulomatous disease, a hereditary disease characterized by a defect in NADPH oxidase, are susceptible to infections caused by catalase-positive microorganisms such as *S. aureus* (Ben Ari et al., 2012). Despite the importance of ROS in host defense against *S. aureus*, the pathogen has high capacity to moderate the effects of ROS (Babior et al., 1975; Hampton et al., 1996; Klebanoff, 1974). For example, catalase and superoxide dismutase protect the pathogen against host oxygen-dependent microbicidal killing mechanisms present in the phagosome (Clements et al., 1999; DeLeo et al., 2009; Valderas and Hart, 2001). Other redundant mechanisms that protect *S. aureus* from oxidant stress include a Mn-dependent and SOD-independent process that is regulated by metal ion homeostasis, alkyl hydroperoxide reductase, bacterioferritin comigratory protein, flavohemoglobin, and thioredoxin (Cosgrove et al., 2007; Horsburgh et al., 2001, 2002; Lu and Holmgren, 2014; Richardson et al., 2006). The staphylococcal golden carotenoid pigment, staphyloxanthin, has also been shown to play a protective role by scavenging oxygen free radicals (Liu et al., 2005). Mutants deficient in staphyloxanthin production and superoxide dismutase are less virulent in murine abscess infection models (Karavolos et al., 2003; Liu et al., 2005). *S. aureus* can also adapt to oxidative stress with a NO-inducible L-lactate dehydrogenase (Richardson et al., 2008) and repair oxidative protein damage with methionine sulfoxide reductase (Singh and Moskovitz, 2003). A microarray study reported global changes in *S. aureus* gene expression after exposure to hydrogen peroxide, HOCl, or azurophilic granule proteins (Palazzolo Ballance et al., 2008). These findings indicated *S. aureus* initiates a survival response following exposure to neutrophil microbicides, a phenomenon that likely contributes to its success as a pathogen.

39.7.7 Exfoliative toxins and superantigens

The production of exfoliative toxins, ETA, ETB, and ETD, is *agr*-regulated (Sheehan et al., 1992). These enzymes are glutamate-specific serine proteases that selectively cleave desmoglein 1 (Dsg1), a host molecule responsible for keratinocyte cell–cell adhesion (Nishifuji et al., 2008). The destruction of host cellular junctions by exfoliative toxins promotes *S. aureus* skin diseases such as bullous impetigo and staphylococcal scalded skin syndrome (Nishifuji et al., 2010).

S. aureus also secretes several toxins classified as superantigens (SAgs), which are characterized by their ability to hyper-stimulate the immune system. A number of staphylococcal SAgs have been identified to date, including the staphylococcal enterotoxins SEA-SEE, SEG-SEJ, and SER-SET, the staphylococcal enterotoxin-like toxins SE*l*K-SE*l*Q and SE*l*U-SE*l*X, and TSST-1 (Holtfreter and Bröker, 2005; Lina et al., 2004; Ono et al., 2008; Proft and Fraser, 2003; Thomas et al., 2007; Wilson et al., 2011). *S. aureus* strains with SAgs genes produce varied levels of the toxins due to the influence of at least four global regulators, *agr*, *sarA*, σ^B, and *saeRS* (Andrey et al., 2010; Kusch et al., 2011; Tseng et al., 2004). Most SAgs interact directly with T-cells by cross-linking T-cell receptor Vβ domains with MHC class II molecules on professional antigen-presenting cells (Fleischer and Schrezenmeier, 1988; Fraser and Proft, 2008). While most conventional peptide antigens stimulate only 1 out of 10^5–10^6 naïve T-cells, SAgs can activate up to 20% of the T-cell pool (Fraser et al., 2000; Proft and Fraser, 2003). The activated T-cells produce high levels of proinflammatory cytokines (Bergdoll et al., 1981; McCormick et al., 2001). The MHC class II interaction with the SAgs also triggers cytokine release by the activated antigen-presenting cells (Proft and Fraser, 2003). Following the massive production of cytokines, these T-cells become anergic and may die, suggesting that the subsequent depletion of T-cells prevents B-cells from mounting an effective antibody response against *S. aureus* (Alderson et al., 1995; Fraser et al., 2000; Rellahan et al., 1990). A recent study shows evidence that *S. aureus* can also block the protective ability of B-cells with protein A, which binds to the Fab regions of the B-cell receptor and instigates apoptosis (Falugi et al., 2013).

39.8 Treatment

Penicillin and related β-lactam antibiotics remain effective for treatment of *S. aureus* infection in areas where the prevalence of methicillin resistance is low (Moellering, 2012). β-lactam antibiotics can be highly effective and they are inexpensive and non-toxic. Oral cephalosporins and clindamycin are also used in first-line treatment of patients with mild disease (Boucher et al., 2010). The dependence on non-β-lactam antibiotics for treatment of *S. aureus* infections has increased dramatically because of the problem with antibiotic-resistant strains – most notably MRSA. Incision and drainage is the preferred treatment for *S. aureus* cutaneous abscesses, since there is evidence that antibiotics are ineffective in most cases (Liu et al., 2011).

S. aureus bacteremia is treated initially by removal of catheters, incision and drainage of abscesses or infective joints, and/or valve replacement for patients with left-sided *S. aureus* endocarditis (Bishara et al., 2001; Cosgrove and Fowler, 2007; Fowler et al., 1998, 2003; Grayson, 2006; John et al., 1998; Vikram et al., 2003). In the past, *S. aureus* bacteremia was treated with 4–6-week courses of antimicrobial therapy, but studies in the 1990s suggested that 10–14-day treatments may be sufficient (Cosgrove and Fowler, 2007; Malanoski et al., 1995; Raad and Sabbagh, 1992; Sullenberger et al., 2005). It is recommended that patients with a high risk of infection undergo >2 weeks of intravenous therapy, as they may experience complications with shorter treatment regimens (Jernigan and Farr, 1993). Suggestions for identifying high-risk patients include follow-up blood cultures, transesophageal echocardiography, and abatement of fever within 72 hours after the start of effective antistaphylococcal therapy (Cosgrove and Fowler, 2007; Fowler et al., 1998, 1999; Mermel et al., 2001; Rosen et al., 1999). Cosgrove and Fowler (2007) suggest patients with catheter-associated *S. aureus* bacteremia undergo treatment for >2 weeks. A comprehensive guide to treatment of MRSA infections is provided by Liu et al. (2011).

39.8.1 Antimicrobial agents

Vancomycin is a glycopeptide antibiotic derived from *Amycolatopsis orientalis* that inhibits cell wall synthesis in Gram-positive bacteria (Levine, 2006). It has been the primary drug used to treat MRSA bacteremia for the past four decades, since it was previously the only antimicrobial agent available with any efficacy against the pathogen. Vancomycin has relatively slow bactericidal activity and causes a number of adverse effects, including "red man" syndrome, neutropenia, and thrombocytopenia (Cosgrove and Fowler, 2007). *S. aureus* strains with reduced susceptibility to vancomycin emerged in the 1990s and are not infrequently isolated from infected individuals, especially those undergoing treatment with vancomycin (and others) for an extended period of time (Howden et al., 2010; Sievert et al., 2008). To address this problem, new antibiotics, such as daptomycin, linezolid, quinupristin–dalfopristin, and tigecycline, have been developed. A few older antibiotics, including trimethoprim–sulfamethoxazole (TMP-SMX) and clindamycin, are widely used in the treatment of CA-MRSA infections (Deresinski, 2009; Liu et al., 2011) (Table 39-3).

Daptomycin is a lipopeptide isolated from *Streptomyces* and acts by disrupting bacterial cell membrane function (Pogliano et al., 2012). Daptomycin has been approved for use in the treatment of *S. aureus* infections, including those caused by MSSA and MRSA, and specifically to treat skin infections, bacteremia, and right-sided endocarditis (Liu et al., 2011). Daptomycin is not recommended for treatment of MRSA pneumonia, because interaction of the drug with pulmonary surfactant inhibits its antibacterial activity (Pertel et al., 2008; Silverman et al., 2005). Linezolid is a synthetic oxazolidinone antibiotic that inhibits bacterial protein synthesis by inhibiting the assembly of the 50S ribosomal subunit (Swaney et al., 1998). Linezolid has been approved for use against skin and soft-tissue infections and pneumonia (Aksoy and Unal, 2008; Liu et al., 2011). This therapeutic agent should be used as a last resort, as it is expensive and prolonged use has several adverse effects, including bone marrow suppression, thrombocytopenia, peripheral neuropathy, optic neuritis, and lactic acidosis (Apodaca and Rakita, 2003; De Vriese et al., 2006; French, 2003; Gerson et al., 2002; Moellering, 2003). Quinupristin and dalfopristin are both streptogramin antibiotics that are delivered intravenously (Cosgrove and Fowler, 2007). Used together, they have synergistic bactericidal activity which inhibits bacterial protein synthesis (Allington and Rivey, 2001). Some adverse effects of quinupristin–dalfopristin are myalgia and arthralgia, pain and inflammation at the infusion site, and thrombophlebitis (Cosgrove and Fowler, 2007). Tigecycline is a glycylcycline that exhibits bacteriostatic action by inhibiting protein translation (Rose and Rybak, 2006). Tigecycline has activity against a number of Gram-negative and Gram-positive organisms, including *S. aureus*, and is used to treat skin and soft-tissue infections and complicated intra-abdominal infections (Rose and Rybak, 2006). Side effects of tigecycline include nausea and vomiting (Cosgrove and Fowler, 2007). TMP-SMX is a sulfonamide antibiotic that is used to treat a wide spectrum of infections, including MRSA (and notably CA-MRSA), and it has a long history of use in healthcare settings. Studies have shown that the combination of these two compounds is rapidly bactericidal against MRSA *in vitro*, making it possible for use in salvage therapies (Cosgrove and Fowler, 2007; Huovinen, 2001; Jemni et al., 1994; Markowitz et al., 1992; Stein et al., 1998). It has become a primary therapeutic agent for the treatment of CA-MRSA skin and soft-tissue infections, and TMP-SMX resistance among prominent CA-MRSA lineages is limited (Liu et al., 2011; Wood et al., 2012). However, there are drug-induced adverse effects, including rashes and hyperkalemia (Ackerman et al., 2013; Nguyen et al., 2013).

A single-agent therapy is usually prescribed for treatment for *S. aureus* osteomyelitis. Benzylpenicillin is usually given for infections caused by penicillin-susceptible strains, while alternatives such as cefazolin, clindamycin, or vancomycin may

Table 39-3 Antibiotics in use and in development for *S. aureus* infections

Antibiotics currently in use		
Antibiotic	**Class**	**Use: type of infection**
Amoxicillin–clavulanate	β-lactam/β-lactamase inhibitor	Penicillin-susceptible strains, SSTI
Ampicillin–sulbactam	β-lactam/β-lactamase inhibitor	Penicillin-susceptible strains, SSTI
Benzylpenicillin	β-lactam	Osteomyelitis, penicillin-susceptible strains
Besifloxacin	Fluoroquinolone	Bacterial conjunctivitis
Cefamandole	Cephalosporin	Osteomyelitis, penicillin-resistant strains
Cefazolin	Cephalosporin	Bacteremia, cellulitis, endocarditis, lung abscess, osteomyelitis, pneumonia, septic arthritis
Ceftaroline	Cephalosporin	Pneumonia, SSTI
Cefuroxime	Cephalosporin	Osteomyelitis, penicillin-resistant strains, SSTI
Cephalexin	Cephalosporin	Bacteremia, cellulitis, lung abscess, osteomyelitis, pneumonia, septic arthritis, SSTI
Cephalothin	Cephalosporin	Bacteremia, cellulitis, lung abscess, osteomyelitis, pneumonia, septic arthritis
Ciprofloxacin	Fluoroquinolone	Bacteremia, cellulitis, endocarditis, osteomyelitis, penicillin-resistant strains
Clindamycin	Lincosamide	Lung abscess, osteomyelitis, pneumonia, septic arthritis, SSTI, TSS
Cloxacillin	β-lactam	Bacteremia, cellulitis, lung abscess, meningitis, osteomyelitis, pneumonia, septic arthritis
Dalbavancin	Glycopeptide	SSTI
Daptomycin	Lipopeptide	Bacteremia, endocarditis, osteomyelitis, septic arthritis, SSTI
Dicloxacillin	β-lactam	Bacteremia, cellulitis, meningitis, SSTI
Doxycycline	Tetracycline	SSTI
Flucloxacillin	β-lactam	Bacteremia, cellulitis, lung abscess, meningitis, osteomyelitis, pneumonia, septic arthritis
Fusidic acid	Steroid	Bacteremia, cellulitis, lung abscess, osteomyelitis, pneumonia, septic arthritis, SSTI
Gentamicin	Aminoglycoside	Bacteremia, endocarditis
Levofloxacin	Fluoroquinolone	Osteomyelitis, penicillin-resistant strains
Linezolid	Oxazolidinone	Brain abscess, CNS infections, epidural abscess, meningitis, osteomyelitis, pneumonia, septic arthritis, SSTI, subdural empyema, TSS
Minocycline	Tetracycline	SSTI
Nafcillin	β-lactam	Bacteremia, brain abscess, cellulitis, endocarditis, epidural abscess, lung abscess, meningitis, osteomyelitis, penicillin-resistant strains, pneumonia, septic arthritis, subdural empyema, TSS
Oxacillin	β-lactam	Bacteremia, brain abscess, cellulitis, endocarditis, epidural abscess, lung abscess, meningitis, osteomyelitis, pneumonia, septic arthritis, subdural empyema, TSS
Piperacillin–tazobactam	β-lactam/β-lactamase inhibitor	Penicillin-susceptible strains
Quinupristin–dalfopristin	Streptogramin	Bacteremia
Retapamulin	Pleuromutilin	Impetigo
Rifampin	Rifamycin	Bacteremia, brain abscess, cellulitis, endocarditis, epidural abscess, lung abscess, meningitis, osteomyelitis, pneumonia, septic arthritis, subdural empyema
Teicoplanin	Glycopeptide	Osteomyelitis
Telavancin	Glycopeptide	SSTI
Tigecycline	Glycylcycline	SSTI
Trimethoprim–sulfamethoxazole (TMP-SMX, Co-trimoxazole)	Sulfonamide	SSTI Brain abscess, CNS infection, epidural abscess, meningitis, osteomyelitis, septic arthritis, subdural empyema
Vancomycin	Glycopeptide	Bacteremia, brain abscess, cellulitis, CNS infections, endocarditis, epidural abscess, lung abscess, meningitis, osteomyelitis, pneumonia, septic arthritis, SSTI, subdural empyema, TSS

(continued)

Table 39-3 (*Continued*)

Antibiotics in development			
Antibiotic	**Class**	**Type of infection**	**Phase**
BC-3205	Pleuromutilin	Pneumonia, SSTI	I
BC-3781	Pleuromutilin	SSTI	II
BC-7013	Pleuromutilin	Pneumonia, SSTI	I
Ceftobiprole	Cephalosporin	Pneumonia, SSTI	III
Cethromycin	Ketolide	Pneumonia	III
Delafloxacin	Quinolone	Pneumonia, SSTI	II
EDP-420	Bicyclolide	Pneumonia	II
Eravacycline	Tetracycline	Intra-abdominal infections	II
F598	hmAb	*S. aureus* infections	I
GSK1322322	PDF inhibitor	Pneumonia, SSTI	II
Iclaprim	DHFR inhibitor	SSTI	I (oral); III (IV)
JNJ-Q2	Fluoroquinolone	SSTI	II
Linopristin–flopristin	Streptogramin	Pneumonia, SSTI	II
Mupirocin	Monoxycarbolic acid	Nasal carriage, SSTI	III
Nemonoxacin	Quinolone	Pneumonia	III
Omadacycline	Tetracycline	Pneumonia, SSTI	III
Oritavancin	Glycopeptide	SSTI	III
Pexiganan (MSI-78)	Antimicrobial peptide	Diabetic foot infections	III
Plazomicin	Aminoglycoside	Intra-abdominal infections, urinary tract infections	II
Radezolid	Oxazolidinone	Pneumonia, SSTI	II
Razupenem	Carbapenem	SSTI	II
Solithromycin	Ketolide	Pneumonia	III
TD-1792	Glycopeptide-cephalosporin	SSTI	II
Tedizolid	Oxazolidinone	SSTI	III
Telithromycin	Ketolide	Pneumonia	IV
Tomopenem	Carbapenem	Pneumonia, SSTI	II
XF-73	Dicationic porphyrin	Nasal carriage	II

Data were gathered from Baddour et al. (2005) Bassetti et al. (2013), Carter and Scott (2010), Cosgrove and Fowler (2007), Covington et al. (2011), DeLeo et al. (2010), Doebbeling et al. (1993), Kurosu et al. (2013), Lew and Waldvogel (2004), Liapikou and Torres (2013), Liu et al. (2011), MacGowan et al. (2011), Naderer et al. (2013), Novak (2011), Oldach et al. (2013), Prince et al. (2013), Rodgers et al. (2013), Rybak et al. (2009), Stevens et al. (2005), Stryjewski et al. (2012), Sugihara et al. (2011), Taylor (2013), Weber (2005), Zhanel et al. (2012). CNS, central nervous system; DHFR, dihydrofolate reductase; hmAb, human monoclonal antibody; IV, intravenous; PDF, peptide deformylase; SSTI, skin and soft-tissue infection; TSS, toxic shock syndrome.

be used. Penicillin-resistant strains can be treated with nafcillin, and alternatively, clindamycin, vancomycin, ciprofloxacin, or levofloxacin with rifampin. Vancomycin and sometimes teicoplanin are recommended for treatment of bone and joint infections caused by MRSA (Liu et al., 2011). Treatments for infections involving prosthetic joints and chronic osteomyelitis also include combination therapy consisting of vancomycin or a β-lactam with rifampin (Liu et al., 2011).

Oral antimicrobial agents often used to treat CA-MRSA infections include clindamycin, tetracyclines (such as doxycycline and minocycline), TMP-SMX, rifampicin, fusidic acid, and linezolid (Liu et al., 2011). Doxycycline and minocycline appear to be effective against MRSA skin and soft-tissue infections, but these agents are not recommended for children younger than 8 years or pregnant women (DeLeo et al., 2010). Fusidic acid is effective against skin and soft-tissue infections, and concomitant use with rifampicin is recommended to treat MRSA infections, especially osteomyelitis (Aboltins et al., 2007; Schöfer and Simonsen, 2010; Wang et al., 2012b). Clindamycin,

a lincosamide antibiotic that inhibits bacterial protein synthesis, has also been used successfully to treat skin and soft-tissue infections – it has antimicrobial activity against more than 80% of CA-MRSA strains tested *in vitro* (Liu et al., 2011; Siberry et al., 2003). Patients that do not respond to oral clindamycin therapy require intravenous treatment, which may use vancomycin, daptomycin, or linezolid (Aksoy and Unal, 2008; Carpenter and Chambers, 2004; Weber, 2005).

39.9 Prevention and control

Rates of *S. aureus* asymptomatic colonization vary globally depending on population (9–58%), and with approximately one-third of the population in the United States colonized by *S. aureus*, nasal carriage is a significant reservoir for the microbe in humans (Gorwitz et al., 2008; Kuehnert et al., 2006; Sollid et al., 2013). Carriers have a higher risk for staphylococcal infections compared with non-carriers, and these individuals are important sources of *S. aureus* transmission. *S. aureus* is usually spread by skin-to-skin contact with a colonized or infected individual or from sharing personal contaminated equipment including towels, washcloths, razors, clothing, and uniforms (CDC, 2013; DeLeo et al., 2010; Holman, 2013; Taylor, 2013). Risk factors for HA-MRSA infections are connected with illness and healthcare settings, while CA-MRSA infection and colonization have been associated with people living or working in close contact with each other (DeLeo et al., 2010). Therefore, general cleanliness and good hygiene are critical for preventing the spread of CA-MRSA.

39.10 Precautionary measures for prevention of infection

The primary defense against transmission is proper hand hygiene. Hands should be washed regularly with soap and water for at least 15 seconds and dried with a disposable towel. If the hands are not soiled, a hand sanitizer composed of at least 60% alcohol should be used. The Centers for Disease Control and Prevention also recommend several precautionary measures for limiting transmission and preventing infections. Healthcare workers should wash hands between patients and wear proper personal protection, such as gloves, gowns, and eye protection, when working with infected patients. Wounds should be kept clean and covered with dry bandages. Covering any infections with pus and avoiding skin-to-skin contact with the wound will prevent the spread of bacteria (Holman, 2013; Taylor, 2013).

Under certain situations, particularly with CA-MRSA infections, contaminated fomites play a role in the transmission of *S. aureus* (Baggett et al., 2004; Miller and Diep, 2008). The bacterium can survive on a variety of items ranging from household items and sports equipment to hospital accoutrements. To avoid spread of infection, sharing of personal items should be avoided and shared equipment should be cleaned after each use. Linens should be washed with hot water and bleach when possible (Holman, 2013).

Another suggested approach to prevention identifies individuals asymptomatically colonized by MRSA by active surveillance. This approach involves obtaining nasal cultures from patients at the time of hospital admission. If a patient is an MRSA carrier, isolation precautions can be implemented to reduce the risk of MRSA transmission to other patients (Diekema and Climo, 2008; Siegel et al., 2007). However, without the addition of isolation procedures or attempts at decolonization and eradication, this screening strategy alone cannot be expected to affect health outcomes. A review of 48 studies examining the efficacy of MRSA active surveillance found that there is currently insufficient evidence that MRSA screening will prevent MRSA transmission (Glick et al., 2013). Many of these observational studies had insufficient controls, and other confounding factors may have influenced the outcomes of MRSA

screening. Thus, future studies should be designed to address these deficiencies in order to determine if active surveillance could have a positive effect on preventing disease transmission.

Decolonization is a preventative strategy predicated on the observation that *S. aureus* infection is often linked to colonization (Wertheim et al., 2005). Decolonization may help reduce the risk of *S. aureus* infection for patients undergoing surgery and other invasive procedures. MSSA decolonization is typically initiated with the application of intranasal mupirocin ointment, which can result in decolonization of up to >90% of individuals, and decolonization can last for several weeks (Doebbeling et al., 1993). Similar results have been reported for decolonization of MRSA carriers (Ammerlaan et al., 2009). However, decolonization is not effective in all cases, especially in trials where topical decolonization was applied but not followed by a regimen for systemic decolonization (Doebbeling et al., 1994; Dryden et al., 2004; Harbarth et al., 1999; Simor et al., 2007). For some patients, short-term eradication of *S. aureus* colonization is successful, but recolonization occurs with the same strain, suggesting that a treatment consisting solely of topical decolonization is insufficient for eliminating *S. aureus* from carriers with non-nasal sites of colonization. Mupirocin resistance is also a concern, as it has been shown to be associated with failure of decolonization (Robicsek et al., 2008; Simor et al., 2007).

39.11 Experimental agents and adjunctive therapy

New drugs developed to combat MRSA infections, including β-lactams with high affinity to PBP2a such as ceftobiprole and ceftaroline, and derivatives of vancomycin such as telavancin, dalbavancin, and oritavancin, have shown high bactericidal activity against the pathogen in animal models and have been approved for clinical use (Micek, 2007; Pan et al., 2008; Stryjewski and Chambers, 2008). Retapamulin is another antibiotic recently approved for use in the United States. It is a pleuromutilin that is used as a topical agent against staphylococcal impetigo (Novak, 2011). Other drugs currently in development for use as potential anti-*S. aureus* agents include tomopenem and razupenem (carbapenems), plazomicin (an aminoglycoside), besifloxacin and nemonoxacin (quinolones), and omadacycline and eravacycline (tetracyclines) (Bassetti et al., 2009, 2013; Chang and Fung, 2010; Liapikou and Torres, 2013; Sutcliffe, 2011; Zhanel et al., 2012) (Table 39-3). It is important to note that few new structural classes of antibiotics have been discovered or are under development, as the "new" drugs are clearly derivatives of older antibiotics. This issue, coupled with the ability of *S. aureus* to easily acquire antibiotic resistance, has led researchers to consider nonconventional approaches for treatment and prevention of MRSA. Some of these new or nonconventional approaches include bacterio-phage therapy, antimicrobial peptides (Lawton et al., 2007), enzymatic molecules active against biofilms such as dispersin B and lysostaphin (Dajcs et al., 2001; Kiedrowski and Horswill, 2011), quorum-sensing inhibitors (Brackman et al., 2011), natural products (Stapleton et al., 2007), and passive and active immu-nization (Schaffer and Lee, 2008; Shinefield et al., 2002; Wardenburg and Schneewind, 2008). Some of the obstacles to development of these approaches include cost, time for research and approval, hyper-sensitivity due to repeated administration of protein products, or unfavorable pharmacological properties such as short half-life and toxicity (DeLeo et al., 2010). Previous attempts to develop vaccines for active and passive immunization have met with failure. This may be due to the conventional strategy of devel-oping a vaccine directed to enhance opsonophago-cytosis, which works for other encapsulated bacteria (Kobayashi and DeLeo, 2011). In this case, all humans have naturally occurring antibodies against *S. aureus*, as normal human serum promotes opsonophagocy-tosis *in vitro*. A key issue with *S. aureus* is that some of the ingested bacteria survive following phagocyto-sis by neutrophils and then ultimately cause cytoly-sis of these host cells (Kobayashi and DeLeo, 2011).

The development of a vaccine based on multivalent antigens has promise, as targeting a single virulence factor is not sufficient to provide protection (Schaffer and Lee, 2009; Spellberg and Daum, 2012). As *S. aureus* causes many diseases and syndromes, a potential vac-cine will likely be targeted for a specific population or syndrome. By comparison, a vaccine that targets multiple types of infections and syndromes will be more difficult to bring to fruition, and such a vac-cine would likely require testing in multiple animal infection models for efficacy (Schaffer and Lee, 2009; Spellberg and Daum, 2012). Other factors that may play a role in determining infection severity include the environment, the immune status of the patient, and genetics. Thus, any future attempts at a success-ful vaccination program would need to consider the target population and aim to bolster the host defense mechanisms other than opsonophagocytosis.

Acknowledgments

The authors are supported by the Intramural Research Program of the National Institute of Allergy and Infec-tious Diseases (NIAID), National Institutes of Health. The authors thank Scott D. Kobayashi (NIAID) for critical review of the article.

References

Aboltins, C.A., Page, M.A., Buising, K.L., Jenney, A.W.J., Daffy, J.R., Choong, P.F.M., and Stanley, P.A. 2007. Treatment of staphylococcal prosthetic joint infections with debridement, prosthesis retention and oral rifampicin and fusidic acid. *Clin. Microbiol. Infect.*, 13, 586–591.

Ackerman, B., Patton, M., Guilday, R., Haith, L., Stair Buch-mann, M., and Reigart, C. 2013. Trimethoprim-induced hyperkalemia in burn patients treated with intravenous or oral trimethoprim sulfamethoxazole for methicillin-resistant *Staphylococcus aureus* and other infections: Nature or nurture? *J. Burn. Care Res.*, 34, 127–132.

Adem, P.V., Montgomery, C.P., Husain, A.N., Koogler, T.K., Arangelovich, V., Humilier, M., Boyle-Vavra, S., and Daum, R.S. 2005. *Staphylococcus aureus* sepsis and the Waterhouse–Friderichsen syndrome in children. *N. Engl. J. Med.*, 353, 1245–1251.

Adler, A., Chmelnitsky, I., Shitrit, P., Sprecher, H., Navon-Venezia, S., Embon, A., Khabra, E., Paitan, Y., Keren, L., Halperin, E., Carmeli, Y., and Schwaber, M.J. 2012. Molecu-lar epidemiology of methicillin-resistant *Staphylococcus aureus* in Israel: Dissemination of global clones and unique features. *J. Clin. Microbiol.*, 50, 134–137.

Aksoy, D.Y., and Unal, S. 2008. New antimicrobial agents for the treatment of Gram-positive bacterial infections. *Clin. Micro-biol. Infect.*, 14, 411–420.

Alderson, M.R., Tough, T.W., Davis-Smith, T., Braddy, S., Falk, B., Schooley, K.A., Goodwin, R.G., Smith, C.A., Ramsdell, F., and Lynch, D.H. 1995. Fas ligand mediates activation-induced cell death in human T lymphocytes. *J. Exp. Med.*, 181, 71–77.

Allington, D.R., and Rivey, M.P. 2001. Quinupristin/dalfopristin: A therapeutic review. *Clin. Ther.*, 23, 24–44.

Alonzo, F., Benson, M., Chen, J., Novick, R., Shopsin, B., and Tor-res, V. 2012. *Staphylococcus aureus* leucocidin ED contributes to systemic infection by targeting neutrophils and promoting bacterial growth in vivo. *Mol. Microbiol.*, 83, 423–435.

Alonzo III, F., Kozhaya, L., Rawlings, S.A., Reyes-Robles, T., DuMont, A.L., Myszka, D.G., Landau, N.R., Unutmaz, D., and Torres, V.J. 2013. CCR5 is a receptor for *Staphylococcus aureus* leukotoxin ED. *Nature*, 493, 51–55.

Aly, R., Shinefield, H.I., Strauss, W.G., and Maibach, H.I. 1977. Bacterial adherence to nasal mucosal cells. *Infect. Immun.*, 17, 546–549.

Ammerlaan, H.S.M., Kluytmans, J.A.J.W., Wertheim, H.F.L., Nouwen, J.L., and Bonten, M.J.M. 2009. Eradication of methicillin-resistant *Staphylococcus aureus* carriage: A system-atic review. *Clin. Infect. Dis.*, 48, 922–930.

Andrey, D.O., Renzoni, A., Monod, A., Lew, D.P., Cheung, A.L., and Kelley, W.L. 2010. Control of the *Staphylococcus aureus* toxic shock *tst* promoter by the global regulator SarA. *J. Bac-teriol.*, 192, 6077–6085.

Apodaca, A.A., and Rakita, R.M. 2003. Linezolid-induced lactic acidosis. *N. Engl. J. Med.*, 348, 86–87.

Arciola, C., Visai, L., Testoni, F., Arciola, S., Campoccia, D., Speziale, P., and Montanaro, L. 2011. Concise survey of *Staphylococcus aureus* virulence factors that promote adhesion and damage to peri-implant tissues. *Int. J. Artif. Organs.*, 34, 771–780.

Armstrong, G.L., Conn, L.A., and Pinner, R.W. 1999. Trends in infectious disease mortality in the United States during the 20th century. *JAMA*, 281, 61–66.

Arnold, S., Elias, D., Buckingham, S., Thomas, E., Novais, E., Arkader, A., and Howard, C. 2006. Changing patterns of acute hematogenous osteomyelitis and septic arthritis: Emer-gence of community-associated methicillin-resistant *Staphy-lococcus aureus*. *J. Pediatr. Orthop.*, 26, 703–708.

Arriola, C.S., Güere, M.E., Larsen, J., Skov, R.L., Gilman, R.H., Gonzalez, A.E., and Silbergeld, E.K. 2011. Presence of methicillin-resistant *Staphylococcus aureus* in pigs in Peru. *PLoS One*, 6, e28529.

Ateba Ngoa, U., Schaumburg, F., Adegnika, A.A., Kösters, K., Möller, T., Fernandes, J.F., Alabi, A., Issifou, S., Becker, K.,

Grobusch, M.P., Kremsner, P.G., and Lell, B. 2012. Epidemiology and population structure of *Staphylococcus aureus* in various population groups from a rural and semi urban area in Gabon, Central Africa. *Acta Trop.*, 124, 42–47.

Athanasopoulos, A.N., Economopoulou, M., Orlova, V.V., Sobke, A., Schneider, D., Weber, H., Augustin, H.G., Eming, S.A., Schubert, U., Linn, T., Nawroth, P.P., Hussain, M., Hammes, H.-P., Herrmann, M., Preissner, K.T., and Chavakis, T. 2006. The extracellular adherence protein (Eap) of *Staphylococcus aureus* inhibits wound healing by interfering with host defense and repair mechanisms. *Blood*, 107, 2720–2727.

Athens, J.W., Haab, O.P., Raab, S.O., Mauer, A.M., Ashenbrucker, H., Cartwright, G.E., and Wintrobe, M.M. 1961. Leukokinetic studies. IV. The total blood, circulating and marginal granulocyte pools and the granulocyte turnover rate in normal subjects. *J. Clin. Invest.*, 40, 989–995.

Atkins, K.L., Burman, J.D., Chamberlain, E.S., Cooper, J.E., Poutrel, B., Bagby, S., Jenkins, A.T.A., Feil, E.J., and van den Elsen, J.M.H. 2008. *S. aureus* IgG-binding proteins SpA and Sbi: Host specificity and mechanisms of immune complex formation. *Mol. Immunol.*, 45, 1600–1611.

Baba, T., Takeuchi, F., Kuroda, M., Yuzawa, H., Aoki, K.-i., Oguchi, A., Nagai, Y., Iwama, N., Asano, K., Naimi, T., Kuroda, H., Cui, L., Yamamoto, K., and Hiramatsu, K. 2002. Genome and virulence determinants of high virulence community-acquired MRSA. *Lancet*, 359, 1819–1827.

Babior, B.M., Curnutte, J.T., and Kipnes, R.S. 1975. Biological defense mechanisms. Evidence for the participation of superoxide in bacterial killing by xanthine oxidase. *J. Lab. Clin. Med.*, 85, 235–244.

Bachert, C., Zhang, N., Holtappels, G., De Lobel, L., van Cauwenberge, P., Liu, S., Lin, P., Bousquet, J., and Van Steen, K. 2010. Presence of IL-5 protein and IgE antibodies to staphylococcal enterotoxins in nasal polyps is associated with comorbid asthma. *J. Allergy. Clin. Immunol.*, 126, 962–968.

Baddour, L.M., Lowrance, C., Albus, A., Lowrance, J.H., Anderson, S.K., and Lee, J.C. 1992. *Staphylococcus aureus* microcapsule expression attenuates bacterial virulence in a rat model of experimental endocarditis. *J. Infect. Dis.*, 165, 749–753.

Baddour, L.M., Wilson, W.R., Bayer, A.S., Fowler, V.G., Bolger, A.F., Levison, M.E., Ferrieri, P., Gerber, M.A., Tani, L.Y., Gewitz, M.H., Tong, D.C., Steckelberg, J.M., Baltimore, R.S., Shulman, S.T., Burns, J.C., Falace, D.A., Newburger, J.W., Pallasch, T.J., Takahashi, M., and Taubert, K.A. 2005. Infective endocarditis: Diagnosis, antimicrobial therapy, and management of complications: A statement for healthcare professionals from the Committee on Rheumatic Fever, Endocarditis, and Kawasaki Disease, Council on Cardiovascular Disease in the Young, and the Councils on Clinical Cardiology, Stroke, and Cardiovascular Surgery and Anesthesia, American Heart Association: Endorsed by the Infectious Diseases Society of America. *Circulation*, 111, e394–e434.

Bae, I.-G., Tonthat, G.T., Stryjewski, M.E., Rude, T.H., Reilly, L.F., Barriere, S.L., Genter, F.C., Corey, G.R., and Fowler, V.G. 2009. Presence of genes encoding the Panton-Valentine leukocidin exotoxin is not the primary determinant of outcome in patients with complicated skin and skin structure infections due to methicillin-resistant *Staphylococcus aureus*: Results of a multinational trial. *J. Clin. Microbiol.*, 47, 3952–3957.

Baggett, H.C., Hennessy, T.W., Rudolph, K., Bruden, D., Reasonover, A., Parkinson, A., Sparks, R., Donlan, R.M., Martinez, P., Mongkolrattanothai, K., and Butler, J.C. 2004. Community-onset methicillin-resistant *Staphylococcus aureus* associated with antibiotic use and the cytotoxin Panton-Valentine leukocidin during a furunculosis outbreak in rural Alaska. *J. Infect. Dis.*, 189, 1565–1573.

Bainton, D.F., Ullyot, J.L., and Farquhar, M.G. 1971. The development of neutrophilic polymorphonuclear leukocytes in human bone marrow. *J. Exp. Med.*, 134, 907–934.

Balk, R.A. 2000. Severe sepsis and septic shock: Definitions, epidemiology, and clinical manifestations. *Crit. Care. Clin.*, 16, 179–192.

Barbu, E.M., Ganesh, V.K., Gurusiddappa, S., Mackenzie, R.C., Foster, T.J., Sudhof, T.C., and Höök, M. 2010. β-neurexin is a ligand for the *Staphylococcus aureus* MSCRAMM SdrC. *PLoS Pathog.*, 6, e1000726.

Barnes, P.J. 2009. Intrinsic asthma: Not so different from allergic asthma but driven by superantigens? *Clin. Exp. Allergy*, 39, 1145–1151.

Bartlett, A.H., Foster, T.J., Hayashida, A., and Park, P.W. 2008. α-Toxin facilitates the generation of CXC chemokine gradients and stimulates neutrophil homing in *Staphylococcus aureus* pneumonia. *J. Infect. Dis.*, 198, 1529–1535.

Bassetti, M., Merelli, M., Temperoni, C., and Astilean, A. 2013. New antibiotics for bad bugs: Where are we? *Ann. Clin. Microbiol. Antimicrob.*, 12, 22–22.

Bassetti, M., Nicolini, L., Esposito, S., Righi, E., and Viscoli, C. 2009. Current status of newer carbapenems. *Curr. Med. Chem.*, 16, 564–575.

Ben Ari, J., Wolach, O., Gavrieli, R., and Wolach, B. 2012. Infections associated with chronic granulomatous disease: Linking genetics to phenotypic expression. *Expert Rev. Anti. Infect. Ther.*, 10, 881–894.

Bera, A., Herbert, S., Jakob, A., Vollmer, W., and Götz, F. 2005. Why are pathogenic staphylococci so lysozyme resistant? The peptidoglycan *O*-acetyltransferase OatA is the major determinant for lysozyme resistance of *Staphylococcus aureus*. *Mol. Microbiol.*, 55, 778–787.

Bergdoll, M.S., Reiser, R.F., Crass, B.A., Robbins, R.N., and Davis, J.P. 1981. A new staphylococcal enterotoxin, enterotoxin F, associated with toxic-shock-syndrome *Staphylococcus aureus* isolates. *Lancet*, 317, 1017–1021.

Bestebroer, J., Poppelier, M.J.J.G., Ulfman, L.H., Lenting, P.J., Denis, C.V., van Kessel, K.P.M., van Strijp, J.A.G., and de Haas, C.J.C. 2007. Staphylococcal superantigen-like 5 binds PSGL-1 and inhibits P-selectin–mediated neutrophil rolling. *Blood*, 109, 2936–2943.

Bhakdi, S., and Tranum-Jensen, J. 1991. Alpha-toxin of *Staphylococcus aureus*. *Microbiol. Rev.*, 55, 733–751.

Bhat, M., Dumortier, C., Taylor, B., Miller, M., Vasquez, G., Yunen, J., Brudney, K., Sánchez-E, J., Rodriguez Taveras, C., Rojas, R., Leon, P., and Lowy, F. 2009. *Staphylococcus aureus* ST398, New York City and Dominican Republic. *Emerg. Infect. Dis.*, 15, 285–287.

Bishara, J., Leibovici, L., Israel, D.G., Sagie, A., Kazakov, A., Miroshnik, E., Ashkenazi, S., and Pitlik, S. 2001. Long-term outcome of infective endocarditis: The impact of early surgical intervention. *Clin. Infect. Dis.*, 33, 1636–1643.

Blevins, J.S., Elasri, M.O., Allmendinger, S.D., Beenken, K.E., Skinner, R.A., Thomas, J.R., and Smeltzer, M.S. 2003. Role of *sarA* in the pathogenesis of *Staphylococcus aureus* musculoskeletal infection. *Infect. Immun.*, 71, 516–523.

Bokarewa, M.I., Jin, T., and Tarkowski, A. 2006. *Staphylococcus aureus*: Staphylokinase. *Int. J. Biochem. Cell. Biol.*, 38, 504–509.

Bone, R.C. 1994. Gram-positive organisms and sepsis. *Arch. Intern. Med.*, 154, 26–34.

Bor, D., Woolhandler, S., Nardin, R., Brusch, J., and Himmelstein, D. 2013. Infective endocarditis in the U.S., 1998–2009: A nationwide study. *PLoS One*, 8, e60033.

Bost, K.L., Ramp, W.K., Nicholson, N.C., Bento, J.L., Marriott, I., and Hudson, M.C. 1999. *Staphylococcus aureus* infection of mouse or human osteoblasts induces high levels of interleukin-6 and interleukin-12 production. *J. Infect. Dis.*, 180, 1912–1920.

Boucher, H., Miller, L.G., and Razonable, R.R. 2010. Serious infections caused by methicillin-resistant *Staphylococcus aureus*. *Clin. Infect. Dis.*, 51, S183–S197.

Boucher, H.W., and Corey, G.R. 2008. Epidemiology of methicillin-resistant *Staphylococcus aureus*. *Clin. Infect. Dis.*, 46, S344–S349.

Boyle-Vavra, S., and Daum, R.S. 2006. Community-acquired methicillin-resistant *Staphylococcus aureus*: The role of Panton-Valentine leukocidin. *Lab. Invest.*, 87, 3–9.

Brackman, G., Cos, P., Maes, L., Nelis, H.J., and Coenye, T. 2011. Quorum sensing inhibitors increase the susceptibility of bacterial biofilms to antibiotics in vitro and in vivo. *Antimicrob. Agents Chemother.*, 55, 2655–2661.

Brauner, J., Hallin, M., Deplano, A., Mendonça, R., Nonhoff, C., Ryck, R., Roisin, S., Struelens, M.J., and Denis, O. 2013. Community-acquired methicillin-resistant *Staphylococcus aureus* clones circulating in Belgium from 2005 to 2009: Changing epidemiology. *Eur. J. Clin. Microbiol. Infect. Dis.*, 32, 613–620.

Brown, E.L., Dumitrescu, O., Thomas, D., Badiou, C., Koers, E.M., Choudhury, P., Vazquez, V., Etienne, J., Lina, G., Vandenesch, F., and Bowden, M.G. 2009. The Panton-Valentine leukocidin vaccine protects mice against lung and skin infections caused by *Staphylococcus aureus* USA300. *Clin. Microbiol. Infect.*, 15, 156–164.

Brumfitt, W., and Hamilton-Miller, J. 1989. Methicillin-resistant *Staphylococcus aureus*. *N. Engl. J. Med.*, 320, 1188–1196.

Brüssow, H., Canchaya, C., and Hardt, W.-D. 2004. Phages and the evolution of bacterial pathogens: From genomic rearrangements to lysogenic conversion. *Microbiol. Mol. Biol. Rev.*, 68, 560–602.

Bubeck Wardenburg, J., Bae, T., Otto, M., DeLeo, F.R., and Schneewind, O. 2007. Poring over pores: α-Hemolysin and Panton-Valentine leukocidin in *Staphylococcus aureus* pneumonia. *Nat. Med.*, 13, 1405–1406.

Buck, J., Como Sabetti, K., Harriman, K., Danila, R., Boxrud, D., Glennen, A., and Lynfield, R. 2005. Community-associated methicillin-resistant *Staphylococcus aureus*, Minnesota, 2000–2003. *Emerg. Infect. Dis.*, 11, 1532–1538.

Bunikowski, R., Mielke, M., Skarabis, H., Herz, U., Bergmann, R.L., Wahn, U., and Renz, H. 1999. Prevalence and role of serum IgE antibodies to the *Staphylococcus aureus*–derived superantigens SEA and SEB in children with atopic dermatitis. *J. Allergy Clin. Immunol.*, 103, 119–124.

Burian, M., Rautenberg, M., Kohler, T., Fritz, M., Krismer, B., Unger, C., Hoffman, W.H., Peschel, A., Wolz, C., and Goerke, C. 2010a. Temporal expression of adhesion factors and activity of global regulators during establishment of *Staphylococcus aureus* nasal colonization. *J. Infect. Dis.*, 201, 1414–1421.

Burian, M., Wolz, C., and Goerke, C. 2010b. Regulatory adaptation of *Staphylococcus aureus* during nasal colonization of humans. *PLoS One*, 5, e10040.

Buyukcangaz, E., Velasco, V., Sherwood, J., Stepan, R., Koslofsky, R., and Logue, C. 2013. Molecular typing of *Staphylococcus aureus* and methicillin-resistant *S. aureus* (MRSA) isolated from animals and retail meat in North Dakota, United States. *Foodborne Pathog. Dis.*, 10, 608–617.

Cameron, D.R., Howden, B.P., and Peleg, A.Y. 2011. The interface between antibiotic resistance and virulence in *Staphylococcus aureus* and its impact upon clinical outcomes. *Clin. Infect. Dis.*, 53, 576–582.

Cañas-Pedrosa, A.M., Vindel, A., Artiles, F., Colino, E., and Lafarga, B. 2012. Antimicrobial resistance and molecular epidemiology of Panton-Valentine leukocidin–positive community-associated methicillin-resistant *Staphylococcus aureus* from Gran Canaria (Canary Islands, Spain). *Diagn. Microbiol. Infect. Dis.*, 74, 432–434.

Carpenter, C.F., and Chambers, H.F. 2004. Daptomycin: Another novel agent for treating infections due to drug-resistant Gram-positive pathogens. *Clin. Infect. Dis.*, 38, 994–1000.

Carter, N.J., and Scott, L.J. 2010. Besifloxacin ophthalmic suspension 0.6%. *Drugs*, 70, 83–97.

Cary, S., Krishnan, M., Marion, T.N., and Silverman, G.J. 1999. The murine clan VH III related 7183, J606 and S107 and DNA4 families commonly encode for binding to a bacterial B cell superantigen. *Mol. Immunol.*, 36, 769–776.

CDC. 2003. Outbreaks of community-associated methicillin-resistant *Staphylococcus aureus* skin infections–Los Angeles County, California, 2002–2003. *MMWR Morb. Mortal. Wkly. Rep.*, 52, 88–88.

CDC. 2013. Methicillin-resistant *Staphylococcus aureus* (MRSA) infections.

Chambers, H.F., and DeLeo, F.R. 2009. Waves of resistance: *Staphylococcus aureus* in the antibiotic era. *Nat. Rev. Micro.*, 7, 629–641.

Chambers, H.F., Korzeniowski, O.M., and Sande, M.A. 1983. *Staphylococcus aureus* endocarditis: Clinical manifestations in addicts and nonaddicts. *Medicine*, 62, 170–177.

Chang, M.H., and Fung, H.B. 2010. Besifloxacin: A topical fluoroquinolone for the treatment of bacterial conjunctivitis. *Clin. Ther.*, 32, 454–471.

Chatterjee, S.S., Chen, L., Joo, H.-S., Cheung, G.Y.C., Kreiswirth, B.N., and Otto, M. 2011. Distribution and regulation of the mobile genetic element-encoded phenol-soluble modulin PSM-*mec* in methicillin-resistant *Staphylococcus aureus*. *PLoS One*, 6, e28781.

Chavakis, T., Hussain, M., Kanse, S.M., Peters, G., Bretzel, R.G., Flock, J.-I., Herrmann, M., and Preissner, K.T. 2002. *Staphylococcus aureus* extracellular adherence protein serves as anti-inflammatory factor by inhibiting the recruitment of host leukocytes. *Nat. Med.*, 8, 687–693.

Chavakis, T., Preissner, K.T., and Herrmann, M. 2007. The anti-inflammatory activities of *Staphylococcus aureus*. *Trends. Immunol.*, 28, 408–418.

Chesney, P., Davis, J.P., Purdy, W.K., Wand, P.J., and Chesney, R.W. 1981. Clinical manifestations of toxic shock syndrome. *JAMA*, 246, 741–748.

Cheung, A.L., Eberhardt, K.J., Chung, E., Yeaman, M.R., Sullam, P.M., Ramos, M., and Bayer, A.S. 1994. Diminished virulence of a sar-/agr- mutant of *Staphylococcus aureus* in the rabbit model of endocarditis. *J. Clin. Invest.*, 94, 1815–1822.

Cheung, G.Y.C., Duong, A.C., and Otto, M. 2012. Direct and synergistic hemolysis caused by Staphylococcus phenol-soluble modulins: Implications for diagnosis and pathogenesis. *Microbes. Infect.*, 14, 380–386.

Cheung, G.Y.C., Rigby, K., Wang, R., Queck, S.Y., Braughton, K.R., Whitney, A.R., Teintze, M., DeLeo, F.R., and Otto, M. 2010. *Staphylococcus epidermidis* strategies to avoid killing by human neutrophils. *PLoS Pathog.*, 6, e1001133.

Cheung, G.Y.C., Wang, R., Khan, B.A., Sturdevant, D.E., and Otto, M. 2011. Role of the accessory gene regulator *agr* in community-associated methicillin-resistant *Staphylococcus aureus* pathogenesis. *Infect. Immun.*, 79, 1927–1935.

Christianson, S., Golding, G.R., Campbell, J., Program, t.C.N.I.S., and Mulvey, M.R. 2007. Comparative genomics of Canadian epidemic lineages of methicillin-resistant *Staphylococcus aureus*. *J. Clin. Microbiol.*, 45, 1904–1911.

Chroboczek, T., Boisset, S., Rasigade, J.-P., Meugnier, H., Akpaka, P., Nicholson, A., Nicolas, M., Olive, C., Bes, M., Vandenesch, F.o., Laurent, F., Etienne, J., and Tristan, A. 2013. Major west indies MRSA clones in human beings: Do they travel with their hosts? *J. Travel. Med.*, 20, 283–288.

Chu, C., Yu, C., Lee, Y., and Su, Y. 2012. Genetically divergent methicillin-resistant *Staphylococcus aureus* and *sec*-dependent mastitis of dairy goats in Taiwan. *BMC Vet. Res.*, 8, 39.

Chua, K., Laurent, F., Coombs, G., Grayson, M.L., and Howden, B.P. 2011a. Not community-associated methicillin-resistant *Staphylococcus aureus* (CA-MRSA)! A clinician's guide to community MRSA – Its evolving antimicrobial resistance and implications for therapy. *Clin. Infect. Dis.*, 52, 99–114.

Chua, K.Y.L., Howden, B.P., Jiang, J.-H., Stinear, T., and Peleg, A.Y. 2014. Population genetics and the evolution of virulence in *Staphylococcus aureus*. *Infect. Genet. Evol.*, 21, 554–562.

Chua, K.Y.L., Seemann, T., Harrison, P.F., Monagle, S., Korman, T.M., Johnson, P.D.R., Coombs, G.W., Howden, B.O., Davies, J.K., Howden, B.P., and Stinear, T.P. 2011b. The dominant Australian community-acquired methicillin-resistant

Staphylococcus aureus clone ST93-IV [2B] Is highly virulent and genetically distinct. *PLoS One*, 6, e25887.

Chuang, Y.-Y., and Huang, Y.-C. 2013. Molecular epidemiology of community-associated meticillin-resistant *Staphylococcus aureus* in Asia. *Lancet Infect. Dis.*, 13, 698–708.

Clarke, S.R. 2010. Phenol-soluble modulins of *Staphylococcus aureus* lure neutrophils into battle. *Cell Host Microbe.*, 7, 423–424.

Clarke, S.R., Andre, G., Walsh, E.J., Dufrêne, Y.F., Foster, T.J., and Foster, S.J. 2009. Iron-regulated surface determinant protein A mediates adhesion of *Staphylococcus aureus* to human corneocyte envelope proteins. *Infect. Immun.*, 77, 2408–2416.

Clarke, S.R., Brummell, K.J., Horsburgh, M.J., McDowell, P.W., Mohamad, S.A.S., Stapleton, M.R., Acevedo, J., Read, R.C., Day, N.P.J., Peacock, S.J., Mond, J.J., Kokai-Kun, J.F., and Foster, S.J. 2006. Identification of in vivo–expressed antigens of *Staphylococcus aureus* and their use in vaccinations for protection against nasal carriage. *J. Infect. Dis.*, 193, 1098–1108.

Clarke, S.R., and Foster, S.J. 2006. Surface adhesins of *Staphylococcus aureus*. In: Robert, K.P., editor. *Advances in Microbial Physiology*. Academic Press. pp. 187–224.

Clements, M.O., Watson, S.P., and Foster, S.J. 1999. Characterization of the major superoxide dismutase of *Staphylococcus aureus* and its role in starvation survival, stress resistance, and pathogenicity. *J. Bacteriol.*, 181, 3898–3903.

Cohen, M.L. 2000. Changing patterns of infectious disease. *Nature*, 406, 762–767.

Coleman, D.C., Sullivan, D.J., Russel, R.J., Arbuthnott, J.P., Carey, B.F., and Pomeroy, H.M. 1989. *Staphylococcus aureus* bacteriophages mediating the simultaneous lysogenic conversion of β-lysin, staphylokinase and enterotoxin A: Molecular mechanism of triple conversion. *J. Gen. Microbiol.*, 135, 1679–1697.

Collins, L.V., Kristian, S.A., Weidenmaier, C., Faigle, M., van Kessel, K.P.M., van Strijp, J.A.G., Götz, F., Neumeister, B., and Peschel, A. 2002. *Staphylococcus aureus* strains lacking D-alanine modifications of teichoic acids are highly susceptible to human neutrophil killing and are virulence attenuated in mice. *J. Infect. Dis.*, 186, 214–219.

Colotta, F., Re, F., Polentarutti, N., Sozzani, S., and Mantovani, A. 1992. Modulation of granulocyte survival and programmed cell death by cytokines and bacterial products. *Blood*, 80, 2012–2020.

Cooke, F.J., and Brown, N.M. 2010. Community-associated methicillin-resistant *Staphylococcus aureus* infections. *Br. Med. Bull.*, 94, 215–227.

Coombs, G.W., Pearson, J.C., Nimmo, G.R., Collignon, P.J., Bell, J.M., McLaws, M.-L., Christiansen, K.J., and Turnidge, J.D. 2013. Antimicrobial susceptibility of *Staphylococcus aureus* and molecular epidemiology of meticillin-resistant *S. aureus* isolated from Australian hospital inpatients: Report from the Australian Group on Antimicrobial Resistance 2011 *Staphylococcus aureus* Surveillance Programme. *J. Global Antimicrob. Resist.*, 1, 149–156.

Cooney, J., Kienle, Z., Foster, T.J., and O'Toole, P.W. 1993. The gamma-hemolysin locus of *Staphylococcus aureus* comprises three linked genes, two of which are identical to the genes for the F and S components of leukocidin. *Infect. Immun.*, 61, 768–771.

Correa de Sa, D.D., Tleyjeh, I., Anavekar, N., Schultz, J., Thomas, J., Lahr, B., Bachuwar, A., Pazdernik, M., Steckelberg, J., Wilson, W., and Baddour, L. 2010. Epidemiological trends of infective endocarditis: A population-based study in Olmsted County, Minnesota. *Mayo Clin. Proc.*, 85, 422–426.

Corrigan, R., Miajlovic, H., and Foster, T. 2009. Surface proteins that promote adherence of *Staphylococcus aureus* to human desquamated nasal epithelial cells. *BMC Microbiol.*, 9, 22.

Cosgrove, K., Coutts, G., Jonsson, I.-M., Tarkowski, A., Kokai-Kun, J.F., Mond, J.J., and Foster, S.J. 2007. Catalase (KatA) and alkyl hydroperoxide reductase (AhpC) have compensatory roles in peroxide stress resistance and are required for

survival, persistence, and nasal colonization in *Staphylococcus aureus*. *J. Bacteriol.*, 189, 1025–1035.

Cosgrove, S., and Fowler, V. 2007. Optimizing therapy for methicillin-resistant *Staphylococcus aureus* bacteremia. *Semin. Respir. Crit. Care. Med.*, 28, 624–631.

Cosgrove, S.E., Qi, Y., Kaye, K.S., Harbarth, S., Karchmer, A.W., and Carmeli, Y. 2005. The impact of methicillin resistance in *Staphylococcus aureus* bacteremia on patient outcomes: Mortality, length of stay, and hospital charges. *Infect. Control Hosp. Epidemiol.*, 26, 166–174.

Covington, P., Davenport, J.M., Andrae, D., O'Riordan, W., Liverman, L., McIntyre, G., and Almenoff, J. 2011. Randomized, double-blind, phase II, multicenter study evaluating the safety/tolerability and efficacy of JNJ-Q2, a novel fluoroquinolone, compared with linezolid for treatment of acute bacterial skin and skin structure infection. *Antimicrob. Agents Chemother.*, 55, 5790–5797.

Cramton, S.E., Gerke, C., Schnell, N.F., Nichols, W.W., and Götz, F. 1999. The intercellular adhesion (*ica*) locus is present in *Staphylococcus aureus* and is required for biofilm formation. *Infect. Immun.*, 67, 5427–5433.

Crémieux, A.-C., Dumitrescu, O., Lina, G., Vallee, C., Côté, J.-F., Muffat-Joly, M., Lilin, T., Etienne, J., Vandenesch, F., and Saleh-Mghir, A. 2009. Panton-Valentine leukocidin enhances the severity of community-associated methicillin-resistant *Staphylococcus aureus* rabbit osteomyelitis. *PLoS One*, 4, e7204.

Cucarella, C., Tormo, M.Á., Knecht, E., Amorena, B., Lasa, Í., Foster, T.J., and Penadés, J.R. 2002. Expression of the biofilm-associated protein interferes with host protein receptors of *Staphylococcus aureus* and alters the infective process. *Infect. Immun.*, 70, 3180–3186.

Cue, D.R., Lei, M.G., and Lee, C. 2012. Genetic regulation of the intercellular adhesion locus in staphylococci. *Front. Cell Infect. Microbiol.*, 2, 38.

Cui, S., Li, J., Hu, C., Jin, S., Li, F., Guo, Y., Ran, L., and Ma, Y. 2009. Isolation and characterization of methicillin-resistant *Staphylococcus aureus* from swine and workers in China. *J. Antimicrob. Chemother.*, 64, 680–683.

Cursino, M.A., Garcia, C.P., Lobo, R.D., Salomão, M.C., Gobara, S., Raymundo, G.F., Kespers, T., Soares, R.E., Mollaco, C.H., Keil, K.G., Malieno, P.B., Krebs, V.L., Gibelli, M.A., Kondo, M.M., Zugaib, M., Costa, S.F., and Levin, A.S. 2012. Performance of surveillance cultures at different body sites to identify asymptomatic *Staphylococcus aureus* carriers. *Diagn. Microbiol. Infect. Dis.*, 74, 343–348.

D'Agata, E.M.C., Webb, G.F., Horn, M.A., Moellering, R.C., and Ruan, S. 2009. Modeling the invasion of community-acquired methicillin-resistant *Staphylococcus aureus* into hospitals. *Clin. Infect. Dis.*, 48, 274–284.

Dajcs, J.J., Thibodeaux, B.A., Hume, E.B., Zheng, X., Sloop, G.D., and O'Callaghan, R.J. 2001. Lysostaphin is effective in treating methicillin-resistant *Staphylococcus aureus* endophthalmitis in the rabbit. *Curr. Eye. Res.*, 22, 451–457.

Dantes, R., Mu, Y., Belflower, R., Aragon, D., Dumyati, G., Harrison, L.H., Lessa, F.C., Lynfield, R., Nadle, J., Petit, S., Ray, S.M., Schaffner, W., Townes, J., Fridkin, S. 2013. Emerging Infections ProgramActive Bacterial Core Surveillance MRSA Surveillance Investigators. National burden of invasive methicillin-resistant *Staphylococcus aureus* infections, United States, 2011. *JAMA Intern. Med.*, 173, 1970–1978.

David, M.Z., and Daum, R.S. 2010. Community-associated methicillin-resistant *Staphylococcus aureus*: Epidemiology and clinical consequences of an emerging epidemic. *Clin. Microbiol. Rev.*, 23, 616–687.

David, M.Z., Taylor, A., Lynfield, R., Boxrud, D.J., Short, G., Zychowski, D., Boyle-Vavra, S., and Daum, R.S. 2013. Comparing pulsed-field gel electrophoresis with multilocus sequence typing, *spa* typing, staphylococcal cassette chromosome *mec* (SCC*mec*) typing, and PCR for Panton-Valentine leukocidin, *arcA*, and *opp3* in methicillin-resistant

Staphylococcus aureus isolates at a U.S. medical center. *J. Clin. Microbiol.*, 51, 814–819.

de Haas, C.J.C., Veldkamp, K.E., Peschel, A., Weerkamp, F., Van Wamel, W.J.B., Heezius, E.C.J.M., Poppelier, M.J.J.G., Van Kessel, K.P.M., and van Strijp, J.A.G. 2004. Chemotaxis inhibitory protein of *Staphylococcus aureus*, a bacterial antiinflammatory agent. *J. Exp. Med.*, 199, 687–695.

De Vriese, A.S., Van Coster, R., Smet, J., Seneca, S., Lovering, A., Van Haute, L.L., Vanopdenbosch, L.J., Martin, J.-J., Ceuterick-de Groote, C., Vandecasteele, S., and Boelaert, J.R. 2006. Linezolid-induced inhibition of mitochondrial protein synthesis. *Clin. Infect. Dis.*, 42, 1111–1117.

DeLeo, F., and Chambers, H. 2009. Reemergence of antibiotic-resistant *Staphylococcus aureus* in the genomics era. *J. Clin. Invest.*, 119, 2464–2474.

DeLeo, F., Kennedy, A., Chen, L., Bubeck Wardenburg, J., Kobayashi, S., Mathema, B., Braughton, K., Whitney, A., Villaruz, A., Martens, C., Porcella, S., McGavin, M., Otto, M., Musser, J., and Kreiswirth, B. 2011. Molecular differentiation of historic phage-type 80/81 and contemporary epidemic *Staphylococcus aureus*. *Proc. Natl. Acad. Sci. USA*, 108, 18091–18096.

DeLeo, F.R., Diep, B.A., and Otto, M. 2009. Host defense and pathogenesis in *Staphylococcus aureus* infections. *Infect. Dis. Clin. North. Am.*, 23, 17–34.

DeLeo, F.R., Otto, M., Kreiswirth, B.N., and Chambers, H.F. 2010. Community-associated meticillin-resistant *Staphylococcus aureus*. *Lancet*, 375, 1557–1568.

Deresinski, S. 2009. Vancomycin in combination with other antibiotics for the treatment of serious methicillin-resistant *Staphylococcus aureus* infections. *Clin. Infect. Dis.*, 49, 1072–1079.

Devriese, L.A., Van Damme, L.R., and Fameree, L. 1972. Methicillin (cloxacillin)-resistant *Staphylococcus aureus* strains isolated from bovine mastitis cases. *Zentralbl. Veterinarmed. B*, 19, 598–605.

Diekema, D.J., and Climo, M. 2008. Preventing MRSA infections: Finding it is not enough. *JAMA*, 299, 1190–1192.

Diekema, D.J., Pfaller, M.A., Schmitz, F.J., Smayevsky, J., Bell, J., Jones, R.N., and Beach, M. 2001. Survey of infections due to Staphylococcus species: Frequency of occurrence and antimicrobial susceptibility of isolates collected in the United States, Canada, Latin America, Europe, and the Western Pacific region for the SENTRY Antimicrobial Surveillance Program, 1997–1999. *Clin. Infect. Dis.*, 32, S114–S132.

Diep, B., Chan, L., Tattevin, P., Kajikawa, O., Martin, T., Basuino, L., Mai, T., Marbach, H., Braughton, K., Whitney, A., Gardner, D., Fan, X., Tseng, C., Liu, G., Badiou, C., Etienne, J., Lina, G., Matthay, M., DeLeo, F., and Chambers, H. 2010. Polymorphonuclear leukocytes mediate *Staphylococcus aureus* Panton-Valentine leukocidin-induced lung inflammation and injury. *Proc. Natl. Acad. Sci. USA*, 107, 5587–5592.

Diep, B.A., Gill, S.R., Chang, R.F., Phan, T.H., Chen, J.H., Davidson, M.G., Lin, F., Lin, J., Carleton, H.A., Mongodin, E.F., Sensabaugh, G.F., and Perdreau-Remington, F. 2006. Complete genome sequence of USA300, an epidemic clone of community-acquired meticillin-resistant *Staphylococcus aureus*. *Lancet*, 367, 731–739.

Diep, B.A., Stone, G.G., Basuino, L., Graber, C.J., Miller, A., des Etages, S.-A., Jones, A., Palazzolo-Ballance, A.M., Perdreau-Remington, F., Sensabaugh, G.F., DeLeo, F.R., and Chambers, H.F. 2008. The arginine catabolic mobile element and staphylococcal chromosomal cassette *mec* linkage: Convergence of virulence and resistance in the USA300 clone of methicillin-resistant *Staphylococcus aureus*. *J. Infect. Dis.*, 197, 1523–1530.

Doebbeling, B.N., Breneman, D.L., Neu, H.C., Aly, R., Yangco, B.G., Holley Jr., H.P., Marsh, R.J., Pfaller, M.A., McGowan Jr., J.E., Scully, B.E., Reagan, D.R., Wenzel, R.P., and Group, T.M.C.S. 1993. Elimination of *Staphylococcus aureus* nasal carriage in health care workers: Analysis of six clinical trials with calcium mupirocin ointment. *Clin. Infect. Dis.*, 17, 466–474.

Doebbeling, B.N., Reagan, D.R., Pfaller, M.A., Houston, A.K., Hollis, R.J., and Wenzel, R.P. 1994. Long-term efficacy of intranasal mupirocin ointment: A prospective cohort study of *Staphylococcus aureus* carriage. *Arch. Intern. Med.*, 154, 1505–1508.

Downer, R., Roche, F., Park, P.W., Mecham, R.P., and Foster, T.J. 2002. The elastin-binding protein of *Staphylococcus aureus* (EbpS) is expressed at the cell surface as an integral membrane protein and not as a cell wall-associated protein. *J. Biol. Chem.*, 277, 243–250.

Dryden, M.S., Dailly, S., and Crouch, M. 2004. A randomized, controlled trial of tea tree topical preparations versus a standard topical regimen for the clearance of MRSA colonization. *J. Hosp. Infect.*, 56, 283–286.

Dryla, A., Gelbmann, D., Von Gabain, A., and Nagy, E. 2003. Identification of a novel iron regulated staphylococcal surface protein with haptoglobin-haemoglobin binding activity. *Mol. Microbiol.*, 49, 37–53.

Dumont, A., Nygaard, T., Watkins, R., Smith, A., Kozhaya, L., Kreiswirth, B., Shopsin, B., Unutmaz, D., Voyich, J., and Torres, V. 2011. Characterization of a new cytotoxin that contributes to *Staphylococcus aureus* pathogenesis. *Mol. Microbiol.*, 79, 814–825.

Edwards, A.M., Bowden, M.G., Brown, E.L., Laabei, M., and Massey, R.C. 2012. *Staphylococcus aureus* extracellular adherence protein triggers TNFα release, promoting attachment to endothelial cells via protein A. *PLoS One*, 7, e43046.

El-Mahdy, T.S., El-Ahmady, M., and Goering, R.V. 2014. Molecular characterization of methicillin-resistant *Staphylococcus aureus* isolated over a 2-year period in a Qatari hospital from multinational patients. *Clin. Microbiol. Infect.*, 20, 169–173.

Ellington, J.K., Harris, M., Hudson, M.C., Vishin, S., Webb, L.X., and Sherertz, R. 2006. Intracellular *Staphylococcus aureus* and antibiotic resistance: Implications for treatment of staphylococcal osteomyelitis. *J. Orthop. Res.*, 24, 87–93.

Enright, M.C., Day, N.P.J., Davies, C.E., Peacock, S.J., and Spratt, B.G. 2000. Multilocus sequence typing for characterization of methicillin-resistant and methicillin-susceptible clones of *Staphylococcus aureus*. *J. Clin. Microbiol.*, 38, 1008–1015.

Enright, M.C., Robinson, D.A., Randle, G., Feil, E.J., Grundmann, H., and Spratt, B.G. 2002. The evolutionary history of methicillin-resistant *Staphylococcus aureus* (MRSA). *Proc. Natl. Acad. Sci. USA*, 99, 7687–7692.

Enright, M.C., and Spratt, B.G. 1999. Multilocus sequence typing. *Trends Microbiol.*, 7, 482–487.

Eriksson, J., Espinosa-Gongora, C., Stamphøj, I., Larsen, A.R., and Guardabassi, L. 2013. Carriage frequency, diversity and methicillin resistance of *Staphylococcus aureus* in Danish small ruminants. *Vet. Microbiol.*, 163, 110–115.

Espadinha, D., Faria, N.A., Miragaia, M., Lito, L.M., Melo-Cristino, J., de Lencastre, H., and Médicos Sentinela, N. 2013. Extensive dissemination of methicillin-resistant *Staphylococcus aureus* (MRSA) between the hospital and the community in a country with a high prevalence of nosocomial MRSA. *PLoS One*, 8, e59960.

Faires, M., Traverse, M., Tater, K., Pearl, D., and Weese, J.S. 2010. Methicillin-resistant and -susceptible *Staphylococcus aureus* infections in dogs. *Emerg. Infect. Dis.*, 16, 69–75.

Falugi, F., Kim, H.K., Missiakas, D.M., and Schneewind, O. 2013. Role of Protein A in the evasion of host adaptive immune responses by *Staphylococcus aureus*. *mBio*, 4, e00575–13.

Fankhauser, C., Schrenzel, J., Francois, P., Renzi, G., Pittet, D., and Harbarth, S. 2013. P052: Molecular epidemiology of methicillin-resistant *Staphylococcus aureus* (MRSA) strains at Geneva University Hospitals (HUG) over a 9 year period. *Antimicrob. Resist. Infect. Control.*, 2, P52.

Feng, Y., Chen, C.-J., Su, L.-H., Hu, S., Yu, J., and Chiu, C.-H. 2008. Evolution and pathogenesis of *Staphylococcus aureus*: Lessons learned from genotyping and comparative genomics. *FEMS Microbiol. Rev.*, 32, 23–37.

Fernandez, S., de Vedia, L., Lopez Furst, M.J., Gardella, N., Di Gregorio, S., Ganaha, M.C., Prieto, S., Carbone, E., Lista, N., Rotrying, F., Stryjewski, M.E., and Mollerach, M. 2013. Methicillin-resistant *Staphylococcus aureus* ST30-SCC*mec* IVc clone as the major cause of community-acquired invasive infections in Argentina. *Infect. Genet. Evol.*, 14, 401–405.

Ferreras, M., Höper, F., Dalla Serra, M., Colin, D.A., Prévost, G., and Menestrina, G. 1998. The interaction of *Staphylococcus aureus* bi-component γ-hemolysins and leucocidins with cells and lipid membranes. *Biochim. Biophys. Acta*, 1414, 108–126.

Fey, P.D., Saïd-Salim, B., Rupp, M.E., Hinrichs, S.H., Boxrud, D.J., Davis, C.C., Kreiswirth, B.N., and Schlievert, P.M. 2003. Comparative molecular analysis of community- or hospital-acquired methicillin-resistant *Staphylococcus aureus*. *Antimicrob. Agents Chemother.*, 47, 196–203.

Fleischer, B., and Schrezenmeier, H. 1988. T cell stimulation by staphylococcal enterotoxins. Clonally variable response and requirement for major histocompatibility complex class II molecules on accessory or target cells. *J. Exp. Med.*, 167, 1697–1707.

Forsgren, A., and Quie, P.G. 1974. Effects of staphylococcal protein a on heat labile opsonins. *J. Immunol.*, 112, 1177–1180.

Foster, A.P. 2012. Staphylococcal skin disease in livestock. *Vet. Dermatol.*, 23, 342–351.

Foster, T. 2005. Immune evasion by staphylococci. *Nat. Rev. Microbiol.*, 3, 948–958.

Fournier, B., and Hooper, D.C. 2000. A new two-component regulatory system involved in adhesion, autolysis, and extracellular proteolytic activity of *Staphylococcus aureus*. *J. Bacteriol.*, 182, 3955–3964.

Fowler, V.G., Sanders, L.L., Kong, L.K., McClelland, R.S., Gottlieb, G.S., Li, J., Ryan, T., Sexton, D.J., Roussakis, G., Harrell, L.J., and Corey, G.R. 1999. Infective endocarditis due to *Staphylococcus aureus*: 59 prospectively identified cases with follow-up. *Clin. Infect. Dis.*, 28, 106–114.

Fowler, V.G., Sanders, L.L., Sexton, D.J., Kong, L., Marr, K.A., Gopal, A.K., Gottlieb, G., McClelland, R.S., and Corey, G.R. 1998. Outcome of *Staphylococcus aureus* bacteremia according to compliance with recommendations of infectious diseases specialists: Experience with 244 patients. *Clin. Infect. Dis.*, 27, 478–486.

Fowler, V.G.J., Olsen, M.K., Corey, G., Woods, C.W., Cabell, C.H., Reller, L.B., Cheng, A.C., Dudley, T., and Oddone, E.Z. 2003. Clinical identifiers of complicated *Staphylococcus aureus* bacteremia. *Arch. Intern. Med.*, 163, 2066–2072.

Francis, J.S., Doherty, M.C., Lopatin, U., Johnston, C.P., Sinha, G., Ross, T., Cai, M., Hansel, N.N., Perl, T., Ticehurst, J.R., Carroll, K., Thomas, D.L., Nuermberger, E., and Bartlett, J.G. 2005. Severe community-onset pneumonia in healthy adults caused by methicillin-resistant *Staphylococcus aureus* carrying the Panton-Valentine leukocidin genes. *Clin. Infect. Dis.*, 40, 100–107.

Fraser, J., Arcus, V., Kong, P., Baker, E., and Proft, T. 2000. Superantigens – Powerful modifiers of the immune system. *Mol. Med. Today*, 6, 125–132.

Fraser, J.D., and Proft, T. 2008. The bacterial superantigen and superantigen-like proteins. *Immunol. Rev.*, 225, 226–243.

French, G. 2003. Safety and tolerability of linezolid. *J. Antimicrob. Chemother.*, 51, ii45–ii53.

Fridkin, S.K., Hageman, J.C., Morrison, M., Sanza, L.T., Como-Sabetti, K., Jernigan, J.A., Harriman, K., Harrison, L.H., Lynfield, R., and Farley, M.M. 2005. Methicillin-resistant *Staphylococcus aureus* disease in three communities. *N. Engl. J. Med.*, 352, 1436–1444.

García, C., Rijnders, M.I.A., Bruggeman, C., Samalvides, F., Stobberingh, E.E., and Jacobs, J. 2012. Antimicrobial resistance and molecular typing of *Staphylococcus aureus* bloodstream isolates from hospitals in Peru. *J. Infect.*, 65, 406–411.

Genestier, A.-L., Michallet, M.-C., Prévost, G., Bellot, G., Chalabreysse, L., Peyrol, S., Thivolet, F., Etienne, J., Lina, G., Vallette, F.M., Vandenesch, F., and Genestier, L. 2005.

Staphylococcus aureus Panton-Valentine leukocidin directly targets mitochondria and induces Bax-independent apoptosis of human neutrophils. *J. Clin. Invest.*, 115, 3117–3127.

Gerson, S.L., Kaplan, S.L., Bruss, J.B., Le, V., Arellano, F.M., Hafkin, B., and Kuter, D.J. 2002. Hematologic effects of linezolid: Summary of clinical experience. *Antimicrob. Agents Chemother.*, 46, 2723–2726.

Gevaert, P., Holtappels, G., Johansson, S.G.O., Cuvelier, C., van Cauwenberge, P., and Bachert, C. 2005. Organization of secondary lymphoid tissue and local IgE formation to *Staphylococcus aureus* enterotoxins in nasal polyp tissue. *Allergy*, 60, 71–79.

Gharsa, H., Ben Slama, K., Lozano, C., Gómez-Sanz, E., Klibi, N., Ben Sallem, R., Gómez, P., Zarazaga, M., Boudabous, A., and Torres, C. 2012. Prevalence, antibiotic resistance, virulence traits and genetic lineages of *Staphylococcus aureus* in healthy sheep in Tunisia. *Vet. Microbiol.*, 156, 367–373.

Gillet, Y., Issartel, B., Vanhems, P., Fournet, J.-C., Lina, G., Bes, M., Vandenesch, F., Piémont, Y., Brousse, N., Floret, D., and Etienne, J. 2002. Association between *Staphylococcus aureus* strains carrying gene for Panton-Valentine leukocidin and highly lethal necrotising pneumonia in young immunocompetent patients. *Lancet*, 359, 753–759.

Ginsburg, I. 2002. Role of lipoteichoic acid in infection and inflammation. *Lancet Infect. Dis.*, 2, 171–179.

Girardin, S.E., Boneca, I.G., Viala, J., Chamaillard, M., Labigne, A., Thomas, G., Philpott, D.J., and Sansonetti, P.J. 2003. Nod2 is a general sensor of peptidoglycan through muramyl dipeptide (MDP) detection. *J. Biol. Chem.*, 278, 8869–8872.

Glick, S.B., Samson, D.J., Huang, E., Vats, V., Weber, S., and Aronson, N. 2013. *Screening for Methicillin-Resistant Staphylococcus aureus (MRSA), AHRQ Comparative Effectiveness Reviews.* Rockville, MD: Agency for Healthcare Research and Quality.

Glikman, D., Davidson, S., Kudinsky, R., Geffen, Y., and Sprecher, H. 2010. First isolation of SCC*mec* IV- and Panton-Valentine leukocidin-positive, sequence type 8, community-associated methicillin-resistant *Staphylococcus aureus* in Israel. *J. Clin. Microbiol.*, 48, 3827–3828.

Goering, R.V., Larsen, A.R., Skov, R., Tenover, F.C., Anderson, K.L., and Dunman, P.M. 2009. Comparative genomic analysis of European and Middle Eastern community-associated methicillin-resistant *Staphylococcus aureus* (CC80:ST80-IV) isolates by high-density microarray. *Clin. Microbiol. Infect.*, 15, 748–755.

Goerke, C., Pantucek, R., Holtfreter, S., Schulte, B., Zink, M., Grumann, D., Bröker, B.M., Doskar, J., and Wolz, C. 2009. Diversity of prophages in dominant *Staphylococcus aureus* clonal lineages. *J. Bacteriol.*, 191, 3462–3468.

Goerke, C., Wirtz, C., Flückiger, U., and Wolz, C. 2006. Extensive phage dynamics in *Staphylococcus aureus* contributes to adaptation to the human host during infection. *Mol. Microbiol.*, 61, 1673–1685.

Goerke, C., and Wolz, C. 2004. Regulatory and genomic plasticity of *Staphylococcus aureus* during persistent colonization and infection. *Int. J. Med. Microbiol.*, 294, 195–202.

González-Domínguez, M., Seral, C., Sáenz, Y., Salvo, S., Gude, M.J., Porres-Osante, N., Torres, C., and Castillo, F.J. 2012. Epidemiological features, resistance genes, and clones among community-onset methicillin-resistant *Staphylococcus aureus* (CO-MRSA) isolates detected in northern Spain. *Int. J. Med. Microbiol.*, 302, 320–326.

Gonzalez, B.E., Hulten, K.G., Dishop, M.K., Lamberth, L.B., Hammerman, W.A., Mason, E.O., and Kaplan, S.L. 2005. Pulmonary manifestations in children with invasive community-acquired *Staphylococcus aureus* infection. *Clin. Infect. Dis.*, 41, 583–590.

Gordon, R.J., and Lowy, F.D. 2008. Pathogenesis of methicillin-resistant *Staphylococcus aureus* infection. *Clin. Infect. Dis.*, 46, S350–S359.

Gorwitz, R., Kruszon Moran, D., McAllister, S., McQuillan, G., McDougal, L., Fosheim, G., Jensen, B., Killgore, G., Tenover, F., and Kuehnert, M. 2008. Changes in the prevalence of nasal colonization with *Staphylococcus aureus* in the United States, 2001–2004. *J. Infect. Dis.*, 197, 1226–1234.

Graham, I.I.I.P.L., Lin, S.X., and Larson, E.L. 2006. A U.S. population-based survey of *Staphylococcus aureus* colonization. *Ann. Intern. Med.*, 144, 318–325.

Graveland, H., Duim, B., van Duijkeren, E., Heederik, D., and Wagenaar, J.A. 2011. Livestock-associated methicillin-resistant *Staphylococcus aureus* in animals and humans. *Int. J. Med. Microbiol.*, 301, 630–634.

Graves, S., Kobayashi, S., Braughton, K., Whitney, A., Sturdevant, D., Rasmussen, D., Kirpotina, L., Quinn, M., and DeLeo, F. 2012. Sublytic concentrations of *Staphylococcus aureus* Panton-Valentine leukocidin alter human PMN gene expression and enhance bactericidal capacity. *J. Leukoc. Biol.*, 92, 361–374.

Gravet, A., Colin, D.A., Keller, D., Giradot, R., Monteil, H., and Prévost, G. 1998. Characterization of a novel structural member, LukE-LukD, of the bi-component staphylococcal leucotoxins family. *FEBS Lett.*, 436, 202–208.

Grayson, M.L. 2006. The treatment triangle for staphylococcal infections. *N. Engl. J. Med.*, 355, 724–727.

Gresham, H.D., Lowrance, J.H., Caver, T.E., Wilson, B.S., Cheung, A.L., and Lindberg, F.P. 2000. Survival of *Staphylococcus aureus* inside neutrophils contributes to infection. *J. Immunol.*, 164, 3713–3722.

Grossman, C. 2008. The first use of penicillin in the United States. *Ann. Intern. Med.*, 149, 135–136.

Grumann, D., Nübel, U., and Bröker, B.M. 2014. *Staphylococcus aureus* toxins – Their functions and genetics. *Infect. Genet. Evol.*, 21, 583–592.

Guardabassi, L., O'Donoghue, M., Moodley, A., Ho, J., and Boost, M. 2009. Novel lineage of methicillin-resistant *Staphylococcus aureus*, Hong Kong. *Emerg. Infect. Dis.*, 15, 1998–2000.

Habib, G., Hoen, B., Tornos, P., Thuny, F., Prendergast, B., Vilacosta, I., Moreillon, P., de Jesus Antunes, M., Thilen, U., Lekakis, J., Lengyel, M., Müller, L., Naber, C., Nihoyannopoulos, P., Moritz, A., and Zamorano, J. 2009. Guidelines on the prevention, diagnosis, and treatment of infective endocarditis (new version 2009): The Task Force on the Prevention, Diagnosis, and Treatment of Infective Endocarditis of the European Society of Cardiology (ESC). Endorsed by the European Society of Clinical Microbiology and Infectious Diseases (ESCMID) and the International Society of Chemotherapy (ISC) for Infection and Cancer. *Eur. Heart. J.*, 30, 2369–2413.

Hageman, J., Uyeki, T., Francis, J., Jernigan, D., Wheeler, J.G., Bridges, C., Barenkamp, S., Sievert, D., Srinivasan, A., Doherty, M., McDougal, L., Killgore, G., Lopatin, U., Coffman, R., MacDonald, J.K., McAllister, S., Fosheim, G., Patel, J., and McDonald, L.C. 2006. Severe community-acquired pneumonia due to *Staphylococcus aureus*, 2003–04 influenza season. *Emerg. Infect. Dis.*, 12, 894–899.

Haggar, A., Ehrnfelt, C., Holgersson, J., Flock, J.-I. 2004. The extracellular adherence protein from *Staphylococcus aureus* inhibits neutrophil binding to endothelial cells. *Infect. Immun.*, 72, 6164–6167.

Haim, M., Trost, A., Maier, C.J., Achatz, G., Feichtner, S., Hintner, H., Bauer, J.W., and Önder, K. 2010. Cytokeratin 8 interacts with clumping factor B: A new possible virulence factor target. *Microbiology*, 156, 3710–3721.

Hajjar, A.M., O'Mahony, D.S., Ozinsky, A., Underhill, D.M., Aderem, A., Klebanoff, S.J., and Wilson, C.B. 2001. Cutting edge: Functional interactions between toll-like receptor (TLR) 2 and TLR1 or TLR6 in response to phenol-soluble modulin. *J. Immunol.*, 166, 15–19.

Hajjeh, R.A., Reingold, A., Weil, A., Shutt, K., Schuchat, A., and Perkins, B.A. 1999. Toxic shock syndrome in the United States: Surveillance update, 1979–1996. *Emerg. Infect. Dis.*, 5, 807–810.

Hall, T.A., Sampath, R., Blyn, L.B., Ranken, R., Ivy, C., Melton, R., Matthews, H., White, N., Li, F., Harpin, V., Ecker, D.J., McDougal, L.K., Limbago, B., Ross, T., Wolk, D.M., Wysocki, V., and Carroll, K.C. 2009. Rapid molecular genotyping and clonal complex assignment of *Staphylococcus aureus* isolates by PCR coupled to electrospray ionization-mass spectrometry. *J. Clin. Microbiol.*, 47, 1733–1741.

Hammel, M., Sfyroera, G., Pyrpassopoulos, S., Ricklin, D., Ramyar, K.X., Pop, M., Jin, Z., Lambris, J.D., and Geisbrecht, B.V. 2007a. Characterization of Ehp, a secreted complement inhibitory protein from *Staphylococcus aureus*. *J. Biol. Chem.*, 282, 30051–30061.

Hammel, M., Sfyroera, G., Ricklin, D., Magotti, P., Lambris, J.D., and Geisbrecht, B.V. 2007b. A structural basis for complement inhibition by *Staphylococcus aureus*. *Nat. Immunol.*, 8, 430–437.

Hampton, M.B., Kettle, A.J., and Winterbourn, C.C. 1996. Involvement of superoxide and myeloperoxidase in oxygen-dependent killing of *Staphylococcus aureus* by neutrophils. *Infect. Immun.*, 64, 3512–3517.

Hamza, T., Dietz, M., Pham, D., Clovis, N., Danley, S., and Li, B. 2013. Intra-cellular *Staphylococcus aureus* alone causes infection in vivo. *Eur. Cell. Mater.*, 25, 341–350.

Hao, H., Dai, M., Wang, Y., Huang, L., and Yuan, Z. 2012. Key genetic elements and regulation systems in methicillin-resistant *Staphylococcus aureus*. *Future Microbiol.*, 7, 1315–1329.

Harbarth, S., Dharan, S., Liassine, N., Herrault, P., Auckenthaler, R., and Pittet, D. 1999. Randomized, placebo-controlled, double-blind trial to evaluate the efficacy of mupirocin for eradicating carriage of methicillin-resistant *Staphylococcus aureus*. *Antimicrob. Agents Chemother.*, 43, 1412–1416.

Harmsen, D., Witte, W., Rothgänger, J., Claus, H., Turnwald, D., and Vogel, U. 2003. Typing of methicillin-resistant *Staphylococcus aureus* in a university hospital setting by using novel software for spa repeat determination and database management. *J. Clin. Microbiol.*, 41, 5442–5448.

Harraghy, N., Hussain, M., Haggar, A., Chavakis, T., Sinha, B., Herrmann, M., and Flock, J.-I. 2003. The adhesive and immunomodulating properties of the multifunctional *Staphylococcus aureus* protein Eap. *Microbiology*, 149, 2701–2707.

Harris, S.R., Cartwright, E.J.P., Török, M.E., Holden, M.T.G., Brown, N.M., Ogilvy-Stuart, A.L., Ellington, M.J., Quail, M.A., Bentley, S.D., Parkhill, J., and Peacock, S.J. 2013. Whole-genome sequencing for analysis of an outbreak of meticillin-resistant *Staphylococcus aureus*: A descriptive study. *Lancet Infect. Dis.*, 13, 130–136.

He, W., Chen, H., Zhao, C., Zhang, F., Li, H., Wang, Q., Wang, X., and Wang, H. 2013. Population structure and characterisation of *Staphylococcus aureus* from bacteraemia at multiple hospitals in China: Association between antimicrobial resistance, toxin genes and genotypes. *Int. J. Antimicrob. Agents*, 42, 211–219.

Heilmann, C. 2011. Adhesion mechanisms of staphylococci. *Adv. Exp. Med. Biol.*, 715, 105–123.

Heilmann, C., Hartleib, J., Hussain, M.S., and Peters, G. 2005. The multifunctional *Staphylococcus aureus* autolysin Aaa mediates adherence to immobilized fibrinogen and fibronectin. *Infect. Immun.*, 73, 4793–4802.

Herold, B.C., Immergluck, L.C., Maranan, M.C., Lauderdale, D.S., Gaskin, R.E., Boyle-Vavra, S., Leitch, C.D., and Daum, R.S. 1998. Community-acquired methicillin-resistant *Staphylococcus aureus* in children with no identified predisposing risk. *JAMA*, 279, 593–598.

Hirschhausen, N., Schlesier, T., Schmidt, M.A., Götz, F., Peters, G., and Heilmann, C. 2010. A novel staphylococcal internalization mechanism involves the major autolysin Atl and heat shock cognate protein Hsc70 as host cell receptor. *Cell. Microbiol.*, 12, 1746–1764.

Ho, P.-L., Chiu, S.S., Chan, M.Y., Gan, Y., Chow, K.-H., Lai, E.L., Lau, Y.-L. 2012. Molecular epidemiology and nasal carriage of *Staphylococcus aureus* and methicillin-resistant *S. aureus* among young children attending day care centers and kindergartens in Hong Kong. *J. Infect.*, 64, 500–506.

Holman, L. 2013. Methicillin-resistant *Staphylococcus aureus*. *Radiol. Technol.*, 84, 307–310.

Holtfreter, S., and Bröker, B. 2005. Staphylococcal superantigens: Do they play a role in sepsis? *Arch. Immunol. Ther. Exp. (Warsz)*, 53, 13–27.

Horsburgh, M.J., Clements, M.O., Crossley, H., Ingham, E., and Foster, S.J. 2001. PerR controls oxidative stress resistance and iron storage proteins and is required for virulence in *Staphylococcus aureus*. *Infect. Immun.*, 69, 3744–3754.

Horsburgh, M.J., Wharton, S.J., Cox, A.G., Ingham, E., Peacock, S., and Foster, S.J. 2002. MntR modulates expression of the PerR regulon and superoxide resistance in *Staphylococcus aureus* through control of manganese uptake. *Mol. Microbiol.*, 44, 1269–1286.

Howden, B.P., Davies, J.K., Johnson, P.D.R., Stinear, T.P., and Grayson, M.L. 2010. Reduced vancomycin susceptibility in *Staphylococcus aureus*, including vancomycin-intermediate and heterogeneous vancomycin-intermediate strains: Resistance mechanisms, laboratory detection, and clinical implications. *Clin. Microbiol. Rev.*, 23, 99–139.

Huang, S.S., Diekema, D.J., Warren, D.K., Zuccotti, G., Winokur, P.L., Tendolkar, S., Boyken, L., Datta, R., Jones, R.M., Ward, M.A., Aubrey, T., Onderdonk, A.B., Garcia, C., and Platt, R. 2008. Strain-relatedness of methicillin-resistant *Staphylococcus aureus* isolates recovered from patients with repeated infection. *Clin. Infect. Dis.*, 46, 1241–1247.

Huang, X.Z., Cash, D.M., Chahine, M.A., Van Horn, G.T., Erwin, D.P., McKay, J.T., Hamilton, L.R., Jerke, K.H., Co, E.M.A., Aldous, W.K., Lesho, E.P., Lindler, L.E., Bowden, R.A., and Nikolich, M.P. 2011. Methicillin-resistant *Staphylococcus aureus* infection in combat support hospitals in three regions of Iraq. *Epidemiol. Infect.*, 139, 994–997.

Hudson, L.O., Murphy, C.R., Spratt, B.G., Enright, M.C., Elkins, K., Nguyen, C., Terpstra, L., Gombosev, A., Kim, D., Hannah, P., Mikhail, L., Alexander, R., Moore, D.F., and Huang, S.S. 2013. Diversity of methicillin-resistant *Staphylococcus aureus* (MRSA) strains isolated from inpatients of 30 hospitals in Orange County, California. *PLoS One*, 8, e62117.

Hudson, M.C., Ramp, W.K., Nicholson, N.C., Williams, A.S., and Nousiainen, M.T. 1995. Internalization of *Staphylococcus aureus* by cultured osteoblasts. *Microb. Pathog.*, 19, 409–419.

Huesca, M., Peralta, R., Sauder, D.N., Simor, A.E., and McGavin, M.J. 2002. Adhesion and virulence properties of epidemic Canadian methicillin-resistant *Staphylococcus aureus* strain 1: Identification of novel adhesion functions associated with plasmin-sensitive surface protein. *J. Infect. Dis.*, 185, 1285–1296.

Huovinen, P. 2001. Resistance to trimethoprim-sulfamethoxazole. *Clin. Infect. Dis.*, 32, 1608–1614.

Hussain, M., Schäfer, D., Juuti, K.M., Peters, G., Haslinger-Löffler, B., Kuusela, P.I., and Sinha, B. 2009. Expression of Pls (plasmin sensitive) in *Staphylococcus aureus* negative for *pls* reduces adherence and cellular invasion and acts by steric hindrance. *J. Infect. Dis.*, 200, 107–117.

Inoshima, I., Inoshima, N., Wilke, G.A., Powers, M.E., Frank, K.M., Wang, Y., and Wardenburg, J.B. 2011. A *Staphylococcus aureus* pore-forming toxin subverts the activity of ADAM10 to cause lethal infection in mice. *Nat. Med.*, 17, 1310–1314.

Ito, T., Katayama, Y., Asada, K., Mori, N., Tsutsumimoto, K., Tiensasitorn, C., and Hiramatsu, K. 2001. Structural comparison of three types of staphylococcal cassette chromosome *mec* integrated in the chromosome in methicillin-resistant *Staphylococcus aureus*. *Antimicrob. Agents Chemother.*, 45, 1323–1336.

Ito, T., Kuwahara-Arai, K., Katayama, Y., Uehara, Y., Han, X., Kondo, Y., and Hiramatsu, K. 2014. Staphylococcal cassette chromosome *mec* (SCC*mec*) analysis of MRSA. In: Ji, Y., editor. *Methicillin-Resistant Staphylococcus Aureus (MRSA) Protocols*. Humana Press. pp. 131–148.

Ito, T., Okuma, K., Ma, X.X., Yuzawa, H., and Hiramatsu, K. 2003. Insights on antibiotic resistance of *Staphylococcus aureus* from its whole genome: Genomic island SCC. *Drug. Resist. Updat.*, 6, 41–52.

IWC-SCC. 2009. Classification of staphylococcal cassette chromosome *mec* (SCC*mec*): Guidelines for reporting novel SCC*mec* elements. *Antimicrob. Agents Chemother.*, 53, 4961–4967.

Jansen van Rensburg, M., Whitelaw, A., and Elisha, B. 2012. Genetic basis of rifampicin resistance in methicillin-resistant *Staphylococcus aureus* suggests clonal expansion in hospitals in Cape Town, South Africa. *BMC Microbiol.*, 12, 46.

Jemni, L., Hmouda, H., and Letaief, A. 1994. Efficacy of trimethoprim-sulfamethoxazole against clinical isolates of methicillin-resistant *Staphylococcus aureus*: A report from Tunisia. *Clin. Infect. Dis.*, 19, 202–203.

Jernigan, J.A., and Farr, B.M. 1993. Short-course therapy of catheter-related *Staphylococcus aureus* bacteremia: Meta-analysis. *Ann. Intern. Med.*, 119, 304–311.

Jevons, M.P. 1961. "Celbenin" – resistant Staphylococci. *BMJ*, 1, 124–125.

Ji, G., Beavis, R.C., and Novick, R.P. 1995. Cell density control of staphylococcal virulence mediated by an octapeptide pheromone. *Proc. Natl. Acad. Sci. USA*, 92, 12055–12059.

John, M.D.V., Hibberd, P.L., Karchmer, A.W., Sleeper, L.A., and Calderwood, S.B. 1998. *Staphylococcus aureus* prosthetic calve endocarditis: Optimal management and risk factors for death. *Clin. Infect. Dis.*, 26, 1302–1309.

Johnson, M., Cockayne, A., and Morrissey, J.A. 2008. Iron-regulated biofilm formation in *Staphylococcus aureus* Newman requires *ica* and the secreted protein. *Emp. Infect. Immun.*, 76, 1756–1765.

Jongerius, I., Köhl, J., Pandey, M.K., Ruyken, M., van Kessel, K.P.M., van Strijp, J.A.G., and Rooijakkers, S.H.M. 2007. Staphylococcal complement evasion by various convertase-blocking molecules. *J. Exp. Med.*, 204, 2461–2471.

Jönsson, K., Signäs, C., Müller, H.-P., and Lindberg, M. 1991. Two different genes encode fibronectin binding proteins in *Staphylococcus aureus*. *Eur. J. Biochem.*, 202, 1041–1048.

Joo, H.-S., Cheung, G.Y.C., and Otto, M. 2011. Antimicrobial activity of community-associated methicillin-resistant *Staphylococcus aureus* is caused by phenol-soluble modulin derivatives. *J. Biol. Chem.*, 286, 8933–8940.

Josefsson, E., Hartford, O., O'Brien, L., Patti, J.M., and Foster, T. 2001. Protection against experimental *Staphylococcus aureus* arthritis by vaccination with clumping factor A, a novel virulence determinant. *J. Infect. Dis.*, 184, 1572–1580.

Karavolos, M.H., Horsburgh, M.J., Ingham, E., and Foster, S.J. 2003. Role and regulation of the superoxide dismutases of *Staphylococcus aureus*. *Microbiology*, 149, 2749–2758.

Katayama, Y., Ito, T., and Hiramatsu, K. 2000. A new class of genetic element, Staphylococcus cassette chromosome *mec*, encodes methicillin resistance in *Staphylococcus aureus*. *Antimicrob. Agents Chemother.*, 44, 1549–1555.

Kawaguchiya, M., Urushibara, N., Ghosh, S., Kuwahara, O., Morimoto, S., Ito, M., Kudo, K., and Kobayashi, N. 2013. Genetic diversity of emerging PVL/ACME-positive ST8-MRSA-IVa and ACME-positive CC5 (ST5/ST764)-MRSA strains in northern Japan. *J. Med. Microbiol.*, 62, 1852–1863.

Kazakova, S., Hageman, J., Matava, M., Srinivasan, A., Phelan, L., Garfinkel, B., Boo, T., McAllister, S., Anderson, J., Jensen, B., Dodson, D., Lonsway, D., McDougal, L., Arduino, M., Fraser, V., Killgore, G., Tenover, F., Cody, S., and Jernigan, D. 2005. A clone of methicillin-resistant *Staphylococcus aureus* among professional football players. *N. Engl. J. Med.*, 352, 468–475.

Kehl-Fie, T.E., Zhang, Y., Moore, J.L., Farrand, A.J., Hood, M.I., Rathi, S., Chazin, W.J., Caprioli, R.M., and Skaar, E.P. 2013. MntABC and MntH contribute to systemic *Staphylococcus aureus* infection by competing with calprotectin for nutrient manganese. *Infect. Immun.*, 81, 3395–3405.

Kennedy, A., Otto, M., Braughton, K., Whitney, A., Chen, L., Mathema, B., Mediavilla, J., Byrne, K., Parkins, L., Tenover, F., Kreiswirth, B., Musser, J., and DeLeo, F. 2008. Epidemic community-associated methicillin-resistant *Staphylococcus aureus*: Recent clonal expansion and diversification. *Proc. Natl. Acad. Sci. USA*, 105, 1327–1332.

Kennedy, A.D., Wardenburg, J.B., Gardner, D.J., Long, D., Whitney, A.R., Braughton, K.R., Schneewind, O., and DeLeo, F.R. 2010. Targeting of alpha-hemolysin by active or passive immunization decreases severity of USA300 skin infection in a mouse model. *J. Infect. Dis.*, 202, 1050–1058.

Kernodle, D.S., Voladri, R.K., Menzies, B.E., Hager, C.C., and Edwards, K.M. 1997. Expression of an antisense hla fragment in *Staphylococcus aureus* reduces alpha-toxin production in vitro and attenuates lethal activity in a murine model. *Infect. Immun.*, 65, 179–184.

Khalil, W., Hashwa, F., Shihabi, A., and Tokajian, S. 2012. Methicillin-resistant *Staphylococcus aureus* ST80-IV clone in children from Jordan. *Diagn. Microbiol. Infect. Dis.*, 73, 228–230.

Kiedrowski, M.R., and Horswill, A.R. 2011. New approaches for treating staphylococcal biofilm infections. *Ann. N. Y. Acad. Sci.*, 1241, 104–121.

Kintarak, S., Whawell, S.A., Speight, P.M., Packer, S., and Nair, S.P. 2004. Internalization of *Staphylococcus aureus* by human keratinocytes. *Infect. Immun.*, 72, 5668–5675.

Klebanoff, S.J. 1968. Myeloperoxidase-halide-hydrogen peroxide antibacterial system. *J. Bacteriol.*, 95, 2131–2138.

Klebanoff, S.J. 1974. Role of the superoxide anion in the myeloperoxidase-mediated antimicrobial system. *J. Biol. Chem.*, 249, 3724–3728.

Klebanoff, S.J. 2005. Myeloperoxidase: Friend and foe. *J. Leukoc. Biol.*, 77, 598–625.

Klevens, R.M., Edwards, J.R., Gaynes, R.P., and System, N.N.I.S. 2008. The impact of antimicrobial-resistant, health care-associated infections on mortality in the United States. *Clin. Infect. Dis.*, 47, 927–930.

Klevens, R.M., Morrison, M., Fridkin, S., Reingold, A., Petit, S., Gershman, K., Ray, S., Harrison, L., Lynfield, R., Dumyati, G., Townes, J., Craig, A., Fosheim, G., McDougal, L., and Tenover, F. 2006. Community-associated methicillin-resistant *Staphylococcus aureus* and healthcare risk factors. *Emerg. Infect. Dis.*, 12, 1991–1993.

Kluytmans, J., van Belkum, A., and Verbrugh, H. 1997. Nasal carriage of *Staphylococcus aureus*: Epidemiology, underlying mechanisms, and associated risks. *Clin. Microbiol. Rev.*, 10, 505–520.

Kobayashi, S., Braughton, K., Palazzolo Ballance, A., Kennedy, A., Sampaio, E., Kristosturyan, E., Whitney, A., Sturdevant, D., Dorward, D., Holland, S., Kreiswirth, B., Musser, J., and DeLeo, F. 2010. Rapid neutrophil destruction following phagocytosis of *Staphylococcus aureus*. *J. Innate. Immun.*, 2, 560–575.

Kobayashi, S., Voyich, J., Burlak, C., and DeLeo, F. 2005. Neutrophils in the innate immune response. *Arch. Immunol. Ther. Exp. (Warsz)*, 53, 505–517.

Kobayashi, S.D., and DeLeo, F.R. 2011. A MRSA-terious enemy among us: Boosting MRSA vaccines. *Nat. Med.*, 17, 168–169.

Kobayashi, S.D., Malachowa, N., Whitney, A.R., Braughton, K.R., Gardner, D.J., Long, D., Wardenburg, J.B., Schneewind, O., Otto, M., and DeLeo, F.R. 2011. Comparative analysis of USA300 virulence determinants in a rabbit model of skin and soft tissue infection. *J. Infect. Dis.*, 204, 937–941.

Koch, T.K., Reuter, M., Barthel, D., Böhm, S., van den Elsen, J., Kraiczy, P., Zipfel, P.F., and Skerka, C. 2012. *Staphylococcus aureus* proteins Sbi and Efb recruit human plasmin to degrade complement C3 and C3b. *PLoS One*, 7, e47638.

Köser, C.U., Holden, M.T.G., Ellington, M.J., Cartwright, E.J.P., Brown, N.M., Ogilvy-Stuart, A.L., Hsu, L.Y., Chewapreecha, C., Croucher, N.J., Harris, S.R., Sanders, M., Enright, M.C., Dougan, G., Bentley, S.D., Parkhill, J., Fraser, L.J., Betley,

J.R., Schulz-Trieglaff, O.B., Smith, G.P., and Peacock, S.J. 2012. Rapid whole-genome sequencing for investigation of a neonatal MRSA outbreak. *N. Engl. J. Med.*, 366, 2267–2275.

Krakauer, T. 1998. Interleukin-8 production by human monocytic cells in response to staphylococcal exotoxins is direct and independent of interleukin-1 and tumor necrosis factor-α. *J. Infect. Dis.*, 178, 573–577.

Kraus, D., Herbert, S., Kristian, S., Khosravi, A., Nizet, V., Gotz, F., and Peschel, A. 2008. The GraRS regulatory system controls *Staphylococcus aureus* susceptibility to antimicrobial host defenses. *BMC Microbiol.*, 8, 85.

Kravitz, G.R., Dries, D.J., Peterson, M.L., and Schlievert, P.M. 2005. Purpura fulminans due to *Staphylococcus aureus*. *Clin. Infect. Dis.*, 40, 941–947.

Kretschmer, D., Gleske, A.-K., Rautenberg, M., Wang, R., Köberle, M., Bohn, E., Schöneberg, T., Rabiet, M.-J., Boulay, F., Klebanoff, S.J., van Kessel, K.A., van Strijp, J.A., Otto, M., and Peschel, A. 2010. Human formyl peptide receptor 2 senses highly pathogenic *Staphylococcus aureus*. *Cell Host Microbe.*, 7, 463–473.

Kristian, S., Drr, M., Van Strijp, J.A.G., Neumeister, B., and Peschel, A. 2003. MprF-mediated lysinylation of phospholipids in *Staphylococcus aureus* leads to protection against oxygen-independent neutrophil killing. *Infect. Immun.*, 71, 546–549.

Kubica, M., Guzik, K., Koziel, J., Zarebski, M., Richter, W., Gajkowska, B., Golda, A., Maciag-Gudowska, A., Brix, K., Shaw, L., Foster, T., and Potempa, J. 2008. A potential new pathway for *Staphylococcus aureus* dissemination: The silent survival of *S. aureus* phagocytosed by human monocyte-derived macrophages. *PLoS One*, 3, e1409.

Kuehnert, M.J., Kruszon-Moran, D., Hill, H.A., McQuillan, G., McAllister, S.K., Fosheim, G., McDougal, L.K., Chaitram, J., Jensen, B., Fridkin, S.K., Killgore, G., and Tenover, F.C. 2006. Prevalence of *Staphylococcus aureus* nasal colonization in the United States, 2001–2002. *J. Infect. Dis.*, 193, 172–179.

Kuroda, M., Kuroda, H., Oshima, T., Takeuchi, F., Mori, H., and Hiramatsu, K. 2003. Two-component system VraSR positively modulates the regulation of cell-wall biosynthesis pathway in *Staphylococcus aureus*. *Mol. Microbiol.*, 49, 807–821.

Kurosu, M., Siricilla, S., and Mitachi, K. 2013. Advances in MRSA drug discovery: Where are we and where do we need to be? *Expert Opin. Drug. Discov.*, 8, 1095–1116.

Kusch, K., Hanke, K., Holtfreter, S., Schmudde, M., Kohler, C., Erck, C., Wehland, J., Hecker, M., Ohlsen, K., Bröker, B., and Engelmann, S. 2011. The influence of SaeRS and σB on the expression of superantigens in different *Staphylococcus aureus* isolates. *Int. J. Med. Microbiol.*, 301, 488–499.

Labandeira-Rey, M., Couzon, F., Boisset, S., Brown, E.L., Bes, M., Benito, Y., Barbu, E.M., Vazquez, V., Höök, M., Etienne, J., Vandenesch, F., and Bowden, M.G. 2007. *Staphylococcus aureus* Panton-Valentine leukocidin causes necrotizing pneumonia. *Science*, 315, 1130–1133.

Laupland, K.B., Ross, T., and Gregson, D.B. 2008. *Staphylococcus aureus* bloodstream infections: Risk factors, outcomes, and the influence of methicillin resistance in Calgary, Canada, 2000–2006. *J. Infect. Dis.*, 198, 336–343.

Lawton, E., Ross, R.P., Hill, C., and Cotter, P. 2007. Two-peptide lantibiotics: A medical perspective. *Mini. Rev. Med. Chem.*, 7, 1236–1247.

Lee, L.Y.L., Höök, M., Haviland, D., Wetsel, R.A., Yonter, E.O., Syribeys, P., Vernachio, J., and Brown, E.L. 2004a. Inhibition of complement activation by a secreted *Staphylococcus aureus* protein. *J. Infect. Dis.*, 190, 571–579.

Lee, L.Y.L., Liang, X., Höök, M., and Brown, E.L. 2004b. Identification and characterization of the C3 binding domain of the *Staphylococcus aureus* extracellular fibrinogen-binding protein (Efb). *J. Biol. Chem.*, 279, 50710–50716.

Levine, D.P. 2006. Vancomycin: A history. *Clin. Infect. Dis.*, 42, S5–S12.

Levy, S.B., and Marshall, B. 2004. Antibacterial resistance worldwide: Causes, challenges and responses. *Nat. Med.*, 10, S122–S129.

Lew, D.P., and Waldvogel, F.A. 2004. Osteomyelitis. *Lancet*, 364, 369–379.

Li, M., Cha, D.J., Lai, Y., Villaruz, A.E., Sturdevant, D.E., and Otto, M. 2007a. The antimicrobial peptide-sensing system *aps* of *Staphylococcus aureus*. *Mol. Microbiol.*, 66, 1136–1147.

Li, M., Cheung, G.Y.C., Hu, J., Wang, D., Joo, H.-S., DeLeo, F., and Otto, M. 2010. Comparative analysis of virulence and toxin expression of global community-associated methicillin-resistant *Staphylococcus aureus* strains. *J. Infect. Dis.*, 202, 1866–1876.

Li, M., Diep, B.A., Villaruz, A.E., Braughton, K.R., Jiang, X., DeLeo, F.R., Chambers, H.F., Lu, Y., and Otto, M. 2009. Evolution of virulence in epidemic community-associated methicillin-resistant *Staphylococcus aureus*. *Proc. Natl. Acad. Sci. USA*, 106, 5883–5888.

Li, M., Du, X., Villaruz, A.E., Diep, B.A., Wang, D., Song, Y., Tian, Y., Hu, J., Yu, F., Lu, Y., and Otto, M. 2012. MRSA epidemic linked to a quickly spreading colonization and virulence determinant. *Nat. Med.*, 18, 816–819.

Li, M., Lai, Y., Villaruz, A.E., Cha, D.J., Sturdevant, D.E., and Otto, M. 2007b. Gram-positive three-component antimicrobial peptide-sensing system. *Proc. Natl. Acad. Sci. USA*, 104, 9469–9474.

Liapikou, A., and Torres, A. 2013. Emerging drugs on methicillin-resistant *Staphylococcus aureus*. *Expert Opin. Emerg. Drugs*, 18, 291–305.

Libman, H., and Arbeit, R.D. 1984. Complications associated with *Staphylococcus aureus* bacteremia. *Arch. Intern. Med.*, 144, 541–545.

Lina, G., Bohach, G.A., Nair, S.P., Hiramatsu, K., Jouvin-Marche, E., and Mariuzza, R. 2004. Standard nomenclature for the superantigens expressed by *Staphylococcus*. *J. Infect. Dis.*, 189, 2334–2336.

Lina, G., Jarraud, S., Ji, G., Greenland, T., Pedraza, A., Etienne, J., Novick, R.P., and Vandenesch, F. 1998. Transmembrane topology and histidine protein kinase activity of AgrC, the agr signal receptor in *Staphylococcus aureus*. *Mol. Microbiol.*, 28, 655–662.

Lindsay, J.A. 2010. Genomic variation and evolution of *Staphylococcus aureus*. *Int. J. Med. Microbiol.*, 300, 98–103.

Lindsay, J.A., and Holden, M.T.G. 2004. *Staphylococcus aureus*: Superbug, super genome? *Trends Microbiol.*, 12, 378–385.

Lindsay, J.A., and Holden, M.T.G. 2006. Understanding the rise of the superbug: Investigation of the evolution and genomic variation of *Staphylococcus aureus*. *Funct. Integr. Genomics*, 6, 186–201.

Lindsay, J.A., Ruzin, A., Ross, H.F., Kurepina, N., and Novick, R.P. 1998. The gene for toxic shock toxin is carried by a family of mobile pathogenicity islands in *Staphylococcus aureus*. *Mol. Microbiol.*, 29, 527–543.

Lipinska, U., Hermans, K., Meulemans, L., Dumitrescu, O., Badiou, C., Duchateau, L., Haesebrouck, F., Etienne, J., and Lina, G. 2011. Panton-Valentine leukocidin does play a role in the early stage of *Staphylococcus aureus* skin infections: A rabbit model. *PLoS One*, 6, e22864.

Liu, C., Bayer, A., Cosgrove, S.E., Daum, R.S., Fridkin, S.K., Gorwitz, R.J., Kaplan, S.L., Karchmer, A.W., Levine, D.P., Murray, B.E., Rybak, M.J., Talan, D.A., and Chambers, H.F. 2011. Clinical practice guidelines by the Infectious Diseases Society of America for the treatment of methicillin-resistant *Staphylococcus aureus* infections in adults and children: Executive summary. *Clin. Infect. Dis.*, 52, 285–292.

Liu, G., Essex, A., Buchanan, J., Datta, V., Hoffman, H., Bastian, J., Fierer, J., and Nizet, V. 2005. *Staphylococcus aureus* golden pigment impairs neutrophil killing and promotes virulence through its antioxidant activity. *J. Exp. Med.*, 202, 209–215.

Lo, W.-T., and Wang, C.-C. 2011. Panton-Valentine leukocidin in the pathogenesis of community-associated methicillin-resistant *Staphylococcus aureus* infection. *Pediatr. Neonatol.*, 52, 59–65.

Lodise, T.P., McKinnon, P.S., Swiderski, L., and Rybak, M.J. 2003. Outcomes analysis of delayed antibiotic treatment for hospital-acquired *Staphylococcus aureus* bacteremia. *Clin. Infect. Dis.*, 36, 1418–1423.

Lowy, F.D. 1998. *Staphylococcus aureus* infections. *N. Engl. J. Med.*, 339, 520–532.

Lu, J., and Holmgren, A. 2014. The thioredoxin antioxidant system. *Free Radic. Biol. Med.*, 66, 75–87.

Luong, T.T., and Lee, C.Y. 2002. Overproduction of type 8 capsular polysaccharide augments *Staphylococcus aureus* virulence. *Infect. Immun.*, 70, 3389–3395.

Lyon, G.J., Mayville, P., Muir, T.W., and Novick, R.P. 2000. Rational design of a global inhibitor of the virulence response in *Staphylococcus aureus*, based in part on localization of the site of inhibition to the receptor-histidine kinase, AgrC. *Proc. Natl. Acad. Sci. USA*, 97, 13330–13335.

MacGowan, A.P., Noel, A., Tomaselli, S., Elliott, H., and Bowker, K. 2011. Pharmacodynamics of razupenem (PZ601) studied in an in vitro pharmacokinetic model of infection. *Antimicrob. Agents Chemother.*, 55, 1436–1442.

Malachowa, N., and DeLeo, F. 2010. Mobile genetic elements of *Staphylococcus aureus*. *Cell. Mol. Life Sci.*, 67, 3057–3071.

Malachowa, N., Kobayashi, S.D., Braughton, K.R., Whitney, A.R., Parnell, M.J., Gardner, D.J., and DeLeo, F.R. 2012. *Staphylococcus aureus* leukotoxin GH promotes inflammation. *J. Infect. Dis.*, 206, 1185–1193.

Malachowa, N., Whitney, A.R., Kobayashi, S.D., Sturdevant, D.E., Kennedy, A.D., Braughton, K.R., Shabb, D.W., Diep, B.A., Chambers, H.F., Otto, M., and DeLeo, F.R. 2011. Global changes in *Staphylococcus aureus* gene expression in human blood. *PLoS One*, 6, e18617.

Malanoski, G.J., Samore, M.H., Pefanis, A., and Karchmer, A.W. 1995. *Staphylococcus aureus* catheter-associated bacteremia: Minimal effective therapy and unusual infectious complications associated with arterial sheath catheters. *Arch. Intern. Med.*, 155, 1161–1166.

Mamo, W., Bodén, M., and Flock, J.-I. 1994. Vaccination with *Staphylococcus aureus* fibrinogen binding proteins (FgBPs) reduces colonisation of *S. aureus* in a mouse mastitis model. *FEMS Immunol. Med. Microbiol.*, 10, 47–53.

Mandell, G.L., Bennett, J.E., and Dolin, R. 2010. *Mandell, Douglas, and Bennett's Principles and Practice of Infectious Diseases*. Churchill Livingstone/Elsevier.

Marcinkiewicz, J. 1997. Neutrophil chloramines: Missing links between innate and acquired immunity. *Immunol. Today*, 18, 577–580.

Markowitz, N., Quinn, E.L., and Saravolatz, L.D. 1992. Trimethoprim-sulfamethoxazole compared with vancomycin for the treatment of *Staphylococcus aureus* infection. *Ann. Intern. Med.*, 117, 390–398.

Mazmanian, S.K., Skaar, E.P., Gaspar, A.H., Humayun, M., Gornicki, P., Jelenska, J., Joachmiak, A., Missiakas, D.M., and Schneewind, O. 2003. Passage of heme-iron across the envelope of *Staphylococcus aureus*. *Science*, 299, 906–909.

McAdam, P.R., Templeton, K.E., Edwards, G.F., Holden, M.T., Feil, E.J., Aanensen, D.M., Bargawi, H.J., Spratt, B.G., Bentley, S.D., Parkhill, J., Enright, M.C., Holmes, A., Girvan, E.K., Godfrey, P.A., Feldgarden, M., Kearns, A.M., Rambaut, A., Robinson, D.A., and Fitzgerald, J.R. 2012. Molecular tracing of the emergence, adaptation, and transmission of hospital-associated methicillin-resistant *Staphylococcus aureus*. *Proc. Natl. Acad. Sci. USA*, 109, 9107–9112.

McAdow, M., Missiakas, D., and Schneewind, O. 2012. *Staphylococcus aureus* secretes coagulase and von Willebrand factor binding protein to modify the coagulation cascade and establish host infections. *J. Innate. Immun.*, 4, 141–148.

McCarthy, A., and Lindsay, J. 2010. Genetic variation in *Staphylococcus aureus* surface and immune evasion genes is lineage associated: Implications for vaccine design and host-pathogen interactions. *BMC Microbiol.*, 10, 173.

McCarthy, A.J., Breathnach, A.S., and Lindsay, J.A. 2012. Detection of mobile-genetic-element variation between colonizing and infecting hospital-associated methicillin-resistant *Staphylococcus aureus* isolates. *J. Clin. Microbiol.*, 50, 1073–1075.

McCormick, J.K., Yarwood, J.M., and Schlievert, P.M. 2001. Toxic shock syndrome and bacterial superantigens: An update. *Annu. Rev. Microbiol.*, 55, 77–104.

McDevitt, D., Nanavaty, T., House-Pompeo, K., Bell, E., Turner, N., McIntire, L., Foster, T., and Höök, M. 1997. Characterization of the interaction between the *Staphylococcus aureus* clumping factor (ClfA) and fibrinogen. *Eur. J. Biochem.*, 247, 416–424.

McDougal, L.K., Fosheim, G.E., Nicholson, A., Bulens, S.N., Limbago, B.M., Shearer, J.E.S., Summers, A.O., and Patel, J.B. 2010. Emergence of resistance among USA300 methicillin-resistant *Staphylococcus aureus* isolates causing invasive disease in the United States. *Antimicrob. Agents Chemother.*, 54, 3804–3811.

McDougal, L.K., Steward, C.D., Killgore, G.E., Chaitram, J.M., McAllister, S.K., and Tenover, F.C. 2003. Pulsed-field gel electrophoresis typing of oxacillin-resistant *Staphylococcus aureus* isolates from the United States: Establishing a national database. *J. Clin. Microbiol.*, 41, 5113–5120.

McLoughlin, R.M., Solinga, R.M., Rich, J., Zaleski, K.J., Cocchiaro, J.L., Risley, A., Tzianabos, A.O., and Lee, J.C. 2006. CD4+ T cells and CXC chemokines modulate the pathogenesis of *Staphylococcus aureus* wound infections. *Proc. Natl. Acad. Sci. USA*, 103, 10408–10413.

Mediavilla, J.R., Chen, L., Mathema, B., and Kreiswirth, B.N. 2012. Global epidemiology of community-associated methicillin resistant *Staphylococcus aureus* (CA-MRSA). *Curr. Opin. Microbiol.*, 15, 588–595.

Medina, G., Egea, A.L., Otth, C., Otth, L., Fernández, H., Bocco, J.L., Wilson, M., and Sola, C. 2013. Molecular epidemiology of hospital-onset methicillin-resistant *Staphylococcus aureus* infections in Southern Chile. *Eur. J. Clin. Microbiol. Infect. Dis.*, 32, 1533–1540.

Mellmann, A., Weniger, T., Berssenbrugge, C., Rothganger, J., Sammeth, M., Stoye, J., and Harmsen, D. 2007. Based Upon Repeat Pattern (BURP): An algorithm to characterize the long-term evolution of *Staphylococcus aureus* populations based on *spa* polymorphisms. *BMC Microbiol.*, 7, 98.

Menegotto, F., González-Cabrero, S., Lorenzo, B., Cubero, Á., Cuervo, W., Gutiérrez, M.P., Simarro, M., Orduña, A., and Bratos, M.Á. 2012. Molecular epidemiology of methicillin-resistant *Staphylococcus aureus* in a Spanish hospital over a 4-year period: Clonal replacement, decreased antimicrobial resistance, and identification of community-acquired and livestock-associated clones. *Diagn. Microbiol. Infect. Dis.*, 74, 332–337.

Menzies, B.E., and Kourteva, I. 1998. Internalization of *Staphylococcus aureus* by endothelial cells induces apoptosis. *Infect. Immun.*, 66, 5994–5998.

Merino, N., Toledo-Arana, A., Vergara-Irigaray, M., Valle, J., Solano, C., Calvo, E., Lopez, J.A., Foster, T.J., Penadés, J.R., and Lasa, I. 2009. Protein A-mediated multicellular behavior in *Staphylococcus aureus*. *J. Bacteriol.*, 191, 832–843.

Mermel, L.A., Cartony, J.M., Covington, P., Maxey, G., and Morse, D. 2011. Methicillin-resistant *Staphylococcus aureus* colonization at different body sites: A prospective, quantitative analysis. *J. Clin. Microbiol.*, 49, 1119–1121.

Mermel, L.A., Farr, B.M., Sherertz, R.J., Raad, I.I., O'Grady, N., Harris, J.S., and Craven, D.E. 2001. Guidelines for the management of intravascular catheter-related infections. *Clin. Infect. Dis.*, 32, 1249–1272.

Meyer, F., Girardot, R., Piémont, Y., Prévost, G., and Colin, D.A. 2009. Analysis of the specificity of Panton-Valentine leucocidin and gamma-hemolysin F component binding. *Infect. Immun.*, 77, 266–273.

Miajlovic, H., Zapotoczna, M., Geoghegan, J.A., Kerrigan, S.W., Speziale, P., and Foster, T.J. 2010. Direct interaction of iron-regulated surface determinant IsdB of *Staphylococcus aureus* with the GPIIb/IIIa receptor on platelets. *Microbiology*, 156, 920–928.

Micek, S.T. 2007. Alternatives to vancomycin for the treatment of methicillin-resistant *Staphylococcus aureus* infections. *Clin. Infect. Dis.*, 45, S184–S190.

Miko, B.A., Uhlemann, A.-C., Gelman, A., Lee, C.J., Hafer, C.A., Sullivan, S.B., Shi, Q., Miller, M., Zenilman, J., and Lowy, F.D. 2012. High prevalence of colonization with *Staphylococcus aureus* clone USA300 at multiple body sites among sexually transmitted disease clinic patients: An unrecognized reservoir. *Microbes. Infect.*, 14, 1040–1043.

Miles, G., Movileanu, L., and Bayley, H. 2002. Subunit composition of a bicomponent toxin: Staphylococcal leukocidin forms an octameric transmembrane pore. *Protein Sci.*, 11, 894–902.

Miller, L., Perdreau Remington, F., Rieg, G., Mehdi, S., Perlroth, J., Bayer, A., Tang, A., Phung, T., and Spellberg, B. 2005. Necrotizing fasciitis caused by community-associated methicillin-resistant *Staphylococcus aureus* in Los Angeles. *N. Engl. J. Med.*, 352, 1445–1453.

Miller, L.G., and Diep, B.A. 2008. Colonization, fomites, and virulence: Rethinking the pathogenesis of community-associated methicillin-resistant *Staphylococcus aureus* infection. *Clin. Infect. Dis.*, 46, 752–760.

Miller, L.G., Remington, F.P., Bayer, A.S., Diep, B., Tan, N., Bharadwa, K., Tsui, J., Perlroth, J., Shay, A., Tagudar, G., Ibebuogu, U., and Spellberg, B. 2007. Clinical and epidemiologic characteristics cannot distinguish community-associated methicillin-resistant *Staphylococcus aureus* infection from methicillin-susceptible *S. aureus* infection: A prospective investigation. *Clin. Infect. Dis.*, 44, 471–482.

Moellering, J.R.C. 2003. Linezolid: The first oxazolidinone antimicrobial. *Ann. Intern. Med.*, 138, 135–142.

Moellering, R.C. 2012. MRSA: The first half century. *J. Antimicrob. Chemother.*, 67, 4–11.

Monecke, S., Coombs, G., Shore, A.C., Coleman, D.C., Akpaka, P., Borg, M., Chow, H., Ip, M., Jatzwauk, L., Jonas, D., Kadlec, K., Kearns, A., Laurent, F., O'Brien, F.G., Pearson, J., Ruppelt, A., Schwarz, S., Scicluna, E., Slickers, P., Tan, H.-L., Weber, S., and Ehricht, R. 2011. A field guide to pandemic, epidemic and sporadic clones of methicillin-resistant *Staphylococcus aureus*. *PLoS One*, 6, e17936.

Monecke, S., Ehricht, R., Slickers, P., Tan, H.L., and Coombs, G. 2009. The molecular epidemiology and evolution of the Panton–Valentine leukocidin-positive, methicillin-resistant *Staphylococcus aureus* strain USA300 in Western Australia. *Clin. Microbiol. Infect.*, 15, 770–776.

Montgomery, C.P., Boyle-Vavra, S., and Daum, R.S. 2010. Importance of the global regulators *agr* and *saeRS* in the pathogenesis of CA-MRSA USA300 infection. *PLoS One*, 5, e15177.

Montoya, M., and Gouaux, E. 2003. β-Barrel membrane protein folding and structure viewed through the lens of α-hemolysin. *Biochim. Biophys. Acta*, 1609, 19–27.

Moran, G.J., Krishnadasan, A., Gorwitz, R.J., Fosheim, G.E., McDougal, L.K., Carey, R.B., and Talan, D.A. 2006. Methicillin-resistant *S. aureus* infections among patients in the emergency department. *N. Engl. J. Med.*, 355, 666–674.

Morikawa, K., Takemura, A.J., Inose, Y., Tsai, M., Nguyen Thi, L.T., Ohta, T., and Msadek, T. 2012. Expression of a cryptic secondary sigma factor gene unveils natural competence for DNA transformation in *Staphylococcus aureus*. *PLoS Pathog.*, 8, e1003003.

Morinaga, N., Kaihou, Y., and Noda, M. 2003. Purification, cloning and characterization of variant LukE-LukD with strong leukocidal activity of staphylococcal bi-component leukotoxin family. *Microbiol. Immunol.*, 47, 81–90.

Murdoch, D.R., Corey, G., Hoen, B., Miró, J.M., Fowler, V.G.J., Bayer, A.S., Karchmer, A.W., Olaison, L., Pappas, P.A., Moreillon, P., Chambers, S.T., Chu, V.H., Falcó, V., Holland, D.J., Jones, P., Klein, J.L., Raymond, N.J., Read, K.M., Tripodi, M.F., Utili, R., Wang, A., Woods, C.W., and Cabell, C.H. 2009. Clinical presentation, etiology, and outcome of infective endocarditis in the 21st century: The international collaboration on endocarditis–prospective cohort study. *Arch. Intern. Med.*, 169, 463–473.

Murray, B.E., and Moellering, R.C. 1978. Patterns and mechanisms of antibiotic resistance. *Med. Clin. North Am.*, 62, 899–923.

Murray, C.K., Griffith, M.E., Mende, K., Guymon, C.H., Ellis, M.W., Beckius, M., Zera, W.C., Yu, X., Co, E.-M.A., Aldous, W., and Hospenthal, D.R. 2010. Methicillin-resistant *Staphylococcus aureus* in wound cultures recovered from a combat support hospital in Iraq. *J. Trauma.*, 69, S102–S108.

Musher, D.M., Lamm, N., Darouiche, R.O., Young, E.J., Hamill, R.J., and Landon, G.C. 1994. The current spectrum of *Staphylococcus aureus* infection in a tertiary care hospital. *Medicine*, 73, 186–208.

Muthukrishnan, G., Quinn, G.A., Lamers, R.P., Diaz, C., Cole, A.L., Chen, S., and Cole, A.M. 2011. Exoproteome of *Staphylococcus aureus* reveals putative determinants of nasal carriage. *J. Proteome. Res.*, 10, 2064–2078.

Naderer, O.J., Jones, L.S., Zhu, J., Kurtinecz, M., and Dumont, E. 2013. Safety, tolerability, and pharmacokinetics of oral and intravenous administration of GSK1322322, a peptide deformylase inhibitor. *J. Clin. Pharmacol.*, 53, 1168–1176.

Naimi, T., LeDell, K., Como Sabetti, K., Borchardt, S., Boxrud, D., Etienne, J., Johnson, S., Vandenesch, F., Fridkin, S., O'Boyle, C., Danila, R., and Lynfield, R. 2003. Comparison of community- and health care-associated methicillin-resistant *Staphylococcus aureus* infection. *JAMA*, 290, 2976–2984.

Nastaly, P., Grinholc, M., and Bielawski, K.P. 2010. Molecular characteristics of community-associated methicillin-resistant *Staphylococcus aureus* strains for clinical medicine. *Arch. Microbiol.*, 192, 603–617.

Newsom, S.W.B. 2008. Ogston's coccus. *J. Hosp. Infect.*, 70, 369–372.

Nguyen, A., Gentry, C., and Furrh, R. 2013. A comparison of adverse drug reactions between high- and standard- dose trimethoprim-sulfamethoxazole in the ambulatory setting. *Curr. Drug. Saf.*, 8, 114–119.

Nguyen, T., Ghebrehiwet, B., and Peerschke, E.I.B. 2000. *Staphylococcus aureus* protein A recognizes platelet gC1qR/p33: A novel mechanism for staphylococcal interactions with platelets. *Infect. Immun.*, 68, 2061–2068.

Ní Eidhin, D., Perkins, S., Francois, P., Vaudaux, P., Höök, M., and Foster, T.J. 1998. Clumping factor B (ClfB), a new surface-located fibrinogen-binding adhesin of *Staphylococcus aureus*. *Mol. Microbiol.*, 30, 245–257.

Nichol, K.A., Adam, H.J., Roscoe, D.L., Golding, G.R., Lagacé-Wiens, P.R.S., Hoban, D.J., Zhanel, G.G., Alliance, o.b.o.t.C.A.R. 2013. Changing epidemiology of methicillin-resistant *Staphylococcus aureus* in Canada. *J. Antimicrob. Chemother.*, 68, i47–i55.

Nilsson, I.-M., Hartford, O., Foster, T., and Tarkowski, A. 1999. Alpha-toxin and gamma-toxin jointly promote *Staphylococcus aureus* virulence in murine septic arthritis. *Infect. Immun.*, 67, 1045–1049.

Nilsson, I.M., Lee, J.C., Bremell, T., Rydén, C., and Tarkowski, A. 1997. The role of staphylococcal polysaccharide microcapsule expression in septicemia and septic arthritis. *Infect. Immun.*, 65, 4216–4221.

Nimmo, G.R. 2012. USA300 abroad: Global spread of a virulent strain of community-associated methicillin-resistant *Staphylococcus aureus*. *Clin. Microbiol. Infect.*, 18, 725–734.

Nimmo, G.R., Bergh, H., Nakos, J., Whiley, D., Marquess, J., Huygens, F., and Paterson, D.L. 2013. Replacement of healthcare-associated MRSA by community-associated MRSA in Queensland: Confirmation by genotyping. *J. Infect.*, 67, 439–447.

Nishifuji, K., Shimizu, A., Ishiko, A., Iwasaki, T., and Amagai, M. 2010. Removal of amino-terminal extracellular domains of desmoglein 1 by staphylococcal exfoliative toxin is sufficient to initiate epidermal blister formation. *J. Dermatol. Sci.*, 59, 184–191.

Nishifuji, K., Sugai, M., and Amagai, M. 2008. Staphylococcal exfoliative toxins: "Molecular scissors" of bacteria that attack the cutaneous defense barrier in mammals. *J. Dermatol. Sci.*, 49, 21–31.

Noble, W.C., Virani, Z., and Cree, R.G.A. 1992. Co-transfer of vancomycin and other resistance genes from *Enterococcus faecalis* NCTC 12201 to *Staphylococcus aureus*. *FEMS Microbiol. Lett.*, 93, 195–198.

Noda, M., Hirayama, T., Kato, I., and Matsuda, F. 1980. Crystallization and properties of staphylococcal leukocidin. *Biochim. Biophys. Acta*, 633, 33–44.

Noskin, G.A., Rubin, R.J., Schentag, J.J., Kluytmans, J., Hedblom, E.C., Jacobson, C., Smulders, M., Gemmen, E., and Bharmal, M. 2007. National trends in *Staphylococcus aureus* infection rates: Impact on economic burden and mortality over a 6-year period (1998–2003). *Clin. Infect. Dis.*, 45, 1132–1140.

Novak, R. 2011. Are pleuromutilin antibiotics finally fit for human use? *Ann. N. Y. Acad. Sci.*, 1241, 71–81.

Novick, R.P. 2003a. Autoinduction and signal transduction in the regulation of staphylococcal virulence. *Mol. Microbiol.*, 48, 1429–1449.

Novick, R.P. 2003b. Mobile genetic elements and bacterial toxinoses: The superantigen-encoding pathogenicity islands of *Staphylococcus aureus*. *Plasmid*, 49, 93–105.

Novick, R.P., Christie, G.E., and Penadés, J.R. 2010. The phage-related chromosomal islands of Gram-positive bacteria. *Nat. Rev. Micro.*, 8, 541–551.

Nygaard, T.K., Pallister, K.B., Ruzevich, P., Griffith, S., Vuong, C., and Voyich, J.M. 2010. SaeR binds a consensus sequence within virulence gene promoters to advance USA300 pathogenesis. *J. Infect. Dis.*, 201, 241–254.

O'Brien, L., Kerrigan, S.W., Kaw, G., Hogan, M., Penadés, J., Litt, D., Fitzgerald, D.J., Foster, T.J., and Cox, D. 2002a. Multiple mechanisms for the activation of human platelet aggregation by *Staphylococcus aureus*: Roles for the clumping factors ClfA and ClfB, the serine–aspartate repeat protein SdrE and protein A. *Mol. Microbiol.*, 44, 1033–1044.

O'Brien, L.M., Walsh, E.J., Massey, R.C., Peacock, S.J., and Foster, T.J. 2002b. *Staphylococcus aureus* clumping factor B (ClfB) promotes adherence to human type I cytokeratin 10: Implications for nasal colonization. *Cell. Microbiol.*, 4, 759–770.

O'Riordan, K., and Lee, J.C. 2004. *Staphylococcus aureus* capsular polysaccharides. *Clin. Microbiol. Rev.*, 17, 218–234.

Okuma, K., Iwakawa, K., Turnidge, J.D., Grubb, W.B., Bell, J.M., O'Brien, F.G., Coombs, G.W., Pearman, J.W., Tenover, F.C., Kapi, M., Tiensasitorn, C., Ito, T., and Hiramatsu, K. 2002. Dissemination of new methicillin-resistant *Staphylococcus aureus* clones in the community. *J. Clin. Microbiol.*, 40, 4289–4294.

Oldach, D., Clark, K., Schranz, J., Das, A., Craft, J.C., Scott, D., Jamieson, B.D., and Fernandes, P. 2013. Randomized, double-blind, multicenter phase 2 study comparing the efficacy and safety of oral solithromycin (CEM-101) to those of oral levofloxacin in the treatment of patients with community-acquired bacterial pneumonia. *Antimicrob. Agents Chemother.*, 57, 2526–2534.

Ono, H.K., Omoe, K., Imanishi, K.i., Iwakabe, Y., Hu, D.-L., Kato, H., Saito, N., Nakane, A., Uchiyama, T., and Shinagawa, K. 2008. Identification and characterization of two novel staphylococcal enterotoxins, types S and T. *Infect. Immun.*, 76, 4999–5005.

Otter, J.A., and French, G.L. 2010. Molecular epidemiology of community-associated meticillin-resistant *Staphylococcus aureus* in Europe. *Lancet Infect. Dis.*, 10, 227–239.

Otter, J.A., and French, G.L. 2011. Community-associated meticillin-resistant *Staphylococcus aureus* strains as a cause of healthcare-associated infection. *J. Hosp. Infect.*, 79, 189–193.

Otter, J.A., and French, G.L. 2012. Community-associated meticillin-resistant *Staphylococcus aureus*: The case for a genotypic definition. *J. Hosp. Infect.*, 81, 143–148.

Otto, M. 2013. Community-associated MRSA: What makes them special? *Int. J. Med. Microbiol.*, 303, 324–330.

Palazzolo Ballance, A., Reniere, M., Braughton, K., Sturdevant, D., Otto, M., Kreiswirth, B., Skaar, E., and DeLeo, F. 2008. Neutrophil microbicides induce a pathogen survival response in community-associated methicillin-resistant *Staphylococcus aureus*. *J. Immunol.*, 180, 500–509.

Palma, M., Nozohoor, S., Schennings, T., Heimdahl, A., and Flock, J.I. 1996. Lack of the extracellular 19-kilodalton fibrinogen-binding protein from *Staphylococcus aureus* decreases virulence in experimental wound infection. *Infect. Immun.*, 64, 5284–5289.

Palmqvist, N., Foster, T., Tarkowski, A., and Josefsson, E. 2002. Protein A is a virulence factor in *Staphylococcus aureus* arthritis and septic death. *Microb. Pathog.*, 33, 239–249.

Palmqvist, N., Patti, J.M., Tarkowski, A., and Josefsson, E. 2004. Expression of staphylococcal clumping factor A impedes macrophage phagocytosis. *Microbes. Infect.*, 6, 188–195.

Pan, A., Lorenzotti, S., and Zoncada, A. 2008. Registered and investigational drugs for the treatment of methicillin-resistant *Staphylococcus aureus* infection. *Recent. Pat. Antiinfect. Drug. Discov.*, 3, 10–33.

Pan, E.S., Diep, B.A., Carleton, H.A., Charlebois, E.D., Sensabaugh, G.F., Haller, B.L., and Remington, F.P. 2003. Increasing prevalence of methicillin-resistant *Staphylococcus aureus* infection in California jails. *Clin. Infect. Dis.*, 37, 1384–1388.

Pan, E.S., Diep, B.A., Charlebois, E.D., Auerswald, C., Carleton, H.A., Sensabaugh, G.F., and Perdreau-Remington, F. 2005. Population dynamics of nasal strains of methicillin-resistant *Staphylococcus aureus*—And their relation to community-associated disease activity. *J. Infect. Dis.*, 192, 811–818.

Pannaraj, P.S., Hulten, K.G., Gonzalez, B.E., Mason, E.O., and Kaplan, S.L. 2006. Infective pyomyositis and myositis in children in the era of community-acquired, methicillin-resistant *Staphylococcus aureus* infection. *Clin. Infect. Dis.*, 43, 953–960.

Panton, P.N., and Valentine, F.C.O. 1932. Staphylococcal toxin. *Lancet*, 219, 506–508.

Pastacaldi, C., Lewis, P., and Howarth, P. 2011. Staphylococci and staphylococcal superantigens in asthma and rhinitis: A systematic review and meta-analysis. *Allergy*, 66, 549–555.

Patel, A.H., Nowlan, P., Weavers, E.D., and Foster, T. 1987. Virulence of protein A-deficient and alpha-toxin-deficient mutants of *Staphylococcus aureus* isolated by allele replacement. *Infect. Immun.*, 55, 3103–3110.

Patti, J.M., House-Pompeo, K., Boles, J.O., Garza, N., Gurusiddappa, S., and Höök, M. 1995. Critical residues in the ligand-binding site of the *Staphylococcus aureus* collagen-binding adhesin (MSCRAMM). *J. Biol. Chem.*, 270, 12005–12011.

Peacock, S.J., Moore, C.E., Justice, A., Kantzanou, M., Story, L., Mackie, K., O'Neill, G., and Day, N.P.J. 2002. Virulent combinations of adhesin and toxin genes in natural populations of *Staphylococcus aureus*. *Infect. Immun.*, 70, 4987–4996.

Perfetto, B., Donnarumma, G., Criscuolo, D., Paoletti, I., Grimaldi, E., Tufano, M.A., and Baroni, A. 2003. Bacterial components induce cytokine and intercellular adhesion molecules-1 and activate transcription factors in dermal fibroblasts. *Res. Microbiol.*, 154, 337–344.

Periasamy, S., Joo, H.-S., Duong, A.C., Bach, T.-H.L., Tan, V.Y., Chatterjee, S.S., Cheung, G.Y.C., and Otto, M. 2012. How *Staphylococcus aureus* biofilms develop their characteristic structure. *Proc. Natl. Acad. Sci. USA*, 109, 1281–1286.

Persoons, D., Van Hoorebeke, S., Hermans, K., Butaye, P., de Kruif, A., Haesebrouck, F., and Dewulf, J. 2009. Methicillin-resistant *Staphylococcus aureus* in poultry. *Emerg. Infect. Dis.*, 15, 452–453.

Pertel, P.E., Bernardo, P., Fogarty, C., Matthews, P., Northland, R., Benvenuto, M., Thorne, G.M., Luperchio, S.A., Arbeit, R.D., and Alder, J. 2008. Effects of prior effective therapy on the efficacy of daptomycin and ceftriaxone for the treatment of community-acquired pneumonia. *Clin. Infect. Dis.*, 46, 1142–1151.

Peschel, A., Jack, R.W., Otto, M., Collins, L.V., Staubitz, P., Nicholson, G., Kalbacher, H., Nieuwenhuizen, W.F., Jung, G., Tarkowski, A., van Kessel, K.P., and van Strijp, J.A. 2001. *Staphylococcus aureus* resistance to human defensins and evasion of neutrophil killing via the novel virulence factor MprF is based on modification of membrane lipids with l-lysine. *J. Exp. Med.*, 193, 1067–1076.

Peschel, A., Otto, M., Jack, R.W., Kalbacher, H., Jung, G., and Götz, F. 1999. Inactivation of the *dlt* operon in *Staphylococcus aureus* confers sensitivity to defensins, protegrins, and other antimicrobial peptides. *J. Biol. Chem.*, 274, 8405–8410.

Pierce, D., Calkins, B., and Thornton, K. 2012. Infectious endocarditis: Diagnosis and treatment. *Am. Fam. Physician.*, 85, 981–986.

Pinchuk, I.V., Beswick, E.J., and Reyes, V.E. 2010. Staphylococcal enterotoxins. *Toxins*, 2, 2177–2197.

Pogliano, J., Pogliano, N., and Silverman, J.A. 2012. Daptomycin-mediated reorganization of membrane architecture causes mislocalization of essential cell division proteins. *J. Bacteriol.*, 194, 4494–4504.

Portillo, B.C., Moreno, J.E., Yomayusa, N., Álvarez, C.A., Cardozo, B.E.C., Pérez, J.A.E., Díaz, P.L., Ibañez, M., Mendez-Alvarez, S., Leal, A.L., and Gómez, N.V. 2013. Molecular epidemiology and characterization of virulence genes of community-acquired and hospital-acquired methicillin-resistant *Staphylococcus aureus* isolates in Colombia. *Int. J. Infect. Dis.*, 17, e744–e749.

Potrich, C., Bastiani, H., Colin, D.A., Huck, S., Prévost, G., and Dalla Serra, M. 2009. The influence of membrane lipids in *Staphylococcus aureus* gamma-hemolysins pore formation. *J. Membr. Biol.*, 227, 13–24.

Powers, M.E., Kim, H.K., Wang, Y., and Bubeck Wardenburg, J. 2012. ADAM10 mediates vascular injury induced by *Staphylococcus aureus* α-hemolysin. *J. Infect. Dis.*, 206, 352–356.

Price, J.R., Didelot, X., Crook, D.W., Llewelyn, M.J., and Paul, J. 2013. Whole genome sequencing in the prevention and control of *Staphylococcus aureus* infection. *J. Hosp. Infect.*, 83, 14–21.

Price, L., Stegger, M., Hasman, H., Aziz, M., Larsen, J., Andersen, P., Pearson, T., Waters, A., Foster, J., Schupp, J., Gillece, J., Driebe, E., Liu, C., Springer, B., Zdovc, I., Battisti, A., Franco, A., Zmudzki, J., Schwarz, S., Butaye, P., Jouy, E., Pomba, C., Porrero, M.C., Ruimy, R., Smith, T., Robinson, D.A., Weese, J.S., Arriola, C., Yu, F., Laurent, F., Keim, P., Skov, R., and Aarestrup, F. 2012. *Staphylococcus aureus* CC398: Host adaptation and emergence of methicillin resistance in livestock. *mBio*, 3.

Prince, W.T., Ivezic-Schoenfeld, Z., Lell, C., Tack, K.J., Novak, R., Obermayr, F., and Talbot, G.H. 2013. Phase II clinical study of BC-3781, a pleuromutilin antibiotic, in treatment of patients with acute bacterial skin and skin structure infections. *Antimicrob. Agents. Chemother.*, 57, 2087–2094.

Proft, T., and Fraser, J.D. 2003. Bacterial superantigens. *Clin. Exp. Immunol.*, 133, 299–306.

Que, Y.-A., and Moreillon, P. 2011. Infective endocarditis. *Nat. Rev. Cardiol.*, 8, 322–336.

Queck, S.Y., Jameson-Lee, M., Villaruz, A.E., Bach, T.-H.L., Khan, B.A., Sturdevant, D.E., Ricklefs, S.M., Li, M., and Otto, M. 2008. RNAIII-independent target gene control by the *agr* quorum-sensing system: Insight into the evolution of virulence regulation in *Staphylococcus aureus*. *Mol. Cell.*, 32, 150–158.

Queck, S.Y., Khan, B.A., Wang, R., Bach, T.-H.L., Kretschmer, D., Chen, L., Kreiswirth, B.N., Peschel, A., DeLeo, F.R., and Otto, M. 2009. Mobile genetic element-encoded cytolysin connects virulence to methicillin resistance in MRSA. *PLoS Pathog.*, 5, e1000533.

Quinn, M., Ammons, M.C.B., and DeLeo, F. 2006. The expanding role of NADPH oxidases in health and disease: No longer just agents of death and destruction. *Clin. Sci. (Lond.)*, 111, 1–20.

Raad, I.I., and Sabbagh, M.F. 1992. Optimal duration of therapy for catheter-related *Staphylococcus aureus* bacteremia: A study of 55 cases and review. *Clin. Infect. Dis.*, 14, 75–82.

Rauch, S., DeDent, A.C., Kim, H.K., Bubeck Wardenburg, J., Missiakas, D.M., and Schneewind, O. 2012. Abscess formation and alpha-hemolysin induced toxicity in a mouse model of *Staphylococcus aureus* peritoneal infection. *Infect. Immun.*, 80, 3721–3732.

Rautenberg, M., Joo, H.-S., Otto, M., and Peschel, A. 2011. Neutrophil responses to staphylococcal pathogens and commensals via the formyl peptide receptor 2 relates to phenol-soluble modulin release and virulence. *FASEB J.*, 25, 1254–1263.

Rellahan, B.L., Jones, L.A., Kruisbeek, A.M., Fry, A.M., and Matis, L.A. 1990. In vivo induction of anergy in peripheral V beta 8+ T cells by staphylococcal enterotoxin B. *J. Exp. Med.*, 172, 1091–1100.

Richardson, A.R., Dunman, P.M., and Fang, F.C. 2006. The nitrosative stress response of *Staphylococcus aureus* is required for resistance to innate immunity. *Mol. Microbiol.*, 61, 927–939.

Richardson, A.R., Libby, S.J., and Fang, F.C. 2008. A nitric oxide-inducible lactate dehydrogenase enables *Staphylococcus aureus* to resist innate immunity. *Science*, 319, 1672–1676.

Richter, A., Sting, R., Popp, C., Rau, J., Tenhagen, B.A., Guerra, B., Hafez, H.M., and Fetsch, A. 2012. Prevalence of types of methicillin-resistant *Staphylococcus aureus* in turkey flocks and personnel attending the animals. *Epidemiol. Infect.*, 140, 2223–2232.

Ricklin, D., Tzekou, A., Garcia, B.L., Hammel, M., McWhorter, W.J., Sfyroera, G., Wu, Y.-Q., Holers, V.M., Herbert, A.P., Barlow, P.N., Geisbrecht, B.V., and Lambris, J.D. 2009. A molecular insight into complement evasion by the staphylococcal complement inhibitor protein family. *J. Immunol.*, 183, 2565–2574.

Rigby, K., and DeLeo, F. 2012. Neutrophils in innate host defense against *Staphylococcus aureus* infections. *Semin. Immunopathol.*, 34, 237–259.

Robicsek, A., Beaumont, J.L., Paule, S.M., Hacek, D.M., Thomson, J.R.B., Kaul, K.L., King, P., and Peterson, L.R. 2008. Universal surveillance for methicillin-resistant *Staphylococcus aureus* in 3 affiliated hospitals. *Ann. Intern. Med.*, 148, 409–418.

Roche, F.M., Downer, R., Keane, F., Speziale, P., Park, P.W., and Foster, T.J. 2004. The N-terminal a domain of fibronectin-binding proteins A and B promotes adhesion of *Staphylococcus aureus* to elastin. *J. Biol. Chem.*, 279, 38433–38440.

Roche, F.M., Meehan, M., and Foster, T.J. 2003. The *Staphylococcus aureus* surface protein SasG and its homologues promote bacterial adherence to human desquamated nasal epithelial cells. *Microbiology*, 149, 2759–2767.

Rodgers, W., Frazier, A.D., and Champney, W.S. 2013. Solithromycin inhibition of protein synthesis and ribosome biogenesis in *Staphylococcus aureus*, *Streptococcus pneumoniae*, and *Haemophilus influenzae*. *Antimicrob. Agents Chemother.*, 57, 1632–1637.

Rodríguez-Noriega, E., Seas, C., Guzmán-Blanco, M., Mejía, C., Alvarez, C., Bavestrello, L., Zurita, J., Labarca, J., Luna, C.M., Salles, M.J.C., and Gotuzzo, E. 2010. Evolution of methicillin-resistant *Staphylococcus aureus* clones in Latin America. *Int. J. Infect. Dis.*, 14, e560–e566.

Rogers, D.E., and Tompsett, R. 1952. The survival of staphylococci within human leukocytes. *J. Exp. Med.*, 95, 209–230.

Rooijakkers, S.H.M., Milder, F.J., Bardoel, B.W., Ruyken, M., van Strijp, J.A.G., and Gros, P. 2007. Staphylococcal complement inhibitor: Structure and active sites. *J. Immunol.*, 179, 2989–2998.

Rooijakkers, S.H.M., Ruyken, M., Roos, A., Daha, M.R., Presanis, J.S., Sim, R.B., van Wamel, W.J.B., van Kessel, K.P.M., and van Strijp, J.A.G. 2005a. Immune evasion by a staphylococcal complement inhibitor that acts on C3 convertases. *Nat. Immunol.*, 6, 920–927.

Rooijakkers, S.H.M., van Wamel, W.J.B., Ruyken, M., van Kessel, K.P.M., and van Strijp, J.A.G. 2005b. Anti-opsonic properties of staphylokinase. *Microbes. Infect.*, 7, 476–484.

Rooijakkers, S.H.M., Wu, J., Ruyken, M., van Domselaar, R., Planken, K.L., Tzekou, A., Ricklin, D., Lambris, J.D., Janssen, B.J.C., van Strijp, J.A.G., and Gros, P. 2009. Structural and functional implications of the alternative complement pathway C3 convertase stabilized by a staphylococcal inhibitor. *Nat. Immunol.*, 10, 721–727.

Rose, W.E., and Rybak, M.J. 2006. Tigecycline: First of a new class of antimicrobial agents. *Pharmacotherapy*, 26, 1099–1110.

Rosen, A.B., Fowler, J.V.G., Corey, G.R., Downs, S.M., Biddle, A.K., Li, J., and Jollis, J.G. 1999. Cost-effectiveness of trans-esophageal echocardiography to determine the duration of therapy for intravascular catheter-associated *Staphylococcus aureus* bacteremia. *Ann. Intern. Med.*, 130, 810–820.

Rosen, H., and Klebanoff, S.J. 1977. Formation of singlet oxygen by the myeloperoxidase-mediated antimicrobial system. *J. Biol. Chem.*, 252, 4803–4810.

Rutland, B., Weese, J.S., Bolin, C., Au, J., and Malani, A. 2009. Human-to-dog transmission of methicillin-resistant *Staphylococcus aureus*. *Emerg. Infect. Dis.*, 15, 1328–1330.

Rybak, M.J., Lomaestro, B.M., Rotscahfer, J.C., Moellering, R.C., Craig, W.A., Billeter, M., Dalovisio, J.R., and Levine, D.P. 2009. Vancomycin therapeutic guidelines: A summary of consensus recommendations from the Infectious Diseases Society of America, the American Society of Health-System Pharmacists, and the Society of Infectious Diseases Pharmacists. *Clin. Infect. Dis.*, 49, 325–327.

Saiman, L., Keefe, M.O., Graham, P.L., Wu, F., Salim, B.S., Kreiswirth, B., LaSala, A., Schlievert, P.M., and Latta, P.D. 2003. Hospital transmission of community-acquired methicillin-resistant *Staphylococcus aureus* among postpartum women. *Clin. Infect. Dis.*, 37, 1313–1319.

Sakamoto, S., Tanaka, Y., Tanaka, I., Takei, T., Yu, J., Kuroda, M., Yao, M., Ohta, T., and Tsumoto, K. 2008. Electron microscopy and computational studies of Ebh, a giant cell-wall-associated protein from *Staphylococcus aureus*. *Biochem. Biophys. Res. Commun.*, 376, 261–266.

Sato'o, Y., Omoe, K., Ono, H.K., Nakane, A., and Hu, D.-L. 2013. A novel comprehensive analysis method for *Staphylococcus aureus* pathogenicity islands. *Microbiol. Immunol.*, 57, 91–99.

Scanvic, A., Denic, L., Gaillon, S., Giry, P., Andremont, A., and Lucet, J.-C. 2001. Duration of colonization by methicillin-resistant *Staphylococcus aureus* after hospital discharge and risk factors for prolonged carriage. *Clin. Infect. Dis.*, 32, 1393–1398.

Schaffer, A.C., and Lee, J.C. 2008. Vaccination and passive immunisation against *Staphylococcus aureus*. *Int. J. Antimicrob. Agents*, 32(Suppl 1), S71–S78.

Schaffer, A.C., and Lee, J.C. 2009. Staphylococcal vaccines and immunotherapies. *Infect. Dis. Clin. North Am.*, 23, 153–171.

Schifferli, J.A., and Peters, D.K. 1983. Immune adherence and staphylococcus protein A binding of soluble immune complexes produced by complement activation. *Clin. Exp. Immunol.*, 54, 827–833.

Schijffelen, M., Boel, C.E., van Strijp, J., and Fluit, A. 2010. Whole genome analysis of a livestock-associated methicillin-resistant *Staphylococcus aureus* ST398 isolate from a case of human endocarditis. *BMC Genomics.*, 11, 376.

Schöfer, H., and Simonsen, L. 2010. Fusidic acid in dermatology: An updated review. *Eur. J. Dermatol.*, 20, 6–15.

Scriba, T.J., Sierro, S., Brown, E.L., Phillips, R.E., Sewell, A.K., and Massey, R.C. 2008. The *Staphyloccous aureus* Eap protein activates expression of proinflammatory cytokines. *Infect. Immun.*, 76, 2164–2168.

Senn, L., Basset, P., Greub, G., Prod'hom, G., Frei, R., Zbinden, R., Gaia, V., Balmelli, C., Pfyffer, G.E., Mühlemann, K.,

Zanetti, G., and Blanc, D.S. 2013. Molecular epidemiology of methicillin-resistant *Staphylococcus aureus* in Switzerland: Sampling only invasive isolates does not allow a representative description of the local diversity of clones. *Clin. Microbiol. Infect.*, 19, E288–E290.

Sharp, J.A., Echague, C.G., Hair, P.S., Ward, M.D., Nyalwidhe, J.O., Geoghegan, J.A., Foster, T.J., and Cunnion, K.M. 2012. *Staphylococcus aureus* surface protein SdrE binds complement regulator factor H as an immune evasion tactic. *PLoS One*, 7, e38407.

Sheehan, B., Foster, T., Dorman, C., Park, S., and Stewart, G.A.B. 1992. Osmotic and growth-phase dependent regulation of the *eta* gene of *Staphylococcus aureus*: A role for DNA supercoiling. *Mol. Gen. Genet.*, 232, 49–57.

Shinefield, H., Black, S., Fattom, A., Horwith, G., Rasgon, S., Ordonez, J., Yeoh, H., Law, D., Robbins, J.B., Schneerson, R., Muenz, L., Fuller, S., Johnson, J., Fireman, B., Alcorn, H., and Naso, R. 2002. Use of a *Staphylococcus aureus* conjugate vaccine in patients receiving hemodialysis. *N. Engl. J. Med.*, 346, 491–496.

Siberry, G.K., Tekle, T., Carroll, K., and Dick, J. 2003. Failure of clindamycin treatment of methicillin-resistant *Staphylococcus aureus* expressing inducible clindamycin resistance in vitro. *Clin. Infect. Dis.*, 37, 1257–1260.

Siboo, I.R., Chaffin, D.O., Rubens, C.E., and Sullam, P.M. 2008. Characterization of the accessory Sec system of *Staphylococcus aureus*. *J. Bacteriol.*, 190, 6188–6196.

Siegel, J.D., Rhinehart, E., Jackson, M., and Chiarello, L. 2007. Management of multidrug-resistant organisms in health care settings, 2006. *Am. J. Infect. Control*, 35, S165–S193.

Sieprawska-Lupa, M., Mydel, P., Krawczyk, K., Wójcik, K., Puklo, M., Lupa, B., Suder, P., Silberring, J., Reed, M., Pohl, J., Shafer, W., McAleese, F., Foster, T., Travis, J., and Potempa, J. 2004. Degradation of human antimicrobial peptide LL-37 by *Staphylococcus aureus*-derived proteinases. *Antimicrob. Agents Chemother.*, 48, 4673–4679.

Sievert, D.M., Rudrik, J.T., Patel, J.B., McDonald, L.C., Wilkins, M.J., and Hageman, J.C. 2008. Vancomycin-resistant *Staphylococcus aureus* in the United States, 2002–2006. *Clin. Infect. Dis.*, 46, 668–674.

Signäs, C., Raucci, G., Jönsson, K., Lindgren, P.E., Anantharamaiah, G.M., Höök, M., and Lindberg, M. 1989. Nucleotide sequence of the gene for a fibronectin-binding protein from *Staphylococcus aureus*: Use of this peptide sequence in the synthesis of biologically active peptides. *Proc. Natl. Acad. Sci. USA*, 86, 699–703.

Silverman, J.A., Mortin, L.I., VanPraagh, A.D.G., Li, T., and Alder, J. 2005. Inhibition of daptomycin by pulmonary surfactant: In vitro modeling and clinical impact. *J. Infect. Dis.*, 191, 2149–2152.

Silversides, J., Lappin, E., and Ferguson, A. 2010. Staphylococcal toxic shock syndrome: Mechanisms and management. *Curr. Infect. Dis. Rep.*, 12, 392–400.

Simor, A., Gilbert, N., Gravel, D., Mulvey, M., Bryce, E., Loeb, M., Matlow, A., McGeer, A., Louie, L., and Campbell, J. 2010. Methicillin-resistant *Staphylococcus aureus* colonization or infection in Canada: National surveillance and changing epidemiology, 1995–2007. *Infect. Control Hosp. Epidemiol.*, 31, 348–356.

Simor, A.E., Phillips, E., McGeer, A., Konvalinka, A., Loeb, M., Devlin, H.R., and Kiss, A. 2007. Randomized controlled trial of chlorhexidine gluconate for washing, intranasal mupirocin, and rifampin and doxycycline versus no treatment for the eradication of methicillin-resistant *Staphylococcus aureus* colonization. *Clin. Infect. Dis.*, 44, 178–185.

Singh, V.K., and Moskovitz, J. 2003. Multiple methionine sulfoxide reductase genes in *Staphylococcus aureus*: Expression of activity and roles in tolerance of oxidative stress. *Microbiology*, 149, 2739–2747.

Sjödahl, J. 1977. Repetitive sequences in protein A from *Staphylococcus aureus*. *Eur. J. Biochem.*, 73, 343–351.

Skinner, D., and Keefer, C.S. 1941. Significance of bacteremia caused by *Staphylococcus aureus*: A study of one hundred and twenty-two cases and a review of the literature concerned with experimental infection in animals. *Arch. Intern. Med.*, 68, 851–875.

Skov, R.L., and Jensen, K.S. 2009. Community-associated meticillin-resistant *Staphylococcus aureus* as a cause of hospital-acquired infections. *J. Hosp. Infect.*, 73, 364–370.

Smith, E., Visai, L., Kerrigan, S., Speziale, P., and Foster, T. 2011. The Sbi protein is a multifunctional immune evasion factor of *Staphylococcus aureus*. *Infect. Immun.*, 79, 3801–3809.

Smith, G. 1982. Ogston's coccus: 102 years and still going strong. *South. Med. J.*, 75, 1559–1562.

Smyth, D.S., and Robinson, D.A. 2009. Integrative and sequence characteristics of a novel genetic element, ICE6013, in *Staphylococcus aureus*. *J. Bacteriol.*, 191, 5964–5975.

Soell, M., Diab, M., Haan-Archipoff, G., Beretz, A., Herbelin, C., Poutrel, B., and Klein, J.P. 1995. Capsular polysaccharide types 5 and 8 of *Staphylococcus aureus* bind specifically to human epithelial (KB) cells, endothelial cells, and monocytes and induce release of cytokines. *Infect. Immun.*, 63, 1380–1386.

Sollid, J.U.E., Furberg, A.S., Hanssen, A.M., and Johannessen, M. 2013. *Staphylococcus aureus*: Determinants of human carriage. *Infect. Genet. Evol.*, 21, 531–541.

Song, L., Hobaugh, M.R., Shustak, C., Cheley, S., Bayley, H., and Gouaux, J.E. 1996. Structure of staphylococcal alpha-hemolysin, a heptameric transmembrane pore. *Science*, 274, 1859–1866.

Sonnevend, Á., Blair, I., Alkaabi, M., Jumaa, P., al Haj, M., Ghazawi, A., Akawi, N., Jouhar, F.S., Hamadeh, M.B., and Pál, T. 2012. Change in meticillin-resistant *Staphylococcus aureus* clones at a tertiary care hospital in the United Arab Emirates over a 5-year period. *J. Clin. Pathol.*, 65, 178–182.

Spellberg, B., and Daum, R. 2012. Development of a vaccine against *Staphylococcus aureus*. *Semin. Immunopathol.*, 34, 335–348.

Speziale, P., Pietrocola, G., Rindi, S., Provenzano, M., Provenza, G., Di Poto, A., Visai, L., and Arciola, C.R. 2009. Structural and functional role of *Staphylococcus aureus* surface components recognizing adhesive matrix molecules of the host. *Future Microbiol.*, 4, 1337–1352.

Standiford, T.J., Arenberg, D.A., Danforth, J.M., Kunkel, S.L., VanOtteren, G.M., and Strieter, R.M. 1994. Lipoteichoic acid induces secretion of interleukin-8 from human blood monocytes: A cellular and molecular analysis. *Infect. Immun.*, 62, 119–125.

Stapleton, P.D., Shah, S., Ehlert, K., Hara, Y., and Taylor, P.W. 2007. The β-lactam-resistance modifier (−)-epicatechin gallate alters the architecture of the cell wall of *Staphylococcus aureus*. *Microbiology*, 153, 2093–2103.

Staubitz, P., Neumann, H., Schneider, T., Wiedemann, I., and Peschel, A. 2004. MprF-mediated biosynthesis of lysylphosphatidylglycerol, an important determinant in staphylococcal defensin resistance. *FEMS Microbiol. Lett.*, 231, 67–71.

Stefani, S., Chung, D.R., Lindsay, J.A., Friedrich, A.W., Kearns, A.M., Westh, H., and MacKenzie, F.M. 2012. Meticillin-resistant *Staphylococcus aureus* (MRSA): Global epidemiology and harmonisation of typing methods. *Int. J. Antimicrob. Agents*, 39, 273–282.

Stein, A., Bataille, J.F., Drancourt, M., Curvale, G., Argenson, J.N., Groulier, P., and Raoult, D. 1998. Ambulatory treatment of multidrug-resistant *Staphylococcus*-infected orthopedic implants with high-dose oral co-trimoxazole (trimethoprim-sulfamethoxazole). *Antimicrob. Agents Chemother.*, 42, 3086–3091.

Stevens, D.L., Bisno, A.L., Chambers, H.F., Everett, E.D., Dellinger, P., Goldstein, E.J.C., Gorbach, S.L., Hirschmann, J.V., Kaplan, E.L., Montoya, J.G., and Wade, J.C. 2005. Practice guidelines for the diagnosis and management of skin and soft-tissue infections. *Clin. Infect. Dis.*, 41, 1373–1406.

Strausbaugh, L.J. 1993. Toxic shock syndrome. Are you recognizing its changing presentations? *Postgrad. Med.*, 94, 107–108, 111.

Strommenger, B., Braulke, C., Heuck, D., Schmidt, C., Pasemann, B., Nübel, U., and Witte, W. 2008. *spa* typing of *Staphylococcus aureus* as a frontline tool in epidemiological typing. *J. Clin. Microbiol.*, 46, 574–581.

Struelens, M.J., Deplano, A., Godard, C., Maes, N., and Serruys, E. 1992. Epidemiologic typing and delineation of genetic relatedness of methicillin-resistant *Staphylococcus aureus* by macrorestriction analysis of genomic DNA by using pulsed-field gel electrophoresis. *J. Clin. Microbiol.*, 30, 2599–2605.

Stryjewski, M.E., and Chambers, H.F. 2008. Skin and soft-tissue infections caused by community-acquired methicillin-resistant *Staphylococcus aureus*. *Clin. Infect. Dis.*, 46, S368–S377.

Stryjewski, M.E., Potgieter, P.D., Li, Y.-P., Barriere, S.L., Churukian, A., Kingsley, J., Corey, G.R., Group TD-1792. 2012. TD-1792 versus vancomycin for treatment of complicated skin and skin structure infections. *Antimicrob. Agents Chemother.*, 56, 5476–5483.

Sugihara, K., Tateda, K., Yamamura, N., Koga, T., Sugihara, C., and Yamaguchi, K. 2011. Efficacy of human-simulated exposures of tomopenem (formerly CS-023) in a murine model of *Pseudomonas aeruginosa* and methicillin-resistant *Staphylococcus aureus* infection. *Antimicrob. Agents Chemother.*, 55, 5004–5009.

Sullenberger, A.L., Avedissian, L., and Kent, S. 2005. Importance of transesophageal echocardiography in the evaluation of *Staphylococcus aureus* bacteremia. *J. Heart Valve. Dis.*, 14, 23–28.

Sung, J.Y., Lee, J., Choi, E.H., and Lee, H.J. 2012. Changes in molecular epidemiology of community-associated and health care-associated methicillin-resistant *Staphylococcus aureus* in Korean children. *Diagn. Microbiol. Infect. Dis.*, 74, 28–33.

Supersac, G., Piémont, Y., Kubina, M., Prévost, G., and Foster, T.J. 1998. Assessment of the role of gamma-toxin in experimental endophthalmitis using a *hlg*-deficient mutant of *Staphylococcus aureus*. *Microb. Pathog.*, 24, 241–251.

Sutcliffe, J.A. 2011. Antibiotics in development targeting protein synthesis. *Ann. N. Y. Acad. Sci.*, 1241, 122–152.

Suzuki, H., Lefebure, T., Bitar, P., and Stanhope, M. 2012. Comparative genomic analysis of the genus *Staphylococcus* including *Staphylococcus aureus* and its newly described sister species *Staphylococcus simiae*. *BMC Genomics.*, 13, 38.

Swaney, S.M., Aoki, H., Ganoza, M.C., and Shinabarger, D.L. 1998. The oxazolidinone linezolid inhibits initiation of protein synthesis in bacteria. *Antimicrob. Agents Chemother.*, 42, 3251–3255.

Szmigielski, S., Sobiczewska, E., Prévost, G., Monteil, H., Colin, D.A., and Jeljaszewicz, J. 1998. Effect of purified staphylococcal leukocidal toxins on isolated blood polymorphonuclear leukocytes and peritoneal macrophages in vitro. *Zentralblatt. für. Bakteriologie*, 288, 383–394.

Tavares, A., Miragaia, M., Rolo, J., Coelho, C., and Lencastre, H. 2013. High prevalence of hospital-associated methicillin-resistant *Staphylococcus aureus* in the community in Portugal: Evidence for the blurring of community-hospital boundaries. *Eur. J. Clin. Microbiol. Infect. Dis.*, 32, 1269–1283.

Taylor, A.R. 2013. Methicillin-resistant *Staphylococcus aureus* infections. *Prim. Care*, 40, 637–654.

Teixeira, M.M., Araújo, M.C., Silva-Carvalho, M.C., Beltrame, C.O., Oliveira, C.C.H.B., Figueiredo, A.M.S., and Oliveira, A.G. 2012. Emergence of clonal complex 5 (CC5) methicillin-resistant *Staphylococcus aureus* (MRSA) isolates susceptible to trimethoprim-sulfamethoxazole in a Brazilian hospital. *Braz. J. Med. Biol. Res.*, 45, 637–643.

Tenover, F.C., and Goering, R.V. 2009. Methicillin-resistant *Staphylococcus aureus* strain USA300: Origin and epidemiology. *J. Antimicrob. Chemother.*, 64, 441–446.

Tenover, F.C., McDougal, L.K., Goering, R.V., Killgore, G., Projan, S.J., Patel, J.B., and Dunman, P.M. 2006. Characterization of a strain of community-associated methicillin-resistant *Staphylococcus aureus* widely disseminated in the United States. *J. Clin. Microbiol.*, 44, 108–118.

Tenover, F.C., Tickler, I.A., Goering, R.V., Kreiswirth, B.N., Mediavilla, J.R., Persing, D.H., MRSA Consortium. 2012. Characterization of nasal and blood culture isolates of methicillin-resistant *Staphylococcus aureus* from patients in United States hospitals. *Antimicrob. Agents Chemother.*, 56, 1324–1330.

Thakker, M., Park, J.S., Carey, V., and Lee, J.C. 1998. *Staphylococcus aureus* serotype 5 capsular polysaccharide is antiphagocytic and enhances bacterial virulence in a murine bacteremia model. *Infect. Immun.*, 66, 5183–5189.

Thomas, D., Chou, S., Dauwalder, O., and Lina, G. 2007. Diversity in *Staphylococcus aureus* enterotoxins. *Chem. Immunol. Allergy*, 93, 24–41.

Thompson, K.M., Abraham, N., and Jefferson, K.K. 2010. *Staphylococcus aureus* extracellular adherence protein contributes to biofilm formation in the presence of serum. *FEMS Microbiol. Lett.*, 305, 143–147.

Thurlow, L.R., Joshi, G.S., Clark, J.R., Spontak, J.S., Neely, C.J., Maile, R., and Richardson, A.R. 2013. Functional modularity of the arginine catabolic mobile element contributes to the success of USA300 methicillin-resistant *Staphylococcus aureus*. *Cell Host Microbe.*, 13, 100–107.

Thurlow, L.R., Joshi, G.S., and Richardson, A.R. 2012. Virulence strategies of the dominant USA300 lineage of community-associated methicillin-resistant *Staphylococcus aureus* (CA-MRSA). *FEMS Immunol. Med. Microbiol.*, 65, 5–22.

Todd, J., Fishaut, M., Kapral, F., and Welch, T. 1978. Toxic-shock syndrome associated with phage-group-I staphylococci. *Lancet*, 312, 1116–1118.

Ton-That, H., Liu, G., Mazmanian, S.K., Faull, K.F., and Schneewind, O. 1999. Purification and characterization of sortase, the transpeptidase that cleaves surface proteins of *Staphylococcus aureus* at the LPXTG motif. *Proc. Natl. Acad. Sci. USA*, 96, 12424–12429.

Tristan, A., Bes, M., Meugnier, H., Lina, G., Bozdogan, B.l., Courvalin, P., Reverdy, M.-E., Enright, M., Vandenesch, F.o., and Etienne, J. 2007. Global distribution of Panton-Valentine leukocidin–positive methicillin-resistant *Staphylococcus aureus*, 2006. *Emerg. Infect. Dis.*, 13, 594–600.

Tseng, C.W., Zhang, S., and Stewart, G.C. 2004. Accessory gene regulator control of staphyloccocal enterotoxin D gene expression. *J. Bacteriol.*, 186, 1793–1801.

Tsuji, B.T., Rybak, M.J., Lau, K.L., and Sakoulas, G. 2007. Evaluation of accessory gene regulator (*agr*) group and function in the proclivity towards vancomycin intermediate resistance in *Staphylococcus aureus*. *Antimicrob. Agents Chemother.*, 51, 1089–1091.

Tufano, M., Ianniello, R., Galdiero, M., and Galdiero, F. 1991. Protein A and other surface components of *Staphylococcus aureus* stimulate production of IL-1 alpha, IL-4, IL-6, TNF and IFN-gamma. *Eur. Cytokine Netw*, 2, 361.

Tung, H.s., Guss, B., Hellman, U., Persson, L., Rubin, K., and Rydén, C. 2000. A bone sialoprotein-binding protein from *Staphylococcus aureus*: A member of the staphylococcal Sdr family. *Biochem. J.*, 345, 611–619.

Turlej, A., Hryniewicz, W., and Empel, J. 2011. Staphylococcal cassette chromosome *mec* (SCC*mec*) classification and typing methods: An overview. *Acta Microbiol. Pol.*, 60, 95–103.

Tzianabos, A.O., Wang, J.Y., and Lee, J.C. 2001. Structural rationale for the modulation of abscess formation by *Staphylococcus aureus* capsular polysaccharides. *Proc. Natl. Acad. Sci. USA*, 98, 9365–9370.

Udo, E.E., Pearman, J.W., and Grubb, W.B. 1993. Genetic analysis of community isolates of methicillin-resistant *Staphylococcus aureus* in Western Australia. *J. Hosp. Infect.*, 25, 97–108.

Uhlemann, A.-C., Hafer, C., Miko, B.A., Sowash, M.G., Sullivan, S.B., Shu, Q., and Lowy, F.D. 2013. Emergence of sequence type 398 as a community- and healthcare-associated methicillin-susceptible *Staphylococcus aureus* in northern Manhattan. *Clin. Infect. Dis.*, 57, 700–703.

Uhlemann, A.-C., Porcella, S., Trivedi, S., Sullivan, S., Hafer, C., Kennedy, A., Barbian, K., McCarthy, A., Street, C., Hirschberg, D., Lipkin, W.I., Lindsay, J., DeLeo, F., and Lowy, F. 2012. Identification of a highly transmissible animal-independent *Staphylococcus aureus* ST398 clone with distinct genomic and cell adhesion properties. *mBio*, 3, e00027–12.

Ulrich, M., Bastian, M., Cramton, S.E., Ziegler, K., Pragman, A.A., Bragonzi, A., Memmi, G., Wolz, C., Schlievert, P.M., Cheung, A., and Döring, G. 2007. The staphylococcal respiratory response regulator SrrAB induces ica gene transcription and polysaccharide intercellular adhesin expression, protecting *Staphylococcus aureus* from neutrophil killing under anaerobic growth conditions. *Mol. Microbiol.*, 65, 1276–1287.

Urushibara, N., Kawaguchiya, M., and Kobayashi, N. 2012. Two novel arginine catabolic mobile elements and staphylococcal chromosome cassette *mec* composite islands in community-acquired methicillin-resistant *Staphylococcus aureus* genotypes ST5-MRSA-V and ST5-MRSA-II. *J. Antimicrob. Chemother.*, 67, 1828–1834.

Valderas, M.W., and Hart, M.E. 2001. Identification and characterization of a second superoxide dismutase gene (*sodM*) from *Staphylococcus aureus*. *J. Bacteriol.*, 183, 3399–3407.

van de Giessen, A.W., van Santen-Verheuvel, M.G., Hengeveld, P.D., Bosch, T., Broens, E.M., and Reusken, C.B.E.M. 2009. Occurrence of methicillin-resistant *Staphylococcus aureus* in rats living on pig farms. *Prev. Vet. Med.*, 91, 270–273.

Van Den Broek, I.V.F., Van Cleef, B.A.G.L., Haenen, A., Broens, E.M., Van Der Wolf, P.J., Van Den Broek, M.J.M., Huijsdens, X.W., Kluytmans, J.A.J.W., Van De Giessen, A.W., and Tiemersma, E.W. 2009. Methicillin-resistant *Staphylococcus aureus* in people living and working in pig farms. *Epidemiol. Infect.*, 137, 700–708.

van Hal, S.J., Jensen, S.O., Vaska, V.L., Espedido, B.A., Paterson, D.L., and Gosbell, I.B. 2012. Predictors of mortality in *Staphylococcus aureus* bacteremia. *Clin. Microbiol. Rev.*, 25, 362–386.

van Wamel, W.J.B., Rooijakkers, S.H.M., Ruyken, M., van Kessel, K.P.M., and van Strijp, J.A.G. 2006. The innate immune modulators staphylococcal complement inhibitor and chemotaxis inhibitory protein of *Staphylococcus aureus* are located on β-hemolysin-converting bacteriophages. *J. Bacteriol.*, 188, 1310–1315.

Vandenbroucke Grauls, C.M., Thijssen, H.M., and Verhoef, J. 1984. Interaction between human polymorphonuclear leucocytes and *Staphylococcus aureus* in the presence and absence of opsonins. *Immunology*, 52, 427–435.

Vandenesch, F., Naimi, T., Enright, M., Lina, G., Nimmo, G., Heffernan, H., Liassine, N., Bes, M.l., Greenland, T., Reverdy, M.-E., and Etienne, J. 2003. Community-acquired methicillin-resistant *Staphylococcus aureus* carrying Panton-Valentine leukocidin genes: Worldwide emergence. *Emerg. Infect. Dis.*, 9, 978–984.

Vardakas, K.Z., Kontopidis, I., Gkegkes, I.D., Rafailidis, P.I., and Falagas, M.E. 2013. Incidence, characteristics, and outcomes of patients with bone and joint infections due to community-associated methicillin-resistant *Staphylococcus aureus*: A systematic review. *Eur. J. Clin. Microbiol. Infect. Dis.*, 32, 711–721.

Vazquez, V., Liang, X., Horndahl, J.K., Ganesh, V.K., Smeds, E., Foster, T.J., and Hook, M. 2011. Fibrinogen is a ligand for the *Staphylococcus aureus* microbial surface components recognizing adhesive matrix molecules (MSCRAMM) bone sialoprotein-binding protein (Bbp). *J. Biol. Chem.*, 286, 29797–29805.

Velasco, C., López-Cortés, L.E., Caballero, F.J., Lepe, J.A., de Cueto, M., Molina, J., Rodríguez, F., Aller, A.I., García Tapia, A.M., Pachón, J., Pascual, Á., and Rodríguez-Baño, J. 2012. Clinical and molecular epidemiology of meticillin-resistant *Staphylococcus aureus* causing bacteraemia in Southern Spain. *J. Hosp. Infect.*, 81, 257–263.

Ventura, C.L., Malachowa, N., Hammer, C.H., Nardone, G.A., Robinson, M.A., Kobayashi, S.D., and DeLeo, F.R. 2010. Identification of a novel *Staphylococcus aureus* two-component leukotoxin using cell surface proteomics. *PLoS One*, 5, e11634.

Verdon, J., Girardin, N., Lacombe, C., Berjeaud, J.-M., and Héchard, Y. 2009. δ-hemolysin, an update on a membrane-interacting peptide. *Peptides*, 30, 817–823.

Vergara-Irigaray, M., Valle, J., Merino, N., Latasa, C., García, B., Ruiz de los Mozos, I., Solano, C., Toledo-Arana, A., Penadés, J.R., and Lasa, I. 2009. Relevant role of fibronectin-binding proteins in *Staphylococcus aureus* biofilm-associated foreign-body infections. *Infect. Immun.*, 77, 3978–3991.

Vergison, A., Machado, A.N., Deplano, A., Doyen, M., Brauner, J., Nonhoff, C., de Mendonça, R., Mascart, G., and Denis, O. 2012. Heterogeneity of disease and clones of community-onset methicillin-resistant *Staphylococcus aureus* in children attending a paediatric hospital in Belgium. *Clin. Microbiol. Infect.*, 18, 769–777.

Verhoef, J., Peterson, P., Kim, Y., Sabath, L.D., and Quie, P.G. 1977. Opsonic requirements for staphylococcal phagocytosis. Heterogeneity among strains. *Immunology*, 33, 191–197.

Vikram, H., Buenconsejo, J., Hasbun, R., and Quagliarello, V. 2003. Impact of valve surgery on 6-month mortality in adults with complicated, left-sided native valve endocarditis: A propensity analysis. *JAMA*, 290, 3207–3214.

Vitale, C., Gross, T.L., and Weese, J.S. 2006. Methicillin-resistant *Staphylococcus aureus* in cat and owner. *Emerg. Infect. Dis.*, 12, 1998–2000.

von Eiff, C., Becker, K., Machka, K., Stammer, H., and Peters, G. 2001. Nasal carriage as a source of *Staphylococcus aureus* bacteremia. *N. Engl. J. Med.*, 344, 11–16.

Voss, A., Loeffen, F., Bakker, J., Klaassen, C., and Wulf, M. 2005. Methicillin-resistant *Staphylococcus aureus* in pig farming. *Emerg. Infect. Dis.*, 11, 1965–1966.

Voyich, J., Braughton, K., Sturdevant, D., Whitney, A., Sad-Salim, B., Porcella, S., Long, R.D., Dorward, D., Gardner, D., Kreiswirth, B., Musser, J., and DeLeo, F. 2005. Insights into mechanisms used by *Staphylococcus aureus* to avoid destruction by human neutrophils. *J. Immunol.*, 175, 3907–3919.

Voyich, J., Otto, M., Mathema, B., Braughton, K., Whitney, A., Welty, D., Long, R.D., Dorward, D., Gardner, D., Lina, G., Kreiswirth, B., and DeLeo, F. 2006. Is Panton-Valentine leukocidin the major virulence determinant in community-associated methicillin-resistant *Staphylococcus aureus* disease? *J. Infect. Dis.*, 194, 1761–1770.

Voyich, J., Vuong, C., DeWald, M., Nygaard, T., Kocianova, S., Griffith, S., Jones, J., Iverson, C., Sturdevant, D., Braughton, K., Whitney, A., Otto, M., and DeLeo, F. 2009. The SaeR/S gene regulatory system is essential for innate immune evasion by *Staphylococcus aureus*. *J. Infect. Dis.*, 199, 1698–1706.

Vuong, C., Voyich, J.M., Fischer, E.R., Braughton, K.R., Whitney, A.R., DeLeo, F.R., and Otto, M. 2004. Polysaccharide intercellular adhesin (PIA) protects *Staphylococcus epidermidis* against major components of the human innate immune system. *Cell. Microbiol.*, 6, 269–275.

Wagenaar, J.A., Yue, H., Pritchard, J., Broekhuizen-Stins, M., Huijsdens, X., Mevius, D.J., Bosch, T., and Van Duijkeren, E. 2009. Unexpected sequence types in livestock associated methicillin-resistant *Staphylococcus aureus* (MRSA): MRSA ST9 and a single locus variant of ST9 in pig farming in China. *Vet. Microbiol.*, 139, 405–409.

Walenkamp, A.M.E., Boer, I.G.J., Bestebroer, J., Rozeveld, D., Timmer Bosscha, H., Hemrika, W., van Strijp, J.A.G., and de Haas, C.J.C. 2009. Staphylococcal superantigen-like 10 inhibits CXCL12-induced human tumor cell migration. *Neoplasia*, 11, 333–344.

Wang, G., Epand, R.F., Mishra, B., Lushnikova, T., Thomas, V.C., Bayles, K.W., and Epand, R.M. 2012a. Decoding the functional roles of cationic side chains of the major antimicrobial region of human cathelicidin LL-37. *Antimicrob. Agents Chemother.*, 56, 845–856.

Wang, J.-L., Tang, H.-J., Hsieh, P.-H., Chiu, F.-Y., Chen, Y.-H., Chang, M.-C., Huang, C.-T., Liu, C.-P., Lau, Y.-J., Hwang, K.-P., Ko, W.-C., Wang, C.-T., Liu, C.-Y., Liu, C.-L., and Hsueh, P.-R. 2012b. Fusidic acid for the treatment of bone and joint infections caused by meticillin-resistant *Staphylococcus aureus*. *Int. J. Antimicrob. Agents.*, 40, 103–107.

Wang, R., Braughton, K.R., Kretschmer, D., Bach, T.-H.L., Queck, S.Y., Li, M., Kennedy, A.D., Dorward, D.W., Klebanoff, S.J., Peschel, A., DeLeo, F.R., and Otto, M. 2007. Identification of novel cytolytic peptides as key virulence determinants for community-associated MRSA. *Nat. Med.*, 13, 1510–1514.

Wang, R., Khan, B.A., Cheung, G.Y.C., Bach, T.-H.L., Jameson-Lee, M., Kong, K.-F., Queck, S.Y., and Otto, M. 2011. *Staphylococcus epidermidis* surfactant peptides promote biofilm maturation and dissemination of biofilm-associated infection in mice. *J. Clin. Invest.*, 121, 238–248.

Wann, E.R., Gurusiddappa, S., and Höök, M. 2000. The fibronectin-binding MSCRAMM FnbpA of *Staphylococcus aureus* is a bifunctional protein that also binds to fibrinogen. *J. Biol. Chem.*, 275, 13863–13871.

Wardenburg, J.B., Patel, R.J., and Schneewind, O. 2007. Surface proteins and exotoxins are required for the pathogenesis of *Staphylococcus aureus* pneumonia. *Infect. Immun.*, 75, 1040–1044.

Wardenburg, J.B., and Schneewind, O. 2008. Vaccine protection against *Staphylococcus aureus* pneumonia. *J. Exp. Med.*, 205, 287–294.

Webb, L., Wagner, W., Carroll, D., Tyler, H., Coldren, F., and Martin, E. 2007. Osteomyelitis and intraosteoblastic *Staphylococcus aureus*. *J. Surg. Orthop. Adv.*, 16, 73–78.

Weber, J.T. 2005. Community-associated methicillin-resistant *Staphylococcus aureus*. *Clin. Infect. Dis.*, 41, S269–S272.

Weese, J.S., Archambault, M., Willey, B.M., Hearn, P., Kreiswirth, B.N., Said Salim, B., McGeer, A., Likhoshvay, Y., Prescott, J.F., and Low, D.E. 2005. Methicillin-resistant *Staphylococcus aureus* in horses and horse personnel, 2000–2002. *Emerg. Infect. Dis.*, 11, 430–435.

Weese, J.S., Caldwell, F., Willey, B.M., Kreiswirth, B.N., McGeer, A., Rousseau, J., and Low, D.E. 2006a. An outbreak of methicillin-resistant *Staphylococcus aureus* skin infections resulting from horse to human transmission in a veterinary hospital. *Vet. Microbiol.*, 114, 160–164.

Weese, J.S., Dick, H., Willey, B.M., McGeer, A., Kreiswirth, B.N., Innis, B., and Low, D.E. 2006b. Suspected transmission of methicillin-resistant *Staphylococcus aureus* between domestic pets and humans in veterinary clinics and in the household. *Vet. Microbiol.*, 115, 148–155.

Weidenmaier, C., Kokai-Kun, J.F., Kulauzovic, E., Kohler, T., Thumm, G., Stoll, H., Götz, F., and Peschel, A. 2008. Differential roles of sortase-anchored surface proteins and wall teichoic acid in *Staphylococcus aureus* nasal colonization. *Int. J. Med. Microbiol.*, 298, 505–513.

Wertheim, H.F.L., Melles, D.C., Vos, M.C., van Leeuwen, W., van Belkum, A., Verbrugh, H.A., and Nouwen, J.L. 2005. The role of nasal carriage in *Staphylococcus aureus* infections. *Lancet Infect. Dis.*, 5, 751–762.

Wertheim, H.F.L., Walsh, E., Choudhurry, R., Melles, D.C., Boelens, H.A.M., Miajlovic, H., Verbrugh, H.A., Foster, T., and van Belkum, A. 2008. Key role for clumping factor B in *Staphylococcus aureus* nasal colonization of humans. *PLoS Med.*, 5, e17.

Whitby, M., McLaws, M.L., and Berry, G. 2001. Risk of death from methicillin-resistant *Staphylococcus aureus* bacteraemia: A meta-analysis. *Med. J. Aust.*, 175, 264–267.

Wilke, G.A., and Wardenburg, J.B. 2010. Role of a disintegrin and metalloprotease 10 in *Staphylococcus aureus* α-hemolysin–mediated cellular injury. *Proc. Natl. Acad. Sci. USA*, 107, 13473–13478.

Williamson, D.A., Roberts, S.A., Ritchie, S.R., Coombs, G.W., Fraser, J.D., and Heffernan, H. 2013. Clinical and molecular epidemiology of methicillin-resistant *Staphylococcus aureus* in

New Zealand: Rapid emergence of sequence type 5 (ST5)-SCC*mec*-IV as the dominant community-associated MRSA clone. *PLoS One*, 8, e62020.

Wilson, C.B., Jacobs, R.F., and Smith, A.L. 1982. Cellular antibiotic pharmacology. *Semin. Perinatol.*, 6, 205–213.

Wilson, G.J., Seo, K.S., Cartwright, R.A., Connelley, T., Chuang-Smith, O.N., Merriman, J.A., Guinane, C.M., Park, J.Y., Bohach, G.A., Schlievert, P.M., Morrison, W.I., and Fitzgerald, J.R. 2011. A novel core genome-encoded superantigen contributes to lethality of community-associated MRSA necrotizing pneumonia. *PLoS Pathog.*, 7, e1002271.

Wirtz, C., Witte, W., Wolz, C., and Goerke, C. 2009. Transcription of the phage-encoded Panton–Valentine leukocidin of *Staphylococcus aureus* is dependent on the phage life-cycle and on the host background. *Microbiology*, 155, 3491–3499.

Wiseman, G.M. 1975. The hemolysins of *Staphylococcus aureus*. *Bacteriol. Rev.*, 39, 317–344.

Witte, W. 2009. Community-acquired methicillin-resistant *Staphylococcus aureus*: What do we need to know? *Clin. Microbiol. Infect.*, 15, 17–25.

Wood, J.B., Smith, D.B., Baker, E.H., Brecher, S.M., and Gupta, K. 2012. Has the emergence of community-associated methicillin-resistant *Staphylococcus aureus* increased trimethoprim-sulfamethoxazole use and resistance?: A 10-year time series analysis. *Antimicrob. Agents Chemother.*, 56, 5655–5660.

Wright, J.S., Jin, R., and Novick, R.P. 2005. Transient interference with staphylococcal quorum sensing blocks abscess formation. *Proc. Natl. Acad. Sci. USA*, 102, 1691–1696.

Wu, K., Zhang, K., McClure, J., Zhang, J., Schrenzel, J., Francois, P., Harbarth, S., and Conly, J. 2013. A correlative analysis of epidemiologic and molecular characteristics of methicillin-resistant *Staphylococcus aureus* clones from diverse geographic locations with virulence measured by a *Caenorhabditis elegans* host model. *Eur. J. Clin. Microbiol. Infect. Dis.*, 32, 33–42.

Wyllie, D.H., Crook, D.W., and Peto, T.E.A. 2006. Mortality after *Staphylococcus aureus* bacteraemia in two hospitals in Oxfordshire, 1997–2003: Cohort study. *BMJ*, 333, 281.

Xiao, M., Wang, H., Zhao, Y., Mao, L.-L., Brown, M., Yu, Y.-S., O'Sullivan, M.V.N., Kong, F., Xu, Y.-C. 2013. National surveillance of methicillin-resistant *Staphylococcus aureus* (MRSA) in China highlights a still evolving epidemiology with fifteen novel emerging multilocus sequence types. *J. Clin. Microbiol.*, 51, 3638–3644.

Xiong, Y.Q., Willard, J., Yeaman, M.R., Cheung, A.L., and Bayer, A.S. 2006. Regulation of *Staphylococcus aureus* α-toxin gene (*hla*) expression by *agr*, *sarA* and *sae* in vitro and in experimental infective endocarditis. *J. Infect. Dis.*, 194, 1267–1275.

Yamashita, K., Kawai, Y., Tanaka, Y., Hirano, N., Kaneko, J., Tomita, N., Ohta, M., Kamio, Y., Yao, M., and Tanaka, I. 2011. Crystal structure of the octameric pore of staphylococcal γ-hemolysin reveals the β-barrel pore formation mechanism by two components. *Proc. Natl. Acad. Sci. USA*, 108, 17314–17319.

Yao, L., Lowy, F.D., and Berman, J.W. 1996. Interleukin-8 gene expression in *Staphylococcus aureus*-infected endothelial cells. *Infect. Immun.*, 64, 3407–3409.

Yin, S., Daum, R.S., and Boyle-Vavra, S. 2006. VraSR two-component regulatory system and its role in induction of *pbp2* and *vraSR* expression by cell wall antimicrobials in *Staphylococcus aureus*. *Antimicrob. Agents Chemother.*, 50, 336–343.

Zecconi, A., and Scali, F. 2013. *Staphylococcus aureus* virulence factors in evasion from innate immune defenses in human and animal diseases. *Immunol. Lett.*, 150, 12–22.

Zhanel, G.G., Lawson, C.D., Zelenitsky, S., Findlay, B., Schweizer, F., Adam, H., Walkty, A., Rubinstein, E., Gin, A.S., Hoban, D.J., Lynch, J.P., and Karlowsky, J.A. 2012. Comparison of the next-generation aminoglycoside plazomicin to gentamicin, tobramycin and amikacin. *Expert Rev. Anti. Infect. Ther.*, 10, 459–473.

Zhang, L., Gray, L., Novick, R.P., and Ji, G. 2002. Transmembrane topology of AgrB, the protein involved in the post-translational modification of AgrD in *Staphylococcus aureus*. *J. Biol. Chem.*, 277, 34736–34742.

Zhang, N., Holtappels, G., Gevaert, P., Patou, J., Dhaliwal, B., Gould, H., and Bachert, C. 2011. Mucosal tissue polyclonal IgE is functional in response to allergen and SEB. *Allergy*, 66, 141–148.

Zollner, T.M., Wichelhaus, T.A., Hartung, A., Von Mallinckrodt, C., Wagner, T.O., Brade, V., and Kaufmann, R. 2000. Colonization with superantigen-producing *Staphylococcus aureus* is associated with increased severity of atopic dermatitis. *Clin. Exp. Allergy*, 30, 994–1000.

Chapter 40

Pathogenesis of *Listeria monocytogenes* in Humans

Hélène Marquis[1], Douglas A. Drevets[2,3], Michael S. Bronze[2], Sophia Kathariou[4], Thaddeus G. Golos[5], and J. Igor Iruretagoyena[6]

[1]Department of Microbiology and Immunology, Cornell University, Ithaca, NY, USA

[2]Department of Internal Medicine, University of Oklahoma Health Sciences Center, Oklahoma City, OK, USA

[3]Department of Veterans Affairs Medical Center, Oklahoma City, OK, USA

[4]Department of Food, Bioprocessing and Nutrition Sciences, North Carolina State University, Raleigh, NC, USA

[5]Department of Comparative Biosciences, Wisconsin National Primate Research Center, Madison, WI, USA

[6]Department of Obstetrics and Gynecology, University of Wisconsin-Madison, Madison, WI, USA

40.1 Pathogen

40.1.1 Introduction to the pathogen

Listeria monocytogenes is the etiologic agent of listeriosis, an invasive disease that affects humans and other animals (mammals, birds). *L. monocytogenes* is the only human pathogen in the genus *Listeria*; the other pathogenic species, *L. ivanovii*, is a pathogen of small ruminants, which are also susceptible to listeriosis due to *L. monocytogenes* (Vazquez-Boland et al., 2001). Even though 13 serotypes are identified in *L. monocytogenes*, most human cases (and a significant portion of animal listeriosis) involve serotypes 1/2a, 1/2b, and 4b (Kathariou, 2002; Swaminathan and Gerner-Smidt, 2007). Serotypes 4a and 4c are relatively uncommon in human disease but noticeably more common in listeriosis of ruminants (Jeffers et al., 2001).

The species *L. monocytogenes* is partitioned in four major genomic divisions (lineages I–IV), of which lineages I and II are predominant. Strains of serotypes 1/2b, 3b, 4d, 4e, 7, and most strains of serotype 4b belong to lineage I, while serotypes 1/2a, 1/2c, 3a, and 3c constitute lineage II. Strains of serotype 4a, 4c, and a minority of serotype 4b belong to lineage III, while strains of lineage IV are rare and their serotype designations remain to be fully determined (Orsi et al., 2011; Ward et al., 2004). Thus, most human listeriosis involves strains of lineage I (i.e., those of serotype 1/2b and 4b) and II (i.e., those of serotype 1/2a).

Most outbreaks of invasive listeriosis have involved strains of serotype 4b, specifically those of three clonal groups, designated epidemic clones (ECs): ECI, ECII, and ECIa (Kathariou, 2002; Ward et al., 2010). However, other ECs, for example, ECIII and ECV are of serotype 1/2a, and the 2011 cantaloupe outbreak (largest to date) involved several different strains of serotypes 1/2a and 1/2b (Laksanalamai et al., 2012; Lomonaco et al., 2013).

40.1.2 Contributions of genomics and other omics tools to the understanding of *L. monocytogenes* as a human pathogen

A tremendous amount of information on the genome content of *L. monocytogenes* has accumulated since the pioneering report on the comparative genomics of *L. monocytogenes* EGD-e (1924 outbreak in rabbits; serotype 1/2a) and the nonpathogenic strain *L. innocua* CLIP 11262 (Glaser et al., 2001). Subsequent studies reported the genome content of two serotype 4b ECI strains (1983–1987 Swiss outbreak involving cheese and 1985 California outbreak, Mexican-style cheese); two serotype 4b ECII strains (1998–1999 hot dog outbreak in the United States and 1999–2000 outbreaks in France) and the 2000 turkey deli meats outbreak (serotype 1/2a, ECIII) (Hain et al., 2012; Nelson et al., 2004; Orsi et al., 2008; Weinmaier et al., 2013). Genome content has also been determined for strains

Human Emerging and Re-emerging Infections: Bacterial & Mycotic Infections, Volume II, First Edition. Edited by Sunit K. Singh.
© 2016 John Wiley & Sons, Inc. Published 2016 by John Wiley & Sons, Inc.

from the 2008 deli meats outbreak in Canada (serotype 1/2a, ECIV) (Gilmour et al., 2010), two strains of ECIa (Scott A, from the Massachusetts outbreak of 1983 and the "Aureli" strain from a 2004 febrile gastroenteritis outbreak) (Briers et al., 2011; den Bakker et al., 2013a), as well as serotype 1/2a and 1/2b strains from the 2011 cantaloupe outbreak (Laksanalamai et al., 2012).

In addition to these epidemic-associated strains, genome sequences of numerous other strains from diverse sources and representing all known different serotypes have been determined (den Bakker et al., 2013a; Kuenne et al., 2013). The majority of these genome sequences are in draft form (incomplete). Besides *L. monocytogenes*, complete or draft genome sequences are available from a small number of strains of nonpathogenic *Listeria* spp., including some (*L. marthii*, *L. fleischmannii*) with novel species designations (Buchrieser et al., 2011; den Bakker et al., 2013a, 2013b; Glaser et al., 2001; Hain et al., 2006; Steinweg et al., 2010).

The novel sequence data have confirmed and expanded those obtained with the early reports (Glaser et al., 2001; Nelson et al., 2004). The genomes are highly conserved in terms of size (ca. 3 Mb and 3000 genes) and GC content (38%) and exhibit pronounced conservation in gene order on the chromosome (synteny) (Buchrieser et al., 2003; den Bakker et al., 2013a; Glaser et al., 2001; Kuenne et al., 2013; Nelson et al., 2004). Species partitioning into the predominant lineages I and II is reflected by genome content diversity at the singe nucleotide polymorphism (SNP) level (den Bakker et al., 2013a; Kuenne et al., 2013; Laksanalamai et al., 2012; Nelson et al., 2004). In spite of the overall conservation, the genome appears to be "open" based on the still only partially catalogued diversity generated by insertions of novel sequences (den Bakker et al., 2010; Kuenne et al., 2013). Comparisons of completely sequenced genome of 16 strains representing all known serotypes indicated that the core genome, that is, sequences conserved among all strains, consisted of 2354 genes, while the pangenome (sum of core sequences and "accessory sequences" harbored by some but not all of the strains) consisted of 4387 genes; it was projected that upon sequencing of 100 genomes the *L. monocytogenes* pangenome would include ca. 6000 genes (Kuenne et al., 2013). Interestingly, and for unknown reasons, accessory sequences are not evenly distributed on the circular chromosomal map; a majority are in the first 65° (starting with the origin of replication), a region that tends to be overall more prone to diversity than the rest of the chromosome (den Bakker et al., 2013a; Kuenne et al., 2013). Recombinational hotspots that exhibit different content in different strains and mobile genomic elements (MGEs), specifically prophages, transposons and putatively genomic islands account for a large portion of the accessory content on the chromosome, with ca. 30% of the accessory content contributed by MGEs (den Bakker et al., 2013a; Kuenne et al., 2013). In addition, plasmids,

harbored by ca. 30% of *L. monocytogenes* strains (Lebrun et al., 1992; McLauchlin et al., 1997) can contribute significantly to the accessory genome (Kuenne et al., 2010; Nelson et al., 2004).

Plasmid content has been typically discussed separately (and infrequently) in analyses of accessory genome content. Nonetheless, plasmids may make significant contributions to *Listeria*'s fitness in foods and food-related environments, and thus to their potential involvement in human disease. Plasmids frequently harbor determinants mediating resistance to cadmium and quaternary ammonium disinfectants (Elhanafi et al., 2010; Kuenne et al., 2010; Lebrun et al., 1992; McLauchlin et al., 1997; Nelson et al., 2004). Although noticeably more common in serotypes 1/2a and 1/2b than in 4b, the large plasmid pLM80 (conferring resistance both to cadmium and to disinfectant) was harbored by the 1998–1999 hot dog outbreak strains (serotype 4b, ECII); a closely related plasmid was also harbored by strains from the 2000 turkey deli meats outbreak (serotype 1/2a, ECIII) (Elhanafi et al., 2010; Kuenne et al., 2010; Nelson et al., 2004).

Key virulence genes of *L. monocytogenes* (to be discussed in subsequent sections) are conserved among different strains. However, genome sequencing has revealed premature stop codons (PMSCs) and other mutations in these genes in several strains (Hain et al., 2012; Nightingale et al., 2007). Many of the sequenced strains have been maintained in the laboratory for long periods and some of these mutations may reflect adaptations to *in vitro* culture conditions.

Comparative genomic analysis and genotyping of multiple strains has yielded two major bodies of data that are highly informative for assessments of food safety risks and human pathogenicity. These concern: (1) the presence (or not) of full-length, wall-anchored internalin A (InlA), a virulence determinant required for enterocyte invasion and encoded by *inlA*; and (2) the special genomic features that differentiate virulence-attenuated strains, especially those of serotype 4a and 4c (members of lineage III which is generally uncommon in human listeriosis) from lineage I and II.

Truncation of InlA is common in virulence-attenuated food and environmental strains of serogroup 1/2. PMSCs that occur prior to the cell wall anchoring LPXTG motif of InlA result in truncated derivatives that can no longer be anchored into the cell wall and interact with the relevant receptor (E-cadherin) on intestinal epithelial cells. Interestingly, and for reasons that remain unclear, PMSCs in *inlA* are commonly (up to 50%) exhibited by food and environmental strains of serogroup 1/2 (serotypes 1/2a, 1/2b, and 1/2c) but not among those of serotype 4b, regardless of origin (Jacquet et al., 2004; Van Stelten et al., 2010; Ward et al., 2010). It is noteworthy that serogroup 1/2 strains are often markedly more common in food and food-processing environments than those of serotype 4b (Kathariou, 2002; Ward et al., 2010). Such PMSCs in *inlA* are

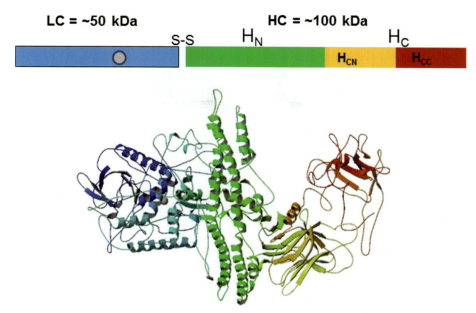

Fig. 43-4 Model and crystal structure of BoNT/A. The light chain (LC) is shown in blue, with the catalytic zinc depicted as a ball in gray. The translocation domain (H_N) of the heavy chain (HC) is shown in green, the N-terminal sub-domain (H_{CN}) of the receptor-binding domain (HC) in yellow, and the C-terminal sub-domain (H_{CC}) in red. The catalytic zinc is depicted as a ball in gray. This model of the crystal structure is derived from PDB ID: 3BTA, and was first published in 1998. With permission from Lacy et al. (1998).

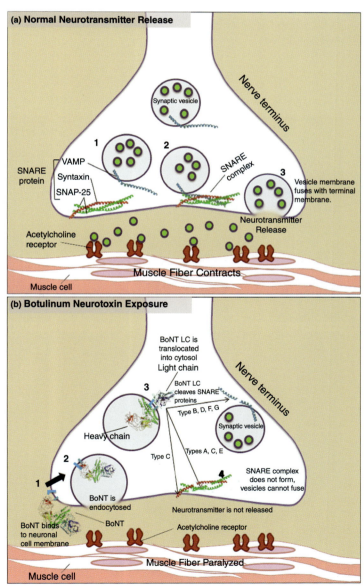

Fig. 43-5 Mechanism of action of Botulinum neurotoxin (BoNT) at the neuromuscular junction: (a) Normal neurotransmitter release: a model of a neuromuscular junction is shown, highlighting the major steps involved in neurotransmitter release. Fusion of the SNARE proteins (SNAP-25, syntaxin, and VAMP/synaptobrevin) to form the SNARE complex mediates fusion of the acetylcholine-filled synaptic vesicle with the neuronal cell membrane. Acetylcholine is released into the synaptic cleft, and binds to the acetylcholine receptor on the muscle cell, causing the muscle cell to contract. (b) BoNT exposure: BoNTs bind to ganglioside and protein receptors on the neuronal cell membrane, leading to endocytosis of the entire toxin. Acidification inside the endocytic vesicle leads to a conformational change and the light chain (LC) is translocated through the endosome membrane into the cells' cytosol. Inside the cytosol, the LC cleaves specific sites on a SNARE protein, preventing the formation of a functional SNARE complex. Thus, the synaptic vesicle cannot fuse with the cell membrane, and no acetylcholine is released. The muscle cell is paralyzed. The BoNT structure and BoNT LC structure in this figure are derived from PDB ID: 3BTA (Lacy et al., 1998) and PDB ID: 4EJ5 (Stura et al., 2012). Figure created by Xiaoyang Serene Hu.

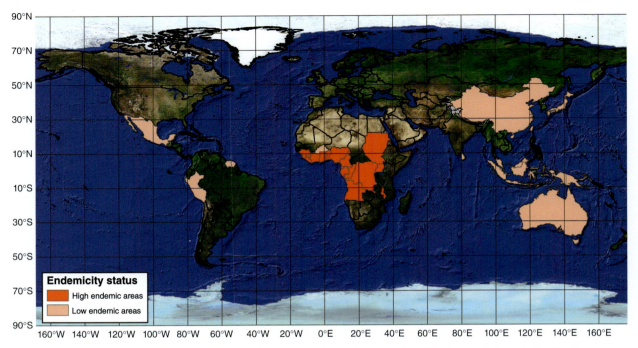

Fig. 44-1 Countries reporting Buruli ulcer. Image and description from Johnson et al. (2005). Acquired through the Creative Commons Attribution License.

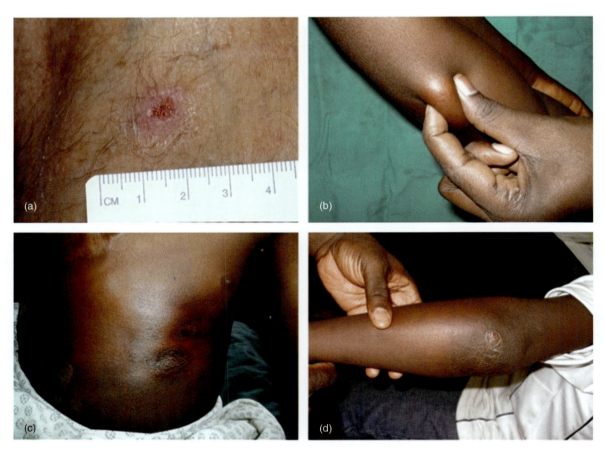

Fig. 44-7 Clinical features of non-ulcerative forms of Buruli ulcer. (a) Papule; (b) nodule; (c) plaque; (d) edema. Images from http://www.who.int/buruli/photos/nonulcerative/en/index.html. Copyright © WHO (2013).

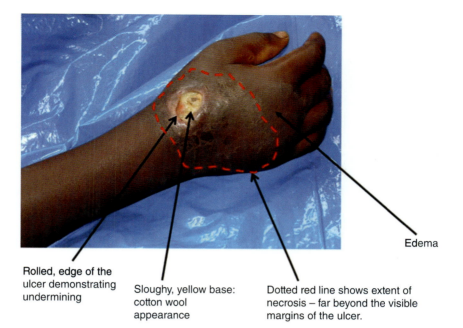

Fig. 44-8 Ulcerative form of Buruli ulcer. Courtesy of Mark Wansbrough-Jones.

Rolled, edge of the ulcer demonstrating undermining

Sloughy, yellow base: cotton wool appearance

Dotted red line shows extent of necrosis – far beyond the visible margins of the ulcer.

Edema

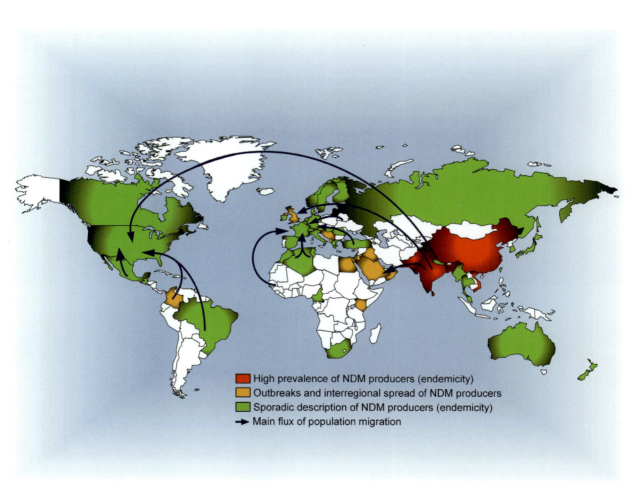

- High prevalence of NDM producers (endemicity)
- Outbreaks and interregional spread of NDM producers
- Sporadic description of NDM producers (endemicity)
- → Main flux of population migration

Fig. 48-1 Geographical spread of NDM producers. The worldwide dissemination of the *bla*$_{NDM}$ genes and the main flux of population migration are represented.

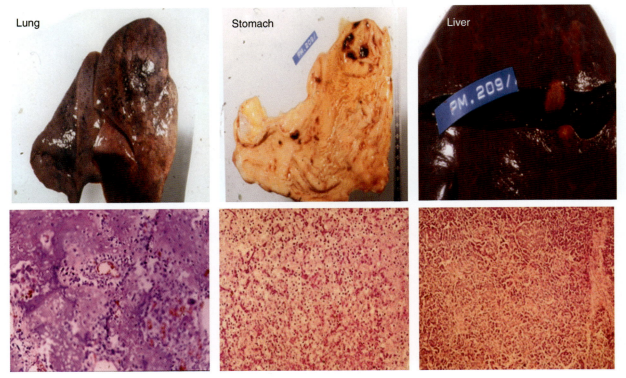

Fig. 52-2 A patient with HIV, admitted in the ICU because of coma and respiratory failure. She received empirically Amphotericin B. She died 48 h later. Autopsy (no whole body) revealed disseminated aspergillosis.

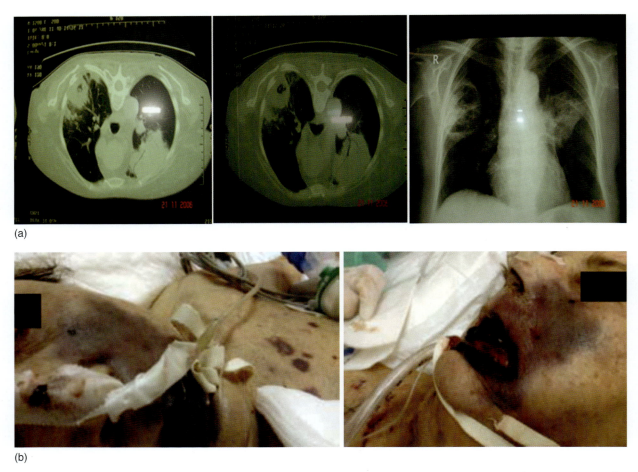

(a)

(b)

Fig. 52-5 (a) A case of pulmonary aspergillosis in a patient with leukemia. The air-crescent sign is apparent. (b) A case of a mixed fungal infection (invasive pulmonary aspergillosis and facial infection due to *Mucor* spp.).

highly uncommon among clinical strains (regardless of serotype or source) (Jacquet et al., 2004; Van Stelten et al., 2010); when present, they may reflect laboratory adaptations. Strains with a truncated InlA are virulence-attenuated in various models (Nightingale et al., 2008). Thus, a substantial portion of serogroup 1/2 strains from foods produce a truncated InlA and may not contribute to human disease. However, it must be kept in mind that serogroup 1/2 strains are relatively abundant in foods, and those (especially 1/2a and 1/2b) with full-length InlA can still be major contributors to human listeriosis.

Virulence-attenuated lineage III strains differ in several genomic attributes from lineage I and II. Lineage III strains are infrequently involved in human disease and some (serotypes 4a, 4c) are virulence-attenuated in certain infection models, while also exhibiting relatively low thermotolerance (De Jesus and Whiting, 2003; Orsi et al., 2011). Lineage III strains are well known for their pronounced genetic diversity which may reflect unusually high propensity for recombination (Orsi et al., 2011; Tsai et al., 2011; Ward et al., 2010). Comparative genomic analysis revealed that these strains differed from lineage I and II in lacking some of the genes coding for virulence-associated surface proteins such as *inlC* (Rajabian et al., 2009), *inlF*, and invasion-associated amidases. An estimated 45 genes (including several encoding metabolic pathway determinants, surface proteins, and transcriptional regulators) conserved in most lineage I and II genomes were absent from those of lineage III (Deng et al., 2010; Doumith et al., 2004; Hain et al., 2012; Kuenne et al., 2013). Apparent loss of such genes in lineage III may account for their infrequent involvement in human disease due to direct impacts on pathogenicity or due to lower fitness in foods and food-related environments.

40.1.3 Can genome sequences determination and other omics-based analyses inform assessments of *Listeria*'s relative potential for human disease or environmental persistence?

Genome content analysis has been critical for elucidating the core and accessory genome of *L. monocytogenes* and in guiding and assisting investigations on *Listeria*'s evolution and adaptations, including pathogenicity. However, with certain notable exceptions (such as discussed in Section 40.1.2 for *inlA* PMSCs and lineage III virulence-attenuated strains) genome content analysis has not yet revealed attributes readily identifiable as associated with the significantly more frequent involvement of certain strains and clonal groups (e.g., ECs) in human listeriosis or in environmental persistence (Briers et al., 2011; Gilmour et al., 2010; Hain et al., 2012; Holch et al., 2013; Laksanalamai et al., 2012; Nelson et al., 2004; Orsi et al., 2008). This may reflect the fact that the functions of a large portion of the accessory

genome (which largely contributes to differences among strains) remain unknown (den Bakker et al., 2013a; Kuenne et al., 2013). Furthermore, human virulence potential and environmental persistence likely reflect the functional outcome of complex networks integrating multiple components of the core and accessory genome (including plasmids) in ways that are currently barely beginning to be understood.

Notable examples of such complex interplays with previously unexpected outcomes include the pronounced upregulation of prophage genes intracellularly and the discovery of major roles of prophage genes in virulence and environmental persistence (Hain et al., 2012; Rabinovich et al., 2012; Toledo-Arana et al., 2009; Verghese et al., 2011); involvement of a transposon-associated cadmium resistance determinant in virulence, first revealed by *in vivo* transcriptomic analysis of strain EGD-e in the murine model (Camejo et al., 2009); the roles of small non-coding RNAs (sRNAs) in virulence, and the identification of numerous sRNAs specifically produced during intracellular growth (Mellin and Cossart, 2012; Mraheil et al., 2011; Toledo-Arana et al., 2009); the role of multidrug efflux systems in innate immune responses (Crimmins et al., 2008; Woodward et al., 2010; Yamamoto et al., 2012); and transcriptomic data suggesting that virulence-attenuated strains tend to overexpress flagellar genes *in vivo* (Hain et al., 2012), possibly facilitating pathogen detection and inflammasome formation in the infected host (Bigot et al., 2005; Sauer et al., 2010; Wu et al., 2010).

Proteomic analysis of *L. monocytogenes* has already contributed to elucidating the composition and functional roles of specific networks, for example, those involving alternative sigma factors (Arous et al., 2004; Mujahid et al., 2013a, 2013b), the wall-associated and secreted proteomes (Garcia-Del Portillo and Pucciarelli, 2012; Renier et al., 2012), the acid tolerance response (Bowman et al., 2012), and those of value in potential source-tracking applications (Dumas et al., 2009). In conjunction with continuing increases in genome sequence data as well as in bioinformatic and proteomic tools, such approaches will be invaluable toward characterizing the responses of *L. monocytogenes* to a wide array of exposures relevant to human disease, ranging from various environmental niches and food to the cells and tissues of an infected individual.

40.2 Epidemiology

40.2.1 Listeriosis: a foodborne disease

Listeria was first described as a zoonotic epidemic infection in 1926 and as a cause of human disease in 1929 (as reviewed) (Cossart, 2007). Over the ensuing decades multiple investigators recognized it as a model pathogen unraveling how the organism resists intracellular killing in macrophages and describing a variety of virulence factors. *L. monocytogenes* has emerged as an important pathogen with multiple, well-described large outbreaks since 1981, when it was first recognized as a cause of foodborne illness, and as a cause of central nervous system (CNS) infection in humans (Schlech, 2000). In fact, approximately 28% of all foodborne-related deaths are attributable to *Listeria* infections (Guevera et al., 2009). It is a ubiquitous organism found widely in nature in soil, decaying vegetation, and in water and as part of the fecal flora of many mammals, including humans (Lorber, 2010; Ramasamy et al., 2007). As a saprophyte, the organism possesses a number of "stress resistance" mechanisms which allow it to live in harsher environments despite being nonsporulating (Freitag et al., 2009). It is an important cause of infections in animals, especially herd animals. Animals excrete the organism in their feces, nasal secretions, and in their milk. *Listeria* infections in animals may result in spontaneous abortions, encephalitis (circling disease), mastitis, infertility, and sepsis (Barbuddhe et al., 2012).

In approximately 5% of healthy adults, the organism can be isolated from stool (Schlech et al., 1983; Swaminathan and Gerner-Smidt, 2007). It may also be cultured from the cervix of women with poor obstetrical histories (Barbuddhe et al., 2012). Human infection is largely transmitted by consumption of contaminated food. Many foods are contaminated with *L. monocytogenes* with recovery rates ranging from 15% to 70% (Farber and Peterkin, 1991). Foods commonly contaminated with *Listeria* and implicated in disease outbreaks include raw vegetables, unpasteurized milk and milk products, meats including those processed in delicatessens. Interestingly, 32% of United States Department of Agriculture (USDA) food recalls between 2002 and 2006 were due to *L. monocytogenes* contamination (Drevets and Bronze, 2008).

In the United States, there are nearly 10 million cases of foodborne illness annually with nearly 56,000 hospitalizations and 1400 deaths (Guevera et al., 2009; Scallan et al., 2011). More than half are due to norovirus, while nearly 30% are collectively due to *Salmonella* and *Campylobacter* species and *Clostridium perfringens* (Scallan et al., 2011). Despite high rates of contamination of foodstuffs, listeriosis is uncommon. The incidence of listeriosis varies by country ranging from 0.1 to 11.3 cases/million, but the treated case fatality rate of *Listeria* meningitis remains quite high ranging from 20% to 30% (Dalton et al., 2011). While the incidence of listeriosis may be increasing in Europe, in the United States, from 1990 to 2005, *Listeria*-related nonperinatal death rates declined from 10.4% annually to 4.26%, although linkage to advanced age, immunosuppressive disease, or medication use and chronic disease remains (Bennion et al., 2008; Munoz et al., 2011). Although not a frequent cause of foodborne illness, patients with listeriosis are more likely to require hospitalization and have a higher mortality rate than those infected with

other causes of foodborne illness (Vrbova et al., 2012; Werber et al., 2013).

Although most human *Listeria* infections are transmitted by consumption of contaminated food, transmission may also occur from mother to child by transplacental or transvaginal routes (as discussed in Section 40.5), by direct contact with infected animals or as nosocomial infections (Barbuddhe et al., 2012). *Listeria* is a very effective foodborne pathogen because it can survive and/or grow at high salt concentrations (10–14%), at low temperatures (−1.5–45°C), or in variable pH environments (4.3–9.1) (CDC, 2011a). Most human *Listeria* infections are due to serotypes 1/2a, 1/2b, or 4b (Dalton et al., 2011; Vrbova et al., 2012). The majority of perinatal infections are due to serotype 4b, whereas foodborne outbreaks are largely due to serotype 1/2 or 4b. Epidemic infections are usually due to serotypes 1/2 and 4b (Ramasamy et al., 2007). Invasive infections are often associated with serotype 4b. The attack rate following exposure to the organism is often quite high due to the consumption of a large inoculum often exceeding 10^5 CFU per gram of food (Ooi and Lorber, 2005). *Listeria* should be suspected in foodborne illness where routine cultures are negative.

A large number of food items have been contaminated with *Listeria* and serve as the transmission sources for multiple outbreaks. Identified food items include coleslaw, milk, soft cheeses (especially if unpasteurized), hot dogs, delicatessen ready-to-eat meats, gravad or cold-smoked fish, pâtés, and fruits and vegetables such as cantaloupes and alfalfa sprouts. Foods identified with higher rates of contamination include fish and seafood, undercooked sausage, soft cheeses, and other milk products and ready-to-eat cooked meats (Barbuddhe et al., 2012). In fact, a major cause of large-scale food recalls is *Listeria* contamination with resultant outbreaks (Pouillot et al., 2012). One of the largest outbreaks occurred in 1997 in which over 1500 school children in Italy developed febrile gastroenteritis due to serotype 4b and attributable to consumption of contaminated cold corn salad (Ooi and Lorber, 2005). Most presented with watery, nonbloody diarrhea and fever and almost 300 required hospitalizations. Several outbreaks have been tied to consumption of Mexican-style soft cheeses (Jackson et al., 2011; Linnan et al., 1988). In 1985, 142 cases of listeriosis were identified in Los Angeles, California, tied to the consumption of unpasteurized Mexican style soft cheese (Linnan et al., 1988). Sixty-five percent of the patients were pregnant women or their neonates and 87% of the maternal–fetal cases were of Hispanic ethnicity. There were 48 reported deaths. Eighty-two percent of the isolates were serotype 4b. In 2009, in the United States, a multistate outbreak tied to pasteurized Mexican-style cheese occurred (Jackson et al., 2011). Contamination occurred during processing after pasteurization and eight pregnant women were infected as well as five neonates. More recently, 146 people were sickened after consuming contaminated cantaloupes in a multi-state outbreak in the United States (http://www.cdc.gov/*Listeria*/outbreaks/cantaloupes-jensen-farms/index.html). Of these cases, 142 required hospitalization and there were 30 deaths. One miscarriage occurred among seven pregnancy-associated infections, and 94% of patients recalled eating cantaloupe.

40.2.2 Prevalence of listeriosis in populations at risk

Although *Listeria* can cause illness in healthy non-pregnant hosts, certain populations of patients are at increased risk for listeriosis, as reviewed (Pouillot et al., 2012). Highest rates of infection are seen in those at extremes of age, pregnant women and their neonates, transplant patients, those immunosuppressed by either disease or medications, and those with chronic illnesses such as diabetes, chronic liver disease, or malignancy, especially hematologic malignancy (CDC, 2011b; Dalton et al., 2011; Guevera et al., 2009; Werber et al., 2013). In one study, incidence of listeriosis in the aggregate at risk population was estimated at 210 cases/100,000 people as compared with 0.7/100,000 in those with no risk factors (Goulet et al., 2012). In a French study, the annual incidence of listeriosis was 0.39 cases per 100,000 residents (Goulet et al., 2012). The incidence increased to 5.6 cases per 100,000 pregnant women and 10–50 cases per 100,000 among patients with hematologic malignancy, whereas the incidence fell to 0.05 cases per 100,000 among those with no chronic illness and under the age of 65.

In a recent US study of 285 patients, factors attributable to risk of infection on multivariate analyses included age ≥70 years, alcoholism, use of corticosteroids, chronic kidney disease, and the presence of a nonhematologic malignancy (Guevera et al., 2009). The presence of an underlying hematologic malignancy, especially chronic lymphocytic leukemia is often cited as a significant risk factor. Advanced age is a significant risk factor, made more alarming due to the increasing age of the world's population (Goulet et al., 2012; Munoz et al., 2011). In fact, nearly 55% of patients from a European review of incidence rates of listeriosis were >65 years of age paralleling an increase in incidence rate for listeriosis in Western Europe (Munoz et al., 2011). Of concern is that the incidence of CNS involvement seems to be higher in the older patients (Munoz et al., 2011). The increased risk of infection in the elderly is thought to be related to senescence of the immune system.

As indicated, most patients with nonperinatal infection have significant medical problems, including malignancy. Safdar and Armstrong (2003b) published a single-center, retrospective review of *Listeria* cases among patients with malignancy. Ninety-four patients were described. Most were actively receiving chemotherapy when the infection was diagnosed and nearly 70% were also receiving corticosteroids.

Bacteremia was the most common manifestation and 36% had CNS infection. Additionally, higher risk is seen among patients with hematologic malignancy (incidence of 10–50 cases/100,000 persons) than with solid tumors such as renal cell carcinoma (incidence of 0.19/100,000 persons) (Goulet et al., 2012).

The use of immune modulating drugs, especially the anti-tumor necrosis factor alpha antagonists heightens the risk for infection due to a variety of organisms, including *L. monocytogenes* (Kelesidis et al., 2010; Salmon-Ceron et al., 2011). Although tuberculosis is the most commonly seen infection (incidence of 144 cases/100,000 persons and 35 cases/100,000 persons for infliximab and etanercept, respectively), the observed incidence of listeriosis was 36/100,000 persons for infliximab and 2/100,000 receiving etanercept (Wallis et al., 2004). A recent review by Bodro and Paeterson (2013) identified 266 cases of listeriosis complicating patients treated with immunomodulating agents. Highest risk was observed in those who received infliximab, while the lowest in those who received either abatacept or golimumab. Most patients received these agents for treatment of rheumatic disease but others were being treated for inflammatory bowel disease or hematologic malignancy. Seventy-three percent of the patients were receiving concomitant immunosuppressant medications and the median time to onset of infection was 184 days. Importantly, mortality ranged between 11.1% and 27.3% (Bodro and Paeterson, 2013).

The risk of listeriosis is markedly increased during pregnancy and intrauterine infection may lead to spontaneous abortion and stillbirth and infection of the neonate (Lamont et al., 2011; Posfay-Barbe and Wald, 2009; Silk et al., 2012). Unlike nonperinatal listeriosis, maternal risk factors for infection other than pregnancy are uncommon. Transmission to the mother is usually foodborne, although spread to the fetus can be transplacental or transvaginal. Most commonly reported food exposures include ice cream, butter, yogurt, soft cheeses, and hot dogs. Recommendations that pregnant women avoid certain high-risk foods and the monitoring of food for contamination with *Listeria* have led to a significant reduction in US cases of perinatal listeriosis (Tappero et al., 1995).

The enhanced risk in pregnancy is thought to be related to a shift from a T_H1 to a T_H2 type of immune response (Posfay-Barbe and Wald, 2009). Pregnancy confers an 18-fold increased risk of *Listeria* infections and 17% of all *Listeria* infections are pregnancy related (Lamont et al., 2011). Maternal infection may be asymptomatic or if symptomatic creating an influenza-like illness or a febrile gastroenteritis. One-third of maternal–fetal infections result in spontaneous abortions or stillbirths, and live births may develop symptoms usually within the first week of life (Allerberger and Wagner, 2010). Jackson and colleagues reported on 128 US cases of pregnancy-associated listeriosis (Jackson et al., 2010). Twenty-eight percent of mothers were asymptomatic while others demonstrated sepsis, influenza-like illnesses, or had febrile gastroenteritis. Eighty-five infants demonstrated invasive infection in which meningitis and sepsis/bacteremia were the most common manifestations. Four neonatal deaths and 26 stillbirths were reported indicating the severity of maternal infection. Maternal outcome following listeriosis is usually favorable.

Fetal and neonatal infection is uncommon estimated at 8.6–17.4 cases/100,000 live births (Drevets and Bronze, 2008; Lamont et al., 2011). Nearly 70% of live births born to mothers with maternal infection will demonstrate listeriosis (Mylonakis et al., 2002). By convention, neonatal listeriosis that occurs within 7 days of delivery is designated as "early onset" and those cases that occur after day 7 as "late onset." Neonates with early-onset infection usually manifest meningitis, bacteremia, and pneumonia, whereas those with late-onset disease usually have meningitis and sepsis. The case fatality rate ranges from 3% to 50% (Posfay-Barbe and Wald, 2009).

Despite effective treatment strategies, mortality rates for *Listeria* infections remain very high. For perinatal infections, the case fatality rate ranges from 1% to 10.3%, whereas in nonperinatal infections it ranges from 9% to 45% depending on the country from which data are reported (Clark et al., 2010).

40.2.3 Clinical syndromes

Human infection with *L. monocytogenes* usually presents with one of three clinical syndromes: febrile gastroenteritis, maternal–fetal/neonatal listeriosis, or nonperinatal infection manifested as bacteremia, CNS infection, or focal infection derived from hematogenous spread. CNS infection may take the form of meningitis, meningoencephalitis, brain abscess, or rarely, rhombencephalitis (Pouillot et al., 2012). A large number of focal infections have been described including peritonitis, cholecystitis, septic arthritis, hepatitis and liver abscess, and endocarditis. Cutaneous listeriosis is uncommon, but may complicate those with chronic skin conditions or those with occupational exposure to infected animals.

Patients with bacteremic *L. monocytogenes* often relate a history of antecedent gastrointestinal (GI) symptoms (Werber et al., 2013). Additionally, there is convincing evidence that *L. monocytogenes* can cause foodborne illness in healthy individuals, as reviewed (Endrikat et al., 2010). Features common to febrile gastroenteritis cases include dose-dependent response in that those who eat multiple servings of the contaminated food are more likely to become ill, ingestion of a large inoculum is required unless the individual is gastric acid suppressed, and attack rates are high (Endrikat et al., 2010). The incubation period is usually less than 24 hours but may range from 6 hours to 10 days. Symptoms are usually self-limited and those commonly reported from outbreaks include fever, diarrhea that is described as nonbloody and watery, vomiting, nausea, and abdominal pain.

Myalgias and arthralgias are usually described in adults, however, bacteremia is uncommon. Recovery occurs within 48 hours and the role of antimicrobial treatment is not clear unless the individual is at high risk for invasive disease.

40.3 Pathogenesis and immunity

40.3.1 Resistance of *L. monocytogenes* to conditions encountered in the gastrointestinal tract

The transition of *L. monocytogenes* from a food item into the human GI tract is associated with a series of environmental stresses that are detrimental to the establishment of infection by a bacterial pathogen. However, *L. monocytogenes* has devised multiple mechanisms to adapt to these stressful conditions and to evade the innate immune mechanisms that prevail in the GI tract.

The pH of the stomach is usually between pH 2 and 4, which is lethal for most bacteria. The ability of *L. monocytogenes* to tolerate acidic conditions encountered in the stomach is mediated by at least two independent systems: the glutamate decarboxylase (GAD) system, which is regulated by the alternative transcription factor sigma B (σ^B) (Cotter et al., 2001; Karatzas et al., 2012; Kazmierczak et al., 2003), and the arginine deiminase (ADI) system, which is regulated by σ^B and the transcriptional regulator ArgR (Ryan et al., 2009). The GAD system works as a glutamate/γ-aminobutyrate antiport in which an extracellular molecule of glutamate is internalized and converted to γ-aminobutyrate in a reaction that consumes a proton. The ADI system converts arginine to ornithine, CO_2, and ammonia, which sequesters a proton to form NH_4^+. Both systems serve to maintain the cytoplasm near neutral pH in an acidic extracellular environment, therefore protecting the bacteria from the potentially lethal effect of the stomach's acidic environment.

Also contributing to the harsh GI environment are bile salts, which are toxic to bacteria due to their effect on the integrity of bacterial membranes and possibly damage to DNA (Merritt and Donaldson, 2009). *L. monocytogenes* has been shown to reside and multiply in the gall bladder and in the GI tract indicating its ability to resist the toxic effect of bile salts (Gahan and Hill, 2005; Hardy et al., 2004). Transcriptional profiling of *L. monocytogenes* has shown that more than 10% of its genes are regulated by bile salts. Among those is the gene coding for MdrT, a multidrug efflux pump for the bile salt cholic acid (Quillin et al., 2011). The *mdrT* gene is repressed by BrtA, a bile-regulated transcription factor. In the presence of cholic acid, BrtA-mediated repression of *mdrT* expression is released, enabling the bacteria to pump out the bile salts. *L. monocytogenes* also possess a bile exclusion system (BilE) (Sleator et al., 2005) and a bile tolerance locus (*btl*) (Begley et al., 2003, 2005), whose specific mechanisms of action are unknown. Another important factor is an intracellular bile salt hydrolase

(BSH) whose activity increases at low oxygen tension, a condition that would be encountered in the GI tract (Dussurget et al., 2002). Overall, *L. monocytogenes* protects itself from bile in multiple ways by preventing entry of bile salts into the cell, pumping them out of the cell, or catabolizing the salts that have entered the cell.

Additional antimicrobial defense comes from Paneth cells, which reside at the base of the crypts in the small intestine and secrete cationic antimicrobial molecules such as defensins (Muller et al., 2005). Defensins are amphipathic peptides that exhibit broad-spectrum antimicrobial activity by binding to the negatively charged bacterial surface and subsequently disrupting membrane integrity (Straus and Hancock, 2006). Consequently, the proton gradient is dissipated potentially leading to autolysis (Forster and Marquis, 2012; Jolliffe et al., 1981; Kemper et al., 1993). *L. monocytogenes* protects itself by modulating ionic charges within the cell envelope to decrease binding of antimicrobial peptides. One mechanism involves the incorporation of D-alanine moieties to negatively charged cell wall teichoic acids and membrane anchored lipoteichoic acids, overall reducing surface charges (Abachin et al., 2002). The four genes involved (*dltA-D*) code for a ligase, a transmembrane protein, a carrier protein, and a membrane-associated protein. A second mechanism used by *L. monocytogenes* involves modifying charges of membrane phospholipids by conjugating a positively charged lysine moiety to diphosphoglycerol-generating lysylphosphatidylglycerol (Thedieck et al., 2006). The enzyme responsible for catalyzing this reaction is MprF, which stands for multiple peptide resistance factor. Thus, modification of charges in the cell wall and cytoplasmic membrane decreases the binding affinity of these cationic antimicrobial peptides and their ability to kill the bacteria.

Paneth cells also secrete lysozyme, a cationic antimicrobial molecule that cleaves the β-(1,4) linkage between the *N*-acetylglucosamine and *N*-acetylmuramic acid residues of peptidoglycan, causing a weakening of the cell wall and osmotic lysis as the intrabacterial milieu is hyperosmotic. The resistance of *L. monocytogenes* to lysozyme results from the deacetylation of *N*-acetylglucosamine residues, preventing the binding of lysozyme to its substrate (Boneca et al., 2007). This modification is mediated by a peptidoglycan *N*-deacetylase encoded by *pgdA*. Therefore, *L. monocytogenes* ability to resist cationic antimicrobial molecules secreted in the small intestine by Paneth cells is, at least in part, due to its ability to enzymatically modify the targets of these molecules.

40.3.2 Bacterial invasion

Surviving within the GI tract is the first step in the pathogenesis of listeriosis. Although the disease has been associated with transient febrile gastroenteritis, listeriosis is primarily an invasive disease. Therefore,

L. monocytogenes needs to cross the intestinal barrier to access internal organs. Uptake of *L. monocytogenes* by M cells at the site of Peyer's patches contributes to invasion of the intestines (Jensen et al., 1998). However, invasion of epithelial cells is the most important pathway used by *L. monocytogenes* to cause invasive disease. Flagella-mediated motility and a bacterial surface molecule called internalin A (InlA) are involved in invasion of the intestinal mucosa. Flagella-mediated motility was shown to enhance *L. monocytogenes* infectivity soon after bacterial ingestion (O'Neil and Marquis, 2006). The flagella do not serve as adhesins to enhance bacterial attachment to epithelial cells, but possibly facilitate the interaction of InlA with its receptor on epithelial cells as InlA distribution at the bacterial surface is polarized (O'Neil and Marquis, 2006; Rafelski and Theriot, 2006). InlA is a leucine-rich repeat surface protein that is covalently attached to the peptidoglycan (Gaillard et al., 1991). It binds specifically to human E-cadherin, a calcium-dependent cell-to-cell adhesion molecule that is critical for the formation of tight junctions between epithelial cells (Lecuit et al., 1999, 2001; Mengaud et al., 1996). Access to E-cadherin is limited since the protein is located at the basolateral surface of polarized cells. However, E-cadherin becomes exposed to the luminal side of the epithelial layer at sites of cell extrusion and around mucus-expelling goblet cells, enabling the binding of *L. monocytogenes* to intestinal cells (Nikitas et al., 2011; Pentecost et al., 2006). After binding to E-cadherin, bacteria are phagocytosed and transcytosed in vacuoles across the intestinal barrier to reach the underlying lamina propria (Nikitas et al., 2011). Once the intestinal barrier has been breached, *L. monocytogenes* can disseminate throughout the body.

Following invasion through the intestinal epithelium, bacteria reach the liver leading to the development of a sub-clinical pyogranulomatous hepatitis (Vazquez-Boland et al., 2001). *L. monocytogenes* invasion of hepatocytes is dependent on internalin B (InlB), a surface protein that is anchored to teichoic acid in a noncovalent manner (Dramsi et al., 1995). The Met receptor tyrosine kinase serves as a receptor for InlB (Shen et al., 2000). Engagement of this receptor leads to the localized activation of type IA phosphoinositide 3-kinase and internalization of bacteria (Ireton et al., 1996, 1999). From this point on, internalized *L. monocytogenes* will begin to proliferate intracellularly and to disseminate to other tissues. This bacterial pathogen has a predilection for the CNS and for the gravid uterus, which is discussed in subsequent sections of this chapter.

40.3.3 The intracellular life cycle

The intracellular life cycle of *L. monocytogenes* was initially described by Tilney and Portnoy (1989). Using a series of electron micrographs from infected mouse macrophages, they observed that *L. monocytogenes* multiplies in the cytosol of host cells and uses an actin-based mechanism of motility to spread from cell to cell without exiting the intracellular milieu. These results served to elucidate an old observation by Mackaness (1962) that protective immunity against *L. monocytogenes* was solely cell-mediated, as following the initial invasion event bacteria remain primarily inside cells. Over the past 25 years, many of the molecular and cellular details of the mechanisms that regulate the intracellular life cycle of *L. monocytogenes* have been revealed. Most of the genes coding for factors that contribute to this life style are under the regulation of PrfA (Chakraborty et al., 1992; Leimeister-Wächter et al., 1990). The *prfA* gene is autoregulated and its 5′-UTR contains a thermosensor riboswitch that prevents translation of the mRNA at 30°C or lower temperature (Bohne et al., 1994; Johansson et al., 2002; Leimeister-Wächter et al., 1992; Xayarath and Freitag, 2012).

Escape from the vacuole is predominantly mediated by a cholesterol-dependent cytolysin called listeriolysin O or LLO. LLO's pore-forming ability is dependent on pH and redox: pore formation is optimal at pH 5.9 and in a reducing environment that is generated by γ-interferon-inducible lysosomal thiol reductase (GILT) (Beauregard et al., 1997; Singh et al., 2008). Mutants that are unable to compartmentalize LLO pore-forming activity to the vacuole are toxic for host cells and avirulent (Decatur and Portnoy, 2000; Glomski et al., 2002). Similarly, LLO-minus mutants of *L. monocytogenes* are completely avirulent in mice (Cossart et al., 1989). In human tissue culture cells, escape from vacuoles occurs in absence of LLO, but at a lower rate of efficiency (Grundling et al., 2003; Marquis et al., 1995; Portnoy et al., 1988). Nevertheless, there has never been a report of human listeriosis caused by a nonhemolytic isolate of *L. monocytogenes* indicating that LLO is essential for virulence in humans.

The efficiency of escape from vacuoles is enhanced by two secreted phospholipases of the C type (PI-PLC and PC-PLC) encoded by *plcA* and *plcB*, respectively (Camilli et al., 1993; Smith et al., 1995a). Studies with single and double gene deletion mutants indicated that the two PLCs have overlapping functions in virulence (Smith et al., 1995a). The activities of the PLCs on phospholipids lead to increased intracellular levels of diacylglycerol and ceramide in infected cells (Camilli et al., 1993; Smith et al., 1995a). PI-PLC mediates calcium and protein kinase C signaling during infection (Goldfine and Wadsworth, 2002; Poussin and Goldfine, 2005; Poussin et al., 2009; Wadsworth and Goldfine, 2002). Inhibition of this activity compromises the efficacy of escape from vacuoles. Similar to LLO, PC-PLC activity is regulated by pH (Marquis et al., 1997). PC-PLC is made as an inactive proenzyme that remains associated with the cell wall until a drop in pH induces autolysis of the metalloprotease of *L. monocytogenes* (Mpl), which mediates the proteolytic activation of PC-PLC (Bitar et al., 2008; Forster et al., 2011; Forster and Marquis, 2012; Marquis

and Hager, 2000; Snyder and Marquis, 2003). A PC-PLC mutant lacking the propeptide is unable to compartmentalize PC-PLC activity to the vacuole causing virulence attenuation (Yeung et al., 2007).

More recently, the DNA uptake competence (Com) system of *L. monocytogenes* has been reported to contribute to escape from vacuoles (Rabinovich et al., 2012). The Com system is interrupted by a prophage, whose excision following cell invasion leads to the production of a functional ComK protein, which is a master regulator of the Com system. It is possible that the Com system serves to translocate bacterial cytoplasmic proteins that contribute to lysis of the vacuole (Chen and Dubnau, 2004; Forster and Marquis, 2012).

L. monocytogenes multiplies in the cytosol of host cells, taking advantage of an abundance of nutrients. The doubling time of *L. monocytogenes* in the cytosol of tissue culture cells can be as fast as 40 min (Marquis et al., 1993), which is close to its doubling time in rich bacterial broth medium, indicating that it is very well adapted to growing in that environment. Even though it is auxotrophic for multiple amino acids, its intracellular growth is supported by its ability to scavenge short peptides, modified lipoyl peptides, and free amino acids from the host cytosol (Keeney et al., 2007; Marquis et al., 1993; O'Riordan et al., 2003). Lipoate-dependent metabolism contributes to the generation of branched-chain fatty acids, which are essential for intracellular growth (Keeney et al., 2009). *L. monocytogenes* also synthesizes a glucose-6-phosphate translocase that enables bacteria to utilize host phosphorylated sugars as a source of carbon (Chico-Calero et al., 2002).

One of the major hallmarks of *L. monocytogenes* is its ability to move from cell to cell without exiting the intracellular milieu (Tilney and Portnoy, 1989). Upon entry in the cytosol, the bacteria synthesize a protein called ActA that is membrane anchored and exposed at the bacterial surface (Kocks et al., 1992). ActA has the ability to polymerize host cell actin, and asymmetric distribution of ActA at the bacterial surface results in the formation of an actin tail (Kocks et al., 1993; Smith et al., 1995b). The Arp2/3 complex, VASP, and profilin are all involved in the nucleation of actin by binding different domains of ActA (Skoble et al., 2000; Welch et al., 1998). The rate of polymerization of actin filaments at the bacterial pole dictates the rate of bacterial movement within the cell, and the ability of *L. monocytogenes* to move directly from cell to cell (Robbins et al., 1999; Theriot et al., 1992). However, membranes of polarized cells are lined with a thick layer of cortical F-actin and myosin that might interfere with cell-to-cell spread. The secreted *Listeria* protein InlC promotes cell-to-cell spread by decreasing cortical tension at membrane junctions (Rajabian et al., 2009). InlC binds to the mammalian protein Tuba, preventing the formation of Tuba–N-WASP complexes at membrane junctions and releasing cortical tension.

In the process of spreading from cell to cell, bacteria become enclosed in double membrane vacuoles from which they must escape to perpetuate the intracellular life cycle of *L. monocytogenes*. Escape from double membrane vacuoles, which are also called secondary vacuoles, is mediated primarily by LLO and PC-PLC (Alberti-Segui et al., 2007; Gedde et al., 2000; Smith et al., 1995a).

40.3.4 The immune response to infection

Despite its efforts to resist killing, *L. monocytogenes* triggers a strong innate and adaptive immune response that effectively clears the infection in most instances. Most of our knowledge about the immune response to *L. monocytogenes* is derived from experiments performed in mice, an excellent animal model to study listeriosis. Following invasion of the liver, the majority of bacteria are rapidly killed presumably by neutrophils and Kupffer cells (Carr et al., 2011; Gregory et al., 1996). However, the ones that survive invade hepatocytes in which they undergo exponential growth for at least 24 hours depending on the infectious dose. In the spleen, bacteria get trapped by CD8α^+ dendritic cells (DCs) that are present in the marginal zone (Edelson et al., 2011). Opsonization of bacteria by platelets and the C3 complement factor facilitate this event (Verschoor et al., 2011). Migration of infected DCs to the T lymphocyte zones is associated with a burst of bacterial growth and lymphocyte apoptosis (Edelson et al., 2011). Infected cells produce interleukin-12 (IL-12) and IL-18 leading to the recruitment and activation of natural killer (NK) cells, which are an important source of interferon gamma (IFN-γ) (Kang et al., 2008). This pro-inflammatory cytokine is essential in controlling the infection as it mediates the activation and maturation of monocytes into tumor necrosis factor alpha (TNFα) and inducible nitric oxide synthase (iNOS) producing DCs (TipDCs) (Serbina et al., 2003). TNFα is a cytokine that increases vascular permeability facilitating the migration of cells to infected tissues, whereas the production of NO empowers cells into controlling intracellular bacterial growth (MacMicking et al., 1995; Ogawa et al., 2001). Moreover, inflammatory monocytes respond to IFN-γ by activating the NADPH oxidase, which produces reactive oxygen species that also contribute to controlling the intracellular growth of bacteria (Myers et al., 2003). Nonetheless, *L. monocytogenes* manages to diminish the effect of IFN-γ on monocytic cells. Bacterial release of c-di-AMP and dsDNA in the cytosol of infected cells induces the production of type I IFN (Leber et al., 2008; Woodward et al., 2010), resulting in the downregulation of IFN-γ receptor, consequently decreasing the ability of IFN-γ to activate inflammatory monocytes (Rayamajhi et al., 2010).

The adaptive immune response is essential for clearing the infection. In SCID mice, which lack adaptive immune responses, the innate immune response to infection is able to control the growth of *L. monocytogenes*, but insufficient to fully clear the infection (Bhardwaj et al., 1998). During systemic listeriosis, bacteria are primarily intracellular, preventing any significant levels of antibody production (Leong et al., 2009). However, *L. monocytogenes* efficiently prime MHC1-restricted CD8$^+$ T cells as a result of bacterial antigen secretion in the cytosol of antigen-presenting cells (Lara-Tejero and Pamer, 2004). The T cell response is dominated by IFN-γ producing T$_H$1 cells and peaks at 5–8 days post-infection (Badovinac et al., 2002). The memory response to infection is very strong, but long-term protection is dependent on bacteria accessing the host cytosolic compartment (Bahjat et al., 2009). Bacteria that cannot escape vacuoles stimulate the production of IL-10, which suppresses the expression of pro-inflammatory cytokines. This observation elucidated why killed or LLO-deficient *L. monocytogenes* are incapable of generating long-term protection.

It is interesting to observe that *L. monocytogenes* has devised multiple mechanisms to evade or control the immune response to infection. In a way, these mechanisms, which only buy time, magnify the generation of a protective and specific memory immune response. On the other hand, the pathogen may win in instances where some aspects of the immune system are dysfunctional.

40.4 Cerebral listeriosis

40.4.1 Clinical features

Clinical presentations of human cerebral listeriosis are quite variable. CNS infection most commonly presents as meningitis or meningoencephalitis, but also can produce cerebritis, encephalitis, brain abscess, and brainstem encephalitis, also known as rhombencephalitis (Drevets and Bronze, 2008; Vazquez-Boland et al., 2001). This spectrum of manifestations contrasts with a monomorphic clinical syndrome of typical meningitis that is produced by more common neuroinvasive bacterial pathogens, such as *Streptococcus pneumoniae* and *Neisseria meningitides*. Host and bacterial factors contribute to the variety of clinical syndromes. From the host perspective, *L. monocytogenes* has a predilection for infecting the elderly and individuals with immune compromising co-morbidities including receiving immunosuppressive pharmaceuticals (Drevets and Bronze, 2008). How immunodeficiency modifies host responses is difficult to quantify yet likely produces the finding that cerebrospinal fluid (CSF) from cases of listeriosis displays lower leukocyte numbers and protein concentrations than does meningitis caused by other bacteria (Amaya-Villar et al., 2010; Mylonakis et al., 1998). Additionally, *L. monocytogenes* frequently evokes a mononuclear pleocytosis rather than a predominance of neutrophils in the CSF (Bartt, 2000; Bouvet et al., 1982).

Principal among bacterial factors is the ability of *L. monocytogenes* to invade and replicate within phagocytic and nonphagocytic mammalian cells. This microbiological feat has profound implications for how *L. monocytogenes* invades the CNS and the subsequent clinical manifestations of CNS infection. Recent work identified specific bone marrow—derived monocytes that harbor *Listeria* and transport it to the brain (Drevets et al., 2004; Join-Lambert et al., 2005). An important corollary is that intracellular *L. monocytogenes* can spread from parasitized monocytes or macrophages to other cells relevant to brain invasion and infection, for example, endothelial cells and neurons (Dramsi et al., 1998; Drevets et al., 1995; Jin et al., 2002). *L. monocytogenes* is also capable of direct invasion of brain microvascular endothelial cells (Greiffenberg et al., 1998; Wilson and Drevets, 1998). The ability to parasitize a wide array of cells enables *L. monocytogenes* to invade CNS by at least three different mechanisms (Drevets and Bronze, 2008). These include (1) a Trojan horse mechanism in which intracellular bacteria are transported to the CNS and across the blood—brain barrier (BBB) or blood—cerebrospinal fluid barrier (BCSFB) within parasitized leukocytes, or (2) a neural route marked by retrograde migration of bacteria into the brain within the axons of cranial nerves, and (3) direct invasion of endothelial cells or cells of the choroid plexus by blood-borne bacteria. In addition, other mechanisms for blood-borne *L. monocytogenes* to bind and invade the CNS, for example, via choroid plexus, are under investigation (Gründler et al., 2013). Studies using organotypic brain slices show *L. monocytogenes* replicates in brain parenchymal cells, predominantly microglia, as well as within axons, astrocytes, and rarely oligodendrocytes (Guldimann et al., 2012). Human studies also show evidence of intercellular spread before and after brain invasion (Antal et al., 2005; Kirk, 1993). The most direct association between microbial virulence factors and clinical syndromes is found in the neural route of CNS invasion and the clinical presentation of rhombencephalitis.

Meningitis and meningoencephalitis together comprise up to 97% of presenting CNS syndromes and are not easily distinguished from meningitis caused by other bacteria (Mylonakis et al., 1998). The majority of patients report fever (temperature ≥38°C), headache, and neck stiffness and essentially all patients report at least one of these symptoms (Mylonakis et al., 1998). The classic triad of fever, stiff neck, and altered mental status are found together in 33–49% of patients, while focal neurological findings and seizures on presentation are observed in 32–43% and 2–9% of patients, respectively, in recent studies (Amaya-Villar et al., 2010; Koopmans et al., 2013). There is a suggestion that typical meningitis is a less common presentation in immunosuppressed patients than in immunocompetent ones, but this is not found in all studies (Amaya-Villar et al., 2010; Skogberg et al., 1992). Similarly, there are not consistent differences between the percentages of patients that present with typical meningitis symptoms due to CNS infection by *L. monocytogenes* compared with CNS infection caused by other bacteria (Amaya-Villar et al., 2010; Mylonakis et al., 1998). Laboratory findings in the periphery are nonspecific and are notable for high white blood cell counts and markers of inflammation, for example, C-reactive protein, and most patients demonstrate hyponatremia. CSF from cases of meningitis is usually reactive with elevated WBC and protein concentrations. In the largest series, median CSF total WBC counts are 550–585 cells/mm^3 with median neutrophil percentages of 70–75% (Amaya-Villar et al., 2010; Mylonakis et al., 1998). Total WBC counts and the percentage of neutrophils, along with protein concentrations, are lower than found in CSF from patients with meningitis caused by other bacterial pathogens (Amaya-Villar et al., 2010; Mylonakis et al., 1998). Additionally, CSF from *L. monocytogenes*—infected patients frequently reveals a predominance of mononuclear leukocytes rather than neutrophils (Bartt, 2000; Bouvet et al., 1982). Gram's stain of the CSF reveals organisms in 5–37% of samples, whereas CSF cultures are positive in 59–100% of cases (Amaya-Villar et al., 2010; Bouvet et al., 1982; Koopmans et al., 2013; Lavetter et al., 1971; Mylonakis et al., 1998). The percentage of positive Gram's stains is notably lower than for other common bacterial CNS pathogens, with the exception of *M. tuberculosis* (Brouwer et al., 2010).

Two less common presentations are encephalitis (cerebritis) and rhombencephalitis. *Listeria* encephalitis, a localized cortical infection that may precede brain abscess, appears as undifferentiated encephalitis not distinguishable on clinical grounds from encephalitis caused by other pathogens (Mailles et al., 2009). In contrast, rhombencephalitis is an unusual and distinct clinical syndrome of inflammation of the brainstem or cerebellum caused by an array of noninfectious and infectious etiologies, of which *L. monocytogenes* is the most common infectious cause (Moragas et al., 2011). Rhombencephalitis accounts for approximately 5% of human cases of *Listeria* CNS infection and likely results from a neural route of infection in which bacteria inoculated into oral mucosa are phagocytosed by tissue macrophages, then invade neurons and enter the brain via their axons (Antal et al., 2001, 2005; Jin et al., 2001). The majority of individuals with *Listeria* rhombencephalitis are not immunosuppressed (Armstrong and Fung, 1993). The classic presentation of *L. monocytogenes* rhombencephalitis is a biphasic illness that begins with a prodromal headache, fever, and nausea/vomiting followed by sudden onset of asymmetrical cranial nerve deficits, cerebellar signs, and hemiparesis (Armstrong and Fung, 1993; Uldry et al., 1993). CSF is usually abnormal, but the magnitude of abnormalities is generally less pronounced than found in meningitis. For example, two series reported median CSF leukocyte counts of 110 and 261 cells/mm^3 with the majority of specimens in both series demonstrating mononuclear predominance (Armstrong and Fung, 1993; Uldry et al., 1993). Collectively in these series, initial Gram's stains and cultures were positive in only 4 of 45 (8.9%) and in 27 of 64 (42.1%) specimens, respectively.

The reported case fatality rates for CNS infections caused by *Listeria* range from 16% to 36% in series representing patients infected over the past half-century (Amaya-Villar et al., 2010; Aouaj et al., 2002; Koopmans et al., 2013; Mylonakis et al., 1998). Although direct comparisons to other bacterial causes are difficult, case fatality rates for *L. monocytogenes* are generally higher than for *N. meningitides* and comparable to *S. pneumoniae* in developed countries (Amaya-Villar et al., 2010; Schuchat et al., 1997; Thigpen et al., 2011; van de Beek et al., 2004, 2006). Age-adjusted mortality rates for all invasive *L. monocytogenes* infections are significantly higher in the elderly and in patients with underlying diseases (Bennion et al., 2008; Goulet et al., 2012). At least one study suggests the same is true for meningitis-associated case fatality rates (Aouaj et al., 2002). Studies from the Netherlands show neurological sequelae are present in 16–30% of survivors (Aouaj et al., 2002; Koopmans et al., 2013). Necropsy findings in the CNS of patients succumbing to fatal *L. monocytogenes* infections include focal hemorrhages and findings characteristic of purulent meningitis with suppurative inflammation and necrosis, microabscess formation, vasculitis with perivascular lymphocytic cuffing, and mononuclear cell infiltration of the vessel walls (Armstrong and Fung, 1993; Gray and Killinger, 1966). Bacteria may be present in necrotic parenchymal lesions rather than in perivascular cuffs. Histological findings in naturally occurring infections of ruminants are also instructive of brainstem encephalitis in humans (Campero et al., 2002; Cordy and Osebold, 1959). In these cases microabscesses are most commonly found near small vessels in the reticular formation of the midbrain, pons, and medulla, but can be present elsewhere. Lesions contain classically activated macrophages expressing iNOS, neutrophils, and few T- or B-lymphocytes (Krueger et al., 1995).

40.4.2 Diagnosis of CNS infection

Diagnosis of CNS infection by *L. monocytogenes* can be challenging due to a lack of pathognomonic clinical features and protean presentations. Thus, while a microbiological diagnosis is being sought, clinical suspicion of neuro-infection in a patient at increased risk for *L. monocytogenes* infection should lead to inclusion of antibiotics active against this organism as part of empiric antimicrobial therapy (Brouwer et al., 2010; Tunkel et al., 2004). Treating clinicians need to possess a high degree of suspicion for *L. monocytogenes* infection, taking clues from a variety of sources. Relevant clues include patient history, for example, consumption of unpasteurized dairy products, underlying conditions that diminish host defenses, for example, malignancies, use of immunosuppressive drugs, chronic illnesses (e.g., liver disease), and extremes of age. In addition, one must have awareness of current foodborne epidemics such as the 2011 cantaloupe-associated outbreak (http://www.cdc.gov/listeria/outbreaks/cantaloup es-jensen-farms/120811/index.html).

Blood cultures before antibiotics should always be performed and are often the only source of a positive culture. Analysis and culture of CSF should be performed as soon as practicable (Straus et al., 2006). However, analysis of CSF leukocytes and chemistries are neither diagnostic for *L. monocytogenes* infection nor are they always indicative of bacterial meningitis (Bouvet et al., 1982; Larsson et al., 1978; Mylonakis et al., 1998; Pollock et al., 1984). Similarly, CSF Gram's stain fails to show organisms in >50% of samples, and even when bacteria are identified they can be reported as cocci or coccobacilli due to bipolar staining, or as Gram negative bacteria due to excessive de-colorization (Amaya-Villar et al., 2010; Bouvet et al., 1982; Koopmans et al., 2013; Lavetter et al., 1971; Mylonakis et al., 1998). Along with other bacterial infections, CSF lactate concentrations are increased in *Listeria* infection and may help differentiate encephalitis/cerebritis caused by *Listeria* from the same syndrome caused by viral pathogens (Cunha et al., 2007). Neuro-imaging using MRI with gadolinium contrast enhancement is a necessary component for diagnosis of rhombencephalitis (Moragas et al., 2011).

The "gold standard" for microbiologic diagnosis is a positive CSF culture which is found in 59–100% of patients in large series (Amaya-Villar et al., 2010; Bouvet et al., 1982; Koopmans et al., 2013; Lavetter et al., 1971; Mylonakis et al., 1998). A presumptive diagnosis for CNS infection can be made if there is a compatible clinical syndrome, reactive CSF, and a positive blood culture for *L. monocytogenes* (Le Monnier et al., 2011). No antigen tests are available and given the low numbers of *L. monocytogenes* in the CSF, this diagnostic approach would likely suffer from poor sensitivity. Serology using Western blot to identify anti-LLO antibodies in patients has been used for presumptive identification of *L. monocytogenes* infection in culture-negative cases of meningitis but is not widely available (Gholizadeh et al., 1997).

Real-time PCR can identify *L. monocytogenes* in CSF, and has been used as part of a multiplex for CNS invasive bacteria or as a single real-time PCR (Chiba et al., 2009; Favaro et al., 2013; Le Monnier et al., 2011). Genes targeted for amplification include the *L. monocytogenes*–specific virulence factors *hly* which encodes LLO (Le Monnier et al., 2011), and *iap* which encodes the p60 invasion-associated protein (Favaro et al., 2013), as well as a *L. monocytogenes*–specific sequence in the 16s rRNA (Chiba et al., 2009). Reported sensitivities are as low as 1 copy, that is, 1 bacterium, for the *hly* PCR and 16 copies for the 16s rRNA target PCR. When tested against a panel of control bacteria, PCR is 100% specific, with the most extensive testing reported for the *hly* and 16s rRNA targets. Importantly, the *hly* product does not amplify closely related hemolysins from *L. ivanovii* or *L. seeligeri* (Le Monnier et al., 2011). Each PCR product correctly identified all *L. monocytogenes* culture-positive samples tested and were negative in non-*Listeria* meningitis cases. In addition, *iap* and *hly* PCRs were positive in some culture-negative specimens suggesting the possibility of greater sensitivity than for culture. Data suggest that amplification of *hly* also correlates with numbers of bacteria present in positive specimens. For example, threshold cycle numbers were significantly higher, indicating fewer bacteria, in patients previously treated with antibiotics ($p = 0.03$) and significantly lower in the presence of a positive Gram's stain ($p = 0.05$) or/and a strong trend for culture ($p = 0.06$) (Le Monnier et al., 2011).

40.4.3 Treatment of CNS infection

L. monocytogenes infections do not occur in sufficient numbers for randomized controlled clinical trials to guide therapy. Thus, current recommendations are based on *in vitro* susceptibility data, animal studies, case series of human patients, and expert opinion (Brouwer et al., 2010; Clauss and Lorber, 2008; Tunkel et al., 2004). Many antibiotics demonstrate *in vitro* activity against *L. monocytogenes*, but most are bacteriostatic *in vivo* and thus are not first-line therapy for CNS infections. In addition, *in vivo* efficacy against intracellular organisms requires that the antibiotic penetrate the intracellular environment as well as the CNS (Hof et al., 1997; Temple and Nahata, 2000). Penicillin-binding protein 3 has been considered the main molecular target of β-lactam antibiotics effective against this organism, however recent data suggest other molecules may be involved (Krawczyk-Balska et al., 2012). *L. monocytogenes* is innately resistant to cephalosporins, clindamycin, and earlier generation fluoroquinolones (i.e., ciprofloxacin) (Marco et al., 2000; Safdar and Armstrong, 2003a). *In vitro* data and animal studies suggest newer fluoroquinolones may be more active, but clinical data are scarce (Michelet et al., 1999; Temple and Nahata, 2000). Thus, ampicillin and penicillin have been considered the mainstays of antimicrobial therapy against *L. monocytogenes* for several decades (Lavetter et al., 1971), with ampicillin used most commonly in clinical practice (Brouwer et al., 2010; Tunkel et al., 2004). Given the potential for relapse of infection, treatment for 3 weeks is recommended for meningitis. Patients with brain abscess or rhombencephalitis are treated for up to 6 weeks with repeated brain imaging used to document resolution of infection prior to cessation of antimicrobial therapy. Trimethoprim/sulfamethoxazole is recommended for penicillin-allergic patients who cannot be desensitized (Brouwer et al., 2010; Michelet et al., 1999; Tunkel et al., 2004). Antimicrobial resistance does not appear to have increased in clinical

isolates of *L. monocytogenes* (Marco et al., 2000; Mylonakis et al., 1998; Safdar and Armstrong, 2003a).

The use of aminoglycosides, typically gentamicin, along with a β-lactam for treating CNS infection with *L. monocytogenes* is controversial. The rationale for adding an aminoglycoside despite their low penetration into the CNS is that these drugs are rapidly bactericidal against extracellular bacteria and could provide synergistic bactericidal activity when combined with a β-lactam (Carryn et al., 2003; Marco et al., 2000; Safdar and Armstrong, 2003a). *In vitro* studies suggest that gentamicin is ineffective against intracellular *L. monocytogenes* in cell lines in which bacteria enter the cytosol rapidly (Carryn et al., 2003), but can potentiate killing by phagocytes able to restrict intracellular bacteria to phagosomes (Drevets et al., 1994). *In vivo* animal experiments show the combination of gentamicin and ampicillin does not produce synergistic bacterial killing in the brain and that gentamicin by itself reduces numbers of bacteria in the bloodstream, but not in the brain (Blanot et al., 1999; Drevets et al., 2001). In patients with CNS infection, Mylonakis et al. (1998) reported 14% mortality in 54 patients treated with ampicillin/penicillin and aminoglycoside compared with 22% mortality in 109 patients treated with ampicillin/penicillin alone. Further statistical analysis was not performed. In contrast, a study by Mitja et al. (2009), including patients with CNS infections and primary bacteremia, reported 33.3% mortality in 33 patients with combined therapy and 14.5% mortality in 69 patients with ampicillin alone (odds ratio (OR), 2.95; 95% confidence interval (CI), 1.10–7.91; $p = 003$). Analysis of predictors of outcome showed renal failure, corticosteroids, and age >65 years significantly contributed to early mortality in univariate and multivariate analyses, whereas combination therapy significantly contributed to mortality in univariate (OR, 8.25; 95% CI, 2.06–33.04; $p = 0.003$) but not multivariate analysis (OR, 3.90; 95% CI, 0.78–19.4; $p = 0.096$). An analysis of patients with meningitis from the Netherlands found that gentamicin use was more common among patients that suffered unfavorable outcomes, but the results did not achieve statistical significance (Koopmans et al., 2013). Similarly, a recent prospective observational study from Spain found use of definitive gentamicin therapy was significantly more common in nonsurvivors (8/12) than in survivors (10/31) (Amaya-Villar et al., 2010). Collectively, the most recent studies either fail to show benefit to combination therapy with an aminoglycoside (Koopmans et al., 2013), or suggest it is harmful (Amaya-Villar et al., 2010; Mitja et al., 2009). Current guidelines advise clinicians to "consider" addition of an aminoglycoside while weighing the risks of renal failure, particularly in patients with underlying renal disease and other co-morbid conditions (Brouwer et al., 2010; Tunkel et al., 2004).

A second controversy centers on the use of corticosteroids as adjunctive anti-inflammatory therapy. Two recent observational studies assessed the impact of steroids on outcomes. Amaya-Villar et al. (2010)

found administration of dexamethasone concomitant to antibiotics was used more frequently in survivors (17/31) than in nonsurvivors (4/12), but the difference was not statistically significant. Koopmans et al. (2013) analyzed two cohorts of patients collected before and after publication of a general recommendation for use of corticosteroids in adults with meningitis. The results showed that although the percentage of patients receiving dexamethasone increased significantly after publication of recommendations, so did unfavorable outcomes, which increased from 27% to 61%. Further analysis showed this was statistically associated with the appearance of a distinct sequence type of *L. monocytogenes* rather than dexamethasone use. Nonetheless, the authors supported the current recommendation that empiric use of dexamethasone should be discontinued once a diagnosis of *L. monocytogenes* has been made (Koopmans et al., 2013; Tunkel et al., 2004).

40.5 Fetal and placental listeriosis

40.5.1 Clinical features

40.5.1.1 Prevalence and impact of listeriosis in pregnancy

Listeriosis in pregnant women is associated with significant neonatal morbidity and mortality, resulting in early pregnancy loss and miscarriage, stillbirth, preterm labor, and neonatal infection (fatality rates of 20–30%) (Lamont et al., 2011; Mylonakis et al., 2002). Mylonakis et al. (2002) reviewed 222 cases of infection in pregnancy, and found abortion or stillbirth in 20% and neonatal sepsis in 68% of surviving neonates. Pregnant women are also disproportionately susceptible to infection: in the large outbreak in Los Angeles County in 1985 associated with contaminated cheese, 65.5% of the cases of listeriosis involved pregnant women or their offspring, and of the nonpregnant adults, 48 of 49 had another predisposing condition (Linnan et al., 1988). The general population incidence of listeriosis is 0.7 per 100,000, which in pregnant women rises to 12 per 100,000 (Southwick and Purich, 1996). It is suggested that this is due to the mild immunosuppression during pregnancy, although it seems possible that adverse outcomes of gestation reveal maternal infection that exists asymptomatically in the nonpregnant population.

During the period from 2009 to 2011, 274 cases of listeriosis in pregnancy (14% of all cases reported) were documented in the United States (CDC, 2013). The relative risk for pregnant women compared with the population as a whole was 10.1, and was 24.1 for pregnant Hispanic women, in agreement with a similar study of 758 listeriosis cases for the period 2004–2007 (Jackson et al., 2010). This study included only listeriosis cases that were documented due to fetal death, or neonatal hospitalization, thus the incidence in pregnancy is almost certainly underestimated. There is reason to expect that the importance of listeriosis in adverse pregnancy outcomes will be increasing. First, *Listeria* is a common contaminant of raw milk (CDC, 2013). As more states consider legalizing raw milk sales to the general public, the possibility for contracting listeriosis during pregnancy will increase. In addition, 46% of the cases of listeriosis in a recent MMWR report (CDC, 2013) were from Hispanic women, most likely due to the consumption of traditional soft (and often homemade) cheese. Given the increasing popularity of unpasteurized dairy products, and the demographics of the US Hispanic population, it seems likely that listeriosis in pregnancy will be an increasing concern in obstetrics. The CDC has recently pointed out that while food safety efforts reduced listeriosis outbreaks in the 1990s, there has not been an improvement in the prevalence since 2001. An improvement in patient education and continued vigilance by regulatory and industry entities is warranted, "from farm to table" (CDC, 2013).

There is a perception that the contribution of *L. monocytogenes* to adverse pregnancy outcomes is underappreciated. Baud and Greub (2011) commented that its role in adverse pregnancy outcomes probably remains underestimated, suggesting that modern diagnostics and rigorous screening may reveal a more significant contribution to adverse outcomes. Lamont et al. (2011) suggested that reluctance to test abortion or miscarriage tissues may contribute to this underestimation. Although it is widely considered that listeriosis in early pregnancy can be catastrophic for the fetus, there have actually been few studies of whether spontaneous miscarriage may be related to *Listeria* infection. The most detailed study of which we are aware (Kaur et al., 2007) indicated that of 61 spontaneous miscarriages analyzed by multiplex PCR, 14% were positive for *Listeria* species, and the confirmed occurrence of *L. monocytogenes* was 3.3%. The latter included both pathogenic and nonpathogenic strains as evaluated with *in vivo* mouse and chick embryo assays. Given the large number of early pregnancy losses of unknown etiology, *Listeria* could have a significant societal as well as economic cost. It has been estimated that the average yearly human exposure to foodborne *L. monocytogenes* may include one episode at $>10^6$ CFU, 4 episodes at up to 10^5 CFU, and nearly 20 exposures to 10^3 CFU (Notermans et al., 1998). It seems likely that a more comprehensive analysis of spontaneous early pregnancy miscarriage tissues would reveal a contribution of listeriosis to early pregnancy loss as well. Finally, the possibility that listeriosis could contribute to a polymicrobial basis of adverse pregnancy outcomes (Cardenas et al., 2011; Czikk et al., 2011) also deserves consideration and experimental investigation.

40.5.1.2 Mechanism and route of placental and fetal infection

The placenta is a known reservoir of *Listeria*, and the organism can be cultured from tissues obtained at term (Janakiraman, 2008). Transplacental transmission of *L. monocytogenes* is likely to be the major route of fetal infection, especially in cases of stillbirth or preterm labor and early-onset perinatal sepsis. Villous disseminated granulomatous lesions and microabscesses, and chorioamnionitis are frequently seen in infected placentas from stillbirths. However, the process by which fetal infection occurs remains incompletely understood. Using a full-term placenta villous explant model, Lecuit et al. (2004) reported that infection of the human syncytiotrophoblast (STB) layer of the chorionic villi, via E-cadherin (CDH1) at the STB apical surface, is the route of bacterial access to the placental villous stroma. Conversely, Robbins et al. (2010) reported that STB do not express surface E-cadherin. Using explants of first-trimester placental villi, these authors concluded that the main route of bacterial access to the fetal circulation is via sporadic infection of the CDH1+ extravillous trophoblasts (EVT) of the trophoblast cell column. It should be noted that CDH1 may not be the sole mechanism by which trophoblast infection takes place. Indeed, mouse CDH1 is not a *L. monocytogenes* receptor, although maternal–fetal infection can take place (Disson et al., 2009; Poulsen et al., 2011). Figure 40-1 summarizes the potential routes by which ingested *Listeria* may infect the fetus.

40.5.1.3 Clinical presentation and outcome of listeriosis in pregnancy

One of the significant challenges in addressing listeriosis during pregnancy is that maternal symptoms may be mild or absent. Even mild symptoms (malaise, nausea, fever) are not readily ascribed to listeriosis,

thus no pathognomonic features can be identified. Maternal fever has been reported to be the most commonly observed symptom (Hasbun et al., 2013; Mylonakis et al., 2002), with clinical presentation typically consisting of nonspecific influenza-like symptoms that may include febrile gastroenteritis. Adverse pregnancy outcomes associated with maternal listeriosis include vaginal bleeding, miscarriage, placental abruption, premature membrane rupture, and preterm labor (Chaudhuri et al., 2012; Jackson et al., 2010). Pregnant women with immunosuppression, corticosteroid use, or comorbidities such as splenectomy, HIV infection, or diabetes are at increased risk for listeriosis (Janakiraman, 2008). Maternal fecal carriage of *Listeria* may be a risk factor for perinatal listeriosis (Gray et al., 1993).

Fetal listeriosis syndrome includes febrile gastroenteritis, pneumonia, and meningitis (including brain abscesses). Bacteremia may or may not be present (Drevets and Bronze, 2008; Mylonakis et al., 2002). The greatest bacterial burden is considered to be in the lung and the gut, suggesting that the fetus is infected through swallowing or inhaling organisms in the amniotic fluid (Baud and Greub, 2011). Fetal infection can result in granulomatosis infantiseptica, with widespread microabscesses and granulomas, particularly in the fetal/neonatal spleen and liver (Baud and Greub, 2011). This presentation is pathognomonic for neonatal *Listeria* infection. Diagnosis of *Listeria* infection should be made from a sterile site, such as amniotic fluid or blood.

The rate of neonatal demise varies among studies. Smith et al. (2009) reported that in 36 cases of listeriosis during pregnancy, 12 (33%) resulted in miscarriage or stillbirth; of the 24 live births, 88% (21) had diagnoses ranging from sepsis to pneumonia and meningitis. The rate of hospitalization in another study (Jackson et al., 2010) was 90.5 % (76/84 infants), with a median hospitalization length of 14 nights (range: 4–50). Overall, the perinatal fatality rate was 14.3%. In addition, neonatal listeriosis is a common cause of neonatal meningitis. In this latter study which evaluated 758 *Listeria* cases during 2004–2007, ethnic minorities, particularly Hispanic women were more than three times more likely to have *Listeria* infection in pregnancy. This finding was consistent with a recent review of cases from 2009 to 2011 (CDC, 2013). A study of 722 pregnant women (McLauchlin, 1990) suspected or confirmed as infected by follow-up culture indicated 10–20% spontaneous abortion, 50% preterm delivery, and a 34% occurrence of fetal distress.

40.5.1.4 Timing of fetoplacental unit susceptibility to infection

It is generally stated that listeriosis in early pregnancy is catastrophic for the fetus. Mylonakis et al. (2002) described case reports in 11 pregnant women, including 9 women who developed symptoms in the third

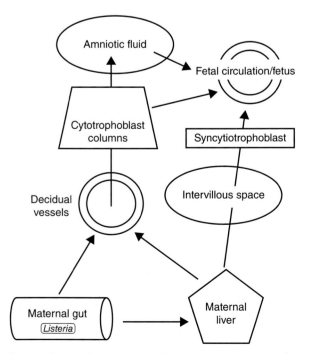

Fig. 40-1 Potential routes by which ingested *Listeria* may infect the fetus.

trimester, and 2 women who developed symptoms in the seventeenth and eighteenth weeks of gestation. These latter two pregnancies ended in spontaneous abortion and death of a preterm baby. Since only one of the third trimester cases resulted in fetal demise, these results suggest that infection earlier in pregnancy may be more deleterious to neonatal health than later infection. Although it was stated (Cherubin et al., 1991) that fetal outcome is more severe when infection occurs earlier in pregnancy, it was not clear in that report exactly when or how maternal infection was determined to have occurred in early gestation. Because sporadic pregnancy loss is not uncommon in early pregnancy, and morning sickness often provides a background level of nausea and malaise, women may be unaware that an early pregnancy loss was associated with a clinically unapparent case of listeriosis.

It is not known whether maternal chronic or pregestational exposure to *L. monocytogenes* is deleterious to pregnancy success. Rappaport et al. (1960) suggested that reproductive tract infection was associated with a history of recurrent abortion, but other studies failed to isolate *Listeria* from the cervix or endometrium of 86 patients with 2 or more fetal losses (Manganiello and Yearke, 1991) or other poor obstetrical outcomes (Ansbacher et al., 1966; Lawler et al., 1964; Macnaughton, 1962). There is some evidence for a higher seroconversion to *L. monocytogenes* in women with a history of spontaneous abortion (Aljicevic et al., 2005), however the sample size in this report was small. Likewise, there are a number of reports (typically individual case studies) (Tridente et al., 1998) or incomplete reports (Romana et al., 1989) suggesting that latent listeriosis may be associated with habitual abortion.

Perinatal infection is referred to as early onset if the neonate demonstrates infection within 5 days of birth, and is thought to be associated with hematogenous (transplacental) infection. It is associated with chorioamnionitis, rapidly presented clinical features including septicemia, pneumonia, or meningitis and a high rate of perinatal death (20–50%) (Boucher and Yonekura, 1986). When maternal diagnosis of listeriosis is made based on placental, blood, or cervical culture, 68% of neonates developed listeriosis, and of that infected group, the mortality rate was 24.5%, with 12.7% exhibiting long-term neurological outcomes, and the remaining 68% having uneventful recovery.

Late-onset neonatal listeriosis is diagnosed 1–4 weeks after delivery and is thought to occur due to colonization of the fetus during passage through the birth canal. In late-onset infection, *Listeria* is thought to spread to the vaginal canal from the lower GI tract (Lamont et al., 2011). A study of early-onset listeriosis noted that high maternal leukocyte count, fetal tachycardia but decreased heart rate variability, and decreased fetal heart rate accelerations intrapartum was associated with a complicated outcome for the neonate. Intrapartum administration of

antibiotics decreased fetal morbidity and mortality (Hasbun et al., 2013).

40.5.2 Diagnosis of maternal infection

Because of the high rate of fetal and neonatal morbidity and mortality, an improvement in early diagnosis and treatment would likely improve neonatal patient outcomes. One of the significant challenges in addressing listeriosis in pregnancy is that maternal symptoms of disease may be mild or absent. Maternal leukocytosis is noted. Culture of maternal fecal samples may be of limited value, with variable recovery rates reported from 1.1% to 12% (Gray et al., 1993; Kampelmacher et al., 1972; Lamont and Postlethwaite, 1986), and relatively low sensitivity to standard microbiological methods (Lamont et al., 2011). Maternal fever has been reported to be the most commonly observed symptom (Hasbun et al., 2013). For this reason, it has been suggested that blood culture should be considered when women present with fever, influenza-like symptoms, or GI distress, recognizing that these are not in and among themselves diagnostic. Correlation with a diet history can be instructive, particularly with a history of consumption of raw vegetables, uncooked meat, or unpasteurized milk products. For prevention, pregnant women should receive counseling on the risk of unpasteurized dairy products, uncooked or ready-to-eat meat products, and other processed foods such as paté or smoked fish.

Diagnostic tests for maternal listeriosis include PCR of fecal or blood samples, serology, and culture of stool or blood. Culture is currently considered the gold standard for diagnosis (Baud and Greub, 2011). Culture of amniotic fluid, placental tissues, meconium, and tracheal aspirates or ear swabs are all used (Lamont et al., 2011). Bacterial isolates may be subjected to further phenotypic analysis, including PCR, for identification at the species level, particularly for identifying possible sources of an outbreak, such as contaminated food.

40.5.3 Diagnosis of fetal infection

The diagnosis of fetal listeriosis in advance of an adverse pregnancy outcome is challenging. Reduced fetal movements, and an abnormal/accelerated heart rate have been described with antenatal fetal listeriosis (Lennon et al., 1984; Quinlivan et al., 1998). A recent study (Hasbun et al., 2013) that evaluated a case of fetal listeriosis reported that the pregnant patient reported reduced fetal movement, and subsequent ultrasound revealed features of bowel inflammation. Evaluation of middle cerebral artery blood flow ruled out fetal anemia, and amniocentesis in this case confirmed the presence of Gram-positive organisms, leading to a diagnosis of chorioamnionitis with intact membranes. Cesarean section allowed initiation of ampicillin treatment of the neonate and neonatal blood culture confirmed fetal listeriosis. These authors concluded that observation of reduced fetal movement and bowel inflammation as markers of intrauterine compromise, coupled with exclusion of fetal anemia and positive infection noted upon amniocentesis, suggest a possible approach for antenatal diagnosis of congenital listeriosis.

40.5.4 Treatment in pregnancy

The treatment of *L. monocytogenes* is challenging in many aspects, beginning with the lack of a prompt diagnosis in pregnant women. After the diagnosis is made, the antibiotic of choice must cross the placenta and reach adequate intracellular concentrations in fetal tissues. Animal studies have shown the presence of *L. monocytogenes* in the fetal compartment as soon as 2 days post-inoculation in guinea pigs (Williams et al., 2011), and as soon as 7 days post-inoculation in cynomolgus monkeys (Golos et al., 2013). Although such data are missing for humans, these animal models suggest rapid infection of the reproductive tract and passage to the fetal compartment, and thus treatment should be instituted in as timely a manner as possible for optimal results.

In vitro, *L. monocytogenes* is resistant to cephalosporins but sensitive to many antibiotics, including penicillin G, ampicillin, erythromycin, sulfamethoxazole, trimethoprim, chloramphenicol, rifampin, tetracycline, and aminoglycosides (Janakiraman, 2008; Lamont et al., 2011; Posfay-Barbe and Wald, 2009). However, erythromycin does not achieve appropriate levels in the fetal compartment (Heikkinen et al., 2000).

As discussed in Section 40.4.3, ampicillin is the drug of choice for treatment. The recommended dose is 6 g/day divided into 2 g every 8 hours. If the patient is allergic to penicillin, then trimethoprim/ sulfamethoxazole is the recommended choice for treatment (Temple and Nahata, 2000). However, only patients with true anaphylactic or angioedema reactions should avoid trimethoprim/sulfamethoxazole (Hof et al., 1997) because, as an anti-folate agent, it is potentially teratogenic in early pregnancy and may cause hyperbilirubinemia if used immediately before delivery and hemolytic anemia in males with glucose-6-phosphate deficiency (Janakiraman, 2008; Lamont et al., 2011). The recommended dose for TMP/SMX

is 200–320 mg for the TMP component (Mylonakis et al., 2002; Temple and Nahata, 2000). Treatment should last a minimum of 14 days, although some studies suggest 21 days depending on the status of the immune system at the time of infection, with longer treatment being appropriate for immunosuppressed patients (Lamont et al., 2011).

40.6 Prevention and control

L. monocytogenes is ubiquitous in nature. It has been isolated from moist decaying vegetation, sewage, river water, manure, and silage among others places (Ivanek et al., 2007; Schlech et al., 1983; Watkins and Sleath, 1981; Welshimer and Donker-Voet, 1971). This bacterium has the ability to survive under a large variety of environmental conditions. Consequently, it is not unusual to find it in the food chain.

Prevention of *Listeria* infections requires careful attention to food-handling practices both in the production phase and in the retail and food service environment. Control of infection is made more difficult because the organism can grow at a wide range of temperature, pH, and salinity. It can also grow well in microaerophilic environments. Pasteurization of milk is effective in controlling risk for infection although cases of infection following ingestion of cheese made from pasteurized milk have been reported implying contamination of the food after processing or a pasteurization failure. Possible mechanisms to control infection in retail and food service environments include addition of bio-preservatives and antimicrobial peptides, maintenance of temperature below 4°C, use of antimicrobial packaging, and thorough reheating of food known to be at increased risk for contamination (Lianou and Sofos, 2007).

Education about high-risk foods, the various sources of contamination, the biology of the pathogen, and basic hygienic rules is one of the premises for reducing the risk of human infection. In particular, pregnant women and immunocompromised individuals should receive counseling about high-risk foods. On the other hand, surveillance programs serve to rapidly detect sources of contaminated food or nascent outbreaks of listeriosis, expediting the application of control measures to curb outbreaks and increasing the awareness of health professionals for early detection of new cases.

References

Abachin, E., Poyart, C., Pellegrini, E., Milohanic, E., Fiedler, F., Berche, P., and Trieu-Cuot, P. 2002. Formation of D-alanyl-lipoteichoic acid is required for adhesion and virulence of *Listeria monocytogenes*. *Mol. Microbiol.*, 43, 1–14.

Alberti-Segui, C., Goeden, K.R., and Higgins, D.E. 2007. Differential function of *Listeria monocytogenes* listeriolysin O and phospholipases C in vacuolar dissolution following cell-to-cell spread. *Cell. Microbiol.*, 9, 179–195.

Aljicevic, M., Beslagic, E., Zvizdic, S., Hamzic, S., and Mahmutovic, S. 2005. *Listeria monocytogenes* in women of reproductive age. *Med. Arh.*, 59, 297–298.

Allerberger, F., and Wagner, M. 2010. Listeriosis: A resurgent foodborne infection. *Clin. Microbiol. Infect.*, 16, 16–23.

Amaya-Villar, R., Garcia-Cabrera, E., Sulleiro-Igual, E., Fernandez-Viladrich, P., Fontanals-Aymerich, D., Catalan-Alonso, P., Rodrigo-Gonzalo de Liria, C., Coloma-Conde, A., Grill-Diaz, F., Guerrero-Espejo, A., Pachon, J., and Prats-Pastor, G. 2010. Three-year multicenter surveillance of community-acquired *Listeria monocytogenes* meningitis in adults. *BMC Infect. Dis.*, 10, 324.

Ansbacher, R., Borchardt, K.A., Hannegan, M.W., and Boyson, W.A. 1966. Clinical investigation of *Listeria monocytogenes* as a possible cause of human fetal wastage. *Am. J. Obstet. Gynecol.*, 94, 386–390.

Antal, E.A., Loberg, E.M., Bracht, P., Melby, K.K., and Maehlen, J. 2001. Evidence for intraaxonal spread of *Listeria monocytogenes* from the periphery to the central nervous system. *Brain Pathol.*, 11, 432–438.

Antal, E.A., Loberg, E.M., Dietrichs, E., and Maehlen, J. 2005. Neuropathological findings in 9 cases of *Listeria monocytogenes* brain stem encephalitis. *Brain Pathol.*, 15, 187–191.

Aouaj, Y., Spanjaard, L., van Leeuwen, N., and Dankert, J. 2002. *Listeria monocytogenes* meningitis: Serotype distribution and patient characteristics in The Netherlands, 1976–95. *Epidemiol. Infect.*, 128, 405–409.

Armstrong, R.W., and Fung, P.C. 1993. Brainstem encephalitis (rhombencephalitis) due to *Listeria monocytogenes*: Case report and review. *Clin. Infect. Dis.*, 16, 689–702.

Arous, S., Buchrieser, C., Folio, P., Glaser, P., Namane, A., Hebraud, M., and Hechard, Y. 2004. Global analysis of gene expression in an *rpoN* mutant of *Listeria monocytogenes*. *Microbiology*, 150, 1581–1590.

Badovinac, V.P., Porter, B.B., and Harty, J.T. 2002. Programmed contraction of CD8(+) T cells after infection. *Nat. Immunol.*, 3, 619–626.

Bahjat, K.S., Meyer-Morse, N., Lemmens, E.E., Shugart, J.A., Dubensky, T.W., Brockstedt, D.G., and Portnoy, D.A. 2009. Suppression of cell-mediated immunity following recognition of phagosome-confined bacteria. *PLoS Pathog.*, 5, e1000568.

Barbuddhe, S.B., Malik, S.V., Kumar, J.A., Kalorey, D.R., and Chakraborty, T. 2012. Epidemiology and risk management of listeriosis in India. *Int. J. Food. Microbiol.*, 154, 113–118.

Bartt, R. 2000. *Listeria* and atypical presentations of *Listeria* in the central nervous system. *Semin. Neurol.*, 20, 361–373.

Baud, D., and Greub, G. 2011. Intracellular bacteria and adverse pregnancy outcomes. *Clin. Microbiol. Infect.*, 17, 1312–1322.

Beauregard, K.E., Lee, K.-D., Collier, R.J., and Swanson, J.A. 1997. pH-dependent perforation of macrophage phagosomes by listeriolysin O from *Listeria monocytogenes*. *J. Exp. Med.*, 186, 1159–1163.

Begley, M., Hill, C., and Gahan, C.G. 2003. Identification and disruption of *btlA*, a locus involved in bile tolerance and general stress resistance in *Listeria monocytogenes*. *FEMS Microbiol. Lett.*, 218, 31–38.

Begley, M., Sleator, R.D., Gahan, C.G., and Hill, C. 2005. Contribution of three bile-associated loci, *bsh*, *pva*, and *btlB*, to gastrointestinal persistence and bile tolerance of *Listeria monocytogenes*. *Infect. Immun.*, 73, 894–904.

Bennion, J.R., Sorvillo, F., Wise, M.E., Krishna, S., and Mascola, L. 2008. Decreasing listeriosis mortality in the United States, 1990–2005. *Clin. Infect. Dis.*, 47, 867–874.

Bhardwaj, V., Kanagawa, O., Swanson, P.E., and Unanue, E.R. 1998. Chronic *Listeria* infection in SCID mice: Requirements for the carrier state and the dual role of T cells in transferring protection or suppression. *J. Immunol.*, 160, 376–384.

Bigot, A., Pagniez, H., Botton, E., Frehel, C., Dubail, I., Jacquet, C., Charbit, A., and Raynaud, C. 2005. Role of FliF and FliI of *Listeria monocytogenes* in flagellar assembly and pathogenicity. *Infect. Immun.*, 73, 5530–5539.

Bitar, A.P., Cao, M., and Marquis, H. 2008. The metalloprotease of *Listeria monocytogenes* is activated by intramolecular autocatalysis. *J. Bacteriol.*, 190, 107–111.

Blanot, S., Boumaila, C., and Berche, P. 1999. Intracerebral activity of antibiotics against *Listeria monocytogenes* during experimental rhombencephalitis. *J. Antimicrob. Chemother.*, 44, 565–568.

Bodro, M., and Paeterson, D.L. 2013. Listeriosis in patients receiving biologic therapies. *Eur. J. Microbiol. Infect. Dis.*, 32, 1225–1230.

Bohne, J., Sokolovic, Z., and Goebel, W. 1994. Transcriptional regulation of *prfA* and PrfA-regulated virulence genes in *Listeria monocytogenes*. *Mol. Microbiol.*, 11, 1141–1150.

Boneca, I.G., Dussurget, O., Cabanes, D., Nahori, M.A., Sousa, S., Lecuit, M., Psylinakis, E., Bouriotis, V., Hugot, J.P., Giovannini, M., Coyle, A., Bertin, J., Namane, A., Rousselle, J.C., Cayet, N., Prevost, M.C., Balloy, V., Chignard, M., Philpott, D.J., Cossart, P., and Girardin, S.E. 2007. A critical role for peptidoglycan N-deacetylation in *Listeria* evasion from the host innate immune system. *Proc. Natl. Acad. Sci. USA*, 104, 997–1002.

Boucher, M., and Yonekura, M.L. 1986. Perinatal listeriosis (early-onset): Correlation of antenatal manifestations and neonatal outcome. *Obstet. Gynecol.*, 68, 593–597.

Bouvet, E., Suter, F., Gibert, C., Witchitz, J.L., Bazin, C., and Vachon, F. 1982. Severe meningitis due to *Listeria monocytogenes*. A review of 40 cases in adults. *Scand. J. Infect. Dis.*, 14, 267–270.

Bowman, J.P., Hages, E., Nilsson, R.E., Kocharunchitt, C., and Ross, T. 2012. Investigation of the *Listeria monocytogenes* Scott A acid tolerance response and associated physiological and phenotypic features via whole proteome analysis. *J. Proteome. Res.*, 11, 2409–2426.

Briers, Y., Klumpp, J., Schuppler, M., and Loessner, M.J. 2011. Genome sequence of *Listeria monocytogenes* Scott A, a clinical isolate from a food-borne listeriosis outbreak. *J. Bacteriol.*, 193, 4284–4285.

Brouwer, M.C., Tunkel, A.R., and van de Beek, D. 2010. Epidemiology, diagnosis, and antimicrobial treatment of acute bacterial meningitis. *Clin. Microbiol. Rev.*, 23, 467–492.

Buchrieser, C., Rusniok, C., Garrido, P., Hain, T., Scortti, M., Lampidis, R., Karst, U., Chakraborty, T., Cossart, P., Kreft, J., Vazquez-Boland, J.A., Goebel, W., and Glaser, P. 2011. Complete genome sequence of the animal pathogen *Listeria ivanovii*, which provides insights into host specificities and evolution of the genus *Listeria*. *J. Bacteriol.*, 193, 6787–6788.

Buchrieser, C., Rusniok, C., Kunst, F., Cossart, P., and Glaser, P. 2003. Comparison of the genome sequences of *Listeria monocytogenes* and *Listeria innocua*: Clues for evolution and pathogenicity. *FEMS Immunol. Med. Microbiol.*, 35, 207–213.

Camejo, A., Buchrieser, C., Couve, E., Carvalho, F., Reis, O., Ferreira, P., Sousa, S., Cossart, P., and Cabanes, D. 2009. In vivo transcriptional profiling of *Listeria monocytogenes* and mutagenesis identify new virulence factors involved in infection. *PLoS Pathog.*, 5, e1000449.

Camilli, A., Tilney, L.G., and Portnoy, D.A. 1993. Dual roles of *plcA* in *Listeria monocytogenes* pathogenesis. *Mol. Microbiol.*, 8, 143–157.

Campero, C.M., Odeon, A.C., Cipolla, A.L., Moore, D.P., Poso, M.A., and Odriozola, E. 2002. Demonstration of *Listeria monocytogenes* by immunohistochemistry in formalin-fixed brain tissues from natural cases of ovine and bovine encephalitis. *J. Vet. Med. B. Infect. Dis. Vet. Public Health*, 49, 379–383.

Cardenas, I., Mor, G., Aldo, P., Lang, S.M., Stabach, P., Sharp, A., Romero, R., Mazaki-Tovi, S., Gervasi, M., and Means, R.E. 2011. Placental viral infection sensitizes to endotoxin-induced pre-term labor: A double hit hypothesis. *Am. J. Reprod. Immunol.*, 65, 110–117.

Carr, K.D., Sieve, A.N., Indramohan, M., Break, T.J., Lee, S., and Berg, R.E. 2011. Specific depletion reveals a novel role for neutrophil-mediated protection in the liver during *Listeria monocytogenes* infection. *Eur. J. Immunol.*, 41, 2666–2676.

Carryn, S., Van Bambeke, F., Mingeot-Leclercq, M.-P., and Tulkens, P.M. 2003. Activity of {beta}-lactams (ampicillin, meropenem), gentamicin, azithromycin and moxifloxacin against intracellular *Listeria monocytogenes* in a 24 h THP-1 human macrophage model. *J. Antimicrob. Chemother.*, 51, 1051–1052.

Centers for Disease Control and Prevention. 2011a. Multistate outbreak of listeriosis associated wiht Jensen Farms cantaloupe–United States, August-September 2011. *Morb. Mortal. Wkly. Rep.*, 60, 1357–1358.

Centers for Disease Control and Prevention. 2011b. Vital signs: Incidence and trends of infection with pathogens transmitted commonly through food-foodborne diseases active surveillance network, 10 U.S. sites, 1996–2010. *Morb. Mortal. Wkly. Rep.*, 60, 749–755.

Centers for Disease Control and Prevention. 2013. Vital signs: *Listeria* illnesses, deaths, and outbreaks–United States, 2009–2011. *MMWR Morb. Mortal. Wkly. Rep.*, 62, 448–452.

Chakraborty, T., Leimeister-Wächter, M., Domann, E., Hartl, M., Goebel, W., Nichterlein, T., and Notermans, S. 1992. Coordinate regulation of virulence genes in *Listeria monocytogenes* requires the product of the *prfA* gene. *J. Bacteriol.*, 174, 568–574.

Chaudhuri, K., Chang, Q.C., Tan, E.K., and Yong, E.L. 2012. Listeriosis in pregnancy with placental abruption. *J. Obstet. Gynaecol.*, 32, 594.

Chen, I., and Dubnau, D. 2004. DNA uptake during bacterial transformation. *Nat. Rev. Microbiol.*, 2, 241–249.

Cherubin, C.E., Appleman, M.D., Heseltine, P.N., Khayr, W., and Stratton, C.W. 1991. Epidemiological spectrum and current treatment of listeriosis. *Rev. Infect. Dis.*, 13, 1108–1114.

Chiba, N., Murayama, S., Morozumi, M., Nakayama, E., Okada, T., Iwata, S., Sunakawa, K., and Ubukata, K. 2009. Rapid detection of eight causative pathogens for the diagnosis of bacterial meningitis by real-time PCR. *J. Infect. Chemother.*, 15, 92–98.

Chico-Calero, I., Suarez, M., Gonzalez-Zorn, B., Scortti, M., Slaghuis, J., and Goebel, W.; European *Listeria* Genome Consortium, and Vazquez-Boland, J.A. 2002. Hpt, a bacterial homolog of the microsomal glucose- 6-phosphate translocase, mediates rapid intracellular proliferation in *Listeria*. *Proc. Natl. Acad. Sci. USA*, 99, 431–436.

Clark, C.G., Farber, J., Pagotto, F., Ciampa, N., Dore, K., Nadon, C., Bernard, K., Ng, L.K., and Cphln, A.T. 2010. Surveillance for *Listeria monocytogenes* and listeriosis, 1995–2004. *Epidemiol. Infect.*, 138, 559–572.

Clauss, H.E., and Lorber, B. 2008. Central nervous system infection with *Listeria monocytogenes*. *Curr. Infect. Dis. Rep.*, 10, 300–306.

Cordy, D.R., and Osebold, J.W. 1959. The neuropathogenesis of *Listeria* encephalomyelitis in sheep and mice. *J. Infect. Dis.*, 104, 164–173.

Cossart, P. 2007. Listeriology (1926–2007): The rise of a model pathogen. *Microbes. Infect.*, 9, 1143–1146.

Cossart, P., Vicente, M.F., Mengaud, J., Baquero, F., Perez-Diaz, J.C., and Berche, P. 1989. Listeriolysin O is essential for virulence of *Listeria monocytogenes*: Direct evidence obtained by gene complementation. *Infect. Immun.*, 57, 3629–3636.

Cotter, P.D., Gahan, C.G., and Hill, C. 2001. A glutamate decarboxylase system protects *Listeria monocytogenes* in gastric fluid. *Mol. Microbiol.*, 40, 465–475.

Crimmins, G.T., Herskovits, A.A., Rehder, K., Sivick, K.E., Lauer, P., Dubensky Jr., T.W., and Portnoy, D.A. 2008. *Listeria monocytogenes* multidrug resistance transporters activate a cytosolic surveillance pathway of innate immunity. *Proc. Natl. Acad. Sci. USA*, 105, 10191–10196.

Cunha, B.A., Fatehpuria, R., and Eisenstein, L.E. 2007. *Listeria monocytogenes* encephalitis mimicking Herpes Simplex virus encephalitis: The differential diagnostic importance of cerebrospinal fluid lactic acid levels. *Heart Lung.*, 36, 226–231.

Czikk, M.J., McCarthy, F.P., and Murphy, K.E. 2011. Chorioamnionitis: From pathogenesis to treatment. *Clin. Microbiol. Infect.*, 17, 1304–1311.

Dalton, C.B., Merritt, T.D., Unicomb, L.E., and Kirk, M.D. 2011. A national case-control study of risk factors for listeriosis. *Epidemiol. Infect.*, 139, 437–445.

De Jesus, A.J., and Whiting, R.C. 2003. Thermal inactivation, growth, and survival studies of *Listeria monocytogenes* strains belonging to three distinct genotypic lineages. *J. Food. Prot.*, 66, 1611–1617.

Decatur, A.L., and Portnoy, D.A. 2000. A PEST-like sequence in listeriolysin O essential for *Listeria monocytogenes* pathogenicity. *Science*, 290, 992–995.

den Bakker, H.C., Cummings, C.A., Ferreira, V., Vatta, P., Orsi, R.H., Degoricija, L., Barker, M., Petrauskene, O., Furtado, M.R., and Wiedmann, M. 2010. Comparative genomics of the bacterial genus *Listeria*: Genome evolution is characterized by limited gene acquisition and limited gene loss. *BMC Genomics.*, 11, 688.

den Bakker, H.C., Desjardins, C.A., Griggs, A.D., Peters, J.E., Zeng, Q., Young, S.K., Kodira, C.D., Yandava, C., Hepburn, T.A., Haas, B.J., Birren, B.W., and Wiedmann, M. 2013a. Evolutionary dynamics of the accessory genome of *Listeria monocytogenes*. *PLoS One*, 8, e67511.

den Bakker, H.C., Manuel, C.S., Fortes, E.D., Wiedmann, M., and Nightingale, K.K. 2013b. Genome sequencing identifies *Listeria fleischmannii* subsp. *coloradensis* subsp. *nov.*, a novel *Listeria fleischmannii* subspecies isolated from a ranch in Colorado. *Int. J. Syst. Evol. Microbiol.*, 63, 3257–3268.

Deng, X., Phillippy, A.M., Li, Z., Salzberg, S.L., and Zhang, W. 2010. Probing the pan-genome of *Listeria monocytogenes*: New insights into intraspecific niche expansion and genomic diversification. *BMC Genomics*, 11, 500.

Disson, O., Nikitas, G., Grayo, S., Dussurget, O., Cossart, P., and Lecuit, M. 2009. Modeling human listeriosis in natural and genetically engineered animals. *Nat. Protoc.*, 4, 799–810.

Doumith, M., Cazalet, C., Simoes, N., Frangeul, L., Jacquet, C., Kunst, F., Martin, P., Cossart, P., Glaser, P., and Buchrieser, C. 2004. New aspects regarding evolution and virulence of *Listeria monocytogenes* revealed by comparative genomics and DNA arrays. *Infect. Immun.*, 72, 1072–1083.

Dramsi, S., Biswas, I., Maguin, E., Braun, L., Mastroeni, P., and Cossart, P. 1995. Entry of *Listeria monocytogenes* into hepatocytes requires expression of InlB, a surface protein of the internalin multigene family. *Mol. Microbiol.*, 16, 251–261.

Dramsi, S., Levi, S., Triller, A., and Cossart, P. 1998. Entry of *Listeria monocytogenes* into neurons occurs by cell-to-cell spread: An in vitro study. *Infect. Immun.*, 66, 4461–4468.

Drevets, D.A., and Bronze, M.S. 2008. *Listeria monocytogenes*: Epidemiology, human disease, and mechanisms of brain invasion. *FEMS Immunol. Med. Microbiol.*, 53, 151–165.

Drevets, D.A., Canono, B.P., Leenen, P.J., and Campbell, P.A. 1994. Gentamicin kills intracellular *Listeria monocytogenes*. *Infect. Immun.*, 62, 2222–2228.

Drevets, D.A., Dillon, M.J., Schawang, J.S., Van Rooijen, N., Ehrchen, J., Sunderkotter, C., and Leenen, P.J. 2004. The Ly-6Chigh monocyte subpopulation transports *Listeria monocytogenes* into the brain during systemic infection of mice. *J. Immunol.*, 172, 4418–4424.

Drevets, D.A., Jelinek, T.A., and Freitag, N.E. 2001. *Listeria monocytogenes*-infected phagocytes can initiate central nervous system infection in mice. *Infect. Immun.*, 69, 1344–1350.

Drevets, D.A., Sawyer, R.T., Potter, T.A., and Campbell, P.A. 1995. *Listeria monocytogenes* infects human endothelial cells by two distinct mechanisms. *Infect. Immun.*, 63, 4268–4276.

Dumas, E., Meunier, B., Berdague, J.L., Chambon, C., Desvaux, M., and Hebraud, M. 2009. The origin of *Listeria monocytogenes* 4b isolates is signified by subproteomic profiling. *Biochim. Biophys. Acta*, 1794, 1530–1536.

Dussurget, O., Cabanes, D., Dehoux, P., Lecuit, M., Buchrieser, C., Glaser, P., and Cossart, P. 2002. *Listeria monocytogenes* bile salt hydrolase is a PrfA-regulated virulence factor involved in the intestinal and hepatic phases of listeriosis. *Mol. Microbiol.*, 45, 1095–1106.

Edelson, B.T., Bradstreet, T.R., Hildner, K., Carrero, J.A., Frederick, K.E., Kc, W., Belizaire, R., Aoshi, T., Schreiber, R.D., Miller, M.J., Murphy, T.L., Unanue, E.R., and Murphy, K.M. 2011. CD8alpha(+) dendritic cells are an obligate cellular entry point for productive infection by *Listeria monocytogenes*. *Immunity*, 35, 236–248.

Elhanafi, D., Dutta, V., and Kathariou, S. 2010. Genetic characterization of plasmid-associated benzalkonium chloride resistance determinants in a *Listeria monocytogenes* strain from the 1998–1999 outbreak. *Appl. Environ. Microbiol.*, 76, 8231–8238.

Endrikat, S., Gallagher, D., Pouillot, R., Hicks Quessenberry, H., Labarre, D., Schroeder, C.M., and Kause, J. 2010. A comparative risk assessment for *Listeria monocytogenes* in prepackaged versus retail-sliced deli meat. *J. Food. Prot.*, 73, 612–619.

Farber, J., and Peterkin, P.I. 1991. *Listeria monocytogenes*, a foodborne pathogen. *Microbiol. Rev.*, 55, 476–511.

Favaro, M., Savini, V., Favalli, C., and Fontana, C. 2013. A multi-target real-time PCR assay for rapid identification of meningitis-associated microorganisms. *Mol. Biotechnol.*, 53, 74–79.

Forster, B.M., Bitar, A.P., Slepkov, E.R., Kota, K.J., Sondermann, H., and Marquis, H. 2011. The metalloprotease of *Listeria monocytogenes* is regulated by pH. *J. Bacteriol.*, 193, 5090–5097.

Forster, B.M., and Marquis, H. 2012. Protein transport across the cell wall of monoderm Gram-positive bacteria. *Mol. Microbiol.*, 84, 405–413.

Freitag, N.E., Port, G.C., and Miner, M.D. 2009. *Listeria monocytogenes*: From saprophyte to intracellular pathogen. *Nat. Rev. Microbiol.*, 7, 623–626.

Gahan, C.G., and Hill, C. 2005. Gastrointestinal phase of *Listeria monocytogenes* infection. *J. Appl. Microbiol.*, 98, 1345–1353.

Gaillard, J.-L., Berche, P., Frehel, C., Gouin, E., and Cossart, P. 1991. Entry of *L. monocytogenes* into cells is mediated by internalin, a repeat protein reminiscent of surface antigens from Gram-positive cocci. *Cell*, 65, 1127–1141.

Garcia-Del Portillo, F., and Pucciarelli, M.G. 2012. Remodeling of the *Listeria monocytogenes* cell wall inside eukaryotic cells. *Commun. Integr. Biol.*, 5, 160–162.

Gedde, M.M., Higgins, D.E., Tilney, L.G., and Portnoy, D.A. 2000. Role of listeriolysin O in cell-to-cell spread of *Listeria monocytogenes*. *Infect. Immun.*, 68, 999–1003.

Gholizadeh, Y., Juvin, M., Beretti, J.L., Berche, P., and Gaillard, J.L. 1997. Culture-negative listeriosis of the central nervous system diagnosed by detection of antibodies to listeriolysin O. *Eur. J. Clin. Microbiol. Infect. Dis.*, 16, 176–178.

Gilmour, M.W., Graham, M., Van Domselaar, G., Tyler, S., Kent, H., Trout-Yakel, K.M., Larios, O., Allen, V., Lee, B., and Nadon, C. 2010. High-throughput genome sequencing of two *Listeria monocytogenes* clinical isolates during a large foodborne outbreak. *BMC Genomics*, 11, 120.

Glaser, P., Frangeul, L., Buchrieser, C., Rusniok, C., Amend, A., Baquero, F., Berche, P., Bloecker, H., Brandt, P., Chakraborty, T., Charbit, A., Chetouani, F., Couve, E., de Daruvar, A., Dehoux, P., Domann, E., Dominguez-Bernal, G., Duchaud, E., Durant, L., Dussurget, O., Entian, K.D., Fsihi, H., Garcia-del Portillo, F., Garrido, P., Gautier, L., Goebel, W., Gomez-Lopez, N., Hain, T., Hauf, J., Jackson, D., Jones, L.M., Kaerst, U., Kreft, J., Kuhn, M., Kunst, F., Kurapkat, G., Madueno, E., Maitournam, A., Vicente, J.M., Ng, E., Nedjari, H., Nordsiek, G., Novella, S., de Pablos, B., Perez-Diaz, J.C., Purcell, R., Remmel, B., Rose, M., Schlueter, T., Simoes, N., Tierrez, A., Vazquez-Boland, J.A., Voss, H., Wehland, J., and Cossart, P. 2001. Comparative genomics of *Listeria* species. *Science*, 294, 849–852.

Glomski, I.J., Gedde, M.M., Tsang, A.W., Swanson, J.A., and Portnoy, D.A. 2002. The *Listeria monocytogenes* hemolysin has an acidic pH optimum to compartmentalize activity and prevent damage to infected host cells. *J. Cell. Biol.*, 156, 1029–1038.

Goldfine, H., and Wadsworth, S.J. 2002. Macrophage intracellular signaling induced by *Listeria monocytogenes*. *Microbes. Infect.*, 4, 1335–1343.

Golos, T.G., Bondarenko, G.I., Durning, M., Faith, N.G., Simmons, H.A., Mejia, A., and Czuprynski, C.J. 2013. Acute fetal demise with early nonhuman primate pregancy infection with *Listeria monocytogenes*. *Placenta*, 34, A43.

Goulet, V., Hebert, M., Hedberg, C., Laurent, E., Vaillant, V., De Valk, H., and Desencio, J.C. 2012. Incidence of listeriosis and related mortality among groups at risk of acquiring listeriosis. *Clin. Infect. Dis.*, 54, 652–660.

Gray, J.W., Barrett, J.F., Pedler, S.J., and Lind, T. 1993. Faecal carriage of *listeria* during pregnancy. *Br. J. Obstet. Gynaecol.*, 100, 873–874.

Gray, M.L., and Killinger, A.H. 1966. *Listeria monocytogenes* and listeric infections. *Bacteriol. Rev.*, 30, 309–382.

Gregory, S.H., Sagnimeni, A.J., and Wing, E.J. 1996. Bacteria in the bloodstream are trapped in the liver and killed by immigrating neutrophils. *J. Immunol.*, 157, 2514–2520.

Greiffenberg, L., Goebel, W., Kim, K.S., Weiglein, I., Bubert, A., Engelbrecht, F., Stins, M., and Kuhn, M. 1998. Interaction of *Listeria monocytogenes* with human brain microvascular endothelial cells: InlB-dependent invasion, long-term intracellular growth, and spread from macrophages to endothelial cells. *Infect. Immun.*, 66, 5260–5267.

Gründler, T., Quednau, N., Stump, C., Orian-Rousseau, V., Ishikawa, H., Wolburg, H., Schroten, H., Tenenbaum, T., and Schwerk, C. 2013. The surface proteins InlA and InlB are interdependently required for polar basolateral invasion by *Listeria monocytogenes* in a human model of the blood–cerebrospinal fluid barrier. *Microbes. Infect.*, 15, 291–301.

Grundling, A., Gonzalez, M.D., and Higgins, D.E. 2003. Requirement of the *Listeria monocytogenes* broad-range phospholipase PC-PLC during infection of human epithelial cells. *J. Bacteriol.*, 185, 6295–6307.

Guevera, R.E., Mascola, L., and Sorvillo, F. 2009. Risk factors for mortality among patients with nonperinatal listeriosis in Los Angeles county; 1992–2004. *Clin. Infect. Dis.*, 48, 1507–1515.

Guldimann, C., Lejeune, B., Hofer, S., Leib, S.L., Frey, J., Zurbriggen, A., Seuberlich, T., and Oevermann, A. 2012. Ruminant organotypic brain-slice cultures as a model for the investigation of CNS listeriosis. *Int. J. Exp. Pathol.*, 93, 259–268.

Hain, T., Ghai, R., Billion, A., Kuenne, C.T., Steinweg, C., Izar, B., Mohamed, W., Mraheil, M.A., Domann, E., Schaffrath, S., Karst, U., Goesmann, A., Oehm, S., Puhler, A., Merkl, R., Vorwerk, S., Glaser, P., Garrido, P., Rusniok, C., Buchrieser, C., Goebel, W., and Chakraborty, T. 2012. Comparative genomics and transcriptomics of lineages I, II, and III strains of *Listeria monocytogenes*. *BMC Genomics*, 13, 144.

Hain, T., Steinweg, C., Kuenne, C.T., Billion, A., Ghai, R., Chatterjee, S.S., Domann, E., Karst, U., Goesmann, A., Bekel, T., Bartels, D., Kaiser, O., Meyer, F., Puhler, A., Weisshaar, B., Wehland, J., Liang, C., Dandekar, T., Lampidis, R., Kreft, J., Goebel, W., and Chakraborty, T. 2006. Whole-genome sequence of *Listeria welshimeri* reveals common steps in genome reduction with *Listeria innocua* as compared to *Listeria monocytogenes*. *J. Bacteriol.*, 188, 7405–7415.

Hardy, J., Francis, K.P., DeBoer, M., Chu, P., Gibbs, K., and Contag, C.H. 2004. Extracellular replication of *Listeria monocytogenes* in the murine gall bladder. *Science*, 303, 851–853.

Hasbun, J., Sepulveda-Martinez, A., Haye, M.T., Astudillo, J., and Parra-Cordero, M. 2013. Chorioamnionitis caused by *Listeria monocytogenes*: A case report of ultrasound features of fetal infection. *Fetal Diagn. Ther.*, 33, 268–271.

Heikkinen, T., Laine, K., Neuvonen, P.J., and Ekblad, U. 2000. The transplacental transfer of the macrolide antibiotics erythromycin, roxithromycin and azithromycin. *BJOG*, 107, 770–775.

Hof, H., Nichterlein, T., and Kretschmar, M. 1997. Management of listeriosis. *Clin. Microbiol. Rev.*, 10, 345–357.

Holch, A., Webb, K., Lukjancenko, O., Ussery, D., Rosenthal, B.M., and Gram, L. 2013. Genome sequencing identifies two nearly unchanged strains of persistent *Listeria monocytogenes* isolated at two different fish processing plants sampled 6 years apart. *Appl. Environ. Microbiol.*, 79, 2944–2951.

Ireton, K., Payrastre, B., Chap, H., Ogawa, W., Sakaue, H., Kasuga, M., and Cossart, P. 1996. A role for phosphoinositide 3-kinase in bacterial invasion. *Science*, 274, 780–782.

Ireton, K., Payrastre, B., and Cossart, P. 1999. The *Listeria monocytogenes* protein InlB is an agonist of mammalian phosphoinositide 3-kinase. *J. Biol. Chem.*, 274, 17025–17032.

Ivanek, R., Grohn, Y.T., Jui-Jung Ho, A., and Wiedmann, M. 2007. Markov chain approach to analyze the dynamics of pathogen fecal shedding–Example of *Listeria monocytogenes* shedding in a herd of dairy cattle. *J. Theor. Biol.*, 245, 44–58.

Jackson, K.A., Biggerstaff, M., Tobin-D'Angelo, M., Sweat, D., Klos, R., Nosari, J., Garrison, O., Boothe, E., Saathoff-Huber, L., Hainstock, L., and Fagan, R.P. 2011. Multistate outbreak of *Listeria monocytogenes* associated with Mexican-style cheese made from pasteurized milk among pregnant, Hispanic women. *J. Food. Prot.*, 74, 949–953.

Jackson, K.A., Iwamoto, M., and Swerdlow, D. 2010. Pregnancy-associated listeriosis. *Epidemiol. Infect.*, 138, 1503–1509.

Jacquet, C., Doumith, M., Gordon, J.I., Martin, P.M., Cossart, P., and Lecuit, M. 2004. A molecular marker for evaluating the pathogenic potential of foodborne *Listeria monocytogenes*. *J. Infect. Dis.*, 189, 2094–2100.

Janakiraman, V. 2008. Listeriosis in pregnancy: Diagnosis, treatment, and prevention. *Rev. Obstet. Gynecol.*, 1, 179–185.

Jeffers, G.T., Bruce, J.L., McDonough, P.L., Scarlett, J., Boor, K.J., and Wiedmann, M. 2001. Comparative genetic characterization of *Listeria monocytogenes* isolates from human and animal listeriosis cases. *Microbiology*, 147, 1095–1104.

Jensen, V.B., Harty, J.T., and Jones, B.D. 1998. Interactions of the invasive pathogens *Salmonella typhimurium*, *Listeria monocytogenes*, and *Shigella flexneri* with M cells and murine Peyer's patches. *Infect. Immun.*, 66, 3758–3766.

Jin, Y., Dons, L., Kristensson, K., and Rottenberg, M.E. 2001. Neural route of cerebral *Listeria monocytogenes* murine infection: Role of immune response mechanisms in controlling bacterial neuroinvasion. *Infect. Immun.*, 69, 1093–1100.

Jin, Y., Dons, L., Kristensson, K., and Rottenberg, M.E. 2002. Colony-stimulating factor 1-dependent cells protect against systemic infection with *Listeria monocytogenes* but facilitate neuroinvasion. *Infect. Immun.*, 70, 4682–4686.

Johansson, J., Mandin, P., Renzoni, A., Chiaruttini, C., Springer, M., and Cossart, P. 2002. An RNA thermosensor controls expression of virulence genes in *Listeria monocytogenes*. *Cell*, 110, 551–561.

Join-Lambert, O.F., Ezine, S., Le Monnier, A., Jaubert, F., Okabe, M., Berche, P., and Kayal, S. 2005. *Listeria monocytogenes*-infected bone marrow myeloid cells promote bacterial invasion of the central nervous system. *Cell. Microbiol.*, 7, 167–180.

Jolliffe, L.K., Doyle, R.J., and Streips, U.N. 1981. The energized membrane and cellular autolysis in *Bacillus subtilis*. *Cell*, 25, 753–763.

Kampelmacher, E.H., Huysinga, W.T., and van Noorle Jansen, L.M. 1972. The presence of *Listeria monocytogenes* in feces of pregnant women and neonates. *Zentralbl. Bakteriol. Orig. A.*, 222, 258–262.

Kang, S.J., Liang, H.E., Reizis, B., and Locksley, R.M. 2008. Regulation of hierarchical clustering and activation of innate immune cells by dendritic cells. *Immunity*, 29, 819–833.

Karatzas, K.A., Suur, L., and O'Byrne, C.P. 2012. Characterization of the intracellular glutamate decarboxylase system: Analysis of its function, transcription, and role in the acid resistance of various strains of *Listeria monocytogenes*. *Appl. Environ. Microbiol.*, 78, 3571–3579.

Kathariou, S. 2002. *Listeria monocytogenes* virulence and pathogenicity, a food safety perspective. *J. Food. Prot.*, 65, 1811–1829.

Kaur, S., Malik, S.V., Vaidya, V.M., and Barbuddhe, S.B. 2007. *Listeria monocytogenes* in spontaneous abortions in humans and its detection by multiplex PCR. *J. Appl. Microbiol.*, 103, 1889–1896.

Kazmierczak, M.J., Mithoe, S.C., Boor, K.J., and Wiedmann, M. 2003. *Listeria monocytogenes* sigma B regulates stress response and virulence functions. *J. Bacteriol.*, 185, 5722–5734.

Keeney, K., Colosi, L., Weber, W., and O'Riordan, M. 2009. Generation of branched-chain fatty acids through lipoate-dependent metabolism facilitates intracellular growth of *Listeria monocytogenes*. *J. Bacteriol.*, 191, 2187–2196.

Keeney, K.M., Stuckey, J.A., and O'Riordan, M.X. 2007. LplA1-dependent utilization of host lipoyl peptides enables *Listeria* cytosolic growth and virulence. *Mol. Microbiol.*, 66, 758–770.

Kelesidis, T., Salhotra, A., Fleisher, J., and Uslan, D.Z. 2010. *Listeria* endocarditis in a patient with psoriatic arthritis on infliximab: Are biologic agents as treatment for inflammatory arthritis increasing the incidence of *Listeria* infections? *J. Infect.*, 60, 386–396.

Kemper, M.A., Urrutia, M.M., Beveridge, T.J., Koch, A.L., and Doyle, R.J. 1993. Proton motive force may regulate cell wall-associated enzymes of *Bacillus subtilis*. *J. Bacteriol.*, 175, 5690–5696.

Kirk, J. 1993. Diagnostic ultrastructure of Listeria monocytogenes in human central nervous tissue. *Ultrastruct. Pathol.*, 17, 583–592.

Kocks, C., Gouin, E., Tabouret, M., Berche, P., Ohayon, H., and Cossart, P. 1992. *L. monocytogenes*-induced actin assembly requires the *actA* gene product, a surface protein. *Cell*, 68, 521–531.

Kocks, C., Hellio, R., Gounon, P., Ohayon, H., and Cossart, P. 1993. Polarized distribution of *Listeria monocytogenes* surface protein ActA at the site of directional actin assembly. *J. Cell. Sci.*, 105, 699–710.

Koopmans, M.M., Brouwer, M.C., Bijlsma, M.W., Bovenkerk, S., Keijzers, W., van der Ende, A., and van de Beek, D. 2013. *Listeria monocytogenes* sequence type 6 and increased rate of unfavorable outcome in meningitis: Epidemiologic cohort study. *Clin. Infect. Dis.*, 57, 247–253.

Krawczyk-Balska, A., Popowska, M., and Markiewicz, Z. 2012. Re-evaluation of the significance of penicillin binding protein 3 in the susceptibility of *Listeria monocytogenes* to beta-lactam antibiotics. *BMC Microbiol.*, 12, 57.

Krueger, N., Low, C., and Donachie, W. 1995. Phenotypic characterization of the cells of the inflammatory response in ovine encephalitic listeriosis. *J. Comp. Pathol.*, 113, 263–275.

Kuenne, C., Billion, A., Mraheil, M.A., Strittmatter, A., Daniel, R., Goesmann, A., Barbuddhe, S., Hain, T., and Chakraborty, T. 2013. Reassessment of the *Listeria monocytogenes* pan-genome reveals dynamic integration hotspots and mobile genetic elements as major components of the accessory genome. *BMC Genomics*, 14, 47.

Kuenne, C., Voget, S., Pischimarov, J., Oehm, S., Goesmann, A., Daniel, R., Hain, T., and Chakraborty, T. 2010. Comparative analysis of plasmids in the genus *Listeria*. *PLoS One*, 5, e12511.

Laksanalamai, P., Jackson, S.A., Mammel, M.K., and Datta, A.R. 2012. High density microarray analysis reveals new insights into genetic footprints of *Listeria monocytogenes* strains involved in listeriosis outbreaks. *PLoS One*, 7, e32896.

Lamont, R.F., Sobel, J., Mazaki-Tovi, S., Kusanovic, J.P., Vaisbuch, E., Kim, S.K., Uldbjerg, N., and Romero, R. 2011. Listeriosis in human pregnancy: A systematic review. *J. Perinat. Med.*, 39, 227–236.

Lamont, R.J., and Postlethwaite, R. 1986. Carriage of *Listeria monocytogenes* and related species in pregnant and non-pregnant women in Aberdeen, Scotland. *J. Infect.*, 13, 187–193.

Lara-Tejero, M., and Pamer, E.G. 2004. T cell responses to *Listeria monocytogenes*. *Curr. Opin. Microbiol.*, 7, 45–50.

Larsson, S., Cronberg, S., and Winblad, S. 1978. Clinical aspects on 64 cases of juvenile and adult listeriosis in Sweden. *Acta Med. Scand.*, 204, 503–508.

Lavetter, A., Leedom, J.M., Mathies, A.W., Ivler, D., and Wehrle, P.F. 1971. Meningitis due to *Listeria monocytogenes*. *N. Engl. J. Med.*, 285, 598–603.

Lawler, F.C., Wood, W.S., King, S., and Metzger, W.I. 1964. *Listeria monocytogenes* as a cause of fetal loss. *Am. J. Obstet. Gynecol.*, 89, 915–923.

Le Monnier, A., Abachin, E., Beretti, J.-L., Berche, P., and Kayal, S. 2011. Diagnosis of *Listeria monocytogenes* meningoencephalitis by real-time PCR for the hly gene. *J. Clin. Microbiol.*, 49, 3917–3923.

Leber, J.H., Crimmins, G.T., Raghavan, S., Meyer-Morse, N.P., Cox, J.S., and Portnoy, D.A. 2008. Distinct TLR- and NLR-mediated transcriptional responses to an intracellular pathogen. *PLoS Pathog.*, 4, e6.

Lebrun, M., Loulergue, J., Chaslus-Dancla, E., and Audurier, A. 1992. Plasmids in *Listeria monocytogenes* in relation to cadmium resistance. *Appl. Environ. Microbiol.*, 58, 3183–3186.

Lecuit, M., Dramsi, S., Gottardi, C., Fedor-Chaiken, M., Gumbiner, B., and Cossart, P. 1999. A single amino acid in E-cadherin responsible for host specificity towards the human pathogen *Listeria monocytogenes*. *EMBO J.*, 18, 3956–3963.

Lecuit, M., Nelson, D.M., Smith, S.D., Khun, H., Huerre, M., Vacher-Lavenu, M.C., Gordon, J.I., and Cossart, P. 2004. Targeting and crossing of the human maternofetal barrier by *Listeria monocytogenes*: Role of internalin interaction with trophoblast E-cadherin. *Proc. Natl. Acad. Sci. USA*, 101, 6152–6157.

Lecuit, M., Vandormael-Pournin, S., Lefort, J., Huerre, M., Gounon, P., Dupuy, C., Babinet, C., and Cossart, P. 2001. A transgenic model for listeriosis: Role of internalin in crossing the intestinal barrier. *Science*, 292, 1722–1725.

Leimeister-Wächter, M., Domann, E., and Chakraborty, T. 1992. The expression of virulence genes in *Listeria monocytogenes* is thermoregulated. *J. Bacteriol.*, 174, 947–952.

Leimeister-Wächter, M., Haffner, C., Domann, E., Goebel, W., and Chakraborty, T. 1990. Identification of a gene that positively regulates expression of listeriolysin, the major virulence factor of *Listeria monocytogenes*. *Proc. Natl. Acad. Sci. USA*, 87, 8336–8340.

Lennon, D., Lewis, B., Mantell, C., Becroft, D., Dove, B., Farmer, K., Tonkin, S., Yeates, N., Stamp, R., and Mickleson, K. 1984. Epidemic perinatal listeriosis. *Pediatr. Infect. Dis.*, 3, 30–34.

Leong, M.L., Hampl, J., Liu, W., Mathur, S., Bahjat, K.S., Luckett, W., Dubensky Jr., T.W., and Brockstedt, D.G. 2009. Impact of preexisting vector-specific immunity on vaccine potency: Characterization of *Listeria monocytogenes*-specific humoral and cellular immunity in humans and modeling studies using recombinant vaccines in mice. *Infect. Immun.*, 77, 3958–3968.

Lianou, A., and Sofos, J.N. 2007. A review of the incidence and transmission of *Listeria monocytogenes* in ready-to-eat products in retail and food service environments. *J. Food. Prot.*, 70, 2172–2198.

Linnan, M.J., Mascola, L., Lou, X.D., Goulet, V., May, S., Salminen, C., Hird, D.W., Yonekura, M.L., Hayes, P., Weaver, R., Audurier, A., Plikaytis, B.D., Fannin, S.L., Kleks, A., and Broome, C.V. 1988. Epidemic listeriosis associated with Mexican-style cheese. *N. Engl. J. Med.*, 319, 823–828.

Lomonaco, S., Verghese, B., Gerner-Smidt, P., Tarr, C., Gladney, L., Joseph, L., Katz, L., Turnsek, M., Frace, M., Chen, Y., Brown, E., Meinersmann, R., Berrang, M., and Knabel, S. 2013. Novel epidemic clones of *Listeria monocytogenes*, United States, 2011. *Emerg. Infect. Dis.*, 19, 147–150.

Lorber, B. 2010. *Listeria monocytogenes*. In: Mandel, G.L., Bennett, J.E., and Dolin, R., editors. *Principles and Practice of Infectious Diseases*, 7th ed. Philadelphia: Churchill Livingstone. pp. 2707–2714.

Mackaness, G.B. 1962. Cellular resistance to infection. *J. Exp. Med.*, 116, 381–406.

MacMicking, J.D., Nathan, C., Hom, G., Chartrain, N., Fletcher, D.S., Trumbauer, M., Stevens, K., Xie, Q.W., Sokol, K., Hutchinson, N., Cen, H., and Mudgett, J.S. 1995. Altered responses to bacterial infection and endotoxic shock in mice lacking inducible nitric oxide synthase. *Cell*, 81, 641–650.

Macnaughton, M.C. 1962. *Listeria monocytogenes* in abortion. *Lancet*, 2, 484.

Mailles, A., Stahl, J.-P., and Steering Committee and Investigators Groupb.o.t.S.Group.I. 2009. Infectious encephalitis in France in 2007: A national prospective study. *Clin. Infect. Dis.*, 49, 1838–1847.

Manganiello, P.D., and Yearke, R.R. 1991. A 10-year prospective study of women with a history of recurrent fetal losses fails to

identify *Listeria monocytogenes* in the genital tract. *Fertil. Steril.*, 56, 781–782.

Marco, F., Almela, M., Nolla-Salas, J., Coll, P., Gasser, I., Ferrer, M.D., and de Simon, M. 2000. In vitro activities of 22 antimicrobial agents against *Listeria monocytogenes* strains isolated in Barcelona, Spain. The Collaborative Study Group of Listeriosis of Barcelona. *Diagn. Microbiol. Infect. Dis.*, 38, 259–261.

Marquis, H., Bouwer, H.G.A., Hinrichs, D.J., and Portnoy, D.A. 1993. Intracytoplasmic growth and virulence of *Listeria monocytogenes* auxotrophic mutants. *Infect. Immun.*, 61, 3756–3760.

Marquis, H., Doshi, V., and Portnoy, D.A. 1995. The broad-range phospholipase C and a metalloprotease mediate listeriolysin O-independent escape of *Listeria monocytogenes* from a primary vacuole in human epithelial cells. *Infect. Immun.*, 63, 4531–4534.

Marquis, H., Goldfine, H., and Portnoy, D.A. 1997. Proteolytic pathways of activation and degradation of a bacterial phospholipase C during intracellular infection by *Listeria monocytogenes*. *J. Cell. Biol.*, 137, 1381–1392.

Marquis, H., and Hager, E.J. 2000. pH-regulated activation and release of a bacteria-associated phospholipase C during intracellular infection by *Listeria monocytogenes*. *Mol. Microbiol.*, 35, 289–298.

McLauchlin, J. 1990. Human listeriosis in Britain, 1967–85, a summary of 722 cases. 1. Listeriosis during pregnancy and in the newborn. *Epidemiol. Infect.*, 104, 181–189.

McLauchlin, J., Hampton, M.D., Shah, S., Threlfall, E.J., Wieneke, A.A., and Curtis, G.D. 1997. Subtyping of *Listeria monocytogenes* on the basis of plasmid profiles and arsenic and cadmium susceptibility. *J. Appl. Microbiol.*, 83, 381–388.

Mellin, J.R., and Cossart, P. 2012. The non-coding RNA world of the bacterial pathogen *Listeria monocytogenes*. *RNA Biol.*, 9, 372–378.

Mengaud, J., Ohayon, H., Gounon, P., Mege, R.-M., and Cossart, P. 1996. E-cadherin is the receptor for internalin, a surface protein required for entry of *L. monocytogenes* into epithelial cells. *Cell*, 84, 923–932.

Merritt, M.E., and Donaldson, J.R. 2009. Effect of bile salts on the DNA and membrane integrity of enteric bacteria. *J. Med. Microbiol.*, 58, 1533–1541.

Michelet, C., Leib, S.L., Bentue-Ferrer, D., and Tauber, M.G. 1999. Comparative efficacies of antibiotics in a rat model of meningoencephalitis due to *Listeria monocytogenes*. *Antimicrob. Agents Chemother.*, 43, 1651–1656.

Mitja, O., Pigrau, C., Ruiz, I., Vidal, X., Almirante, B., Planes, A.-M., Molina, I., Rodriguez, D., and Pahissa, A. 2009. Predictors of mortality and impact of aminoglycosides on outcome in listeriosis in a retrospective cohort study. *J. Antimicrob. Chemother.*, 64, 416–423.

Moragas, M., Martinez-Yelamos, S., Majos, C., Fernandez-Viladrich, P., Rubio, F., and Arbizu, T. 2011. Rhombencephalitis: A series of 97 patients. *Medicine (Baltimore)*, 90, 256–261.

Mraheil, M.A., Billion, A., Mohamed, W., Mukherjee, K., Kuenne, C., Pischimarov, J., Krawitz, C., Retey, J., Hartsch, T., Chakraborty, T., and Hain, T. 2011. The intracellular sRNA transcriptome of *Listeria monocytogenes* during growth in macrophages. *Nucleic Acids Res.*, 39, 4235–4248.

Mujahid, S., Orsi, R.H., Boor, K.J., and Wiedmann, M. 2013a. Protein level identification of the *Listeria monocytogenes* sigma H, sigma L, and sigma C regulons. *BMC Microbiol.*, 13, 156.

Mujahid, S., Orsi, R.H., Vangay, P., Boor, K.J., and Wiedmann, M. 2013b. Refinement of the *Listeria monocytogenes* sigmaB regulon through quantitative proteomic analysis. *Microbiology*, 159, 1109–1119.

Muller, C.A., Autenrieth, I.B., and Peschel, A. 2005. Innate defenses of the intestinal epithelial barrier. *Cell. Mol. Life Sci.*, 62, 1297–1307.

Munoz, P., Rojas, L., Bunsow, E., Saez, E., Sanchez-Cambronero, L., Alcala, L., Rodríguez-Creixems, M., and Bouza, E. 2011. Listeriosis: An emerging public health problem especially among the elderly. *J. Infect.*, 64, 19–33.

Myers, J.T., Tsang, A.W., and Swanson, J.A. 2003. Localized reactive oxygen and nitrogen intermediates inhibit escape of *Listeria monocytogenes* from vacuoles in activated macrophages. *J. Immunol.*, 171, 5447–5453.

Mylonakis, E., Hohmann, E.L., and Calderwood, S.B. 1998. Central nervous system infection with *Listeria monocytogenes*. 33 years' experience at a general hospital and review of 776 episodes from the literature. *Medicine (Baltimore)*, 77, 313–336.

Mylonakis, E., Paliou, M., Hohmann, E.L., and Calderwood, S.B. 2002. Listeriosis during pregnancy: A case series and review of 222 cases. *Medicine (Baltimore)*, 81, 260–269.

Nelson, K.E., Fouts, D.E., Mongodin, E.F., Ravel, J., DeBoy, R.T., Kolonay, J.F., Rasko, D.A., Angiuoli, S.V., Gill, S.R., Paulsen, I.T., Peterson, J., White, O., Nelson, W.C., Nierman, W., Beanan, M.J., Brinkac, L.M., Daugherty, S.C., Dodson, R.J., Durkin, A.S., Madupu, R., Haft, D.H., Selengut, J., Van Aken, S., Khouri, H., Fedorova, N., Forberger, H., Tran, B., Kathariou, S., Wonderling, L.D., Uhlich, G.A., Bayles, D.O., Luchansky, J.B., and Fraser, C.M. 2004. Whole genome comparisons of serotype 4b and 1/2a strains of the food-borne pathogen *Listeria monocytogenes* reveal new insights into the core genome components of this species. *Nucleic Acids Res.*, 32, 2386–2395.

Nightingale, K.K., Ivy, R.A., Ho, A.J., Fortes, E.D., Njaa, B.L., Peters, R.M., and Wiedmann, M. 2008. *inlA* premature stop codons are common among *Listeria monocytogenes* isolates from foods and yield virulence-attenuated strains that confer protection against fully virulent strains. *Appl. Environ. Microbiol.*, 74, 6570–6583.

Nightingale, K.K., Milillo, S.R., Ivy, R.A., Ho, A.J., Oliver, H.F., and Wiedmann, M. 2007. *Listeria monocytogenes* F2365 carries several authentic mutations potentially leading to truncated gene products, including *inlB*, and demonstrates atypical phenotypic characteristics. *J. Food. Prot.*, 70, 482–488.

Nikitas, G., Deschamps, C., Disson, O., Niault, T., Cossart, P., and Lecuit, M. 2011. Transcytosis of *Listeria monocytogenes* across the intestinal barrier upon specific targeting of goblet cell accessible E-cadherin. *J. Exp. Med.*, 208, 2263–2277.

Notermans, S., Dufrenne, J., Teunis, P., and Chackraborty, T. 1998. Studies on the risk assessment of *Listeria monocytogenes*. *J. Food. Prot.*, 61, 244–248.

O'Neil, H.S., and Marquis, H. 2006. *Listeria monocytogenes* flagella are used for motility, not as adhesins, to increase host cell invasion. *Infect. Immun.*, 74, 6675–6681.

O'Riordan, M., Moors, M.A., and Portnoy, D.A. 2003. *Listeria* intracellular growth and virulence require host-derived lipoic acid. *Science*, 302, 462–464.

Ogawa, R., Pacelli, R., Espey, M.G., Miranda, K.M., Friedman, N., Kim, S.M., Cox, G., Mitchell, J.B., Wink, D.A., and Russo, A. 2001. Comparison of control of *Listeria* by nitric oxide redox chemistry from murine macrophages and NO donors: Insights into listeriocidal activity of oxidative and nitrosative stress. *Free Radic Biol. Med.*, 30, 268–276.

Ooi, S.T., and Lorber, B. 2005. Gastroenteritis due to *Listeria monocytogenes*. *Clin. Infect. Dis.*, 40, 1327–1332.

Orsi, R.H., Borowsky, M.L., Lauer, P., Young, S.K., Nusbaum, C., Galagan, J.E., Birren, B.W., Ivy, R.A., Sun, Q., Graves, L.M., Swaminathan, B., and Wiedmann, M. 2008. Short-term genome evolution of *Listeria monocytogenes* in a non-controlled environment. *BMC Genomics*, 9, 539.

Orsi, R.H., den Bakker, H.C., and Wiedmann, M. 2011. *Listeria monocytogenes* lineages: Genomics, evolution, ecology, and phenotypic characteristics. *Int. J. Med. Microbiol.*, 301, 79–96.

Pentecost, M., Otto, G., Theriot, J.A., and Amieva, M.R. 2006. *Listeria monocytogenes* invades the epithelial junctions at sites of cell extrusion. *PLoS Pathog.*, 2, e3.

Pollock, S.S., Pollock, T.M., and Harrison, M.J. 1984. Infection of the central nervous system by *Listeria monocytogenes*: A review of 54 adult and juvenile cases. *Q. J. Med.*, 53, 331–340.

Portnoy, D.A., Jacks, P.S., and Hinrichs, D.J. 1988. Role of hemolysin for the intracellular growth of *Listeria monocytogenes*. *J. Exp. Med.*, 167, 1459–1471.

Posfay-Barbe, K.M., and Wald, E.R. 2009. Listeriosis. *Semin. Fetal. Neonatal. Med.*, 14, 228–233.

Pouillot, R., Hoelzer, K., Jackson, K.A., Henao, O.L., and Silk, B.J. 2012. Relative risk of listeriosis in foodborne diseases active surveillance network (FoodNet) sites according to age, pregnancy and ethnicity. *Clin. Infect. Dis.*, 54, S405–S410.

Poulsen, K.P., Faith, N.G., Steinberg, H., and Czuprynski, C.J. 2011. Pregnancy reduces the genetic resistance of C57BL/6 mice to *Listeria monocytogenes* infection by intragastric inoculation. *Microb. Pathog.*, 50, 360–366.

Poussin, M.A., and Goldfine, H. 2005. Involvement of *Listeria monocytogenes* phosphatidylinositol-specific phospholipase C and host protein kinase C in permeabilization of the macrophage phagosome. *Infect. Immun.*, 73, 4410–4413.

Poussin, M.A., Leitges, M., and Goldfine, H. 2009. The ability of *Listeria monocytogenes* PI-PLC to facilitate escape from the macrophage phagosome is dependent on host PKCbeta. *Microb. Pathog.*, 46, 1–5.

Quillin, S.J., Schwartz, K.T., and Leber, J.H. 2011. The novel *Listeria monocytogenes* bile sensor BrtA controls expression of the cholic acid efflux pump MdrT. *Mol. Microbiol.*, 81, 129–142.

Quinlivan, J.A., Newnham, J.P., and Dickinson, J.E. 1998. Ultrasound features of congenital listeriosis–a case report. *Prenat. Diagn.*, 18, 1075–1078.

Rabinovich, L., Sigal, N., Borovok, I., Nir-Paz, R., and Herskovits, A.A. 2012. Prophage excision activates *Listeria* competence genes that promote phagosomal escape and virulence. *Cell*, 150, 792–802.

Rafelski, S.M., and Theriot, J.A. 2006. Mechanism of polarization of *Listeria monocytogenes* surface protein ActA. *Mol. Microbiol.*, 59, 1262–1279.

Rajabian, T., Gavicherla, B., Heisig, M., Muller-Altrock, S., Goebel, W., Gray-Owen, S.D., and Ireton, K. 2009. The bacterial virulence factor InlC perturbs apical cell junctions and promotes cell-to-cell spread of *Listeria*. *Nat. Cell. Biol.*, 11, 1212–1218.

Ramasamy, V., Cresence, V.M., Rejitha, J.S., Lekshmi, M.U., Dharsana, K.S., Prasad, S.P., and Vijila, H.M. 2007. *Listeria*–Review of epidemiology and pathogenesis. *J. Microbiol. Immunol. Infect.*, 40, 4–13.

Rappaport, F., Rabinovitz, M., Toaff, R., and Krochik, N. 1960. Genital listeriosis as a cause of repeated abortion. *Lancet*, 1, 1273–1275.

Rayamajhi, M., Humann, J., Penheiter, K., Andreasen, K., and Lenz, L.L. 2010. Induction of IFN-alphabeta enables *Listeria monocytogenes* to suppress macrophage activation by IFN-gamma. *J. Exp. Med.*, 207, 327–337.

Renier, S., Micheau, P., Talon, R., Hebraud, M., and Desvaux, M. 2012. Subcellular localization of extracytoplasmic proteins in monoderm bacteria: Rational secretomics-based strategy for genomic and proteomic analyses. *PLoS One*, 7, e42982.

Robbins, J.R., Barth, A.I., Marquis, H., de Hostos, E.L., Nelson, W.J., and Theriot, J.A. 1999. *Listeria monocytogenes* exploits normal host cell processes to spread from cell to cell. *J. Cell. Biol.*, 146, 1333–1349.

Robbins, J.R., Skrzypczynska, K.M., Zeldovich, V.B., Kapidzic, M., and Bakardjiev, A.I. 2010. Placental syncytiotrophoblast constitutes a major barrier to vertical transmission of *Listeria monocytogenes*. *PLoS Pathog.*, 6, e1000732.

Romana, C., Salleras, L., and Sage, M. 1989. Latent listeriosis may cause habitual abortion intrauterine deaths, fetal malformations. When diagnosed and treated adequately normal children will be born. *Acta Microbiol. Hung.*, 36, 171–172.

Ryan, S., Begley, M., Gahan, C.G., and Hill, C. 2009. Molecular characterization of the arginine deiminase system in *Listeria monocytogenes*: Regulation and role in acid tolerance. *Environ. Microbiol.*, 11, 432–445.

Safdar, A., and Armstrong, D. 2003a. Antimicrobial activities against 84 *Listeria monocytogenes* isolates from patients with systemic listeriosis at a comprehensive cancer center (1955–1997). *J. Clin. Microbiol.*, 41, 483–485.

Safdar, A., and Armstrong, D. 2003b. Listeriosis in patients at a comprehensive cancer center, 1955–1997. *Clin. Infect. Dis.*, 37, 359–364.

Salmon-Ceron, D., Tubach, F., Lortholary, O., Chosidow, O., Bretagne, S.l., Nicolas, N., Cuillerier, E., Fautrel, B., Michelet, C., Morel, J., Puéchal, X., Wendling, D., Lemann, M., Ravaud, P., Mariette, X., and RATIO group. 2011. Drug-specific risk of non-tuberculosis opportunistic infections in patients receiving anti-TNF therapy reported to the 3-year prospective French RATIO registry. *Ann. Rheum. Dis.*, 70, 616–623.

Sauer, J.D., Witte, C.E., Zemansky, J., Hanson, B., Lauer, P., and Portnoy, D.A. 2010. *Listeria monocytogenes* triggers AIM2-mediated pyroptosis upon infrequent bacteriolysis in the macrophage cytosol. *Cell. Host. Microbe.*, 7, 412–419.

Scallan, E., Hoekstra, R.M., Angulo, F.J., Tauxe, R.V., Widdowson, M.A., Roy, S.L., Jones, J.L., and Griffin, P.M. 2011. Foodborne illness acquired in the United States–Major pathogens. *Emerg. Infect. Dis.*, 17, 7–15.

Schlech 3rd, W.F. 2000. Foodborne listeriosis. *Clin. Infect. Dis.*, 31, 770–775.

Schlech 3rd, W.F., Lavigne, P.M., Bortolussi, R.A., Allen, A.C., Haldane, E.V., Wort, A.J., Hightower, A.W., Johnson, S.E., King, S.H., Nicholls, E.S., and Broome, C.V. 1983. Epidemic listeriosis–Evidence for transmission by food. *N. Engl. J. Med.*, 308, 203–206.

Schuchat, A., Robinson, K., Wenger, J.D., Harrison, L.H., Farley, M., Reingold, A.L., Lefkowitz, L., and Perkins, B.A. 1997. Bacterial meningitis in the United States in 1995. Active Surveillance Team. *N. Engl. J. Med.*, 337, 970–976.

Serbina, N.V., Salazar-Mather, T.P., Biron, C.A., Kuziel, W.A., and Pamer, E.G. 2003. TNF/iNOS-producing dendritic cells mediate innate immune defense against bacterial infection. *Immunity*, 19, 59–70.

Shen, Y., Naujokas, M., Park, M., and Ireton, K. 2000. InlB-dependent internalization of *Listeria* is mediated by the met receptor tyrosine kinase. *Cell*, 103, 501–510.

Silk, B.J., Date, K.A., Jackson, K.A., Pouillot, R., Holt, G., Graves, L.M., Ong, K.L., Hurd, S., Meyer, R., Marcus, R., Shiferaw, B., Norton, D.M., Medus, C., Zansky, S.M., Cronquist, A.B., Henao, O.L., Jones, T.F., Vugia, D.J., Farley, M.M., and Mahon, B.E. 2012. Invasive listeriosis in the foodborne diseases active surveillance network (FoodNet), 2004–2009: Further targeted prevention needed for higher risk groups. *Clin. Infect. Dis.*, 54, S396-S404.

Singh, R., Jamieson, A., and Cresswell, P. 2008. GILT is a critical host factor for *Listeria monocytogenes* infection. *Nature*, 455, 1244–1247.

Skoble, J., Portnoy, D.A., and Welch, M.D. 2000. Three regions within ActA promote Arp2/3 complex-mediated actin nucleation and *Listeria monocytogenes* motility. *J. Cell. Biol.*, 150, 527–538.

Skogberg, K., Syrjanen, J., Jahkola, M., Renkonen, O.V., Paavonen, J., Ahonen, J., Kontiainen, S., Ruutu, P., and Valtonen, V. 1992. Clinical presentation and outcome of listeriosis in patients with and without immunosuppressive therapy. *Clin. Infect. Dis.*, 14, 815–821.

Sleator, R.D., Wemekamp-Kamphuis, H.H., Gahan, C.G., Abee, T., and Hill, C. 2005. A PrfA-regulated bile exclusion system (BilE) is a novel virulence factor in *Listeria monocytogenes*. *Mol. Microbiol.*, 55, 1183–1195.

Smith, B., Kemp, M., Ethelberg, S., Schiellerup, P., Bruun, B.G., Gerner-Smidt, P., and Christensen, J.J. 2009. *Listeria monocytogenes*: Maternal-foetal infections in Denmark 1994–2005. *Scand. J. Infect. Dis.*, 41, 21–25.

Smith, G.A., Marquis, H., Jones, S., Johnston, N.C., Portnoy, D.A., and Goldfine, H. 1995a. The two distinct phospholipases C of *Listeria monocytogenes* have overlapping roles in escape from a vacuole and cell-to-cell spread. *Infect. Immun.*, 63, 4231–4237.

Smith, G.A., Portnoy, D.A., and Theriot, J.A. 1995b. Asymmetric distribution of the *Listeria monocytogenes* ActA protein is required and sufficient to direct actin-based motility. *Mol. Microbiol.*, 17, 945–951.

Snyder, A., and Marquis, H. 2003. Restricted translocation across the cell wall regulates secretion of the broad-range phospholipase C of *Listeria monocytogenes*. *J. Bacteriol.*, 185, 5953–5958.

Southwick, F.S., and Purich, D.L. 1996. Intracellular pathogenesis of listeriosis. *N. Engl. J. Med.*, 334, 770–776.

Steinweg, C., Kuenne, C.T., Billion, A., Mraheil, M.A., Domann, E., Ghai, R., Barbuddhe, S.B., Karst, U., Goesmann, A., Puhler, A., Weisshaar, B., Wehland, J., Lampidis, R., Kreft, J., Goebel, W., Chakraborty, T., and Hain, T. 2010. Complete genome sequence of *Listeria seeligeri*, a nonpathogenic member of the genus *Listeria*. *J. Bacteriol.*, 192, 1473–1474.

Straus, S.E., Thorpe, K.E., and Holroyd-Leduc, J. 2006. How do I perform a lumbar puncture and analyze the results to diagnose bacterial meningitis? *JAMA*, 296, 2012–2022.

Straus, S.K., and Hancock, R.E. 2006. Mode of action of the new antibiotic for Gram-positive pathogens daptomycin: Comparison with cationic antimicrobial peptides and lipopeptides. *Biochim. Biophys. Acta*, 1758, 1215–1223.

Swaminathan, B., and Gerner-Smidt, P. 2007. The epidemiology of human listeriosis. *Microbes. Infect.*, 9, 1236–1243.

Tappero, J.W., Schuchat, A., Deaver, K.A., Mascola, L., and Wenger, J.D. 1995. Reduction in the incidence of human listeriosis in the United States. Effectiveness of prevention efforts? The Listeriosis Study Group. *JAMA*, 273, 1118–1122.

Temple, M.E., and Nahata, M.C. 2000. Treatment of listeriosis. *Ann. Pharmacother.*, 34, 656–661.

Thedieck, K., Hain, T., Mohamed, W., Tindall, B.J., Nimtz, M., Chakraborty, T., Wehland, J., and Jansch, L. 2006. The MprF protein is required for lysinylation of phospholipids in listerial membranes and confers resistance to cationic antimicrobial peptides (CAMPs) on *Listeria monocytogenes*. *Mol. Microbiol.*, 62, 1325–1339.

Theriot, J.A., Mitchison, T.J., Tilney, L.G., and Portnoy, D.A. 1992. The rate of actin-based motility of intracellular *Listeria monocytogenes* equals the rate of actin polymerization. *Nature*, 357, 257–260.

Thigpen, M.C., Whitney, C.G., Messonnier, N.E., Zell, E.R., Lynfield, R., Hadler, J.L., Harrison, L.H., Farley, M.M., Reingold, A., Bennett, N.M., Craig, A.S., Schaffner, W., Thomas, A., Lewis, M.M., Scallan, E., and Schuchat, A. 2011. Bacterial meningitis in the United States, 1998–2007. *N. Engl. J. Med.*, 364, 2016–2025.

Tilney, L.G., and Portnoy, D.A. 1989. Actin filaments and the growth, movement, and spread of the intracellular bacterial parasite, *Listeria monocytogenes*. *J. Cell. Biol.*, 109, 1597–1608.

Toledo-Arana, A., Dussurget, O., Nikitas, G., Sesto, N., Guet-Revillet, H., Balestrino, D., Loh, E., Gripenland, J., Tiensuu, T., Vaitkevicius, K., Barthelemy, M., Vergassola, M., Nahori, M.A., Soubigou, G., Regnault, B., Coppee, J.Y., Lecuit, M., Johansson, J., and Cossart, P. 2009. The *Listeria* transcriptional landscape from saprophytism to virulence. *Nature*, 459, 950–956.

Tridente, V., Cataldi, U.M., Mossa, B., Morini, F., Bonessio, L., Ciardo, A., and Salvi, M. 1998. A case of maternal and neonatal infection due to *Listeria monocytogenes*. *Clin. Ter.*, 149, 307–311.

Tsai, Y.H., Maron, S.B., McGann, P., Nightingale, K.K., Wiedmann, M., and Orsi, R.H. 2011. Recombination and positive selection contributed to the evolution of *Listeria monocytogenes* lineages III and IV, two distinct and well supported uncommon *L. monocytogenes* lineages. *Infect. Genet. Evol.*, 11, 1881–1890.

Tunkel, A.R., Hartman, B.J., Kaplan, S.L., Kaufman, B.A., Roos, K.L., Scheld, W.M., and Whitley, R.J. 2004. Practice guidelines for the management of bacterial meningitis. *Clin. Infect. Dis.*, 39, 1267–1284.

Uldry, P.A., Kuntzer, T., Bogousslavsky, J., Regli, F., Miklossy, J., Bille, J., Francioli, P., and Janzer, R. 1993. Early symptoms and outcome of *Listeria monocytogenes* rhombencephalitis: 14 adult cases. *J. Neurol.*, 240, 235–242.

van de Beek, D., de Gans, J., Spanjaard, L., Weisfelt, M., Reitsma, J.B., and Vermeulen, M. 2004. Clinical features and prognostic factors in adults with bacterial meningitis. *N. Engl. J. Med.*, 351, 1849–1859.

van de Beek, D., de Gans, J., Tunkel, A.R., and Wijdicks, E.F.M. 2006. Community-acquired bacterial meningitis in adults. *N. Engl. J. Med.*, 354, 44–53.

Van Stelten, A., Simpson, J.M., Ward, T.J., and Nightingale, K.K. 2010. Revelation by single-nucleotide polymorphism genotyping that mutations leading to a premature stop codon in *inlA* are common among *Listeria monocytogenes* isolates from ready-to-eat foods but not human listeriosis cases. *Appl. Environ. Microbiol.*, 76, 2783–2790.

Vazquez-Boland, J.A., Kuhn, M., Berche, P., Chakraborty, T., Dominguez-Bernal, G., Goebel, W., Gonzalez-Zorn, B., Wehland, J., and Kreft, J. 2001. *Listeria* pathogenesis and molecular virulence determinants. *Clin. Microbiol. Rev.*, 14, 584–640.

Verghese, B., Lok, M., Wen, J., Alessandria, V., Chen, Y., Kathariou, S., and Knabel, S. 2011. *comK* prophage junction fragments as markers for *Listeria monocytogenes* genotypes unique to individual meat and poultry processing plants and a model for rapid niche-specific adaptation, biofilm formation, and persistence. *Appl. Environ. Microbiol.*, 77, 3279–3292.

Verschoor, A., Neuenhahn, M., Navarini, A.A., Graef, P., Plaumann, A., Seidlmeier, A., Nieswandt, B., Massberg, S., Zinkernagel, R.M., Hengartner, H., and Busch, D.H. 2011. A platelet-mediated system for shuttling blood-borne bacteria to CD8alpha+ dendritic cells depends on glycoprotein GPIb and complement C3. *Nat. Immunol.*, 12, 1194–1201.

Vrbova, L., Johnson, K., Whitfield, Y., and MIddleton, D. 2012. A descriptive study of reportable gastrointestinal illnesses in Ontario, Canada, from 2007–2009. *BMC Public Health*, 12, 970–980.

Wadsworth, S.J., and Goldfine, H. 2002. Mobilization of protein kinase C in macrophages induced by *Listeria monocytogenes* affects its internalization and escape from the phagosome. *Infect. Immun.*, 70, 4650–4660.

Wallis, R.S., Broder, M.S., Wong, J.Y., Hanson, M.E., and Beenhouwer, D.O. 2004. Granulomatous infectious diseases associated with tumor necrosis factor antagonists. *Clin. Infect. Dis.*, 38, 1261–1265.

Ward, T.J., Gorski, L., Borucki, M.K., Mandrell, R.E., Hutchins, J., and Pupedis, K. 2004. Intraspecific phylogeny and lineage group identification based on the *prfA* virulence gene cluster of *Listeria monocytogenes*. *J. Bacteriol.*, 186, 4994–5002.

Ward, T.J., Usgaard, T., and Evans, P. 2010. A targeted multilocus genotyping assay for lineage, serogroup, and epidemic clone typing of *Listeria monocytogenes*. *Appl. Environ. Microbiol.*, 76, 6680–6684.

Watkins, J., and Sleath, K.P. 1981. Isolation and enumeration of *Listeria monocytogenes* from sewage, sewage sludge and river water. *J. Appl. Bacteriol.*, 50, 1–9.

Weinmaier, T., Riesing, M., Rattei, T., Bille, J., Arguedas-Villa, C., Stephan, R., and Tasara, T. 2013. Complete genome sequence of *Listeria monocytogenes* LL195, a serotype 4b strain from the 1983–1987 listeriosis epidemic in Switzerland. *Genome Announc*, 1.

Welch, M.D., Rosenblatt, J., Skoble, J., Portnoy, D.A., and Mitchison, T.J. 1998. Interaction of human Arp2/3 complex and the *Listeria monocytogenes* ActA protein in actin filament nucleation. *Science*, 281, 105–108.

Welshimer, H.J., and Donker-Voet, J. 1971. *Listeria monocytogenes* in nature. *Appl. Microbiol.*, 21, 516–519.

Werber, D., Hille, K., Frank, C., Dehnert, M., Altmann, D., Mueller, J., Koch, J., and Stark, K. 2013. Years of potential life lost for six major enteric pathogens, Germany, 2004–2008. *Epidemiol. Infect.*, 141, 961–968.

Williams, D., Dunn, S., Richardson, A., Frank, J.F., and Smith, M.A. 2011. Time course of fetal tissue invasion by *Listeria monocytogenes* following an oral inoculation in pregnant guinea pigs. *J. Food. Prot.*, 74, 248–253.

Wilson, S.L., and Drevets, D.A. 1998. *Listeria monocytogenes* infection and activation of human brain microvascular endothelial cells. *J. Infect. Dis.*, 178, 1658–1666.

Woodward, J.J., Iavarone, A.T., and Portnoy, D.A. 2010. c-di-AMP secreted by intracellular *Listeria monocytogenes* activates a host type I interferon response. *Science*, 328, 1703–1705.

Wu, J., Fernandes-Alnemri, T., and Alnemri, E.S. 2010. Involvement of the AIM2, NLRC4, and NLRP3 inflammasomes in caspase-1 activation by *Listeria monocytogenes*. *J. Clin. Immunol.*, 30, 693–702.

Xayarath, B., and Freitag, N.E. 2012. Optimizing the balance between host and environmental survival skills: Lessons learned from *Listeria monocytogenes*. *Future Microbiol.*, 7, 839–852.

Yamamoto, T., Hara, H., Tsuchiya, K., Sakai, S., Fang, R., Matsuura, M., Nomura, T., Sato, F., Mitsuyama, M., and Kawamura, I. 2012. *Listeria monocytogenes* strain-specific impairment of the TetR regulator underlies the drastic increase in cyclic di-AMP secretion and beta interferon-inducing ability. *Infect. Immun.*, 80, 2323–2332.

Yeung, P.S., Na, Y., Kreuder, A.J., and Marquis, H. 2007. Compartmentalization of the broad-range phospholipase C activity to the spreading vacuole is critical for *Listeria monocytogenes* virulence. *Infect. Immun.*, 75, 44–51.

Chapter 41

Anthrax Infection

Kenneth E. Remy[1], Caitlin Hicks[2], and Peter Q. Eichacker[1]

[1]Critical Care Medicine Department, Clinical Center, National Institutes of Health, Bethesda, MD, USA

[2]Johns Hopkins Hospital, Department of Surgery, Baltimore, MD, USA

41.1 Introduction

Bacillus anthracis (*B. anthracis*) is a bacterium that exists in either a dormant spore form or an actively replicating vegetative rod form. *B. anthracis* infection is often referred to as anthrax due to the characteristic black color of the skin lesion that occurs with cutaneous disease. Naturally occurring anthrax may have first been noted in Egypt at the time of Moses (Dirckx, 1981; Stern et al., 2008). Inhalational anthrax was formally described in the mid-1800s in British textile workers sorting wool and was termed woolsorters' disease (Metcalfe, 2004; Sidel et al., 2002). Close to this time, Robert Koch isolated and described *B. anthracis* in the laboratory, demonstrated its ability to form spores, and produced experimental anthrax by injecting it into animals (Brachman, 1980; Cohn, 1875).

B. anthracis spores are found in soil around the world but natural disease precipitated by the vegetative form occurs predominantly in warmer regions (Mock and Fouet, 2001). *B. anthracis* infection normally develops in herbivore mammals (e.g., cattle, sheep, goats, camels, and antelopes) that ingest spores when grazing in contaminated areas. However, contact with contaminated animal products and internalization of spores can produce disease in humans.

B. anthracis has previously been associated with three forms of clinical infection depending on its route of entry: cutaneous, gastrointestinal (GI), or inhalational (Kiple, 1993). However, a fourth form of the disease related to injection of *B. anthracis* contaminated heroin has recently been recognized and is now referred to as injectional anthrax (Booth et al., 2010; Hicks et al., 2012; Ramsay et al., 2010).

Over the past century, *B. anthracis* infection has generally been associated with less developed regions of the world where widespread measures to prevent infection in domesticated animals and at-risk individuals have not been possible. However, recent outbreaks of inhalational, GI, and injectional anthrax in the United States and Europe, as well as the potential weaponization of *B. anthracis* for bioterrorist use has increased concern for the disease in the developed world (Inglesby et al., 2002; Kyriacou et al., 2006). Important aspects of *B. anthracis* infection are reviewed here.

41.2 Occurrence, geographic distribution, and mortality rates

The occurrence, geographic distribution, and mortality rate of *B. anthracis* infection in humans vary based on the form of disease. Generally, the characteristic skin lesion occurring with cutaneous anthrax and its association with particular occupational exposures contributes to a relatively high rate of early and successful diagnosis and treatment. Cutaneous anthrax is the most frequently reported form and occurs predominantly in underdeveloped regions of the world following contact with infected animals or contaminated animal products. The largest reported epidemic included more than 10,000 cases in Zimbabwe between 1979 and 1985, while the most recent outbreak appears to have occurred in July and August of 2011 in a village in India (Davies, 1982, 1983; Mwenye et al., 1996; Nass, 1992; Reddy et al., 2012). Table 41-1 summarizes several recent (over the past 50 years) outbreaks of cutaneous anthrax occurring outside of the United States (Chakraborty et al., 2012b; Demirdag et al., 2003; Doganay and Metan, 2009; Doganay et al., 2010; Heyworth et al., 1975; Lakshmi and Kumar, 1992; Ndyabahinduka et al., 1984; Peck and Fitzgerald, 2007; Phonboon, 1984; Ray et al., 2009; Reddy et al., 2012; Singh et al., 1992; Sirisanthana et al., 1984). In the United States, between 1944 and 1994, there were 224 cases of cutaneous anthrax reported (Inglesby et al., 2002). However, in the 2001 US anthrax outbreak, 12 of the 23 cases were cutaneous (Jernigan et al., 2001; Tutrone et al., 2002). There

Human Emerging and Re-emerging Infections: Bacterial & Mycotic Infections, Volume II, First Edition. Edited by Sunit K. Singh.
© 2016 John Wiley & Sons, Inc. Published 2016 by John Wiley & Sons, Inc.

Table 41-1 Selected outbreaks of cutaneous *B. anthracis* infection outside the United States over the past 50 years

Country of outbreak	Initial year of outbreak	Number of cases	Infected sources	Antibiotics administered	Mortality nonsurvivors/total (%)
Gambia	1970	448	Cattle, sheep, or goats	NR	10/443 (4)
Zimbabwe	1976	10738	Cattle	NR	182/10,738 (2)
Thailand	1982	28	Water buffalo or cattle	Penicillin	0/22 (0)
Thailand	1982	52	Water buffalo or cattle	NR	0/52 (0)
India	1989	10	Cattle or sheep	NR	0/10 (0)
Turkey	1990	413	Cattle or sheep	364 patients: penicillin alone, 14 patients: penicillin and ciprofloxacin	4/413 (1)
India	1992	18	Cattle	NR	0/18 (0)
Haiti	1992	87	Cattle or goats	All patients: penicillin	7/87 (8)
Turkey	2002	22	Cattle or sheep	15 patients: penicillin, 4 patients: amoxicillin, 1 patient: ciprofloxacin, 1 patient: doxycycline, 1 patient: clindamycin	0/22 (0)
India	2007	89	Cattle	NR	2/89 (2)
Bangladesh	2009	273	Cattle, goats, or sheep	Ciprofloxacin	0/273 (0)
India	2011	9	Cattle, goats, and buffalo	Ciprofloxacin	0/9 (0)

NR, not reported.

have subsequently been two additional US cases in 2007 related to drum making with infected goat hides from West Africa (CDC, 2008). Worldwide, mortality rates for patients with cutaneous anthrax not treated with antibiotics have been reported to be as high as 20%, but with antibiotics they are much lower (Table 41-1) (Hicks et al., 2012; Tutrone et al., 2002). The recent US cases all survived.

GI anthrax also occurs primarily in underdeveloped regions of the world, especially in Africa and Asia, and is typically related to the ingestion of inadequately cooked contaminated meat (Beatty et al., 2003; Plotkin and Orenstein, 1999; Sirisanthana et al., 1988). Due to its nonspecific symptomatology, it is speculated that many cases of GI anthrax are not identified. Table 41-2 summarizes recent (over the past 50 years) outbreaks of GI anthrax occurring outside of the United States (Chakraborty et al., 2012a; Doganay et al., 1986; Doganay and Metan, 2009; Kanafani et al., 2003; Ndyabahinduka et al., 1984; Phonboon, 1984; Sirisanthana et al., 1984). Several of these outbreaks were among individuals known to

Table 41-2 Selected outbreaks of gastrointestinal *B. anthracis* infection outside the United States over the past 50 years

Country of outbreak	Initial year of outbreak	Number of cases	Infected sources	Antibiotics administered	Mortality nonsurvivors/total (%)
Lebanon	1960	100	Goats	NR	NR
Gambia	1970	2	Cattle, sheep, or goats	NR	2/2 (100)
Thailand	1982	74	Water buffalo and cattle	Penicillin	3/74 (4)
Thailand	1982	24	Water buffalo or cattle	22 patients: penicillin alone or in combination with other antibiotics, 1 patient: streptomycin	2/24 (12)
Uganda	1984	143	Oxen	Penicillin or tetracycline	9/143 (6)
Turkey	1986	6	Cattle or sheep	NR	3/6 (50)
India	1989	20	Cattle or sheep	NR	5/20 (25)
Turkey	1990	8	Cattle or sheep	Penicillin	3/8 (38)
India	1992	7	Cattle	NR	2/7 (29)
India	2004	NR	Deer	NR	4/NR
India	2009	75	Cattle	NR	9/75 (12)
Bangladesh	2009	25	Cattle, goats, or sheep	Ciprofloxacin	0/25 (0)

NR, not reported.

have been exposed to the same contaminated food source (Alizad et al., 1995; Dutz and Kohout-Dutz, 1981; Heyworth et al., 1975; Lakshmi and Kumar, 1992; Navacharoen et al., 1985; Ndyabahinduka et al., 1984; Sekhar et al., 1990; Sirisanthana and Brown, 2002; Sirisanthana et al., 1984). The largest reported outbreak occurred in Uganda in 1984 among 143 people who had ingested infected oxen meat (Ndyabahinduka et al., 1984). The most recent non-US outbreak occurred in 2009 in Bangladesh (Chakraborty et al., 2012a). On the contrary, the number of cases of GI anthrax reported in the United States is very low. In 2000, there were two US cases related to the ingestion of contaminated meat in Minnesota, whereas the most recent case was a patient who likely inhaled and then ingested spores from a contaminated animal-hide drum (CDC, 2000; Klempner et al., 2010).

The reported mortality rate with intestinal anthrax varies between 15% and 60% but may be less than 40% with correct diagnosis and antibiotic treatment (Beatty et al., 2003; Brook, 2002; Inglesby et al., 1999; Kamal et al., 2011). In one outbreak in 2004 in Mysore, India, all four cases died after consuming contaminated deer meat (Ichhpujani et al., 2004). However, in a Ugandan outbreak in 1984, while 143 villagers consuming contaminated bovine meat appeared to develop GI anthrax, only 9 (6%) died and these were all children (Ndyabahinduka et al., 1984). Each of the three US cases were treated with timely antibiotics and survived (CDC, 2000, 2010; Klempner et al., 2010).

Inhalational anthrax was formally described in the mid–1800s in British textile workers sorting wool (i.e., woolsorters' disease) (Metcalfe, 2004; Sidel et al., 2002). Substantial reductions in the incidence of occupational inhalational anthrax have occurred with improved industrial hygiene, reduced use of contaminated animal products, and programs that immunized at-risk workers (Bell, 1880; Bell et al., 2002). A review of the literature from 1900 to 2005 identified 82 confirmed cases of inhalational anthrax. This series included 37 cases from the United States, 26 from 1900 to 1976 and 11 from the 2001 outbreak related to contaminated mail. Overall, 43 of the 82 cases were related to exposure to infected animal products (including all US cases from 1976 and earlier), 2 were in laboratory workers, 11 were from the 2001 US outbreak, and 26 had an unknown source. The report noted that it did not include cases from an outbreak in Sverdlovsk, Russia, in 1979 (Holty et al., 2006a). More recently, three cases of inhalational anthrax were described in drum-makers using hides from infected animals; one in the United States and two in the United Kingdom (Anaraki et al., 2008; Chan, 2006; Euro Surveillance Team, 2006; Nguyen et al., 2010). The most recent US case occurred in 2011 in Minnesota after an individual traveling in the southwest reportedly contracted the disease from the natural environment (Minnesota Department of Health, 2011).

Of the 82 cases identified in the world's literature from 1900 to 2005 (excluding the cases from Sverdlovsk), the overall mortality rate was 85% (Holty et al., 2006a). Mortality rates were 92% without antibiotics and 75% with antibiotics (Holty et al., 2006a). However, in the 2001 US outbreak where patients received early antibiotics and aggressive support, the mortality rate was only 45%. In the recent cases in drum-makers and the traveler in the US southwest, one patient (a drum-maker) died.

Two outbreaks of inhalational anthrax demonstrate its potential risk as a biologic weapon. The first occurred in 1979 in the city of Sverdlovsk in the former Soviet Union and was ultimately attributed to an accidental release of spores from a military microbiology facility (Christopher et al., 1997). Although this is the largest reported outbreak of human inhalational anthrax, the precise number of individuals infected is unknown. While one account suggested that there were up to 96 cases with 64 deaths, a comprehensive review of this outbreak noted 77 cases with 66 deaths (Meselson et al., 1994). The other outbreak was the one that occurred in the United States in 2001 related to the mailing of spore-containing envelopes (Jernigan et al., 2002). There were 11 reported cases of inhalational disease with 5 of these patients developing shock and subsequently dying. The specific source of the contaminated envelopes may have been a US biodefense laboratory (Rasko et al., 2011; Welsh, 2011).

Injectional anthrax was first described in a heroin user from Norway in 2000 who died with *B. anthracis* soft tissue infection following a contaminated injection (Ringertz et al., 2000). Then, between 2009 and 2010, 54 confirmed *B. anthracis* soft tissue infections related to injection with contaminated heroin were reported in Scotland, England, and Germany (47, 5, and 2 cases, respectively) of which 18 patients died despite receiving aggressive supportive therapy (Health Protection Scotland (HPS), 2011; Sweeney et al., 2011). During this outbreak in Scotland, there were also 72 patients who reportedly had either probable (35) or possible (37) infection based on HPS (2011) criteria. Although that outbreak was declared over in late 2010, there have been additional injectional anthrax cases reported over the past 2 years. This very recent outbreak began in 2012 with 14 cases occurring in Germany, France, Denmark, and the United Kingdom, with 6 deaths overall (Grunow et al., 2012, 2013; Guidi-Rontani, 2002). Cases presenting from 2009 on have all been attributed to a similar strain of *B. anthracis* that may have contaminated heroin during the drug's shipment through Afghanistan and Turkey (Grunow et al., 2013).

41.3 Microbiology and pathogenesis

B. anthracis is part of the *B. cereus* group of bacilli (Murray and Baron, 2007). *B. anthracis* is a Gram-positive, nonmotile, aerobic, facultative anaerobic, large (1–8 μm long, 1–1.5 μm wide) rod-shaped bacterium that forms spores (Mock and Fouet, 2001). The spore form is approximately 1 μm in size and dormant (Inglesby et al., 2002). However, with increased temperature and CO_2 levels, the spore transforms into the rod-shaped vegetative form. This form is capable of replication and the production of virulence factors, and ultimately disease. When the vegetative form is exposed to O_2 and reduced temperature, such as after the death and decomposition of an infected animal, the bacterium sporulates (Mock and Fouet, 2001).

B. anthracis survives or grows in three distinct environments: soil, animals (herbivores primarily), and humans. Spores can persist in the soil for prolonged periods (Titball et al., 1991). However, after ingestion by herbivores, spores germinate within the host producing the vegetative forms. These vegetative forms then replicate within the host, express their virulence factors, and if untreated, can kill the host. Raw materials from animals dying of *B. anthracis* infection, such as meat, hide, hair, or bone, can be heavily contaminated with anthrax spores and serve as a source for human infection (Brachman, 1970; Mock and Fouet, 2001; World Health Organization et al., 2008).

B. anthracis infection in humans occurs when spores enter the body through the skin, are ingested with contaminated food, are inhaled, or are injected into the soft tissue resulting in cutaneous, GI, inhalational, or injectional anthrax, respectively (Hicks et al., 2012; Holty et al., 2006b; Zakowska et al., 2012). Human-to-human transmission of anthrax has not been reported to occur (Dixon et al., 1999). Whether the progression of infection from initial spore entry to dissemination of vegetative forms in the blood stream and lymphatic system is similar for all four forms of disease is unclear (Corre et al., 2013; Glomski et al., 2007; Weiner and Glomski, 2012). For cutaneous and GI anthrax, germination is described as occurring in macrophages after delivery of spores to subepithelial layers (Pile et al., 1998; World Health Organization et al., 2008). This is associated with local edema and necrosis. Early vegetative forms then localize at regional lymph nodes where they replicate and release toxin and other virulence factors. Dependent on the host's response, infection may be gradually controlled. However, if it is not, vegetative forms proliferate further and spread into the lymphatic system or blood stream. Soft tissue infection and systemic spread with injectional anthrax may follow similar steps. In the case of GI anthrax, it has also been suggested that vegetative forms develop in partially cooked meat and are then delivered to the GI tract during eating (Inglesby et al., 2002).

There is debate regarding steps in the progression of inhalational anthrax. It was believed for some time that spores deposited in the alveolar spaces were phagocytized by macrophages and transported by these cells to mediastinal lymph nodes (Guidi-Rontani, 2002; Weiner and Glomski, 2012; Young and Zelle, 1946). Germination occurred either in macrophages or after release of spores in the mediastinum. Vegetative forms then replicated, proliferated, and were subsequently released into the lymphatic system and blood stream producing systemic disease. However, more recent data from bioluminescent imaging in mouse models have raised the possibility that more analogous to cutaneous and GI anthrax, inhalational disease arises after direct spread of spores to lymphatic tissue in the upper airways where germination and the subsequent spread of vegetative forms occurs (Glomski et al., 2008; Weiner and Glomski, 2012).

Three factors have classically been associated with the pathogenesis of *B. anthracis*: a phagocytic-resistant capsule; the production of two exotoxins, lethal and edema toxin (LT and ET, respectively); and the bacteria's ability to replicate and produce high bacterial loads in infected hosts (Friedlander, 2001). However, growing research over the past 10–15 years has begun to identify additional potentially important factors including biologically active components in the vegetative bacterium's cell wall and metalloproteases other than LT (Sweeney et al., 2011).

Fully virulent *B. anthracis* strains carry two large plasmids, pXO1 and pXO2, which encode for capsule and toxin components. Vegetative forms of the bacteria express these virulence genes (Dixon et al., 1999; Mock and Fouet, 2001). The transcriptional activator AtxA (DAI) regulates expression of the toxin and capsule genes. Host-specific factors such as elevated temperature (>37°C) and carbon dioxide concentration (>5%) and the presence of serum components appear to control AtxA activity (Dale et al., 2012; Fouet, 2010; Hammerstrom et al., 2011; Makino et al., 1988, 1989).

The capsule-bearing plasmid of *B. anthracis*, pXO2, is 95.3 kbp in size and includes three genes (capB, capC, and capA) that participate in the synthesis of a poly-D-glutamic acid capsule and is an important mechanism for immune cell evasion by the bacteria (Makino et al., 1989). The capsule is weakly immunogenic, resists phagocytosis by macrophages and neutrophils, and may enhance the lethal effects of LT (Jang et al., 2011, 2013; Scorpio et al., 2007, 2010).

The toxin-bearing plasmid, pXO1, is 184.5 kbp in size and includes genes producing the three proteins necessary for toxin formation: protective antigen (PA), lethal factor (LF), and edema factor (EF). The toxins are binary or A-B type toxins and include LT, made up of LF and PA, and ET, made up of EF and PA (Moayeri and Leppla, 2009; Muehlbauer et al., 2007; Pezard et al., 1991; Welsh, 2011). PA is the B component that initiates host cell binding and uptake of LF or EF, the two toxigenic or A components (Figure 41-1). The crystal structures of PA and LF have been described (Gogol et al., 2013; Katayama et al., 2010; Liddington et al., 1999; Pannifer et al., 2001; Petosa et al., 1997).

LF is a zinc metalloprotease that inactivates mitogen-activated protein kinase kinase pathways and causes lysis of macrophages *in vitro* (Duesbery and Vande Woude, 1999; Pellizzari et al., 1999; Vitale

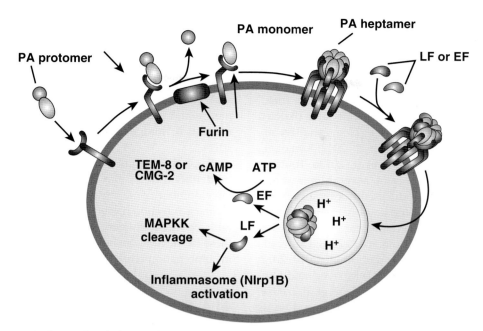

Fig. 41-1 Key events in the uptake of edema or lethal toxin by host cells as well as the potential effects of edema (EF) and lethal factor (LF). During infection, circulating protective antigen (PA) binds to one of at least two host cellular receptors (anthrax toxin receptors, ATR-1 and ATR-2) encoded by the tumor endothelial marker 8 (TEM-8) gene or another receptor encoded by the capillary morphogenesis gene 2 (CMG-2), both of which are present on many tissues. The bound PA precursor molecule undergoes furin cleavage with release of an unbound subunit. Bound PA monomers form a heptamer that 1–3 circulating LF or EF proteins competitively bind to. This complex undergoes endocytosis and the toxic factors are then released into the cytoplasm. EF has calmodulin-dependent adenyl cyclase activity and increases intracellular cAMP levels. LF inhibits mitogen-activated protein kinase 1–4, 6, and 7 and also results in inflammasome (Nlrp1B) activation.

et al., 1998, 1999). LF has been demonstrated to inhibit a variety of important host cell functions (Agrawal et al., 2003; Chavarria-Smith and Vance, 2013; Dang et al., 2004; Moayeri et al., 2012; Park et al., 2002; Popov et al., 2002). However, growing evidence also shows that the proteolytic activity of LF can stimulate host cell Nlrp1b inflammasome formation (Levinsohn et al., 2012; Moayeri et al., 2012; Newman et al., 2010). This occurs through constitutive p38 or AKT activation in macrophages and dendritic cells, resulting in IL-1β and IL-18 maturation and rapid cell death (Ali et al., 2011; Chavarria-Smith and Vance, 2013; Frew et al., 2012; Levinsohn et al., 2012; Liao and Mogridge, 2009; Moayeri et al., 2010, 2012).

EF is a calmodulin-dependent adenyl cyclase that increases intracellular cAMP levels to high levels and can impair host defenses, including phagocytosis and cytokine production by macrophages (Hoover et al., 1994; Leppla, 1982). It causes edema when injected subcutaneously into experimental animals and has been implicated in the massive extravascular fluid collection sometimes noted in patients with cutaneous anthrax (Tippetts and Robertson, 1988). While this edema formation suggests that ET may alter endothelial barrier function, evidence also indicates that ET produces arterial and venous relaxation (Hicks et al., 2011a, 2011b; Li et al., 2013a, 2013b).

During infection all three toxin components are released by vegetative bacterial forms (Baldari et al., 2006; Tournier et al., 2009a; Welsh, 2011). Based on studies in rabbits, LF may be produced in greater proportions than EF (Molin et al., 2008). Host cells

produce at least two anthrax toxin receptors: tumor endothelial marker 8 (TEM-8 or ATR-1) or capillary morphogenesis gene 2 (CMG-2 or ATR-2) (Bradley et al., 2001; Cryan and Rogers, 2011; Liu et al., 2013; Scobie et al., 2003). Circulating PA can recognize and bind to each of these receptors, although CMG-2 may have more influence during the development of disease (Liu et al., 2010; Moayeri and Leppla, 2009). A range of tissues have been shown to express both receptors including heart, lung, small intestine, spleen liver, kidney, skeletal muscle, and skin (Bonuccelli et al., 2005; Hicks et al., 2011a). After binding, the PA precursor molecule is cleaved by a furin protease and the bound subunits form a heptamer (Abrami et al., 2010; Gordon et al., 1997; Guichard et al., 2012; Klimpel et al., 1994; Mogridge et al., 2002). Then 1–3 circulating LF and EF proteins competitively bind to this heptamer. The resulting complex combines with lipid rafts on the host cell surface and undergoes endocytosis and internalization. After progressive acidification of the endocyst, LF and EF are released into the host cell's cytoplasm. Substantial work has been done defining the events that result in the internalization and release of LF and EF (Elliott et al., 2000; Young and Collier, 2007). Data also suggest that EF may facilitate uptake of both toxic factors by increasing cell surface receptor expression (Larabee et al., 2011).

While LT appears more lethal than ET on a weight basis in some species, both toxins can produce shock and lethality (Cui et al., 2007; Firoved et al., 2005; Sweeney et al., 2010). The mechanisms underlying these actions are not entirely clear. Neither toxin produces the excessive inflammatory response typically associated with bacterial sepsis and septic shock. In fact, both LT and ET appear to inhibit elements in the innate and adaptive immune responses and may contribute to the bacteria's ability to produce infection initially and high microbial loads later (Baldari et al., 2006; Chung et al., 2011b; Coggeshall et al., 2013; Comer et al., 2005; Hicks et al., 2012; Kiel et al., 2002; Pickering et al., 2004; Qiu et al., 2013a; Tournier et al., 2009a, 2009b). At present, several lines of evidence suggest that LT produces shock by either disrupting the integrity of arterial and venous endothelium or by producing direct myocardial depression (Barochia et al., 2012; Bolcome et al., 2008; Cui et al., 2004; Frankel et al., 2009; Kandadi et al., 2010, 2012; Kirby, 2004; Lawrence et al., 2011; Moayeri and Leppla, 2009; Sweeney et al., 2010; Watson et al., 2007a, 2007b). While ET was originally named based on its ability to produce local edema after subcutaneous injection, it is unclear whether this action underlies development of shock. A large body of data suggests that increases in intracellular cAMP actually promotes vascular integrity and mitigates against edema formation (Guichard et al., 2012; Nguyen et al., 2012; Sayner, 2011; Sayner et al., 2011). On the other hand, cAMP increases in vascular smooth muscle are well recognized to cause arterial and vascular relaxation (Li et al., 2013). Thus, shock and organ injury with ET may relate to excessive vascular dilation that in turn causes edema formation (Hicks et al., 2011a; Li et al., 2013; Sweeney et al., 2010).

While LT or ET do not produce an excessive inflammatory response in animal models, live *B. anthracis* challenge does and as with other types of pathogenic Gram-positive bacteria, this may contribute to shock and organ injury (Chakrabarty et al., 2007; Stearns-Kurosawa et al., 2006). Studies show that the *B. anthracis* cell wall and its peptidoglycan component can produce a robust inflammatory response via the stimulation of several different host pathogen pattern recognition proteins including TLR2/6 heterodimers (Hsu et al., 2008; Langer et al., 2008; Triantafilou et al., 2007). Challenge in rats with whole anthrax cell wall as well as purified peptidoglycan alone produced lethality; hypotension and organ injury; increases in circulating inflammatory cytokines, chemokines, and nitric oxide; and coagulopathy and thrombocytopenia (Cui et al., 2010; Qiu et al., 2013a). Interestingly, LT and ET, while not only contributing to hemodynamic dysfunction directly, may also interfere with host defense mechanisms and promote bacterial growth and the dissemination of toxic cell wall components (Cui et al., 2006; Kau et al., 2010).

Growing evidence also suggests that *B. anthracis* produces metalloproteases other than LF which can produce shock and tissue injury. These include proteases belonging to the M4 thermolysin and M9 bacterial collagenase families, and Npr599 and InhA. These proteases cleave proteins important in endothelial function including plasma ADAMTS13, von Willebrand factor (VWF) substrate FRETS-VWF73, and VWF itself (Chung et al., 2008, 2009, 2011a, 2011b, 2013; Popov et al., 2004). InhA also inactivates plasminogen activator inhibitor (PAI)-1 in mouse liver and increases blood–brain barrier permeability in mouse and human brain microvasculature endothelial cells (Chung et al., 2009, 2011a; Mukherjee et al., 2011).

These diverse pathogenic factors, as well as other unidentified ones, may provide a basis for why the development of shock and organ injury during *B. anthracis* infection has been observed to be associated with a particularly poor prognosis (Figure 41-2). How these factors interact however and whether they work in combination requires further study.

41.4 Clinical disease

41.4.1 Cutaneous anthrax

Cutaneous anthrax occurs when spores are introduced subcutaneously via a laceration or abrasion. Excoriation of the skin can promote this entry. Uncovered areas of the skin, such as the head, neck, and distal extremities are most commonly infected (Doganay et al., 2010; Inglesby et al., 2002; Kayabas et al., 2012). In a recent review of 58 cases of cutaneous anthrax cases of which 62% were related to the butchering of infected animals, the affected sites were hands (39%), fingers (29%), forearms (12%), eyelids (7%), and neck (3%) (Baykam et al., 2009). Reports of other outbreaks have described similar findings (Doganay and Metan, 2009). Spores introduced locally germinate resulting in vegetative forms that produce local edema. The primary skin lesion is usually a poorly defined, painless, pruritic papule developing 3–5 days following spore introduction. After 1–2 additional days, vesicular lesions develop that undergo central necrosis and drying, leaving a characteristic black eschar surrounded by edema and several purplish vesicles. The eschar is subsequently shed after 1–2 weeks (Kutluk et al., 1987). Cutaneous lesions generally demonstrate necrosis and edema with lymphocytic infiltrates (Mallon and McKee, 1997). Hemorrhage and thrombosis occur in the absence of tissue liquefaction or abscess formation.

Although cutaneous anthrax is generally self-limited, antibiotic treatment is recommended to decrease the potential for systemic or severe disease. However, severe disease can occur. In a recent review of 20 patients from Turkey, 45% of cases were described as severe with 9% having septic shock (Doganay et al., 2010). When severe disease with cutaneous anthrax does develop, it can be characterized by extensive edema and shock. When the neck and chest are involved, airway compromise can occur (Doganay et al., 2010; Karahocagil et al., 2008). However, with appropriate antibiotic treatment, death with cutaneous anthrax is rare (Tutrone et al., 2002).

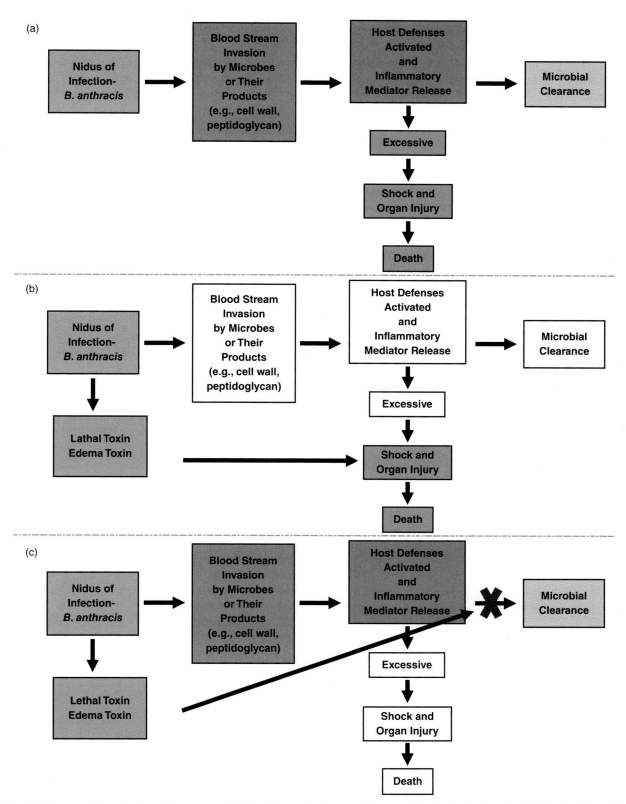

Fig. 41-2 An overview of basic pathways potentially leading to shock, organ injury, and death during *B. anthracis* infection. (Panel a) Gram-positive bacteria, *B. anthracis* and its products (e.g., cell wall and peptidoglycan) activate host defenses and inflammatory mediator release which are necessary for microbial clearance. However, if this response is excessive it may result in the development of shock, organ failure, and death. (Panel b) *B. anthracis* also produces two exotoxins: lethal and edema toxins, which are capable of contributing directly to shock, organ injury, and death via diverse mechanisms. (Panel c) Lethal and edema toxin also appear capable of subverting critical host defense systems and contributing to the pathogenesis of shock, organ injury, and death by limiting microbial clearance. Other mechanisms not depicted in this figure such as the activation of metalloproteases other than LF may contribute to shock and organ injury with *B. anthracis* as well (see text).

41.4.2 Gastrointestinal anthrax

GI anthrax occurs in one of two forms: oral-pharyngeal or lower GI. The oral-pharyngeal form starts with the development of an oral or esophageal ulcer followed by regional lymphadenopathy, edema, and sepsis (Sirisanthana et al., 1984). The lower GI form is associated with intestinal lesions occurring predominantly in the terminal ileum or cecum. Patients may present initially with nausea, vomiting, and malaise; which is followed by a severe phase with bloody diarrhea, evidence of peritonitis, or sepsis (Beatty et al., 2003; Khoddami et al., 2010). If disease occurs high enough in the GI tract, hematemesis may be evident. About 2–4 days after the onset of symptoms, ascites can develop which ranges from clear to purulent in appearance and yields *B. anthracis* in culture (Dutz et al., 1970; Hatami et al., 2010; Sirisanthana and Brown, 2002). Pathologic examination of the GI tract demonstrates necrosis and edema of infected tissues with mesenteric lymphadenitis and large Gram-positive bacilli associated with inflammatory infiltrates in the mucosal, submucosal, and lymphatic tissues (Khoddami et al., 2010; Klempner et al., 2010; Sirisanthana and Brown, 2002).

In the last confirmed US case of GI anthrax, the patient initially experienced flu-like symptoms followed by nausea, vomiting, and abdominal cramps and then hypotension (CDC, 2010; Klempner et al., 2010). Laboratory data showed leukocytosis and hemoconcentration while an abdominal CT demonstrated prominent ascites, thickened small bowel segments, and retroperitoneal lymphadenopathy. Exploratory laparotomy demonstrated nodular hemorrhagic lesions in the mesentery and two areas of necrotic small bowel which required resection. Hypotension and respiratory failure later worsened and repeat laparotomy showed a retroperitoneal hematoma that was removed. Following the second surgical procedure, the patient slowly recovered (Klempner et al., 2010).

41.4.3 Inhalational anthrax

As previously described, it has long been thought that inhalational anthrax occurred when anthrax spores less than 5 μm in size were inhaled and reached the lower respiratory tract (Druett et al., 1953). Alveolar macrophages then phagocytized and transported the spores to the hilar and mediastinal lymph nodes, where they germinated, proliferated, and spread systemically (Guidi-Rontani, 2002; Inglesby et al., 2002; Weiner and Glomski, 2012). However, new evidence has raised the possibility that spores gain entry to subepithelial and lymphatic tissue in the upper airways where germination occurs and where vegetative forms can spread (Weiner and Glomski, 2012).

Regardless of the site of entry, the clinical presentation of inhalational anthrax appears to occur in two stages. Patients initially develop a prodromal syndrome with flu-like symptoms including cough, fever, and fatigue (Brachman, 1980; Holty et al., 2006a). This prodromal stage can last from hours to a few days and may be followed by a brief period of apparent recovery. The second stage is a fulminant one with rising temperature, worsening respiratory distress, and the onset of shock. Patients presenting in the 2001 outbreak in the United States appeared to follow this course as have recent isolated cases related either to infected drums or the case from Minnesota (Jernigan et al., 2001; Minnesota Department of Health, 2011). It is reported that 38–76% of patients with inhalational anthrax develop meningoencephalitis with associated symptoms and up to 50% of these cases have a hemorrhagic component (Abramova et al., 1993; Brachman, 1980; Gordon et al., 1997; Holty et al., 2006a, 2006b; Inglesby et al., 1999; Lanska, 2002). In such cases, cyanosis and hypotension progress rapidly and death occurs within hours. Inhalational anthrax is also associated with mediastinal changes that can be detected on chest radiograph relatively early in the course of disease (Vessal et al., 1975; Wood et al., 2003). It is frequently associated with pleural effusions which may be hemorrhagic (Holty et al., 2006a). Autopsy examination can show a focus of necrotizing hemorrhagic pneumonitis possibly representing a portal of infection as well as hemorrhage and necrosis of peribronchial and mediastinal lymph nodes (Abramova et al., 1993). GI and leptomeningeal lesions as a result of hematogenous spread are also often noted in patients at autopsy.

Based on findings from the Sverdlovsk outbreak, the modal incubation time for inhalational anthrax appears to be about 10 days (Dixon et al., 1999; Meselson et al., 1994). However, in some patients the onset of illness occurred up to 6 weeks after the reported time of exposure, suggesting that anthrax spores may remain viable in the lung for prolonged periods. Due to its nonspecific symptoms, diagnosis of early inhalational anthrax requires a high index of suspicion. Analysis and modeling of known cases of inhalational anthrax from the past 100 years have suggested that age, duration of disease prior to treatment, the use of antibiotics or antiserum, and possibly repetitive pleural drainage influence mortality (Bravata et al., 2007; Holty et al., 2006a). Widened mediastinum or pleural effusion on radiography, hemoconcentration, and hyponatremia on laboratory exam and tachycardia may differentiate pulmonary disease with anthrax from other types of community-acquired pneumonia (Abrami et al., 2003; Ingram et al., 2010; Jernigan et al., 2001; Kuehnert et al., 2003; Mayer et al., 2001; Mytle et al., 2013).

41.4.4 Injectional anthrax

Due to the relatively recent recognition of injectional anthrax, a clear clinical picture of this form of disease has not yet developed. The first known case was a 49-year-old heroin injection user who presented with a 4-day history of an infection in his gluteal region (Ringertz et al., 2000). The infected area was erythematous but did not demonstrate pus or an eschar. The patient was treated with oral dicloxacillin but re-presented 4 days later in shock and coma and with evidence of meningitis on lumbar puncture. The gluteal region, thigh, and lower abdominal wall were erythematous but still without eschar formation. Surgical exploration demonstrated extensive edema of muscles and subcutaneous tissue but no pus or necrosis. Despite treatment with high-dose penicillin, chloramphenicol, and dexamethasone, the patient died several days after admission. Cerebrospinal fluid (CSF) and wound samples grew *B. anthracis*. The absence of an eschar and the similarity of the case to reports of soft tissue anthrax infection after injection with anthrax-contaminated antibiotic in India as well as disease in chimpanzees after subcutaneous injection with the bacteria led to the term injectional anthrax for this form of infection (Berdjis and Gleiser, 1964; Lalitha et al., 1988).

The next reported cases of injectional anthrax occurred in December 2009, when two hospitalized injection drug users from Glasgow in the United Kingdom were found to have blood cultures positive for *B. anthracis* (Booth et al., 2010; Ramsay et al., 2010). Additional cases subsequently presented and by the end of the outbreak in December 2010, there were 47 confirmed cases with 13 deaths in Scotland, 5 cases with 4 deaths in England, and 2 cases with 1 death in Germany (HPS, 2011). At least 10 of these cases have now been reported on individually (Beaumont, 2010; Jallali et al., 2011; Parcell et al., 2011; Powell et al., 2011; Radun et al., 2010). Nine patients presented with tissue swelling and evidence of soft tissue infection 1–10 days after the injection of heroin. Skin changes, noted in some but not all of these cases, were not distinctive and appeared similar to other skin manifestations of injection drug use. One patient presented with abdominal symptoms and evidence of peritonitis following groin injection. Confirmed diagnosis was based on positive blood or tissue cultures and/or PCR analysis. Nine cases were reported to require surgical debridement and/or fasciotomy. Laparotomy was performed in the patient with abdominal symptoms. Excessive bleeding and edema were frequently noted at surgery. While this initial outbreak appeared to end in December 2010, new cases of injectional anthrax have been reported in the United Kingdom and Europe and some have been reported in Grunow et al. (2012, 2013) and Hicks et al. (2012).

41.5 Diagnostic methods

Based on the Center for Disease Control (CDC) (CDC, 2001b) recommendations, a confirmed anthrax case is defined as a clinically compatible one in combination with isolation of *B. anthracis* itself or with at least two positive supportive tests. Routine culture with confirmation by immunohistochemical staining or real-time PCR is most frequently employed (CDC, 2001b). A variety of samples allow isolation of *B. anthracis* including blood, surgical tissue samples, exudate from skin lesions, CSF, pleural fluid, sputum, and feces. In systemic infections, while organisms can easily be cultured from early blood samples, this is not the case after antimicrobial therapy has been administered. Anthrax should immediately be considered if specimens demonstrate Gram-positive bacilli in chains, especially if these are in high concentrations (CDC, 2001b).

Because infection can progress rapidly, suspicion of anthrax should be immediately followed by testing and initiation of antibiotic therapy (Sejvar et al., 2005). In 1999, the Laboratory Response Network (LRN) was established with input from the CDC, the Association of Public Health Laboratories, the Federal Bureau of Investigation, and the United States Army Research Institute of Infectious Diseases (Morse et al., 2003). Suspected *B. anthracis* isolates should be sent to a LRN reference laboratory for testing, and local or state health departments should be notified (Morse et al., 2003). Most clinical laboratories do not have the testing capabilities required to confirm the diagnosis of *B. anthracis* but LRN laboratories do. Such testing includes susceptibility to gamma phage lysis, an LRN real-time PCR assay, a direct fluorescent assay (DFA), and time-resolved fluorescent assay (TRF) for specific detection of *B. anthracis* antigens (Brown and Cherry, 1955; De et al., 2002; Hoffmaster et al., 2002). A commercially available immunochromatographic test that uses nonhemolytic *Bacillus* isolates cultured on sheep blood agar (Redline Alert, Tetracore, Inc., Rockville, MD, USA) has also been approved by the US Food and Drug Administration (FDA) for the identification of *B. anthracis* isolates (HPS, 2011; FDA, 2004). Supportive laboratory tests include the LRN PCR assay, immuno-histochemical staining (IHC) of tissues, and an anti-PA IgG detected by an enzyme-linked immunosorbent assay (ELISA) (CDC, 2001a). The LRN PCR targets three loci on the *B. anthracis* chromosome and each of the two virulence plasmids (pXO1 and pXO2). It has a limit of detection of 1 pg of DNA (approximately 167 cells) (Hoffmaster et al., 2002). Immunohistochemical assays permit specific detection of *B. anthracis* cells in formalin-fixed tissues with antibodies specific for a *B. anthracis* cell wall or capsule antigen (Shieh et al., 2003). Serologic assays for the detection of an antibody response against the PA anthrax toxin protein include the quantitative human anti-PA IgG ELISA and a commercially available qualitative kit (QuickELISA™ Anthrax-PA Kit; Immunetics, Boston, MA) (Quinn et al., 2002). Several cases in the outbreak of injectional anthrax in the United Kingdom between 2009 and 2010 were diagnosed based on measurable increases in concentrations of serum antibody to PA and LF from paired samples drawn at the time of initial hospitalization and then later (HPS, 2011). Table 41-3 summarizes diagnostic tests for *B. anthracis* infection.

Table 41-3 Laboratory response network (LRN) approved tests for *B. anthracis*

Test	Test summary
Gram stain	Identifies *Bacillus* genus
Culture with immunohistochemical staining	Directly identifies *B. anthracis* and offers susceptibility testing on 5% sheep blood or chocolate agar plates; *B. anthracis* appearance: colonies will be 2–5 mm in diameter with irregularly round borders and a ground glass appearance; nonhemolytic and nonmotile; sporulate 18–24 h after incubation at 35–37°C in a non-CO_2 environment
Gamma phage lysis susceptibility	Gamma phage specifically lyses *B. anthracis* vegetative cells and can be used to confirm diagnosis for isolates with a concomitant positive capsule stain
Real-time polymerase chain reaction (PCR)	Targets three distinct loci on the *B. anthracis* chromosome and both virulent plasmids
Direct fluorescent assay (DFA)	Two-component assay that uses fluorescein-labeled monoclonal antibodies to detect the galactose/N-acetylglucosamine cell-wall-associated polysaccharide and capsule produced by *B. anthracis* vegetative cells either in culture or directly in clinical specimens
Time-resolved fluorescent assay	Anti-PA assay similar to ELISA, but that uses detector antibodies labeled with fluorescing lanthanide chelates instead of enzymes and pulses of excitation energy to measure fluorescence
Immunochromatography; Redline Alert	A lateral flow immunoassay containing a monoclonal antibody that is specific for the presence of a cell-surface protein found in *B. anthracis* vegetative cells and used for rapid presumptive identification of *B. anthracis* from nonhemolytic Bacillus colonies cultured on sheep blood agar plates
Anti-PA IgG detection by enzyme-linked immunosorbent assay (ELISA)	Enzyme-based colorimetric detection of antibody response against the protective antigen (PA) anthrax toxin protein
Molecular characterization	Multi-locus variable-number tandem repeat analysis (MLVA) and sequencing of genes coding for 16S ribosomal RNA may be conducted for species identification and molecular characterization of *B. anthracis* isolates
Electrophoretic immunotransblot (EITB) reaction	Measures antibody protective antigen and/or lethal factor (LF) bands in serum from patients with suspected *B. anthracis* infection
Fluorescence resonance energy transfer (FRET) assay	Fluorogenic peptide-based assay used to screen for *B. anthracis* LF protease activity
Capsule staining	India ink, McFadyean staining, or DFA may be used to visualize encapsulated *B. anthracis* either in culture or directly in clinical specimens
Europium nanoparticle-based immunoassay (ENIA)	Nanoparticle-based detection of PA antibody response

41.6 Management

The CDC has recently revised the clinical guidelines for the treatment of anthrax in adults based on findings from an expert panel that met in 2011 and 2012. These guidelines are currently under revision for publication and this section on the management of *B. anthracis* infection reflects present recommendations (Stern et al., 2008). Treatment of *B. anthracis* infection must be considered in the context of those patients with known active disease and of those who have potentially been exposed to but do not yet demonstrate clinical symptoms. Given the possible severity of *B. anthracis* infection, the first suspicion of disease must prompt initiation of antibiotic treatment pending confirmed diagnosis. Immediate notification of local or state public health departments and laboratories should also occur (Inglesby et al., 2002). *B. anthracis* is very susceptible to a variety of antimicrobial agents including penicillin, chloramphenicol, tetracycline, erythromycin, streptomycin, and fluoroquinolones (Bakici et al., 2002; Turnbull et al., 2004). It is not susceptible to cephalosporins or trimethoprim–sulfamethoxazole and these agents should not be used for treatment of active disease (Kyriacou et al., 2006).

41.6.1 Active infection

For patients diagnosed with active *B. anthracis* infection, treatment recommendations differ depending on which stage of the disease is present (i.e., stage 1 (mild localized disease) vs. stage 2 (systemic disease manifested by shock, respiratory failure, or other systemic manifestations)) (Table 41-4). For stage 1 active inhalational anthrax, the CDC recommends intravenous (IV) treatment with ciprofloxacin or doxycycline and one or two additional antimicrobials (CDC, 2001c). Initial therapy may be altered based on the clinical course of the patient; one or two oral antimicrobials may be adequate as the patient improves. Oral antimicrobial treatment should be continued to complete a total course of antibiotic treatment of not less than 60 days. The treatment regimens for stage 1 GI or oropharyngeal anthrax are the same as those recommended for stage 1 inhalational anthrax (CDC, 2001c).

For patients affected with stage 2 active inhalational anthrax, IV ciprofloxacin is recommended over doxycycline as the primary antimicrobial agent unless ciprofloxacin use is contraindicated (Stern et al., 2008). This is mainly because fluoroquinolones are bactericidal, whereas tetracyclines are bacteriostatic. Ciprofloxacin is also favored over doxycycline

Table 41-4 Brief summary of recommended antibiotic prophylaxis or treatment for *B. anthracis* exposure or active infection[a,b]

	Exposure without active infection		Active infection
	Prophylaxis	Treatment for cutaneous anthrax lacking systemic involvement	Treatment for systemic anthrax (gastrointestinal, inhalational, injectional, or severe cutaneous)
	Ciprofloxacin 500 mg every 12 h orally	Ciprofloxacin 500 mg every 12 h orally	Ciprofloxacin 400 mg every 8 h *(preferred)* intravenously (IV)
	OR	OR	OR
	Doxycycline 100 mg every 12 h orally	Doxycycline 100 mg every 12 h orally	Doxycycline 100 mg every 12 h IV PLUS
			Meropenem 1 gm every 6–8 h IV OR
			Penicillin G 4 MU every 4–6 h IV PLUS
			Rifampin 300 mg every 12 h orally OR
			Clindamycin 600 mg every 8 h IV
Duration of therapy	60 days	60 days; if systemic disease or extensive facial/neck edema, treat with regimen for systemic involvement	IV therapy: 2–3 weeks or greater; at completion of IV therapy, an oral regimen should be continued to complete 60 days of total coverage

[a]Refer to text for further explanation.
[b]Revised Center for Disease Control (CDC) guidelines for treatment soon to be published.

because meningeal involvement is likely in systemic anthrax cases, and CNS penetration of ciprofloxacin in the presence of meningeal inflammation is higher than for doxycycline (Sejvar et al., 2005). In addition to ciprofloxacin, at least one or more additional agents with adequate CNS penetration and *in vitro* activity against *B. anthracis* (e.g., ampicillin or penicillin, meropenem, rifampin, or vancomycin) should be used in the treatment of systemic cases of anthrax regardless of clinical suspicion of meningeal involvement. Clindamycin or another agent capable of inhibiting protein synthesis is also strongly recommended for inclusion in the regimen because of their potential ability to reduce exotoxin production (Stern et al., 2008). As with stage 1, patients affected with stage 2 inhalational anthrax should follow a 60-day course of antimicrobial therapy, with adjustment of the regimen to treatment with oral antimicrobials based on the patient's clinical course (Stern et al., 2008). The treatment regimens for stage-2 GI anthrax are the same as those recommended for stage-2 inhalational anthrax.

Patients affected with localized or uncomplicated cases of naturally acquired cutaneous anthrax should be treated with oral ciprofloxacin or doxycycline for 60 days. Although previous guidelines have suggested treating cutaneous anthrax for 7–10 days, 60 days is presently recommended given the potential for concurrent exposure to aerosolized *B. anthracis* (CDC, 2001c; Inglesby et al., 1999). Oral amoxicillin is an option for completion of therapy with susceptible strains following clinical improvement, since penicillin has been shown to render cutaneous anthrax lesions culture-negative within 24 hours, and has long

been the treatment of choice in many parts of the world (Stern et al., 2008). For severe cases of cutaneous anthrax with signs of systemic involvement, extensive edema, or lesions on the head or neck, IV therapy using a multidrug approach is recommended.

Injectional anthrax treatment requires early antimicrobial management, appropriate treatment for shock, and source control. According to recommendations from HPS, management should include both prompt antimicrobial therapy and immediate surgical evaluation (Karahocagil et al., 2008). With soft tissue infection, surgical exploration and debridement are necessary for treatment as well as for possible diagnosis. Antimicrobial therapy includes empiric antibiotic treatment with ciprofloxacin and clindamycin intravenously in combination with other antibiotics such as penicillin, flucloxacillin, and metronidazole. Based on HPS recommendations, this therapy should be continued for 10–14 days therapy, followed by a consideration of either continuing this regimen, modifying it (including route of administration), or stopping therapy depending on the clinical course of the individual patient. If antimicrobial therapy is stopped at this point, patients must be monitored carefully for worsening of symptoms (Karahocagil et al., 2008). These recommendations from HPS follow the 2008 World Health Organization (WHO) recommendations for systemic or life-threatening disease which suggest a minimum of 10–14 days therapy with three agents; ciprofloxacin (IV at first) with at least one more additional agent with adequate CNS penetration (ampicillin or penicillin, meropenem, rifampicin, or vancomycin) (Stern et al., 2008) and the HPS suggests adding clindamycin to this regimen based on its potential ability to reduce toxin production (Karahocagil et al., 2008).

Prior to the advent of antimicrobial therapy, passive immunization with antiserum from horses previously immunized with the Sterne strain of *B. anthracis* (a cell-free, non-encapsulated relatively avirulent strain) was used for the treatment of anthrax with some success in both animals and patients. A systematic review of inhalational anthrax cases between 1900 and 2005 suggested that mortality was lower among patients who received antiserum (Holty et al., 2006a). Based on this past experience and the contribution, toxin production is believed to have in the pathogenesis of infection, there is increasing interest in immunoglobulin-based therapies for anthrax. Two such preparations have now been included in the U.S. Strategic National Stockpile. One of these preparations, termed anthrax immune globulin (AIG), is a polyclonal antibody developed from the serum of individuals previously vaccinated with AVA vaccine (Emergent BioSolutions, Inc., Rockville, MD; and Cangene, Inc., Frederick, MD).

AIG has toxin-neutralizing activity *in vitro* and protects mice, rabbits, and nonhuman primates (NHP) from lethal spore challenge with *B. anthracis* Sterne strain (mice) and Ames Strain (rabbits and NHP) (Mytle et al., 2013; Schneemann and Manchester, 2009). AIG has now been used in several cases of inhalational and injectional anthrax and one case of GI disease (Anaraki et al., 2008; Biagini et al., 2006; Walsh et al., 2007). Emergent BioSolutions, Inc., has evaluated the safety and pharmacokinetics of their AIG preparation in 174 healthy adult volunteers (Croasdell and Gale, 2009; Mytle et al., 2013). At least 19 patients have received supplemental AIG to antimicrobials during active infection (3 with inhalation anthrax, 15 with injectional anthrax, and 1 with GI anthrax) with a total mortality of 32% (Booth et al., 2013; HPS, 2011; Klempner et al., 2010; Minnesota Department of Health, 2011; Walsh et al., 2007). However, it is currently unclear whether AIG improves the outcome of patients with anthrax treated aggressively with conventional treatment (Booth et al., 2013). AIG is available from the CDC under an Investigational New Drug protocol for the treatment of patients with confirmed life-threatening anthrax.

The other immunoglobulin preparation now included in the U.S. Strategic National Stockpile is a human monoclonal antibody directed against recombinant PA termed raxibacumab (ABthrax) developed by Human Genome Sciences (HGS; Rockville, MD). HGS was subsequently purchased by GlaxoSmithKline (GSK, London, United Kingdom). Raxibacumab was shown to improve survival when administered before or after challenge with anthrax spore challenge in rabbits and NHP not treated with antibiotics as well as in rabbits receiving antibiotics (Corey et al., 2013; Migone et al., 2009). In canines challenged with 24 h infusions of either LT or ET in doses producing progressive shock and lethality, raxibacumab added to the beneficial effects of fluid and vasopressor hemodynamic support and improved survival (Barochia et al., 2012; Remy et al., 2013). Although raxibacumab has been shown to be safe in uninfected humans, it has not yet been administered to infected patients (Migone et al., 2009). It was approved by the FDA in 2013 (Kaye, 2013).

Other antibody- and non-antibody-based preparations directed against PA, LF, and EF have also been proposed as anthrax treatments and are under study (Albrecht et al., 2007; Altaweel et al., 2011; Artenstein and Opal, 2012; Chen et al., 2009a, 2009b, 2011; Guidi-Rontani, 2002; Hull et al., 2005; Li et al., 2009; Lim et al., 2005; Vitale et al., 2006; Wang et al., 2004; Wild et al., 2007; Zhao et al., 2003). Examples of the latter include the soluble forms of ANTXR2 (sATR/TEM8 and sCMG-2, respectively) that bind to and inhibit circulating PA; and Auranofin, a gold complex that blocks LT-induced caspase-1 activation (Artenstein and Opal, 2012; Cai et al., 2011; Manayani et al., 2007; Newman et al., 2011; Schneemann and Manchester, 2009; Scobie et al., 2005).

Inhalational anthrax is characterized by the development of pleural effusions which frequently re-accumulate after drainage. Evidence suggests that continued or intermittent drainage of these effusions may be important in the treatment of this form of anthrax. Such drainage may have improved outcome both in the 2001 US outbreak as well as in several isolated cases since (Holty et al., 2006a; Jernigan et al., 2001; Walsh et al., 2007). In one of these cases, analysis showed high pleural fluid lethal toxin levels, and the case's positive outcome was attributed in part to the beneficial effects pleural fluid drainage on respiratory mechanics as well as reductions in lethal toxin levels (Walsh et al., 2007).

Based on the use of glucocorticoids in other types of infection as well as in prior cases of anthrax, they should be considered in patients with meningoencephalitis and in cutaneous disease with extensive involvement of the neck and chest (Doust et al., 1968; Lanska, 2002; Sejvar et al., 2005). However, the effectiveness of glucocorticoids for anthrax infection is unclear. On one hand, while LT represses glucocorticoid receptor transactivation, dexamethasone did not improve outcome in a toxin-challenged mouse model (Moayeri et al., 2005; Webster et al., 2003, 2004). On the other hand, emerging data regarding the potential role *B. anthracis* cell wall components have on eliciting an injurious host inflammatory response does provide a potential rational for the anti-inflammatory effects of corticosteroids (Cui et al., 2010; Qiu et al., 2013a).

Patients presenting with *B. anthracis* infection and shock should be treated aggressively with the same type of hemodynamic support utilized for septic shock due to other types of bacterial infection (Inglesby et al., 2002). Studies in a canine model have demonstrated that conventional hemodynamic support with fluids and vasopressors titrated to pulmonary and systemic arterial pressures was beneficial during the development of shock related to LT challenge (Barochia et al., 2012).

41.6.2 Post-exposure prophylaxis

Besides patients with active *B. anthracis* disease, there are also those who may have been exposed to infection but are asymptomatic. Such patients must be considered for post-exposure prophylaxis (PEP). Although penicillin has been used historically for the treatment of anthrax, resistance has been found in some naturally occurring strains possibly because of relatively low antibiotic levels in pulmonary tissues related to oral agents (Cook et al., 1994). In 2008, the CDC recommended that for anthrax PEP, ciprofloxacin or doxycycline be considered as equivalent first-line antimicrobial agents (Stern et al., 2008). During follow-up of the 2001 US anthrax outbreak, there were no differences in self-reported symptoms when either drug was used for PEP; and no serious adverse events appeared related to either (Tierney et al., 2003).

In addition to antibiotics for PEP therapy, the CDC has recommended treatment with three doses of anthrax vaccine adsorbed (AVA) (BioThrax; Bio-Port Corporation, Lansing, MI), with the first dose administered at time zero and then 2 and 4 weeks after exposure (CDC, 2002; Stern et al., 2008; Wright et al., 2010). AVA is an aluminum hydroxide–precipitated preparation of PA from attenuated, non-encapsulated *B. anthracis* cultures of the Sterne strain that was developed for use in the United States (Gu et al., 2007; Hannesschlager, 2007). The United Kingdom employs a similar vaccine (anthrax vaccine precipitated, AVP) produced by the Health Protection Agency (HPA; Porton Down, UK) also composed of alum-precipitated PA and LF from the Sterne strain (Baillie et al., 2004). The rationale for the use of AVA with antibiotics was based on studies in NHP showing that anthrax spores persisted for up to 100 days after inhalational exposure (Friedlander et al., 1993; Henderson et al., 1956). The development of infection in some patients several weeks after the release of spores in the Sverdlovsk outbreak was consistent with these experimental findings (Meselson et al., 1994). Since disease can develop long after exposure to spores, animal studies have examined the use of post-exposure vaccination in combination with antimicrobial agents. In a study of NHP challenged with a high dose of aerosolized *B. anthracis* spores, post-exposure vaccination with AVA enhanced protection afforded by 14 days of antibiotic prophylaxis alone and completely protected animals (Vietri et al., 2006). Similar studies with guinea pigs have supported these findings (Altboum et al., 2002; Chitlaru et al., 2011). Thus, the simultaneous use of vaccine with antimicrobial therapy may confer prolonged immunity after the initial anthrax exposure and provide protection from persistent viable spores. At present, AVA is recommended by the CDC as adjunctive therapy with antibiotics for post-exposure

prophylaxis but not with active disease (Stern et al., 2008; Wright et al., 2010). Although AVA is not FDA-approved for PEP, it is available under an Investigational New Drug protocol. While the CDC has recommended use of AVA for PEP in pregnant women exposed to aerosolized *B. anthracis* spores, at present additional safety data are needed to support such a decision for children (Stern et al., 2008; Wright et al., 2010).

New vaccines with improved efficacy and the ability to meet current licensing criteria are being developed (Baillie, 2001). A recombinant protective antigen (rPA) vaccine has demonstrated efficacy in animal models (Ivins et al., 1998). In phase 1 studies, the immune response with rPA and adjuvant was not statistically different from AVA (Campbell et al., 2007; Gorse et al., 2006). However, further testing of rPA has been delayed due to problems with formulation of the vaccine.

41.6.3 Pre-exposure prophylaxis

Since 1970, the FDA has licensed AVA for the pre-exposure prophylaxis of persons at occupational risk of acquiring *B. anthracis* infection including US armed forces personnel, wool mill and laboratory workers, and veterinarians (Bartlett et al., 2002). Subcutaneous vaccination is given to healthy individuals aged 18–65 years at 0, 2, and 4 weeks and 6, 12, and 18 months, followed by annual boosters (Friedlander et al., 1999). In 2010, the CDC updated recommendation regarding pre-exposure prophylaxis which included reducing the number of doses of AVA required from 6 to 5 doses and administering doses intramuscularly rather than subcutaneously (Wright et al., 2010).

41.7 Hospital and ICU considerations

Since no data exist suggesting that person-to-person transmission of anthrax occurs, standard barrier precautions are recommended for hospitalized patients with anthrax and airborne protection is not indicated (Heymann, 2008; Pile et al., 1998). Contact isolation precautions should be used for patients with draining anthrax lesions (Mandell et al., 2005). There is also no need to immunize or provide PEP to healthcare workers of household contacts unless those persons were exposed to an aerosol or surface contamination (Inglesby et al., 2002).

Dressings from draining anthrax lesions should be treated and disposed of as biohazard waste. Individuals directly contacting *B. anthracis*-containing substances should wash their exposed skin and clothing thoroughly with soap and water. For contaminated environmental surfaces, a standard hospital disinfectant (e.g., hypochlorite, aqueous chlorine dioxide, sodium hypochlorite, hydrogen peroxide, and peroxyacetic acid combined) is sufficient for cleaning (OSHA, 2013; Kyriacou et al., 2006). *B.*

anthracis–contaminated human and animal remains should be cremated and not buried to prevent further disease spread (Inglesby et al., 2002).

41.8 Bioterrorism potential

B. anthracis is a bioterrorist threat because it is relatively easy to stimulate vegetative forms to first replicate to high numbers and to then sporulate. Once in the spore form, the bacteria can be stored, transported, and then dispersed widely as an aerosol. Events in Sverdlovsk where the airflow of a fan was inadvertently directed from the inside of a chamber holding spores to the external environment, and the subsequent infection of a large number of cases downwind of this facility, supports such a scenario. The WHO estimated that 50 kg of *B. anthracis* spores dispersed over an urban population of 5 million would sicken 250,000 and kill 10,000 (Bossi et al., 2004). A US Congressional Office of Technology analysis estimated that 100 kg of *B. anthracis* could produce between 130,000 and 3 million deaths (Inglesby et al., 2002). Although most national offensive bioweapons programs were terminated following ratification of the Biological Weapons Convention in the early 1970s, some nations did continue bioweapons development programs (Christopher et al., 1997). In 1995, Iraq acknowledged to the United Nations Special Commission that it had weaponized *B. anthracis* (Zilinskas, 1997). The former Soviet Union also had a large *B. anthracis* production program (Alibek and Handelman, 1999). Reports from several nongovernmental sources suggest that biological weapon programs may still exist in at least 13 countries (Haugen, 2001).

The US Department of Defense has shown that as few as three individuals with technical skills but no expert bioweapons knowledge could produce a *B anthracis*-based weapon at a cost of approximately 1 million dollars (Miller, 2001). Therefore, an adequately funded group could purchase a manufactured weaponized biological agent or acquire the expertise and resources to produce one (Inglesby et al., 2002; Miller, 2001). The cult responsible for dispersing Sarin nerve gas in a Tokyo subway station in 1995 also released aerosols of anthrax and botulism in this city at least eight times, although these efforts did not prove effective (Keim et al., 2001). The terrorist organization Al Qaeda reportedly sought to obtain bioweapons (Inglesby et al., 2002). Thus, use of *B. anthracis* as an agent of bioterrorism appears to be a legitimate concern (Dixon et al., 1999).

41.9 Conclusion

Over 125 years ago, anthrax emerged as a damaging disease of domesticated animals and the farming economy (Brachman, 1980). In the early 1900s, improved industrial and animal husbandry hygiene, decreased use of potentially contaminated animal products, and vaccination of those at risk of occupational exposure resulted in a steady and significant decline in the number of cases of anthrax reported in the developed world (Kyriacou et al., 2006). However, the 2001 US outbreak of inhalational disease, recent isolated cases of inhalational and GI disease in the United States and Europe, and the emergence of injectional anthrax in Europe has renewed interest in this potentially deadly infection in developed countries. This interest has stimulated a remarkable body of research directed at the pathogenesis and management *B. anthracis* infection which will likely continue to grow.

References

Abrami, L., Bischofberger, M., Kunz, B., Groux, R., and van der Goot, F.G. 2010. Endocytosis of the anthrax toxin is mediated by clathrin, actin and unconventional adaptors. *PLoS Pathog.*, 6, e1000792.

Abrami, L., Liu, S., Cosson, P., Leppla, S.H., and van der Goot, F.G. 2003. Anthrax toxin triggers endocytosis of its receptor via a lipid raft-mediated clathrin-dependent process. *J. Cell Biol.*, 160, 321–328.

Abramova, F.A., Grinberg, L.M., Yampolskaya, O.V., and Walker, D.H. 1993. Pathology of inhalational anthrax in 42 cases from the Sverdlovsk outbreak of 1979. *Proc. Natl. Acad. Sci. USA*, 90, 2291–2294.

Agrawal, A., Lingappa, J., Leppla, S.H., Agrawal, S., Jabbar, A., Quinn, C., and Pulendran, B. 2003. Impairment of dendritic cells and adaptive immunity by anthrax lethal toxin. *Nature*, 424, 329–334.

Albrecht, M.T., Li, H., Williamson, E.D., LeButt, C.S., Flick-Smith, H.C., Quinn, C.P., Westra, H., Galloway, D., Mateczun, A., Goldman, S., Groen, H., and Baillie, L.W. 2007. Human monoclonal antibodies against anthrax lethal factor and protective antigen act independently to protect against *Bacillus anthracis* infection and enhance endogenous immunity to anthrax. *Infect. Immun.*, 75, 5425–5433.

Ali, S.R., Timmer, A.M., Bilgrami, S., Park, E.J., Eckmann, L., Nizet, V., and Karin, M. 2011. Anthrax toxin induces macrophage death by p38 MAPK inhibition but leads to inflammasome activation via ATP leakage. *Immunity*, 35, 34–44.

Alibek, K., and Handelman, S. 1999. *Biohazard: The Chilling True Story of the Large Covert Biological Weapons Prgram in the World, Told from the Inside by the Man Who Ran it*, 1st ed. New York: Random House.

Alizad, A., Ayoub, E.M., and Makki, N. 1995. Intestinal anthrax in a two-year-old child. *Pediatr. Infect. Dis. J.*, 14, 394–395.

Altaweel, L., Chen, Z., Moayeri, M., Cui, X., Li, Y., Su, J., Fitz, Y., Johnson, S., Leppla, S.H., Purcell, R., and Eichacker, P.Q. 2011. Delayed treatment with W1-mAb, a chimpanzee-derived monoclonal antibody against protective antigen, reduces mortality from challenges with anthrax edema or lethal toxin in rats and with anthrax spores in mice. *Crit. Care Med.*, 39, 1439–1447.

Altboum, Z., Gozes, Y., Barnea, A., Pass, A., White, M., and Kobiler, D. 2002. Postexposure prophylaxis against anthrax: Evaluation of various treatment regimens in intranasally infected guinea pigs. *Infect. Immun.*, 70, 6231–6241.

Anaraki, S., Addiman, S., Nixon, G., Krahe, D., Ghosh, R., Brooks, T., Lloyd, G., Spencer, R., Walsh, A., McCloskey, B., and Lightfoot, N. 2008. Investigations and control measures following a case of inhalation anthrax in East London in a drum maker and drummer, October 2008. *Euro Surveill.*, 13.

Artenstein, A.W., and Opal, S.M. 2012. Novel approaches to the treatment of systemic anthrax. *Clin. Infect. Dis.*, 54, 1148–1161.

Baillie, L. 2001. The development of new vaccines against *Bacillus anthracis*. *J. Appl. Microbiol.*, 91, 609–613.

Baillie, L., Townend, T., Walker, N., Eriksson, U., and Williamson, D. 2004. Characterization of the human immune response to the UK anthrax vaccine. *FEMS Immunol. Med. Microbiol.*, 42, 267–270.

Bakici, M.Z., Elaldi, N., Bakir, M., Dokmetas, I., Erandac, M., and Turan, M. 2002. Antimicrobial susceptibility of *Bacillus anthracis* in an endemic area. *Scand. J. Infect. Dis.*, 34, 564–566.

Baldari, C.T., Tonello, F., Paccani, S.R., and Montecucco, C. 2006. Anthrax toxins: A paradigm of bacterial immune suppression. *Trends Immunol.*, 27, 434–440.

Barochia, A.V., Cui, X., Sun, J., Li, Y., Solomon, S.B., Migone, T.S., Subramanian, G.M., Bolmer, S.D., and Eichacker, P.Q. 2012. Protective antigen antibody augments hemodynamic support in anthrax lethal toxin shock in canines. *J. Infect. Dis.*, 205, 818–829.

Bartlett, J.G., Inglesby, T.V., Jr., and Borio, L. 2002. Management of anthrax. *Clin. Infect. Dis.*, 35, 851–858.

Baykam, N., Ergonul, O., Ulu, A., Eren, S., Celikbas, A., Eroglu, M., and Dokuzoguz, B. 2009. Characteristics of cutaneous anthrax in Turkey. *J. Infect. Dev. Countr.*, 3, 599–603.

Beatty, M.E., Ashford, D.A., Griffin, P.M., Tauxe, R.V., and Sobel, J. 2003. Gastrointestinal anthrax: Review of the literature. *Arch. Intern. Med.*, 163, 2527–2531.

Beaumont, G. 2010. Anthrax in a Scottish intravenous drug user. *J. Forensic Leg. Med.*, 17, 443–445.

Bell, J.H. 1880. On anthrax and anthracaemia in wool-sorters, heifers, and sheep. *Br. Med. J.*, 2, 656–657.

Bell, J.H., Fee, E., and Brown, T.M. 2002. Anthrax and the wool trade. 1902. *Am. J. Public Health.*, 92, 754–757.

Berdjis, C.C., and Gleiser, C.A. 1964. Experimental subcutaneous anthrax in chimpanzees. *Exp. Mol. Pathol.*, 33, 63–75.

Biagini, R.E., Sammons, D.L., Smith, J.P., MacKenzie, B.A., Striley, C.A.F., Snawder, J.E., Robertson, S.A., and Quinn, C.P. 2006. Rapid, sensitive, and specific lateral-flow immunochromatographic device to measure anti-anthrax protective antigen immunoglobulin G in serum and whole blood. *Clin. Vaccine Immunol.*, 13, 541–546.

Bolcome 3rd, R.E., Sullivan, S.E., Zeller, R., Barker, A.P., Collier, R.J., and Chan, J. 2008. Anthrax lethal toxin induces cell death-independent permeability in zebrafish vasculature. *Proc. Natl. Acad. Sci. USA*, 105, 2439–2444.

Bonuccelli, G., Sotgia, F., Frank, P.G., Williams, T.M., de Almeida, C.J., Tanowitz, H.B., Scherer, P.E., Hotchkiss, K.A., Terman, B.I., Rollman, B., Alileche, A., Brojatsch, J., and Lisanti, M.P. 2005. ATR/TEM8 is highly expressed in epithelial cells lining *Bacillus anthracis'* three sites of entry: Implications for the pathogenesis of anthrax infection. *Am. J. Physiol. Cell Physiol.*, 288, C1402–C1410.

Booth, M., Donaldson, L., Xizhong, C., Junfeng, S., and Eichacker, P. 2013. Comparison of Survivors and Nonsurvivors from 27 Confirmed Injectional Anthrax Cases from the 2009 Outbreak in Scotland. *Critical Care*, 17, P72.

Booth, M.G., Hood, J., Brooks, T.J., and Hart, A. 2010. Anthrax infection in drug users. *Lancet*, 375, 1345–1346.

Bossi, P., Tegnell, A., Baka, A., Van Loock, F., Hendriks, J., Werner, A., Maidhof, H., and Gouvras, G. 2004. Bichat guidelines for the clinical management of anthrax and bioterrorism-related anthrax. *Euro Surveill.*, 9, E3–E4.

Brachman, P.S. 1970. Anthrax. *Ann. N. Y. Acad. Sci.*, 174, 577–582.

Brachman, P.S. 1980. Inhalation anthrax. *Ann. N. Y. Acad. Sci.*, 353, 83–93.

Bradley, K.A., Mogridge, J., Mourez, M., Collier, R.J., and Young, J.A. 2001. Identification of the cellular receptor for anthrax toxin. *Nature*, 414, 225–229.

Bravata, D.M., Holty, J.E., Wang, E., Lewis, R., Wise, P.H., McDonald, K.M., and Owens, D.K. 2007. Inhalational, gastrointestinal, and cutaneous anthrax in children: A systematic review of cases: 1900 to 2005. *Arch. Pediatr. Adolesc. Med.*, 161, 896–905.

Brook, I. 2002. The prophylaxis and treatment of anthrax. *Int. J. Antimicrob. Agents*, 20, 320–325.

Brown, E.R., and Cherry, W.B. 1955. Specific identification of *Bacillus anthracis* by means of a variant bacteriophage. *J. Infect. Dis.*, 96, 34–39.

Cai, C., Che, J., Xu, L., Guo, Q., Kong, Y., Fu, L., Xu, J., Cheng, Y., and Chen, W. 2011. Tumor endothelium marker-8 based decoys exhibit superiority over capillary morphogenesis protein-2 based decoys as anthrax toxin inhibitors. *PloS One*, 6, e20646.

Campbell, J.D., Clement, K.H., Wasserman, S.S., Donegan, S., Chrisley, L., and Kotloff, K.L. 2007. Safety, reactogenicity and immunogenicity of a recombinant protective antigen anthrax vaccine given to healthy adults. *Hum. Vaccin.*, 3, 205–211.

Centers for Disease Control and Prevention (CDC). 2000. Human ingestion of *Bacillus anthracis*-contaminated meat–Minnesota, August 2000. *JAMA*, 284, 1644–1646.

Centers for Disease Control and Prevention (CDC). 2001a. Update: Investigation of anthrax associated with intentional exposure and interim public health guidelines, October 2001. *MMWR Morb. Mortal. Wkly. Rep.*, 50, 889–893.

Centers for Disease Control and Prevention (CDC). 2001b. Update: Investigation of bioterrorism-related anthrax and interim guidelines for clinical evaluation of persons with possible anthrax. *MMWR Morb. Mortal. Wkly. Rep.*, 50, 941–948.

Centers for Disease Control and Prevention (CDC). 2001c. Update: Investigation of bioterrorism-related anthrax and interim guidelines for exposure management and antimicrobial therapy, October 2001. *MMWR Morb. Mortal. Wkly. Rep.*, 50, 909–919.

Centers for Disease Control and Prevention (CDC). 2002. Use of anthrax vaccine in response to terrorism: Supplemental recommendations of the Advisory Committee on Immunization Practices. *JAMA*, 288, 2681–2682.

Centers for Disease Control and Prevention (CDC). 2008. Cutaneous anthrax associated with drum making using goat hides from West Africa–Connecticut, 2007. *MMWR Morb. Mortal. Wkly. Rep.*, 57, 628–631.

Centers for Disease Control and Prevention (CDC). 2010. Gastrointestinal anthrax after an animal-hide drumming event - New Hampshire and Massachusetts, 2009. *MMWR Morb. Mortal. Wkly. Rep.*, 59, 872–877.

Chakrabarty, K., Wu, W., Booth, J.L., Duggan, E.S., Nagle, N.N., Coggeshall, K.M., and Metcalf, J.P. 2007. Human lung innate immune response to *Bacillus anthracis* spore infection. *Infect. Immun.*, 75, 3729–3738.

Chakraborty, A., Khan, S.U., Hasnat, M.A., Parveen, S., Islam, M.S., Mikolon, A., Chakraborty, R.K., Ahmed, B.N., Ara, K., Haider, N., Zaki, S.R., Hoffmaster, A.R., Rahman, M., Luby, S.P., and Hossain, M.J. 2012a. Anthrax outbreaks in Bangladesh, 2009–2010. *Am. J. Trop. Med. Hyg.*, 86, 703–710.

Chakraborty, P.P., Thakurt, S.G., Satpathi, P.S., Hansda, S., Sit, S., Achar, A., and Banerjee, D. 2012b. Outbreak of cutaneous anthrax in a tribal village: A clinico-epidemiological study. *J. Assoc. Physicians India*, 60, 89–93.

Chan, S. 2006. An Apartment in Brooklyn Is Cleared in Anthrax Test, *The New York Times*.

Chavarria-Smith, J., and Vance, R.E. 2013. Direct proteolytic cleavage of NLRP1B is necessary and sufficient for inflammasome activation by anthrax lethal factor. *PLoS Pathog.*, 9, e1003452.

Chen, Z., Moayeri, M., Crown, D., Emerson, S., Gorshkova, I., Schuck, P., Leppla, S.H., and Purcell, R.H. 2009a. Novel chimpanzee/human monoclonal antibodies that neutralize anthrax lethal factor, and evidence for possible synergy with anti-protective antigen antibody. *Infect. Immun.*, 77, 3902–3908.

Chen, Z., Moayeri, M., and Purcell, R. 2011. Monoclonal antibody therapies against anthrax. *Toxins*, 3, 1004–1019.

Chen, Z., Moayeri, M., Zhao, H., Crown, D., Leppla, S.H., and Purcell, R.H. 2009b. Potent neutralization of anthrax edema toxin by a humanized monoclonal antibody that competes with calmodulin for edema factor binding. *Proc. Natl. Acad. Sci. USA*, 106, 13487–13492.

Chitlaru, T., Altboum, Z., Reuveny, S., and Shafferman, A. 2011. Progress and novel strategies in vaccine development and treatment of anthrax. *Immunol. Rev.*, 239, 221–236.

Christopher, G.W., Cieslak, T.J., Pavlin, J.A., and Eitzen Jr., E.M. 1997. Biological warfare. A historical perspective. *JAMA* 278, 412–417.

Chung, M.C., Jorgensen, S.C., Popova, T.G., Tonry, J.H., Bailey, C.L., and Popov, S.G. 2009. Activation of plasminogen activator inhibitor implicates protease InhA in the acute-phase response to *Bacillus anthracis* infection. *J. Med. Microbiol.*, 58, 737–744.

Chung, M.C., Jorgensen, S.C., Tonry, J.H., Kashanchi, F., Bailey, C., and Popov, S. 2011a. Secreted *Bacillus anthracis* proteases target the host fibrinolytic system. *FEMS Immunol. Med. Microbiol.*, 62, 173–181.

Chung, M.C., Narayanan, A., Popova, T.G., Kashanchi, F., Bailey, C.L., and Popov, S.G. 2013. *Bacillus anthracis*-derived nitric oxide induces protein S-nitrosylation contributing to macrophage death. *Biochem. Biophys. Res. Commun.*, 430, 125–130.

Chung, M.C., Popova, T.G., Jorgensen, S.C., Dong, L., Chandhoke, V., Bailey, C.L., and Popov, S.G. 2008. Degradation of circulating von Willebrand factor and its regulator ADAMTS13 implicates secreted *Bacillus anthracis* metalloproteases in anthrax consumptive coagulopathy. *J. Biol. Chem.*, 283, 9531–9542.

Chung, M.C., Tonry, J.H., Narayanan, A., Manes, N.P., Mackie, R.S., Gutting, B., Mukherjee, D.V., Popova, T.G., Kashanchi, F., Bailey, C.L., and Popov, S.G. 2011b. *Bacillus anthracis* interacts with plasmin(ogen) to evade C3b-dependent innate immunity. *PloS One*, 6, e18119.

Coggeshall, K.M., Lupu, F., Ballard, J., Metcalf, J.P., James, J.A., Farris, D., and Kurosawa, S. 2013. The sepsis model: An emerging hypothesis for the lethality of inhalation anthrax. *J. Cell. Mol. Med.*, 17, 914–920.

Cohn, F. 1875. *Beiträge zur Biologie der Pflanzen*. Breslau: J.U. Kern's Verlag (Max Müller). p. v.

Comer, J.E., Chopra, A.K., Peterson, J.W., and Konig, R. 2005. Direct inhibition of T-lymphocyte activation by anthrax toxins in vivo. *Infect. Immun.*, 73, 8275–8281.

Cook, P.J., Andrews, J.M., Woodcock, J., Wise, R., and Honeybourne, D. 1994. Concentration of amoxycillin and clavulanate in lung compartments in adults without pulmonary infection. *Thorax*, 49, 1134–1138.

Corey, A., Migone, T.S., Bolmer, S., Fiscella, M., Ward, C., Chen, C., and Meister, G. 2013. *Bacillus anthracis* protective antigen kinetics in inhalation spore-challenged untreated or levofloxacin/raxibacumab-treated New Zealand white rabbits. *Toxins*, 5, 120–138.

Corre, J.P., Piris-Gimenez, A., Moya-Nilges, M., Jouvion, G., Fouet, A., Glomski, I.J., Mock, M., Sirard, J.C., and Goossens, P.L. 2013. In vivo germination of *Bacillus anthracis* spores during murine cutaneous infection. *J. Infect. Dis.*, 207, 450–457.

Croasdell, G., and Gale, S. 2009. 27th Annual JPMorgan Healthcare Conference–BioCryst and emergent biosolutions. *IDrugs*, 12, 149–151.

Cryan, L.M., and Rogers, M.S. 2011. Targeting the anthrax receptors, TEM-8 and CMG-2, for anti-angiogenic therapy. *Front. Biosci. (Landmark Ed)*, 16, 1574–1588.

Cui, X., Li, Y., Li, X., Haley, M., Moayeri, M., Fitz, Y., Leppla, S.H., and Eichacker, P.Q. 2006. Sublethal doses of *Bacillus anthracis* lethal toxin inhibit inflammation with lipopolysaccharide and Escherichia coli challenge but have opposite effects on survival. *J. Infect. Dis.*, 193, 829–840.

Cui, X., Li, Y., Li, X., Laird, M.W., Subramanian, M., Moayeri, M., Leppla, S.H., Fitz, Y., Su, J., Sherer, K., and Eichacker, P.Q. 2007. *Bacillus anthracis* edema and lethal toxin have different hemodynamic effects but function together to worsen shock and outcome in a rat model. *J. Infect. Dis.*, 195, 572–580.

Cui, X., Moayeri, M., Li, Y., Li, X., Haley, M., Fitz, Y., Correa-Araujo, R., Banks, S.M., Leppla, S.H., and Eichacker, P.Q. 2004. Lethality during continuous anthrax lethal toxin infusion is associated with circulatory shock but not inflammatory cytokine or nitric oxide release in rats. *Am. J. Physiol. Regul. Integr. Comp. Physiol.*, 286, R699–R709.

Cui, X., Su, J., Li, Y., Shiloach, J., Solomon, S., Kaufman, J.B., Mani, H., Fitz, Y., Weng, J., Altaweel, L., Besch, V., and Eichacker, P.Q. 2010. *Bacillus anthracis* cell wall produces injurious inflammation but paradoxically decreases the lethality of anthrax lethal toxin in a rat model. *Intensive Care Med.*, 36, 148–156.

Dale, J.L., Raynor, M.J., Dwivedi, P., and Koehler, T.M. 2012. cis-Acting elements that control expression of the master virulence regulatory gene atxA in *Bacillus anthracis*. *J. Bacteriol.*, 194, 4069–4079.

Dang, O., Navarro, L., Anderson, K., and David, M. 2004. Cutting edge: Anthrax lethal toxin inhibits activation of IFN-regulatory factor 3 by lipopolysaccharide. *J. Immunol.*, 172, 747–751.

Davies, J.C. 1982. A major epidemic of anthrax in Zimbabwe. *Cent. Afr. J. Med.*, 28, 291–298.

Davies, J.C. 1983. A major epidemic of anthrax in Zimbabwe. Part II. *Cent. Afr. J. Med.*, 29, 8–12.

De, B.K., Bragg, S.L., Sanden, G.N., Wilson, K.E., Diem, L.A., Marston, C.K., Hoffmaster, A.R., Barnett, G.A., Weyant, R.S., Abshire, T.G., Ezzell, J.W., and Popovic, T. 2002. A two-component direct fluorescent-antibody assay for rapid identification of *Bacillus anthracis*. *Emerg. Infect. Dis.*, 8, 1060–1065.

Demirdag, K., Ozden, M., Saral, Y., Kalkan, A., Kilic, S.S., and Ozdarendeli, A. 2003. Cutaneous anthrax in adults: A review of 25 cases in the eastern Anatolian region of Turkey. *Infection*, 31, 327–330.

Dirckx, J.H. 1981. Virgil on anthrax. *Am. J. Dermatopathol.*, 3, 191–195.

Dixon, T.C., Meselson, M., Guillemin, J., and Hanna, P.C. 1999. Anthrax. *N. Engl. J. Med.*, 341, 815–826.

Doganay, M., Almac, A., and Hanagasi, R. 1986. Primary throat anthrax. A report of six cases. *Scand. J. Infect. Dis.*, 18, 415–419.

Doganay, M., and Metan, G. 2009. Human anthrax in Turkey from 1990 to 2007. *Vector Borne Zoonotic Dis.*, 9, 131–140.

Doganay, M., Metan, G., and Alp, E. 2010. A review of cutaneous anthrax and its outcome. *J. Infect. Public Health*, 3, 98–105.

Doust, J.Y., Sarkarzadeh, A., and Kavoossi, K. 1968. Corticosteroid in treatment of malignant edema of chest wall and neck (anthrax). *Dis. Chest*, 53, 773–774.

Druett, H.A., Henderson, D.W., Packman, L., and Peacock, S. 1953. Studies on respiratory infection. I. The influence of particle size on respiratory infection with anthrax spores. *J. Hyg. (Lond)*, 51, 359–371.

Duesbery, N.S., and Vande Woude, G.F. 1999. Anthrax lethal factor causes proteolytic inactivation of mitogen-activated protein kinase kinase. *J. Appl. Microbiol.*, 87, 289–293.

Dutz, W., and Kohout-Dutz, E. 1981. Anthrax. *Int. J. Dermatol.*, 20, 203–206.

Dutz, W., Saidi, F., and Kohout, E. 1970. Gastric anthrax with massive ascites. *Gut*, 11, 352–354.

Elliott, J.L., Mogridge, J., and Collier, R.J. 2000. A quantitative study of the interactions of *Bacillus anthracis* edema factor and lethal factor with activated protective antigen. *Biochemistry*, 39, 6706–6713.

Euro Surveillance Team. 2006. Probable human anthrax death in Scotland. *Euro Surveill.*, 11.

Firoved, A.M., Miller, G.F., Moayeri, M., Kakkar, R., Shen, Y., Wiggins, J.F., McNally, E.M., Tang, W.J., and Leppla, S.H. 2005. *Bacillus anthracis* edema toxin causes extensive tissue lesions and rapid lethality in mice. *Am. J. Pathol.*, 167, 1309–1320.

Fouet, A. 2010. AtxA, a *Bacillus anthracis* global virulence regulator. *Res. Microbiol.*, 161, 735–742.

Frankel, A.E., Kuo, S.R., Dostal, D., Watson, L., Duesbery, N.S., Cheng, C.P., Cheng, H.J., and Leppla, S.H. 2009. Pathophysiology of anthrax. *Front. Biosci.*, 14, 4516–4524.

Frew, B.C., Joag, V.R., and Mogridge, J. 2012. Proteolytic processing of Nlrp1b is required for inflammasome activity. *PLoS Pathog.*, 8, e1002659.

Friedlander, A.M. 2001. Tackling anthrax. *Nature*, 414, 160–161.

Friedlander, A.M., Pittman, P.R., and Parker, G.W. 1999. Anthrax vaccine: Evidence for safety and efficacy against inhalational anthrax. *JAMA* 282, 2104–2106.

Friedlander, A.M., Welkos, S.L., Pitt, M.L., Ezzell, J.W., Worsham, P.L., Rose, K.J., Ivins, B.E., Lowe, J.R., Howe, G.B., Mikesell, P., and Lawrence, W.B. 1993. Postexposure prophylaxis against experimental inhalation anthrax. *J. Infect. Dis.*, 167, 1239–1243.

Glomski, I.J., Corre, J.P., Mock, M., and Goossens, P.L. 2007. Noncapsulated toxinogenic *Bacillus anthracis* presents a specific growth and dissemination pattern in naive and protective antigen-immune mice. *Infect. Immun.*, 75, 4754–4761.

Glomski, I.J., Dumetz, F., Jouvion, G., Huerre, M.R., Mock, M., and Goossens, P.L. 2008. Inhaled non-capsulated *Bacillus anthracis* in A/J mice: Nasopharynx and alveolar space as dual portals of entry, delayed dissemination, and specific organ targeting. *Microbes Infect.*, 10, 1398–1404.

Gogol, E.P., Akkaladevi, N., Szerszen, L., Mukherjee, S., Chollet-Hinton, L., Katayama, H., Pentelute, B.L., Collier, R.J., and Fisher, M.T. 2013. Three dimensional structure of the anthrax toxin translocon-lethal factor complex by cryo-electron microscopy. *Protein Sci.*, 22, 586–594.

Gordon, V.M., Rehemtulla, A., and Leppla, S.H. 1997. A role for PACE4 in the proteolytic activation of anthrax toxin protective antigen. *Infect. Immun.*, 65, 3370–3375.

Gorse, G.J., Keitel, W., Keyserling, H., Taylor, D.N., Lock, M., Alves, K., Kenner, J., Deans, L., and Gurwith, M. 2006. Immunogenicity and tolerance of ascending doses of a recombinant protective antigen (rPA102) anthrax vaccine: A randomized, double-blinded, controlled, multicenter trial. *Vaccine*, 24, 5950–5959.

Grunow, R., Klee, S.R., Beyer, W., George, M., Grunow, D., Barduhn, A., Klar, S., Jacob, D., Elschner, M., Sandven, P., Kjerulf, A., Jensen, J.S., Cai, W., Zimmermann, R., and Schaade, L. 2013. Anthrax among heroin users in Europe possibly caused by same *Bacillus anthracis* strain since 2000. *Euro Surveill.*, 18.

Grunow, R., Verbeek, L., Jacob, D., Holzmann, T., Birkenfeld, G., Wiens, D., von Eichel-Streiber, L., Grass, G., and Reischl, U. 2012. Injection anthrax–a new outbreak in heroin users. *Dtsch. Arztebl. Int.*, 109, 843–848.

Gu, M., Hine, P.M., James Jackson, W., Giri, L., and Nabors, G.S. 2007. Increased potency of BioThrax anthrax vaccine with the addition of the C-class CpG oligonucleotide adjuvant CPG 10109. *Vaccine*, 25, 526–534.

Guichard, A., Nizet, V., and Bier, E. 2012. New insights into the biological effects of anthrax toxins: Linking cellular to organismal responses. *Microbes Infect.*, 14, 97–118.

Guidi-Rontani, C. 2002. The alveolar macrophage: The Trojan horse of *Bacillus anthracis*. *Trends Microbiol.*, 10, 405–409.

Hammerstrom, T.G., Roh, J.H., Nikonowicz, E.P., and Koehler, T.M. 2011. *Bacillus anthracis* virulence regulator AtxA: Oligomeric state, function and CO(2)-signalling. *Mol. Microbiol.*, 82, 634–647.

Hannesschlager, J.R. 2007. Results of a safety and immunogenicity study of an early-stage investigational rPA102 vaccine as compared to the FDA licensed anthrax vaccine (anthrax vaccine adsorbed (AVA. BioThrax)). *Vaccine*, 25, 3247.

Hatami, H., Ramazankhani, A., and Mansoori, F. 2010. Two cases of gastrointestinal anthrax with an unusual presentation from Kermanshah (western Iran). *Arch. Iran. Med.*, 13, 156–159.

Haugen, D.M. 2001. *Biological and Chemical Weapons*. San Diego, CA: Greenhaven Press.

Health Protection Scotland (HPS). 2011. A report on behalf of the National Anthrax Outbreak Control Team. An Outbreak of Anthrax Among Drug Users in Scotland, December 2009 to December 2010. http://www.documents.hps.scot.nhs.uk/giz/anthrax-outbreak/anthrax-outbreak-report-2011-12.pdf

Henderson, D.W., Peacock, S., and Belton, F.C. 1956. Observations on the prophylaxis of experimental pulmonary anthrax in the monkey. *J. Hyg. (Lond)*, 54, 28–36.

Heymann, D. 2008. *Control of Communicable Diseases Manual*, 19th ed. Washington, DC: American Public Health Association.

Heyworth, B., Ropp, M.E., Voos, U.G., Meinel, H.I., and Darlow, H.M. 1975. Anthrax in the Gambia: An epidemiological study. *Br. Med. J.*, 4, 79–82.

Hicks, C.W., Cui, X., Sweeney, D.A., Li, Y., Barochia, A., and Eichacker, P.Q. 2011a. The potential contributions of lethal and edema toxins to the pathogenesis of anthrax associated shock. *Toxins (Basel)*, 3, 1185–1202.

Hicks, C.W., Li, Y., Okugawa, S., Solomon, S.B., Moayeri, M., Leppla, S.H., Mohanty, A., Subramanian, G.M., Mignone, T.S., Fitz, Y., Cui, X., and Eichacker, P.Q. 2011b. Anthrax edema toxin has cAMP-mediated stimulatory effects and high-dose lethal toxin has depressant effects in an isolated perfused rat heart model. *Am. J. Physiol. Heart Circ. Physiol.*, 300, H1108–H1118.

Hicks, C.W., Sweeney, D.A., Cui, X., Li, Y., and Eichacker, P.Q. 2012. An overview of anthrax infection including the recently identified form of disease in injection drug users. *Intensive Care Med.*, 38, 1092–1104.

Hoffmaster, A.R., Meyer, R.F., Bowen, M.D., Marston, C.K., Weyant, R.S., Thurman, K., Messenger, S.L., Minor, E.E., Winchell, J.M., Rassmussen, M.V., Newton, B.R., Parker, J.T., Morrill, W.E., McKinney, N., Barnett, G.A., Sejvar, J.J., Jernigan, J.A., Perkins, B.A., and Popovic, T. 2002. Evaluation and validation of a real-time polymerase chain reaction assay for rapid identification of *Bacillus anthracis*. *Emerg. Infect. Dis.*, 8, 1178–1182.

Holty, J.E., Bravata, D.M., Liu, H., Olshen, R.A., McDonald, K.M., and Owens, D.K. 2006a. Systematic review: A century of inhalational anthrax cases from 1900 to 2005. *Ann. Intern. Med.*, 144, 270–280.

Holty, J.E., Kim, R.Y., and Bravata, D.M. 2006b. Anthrax: A systematic review of atypical presentations. *Ann. Emerg. Med.*, 48, 200–211.

Hoover, D.L., Friedlander, A.M., Rogers, L.C., Yoon, I.K., Warren, R.L., and Cross, A.S. 1994. Anthrax edema toxin differentially regulates lipopolysaccharide-induced monocyte production of tumor necrosis factor alpha and interleukin-6 by increasing intracellular cyclic AMP. *Infect. Immun.*, 62, 4432–4439.

Hsu, L.C., Ali, S.R., McGillivray, S., Tseng, P.H., Mariathasan, S., Humke, E.W., Eckmann, L., Powell, J.J., Nizet, V., Dixit, V.M., and Karin, M. 2008. A NOD2-NALP1 complex mediates caspase-1-dependent IL-1beta secretion in response to *Bacillus anthracis* infection and muramyl dipeptide. *Proc. Natl. Acad. Sci. USA*, 105, 7803–7808.

Hull, A.K., Criscuolo, C.J., Mett, V., Groen, H., Steeman, W., Westra, H., Chapman, G., Legutki, B., Baillie, L., and Yusibov, V. 2005. Human-derived, plant-produced monoclonal antibody for the treatment of anthrax. *Vaccine*, 23, 2082–2086.

Ichhpujani, R.L., Rajagopal, V., Bhattacharya, D., Rana, U.V., Mittal, V., Rai, A., Ravishankar, A.G., Pasha, S.T., Sokhey, J., and Biswas, S. 2004. An outbreak of human anthrax in Mysore (India). *J. Commun. Dis.*, 36, 199–204.

Inglesby, T.V., Henderson, D.A., Bartlett, J.G., Ascher, M.S., Eitzen, E., Friedlander, A.M., Hauer, J., McDade, J., Osterholm, M.T., O'Toole, T., Parker, G., Perl, T.M., Russell, P.K., and Tonat, K. 1999. Anthrax as a biological weapon: Medical and public health management. Working Group on Civilian Biodefense. *JAMA* 281, 1735–1745.

Inglesby, T.V., O'Toole, T., Henderson, D.A., Bartlett, J.G., Ascher, M.S., Eitzen, E., Friedlander, A.M., Gerberding, J., Hauer, J., Hughes, J., McDade, J., Osterholm, M.T., Parker, G., Perl, T.M., Russell, P.K., and Tonat, K. 2002. Anthrax as a biological weapon, 2002: Updated recommendations for management. *JAMA* 287, 2236–2252.

Ingram, R.J., Metan, G., Maillere, B., Doganay, M., Ozkul, Y., Kim, L.U., Baillie, L., Dyson, H., Williamson, E.D., Chu, K.K., Ascough, S., Moore, S., Huwar, T.B., Robinson, J.H., Sriskandan, S., and Altmann, D.M. 2010. Natural exposure to cutaneous anthrax gives long-lasting T cell immunity encompassing infection-specific epitopes. *J. Immunol.*, 184, 3814–3821.

Ivins, B.E., Pitt, M.L., Fellows, P.F., Farchaus, J.W., Benner, G.E., Waag, D.M., Little, S.F., Anderson Jr., G.W., Gibbs, P.H., and Friedlander, A.M. 1998. Comparative efficacy of experimental anthrax vaccine candidates against inhalation anthrax in rhesus macaques. *Vaccine*, 16, 1141–1148.

Jallali, N., Hettiaratchy, S., Gordon, A.C., and Jain, A. 2011. The surgical management of injectional anthrax. *J. Plast. Reconstr. Aesthet. Surg.*, 64, 276–277.

Jang, J., Cho, M., Chun, J.H., Cho, M.H., Park, J., Oh, H.B., Yoo, C.K., and Rhie, G.E. 2011. The poly-gamma-D-glutamic acid capsule of *Bacillus anthracis* enhances lethal toxin activity. *Infect. Immun.*, 79, 3846–3854.

Jang, J., Cho, M., Lee, H.R., Cha, K., Chun, J.H., Hong, K.J., Park, J., and Rhie, G.E. 2013. Monoclonal antibody against the poly-gamma-D-glutamic acid capsule of *Bacillus anthracis* protects mice from enhanced lethal toxin activity due to capsule and anthrax spore challenge. *Biochim. Biophys. Acta*, 1830, 2804–2812.

Jernigan, D.B., Raghunathan, P.L., Bell, B.P., Brechner, R., Bresnitz, E.A., Butler, J.C., Cetron, M., Cohen, M., Doyle, T., Fischer, M., Greene, C., Griffith, K.S., Guarner, J., Hadler, J.L., Hayslett, J.A., Meyer, R., Petersen, L.R., Phillips, M., Pinner, R., Popovic, T., Quinn, C.P., Reefhuis, J., Reissman, D., Rosenstein, N., Schuchat, A., Shieh, W.J., Siegal, L., Swerdlow, D.L., Tenover, F.C., Traeger, M., Ward, J.W., Weisfuse, I., Wiersma, S., Yeskey, K., Zaki, S., Ashford, D.A., Perkins, B.A., Ostroff, S., Hughes, J., Fleming, D., Koplan, J.P., and Gerberding, J.L. 2002. Investigation of bioterrorism-related anthrax, United States, 2001: Epidemiologic findings. *Emerg. Infect. Dis.*, 8, 1019–1028.

Jernigan, J.A., Stephens, D.S., Ashford, D.A., Omenaca, C., Topiel, M.S., Galbraith, M., Tapper, M., Fisk, T.L., Zaki, S., Popovic, T., Meyer, R.F., Quinn, C.P., Harper, S.A., Fridkin, S.K., Sejvar, J.J., Shepard, C.W., McConnell, M., Guarner, J., Shieh, W.J., Malecki, J.M., Gerberding, J.L., Hughes, J.M., and Perkins, B.A. 2001. Bioterrorism-related inhalational anthrax: The first 10 cases reported in the United States. *Emerg. Infect. Dis.*, 7, 933–944.

Kamal, S., Rashid, A.M., Bakar, M., and Ahad, M. 2011. Anthrax: An update. *Asian Pac. J. Trop. Biomed.*, 1, 496–501.

Kanafani, Z.A., Ghossain, A., Sharara, A.I., Hatem, J.M., and Kanj, S.S. 2003. Endemic gastrointestinal anthrax in 1960s Lebanon: Clinical manifestations and surgical findings. *Emerg. Infect. Dis.*, 9, 520–525.

Kandadi, M.R., Frankel, A.E., and Ren, J. 2012. Toll-like receptor 4 knockout protects against anthrax lethal toxin-induced cardiac contractile dysfunction: Role of autophagy. *Br. J. Pharmacol.*, 167, 612–626.

Kandadi, M.R., Hua, Y., Ma, H., Li, Q., Kuo, S.R., Frankel, A.E., and Ren, J. 2010. Anthrax lethal toxin suppresses murine cardiomyocyte contractile function and intracellular Ca2 +handling via a NADPH oxidase-dependent mechanism. *PloS One*, 5, e13335.

Karahocagil, M.K., Akdeniz, N., Akdeniz, H., Calka, O., Karsen, H., Bilici, A., Bilgili, S.G., and Evirgen, O. 2008. Cutaneous anthrax in Eastern Turkey: A review of 85 cases. *Clin. Exp. Dermatol.*, 33, 406–411.

Katayama, H., Wang, J., Tama, F., Chollet, L., Gogol, E.P., Collier, R.J., and Fisher, M.T. 2010. Three-dimensional structure of the anthrax toxin pore inserted into lipid nanodiscs and lipid vesicles. *Proc. Natl. Acad. Sci. USA*, 107, 3453–3457.

Kau, J.H., Sun, D.S., Huang, H.S., Lien, T.S., Huang, H.H., Lin, H.C., and Chang, H.H. 2010. Sublethal doses of anthrax lethal toxin on the suppression of macrophage phagocytosis. *PloS One*, 5, e14289.

Kayabas, U., Karahocagil, M.K., Ozkurt, Z., Metan, G., Parlak, E., Bayindir, Y., Kalkan, A., Akdeniz, H., Parlak, M., Simpson, A.J., and Doganay, M. 2012. Naturally occurring cutaneous anthrax: Antibiotic treatment and outcome. *Chemotherapy*, 58, 34–43.

Kaye, D. 2013. FDA approves raxibacumab to treat inhalational anthrax. *Clin. Infect. Dis.*, 56, I–I.

Keim, P., Smith, K.L., Keys, C., Takahashi, H., Kurata, T., and Kaufmann, A. 2001. Molecular investigation of the Aum Shinrikyo anthrax release in Kameido, Japan. *J. Clin. Microbiol.*, 39, 4566–4567.

Khoddami, M., Shirvani, F., Esmaeili, J., and Beladimogaddam, N. 2010. Two rare presentations of fatal anthrax: Meningeal and intestinal. *Arch. Iran. Med.*, 13, 432–435.

Kiel, J.L., Parker, J.E., Gifford, H., Stribling, L.J., Alls, J.L., Meltz, M.L., McCreary, R.P., and Holwitt, E.A. 2002. Basis for the extraordinary genetic stability of anthrax. *Ann. N. Y. Acad. Sci.*, 969, 112–118.

Kiple, K.F. 1993. *The Cambridge World History of Human Disease.* Cambridge, NY: Cambridge University Press.

Kirby, J.E. 2004. Anthrax lethal toxin induces human endothelial cell apoptosis. *Infect. Immun.*, 72, 430–439.

Klempner, M.S., Talbot, E.A., Lee, S.I., Zaki, S., and Ferraro, M.J. 2010. Case records of the Massachusetts General Hospital. Case 25–2010. A 24-year-old woman with abdominal pain and shock. *N. Engl. J. Med.*, 363, 766–777.

Klimpel, K.R., Arora, N., and Leppla, S.H. 1994. Anthrax toxin lethal factor contains a zinc metalloprotease consensus sequence which is required for lethal toxin activity. *Mol. Microbiol.*, 13, 1093–1100.

Kuehnert, M.J., Doyle, T.J., Hill, H.A., Bridges, C.B., Jernigan, J.A., Dull, P.M., Reissman, D.B., Ashford, D.A., and Jernigan, D.B. 2003. Clinical features that discriminate inhalational anthrax from other acute respiratory illnesses. *Clin. Infect. Dis.*, 36, 328–336.

Kutluk, M.T., Secmeer, G., Kanra, G., Celiker, A., and Aksoyek, H. 1987. Cutaneous anthrax. *Cutis*, 40, 117–118.

Kyriacou, D.N., Adamski, A., and Khardori, N. 2006. Anthrax: From antiquity and obscurity to a front-runner in bioterrorism. *Infect. Dis. Clin. North Am.*, 20, 227–251.

Lakshmi, N., and Kumar, A.G. 1992. An epidemic of human anthrax–A study. *Indian J. Pathol. Microbiol.*, 35, 1–4.

Lalitha, M.K., Anandi, V., Walter, N., Devadatta, J.O., and Pulimood, B.M. 1988. Primary anthrax presenting as an injection "abscess". *Indian J. Pathol. Microbiol.*, 31, 254–256.

Langer, M., Malykhin, A., Maeda, K., Chakrabarty, K., Williamson, K.S., Feasley, C.L., West, C.M., Metcalf, J.P., and Coggeshall, K.M. 2008. *Bacillus anthracis* peptidoglycan stimulates an inflammatory response in monocytes through the p38 mitogen-activated protein kinase pathway. *PloS One*, 3, e3706.

Lanska, D.J. 2002. Anthrax meningoencephalitis. *Neurology*, 59, 327–334.

Larabee, J.L., Maldonado-Arocho, F.J., Pacheco, S., France, B., DeGiusti, K., Shakir, S.M., Bradley, K.A., and Ballard, J.D. 2011. Glycogen synthase kinase 3 activation is important for anthrax edema toxin-induced dendritic cell maturation and anthrax toxin receptor 2 expression in macrophages. *Infect. Immun.*, 79, 3302–3308.

Lawrence, W.S., Marshall, J.R., Zavala, D.L., Weaver, L.E., Baze, W.B., Moen, S.T., Whorton, E.B., Gourley, R.L., and Peterson, J.W. 2011. Hemodynamic effects of anthrax toxins in the rabbit model and the cardiac pathology induced by lethal toxin. *Toxins*, 3, 721–736.

Leppla, S.H. 1982. Anthrax toxin edema factor: A bacterial adenylate cyclase that increases cyclic AMP concentrations of eukaryotic cells. *Proc. Natl. Acad. Sci. USA*, 79, 3162–3166.

Levinsohn, J.L., Newman, Z.L., Hellmich, K.A., Fattah, R., Getz, M.A., Liu, S., Sastalla, I., Leppla, S.H., and Moayeri, M. 2012. Anthrax lethal factor cleavage of Nlrp1 is required for activation of the inflammasome. *PLoS Pathog.*, 8, e1002638.

Li, G., Qu, Y., Cai, C., Kong, Y., Liu, S., Zhang, J., Zhao, J., Fu, L., Xu, J., and Chen, W. 2009. The inhibition of the interaction between the anthrax toxin and its cellular receptor by an anti-receptor monoclonal antibody. *Biochem. Biophys. Res. Commun.*, 385, 591–595.

Li, Y., Cui, X., Solomon, S.B., Remy, K., Fitz, Y., and Eichacker, P.Q. 2013a. *B. anthracis* edema toxin increases cAMP levels and inhibits phenylephrine stimulated contraction in a rat aortic ring model. *Am. J. Physiol. Heart Circ. Physiol.*, 305, H238–H250.

Li, Y., Cui, X., Fitz, Y., and Eichacker, P.Q. 2013b. Endothelium may contribute to the arterial relaxant effects of anthrax edema toxin. *Am. J. Respir. Crit. Care Med.*, 187, A3050 [Abstract #39149].

Liao, K.C., and Mogridge, J. 2009. Expression of Nlrp1b inflammasome components in human fibroblasts confers susceptibility to anthrax lethal toxin. *Infect. Immun.*, 77, 4455–4462.

Liddington, R., Pannifer, A., Hanna, P., Leppla, S., and Collier, R.J. 1999. Crystallographic studies of the anthrax lethal toxin. *J. Appl. Microbiol.*, 87, 282.

Lim, N.K., Kim, J.H., Oh, M.S., Lee, S., Kim, S.Y., Kim, K.S., Kang, H.J., Hong, H.J., and Inn, K.S. 2005. An anthrax lethal factor-neutralizing monoclonal antibody protects rats before and after challenge with anthrax toxin. *Infect. Immun.*, 73, 6547–6551.

Liu, S., Miller-Randolph, S., Crown, D., Moayeri, M., Sastalla, I., Okugawa, S., and Leppla, S.H. 2010. Anthrax toxin targeting of myeloid cells through the CMG2 receptor is essential for establishment of *Bacillus anthracis* infections in mice. *Cell Host Microbe*, 8, 455–462.

Liu, S., Zhang, Y., Hoover, B., and Leppla, S.H. 2013. The receptors that mediate the direct lethality of anthrax toxin. *Toxins*, 5, 1–8.

Makino, S., Sasakawa, C., Uchida, I., Terakado, N., and Yoshikawa, M. 1988. Cloning and CO2-dependent expression of the genetic region for encapsulation from *Bacillus anthracis*. *Mol. Microbiol.*, 2, 371–376.

Makino, S., Uchida, I., Terakado, N., Sasakawa, C., and Yoshikawa, M. 1989. Molecular characterization and protein analysis of the cap region, which is essential for encapsulation in *Bacillus anthracis*. *J. Bacteriol.*, 171, 722–730.

Mallon, E., and McKee, P.H. 1997. Extraordinary case report: Cutaneous anthrax. *Am. J. Dermatopathol.*, 19, 79–82.

Manayani, D.J., Thomas, D., Dryden, K.A., Reddy, V., Siladi, M.E., Marlett, J.M., Rainey, G.J., Pique, M.E., Scobie, H.M., Yeager, M., Young, J.A., Manchester, M., and Schneemann, A. 2007. A viral nanoparticle with dual function as an anthrax antitoxin and vaccine. *PLoS Pathog.*, 3, 1422–1431.

Mandell, G.L., Douglas, R.G., Bennett, J.E., and Dolin, R. 2005. *Mandell, Douglas, and Bennett's Principles and Practice of Infectious Diseases*, 6th ed. New York: Elsevier/Churchill Livingstone, pp. 2 v. (xxxviii, 3661, cxxx p.) ill. (some col.) 3629 cm. + 3661 CD-ROM (3664 3663/3664 in.). http://www.e-streams.com/es0806_7/es0867_4196.html.

Mayer, T.A., Bersoff-Matcha, S., Murphy, C., Earls, J., Harper, S., Pauze, D., Nguyen, M., Rosenthal, J., Cerva, D., Druckenbrod, G., Hanfling, D., Fatteh, N., Napoli, A., Nayyar, A., and Berman, E.L. 2001. Clinical presentation of inhalational anthrax following bioterrorism exposure - Report of 2 surviving patients. *JAMA*, 286, 2549–2553.

Meselson, M., Guillemin, J., Hugh-Jones, M., Langmuir, A., Popova, I., Shelokov, A., and Yampolskaya, O. 1994. The Sverdlovsk anthrax outbreak of 1979. *Science*, 266, 1202–1208.

Metcalfe, N. 2004. The history of woolsorters' disease: A Yorkshire beginning with an international future? *Occup. Med. (Lond.)*, 54, 489–493.

Migone, T.S., Subramanian, G.M., Zhong, J., Healey, L.M., Corey, A., Devalaraja, M., Lo, L., Ullrich, S., Zimmerman, J., Chen, A., Lewis, M., Meister, G., Gillum, K., Sanford, D., Mott, J., and Bolmer, S.D. 2009. Raxibacumab for the treatment of inhalational anthrax. *N. Engl. J. Med.*, 361, 135–144.

Miller, J. 2001. A Germ-Making Plant, *The New York Times*, p. A1.

Minnesota Department of Health. 2011. Health officials investigate case of inhalational anthrax from suspected natural environmental exposure (News Release). http://www.health.state.mn.us/news/pressrel/2011/anthrax080911.html.

Moayeri, M., Crown, D., Newman, Z.L., Okugawa, S., Eckhaus, M., Cataisson, C., Liu, S., Sastalla, I., and Leppla, S.H. 2010. Inflammasome sensor Nlrp1b-dependent resistance to anthrax is mediated by caspase-1, IL-1 signaling and neutrophil recruitment. *PLoS Pathog.*, 6, e1001222.

Moayeri, M., and Leppla, S.H. 2009. Cellular and systemic effects of anthrax lethal toxin and edema toxin. *Mol. Aspects Med.*, 30, 439–455.

Moayeri, M., Sastalla, I., and Leppla, S.H. 2012. Anthrax and the inflammasome. *Microbes Infect.*, 14, 392–400.

Moayeri, M., Webster, J.I., Wiggins, J.F., Leppla, S.H., and Sternberg, E.M. 2005. Endocrine perturbation increases susceptibility of mice to anthrax lethal toxin. *Infect. Immun.*, 73, 4238–4244.

Mock, M., and Fouet, A. 2001. Anthrax. *Annu. Rev. Microbiol.*, 55, 647–671.

Mogridge, J., Cunningham, K., and Collier, R.J. 2002. Stoichiometry of anthrax toxin complexes. *Biochemistry*, 41, 1079–1082.

Molin, F.D., Fasanella, A., Simonato, M., Garofolo, G., Montecucco, C., and Tonello, F. 2008. Ratio of lethal and edema factors in rabbit systemic anthrax. *Toxicon*, 52, 824–828.

Morse, S., Kellogg, R.B., and Perry, S. 2003. Detecting biothreat agents: The laboratory response network. *ASM News*, 69, 433.

Muehlbauer, S.M., Evering, T.H., Bonuccelli, G., Squires, R.C., Ashton, A.W., Porcelli, S.A., Lisanti, M.P., and Brojatsch, J. 2007. Anthrax lethal toxin kills macrophages in a strain-specific manner by apoptosis or caspase-1-mediated necrosis. *Cell Cycle*, 6, 758–766.

Mukherjee, D.V., Tonry, J.H., Kim, K.S., Ramarao, N., Popova, T.G., Bailey, C., Popov, S., and Chung, M.C. 2011. *Bacillus anthracis* protease InhA increases blood-brain barrier permeability and contributes to cerebral hemorrhages. *PLoS One*, 6, e17921.

Murray, P.R., and Baron, E.J. 2007. *Manual of Clinical Microbiology*, 9th ed. Washington, DC: ASM Press.

Mwenye, K.S., Siziya, S., and Peterson, D. 1996. Factors associated with human anthrax outbreak in the Chikupo and Ngandu villages of Murewa district in Mashonaland East Province, Zimbabwe. *Cent. Afr. J. Med.*, 42, 312–315.

Mytle, N., Hopkins, R.J., Malkevich, N.V., Basu, S., Meister, G.T., Sanford, D.C., Comer, J.E., Van Zandt, K.E., Al-Ibrahim, M., Kramer, W.G., Howard, C., Daczkowski, N., Chakrabarti, A.C., Ionin, B., Nabors, G.S., and Skiadopoulos, M.H. 2013. Evaluation of anthrax immune globulin intravenous for treatment of inhalation anthrax. *Antimicrob. Agents Chemother.*, 57, 5684–5692.

Nass, M. 1992. Anthrax epizootic in Zimbabwe, 1978–1980: Due to deliberate spread? *PSR Quarterly*, 2, 198–209.

Navacharoen, N., Sirisanthana, T., Navacharoen, W., and Ruckphaopunt, K. 1985. Oropharyngeal anthrax. *J. Laryngol. Otol.*, 99, 1293–1295.

Ndyabahinduka, D.G., Chu, I.H., Abdou, A.H., and Gaifuba, J.K. 1984. An outbreak of human gastrointestinal anthrax. *Ann Ist. Super. Sanita*, 20, 205–208.

Newman, Z.L., Crown, D., Leppla, S.H., and Moayeri, M. 2010. Anthrax lethal toxin activates the inflammasome in sensitive rat macrophages. *Biochem. Biophys. Res. Commun.*, 398, 785–789.

Newman, Z.L., Sirianni, N., Mawhinney, C., Lee, M.S., Leppla, S.H., Moayeri, M., and Johansen, L.M. 2011. Auranofin protects against anthrax lethal toxin-induced activation of the Nlrp1b inflammasome. *Antimicrob. Agents Chemother.*, 55, 1028–1035.

Nguyen, C., Feng, C., Zhan, M., Cross, A.S., and Goldblum, S.E. 2012. *Bacillus anthracis*-derived edema toxin (ET) counter-regulates movement of neutrophils and macromolecules through the endothelial paracellular pathway. *BMC Microbiol.*, 12, 2.

Nguyen, T.Q., Clark, N., Karpati, A., Goldberg, A., Paykin, A., Tucker, A., Baker, A., Almiroudis, A., Fine, A., Tsoi, B., Aston, C., Berg, D., Weiss, D., Connelly, E., Beaudry, G., Weisfuse, I., Durrah, J.C., Prudhomme, J., Leighton, J., Ackelsberg, J., Mahoney, K., Van Vynck, L., Lee, L., Moskin, L., Layton, M., Wong, M., Raphael, M., Robinson, M., Phillips, M., Jones, M., Jeffery, N., Nieves, R., Slavinski, S., Mullin, S., Beatrice, S.T., Balter, S., Blank, S., Frieden, T., Keifer, M., Rosenstein, N., Diaz, P., Clark, T., Compton, H., Daloia, J., Cardarelli, J., Norrell, N., Horn, E., Jackling, S., Bacon, C., Glasgow, E., Gomez, T., Baltzersen, R.A., Kammerdener, C., Margo-Zavazky, D., Colgan, J., and Pulaski, P. 2010. Public health and environmental response to the first case of naturally acquired inhalational anthrax in the United States in 30 years: Infection of a New York City resident who worked with dried animal hides. *J. Public Health Manag. Pract.*, 16, 189–200.

Occupational Safety and Health Administration (OSHA). 2013. How should I decontaminate during response actions? United States Department of Labor. www.osha.gov/SLTC/etools/anthrax/decon.html.

Pannifer, A.D., Wong, T.Y., Schwarzenbacher, R., Renatus, M., Petosa, C., Bienkowska, J., Lacy, D.B., Collier, R.J., Park, S., Leppla, S.H., Hanna, P., and Liddington, R.C. 2001. Crystal structure of the anthrax lethal factor. *Nature*, 414, 229–233.

Parcell, B.J., Wilmshurst, A.D., France, A.J., Motta, L., Brooks, T., and Olver, W.J. 2011. Injection anthrax causing compartment syndrome and necrotising fasciitis. *J. Clin. Pathol.*, 64, 95–96.

Park, J.M., Greten, F.R., Li, Z.W., and Karin, M. 2002. Macrophage apoptosis by anthrax lethal factor through p38 MAP kinase inhibition. *Science*, 297, 2048–2051.

Peck, R.N., and Fitzgerald, D.W. 2007. Cutaneous anthrax in the Artibonite Valley of Haiti: 1992–2002. *Am. J. Trop. Med. Hyg.*, 77, 806–811.

Pellizzari, R., Guidi-Rontani, C., Vitale, G., Mock, M., and Montecucco, C. 1999. Anthrax lethal factor cleaves MKK3 in macrophages and inhibits the LPS/IFNgamma-induced release of NO and TNFalpha. *FEBS Lett.*, 462, 199–204.

Petosa, C., Collier, R.J., Klimpel, K.R., Leppla, S.H., and Liddington, R.C. 1997. Crystal structure of the anthrax toxin protective antigen. *Nature*, 385, 833–838.

Pezard, C., Berche, P., and Mock, M. 1991. Contribution of individual toxin components to virulence of *Bacillus anthracis*. *Infect. Immun.*, 59, 3472–3477.

Phonboon, K., Ratanastri, P., Peeraprakorn, S., Choomkasien, P., Chongcharoen, P., and Sritasoi, S. 1984. Anthrax outbreak in Udon Thani. *Commun. Dis. J.*, 10, 207–220.

Pickering, A.K., Osorio, M., Lee, G.M., Grippe, V.K., Bray, M., and Merkel, T.J. 2004. Cytokine response to infection with *Bacillus anthracis* spores. *Infect. Immun.*, 72, 6382–6389.

Pile, J.C., Malone, J.D., Eitzen, E.M., and Friedlander, A.M. 1998. Anthrax as a potential biological warfare agent. *Arch. Intern. Med.*, 158, 429–434.

Plotkin, S.A., and Orenstein, W.A. 1999. *Vaccines*, 3rd ed. Philadelphia, PA: W.B. Saunders Co.

Popov, S.G., Popova, T.G., Grene, E., Klotz, F., Cardwell, J., Bradburne, C., Jama, Y., Maland, M., Wells, J., Nalca, A., Voss, T., Bailey, C., and Alibek, K. 2004. Systemic cytokine response in murine anthrax. *Cell. Microbiol.*, 6, 225–233.

Popov, S.G., Villasmil, R., Bernardi, J., Grene, E., Cardwell, J., Popova, T., Wu, A., Alibek, D., Bailey, C., and Alibek, K. 2002. Effect of *Bacillus anthracis* lethal toxin on human peripheral blood mononuclear cells. *FEBS Lett.*, 527, 211–215.

Powell, A.G., Crozier, J.E., Hodgson, H., and Galloway, D.J. 2011. A case of septicaemic anthrax in an intravenous drug user. *BMC Infect. Dis.*, 11, 21.

Qiu, P., Li, Y., Shiloach, J., Cui, X., Sun, J., Trinh, L., Kubler-Kielb, J., Vinogradov, E., Mani, H., Al-Hamad, M., Fitz, Y., and Eichacker, P.Q. 2013a. *B. anthracis* cell wall peptidoglycan but not Lethal or Edema toxins produces changes consistent with disseminated intravascular coagulation in a rat model. *J. Infect. Dis.*, 208, 978–989.

Quinn, C.P., Semenova, V.A., Elie, C.M., Romero-Steiner, S., Greene, C., Li, H., Stamey, K., Steward-Clark, E., Schmidt, D.S., Mothershed, E., Pruckler, J., Schwartz, S., Benson, R.F., Helsel, L.O., Holder, P.F., Johnson, S.E., Kellum, M., Messmer, T., Thacker, W.L., Besser, L., Plikaytis, B.D., Taylor Jr., T.H., Freeman, A.E., Wallace, K.J., Dull, P., Sejvar, J., Bruce, E., Moreno, R., Schuchat, A., Lingappa, J.R., Martin, S.K., Walls, J., Bronsdon, M., Carlone, G.M., Bajani-Ari, M., Ashford, D.A., Stephens, D.S., and Perkins, B.A. 2002. Specific, sensitive, and quantitative enzyme-linked immunosorbent assay for human immunoglobulin G antibodies to anthrax toxin protective antigen. *Emerg. Infect. Dis.*, 8, 1103–1110.

Radun, D., Bernard, H., Altmann, M., Schoneberg, I., Bochat, V., van Treeck, U., Rippe, R.M., Grunow, R., Elschner, M., Biederbick, W., and Krause, G. 2010. Preliminary case report of fatal anthrax in an injecting drug user in North-Rhine-Westphalia, Germany, December 2009. *Euro Surveill.*, 15.

Ramsay, C.N., Stirling, A., Smith, J., Hawkins, G., Brooks, T., Hood, J., Penrice, G., Browning, L.M., and Ahmed, S. 2010. An outbreak of infection with *Bacillus anthracis* in injecting drug users in Scotland. *Euro Surveill.*, 15.

Rasko, D.A., Worsham, P.L., Abshire, T.G., Stanley, S.T., Bannan, J.D., Wilson, M.R., Langham, R.J., Decker, R.S., Jiang, L., Read, T.D., Phillippy, A.M., Salzberg, S.L., Pop, M., Van Ert, M.N., Kenefic, L.J., Keim, P.S., Fraser-Liggett, C.M., and Ravel, J. 2011. *Bacillus anthracis* comparative genome analysis in support of the Amerithrax investigation. *Proc. Natl. Acad. Sci. USA*, 108, 5027–5032.

Ray, T.K., Hutin, Y.J., and Murhekar, M.V. 2009. Cutaneous anthrax, West Bengal, India, 2007. *Emerg. Infect. Dis.*, 15, 497–499.

Reddy, R., Parasadini, G., Rao, P., Uthappa, C.K., and Murhekar, M.V. 2012. Outbreak of cutaneous anthrax in Musalimadugu village, Chittoor district, Andhra Pradesh, India, July–August 2011. *J. Infect. Dev. Countr.*, 6, 695–699.

Remy, K.E., Cui, X., Solomon, S.D., Sun, J., Migone, T.-S., Bolmer, S.D., Al Hamad, M., Li, Y., Fitz, Y., and Eichacker, P.Q. 2013. Hemodynamic support and early treatment with protective antigen directed monoclonal together antibody improve survival in a canine model of anthrax edema toxin induced shock. *Am. J. Respir. Crit. Care Med.*, 187, A3055.

Ringertz, S.H., Hoiby, E.A., Jensenius, M., Maehlen, J., Caugant, D.A., Myklebust, A., and Fossum, K. 2000. Injectional anthrax in a heroin skin-popper. *Lancet*, 356, 1574–1575.

Sayner, S.L. 2011. Emerging themes of cAMP regulation of the pulmonary endothelial barrier. *Am. J. Physiol. Lung Cell. Mol. Physiol.*, 300, L667–L678.

Sayner, S.L., Balczon, R., Frank, D.W., Cooper, D.M., and Stevens, T. 2011. Filamin A is a phosphorylation target of membrane but not cytosolic adenylyl cyclase activity. *Am. J. Physiol. Lung Cell. Mol. Physiol.*, 301, L117–L124.

Schneemann, A., and Manchester, M. 2009. Anti-toxin antibodies in prophylaxis and treatment of inhalation anthrax. *Future Microbiol.*, 4, 35–43.

Scobie, H.M., Rainey, G.J., Bradley, K.A., and Young, J.A. 2003. Human capillary morphogenesis protein 2 functions as an anthrax toxin receptor. *Proc. Natl. Acad. Sci. USA*, 100, 5170–5174.

Scobie, H.M., Thomas, D., Marlett, J.M., Destito, G., Wigelsworth, D.J., Collier, R.J., Young, J.A., and Manchester, M. 2005. A soluble receptor decoy protects rats against anthrax lethal toxin challenge. *J. Infect. Dis.*, 192, 1047–1051.

Scorpio, A., Chabot, D.J., Day, W.A., Hoover, T.A., and Friedlander, A.M. 2010. Capsule depolymerase overexpression reduces *Bacillus anthracis* virulence. *Microbiology*, 156, 1459–1467.

Scorpio, A., Chabot, D.J., Day, W.A., O'Brien, D.K., Vietri, N.J., Itoh, Y., Mohamadzadeh, M., and Friedlander, A.M. 2007. Poly-gamma-glutamate capsule-degrading enzyme treatment enhances phagocytosis and killing of encapsulated *Bacillus anthracis*. *Antimicrob. Agents Chemother.*, 51, 215–222.

Sejvar, J.J., Tenover, F.C., and Stephens, D.S. 2005. Management of anthrax meningitis. *Lancet Infect. Dis.*, 5, 287–295.

Sekhar, P.C., Singh, R.S., Sridhar, M.S., Bhaskar, C.J., and Rao, Y.S. 1990. Outbreak of human anthrax in Ramabhadrapuram village of Chittoor district in Andhra Pradesh. *Indian J. Med. Res.*, 91, 448–452.

Shieh, W.J., Guarner, J., Paddock, C., Greer, P., Tatti, K., Fischer, M., Layton, M., Philips, M., Bresnitz, E., Quinn, C.P., Popovic, T., Perkins, B.A., and Zaki, S.R. 2003. The critical role of pathology in the investigation of bioterrorism-related cutaneous anthrax. *Am. J. Pathol.*, 163, 1901–1910.

Sidel, V., Cohen, H.W., and Gould, R.M. 2002. From woolsorters to mail sorters: Anthrax past, present, and future. *Am. J. Public Health.*, 92, 705–706.

Singh, R.S., Sridhar, M.S., Sekhar, P.C., and Bhaskar, C.J. 1992. Cutaneous anthrax–a report of ten cases. *J. Assoc. Physicians India*, 40, 46–49.

Sirisanthana, T., and Brown, A.E. 2002. Anthrax of the gastrointestinal tract. *Emerg. Infect. Dis.*, 8, 649–651.

Sirisanthana, T., Navachareon, N., Tharavichitkul, P., Sirisanthana, V., and Brown, A.E. 1984. Outbreak of oral-oropharyngeal anthrax: An unusual manifestation of human infection with *Bacillus anthracis*. *Am. J. Trop. Med. Hyg.*, 33, 144–150.

Sirisanthana, T., Nelson, K.E., Ezzell, J.W., and Abshire, T.G. 1988. Serological studies of patients with cutaneous and oral-oropharyngeal anthrax from northern Thailand. *Am. J. Trop. Med. Hyg.*, 39, 575–581.

Stearns-Kurosawa, D.J., Lupu, F., Taylor Jr., F.B., Kinasewitz, G., and Kurosawa, S. 2006. Sepsis and pathophysiology of anthrax in a nonhuman primate model. *Am. J. Pathol.*, 169, 433–444.

Stern, E.J., Uhde, K.B., Shadomy, S.V., and Messonnier, N. 2008. Conference report on public health and clinical guidelines for anthrax. *Emerg. Infect. Dis.*, 14.

Sweeney, D.A., Cui, X., Solomon, S.B., Vitberg, D.A., Migone, T.S., Scher, D., Danner, R.L., Natanson, C., Subramanian, G.M., and Eichacker, P.Q. 2010. Anthrax lethal and edema toxins produce different patterns of cardiovascular and renal dysfunction and synergistically decrease survival in canines. *J. Infect. Dis.*, 202, 1885–1896.

Sweeney, D.A., Hicks, C.W., Cui, X., Li, Y., and Eichacker, P.Q. 2011. Anthrax infection. *Am. J. Respir. Crit. Care Med.*, 184, 1333–1341.

Tierney, B.C., Martin, S.W., Franzke, L.H., Marano, N., Reissman, D.B., Louchart, R.D., Goff, J.A., Rosenstein, N.E., Sever, J.L., and McNeil, M.M. 2003. Serious adverse events among participants in the Centers for Disease Control and Prevention's Anthrax Vaccine and Antimicrobial Availability Program for persons at risk for bioterrorism-related inhalational anthrax. *Clin. Infect. Dis.*, 37, 905–911.

Tippetts, M.T., and Robertson, D.L. 1988. Molecular cloning and expression of the *Bacillus anthracis* edema factor toxin gene:

A calmodulin-dependent adenylate cyclase. *J. Bacteriol.*, 170, 2263–2266.

Titball, R.W., Turnbull, P.C., and Hutson, R.A. 1991. The monitoring and detection of *Bacillus anthracis* in the environment. *Soc. Appl. Bacteriol. Symp. Ser.*, 20, 9S–18S.

Tournier, J.N., Rossi Paccani, S., Quesnel-Hellmann, A., and Baldari, C.T. 2009a. Anthrax toxins: A weapon to systematically dismantle the host immune defenses. *Mol. Aspects Med.*, 30, 456–466.

Tournier, J.N., Ulrich, R.G., Quesnel-Hellmann, A., Mohamadzadeh, M., and Stiles, B.G. 2009b. Anthrax, toxins and vaccines: A 125-year journey targeting *Bacillus anthracis*. *Expert Rev. Anti. Infect. Ther.*, 7, 219–236.

Triantafilou, M., Uddin, A., Maher, S., Charalambous, N., Hamm, T.S., Alsumaiti, A., and Triantafilou, K. 2007. Anthrax toxin evades Toll-like receptor recognition, whereas its cell wall components trigger activation via TLR2/6 heterodimers. *Cell. Microbiol.*, 9, 2880–2892.

Turnbull, P.C., Tindall, B.W., Coetzee, J.D., Conradie, C.M., Bull, R.L., Lindeque, P.M., and Huebschle, O.J. 2004. Vaccine-induced protection against anthrax in cheetah (*Acinonyx jubatus*) and black rhinoceros (*Diceros bicornis*). *Vaccine*, 22, 3340–3347.

Tutrone, W.D., Scheinfeld, N.S., and Weinberg, J.M. 2002. Cutaneous anthrax: A concise review. *Cutis*, 69, 27–33.

U.S Food and Drug Administration (FDA). 2004. New Lab Test for Anthrax. FDA Patient Safety News. http://www.accessdata.fda.gov/psn/printer.cfm?id=203.

Vessal, K., Yeganehdoust, J., Dutz, W., and Kohout, E. 1975. Radiological changes in inhalation anthrax. A report of radiological and pathological correlation in two cases. *Clin. Radiol.*, 26, 471–474.

Vietri, N.J., Purcell, B.K., Lawler, J.V., Leffel, E.K., Rico, P., Gamble, C.S., Twenhafel, N.A., Ivins, B.E., Heine, H.S., Sheeler, R., Wright, M.E., and Friedlander, A.M. 2006. Short-course postexposure antibiotic prophylaxis combined with vaccination protects against experimental inhalational anthrax. *Proc. Natl. Acad. Sci. USA*, 103, 7813–7816.

Vitale, G., Pellizzari, R., Recchi, C., Napolitani, G., Mock, M., and Montecucco, C. 1998. Anthrax lethal factor cleaves the N-terminus of MAPKKs and induces tyrosine/threonine phosphorylation of MAPKs in cultured macrophages. *Biochem. Biophys. Res. Commun.*, 248, 706–711.

Vitale, G., Pellizzari, R., Recchi, C., Napolitani, G., Mock, M., and Montecucco, C. 1999. Anthrax lethal factor cleaves the N-terminus of MAPKKS and induces tyrosine/threonine phosphorylation of MAPKS in cultured macrophages. *J. Appl. Microbiol.*, 87, 288.

Vitale, L., Blanset, D., Lowy, I., O'Neill, T., Goldstein, J., Little, S.F., Andrews, G.P., Dorough, G., Taylor, R.K., and Keler, T. 2006. Prophylaxis and therapy of inhalational anthrax by a novel monoclonal antibody to protective antigen that mimics vaccine-induced immunity. *Infect. Immun.*, 74, 5840–5847.

Walsh, J.J., Pesik, N., Quinn, C.P., Urdaneta, V., Dykewicz, C.A., Boyer, A.E., Guarner, J., Wilkins, P., Norville, K.J., Barr, J.R., Zaki, S.R., Patel, J.B., Reagan, S.P., Pirkle, J.L., Treadwell, T.A., Messonnier, N.R., Rotz, L.D., Meyer, R.F., and Stephens, D.S. 2007. A case of naturally acquired inhalation anthrax: Clinical care and analyses of anti-protective antigen immunoglobulin G and lethal factor. *Clin. Infect. Dis.*, 44, 968–971.

Wang, F., Ruther, P., Jiang, I., Sawada-Hirai, R., Sun, S.M., Nedellec, R., Morrow, P.R., and Kang, A.S. 2004. Human monoclonal antibodies that neutralize anthrax toxin by inhibiting heptamer assembly. *Hum. Antibodies*, 13, 105–110.

Watson, L.E., Kuo, S.R., Katki, K., Dang, T., Park, S.K., Dostal, D.E., Tang, W.J., Leppla, S.H., and Frankel, A.E. 2007a. Anthrax toxins induce shock in rats by depressed cardiac ventricular function. *PloS One*, 2, e466.

Watson, L.E., Mock, J., Lal, H., Lu, G., Bourdeau, R.W., Tang, W.J., Leppla, S.H., Dostal, D.E., and Frankel, A.E. 2007b. Lethal and edema toxins of anthrax induce distinct hemodynamic dysfunction. *Front. Biosci.*, 12, 4670–4675.

Webster, J.I., Moayeri, M., and Sternberg, E.M. 2004. Novel repression of the glucocorticoid receptor by anthrax lethal toxin. *Ann. N. Y. Acad. Sci.*, 1024, 9–23.

Webster, J.I., Tonelli, L.H., Moayeri, M., Simons Jr., S.S., Leppla, S.H., and Sternberg, E.M. 2003. Anthrax lethal factor represses glucocorticoid and progesterone receptor activity. *Proc. Natl. Acad. Sci. USA*, 100, 5706–5711.

Weiner, Z.P., and Glomski, I.J. 2012. Updating perspectives on the initiation of *Bacillus anthracis* growth and dissemination through its host. *Infect. Immun.*, 80, 1626–1633.

Welsh, J. 2011. Anthrax in 2001 letters was traced to Maryland by genetic mutations. *Live Sci.*, http://www.livescience.com/13229-anthrax-attacks-2001-genetics-110314.html.

Wild, M.A., Kumor, K., Nolan, M.J., Lockman, H., and Bowdish, K.S. 2007. A human antibody against anthrax protective antigen protects rabbits from lethal infection with aerosolized spores. *Hum. Antibodies*, 16, 99–105.

Wood, B.J., DeFranco, B., Ripple, M., Topiel, M., Chiriboga, C., Mani, V., Barry, K., Fowler, D., Masur, H., and Borio, L. 2003. Inhalational anthrax: Radiologic and pathologic findings in two cases. *AJR Am. J. Roentgenol.*, 181, 1071–1078.

World Health Organization, International Office of Epizootics, and Food and Agriculture Organization of the United Nations. 2008. *Anthrax in Humans and Animals*, 4th ed. Geneva, Switzerland: World Health Organization.

Wright, J.G., Quinn, C.P., Shadomy, S., and Messonnier, N. 2010. Use of anthrax vaccine in the United States: Recommendations of the Advisory Committee on Immunization Practices (ACIP), 2009. *MMWR*, 59, 1–30.

Young Jr., G.A., and Zelle, M.R. 1946. Respiratory pathogenicity of *Bacillus anthracis* spores; chemical–biological synergisms. *J. Infect. Dis.*, 79, 266–271.

Young, J.A., and Collier, R.J. 2007. Anthrax toxin: Receptor binding, internalization, pore formation, and translocation. *Annu. Rev. Biochem.*, 76, 243–265.

Zakowska, D., Bartoszcze, M., Niemcewicz, M., Bielawska-Drozd, A., and Kocik, J. 2012. New aspects of the infection mechanisms of *Bacillus anthracis*. *Ann. Agric. Environ. Med.*, 19, 613–618.

Zhao, J., Roy, S.A., and Nelson, D.J. 2003. MD simulations of anthrax edema factor: Calmodulin complexes with mutations in the edema factor "switch a" region and docking of 3'-deoxy ATP into the adenylyl cyclase active site of wild-type and mutant edema factor variants. *J. Biomol. Struct. Dyn.*, 21, 159–170.

Zilinskas, R.A. 1997. Iraq's biological weapons. The past as future? *JAMA*, 278, 418–424.

Chapter 42

Pathogenesis of *Streptococcus* in Humans

Roberta Creti[1], Giovanni Gherardi[2], Alberto Berardi[3], and Lucilla Baldassarri[1]

[1]Department of Infectious, Parasitic and Immune-Mediated Diseases, Istituto Superiore di Sanità, Rome, Italy

[2]Centro Integrato di Ricerche, Laboratory of Microbiology, University Campus Biomedico, Rome, Italy

[3]Neonatal Intensive Care Unit, Polyclinic University Hospital, Modena, Italy

42.1 Introduction

Streptococcal disease in human defines a plethora of clinical presentations, gravity of symptoms, outcome, and sequelae that constitute a heavy burden at all ages and worldwide.

β-Hemolytic streptococci, in particular, constitute a very heterogeneous taxon classifiable into serological groups by the capital letters A, B, C, G, according to the scheme that was developed by Rebecca Lancefield in the 1930s based on the different antigenic properties of the carbohydrate on their cell surface.

Of these, the major pathogens for man are the group A streptococcus (GAS; *Streptococcus pyogenes*) and group B streptococcus (GBS; *Streptococcus agalactiae*).

GAS is an exclusively human pathogen, whereas GBS was first identified in bovine mastitis and emerged as responsible for invasive neonatal infection in the developed countries in the 1970s.

This chapter intends to give an updated overview on aspects regarding the management of human infections by these two streptococcal species (clinical features, diagnosis, treatment, prevention, and control) as well as on how a decade of genomic studies has driven new perspectives on the GBS and GAS genetic population structure and pathogenesis research.

42.2 Clinical features

42.2.1 *Streptococcus pyogenes* (GAS)

S. pyogenes can cause superficial or deep infections due to toxin-mediated and immunologically mediated mechanisms of disease (Table 42-1).

S. pyogenes colonizes the human throat and skin and has developed complex virulence mechanisms to avoid host defenses (Cunningham, 2000; Efstratiou, 2000). *S. pyogenes* is the most common cause of bacterial pharyngitis primarily affecting school-age children in developed countries and of impetigo (pyoderma), most commonly occurring in economically disadvantaged children living in tropical or subtropical climates (Valery, 2008). Pharyngitis is spread by person-to-person contact, presumably via nasal secretions or droplets of saliva (Bisno and Stevens, 2010; Wessels, 2011). Epidemics are common in military training facilities or other places of crowding such as schools. Clinical symptoms of GAS pharyngitis include sore throat, malaise, fever, headache, nausea, and vomiting (Choby, 2009; Wessels, 2011). It is generally a self-limiting infection; prompt treatment of pharyngitis is, however, important in decreasing the likelihood of dissemination of the pathogen and to prevent post-infection sequelae such as acute rheumatic fever (ARF), an autoimmune disease caused by antibodies directed to *S. pyogenes* that also react with human tissue (Dmitriev and Chaussee, 2010). Impetigo manifests as a papule that quickly evolves into a vesicle surrounded by redness, and then turns into pustules (Bisno and Stevens, 2010). Often, both *Staphylococcus aureus* and GAS are isolated from impetigo lesions. Other superficial infections by GAS can result in a number of conditions including erysipelas, vaginitis, and postpartum infections. Erysipelas is a superficial cutaneous cellulitis, typically characterized by a localized erythematous indurated that most commonly occurs in old and immunocompromised individuals as well as in neonates and small children (Celestin et al., 2007).

Human Emerging and Re-emerging Infections: Bacterial & Mycotic Infections, Volume II, First Edition. Edited by Sunit K. Singh.
© 2016 John Wiley & Sons, Inc. Published 2016 by John Wiley & Sons, Inc.

Table 42-1 Most common diseases caused by *Streptococcus pyogenes*

Superficial infections	Pharyngitis (streptococcus sore throat)
	Impetigo (blistering skin infections)
	Erysipelas (superficial cutaneous *cellulitis*)
	Vaginitis
	Postpartum infections
Severe systemic and invasive infections	Bacteremia
	Sepsis
	Deep soft-tissue infections
	Cellulitis
	Bacteremia
	Sepsis (including puerperal sepsis)
	Necrotizing fasciitis
	Scarlet fever (exotoxin-mediated)
	Streptococcal toxic shock syndrome (STSS) (exotoxin-mediated)
	Myositis (rare)
	Osteomyelitis (rare)
	Septic arthritis (rare)
	Pneumonia (rare)
	Meningitis (rare)
	Endocarditis (rare)
	Pericarditis (rare)
	Severe neonatal infections (rare)
Non-suppurative poststreptococcal infections	Acute rheumatic fever (ARF)
	Rheumatic heart disease (RHD)
	Acute poststreptococcal glomerulonephritis (GN)
	PANDAS

With early diagnosis and treatment, patients generally make a full recovery. However, in rare cases, infection can spread to deeper levels of the skin and soft tissues resulting in complications such as abscess formation, necrotizing fasciitis, and septicemia (Bisno and Stevens, 2010).

S. pyogenes is also an important contributor to the so-called excessive mortality associated with influenza, which occurs when influenza infections are succeeded by secondary bacterial infections, as was observed during both the influenza pandemic of 1918–1919 and the 2009 H1N1 swine-origin influenza pandemic (Dmitriev and Chaussee, 2010).

42.2.1.1 Invasive GAS infections

Invasive GAS infections are complex infections associating septic, toxic, and immunological disorders. In the past, *S. pyogenes* was a common cause of childbed fever and puerperal sepsis. Since the late 1980s, a resurgence of severe invasive infections due to *S. pyogenes* has been reported worldwide (Johansson et al., 2010; Stevens, 1999). The majority of invasive GAS infections are sporadic and community-based, although outbreaks of severe GAS infections may occur among patients in acute care hospitals and residents in long-term care facilities (Deutscher et al., 2011). Cellulitis and bacteremia are the most common GAS invasive diseases, each accounting for about

20–40% of invasive GAS disease in published epidemiological reports (Lamagni et al., 2008; O'Loughlin et al., 2007). Clinically, cellulitis is characterized by redness and inflammation of the skin, with associated pain and swelling (Bernard, 2008). Bacteremia causes a rapid and robust pro-inflammatory cytokine response that results in high fever, nausea, and vomiting (Sriskandan and Altmann, 2008). While a serious health concern in their own right, cellulitis and bacteremia may also precede the more life-threatening invasive diseases necrotizing fasciitis and streptococcal toxic shock syndrome (STSS). The first, termed also "flesh-eating disease" is a devastating GAS infection involving the skin, subcutaneous and deep soft tissue, and muscle (Olsen et al., 2009). Pathogenesis of necrotizing fasciitis is complex and poorly understood at the molecular level. However, the rapid tissue destruction and bacterial spread are thought to involve several host and bacterial factors (Cole et al., 2011; Johansson et al., 2008; Olsen et al., 2009; Walker et al., 2005). Invasive GAS diseases may also result in the development of STSS: a "cytokine avalanche" produced in response to GAS superantigens that substantially increases the risk of death (Lamagni et al., 2008; Lappin and Ferguson 2009; O'Loughlin et al., 2007). Less common presentations include myositis, osteomyelitis, septic arthritis, pneumonia, meningitis, endocarditis, pericarditis, and severe neonatal infections following intrapartum transmission.

Scarlet fever, which results from infection (usually pharyngeal) with a GAS strain that secretes one or more pyrogenic exotoxins, is characterized by a deep red rash, fever, "strawberry tongue," and exudative pharyngitis (Bisno and Stevens, 2010). A severe form of scarlet fever, septic scarlet fever, is characterized by the spread of GAS into the bloodstream. Although scarlet fever was in the past a significant cause of childhood morbidity and mortality, since the advent of antibiotics it is considered a rare disease. While systemic toxic effects occur rarely with scarlet fever, severe clinical manifestations in STSS may result from massive superantigen-induced cytokine and lymphokine production.

42.2.1.2 Nonsuppurative sequelae

Nonsuppurative complications include poststreptococcal glomerulonephritis (GN) and ARF. While both GN and ARF may follow pharyngitis, only GN is linked with skin infections due to *S. pyogenes*. However, in the Northern Territory of Australia where ARF occurs at endemic levels (Carapetis et al., 2000), GAS throat colonization and pharyngitis are rare. Rather, the skin is the primary tissue reservoir of GAS and impetigo is the main form of superficial GAS infection in these populations (McDonald et al., 2004). ARF can appear as an inflammation of the joints (arthritis), heart (carditis), central nervous system (chorea), skin (erythema marginatum), or subcutaneous nodules

(Cunningham, 2008). Rheumatic heart disease (RHD) can occur following repeated cases of ARF (Carapetis et al., 2005a), often resulting in long-term damage to the heart valves. ARF is the most common cause of pediatric heart disease worldwide (Stollerman, 1997a, 1997b). ARF is triggered when antibodies generated against proteins and molecules on the GAS cell surface immunologically cross-react with proteins in human tissue, such as cardiac myosin, resulting in autoimmunity (McNamara et al., 2008). Host susceptibility has also been reported as a possible factor in the development of ARF and RHD (Guedez et al., 1999). Acute poststreptococcal GN is characterized by symptoms such as edema, hypertension, urinary sediment abnormalities, and decreased levels of complement components in serum (Bisno and Stevens, 2010). GN mainly occurs in children and teenagers, with predominately more males affected than females. Factors such as crowding, poor hygiene, and poverty are associated with GN outbreaks (Marshall et al., 2011). Unlike ARF, outbreaks of GN continue to decline (Cunningham, 2008). It is thought that immune complex deposition (Friedman et al., 1984), direct activation of complement due to the presence of streptococcal products in the glomeruli (Cunningham, 2000), and cross-reactivity of antibodies reactive with both streptococcal antigens and glomerular tissue (Goroncy-Bermes et al., 1987) may play a role in the pathogenesis of GN.

Besides chorea, other pediatric movement disorders have been associated to autoimmunity mechanisms triggered by GAS. The term PANDAS (Pediatric Autoimmune Neuropsychiatric Disorders Associated with Streptococcal infections) has been proposed following the description of a rare group of patients with Tourette's syndrome in whom recrudescence of tics and worsening of behavioral features were associated to variations in the immunological response to GAS (Swedo et al., 1997). Although the precise role *S. pyogenes* plays in these manifestations remains unclear (Leckman et al., 2011), some studies indicated that tic patient sera exhibit immunological profiles typical of individuals who elicited a broad, specific, and strong immune response against GAS (Bombaci et al., 2009; Cardona and Orefici, 2001).

42.2.2 *Streptococcus agalactiae* (GBS)

Despite advances in intensive care supports, GBS infections remain associated with high mortality and morbidity in infancy, particularly among neonates born preterm or those with meningitis. The case fatality ratio is inversely related to gestational age, and is up to eight times higher in early preterm with respect to full-term neonates. Reasons for increased susceptibility to serious infections of preterm neonates include immature immune responses and the many invasive

Table 42-2 Most common diseases caused by *Streptococcus agalactiae*

Early-onset disease	By 7 days of age (pneumonia, meningitis, bacteremia)
Late-onset disease	From 8 days to 3 months (meningitis and bacteremia)
Pregnant women infections	Postpartum endometritis
	Wound infections
	Urinary tract infections (UTIs)
	Bacteremia
Other adults infections	Bacteremia
	Pneumonia
	Joint infections
	Arthritis
	Skin and soft-tissue infections

devices used to provide life-supporting care (Baker, 1997).

GBS may also cause significant morbidity in pregnant women, and remains a prominent cause of morbidity and mortality in the elderly with underlying conditions (malignancy, diabetes mellitus, heart disease) or in immunocompromised hosts (Schuchat, 1998). The disease burden in adults is substantial and increasing rates are reported in some studies (Phares et al., 2008) (Table 42-2).

42.2.2.1 GBS infections in neonates and infants

GBS infections may present as fulminating septicemic illness or with subtle and unspecific early signs that overlap with non-infectious diseases (respiratory, metabolic, neurological disorders, or congenital heart diseases). Without a prompt therapy, GBS infections may lead to a rapid clinical deterioration, septic shock, multi-organ failure, disseminated intravascular coagulopathy, pulmonary and intracranial hemorrhages.

Commonly, two patterns of neonatal clinical syndromes have been identified: early-onset disease (EOD) presenting at age 0–6 days, and late-onset disease (LOD) affecting infants aged 7–89 days. Cases of infection occurring after the third month of life (ultra late onset disease, ULOD) (Hussain et al., 1995) have been recently defined with more details (Guilbert et al., 2010; Wu et al., 2004).

Some decades ago, before the era of intrapartum antibiotic prophylaxis (IAP), EOD accounted for approximately 75–80% of GBS cases (Schrag et al., 2000). A recent systematic review and meta-analysis by region, conducted on all continents for the years 2000–2011, found that the incidence of EOD was two times higher than that of LOD (Edmond et al., 2012). However, countries having effective strategies for prevention of EOD now report EOD incidence rates comparable to (Verani et al., 2010) or lower than LOD (Berardi et al., 2013a, 2013b). The declining rates of EOD have not been associated with changes in the rates of LOD.

Mortality rates vary in infants according to gestational age at diagnosis. A mean EOD case fatality of 12.1% was found among infants aged 0–89 days; death for EOD was twice as high as LOD. Case fatality ratio was approximately three times higher in low- (12.6%) than in high-income countries (4.6%) (Edmond et al., 2012).

42.2.2.2 Early-onset disease

Up to 30% of pregnant women can be asymptomatically colonized with GBS in the vagina and/or rectum and represent the primary risk factor for the vertical transmission during labor. In the absence of any prevention strategies, about 50% of neonates born to GBS colonized mothers are colonized at the mucosal and skin sites and an estimated 1–2% develop early-onset GBS invasive infections (Boyer et al., 1983; Boyer and Gotoff, 1985).

EOD is acquired at delivery via vertical transmission (CDC, 1996). Most infections are acquired *in utero*, and up to 90% of cases present with signs of systemic infection at birth (low Apgar score) or within the first 12–24 h of age (Berardi et al., 2010; RCOG, 2012; Verani et al., 2010). Exposure to intrapartum antibiotics do not delay disease presentation and do not change clinical manifestations of EOD (Berardi et al., 2011; Bromberger et al., 2000).

Most cases of EOD present with bacteremia without focus where respiratory signs (including cyanosis, apnea, tachypnea, grunting, and retractions), lethargy (or irritability), and cardiovascular instability are the main initial signs. Pneumonia affects ~10% of neonates (Berardi et al., 2010; Phares et al., 2008). The radiological features of pneumonia may sometime be identical to hyaline membrane disease of preterm neonates that is not complicated by infection. Persistent fetal circulation may complicate the course of infection.

More severe cases of EOD may be characterized by shock and severe respiratory failure at birth. Clinical features may be identical to those found in intrapartum asphyxia (Keogh et al., 1999).

Fulminant course with neutropenia (Lannering et al., 1983), disseminated intravascular coagulation, and hemodynamic abnormalities (including tachycardia, acidosis, and shock) may sometime develop.

Payne and coworkers studied clinical characteristics of 69 neonates with EOD. They found that features adequately predicting fatal outcome were birth weight less than 2500 g, absolute neutrophil count less than 1500 cells/mm^3, hypotension, apnea, and a pleural effusion on the initial chest radiographs. With these five variables and an initial blood pH less than 7.25, a clinical score correctly predicted outcome in 93% of patients in this study (87% sensitivity, 95% specificity) (Payne et al., 1988).

A few EOD cases present with mild, unspecific symptoms, that is, jaundice or poor feeding (Howard and McCracken, 1974). Currently some cases of EOD

are diagnosed in neonates who remain asymptomatic despite a positive blood culture (obtained at birth, because of maternal risk factors for EOD) (Berardi et al., 2013a).

Early studies reported high case fatality ratio (up to 50% of neonatal cases) (Baker and Barrett, 1973). However, early recognition, prompt treatment, and intensive care support led to decreased mortality. Currently EOD mortality in some studies affects almost exclusively preterm neonates (Berardi et al., 2013a; Weston et al., 2011).

42.2.2.3 Late-onset disease

The predominant routes of LOD transmission are poorly understood and the pathogenesis of LOD is less well known than that of EOD. Maternal GBS carriage and preterm birth are risk factors for LOD (Berardi et al., 2013a, 2013b; Verani et al., 2010). In a case–control study of 122 infants (Lin et al., 2003), the risk of LOD increased by a factor of 1.34 (95% CI, 1.15–1.56) for each week of decreasing gestation. This is probably due in part to low levels of maternal protective antibodies, which cross the placenta in the third trimester of pregnancy.

Nosocomial transmission of GBS through the hands of healthcare workers (frequent some decades ago, when postparturient mothers and their infants typically remained in the hospital ≥1 week) (Anthony et al., 1979; Paredes et al., 1977) is probably less common in recent years. Present data show that outbreaks of nosocomial transmission occur more rarely and predominantly affect preterm neonates (Barbadoro et al., 2011; Berardi et al., 2013b; MacFarquhar et al., 2010).

Anecdotal case reports have also suggested breast milk as a possible source of LOD (Filleron et al., 2014), but the relationship between contaminated milk and LOD is still not clear (Berardi et al., 2013c).

In early studies, LOD was uncommon in preterm neonates but increased survival of very low birth weight, early-preterm neonates led to high rates of premature newborns suffering from LOD (Edmond et al., 2012). The mean age at diagnosis for infants with LOD is 1 month, but preterm neonates may present LOD at older age (Berardi et al., 2013b).

The clinical spectrum of LOD in infancy is wider than that of EOD; most cases present as bacteremia without focus or meningitis. More rarely GBS infections may affect bone, joint, soft tissue, or urinary tract (focal infections). Cellulitis and adenitis are often localized to the submandibular or parotid region and case reports have associated infections in these sites to breastfeeding (Bertini and Dani, 2008; Fluegge et al., 2003; Lanari et al., 2007). Rarely GBS may lead to endocarditis and pericarditis.

Most cases present as full-term infants (2–4 weeks old) and admitted to hospital because of fever, irritability (or lethargy), poor feeding, and reduced perfusion. Route of GBS transmission as well as risk factors remain often unclear, but frequently mothers are found to carry GBS at recto-vaginal site. In one study, approximately two-third of mothers of neonates with LOD were GBS positive (vaginal site) at prenatal screening and/or at the time when LOD was diagnosed in the neonate (Berardi et al., 2013c).

Some cases of LOD may have a fulminant course, and death may occur despite maximal doses of antibiotics and intensive care support. The presence of shock at admission as well as a peripheral leucocyte count of less than $4 \times 10^3/L$ are poor prognostic signs. Meningitis is more common in LOD than in EOD. Early symptoms of meningitis may be unspecific and overlap with sepsis; typical signs of CNS involvement, such as bulging fontanel, nuchal rigidity, and seizures are not common at this stage, underscoring the need to evaluate with a lumbar puncture all neonates with suspected sepsis for excluding meningeal involvement.

Unlike outpatient full-term neonates, in hospital early preterm neonates present as a septic newborn aged 4–8 weeks or more; intensive care support during the weeks before LOD presentation and prolonged courses of postpartum antibiotics are common findings. Early preterm neonates often suffer from more severe LOD and have highest mortality (Berardi et al., 2013b). In a study from Active Bacterial Core Surveillance of USA (1990–2005), the case fatality ratio was 1% in full-term infants, 3% in infants with a gestational age of 29–36 weeks, and 8% in infants born at less than 29 weeks gestational age (Jordan et al., 2008).

LOD may often present as a less severe disease in more mature (>4–6 weeks old), full-term or late-preterm infants admitted to hospital because of fever; they have often nontoxic appearance with no apparent clinical or laboratory evidence of focal infection. This milder LOD (so-called occult bacteremia) is more common among outpatient older infants (Peña et al., 1998). However, if untreated, these infections may still progress to severe complications.

42.2.2.4 Ultra late onset disease

About 10–20% of LOD cases present in infants aged 90 or more days (ULOD). More than half cases of ULOD suffer from bacteremia without focus, whereas meningitis is less common in ULOD than in LOD. Bone, joint, or soft tissues are rarely involved.

ULOD cases often affect infants born preterm with a prolonged hospital stay. In a French study of 242 children hospitalized between 2001 and 2006 for GBS meningitis (220 in the LOD group and 22 in the ULOD group), early preterm birth was the only factor associated to ultra late onset meningitis with respect to late-onset meningitis (Guilbert et al., 2010).

Underlying conditions are not common and include HIV infection (Hussain, 1995), neurologic disorders, immunosuppressive therapies, asthma, malignancy, and renal disease (Phares et al., 2008).

42.2.2.5 Recurrent GBS neonatal infections

The recurrence of one or more episodes of GBS infection affects 0.5–3.0% of newborn infants who have had a first episode of invasive disease (Edwards and Nizet, 2011). Frequently the infant presents with the same clinical manifestations, and isolates from initial and second episodes are often found to have identical genotypes. The recurrence may result from a not drained focus of infection or may occur days or months after a full recovery from the initial episode.

The pathogenesis of recurrent infections is poorly understood. Infants receiving treatment for invasive GBS infection frequently remain colonized at mucous membranes, and reinvasion from persistently colonized mucous membranes or re-exposure to a household carrier has been proposed. Breast milk has been suggested as a source of recurrent infection. The GBS-colonized oropharynx of the infant and mother's GBS-contaminated milk might result in persistent exposure/colonization of both the infant and the mother. Recent data show that heavy neonatal GBS colonization may be associated with milk contamination (Berardi et al., 2013c). In some cases, the infected neonate (colonized in the oropharynx) might be the source of contaminated milk. This would explain cases of recurrent infection having no benefits from discontinuing infected breast milk (Soukka et al., 2010).

There are no recommendations for managing neonates presenting a GBS recurrence. The administration of rifampin is effective in eradicating neonatal GBS colonization in 50% of cases only (Fernandez et al., 2001).

42.2.2.6 Outcome in infants

Data concerning long-term outcome of infants with GBS infection come mostly from studies in the 1970s and 1980s (Edwards et al., 1985; Haslam et al., 1977; Wald et al., 1986), whereas contemporary data are limited. The outcome of GBS infections is closely related to the site of infection and to the severity of the disease. Early recognition and prompt therapy have improved survival rates in past decades, but long-term sequelae, particularly from meningitis, represent a major morbidity. However, even in the absence of meningitis, severe complications (such as periventricular leukomalacia or severe developmental delay) may result from serious disease leading to septic shock.

Sequelae of varying severity are reported in 25–50% of infants surviving LOD and include developmental delay, spastic quadriplegia, microcephaly or hydrocephalus, cortical blindness, deafness, seizures disorders, and retarded speech and language.

In the era of intrapartum chemoprophylaxis, meningitis following EOD has become less frequent (10–15% of cases), but it remains a frequent complication of LOD (up to 50% of cases). An Italian study analyzed 100 cases of LOD. Four percent of infants (mostly born early preterm) died; approximately one-third of infants (of which 80% were born full-term) suffered from meningitis; and more than 90% of neonates with brain lesions at discharge had meningitis (Berardi et al., 2013b).

A few contemporary data show that survivors of GBS meningitis continue to have substantial long-term morbidity. Ninety children (born full- or near-term) were evaluated in the United States at age ≥3 years. About 56% of them demonstrated age-appropriate development, whereas 25% had mild-to-moderate impairment, and ~20% had severe impairment. Death after hospital discharge or severe impairment were associated with specific features at admission, including coma or semicoma, lethargy, seizures, bulging fontanel, respiratory distress, leukopenia, acidosis, cerebrospinal fluid (CSF) protein 0.300 mg/dL, CSF glucose <20 mg/dL, and need for ventilator or pressor support (Libster et al., 2012).

42.2.2.7 GBS infections in pregnant woman and in nonpregnant adults

Less than half of GBS cases in pregnant women present as bacteremia without focus. Most cases are associated with infection of the upper genital tract, placenta, or amniotic sac (endometritis, chorioamnionitis). Pneumonia, puerperal sepsis, skin and soft-tissue infections, or meningitis are rare in the pregnant woman. Mortality is uncommon, but frequently these infections may lead to complications of pregnancy, as spontaneous abortion or still-born infants.

Common clinical syndromes of GBS invasive disease in nonpregnant adult include bacteremia without focus, skin and/or soft-tissue infection, pneumonia, osteomyelitis, and joint infection.

Up to ~90% of adult cases may have underlying condition, mostly diabetes mellitus, heart disease, and malignancy (Phares et al., 2008; Tazi et al., 2011).

In one study from the U.S. surveillance areas of the Centers for Disease Control and Prevention, cases of disease in nonpregnant adult (aged from 18 to ≥65 years) presented primarily as bacteremia without focus (nearly 40% of cases), skin and/or soft-tissue infection (25%), osteomyelitis and joint infection (17%), and pneumonia (12%). Case fatality was 7.5% in all age groups, but mortality was higher (10%) among people aged ≥65 years (Skoff et al., 2009).

42.3 Diagnosis

42.3.1 Specimen collection

42.3.1.1 Group A streptococcus

The clinical diagnosis of streptococcal pharyngitis is notoriously unreliable (Centor et al., 1981; Poses et al., 1985). Symptoms and signs are variable, and the severity of illness ranges from mild throat discomfort alone to classic exudative pharyngitis with high fever and prostration. Because the presentation is nonspecific, the diagnosis of streptococcal pharyngitis should be based on the results of a specific test to detect the presence of the organism, such as a throat culture or a rapid antigen-detection test (RADT) of a throat-swab specimen. Swabbing the posterior pharynx and tonsils and not the tongue, lips, or buccal mucosa increases the sensitivity of both the culture and RADT (Fox et al., 2006).

Diagnostic studies for GAS pharyngitis are not indicated for children less than 3 years old because ARF and the incidence of streptococcal pharyngitis are rare in this population (Shulman et al., 2012).

Contrarily to many streptococcal species that loose viability fairly quickly, *S. pyogenes* can easily and safely be transported on dry swabs, since desiccation enhances recovery from mixed cultures by inhibition of the accompanying saprophytic flora (Martin et al., 1977). Microscopic examination of group A and B streptococci is most useful in the case of clinical specimens from sterile body sites, such as blood. In specimens from not sterile body sites, such as vaginal swabs for GBS identification or pharyngeal swabs for GAS identification, the interpretation of Gram stain is difficult due to the abundant residential microbiota of different streptococcal species.

42.3.1.2 Group B streptococcus

As pointed out in the Section 42.2.2.2, approximately 10–30% of pregnant women are colonized with GBS in the vagina or rectum. The gastrointestinal tract serves as the primary reservoir for GBS and is the likely source of vaginal colonization. Because GBS colonization status can change over the course of a pregnancy, the timing of specimen collection as well as the laboratory method used for specimen processing are very important. GBS cultures performed until less than 5 weeks before delivery are considered reliable (Verani et al., 2010); nevertheless, the positive predictive value of prenatal cultures for GBS carriage during labor varies through the literature and may sometimes be lower (Lin et al., 2011; Valkenburg-van den Berg et al., 2010).

Swabbing the lower vagina, followed by the rectum increases the culture yield substantially compared with sampling the cervix or the vagina without also swabbing the rectum (Dillon et al., 1982).

The use of appropriate transport media can help sustain the viability of GBS in settings where immediate laboratory processing is not possible (Teese et al., 2003). GBS isolates can remain viable in transport media for several days at room temperature; however, the recovery of isolates declines over 1–4 days, especially at elevated temperatures, which can lead to false-negative results. Even when appropriate transport media are used, the sensitivity of culture is greatest when the specimen is stored at 4°C before culture and processed within 24 h of collection (Rosa-Fraile et al., 2005).

42.3.2 Isolation procedures and colony morphology

Generally, all streptococci are usually grown on blood agar media because the assessment of the β-hemolytic reaction is important for identification. The optimal incubation temperature for most streptococcal species ranges between 35°C and 37°C, and supplemental carbon dioxide (5% CO_2) or anaerobic conditions seem to enhance the growth of many streptococcal species.

After 18–24 h of incubation, culture plates should be examined for growth of β-hemolytic colonies. Negative cultures should be reexamined after an additional 24 h incubation period.

Colony size varies among the different β-hemolytic species and helps to distinguish groups of streptococci. β-hemolytic streptococci, such as *S. pyogenes* and *S. agalactiae*, form colonies of >0.5 mm after 24 h of incubation, in contrast to the β-hemolytic strains belonging to *S. anginosus* group, formerly named as *S. milleri* group, which present with pin-point colonies of <0.5 mm after the same incubation period. Within *S. pyogenes* species, dry colonies or large and mucoid colonies are also rarely encountered. *S. agalactiae* produces the largest colonies with a relatively small zone of β-hemolysis. Nonhemolytic *S. agalactiae* strains do also occur (1–4%) and resemble to enterococci.

For the diagnosis of *S. pyogenes* acute pharyngitis, the throat swab is directly streaked on blood agar plates. The isolation of only a few colonies of *S. pyogenes* may reflect inadequate specimen collection and, for this reason, it does not allow the differentiation between a carrier and an acutely infected individual (Bisno, 2001).

In the case of the processing of vaginal-rectal swab for the identification of GBS, the swab should be inoculated in a recommended selective broth medium, such as Todd-Hewitt broth supplemented with either gentamicin (8 µg/mL) and nalidixic acid (15 µg/mL) (TransVag broth), or with colistin (10 µg/mL) and nalidixic acid (15 µg/mL) (Lim broth). The use of a selective enrichment broth improves detection substantially; when direct agar plating is used instead of selective enrichment broth, only heavy colonization is detected and as many as 50% of women who are GBS carriers have false-negative culture results (Platt et al., 1995).

After an incubation for 18–24 h at 35–37°C in ambient air or 5% CO_2, subcultures to an appropriate agar plate (e.g., tryptic soy agar with 5% defibrinated sheep blood, Colombia agar with colistin and nalidixic acid) should follow for the identification of β-hemolytic colonies. More recently, chromogenic agars that undergo color change in the presence of β-hemolytic colonies of GBS have become available (Tazi et al., 2008). These chromogenic agars can facilitate detection of β-hemolytic GBS, but the majority will not detect nonhemolytic strains.

42.3.3 Phenotypic tests for identification of *S. pyogenes* and *S. agalactiae*

All β-hemolytic streptococci do not produce catalase, and this reaction is essential to distinguish between Streptococci and Staphylococci that are catalase-positive. Identification of some β-hemolytic streptococci, including *S. pyogenes* and *S. agalactiae*, can be achieved by commercially available latex agglutination immunological assays. The presence of the Lancefield group B antigen in β-hemolytic isolates from human clinical specimens correlates with the species *S. agalactiae*. The presence of Lancefield group A streptococcal antigen is usually found among *S. pyogenes* isolates but it necessitates of further identification tests, since it can be also associated with other β-hemolytic streptococcal species, such as those belonging to *S. anginosus* group or *S. dysgalactiae* subsp. *equisimilis* (SDSE) (Facklam, 2002).

The definitive identification of β-hemolytic streptococci is achieved by phenotypic tests. These tests include the combined use of PYR test and bacitracin susceptibility for *S. pyogenes*, and the combined use of CAMP test and the hippurate hydrolysis test for *S. agalactiae*.

Over the years, identification methods miniaturized discriminating biochemical reactions to create metabolic profiles that are compared with established databases. These systems can include manual or automated systems. Most recently, proteomic profiles by mass spectrometry (e.g., MALDI-TOF) have emerged as useful technique for the identification of microorganisms (Freiwald and Sauer 2009; Russell, 2009). MALDI-TOF produces reproducible, species-specific spectral patterns including streptococci.

42.3.4 Direct antigen detection of *S. pyogenes* from throat specimens

If diagnosis of acute pharyngitis caused by *S. pyogenes* can be provided rapidly, antibiotic therapy can be initiated promptly to relieve symptoms, to avoid sequelae, and to reduce transmission. Numerous assays for the qualitative detection of the Lancefield group A–specific carbohydrate antigen directly from throat specimen by agglutination methods or chromatographic immunoassays also referred to as RADT are commercially available (Gerber and Shulman, 2004). These tests are based on acid-extraction of cell-wall carbohydrate antigen and detection of the antigen with the use of a specific antibody. A wide range of sensitivity (generally, 70–90%) has been reported for currently available RADT, dependent on the clinical likelihood of streptococcal infection in the test population (Edmonson and Farwell, 2005; Tanz et al., 2009). The specificity of RADT is 95% or greater, and thus a positive result can be considered to be definitive avoiding the need for culture. Although the specificity is generally high, false-positive antigen results have been described particularly from patients with previous diagnosis and/or treatment for *S. pyogenes* infection (Chapin et al., 2002). A RADT is less sensitive than culture, so most guidelines recommend obtaining a throat culture if the RADT is negative. Although these tests provide rapid results and allow early treatment decisions, the throat culture remains the gold standard.

42.3.5 Serologic tests for *S. pyogenes*

Anti-streptococcal antibody titers are not recommended in the routine diagnosis of acute pharyngitis as they reflect past, not current events. Thus, determination of streptococcal antibodies is indicated for the diagnosis of poststreptococcal diseases, such as ARF or GN (Shet and Kaplan, 2002). A fourfold rise in antibody titer is regarded as definitive proof of an antecedent streptococcal infection. Multiple variables, including site of infection, age, the background prevalence of streptococcal infections, antimicrobial therapy, and other comorbidities, influence antibody levels (Ayoub et al., 2003). The most widely tested antibodies are anti-streptolysin O (ASO) and anti-DNase B.

Due to frequent exposure to *S. pyogenes*, ASO and anti-DNase B titers are higher in children (Kaplan et al., 1998). Prompt antibiotic treatment of streptococcal infections can reduce the titer but does not abolish antibody production. Streptococcal carriers do not experience a rise in streptococcal antibody titers.

42.3.6 Nucleic acid detection technologies

Nucleic acid–based identification tests rely on the rapid identification of species-specific DNA sequences by means of hybridization with a chemiluminescent probe or broad and specific nucleic acid amplification assays from culture, biological fluid, or isolated colonies with high rates of sensitivity and specificity (Chapin et al., 2002; Pokorski et al., 1994; Uhl et al., 2003).

Rapid molecular diagnostic assays are of particular importance for the detection of *S. agalactiae* genital colonization in laboring women. These molecular tests are based on the detection of *S. agalactiae*–specific genes by real-time PCR assay (Davies et al., 2004; Gavino and Wang, 2007; Honest et al., 2006). Even if the tests costs are exceedingly higher than selective culture, the major advantages are that they are performed directly on clinical samples and results may be available within a short time frame allowing the administration of the IAP in a more proper way.

42.4 Treatment

42.4.1 Group A streptococcus

Since streptococcal pharyngitis is a self-limited illness in the vast majority of cases, a reasonable question is whether it is worthwhile to pursue diagnostic testing and to offer antibiotic treatment for suspected or confirmed cases. Although poststreptococcal GN does not appear to be prevented by antibiotic treatment of streptococcal pharyngitis, several other potential benefits have been suggested to justify treatment. Antibiotic therapy reduces the risk of subsequent development of ARF (Chamovitz et al., 1954; Denny et al., 1950) and also it reduces the risk of suppurative complications of streptococcal infection, such as acute otitis media and peritonsillar abscess. Antibiotic therapy also reduces the duration of streptococcal symptoms. Follow-up post-treatment throat cultures or RADT are not recommended routinely but they may be considered in special circumstances.

The 2012 updated guidelines of the Infectious Diseases Society of America (IDSA) indicated that penicillin or amoxicillin remain the treatment of choice for streptococcal pharyngitis. As a matter of fact, GAS remains uniformly susceptible to penicillin; reports about reduced susceptibility to penicillin in strains of *S. pyogenes* have not ever been confirmed by reference laboratories. Antibiotic treatment for penicillin-allergic patients include clindamycin, along with first-generation cephalosporin (for those not anaphylactically sensitive), and macrolides (Shulman et al., 2012).

Fluoroquinolones (FQs) can constitute a useful therapeutic alternative when resistance to both macrolide and lincosamides is detected in penicillin-allergic patients. The widespread belief that trimethoprim–sulfamethoxazole (SXT) is ineffective for *S. pyogenes* infections because of inherent antimicrobial resistance is a fallacy due to technical limitations (Bowen et al., 2012). Nevertheless, there are currently no good clinical trial data to support the use of SXT for the treatment of *S. pyogenes* infections (Gelfand et al., 2013).

Tetracyclines should not be used because of the high prevalence of resistant strains, while resistance was not detected for daptomycin, linezolid, or quinupristin–dalfopristin (Brown et al., 2008; Jones et al., 2008).

Treatment for invasive GAS infections has to be effective on both the etiologic agent and its toxins, due to the severity of the disease (associated to the spread of highly virulent bacterial clones).

In the case of toxic shock syndrome, it is often difficult to distinguish between streptococcal and staphylococcal infection before cultures become available and so antibacterial choice must include coverage of both of these organisms. In addition, clindamycin is an important adjunctive antibacterial because of its anti-toxin effects and excellent tissue penetration. Early institution of intravenous immunoglobulin therapy should be considered in cases of toxic shock syndrome and severe invasive infection, including necrotizing fasciitis. Early surgical debridement of necrotic tissue is also an important part of management in cases of necrotizing fasciitis (Steer et al., 2012).

42.4.2 Group B streptococcus

42.4.2.1 Treatment of the newborn infant

Because sepsis in the newborn infant can be rapidly fatal if left untreated, antibiotics should be promptly intravenously administered, at the highest clinically validated doses. Currently, national pediatric associations recommend as empirical treatment of neonatal sepsis a combination of penicillin/ampicillin and gentamicin, or third-generation cephalosporins (e.g., ceftriaxone or cefotaxime) for 10–14 days. These antibiotics are safe and retain efficacy when administered at extended intervals (e.g., twice daily or daily dosing). This combination is effective against most Gram-positive and Gram-negative pathogens. Once cultures confirm a GBS invasive disease and antibiotic susceptibility is available, therapy should be reassessed, and penicillin G alone can be given.

A 10-day course of penicillin G is generally administered in GBS bacteremia without focus.

The administration of antibiotics both to the mother and the baby has been suggested as a means of eradicating GBS colonization in cases of recurrent infection. Rifampicin has proved to be effective in eradicating colonization only in half of the cases (Atkins et al., 1998).

In the case of neonatal meningitis, treatment must be aggressive as the goal is to achieve bactericidal concentration and to sterilize CSF as soon as possible (Heath et al., 2011; Tunkel et al., 2004). This depends on the antibiotic lipophilic properties (which enable the diffusion even in the absence of meningeal inflammation), the low molecular weight and relatively simple structure, the reduced protein binding in serum, and the presence of an active system for transporting the drug from blood to CSF. β-Lactams have decreased penetration into CSF in the absence of meningeal inflammation. Because GBS presents a minimum bactericidal concentration (MBC) that is 10-fold higher than GAS, and the bacterial density in the CSF of neonates with meningitis is generally much higher than that of children with meningitis, it is recommended that large doses of antibiotics are administered in neonates and young infants. After an initial 48–72 h of combined therapy of β-lactams with an aminoglycoside or a third-generation cephalosporin, penicillin G or ampicillin alone are usually satisfactory. However, because there is synergy for ampicillin and gentamycin in the treatment of group B streptococcal infections, many experts have recommended combined therapy at least until the CSF is sterilized.

The duration of treatment for neonatal meningitis depends on the clinical response and duration of positive CSF cultures after treatment is started. Fourteen days is usually satisfactory. While steroids are extensively used for the treatment of pediatric meningitis, there is no evidence to support their use in the treatment of neonates.

Infants with positive CSF cultures after initiation of appropriate therapy are at risk of a poor outcome as well as complications (subdural empyema, obstructive ventriculitis, or multiple small vessel thrombi). A cranial CT scan or MRI should be undertaken during treatment to ascertain that intracranial complications have not occurred. Lumbar puncture is not required if clinical course is not complicated and the infant has rapidly improved (Schaad et al., 1981).

42.4.3 Antimicrobial susceptibility testing

Due to the uniform susceptibility of *S. pyogenes* to penicillin, resistance testing to penicillins or other β-lactam antibiotics approved for treatment of GAS and GBS infections is not necessary for clinical purposes. So far, the identification of rare isolates of GBS with reduced susceptibility to penicillin in Asia and the United States (Dahesh et al., 2008; Kimura et al., 2008) has not resulted in a change of recommendations, unless increasing numbers of such strains are encountered and reported.

Macrolide resistance rates among isolates of *S. pyogenes* have been increasing worldwide (Desjardins et al., 2004). Resistance rates correlate with the use of macrolides in clinical practice, and geographic differences in resistance rates are often due to differences in macrolide use. Some recent European studies have however reported a decline in macrolide resistance among *S. pyogenes* isolates likely associated with a fall in macrolide consumption (d'Humières et al., 2012; Silva-Costa et al., 2012).

Susceptibility testing for macrolides should be performed by using erythromycin, since resistance and susceptibility to azithromycin, clarithromycin, and dirithromycin can be predicted by testing erythromycin. Mechanisms of action (inhibition of protein synthesis) and mechanisms of resistance for macrolide and clindamycin are very similar. There are two main mechanisms of macrolide resistance. One major mechanism is an efflux pump that affects only macrolides (designated M-type for macrolide), and it is mediated in streptococci by *mef* genes. The second most important erythromycin resistance mechanism is a methylase that alters the ribosomal binding and confers resistance to macrolides, lincosamides, and streptogramin B agents (denominated MLS_B-type, for macrolide–lincosamide–streptogramin B), and it is mediated by an *erm* gene. The MLS_B-type resistance in β-hemolytic streptococci, including group A and B streptococci, may be either inducible or constitutive (Varaldo et al., 2009). This means that phenotypically, an isolate with MLS_B-type resistance is erythromycin resistant but may be susceptible or resistant to clindamycin because it is either constitutive or inducible to that drug. If the resistance is inducible, the strains will be resistant to erythromycin (and the other 14- and 15-membered macrolides) but appear to be susceptible to clindamycin unless resistance is induced.

If erythromycin is being used to treat infections caused by group A streptococci and treatment failure is suspected, testing might be considered. If clindamycin is being considered for treating infections by β-hemolytic streptococci that become clindamycin susceptible and erythromycin resistant, the determination of the inducible clindamycin resistance should be also considered.

In the past few years, *S. pyogenes* isolates with low-level resistance to FQs have been reported by several authors (Carapetis et al., 2005b), but isolates with a high level of resistance have been detected very infrequently (Malhotra-Kumar et al., 2009; Reinert et al., 2004).

42.5 Prevention and control

42.5.1 Group A streptococcus

Although reliable data from epidemiological studies are not available from many countries in the world because systematic reporting is not mandatory, evidence suggests that there has been no decrease in the incidence of group A streptococcal infections, even during the antibiotic era. While the incidence of nonsupporative sequelae as rheumatic fever has fallen significantly in some parts of the world, GAS infections remain one of the most common causes of illness in children (Steer, 2012).

In the past decades, upsurges of severe GAS infections has been witnessed in Europe and beyond, with several cases found to be healthcare-associated, with most of them being postsurgical infections. Moreover, severe GAS infections have been described to be associated with recent childbirth, with an increasing rate of maternal deaths associated with maternal GAS infections. For this reason, hospital outbreaks of group A streptococcal (GAS) infection deserve particular awareness, since they can be devastating and occasionally result in the death of previously well patients. Recently, guidelines for prevention and control of group A streptococcal infection in the United Kingdom produced by a multidisciplinary working group provided an evidence-based systematic approach to the investigation of single cases or outbreaks of healthcare-associated GAS infection in acute care or maternity settings (Steer et al., 2012). In particular, they underlined the importance of successful management of every case of healthcare-associated GAS infection, not only to prevent spread and possible serious infections, but also to investigate if transmission is occurring from an ongoing and preventable source. Healthcare workers, the environment, and other patients are possible sources of transmission. Prospective and retrospective surveillances, implementation of barriers such as patient isolation and personal protective equipment, hand hygiene, environmental cleaning, linen and waste handled as hazardous, and the transfer of infected patients to another healthcare facility represent essential preventive strategies. Communal facilities, such as baths, bidets, and showers, should be cleaned and decontaminated between all patients, especially on delivery suites, postnatal wards, and other high-risk areas. Continuous surveillance is required to identify outbreaks which arise over long periods of time. Moreover, formation of outbreak control teams, screening of healthcare workers and of patients, and good and efficient communication strategies are also important steps to be pursued. Indeed, GAS isolates from in-patients, peri-partum patients, neonates, and postoperative wounds should be saved for 6 months to facilitate outbreak investigation (Steer et al., 2012).

Antimicrobial therapy is not indicated for the large majority of chronic streptococcal carriers. However, there are special situations in which eradication of carriage may be desirable, including a community outbreak of ARF, acute poststreptococcal GN, or invasive GAS infection. A number of antimicrobial schedules (e.g., amoxicillin–clavulanic acid) have been demonstrated to be substantially more effective than penicillin or amoxicillin in eliminating chronic streptococcal carriage (Shulman et al., 2012).

42.5.2 Group B streptococcus

Present strategies to prevent the invasive GBS neonatal disease are aimed to minimize the risk for the vertical transmission from the mother to the newborn and the early-onset GBS disease by the use of an IAP (Melin and Efstratiou, 2013). It consists in the intravenous administration of β-lactam antibiotics (penicillin G, ampicillin) or clindamycin, in the case of serious penicillin allergy, for >4 h to labor women.

First proposed in 1976 (Ablow et al., 1976), clinical trials in the late 1980s demonstrated that IAP could reduce mother to newborn bacterial transmission (Boyer and Gotoff, 1986; Matorras et al., 1991; Tuppurainen and Hallman, 1989). IAP has not, however, been studied in adequately sized double-blind controlled trials, and a Cochrane review states that the actual effectiveness of this strategy is unclear (Ohlsson and Shah, 2009). Guidelines recommend the use of one of two approaches for identifying women who should receive IAP: a risk-based approach or a culture-based screening approach. The risk-based method identifies candidates for intrapartum chemoprophylaxis according to the presence of any of the following intrapartum risk factors: delivery at <37 weeks' gestation, intrapartum temperature ≥38.0°C, or rupture of membranes for ≥18 h. The culture-based screening method consists the screening of all pregnant women for vaginal and rectal GBS colonization between 35 and 37 week's gestation. Colonized women are offered intrapartum antibiotics at the time of labor onset or rupture of membranes if before labor.

Under both strategies, IAP is recommended for women with GBS bacteriuria at any time during their current pregnancy or for women who had given birth previously to an infant with invasive early-onset GBS disease. IAP is not indicated if a cesarean delivery is performed before onset of labor on a woman with intact amniotic membranes, regardless of GBS colonization status or gestational age (CDC, 2010).

One of the most compelling evidence of the effectiveness of IAP is the obvious reduction in the incidence rates of EOD observed in countries where a prevention (based on universal antenatal screening) has been adequately implemented. The incidence rates of EOD have declined in the United States from 1.7 cases per 1000 live births in the early 1990s to 0.25 cases per 1000 live births in 2010 (CDC, 2009; Weston et al., 2011). A decrease in EOD incidence (from 1.45 to 0.45 per 1000 live births) has been observed in a Spanish hospital network (López Sastre et al., 2005).

Antenatal screening and IAP have not, however, changed the incidence of LOD.

In Europe, routine bacteriological screening of all pregnant women for antenatal GBS carriage to optimize the identification of women who should receive IAP is part of national guidelines in Spain (1998), France (2001), Belgium (2003), Germany (1996), Czech Republic (2008), Poland (2008), Switzerland (2007), but not in the United Kingdom, Holland, and Norway that opted for the risk-based approach (Berardi et al., 2008; Melin, 2011).

The emerging aspect in the post-IAP era is that about 35–40% of EOD are from negative antenatal screened mothers (Berardi et al., 2010, 2013a; Pulver et al., 2009; Puopolo et al., 2005) and it constitutes, nowadays, the prominent failure in the prevention strategies. Maternal colonization at the vaginal level by GBS can be heavy, light, and transient, possibly giving false-negative microbiological results at the time of antenatal screening. The recommended combined rectal-vaginal swabbing increased sensitivity, but the relevant proportion of perinatal infections from negative mothers strongly suggests new solutions should be proposed. Rapid molecular diagnostic assays to be used at the time of delivery could partly contribute to identify carrier women are on the road.

42.5.3 Vaccines

Even perfect adherence to the recommended prevention guidelines would not prevent GBS neonatal disease (Schrag and Verani, 2013). As reported in the Section 42.5.2, in those countries where universal antenatal screening is applied, false-negative results are reported. Other limitations include precipitous deliveries that can impede the ability to provide an adequate duration of IAP before delivery, often including preterm deliveries where babies are not protected because of incomplete transfer of maternal antibody.

In addition, a prevention strategy reliant on antibiotic prophylaxis is vulnerable to the emergence of antimicrobial resistance among GBS or other newborn pathogens. Finally, IAP is not effective in preventing LOD that is now higher in occurrence than EOD in the United States.

Vaccination of pregnant women is an important strategy that has the potential to improve further on existing protocols and the optimal strategy to ensure that protective levels of antibody are present at delivery (Chen et al., 2013). A three-valent conjugate vaccine is, at present, in phase II trial and it is effective in preventing maternal colonization (Hillier et al., 2009). Its efficacy will be established in the forthcoming phase III clinical trials. By a reverse vaccinology approach (Rappuoli, 2000), novel, highly conserved GBS surface protein antigens have been identified as possible candidates to be used as carriers, promising a more specific protection of glycoconjugate vaccines and a broader coverage of GBS disease (Black et al., 2013; Maione et al., 2005).

The development of GAS vaccines faces obstacles such as the occurrence of many unique serotypes (more than 150 M types), antigenic variation within the same serotype, differences in geographical distribution of serotypes, and the production of cross-reactive antibodies with human tissue which can lead to host autoimmune disease. Early studies involved the vaccination of human volunteers with crude M protein preparations followed by the administration of live GAS to the pharynx, although an increased incidence of ARF in some of the vaccinated children was observed (Fox et al., 1973). In an attempt to avoid the induction of autoimmune disease, studies have focused on peptides derived from the serotype-specific hypervariable N-terminal A-repeats of the M protein and two such preparations have reached clinical trials: a hexavalent preparation (Dale, 1999; Hall et al., 2004; Kotloff et al., 2005) and a multivalent preparation containing 26 N-terminal M protein fragments (Hu et al., 2002). Even though the 26-valent preparation was observed to be safe and immunogenic in phase I human clinical trials (McNeil et al., 2005), the serotype coverage appears to be quite disparate between different geographical regions.

42.6 Pathogenesis

Infections caused by either group A or group B streptococci share several aspects as far as pathogenic mechanisms and type of virulence factors are concerned.

Indeed, both pathogens colonize mucosal surfaces of the host where they proliferate and from which they can spread to different body sites. In this view, surface structures enabling them to adhere and survive, often protected from the clearance mechanism operated by the host, are of utmost importance for both pathogens.

For example, among such structures, pili represent the most recently detected in both *S. pyogenes* and *S. agalactiae*, after having been known for decades for other streptococcal species (Kline et al., 2010). They represent a new class of virulence factors with adhesive and matrix protein-binding activity. Pili and/or constituents parts are being considered as important components of protein-based vaccines against streptococci, particularly against GBS (Lauer et al., 2005; Mora et al., 2005; Rosini et al., 2006). Indeed, these surface structures play a role in many aspects involved in pathogenesis such as host cell and tissue adherence, paracellular translocation, and biofilm formation (reviewed by Kreikemeyer, 2012), and preliminary studies have shown a protective activity of highly conserved region of pili against several GBS serogroups (Nuccitelli et al., 2011).

A peculiar characteristic of GAS is the ability to cause a large variety of infections, though only in humans, in the most diverse host compartments. Such strict link to humans is probably related to the adaptation to the human host and particularly to the human saliva and pharynx environment (Lefébvre et al., 2012).

Infections with specific serotypes (M types) are more likely to result in severe reactions and specific clinical outcomes (Bessen et al., 1989; Bisno, 1980; Green et al., 2005; Luca-Harari et al., 2009; Stollerman, 1997a, 1997b), indicating M proteins as primary virulence factors for GAS strains. M proteins possess a series of characteristics that enable them to play a role in pathogenesis. As defense against the host response, M proteins provide immunologically distinct surface coats; as these rarely cross-react, novel serotypes can avoid antibodies raised by hosts in response to previous infections (Musser and Shelburne, 2009). As virulence factors, M proteins confer resistance to phagocytosis both by binding either to the inhibitory regulators of the complement system (Horstmann et al., 1988) or fibrinogen (Courtney et al., 2006; Horstmann et al., 1992; Whitnack and Beachey, 1982) (thus evading both the classic and the alternative pathways of complement activation), and by inhibiting the non-immune binding of the IgG's Fc region. Immune evasion is part also of the pathogenic mechanism of GBS, with the sialic acid–rich polysaccharide capsule preventing complement factor C3 deposition and phagocytosis (Campbell et al., 1991; Marques et al., 1992).

M proteins may also act as adhesins for GAS, together with F/Sfb protein, fibronectin, vitronectin, and collagen-binding proteins (reviewed in Cunningham, 2000), all responsible for the ability of GAS to adhere to, colonize, and invade human skin and mucous membranes under varying environmental conditions. Fibrinogen- (Gutenkunst et al., 2004; Jacobsson, 2003) and laminin- (Spellerberg et al., 1999; Tenenbaum et al., 2007) binding proteins are also part of the virulence factors repertoire of GBS, mediating adherence to extracellular matrix proteins and cells and cell invasion.

Given the relevant role in so many aspects of GAS biology, M protein, or more specifically the highly conserved carboxyl (C) terminus of the M protein, is central to all the vaccines preparation currently under study (Cole et al., 2008; Dale et al., 1996; Zaman, 2014).

Also lipoteichoic acid (LTA), the major component of Gram-positives' cell wall, has been suggested involved in virulence, as modifications in the D-alanylation of LTA result in decreased virulence (Cox et al., 2009; Kristian et al., 2005) and increased susceptibility to antimicrobial peptides (Fabretti et al., 2006).

A recent work (Flores et al., 2012) contrasted the general concept of the hyaluronic acid capsule in GAS as a major virulence factor contributing to human pharyngeal and invasive infection. Instead, the ability to form biofilm, which has long been subject of debate whether it was only a different expression of a capsule, has been suggested to be responsible for therapy failure and recurrent GAS infections (Baldassarri et al., 2006; Conley et al., 2003).

The GAS streptococcal pyrogenic toxins function as superantigens interacting with antigen-presenting cells and T cells to induce T-cell proliferation and massive cytokine production, cause of the major symptoms of toxic shock syndrome (Kotb, 1995). The speA gene (exotoxin A) has been associated with specific serotypes of *S. pyogenes* such as M1. SpeB encodes the pyrogenic exotoxin B, a cysteine protease that is implicated in degradation of host extracellular matrix proteins, antibacterial peptides, and immunoglobulins, and is of importance particularly in the pathogenesis of necrotizing fasciitis (Thulin, 2006). Chatellier and colleagues (2000), however, did not observe a correlation between the amount of Spe expressed and disease severity, underscoring the importance of host factors in the outcome of GAS infections.

In the matter of extracellular enzymes, C5a peptidase produced by GBS enables the bacterium to evade the immune response by impairing neutrophil recruitment (Takahashi et al., 1995). This same protein, as well as serine proteases and others, also facilitates binding to human epithelial cells and to extracellular matrix protein (Beckmann et al., 2002; Cheng et al., 2002; Harris et al., 2003).

The aforementioned properties are only examples of a large repertoire of virulence factors of both GAS and GBS that may contribute to pathogenesis and that have been largely and thoroughly reviewed elsewhere (Maisey et al., 2008; Nitsche-Schmitz and Chhatwal, 2013; Steer et al., 2012).

More recently, a large number of studies using DNA microarrays to characterize, both qualitatively and quantitatively, the transcriptome have led to significant insights into the pathogen's response to changes in its environment and a realization that the genomic diversity observed in the species is compounded by strain-associated variation in regulatory networks. Proteomic approaches illustrated that the complexity generated at protein level likely contributes to phenotypic variation within the species and, at least to some extent, the diverse clinical outcomes associated with infection. From here, increasing attention has been given in the most recent years to the regulatory processes controlling the expression of the aforementioned factors in response to the different stimuli rather than to the factors themselves. Again for both pathogens considered, two component global regulatory systems have been pointed at for their important role. The CovR/CovS appear to be active in both group A and group B streptococci, regulating the expression of almost the totality of factors, that is, speB and biofilm formation in GAS, pore-forming toxins and C5a peptidase in GBS, only to cite a few (Patenge et al., 2013; Rajagopal, 2009). Also standalone regulators are known (Mga, RovS, RogB) (Patenge et al., 2013; Rajagopal, 2009) modulating, for example, pili expression and other adherence factors in GBS, though the specific mechanism(s) is still unclear (Dramsi et al., 2006).

42.7 Genomics and genetic population structure

42.7.1 S. *pyogenes* and S. *agalactiae* genomes

Since the publication of the complete sequenced genome of the GAS strain SF370 (Ferretti et al., 2001), 19 complete gap-free GAS genomes are available at present (refs in Table 42-3). The repertoire comprises different serotypes (based on the polymorphism of the surface M protein), source of infection/isolation, and geographical origin.

Only three complete gap-free GBS genomes isolated from neonatal sepsis have been published; nevertheless, other three complete and eight permanent draft genomes, including nonhuman isolates, are publicly available (references in Table 42-4).

This temporal progression of genome-scale studies of S. *pyogenes* and S. *agalactiae* has greatly enhanced our appreciation of the genetic diversity of these pathogens. In particular, the studies on GBS genomes introduced to the concept of "pan genome," proposed by Tettelin and colleagues (2005) who noted that each new GBS genome had an average of 30 genes that were not present in any of the previously sequenced genomes; this suggested that the number of genes associated with this species could, theoretically, be unlimited.

The pan genome can be divided into three elements: the "core" genome comprising genes present in all isolates, the "dispensable" genome that refers to those portions present in only a subset of isolates, and the "strain-specific" genome, unique to each isolate. The core genome encodes the basic aspects of the bacterial species biology and the dispensable and strain-specific genes are typically associated with horizontally transmitted genetic elements such as temperate bacteriophage, transposons, and insertion elements (Medini et al., 2005).

S. *agalactiae* has a large pan genome (in excess of 2800 genes), whereas S. *pyogenes* has a smaller pan genome (about 2500 genes), perhaps reflecting the broader habitat range and gene pool for GBS compared with S. *pyogenes*.

Core genome is, however, not synonymous of stable genome: GBS has around 18% recombination in its core genome, whereas as much as 35% of the core genome of GAS is recombinant. Recombination is, then, an important evolutionary force in shaping *Streptococcus* genomes, not only in the acquisition of significant portions of the genome as lineage-specific loci, but also in facilitating rapid evolution of the core genome (Bessen et al., 2011; Lefébure and Stanhope, 2007).

Comparative genomics have demonstrated that additive transfer events are enriched in the dispensable genome and homologous recombination occurs at higher rate in the core genome.

GBS genomes revealed a mosaic organization with a conserved genomic backbone and multiple interspersed genomic islands representing about one quarter of the genome (Lachenauer et al., 2000; Tettelin et al., 2005). These genomic islands are formed by integrative and conjugative elements (ICE) or derivatives of ICEs and phages (Brochet et al., 2008). So, more than phage-mediated transduction that involves the transfer of small pieces of genomes, large conjugal exchanges have contributed significantly to the genome dynamics of GBS.

GAS genomes are instead poly-lysogenic, where prophages constitute about 10% of the total genome of sequenced strains (Brüssow et al., 2004) and they are apparently hotspots for genetic recombination. The prophages encode a wide variety of putative and established virulence factors, contributing to the emergence of virulent subclones.

This could seemingly constitute a contradiction with the presence of clustered regularly interspaced short palindromic repeats (CRISPR)–Cas systems in GAS (Nozawa et al., 2011) but also in GBS genomes (Lopez-Sanchez et al., 2012).

CRISPR, together with associated cas protein genes, form the CRISPR–cas adaptive immune system, which provides resistance to viruses and plasmids in bacteria and archaea (Makarova et al., 2011).

When bacteria are exposed to phages, short fragments derived from phage DNA are integrated into clusters of the CRISPR element as spacers. Indeed, two distinct CRISPR loci have been identified in GAS (Nozawa et al., 2011). Nevertheless, despite the fact that GAS CRISPR appear to be active in spacer acquisition, the number of spacers is small and a significant inverse correlation of GAS prophages and the number of spacers in a CRISPR locus has been noted. The mean number of CRISPR spacers (about four) is less than for other streptococci as food industry streptococci, that is, S. *thermophilus* or other β-hemolytic human streptococci as SDSE, where the high number of spacers suggested that the prophage infection of SDSE is somewhat restricted, as confirmed by the absence of prophagic virulence factors (Shimomura et al., 2011). In addition, some GAS strains lack CRISPR and, indeed, they possess significantly more prophages than CRISPR harboring strains.

Also in the case of GBS, two CRISPR–Cas systems have been identified: one is ubiquitous and the other is present in few strains; similarly to GAS, in GBS the spacers mainly target prophages and ICEs that are widespread among S. *agalactiae* strains. It has been proposed that the CRISPR–Cas system contributes to the diversity of natural GBS population modulating

Table 42-3 Description of complete gap-free GAS genomes

Organism name	Strain	EMM/MLST types	Size (kb)	Prophages	Culture collection	Publication year	Isolation	Geographic location/year
Streptococcus pyogenes	GAS SF370	emm1/ST28	1852	4	ATCC 700294	Ferretti et al. (2001)	Wound infection	USA/not available
	MGAS315	emm3/ST15	1900	6	ATCC BAA-595	Beres et al. (2002)	STSS	USA/1980s
	MGAS8232	emm18/ST42	1895	5	ATCC BAA-572	Smoot et al. (2002)	ARF	USA/1987
	SSI-1	emm3/ST15	1894	6	Not available	Nakagawa et al. (2003)	STSS	Japan/1994
	MGAS10394	emm6/ST382	1899	8	ATCC BAA-946	Banks et al. (2004)	Pharyngitis	USA/2001
	MGAS5005	emm1/ST28	1838	3	ATCC BAA-947	Sumby et al. (2005)	Invasive case	Ontario, Canada/1996
	MGAS6180	emm28/ST52	1897	4	ATCC BAA-1064	Green et al. (2005)	Puerperal sepsis	Texas, USA/1998
	MGAS9429	emm12/ST36	1836	3	ATCC BAA-1315	Beres et al. (2006)	Pharyngeal swab, animal	USA/2001
	MGAS10270	emm2/ST55	1928	5	ATCC BAA-1063	Beres et al. (2006)	Pharyngitis	USA/1990s
	MGAS2096	emm12/ST36	1860	2	ATCC BAA-1065	Beres et al. (2006)	Not available	1993
	MGAS10750	emm4/ST39	1937	4	ATCC BAA-1066	Beres et al. (2006)	Pharyngitis	Florida, USA/2002
	Manfredo	emm5/ST99	1840	5	Not available	Holden et al. (2007)	ARF	Chicago, USA/1950s
	NZ131	emm49/ST30	1815	3	ATCC BAA-1633	McShan et al. (2008)	Acute glomerulonephritis	New Zealand/1990s
	Alab49	emm53/ST11	1827	4	ATCC BAA-1323	Bessen et al. (2011)	Impetigo lesion	USA/1986
	MGAS15252	emm59/ST172	1750	Not available	NR-33709	Fittipaldi et al. (2012)	Soft-tissue infection	Canada/2008
	476	emm1/not available	1813	5	Not available	Miyoshi-Akiyama et al. (2012)	Toxic shock syndrome	Japan/1994
	A20	emm1/ST28	1837	3	Not available	Zheng et al. (2013)	Necrotizing fasciitis	Taiwan
	HSC5	emm14/not available	1818	Not available	Not available	Port et al. (2013)	Not available	USA/not available
	MGAS1882	emm59/ST172	1781	Not available	NR-33708	Unpublished	Not available	USA/1960s

Table 42-4 Description of complete gap-free GBS genomes isolated from human

Organism name	Strain	Serotype/ MLST types	Size (kb)	Alpha-like protein gene	Culture collection	Publication year	Isolation	Geographic location/year
Streptococcus agalactiae	2603	V/ST110	2160	rib	ATCC BAA-611	Tettelin et al. (2002)	Neonatal sepsis	Italy/1993
	A909	Ia/ST7	2128	alpha	ATCC 27591	Tettelin et al. (2005)	Neonatal sepsis	USA/1970s
	NEM316	III/ST23	2211	alp2	ATCC 12403	Glaser et al. (2002)	Neonatal sepsis	USA/1970s

the fitness in diverse environments (Lopez-Sanchez et al., 2012).

42.7.2 Epidemiology

Typing methods are major tools for the epidemiological characterization of bacterial pathogens, allowing the determination of the clonal relationships between isolates based on their genotypic or phenotypic characteristics. Typing applies "distinct labels" to bacterial isolates contributing to investigations of infectious disease pathogenesis and bacterial population structures.

The advent of genomics led to new experimental approaches for assessing bacterial diversity; in particular, the availability of different genome sequences within the same bacterial species have enabled the identification of polymorphic sequences to be used in molecular epidemiology studies. These have underlined that the variation in gene content among strains of GBS and GAS contributes to biomedically relevant difference in phenotypes as virulence, antimicrobial resistance, and disease manifestation, and have facilitated the classification of strains by source or pathogenicity and helped to identify pathogen-specific genes in pathogenic lineages that can lead to strain emergence (Beres and Musser, 2007).

On the patient side, the importance of having a bacterial isolate can be useless on the immediate but the possibility to have bacterial collections to be typed and compared on global scale can be prospectively beneficial in terms of a better understanding of pathogenic mechanisms, diffusion and transmission of antimicrobial resistance determinants, and association of particular clones with clinical outcomes.

42.7.2.1 Group A streptococcus

For decades, the serological typing scheme for GAS has been based on the polymorphism of the M protein, a fibrillar surface protein of which more than 120 types are recognized, constituting the most polymorphic bacterial protein. Additional serological markers included the determination of T antigen type and the presence of the opacity factor (OF).

Nowadays, the classical serotyping methods have been replaced by the polymorphism analysis of the corresponding protein coding genes, with *emm* typing

(encoding M protein) (Beall et al., 1996) as the most widely used single-locus GAS genotyping method.

The determinants of serological M type lie at the hypervariable aminoterminal portion and the *emm* typing method consists in the sequence analysis of the portion of the emm gene that encodes M serospecificity. This enables the assignment of an *emm* type and, more recently, also of an *emm*-subtype. *emm* typing remains, at present, the most directly informative and well-documented method for GAS typing in that it is a universal typing marker and it is relatively easy to perform on large number of strains. Often coupled with other epidemiological markers as superantigen gene profile, *emm* typing applies "distinct labels" for tracking outbreaks of GAS and measuring the general threat presented by GAS at any given time and place (Friães et al., 2013a, 2013b; Metzgar and Zampolli, 2011).

M protein and other intrinsic virulence factors of *S. pyogenes* are encoded on an ancient pathogenicity island, acquired before speciation (Panchaud et al., 2009).

Importantly, epidemiologic studies conducted over many decades have repeatedly found that certain M protein types are nonrandomly associated with particular human infections. Infection with specific serotypes (M types) is more likely to result in severe reactions, rheumatic sequelae, and specific clinical outcomes. For example, invasive infection are commonly caused by serotype M1 and M3 GAS strains, M18 is often linked to ARF, and M28 GAS strains are significantly overrepresented among puerperal sepsis and neonatal GAS infections. Despite genetic changes in these serotypes has occurred in time including also antibiotic resistance and virulence factors, these associations generally remain stable (Creti et al., 2007; Metzgar and Zampolli, 2011).

Another typing method, called *emm* pattern-typing, distinguishes distinct chromosomal architectures (patterns A–C, D, and E) based on the presence and arrangement of *emm* and *emm*-like genes within the GAS genome (Bessen, 2010). Specific *emm* types correlate well with specific *emm* patterns, which, in turn, correlate with tissue tropism. GAS strains with the *emm* pattern A–C genotype are strongly associated with throat infections (throat specialists); pattern D strains are mainly recovered from impetigo infections (skin specialists), whereas pattern E represents a "generalist" group associated with both tissue sites (Bessen, 2010; McMillan et al., 2013).

The tissue tropism reflects also a different geographical prevalence and variability of GAS serotypes. *emm*-type-based surveillance studies showed that the diversity of serotype strains circulating in developing countries far exceeds that in Western countries, mirroring the broader variability of the pattern D strains, responsible for the skin infections that predominate in warm and tropical climate, with respect to the more restricted spectrum of the pattern A–C strains, responsible for the upper respiratory tract infections that predominate in temperate regions (Bessen, 2010).

Whole genome typing techniques, commonly used to study bacterial diversity, such as pulsed field gel electrophoresis (PFGE) and multi-locus sequence typing (MLST), are valuable tools also for GAS genotyping (Carriço et al., 2006; Enright et al., 2001, McGregor et al., 2004; Silva-Costa et al., 2006). More recently additional sequence-based methods to study molecular epidemiology of GAS have been developed (Obszańska et al., 2011; Pittiglio et al., 2010).

It has been noted that relationships between *emm* type and genetic background differ among the three host tissue–related groups and the selection pressures acting on *emm* appear to be strongest for the throat specialists (Bessen, 2010): *emm* typing alone, therefore, as a single-locus typing, is a poor marker of clone definition to study population genetic structure. The *emm* gene also undergoes horizontal transfer, and therefore the utility of *emm* type as a strict definition for clone is somewhat limited. Yet, against a highly random genetic background, adaptations to key ecological niches appear to be highly stable. The genetic basis for the ecological phenotypes for preferred sites of infection most likely lie in a unique combination of genes, small indels, point mutations, or gene arrangements clustered in the accessory gene regions and strongly associated with different *emm* pattern strains (Bessen et al., 2011).

Periodic resurgences in invasive group A streptococcal infections in industrialized countries have been reported from the 1980s onward. The availability of multiple GAS genomes facilitated the development of functional genomics, including DNA microarray and single-nucleotide polymorphism analyses, currently used to assess strain variation. These studies have demonstrated that each epidemic was caused by a shift in GAS population dynamics resulting in the emergence of a subpopulation of strains within the previously immune community (Beres et al., 2010).

The recently emerged unusual virulent subclones of M3 *S. pyogenes* strains are the result of sequential acquisition of three prophages and their associated virulence genes over the last 80 years (Lynskey et al., 2011). The hyperinvasive M1T1 clone, first detected in the mid-1980s in the United States, has since disseminated worldwide and remains a major cause of severe invasive human infections. A key feature of the M1T1 clone is its ability to switch from a noninvasive to an invasive phenotype by a single point mutation of the regulatory systems CovR/S (control of virulence) which, in turn, limits the autodegradation of a number of major virulence genes. These virulence factors have been gradually acquired and consist of a 36 kb genome segment from serotype M12 GAS and the bacteriophage-encoded DNase *Sda*1. The more recent acquisition of the phage-encoded superantigen SpeA is likely to have provided selection advantage for the global dissemination of the M1T1 clone (Maamary et al., 2012; Sumby et al., 2005).

New manifestations of GAS infection reported from Europe and North America in the 1980–1990s comprised toxin-mediated disease (Hoge et al., 1993; Lamagni et al., 2005). These led to clinical case definition for STSS and necrotizing fasciitis (Stevens, 1995).

On the wave of fears for the new epidemic of "flesh-eating bacteria," enhanced surveillance programs were launched. Strep-Euro program gave valuable insight on the epidemiology and antibiotic resistance of GAS strains in Europe in the years 2002–2005 (Lamagni et al., 2008; Luca-Harari et al., 2009).

Unfortunately, this did not produce a systematic reporting for GAS disease onward and disease from β-hemolytic streptococci is still not in the list of European Centre for Prevention and Control (ECDC) Disease Networks/Specific Surveillances. Thus, at present, the lack of systematic reporting for GAS disease makes the assessment of national and global disease burdens very scattered. Only few European countries, such as the United Kingdom, recently included invasive GAS disease among the reportable diseases (Lamagni et al., 2009). This enabled the detection of an increase in invasive GAS infections beyond the seasonally expected, in winter 2010–2011 (Zakikhany et al., 2011).

42.7.2.2 Group B streptococcus

GBS is traditionally differentiated by type-specific capsular polysaccharides and protein antigens by using both serological and molecular methods (Creti et al., 2004; Imperi et al., 2010; Manning et al., 2005Poyart et al., 2007; Yao et al., 2013; Zeng et al., 2006). A total of 10 different GBS capsular polysaccharides antigenes (Ia, Ib, II–IX) have been described so far.

A study on global serotype estimation showed an overall prevalence of five serotypes: Ia, Ib, II, III, and V; these accounted for more than 85% of serotypes in all global regions with serotype data (Americas 96%, Europe 93%, Western Pacific 89%) (Edmond et al., 2012). Other serotypes have been reported mainly in colonization studies with a relevant incidences of serotypes VI and VIII in Japan and serotype IV in the United Arab Emirates and the United States (Amin et al., 2002; Diedrick et al., 2010; Ferrieri et al., 2013; Kimura et al., 2013).

In Europe, surveillance of invasive neonatal infections was undertaken during years 2008 –2010 in eight countries (Belgium, Bulgaria, Czech Republic, Denmark, Germany, Spain, Italy, the United Kingdom) under the auspices of a Pan-European study "DEVANI" (Rodriguez-Granger et al., 2012). During the study, more than 25,000 pregnancies were followed and 188 cases of neonatal disease documented; 61% comprised EOD cases and 39% LOD (Melin and Efstratiou, 2013).

MLST showed that most isolates clustered in well-defined clonal complexes (CC) (Jones, 2003; Springman, 2009). The association between clonal cluster and serotype is quite conserved but prevalence among colonization and invasive disease in adults and infants is different (Gherardi et al., 2007; Imperi et al., 2011; Poyart et al., 2008; Tazi et al., 2011).

In the case of neonatal invasive infections, despite the introduction of prophylaxis which resulted in a significant decrease in EOD, epidemiological studies revealed a very stable clonal structure of GBS in different countries (Martins, 2011). EOD is generally associated with Ia, Ib, II, III, and V serotypes; this reflects the serotype distribution of colonized women, confirming that the vertical transmission from mother to baby is the principal cause of disease. LOD, on the contrary, is associated primarily with serotype III.

Indeed, serotype III is split into two clonal complexes, CC17 and CC19, and the CC17 lineage is the very homogenous group that is overrepresented among isolates responsible for invasive infections in neonates. Among the major evolutionary lineages revealed by population genetics analysis, the GBS CC17 hypervirulent lineage is the prototype of an emerging clone, well adapted to the human neonate and which has successfully spread (Singh et al., 2012). It probably accounts for the increased prevalence of neonatal infection observed after 1960 and, at present, it is responsible for the majority (up to 90%) of LOD cases (Imperi et al., 2010; Poyart et al., 2008).

The capsular switching from serotype III to serotype IV within the homogenous clonal complex CC17 by exchange of the entire polysaccharide operon has been reported (Bellais et al., 2012; Ferrieri et al., 2013; Florindo et al., 2014; Martins et al., 2011; Meehan et al., 2014; Teatero et al., 2015). This mechanism was also responsible for the natural loss of the capsular locus in a GBS strain isolated from a pregnant woman (Creti et al., 2012).

Provided these are few reports, they leave these organisms unrecognizable by the vaccines currently under development, deserving constant attention and, hopefully, routine surveillance.

References

Ablow, R.C., Driscoll, S.G., Effmann, E.L., Gross, I., Jolles, C.J., Uauy, R., and Warshaw, J.B. 1976. A comparison of early-onset group B streptococcal neonatal infection and the respiratory-distress syndrome of the newborn. *N. Engl. J. Med.*, 294(2), 65–70.

Amin, A., Abdulrazzaq, Y.M., and Uduman, S. 2002. Group B streptococcal serotype distribution of isolates from colonized pregnant women at the time of delivery in United Arab Emirates. *J. Infect.*, 45(1), 42–46.

Anthony, B.F., Okada, D.M., and Hobel, C.J. 1979. Epidemiology of the group B streptococcus: Maternal and nosocomial sources for infant acquisitions. *J. Pediatr.*, 95(3), 431–436.

Atkins, J.T., Heresi, G.P., Coque, T.M., and Baker, C.J. 1998. Recurrent group B streptococcal disease in infants: Who should receive rifampin? *J. Pediatr.*, 132(3 Pt 1), 537–539.

Ayoub, E.M., Nelson, B., Shulman, S.T., Barrett, D.J., Campbell, J.D., Armstrong, G., Lovejoy, J., Angoff, G.H., and Rockenmacher, S. 2003. Group A streptococcal antibodies in subjects with or without rheumatic fever in areas with high or low incidences of rheumatic fever. *Clin. Diagn. Lab. Immunol.*, 10(5), 886–890.

Baker, C.J. 1997. Group B streptococcal infections. *Clin. Perinatol.*, 24(1), 59–70.

Baker, C.J., and Barrett, F.F. 1973. Transmission of group B streptococci among parturient women and their neonates. *J. Pediatr.*, 83(6), 919–925.

Baldassarri, L., Creti, C., Recchia, S., Imperi, M., Facinelli, B., Giovanetti, E., Pataracchia, M., Alfarone, G., and Orefici, G. 2006. Therapeutic failures of antibiotics used to treat macrolide-susceptible *Streptococcus pyogenes* infections may be due to biofilm formation. *J. Clin. Microbiol.*, 44(8), 2721–2727.

Banks, D.J., Porcella, S.F., Barbian, K.D., Beres, S.B., Philips, L.E., Voyich, J.M., DeLeo, F.R., Martin, J.M., Somerville, G.A.,

and Musser, J.M. 2004. Progress toward characterization of the group A streptococcus metagenome: Complete genome sequence of a macrolide-resistant serotype M6 strain. *J. Infect. Dis.*, 190(4), 727–738.

Barbadoro, P., Marigliano, A., Savini, S., D'Errico, M.M., and Prospero, E. 2011. Group B streptococcal sepsis: An old or ongoing threat? *Am. J. Infect. Control*, 39(8), e45–e48.

Beall, B., Facklam, R., and Thompson, T. 1996. Sequencing emm-specific PCR products for routine and accurate typing of group A streptococci. *J. Clin. Microbiol.*, 34(4), 953–958.

Beckmann, C., Waggoner, J.D., Harris, T.O., Tamura, G.S., and Rubens, C.E. 2002. Identification of a novel adhesions from group B streptococci by use of phage display reveals that C5a peptidase mediates fibronectin binding. *Infect. Immun.*, 70(6), 2869–2876.

Bellais, S., Six, A., Fouet, A., Longo, M., Dmytruk, N., Glaser, P., Trieu-Cuot, P., and Poyart, C. 2012. Capsular switching in group B streptococcus CC17 hypervirulent clone: A future challenge for polysaccharide vaccine development. *J. Infect. Dis.*, 206(11), 1745–1752.

Berardi, A., Lugli, L., Baronciani, D., Rossi, C., Ciccia, M., Creti, R., Gambini, L., Mariani, S., Papa, I., Tridapalli, E., Vagnarelli, F., Ferrari, F., and GBS Prevention Working Group of Emilia-Romagna. 2010. Group B streptococcus early-onset disease in Emilia-romagna: Review after introduction of a screening-based approach. *Pediatr. Infect. Dis. J.*, 29(2), 115–121.

Berardi, A., Lugli, L., Rossi, C., China, M., Chiossi, C., Gambini, L., Guidi, B., Pedna, M.F., Piepoli, M., Simoni, A., and Ferrari, F. 2011. Intrapartum antibiotic prophylaxis failure and group-B streptococcus early-onset disease. *J. Matern. Fetal. Neonatal. Med.*, 24(10), 1221–1224.

Berardi, A., Lugli, L., Rossi, C., Guidotti, I., Lanari, M., Creti, R., Perrone, E., Biasini, A., Sandri, F., Volta, A., China, M., Sabatini, L., Baldassarri, L., Vagnarelli, F., Ferrari, F., and GBS

Prevention Working Group, Emilia-Romagna. 2013a. Impact of perinatal practices for early-onset group B streptococcal disease prevention. *Pediatr. Infect. Dis. J.*, 32(7), e265–e271.

Berardi, A., Lugli, L., Rossi, C., Morini, M.S., Vagnarelli, F., and Ferrari, F. 2008. Group B streptococcus and preventive strategies in Europe. *Arch. Dis. Child. Fetal Neonatal Ed.*, 93(3), F249.

Berardi, A., Rossi, C., Creti, R., China, M., Gherardi, G., Venturelli, C., Rumpianesi, F., and Ferrari, F. 2013c. Group B streptococcal colonization in 160 mother–baby pairs: A prospective cohort study. *J. Pediatr.*, 163(4), 1099–1104.

Berardi, A., Rossi, C., Lugli, L., Creti, R., Bacchi Reggiani, M.L., Lanari, M., Memo, L., Pedna, M.F., Venturelli, C., Perrone, E., Ciccia, M., Tridapalli, E., Piepoli, M., Contiero, R., and Ferrari, F., GBS Prevention Working Group, Emilia-Romagna. 2013b. Group B streptococcus late-onset disease: 2003–2010. *Pediatrics*, 131(2), e361–e368.

Beres, S.B., Carroll, R.K., Shea, P.R., Sitkiewicz, I., Martinez-Gutierrez, J.C., Low, D.E., McGeer, A., Willey, B.M., Green, K., Tyrrell, G.J., Goldman, T.D., Feldgarden, M., Birren, B.W., Fofanov, Y., Boos, J., Wheaton, W.D., Honisch, C., and Musser, J.M. 2010. Molecular complexity of successive bacterial epidemics deconvoluted by comparative pathogenomics. *Proc. Natl. Acad. Sci. USA*, 107(9), 4371–4376.

Beres, S.B., and Musser, J.M. 2007. Contribution of exogenous genetic elements to the group A streptococcus metagenome. *PLoS One*, 2(8), e800.

Beres, S.B., Richter, E.W., Nagiec, M.J., Sumby, P., Porcella, S.F., DeLeo, F.R., and Musser, J.M. 2006. Molecular genetic anatomy of inter- and intraserotype variation in the human bacterial pathogen group A streptococcus. *Proc. Natl. Acad. Sci. USA*, 103(18), 7059–7064.

Beres, S.B., Sylva, G.L., Barbian, K.D., Lei, B., Hoff, J.S., Mammarella, N.D., Liu, M,Y., Smoot, J.C., Porcella, S.F., Parkins, L.D., Campbell, D.S., Smith, T.M., McCormick, J.K., Leung, D.Y., Schlievert, P.M., and Musser, J.M. 2002. Genome sequence of a serotype M3 strain of group A streptococcus: Phage-encoded toxins, the high-virulence phenotype, and clone emergence. *Proc. Natl. Acad. Sci. USA*, 99(15), 10078–10083.

Bernard, P. 2008. Management of common bacterial infections of the skin. *Curr. Opin. Infect. Dis.*, 21(2), 122–128.

Bertini, G., and Dani, C. 2008. Group B streptococcal late-onset sepsis with submandibular phlegmon in a premature infant after beginning of breast-feeding. *J. Matern. Fetal Neonatal Med.*, 21(3), 213–215.

Bessen, D., Jones, K.F., and Fischetti, V.A. 1989. Evidence for two distinct classes of streptococcal M protein and their relationship to rheumatic fever. *J. Exp. Med.*, 169(1), 269–283.

Bessen, D.E. 2010. Tissue tropism in group A streptococcal infections. *Future Microbiol.*, 5(4), 623–638.

Bessen, D.E., Kumar, N., Hall, G.S., Riley, D.R., Luo, F., Lizano, S., Ford, C.N., McShan, W.M., Nguyen, S.V., Dunning Hotopp, J.C., and Tettelin, H. 2011. Whole-genome association study on tissue tropism phenotypes in group A streptococcus. *J. Bacteriol.*, 193(23), 6651–6663.

Bisno, A.L. 1980. The concept of rheumatogenic and non-rheumatogenic group A streptococci. In: Read, S.E., and Zabriskie, J.B., editors. *Streptococcal Diseases and the Immune Response*. New York, NY: Academic Press Inc. pp. 789–903.

Bisno, A.L. 2001. Acute pharyngitis. *N. Engl. J. Med.*, 344(3), 205–211.

Bisno, A.L., and Stevens D.L. 2010. *Streptococcus pyogenes*. In: Mandell, G.L., Bennet, J.E., and Dolin, R., editors. *Principles and Practice of Infectious Diseases*, 7th ed. Philadelphia, PA: Livingstone Elsevier. pp. 2593–2610.

Black, S., Margarit, I., and Rappuoli, R. 2013. Preventing newborn infection with maternal immunization. *Sci. Transl. Med.*, 5(195), 195ps11.

Bombaci, M., Grifantini, R., Mora, M., Reguzzi, V., Petracca, R., Meoni, E., Balloni, S., Zingaretti, C., Falugi, F., Manetti, A.G., Margarit, I., Musser, J.M., Cardona, F., Orefici, G., Grandi, G., and Bensi, G. 2009. Protein array profiling of tic patient sera reveals a broad range and enhanced immune response against group A streptococcus antigens. *PLoS One*, 4(7), e6332.

Bowen, A.C., Lilliebridge, R.A., Tong, S.Y.C., Baird, R.W., Ward, P., McDonald, M.I., Currie, B.J., and Carapetis, J.R. 2012. Is *Streptococcus pyogenes* resistant or susceptible to trimethoprim-sulfamethoxazole? *J. Clin. Microbiol.*, 50(12), 4067–4072.

Boyer, K.M., Gadzala, C.A., Kelly, P.D., and Gotoff, S.P. 1983. Selective intrapartum chemoprophylaxis of neonatal group B streptococcal early-onset disease. III. Interruption of mother-to-infant transmission. *J. Infect. Dis.*, 148(5), 810–816.

Boyer, K.M., and Gotoff, S.P. 1985. Strategies for chemoprophylaxis of GBS early-onset infections. *Antibiot. Chemother.*, 35, 267–280.

Boyer, K.M., and Gotoff, S.P. 1986. Prevention of early-onset neonatal group B streptococcal disease with selective intrapartum chemoprophylaxis. *N. Engl. J. Med.*, 314(26), 1665–1669.

Brochet, M., Rusniok, C., Couvé, E., Dramsi, S., Poyart, C., Trieu-Cuot, P., Kunst, F., and Glaser, P. 2008. Shaping a bacterial genome by large chromosomal replacements, the evolutionary history of *Streptococcus agalactiae*. *Proc. Natl. Acad. Sci. USA*, 105(41), 15961–15966.

Bromberger, P., Lawrence, J.M., Braun, D., Saunders, B., Contreras, R., and Petitti, D.B. 2000. The influence of intrapartum antibiotics on the clinical spectrum of early-onset group B streptococcal infection in term infants. *Pediatrics*, 106(2 Pt 1), 244–250.

Brown, D.F., Hope, R., Livermore, D.M., Brick, G., Broughton, K., George, R.C., Reynolds, R., and BSAC Working Parties on Resistance Surveillance. 2008. Non-susceptibility trends among enterococci and non-pneumococcal streptococci from bacteraemias in the UK and Ireland, 2001–06. *J. Antimicrob. Chemother.*, 62(Suppl 2), ii75–ii85.

Brüssow, H., Canchaya, C., and Hardt, W.D. 2004. Phages and the evolution of bacterial pathogens: From genomic rearrangements to lysogenic conversion. *Microbiol. Mol. Biol. Rev.*, 68(3), 560–602.

Campbell, J.R., Baker, C.J., and Edwards, M.S. 1991 Deposition and degradation of C3 on type III group B streptococci. *Infect. Immun.*, 59(6), 1978–1983.

Carapetis, J.R., Currie, B.J., and Mathews, J.D. 2000. Cumulative incidence of rheumatic fever in an endemic region: A guide to the susceptibility of the population? *Epidemiol. Infect.*, 124(2), 239–244.

Carapetis, J.R., Mcdonald, M., and Wilson, N.J. 2005a. Acute rheumatic fever. *Lancet*, 366(9480), 155–168.

Carapetis, J.R., Steer, A.C., Mulholland, E.K., and Weber, M. 2005b. The global burden of group A streptococcal diseases. *Lancet Infect. Dis.*, 5(11), 685–694.

Cardona, F., and Orefici, G. 2001. Group A streptococcal infections and tic disorders in an Italian pediatric population. *J. Pediatr.*, 138(1), 71–75.

Carriço, J.A., Silva-Costa, C., Melo-Cristino, J., Pinto, F.R., de Lencastre, H., Almeida, J.S., and Ramirez, M. 2006. Illustration of a common framework for relating multiple typing methods by application to macrolide-resistant *Streptococcus pyogenes*. *J. Clin. Microbiol.*, 44(7), 2524–2532.

Celestin R., Brown, J., Khiczak, G., and Schwartz, R.A. 2007. Erysipelas: A common potentially dangerous infection. *Acta Dermatovenerol. Alp. Panonica Adriat.*, 16(3), 123–127.

Centers for Disease Control and Prevention (CDC). 1996. Prevention of perinatal group B streptococcal disease: A public health perspective. *MMWR Recomm. Rep.*, 45(31), 1–24.

Centers for Disease Control and Prevention (CDC). 2009. Trends in perinatal group B streptococcal disease – United States, 2000–2006. *MMWR Morb. Mortal. Wkly. Rep.*, 58(5), 109–112.

Centers for Disease Control and Prevention (CDC). 2010. Active Bacterial Core Surveillance Report, Emerging Infections Program Network, Group B Streptococcus, provisional–2010. http://www.cdc.gov/abcs/reports-findings/survreports/gbs10.pdf. Accessed March 7, 2013.

Centor, R.M., Witherspoon, J.M., Dalton, H.P., Brody, C.E., and Link, K. 1981. The diagnosis of strep throat in adults in the emergency room. *Med. Decis. Making*, 1(3), 239–246.

Chamovitz, R., Catanzaro, F.J., Stetson, C.A., and Rammelkamp Jr., C.H. 1954. Prevention of rheumatic fever by treatment of previous streptococcal infections. I. Evaluation of benzathine penicillin G. *N. Engl. J. Med.*, 251(12), 466–471.

Chapin, K.C., Blake, P., and Wilson, C.D. 2002. Performance characteristics and utilization of rapid antigen test, DNA probe, and culture for detection of group A streptococci in an acute care clinic. *J. Clin. Microbiol.*, 40(11), 4207–4210.

Chatellier, S.N., Ihendyane, I., Kansal, R.G., Khambaty, F., Basma, H., Norrby-Teglund, A., Low, D.E., McGeer, A., and Kotb, M. 2000. Genetic relatedness and superantigen expression in group A streptococcus serotype M1 isolates from patients with severe and nonsevere invasive diseases. *Infect. Immun.*, 68(6), 3523–3534.

Chen, V.L., Avci, F.Y., and Kasper, D.L. 2013. A maternal vaccine against group B streptococcus: Past, present, and future. *Vaccine*, 31(Suppl 4), D13–D19.

Cheng, Q., Stafslien, D., Purushothaman, S.S., and Cleary, P. 2002. The group B streptococcal C5a peptidase is both a specific protease and an invasion. *Infect. Immun.*, 70(5), 2408–2413.

Choby, B.A. 2009. Diagnosis and treatment of streptococcal pharyngitis. *Am. Fam. Physician*, 79(5), 383–390.

Cole, J.N., Barnett, T.C., Nizet, V., and Walker, M.J. 2011. Molecular insight into invasive group A streptococcal disease. *Nat. Rev. Microbiol.*, 9(10), 724–736.

Cole, J.N., Henningham, A., Gillen, C.M., Ramachandran, V., and Walker, M.J. 2008. Human pathogenic streptococcal proteomics and vaccine development. *Proteomics Clin. Appl.*, 2(3), 387–410.

Conley, J., Olson, M.E., Cook, L.S., Ceri, H., Phan, V., and Davies, H.D. 2003. Biofilm formation by group a streptococci: Is there a relationship with treatment failure? *J. Clin. Microbiol.*, 41(9), 4043–4048.

Courtney, H.S., Hasty, D.L., and Dale, J.B. 2006. Anti-phagocytic mechanisms of *Streptococcus pyogenes*: Binding of fibrinogen to M-related protein. *Mol. Microbiol.*, 59(3), 936–947.

Cox, K.H., Ruiz-Bustos, E., Courtney, H.S., Dale, J.B., Pence, M.A., Nizet, V., Aziz, R.K., Gerling, I., Price, S.M., and Hasty, D.L. 2009. Inactivation of DltA modulates virulence factor expression in *Streptococcus pyogenes*. *PLoS One*, 4(4), e5366.

Creti, R., Fabretti, F., Orefici, G., and von Hunolstein, C. 2004. Multiplex PCR assay for direct identification of group B streptococcal alpha-protein-like protein genes. *J. Clin. Microbiol.*, 42(3), 1326–1329.

Creti, R., Imperi, M., Baldassarri, L., Pataracchia, M., Recchia, S., Alfarone, G., and Orefici G. 2007. Emm types, virulence factors, and antibiotic resistance of invasive *Streptococcus pyogenes* isolates from Italy: What has changed in 11 years? *J. Clin. Microbiol.*, 45(7), 2249–2256.

Creti, R., Imperi, M., Pataracchia, M., Alfarone, G., Recchia, S., and Baldassarri, L. 2012. Identification and molecular characterization of a S. agalactiae strain lacking the capsular locus. *Eur. J. Clin. Microbiol. Infect. Dis.*, 31(3), 233–235.

Cunningham, M.W. 2000. Pathogenesis of group A streptococcal infections. *Clin. Microbiol. Rev.*, 13(3), 470–511.

Cunningham, M. 2008. Pathogenesis of group A streptococcal infections and their sequelae. *Adv. Exp. Med. Biol.*, 609, 29–42.

d'Humières, C., Cohen, R., Levy, C., Bidet, P., Thollot, F., Wollner, A., and Bingen, E. 2012. Decline in macrolide-resistant *Streptococcus pyogenes* isolates from French children. *Int. J. Med. Microbiol.*, 302(7–8), 300–303.

Dahesh, S., Hensler, M.E., Van Sorge, N.M., Gertz, R.E., Schrag, S., Nizet, V., and Beall, B.W. 2008. Point mutation in the group B streptococcal *pbp2x* gene conferring decreased susceptibility to beta-lactam antibiotics. *Antimicrob. Agents Chemother.*, 52(8), 2915–2918.

Dale, J.B. 1999. Multivalent group A streptococcal vaccine designed to optimize the immunogenicity of six tandem M protein fragments. *Vaccine*, 17(2), 193–200.

Dale, J.B., Simmons, M., Chiang, E.C., and Chiang, E.Y. 1996. Recombinant octavalent group A streptococcal M protein vaccine. *Vaccine*, 14(10), 944–948.

Davies, H.D., Miller, M.A., Faro, S., Gregson, D., Kehl, S.C., and Jordan, J.A. 2004. Multicenter study of a rapid molecular-based assay for the diagnosis of group B streptococcus colonization in pregnant women. *Clin. Infect. Dis.*, 39(8), 1129–1135.

Denny, F.W., Wannamaker, L.W., Brink, W.R., Rammelkamp Jr., C.H., and Custer, E.A. 1950. Prevention of rheumatic fever; treatment of the preceding streptococci infection. *J. Am. Med. Assoc.*, 143(2), 151–153.

Desjardins, M., Delgaty, K.L., Ramotar, K., Seetaram, C., and Toye, B. 2004. Prevalence and mechanisms of erythromycin resistance in group A and group B streptococcus: Implications for reporting susceptibility results. *J. Clin. Microbiol.*, 42(12), 5620–5623.

Deutscher, M., Schillie, S., Gould, C., Baumbach, J., Mueller, M., Avery, C., and Van Beneden, C.A. 2011. Investigation of a group a streptococcal outbreak among residents of a long-term acute care hospital. *Clin. Infect. Dis.*, 52(8), 988–994.

Diedrick, M.J., Flores, A.E., Hillier, S.L., Creti, R., and Ferrieri, P. 2010. Clonal analysis of colonizing group B streptococcus, serotype IV, an emerging pathogen in the United States. *J. Clin. Microbiol.*, 48(9), 3100–3104.

Dillon Jr., H.C., Gray, E., Pass, M.A., and Gray, B.M. 1982. Anorectal and vaginal carriage of group B streptococci during pregnancy. *J. Infect. Dis.*, 145(6), 794–799.

Dmitriev, A.V., and Chaussee, M.S. 2010. The *Streptococcus pyogenes* proteome: Maps, virulence factors and vaccine candidates. *Future Microbiol.*, 5(10), 1539–1551.

Dramsi, S., Caliot, E., Bonne, I., Guadagnini, S., Prévost, M.C., Kojadinovic, M., Lalioui, L., Poyart, C., and Trieu-Cuot, P. 2006. Assembly and role of pili in group B streptococci. *Mol. Microbiol.*, 60(6), 1401–1413.

Edmond, K.M., Kortsalioudaki, C., Scott, S., Schrag, S.J., Zaidi, A.K., Cousens, S., and Heath, P.T. 2012. Group B streptococcal disease in infants aged younger than 3 months: Systematic review and meta-analysis. *Lancet*, 379(9815), 547–556.

Edmonson, M.B., and Farwell, K.R. 2005. Relationship between the clinical likelihood of group A streptococcal pharyngitis and the sensitivity of a rapid antigen-detection test in a pediatric practice. *Pediatrics*, 115(2), 280–285.

Edwards, M.S., and Nizet, V. 2011. Group B streptococcal infections. In: Remington, J.S., and Klein, J.O., editors. *Remington and Klein, Infectious Diseases of the Fetus and Newborn Infant*. 7th ed. Philadelphia, PA: Elsevier Saunders. pp 419–469.

Edwards, M.S., Rench, M.A., Haffar, A.A., Murphy, M.A., Desmond, M.M., and Baker, C.J. 1985. Long-term sequelae of group B streptococcal meningitis in infants. *J. Pediatr.*, 106(5), 717–722.

Efstratiou, A. 2000. Group A streptococci in the 1990s. *J. Antimicrob. Chemother.*, 45(suppl), 3–12.

Enright, M.C., Spratt, B.G., Kalia, A., Cross, J.H., and Bessen, D.E. 2001. Multilocus sequence typing of *Streptococcus pyogenes* and the relationships between emm type and clone. *Infect. Immun.*, 69(4), 2416–2427.

Fabretti, F., Theilacker, C., Baldassarri, L., Kaczynski, Z., Kropec, A., Olst, O., and Huebner, J. 2006. Alanine esters of enterococcal lipoteichoic acid play a role in biofilm formation and resistance to antimicrobial peptides. *Infect. Immun.*, 74(7), 4164–4171.

Facklam, R.R. 2002. What happened to the streptococci: Overview of taxonomic and nomenclature changes. *Clin. Microbiol. Rev.*, 15(4), 613–630.

Fernandez, M., Rench, M.A., Albanyan, E.A., Edwards, M.S., and Baker, C.J. 2001. Failure of rifampin to eradicate group B streptococcal colonization in infants. *Pediatr. Infect. Dis. J.,* 20(4), 371–376.

Ferretti, J.J., McShan, W.M., Ajdic, D., Savic, D.J., Savic, G., Lyon, K., Primeaux, C., Sezate, S., Suvorov, A.N., Kenton, S., Lai, H.S., Lin, S.P., Qian, Y., Jia, H.G., Najar, F.Z., Ren, Q., Zhu, H., Song, L., White, J., Yuan, X., Clifton, S.W., Roe, B.A., and McLaughlin, R. 2001. Complete genome sequence of an M1 strain of *Streptococcus pyogenes. Proc. Natl. Acad. Sci. USA,* 98(8), 4658–4663.

Ferrieri, P., Lynfield, R., Creti, R., and Flores, A.E. 2013. Serotype IV and invasive group B streptococcus disease in neonates, Minnesota, USA, 2000–2010. *Emerg. Infect. Dis.,* 19(4), 551–558.

Filleron, A., Lombard, F., Jacquot, A., Jumas-Bilak, E., Rodière, M., Cambonie, G., and Marchandin, H. 2014. Group B streptococci in milk and late neonatal infections: An analysis of cases in the literature. *Arch. Dis. Child Fetal Neonatal Ed.,* 99(1), F41–F47.

Fittipaldi, N., Beres, S.B., Olsen, R.J., Kapur, V., Shea, P.R., Watkins, M.E., Cantu, C.C., Laucirica, D.R., Jenkins, L., Flores, A.R., Lovgren, M., Ardanuy, C., Liñares, J., Low, D.E., Tyrrell, G.J., and Musser, J.M. 2012. Full-genome dissection of an epidemic of severe invasive disease caused by a hypervirulent, recently emerged clone of group A streptococcus. *Am. J. Pathol.,* 180(4), 1522–1534.

Flores, R., Jewell, E.B., Fittipaldi, N., Beres, S.B., and Musser, J.M. 2012. Human disease isolates of serotype M4 and M22 group A streptococcus lack genes required for hyaluronic acid capsule biosynthesis. *Mbio,* 3(6), e00413-12.

Florindo, C., Damiao, V., Silvestre, I., Farinha, C., Rodrigues, F., Nogueira, F., Martins-Pereira, F., Castro, R., Borrego, M.J., Santos-Sanches, I., and Group for the Prevention of Neonatal GBS Infection. 2014. Epidemiological surveillance of colonising group B Streptococcus epidemiology in the Lisbon and Tagus Valley regions, Portugal (2005 to 2012): emergence of a new epidemic type IV/clonal complex 17 clone. *Euro Surveill.,* 19(23), pii:20825.

Fluegge, K., Greiner, P., and Berner, R. 2003. Late onset group B streptococcal disease manifested by isolated cervical lymphadenitis. *Arch. Dis. Child.,* 88(11), 1019–1020.

Fox, E.N., Waldman, R.H., Wittner, M.K., Mauceri, A.A., and Dorfman, A. 1973. Protective study with a group A streptococcal M protein vaccine. Infectivity challenge of human volunteers. *J. Clin. Invest.,* 52(8), 1885–1892.

Fox, J.W., Marcon, M.J., and Bonsu, B.K. 2006. Diagnosis of streptococcal pharyngitis by detection of *Streptococcus pyogenes* in posterior pharyngeal versus oral cavity specimens. *J. Clin. Microbiol.,* 44(7), 2593–2594.

Freiwald, A., and Sauer, S. 2009. Phylogenetic classification and identification of bacteria by mass spectrometry. *Nat. Protocols,* 4(5), 732–742.

Friães, A., Lopes, J.P., Melo-Cristino, J., Ramirez, M., and The Portuguese Group for the Study of Streptococcal Infections. 2013b. Changes in *Streptococcus pyogenes* causing invasive disease in Portugal: Evidence for superantigen gene loss and acquisition. *Int. J. Med. Microbiol.,* 303(8), 505–513.

Friães, A., Pinto, F.R., Silva-Costa, C., Ramirez, M., and Melo-Cristino, J. 2013a. Superantigen gene complement of *Streptococcus pyogenes*–Relationship with other typing methods and short-term stability. *Eur. J. Clin. Microbiol. Infect. Dis.,* 32(1), 115–125.

Friedman, J., Van De Rijn, I., Ohkuni, H., Fischetti, V.A., and Zabriskie, J.B. 1984. Immunological studies of post-streptococcal sequelae. Evidence for presence of streptococcal antigens in circulating immune complexes. *J. Clin. Invest.,* 74(3), 1027–1034.

Gavino, M., and Wang, E. 2007. A comparison of a new rapid real-time polymerase chain reaction system to traditional culture in determining group B streptococcus colonization. *Am. J. Obstet. Gynecol.,* 197(4), 388.e1-4.

Gelfand, M.S., Cleveland, K.O., and Ketterer, D.C. 2013. Susceptibility of *Streptococcus pyogenes* to trimethoprim-sulfamethoxazole. *J. Clin. Microbiol.,* 51(4), 1350.

Gerber, M.A., and Shulman, S.T. 2004. Rapid diagnosis of pharyngitis caused by group A streptococci. *Clin. Microbiol. Rev.,* 17(3), 571–580.

Gherardi, G., Imperi, M., Baldassarri, L., Pataracchia, M., Alfarone, G., Recchia, S., Orefici, G., Dicuonzo, G., and Creti, R. 2007. Molecular epidemiology and distribution of serotypes, surface proteins, and antibiotic resistance among group B streptococci in Italy. *J. Clin. Microbiol.,* 45(9), 2909–2916.

Glaser, P., Rusniok, C., Buchrieser, C., Chevalier, F., Frangeul, L., Msadek, T., Zouine, M., Couvé, E., Lalioui, L., Poyart, C., Trieu-Cuot, P., and Kunst, F. 2002. Genome sequence of Streptococcus agalactiae, a pathogen causing invasive neonatal disease. *Mol. Microbiol.,* 45(6), 1499–1513.

Goroncy-Bermes, P., Dale, J.B., Beachey, E.H., and Opferkuch, W. 1987. Monoclonal antibody to human renal glomeruli cross-reacts with streptococcal M protein. *Infect. Immun.,* 55(10), 2416–2419.

Green, N.M., Zhang, S., Porcella, S.F., Nagiec, M.J., Barbian, K.D., Beres, S.B., LeFebvre, R.B., and Musser, J.M. 2005. Genome sequence of a serotype M28 strain of group a streptococcus: Potential new insights into puerperal sepsis and bacterial disease specificity. *J. Infect. Dis.,* 192(5), 760–770.

Guedez, Y., Kotby, A., El-Demellawy, M., Galai, A., Thomson, G., Zaher, S., and Kassem, S. 1999. HLA class II associations with rheumatic heart disease are more evident and consistent among clinically homogeneous patients. *Circulation,* 99(21), 2784–2790.

Guilbert, J., Levy, C., Cohen, R., Bacterial Meningitis Group, Delacourt, C., Renolleau, S., and Flamant, C. 2010. Late and ultra-late onset streptococcus B meningitis: Clinical and bacteriological data over 6 years in France. *Acta Paediatr.,* 99(1), 47–51.

Gutenkunst, H., Eikmanns, B.J., and Reinscheid, D.J. 2004. The novel fibrinogen binding protein FbsB promotes *Streptococcus agalactiae* invasion into epithelial cells. *Infect. Immun.,* 72(6), 3495–3504.

Hall, M.A., Stroop, S.D., Hu, M.C., Walls, M.A., Reddish, M.A., Burt, D.S., Lowell, G.H., and Dale, J.B. 2004. Intranasal immunization with multivalent group A streptococcal vaccines protects mice against intranasal challenge infections. *Infect. Immun.,* 72(5), 2507–2512.

Harris, T.O., Shelver, D.W., Bohnsack, J.F., and Rubens, C.E. 2003. A novel streptococcal surface protease promotes virulence, resistance to opsonophagocytosis, and cleavage of human fibrinogen. *J. Clin. Invest.,* 111(1), 61–70.

Haslam, R.H., Allen, J.R., Dorsen, M.M., Kanofsky, D.L., Mellitus, E.D., and Norris, D.A. 1977. The sequelae of group B beta-hemolytic streptococcal meningitis in early infancy. *Am. J. Dis. Child,* 131(8), 845–849.

Heath, P.T., Okike, I.O., and Oeser, C. 2011. Neonatal meningitis: Can we do better? *Adv. Exp. Med. Biol.,* 719, 11–24.

Hillier, S., Ferris, D., Fine, D., Ferrieri, P., Edwards, M.S.C., Ewell, M., et al. 2009. *Women Receiving Group B Streptococcus Serotype III Tetanus Toxoid (GBS III-TT) Vaccine Have Reduced Vaginal and Rectal Acquisition of GBS Type III.* Philadelphia, PA: Infectious Diseases Society of America.

Hoge, C.W., Schwartz, B., Talkington, D.F., Breiman, R.F., MacNeill, E.M., and Englender, S.J. 1993. The changing epidemiology of invasive group A streptococcal infections and the emergence of streptococcal toxic shock-like syndrome. A retrospective population-based study. *JAMA,* 269(3), 384–389.

Holden, M.T., Scott, A., Cherevach, I., Chillingworth, T., Churcher, C., Cronin, A., Dowd, L., Feltwell, T., Hamlin, N., Holroyd, S., Jagels, K., Moule, S., Mungall, K., Quail, M.A., Price, C., Rabbinowitsch, E., Sharp, S., Skelton, J., Whitehead, S., Barrell, B.G., Kehoe, M., and Parkhill, J. 2007. Complete genome of acute rheumatic fever-associated serotype

M5 *Streptococcus pyogenes* strain Manfredo. *J. Bacteriol.*, 189(4), 1473–1477.

Honest, H., Sharma, S., and Khan, K.S. 2006. Rapid tests for group B streptococcus colonization in laboring women: A systematic review. *Pediatrics*, 117(4), 1055–1066.

Horstmann, R.D., Sievertsen, H.J., Knobloch, J., and Fischetti, V.A. 1988. Antiphagocytic activity of streptococcal M protein: Selective binding of complement control protein factor H. *Proc. Natl. Acad. Sci. USA*, 85(5), 1657–1661.

Horstmann, R.D., Sievertsen, H.J., Leippe, M., and Fischetti, V.A. 1992. Role of fibrinogen in complement inhibition by streptococcal M protein. *Infect. Immun.*, 60(12), 5036–5041.

Howard, J.B., and McCracken Jr., G.H. 1974. The spectrum of group B streptococcal infections in infancy. *Am. J. Dis. Child.*, 128(6), 815–818.

Hu, M.C., Walls, M.A., Stroop, S.D., Reddish, M.A., Beall, B., and Dale, J.B. 2002. Immunogenicity of a 26-valent group A streptococcal vaccine. *Infect. Immun.*, 70(4), 2171–2177.

Hussain, S.M., Luedtke, G.S., Baker, C.J., Schlievert, P.M., and Leggiadro, R.J. 1995. Invasive group B streptococcal disease in children beyond early infancy. *Pediatr. Infect. Dis. J.*, 14(4), 278–288.

Imperi, M., Gherardi, G., Berardi, A., Baldassarri, L., Pataracchia, M., Dicuonzo, G., Orefici, G., and Creti, R. 2011. Invasive neonatal GBS infections from an area-based surveillance study in Italy. *Clin. Microbiol. Infect.*, 17(12), 1834–1839.

Imperi, M., Pataracchia, M., Alfarone, G., Baldassarri, L., Orefici, G., and Creti, R. 2010. A multiplex PCR assay for the direct identification of the capsular type (Ia to IX) of *Streptococcus agalactiae*. *J. Microbiol. Methods*, 80(2), 212–214.

Jacobsson, K. 2003. A novel family of fibrinogen binding proteins in *Streptococcus agalactiae*. *Vet. Microbiol.*, 96(1), 103–113.

Johansson, L., Thulin, P., Low, D.E., and Norrby-Teglund, A. 2010. Getting under the skin: The immunopathogenesis of *Streptococcus pyogenes* deep tissue infections. *Clin. Infect. Dis.*, 51(1), 58–65.

Johansson, L., Thulin, P., Sendi, P., Hertzén, E., Linder, A., Akesson, P., Low, D.E., Agerberth, B., and Norrby-Teglund, A. 2008. Cathelicidin LL-37 in severe Streptococcus pyogenes soft tissue infections in humans. *Infect. Immun.*, 76(8), 3399–3404.

Jones, N., Bohnsack, J.F., Takahashi, S., Oliver, K.A., Chan, M.S., Kunst, F., Glaser, P., Rusniok, C., Crook, D.W., Harding, R.M., Bisharat, N., and Spratt, B.G. 2003. Multilocus sequence typing system for group B streptococcus. *J. Clin. Microbiol.*, 41(6), 2530–2536.

Jones, R.N., Ross, J.E., Castanheira, M., and Mendes, R.E. 2008. United States resistance surveillance results for linezolid (LEADER Program for 2007). *Diagn. Microbiol. Infect. Dis.*, 62(4), 416–426.

Jordan, H.T., Farley, M.M., Craig, A., Mohle-Boetani, J., Harrison, L.H., Petit, S., Lynfield, R., Thomas, A., Zansky, S., Gershman, K., Albanese, B.A., Schaffner, W., Schrag, S.J., and Active Bacterial Core Surveillance (ABCs)/Emerging Infections Program Network, CDC. 2008. Revisiting the need for vaccine prevention of late-onset neonatal group B streptococcal disease: A multistate, population-based analysis. *Pediatr. Infect. Dis. J.*, 27(12), 1057–1064.

Kaplan, E.L., Rothermel, C.D., and Johnson, D.R. 1998. Antistreptolysin O and anti-deoxyribonuclease B titers: Normal values for children ages 2 to 12 in the United States. *Pediatrics*, 101(1 Pt 1), 86–88.

Keogh, J.M., Badawi, N., Kurinczuk, J.J., Pemberton, P.J., and Stanley, F.J. 1999. Group B streptococcus infection, not birth asphyxia. *Aust. N. Z. J. Obstet. Gynaecol.*, 39(1), 108–110.

Kimura, K., Matsubara, K., Yamamoto, G., Shibayama, K., and Arakawa, Y. 2013. Active screening of group B streptococci with reduced penicillin susceptibility and altered serotype distribution isolated from pregnant women in Kobe, Japan. *Jpn. J. Infect. Dis.*, 66(2), 158–160.

Kimura, K., Suzuki, S., Wachino, J., Kurokawa, H., Yamane, K., Shibata, N., Nagano, N., Kato, H., Shibayama, K., and Arakawa, Y. 2008. First molecular characterization of group B streptococci with reduced penicillin susceptibility. *Antimicrob. Agents Chemother.*, 52(8), 2890–2897.

Kline, K.A., Dodson, K.W., Caparon, M.G., and Hultgren, S.J. 2010. A tale of two pili: Assembly and function of pili in bacteria. *Trends Microbiol.*, 18(5), 224–232.

Kotb, M. 1995. Bacterial pyrogenic exotoxins as superantigen genes. *Clin. Microbiol. Rev.*, 8(3), 411–426.

Kotloff, K.L., Wasserman, S.S., Jones, K.F., Livio, S., Hruby, D.E., Franke, C.A., and Fischetti, V.A. 2005. Clinical and microbiological responses of volunteers to combined intranasal and oral inoculation with a *Streptococcus gordonii* carrier strain intended for future use as a group A streptococcus vaccine. *Infect. Immun.*, 73(4), 2360–2366.

Kreikemeyer, B., Gámez, G., Margarit, I., Giard, J.C., Hammerschmidt, S., Hartke, A., and Podbielski, A. 2012. Genomic organization, structure, regulation and pathogenic role of pilus constituents in major pathogenic Streptococci and Enterococci. *Int. J. Med. Microbiol.*, 301(3), 240–251.

Kristian, S.A., Datta, V., Weidenmaier, C., Kansal, R., Fedtke, I., Peschel, A., Gallo, R.L., and Nizet, V. 2005. D-alanylation of teichoic acids promotes group A streptococcus antimicrobial peptide resistance, neutrophil survival, and epithelial cell invasion. *J. Bacteriol.*, 187(19), 6719–6725.

Lachenauer, C.S., Creti, R., Michel, J.L., and Madoff, L.C. 2000. Mosaicism in the alpha-like protein genes of group B streptococci. *Proc. Natl. Acad. Sci. USA*, 97(17), 9630–9635.

Lamagni, T.L., Darenberg, J., Luca-Harari, B., Siljander, T., Efstratiou, A., Henriques-Normark, B., Vuopio-Varkila, J., Bouvet, A., Creti, R., Ekelund, K., Koliou, M., Reinert, R.R., Stathi, A., Strakova, L., Ungureanu, V., Schalén, C., Strep-EURO Study Group, and Jasir, A. 2008. Epidemiology of severe *Streptococcus pyogenes* disease in Europe. *J. Clin. Microbiol.*, 46(7), 2359–2367.

Lamagni, T.L., Efstratiou, A., Dennis, J., Nair, P., Kearney, J., George, R., and National Incident Management Team. 2009. Increase in invasive group A streptococcal infections in England, Wales and Northern Ireland, 2008–9. *Euro Surveill.*, 14(5).

Lamagni, T.L., Efstratiou, A., Vuopio-Varkila, J., Jasir, A., Schalen, C., and Strep-Euro. 2005. The epidemiology of severe *Streptococcus pyogenes* associated disease in Europe. *Euro Surveill.*, 10(9), 179–184.

Lanari, M., Serra, L., Cavrini, F., Liguori, G., and Sambri, V. 2007. Late-onset group B streptococcal disease by infected mother's milk detected by polymerase chain reaction. *New Microbiol.*, 30(3), 253–254.

Lannering, B., Larsson, L.E., Rojas, J., and Stahlman, M.T. 1983. Early onset group B streptococcal disease. Seven year experience and clinical scoring system. *Acta Paediatr. Scand.*, 72(4), 597–602.

Lappin, E., and Ferguson, A.J. 2009. Gram-positive toxic shock syndromes. *Lancet Infect. Dis.*, 9(5), 281–290.

Lauer, P., Rinaudo, C.D., Soriani, M., Margarit, I., Maione, D., Rosini, R., Taddei, A.R., Mora, M., Rappuoli, R., Grandi, G., and Telford, J.L. 2005. Genome analysis reveals pili in group B streptococcus. *Science*, 309(5731), 105.

Leckman, J.F., King, R.A., Gilbert, D.L., Coffey, B.J., Singer, H.S., Dure, L.S., Grantz, H., Katsovich, L., Lin, H., Lombroso, P.J., Kawikova, I., Johnson, D.R., Kurlan, R.M., and Kaplan, E.L. 2011. Streptococcal upper respiratory tract infections and exacerbations of tic and obsessive-compulsive symptoms: A prospective longitudinal study. *J. Am. Acad. Child Adolesc. Psychiatry*, 50(2), 108–118.

Lefébure, T., and Stanhope, M.J. 2007. Evolution of the core and pan-genome of streptococcus: Positive selection, recombination, and genome composition. *Genome Biol.*, 8(5), R71.

Lefébure, T., Richards, V.P., Lang, P., Pavinski-Bitar, P., and Stanhope, M.J. 2012. Gene repertoire evolution of *Streptococcus pyogenes* inferred from phylogenomic analysis with *Streptococcus canis* and *Streptococcus dysgalactiae*. *PLoS One*, 7(5), e37607.

Libster, R., Edwards, K.M., Levent, F., Edwards, M.S., Rench, M.A., Castagnini, L.A., Cooper, T., Sparks, R.C., Baker, C.J., and Shah, P.E. 2012. Long-term outcomes of group B streptococcal meningitis. *Pediatrics*, 130(1), e8–e15.

Lin, F.Y., Weisman, L.E., Azimi, P., Young, A.E., Chang, K., Cielo, M., Moyer, P., Troendle, J.F., Schneerson, R., and Robbins, J.B. 2011. Assessment of intrapartum antibiotic prophylaxis for the prevention of early-onset group B Streptococcal disease. *Pediatr. Infect. Dis. J.*, 30(9), 759–763.

Lin, F.Y., Weisman, L.E., Troendle, J., and Adams, K. 2003. Prematurity is the major risk factor for late-onset group B streptococcus disease. *J. Infect. Dis.*, 188(2), 267–271.

López Sastre, J.B., Fernández Colomer, B., Coto Cotallo, G.D., Ramos Aparicio, A., and Grupo de Hospitales Castrillo. 2005. Trends in the epidemiology of neonatal sepsis of vertical transmission in the era of group B streptococcal prevention. *Acta Paediatr.*, 94(4), 451–457.

Lopez-Sanchez, M.J., Sauvage, E., Da Cunha, V., Clermont, D., Ratsima Hariniaina, E., Gonzalez-Zorn, B., Poyart, C., Rosinski-Chupin, I., and Glaser, P. 2012. The highly dynamic CRISPR1 system of *Streptococcus agalactiae* controls the diversity of its mobilome. *Mol. Microbiol.*, 85(6), 1057–1071.

Luca-Harari, B., Darenberg, J., Neal, S., Siljander, T., Strakova, L., Tanna, A., Creti, R., Ekelund, K., Koliou, M., Tassios, P.T., van der Linden, M., Straut, M., Vuopio-Varkila, J., Bouvet, A., Efstratiou, A., Schalén, C., Henriques-Normark, B., Strep-EURO Study Group, and Jasir, A. 2009. Clinical and microbiological characteristics of severe *Streptococcus pyogenes* disease in Europe. *J. Clin. Microbiol.*, 47(4), 1155–1165.

Lynskey, N.N., Lawrenson, R.A., and Sriskandan, S. 2011. New understandings in *Streptococcus pyogenes*. *Curr. Opin. Infect. Dis.*, 24(3), 196–202.

Maamary, P.G., Ben Zakour, N.L., Cole, J.N., Hollands, A., Aziz, R.K., Barnett, T.C., Cork, A.J., Henningham, A., Sanderson-Smith, M., McArthur, J.D., Venturini, C., Gillen, C.M., Kirk, J.K., Johnson, D.R., Taylor, W.L., Kaplan, E.L., Kotb, M., Nizet, V., Beatson, S.A., and Walker, M.J. 2012. Tracing the evolutionary history of the pandemic group A streptococcal M1T1 clone. *FASEB J.*, 26(11), 4675–4684.

MacFarquhar, J.K., Jones, T.F., Woron, A.M., Kainer, M.A., Whitney, C.G., Beall, B., Schrag, S.J., and Schaffner, W. 2010. Outbreak of late-onset group B streptococcus in a neonatal intensive care unit. *Am. J. Infect. Control.*, 38(4), 283–288.

Maione, D., Margarit, I., Rinaudo, C.D., Masignani, V., Mora, M., Scarselli, M., Tettelin, H., Brettoni, C., Iacobini, E.T., Rosini, R., D'Agostino, N., Miorin, L., Buccato, S., Mariani, M., Galli, G., Nogarotto, R., Nardi-Dei, V., Vegni, F., Fraser, C., Mancuso, G., Teti, G., Madoff, L.C., Paoletti, L.C., Rappuoli, R., Kasper, D.L., Telford, J.L., and Grandi, G. 2005. Identification of a universal group B streptococcus vaccine by multiple genome screen. *Science*, 309(5731), 148–150.

Maisey, H.C., Doran, K.S., and Nizet, V. 2008. Recent advances in understanding the molecular basis of group B streptococcus virulence. *Expert Rev. Mol. Med.*, 10, e27.

Makarova, K.S., Haft, D.H., Barrangou, R., Brouns, S.J., Charpentier, E., Horvath, P., Moineau, S., Mojica, F.J., Wolf, Y.I., Yakunin, A.F., van der Oost, J., and Koonin, E.V. 2011. Evolution and classification of the CRISPR-Cas systems. *Nat. Rev. Microbiol.*, 9(6), 467–477.

Malhotra-Kumar, S., Van Heirstraeten, L., Lammens, C., Chapelle, S., and Goossens, H. 2009. Emergence of high-level fluoroquinolone resistance in emm6 *Streptococcus pyogenes* and *in vitro* resistance selection with ciprofloxacin, levofloxacin and moxifloxacin. *J. Antimicrob. Chemother.*, 63(5), 886–894.

Manning, S.D., Lacher, D.W., Davies, H.D., Foxman, B., and Whittam, T.S. 2005. DNA polymorphism and molecular subtyping of the capsular gene cluster of group B streptococcus. *J. Clin. Microbiol.*, 43(12), 6113–6116.

Marques, M.B., Kasper, D.L., Pangburn, M.K., and Wessels, M.R. 1992. Prevention of C3 deposition by capsular polysaccharide is a virulence mechanism of type III group B streptococci. *Infect. Immun.*, 60(10), 3986–3993.

Marshall, C.S., Cheng, A.C., Markey, P.G., Towers, R.J., Richardson, L.J., Fagan, P.K., Scott, L., Krause, V.L., and Currie, B.J. 2011. Acute post-streptococcal glomerulonephritis in the Northern Territory of Australia: A review of 16 years data and comparison with the literature. *Am. J. Trop. Med. Hyg.*, 85(4), 703–710.

Martin, D.R., Stanhope, J.M., and Finch, L.A. 1977. Delayed culture of group-A streptococci: An evaluation of variables in methods of examining throat swabs. *J. Med. Microbiol.*, 10(2), 249–253.

Martins, E.R., Andreu, A., Correia, P., Juncosa, T., Bosch, J., Ramirez, M., Melo-Cristino, J., and Microbiologist Group for the Study of Vertical Transmission Infections from the Catalan Society for Clinical Microbiology and Infectious Diseases. 2011. Group B streptococci causing neonatal infections in Barcelona are a stable clonal population: 18-year surveillance. *J. Clin. Microbiol.*, 49(8), 2911–2918.

Matorras, R., García-Perea, A., Omeñaca, F., Diez-Enciso, M., Madero, R., and Usandizaga, J.A. 1991. Intrapartum chemoprophylaxis of early-onset group B streptococcal disease. *Eur. J. Obstet. Gynecol. Reprod. Biol.*, 40(1), 57–62.

McDonald, M., Currie, B.J., and Carapetis, J.R. 2004. Acute rheumatic fever: A link in the chain that links the heart to the throat? *Lancet Infect. Dis.*, 4(4), 240–245.

McGregor, K.F., Spratt, B.G., Kalia, A., Bennett, A., Bilek, N., Beall, B., and Bessen, D.E. 2004. Multilocus sequence typing of *Streptococcus pyogenes* representing most known emm types and distinctions among subpopulation genetic structures. *J. Bacteriol.*, 186(13), 4285–4294.

McMillan, D.J., Sanderson-Smith, M.L., Smeesters, P.R., and Sriprakash, K.S. 2013. Molecular markers for the study of streptococcal epidemiology. *Curr. Top. Microbiol. Immunol.*, 368, 29–48.

McNamara, C., Zinkernagel, A.S., Macheboeuf, P., Cunningham, M.W., Nizet, V., and Ghosh, P. 2008. Coiled-coil irregularities and instabilities in group A streptococcus M1 are required for virulence. *Science*, 319(5868), 1405–1408.

McNeil, S.A., Halperin, S.A., Langley, J.M., Smith, B., Warren, A., Sharratt, G.P., Baxendale, D.M., Reddish, M.A., Hu, M.C., Stroop, S.D., Linden, J., Fries, L.F., Vink, P.E., and Dale, J.B. 2005. Safety and immunogenicity of 26-valent group a streptococcus vaccine in healthy adult volunteers. *Clin. Infect. Dis.*, 41(8), 1114–1122.

McShan, W.M., Ferretti, J.J., Karasawa, T., Suvorov, A.N., Lin, S., Qin, B., Jia, H., Kenton, S., Najar, F., Wu, H., Scott, J., Roe, B.A., and Savic, D.J. 2008. Genome sequence of a nephritogenic and highly transformable M49 strain of *Streptococcus pyogenes*. *J. Bacteriol.*, 190(23), 7773–7785.

Medini, D., Donati, C., Tettelin, H., Masignani, V., and Rappuoli, R. 2005. The microbial pan-genome. *Curr. Opin. Genet. Dev.*, 15(6), 589–594.

Meehan, M., Cunney, R., and Cafferkey, M. 2014. Molecular epidemiology of group B streptococci in Ireland reveals a diverse population with evidence of capsular switching. *Eur. J. Clin. Microbiol. Infect. Dis.*, 33(7), 1155–1162.

Melin, P. 2011. Neonatal group B streptococcal disease: From pathogenesis to preventive strategies. *Clin. Microbiol. Infect.*, 17(9), 1294–1303.

Melin, P., and Efstratiou, A. 2013. Group B streptococcal epidemiology and vaccine needs in developed countries. *Vaccine*, 31(Suppl 4), D31–D42.

Metzgar, D., and Zampolli, A. 2011. The M protein of group A streptococcus is a key virulence factor and a clinically relevant strain identification marker. *Virulence*, 2(5), 402–412.

Miyoshi-Akiyama, T., Watanabe, S., and Kirikae, T. 2012. Complete genome sequence of Streptococcus pyogenes M1 476, isolated from a patient with streptococcal toxic shock syndrome. *J. Bacteriol.*, 194(19), 5466.

Mora, M., Bensi, G., Capo, S., Falugi, F., Zingaretti, C., Manetti, A.G., Maggi, T., Taddei, A.R., Grandi, G., and Telford, J.L. 2005. Group A streptococcus produce pilus-like structures containing protective antigens and Lancefield T antigens. *Proc. Natl. Acad. Sci. USA*, 102(43), 15641–15646.

Musser, J.M., and Shelburne, S.A. 2009. A decade of molecular pathogenomic analysis of group A streptococcus. *J. Clin. Invest.*, 119(9), 2455–2463.

Nakagawa, I., Kurokawa, K., Yamashita, A., Nakata, M., Tomiyasu, Y., Okahashi, N., Kawabata, S., Yamazaki, K., Shiba, T., Yasunaga, T., Hayashi, H., Hattori, M., and Hamada, S. 2003. Genome sequence of an M3 strain of *Streptococcus pyogenes* reveals a large-scale genomic rearrangement in invasive strains and new insights into phage evolution. *Genome Res.*, 13(6A), 1042–1055.

Nitsche-Schmitz, D.P., and Chhatwal, G.S. 2013. Host-pathogen interactions in streptococcal immune sequelae. *Curr. Top. Microbiol. Immunol.*, 368, 155–171.

Nozawa, T., Furukawa, N., Aikawa, C., Watanabe, T., Haobam, B., Kurokawa, K., Maruyama, F., and Nakagawa, I. 2011. CRISPR inhibition of prophage acquisition in *Streptococcus pyogenes*. *PLoS One*, 6(5), e19543.

Nuccitelli, R., Cozzi, L.J., Gourlay, D., Donnarumma, F., Necchi, N., Norais, J.L., Telford, R., Rappuoli, M., Bolognesi, D., Maione, G., Grandi, G., and Rinaudo, C.D. 2011. Structure-based approach to rationally design a chimeric protein for an effective vaccine against Group B streptococcus infections. *Proc. Natl. Acad. Sci. USA*, 108, 10278–10283.

Obszańska, K., Borek, A.L., Izdebski, R., Hryniewicz, W., and Sitkiewicz, I. 2011. Multilocus variable number tandem repeat analysis (MLVA) of *Streptococcus pyogenes*. *J. Microbiol. Methods*, 87(2), 143–149.

Ohlsson, A., and Shah, V.S. 2009. Intrapartum antibiotics for known maternal Group B streptococcal colonization. *Cochrane Database Syst. Rev.*, (3), art. no. CD007467.

O'Loughlin, R.E., Roberson, A., Cieslak, P.R., Lynfield, R., Gershman, K., Craig, A., Albanese, B.A., Farley, M.M., Barrett, N.L., Spina, N.L., Beall, B., Harrison, L.H., Reingold, A., Van Beneden, C., and Team ABCS. 2007. The epidemiology of invasive group A streptococcal infection and potential vaccine implications: United States, 2000–2004. *Clin. Infect. Dis.*, 45(7), 853–862.

Olsen, R.J., Shelburne, S.A., and Musser, J.M. 2009. Molecular mechanisms underlying group A streptococcal pathogenesis. *Cell. Microbiol.*, 11(1), 1–12.

Panchaud, A., Guy, L., Collyn, F., Haenni, M., Nakata, M., Podbielski, A., Moreillon, P., and Roten, C.A. 2009. M-protein and other intrinsic virulence factors of *Streptococcus pyogenes* are encoded on an ancient pathogenicity island. *BMC Genomics*, (10), 198.

Paredes, A., Wong, P., Mason Jr., E.O., Taber, L.H., and Barrett, F.F. 1977. Nosocomial transmission of group B Streptococci in a newborn nursery. *Pediatrics*, 59(5), 679–682.

Patenge, N., Fiedler, T., and Kreikemeyer, B. 2013. Common regulators of virulence in streptococci. *Curr. Top Microbiol. Immunol.*, 368, 111–153.

Payne, N.R., Burke, B.A., Day, D.L., Christenson, P.D., Thompson, T.R., and Ferrieri, P. 1988. Correlation of clinical and pathologic findings in early onset neonatal group B streptococcal infection with disease severity and prediction of outcome. *Pediatr. Infect. Dis. J.*, 7, 836–847.

Peña, B.M., Harper, M.B., and Fleisher, G.R. 1998. Occult bacteremia with group B streptococci in an outpatient setting. *Pediatrics*, 102(1 Pt 1), 67–72.

Phares, C.R., Lynfield, R., Farley, M.M., Mohle-Boetani, J., Harrison, L.H., Petit, S., Craig, A.S., Schaffner, W., Zansky, S.M., Gershman, K., Stefonek, K.R., Albanese, B.A., Zell, E.R., Schuchat, A., Schrag, S.J., and Active Bacterial Core surveillance/Emerging Infections Program Network. 2008. Epidemiology of invasive group B streptococcal disease in the United States, 1999–2005. *JAMA*, 299(17), 2056–2065.

Pittiglio, V., Ciammaruconi, A., D'Avenio, G., Gherardi, G., Pourcel, C., and Creti, R. 2010. A MLVA assay for genotyping of *Streptococcus pyogenes*. Tenth International Conference on Molecular Epidemiology and Evolutionary Genetics of Infectious Diseases, 3–5 November, at Amsterdam, The Netherlands.

Platt, M.W., McLaughlin, J.C., Gilson, G.J., Wellhoner, M.F., and Nims, L.J. 1995. Increased recovery of group B streptococcus by the inclusion of rectal culturing and enrichment. *Diagn. Microbiol. Infect. Dis.*, 21(2), 65–68.

Pokorski, S.J., Vetter, E.A., Wollan, P.C., and Cockerill 3rd., F.R. 1994. Comparison of gen-probe group A streptococcus direct test with culture for diagnosing streptococcal pharyngitis. *J. Clin. Microbiol.*, 32(6), 1440–1443.

Port, G.C., Paluscio, E., and Caparon, M.G. 2013. Complete genome sequence of emm type 14 *Streptococcus pyogenes* strain HSC5. *Genome Announc.*, 1(4), e00612-13.

Poses, R.M., Cebul, R.D., Collins, M., and Fager, S.S. 1985. The accuracy of experienced physicians' probability estimates for patients with sore throats: Implications for decision making. *JAMA*, 254(7), 925–929.

Poyart, C., Réglier-Poupet, H., Tazi, A., Billoët, A., Dmytruk, N., Bidet, P., Bingen, E., Raymond, J., and Trieu-Cuot, P. 2008. Invasive group B streptococcal infections in infants, France. *Emerg. Infect. Dis.*, 14(10), 1647–1649.

Poyart, C., Tazi, A., Réglier-Poupet, H., Billoët, A., Tavares, N., Raymond, J., and Trieu-Cuot, P. 2007. Multiplex PCR assay for rapid and accurate capsular typing of group B streptococci. *J. Clin. Microbiol.*, 45(6), 1985–1988.

Pulver, L.S., Hopfenbeck, M.M., Young, P.C., Stoddard, G.J., Korgenski, K., Daly, J., and Byington, C.L. 2009. Continued early onset group B streptococcal infections in the era of intrapartum prophylaxis. *J. Perinatol.*, 29(1), 20–25.

Puopolo, K.M., Madoff, L.C., and Eichenwald, E.C. 2005. Early-onset group B streptococcal disease in the era of maternal screening. *Pediatrics*, 115(5), 1240–1246.

Rajagopal, L. 2009. Understanding the regulation of group B streptococcal virulence factors. *Rev. Future Microbiol.*, 4(2), 201–221.

Rappuoli, R. 2000. Reverse vaccinology. *Curr. Opin. Microbiol.*, 3(5), 445–450.

Reinert, R.R., Lutticken, R., and Al-Lahham, A. 2004. High-level fluoroquinolone resistance in a clinical *Streptococcus pyogenes* isolate in Germany. *Clin. Microbiol. Infect.*, 10(7), 659–662.

Rodriguez-Granger, J., Alvargonzalez, J.C., Berardi, A., Berner, R., Kunze, M., Hufnagel, M., Melin, P., Decheva, A., Orefici, G., Poyart, C., Telford, J., Efstratiou, A., Killian, M., Krizova, P., Baldassarri, L., Spellerberg, B., Puertas, A., and Rosa-Fraile, M. 2012. Prevention of group B streptococcal neonatal disease revisited. The DEVANI European Project. *Eur. J. Clin. Microbiol. Infect. Dis.*, 31(9), 2097–2104.

Rosa-Fraile, M., Camacho-Muñoz, E., Rodríguez-Granger, J., and Liébana-Martos, C. 2005. Specimen storage in transport medium and detection of group B streptococci by culture. *J. Clin. Microbiol.*, 43(2), 928–930.

Rosini, R., Rinaudo, C.D., Soriani, M., Lauer, P., Mora, M., Maione, D., Taddei, A., Santi, I., Ghezzo, C., Brettoni, C., Buccato, S., Margarit, I., Grandi, G., and Telford, J.L. 2006. Identification of novel genomic islands coding for antigenic pilus-like structures in *Streptococcus agalactiae*. *Mol. Microbiol.*, 61(1), 126–141.

Royal College of Obstetricians and Gynaecologists: Guidelines. 2012. The prevention of early onset neonatal group B streptococcal disease. Green-top Guideline No. 36. 2nd ed. pp. 1–13 [http://www.rcog.org.uk/files/rcog-corp/GTG36_GBS.pdf].

Russell, S.C. 2009. Microorganism characterization by single particle mass spectrometry. *Mass Spectrom. Rev.*, 28(2), 376–387.

Schaad, U.B., Nelson, J.D., and McCracken Jr., G.H. 1981. Recrudescence and relapse in bacterial meningitis of childhood. *Pediatrics*, 67, 188–195.

Schrag, S.J., and Verani, J.R. 2013. Intrapartum antibiotic prophylaxis for the prevention of perinatal group B streptococcal disease: Experience in the United States and implications for a potential group B streptococcal vaccine. *Vaccine*, 31 (Suppl 4), D20–D26.

Schrag, S.J., Zywicki, S., Farley, M.M., Reingold, A.L., Harrison, L.H., Lefkowitz, L.B., Hadler, J.L., Danila, R., Cieslak, P.R., and Schuchat, A. 2000. Group B streptococcal disease in the era of intrapartum antibiotic prophylaxis. *N. Engl. J. Med.*, 342(1), 15–20.

Schuchat, A. 1998. Epidemiology of group B streptococcal disease in the United States: Shifting paradigms. *Clin. Microbiol. Rev.*, 11(3), 497–513.

Shet, A., and Kaplan, E.L. 2002. Clinical use and interpretation of group A streptococcal antibody tests: A practical approach for the pediatrician or primary care physician. *Pediatr. Infect. Dis. J.*, 21(5), 420–426, quiz 427–430.

Shimomura, Y., Okumura, K., Murayama, S.Y., Yagi, J., Ubukata, K., Kirikae, T., and Miyoshi-Akiyama, T. 2011. Complete genome sequencing and analysis of a Lancefield group G *Streptococcus dysgalactiae* subsp. *equisimilis* strain causing streptococcal toxic shock syndrome (STSS). *BMC Genomics*, 11(12), 17.

Shulman, S.T., Bisno, A.L., Clegg, H.W., Gerber, M.A., Kaplan, E.L., Lee, G., Martin, J.M., and Van Beneden, C. 2012. Clinical practice guideline for the diagnosis and management of group a streptococcal pharyngitis: 2012 Update by the Infectious Diseases Society of America. *Clin. Infect. Dis.*, 55(10), 1279–1282.

Silva-Costa, C., Friães, A., Ramirez, M., Melo-Cristino, J., and the Portuguese Group for the Study of Streptococcal Infections. 2012. Differences between macrolide-resistant and -susceptible *Streptococcus pyogenes*: Importance of clonal properties in addition to antibiotic consumption. *Antimicrob. Agents Chemother.*, 56(11), 5661–5666.

Silva-Costa, C., Ramirez, M., and Melo-Cristino, J. 2006. Identification of macrolide-resistant clones of *Streptococcus pyogenes* in Portugal. *Clin. Microbiol. Infect.*, 12(6), 513–518.

Singh, P., Springman, A.C., Davies, H.D., and Manning, S.D. 2012. Whole-genome shotgun sequencing of a colonizing multilocus sequence type 17 *Streptococcus agalactiae* strain. *J. Bacteriol.*, 194(21), 6005.

Skoff, T.H., Farley, M.M., Petit, S., Craig, A.S., Schaffner, W., Gershman, K., Harrison, L.H., Lynfield, R., Mohle-Boetani, J., Zansky, S., Albanese, B.A., Stefonek, K., Zell, E.R., Jackson, D., Thompson, T., and Schrag, S.J. 2009. Increasing burden of invasive group B streptococcal disease in nonpregnant adults, 1990–2007. *Clin. Infect. Dis.*, 49(1), 85–92.

Smoot, J.C., Barbian, K.D., Van Gompel, J.J., Smoot, L.M., Chaussee, M.S., Sylva, G.L., Sturdevant, D.E., Ricklefs, S.M., Porcella, S.F., Parkins, L.D., Beres, S.B., Campbell, D.S., Smith, T.M., Zhang, Q., Kapur, V., Daly, J.A., Veasy, L.G., and Musser, J.M. 2002. Genome sequence and comparative microarray analysis of serotype M18 group A streptococcus strains associated with acute rheumatic fever outbreaks. *Proc. Natl. Acad. Sci. USA*, 99(7), 4668–4673.

Soukka, H., Rantakokko-Jalava, K., Vähäkuopus, S., and Ruuskanen, O. 2010. Three distinct episodes of GBS septicemia in a healthy newborn during the first month of life. *Eur. J. Pediatr.*, 169(10), 1275–1277.

Spellerberg, B., Rodzinski, E., and Martins, S. 1999. Lmb, a protein with similarities to the LrsI adhesion family, mediates attachment of *Streptococcus agalactiae* to human laminin. *Infect. Immun.*, 67(12), 871–878.

Springman, A.C., Lacher, D.W., Wu, G., Milton, N., Whittam, T.S., Davies, H.D., and Manning, S.D. 2009. Selection, recombination, and virulence gene diversity among group B streptococcal genotypes. *J. Bacteriol.*, 191(17), 5419–5427.

Sriskandan, S., and Altmann, D.M. 2008. The immunology of sepsis. *J. Pathol.*, 214(2), 211–223.

Steer, A.C., Lamagni, T., Curtis, N., and Carapetis, J.R. 2012. Invasive group a streptococcal disease: Epidemiology, pathogenesis and management. *Drugs*, 72(9), 1213–1227.

Stevens, D.L. 1995. Streptococcal toxic-shock syndrome: Spectrum of disease, pathogenesis, and new concepts in treatment. *Emerg. Infect. Dis.*, 1(3), 69–78.

Stevens, D.L. 1999. The flesh-eating bacterium: What's next? *J. Infect. Dis.*, 179(Suppl 2), S366–S374.

Stollerman, G.H. 1997a. Changing streptococci and prospects for the global eradication of rheumatic fever. *Perspect. Biol. Med.*, 40(2), 165–189.

Stollerman, G.H. 1997b. Rheumatic fever. *Lancet*, (349), 935–942.

Sumby, P., Porcella, S.F., Madrigal, A.G., Barbian, K.D., Virtaneva, K., Ricklefs, S.M., Sturdevant, D.E., Graham, M.R., Vuopio-Varkila, J., Hoe, N.P., and Musser, J.M. 2005. Evolutionary origin and emergence of a highly successful clone of serotype M1 group A streptococcus involved multiple horizontal gene transfer events. *J. Infect. Dis.*, 192(5), 771–782.

Swedo, S.E., Leonard, H.L., Mittleman, B.B., Allen, A.J., Rapoport, J.L., Dow, S.P., Kanter, M.E., Chapman, F., and Zabriskie, J. 1997. Identification of children with pediatric autoimmune neuropsychiatric disorders associated with streptococcal infections by a marker associated with rheumatic fever. *Am. J. Psychiatry*, 154(1), 110–112.

Takahashi, S., Nagano, Y., Nagano, N., Hayashy, O., Taguchi, F., and Okuwaki, Y. 1995. Role of C5a-ase in group B streptococcal resistance to opsonophagocytic killing. *Infect. Immun.*, 63(12), 4764–4769.

Tanz, R.R., Gerber, M.A., Kabat, W., Rippe, J., Seshadri, R., and Shulman, S.T. 2009. Performance of a rapid antigen-detection test and throat culture in community pediatric offices: Implications for management of pharyngitis. *Pediatrics*, 123(2), 437–444.

Tazi, A., Morand, P.C., Réglier-Poupet, H., Dmytruk, N., Billoët, A., Antona, D., Trieu-Cuot, P., and Poyart, C. 2011. Invasive group B streptococcal infections in adults, France (2007–2010). *Clin. Microbiol. Infect.*, 17(10), 1587–1589.

Tazi, A., Réglier-Poupet, H., Dautezac, F., Raymond, J., and Poyart, C. 2008. Comparative evaluation of Strepto B ID chromogenic medium and Granada media for the detection of group B streptococcus from vaginal samples of pregnant women. *J. Microbiol. Methods*, 73(3), 263–265.

Teatero, S., McGeer, A., Li, A., Gomes, J., Seah, C., Demczuk, W., Martin, I., Wasserscheid, J., Dewar, K., Melano, R.G., and Fittipaldi, N. 2015. Population structure and antimicrobial resistance of invasive serotype IV group B Streptococcus, Toronto, Ontario, Canada. *Emerg. Infect. Dis.*, 21(4), 585–591.

Teese, N., Henessey, D., Pearce, C., Kelly, N., and Garland, S. 2003. Screening protocols for group B streptococcus: Are transport media appropriate? *Infect. Dis. Obstet. Gynecol.*, 11(4), 199–202.

Tenenbaum, T., Spellerberg, B., Adam, R., Vogel, M., Kim, K.S., and Schroten, H. 2007. *Streptococcus agalctiae* invasion of human brain microvascular endothelial cells is promoted by the laminin binding protein Lmb. *Microbes Infect.*, 9(6), 714–720.

Tettelin, H., Masignani, V., Cieslewicz, M.J., Eisen, J.A., Peterson, S., Wessels, M.R., Paulsen, I.T., Nelson, K.E., Margarit, I., Read, T.D., Madoff, L.C., Wolf, A.M., Beanan, M.J., Brinkac, L.M., Daugherty, S.C., DeBoy, R.T., Durkin, A.S., Kolonay, J.F., Madupu, R., Lewis, M.R., Radune, D., Fedorova, N.B., Scanlan, D., Khouri, H., Mulligan, S., Carty, H.A., Cline, R.T., Van Aken, S.E., Gill, J., Scarselli, M., Mora, M., Iacobini, E.T., Brettoni, C., Galli, G., Mariani, M., Vegni, F., Maione, D., Rinaudo, D., Rappuoli, R., Telford, J.L., Kasper, D.L., Grandi, G., and Fraser, C.M. 2002. Complete genome sequence and comparative genomic analysis of an emerging human pathogen, serotype V Streptococcus agalactiae. *Proc. Natl. Acad. Sci. USA*, 99(19), 12391–12396.

Tettelin, H., Masignani, V., Cieslewicz, M.J., Donati, C., Medini, D., Ward, N.L., Angiuoli, S.V., Crabtree, J., Jones, A.L.,

Durkin, A.S., Deboy, R.T., Davidsen, T.M., Mora, M., Scarselli, M., Margarit y Ros, I., Peterson, J.D., Hauser, C.R., Sundaram, J.P., Nelson, W.C., Madupu, R., Brinkac, L.M., Dodson, R.J., Rosovitz, M.J., Sullivan, S.A., Daugherty, S.C., Haft, D.H., Selengut, J., Gwinn, M.L., Zhou, L., Zafar, N., Khouri, H., Radune, D., Dimitrov, G., Watkins, K., O'Connor, K.J., Smith, S., Utterback, T.R., White, O., Rubens, C.E., Grandi, G., Madoff, L.C., Kasper, D.L., Telford, J.L., Wessels, M.R., Rappuoli, R., and Fraser, C.M. 2005. Genome analysis of multiple pathogenic isolates of *Streptococcus agalactiae*: Implications for the microbial "pan-genome". *Proc. Natl. Acad. Sci. USA*, 102(39), 13950–13955.

Thulin, P., Johansson, L., Low, D.E., Gan, B.S., Kotb, M., McGeer, A., and Norrby-Teglund, A. 2006. Viable group A streptococci in macrophages during acute soft tissue infection. *PLoS Med.*, 3(3), e53.

Tunkel, A.R., Hartman, B.J., Kaplan, S.L., Kaufman, B.A., Roos, K.L., Scheld, W.M., and Whitley, R.J. 2004. Practice guidelines for the management of bacterial meningitis. *Clin. Infect. Dis.*, 39(9), 1267–1284.

Tuppurainen, N., and Hallman, M. 1989. Prevention of neonatal group B streptococcal disease: Intrapartum detection and chemoprophylaxis of heavily colonized parturients. *Obstet. Gynecol.*, 73(4), 583–587.

Uhl, J.R., Adamson, S.C., Vetter, E.A., Schleck, C.D., Harmsen, W.S., Iverson, L.K., Santrach, P.J., Henry, N.K., and Cockerill, F.R. 2003. Comparison of LightCycler PCR, rapid antigen immunoassay, and culture for detection of group A streptococci from throat swabs. *J. Clin. Microbiol.*, 41(1), 242–249.

Valery, P.C., Wenitong, M., Clements, V., Sheel, M., Mcmillan, D., Stirling, J., Sriprakash, K.S., Batzloff, M., Vohra, R., and Mccarthy, J.S. 2008. Skin infections among indigenous Australians in an urban setting in far North Queensland. *Epidemiol. Infect.*, 136(8), 1103–1108.

Valkenburg-van den Berg, A.W., Houtman-Roelofsen, R.L., Oostvogel, P.M., Dekker, F.W., Dörr, P.J., and Sprij, A.J. 2010. Timing of group B streptococcus screening in pregnancy: A systematic review. *Gynecol. Obstet. Invest.*, 69(3), 174–183.

Varaldo, P.E., Montanari, M.P., and Giovanetti, E. 2009. Genetic elements responsible for erythromycin resistance in streptococci. *Antimicrob. Agents Chemother.*, 53(2), 343–353.

Verani, J.R., McGee, L., Schrag, S.J., and Division of Bacterial Diseases, National Center for Immunization and Respiratory Diseases, Centres for Disease Control and Prevention (CDC). 2010. Prevention of perinatal group B streptococcal disease—revised guidelines from CDC, 2010. *MMWR Recomm. Rep.*, 59(RR-10), 1–36.

Wald, E.R., Bergman, I., Taylor, H.G., Chiponis, D., Porter, C., and Kubek, K. 1986. Long-term outcome of group B streptococcal meningitis. *Pediatrics*, 77(2), 217–221.

Walker, M.J., Mcarthur, J.D., Mckay, F., and Ranson, M. 2005. Is plasminogen deployed as a Streptococcus pyogenes virulence factor? *Trends Microbiol.*, 13(7), 308–313.

Wessels, M.R. 2011. Streptococcal pharyngitis. *N. Engl. J. Med.*, 364(7), 648–655.

Weston, E.J., Pondo, T., Lewis, M.M., Martell-Cleary, P., Morin, C., Jewell, B., Daily, P., Apostol, M., Petit, S., Farley, M., Lynfield, R., Reingold, A., Hansen, N.I., Stoll, B.J., Shane, A.J., Zell, E., and Schrag, S.J. 2011. The burden of invasive early-onset neonatal sepsis in the United States, 2005–2008. *Pediatr. Infect. Dis. J.*, 30(11), 937–941.

Whitnack, E., and Beachey, E.H. 1982. Antiopsonic activity of fibrinogen bound to M protein on the surface of group A streptococci. *J. Clin. Invest.*, 69(4), 1042–1051.

Wu, C.S., Wang, S.M., Ko, W.C., Wu, J.J., Yang, YJ., and Liu, C.C. 2004 Group B streptococcal infections in children in a tertiary care hospital in southern Taiwan. *J. Microbiol. Immunol. Infect*, 37(3), 169–175.

Yao, K., Poulsen, K., Maione, D., Rinaudo, C.D., Baldassarri, L., Telford, J.L., Sørensen, U.B., Members of the DEVANI Study Group, and Kilian, M. 2013. Capsular gene typing of *Streptococcus agalactiae* compared to serotyping by latex agglutination. *J. Clin. Microbiol.*, 51(2), 503–507.

Zakikhany, K., Degail, M.A., Lamagni, T., Waight, P., Guy, R., Zhao, H., Efstratiou, A., Pebody, R., George, R., and Ramsay, M. 2011. Increase in invasive *Streptococcus pyogenes* and *Streptococcus pneumoniae* infections in England, December 2010 to January 2011. *Euro Surveill.*, 16(5), 1–4.

Zaman, M., Chandrudu, S., Giddam, A.K., Reiman, J., Skwarczynski, M., McPhun, V., Moyle, P.M., Batzloff, M.R., Good, M.F., and Toth, I. 2014. Group A Streptococcal vaccine candidate: contribution of epitope to size, antigen presenting cell interaction and immunogenicity. *Nanomedicine*, 9(17), 2613–2624.

Zeng, X., Kong, F., Morgan, J., and Gilbert, G.L. 2006. Evaluation of a multiplex PCR-based reverse line blot-hybridization assay for identification of serotype and surface protein antigens of *Streptococcus agalactiae*. *J. Clin. Microbiol.*, 44(10), 3822–3825.

Zheng, P.X., Chung, K.T., Chiang-Ni, C., Wang, S.Y., Tsai, P.J., Chuang, W.J., Lin, Y.S., Liu, C.C., and Wu, J.J. 2013. Complete genome sequence of emm1 *Streptococcus pyogenes* A20, a strain with an intact two-component system, CovRS, isolated from a patient with necrotizing fasciitis. *Genome Announc.*, 1(1).

Chapter 43

Pathogenesis of *Clostridium botulinum* in Humans

Sabine Pellett

Department of Bacteriology, University of Wisconsin-Madison, Madison, WI, USA

43.1 Introduction

Botulinum neurotoxins (BoNTs) are the most potent toxins known to humankind, and are the causative agent of the severe and potentially fatal neuroparalytic disease botulism. Even though botulism is a rare disease in humans, the severity and duration of symptoms, the great potency of the toxins, the lack of effective countermeasures, the high mortality if untreated, and the high cost involved in medical care of patients make botulism a serious concern. Botulinum toxins are produced and secreted by *Clostridium botulinum* (*C. botulinum*) during vegetative growth. *C. botulinum* comprises a heterogeneous group of anaerobic, spore-forming, Gram-positive bacteria that are related only by being clostridia and having the ability to produce BoNT (Peck, 2009). While the vast majority of BoNT-producing bacteria are classified into this group, a few otherwise nontoxic *Clostridium* strains, including *C. baratii*, *C. butyricum*, and *C. sporogenes*, have also been found to produce BoNT. The *Clostridia* are soil-dwelling organisms, and due to their ability to form spores that are resistant to environmental insults, these organisms are widespread in soil and water throughout the world. While the spores themselves are harmless, germination of the spores under favorable conditions results in vegetative growth of the bacteria accompanied by BoNT production. Entry of BoNTs into the body causes human and animal botulism. There are three naturally occurring forms of botulism, infant or intestinal botulism, foodborne botulism, and wound botulism. In all cases, the pathogenesis of botulism is due to the toxin, not the bacteria themselves. Infant botulism is caused by colonization of the infant's immature intestine by *C. botulinum*, where the bacteria produce BoNT which penetrates the intestinal wall and is taken up into the circulation. In very rare instances, the intestines of adults or children older than 1 year can also be colonized. Foodborne botulism, which is a perennial concern in the food industry, is the result of ingestion of BoNT that has been produced by *C. botulinum* growing in contaminated foods. The anaerobic environment of canned foods or some bottled juices allows germination of contaminating spores in the foods, accompanied by toxin production. If the contaminated food is not properly heated to deactivate the toxin, ingestion of it causes foodborne botulism. Wound botulism is the result of infection of a wound with *C. botulinum* and toxin production by the bacteria growing in the wound. Additional rare botulism cases have been reported due to accidental exposure in a clinical or laboratory setting. BoNTs are categorized into seven immunologically distinct serotypes (denoted A–G), and within most serotypes several subtypes are distinguished based on amino acid differences, with a total of over 35 different BoNTs described to date (Hill and Smith, 2013). Recently, a new BoNT isoform has been described based on distinct immunological features of culture supernatant and by DNA sequencing, and has been proposed as the eighth serotype, named BoNT/H (Barash and Arnon, 2013; Dover et al., 2013). While the general mechanism of pathogenesis is the same for all BoNT serotypes, there are important differences between the serotypes in species specificity, severity and duration of disease, and immunologic characteristics. Four of the BoNT serotypes (A, B, E, and F) cause human botulism, while BoNT/C and D cause botulism in birds and mammals. Only one outbreak of human botulism has been associated with BoNT/G (Sonnabend et al., 1981).

In spite of many control measures and education efforts, sporadic outbreaks of botulism continue to occur. Currently, these outbreaks are generally small and rare, with about 150 cases per year being reported in the United States. However, the potency of BoNTs, the prevalence of *C. botulinum* spores in the environment worldwide, and the severity of human botulism are of great concern to both regulatory agencies in the food industry and in biodefense.

43.2 Epidemiology

Botulism was first described following a series of large and especially frequent outbreaks in Germany in the late 18th and early 19th century. The disease was a significant concern because of the severity of resulting paralysis and a mortality rate of about 60%, prompting a detailed analysis and description by the German physician Justinus Kerner of a total of over 250 cases (Kerner, 1817, 1820, 1822). The disease usually occurred after ingestion of sausage, leading Kerner to term it botulism, derived from the Latin word for sausage, "botulus." It was not until 1897 that botulism was found to be associated with bacterial growth in foods, and the first strain of *C. botulinum* was isolated and cultured by Emile Pierre van Ermengem (van Ermengem, 1897, 1979). Van Ermengem also recognized at the time that it was not the bacteria themselves, but a toxin produced by them, that caused the disease.

43.2.1 Human botulism

Today botulism is a rare disease in developed countries, with about 150 cases per year in the United States. Mortality has now been reduced to below 10%, mainly due to improved diagnosis and medical equipment such as ventilators as well as antitoxin. The majority of cases in the United States (about 60–70%) are infant botulism, about 10–20% are foodborne botulism, and about 10–25% are wound botulism (Figure 43-1). Every year, very few cases of

other or unknown etiology are also reported. Since the Centers for Disease Control began monitoring national botulism cases starting in 1973, the total number of reported cases per year has remained relatively steady at around 110–150. However, it is likely that additional non-reported or misdiagnosed cases occur, which is probably also the reason for very rare reports of botulism in non-developed countries. Almost all human botulism is caused by BoNT/A, B, E, or F, with the vast majority being due to BoNT/A and B (CDC, 2011b; Smith, 2009). Unfortunately, BoNT/A and B are also the BoNT serotypes that cause the most severe and longest lasting disease (3–6 months), with botulism caused by BoNT/A being more severe and having a longer duration than BoNT/B. It is, however, possible that due to the much shorter duration of action of BoNT/E and F (1 day to 1 month), botulism caused by these serotypes is underreported.

Infant botulism usually occurs as single cases, due to infants receiving food contaminated with *C. botulinum* spores or breathing in dust containing *C. botulinum* spores (Fenicia and Anniballi, 2009; Grant et al., 2013). However, several clusters of infant botulism have also been reported (Arnon et al., 1979; Istre et al., 1986; Long, 1985). Risk factors for infant botulism include living in rural areas or having a parent who works with soil, because *C. botulinum* spores are prevalent in soils (Fox et al., 2005). Additionally, consumption of honey is another potential source of spores. Although there is some controversy surrounding whether consumption of honey by infants can cause botulism, some cases of infant botulism have been associated with prior consumption of honey (Arnon et al., 1979; Aureli et al., 2002; Grant et al., 2013). Due to a successful public awareness campaign in the United States, such cases have now decreased to below 20% of all cases, compared with 59% of cases in Europe (Aureli et al., 2002).

Wound botulism is a concern for individuals sustaining deep injuries that may come in contact with *C. botulinum* spores present in the soil, such as troops injured in the field. However, currently, most wound botulism cases in the United States are reported in drug users as a result of using dirty needles, injecting into dirty skin, or using contaminated drugs. Most wound botulism cases are associated with injection of black-tar heroin (Palmateer et al., 2013; Yuan et al., 2011). Wound botulism may occur in clusters.

Food-borne botulism almost always occurs in relatively small outbreaks. Home-canned products are the leading cause of foodborne botulism, but rarely commercially produced products may also cause an outbreak (Date et al., 2011). Other outbreaks include cases due to prison-made illicit alcohol drinks (Pruno) (CDF, 2012, 2013; Vugia et al., 2009) as well commercial food products such as salted fish, cheese, canned foods, and meat products (Aureli et al., 2000; Juliao et al., 2013; Leclair et al., 2013; Walton et al., 2014; Zhang et al., 2010). Foodborne botulism outbreaks

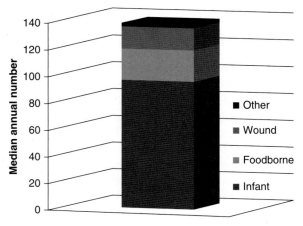

Fig. 43-1 Median number of confirmed botulism cases by botulism type in the United States, 2000–2012. Data obtained from Center for Disease Control National Botulism Surveillance System: http://www.cdc.gov/nationalsurveillance/botulism_surveillance.htm.

and associated mortality rates are presumably somewhat greater in developing countries; however, due to inconsistent reporting, it is difficult to estimate the incidence (Akhtar et al., 2012). The incidence of foodborne botulism varies widely by geographical region within the United States and worldwide, and correlates with the prevalence and serotype of spores present in soil as well as with local food preservation methods (Hatheway, 1995; Peck, 2009; Sobel, 2005).

Other forms of botulism are very rare and include adult intestinal botulism, accidental exposure, and iatrogenic botulism. Only a few cases of adult intestinal botulism have been reported (Bartlett, 1986; Freedman et al., 1986; McCroskey and Hatheway, 1988; Sheppard et al., 2012). While *C. botulinum* spores do not colonize a healthy, adult, human intestine after ingestion, such colonization is possible if the normal intestinal flora is severely interrupted due to surgery, disease, or long-term broad spectrum antibiotic use (Bartlett, 1986; Cherington, 1998; McCroskey and Hatheway, 1988; Sheppard et al., 2012). Only one accidental laboratory exposure involving three cases has been reported, which also presents the only known cases of human inhalation botulism (Holzer, 1962). Iatrogenic botulism is inadvertent botulism caused by injection of BoNT for therapeutic purposes (Chertow et al., 2006; Coban et al., 2010; Crowner et al., 2007; Ghasemi et al., 2012; Partikian and Mitchell, 2007). Such cases are very rare, and at least in some instances have been attributed to severe overdose of the toxin (Chertow et al., 2006; Crowner et al., 2007).

43.2.2 Animal botulism

Many animal species including several mammals, birds, and fish can also be affected by botulism. While animal botulism is not spread to humans, it can affect humans in several ways as discussed below, and much can be learned with regard to human botulism from animal cases. Botulism in animals occurs in large outbreaks with high mortality rates and thus is an important environmental and economic concern. *C. botulinum* spores are ubiquitous in soil and water, and germination of the spores occurs frequently in the environment when in contact with decaying organic matter or animal carcasses. The vegetative cells produce BoNTs as they grow, and ingestion of the toxin along with organic matter by animals leads to botulism. In addition, ingestion of large numbers of *C. botulinum* spores by animals or infection of a wound can lead to toxico-infections.

Periodic and at times large outbreaks in cattle herds are reported throughout the world, with a high mortality rate of above 80% and large associated economic losses (Lindstrom et al., 2010). *C. botulinum* spores have been identified in feed as well as in the intestines of cows and rarely in milk (Lindstrom et al., 2010), which led to a concern of potential secondary human exposure. In fact, recent reports attribute a chronic disease of farmers, which is characterized by similar but milder symptoms as botulism, to repeated BoNT exposure by the farmers through contact with animal feed (Kruger et al., 2012; Rodloff and Kruger, 2012). A botulisms vaccine for cattle and sheep is available to protect against the disease (Steinman et al., 2006). Mink and foxes are known to be very susceptible to botulism, which they acquire by ingestion of BoNT in contaminated meats. An effective botulism vaccine is available for mink to avoid large economic losses by mink farmers (Kolbe and Coe Clough, 2008). Horses can develop botulism after ingestion of BoNTs in contaminated feed, as well as rarely wound botulism or intestinal botulism in foals (Galey, 2001). The disease is severe and devastating in horses, with mortality rates estimated as high as 85–95%. While a vaccine for horses is also available, periodic outbreaks are reported worldwide which lead to loss of horses every year due to botulism (Galey, 2001). Periodic outbreaks in many species of birds are reported, including commercially raised chicken, looms, gulls, waterfowl, herons, and songbirds (Rocke and Friend, 1999). These outbreaks can lead to loss of large numbers of birds, with large outbreaks killing up to a million birds at a time. Birds may ingest BoNTs directly, usually by eating decomposing animal or fish carcasses that are contaminated with *C. botulinum*, or by eating toxin-laden invertebrates such as maggots. *C. botulinum* spores have also been isolated from bird feces in many cases, suggesting infection of the birds with the spores. Mortality rates in chicken is estimated at around 40%, but vaccination can protect against the disease (Dohms et al., 1982). In addition, removal of decaying organic matter and animal carcasses plays a large role in preventing avian botulism (Rocke and Friend, 1999). Cats, dogs, and pigs are relatively resistant to botulism, and only very few cases have been reported (Borst et al., 1986; Elad et al., 2004; Uriarte et al., 2010; van Nes and van der Most van Spijk, 1986).

43.3 *Clostridium botulinum*

The species *C. botulinum* comprises a large and diverse group of *Clostridia* classified as *C. botulinum* based solely on their ability to produce BoNTs. *C. botulinum* are anaerobic, Gram-positive, rod-shaped (arranged in singles, pairs, or chains), and spore-forming bacteria (Figure 43-2). The dormant endospores are oval-shaped and resistant to many extreme environmental conditions that kill most bacteria, providing a means of survival in the environment under unfavorable conditions. The spores can withstand temperatures over 100°C (the temperature of boiling water at sea level), UV light, acidic environments, desiccation, high pressure, aerobic conditions, and many chemicals. Killing the spores requires temperatures of above 121°C at increased pressure for several minutes. Heating canned products at 121°C for at least 3 minutes in a pressure cooker has been termed the "botulinum cook" and is a required procedure for commercial canned products in the United States

(a)

(b)

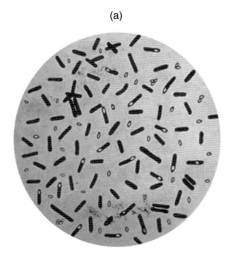

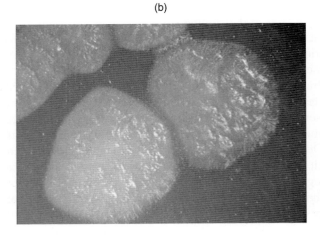

Fig. 43-2 *Clostridium botulinum.* (a) Photomicrographic view of a gentian violet-stained culture specimen revealing the presence of numerous Gram-positive *C. botulinum* bacteria and bacterial endospores. (b) *C. botulinum* type E colonies grown on an egg yolk agar plate for 48 hours, magnified 1.9×. This medium is from the Centers for Disease Control and Prevention Public Health Image Library (PHIL), with identification numbers #2107 and #3871.

and recommended for home-canning (Stumbo, 1973). Other methods of killing *C. botulinum* spores involve extreme acidity, very low pH, or harsh chemical treatments such as, for example, sodium hypochlorite treatment, which is not conducive to food treatments.

C. botulinum has been categorized into four genetically and physiologically distinct groups, denoted as Group I–IV (Peck, 2006). In addition, a few strains of the otherwise nontoxic species of *C. baratii* and *C. butyricum* also produce BoNT and are referred to as Group V and VI. *C. botulinum* Group I are proteolytic (can degrade protein) and produce BoNT serotypes A, B, and F. *C. botulinum* Group II are non-proteolytic (do not degrade protein) and saccharolytic (can metabolize sugars) and produce B, E, and F toxins. These two groups are responsible for almost all human botulism cases. *C. botulinum* Group III produce BoNTs C and D, which cause avian and animal botulism, and Group IV (also called *C. argentinense*) produces type G toxin, which has only been associated with one botulism outbreak (Sonnabend et al., 1981). Each group of *C. botulinum* has distinct optimal growth conditions and survival capabilities (Table 43-1). The spores formed by Group I are the most heat resistant withstanding temperatures up to 120°C, while Group II *C. botulinum*

Table 43-1 Characteristics of *C. botulinum*

	C. botulinum Group I	*C. botulinum* Group II	*C. botulinum* Group III	*C. botulinum* Group IV	Toxigenic *C. baratii*	Toxigenic *C. butyricum*
BoNT serotypes	A, B, F	B, E, F	C, D	G	F	E
Botulism in	Humans	Humans	Birds, mammals	–	Humans	Humans
Minimum temperature for growth	10–12°C	2.5°C	15°C		10°C	12°C
Optimal growth temperature	37°C	25°C	40°C	37°C	30–45°C	30–37°C
Minimum pH required for growth	4.6	5.0	5.1			4.8
Spore heat resistance in phosphate buffer (*D* = decimal reduction time)	$D_{121°C} = 0.21$ min	$D_{82.2°C} = 2.4$ min (without lysozyme) $D_{82.2°C} = 231$ min (with lysozyme)	$D_{104°C} = 0.9$ min	$D_{104°C} = 1.1$ min		$D_{100°C} < 0.1$ min
Toxin genes location	Chromosome or large plasmid	Chromosome or plasmid	Bacteriophage	Large plasmid		
NaCl concentration preventing growth	10%	5%		6.5%		
Ferment glucose	+	+	+	–	+	+
Ferment fructose	+/–	+	+/–	–	+	+
Ferment maltose	+/–	+	+	–	+	+
Ferment sucrose	–	+	–	–	+	+

Sources: Hill and Smith (2013); Peck (2006), (2009), Peck et al. (2011), and Stringer et al. (2013).
Empty fields did not contain information in the literature.

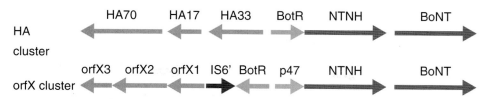

Fig. 43-3 Botulinum neurotoxin gene clusters. The BoNT genes are aligned into one of two conserved gene clusters. Proteins encoded by the genes are indicated in black. BoNT, botulinum neurotoxin; HA, hemagglutinin; BotR, regulator of botulinum neurotoxin; NTNH, non-hemagglutinin; orfX, open reading frame of unknown function; IS, insertional element; p47, p47 protein.

are somewhat less resistant to heat. It is interesting to note, though, that even though the spores of non-proteolytic Group II *C. botulinum* are effectively inactivated by heating to 80–85°C for a few minutes, lysozyme, which is present in many foods, has a protective effect on the spores such that they will survive this heat inactivation (Peck et al., 2011). Thus, all canned foods should always be prepared using the "botulinum cook" of heating at 121°C for at least 3 minutes.

Several *C. botulinum* strains have been sequenced to date. The genome of proteolytic *C. botulinum* is 3.76–4.26 Mb with a low GC content of 28.1–28.5%. The genomes of non-proteolytic *C. botulinum* members are slightly smaller with a slightly lower GC content, and have no synteny with genomes of proteolytic *C. botulinum* strains (Peck et al., 2011). Many strains contain extra-chromosomal plasmids or phages, and the genes encoding BoNTs can be located on the chromosomes or on plasmids or phages (Table 43-1). The BoNT genes are always localized within one of two conserved gene clusters, either the ha cluster or the orf-x cluster (Figure 43-3), which encode non-toxic proteins that associate with the toxin protein, forming the toxin complex (Hill and Smith, 2013). Most *C. botulinum* strains produce only one BoNT serotype, but a few strains contain toxin genes for two or even three different BoNT serotypes, and in some cases two or three BoNTs are expressed in these strains (dual toxin producing strains), usually with one serotype being predominantly expressed (Hill and Smith, 2013). Table 43-1 summarizes the characteristics of the six groups of BoNT-producing *Clostridia*.

43.4 Botulinum neurotoxins

BoNTs are the most potent neurotoxins known to humankind. As little as 1–2 ng per kg of body weight are estimated to be lethal to humans if injected intravenously, 10–20 ng/kg if inhaled, and about 1–2 μg/kg if ingested (Arnon et al., 2001). In contrast to *C. botulinum* bacteria, though, which form spores that are resistant to extreme environmental conditions, BoNTs are easily inactivated by heating to 85°C, by agitation (unless stabilizers are added), and by many chemical treatments such as ethanol or bleach. BoNTs are categorized into seven immunologically distinct serotypes, named A, B, C, D, E, F, and G

(Gimenez and Gimenez, 1995). Monovalent antitoxins that completely neutralize the effects of each toxin serotype in mice have been developed against all seven serotypes and are being distributed by the CDC for serotype identification purposes (Hatheway, 1995). Most serotypes are further categorized into several subtypes based on differences in their amino acid sequence which can vary from 0.9 to 36% within one serotype (Hill and Smith, 2013). The subtypes are denoted by a number following the serotype letter designation, such as A1 and A2. To date, a total of at least 36 different BoNTs has been described, and the list is growing. Most recently, a new *C. botulinum* strain was isolated from an infant botulism case, and none of the currently available antitoxins or combinations thereof was able to completely neutralize the effects of the toxin(s) produced by this strain (Barash and Arnon, 2013). While sequence analysis of the strain indicated that this strain contained two BoNT genes, one encoding BoNT/B and another encoding an as yet unidentified hybrid toxin consisting of a combination of BoNT/A and F (Dover et al., 2013), the lack of an effective antitoxin prompted the suggestion to denote the A/F hybrid toxin as a new serotype, H. Further research will be required to isolate and characterize the new protein toxin.

BoNTs are 150-kDa proteins that consist of a 100-kDa heavy chain (HC) linked via a disulfide bond and a belt loop to a 50-kDa light chain (LC) (Montal, 2010). The BoNTs are produced as a largely inactive 150-kDa polypeptide chain, which is then processed into the active di-chain molecule by cleavage between the HC and LC. This cleavage is conferred by either clostridial proteases (in proteolytic strains) or by tissue proteases present in host organisms or foods. Thus, the activated BoNT consists of three functional modules. The N-terminal 50-kDa LC is a zinc-dependent endopeptidase that, once internalized into neuronal cells, cleaves one of the soluble *N*-ethylmaleimide-sensitive-factor attachment protein receptor (SNARE) proteins (Montecucco and Schiavo, 1994; Schiavo et al., 1995). The C-terminal HC is divided into the N-terminal 50-kDa translocation domain (H_N or HCN), which is involved in cytosolic entry of the LC, and the C-terminal 50-kDa receptor binding domain (H_C or HCR). The H_C domain is further sub-divided into the β-sheet jelly roll fold (H_{CN}) of as yet unknown function and the β-tree foil fold (H_{CC}) that contains the receptor binding sites for specific binding to neuronal cells (Figure 43-4) (Montal, 2010).

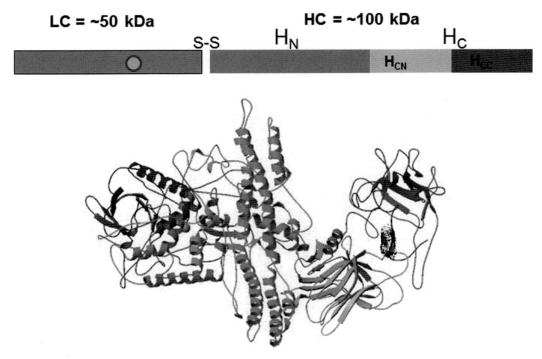

Fig. 43-4 Model and crystal structure of BoNT/A. The light chain (LC) is shown in blue, with the catalytic zinc depicted as a ball in gray. The translocation domain (H_N) of the heavy chain (HC) is shown in green, the N-terminal sub-domain (H_{CN}) of the receptor-binding domain (HC) in yellow, and the C-terminal sub-domain (H_{CC}) in red. The catalytic zinc is depicted as a ball in gray. This model of the crystal structure is derived from PDB ID: 3BTA, and was first published in 1998. With permission from Lacy et al. (1998). For a color version of this figure, see the color plate section.

While all BoNTs seem to share these structural and functional features, there are important differences in the characteristics of the BoNT serotypes and even between subtypes within one serotype. The LC of the seven BoNT serotypes each has their own unique and specific SNARE target and cleavage site. BoNT/A, E, and C cleave the t-SNARE synaptosomal-associated protein 25 (SNAP-25) at distinct sites, while BoNT/B, D, F, and G cleave the v-SNARE vesicle-associated membrane protein (VAMP, also called synaptobrevin) at distinct sites (Schiavo et al., 1995, 2000). BoNT/C is unique in that it also cleaves the t-SNARE syntaxin in addition to cleavage of SNAP-25. The HCs of the serotypes associate with different neuronal cell receptors. In addition to binding to gangliosides, BoNT/A, D, E, and F associate with SV2 receptors with different affinities (Dong et al., 2006, 2008; Peng et al., 2011; Rummel et al., 2009), whereas BoNT/B and G associate with synaptotagmin I and II (Dong et al., 2003; Nishiki et al., 1994; Rummel et al., 2004). There are also differences in ganglioside binding. For example, BoNT/B and C have been shown to bind gangliosides with greater affinity than BoNT/A (Rummel et al., 2011).

In addition to immunogenic differences and distinct SNARE cleavage sites, the serotypes differ in other important characteristics such as species specificity and longevity of action. Most human botulism cases are caused by BoNT serotypes A, B, E, and rarely F, and serotypes C and D as well as A, B, and E cause botulism in birds and some mammals. Botulism caused by BoNT/G is rare and has only been identified postmortem in one small cluster of human bodies (Sonnabend et al., 1981). Human botulism caused by serotypes C and D has also been described in isolated cases (Demarchi et al., 1958; Oguma et al., 1990). Importantly, the severity and characteristics of human botulism symptoms vary with the serotype of BoNT. BoNT/A causes the most severe botulism with the longest lasting paralysis (3–6 months and up to 1 year) and the highest death rate of about 8% (Varma et al., 2004), whereas botulism caused by BoNT/B and E results in faster recovery (3 months and ~1 month, respectively) and a reduced death rate of 1–3% (Boyer et al., 2001; Gottlieb et al., 2007). The fastest recovery of symptoms occurs with BoNT/F (a few days to 1 month) (Gupta et al., 2005), and no deaths have been reported after intoxication with BoNT/F. While most symptoms of botulism are due to neuromuscular paralysis consistent with a disabling of the neuromuscular junction, autonomic nerve dysfunction has also been described in a few cases, more frequently for BoNT/B and E poisoning than for BoNT/A poisoning (Johnson and Montecucco, 2008).

Furthermore, several cell-based and *in vivo* studies using BoNT serotypes or subtypes have indicated that significant differences exist between the biological characteristics of the BoNT sero- and subtypes. Neuronal cells in culture recover at different speeds after intoxication with BoNT serotypes, with recovery half-lives ranging from 0.2 to several days for BoNT/F and E, to about 10 days for BoNT/B, and to several months for BoNT/C and A (Foran et al., 2003; Keller, 2006). Correspondingly, local injection studies in humans have shown that BoNT serotypes A, B, C, E, and F cause differing paralysis duration of 3–6, 3, 2–4, 1–3, and 1–2 months, respectively (Eleopra et al., 1998, 2004). Local injection studies in mice using BoNT/A, B, or E have similarly shown clear differences in the onset and duration of symptoms as well as in the onset and rate of recovery (Keller, 2006). Only little is known about the response in humans to the BoNT serotypes that typically do not cause human botulism, such as C, D, and G. While one study indicates that BoNT/C may be cytotoxic to neurons *in vitro* (Peng et al., 2013), data from a small trial study in humans indicate that injection of low concentrations of BoNT/C is as safe as BoNT/A injections (Eleopra et al., 2004; Sloop et al., 1997). Local injection of up to 10 U of BoNT/D in humans failed to elicit a significant paralytic response (Eleopra et al., 2013), although the attribution of one outbreak of foodborne botulism to BoNT/D (Demarchi et al., 1958) would indicate that high doses of BoNT/D can also cause paralysis in humans. Recent studies comparing the characteristics of BoNT/A subtypes 1–5 *in vitro* and *in vivo* indicate distinct symptoms and *in vitro* characteristics of these subtypes (Akaike et al., 2013; Henkel et al., 2009; Ma et al., 2012; Pier et al., 2011; Tepp et al., 2012; Torii et al., 2011, 2010; Whitemarsh et al., 2013). The significance of the different characteristics of these subtypes to the pathology of botulism is currently unknown.

43.5 Pathogenesis

While botulism usually occurs in small outbreaks, it is not a contagious disease, and develops only after entry of BoNTs into the human body. BoNTs can gain entry into the human (or animal) body through ingestion of the toxin from foods contaminated with *C. botulinum*, through inhalation, or by production of *C. botulinum* that have colonized the body either in the intestine (infant botulism or adult intestinal colonization) or in wounds (wound botulism). The toxins cannot penetrate through intact skin. Irrespective of the route of entry, the toxins penetrate through a mucosal surface (gut, lungs, or a wound), enter the circulation, and distribute to the peripheral nervous system (Cherington, 1998; Hatheway, 1988). The BoNTs, then, enter the neurons and selectively block neurotransmitter release of peripheral motor neurons, thereby triggering the flaccid paralysis that is a hallmark of botulism. While the related tetanus toxin (TeNT), which is produced by *C. tetani*, also enters neurons of the peripheral nervous system, TeNT uses a distinct entry mechanism and is transported into the central nervous system by retrograde transport and transcytosis (Lalli et al., 2003a, 2003b). TeNT is activated in inhibitory neurons, where it, similar to BoNTs, cleaves VAMP and blocks neurotransmitter release (Schiavo et al., 1992). However, because the

transmitter release is selectively blocked in inhibitory neurons, as opposed to the motor neurons which are affected by BoNTs, TeNT causes over-activity of the musculature leading to spastic paralysis known as tetanus versus the flaccid paralysis of botulism (Montecucco and Schiavo, 1994; Schiavo et al., 2000).

The neuronal cell entry by BoNTs proceeds via a series of consecutive and essential steps (Figure 43-5) (Montal, 2010). To initiate cell intoxication, BoNT binds to specific protein and ganglioside receptors on the cell surface via the HCC domain, leading to endocytosis. All BoNTs associate with various gangliosides on the cell membrane, and specific protein receptors have been identified for several BoNTs (synaptotagmin I and II for BoNT/B and G, and SV2 for BoNT/A, D, and E) (Rummel, 2013). Recently, Fibroblast Growth Factor Receptor 3 (FGFR3) was suggested as an additional or alternative protein receptor for BoNT/A, and overexpression of FGFR3 in neurons correlated with increased BoNT/A1 entry (Jacky et al., 2013). In the dual receptor model proposed for BoNT, the toxins first associate with gangliosides, which brings the toxins in close contact with the protein receptor for subsequent binding (Montecucco, 1986; Rummel, 2013; Stenmark et al., 2008). This association then leads to endocytosis of the entire BoNT molecule. In the endocytic vesicle, protonation causes a conformational shift in the BoNT protein, ultimately resulting in the incorporation of the HC into the endocytic vesicle membrane to form a channel through which the LC is translocated (Fischer and Montal, 2007; Fischer et al., 2009; Montal, 2010). The disulfide bond between HC and LC is cleaved and the LC is released into the cytosol, where it is refolded to its enzymatically active conformation. The endoprotease function of the LC then specifically cleaves a cytosolic SNARE protein, thereby preventing SNARE-mediated protein transport and transmitter release (Montecucco and Schiavo, 1994). SNARE proteins are a large family of proteins that mediate vesicle fusion or exocytosis. In neurons, three SNAREs are required for synaptic vesicle fusion with the presynaptic membrane to release neurotransmitter at the neuronal synapse. These SNAREs include VAMP, which is anchored to the synaptic vesicle membrane, SNAP-25, which is associated with the plasma membrane, and syntaxin, which is anchored in the plasma membrane. When an action potential depolarizes the presynaptic membrane of a neuron, Ca^{2+}-channels open, leading to an influx of Ca^{2+} into the presynaptic nerve terminal. This Ca^{2+} influx triggers the formation of the SNARE complex, which consists of a four α-helix bundle, where one α-helix is contributed by syntaxin-1, one by synaptobrevin, and two by SNAP-25. The formation of the SNARE complex leads to synaptic vesicle fusion and neurotransmitter release into the synaptic cleft. The released neurotransmitter then binds to the respective receptor on the post-synaptic membrane, thereby transmitting the signal from the neuron.

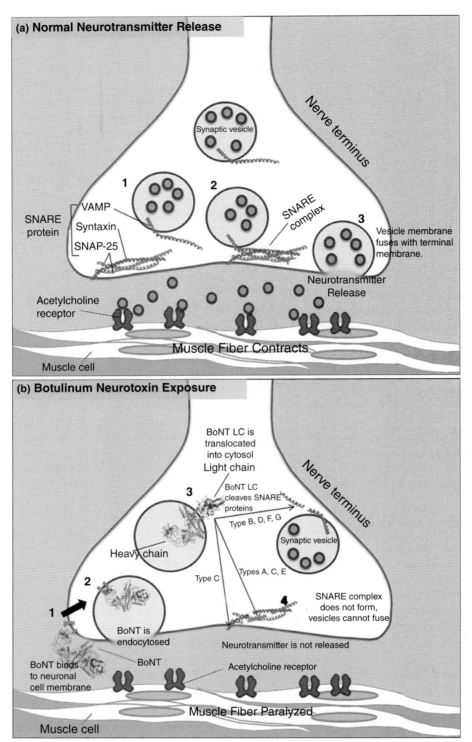

Fig. 43-5 Mechanism of action of Botulinum neurotoxin (BoNT) at the neuromuscular junction: (a) Normal neurotransmitter release: a model of a neuromuscular junction is shown, highlighting the major steps involved in neurotransmitter release. Fusion of the SNARE proteins (SNAP-25, syntaxin, and VAMP/synaptobrevin) to form the SNARE complex mediates fusion of the acetylcholine-filled synaptic vesicle with the neuronal cell membrane. Acetylcholine is released into the synaptic cleft, and binds to the acetylcholine receptor on the muscle cell, causing the muscle cell to contract. (b) BoNT exposure: BoNTs bind to ganglioside and protein receptors on the neuronal cell membrane, leading to endocytosis of the entire toxin. Acidification inside the endocytic vesicle leads to a conformational change and the light chain (LC) is translocated through the endosome membrane into the cells' cytosol. Inside the cytosol, the LC cleaves specific sites on a SNARE protein, preventing the formation of a functional SNARE complex. Thus, the synaptic vesicle cannot fuse with the cell membrane, and no acetylcholine is released. The muscle cell is paralyzed. The BoNT structure and BoNT LC structure in this figure are derived from PDB ID: 3BTA (Lacy et al., 1998) and PDB ID: 4EJ5 (Stura et al., 2012). Figure created by Xiaoyang Serene Hu. For a color version of this figure, see the color plate section.

Once a BoNT LC has gained entry into the cell, its endoprotease function cleaves one of the SNARE proteins at a specific site. BoNT/A and E cleave SNAP-25, BoNT/B, D, F, and G cleave VAMP, and BoNT/C cleaves both SNAP-25 and syntaxin, each at distinct cleavage sites. Irrespective of which SNARE protein is cleaved and where it is cleaved, this cleavage always causes a disruption in the SNARE complex formation and thereby prevents vesicle fusion and neurotransmitter release (Figure 43-5). This blockage of neurotransmitter release primarily at the neuromuscular junction leads to the flaccid paralysis, which is the hallmark symptom of botulism.

43.6 Clinical features

The symptoms of botulism were first described in the early 19th century by Justinus Kerner in great detail (Kerner, 1822; Pellett, 2012). Kerner accurately observed that the symptoms of botulism were largely restricted to motor neurons of the peripheral nervous system, whereas the brain and sensory neurons seemed largely unchanged. In spite of contrary advice by colleagues, Kerner conducted a risky experiment on himself, which was a fairly common practice among scientists at the time. He placed a few drops of botulinum toxin containing sausage extract on his tongue and fortunately lived to record his body's response. In direct translation, he reported: "A few drops of this acid placed onto the tongue cause, to a greater degree than the impure diluted sausage acid, this feeling of contraction and choking in the area of the larynx, a feeling that one also notices after consumption of meconic acid or iodine. After stronger administration, a feeling of listlessness and tension in the eyelids results, the eyes go stupid, one feels a slight burning in the urea, dull pain in the stomach, constipation, and very dry palms and soles of the feet. The feeling of drying of the throat disappears intermittently, but then again comes back more frequently together with that feeling of tension in the eyelids" (Häfner, 2009; Pellett, 2012). This accurately describes the first symptoms of botulism.

Botulism is an intoxication characterized by (usually) symmetric descending flaccid paralysis, beginning with cranial paralysis and followed by progressive muscle paralysis of the entire body (Arnon et al., 2001; Cherington, 1998, 2004; Sobel, 2005). These characteristic symptoms are due primarily to a block of neurotransmitter release at the neuromuscular junction of skeletal muscle and the resulting flaccid paralysis. The onset of the symptoms of foodborne botulism usually is from 12 to 36 hours post-ingestion of the toxins, but can be as early as 6 hours or as late as 10 days post-ingestion. The first symptoms of foodborne or wound botulism include bulbar palsies such as diplopia (double vision), dysarthria (difficulty with speech), dysphonia (altered voice), and dysphagia (difficulty swallowing) (Arnon et al., 2001). Dry mouth, difficulty swallowing, fixed pupils not

reactive to light, nasal regurgitation, and a sensation of suffocating are common. Abdominal pain, nausea, vomiting, and diarrhea may be present as early symptoms in foodborne botulism, but it is unclear whether these symptoms are due to the action of BoNTs or other constituents of the contaminated food that was ingested. The face often is expressionless and sagging and eyelids are droopy due to paralysis of the facial muscles. This is followed by a progressive, usually symmetrical, descending, flaccid paralysis of the arms, and then the legs. Most patients eventually suffer from severe constipation and urinary retention. As the disease progresses, patients become weaker and fatigued, and in severe cases paralysis of the respiratory musculature develops requiring mechanical ventilation. Death in untreated cases is usually due to paralysis of the respiratory musculature. The typical symptoms of botulism are listed in Table 43-2. It has long been believed that the BoNTs cannot penetrate the blood–brain barrier, and thus brain function is unaffected and the patients remain fully alert and show no signs of cognitive or sensory abnormalities. However, recent evidence from multiple laboratories indicates that some degree of retrograde transport occurs with at least some BoNT isotypes (Alexiades-Armenakas, 2008; Antonucci et al., 2008; Bach-Rojecky and Lackovic, 2009; Filipovic et al., 2012; Marchand-Pauvert et al., 2013; Marinelli et al., 2012; Matak et al., 2012; Restani et al., 2012a, 2012b), and it is currently unclear whether and to which extend CNS function and plasticity is affected by BoNTs.

The duration of symptoms depends on the BoNT serotype and dose, and can range from a few days (BoNT/F), or a few days or a few weeks (BoNT/E) to 3 months (BoNT/B) or 6 months (BoNT/A) (Eleopra et al., 2004; Pellett, 2012; Sloop et al., 1997). Recovery is gradual and some symptoms such as dry mouth, constipation, and impotence may remain for up to 1–2 years, but patients eventually fully recover. While botulism itself does not cause depression, the utter helplessness experienced by patients suffering from severe botulism, especially if prolonged respiration is required, while being fully aware of their situation, may lead to secondary frustration or depression.

Infant botulism has first been reported in 1976 (Pickett, 1982; Pickett et al., 1976), and since 1990 has become the most frequently reported form of botulism each year to the Centers for Disease Control and Prevention (CDC). Infant botulism is due to colonization of the infant's intestine by *C. botulinum*, which has not yet developed a mature intestinal flora. In healthy adults, ingested *C. botulinum* spores are incapable of colonizing the intestine and cause no harm. However, in the immature intestines of infants, ingested *C. botulinum* spores are able to colonize, germinate, grow, and produce toxin within the large intestine (Arnon, 1980; Cherington, 1998; Fenicia and Anniballi, 2009). Most cases occur before the age of 6 months, but may occur up to 1 year of age. Several days to weeks after ingestion of the spores, symptoms develop. The first

Table 43-2 Signs and symptoms of botulism

	Foodborne/wound botulism	Infant botulism
Eyes	• Double vision • Blurred vision • Fixed or dilated pupils • Ptosis • Nystagmus • Drooping eyelids	• Fixed or dilated pupils • Ptosis • Ocular weakness • Drooping eyelids • Dry mouth
Face	• Expressionless • Sagging, weak muscle tone	• Hypotonia
Mouth/throat	• Dysphagia • Dry mouth • Dysarthria • Sore throat • Diminished gag reflex • Tongue weakness	• Weak suck • Poor feeding • Weak cry
Respiratory system	• Dyspnea • Inability to breath in severe cases	• Dyspnea • Inability to breath in severe cases
Bladder	• Urinary retention	• Neurogenic bladder
Gastrointestinal tract	• Constipation • Nausea } May be early • Vomiting } symptoms in • Abdominal cramps } mild cases • Diarrhea	• Constipation
Musculoskeletal system	• Arm weakness • Leg weakness • Paresthesia • Hyporeflexia or areflexia • Ataxia • Descending symmetric flaccid paralysis	• Hypotonia • Loss of head control • Diminished spontaneous movement • Weakness of arms and legs • Descending symmetric flaccid paralysis
General	• Fatigue • Dizziness	• Lethargy
Central nervous system	• Mental alertness • No cognitive or sensory abnormalities	

Sources: Arnon et al. (2001), CDC (2009), Cherington (1998), and Cherington (2004).

signs of infant botulism are severe constipation, followed by ocular weakness, ptosis, a weak cry, poor feeding, and diminished suckling. More severe cases lead to lethargy and loss of head control due to neck weakness, hypotonia, and decrease in spontaneous movements. When lifted, a baby suffering from botulism will not be able to lift its head or limbs, and is commonly referred to as "floppy baby." Severe cases progress to respiratory failure, often leading to a requirement for prolonged mechanical ventilation. With adequate support, most infants recover fully without long-term effects other than the developmental time period that was missed. However, mortality rates of hospitalized patients up to 3–5% have been reported (Cherington, 1998; Fox et al., 2005), and recovery may take several weeks to months. In some cases, infant botulism develops rapidly over a few hours, followed by fast recovery (Fox et al., 2005).

Adult intestinal botulism is very rare, and only a few cases have been reported. *C. botulinum* spores are incapable of colonizing a healthy adult intestine. However, a significant interruption in the intestinal flora, such as after intensive broad-spectrum antibiotic therapy, allows a niche for *C. botulinum* colonization and germination in children over the age of 1 or in adults. The symptoms of adult intestinal botulism resemble those of foodborne and wound botulism.

Even though infant botulism, wound botulism, and adult intestinal botulism involve the colonization of the intestine or a wound with *C. botulinum* bacteria and thus could be considered an infection, all pathogenesis is due to the botulinum toxins produced by the organism and not to the bacteria. Thus, botulism is always an intoxication.

43.7 Immunity

Patients who recover from botulism generally do not develop immunity against the toxin or the bacteria, even after severe disease. Thus, an individual who has recovered from botulism is not immune to repeat intoxication and may acquire botulism again. However, passive immunity to botulism can be provided by intravenous administration of equine botulinum antitoxin (CDC, 2010) or, for infants only, specific human hyperimmune globulin (Arnon et al., 2006). In addition, endogenous immunity may be achieved by vaccination of individuals with inactive BoNT toxoid (see Section 43.9). Thus, it is possible for humans to mount a protective immune response to BoNTs. The lack of an immune response in naturally occurring botulism cases is most likely due to the high potency and characteristics of BoNTs. Even though BoNTs enter the circulation, the toxins enter neuronal cells shortly thereafter and within a few to 72 hours after intoxication. Once inside neuronal cells, the BoNTs are protected from the immune system. Because very small amounts of BoNTs are sufficient to cause botulism, there is usually not enough protein present in the circulation to mount a robust immune response. However, passive immunization can be achieved by administration of antitoxin after the intoxication and before the BoNTs have entered neurons (discussed in more detail in Section 43.9).

Immunization with a toxoid form or non-toxic portions of BoNTs imparts immunity in animals (Dohms et al., 1982; Galey, 2001; Kolbe and Coe Clough, 2008; Steinman et al., 2006), and has been used in humans on an investigational basis to protect at-risk groups (Baldwin et al., 2008; CDC, 2011a; Pier et al., 2008; Smith, 2009; Webb and Smith, 2013). Immunization of the entire human population is impractical and undesirable, because BoNTs are also extensively used for medical purposes. Thus, even though natural botulism does not result in immunity, it is possible to achieve immunity by vaccination. An undesired side effect of the ability of humans to mount a neutralizing immune response to BoNTs is the formation of such a response in a subset of patients receiving repeat injections of active BoNT preparations for medical purposes (see Section 43.10). These patients become refractory to further treatments.

43.8 Diagnosis

The initial diagnosis of botulism is based on the characteristic clinical findings (Cherington, 1998). Botulism may be recognized by a combination of three main characteristics: (1) symmetric, descending flaccid paralysis with prominent bulbar palsies, (2) an afebrile patient, and (3) absence of cognitive and sensory abnormalities (Arnon et al., 2001). This early clinical diagnosis is important to enable immediate treatment with antitoxin. The antitoxin can be acquired from the CDC and is a critical factor in preventing mortality and reducing severity of the disease (see Section 43.9), and thus all treatment and management decisions should be based on this early clinical diagnosis. The clinical diagnosis is then followed by confirmation in the laboratory. Definitive laboratory

diagnosis requires demonstration of the presence of BoNT in serum, stool or vomitus, or food, or alternatively culturing of *C. botulinum* or a botulinum toxin-producing *Clostridium* sp. strain from stool, wound, or food (Hatheway, 1988, 1995). All suspected and confirmed diagnoses of botulism should be reported to the CDC. The CDC provides and maintains a webpage containing information and guidance for clinicians (CDC, 2009). This website contains information on the background of botulism as well as the most important clinical symptoms, criteria for diagnosis, treatment options, control measures, and patient education. The website further states that "Every case of foodborne botulism is treated as a public health emergency, because the food source, whether homemade or commercial, might still be available for consumption and could make unsuspecting persons ill" and "Health-care providers should be alert to illness patterns and diagnostic clues that might indicate an unusual infectious disease outbreak associated with intentional release of a biologic agent and should report any clusters or findings to their local or state health department."

43.8.1 Differential diagnosis

The early clinical diagnosis of botulism is a critical factor in treatment outcome and prognosis of botulism. Unfortunately, botulism is frequently misdiagnosed leading to a delay in antitoxin administration during the critical time window (up to 72 hours after intoxication). The most common disorders that botulism is mistaken for are polyradiculoneuropathy (Guillain–Barré or Miller–Fisher syndrome), myasthenia gravis, or other diseases of the central nervous system (CDC, 2009). Table 43-3 provides a comprehensive table of nervous system disorders that botulism may be misdiagnosed as. Botulism can be differentiated from other causes of flaccid paralyses in that it always manifests initially with prominent cranial paralysis evident by bulbar palsies (double vision, droopy eyelids,

Table 43-3 Differential diagnosis for botulism

In adults	In infants
Guillain–Barré syndrome	Sepsis, infections
Myasthenia gravis	Meningitis
Cerebrovascular accident (CVA)	Electrolyte–mineral imbalance
Bacterial and/or chemical food poisoning	Reye's syndrome
Tick paralysis	Congenital myopathy
Chemical intoxication (e.g., carbon monoxide)	Werdnig–Hoffman disease
Mushroom poisoning	Metabolic disorders (Leigh disease, Maple syrup urine disease, glutaric aciduria type I)
Poliomyelitis	Spinal muscular atrophy (SMA)
Psychiatric illness	

Sources: CDC (2009) and Francisco and Arnon (2007).

difficulty swallowing and with speech, altered voice) and the invariable descending progression of flaccid paralysis in a (usually) symmetric fashion, and in its apparent absence of sensory nerve damage. In addition, botulism is more likely to occur in a small cluster of cases of acute flaccid paralysis than Guillain–Barré syndrome, chemical poisoning, or poliomyelitis (CDC, 2009).

Routine lab tests such as complete blood counts, electrolytes analysis, urinalysis, and liver function tests generally show no abnormalities. Cerebrospinal fluid analysis is usually normal but may show borderline elevated protein levels. A tensilon test, in which the patient gets injected with tensilon (edrophonium) followed by observation of improvements in muscle strength, may be used to differentiate botulism from myasthenia gravis (Benatar, 2006; Meriggioli and Sanders, 2005). However, borderline positive tests may occur in botulism patients (CDC, 2009). Normal Computed Tomography (CT) and Magnetic Resonance Imaging (MRI) results help to rule out a cerebrovascular accident (CVA) (CDC, 2009). Results from electrophysiology tests such as repetitive nerve stimulation (RNS) may help to diagnose botulism as the cause of flaccid paralysis (Meriggioli and Sanders, 2005; Witoonpanich et al., 2009). However, the most important aspect of early clinical diagnosis remains recognition of the characteristic symptoms of botulism (Table 43-2).

In infants, the two most common disorders mistaken as botulism are spinal muscular dystrophy (SMA) type I and metabolic disorders (Table 43-3) (Francisco and Arnon, 2007). SMA can be differentiated from botulism by the longer history of generalized weakness versus the acute to subacute onset of botulism symptoms. In addition, infants suffering from botulism typically present with paralysis of the ocular muscles and decrease in anal sphincter tone, which is usually absent in SMA patients. Metabolic disorders can be excluded by the corresponding laboratory tests. Cerebrospinal fluid analysis, nerve conduction studies, and electromyography help to distinguish botulism from Guillain–Barré syndrome (Francisco and Arnon, 2007). Most botulism cases occur in infants younger than 6 months, which also helps in differential diagnosis. As with adult botulism, the best practice for differential diagnosis of botulism is a careful history and neurologic examination (see Table 43-2 for characteristic signs and symptoms of infant botulism) (Francisco and Arnon, 2007).

43.8.2 Laboratory confirmation

Definitive diagnosis of botulism requires laboratory confirmation of the presence of BoNTs in serum, stool, enema liquid, vomitus, or food or other potential environmental sources, or the isolation of *C. botulinum* from stool, enema, tissue samples or exudates, a wound, food, or other environmental source (CDC, 2009; Hatheway, 1988). However, this testing may take

several days, and thus treatment should be based on the clinical diagnosis and not be delayed to wait for laboratory test results. Laboratory testing of botulism requires trained staff and approval of laboratories and staff to handle potentially toxic compounds and cultures. Most laboratories able to conduct these studies are at the federal or state level, and only very few county- or city-level laboratories can perform this testing. The Sentinal Laboratory Guidelines for Suspected Agents of Bioterrorism developed and disseminated by the American Society of Microbiology provides guidelines for proper collection and shipping of samples to be tested (American Society of Microbiology, 2003). The Bacteriological Analytical Manual Online published by the Food and Drug Administration (FDA) outlines the protocols for BoNT detection and *C. botulinum* isolation (Solomon and Lilly, 2001).

Currently, the only method for BoNT detection accepted by the FDA is the mouse bioassay, which has been used for BoNT detection for decades (Hatheway, 1988; Schantz and Kautter, 1978). In this assay, mice are injected with the test substance and observed for characteristic signs of botulism, including ruffled fur, labored breathing, weakness of limbs, and finally total paralysis with gasping for breath and death due to respiratory failure (Solomon and Lilly, 2001). The sample is then pre-mixed with serotype-specific antisera prior to injection into mice. The antisera that neutralizes the toxic effects of the testing sample identifies the BoNT serotype, and confirms the presence of BoNTs (Solomon and Lilly, 2001). The testing procedure should closely follow the guidelines described in the Bacteriological Analytical Manual (Solomon and Lilly, 2001) to avoid false positives and false negatives. It is important to use several dilutions of the testing material, prepared in an appropriate buffer (i.e. gel-phosphate buffer) and sterile filtered prior to injection to minimizing non-specific deaths. In addition, some BoNT types may need activation by trypsin, and thus half of the sample should always be treated with trypsin prior to dilution and injection into mice. Assays testing for neutralization by monovalent antisera should always be conducted to confirm that mouse deaths were specific to BoNT and to determine the serotype. Identification of the serotype is important as the progression and severity of botulism depends on the serotype.

Procedures for the isolation of *C. botulinum* are also described in the Bacteriological Analytical Manual (Solomon and Lilly, 2001). This procedure involves growing an enrichment culture from the testing sample in trypticase-peptone-glucose-yeast extract broth (TPGY) broth under strictly anaerobic conditions. The culture is incubated for 5–10 days and examined by microscopy for the presence of cells that match the morphology of *C. botulinum* (Figure 43-1). To isolate pure cultures, the enrichment culture, which now contains spores, is treated with 50% ethanol to kill any non-sporulated cells. The remaining spores can then be streaked out onto liver–veal–egg yolk agar and/or anaerobic egg yolk agar and incubated under anaerobic conditions for 2 days to obtain isolated colonies. Colonies that resemble *C. botulinum* (Figure 43-1) are then transferred to fresh TPGY broth or cooked meat medium. Cultures may be re-streaked for further isolation, and grown under both anaerobic conditions and aerobic conditions to distinguish from other bacteria that are capable of growing aerobically. Finally, culture supernatant from the broth culture grown for 5 days is tested for the presence of BoNT and the serotype is determined as described above (Solomon and Lilly, 2001).

It should be noted that many alternative BoNT detection and sero- or subtype identification methods have been developed, some of which are as or more sensitive than the mouse bioassay. These include mostly *in vitro* and cell-based assays, and some alternative *in vivo* assays (Dorner et al., 2013; Pellett, 2013). However, currently, the mouse bioassay is the only assay that measures all function of the BoNTs and is robust enough to detect BoNTs in many different samples. While the mouse bioassay may lead to false positives due to non-specific deaths or to false negatives due to excessive dilution or careless handling of the testing material, most of these can be eliminated by using proper procedure and including neutralization assays with antisera, and appropriate training of staff conducting the assay. The *in vitro* assays, on the other hand, measure only one or in some cases two functions of the BoNTs such as catalytic activity or antibody binding, and thus are prone to false positives or negatives. The cell-based assays measure all aspects of cellular intoxication by BoNTs, are specific, and can be more sensitive than the mouse bioassay. Thus cell-based assay offers a potential alternative to the mouse bioassay, but robustness to different testing samples remains to be investigated before such assays can be implemented.

43.9 Treatment, prevention, and control

There is no cure for botulism, but the severity and duration of the disease as well as mortality can be decreased significantly with proper treatment. Additionally, most botulism cases can be prevented by simple measures.

43.9.1 Treatment

Treatment for botulism is restricted to early administration of antitoxin, which is effective only if administered within 24–72 hours after intoxication. The currently available antitoxin is a heptavalent equine antiserum called HBAT, produced by Cangene Corporation, and distributed by the CDC (2010). While this antitoxin was first used on an investigational basis, it has been approved by the FDA in early 2013 for commercial distribution, and is now replacing the previously available trivalent (A, B, E) antitoxin. HBAT can neutralize all seven serotypes of BoNTs, but is

only effective in the time period before BoNTs enter neurons (up to 72 hours post-intoxication). Once the time window for antitoxin treatment has elapsed, the only available treatment option is supportive therapy, which in severe cases includes mechanical respiration. Thus, early clinical diagnosis and immediate reporting to the state health department is critical. The state health department then immediately contacts CDC to request antitoxin and possibly consult on the diagnosis. HBAT is stored at CDC Quarantine Stations located in major airports around the nation, ensuring delivery to any location within the United States within hours. The earlier the antitoxin is administered, the more BoNTs will be neutralized and cleared from the body and thereby prevented from entering neurons, which will lead to less severe symptoms and faster recovery. Side effects and contraindications to the antitoxin include hypersensitivity to equine serum, including the potential to develop anaphylaxis and serum sickness (in about 6% of patients), and older literature contains controversy as to whether antitoxin treatment is beneficial (Cherington, 1998). However, it has now been shown that early antitoxin administration reduces morbidity and deaths and predicts the need for mechanical respiration (Sandrock and Murin, 2001; Tacket et al., 1984). Current efforts are attempting to develop humanized immunoglobulin treatments, but in the meantime early administration of HBAT is recommended.

To reduce the risks associated with equine antitoxin in infants, to avoid lifelong sensitization to equine proteins, and because of the relatively short half-life of the equine antitoxin (CDC, 2010), a human immune globulin named BabyBIG® (or BIG-IV) was developed specifically to treat infant botulism (Arnon et al., 2006). BabyBIG® is derived from plasma donations of humans previously immunized with a pentavalent (BoNT/A/B/C/D/E) toxoid vaccine that was available on an investigational basis for individuals at risk of BoNT exposure such as laboratory staff, health care workers, and military personnel. This pentavalent vaccine has since been discontinued due to a loss in efficacy and increased side effects (CDC, 2011a). Since maintenance of neutralizing antibodies to BoNTs requires annual boosters of the vaccine, previously immunized individuals can no longer donate plasma for production of BabyBIG®. However, new vaccines are being developed to replace the pentavalent vaccine in the future (Karalewitz and Barbieri, 2012; Webb and Smith, 2013), which will allow for continuation of the BabyBIG® program. As with adult botulism, treatment of infant botulism is restricted to antitoxin (BabyBIG®) treatment and supportive care. However, because of the continuous toxin production by *C. botulinum* that have colonized the infant intestines, the time window for potential benefits from antitoxin treatment is greater. BabyBIG® remains in the circulation for at least 6 months where it can neutralize any newly produced BoNT (Arnon et al., 2006). Prompt treatment of infants with BabyBIG®

significantly reduces the length of the hospital stay and severity of the disease (Arnon et al., 2006). Due to the limited supply of BabyBIG®, this product is restricted to treatments of infant botulism only and is available through the California Infant Botulism Treatment and Prevention Program.

Wound botulism requires additional treatment of the wound. The wound is usually cleaned surgically to remove all invading bacteria, and appropriate antibiotics such as vancomycin or metronidazole may be administered. However, it should be noted that treatment with some antibiotics such as aminoglycosides has been associated with increased severity of botulism (L'Hommedieu et al., 1979; Schwartz and Eng, 1982; Wang et al., 1984), and antibiotic treatments for foodborne or infant botulism are not suggested.

Death from botulism is usually the result of respiratory failure due to paralysis of the respiratory musculature (Arnon, 1980; Arnon et al., 2001; CDC, 2009; Woodruff et al., 1992), although cardiac arrest has also been reported as the cause of death in botulism patients (Tacket et al., 1984). Respiratory support by mechanical ventilation is the most important supportive treatment for severe botulism cases that involve weakening of the respiratory musculature and should be initiated before respiratory arrest sets in (Cherington, 1998). The requirement for respiratory support may be prolonged, usually several weeks to months (Arnon et al., 2006; Chalk et al., 2011; Cherington, 1998, 2004). Additional intensive nursing care is usually required including tube feedings and digestive support. No other treatments have been shown to be advantageous.

Most patients who recover from botulism recover fully within several weeks or in severe cases up to a year with no permanent weakness of neurologic abnormalities. However, secondary infections such as pneumonia or other complications such as being paralyzed for weeks to months may occur and can potentially lead to morbidity or death. Some patients recovering from very severe botulism may remain fatigued and weak for several years and may require a long-term therapy (CDC, 2011b).

43.9.2 Prevention and control

The best treatment of botulism is prevention. Many botulism cases are preventable. Foodborne botulism can be prevented by proper handling during manufacture, at retail, or by consumers. While C. botulinum spores are widespread and can easily contaminate food sources, simple measures such as including the "botulinum cook" (Stumbo, 1973) during canning procedures to kill C. botulinum spores will protect the food source. Foodborne botulism has most often been caused by home-canned foods with low acid content such as asparagus, green beans, beets, and corn that was not properly canned. Other seemingly unlikely or unusual sources include chopped garlic in oil, canned cheese sauce, chili peppers, tomatoes, carrot juice, and baked potatoes wrapped in foil, fermented fish, and other aquatic game foods (CDC, 2011b). All of the contaminated food products have in common that they were handled or packaged improperly. Unlike the C. botulinum spores, the BoNTs are easily destroyed by normal cooking temperatures (85°C), and thus properly heating canned and preserved food products ensures safety.

Wound botulism can be prevented by not using injectable street drugs, especially black tar heroin, and by seeking medical attention for infected wounds. Infant botulism cases may be caused by spores present in soil or house dust, which cannot easily be avoided. However, honey and herbal teas have been a suspected food for infant botulism. About 25% of honey has been found to contain C. botulinum spores (Arnon et al., 1979; Aureli et al., 2002; Bianco et al., 2008, 2009; Grant et al., 2013). Thus, infants under 1 year of age should not be fed honey or honey-containing products.

A vaccine to botulism was available on an investigation basis until 2011, and at-risk individuals such as laboratory workers, health care workers, and military personnel were vaccinated. Maintenance of neutralizing antibody titers in vaccinated individuals required yearly booster shots (Smith, 2009). However, in 2011, the CDC determined that the vaccine lost effectiveness and the incidence in side effects increased, leading to discontinuation of the product (Centers for Disease Control and Prevention, 2011). Currently, there is no vaccine available, but efforts are underway to develop a new vaccine for protection of at-risk populations. Vaccination of humans, in general, is undesirable because of the many benefits that medical uses of BoNTs can impart.

Public education campaigns about botulism and its prevention has been extensive and is responsible for a large reduction of foodborne botulism in the past few decades and reduced mortality rates of botulism cases, and it still is an ongoing activity (CDC, 2009, 2011b). Trained staff are available at state health departments and at the CDC to physicians 24 hours a day, and the CDC is providing online educational material for clinicians (CDC, 2009).

43.10 Botulinum neurotoxins in medicine and as potential weapons

Due to the high potency of BoNTs, the relative ease of production and transport, the potential for weaponization, and a lack of effective countermeasures, the CDC has categorized BoNTs as a Tier 1, Category A, Select Agent (the highest threat level to be misused as a weapon of bioterrorism). The consequences of a moderate- to large-scale human intoxication would be severe in that it would quickly overwhelm the health care system due to limited availability of respirators and the required long-term intensive nursing care, leading to high costs associated with patient care and a high death rate (Arnon et al., 2001; Balali-Mood et al., 2013; Rega et al., 2009). In the past, a few attempts to use BoNTs as a biological warfare agent or for bioterrorism purposes have been documented, but fortunately all were unsuccessful. Three active attempts to deploy BoNTs as a weapon in Japan between 1990 and 1995 are documented (Arnon et al., 2001). In addition, after the 1991 Persian Gulf War, Iraq admitted to United Nations inspectors that they had produced 19,000 L of botulinum toxin, with 10,000 L loaded into SCUD missile heads. The toxin contained in the 19,000 L was approximately three times the amount of BoNT needed to kill the entire human population by inhalation (Arnon et al., 2001). Most but not all of the toxin has been accounted for (Zilinskas, 1997). Possible routes of exposure used in a bioterrorism event might be contamination of a common food source ingested by many people or aerosolization of the toxin. While inhalation botulism does not occur naturally, the one human outbreak recorded in three laboratory workers accidentally exposed to BoNTs through accidental inhalation indicates that very small amounts of toxin would be sufficient to cause inhalation botulism (Holzer, 1962). In addition, animal studies indicate that inhalation of minute amounts of aerosolized BoNTs results in botulism (Al-Saleem et al., 2012; Arnon et al., 2001; Franz et al., 1993; Park and Simpson, 2003).

Conversely, the same properties that render BoNTs a threat for abuse also allowed their introduction into medicine as a unique and important therapeutic in 1989. The BoNTs' high potency and neuronal specificity is now utilized extensively to treat many spastic muscle disorders, for aesthetic purposes, for treatment of migraine headaches, and for other neuronal disorders of the peripheral nervous system such as hyperhidrosis (Brashear, 2010; Dressler, 2012). New medical applications are developed continuously, and recently the utility of BoNTs for treatment of pain disorders is being investigated (Pavone and Luvisetto, 2010; Ranoux et al., 2008; Rawicki et al., 2010; Vacca et al., 2013; Xiao et al., 2010; Yuan et al., 2009). Pharmaceutical BoNTs have proven to be a very safe drug with only few and minor side effects if used properly and by trained physicians. Only very few cases of iatrogenic botulism (inadvertent botulism caused by injection of BoNT for therapeutic purposes) have

been reported (Chertow et al., 2006; Coban et al., 2010; Crowner et al., 2007; Ghasemi et al., 2012; Partikian and Mitchell, 2007), and at least in some cases have been attributed to severe overdose of the toxin (Chertow et al., 2006; Crowner et al., 2007).

43.11 Conclusion

Botulism is a rare but severe and potentially deadly disease. Sporadic outbreaks of botulism continue to occur, and prevention of *C. botulinum* or BoNT contamination of foods is a perennial concern of the food industry. Most outbreaks of botulism are preventable by using proper food preservation and preparation techniques, and by avoiding wound infections, especially due to drug use. Any case of botulism is always treated as a public health emergency, because the offending source might still be available to unsuspecting persons, and because of the constant danger of an intentional release of BoNTs with potentially severe consequences. On the other hand, BoNTs are also unique and important pharmaceuticals to treat many neurologic disorders that are otherwise difficult or impossible to treat. The large number of different BoNT isotypes and their genetic relatedness within the background of a diverse population of *Clostridia* sp. is intriguing and suggests that recombination occurs among BoNT-producing bacteria, and that the toxin gene clusters may be passed between different bacteria. This is an important consideration when designing countermeasure to botulism and counter-terrorism strategies, and emphasizes the need for future research in this area. Most research on BoNTs so far has focused on the two isotypes used as therapeutics (BoNT/A1 and B1). Future efforts investigating the commonalities and differences between BoNT sero- and subtypes are important to develop novel therapeutics as well as effective botulism countermeasures.

Acknowledgments

I would like to thank Xiaoyang Serene Hu for preparation of Figure 43-5. Funding for this work was provided by National Institute of Allergy and Infectious Diseases (R01AI095274 and R01AI093504).

References

Akaike, N., Shin, M.C., Wakita, M., Torii, Y., Harakawa, T., Ginnaga, A., Kato, K., Kaji, R., and Kozaki, S. 2013. Transsynaptic inhibition of spinal transmission by A2 botulinum toxin. *J. Physiol.*, 591, 1031–1043.

Akhtar, S., Sarker, M.R., and Hossain, A. 2012. Microbiological food safety: A dilemma of developing societies. *Crit. Rev. Microbiol.*, 40, 348–359.

Alexiades-Armenakas, M. 2008. Retrograde transport and transcytosis of botulinum toxin serotypes to the brain: Analysis of potential neurotoxicity. *J. Drugs Dermatol.*, 7, 1006–1007.

Al-Saleem, F.H., Ancharski, D.M., Joshi, S.G., Elias, M., Singh, A., Nasser, Z., and Simpson, L.L. 2012. Analysis of the mechanisms that underlie absorption of botulinum toxin by the inhalation route. *Infect. Immun.*, 80, 4133–4142.

American Society of Microbiology. 2003. Sentinel laboratory guidelines for suspected agents of bioterrorism. In: Snyder, J., editor. *Botulinum Toxin*. 22: http://www.ok.gov/health2/documents/Botulism_toxin%202-15-13.pdf

Antonucci, F., Rossi, C., Gianfranceschi, L., Rossetto, O., and Caleo, M. 2008. Long-distance retrograde effects of botulinum neurotoxin A. *J. Neurosci.*, 28, 3689–3696.

Arnon, S.S. 1980. Infant botulism. *Annu. Rev. Med.*, 31, 541–560.

Arnon, S.S., Midura, T.F., Damus, K., Thompson, B., Wood, R.M., and Chin, J. 1979. Honey and other environmental risk factors for infant botulism. *J. Pediatr.*, 94, 331–336.

Arnon, S.S., Schechter, R., Inglesby, T.V., Henderson, D.A., Bartlett, J.G., Ascher, M.S., Eitzen, E., Fine, A.D., Hauer, J., Layton, M., Lillibridge, S., Osterholm, M.T., O'Toole, T., Parker, G., Perl, T.M., Russell, P.K., Swerdlow, D.L., Tonat, K., and Working Group on Civilian Biodefense. 2001. Botulinum toxin as a biological weapon: Medical and public health management. *J. Am. Med. Assoc.*, 285, 1059–1070.

Arnon, S.S., Schechter, R., Maslanka, S.E., Jewell, N.P., and Hatheway, C.L. 2006. Human botulism immune globulin for the treatment of infant botulism. *N. Engl. J. Med.*, 354, 462–471.

Aureli, P., Di Cunto, M., Maffei, A., De Chiara, G., Franciosa, G., Accorinti, L., Gambardella, A.M., and Greco, D. 2000. An outbreak in Italy of botulism associated with a dessert made with mascarpone cream cheese. *Eur. J. Epidemiol.*, 16, 913–918.

Aureli, P., Franciosa, G., and Fenicia, L. 2002. Infant botulism and honey in Europe: A commentary. *Pediatr. Infect. Dis. J.*, 21, 866–868.

Bach-Rojecky, L., and Lackovic, Z. 2009. Central origin of the antinociceptive action of botulinum toxin type A. *Pharmacol. Biochem. Behav.*, 94, 234–238.

Balali-Mood, M., Moshiri, M., and Etemad, L. 2013. Medical aspects of bio-terrorism. *Toxicon*, 69, 131–142.

Baldwin, M.R., Tepp, W.H., Przedpelski, A., Pier, C.L., Bradshaw, M., Johnson, E.A., and Barbieri, J.T. 2008. Subunit vaccine against the seven serotypes of botulism. *Infect. Immun.*, 76, 1314–1318.

Barash, J.R., and Arnon, S.S. 2013. A novel strain of *Clostridium botulinum* that produces type B and type H botulinum toxins. *J. Infect. Dis.*, 209, 183–191.

Bartlett, J.C. 1986. Infant botulism in adults. *N. Engl. J. Med.*, 315, 254–255.

Benatar, M. 2006. A systematic review of diagnostic studies in myasthenia gravis. *Neuromuscul. Disord.*, 16, 459–467.

Bianco, M.I., Luquez, C., de Jong, L.I., and Fernandez, R.A. 2008. Presence of *Clostridium botulinum* spores in *Matricaria chamomilla* (chamomile) and its relationship with infant botulism. *Int. J. Food. Microbiol.*, 121, 357–360.

Bianco, M.I., Luquez, C., De Jong, L.I., and Fernandez, R.A. 2009. Linden flower (*Tilia* spp.) as potential vehicle of *Clostridium botulinum* spores in the transmission of infant botulism. *Rev. Argent. Microbiol.*, 41, 232–236.

Borst, G.H., Lambers, G.M., and Haagsma, J. 1986. Type-C botulism in dogs. *Tijdschr. Diergeneeskd.*, 111, 1104–1105.

Boyer, A., Girault, C., Bauer, F., Korach, J.M., Salomon, J., Moirot, E., Leroy, J., and Bonmarchand, G. 2001. Two cases of foodborne botulism type E and review of epidemiology in France. *Eur. J. Clin. Microbiol. Infect. Dis.*, 20, 192–195.

Brashear, A. 2010. Botulinum toxin type A: Exploring new indications. *Drugs Today (Barc)*, 46, 671–682.

CDC. 2009. Botulism: Information and guidance for clinicians. In: Centers for Disease Control and Prevention, editor. *Prevention*. Atlanta, GA: Office of the Associate Director for Communication, Division of News and Electronic Media. http://www.bt.cdc.gov/agent/botulism/clinicians/index.asp

CDC. 2010. Investigational heptavalent botulinum antitoxin (HBAT) to replace licensed botulinum antitoxin AB and investigational botulinum antitoxin E. *MMWR Morb. Mortal. Wkly. Rep.*, 59, 299.

CDC. 2011a. Notice of CDC's discontinuation of investigational pentavalent (ABCDE) botulinum toxoid vaccine for workers at risk for occupational exposure to botulinum toxins. *MMWR Morb. Mortal. Wkly. Rep.*, 60, 1454–1455.

CDC. 2011b. *Botulism*. Atlanta, GA: Center for Disease Control and Prevention.

CDF. 2012. Botulism from drinking prison-made illicit alcohol – Utah 2011. *MMWR Morb. Mortal. Wkly. Rep.*, 61, 782–784.

CDF. 2013. Notes from the field: Botulism from drinking prison-made illicit alcohol – Arizona, 2012. *MMWR. Morb. Mortal. Wkly. Rep.*, 62, 88.

Chalk, C., Benstead, T.J., and Keezer, M. 2011. Medical treatment for botulism. *Cochrane Database Syst. Rev.*, 2, CD008123.

Cherington, M. 1998. Clinical spectrum of botulism. *Muscle Nerve*, 21, 701–710.

Cherington, M. 2004. Botulism: Update and review. *Semin. Neurol.*, 24, 155–163.

Chertow, D.S., Tan, E.T., Maslanka, S.E., Schulte, J., Bresnitz, E.A., Weisman, R.S., Bernstein, J., Marcus, S.M., Kumar, S., Malecki, J., Sobel, J., and Braden, C.R. 2006. Botulism in 4 adults following cosmetic injections with an unlicensed, highly concentrated botulinum preparation. *J. Am. Med. Assoc.*, 296, 2476–2479.

Coban, A., Matur, Z., Hanagasi, H.A., and Parman, Y. 2010. Iatrogenic botulism after botulinum toxin type A injections. *Clin. Neuropharmacol.*, 33, 158–160.

Crowner, B.E., Brunstrom, J.E., and Racette, B.A. 2007. Iatrogenic botulism due to therapeutic botulinum toxin a injection in a pediatric patient. *Clin. Neuropharmacol.*, 30, 310–313.

Date, K., Fagan, R., Crossland, S., Maceachern, D., Pyper, B., Bokanyi, R., Houze, Y., Andress, E., and Tauxe, R. 2011. Three outbreaks of foodborne botulism caused by unsafe home canning of vegetables – Ohio and Washington, 2008 and 2009. *J. Food Prot.*, 74, 2090–2096.

Demarchi, J., Mourgues, C., Orio, J., and Prevot, A.R. 1958. Existence of type D botulism in man. *Bull. Acad. Natl. Med.*, 142, 580–582.

Dohms, J.E., Allen, P.H., and Cloud, S.S. 1982. The immunization of broiler chickens against type C botulism. *Avian Dis.*, 26, 340–345.

Dong, M., Liu, H., Tepp, W.H., Johnson, E.A., Janz, R., and Chapman, E.R. 2008. Glycosylated SV2A and SV2B mediate the entry of botulinum neurotoxin E into neurons. *Mol. Biol. Cell*, 19, 5226–5237.

Dong, M., Richards, D.A., Goodnough, M.C., Tepp, W.H., Johnson, E.A., and Chapman, E.R. 2003. Synaptotagmins I and II mediate entry of botulinum neurotoxin B into cells. *J. Cell Biol.*, 162, 1293–1303.

Dong, M., Yeh, F., Tepp, W.H., Dean, C., Johnson, E.A., Janz, R., and Chapman, E.R. 2006. SV2 is the protein receptor for botulinum neurotoxin A. *Science*, 312, 592–596.

Dorner, M.B., Schulz, K.M., Kull, S., and Dorner, B.G. 2013. Complexity of botulinum neurotoxins: Challenges for detection technology. *Curr. Top. Microbiol. Immunol.*, 364, 219–255.

Dover, N., Barash, J.R., Hill, K.K., Xie, G., and Arnon, S.S. 2013. Molecular characterization of a novel botulinum neurotoxin type H gene. *J. Infect. Dis.*, 209, 192–202.

Dressler, D. 2012. Clinical applications of botulinum toxin. *Curr. Opin. Microbiol. (Engl.)*, 15, 325–336.

Elad, D., Yas-Natan, E., Aroch, I., Shamir, M.H., Kleinbart, S., Hadash, D., Chaffer, M., Greenberg, K., and Shlosberg, A. 2004. Natural *Clostridium botulinum* type C toxicosis in a group of cats. *J. Clin. Microbiol.*, 42, 5406–5408.

Eleopra, R., Tugnoli, V., Rossetto, O., De Grandis, D., and Montecucco, C. 1998. Different time courses of recovery after poisoning with botulinum neurotoxin serotypes A and E in humans. *Neurosci. Lett.*, 256, 135–138.

Eleopra, R., Tugnoli, V., Quatrale, R., Rossetto, O., and Montecucco, C. 2004. Different types of botulinum toxin in humans. *Mov. Disord.*, 19(suppl. 8), S53–S59.

Eleopra, R., Montecucco, C., Devigili, G., Lettieri, C., Rinaldo, S., Verriello, L., Pirazzini, M., Caccin, P., and Rossetto, O. 2013. Botulinum neurotoxin serotype D is poorly effective in humans: An in vivo electrophysiological study. *Clin. Neurophysiol.*, 124, 999–1004.

Fenicia, L., and Anniballi, F. 2009. Infant botulism. *Ann. Ist. Super. Sanita*, 45, 134–146.

Filipovic, B., Matak, I., Bach-Rojecky, L., and Lackovic, Z. 2012. Central action of peripherally applied botulinum toxin type a on pain and dural protein extravasation in rat model of trigeminal neuropathy. *PloS One*, 7, e29803.

Fischer, A., and Montal, M. 2007. Single molecule detection of intermediates during botulinum neurotoxin translocation across membranes. *Proc. Natl. Acad. Sci. USA*, 104, 10447–10452.

Fischer, A., Nakai, Y., Eubanks, L.M., Clancy, C.M., Tepp, W.H., Pellett, S., Dickerson, T.J., Johnson, E.A., Janda, K.D., and Montal, M. 2009. Bimodal modulation of the botulinum neurotoxin protein-conducting channel. *Proc. Natl. Acad. Sci. USA*, 106, 1330–1335.

Foran, P.G., Mohammed, N., Lisk, G.O., Nagwaney, S., Lawrence, G.W., Johnson, E., Smith, L., Aoki, K.R., and Dolly, J.O. 2003. Evaluation of the therapeutic usefulness of botulinum neurotoxin B, C1, E, and F compared with the long lasting type A. Basis for distinct durations of inhibition of exocytosis in central neurons. *J. Biol. Chem.*, 278, 1363–1371.

Fox, C.K., Keet, C.A., and Strober, J.B. 2005. Recent advances in infant botulism. *Pediatr. Neurol.*, 32, 149–154.

Francisco, A.M., and Arnon, S.S. 2007. Clinical mimics of infant botulism. *Pediatrics*, 119, 826–828.

Franz, D.R., Pitt, L.M., Clayton, M.A., Hanes, M.A., and Rose, K.J. 1993. Efficacy of prophylactic and therapeutic administration of antitoxin for inhalation botulism. In: Das Gupta, B., editor. *Botulinum and Tetanus Neurotoxins*. New York, NY: Plenum Press. pp. 473–476.

Freedman, M., Armstrong, R.M., Killian, J.M., and Boland, D. 1986. Botulism in a patient with jejunoileal bypass. *Ann. Neurol.*, 20, 641–643.

Galey, F.D. 2001. Botulism in the horse. *Vet. Clin. North Am. Equine Pract.*, 17, 579–588.

Ghasemi, M., Norouzi, R., Salari, M., and Asadi, B. 2012. Iatrogenic botulism after the therapeutic use of botulinum toxin-A: A case report and review of the literature. *Clin. Neuropharmacol.*, 35, 254–257.

Gimenez, D.F., and Gimenez, J.A. 1995. The typing of botulinal neurotoxins. *Int. J. Food Microbiol.*, 27, 1–9.

Gottlieb, S.L., Kretsinger, K., Tarkhashvili, N., Chakvetadze, N., Chokheli, M., Chubinidze, M., Michael Hoekstra, R., Jhorjholiani, E., Mirtskhulava, M., Moistsrapishvili, M., Sikharulidze, M., Zardiashvili, T., Imnadze, P., and Sobel, J. 2007. Long-term outcomes of 217 botulism cases in the Republic of Georgia. *Clin. infect. Dis.*, 45, 174–180.

Grant, K.A., McLauchlin, J., and Amar, C. 2013. Infant botulism: Advice on avoiding feeding honey to babies and other possible risk factors. *Community Pract.*, 86, 44–46.

Gupta, A., Sumner, C.J., Castor, M., Maslanka, S., and Sobel, J. 2005. Adult botulism type F in the United States, 1981–2002. *Neurology*, 65, 1694–1700.

Häfner, S. 2009. Justinus Kerner's heroic experiments with botulinum toxin. *Aktuelle Neurologie*, 36, 412–416.

Hatheway, C.L. 1988. Botulism. In: Balows, A., Hausler, W.H., Ohashi, M., and Turano, M.A., editors. *Laboratory Diagnosis of Infectious Diseases: Principles and Practice*. New York: Springer-Verlag. pp. 111–133.

Hatheway, C.L. 1995. Botulism: The present status of the disease. *Curr. Top. Microbiol. Immunol.*, 195, 55–75.

Henkel, J.S., Jacobson, M., Tepp, W., Pier, C., Johnson, E.A., and Barbieri, J.T. 2009. Catalytic properties of botulinum neurotoxin subtypes A3 and A4 (dagger). *Biochemistry*, 48, 2522.

Hill, K.K., and Smith, T.J. 2013. Genetic diversity within *Clostridium botulinum* serotypes, botulinum neurotoxin gene clusters and toxin subtypes. *Curr. Top. Microbiol. Immunol.*, 364, 1–20.

Holzer, E. 1962. Botulism caused by inhalation. *Med. Klin.*, 57, 1735–1738.

Istre, G.R., Compton, R., Novotny, T., Young, J.E., Hatheway, C.L., and Hopkins, R.S. 1986. Infant botulism. Three cases in a small town. *Am. J. Dis. Child*, 140, 1013–1014.

Jacky, B.P., Garay, P.E., Dupuy, J., Nelson, J.B., Cai, B., Molina, Y., Wang, J., Steward, L.E., Broide, R.S., Francis, J., Aoki, K.R.,

Stevens, R.C., and Fernandez-Salas, E. 2013. Identification of Fibroblast Growth Factor Receptor 3 (FGFR3) as a protein receptor for botulinum neurotoxin Serotype A (BoNT/A). *PLoS Pathog.*, 9, e1003369.

Johnson, E.A., and Montecucco, C. 2008. Botulism, Chapter 11. In: Andrew, G.E., editor. *Handbook of Clinical Neurology*. Netherlands: Elsevier. pp. 333–368.

Juliao, P.C., Maslanka, S., Dykes, J., Gaul, L., Bagdure, S., Granzow-Kibiger, L., Salehi, E., Zink, D., Neligan, R.P., Barton-Behravesh, C., Luquez, C., Biggerstaff, M., Lynch, M., Olson, C., Williams, I., and Barzilay, E.J. 2013. National outbreak of type a foodborne botulism associated with a widely distributed commercially canned hot dog chili sauce. *Clin. Infect. Dis.*, 56, 376–382.

Karalewitz, A.P., and Barbieri, J.T. 2012. Vaccines against botulism. *Curr. Opin. Microbiol.*, 15, 317–324.

Keller, J.E. 2006. Recovery from botulinum neurotoxin poisoning in vivo. *Neuroscience*, 139, 629–637.

Kerner, J. 1817. Vergiftung durch verdorbene Würste. *Tübinger Blätter für Naturwissenschaften und Arzneykunde*, 3, 1–25.

Kerner, J. 1820. *Neue Beobachtungen über die in Würtemberg so häufig vorfallenden tödtlichen Vergiftungen durch den Genuß geräucherter Würste*. Tübingen: Osiander.

Kerner, J. 1822. *Das Fettgift oder die Fettsäure und ihre Wirkungen auf den thierischen Organismus, ein Beytrag zur Untersuchung des in verdorbenen Würsten giftig wirkenden Stoffes*. Stuttgart, Tübingen: Cotta.

Kolbe, D.R., and Coe Clough, N.E. 2008. Correlation of *Clostridium botulinum* type C antitoxin titers in mink and guinea pigs to protection against type C intoxication in mink. *Anaerobe*, 14, 128–130.

Kruger, M., Grosse-Herrenthey, A., Schrodl, W., Gerlach, A., and Rodloff, A. 2012. Visceral botulism at dairy farms in Schleswig Holstein, Germany: Prevalence of *Clostridium botulinum* in feces of cows, in animal feeds, in feces of the farmers, and in house dust. *Anaerobe*, 18, 221–223.

L'Hommedieu, C., Stough, R., Brown, L., Kettrick, R., and Polin, R. 1979. Potentiation of neuromuscular weakness in infant botulism by aminoglycosides. *J. Pediatr.*, 95, 1065–1070.

Lacy, D.B., Tepp, W., Cohen, A.C., DasGupta, B.R., and Stevens, R.C. 1998. Crystal structure of botulinum neurotoxin type A and implications for toxicity. *Nat. Struct. Biol.*, 5, 898–902.

Lalli, G., Bohnert, S., Deinhardt, K., Verastegui, C., and Schiavo, G. 2003a. The journey of tetanus and botulinum neurotoxins in neurons. *Trends Microbiol.*, 11, 431–437.

Lalli, G., Gschmeissner, S., and Schiavo, G. 2003b. Myosin Va and microtubule-based motors are required for fast axonal retrograde transport of tetanus toxin in motor neurons. *J. Cell Sci.*, 116, 4639–4650.

Leclair, D., Fung, J., Isaac-Renton, J.L., Proulx, J.F., May-Hadford, J., Ellis, A., Ashton, E., Bekal, S., Farber, J.M., Blanchfield, B., and Austin, J.W. 2013. Foodborne botulism in Canada, 1985–2005. *Emerg. Infect. Dis.*, 19, 961–968.

Lindstrom, M., Myllykoski, J., Sivela, S., and Korkeala, H. 2010. *Clostridium botulinum* in cattle and dairy products. *Crit. Rev. Food. Sci. Nutr.*, 50, 281–304.

Long, S.S. 1985. Epidemiologic study of infant botulism in Pennsylvania: Report of the Infant Botulism Study group. *Pediatrics*, 75, 928–934.

Ma, L., Nagai, J., Sekino, Y., Goto, Y., Nakahira, S., and Ueda, H. 2012. Single application of A2 NTX, a botulinum toxin A2 subunit, prevents chronic pain over long periods in both diabetic and spinal cord injury-induced neuropathic pain models. *J. Pharmacol. Sci.*, 119, 282–286.

Marchand-Pauvert, V., Aymard, C., Giboin, L.S., Dominici, F., Rossi, A., and Mazzocchio, R. 2013. Beyond muscular effects: Depression of spinal recurrent inhibition after botulinum neurotoxin A. *J. Physiol.*, 591, 1017–1029.

Marinelli, S., Vacca, V., Ricordy, R., Uggenti, C., Tata, A.M., Luvisetto, S., and Pavone, F. 2012. The analgesic effect on neuropathic pain of retrogradely transported botulinum neurotoxin A involves Schwann cells and astrocytes. *PLoS One*, 7, e47977.

Matak, I., Riederer, P., and Lackovic, Z. 2012. Botulinum toxin's axonal transport from periphery to the spinal cord. *Neurochem. Int.*, 61, 236–239.

McCroskey, L.M., and Hatheway, C.L. 1988. Laboratory findings in four cases of adult botulism suggest colonization of the intestinal tract. *J. Clin. Microbiol.*, 26, 1052–1054.

Meriggioli, M.N., and Sanders, D.B. 2005. Advances in the diagnosis of neuromuscular junction disorders. *Am. J. Phys. Med. Rehabil.*, 84, 627–638.

Montal, M. 2010. Botulinum neurotoxin: A marvel of protein design. *Annu. Rev. Biochem.*, 79, 591–617.

Montecucco, C. 1986. How do tetanus and botulinum toxins bind to neuronal membranes? *Trends Biochem. Sci.*, 11, 314–317.

Montecucco, C., and Schiavo, G. 1994. Mechanism of action of tetanus and botulinum neurotoxins. *Mol. Microbiol.*, 13, 1–8.

Nishiki, T., Kamata, Y., Nemoto, Y., Omori, A., Ito, T., Takahashi, M., and Kozaki, S. 1994. Identification of protein receptor for Clostridium botulinum type B neurotoxin in rat brain synaptosomes. *J. Biol. Chem.*, 269, 10498–10503.

Oguma, K., Yokota, K., Hayashi, S., Takeshi, K., Kumagai, M., Itoh, N., Tachi, N., and Chiba, S. 1990. Infant botulism due to *Clostridium botulinum* type C toxin. *Lancet*, 336, 1449–1450.

Palmateer, N.E., Hope, V.D., Roy, K., Marongiu, A., White, J.M., Grant, K.A., Ramsay, C.N., Goldberg, D.J., and Ncube, F. 2013. Infections with spore-forming bacteria in persons who inject drugs, 2000–2009. *Emerg. Infect. Dis.*, 19, 29–34.

Park, J.B., and Simpson, L.L. 2003. Inhalational poisoning by botulinum toxin and inhalation vaccination with its heavy-chain component. *Infect. Immun.*, 71, 1147–1154.

Partikian, A., and Mitchell, W.G. 2007. Iatrogenic botulism in a child with spastic quadriparesis. *J. Child. Neurol.*, 22, 1235–1237.

Pavone, F., and Luvisetto, S. 2010. Botulinum neurotoxin for pain management: Insights from animal models. *Toxins (Basel)*, 2, 2890–2913.

Peck, M.W. 2006. *Clostridium botulinum* and the safety of minimally heated, chilled foods: An emerging issue? *J. Appl. Microbiol.*, 101, 556–570.

Peck, M.W. 2009. Biology and genomic analysis of *Clostridium botulinum*. *Adv. Microb. Physiol.*, 55, 183–265.

Peck, M.W., Stringer, S.C., and Carter, A.T. 2011. *Clostridium botulinum* in the post-genomic era. *Food Microbiol.*, 28, 183–191.

Pellett, S. 2012. Learning from the past: Historical aspects of bacterial toxins as pharmaceuticals. *Curr. Opin. Microbiol.*, 15, 292–299.

Pellett, S. 2013. Progress in cell based assays for botulinum neurotoxin detection. *Curr. Top. Microbiol. Immunol.*, 364, 257–285.

Peng, L., Liu, H., Ruan, H., Tepp, W.H., Stoothoff, W.H., Brown, R.H., Johnson, E.A., Yao, W.D., Zhang, S.C., and Dong, M. 2013. Cytotoxicity of botulinum neurotoxins reveals a direct role of syntaxin 1 and SNAP-25 in neuron survival. *Nat. Commun.*, 4, 1472.

Peng, L., Tepp, W.H., Johnson, E.A., and Dong, M. 2011. Botulinum Neurotoxin D Uses Synaptic Vesicle Protein SV2 and Gangliosides as Receptors. *PLoS pathog.*, 7, e1002008.

Pickett, J.B. 1982. Infant botulism – the first five years. *Muscle Nerve*, 5, S26–S27.

Pickett, J., Berg, B., Chaplin, E., and Brunstetter-Shafer, M.A. 1976. Syndrome of botulism in infancy: Clinical and electrophysiologic study. *N. Engl. J. Med.*, 295, 770–772.

Pier, C.L., Tepp, W.H., Bradshaw, M., Johnson, E.A., Barbieri, J.T., and Baldwin, M.R. 2008. Recombinant holotoxoid vaccine against botulism. *Infect. Immun.*, 76, 437–442.

Pier, C.L., Chen, C., Tepp, W.H., Lin, G., Janda, K.D., Barbieri, J.T., Pellett, S., and Johnson, E.A. 2011. Botulinum neurotoxin subtype A2 enters neuronal cells faster than subtype A1. *FEBS Lett.*, 585, 199–206.

Ranoux, D., Attal, N., Morain, F., and Bouhassira, D. 2008. Botulinum toxin type A induces direct analgesic effects in chronic neuropathic pain. *Ann. Neurol.*, 64, 274–283.

Rawicki, B., Sheean, G., Fung, V.S., Goldsmith, S., Morgan, C., Novak, I., and Cerebral Palsy Institute. 2010. Botulinum toxin assessment, intervention and aftercare for paediatric and adult niche indications including pain: International consensus statement. *Eur. J. Neurol.*, 17(suppl. 2), 122–134.

Rega, P., Burkholder-Allen, K., and Bork, C. 2009. An algorithm for the evaluation and management of red, yellow, and green zone patients during a botulism mass casualty incident. *Am. J. Disaster. Med.*, 4, 192–198.

Restani, L., Giribaldi, F., Manich, M., Bercsenyi, K., Menendez, G., Rossetto, O., Caleo, M., and Schiavo, G. 2012a. Botulinum neurotoxins a and e undergo retrograde axonal transport in primary motor neurons. *PLoS pathog.*, 8, e1003087.

Restani, L., Novelli, E., Bottari, D., Leone, P., Barone, I., Galli-Resta, L., Strettoi, E., and Caleo, M. 2012b. Botulinum neurotoxin A impairs neurotransmission following retrograde transsynaptic transport. *Traffic*, 13, 1083–1089.

Rocke, T.E., and Friend, M. 1999. Avian botulism, Information and Technology Report, Reston, VA. pp. 271–281.

Rodloff, A.C., and Kruger, M. 2012. Chronic *Clostridium botulinum* infections in farmers. *Anaerobe*, 18, 226–228.

Rummel, A. 2013. Double receptor anchorage of botulinum neurotoxins accounts for their exquisite neurospecificity. *Curr. Top. Microbiol. Immunol.*, 364, 61–90.

Rummel, A., Hafner, K., Mahrhold, S., Darashchonak, N., Holt, M., Jahn, R., Beermann, S., Karnath, T., Bigalke, H., and Binz, T. 2009. Botulinum neurotoxins C, E and F bind gangliosides via a conserved binding site prior to stimulation-dependent uptake with botulinum neurotoxin F utilising the three isoforms of SV2 as second receptor. *J. Neurochem.*, 110, 1942–1954.

Rummel, A., Karnath, T., Henke, T., Bigalke, H., and Binz, T. 2004. Synaptotagmins I and II act as nerve cell receptors for botulinum neurotoxin G. *J. Biol. Chem.*, 279, 30865–30870.

Rummel, A., Mahrhold, S., Bigalke, H., and Binz, T. 2011. Exchange of the H(CC) domain mediating double receptor recognition improves the pharmacodynamic properties of botulinum neurotoxin. *FEBS J.*, 278, 4506–4515.

Sandrock, C.E., and Murin, S. 2001. Clinical predictors of respiratory failure and long-term outcome in black tar heroin-associated wound botulism. *Chest*, 120, 562–566.

Schantz, E.J., and Kautter, D.A. 1978. Standardized assay for *Clostridium botulinum* toxins. *J. Assoc. off. Anal. Chem.*, 61, 96–99.

Schiavo, G., Benfenati, F., Poulain, B., Rossetto, O., Polverino de Laureto, P., DasGupta, B.R., and Montecucco, C. 1992. Tetanus and botulinum-B neurotoxins block neurotransmitter release by proteolytic cleavage of synaptobrevin. *Nature*, 359, 832–835.

Schiavo, G., Rossetto, O., Tonello, F., and Montecucco, C. 1995. Intracellular targets and metalloprotease activity of tetanus and botulism neurotoxins. *Curr. Top. Microbiol. Immunol.*, 195, 257–274.

Schiavo, G., Matteoli, M., and Montecucco, C. 2000. Neurotoxins affecting neuroexocytosis. *Physiol. Rev.*, 80, 717–766.

Schwartz, R.H., and Eng, G. 1982. Infant botulism: Exacerbation by aminoglycosides. *Am. J. Dis. Child*, 136, 952.

Sheppard, Y.D., Middleton, D., Whitfield, Y., Tyndel, F., Haider, S., Spiegelman, J., Swartz, R.H., Nelder, M.P., Baker, S.L., Landry, L., Maceachern, R., Deamond, S., Ross, L., Peters, G., Baird, M., Rose, D., Sanders, G., and Austin, J.W. 2012. Intestinal toxemia botulism in 3 adults, Ontario, Canada, 2006–2008. *Emerg. Infect. Dis.*, 18, 1–6.

Sloop, R.R., Cole, B.A., and Escutin, R.O. 1997. Human response to botulinum toxin injection: Type B compared with type A. *Neurology*, 49, 189–194.

Smith, L.A. 2009. *Botulism and vaccines for its prevention*. *Vaccine (Netherlands)*, 27(suppl 4), D33–D39.

Sobel, J. 2005. Botulism. *Clin. Infect. Dis.*, 41, 1167–1173.

Solomon Jr., H.M., and Lilly, T. 2001. *Clostridium botulinum. BAM Bacteriological Analytical Manual*. Chapter 17. U.S. Food and Drug Administration.

Sonnabend, O., Sonnabend, W., Heinzle, R., Sigrist, T., Dirnhofer, R., and Krech, U. 1981. Isolation of *Clostridium botulinum* type G and identification of type G botulinal toxin in humans: Report of five sudden unexpected deaths. *J. Infect. Dis.*, 143, 22–27.

Steinman, A., Chaffer, M., Elad, D., and Shpigel, N.Y. 2006. Quantitative analysis of levels of serum immunoglobulin G against botulinum neurotoxin type D and association with protection in natural outbreaks of cattle botulism. *Clin. Vaccine Immunol.*, 13, 862–868.

Stenmark, P., Dupuy, J., Imamura, A., Kiso, M., and Stevens, R.C. 2008. Crystal structure of botulinum neurotoxin type A in complex with the cell surface co-receptor GT1b-insight into the toxin-neuron interaction. *PLoS Pathog.*, 4, e1000129.

Stringer, S.C., Carter, A.T., Webb, M.D., Wachnicka, E., Crossman, L.C., Sebaihia, M., and Peck, M.W. 2013. Genomic and physiological variability within Group II (non-proteolytic) *Clostridium botulinum*. *BMC Genomics*, 14, 333.

Stumbo, C.R. 1973. *Thermobacteriology in Food Processing*. New York and London: Academic Press.

Stura, E.A., Le Roux, L., Guitot, K., Garcia, S., Bregant, S., Beau, F., Vera, L., Collet, G., Ptchelkine, D., Bakirci, H., and Dive, V. 2012. Structural framework for covalent inhibition of *Clostridium botulinum* neurotoxin A by targeting Cys165. *J. Biol. Chem.*, 287, 33607–33614.

Tacket, C.O., Shandera, W.X., Mann, J.M., Hargrett, N.T., and Blake, P.A. 1984. Equine antitoxin use and other factors that predict outcome in type A foodborne botulism. *Am. J. Med.*, 76, 794–798.

Tepp, W.H., Lin, G., and Johnson, E.A. 2012. Purification and characterization of a novel subtype a3 botulinum neurotoxin. *Appl. Environ. Microbiol.*, 78, 3108–3113.

Torii, Y., Kiyota, N., Sugimoto, N., Mori, Y., Goto, Y., Harakawa, T., Nakahira, S., Kaji, R., Kozaki, S., and Ginnaga, A. 2010. Comparison of effects of botulinum toxin subtype A1 and A2 using twitch tension assay and rat grip strength test. *Toxicon*, 57, 93–99.

Torii, Y., Akaike, N., Harakawa, T., Kato, K., Sugimoto, N., Goto, Y., Nakahira, S., Kohda, T., Kozaki, S., Kaji, R., and Ginnaga, A. 2011. Type A1 but not type A2 botulinum toxin decreases the grip strength of the contralateral foreleg through axonal transport from the toxin-treated foreleg of rats. *J. Pharmacol. Sci.*, 117, 275–285.

Uriarte, A., Thibaud, J.L., and Blot, S. 2010. Botulism in 2 urban dogs. *Can. Vet. J.*, 51, 1139–1142.

Vacca, V., Marinelli, S., Luvisetto, S., and Pavone, F. 2013. Botulinum toxin A increases analgesic effects of morphine, counters development of morphine tolerance and modulates glia activation and μ opioid receptor expression in neuropathic mice. *Brain Behav. Immun.*, 32, 40–50.

van Ermengem, E. 1897. Ueber einen neuen anaëroben Bacillus und seine Beziehungen zum Botulismus. *Med. Microbiol. Immunol. (Berl.)*, 26, 1–56.

van Ermengem, E. 1979. Classics in infectious diseases. A new anaerobic bacillus and its relation to botulism. E. van Ermengem. Originally published as "Ueber einen neuen anaeroben Bacillus und seine Beziehungen zum Botulismus" in Zeitschrift fur Hygiene und Infektionskrankheiten 26: 1–56, 1897. *Rev. Infect. Dis.*, 1, 701–719.

van Nes, J.J., and van der Most van Spijk, D. 1986. Electrophysiological evidence of peripheral nerve dysfunction in six dogs with botulism type C. *Res. Vet. Sci.*, 40, 372–376.

Varma, J.K., Katsitadze, G., Moiscrafishvili, M., Zardiashvili, T., Chokheli, M., Tarkhashvili, N., Jhorjholiani, E., Chubinidze, M., Kukhalashvili, T., Khmaladze, I., Chakvetadze, N., Imnadze, P., Hoekstra, M., and Sobel, J. 2004. Signs and symptoms predictive of death in patients with foodborne botulism – Republic of Georgia, 1980–2002. *Clin. infect. Dis.*, 39, 357–362.

Vugia, D.J., Mase, S.R., Cole, B., Stiles, J., Rosenberg, J., Velasquez, L., Radner, A., and Inami, G. 2009. Botulism from drinking pruno. *Emerg. Infect. Dis.*, 15, 69–71.

Walton, R.N., Clemens, A., Chung, J., Moore, S., Wharton, D., Haydu, L., de Villa, E., Sanders, G., Bussey, J., Richardson, D., and Austin, J.W. 2014. Outbreak of type E foodborne botulism linked to traditionally prepared salted fish in Ontario, Canada. *Foodborne Pathog. Dis.*, 11, 830–834.

Wang, Y.C., Burr, D.H., Korthals, G.J., and Sugiyama, H. 1984. Acute toxicity of aminoglycoside antibiotics as an aid in detecting botulism. *Appl. Environ. Microbiol.*, 48, 951–955.

Webb, R.P., and Smith, L.A. 2013. What next for botulism vaccine development? *Expert Rev. Vaccines*, 12, 481–492.

Whitemarsh, R.C., Tepp, W.H., Bradshaw, M., Lin, G., Pier, C.L., Scherf, J.M., Johnson, E.A., and Pellett, S. 2013. Characterization of botulinum neurotoxin A subtypes 1 through 5 by investigation of activities in mice, neuronal cell cultures, and in vitro. *Infect. Immun.*, 81, 3894–3902.

Witoonpanich, R., Vichayanrat, E., Tantisiriwit, K., Rattanasiri, S., and Ingsathit, A. 2009. Electrodiagnosis of botulism and clinico-electrophysiological correlation. *Clin. Neurophysiol.*, 120, 1135–1138.

Woodruff, B.A., Griffin, P.M., McCroskey, L.M., Smart, J.F., Wainwright, R.B., Bryant, R.G., Hutwagner, L.C., and Hatheway, C.L. 1992. Clinical and laboratory comparison of botulism from toxin types A, B, and E in the United States, 1975–1988. *J. Infect. Dis.*, 166, 1281–1286.

Xiao, L., Mackey, S., Hui, H., Xong, D., Zhang, Q., and Zhang, D. 2010. Subcutaneous injection of botulinum toxin a is beneficial in postherpetic neuralgia. *Pain Med.*, 11, 1827–1833.

Yuan, R.Y., Sheu, J.J., Yu, J.M., Chen, W.T., Tseng, I.J., Chang, H.H., and Hu, C.J. 2009. Botulinum toxin for diabetic neuropathic pain: A randomized double-blind crossover trial. *Neurology*, 72, 1473–1478.

Yuan, J., Inami, G., Mohle-Boetani, J., and Vugia, D.J. 2011. Recurrent wound botulism among injection drug users in California. *Clin. Infect. Dis.*, 52, 862–866.

Zhang, S., Wang, Y., Qiu, S., Dong, Y., Xu, Y., Jiang, D., Fu, X., Zhang, J., He, J., Jia, L., Wang, L., Zhang, C., Sun, Y., and Song, H. 2010. Multilocus outbreak of foodborne botulism linked to contaminated sausage in Hebei province, China. *Clin. Infect. Dis.*, 51, 322–325.

Zilinskas, R.A. 1997. Iraq's biological weapons. The past as future? *J. Am. Med. Assoc.*, 278, 418–424.

Chapter 44

Mycobacterium ulcerans and Buruli Ulcer

Mark Eric Benbow[1], Belinda Hall[2], Lydia Mosi[3], Sophie Roberts[4], Rachel Simmonds[2], and Heather Williamson Jordan[5]

[1]Department of Entomology and Department of Osteopathic Medical Specialties, Michigan State University, MI, USA

[2]Department of Microbial and Cellular Sciences and School of Biosciences and Medicine, University of Surrey, Guildford, UK

[3]Department of Biochemistry, Cell and Molecular Biology; West African Center for Cell Biology of Infectious Pathogens, University of Ghana, Legon

[4]Royal Liverpool Hospital, Liverpool, United Kingdom

[5]Department of Biological Sciences, Mississippi State University, MS, USA

44.1 Introduction

Buruli ulcer (BU) is a tropical skin disease with a significant public health impact to affected populations, primarily in rural West Africa (Walsh et al., 2010). The disease begins as a painless nodule or papule that, if left untreated, can lead to ulceration that can cover large portions of the body. Infection is due to an environmental pathogen, *Mycobacterium ulcerans*, whose pathogenesis is mediated through the production of an immunosuppressive lipid toxin, mycolactone, that causes the necrosis characteristic of the disease. Though mortality is low, morbidity is extremely high in rural populations most affected because disability and disfigurement are often disease outcomes (World Health Organization, 2013a). The WHO declared BU an emerging disease in 1998 and the Global BU Initiative was established to develop programs for addressing the severe morbidity (Johnson et al., 2005). The initiative has sought to increase surveillance, promote awareness through education, and improve treatment by teaming academic institutions with non-government organizations and health leaders of endemic countries. Pooling resources and experience from microbiologists, sociologists, epidemiologists, funding agencies, and medical doctors have resulted in significant advances in understanding and preventing morbidity due to BU.

44.2 Disease burden

BU is the third most common mycobacterium disease following only tuberculosis and leprosy (Johnson et al., 2005). BU is endemic in at least 30 countries (Figure 44-1). The highest number of reported cases is from West Africa (2004). In Cote d'Ivoire, over 24,000 cases were reported from 1978 to 2006 (Anonymous, 2004). Prevalence in Ghana has been reported to be 11,000 cases since 1993. However, the disease is underreported due to cultural reasons and few countries conduct active surveillance. As a result, official figures on BU seriously underestimate the true burden of disease. Some patients seek medical attention from traditional healers or attempt to treat the disease themselves, and most of these instances are not reported to the ministries of health. In addition, there is often a stigma attached to the disease at least partially because it is often considered to be due to witchcraft. Despite efforts by WHO to obtain accurate epidemiological data, BU remains a non-notifiable disease in at least two dozen countries.

44.3 Transmission

Though BU is thought to be associated with large- and small-scale disturbances to the landscape and bodies of water frequented by human populations (Merritt et al., 2010; van Ravensway et al., 2012), primary prevention of BU is difficult because the mode of transmission is not known. Education regarding risk factors is key, especially in endemic areas for individuals to identify personal risks and take protective actions to prevent disease. The epidemiology of BU was recently reviewed by White et al. (2013) and Walsh et al. (2011). These reviews provided

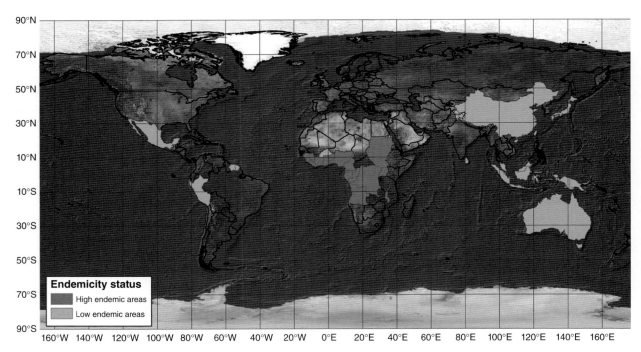

Fig. 44-1 Countries reporting Buruli ulcer. Image and description from Johnson et al. (2005). Acquired through the Creative Commons Attribution License. For a color version of this figure, see the color plate section.

background into the epidemiology of BU and also evaluated the risk factors associated with this disease from areas all over the globe. Furthermore, understanding and reporting consistent epidemiological risk factors of BU have been complex, since the mode of transmission is still not fully understood. Merritt et al. (2010) provided a global systematic review of the transmission of *M. ulcerans* in a way to facilitate future epidemiological studies of BU. Here, we provide a short review of the most common environmental risk factors for BU and recent research into understanding its transmission. Understanding both the epidemiology and transmission will facilitate the control and management of this neglected tropical disease, as this information can then be included into disease risk mapping at local, regional, and global scales (Stanford et al., 1975).

While broad-scale risk factors were reported by White et al. (2013), they also recognized in their review that many reported risk factors were not consistently found among specific studies, frustrating attempts to identify more specific human activity patterns important to disease risk. They argued that these inconsistent associations were the result of variation in study locations, analyses, and how cases and control populations were defined. Some of these inconsistent risk factors were the extent and specific activity within or near water, the use of bed nets, and the reporting of mosquito or other insect bites (see Table 1 in White et al. (2013)). Much of these inconsistencies are probably explained by not understanding the mode of transmission of *M. ulcerans* to humans and possibly the causal factors that change both in space and time (e.g., rainfall patterns that affect habitat abundance of possible reservoirs or vectors).

The most consistent risk factor associated with BU is residence and/or working near aquatic habitats (Jacobsen and Padgett, 2010). Associations with water include bathing in rivers, and time spent at the water source. However, fishing, swimming, aquaculture, and washing clothes were not strongly associated. In Africa, farming of dry land crops such as cassava, banana, beans, and maize was associated with a decreased risk. In Australia, overland activities such as bird watching, cycling, golf, and gardening were also negatively associated (Quek et al., 2007). In several studies, failure to wear protective clothing (e.g., long pants and long sleeve shirts) and poor wound care were also identified as significant risk factors for BU (White et al., 2013).

The association with aquatic habitats has been a recognized and accepted risk factor of BU for several decades (Merritt et al., 2010), but only recently have there been quantitative studies to statistically associate BU cases with low-lying topographic areas that are commonly flooded during prolonged periods of rainfall (van Ravensway et al., 2012; Wagner et al., 2008a, 2008b). These studies of landscape epidemiology were of large scale (i.e., region scale), and though of value, they have not provided adequate resolution for assessing the association between landscape features and disease at the village level. Because of the focal distribution of BU cases within a small geographical area, more detailed landscape studies could add significantly to an understanding of the ecology of this disease. A step in the right direction has been provided in a recent study by Williamson et al. (2012), where a significant association of *M. ulcerans* from environmental samples with BU prevalence provided evidence for a close association between human activity in aquatic environments and the risk of disease.

To better understand transmission, the basic ecology of *M. ulcerans* must be understood, including the reservoir and distribution of the bacteria in the natural environment. There have been several studies indicating that physical and chemical changes in water body condition due to human activities (e.g., sedimentation, eutrophication) lead to changes in water quality that positively influence *M. ulcerans* communities (MacCallum et al., 1948; Merritt et al., 2005), and that *M. ulcerans* can be found in samples taken from highly impacted water bodies (Williamson et al., 2008).

Water quality affects the aquatic invertebrate communities of both standing and flowing water bodies. These communities are represented by several taxa hypothesized to be part of the transmission cycle of *M. ulcerans* (Merritt et al., 2010). These taxa include biting water bugs (Hemiptera) of the Naucoridae and Belostomatidae families and mosquitoes (Diptera: Culicidae) (Agbenorku et al., 2001; Lavender et al., 2011; Marsollier et al., 2002; Portaels et al., 1999; Sopoh et al., 2010; World Health Organization, 2013a, 2013c). However, while laboratory and small-scale field studies (e.g., sampling from only 1–4 sites) have facilitated these hypotheses, there have been very few studies conducted using standardized and statistically defendable methodology, including the evaluation of replicate water bodies at a large spatial scale (i.e., more than 10 sites and over an entire country covering thousands of square kilometers). The studies of transmission have been locally focused, which is different from the studies addressing large-scale landscape risk patterns mentioned above (van Ravensway et al., 2012; Wagner et al., 2008a, 2008b).

For example, Marion et al. (2010) took aquatic samples of Hemiptera during October, November, December, January, April, and July from one endemic village site and in April and July from a non-endemic village site and reported *M. ulcerans* colonization of hemipteran tissue to range from 1.4 to 33.9% (depending on time of year) in the endemic site, but no pooled samples were positive for *M. ulcerans* DNA in the non-endemic site. They tested 616 pools of hemiptera from the endemic site compared to only 80 pools from the non-endemic site, suggesting that the analysis effort was not equal between the two sites. A further limitation of this study was that analysis was biased. Although a large group of insect taxa were collected, none but the Hemiptera were analyzed for *M. ulcerans* DNA. This in an important omission because other invertebrate taxa such as mosquito (Diptera: Culicidae) or the highly abundant midge (Diptera: Chironomidae) larvae have been positive for *M. ulcerans* DNA in other studies (Benbow et al., 2008; Johnson et al., 2007; Lavender et al., 2011; Williamson et al., 2008). Finally, sampling did not include other potential environmental sources such as plants, plant biofilms, or other aquatic organisms (Marsollier et al., 2004b, 2007).

Laboratory experiments have demonstrated that *M. ulcerans* could be passed through trophic levels in mollusks and aquatic insects, suggesting that *M. ulcerans* is an integral part of the aquatic environment, including the food web (Marsollier et al., 2002, 2004a; Mosi et al., 2008; Wallace et al., 2010); however, little is known about the between-habitat (e.g., flowing vs. non-flowing water) or micro-habitat (e.g., emergent vs. submerged plants) distribution of this environmental pathogen. The findings that *M. ulcerans* is found associated with many aquatic invertebrate taxa, can be incorporated into the food web, and is associated with aquatic habitats in general suggest that additional research is needed to more thoroughly evaluate the role of biting water bugs in transmission. This would be useful for developing competing hypotheses regarding direct transmission from the water with no insect vectoring to correlations of BU and *M. ulcerans* positivity in adult mosquitoes that have been reported from southeast Australia (Fyfe et al., 2007; Johnson et al., 2007).

Most recently, there was a reported correlation between the detection of *M. ulcerans* in adult mosquitoes and BU incidence, suggesting a possible transmission cycle involving mammals and mosquitoes in Australia (Fyfe et al., 2010); however, this was only a correlation and deserves additional study to meet Koch's postulates (Merritt et al., 2010; Plowright et al., 2008). Additionally, in Australia, *M. ulcerans* has been detected from terrestrial animals including koalas (Mitchell et al., 1984), possums (Adjei et al., 2001; Fyfe et al., 2010), horses (Ameixa and Friedland, 2001), a cat (Ameixa and Friedland, 2001), and adult mosquitoes (Liaw et al., 2002; Sopoh et al., 2010; Teixeira et al., 2010). The apparent broad distribution of *M. ulcerans* within the environment and among many different plant and animal taxa had made the identification of specific ecological relationships and modes of transmission difficult until Fyfe et al. (2010) found that *M. ulcerans* was in high abundance (using quantitative polymerase chain reaction (qPCR)) in possum wounds, internal organs, and feces, suggesting an important role of terrestrial mammals in the ecology of this disease. However, studies of mosquitoes in Africa have not been completed and there have been no positive *M. ulcerans* findings associated with synanthropic mammals in the few such studies from Africa (Vandelannoote et al., 2010). All of these findings in both Africa and Australia also suggest that there may be different disease systems and modes of transmission between the continents, and this could be the reason that developing a comprehensive understanding of transmission and BU epidemiology has been challenging and complex.

44.4 Microbiology of *M. ulcerans*

M. ulcerans is a slow growing, acid-fast bacterium (Figure 44-2) with a temperature-restricted growth pattern. *M. ulcerans* grows optimally at 32°C with a

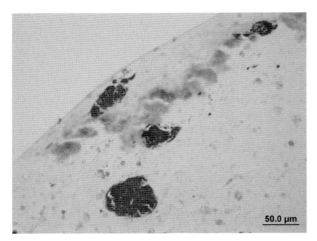

Fig. 44-2 *M. ulcerans* in infected tissue. *M. ulcerans* are acid-fast bacilli that are primarily extracellular.

doubling time ranging from 24 to 84 hours under laboratory conditions on standard mycobacterial media (Mve-Obiang et al., 2003). Because of the organism's slow growth, it is difficult to maintain samples for culture in areas where resources are limited. The temperature-restricted growth is consistent with the organism's inability to cause systemic infection in humans since the bacterium cannot grow at 37°C.

44.5 Molecular biology of *M. ulcerans*

Sequencing shows that *M. ulcerans* is very closely related to the fish pathogen *M. marinum* that causes an intracellular, granulomatous skin infection in humans. In fact, 90% of *M. ulcerans* coding DNA sequences (CDS) have orthologs in *M. marinum* with 98% DNA identity, although *M. ulcerans* was found to have a reduced genome (Stinear et al., 2007).

M. ulcerans is composed of a 5,631,606-bp chromosome that harbors 4160 protein-coding genes (compared to *M. marinum* with 5426) and 771 pseudogenes (Figure 44-3). Despite the close identity, a few differences distinguish *M. ulcerans* from *M. marinum*. The first is the presence of two mobile genetic elements, IS2404 and IS2606 found in high copy (209 copies and 83 copies, respectively) on the *M. ulcerans* chromosome, but absent from *M. marinum* genomes. The second is the presence of a large, 174 kb circular plasmid in *M. ulcerans* (Figure 44-4). This is the first virulence plasmid identified in a pathogenic mycobacterial species. Sixty per cent of the plasmid is composed of three polyketide synthase genes and three accessory genes that are responsible for mycolactone production (Stinear et al., 2005). There are also 26 copies of insertion sequences including IS2404 and IS2606 located on the plasmid. Additionally, 47 CDS have been identified that are potentially unique to *M. ulcerans* (Stinear et al., 2007).

Genome comparisons of *M. ulcerans* and *M. marinum* reveal 157 regions accounting for 1232 bp present in *M. marinum*, but absent in *M. ulcerans*. The majority of the regions of difference (RDs) are from

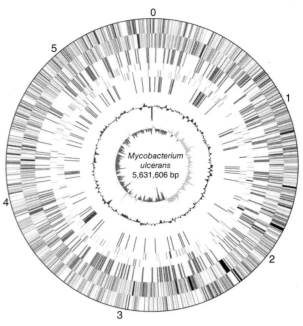

Fig. 44-3 Circular representation of the *M. ulcerans* chromosome. The scale is shown in megabases in the outer black circle. Moving inward, the next two circles show forward and reverse strand CDS, respectively, with colors representing the functional classification (red, replication; light blue, regulation; light green, hypothetical protein; dark green, cell wall and cell processes; orange, conserved hypothetical protein; cyan, IS elements; yellow, intermediate metabolism; gray, lipid metabolism; purple, PE/PPE). The location of each copy of IS2404 and IS2606 is then shown (cyan). The following two circles show forward and reverse strand pseudogenes (colors represent the functional classification), followed by the G+C content and finally the GC skew (G−C)/(G+C) using a 20-kb window. Image and description from Stinear et al. (2007). Copyright © Cold Spring Harbor Laboratory Press.

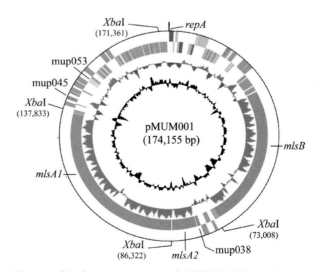

Fig. 44-4 Circular representation of pMUM001. The scale is shown in kilobases by the outer black circle. Moving in from the outside, the next two circles show forward and reverse strand protein-coding DNA sequences, respectively. These circles are followed by the GC skew (G–C)/(G+C) and finally the G+C content by using a 1-kb window. The arrangement of the mycolactone biosynthetic cluster has been highlighted, and the locations of all *Xba*I sites are indicated. Image and description is from Stinear et al. (2004). Copyright © National Academy of Sciences, USA.

deletions in *M. ulcerans*, accounting for 1064 kb as well as many DNA rearrangements (Stinear et al., 2007). Noteworthy deletions in *M. ulcerans* include the ESX loci. There are five complete ESX systems in *M. marinum*, but only three remain in *M. ulcerans*. In *M. tuberculosis*, the *esx1* locus encodes a novel protein secretion apparatus that contributes to its virulence and immunogenicity (Brodin et al., 2004). Effector proteins secreted by ESX systems belong to the ESAT-6 or the EspA families and 13 ESAT-6 proteins exist in *M. ulcerans*. But the *espA* gene, encoding a putative substrate of ESX-1, has been deleted. Loss of these systems, which trigger granuloma formation by *M. marinum* and *M. tuberculosis*, has been speculated to contribute to the primarily extracellular nature of *M. ulcerans* in infected tissue (Huber et al., 2008).

The presence of insertion sequences has led to genomic mutations in the form of deletions, insertions, inversions, and replicon fusions, often resulting in pseudogenes (Table 44-1). Inactive genes most represented among the pseudogenes are the Proline-Glutamate (PE) and Proline-Proline-Glutamate (PPE) multigenes. Members of these protein families contain mycobacteria conserved PE or PPE residues that are surface-associated cell wall proteins believed to provide adverse antigenic profiles and thus may play a role in immune invasion. This loss suggests a reductive evolution and move toward a particular niche, similar to that found in *M. leprae* that has lost almost all of its genes in this family (Cole et al., 2001).

Other gene inactivations have been found in *M. ulcerans* compared to those active in *M. marinum*. Some of these provide hints toward the *M. ulcerans* specialized environments. For instance, though intact in *M. marinum*, multiple mutations in the anaerobic respiration pathway suggest that *M. ulcerans* is incapable of growth under anaerobic conditions (Stinear et al., 2007). Growth of mycobacteria under

microaerophilic conditions involves the upregulation of a high oxygen affinity cytochrome *bd* oxidase (*cydABCD*). Though this locus is present in *M. ulcerans*, the *cydA* ortholog is a pseudogene. A knockout mutation of *cydA* in *M. smegmatis* leads to a competitive growth disadvantage under microaerophilic conditions (Kana et al., 2001), and suggests the same would occur with *M. ulcerans* grown under microaerophilic conditions. *M. ulcerans* has, however, maintained many (greater than 400 while *M. marinum* has greater than 590) enzymes involved in aerobic respiration, suggesting that, under aerobic conditions, a robust and complex respiratory potential is present.

Data from the *M. ulcerans* genome studies suggest recent evolution of *M. ulcerans* from a *M. marinum* progenitor and adaptation of *M. ulcerans* to a particular niche (Stinear et al., 2007). Data based upon RDs among geographically diverse *M. ulcerans* strains show further evolution resulting in five insertion–deletion haplotypes that separate into two distinct *M. ulcerans* lineages (Figure 44-5) (Kaser et al., 2007). Members of the ancestral lineage are more genetically related to the *M. marinum* progenitor in sequence composition and orientation showing only minor changes such as single nucleotide polymorphisms (SNPs), small deletions, or sequence variation. Members of the ancestral lineage include human isolates from Asia, South America, and Mexico, as well as isolates from mycolactone-producing mycobacteria (MPMs) causing disease in ectotherms such as fish and frogs (Kaser et al., 2007). The classical lineage comprises isolates from Africa, Australia, Malaysia, and Papua New Guinea. These isolates are among the most pathogenic and show major genomic polymorphism, including rearrangements caused by IS elements when compared to the *M. marinum* progenitor.

Though methods such as the detection of amplified fragment length polymorphisms (AFLP), multilocus

Table 44-1 Representative pseudogenes found in *M. ulcerans* genome, their location and function, and type of mutation rendering them inactive.

Gene/locus name	Location in genome	Function	Type of mutation
cydA	173,992–174,1378	Integral membrane cytochrome D ubiquinol oxidase: involved in growth of mycobacteria under microaerophilic conditions	N-term disrupted by internal stop codon
emrB	78,290–79,917	Translocase involved in export of multiple drugs. Thought to be involved in multidrug resistance	Frame shift mutation
esxS	2,496,251–2,496,546	ESAT-6 family protein thought to be involved in immunogenicity and virulence	Frame shift mutation
arsC	3,661,459–3,663,077	Involved in arsenic export	Frame shift mutation
arsB1	3,696,735–3,698,021	Involved in arsenic export	Stop codon
nark1		Involved in nitrate reduction	C-term truncated (missing 437 aa)
cstA	3,790,554–3,792,832	Peptide utilization during carbon starvation	Frame shift has removed 16 aa from N-term
MUL_4267	4,736,890–4,737,470	Formate dehydrogenase α subunit: potential role in growth under anaerobic conditions	Frame shift mutation
MUL_4268	4,737,567–4,740,083	Formate dehydrobenase β subunit: potential role in growth under anaerobic conditions	Frame shift mutation

Data are from *Mycobacterium ulcerans* Agy99, complete genome Genbank CP000325.1. http://www.ncbi.nlm.nih.gov/nuccore/CP000325.1.

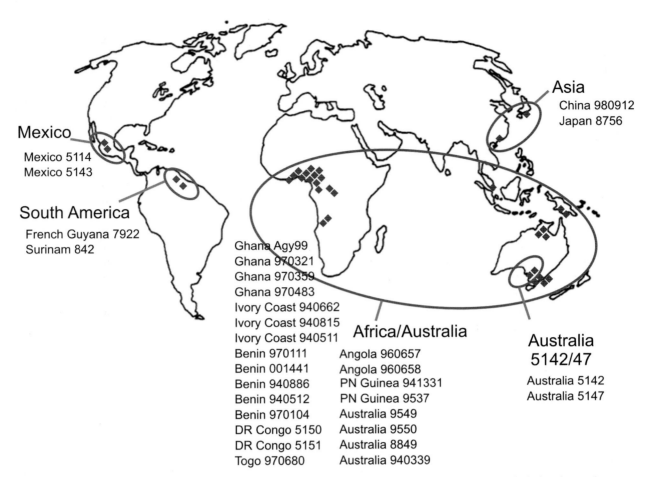

Fig. 44-5 Geographical distribution of the five *M. ulcerans* haplotypes. The origin of *M. ulcerans* strains included in this study is shown in the world map, with each dot representing one patient isolate. The five InDel haplotypes are encircled. Image from Kaser et al. (2007).

sequence analysis, and PCR of the intragenic regions between insertion sequences IS*2404* and IS*2606* have detected genetic variability among *M. ulcerans* strains, these methods could not distinguish between different strains from the same geographical region (Chemlal et al., 2001; Jackson et al., 1995; Stinear et al., 2000). Strain discrimination has been achieved using variable-number tandem repeat (VNTR) analysis, and more recently, detection of SNPs. Targeting two VNTR loci, three allele patterns were found from *M. ulcerans* clinical isolates from Ghana (Hilty et al., 2006). Recently, SNP analysis was used, which provided enough resolution to identify ten *M. ulcerans* haplotypes within a relatively small and highly endemic region of Ghana (Roltgen et al., 2010). Those results suggest great promise for the use of SNP analysis for future micro-epidemiological studies of *M. ulcerans* transmission.

44.6 *M. ulcerans* ecovars and the *M. ulcerans M. marinum* complex

Recently, other closely related MPMs have been isolated from aquatic environmental sources. *M. liflandii* was isolated from *Xenopus tropicalis* and *X. laevis* frogs imported into the United States from West Africa. *M. liflandii* produces a unique mycolactone,

mycolactone E (Figure 44-6) (Godfrey et al., 2007; Mve-Obiang et al., 2005; Trott et al., 2004). *M. pseudoshottsii*, isolated from striped bass in the Chesapeake Bay, produces mycolactone F (Rhodes et al., 2005) as do mycolactone-producing *M. marinum* strains isolated from diseased fish in both the Red and Mediterranean Seas (Ranger et al., 2006; Ucko and Colorni, 2005) (Figure 44-6). These organisms grow at lower temperatures than do *M. ulcerans* and are not known to infect humans.

The close phenotypic and genotypic relationship between *M. marinum*, *M. ulcerans*, and other closely related MPMs has prompted researchers in the field to define these as members of a *M. ulcerans*–*M. marinum* complex, and MPMs are now further characterized as *M. ulcerans* ecovars (Doig et al., 2012; Pidot et al., 2010a; Tobias et al., 2013).

44.7 Pathogenesis and immunity

There are several factors that potentially influence the pathogenicity of *M. ulcerans*. The first is the slow growth rate. Most pathogenic mycobacteria are characterized by their slow growth, including M. tuberculosis, M. leprae, M. bovis, M. avium, and *M. ulcerans*. Incubation of these organisms takes weeks to months for visible colonies to appear, though the exact link

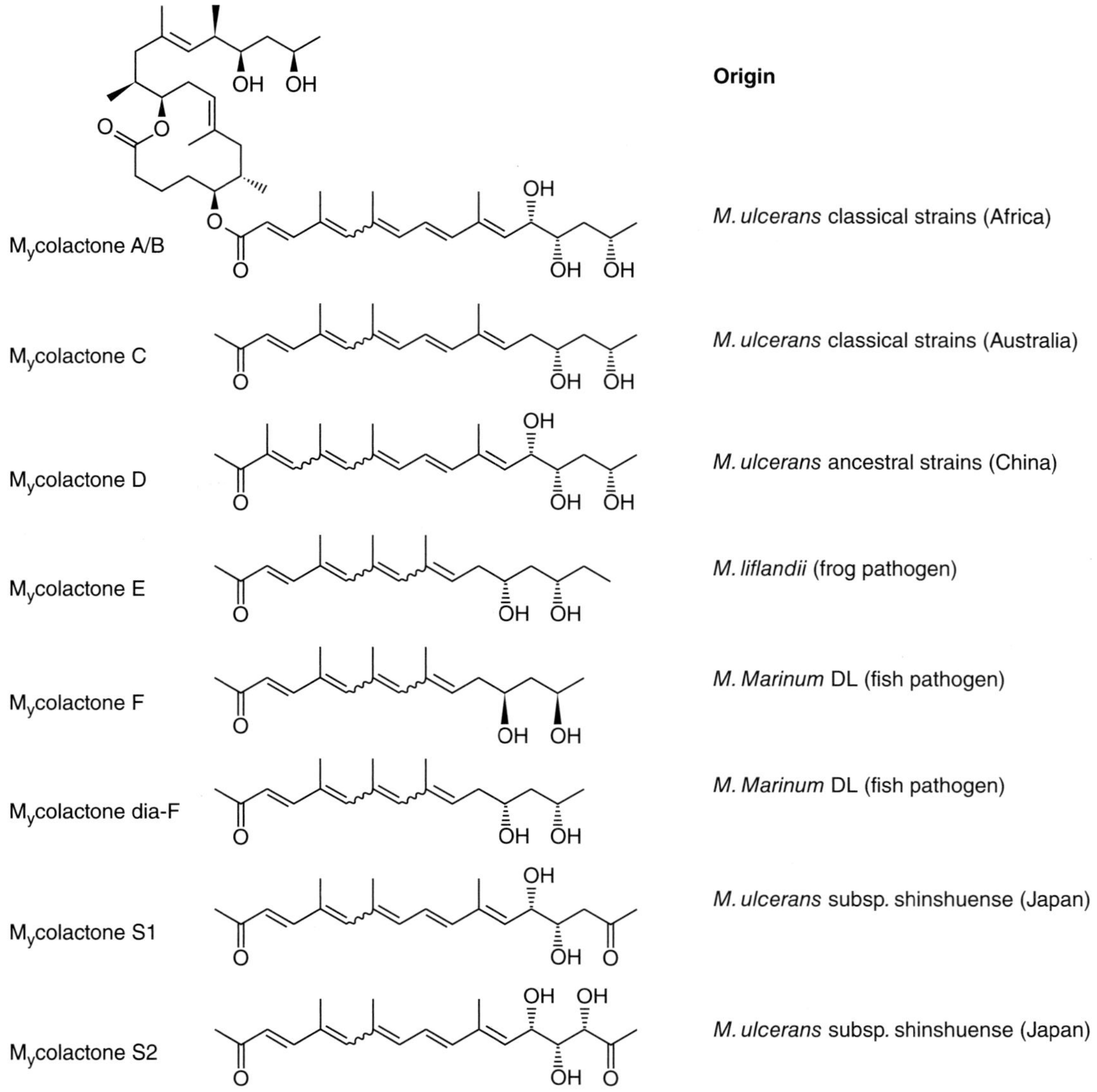

Fig. 44-6 The structure of naturally occurring mycolactones. The complete structure of mycolactone A/B is shown, which exists in nature as 3:2 ratio of Z-/E-isomers of the C4–C5 bond in the long polyketide side chains. Since it is this long chain that varies between geographically diverse clinical isolates (mycolactones A–D) and other non-pathogenic MPM (mycolactones E, F and S1, S2), these chains alone are shown.
Source: Hall and Simmonds (2014). Copyright © The Biochemical Society.

to pathogenesis is not clear (Hett and Rubin, 2008). Though the incubation period is not clearly defined, the slow growth of *M. ulcerans* leads to several months of incubation, depending on infective dose, before the presentation of clinical signs of disease, as well as a resulting slow progression of lesions (van der Werf et al., 2005). The second is the low growth temperature that restricts *M. ulcerans* growth to primarily cooler parts of the body, particularly the skin, and limits systemic infection (Anonymous, 1970; Johnson et al., 2005). The third and major pathogenicity factor of *M. ulcerans* is the presence of the lipid toxin mycolactone. Details of mycolactone are discussed in the next section.

44.8 Mycolactone: the sole virulence determinant

The existence of a factor that contributed to the virulence of *M. ulcerans* was postulated as long ago as

1965 (Connor and Lunn, 1965), but was first assumed to be a secreted protein toxin. Sterile filtrates of liquid cultures were shown to induce rounding up, detachment, and death of L929 fibroblasts in culture (Hockmeyer et al., 1978), and also to inhibit the proliferation of T cells and phagocytosis by murine macrophages (Pimsler et al., 1988). The cytotoxic activity and an ability to suppress cytokine production were, subsequently, shown to reside in the alcohol-soluble lipids (ASLs) extracted from *M. ulcerans* culture supernatants (George et al., 1998; Pahlevan et al., 1999). But it was not until 1999 that the toxin was isolated to purity by Pamela Small's group, showing that *M. ulcerans* produces a toxic macrolide, called mycolactone, that diffuses through infected tissue resulting in pathology extending further than the site of bacterial colonization (George et al., 2000). Furthermore, these studies showed that purified mycolactone injected into guinea pigs is sufficient to cause lesions similar to wild-type

M. ulcerans (George et al., 1999), while a mycolactone-negative *M. ulcerans* mutant is avirulent (George et al., 1999). A later study showed that chemical complementation of avirulent, mycolactone-negative *M. ulcerans* mutants restores wild-type *M. ulcerans* pathology, suggesting that mycolactone was the sole virulence determinant (Adusumilli et al., 2005).

Mycolactone is a highly hydrophobic, polyketide-derived lipid toxin whose structure consists of a 12-membered macrolide core structure and two fatty acid side chains (George et al., 1999). Most strains of *M. ulcerans* make a range of mycolactone-like molecules (Mve-Obiang et al., 2003). Though the core structure is invariant in all isolates of *M. ulcerans*, variation in the side chain and the location of hydroxyl groups and double bonds within lead to differing abilities to induce cellular cytotoxicity and suppress cytokine production (Hong et al., 2008b; Kishi, 2011; Mve-Obiang et al., 2003). A recent study where synthetic mycolactones A/B, C, and F had differing LC_{50} values showed that loss of the long polyketide side chain renders the compound effectively inert (Scherr et al., 2013). On the other hand, modifications of the short polyketide chain had a less profound influence over cytotoxic activity (Scherr et al., 2013).

The most virulent strains of *M. ulcerans* make a 3:2 ratio of mycolactone A/B (*Z*-/*E*-isomers of the C4–C5 bond in the long polyketide side chains) and most research into the biological functions of mycolactone has focused on this. A number of naturally occurring variants of mycolactone are produced by clinical isolates from geographically diverse areas as well as from other non-pathogenic MPM (Figure 44-6). Isolates from Africa and Australia produce mycolactone A/B and C, respectively, and have been found to be the most cytotoxic to cells (Mve-Obiang et al., 2003). Isolates from Asia, which produce mycolactone D, are less cytotoxic as compared to mycolactones A/B and C producing strains (Hong et al., 2008a; Kishi, 2011; Mve-Obiang et al., 2003). Mycolactone E is produced by *M. liflandii* and *M. pseudoshottsii* strains, and mycolactone E by *M. marinum* strains. These mycolactones have been found to be cytotoxic to human cells *in vitro*, though these *M. ulcerans* ecovars are not known to cause infections in humans (Hong et al., 2008b; Ranger et al., 2006).

Genes for mycolactone synthesis form a 110-kb cluster on a large 174-kb plasmid (pMUM; Figure 44-4) (Stinear et al., 2005). The *mls* genes encoded in pMUM have been reported to have a strong sigA-type promoter that is constitutively expressed during planktonic monoculture under laboratory conditions (Tobias et al., 2009). But how this relates to expression by *M. ulcerans in* its environmental niche (itself not clearly defined) or during different stages of clinical ulcers is not known. It is possible that mycolactone production is regulated *in vivo*. Though mycolactone has been successfully isolated from ulcer exudates of patients who have completed a sterilizing course of antibiotic therapy, treatment of patients with advanced ulcers leads to a switch

from a non-inflammatory immunohistochemistry to a granulomatous pattern more similar to that seen in other mycobacterial infections (such as *M. tuberculosis*, *M. leprae*, and *M. marinum*). This suggests the possibility that rifampicin/streptomycin are able to switch off mycolactone production as part of the mycobacteriocidal activity (Schutte and Pluschke, 2009).

Mycolactone largely resides in the extracellular matrix that forms around growing bacterial colonies/aggregates (biofilms) (Marsollier et al., 2007). It is assumed that mycolactone accesses mammalian cells by means of passive diffusion due to its hydrophobic nature. Bodipy-labeled mycolactone has been detected in the cytosol of cells within 2 minutes of exposure, but interestingly is excluded from the nucleus (Chany et al., 2011; Snyder and Small, 2003). Other *in vivo* experiments have demonstrated that mycolactone can diffuse away from the site of injection and can access all bodily compartments with the exception of the brain (Hong et al., 2008a). In a patient study, mycolactone could be detected in the serum of three out of six BU patients prior to starting antibiotic therapy (Sarfo et al., 2011). Thus, it seems that mycolactone can have wide-reaching systemic, yet not universal, effects.

44.8.1 Cytotoxic activity

At high doses (usually considered to be above 1 μg/mL, around 1.3 μM) and/or prolonged exposure times (>24–48 hours), mycolactone is generally cytotoxic to mammalian cells. The most sensitive cells identified to date are L929 murine fibroblasts, with LD_{50} in a low nanomolar range, and these are frequently used as reporter cells. DNA content analysis showed that mycolactone induces cell cycle arrest at G1/G0 (George et al., 1999). The cells also undergo cytoskeletal rearrangements and become apoptotic. Interestingly, different types of cells show a wide range in different susceptibility to mycolactone-induced cytotoxicity. For instance, purified human monocytes (Simmonds et al., 2009) show no significant reduction in viability after 24 hours exposure, whereas immature dendritic cells (DCs) are much more sensitive, but only after 48 hours (Coutanceau et al., 2007). Even prolonged (>24 hours) exposure of PBMCs (Boulkroun et al., 2010), mature DCs (Coutanceau et al., 2007), Jurkat T cells (Boulkroun et al., 2010) and myeloid cell lines, such as J774 (human macrophage-like) (George et al., 2000), and CLC (adherent carp monocytic) cells (Mosi et al., 2012) causes only minimal cytotoxicity, growth inhibition, or apoptosis.

Myocytes, keratinocytes, adipocytes, fibroblasts, and Schwann cells are all sensitive to mycolactone (Bozzo et al., 2010; Dobos et al., 2001; En et al., 2008; Houngbedji et al., 2011), although adipocytes in particular seem to require extended exposure times for induction of cell death (Dobos et al., 2001). The neurotoxicity has been a suggested mechanism for the painlessness of BU ulcers although the connection between this and the cell cycle arrest seen in actively growing cell populations is unclear (En et al., 2008). Even though the keratinocytes at the basal layer of epidermis of human skin are rapidly dividing and are reportedly sensitive to mycolactone, the epidermis is often observed to be undergoing hyperplasia during BU (Guarner et al., 2003). On the other hand, the relatively resistant adipocytes do not divide beyond puberty but are nevertheless killed in BU, providing the classic "ghost-cell" appearance of coagulative necrosis seen in histological specimens and indicating that they must be susceptible to mycolactone's effects, either directly or indirectly.

The mechanism of cell death as a consequence of mycolactone exposure in *M. ulcerans* infection appears to be more complex than simply a matter of cell cycle arrest. In fact, recent work by Caroline Demangel's group has shed light on a mechanism of mycolactone-

induced cell death (Guenin-Mace et al., 2013). Here, low doses of mycolactone were shown to cause cells to rapidly form filopodia, clusters of actin bundles associated with wound healing and cell–cell interactions. This was found to be due to mycolactone-dependent enhancement of actin polymerization following its binding to the GTPase domain of Wiskott–Aldrich syndrome protein (WASP) family proteins. The inappropriate activation of WASP and subsequent relocalization of the actin nucleating complex Arp2/3 offer an explanation for the well-characterized cytoskeletal rearrangement in mycolactone-treated cells. These were associated with a mycolactone-induced loss of cell adhesion and e-cadherin-dependent tight junctions, which could be at least partially restored by the addition of the WASP inhibitor, wiskostatin. Interestingly, e-cadherin itself appears to be downregulated by mycolactone (Guenin-Mace et al., 2013; Table 44-2). It therefore appears that, at least *in vitro*,

Table 44-2 Proteins shown to be downregulated by mycolactone

Protein downregulated	Cell type	Function	References
Membrane proteins			
CD3	T cell	TCR signal transduction	Boulkroun et al. (2010)
CD25	DC	IL-2 receptor α chain, DC maturation marker	Coutanceau et al. (2007)
CD40 (partial)	DC	Co-stimulatory protein	Coutanceau et al. (2007)
CD62-L	T cell	Adhesion, T cell homing	Guenin-Mace et al. (2011)
CD80 (partial)	DC	Co-stimulatory protein	Coutanceau et al. (2007)
CD83	DC	Dendritic maturation marker	Coutanceau et al. (2007)
CD86	DC	Co-stimulatory protein	Coutanceau et al. (2007)
e-Cadherin	HeLa	Adhesion molecule	Guenin-Mace et al. (2013)
LFA-1	T cell	Adhesion, T cell homing	Guenin-Mace et al. (2011)
MHCII	DC	Antigen presentation molecule	Coutanceau et al. (2007)
T cell receptor	T cell	Antigen recognition	Boulkroun et al. (2010)
Cytokines			
IL-1β (partial)	Mo Mφ DC	Cytokine (proinflammatory)	Coutanceau et al. (2007), Simmonds et al. (2009)
IL-2	T cell	Cytokine (T cell proliferation and differentiation)	Boulkroun et al. (2010), Hong et al. (2008a), Pahlevan et al. (1999), Phillips et al. (2009)
IL-4	T cell	Cytokine (Th2 responses)	Phillips et al. (2009)
IL-6	Mφ Mo DC	Cytokine (proinflammatory)	Coutanceau et al. (2007), Simmonds et al. (2009)
IL-10	T cell Mo Mφ	Cytokine (Anti-inflammatory, Th2 responses)	Pahlevan et al. (1999), Simmonds et al. (2009)
IL-12 (partial)	DC	Cytokine	Coutanceau et al. (2007)
IL-17	T cell	Cytokine (proinflammatory)	Phillips et al. (2009)
Interferon-γ	T cell	Cytokine (proinflammatory)	Phillips et al. (2009)
TNF-α	T cell Mo Mφ	Cytokine (proinflammatory)	Pahlevan et al. (1999), Phillips et al. (2009), Simmonds et al. (2009), Torrado et al. (2007a)
Chemokines			
IL-8	T cell Mo Mφ	Chemokine	Phillips et al. (2009), Simmonds et al. (2009)
IP-10	Mo Mφ DC	Chemokine (reporter for IFN-α/β production)	Coutanceau et al. (2007), Simmonds et al. (2009)
MCP-1	DC	Chemokine	Coutanceau et al. (2007)
MIP-1α	DC	Activates granulocytes, chemokine	Coutanceau et al. (2007)
MIP-1β	T cell	Activates granulocytes, chemokine	Phillips et al. (2009)
RANTES	DC	Chemokine	Coutanceau et al. (2007)
Intracellular			
COX-2	Mo Mφ	Prostaglandin synthesis	Simmonds et al. (2009)

CD, cluster of differentiation; COX-2, cycolooxygenase-2; DC, dendritic cell; IL, interleukin; MCP-1, monocyte chemotactic protein-1; MIP-1α, macrophage inflammatory protein-1α; Mo, monocytes; Mφ, macrophage; RANTES, regulated upon activation normal T cell expressed and presumably secreted; TNF-α, tumour necrosis factor α.
Source: Hall and Simmonds (2014). Copyright © The Biochemical Society.

cell death may be primarily related to anoikis, that is, apoptosis due to detachment.

44.8.2 Immunosuppressive activity

Histopathology of BU sections shows abundant foci of extracellular acid-fast bacilli in the absence of an inflammatory infiltrate. Instead, mononuclear cells are mostly found in the periphery of the infection rather than the regions close to the bacilli, suggesting that the signals required to access the infection are missing (Hong et al., 2008a; Oliveira et al., 2005; Torrado et al., 2007b). It is now generally accepted that mycolactone is the responsible immunosuppressive factor. Immunosuppressive doses of mycolactone are generally lower than cytotoxic doses on the same cell type, and are certainly faster acting. For instance, primary human monocytes begin to produce tumor necrosis factor (TNF) within 40 minutes of exposure to an inflammatory agent and mycolactone effectively obliterates this at a tenth of the cytotoxic dose (Simmonds et al., 2009).

Mycolactone can suppress both innate and adaptive immune responses in patients and in experimentally infected mice. Innate responses are suppressed by restricting many of the functions of tissue-resident macrophages. Like other mycobacterial infections, *M. ulcerans* can be phagocytosed by macrophages in the early stages of infection (Coutanceau et al., 2005), presumably escaping digestion by preventing phagolysosome formation similarly to *M. tuberculosis*. Over short time periods and at low multiplicity of infections (MOIs), it can proliferate inside murine macrophages, until mycolactone causes the apoptosis/necrosis of the macrophages and the release of the bacilli into the intercellular space (Torrado et al., 2007b). However, it has also been shown *in vitro* that wild-type strains of *M. ulcerans* are not phagocytosed as effectively as mycolactone-negative strains and exogenous mycolactone has the ability to complement this activity (Adusumilli et al., 2005; Coutanceau et al., 2005).

Monocytes, macrophages, T cells, and DCs exposed to mycolactone-producing strains of *M. ulcerans* also secrete dramatically less TNF than those encountering mycolactone-negative strains (Pahlevan et al., 1999; Phillips et al., 2009; Simmonds et al., 2009; Torrado et al., 2007a). Stimulation of production of other cytokines and chemokines (including IL-6, IL-8, IL-10, and IP-10) is also substantially reduced in both primary human monocytes and macrophages, as is expression of the pro-inflammatory mediator cyclooxygenase-2 (COX-2) (Table 44-1; Simmonds et al., 2009). Failure to produce cytokines and chemokines at the site of infection is considered a likely explanation for the lack of an inflammatory infiltrate close to bacilli (Silva et al., 2009); however, the technical expertise to measure or detect such low levels of mycolactone *in situ* is still lacking.

In addition to inhibiting macrophage function, mycolactone also inhibits DC maturation, and suppresses DC chemokine production (Coutanceau et al., 2007), possibly explaining the anergy that has been reported in patients, where their T cells no longer respond to mycobacterial antigens after exposure (Gooding et al., 2003). This mycolactone-mediated inhibition could also contribute to the reduced Th1/Th2/Th17 responses seen in whole blood restimulations (Phillips et al., 2009). T cells themselves are also directly sensitive, as low doses of mycolactone directly block the mitogen- and CD3/CD28-depedent production of a range of cytokines including IL-2, IL-4, IL-8, IL-10, IL-17, IFN-γ, and MIP-1β from CD4+ T cells (Phillips et al., 2009). Furthermore, lymphocyte homing to peripheral lymph nodes is inhibited due to a downregulation of L-selectin and LFA-1 on the surface of T cells (Guenin-Mace et al., 2011).

44.8.3 Mechanism underlying mycolactone-dependent inhibition of protein production

It is predicted that the proteins affected by mycolactone may share an underlying mechanism of production that could explain their co-regulation. Early work on the mechanism of suppression by mycolactone was carried out in Jurkat T cells using ASLs and focused on the suppression of IL-2 production (Pahlevan et al., 1999). In this cell line, mycolactone strongly inhibited the activation of NFκB following TNF stimulation, which led to the initial assumption that this was a generalizable mechanism. However, these authors found no comparable effect on NFκB activation in other model systems (phorbol 12-myristate 13-acetate (PMA) induced in Jurkat cells, IL-1 induced in HeLa cells, and lipopolysaccharide (LPS) induced in RAW264.7 cells) (Pahlevan et al., 1999). Indeed, the effect of mycolactone on IL-2 mRNA levels turns out to be dependent on the stimulus, being substantially reduced during the CD3/CD28 response and enhanced for PMA/ionomycin (Boulkroun et al., 2010). More recently, transcriptional profiling of primary human T cells treated with mycolactone found no significant changes in mRNA level (Guenin-Mace et al., 2011). The situation is similar in myeloid cells since a detailed investigation of the effect of mycolactone on the LPS-dependent upregulation of TNF, IL-6, and COX-2 showed that while protein production is completely ablated by low immunosuppressive (non-cytotoxic) doses of mycolactone, the induced mRNA levels are not affected or are even enhanced in the case of TNF (Simmonds et al., 2009). This is true for primary human monocytes (Simmonds et al., 2009), RAW264.7 cells (Coutanceau et al., 2005), primary human macrophages, and HeLa cells. Thus, it seems that, with the exception of the transformed Jurkat T cell line, transcriptional regulation cannot explain the immunosuppression seen in mycolactone-exposed cells.

Detailed analyses of the signaling pathways in primary human monocytes led to the first proposal of a post-transcriptional mechanism for mycolactone

Table 44-3 Proteins known to be unaffected or upregulated by mycolactone

Protein unaffected	Cell type	Function	References
Actin	Mo, HeLa	Cytoskeleton	Guenin-Mace et al. (2013), Simmonds et al. (2009)
CD28	T cell	Costimulatory receptor	Boulkroun et al. (2010)
eIF2α	T cell, Mo	Translation initiation	Boulkroun et al. (2010), Simmonds et al. (2009)
fyn	T cell	Signaling molecule	Boulkroun et al. (2010)
IκBα	Mo	Signaling molecule (inhibition of NFκB)	Simmonds et al. (2009)
LAT	T cell	Signaling molecule	Boulkroun et al. (2010)
lck	T cell	Signaling molecule	Boulkroun et al. (2010)
p34 Arp2/3	HeLa	Cytoskeletal regulator	Guenin-Mace et al. (2013)
p38 MAPK	T cell, Mo	Signaling molecule	Boulkroun et al. (2010), Simmonds et al. (2009)
PLCγ1	T cell	Signaling molecule	Boulkroun et al. (2010)
N-WASP	HeLa	Cytoskeletal regulator	Guenin-Mace et al. (2013)
ZAP70	T cell	Signaling molecule	Boulkroun et al. (2010)
CCR7	T cell	Chemokine receptor	Boulkroun et al. (2010)
MuRF-1	Muscle	Muscle-specific E3 ubiquitin ligase	Houngbedji et al. (2011)
Atrogin-1	Muscle	Muscle-specific E3 ubiquitin ligase	Houngbedji et al. (2011)

CD, cluster of differentiation; DC, dendritic cell; Mo, monocytes; Mφ, macrophage.

Source: Hall and Simmonds (2014). Copyright © The Biochemical Society.

function, probably at the level of translation (Simmonds et al., 2009). This study revealed that neither mitogen-activated protein kinase (MAPK) nor nuclear factor kappa-light-chain-enhancer of activated B cells (NFκB) signaling was affected in these cells, nor did the induced proteins appear to be trapped inside the cells. The enhanced or unaffected transcription would also rule out an effect on mRNA stability – indeed, even inhibitors such as SB203080, which targets p38 MAPK, and which regulates pro-inflammatory protein production by stabilizing AU-rich element containing mRNAs, cause only partial inhibition of TNF production (Brook et al., 2000), in contrast to the complete block seen following mycolactone exposure. While it is clear that mycolactone is acting post-transcriptionally, mycolactone is not a general inhibitor of protein production in monocytes, since it does not suppress total protein synthesis. Total incorporation of ^{35}S-labeled Met/Cys is not altered by the presence of mycolactone (Simmonds et al., 2009) and neither are the general pathways known to control translation: mycolactone does not affect phosphorylation of the mammalian target of rapamycin complex 1 (MTORC1) targets eIF4E or p70s6 kinase. Neither is phosphorylation of the endoplasmic reticulum stress response marker eIF2α altered in mycolactone-treated monocytes (Simmonds et al., 2009). These findings, later confirmed in T cells (Boulkroun et al., 2010), suggest that mycolactone is selectively targeting a subset of genes for control by a novel mechanism. A small yet significant decrease in let-7b levels has been observed in T cells exposed to mycolactone. Surprisingly, inhibition of let-7b caused a decrease in L-selectin mRNA and overexpression increased it, suggesting that L-selectin is not a direct target of let-7b (since the opposite pattern would have been expected). However, the findings were not confirmed at the protein level.

Taken together, the loss of key proteins involved in the immune response and cellular adhesion would appear to be driving the both immunosuppression and tissue necrosis in BU (Table 44-2). Indeed, when all available literature are surveyed, only three proteins could be identified that are upregulated by mycolactone (Table 44-3); the chemokine receptor CCR7 (Guenin-Mace et al., 2011) and two muscle-specific E3 ubiquitin ligases (MuRF-1 and Atrogin-1) (Houngbedji et al., 2011). Most proteins studied appear to escape mycolactone-dependent regulation (notably constitutive expression of actin, CD28, p38 MAPK, and eIF2α, and also the LPS-dependent resynthesis of IκBα (Table 44-2)) and total cellular protein production rates are not suppressed (Simmonds et al., 2009). However, it is currently unclear how the suppression of constitutively expressed proteins such as e-cadherin and CD3 fits with our current models of mycolactone function. While the recent finding of a direct interaction between mycolactone and WASP is exciting, it may not explain the inhibition of cytokine production. Instead, the cellular compartments of the sensitive and insensitive proteins may be informative when considering a unifying mechanism for this fascinating pleiotropic molecule.

44.9 Immunity

44.9.1 Cell-mediated immunity

Despite the immunosuppressive properties of mycolactone, several studies have been conducted that point to a mounted and modulated cell-mediated response to *M. ulcerans* infection. In humans, histopathological examination of nodules in the early stages of disease shows extensive necrosis with many aggregates of extracellular, acid-fast bacilli and very little inflammation. However, in older lesions, acid-fast bacilli are scarce, and both acute and chronic inflammations are possible, sometimes including granuloma formation (Dodge and Lunn, 1962;

Guarner et al., 2003; Hayman, 1993), suggesting the involvement of cell-mediated immunity in healing.

Pro-inflammatory cytokines are expressed at lower levels compared to healthy controls at various stages of the infection, suggesting an immune response similar to that seen from other mycobacteria (Appelberg, 2006; Ciaramella et al., 2002; Krutzik et al., 2003). Both IFN-γ and IL-10 are expressed in active BU nodules and ulcers. IFN-γ expression is modulated during BU progression, and is highest in patients with ulcerative and healed lesions compared to those with early disease (Phillips et al., 2006a, 2006b; Westenbrink et al., 2005; Yeboah-Manu et al., 2006; Zavattaro et al., 2010). In some cases, IFN-γ has been found expressed in nodules during early infection and correlated histologically with the presence of neutrophilia as a marker of an acute inflammatory response. This suggests an initial mounting of an early TH1 response (Phillips et al., 2006b), prior to mycolactone-mediated inhibition during the active phase of disease. *M. ulcerans* lesions with granulomas express significantly high levels of IFN-γ and IL-8. The expression of these cytokines has frequently been associated with granuloma formation in other mycobacterial diseases such as tuberculosis and leprosy (Ameixa and Friedland, 2002; Cavalcanti et al., 2012; Wang et al., 2013). IL-8 expression is also significantly associated with ulcers showing neutrophilia, and has been found to be the strongest among patients who had received antibiotic therapy, suggesting the development of specific immune responses once therapeutic treatments have limited the production of the toxic mycolactone. TNF expression was also detected at low levels in *M. ulcerans*-infected tissues (Phillips et al., 2006b; Westenbrink et al., 2005; Zavattaro et al., 2010). TNF is an effector cytokine produced mainly by macrophages and has an autocrine effect on the activation of the macrophage's microbicidal activity against intracellular parasites (Nacy et al., 1991). These results suggest that the patients were able to mount an early Th1 response to *M. ulcerans* despite the presence of mycolactone that is known to influence cytokine production *in vivo*.

There is also evidence suggesting delayed-type hypersensitivity in BU patients. Intradermal injection of *M. ulcerans* sonicate (the burulin skin test) showed that those patients with early disease did not react, whereas a positive response was elicited from 92% of patients with lesions, indicating a T-cell sensitization (Stanford et al., 1975). Also, some patients do spontaneously heal and this is associated with a delayed-type hypersensitivity response (Dobos et al., 2000).

44.9.2 Antibody-mediated immunity

It is not clear whether antibodies play a protective role against BU. The precise antigens that induce a protective immune response are poorly defined. Antibody responses against *M. ulcerans hsp* (homologous to *M. leprae* 18-kDa heat shock protein) can be used as a serological marker for exposure to *M. ulcerans* (Diaz et al., 2006; Pidot et al., 2010b, 2010c). Though *hsp* is expressed in considerable amounts in *M. ulcerans*, and is immunogenic for both B and T cells in mice, vaccination limited only weakly the progression of experimental infection. IgM antibody responses against *M. ulcerans* culture filtrate proteins can be detected from sera in 85% of BU patients, and only in a small proportion of healthy controls (Okenu et al., 2004).

Antigen 85 is a major secreted component in the culture filtrate of many mycobacteria. In *M. tuberculosis*, it is involved in cell wall synthesis and is one of the most promising tuberculosis vaccine candidates (White et al., 2013). Vaccination using *M. ulcerans* 85A followed by boosting was effective in reducing 100-fold the bacterial number and offered more protection than BCG vaccination (Tanghe et al., 2001, 2008). But the protection was not sterilizing and eventually all mice developed footpad swelling. There was a reduced or delayed mycolactone production, but mycolactone was not completely blocked (Tanghe et al., 2008).

Inoculation of small doses of *M. bovis* BCG protects to some extent against infection with *M. ulcerans*, but a booster does not increase the protective effect (Tanghe et al., 2007). Protection is also short lived, but does seem to exert a significant protection against osteomyelitis. DNA vaccination induces a broader T cell repertoire where more protein epitopes are recognized (Tanghe et al., 2008). Cross reactivity was observed with partial specificity of some epitopes. Combination with vaccines targeting mycolactone or with vaccines targeting enzymes involved in mycolactone synthesis may be a way to strengthen its efficacy.

44.9.3 Innate susceptibility

Factors such as genetics, family relationships, and joint human activities are just beginning to be explored, and genetic host susceptibility factors such as those found for tuberculosis and leprosy have been hypothesized to play an important role (Stienstra et al., 2001). A large scale, cross-sectional study with matched controls in Ghana found that genetic polymorphisms in the macrophage protein gene SLC11A1 (NRAMP1) provided a 13% population attributable risk to BU (Stienstra et al., 2006). Polymorphisms in this gene are also positively associated with tuberculosis and leprosy (Liaw et al., 2002; Teixeira et al., 2010). A study in Benin found risk three times higher in cases than in controls with daily contact with natural water sources and five times higher for those who had a history of BU in their family. It was unclear, however, from this study whether genetics or behavior played a larger role in acquiring the disease (Sopoh et al., 2010).

44.10 Clinical presentation

In infected humans, the organism is found primarily in necrotic cutaneous tissues devoid of inflammatory cells (Adusumilli et al., 2005). Though recent studies have shown phases of both intracellular and extracellular survival and reproduction (Coutanceau et al., 2005; Torrado et al., 2007b), the infection is primarily extracellular in contrast to other mycobacterial infections that are primarily intracellular (Figure 44-2). Cross reactivity with mycobacterial antigens from environmental mycobacteria as well as with *M. tuberculosis* makes the creation of serological tests for *M. ulcerans* difficult where tuberculosis is also endemic (Diaz et al., 2006).

The clinical features of BU depend on disease stage. Early infection presents as one of the following non-ulcerative forms depending upon geographical location of the patient: papule, nodule, plaque, and edema. In Australia, a papule usually indicates the first sign of disease, whereas in Africa, the first sign is usually a nodule (van der Werf et al., 2005). Late-stage infection is signified by the presence of an ulcer.

Non-ulcerative forms can be identified by the following clinical features (Figure 44-7) (Adjei et al., 2001; World Health Organization, 2013a, 2013c):

- **Papule**: painless, raised skin lesion, usually <1 cm in diameter. Surrounding skin is commonly erythematous, with no apparent involvement of subcutaneous fat.
- **Nodule**: Both the skin and subcutaneous tissue are involved. It appears as a discrete painless, raised lump, approximately 1–2 cm in diameter. Commonly it has a central punctuate mark. Again the overlying skin is erythematous secondary to inflammation.
- **Plaque**: a firm, discrete area of induration, greater than 2 cm in diameter. The overlying skin is often discolored or erythematous in response to infection.
- **Edema**: diffuse, ill-defined, extensive, firm, usually non-pitting swelling of all or part of the limb affected by *M. ulcerans*. There may be associated changes in color in the overlying skin (unlike other non-ulcerative forms, edema can persist in the presence of an ulcer).

M. ulcerans releases mycolactone toxin, which preferentially necroses fat. As infection progresses, mycolactone necroses subcutaneous fat resulting in destruction of the overlying skin. In this way, the aforementioned non-ulcerative lesions break down and ulcerate.

Resultant ulcers have a characteristic rolled-edged appearance with a yellow-white, "cotton-wool-like" sloughy base (Figure 44-8) (Adjei et al., 2001; Agbenorku et al., 2001). The surrounding skin is

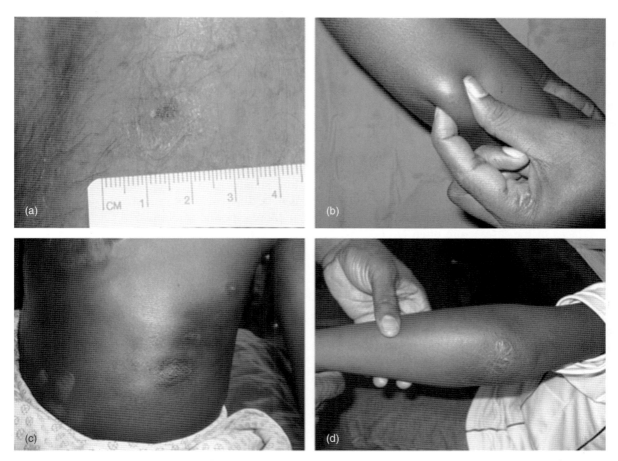

Fig. 44-7 Clinical features of non-ulcerative forms of Buruli ulcer. (a) Papule; (b) nodule; (c) plaque; (d) edema. Images from http://www.who.int/buruli/photos/nonulcerative/en/index.html. Copyright © WHO (2013). For a color version of this figure, see the color plate section.

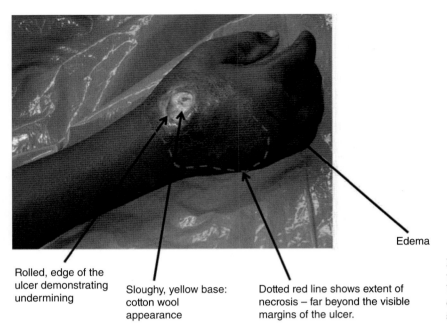

Edema

Rolled, edge of the ulcer demonstrating undermining

Sloughy, yellow base: cotton wool appearance

Dotted red line shows extent of necrosis – far beyond the visible margins of the ulcer.

Fig. 44-8 Ulcerative form of Buruli ulcer. Courtesy of Mark Wansbrough-Jones. For a color version of this figure, see the color plate section.

frequently discolored owing to extensive induration. Induration results from subcutaneous fat necrosis by mycolactone. Necrosis extends far beyond visible ulcer edges, accounting for the "rolled-edge" appearance (Figure 44-8) (Adjei et al., 2001; Agbenorku et al., 2001).

Both non-ulcerative and ulcerative lesions are described as painless. This can be attributed to the immunosuppressant properties of mycolactone, which prevents a local inflammatory response. Pain ensues only in the presence of secondary wound infection, or in the event of osteomyelitis, a rare feature of *M. ulcerans* infection. Osteomyelitis may be present even when the overlying skin is intact. With advancing osteomyelitis, fistula formation may occur with necrotic tissue being exuded onto the surface of the overlying skin (Pszolla et al., 2003).

44.11 Diagnosis

Due to the painless nature of the disease and the lack of epidemiological data on the incubation period, early stages of the disease are often ignored. Thus, most affected individuals report to health centers only after development of a large persistent ulcer. Diagnosis of the infection has been based on the presentation of the symptoms of the disease at a health facility in endemic areas or active case search usually undertaken by the Buruli Ulcer Control Programs in endemic areas. This has been quite challenging due to the vast number of other skin infections or conditions that may exhibit symptoms that are similar to that known for BU. Thus, the WHO has directed that all clinically diagnosed or suspected cases of BU be confirmed. Confirmation can, however, only be accomplished in reference laboratories leaving a gap between accurate diagnosis and treatment.

Laboratory confirmation is difficult to undertake at the point of care. *M. ulcerans* strains are shown to

be acid fast with the Ziehl–Neelsen staining procedure but this is not specific as there may be false positives from other environmental mycobacteria. Subsequently, swabs taken from open lesions do not always show acid-fast bacilli (AFB) by microscopic examination due to the disparity in the distribution of the bacilli in the lesion. Histopathology, although a robust confirmatory test has a low detection rate for AFB, trained medical personnel will be able to confirm *M. ulcerans* infection due to the lack of inflammatory cells and general necrosis of the specimen. Culture of *M. ulcerans* is the gold standard; however, it has a sensitivity of only 35–50% (Phillips et al., 2005). *M. ulcerans* is a notoriously slow-growing mycobacterium, thus decontamination methods are employed to prevent faster growing species (Yeboah-Manu et al., 2011) with some loss of viability. Also, the time required to observe growth is too long, making it unfavorable for case confirmation prior to the onset of treatment. Culture, staining, and histology also fail to recognize any genetic differences between strains.

More specific molecular PCR methods have been developed based on probing for an insertion sequence, IS2404, found on the chromosome as well as on the plasmid (Stinear et al., 2004). However, due to the very technical and expensive nature of these techniques, they are confined only to reference laboratories. In addition to the technical nature of these methods, the IS2404 methods are not very specific to *M. ulcerans* detection and may amplify other related mycobacteria (Williamson et al., 2008). IS2404-PCR has been performed using swab samples taken from skin exudates or punch biopsies of patients with suspected BU and has become the gold standard for diagnosis (Beissner et al., 2010; World Health Organization, 2013a).

More recently, the collection of swabs for open lesions and fine needle aspirations from unbroken skin are replacing punch biopsies as the preferred method of obtaining samples for diagnosis. It has been shown that the sensitivity of these newer methods is an improvement on the older method, albeit less invasive (Phillips et al., 2009). Nonetheless, these tests can be fairly accurate. But, there is no simple, rapid test that is appropriate for early diagnosis and use in the low-resource settings in West Africa where the disease is most prevalent. This represents a huge unmet need, because early diagnosis (before ulcers and deformities occur) is critical, and therefore there is the need for, effective, simple, and inexpensive tools that can be easily applied in rural settings for a quick and reliable diagnosis of BU disease.

More recently, the loop-mediated isothermal amplification (LAMP) has been investigated as a possible point-of-care diagnostic (Abolordey et al., 2012; de Souza et al., 2012). This method requires minimal manipulation to extract DNA from a specimen and there is minimal equipment involved in the amplification of the DNA. More testing is required, however, to improve the specificity of this method as well as to further reduce the cost of reagents. There are efforts also to employ the direct detection of mycolactone in tissues using solvent extraction followed by UV and chromatographic detection (Sarfo et al., 2010).

44.12 Treatment

BU treatment is constantly evolving. A series of randomized control trials demonstrated that *M. ulcerans* infection could be effectively treated using a combination of rifampicin and streptomycin (Ji et al., 2007). In 2004, the WHO introduced the first antibiotic treatment regime, reducing the need for extensive surgical debridement and resultant post-operative complications.

Prior to 2004, surgical excision/debridement had been the mainstay of treatment (World Health Organization, 2004), irrespective of the stage of disease presentation. Nodules were excised, leaving scars of approximately 1–2 cm, while papules, plaques, and ulcers were extensively debrided.

Since then, clinical trials indicate that antibiotics are potentially curative if administered during early, non-ulcerative stages of infection, or in those with ulcers of <3 cm diameter (World Health Organization, 2013a, 2013c). Antibiotics administered to patients with advanced infection (lesions >3 cm), reduced lesion size, thus reducing the need for extensive debridement. In this way, antibiotics were proven to be curative or of benefit as an adjuvant therapy to surgery.

44.12.1 Antibiotics

Currently, *M. ulcerans* infection is treated using antibiotics or a combination of antibiotic therapy and surgical debridement (Friedman et al., 2013; O'Brien et al., 2012; World Health Organization, 2013a, 2013c). The

WHO recommends a variety of antibiotic treatment regimens (see listed), though efficacy of rifampicin in combination with clarithromycin or moxifloxacin has not yet been proven by randomized control trial. Treatment is based upon size of the lesion and, for the purposes of guidance, the WHO recommends patients be divided into three size categories (World Health Organization, 2013b). Category I includes small, early lesions such as nodules/papules, and ulcers less than 5 cm in diameter. Category II includes non-ulcerative and ulcerative plaque, edematous forms, and ulcers larger than 5 cm in diameter. All lesions found in the head and neck are also included in category II. Disseminated forms of the disease including osteomyelitis and with joint involvement are included in category III (World Health Organization, 2013b).

In all cases, antibiotics are administered under directly observed therapy (DOT) for a minimum of 8 weeks (World Health Organization, 2013a):

- Combined use of rifampicin (10 mg/kg once daily) and streptomycin (15 mg/kg once daily); or
- Combined use of rifampicin (10 mg/kg once daily) and clarithromycin (7.5 mg/kg twice daily); or
- Combined use of rifampicin (10 mg/kg once daily) and moxifloxacin (400 mg once daily)

Most commonly, rifampicin and streptomycin are used in combination. However, antibiotic regimen choice may be determined by wound size at diagnosis, response to antibiotics – qualified as "healing time" patient factors such as allergy and pregnancy. (Streptomycin use is contraindicated in pregnancy (World Health Organization, 2004) and so clarithromycin is preferentially prescribed).

In late-stage infection, when lesions are large, antibiotics can be used as an adjuvant to surgery. Antibiotic treatment decreases wound size, reducing the need for extensive surgical debridement. Post-operative outcomes have improved significantly with the administration of antibiotics (O'Brien et al., 2012).

44.12.2 Surgery as an adjunct to antibiotic treatment

Surgical debridement may be necessary in the event of extensive ulceration (World Health Organization, 2013a), poor response to antibiotics, or osteomyelitis. Margins for surgical excision often extend several inches beyond the visible lesion edges. Owing to the necrotizing properties of mycolactone, extensive debridement is essential to ensure successful removal of all infective tissues. Debridement exposes uninfected skin, fat, fascia, and muscle, resulting in a painful wound. For this reason, good post-operative care is imperative, to avoid possible surgical sequelae: limb contractures and resultant loss of functionality.

44.13 The multidisciplinary team

A multidisciplinary approach to treatment and patient care is essential for optimizing treatment outcomes. Physiotherapy is paramount minimizing and/or preventing disabilities.

Post-operative wounds are painful. Physiotherapy must coincide with the use of analgesia. Analgesia must be administered prior to commencing physiotherapy, to ensure that the patient is able to optimize limb use during exercise. Where excision followed by skin grafting involving a joint has taken place, the physiotherapist must work with the patient to mobilize the joint through its full range of movement to prevent contracture formation. Often patients are young, requiring constant supervision and encouragement regarding practice and implementation of physiotherapy advice.

44.13.1 Managing patients with co-morbidities

In all cases, patients should be investigated and treated for co-morbidities that may result in immune-suppression. Co-infection with HIV, TB, and/or enteric parasites may prolong or prevent BU healing. In all cases, HIV counseling should be provided and in the case of confirmed HIV infection, early retroviral treatment is beneficial (Johnson et al., 2008; Toll et al., 2005).

44.13.2 BU control

In the absence of a clear path for primary prevention, control is necessary to prevent progression from early lesions to ulcers through education and treatment. One of the most significant activities impacting awareness is active case detection. Active case finding followed by BU sensitization campaigns has contributed to the dissemination of information of BU and has led to a significant increase in early lesions and the simple ulcerative forms, and a decrease in category III ulcers. This has also increased the proportion of healing with reduced complications, as help-seeking behavior is influenced by the perceived effectiveness of treatment (Ackumey et al., 2011; Agbenorku et al., 2011; Phanzu et al., 2011; Porten et al., 2009). These works advocate the cooperation from National Control programs, municipal executives, health staff, teachers, school children, community leaders, and community health volunteers to overcome difficulties associated with health care costs, accommodations, and adequate infrastructure for surgery, wound treatment, and aftercare.

Acknowledgments

The authors are grateful to Dr. Pamela Small and Dr. Mark Wansbrough-Jones for critical and insightful comments and advice regarding manuscript content.

References

Ablordey, A., Amissah, D.A., Aboagye, I.F., Hatano B., Yamazaki, T., Sata, T., Ishikawa, K., and Katano, H. 2012. Detection of Mycobacterium ulcerans by the loop mediated isothermal amplification method. *PLoS Negl. Trop. Dis.*, 6, e1590.

Ackumey, M.M., Gyapong, M., Pappoe, M., and Weiss, M.G. 2011. Help-seeking for pre-ulcer and ulcer conditions of *Mycobacterium ulcerans* disease (Buruli ulcer) in Ghana. *Am. J. Trop. Med. Hyg.*, 85, 1106–1113.

Adjei, O., Carbonnelle, B., Mensah, P., Johnson, P., Kouakou, H., Meyers, W.M., Portaels, F., and Asiedu, K. 2001. *Buruli ulcer: Diagnosis of Mycobacterium ulcerans disease. A manual for health care providers.* World Health Organization, Geneva, Switzerland.

Adusumilli, S., Mve-Obiang, A., Sparer, T., Meyers, W., Hayman, J., and Small, P.L. 2005. *Mycobacterium ulcerans* toxic macrolide, mycolactone modulates the host immune response and cellular location of *M. ulcerans* in vitro and in vivo. *Cell. Microbiol.*, 7, 1295–1304.

Agbenorku, P., Agbenorku, M., Amankwa, A., Tuuli, L., and Saunderson, P. 2011. Factors enhancing the control of Buruli ulcer in the Bomfa communities, Ghana. *Trans. R. Soc. Trop. Med. Hyg.*, 105, 459–465.

Agbenorku, P., Aguiar, J., Asse, H., Buntine, J., Crofts, K., Oberlin, C., Priuli, G.B., Smith, M., Steffen, C., Asiedu, K., Eklund, A., Goerdt, A., and Emmanuel, J. 2001. *Buruli Ulcer: Management of Mycobacterium ulcerans disease. A manual for health care providers.* World Health Organization, Geneva, Switzerland.

Ameixa, C., and Friedland, J.S. 2001. Down-regulation of interleukin-8 secretion from *Mycobacterium tuberculosis*-infected monocytes by interleukin-4 and -10 but not by interleukin-13. *Infect. Immun.*, 69, 2470–2476.

Ameixa, C., and Friedland, J.S. 2002. Interleukin-8 secretion from *Mycobacterium tuberculosis*-infected monocytes is regulated by protein tyrosine kinases but not by ERK1/2 or p38 mitogen-activated protein kinases. *Infect. Immun.*, 70, 4743–4746.

Anonymous. 1970. Clinical features and treatment of pre-ulcerative Buruli lesions (*Mycobacterium ulcerans* infection). Report II of the Uganda Buruli Group. *Br. Med. J.*, 2, 390–393.

Anonymous. 2004. Buruli ulcer disease. *Wkly Epidemiol. Rec.* 79, 194–199.

Appelberg, R. 2006. Pathogenesis of *Mycobacterium avium* infection: Typical responses to an atypical mycobacterium? *Immunol. Res.*, 35, 179–190.

Beissner, M., Herbinger, K.H., and Bretzel, G. 2010. Laboratory diagnosis of Buruli ulcer disease. *Future Microbiol.*, 5, 363–370.

Benbow, M., Williamson, H., Kimbirauskus, R., McIntosh, M., Kolar, R., Quaye, C., Boakye, D., Small, P., and Merritt, R. 2008. Aquatic invertebrates as unlikely vectors of Buruli ulcer disease. *Emerg. Infect. Dis.*, 14, 1247–1254.

Boulkroun, S., Guenin-Mace, L., Thoulouze, M.I., Monot, M., Merckx, A., Langsley, G., Bismuth, G., Di Bartolo, V., and Demangel, C. 2010. Mycolactone suppresses T cell responsiveness by altering both early signaling and posttranslational events. *J. Immunol.*, 184, 1436–1444.

Bozzo, C., Tiberio, R., Graziola, F., Pertusi, G., Valente, G., Colombo, E., Small, P.L., and Leigheb, G. 2010. A *Mycobacterium ulcerans* toxin, mycolactone, induces apoptosis in primary human keratinocytes and in HaCaT cells. *Microbes Infect.*, 12, 1258–1263.

Brodin, P., Majlessi, L., Brosch, R., Smith, D., Bancroft, G., Clark, S., Williams, A., Leclerc, C., and Cole, S.T. 2004. Enhanced protection against tuberculosis by vaccination with recombinant *Mycobacterium microti* vaccine that induces T cell immunity against region of difference 1 antigens. *J. Infect. Dis.*, 190, 115–122.

Brook, M., Sully, G., Clark, A.R., and Saklatvala, J. 2000. Regulation of tumour necrosis factor alpha mRNA stability by the mitogen-activated protein kinase p38 signalling cascade. *FEBS Lett.*, 483, 57–61.

Cavalcanti, Y.V., Brelaz, M.C., Neves, J.K., Ferraz, J.C., and Pereira, V.R. 2012. Role of TNF-alpha, IFN-gamma, and IL-10 in the development of pulmonary tuberculosis. *Pulm. Med.*, 2012, 745483.

Chany, A.C., Casarotto, V., Schmitt, M., Tarnus, C., Guenin-Mace, L., Demangel, C., Mirguet, O., Eustache, J., and Blanchard, N. 2011. A diverted total synthesis of mycolactone analogues: An insight into Buruli ulcer toxins. *Chemistry*, 17, 14413–14419.

Chemlal, K., Huys, G., Fonteyne, P.A., Vincent, V., Lopez, A.G., Rigouts, L., Swings, J., Meyers, W.M., and Portaels, F. 2001. Evaluation of PCR-restriction profile analysis and IS2404 restriction fragment length polymorphism and amplified fragment length polymorphism fingerprinting for identification and typing of *Mycobacterium ulcerans* and *M. marinum*. *J. Clin. Microbiol.*, 39, 3272–3278.

Ciaramella, A., Cavone, A., Santucci, M.B., Amicosante, M., Martino, A., Auricchio, G., Pucillo, L.P., Colizzi, V., and Fraziano, M. 2002. Proinflammatory cytokines in the course of *Mycobacterium tuberculosis*-induced apoptosis in monocytes/macrophages. *J. Infect. Dis.*, 186, 1277–1282.

Cole, S.T., Eiglmeier, K., Parkhill, J., James, K.D., Thomson, N.R., Wheeler, P.R., Honore, N., Garnier, T., Churcher, C., Harris, D., Mungall, K., Basham, D., Brown, D., Chillingworth, T., Connor, R., Davies, R.M., Devlin, K., Duthoy, S., Feltwell, T., Fraser, A., Hamlin, N., Holroyd, S., Hornsby, T., Jagels, K., Lacroix, C., Maclean, J., Moule, S., Murphy, L., Oliver, K., Quail, M.A., Rajandream, M.A., Rutherford, K.M., Rutter, S., Seeger, K., Simon, S., Simmonds, M., Skelton, J., Squares, R., Squares, S., Stevens, K., Taylor, K., Whitehead, S., Woodward, J.R., and Barrell, B.G. 2001. Massive gene decay in the leprosy bacillus. *Nature*, 409, 1007–1011.

Connor, D.H., and Lunn, H.F. 1965. *Mycobacterium ulcerans* infection (with comments on pathogenesis). *Int. J. Lepr.*, 33(suppl.), 698–709.

Coutanceau, E., Decalf, J., Martino, A., Babon, A., Winter, N., Cole, S.T., Albert, M.L., and Demangel, C. 2007. Selective suppression of dendritic cell functions by *Mycobacterium ulcerans* toxin mycolactone. *J. Exp. Med.*, 204, 1395–1403.

Coutanceau, E., Marsollier, L., Brosch, R., Perret, E., Goossens, P., Tanguy, M., Cole, S.T., Small, P.L., and Demangel, C. 2005. Modulation of the host immune response by a transient intracellular stage of *Mycobacterium ulcerans*: The contribution of endogenous mycolactone toxin. *Cell. Microbiol.*, 7, 1187–1196.

de Souza, D.K., Quaye, C., Mosi, L., Addo, P., and Boakye, D.A. 2012. A quik and cost effective method for the diagnosis of Mycobacterium ulcerans infection. *BMC Infect. Dis.*, 12, doi: 10.1186.

Diaz, D., Dobeli, H., Yeboah-Manu, D., Mensah-Quainoo, E., Friedlein, A., Soder, N., Rondini, S., Bodmer, T., and Pluschke, G. 2006. Use of the immunodominant 18-kilo Dalton small heat shock protein as a serological marker for exposure to *Mycobacterium ulcerans*. *Clin. Vaccine Immunol.*, 13, 1314–1321.

Dobos, K.M., Small, P.L., Deslauriers, M., Quinn, F.D., and King, C.H. 2001. *Mycobacterium ulcerans* cytotoxicity in an adipose cell model. *Infect. Immun.*, 69, 7182–7186.

Dobos, K.M., Spotts, E.A., Marston, B.J., Horsburgh Jr., C.R., and King, C.H. 2000. Serologic response to culture filtrate antigens of *Mycobacterium ulcerans* during Buruli ulcer disease. *Emerg. Infect. Dis.*, 6, 158–164.

Dodge, O.G., and Lunn, H.F. 1962. Buruli ulcer: A mycobacterial skin ulcer in a Uganda child. *J. Trop. Med. Hyg.*, 65, 139–142.

Doig, K.D., Holt, K.E., Fyfe, J.A., Lavender, C.J., Eddyani, M., Portaels, F., Yeboah-Manu, D., Pluschke, G., Seemann, T., and Stinear, T.P. 2012. On the origin of *Mycobacterium ulcerans*, the causative agent of Buruli ulcer. *BMC Genomics*, 13, 258.

En, J., Goto, M., Nakanaga, K., Higashi, M., Ishii, N., Saito, H., Yonezawa, S., Hamada, H., and Small, P.L. 2008. Mycolactone is responsible for the painlessness of *Mycobacterium ulcerans* infection (buruli ulcer) in a murine study. *Infect. Immun.*, 76, 2002–2007.

Friedman, N.D., Athan, E., Hughes, A.J., Khajehnoori, M., McDonald, A., Callan, P., Rahdon, R., and O'Brien, D.P. 2013. *Mycobacterium ulcerans* disease: Experience with primary oral medical therapy in an Australian cohort. *PLoS Negl. Trop. Dis.*, 7, e2315.

Fyfe, J.A., Lavender, C.J., Johnson, P.D., Globan, M., Sievers, A., Azuolas, J., and Stinear, T.P. 2007. Development and application of two multiplex real-time PCR assays for the detection of *Mycobacterium ulcerans* in clinical and environmental samples. *Appl. Environ. Microbiol.*, 73, 4733–4740.

Fyfe, J.A.M., Lavender, C.J., Handasyde, K.A., Legione, A.R., O'Brien, C.R., Stinear, T.P., Pidot, S.J., Seemann, T., Benbow, M.E., Wallace, J.R., McCowan, C., and Johnson, P.D.R. 2010. A major role for mammals in the ecology of *Mycobacterium ulcerans*. *PLoS Negl. Trop. Dis.*, 4, e791.

George, K.M., Barker, L.P., Welty, D.M., and Small, P.L. 1998. Partial purification and characterization of biological effects of a lipid toxin produced by *Mycobacterium ulcerans*. *Infect. Immun.*, 66, 587–593.

George, K.M., Chatterjee, D., Gunawardana, G., Welty, D., Hayman, J., Lee, R., and Small, P.L. 1999. Mycolactone: A polyketide toxin from *Mycobacterium ulcerans* required for virulence. *Science*, 283, 854–857.

George, K.M., Pascopella, L., Welty, D.M., and Small, P.L. 2000. A *Mycobacterium ulcerans* toxin, mycolactone, causes apoptosis in guinea pig ulcers and tissue culture cells. *Infect. Immun.*, 68, 877–883.

Godfrey, D., Williamson, H., Silverman, J., and Small, P.L. 2007. Newly identified *Mycobacterium* species in a *Xenopus laevis* colony. *Comp. Med.*, 57, 97–104.

Gooding, T.M., Kemp, A.S., Robins-Browne, R.M., Smith, M., and Johnson, P.D. 2003. Acquired T-helper 1 lymphocyte anergy following infection with *Mycobacterium ulcerans*. *Clin. Infect. Dis.*, 36, 1076–1077.

Guarner, J., Bartlett, J., Whitney, E.A., Raghunathan, P.L., Stienstra, Y., Asamoa, K., Etuaful, S., Klutse, E., Quarshie, E., van der Werf, T.S., van der Graaf, W.T., King, C.H., and Ashford, D.A. 2003. Histopathologic features of *Mycobacterium ulcerans* infection. *Emerg. Infect. Dis.*, 9, 651–656.

Guenin-Mace, L., Carrette, F., Asperti-Boursin, F., Le Bon, A., Caleechurn, L., Di Bartolo, V., Fontanet, A., Bismuth, G., and Demangel, C. 2011. Mycolactone impairs T cell homing by suppressing microRNA control of L-selectin expression. *Proc. Natl. Acad. Sci. USA*, 108, 12833–12838.

Guenin-Mace, L., Veyron-Churlet, R., Thoulouze, M.I., Romet-Lemonne, G., Hong, H., Leadlay, P.F., Danckaert, A., Ruf, M.T., Mostowy, S., Zurzolo, C., Bousso, P., Chretien, F., Carlier, M.F., and Demangel, C. 2013. Mycolactone activation of Wiskott–Aldrich syndrome proteins underpins Buruli ulcer formation. *J. Clin. Invest.*, 123, 1501–1512.

Hall, B., and Simmonds, R. 2014. Pleotropic molecular effects of the *Mycobacterium ulcerans* virulence factor mycolactone underlying the cell death and immunosuppression seen in Buruli ulcer. *Biochem. Soc. Trans.*, 42, 177–183.

Hayman, J. 1993. Out of Africa: Observations on the histopathology of *Mycobacterium ulcerans* infection. *J. Clin. Pathol.*, 46, 5–9.

Hett, E.C., and Rubin, E.J. 2008. Bacterial growth and cell division: A mycobacterial perspective. *Microbiol. Mol. Biol. Rev.*, 72, 126–156, table of contents.

Hilty, M., Yeboah-Manu, D., Boakye, D., Mensah-Quainoo, E., Rondini, S., Schelling, E., Ofori-Adjei, D., Portaels, F.,

Zinsstag, J., and Pluschke, G. 2006. Genetic diversity in *Mycobacterium ulcerans* isolates from Ghana revealed by a newly identified locus containing a variable number of tandem repeats. *J. Bacteriol.*, 188, 1462–1465.

Hockmeyer, W.T., Krieg, R.E., Reich, M., and Johnson, R.D. 1978. Further characterization of *Mycobacterium ulcerans* toxin. *Infect. Immun.*, 21, 124–128.

Hong, H., Coutanceau, E., Leclerc, M., Caleechurn, L., Leadlay, P.F., and Demangel, C. 2008a. Mycolactone diffuses from *Mycobacterium ulcerans*-infected tissues and targets mononuclear cells in peripheral blood and lymphoid organs. *PLoS Negl. Trop. Dis.*, 2, e325.

Hong, H., Demangel, C., Pidot, S.J., Leadlay, P.F., and Stinear, T. 2008b. Mycolactones: Immunosuppressive and cytotoxic polyketides produced by aquatic mycobacteria. *Nat. Prod. Rep.*, 25, 447–454.

Houngbedji, G.M., Bouchard, P., and Frenette, J. 2011. *Mycobacterium ulcerans* infections cause progressive muscle atrophy and dysfunction, and mycolactone impairs satellite cell proliferation. *Am. J. Physiol. Regul. Integr. Comp. Physiol.*, 300, R724–R732.

Huber, C.A., Ruf, M.T., Pluschke, G., and Kaser, M. 2008. Independent loss of immunogenic proteins in *Mycobacterium ulcerans* suggests immune evasion. *Clin. Vaccine Immunol.*, 15, 598–606.

Jackson, K., Edwards, R., Leslie, D.E., and Hayman, J. 1995. Molecular method for typing *Mycobacterium ulcerans*. *J. Clin. Microbiol.*, 33, 2250–2253.

Jacobsen, K.H., and Padgett, J.J. 2010. Risk factors for *Mycobacterium ulcerans* infection. *Int. J. Infect. Dis.*, 14, e677–e681.

Ji, B., Chauffour, A., Robert, J., Lefrancois, S., and Jarlier, V. 2007. Orally administered combined regimens for treatment of *Mycobacterium ulcerans* infection in mice. *Antimicrob. Agents Chemother.*, 51, 3737–3739.

Johnson, P.D., Stinear, T., Small, P.L., Pluschke, G., Merritt, R.W., Portaels, F., Huygen, K., Hayman, J.A., and Asiedu, K. 2005. Buruli ulcer (*M. ulcerans* infection): New insights, new hope for disease control. *PLoS Med.*, 2, e108.

Johnson, P.D., Azuolas, J., Lavender, C.J., Wishart, E., Stinear, T.P., Hayman, J.A., Brown, L., Jenkin, G.A., and Fyfe, J.A. 2007. *Mycobacterium ulcerans* in mosquitoes captured during outbreak of Buruli ulcer, southeastern Australia. *Emerg. Infect. Dis.*, 13, 1653–1660.

Johnson, R.C., Nackers, F., Glynn, J.R., de Biurrun Bakedano, E., Zinsou, C., Aguiar, J., Tonglet, R., and Portaels, F. 2008. Association of HIV infection and *Mycobacterium ulcerans* disease in Benin. *Aids*, 22, 901–903.

Kana, B.D., Weinstein, E.A., Avarbock, D., Dawes, S.S., Rubin, H., and Mizrahi, V. 2001. Characterization of the cydAB-encoded cytochrome bd oxidase from *Mycobacterium smegmatis*. *J. Bacteriol.*, 183, 7076–7086.

Kaser, M., Rondini, S., Naegeli, M., Stinear, T., Portaels, F., Certa, U., and Pluschke, G. 2007. Evolution of two distinct phylogenetic lineages of the emerging human pathogen *Mycobacterium ulcerans*. *BMC Evol. Biol.*, 7, 177.

Kishi, Y. 2011. Chemistry of mycolactones, the causative toxins of Buruli ulcer. *Proc. Natl. Acad. Sci. USA*, 108, 6703–6708.

Krutzik, S.R., Ochoa, M.T., Sieling, P.A., Uematsu, S., Ng, Y.W., Legaspi, A., Liu, P.T., Cole, S.T., Godowski, P.J., Maeda, Y., Sarno, E.N., Norgard, M.V., Brennan, P.J., Akira, S., Rea, T.H., and Modlin, R.L. 2003. Activation and regulation of Toll-like receptors 2 and 1 in human leprosy. *Nat. Med.*, 9, 525–532.

Lavender, C.J., Fyfe, J.A., Azuolas, J., Brown, K., Evans, R.N., Ray, L.R., and Johnson, P.D. 2011. Risk of Buruli ulcer and detection of *Mycobacterium ulcerans* in mosquitoes in southeastern Australia. *PLoS Negl. Trop. Dis.*, 5, e1305.

Liaw, Y.S., Tsai-Wu, J.J., Wu, C.H., Hung, C.C., Lee, C.N., Yang, P.C., Luh, K.T., and Kuo, S.H. 2002. Variations in the NRAMP1 gene and susceptibility of tuberculosis in Taiwanese. *Int. J. Tuberc. Lung Dis.*, 6, 454–460.

MacCallum, P., Tolhurst, J.C., Buckle, G., and Sissons, H.A. 1948. A new mycobacterial infection in man. *J. Pathol. Bacteriol.*, 60, 93–122.

Marion, E., Eyangoh, S., Yeramian, E., Doannio, J., Landier, J., Aubry, J., Fontanet, A., Rogier, C., Cassisa, V., Cottin, J., Marot, A., Eveillard, M., Kamdem, Y., Legras, P., Deshayes, C., Saint-André, J.-P., and Marsollier, L. 2010. Seasonal and regional dynamics of *M. ulcerans* transmission in environmental context: Deciphering the role of water bugs as hosts and vectors. *PLoS Negl. Trop. Dis.*, 4, e731.

Marsollier, L., Robert, R., Aubry, J., Saint Andre, J.P., Kouakou, H., Legras, P., Manceau, A.L., Mahaza, C., and Carbonnelle, B. 2002. Aquatic insects as a vector for *Mycobacterium ulcerans*. *Appl. Environ. Microbiol.*, 68, 4623–4628.

Marsollier, L., Severin, T., Aubry, J., Merritt, R.W., Saint Andre, J.P., Legras, P., Manceau, A.L., Chauty, A., Carbonnelle, B., and Cole, S.T. 2004a. Aquatic snails, passive hosts of *Mycobacterium ulcerans*. *Appl. Environ. Microbiol.*, 70, 6296–6298.

Marsollier, L., Stinear, T., Aubry, J., Saint Andre, J.P., Robert, R., Legras, P., Manceau, A.L., Audrain, C., Bourdon, S., Kouakou, H., and Carbonnelle, B. 2004b. Aquatic plants stimulate the growth of and biofilm formation by *Mycobacterium ulcerans* in axenic culture and harbor these bacteria in the environment. *Appl. Environ. Microbiol.*, 70, 1097–1103.

Marsollier, L., Brodin, P., Jackson, M., Kordulakova, J., Tafelmeyer, P., Carbonnelle, E., Aubry, J., Milon, G., Legras, P., Andre, J.P., Leroy, C., Cottin, J., Guillou, M.L., Reysset, G., and Cole, S.T. 2007. Impact of *Mycobacterium ulcerans* biofilm on transmissibility to ecological niches and Buruli ulcer pathogenesis. *PLoS Pathog.*, 3, e62.

Merritt, R.W., Benbow, M.E., and Small, P.L.C. 2005. Unraveling an emerging disease associates with disturbed aquatic environments: The case of Buruli ulcer. *Front. Ecol. Environ.*, 3, 323–331.

Merritt, R.W., Walker, E.D., Small, P.L.C., Wallace, J.R., Johnson, P.D.R., Benbow, M.E., and Boakye, D.A. 2010. Ecology and transmission of Buruli ulcer disease: A systematic review. *PLoS Negl. Trop. Dis.*, 4, e911.

Mitchell, P.J., Jerrett, I.V., and Slee, K.J. 1984. Skin ulcers caused by *Mycobacterium ulcerans* in Koalas near Bairnsdale, Australia. *Pathology*, 16, 256–260.

Mosi, L., Williamson, H., Wallace, J.R., Merritt, R.W., and Small, P.L. 2008. Persistent association of *Mycobacterium ulcerans* with West African predaceous insects of the family belostomatidae. *Appl. Environ. Microbiol.*, 74, 7036–7042.

Mosi, L., Mutoji, N.K., Basile, F.A., Donnell, R., Jackson, K.L., Spangenberg, T., Kishi, Y., Ennis, D.G., and Small, P.L. 2012. *Mycobacterium ulcerans* causes minimal pathogenesis and colonization in medaka (*Oryzias latipes*): An experimental fish model of disease transmission. *Microbes Infect.*, 14, 719–729.

Mve-Obiang, A., Lee, R.E., Portaels, F., and Small, P.L. 2003. Heterogeneity of mycolactones produced by clinical isolates of *Mycobacterium ulcerans*: Implications for virulence. *Infect. Immun.*, 71, 774–783.

Mve-Obiang, A., Lee, R.E., Umstot, E.S., Trott, K.A., Grammer, T.C., Parker, J.M., Ranger, B.S., Grainger, R., Mahrous, E.A., and Small, P.L. 2005. A newly discovered mycobacterial pathogen isolated from laboratory colonies of *Xenopus* species with lethal infections produces a novel form of mycolactone, the *Mycobacterium ulcerans* macrolide toxin. *Infect. Immun.*, 73, 3307–3312.

Nacy, C.A., Meierovics, A.I., Belosevic, M., and Green, S.J. 1991. Tumor necrosis factor-alpha: Central regulatory cytokine in the induction of macrophage antimicrobial activities. *Pathobiology*, 59, 182–184.

O'Brien, D.P., McDonald, A., Callan, P., Robson, M., Friedman, N.D., Hughes, A., Holten, I., Walton, A., and Athan, E. 2012. Successful outcomes with oral fluoroquinolones combined with rifampicin in the treatment of *Mycobacterium ulcerans*: An observational cohort study. *PLoS Negl. Trop. Dis.*, 6, e1473.

Okenu, D.M., Ofielu, L.O., Easley, K.A., Guarner, J., Spotts Whitney, E.A., Raghunathan, P.L., Stienstra, Y., Asamoa, K., van der Werf, T.S., van der Graaf, W.T., Tappero, J.W., Ashford, D.A., and King, C.H. 2004. Immunoglobulin M antibody responses to *Mycobacterium ulcerans* allow discrimination between cases of active Buruli ulcer disease and matched family controls in areas where the disease is endemic. *Clin. Diagn. Lab. Immunol.*, 11, 387–391.

Oliveira, M.S., Fraga, A.G., Torrado, E., Castro, A.G., Pereira, J.P., Filho, A.L., Milanezi, F., Schmitt, F.C., Meyers, W.M., Portaels, F., Silva, M.T., and Pedrosa, J. 2005. Infection with *Mycobacterium ulcerans* induces persistent inflammatory responses in mice. *Infect. Immun.*, 73, 6299–6310.

Pahlevan, A.A., Wright, D.J., Andrews, C., George, K.M., Small, P.L., and Foxwell, B.M. 1999. The inhibitory action of *Mycobacterium ulcerans* soluble factor on monocyte/T cell cytokine production and NF-kappa B function. *J. Immunol.*, 163, 3928–3935.

Phanzu, D.M., Suykerbuyk, P., Imposo, D.B., Lukanu, P.N., Minuku, J.B., Lehman, L.F., Saunderson, P., de Jong, B.C., Lutumba, P.T., Portaels, F., and Boelaert, M. 2011. Effect of a control project on clinical profiles and outcomes in buruli ulcer: A before/after study in Bas-Congo, Democratic Republic of Congo. *PLoS Negl. Trop. Dis.*, 5, e1402.

Phillips, R., Horsfield, C., Kuijper, S., Lartey, A., Tetteh, I., Etuaful, S., Nyamekye, B., Awuah, P., Nyarko, K.M., Osei-Sarpong, F., Lucas, S., Kolk, A.H., and Wansbrough-Jones, M. 2005. Sensitivity of PCR targeting the IS2404 insertion sequence of *Mycobacterium ulcerans* in an assay using punch biopsy specimens for diagnosis of Buruli ulcer. *J. Clin. Microbiol.*, 43, 3650–3656.

Phillips, R., Horsfield, C., Kuijper, S., Sarfo, S.F., Obeng-Baah, J., Etuaful, S., Nyamekye, B., Awuah, P., Nyarko, K.M., Osei-Sarpong, F., Lucas, S., Kolk, A.H., and Wansbrough-Jones, M. 2006a. Cytokine response to antigen stimulation of whole blood from patients with *Mycobacterium ulcerans* disease compared to that from patients with tuberculosis. *Clin. Vaccine Immunol.*, 13, 253–257.

Phillips, R., Horsfield, C., Mangan, J., Laing, K., Etuaful, S., Awuah, P., Nyarko, K., Osei-Sarpong, F., Butcher, P., Lucas, S., and Wansbrough-Jones, M. 2006b. Cytokine mRNA expression in *Mycobacterium ulcerans*-infected human skin and correlation with local inflammatory response. *Infect. Immun.*, 74, 2917–2924.

Phillips, R., Sarfo, F.S., Guenin-Mace, L., Decalf, J., Wansbrough-Jones, M., Albert, M.L., and Demangel, C. 2009. Immunosuppressive signature of cutaneous *Mycobacterium ulcerans* infection in the peripheral blood of patients with buruli ulcer disease. *J. Infect. Dis.*, 200, 1675–1684.

Pidot, S.J., Asiedu, K., Kaser, M., Fyfe, J.A., and Stinear, T.P. 2010a. *Mycobacterium ulcerans* and other mycolactone-producing mycobacteria should be considered a single species. *PLoS Negl. Trop. Dis.*, 4, e663.

Pidot, S.J., Porter, J.L., Marsollier, L., Chauty, A., Migot-Nabias, F., Badaut, C., Benard, A., Ruf, M.T., Seemann, T., Johnson, P.D., Davies, J.K., Jenkin, G.A., Pluschke, G., and Stinear, T.P. 2010b. Serological evaluation of *Mycobacterium ulcerans* antigens identified by comparative genomics. *PLoS Negl. Trop. Dis.*, 4, e872.

Pidot, S.J., Porter, J.L., Tobias, N.J., Anderson, J., Catmull, D., Seemann, T., Kidd, S., Davies, J.K., Reynolds, E., Dashper, S., and Stinear, T.P. 2010c. Regulation of the 18 kDa heat shock protein in *Mycobacterium ulcerans*: An alpha-crystallin orthologue that promotes biofilm formation. *Mol. Microbiol.*, 78, 1216–1231.

Pimsler, M., Sponsler, T.A., and Meyers, W.M. 1988. Immunosuppressive properties of the soluble toxin from *Mycobacterium ulcerans*. *J. Infect. Dis.*, 157, 577–580.

Plowright, R.K., Sokolow, S.H., Gorman, M.E., Daszak, P., and Foley, J.E. 2008. Causal inference in disease ecology: Investigating ecological drivers of disease ecology: Investigating ecological drivers of disease emergence. *Front. Ecol. Environ.*, 6, 420–429.

Portaels, F., Elsen, P., Guimaraes-Peres, A., Fonteyne, P., and Meyers, W.M. 1999. Insects in the transmission of *Mycobacterium ulcerans* infection. *Lancet*, 353, 986.

Porten, K., Sailor, K., Comte, E., Njikap, A., Sobry, A., Sihom, F., Meva'a, A., Eyangoh, S., Myatt, M., Nackers, F., and Grais, R.F. 2009. Prevalence of Buruli ulcer in Akonolinga health district, Cameroon: Results of a cross sectional survey. *PLoS Negl. Trop. Dis.*, 3, e466.

Pszolla, N., Sarkar, M.R., Strecker, W., Kern, P., Kinzl, L., Meyers, W.M., and Portaels, F. 2003. Buruli ulcer: A systemic disease. *Clin. Infect. Dis.*, 37, e78–e82.

Quek, T.Y., Athan, E., Henry, M.J., Pasco, J.A., Redden-Hoare, J., Hughes, A., and Johnson, P.D. 2007. Risk factors for *Mycobacterium ulcerans* infection, southeastern Australia. *Emerg. Infect. Dis.*, 13, 1661–1666.

Ranger, B.S., Mahrous, E.A., Mosi, L., Adusumilli, S., Lee, R.E., Colorni, A., Rhodes, M., and Small, P.L. 2006. Globally distributed mycobacterial fish pathogens produce a novel plasmid-encoded toxic macrolide, mycolactone F. *Infect. Immun.*, 74, 6037–6045.

Rhodes, M.W., Kator, H., McNabb, A., Deshayes, C., Reyrat, J.M., Brown-Elliott, B.A., Wallace Jr., R., Trott, K.A., Parker, J.M., Lifland, B., Osterhout, G., Kaattari, I., Reece, K., Vogelbein, W., and Ottinger, C.A. 2005. *Mycobacterium pseudoshottsii* sp. nov., a slowly growing chromogenic species isolated from Chesapeake Bay striped bass (*Morone saxatilis*). *Int. J. Syst. Evol. Microbiol.*, 55, 1139–1147.

Roltgen, K., Qi, W., Ruf, M.T., Mensah-Quainoo, E., Pidot, S.J., Seemann, T., Stinear, T.P., Kaser, M., Yeboah-Manu, D., and Pluschke, G. 2010. Single nucleotide polymorphism typing of *Mycobacterium ulcerans* reveals focal transmission of buruli ulcer in a highly endemic region of Ghana. *PLoS Negl. Trop. Dis.*, 4, e751.

Sarfo, F.S., Le Chevalier, F., Aka, N., Phillips, R.O., Amoako, Y., Boneca, I.G., Lenormand, P., Dosso, M., Wansbrough-Jones, M., Veyron-Churlet, R., Guenin-Mace, L., and Demangel, C. 2011. Mycolactone diffuses into the peripheral blood of Buruli ulcer patients – implications for diagnosis and disease monitoring. *PLoS Negl. Trop. Dis.*, 5, e1237.

Sarfo, F.S., Phillips, R.O., Rangers, B., Mahrous, E.A., Lee, R.E., et al. 2010. Detection of Mycolactone A/B in *Mycobacterium ulcerans*-Infected Human Tissue. *PLoS Negl. Trop. Dis.*, 4(1), e577.

Scherr, N., Gersbach, P., Dangy, J.P., Bomio, C., Li, J., Altmann, K.H., and Pluschke, G. 2013. Structure–activity relationship studies on the macrolide exotoxin mycolactone of *Mycobacterium ulcerans*. *PLoS Negl. Trop. Dis.*, 7, e2143.

Schutte, D., and Pluschke, G. 2009. Immunosuppression and treatment-associated inflammatory response in patients with *Mycobacterium ulcerans* infection (Buruli ulcer). *Expert. Opin. Biol. Ther.*, 9, 187–200.

Silva, M.T., Portaels, F., and Pedrosa, J. 2009. Pathogenetic mechanisms of the intracellular parasite *Mycobacterium ulcerans* leading to Buruli ulcer. *Lancet Infect. Dis.*, 9, 699–710.

Simmonds, R.E., Lali, F.V., Smallie, T., Small, P.L., and Foxwell, B.M. 2009. Mycolactone inhibits monocyte cytokine production by a posttranscriptional mechanism. *J. Immunol.*, 182, 2194–2202.

Snyder, D.S., and Small, P.L. 2003. Uptake and cellular actions of mycolactone, a virulence determinant for *Mycobacterium ulcerans*. *Microb. Pathog.*, 34, 91–101.

Sopoh, G.E., Barogui, Y.T., Johnson, R.C., Dossou, A.D., Makoutode, M., Anagonou, S.Y., Kestens, L., and Portaels, F. 2010. Family relationship, water contact and occurrence of Buruli ulcer in Benin. *PLoS Negl. Trop. Dis.*, 4, e746.

Stanford, J.L., Revill, W.D., Gunthorpe, W.J., and Grange, J.M. 1975. The production and preliminary investigation of Burulin, a new skin test reagent for *Mycobacterium ulcerans* infection. *J. Hyg. (Lond.)*, 74, 7–16.

Stienstra, Y., van der Graaf, W.T., te Meerman, G.J., The, T.H., de Leij, L.F., and van der Werf, T.S. 2001. Susceptibility to development of *Mycobacterium ulcerans* disease: Review of possible risk factors. *Trop. Med. Int. Health*, 6, 554–562.

Stienstra, Y., van der Werf, T.S., Oosterom, E., Nolte, I.M., van der Graaf, W.T., Etuaful, S., Raghunathan, P.L., Whitney, E.A., Ampadu, E.O., Asamoa, K., Klutse, E.Y., te Meerman, G.J., Tappero, J.W., Ashford, D.A., and van der Steege, G. 2006. Susceptibility to Buruli ulcer is associated with the SLC11A1 (NRAMP1) D543N polymorphism. *Genes Immun.*, 7, 185–189.

Stinear, T., Davies, J.K., Jenkin, G.A., Portaels, F., Ross, B.C., Oppedisano, F., Purcell, M., Hayman, J.A., and Johnson, P.D. 2000. A simple PCR method for rapid genotype analysis of *Mycobacterium ulcerans*. *J. Clin. Microbiol.*, 38, 1482–1487.

Stinear, T.P., Mve-Obiang, A., Small, P.L., Frigui, W., Pryor, M.J., Brosch, R., Jenkin, G.A., Johnson, P.D., Davies, J.K., Lee, R.E., Adusumilli, S., Garnier, T., Haydock, S.F., Leadlay, P.F., Cole, S.T. 2004. Giant plasmid-encoded polyketide synthases produce the macrolide toxin of *Mycobacterium ulcerans*. *Proc. Natl. Acad. Sci. USA*, 101, 1345–1349.

Stinear, T.P., Pryor, M.J., Porter, J.L., and Cole, S.T. 2005. Functional analysis and annotation of the virulence plasmid pMUM001 from *Mycobacterium ulcerans*. *Microbiology*, 151, 683–692.

Stinear, T.P., Seemann, T., Pidot, S., Frigui, W., Reysset, G., Garnier, T., Meurice, G., Simon, D., Bouchier, C., Ma, L., Tichit, M., Porter, J.L., Ryan, J., Johnson, P.D., Davies, J.K., Jenkin, G.A., Small, P.L., Jones, L.M., Tekaia, F., Laval, F., Daffe, M., Parkhill, J., and Cole, S.T. 2007. Reductive evolution and niche adaptation inferred from the genome of *Mycobacterium ulcerans*, the causative agent of Buruli ulcer. *Genome Res.*, 17, 192–200.

Tanghe, A., Content, J., Van Vooren, J.P., Portaels, F., and Huygen, K. 2001. Protective efficacy of a DNA vaccine encoding antigen 85A from *Mycobacterium bovis* BCG against Buruli ulcer. *Infect. Immun.*, 69, 5403–5411.

Tanghe, A., Adnet, P.Y., Gartner, T., and Huygen, K. 2007. A booster vaccination with Mycobacterium bovis BCG does not increase the protective effect of the vaccine against experimental *Mycobacterium ulcerans* infection in mice. *Infect. Immun.*, 75, 2642–2644.

Tanghe, A., Dangy, J.P., Pluschke, G., and Huygen, K. 2008. Improved protective efficacy of a species-specific DNA vaccine encoding mycolyl-transferase Ag85A from *Mycobacterium ulcerans* by homologous protein boosting. *PLoS Negl. Trop. Dis.*, 2, e199.

Teixeira, M.A., Silva, N.L., Ramos Ade, L., Hatagima, A., and Magalhaes, V. 2010. NRAMP1 gene polymorphisms in individuals with leprosy reactions attended at two reference centers in Recife, northeastern Brazil. *Rev. Soc. Bras. Med. Trop.*, 43, 281–286.

Tobias, N.J., Seemann, T., Pidot, S.J., Porter, J.L., Marsollier, L., Marion, E., Letournel, F., Zakir, T., Azuolas, J., Wallace, J.R., Hong, H., Davies, J.K., Howden, B.P., Johnson, P.D., Jenkin, G.A., and Stinear, T.P. 2009. Mycolactone gene expression is controlled by strong SigA-like promoters with utility in studies of *Mycobacterium ulcerans* and buruli ulcer. *PLoS Negl. Trop. Dis.*, 3, e553.

Tobias, N.J., Doig, K.D., Medema, M.H., Chen, H., Haring, V., Moore, R., Seemann, T., and Stinear, T.P. 2013. Complete genome sequence of the frog pathogen *Mycobacterium ulcerans* ecovar Liflandii. *J. Bacteriol.*, 195, 556–564.

Toll, A., Gallardo, F., Ferran, M., Gilaberte, M., Iglesias, M., Gimeno, J.L., Rondini, S., and Pujol, R.M. 2005. Aggressive multifocal Buruli ulcer with associated osteomyelitis in an HIV-positive patient. *Clin. Exp. Dermatol.*, 30, 649–651.

Torrado, E., Adusumilli, S., Fraga, A.G., Small, P.L., Castro, A.G., and Pedrosa, J. 2007a. Mycolactone-mediated inhibition of tumor necrosis factor production by macrophages infected with *Mycobacterium ulcerans* has implications for the control of infection. *Infect. Immun.*, 75, 3979–3988.

Torrado, E., Fraga, A.G., Castro, A.G., Stragier, P., Meyers, W.M., Portaels, F., Silva, M.T., and Pedrosa, J. 2007b. Evidence for an intramacrophage growth phase of *Mycobacterium ulcerans*. *Infect. Immun.*, 75, 977–987.

Trott, K.A., Stacy, B.A., Lifland, B.D., Diggs, H.E., Harland, R.M., Khokha, M.K., Grammer, T.C., and Parker, J.M. 2004. Characterization of a *Mycobacterium ulcerans*-like infection in a colony of African tropical clawed frogs (*Xenopus tropicalis*). *Comp. Med.*, 54, 309–317.

Ucko, M., and Colorni, A. 2005. *Mycobacterium* marinum infections in fish and humans in Israel. *J. Clin. Microbiol.*, 43, 892–895.

van der Werf, T.S., Stienstra, Y., Johnson, R.C., Phillips, R., Adjei, O., Fleischer, B., Wansbrough-Jones, M.H., Johnson, P.D., Portaels, F., van der Graaf, W.T., and Asiedu, K. 2005. *Mycobacterium ulcerans* disease. *Bull. World Health Organ.*, 83, 785–791.

Vandelannoote, K., Durnez, L., Amissah, D., Gryseels, S., Dodoo, A., Yeboah, S., Addo, P., Eddyani, M., Leirs, H., Ablordey, A., and Portaels, F. 2010. Application of real-time PCR in Ghana, a Buruli ulcer-endemic country, confirms the presence of *Mycobacterium ulcerans* in the environment. *FEMS Microbiol. Lett.*, 304, 191–194.

van Ravensway, J., Eric Benbow, M.E., Tsonis, A.A., Pierce, S.J., Campbell, L.P., Fyfe, J.A.M., Hayman, J.A., Johnson, P.D.R., Wallace, J.R., and Qi, J. 2012. Climate and landscape factors associated with Buruli ulcer incidence in Victoria, Australia. *PloS One*, 7, e51074.

Wagner, T., Benbow, M.E., Brenden, T., Qi, J., and Johnson, R.C. 2008a. Buruli ulcer disease prevalence in Benin, West Africa: Associations with land use/cover and the identification of disease clusters. *Int. J. Health Geogr.*, 7, 25.

Wagner, T., Benbow, M.E., Burns, M., Johnson, R.C., Merritt, R., Qi, J., and Small, P. 2008b. A landscape-based model for predicting *Mycobacterium ulcerans* infection (Buruli ulcer disease) presence in Benin, West Africa. *EcoHealth*, 5, 69–79.

Wallace, J.R., Gordon, M.C., Hartsell, L., Mosi, L., Benbow, M.E., Merritt, R.W., and Small, P.L. 2010. Interaction of *Mycobacterium ulcerans* with mosquito species: Implications for transmission and trophic relationships. *Appl. Environ. Microbiol.*, 76, 6215–6222.

Walsh, D.S., Portaels, F., and Meyers, W.M. 2010. Recent advances in leprosy and Buruli ulcer (*Mycobacterium ulcerans* infection). *Curr. Opin. Infect. Dis.*, 23, 445–455.

Walsh, D.S., Portaels, F., and Meyers, W.M. 2011. Buruli ulcer: Advances in understanding *Mycobacterium ulcerans* infection. *Dermatol. Clin.*, 29, 1–8.

Wang, H., Maeda, Y., Fukutomi, Y., and Makino, M. 2013. An in vitro model of *Mycobacterium leprae* induced granuloma formation. *BMC Infect. Dis.*, 13, 279.

Westenbrink, B.D., Stienstra, Y., Huitema, M.G., Thompson, W.A., Klutse, E.O., Ampadu, E.O., Boezen, H.M., Limburg, P.C., and van der Werf, T.S. 2005. Cytokine responses to stimulation of whole blood from patients with Buruli ulcer disease in Ghana. *Clin. Diagn. Lab. Immunol.*, 12, 125–129.

White, A.D., Sibley, L., Dennis, M.J., Gooch, K., Betts, G., Edwards, N., Reyes-Sandoval, A., Carroll, M.W., Williams, A., Marsh, P.D., McShane, H., and Sharpe, S.A. 2013. Evaluation of the safety and immunogenicity of a candidate tuberculosis vaccine, MVA85A, delivered by aerosol to the lungs of macaques. *Clin. Vaccine Immunol.*, 20, 663–672.

Williamson, H.R., Benbow, M.E., Nguyen, K.D., Beachboard, D.C., Kimbirauskas, R.K., McIntosh, M.D., Quaye, C., Ampadu, E.O., Boakye, D., Merritt, R.W., and Small, P.L.C. 2008. Distribution of *Mycobacterium ulcerans* in Buruli ulcer endemic and non-endemic aquatic sites in Ghana. *PLoS Negl. Trop. Dis.*, 2, e205.

Williamson, H.R., Benbow, M.E., Campbell, L.P., Johnson, C.R., Sopoh, G., Barogui, Y., Merritt, R.W., and Small, P.L.C. 2012. Detection of *Mycobacterium ulcerans* in the environment predicts prevalence of Buruli ulcer in Benin. *PLoS Negl. Trop. Dis.*, 6, e1506.

World Health Organization. 2013a. *Buruli Ulcer Fact Sheet Number 199*. World Health Organization, Geneva, Switzerland.

World Health Organization. 2013b. *Buruli Ulcer: Provisional Guidance on the Role of Specific Antibiotics in the Management of Mycobacterium ulcerans Disease (Buruli Ulcer)*. World Health Organization, Geneva, Switzerland.

World Health Organization. 2013c. *Mycobacterium ulcerans Infection Online Factsheet*. World Health Organization, Geneva, Switzerland.

World Health Organization. Global Buruli Ulcer Initiative. 2004. *Provisional Guidance on the Roles of Specific Antibiotics in the Management of Mycobacterium ulcerans Disease (Buruli ulcer)*. World Health Organization, Geneva, Switzerland.

Yeboah-Manu, Danso, E., Ampah, K., Asante-Poku, A., Nakobu, Z., and Pluschke, G. 2011. Isolation of *Mycobacterium ulcerans* from swab and fine-needle-aspiration specimens. *J. Clin. Microbiol.*, 49(5), 1997–1999.

Yeboah-Manu, D., Peduzzi, E., Mensah-Quainoo, E., Asante-Poku, A., Ofori-Adjei, D., Pluschke, G., and Daubenberger, C.A. 2006. Systemic suppression of interferon-gamma responses in Buruli ulcer patients resolves after surgical excision of the lesions caused by the extracellular pathogen *Mycobacterium ulcerans*. *J. Leukoc. Biol.*, 79, 1150–1156.

Zavattaro, E., Mesturini, R., Dossou, A., Melensi, M., Johnson, R.C., Sopoh, G., Dianzani, U., and Leigheb, G. 2010. Serum cytokine profile during *Mycobacterium ulcerans* infection (Buruli ulcer). *Int. J. Dermatol.*, 49, 1297–1302.

Chapter 45

Challenges Associated with Diagnostics, Drug Resistance, and Pathogenesis of *Mycobacterium tuberculosis*

Juan Carlos Palomino and Anandi Martin

Laboratory of Microbiology, Department of Biochemistry and Microbiology, Ghent University, Ghent, Belgium

45.1 Introduction

Tuberculosis (TB) still remains as an important public health problem at the global level. According to the latest report published by the World Health Organization (WHO), there were, in 2011, 8.7 million incident cases of TB and 1.4 million deaths were attributed to the disease (World Health Organization, 2012). The same report also estimates the occurrence of 630,000 cases of multidrug-resistant (MDR)-TB. MDR-TB is defined as cases of TB caused by strains of *Mycobacterium tuberculosis* that are resistant to at least rifampicin (RIF) and isoniazid (INH), two key drugs in the treatment of the disease. Even more worrying is the emergence of extensively drug-resistant (XDR)-TB, caused by strains of *M. tuberculosis* that, in addition to being MDR, are resistant to any fluoroquinolone and to at least one of the second-line injectable drugs, kanamycin (KAN), capreomycin (CAP), or amikacin (AMK) (Almeida da Silva and Palomino, 2011). It is estimated that 9% of MDR-TB cases are also XDR-TB (World Health Organization, 2012). To make matters worse, there have been recent reports of strains of *M. tuberculosis* that were practically resistant to all antibiotics that were tested, a situation that has been termed as totally drug-resistant (TDR)-TB (Migliori et al., 2012; Udwadia et al., 2012; Velayati et al., 2012). The latter denomination, though, has not been formally agreed.

Important progress has been made in the last years for a better control of the disease. Globally, there is a trend to achieve the Millennium Development Goal target of a 50% reduction in TB burden by 2015. However, global TB burden still remains high and only one in five cases of MDR-TB are estimated to be detected. In the absence of a new and more effective TB vaccine and newer and improved treatment options, there is an urgent need to develop accurate, rapid, and affordable diagnostics and innovative technologies for the timely detection of drug-resistant TB (Palomino, 2012; Young et al., 2008).

45.2 The classical route

45.2.1 Direct microscopy

Detection of *M. tuberculosis* by direct microscopy examination of Ziehl–Neelsen (Z–N)-stained sputum smears is the simplest way of presumptively diagnosing TB and it stands as the sole method of diagnosis in many low-resource settings. However, sensitivity of sputum smear microscopy remains low and 5000–10,000 bacilli/mL of sputum are necessary to be spotted as positive (Yeager et al., 1967). An increase of sensitivity can be obtained by using fluorescence microscopy of auramine-stained sputum smears. The use of auramine was introduced many years ago, and later on re-assessed using the combination auramine-O/rhodamine (Ba and Rieder, 1999). This method is associated with higher detection rates since the slides are examined at lower magnifications. It is, then, usually accepted that the fluorescence method should be given preference over Z–N, especially in laboratories with a large number of specimens.

Human Emerging and Re-emerging Infections: Bacterial & Mycotic Infections, Volume II, First Edition. Edited by Sunit K. Singh.
© 2016 John Wiley & Sons, Inc. Published 2016 by John Wiley & Sons, Inc.

Fluorescence microscopy has been recently boosted by the introduction of light-emitting diode (LED) microscopes that do away with the need of replacing old and costly mercury vapor lamps (Anthony et al., 2006). LED microscopy appears to be more sensitive and faster than classical fluorescence microscopy or Z–N staining (Marais et al., 2008). These results have been corroborated in larger and more recent evaluations of the implementation of LED microscopy in different settings (Cuevas et al., 2011; Minion et al., 2011). Based on this evidence, the WHO has recently recommended that fluorescence microscopy should be replaced by LED microscopy using auramine and that the latter should be phased in as an alternative to the classical Z–N light microscopy (World Health Organization, 2011).

Several efforts have also been made in the past to increase the sensitivity of Z–N microscopy. The use of sodium hypochlorite (household bleach) for liquefying the sputum followed by concentration by centrifugation was proposed. An earlier evaluation of the studies performed found a statistically significant improvement in the proportion of positive smears or sensitivity (Angeby et al., 2004). More recent evaluations, however, have found only small increases in sensitivity with concomitant small decreases in specificity (Cattamanchi et al., 2010). Combined application of LED microscopy and sodium hypochlorite sedimentation has not improved further the detection rates either (Bonnet et al., 2011; Habtamu et al., 2012).

Being simple and inexpensive, microscopy examination of sputum-stained smears is still in need of more innovative developments to turn it into a true point-of-care (POC) test for TB diagnosis.

45.2.2 Cultivation of *M. tuberculosis*

The gold standard for TB diagnosis in the laboratory remains the isolation of *M. tuberculosis* in culture from a clinical specimen. Since many years, this has been accomplished by using agar-based or egg-based culture media; the most commonly used being the Middlebrook 7H10 or 7H11 agars and the Löwenstein–Jensen (L–J) medium (Heifets, 1997). In contrast to sputum smear microscopy, culture techniques can detect up to 100 bacilli/mL of sample (Tiruviluamala and Reichman, 2002) and they provide material for further identification and drug susceptibility testing. Following decontamination and liquefaction procedures, specimens are inoculated and incubated to detect morphological growth, which usually occurs after several weeks of incubation. Identification as *M. tuberculosis* was traditionally done by performing various biochemical tests (Metchock et al., 1999). Nevertheless, these classical procedures are laborious and time-consuming requiring from 3 to 8 weeks to provide results.

Faster and reliable liquid culture-based technologies became available some time ago. The first generation of these newer cultivation techniques is represented by the radiometric culture method such as the BACTEC TB-460 system (Becton Dickinson, Sparks, MD, USA). This commercial system was kind of a breakthrough at the time, since it allowed detection of *M. tuberculosis* in a matter of days compared to weeks needed with the conventional culture media (Roberts et al., 1983). However, the use of isotope-labeled culture media and the cost of the equipment precluded its use on a routine basis, with the exception of reference laboratories predominantly in developed countries. A newer generation of non-radiometric, semi-automated liquid culture-based systems became available recently (Palomino, 2005). Three such systems became commercially available: the Mycobacterium Growth Indicator Tube (MGIT), the BacT/Alert system, and the Versa TREK system (Palomino et al., 2008). The most commonly used of the three is the MGIT system. This system, initially launched as a manual method, is now available as the MGIT960 and MGIT320 systems. The method is based on fluorescence detection that correlates with mycobacterial growth. MGIT culture vials contain an enriched Middlebrook 7H9 broth and a fluorescence quenching-based oxygen sensor embedded at the bottom of the tube. Consumption of the dissolved oxygen in the medium by the growing mycobacteria allows detecting the fluorescence when interrogated by a UV light. This was accomplished with an UV lamp in the original manual system, but is now performed inside the machine in the more recent automated systems. Both manual and automated systems perform similarly with results comparable to those obtained with the former radiometric BACTEC method. The time to detection varies between 5 and 12 days (Palomino, 2005). The MGIT960 system has been widely evaluated for detection and identification of *M. tuberculosis* including in high-endemic low-resource countries (Rodrigues et al., 2009). Based on the evidence generated in multiple studies, the WHO has endorsed the use of liquid culture assays and rapid identification to the species level according to country-specific plans of laboratory capacity strengthening (World Health Organization, 2007).

45.2.3 Immunological diagnosis of tuberculosis

Until today there is no reliable and accurate way of diagnosing TB by immunological methods and this remains as an unfinished task. In contrast to other bacterial infections, the role of serological tests in TB diagnosis has not been fully demonstrated. Such serological tests would normally detect the antibody response in serum to one or more mycobacterial antigens common to *M. tuberculosis* but absent from non-tuberculous mycobacteria (NTM) and other microorganisms. Several types of blood tests have been proposed, but until now, none of them have shown to be predictive enough to warrant their use as a diagnostic TB test (Steingart et al., 2007). Moreover, the WHO has recently given a negative recommendation for these serological tests based on the evidence available until now (Morris, 2011). But considering that one-third of the world's population is estimated to be latently infected with *M. tuberculosis*, it would be desirable to have a way to predict who, among the latently infected persons, will develop active TB. Diagnosis of latent TB infection, however, remains a difficult endeavor. Since long, the only available tool to assess TB infection was the tuberculin skin test (TST) or Mantoux test, which was considered as the gold standard. The TST measures cell-mediated response to a crude mixture of mycobacterial proteins prepared from cultures of *M. tuberculosis*. Unfortunately, TST lacks sensitivity, especially in immuno-compromised persons and is not specific (Cobelens et al., 2006). False-positive results are relatively common following exposure to *Mycobacterium bovis* BCG and environmental NTM (Farhat et al., 2006).

45.2.4 Interferon gamma release assays

In the last decade, newer *in vitro* T cell-based assays were proposed. The interferon gamma release assays (IGRAs) gained much attention for the possibility of differentiating latent from active TB. It is well known that when T cells are sensitized with *M. tuberculosis*, they produce interferon gamma when re-exposed to mycobacterial antigens. A high production of interferon gamma is then assumed to correlate with TB infection (Pai et al., 2004). The first versions of IGRAs used purified protein derivative (PPD) as the antigen for stimulation; subsequent versions use *M. tuberculosis*-specific antigens, such as the early secretory antigen target 6 (ESAT6) and the culture filtrate protein 10 (CFP10). These antigens are coded by RD1 genes in the *M. tuberculosis* genome and are absent in *M. bovis* BCG and most NTM, with the exception of *M. marinum*, *M. sulzgai*, and *M. kansasii* (Dinnes et al., 2007). Two commercially available IGRA tests are in the market, the QuantiFERON-TB Gold In-Tube assay (Cellestis Ltd., Carnegie, Australia) and the T-SPOT.TB test (Oxford Immunotec, Oxford, UK). These tests measure production of interferon gamma in response to *M. tuberculosis* antigens by ELISA and enzyme-linked immunospot, respectively.

The QuantiFERON®-TB Gold In-Tube test uses whole blood for measuring the concentration of secreted interferon gamma. The T-SPOT TB assay uses peripheral blood mononuclear cells to measure the number of interferon gamma-producing T-cells (Lalvani and Pareek, 2004). Published studies evaluating the performance of IGRAs have shown advantages over TST, higher specificity, better correlation with pre-exposure to *M. tuberculosis*, and low cross-reactivity due to BCG vaccination or previous exposure to NTM (Dinnes et al., 2007; Lalvani and Pareek, 2004). According to published evidence, QuantiFERON-TB Gold In-Tube test

has apparently similar sensitivity to TST, while T-SPOT.TB assay sensitivity appears to be higher (Lalvani and Pareek, 2004). There are several published reviews that assess the performance of IGRAs in different settings and study populations (Diel et al., 2010; Pai et al., 2008).

The current consensus, summarized in a recent European Center for Disease Control (ECDC) Guidance document, is that, based on the available evidence, IGRAs should not replace the current diagnostic methods for active TB, and in relation to the diagnosis of latent TB infection, IGRAs should only be used with an overall risk assessment to identify individuals whom might benefit from preventive treatment (ECDC, 2011). In conclusion, neither method is capable by itself of diagnosing TB or differentiating active disease from latent infection. Immunological diagnosis of TB directly from clinical samples and the differentiation of active TB disease from latent TB infection remains, thus, the Holy Grail in the field of TB diagnosis.

45.3 Detecting drug resistance

Since many years ago, detection of drug resistance in *M. tuberculosis* has been done by performing classical bacteriological procedures based on the growth of the bacteria in the presence or absence of antibiotics. Procedures such as the proportion method, the resistance ratio method, and the absolute concentration method, all developed almost 50 years ago, are still in use in many TB diagnostic laboratories that perform drug susceptibility testing. Moreover, they are still considered as the gold standard for detecting drug resistance in TB. Although mostly reliable, they suffer from the same disadvantage as most bacteriological-based methods in TB, namely, their delay in giving results due to the slow growth of the tubercle bacillus.

The advent of the radiometric-based BACTEC TB-460 in the 1980s and the other liquid culture-based systems in the last decade offered the possibility to improve turnaround time for drug resistance testing, giving results in a matter of several days instead of few weeks (Palomino et al., 2008). The MGIT960 system is also considered nowadays as a gold standard method for detecting drug resistance.

45.4 New avenues for TB diagnosis and detection of drug resistance

Microscopy techniques and culture-based methods are currently the cornerstone for TB diagnosis and drug resistance detection in many parts of the world. However, the limitations of these techniques and the alarming rates of TB burden and drug resistance in many settings call for improved and faster technologies to address this problem. The following sections will present technologies and methodologies of more recent introduction for TB diagnosis and for the detection of drug resistance. Some of them are already incorporated into the routine diagnostic algorithms for TB in many countries. Other more recent developments are still undergoing evaluation.

45.5 Molecular-based technologies

The sequencing of the genome of *M. tuberculosis* in 1998 was a breakthrough at the time (Cole et al., 1998). This, linked to other developments in nucleic acid amplification techniques after the introduction of the polymerase chain reaction (PCR) method in the previous decade, opened many possibilities for the rapid detection of *M. tuberculosis* by molecular techniques (Pai et al., 2006; Palomino, 2009). Nucleic acid amplification tests (NAATs) depend on the enzymatic amplification of short and specific sequences of DNA or RNA of *M. tuberculosis* by PCR, which are then detected in agarose or acrylamide gels by electrophoresis, or by hybridization using various formats. The most frequently used target for *M. tuberculosis* detection has been the insertion sequence IS*6110*. This marker together with the use of a nested-PCR approach was associated with higher accuracy values in an analysis of several previous studies (Flores et al., 2005). As a general rule, sensitivity and specificity of NAATs have been higher for respiratory specimens compared with non-respiratory samples. Many "in-house" PCR assays were developed and are thoroughly described in the literature (Greco et al., 2009). As with other NAATs when applied directly to clinical samples, PCR can be affected by the presence of inhibitors in the samples.

Commercially available molecular tests for TB diagnosis have gained increased acceptance. Some of them have been developed with the main purpose of detecting the presence of *M. tuberculosis*, while others can also be applied for the detection of drug resistance. The following sections will describe some of them.

45.5.1 Molecular TB diagnosis: the pioneers

Arguably, there are two or three molecular tests that were the pioneers in proposing rapid and specific detection of *M. tuberculosis* in sputum samples. These were the Amplicor Mycobacterium Tuberculosis Test (MTB; Amplicor, Roche Diagnostic Systems Inc., New Jersey, NJ, USA) and the Amplified *Mycobacterium tuberculosis* Direct Test (MTD; Gen-Probe Inc., San Diego, CA, USA). Both tests were approved by the US Food and Drug Administration (FDA) many years ago. The Amplicor MTB test is a DNA-based assay that involves amplification of a fragment of the 16S rRNA gene with genus-specific primers. Following hybridization to oligonucleotide probes, the detection is performed in a colorimetric reaction performed in a microplate (Dalovisio et al., 1996). Together with the COBAS Amplicor analyzer (Roche Diagnostics, Switzerland) subsequently introduced, the COBAS Amplicor MTB test allowed amplification and detection to be performed automatically in one system.

A new version of the system, the COBAS TaqMan® MTB test, makes use of real-time PCR performed in a COBAS TaqMan 48 analyzer, which can run up to 48 samples in approximately 2.5 hours. The Amplicor assays were intended to be used with decontaminated and concentrated smear-positive respiratory samples of patients without previous treatment (Michos et al., 2006). The reported overall sensitivity in those studies was 83.0–92.4% for respiratory samples, 90–100% for smear-positive samples, and 50.0–95.9% for smear-negative samples. In extra-pulmonary samples, the sensitivity reported was persistently lower. Overall specificity ranged from 91.3 to 100% (Ozkutuk et al., 2006). A recent thorough comparison of both systems, Amplicor MTB PCR and TaqMan MTB PCR, shows further improvement in specificity (close to 100%), while the sensitivity remained almost the same at 75–77% (Tortoli et al., 2012). The need for more effective DNA extraction procedures and a better lysis of the mycobacterial cell wall have been signaled as issues for improvement.

The Amplified MTD, on the other hand, is based on isothermal amplification of 16S ribosomal transcripts, which are detected by hybridization in an assay with acridinium ester-labeled DNA probes, specific for *M. tuberculosis* complex (Abe et al., 1993). The test was FDA approved for direct detection of *M. tuberculosis* in smear-positive and smear-negative respiratory samples, requiring a luminometer for interpretation (Coll et al., 2003). Previous evaluations have reported an overall sensitivity in the range of 77–100%, with 90–100% in smear-positive and 63–100% in smear-negative samples (Piersimoni and Scarparo, 2003).

The third molecular assay launched some years ago was the BD ProbeTec MTB Test (Becton Dickinson, Sparks, MD, USA) as a semi-automated system for the rapid diagnosis of TB (Bergmann and Woods, 1998). This test was based on the strand-displacement amplification format, which uses enzymatic replication of target sequences in IS*6110* and the 16S rRNA gene. The amplicons are then detected in a luminometer. In evaluations with respiratory samples, the sensitivity was 100% for smear-positive specimens and 92–100% in smear-negative samples. Overall specificity was reported as 96–99% (Pfyffer et al., 1999). One disadvantage of this version of the test was the time needed for sample preparation (~2 hours). A new version of the system, the BDProbe Tec ET that includes an internal amplification control to rule out the presence of inhibitors, was later introduced. Again, higher sensitivity and specificity values were found in respiratory smear-positive samples (Rusch-Gerdes and Richter, 2004).

45.5.2 Line probe assays

Molecular detection of *M. tuberculosis* by line probe assays (LPAs) was already introduced in 1995. The format also allowed rapid detection of drug resistance by looking for drug resistance-associated mutations, although this was initially restricted to RIF (De Beenhouwer et al., 1995). Since then, DNA sequencing became the reference standard for detection of drug resistance by molecular methods in *M. tuberculosis* (Garcia de Viedma, 2003). Yet, for practical, economical, and logistical reasons, it would be extremely difficult for the time being to organize routine laboratory facilities for resistance detection to all major anti-TB drugs by DNA sequencing alone. Furthermore, not all mechanisms of drug resistance of *M. tuberculosis* are fully known for all the drugs involved.

LPAs, also known as solid-phase hybridization assays, are based on the reverse hybridization of oligonucleotides on nitrocellulose strips to which specific probes have been immobilized. Amplified products from the samples are then bound to specific probes, and hybridization is visualized by the development of a colored band on the strip. There are now three LPAs for the rapid detection of *M. tuberculosis* and its associated drug resistance commercially available: the Line Probe Assay (LiPA) (INNO-LiPA Rif TB Assay, Innogenetics, Belgium), the GenoType MTB-DRPlus and GenoType MTBDRsl (Hain Lifesciences, Germany), and the AID TB Resistance Modules (AID Diagnostika, GmbH, Strassberg, Germany).

The INNO-LiPA Rif TB Assay was the first to appear some years go. It relies on the reverse hybridization of amplified DNA fragments from *M. tuberculosis* to ten probes covering the core region of the *rpoB* gene immobilized on a nitrocellulose strip. The hybridization pattern obtained reveals the presence or absence of mutated or wild regions that are seen as colored bands on the strip. The sample is, thus, considered as resistant or susceptible to RIF (De Beenhouwer et al., 1995). The INNO-LiPA Rif TB assay can be performed on *M. tuberculosis* isolates or directly on sputum samples (Tortoli and Marcelli, 2007; Traore et al., 2006). Since detection of RIF resistance is considered as a good surrogate indicator of MDR-TB, several studies evaluated the application of the test for MDR-TB detection. The study by Traore et al. analyzed 420 sputum samples from different countries and found 99.6% agreement between INNO-LiPA Rif TB and culture.

The GenoType MTBDR*plus* and GenoType MTBDR*sl* are two versions of the same LPA developed for the rapid detection of *M. tuberculosis* and its associated resistance to several anti-TB drugs. GenoType MTBDR*plus* followed the original Geno-Type MTBDR incorporating detection of resistance to INH and RIF, both in clinical isolates and sputum samples. It is based on the detection of the most common mutations in the *katG*, *inhA*, and *rpoB* genes (Hillemann et al., 2007). Like the INNO-LiPA Rif TB assay, GenoType MTBDR*plus* is based on PCR and reverse-hybridization of probes immobilized on a plastic strip. A large evaluation of the implementation of GenoType MTBDR*plus* in a high-volume public health laboratory for MDR-TB screening found that,

with smear-positive samples, 97% of results were available within 1–2 days. Sensitivity for RIF resistance detection was 98.9% while for INH resistance was 94.2%. Specificity values were 99.4% and 99.7%, respectively. For MDR-TB detection, sensitivity and specificity was 98.8% and 100%, respectively, compared with conventional drug susceptibility testing (Barnard et al., 2008). The application of the GenoType MTBDR*plus* in other TB endemic settings has been recently further confirmed (Imperiale et al., 2012; Maschmann et al., 2013). The other version, GenoType MTBDR*sl* (Hain Life Sciences, Germany), allows detecting resistance to second-line drugs: fluoroquinolones, AMK, KAN, CAP, and also ethambutol, by looking for the most common mutations in the genes *gyrA*, *rrs*, and *embB* (Hillemann et al., 2009). Initial evaluations of this version of the test have given varied results (Brossier et al., 2010). A recent meta-analysis to assess the diagnostic accuracy of GenoType MTBDR*sl*, based on the several published studies, concluded that the test had good accuracy for detecting resistance to fluoroquinolones, AMK, and CAP. The test was not optimal, however, to detect resistance to KAN and ethambutol (Feng et al., 2013). Variation in results can be explained by the fact that mutations not included in the test, such as those present in *gyrB* or resistance mechanisms not yet fully described (e.g., for ethambutol), are obviously not detected by the test. This reinforces the current recommendation that LPAs should be used as a first screening tool followed by culture-based drug susceptibility testing methods.

The AID TB Resistance Modules (AID Diagnostika, GmbH, Strassberg, Germany) have been recently introduced. There are three separate modules, TB Resistance Module Isoniazid/Rifampicin, TB Resistance Module Aminoglycosides, and TB Resistance Module Fluoroquinolones. They look for the most common mutations in genes *rpoB*, *katG*, *inhA*, *rrs*, *rpsL*, *gyrA*, and *embB*.

In 2008, the WHO endorsed the use of molecular LPAs for rapid screening of patients at risk of MDR-TB (World Health Organization, 2008).

Additionally, there are several other "home-brew" formats of reverse-hybridization tests developed for TB diagnosis and detection of drug resistance that will not be discussed in more detail in this chapter (Giannoni et al., 2005; Hernández-Neuta et al., 2010; Senna et al., 2006).

45.5.2.1 Newer real-time PCR-based methods

In addition to the COBAS TaqMan MTB test mentioned in Section 45.5.1, several other methods for rapid diagnosis of *M. tuberculosis* in clinical samples and for the detection of RIF resistance using real-time PCR have been introduced (Varma-Basil et al., 2004). Real-time PCR-based methods use fluorescent-labeled probes with the reaction monitored inside a thermal cycler. The increase in fluorescence is directly proportional to the amount of amplified product in the reaction tube. The major advantage of real-time PCR is the speed to obtain results and minimal contamination rates since reaction and detection occur in a single vial.

45.5.3 The GeneXpert technology

The real-time PCR-based technology which has recently acquired greater visibility is undoubtedly the GeneXpert system (Cepheid, Sunnyvale, CA, USA), based on a single-use sample-processing cartridge that integrates real-time PCR technology to detect the target (Raja et al., 2005). In the original description of the prototype instrument, the GeneXpert was presented as a system to perform automated nucleic acid isolation, reverse transcription, and quantitative PCR in about half an hour. The system was applied for the analysis of different samples to detect cancer markers. More recently, it has been proposed as a rapid on-demand near-patient technology for detection of *M. tuberculosis* and resistance to RIF (Helb et al., 2010). The GeneXpert system comprises a self-contained, integrated, and automated machine that can be operated with minimal technical skills. It combines onboard sample preparation and automated real-time PCR and detection functions. The cartridge-based system contains microfluidics components pre-loaded with buffers and lyophilized reagents used for the automated nucleic acid purification, concentration, detection, and identification of targeted sequences directly from clinical samples (Lawn and Nicol, 2011).

Currently known as the Xpert MTB/RIF assay, the test is based on a previous prototype assay that used molecular beacons (El-Hajj et al., 2001; Tyagi and Kramer, 1996). This approach uses five different probes in one multiplex reaction. The probes are complementary to target sequences spanning the rifampicin-resistance-determining-region of the *rpoB* gene of a wild-type *M. tuberculosis*. Each probe is labeled with a different fluorescent color. The presence of a mutation within each region will affect the hybridization, causing partial or complete inhibition of fluorescence in the corresponding molecular beacon. Sputum samples are pre-treated with a "sample reagent" containing sodium hydroxide and isopropanol for 15 minutes. The treated sample is then manually loaded into the cartridge that is entered in the GeneXpert machine. For the identification of *M. tuberculosis* (TB diagnosis), at least two of the five probes must give a positive signal. The machine indicates the presence (or absence) of *M. tuberculosis* and the occurrence of RIF resistance based on the fluorescence signals after hybridization of the probes. Results are generated within 2 hours.

Initial laboratory evaluation of the Xpert MTB/RIF assay reported a limit of detection of 4.5 genome copies and over 100 bacilli/mL in spiked sputum samples (Helb et al., 2010), with very good specificity both for *M. tuberculosis* identification and RIF resistance detection. The first large clinical evaluation of the Xpert MTB/RIF assay was published in 2010 (Boehme et al., 2010). This study found that among culture-positive TB patients, Xpert MTB/RIF detected 98.2% and 72.5% of patients with smear-positive and smear-negative TB, respectively. The specificity was 99.2% in this study. The sensitivity in smear-negative samples could be increased if a second or third test was performed on the same sample. RIF resistance was detected with 97.6% sensitivity and 98.1% specificity compared to conventional drug detection methods.

Since the initial description of the test and the large evaluation performed in several countries, numerous studies have evaluated the accuracy of Xpert MTB/RIF for the diagnosis of TB and detection of RIF resistance. A recent meta-analysis that included 18 published studies and more than 10,000 suspected clinical samples reports an overall sensitivity of 90.4% and specificity of 98.4%. Detection of RIF resistance was 94.1% sensitive and 97% specific. These values were found to be lower in extra-pulmonary TB and in smear-negative samples (Chang et al., 2012). Currently ongoing and future studies will better document the accuracy of this test in the pediatric population and in HIV co-infected patients.

45.5.4 Loop-mediated isothermal amplification

A newer development for TB diagnosis is the loop-mediated isothermal amplification (LAMP) assay, originally described by Notomi et al. some years ago (Mori et al., 2001; Notomi et al., 2000). The main advantages of LAMP include: the capacity to amplify the target DNA under isothermal conditions (60–65°C) obviating the need for a thermocycler; the increased specificity, since it uses several primers to amplify different regions of the target sequence; the ability to yield an amplification of 10 × 9 copies of the target sequence, comparing very favorably in terms of sensitivity to classical PCR. In the original description of the assay, the amount of amplified DNA was quantified by measuring the pyrophosphate formed as a by-product of the reaction, seen as a white precipitate. This was later improved by the addition of different fluorescent dyes, such as SYBR Green I. A further modification has been the development of a fully colorimetric format by incorporation of hydroxy-naphthol blue with the possibility to perform many tests in a microwell plate format (Goto et al., 2009).

A first performance evaluation of the LAMP assay for diagnosis of pulmonary TB was conducted in three microscopy centers in low-resource countries (Boehme et al., 2007). In this initial study, the sensitivity for smear-positive and culture-positive sputum samples was 97.7%, while in smear-negative, culture-positive samples, the sensitivity was only 48.8%. An improved LAMP assay has been developed recently as a test kit (LAMP, Eiken Chemical Co. Ltd., Tokyo, Japan). A performance evaluation of this kit has been done in sputum samples with somewhat better results than in the previous study (98.2% and 55.6% sensitivity, respectively). Several studies have also been performed to test different targets of *M. tuberculosis* to improve performance (Pandey et al., 2008; Zhu et al., 2009).

Two very recent developments that make use of the LAMP technology for TB diagnostics in new formats are worth mentioning. The first is a high-throughput microfluidics system that integrates LAMP analysis (Liu et al., 2013). The system makes use of a capillary array with droplet technology and magnetic beads. It is able to process ten samples simultaneously and has a limit of detection of 10 bacteria with readout available within 50 minutes. The second development combines LAMP with lateral flow dipstick format for detection of *M. tuberculosis* in clinical samples (Kaewphinit et al., 2013). The test is able to detect *M. tuberculosis* in approximately 1 hour. A preliminary testing with sputum samples rendered a sensitivity of 98.9% and 100% specificity. This format has the potential to become a near-the-patient test.

One study has proposed to apply LAMP assay coupled with restriction endonuclease digestion and ELISA to detect drug resistance (Lee et al., 2010). This modified assay was able to detect resistance to INH, ethambutol, and streptomycin indirectly on *M. tuberculosis* isolates. No studies have been done yet directly on clinical samples.

45.6 Alternative culture-based methods

Several alternative new culture-based approaches have been described for a faster and improved diagnosis of TB, which are also applied for the detection of drug resistance. There are three newer noncommercial methods for this purpose that have been recently endorsed by the WHO for the monitoring of patients at risk of MDR-TB (World Health Organization, 2009). They are the colorimetric redox indicator (CRI) methods, the nitrate reduction assay (NRA), and the microscopic observation broth drug susceptibility (MODS) assay.

The CRI methods, instead of looking for mycobacterial colony growth, detect the metabolic activity of the bacteria measured in a colored reaction in liquid medium. The most commonly used indicators are resazurin, used in the REMA assay, and 3-(4,5-dimethylthiazol-2-yl)-2,5-diphenyl tetrazolium bromide (MTT) (Martin et al., 2005a; Mshana et al., 1998; Palomino et al., 2002). Numerous evaluations of the performance and applicability of the CRI methods for detection of drug resistance have shown good accuracy and reliability (Martin et al., 2007; Montoro et al., 2005). The CRI methods have also been used for rapid direct detection of *M. tuberculosis* and RIF resistance from sputum samples (Abate et al., 2004; Boum et al., 2013).

The nitrate reductase assay (NRA) is based on the ability of *M. tuberculosis* to reduce nitrate to nitrite, which is then detected in a colored reaction with the Griess reagent (Angeby et al., 2002). As originally proposed, the test is performed in the L–J culture medium, which facilitates its integration into the routine tests of TB diagnostic laboratories. In an effort to obtain faster results, a liquid culture format has also been proposed (Syre et al., 2003). Numerous studies have shown the ease of use and good accuracy of the NRA for detecting drug resistance (Martin et al., 2005b, 2008). NRA has also been applied as a direct diagnostic method in sputum (Musa et al., 2005; Shikama et al., 2009).

The MODS assay looks into micro-colonies of *M. tuberculosis* instead of waiting for normal bacterial colonies to grow, which usually takes 3–6 weeks (Caviedes et al., 2000; Palomino et al., 2008). This is based on the characteristic growth of *M. tuberculosis in vitro*, which tends to form "cords" that can be distinguished by a trained observer, allowing presumptive identification of the TB bacillus. MODS is capable of detecting growth within 7–11 days, while also allowing presumptive identification of *M. tuberculosis*. The test is performed in liquid medium and observation of micro-colonies is done with an inverted microscope. In a large operational study conducted in Peru, the performance of MODS was evaluated against two reference methods for culture and drug susceptibility testing of *M. tuberculosis* (Moore et al., 2006). MODS showed a sensitivity of 98.7% for detection with a median time to culture positivity of 7 days. A recent meta-analysis that assessed the accuracy of MODS for detection of drug resistance yielded a sensitivity of 98% and specificity of 99.4% for RIF, and 97.7% and 95.8% of sensitivity and specificity, respectively, for INH (Minion et al., 2010).

A very similar assay, the thin-layer agar (TLA) method, looks also for micro-colonies of *M. tuberculosis* but the test is performed in solid culture media (Mejia et al., 1999). TLA has also been evaluated for performance in different settings giving similar good results (Martin et al., 2009; Robledo et al., 2006).

Phage-based methods were also proposed some time ago as tools for rapid TB diagnosis and drug resistance detection. The originally described approaches did not perform, however, well under field conditions and did not continue to be implemented widely as initially thought (Biswas et al., 2008; Mole et al., 2007). An interesting variant with fluorescent mycobacteriophages has been recently proposed (Piuri et al., 2009). In this format, fluorescent mycobacteriophages deliver a green fluorescent protein (GFP) or ZsYellow fluorescent marker gene to *M. tuberculosis*, which is then detected by fluorescence microscopy or flow cytometry. One advantage of this assay is that the fluorescence readout resists fixation with paraformaldehyde, providing a biosafety advantage when testing drug-resistant strains by microscopy or flow cytometry. In preliminary

Table 45-1 Main current technologies for the diagnosis of tuberculosis and detection of drug resistance

Technique/methodology	Aim	Requirements	Cost
Light microscopy	Diagnosis	Low	Low
Fluorescence microscopy	Diagnosis	Low/medium	Low
Conventional culture (L–J, 7H10)	Diagnosis/drug resistance detection	Low/medium	Moderate
Liquid culture systems	Diagnosis/drug resistance detection	High	High
Molecular PCR based	Diagnosis	High	High
Molecular isothermal	Diagnosis	Medium	Medium
Molecular real-time PCR based	Diagnosis/RIF resistance	High	High
Alternative culture methods (CRI, NRA, MODS)	Diagnosis/drug resistance detection	Low/medium	Moderate
Immunological (IGRAs)	Diagnosis	Low/medium	Moderate

evaluations, resistance to RIF or streptomycin was detected, as lack of fluorescence, in less than 24 hours. A more recent evaluation of this assay with clinical isolates of *M. tuberculosis* has confirmed these results (Rondón et al., 2011). Table 45-1 shows some of the main technologies currently available for TB diagnosis and for detection of drug resistance.

45.7 Pathogenesis of tuberculosis

The granuloma formation is the characteristic sign of TB pathogenesis. These organized structures consist of infected and non-infected macrophages, foamy macrophages, epithelioid cells, and multi-nucleated giant cells, known as Langerhans cells, all surrounded by lymphocytes. The classical view states that the granuloma mediates the long-term containment of the mycobacteria in the lung (Saunders and Cooper, 2000). The formation of the granuloma comprises the interrelated production of several chemokines and cytokines as well as the upregulation of their corresponding receptors for the recruitment and migration of these cells to the granuloma. Recently, the role of the Th17 immune response and how it regulates the formation of the granuloma and its influence in the anti-mycobacterial immunity and pathology of TB have also been recognized (Torrado and Cooper, 2010).

The route of infection in TB is through the airborne channel. Following inhalation, the bacilli are phagocytosed by macrophages, which lead into the interstitial space and eventually come into contact with dendritic cells. These become activated migrating to the draining lymph nodes and then stimulate naïve T cells. Under normal circumstances, concerted antigen presentation, macrophage and T-cell activation, all together contribute to an efficient containment of bacillary growth and control of the infection.

However, complete sterilization is hardly achieved and persons infected with *M. tuberculosis* remain responsive to mycobacterial antigens. The tubercle bacillus, thus, enters into a dormant state and this is known as latent TB infection (LTBI).

It is well known that only a minor proportion of infected individuals (around 5–10%) develop active disease. The risk of developing active TB after infection has been estimated as 5–10% during a lifetime, which can increase to 10% annually in case of immunocompromised individuals, such as HIV co-infected patients. But how this latent state is maintained and at what time and for what reason bacilli that have been dormant, even for decades, can "wake-up" and cause overt disease and pathology? Why this occurs in certain individuals and not in others? These are important and unanswered questions at the center of TB pathogenesis and many efforts are being made to address this issue. One key element for the control of the disease would be to be able to predict who among the TB-infected persons will develop active disease, so this can be prevented in time.

New knowledge is being generated in our understanding of the formation and role of the TB granuloma. For example, recent studies in the zebrafish model using the surrogate bacteria *Mycobacterium marinum* suggests that the granuloma, instead, might be contributing to the bacterial growth at the early steps of infection (Davis and Ramakrishnan, 2009). By applying quantitative intravital microscopy, these researchers showed that *M. marinum* utilized ESX-1/RD1, a known virulence locus, to promote the recruitment of new macrophages to the granuloma. These new macrophages then phagocytosed infected macrophages and underwent apoptosis, in this way producing an expansion of the infected macrophages and at the same time of the bacterial load. These results, the authors concluded, might be signaling at how the bacteria exploit the granuloma during the innate immune response for their own benefit.

Nevertheless, the zebrafish model has its detractors, who argue that since they, actually, do not have lungs, it could be difficult to extrapolate such results to what really happens in the human host.

Studying the human TB granuloma presents its challenges and it is not an easy task. There are some few established models all of them being surrogates of the human TB granuloma (Guirado and Schlesinger, 2013). Among the animal models that can reproduce some of the events occurring in humans, the most commonly used are mice, guinea pigs, rabbits, non-human primates, the zebrafish, and minipigs. The mouse model is undoubtedly the most commonly used, although it resembles more a chronic rather than a latent infection as in humans. Moreover, the granuloma that forms in the mouse model does not resemble the human TB granuloma. Some newer mouse models are being proposed that resemble more closely the necrotic nature of the human granuloma (Driver et al., 2012; Harper et al., 2012). The non-human primate model, on the other hand, is the one that probably better resembles the course of infection and pathogenesis that occurs in humans. Unfortunately, several practical and ethical issues severely restrict the use of non-human primates for trials. A minipig model has been recently introduced that apparently have similar pulmonary structure as those in humans and reproduce fairly well the lesions as those appearing in humans (Gil et al., 2010).

Study of the TB granuloma has also been attempted by models *in vitro*. The so-called *"in vitro* granuloma" has been proposed as an alternative option to study early granuloma formation and maintenance. These models are based on the use of peripheral blood mononuclear cells and some kind of support that could be matrix gels or agarose beads that can be treated with mycobacterial antigens (Birkness et al., 2007; Seitzer and Gerdes, 2003).

More recently, mathematical and computational modeling have also started to be explored in the study of host–pathogen interactions, including the immune response to TB and the formation of the granuloma (Fallahi-Sichani et al., 2011; Marino et al., 2011). These approaches could well complement the knowledge obtained by the other conventional research tools.

45.7.1 The role of virulence factors

Virulence factors play an important role for the survival of *M. tuberculosis* inside the macrophages. The main survival mechanism of the bacteria in the alveolar macrophage is by inhibiting phagosome acidification and maturation. In other pathogens, the role of secreted proteins, such as phosphatases, in modulating the immune response of the host has been thoroughly demonstrated (Heneberg, 2012). In this context, several previously described virulence factors seem to play an important role in the pathogenesis of the TB bacillus, by interfering important functions and defense mechanisms of the host (Chao et al., 2010; Wong et al., 2013).

For example, the protein tyrosine phosphatase A (PtpA), a low-molecular weight protein secreted by *M. tuberculosis*, blocks the phagosome–lysosome fusion by acting on the regulation of membrane fusion and is essential for the intracellular survival of the bacteria (Bach et al., 2008). Another phosphatase, PtpB, also acts by interfering with the normal immune response of the host, although its specific target is still unknown (Singh et al., 2003). A third important virulence factor described in *M. tuberculosis* is the lipid phosphatase, secreted acid phosphatase (SapM), which inhibits the production of phosphatidylinositol

3-phosphate, essential in the biogenesis of phagolysosomes (Vergne et al., 2005). A recent report using *M. bovis* BCG SapM mutants in vaccination studies has also shown that it inhibits the recruitment of dendritic cells and their activation at the vaccination site (Festjens et al., 2011).

Being important factors of mycobacterial virulence and mechanisms of defense to escape the immune response of the host, these proteins have been recently considered as possible targets for antimycobacterial action. Since PtpA and PtpB are proteins secreted to the extracellular medium *in vivo*, this might be an advantage to bypass the intrinsic impermeability of the mycobacterial cell wall to common antibiotics, rendering them more vulnerable to the action of inhibitors (Barczak and Hung, 2009). To exploit this in a clinically relevant mode, however, specificity issues would have to be first solved, since, especially PtpA, shares an important identity with its human ortholog, but this is not the case for PtpB (Grundner et al., 2005; Stehle et al., 2012).

Translating all this knowledge into clinical practice, however, would not be easy. Clinical strains of *M. tuberculosis* are genetically diverse with several clades and families already identified (Gagneux and Small, 2007). Recent studies performed with macrophages and infected mice have shown that the pathogenicity of the bacteria depended on the genetic lineage of the infecting strain (Reiling et al., 2013). This represents an additional issue to be solved if we want to develop tools to counter *M. tuberculosis* pathogenicity in humans. In summary, we need a much better understanding of all the events that take place inside the granuloma if we want to better control the pathogenicity of the bacteria and try to understand and eventually predict when a latent infection would turn into active disease.

45.8 Conclusion

A better control and eventual eradication of TB will depend on several factors related to the pathogen as well as the host. As we have seen in this chapter, our knowledge of the physiology and biochemistry of the bacteria is still insufficient and this has consequences in the way we can successfully detect and combat this pathogen. Diagnosis of TB in many low-resource countries is still based on a microscopy technique more than a hundred years old that lacks specificity. Confirmation of diagnosis in the laboratory is still based on cultivating the bacteria in artificial media that do not reproduce well the environment the pathogen encounters in the host. This, compounded by the slow generation time of the TB bacillus, makes the isolation of the bacteria *in vitro* extremely slow needing 4–6 weeks to become positive. But not only diagnosis of the disease is slow in the laboratory, also detection of drug resistance that has been traditionally been based on similar cultivation techniques that give results too late to properly manage TB-diagnosed patients. Undeniably much progress has been made in the last years and several innovative approaches have been developed to improve the diagnosis and detection of drug resistance. Liquid culture-based methods and semi-automated systems have reduced the time needed to confirm the diagnosis of TB in the laboratory. Molecular-based methods have added new tools not only to speed the diagnosis but also to detect drug resistance-associated mutations that give an early signal of the resistance profile of the infecting strain. Microfluidics and self-contained modules have recently improved our possibilities to rapidly detect the presence of the bacteria in clinical samples.

So what is missing? Several issues still need improvement so we can make a real advance in the way TB is diagnosed, drug resistance is detected, and pathogenesis is properly addressed. For example, microscopy, the simplest of the diagnostic techniques mentioned in the chapter suffers from low sensitivity. Even though newer LED-based microscopes have improved microscopical diagnosis significantly, they have not really produced a considerable increase in sensitivity of diagnosis. There is some room for improvement here, with new concentration approaches to be explored or new and reliable markers to be discovered that could give microscopy a real boost as a simple diagnostic tool for TB. Culture-based techniques have probably reached their limit in terms of time to detection of growth. The latest liquid culture-based semi-automated systems are able to detect growth of *M. tuberculosis* in 5–7 days from a highly positive clinical specimen. If we compare this to what was possible with the classical culture approaches, which need 3–6 weeks to give results, we can consider that we have made a great improvement. Do we need faster methods? Is a real difference to be able to detect the presence of *M. tuberculosis* in 5 days or in 1–2 days? In some circumstances, probably yes, when it has a direct impact on how a patient is managed and treated. In some other cases, probably is not of much difference. However, early diagnosis of TB is always a priority to avoid transmission of the bacilli to uninfected subjects. What we are missing is probably better markers of *M. tuberculosis* presence and activity that would allow us to differentiate active disease from latent infection. In this context, the availability of a POC diagnostic test is badly needed. As we have seen in Section 45.7 in the chapter, the development of the TB granuloma is the hallmark of the disease. How successful is the host in containing the bacteria or how successful is the pathogen in evading the immune system will determine the ultimate fate of the infection. If we succeed in containing the bacilli, we will have a latent infection, otherwise we will develop the active disease. A diagnostic test able to reliably differentiate these two states, and, ultimately, to accurately predict who among those latently infected, one-third of the world's human population, will develop active disease constitutes our current challenge. How successfully we are in addressing these issues will have an impact on the perspectives for controlling the disease in the near future.

References

Abate, G., Aseffa, A., Selassie, A., Goshu, S., Fekade, B., WoldeMeskal, D., and Miörner, H. 2004. Direct colorimetric assay for rapid detection of rifampin-resistant *Mycobacterium tuberculosis*. *J. Clin. Microbiol.*, 42, 871–873.

Abe, C., Hirano, K., Wada, M., Kazumi, Y., Takahashi, M., Fukasawa, Y., Yoshimura, T., Miyagi, C., and Goto, S. 1993. Detection of *Mycobacterium tuberculosis* in clinical specimens by polymerase chain reaction and Gen-Probe Amplified Mycobacterium Tuberculosis Direct Test. *J. Clin. Microbiol.*, 31, 3270–3274.

Almeida da Silva, P.E., and Palomino, J.C. 2011. Molecular basis and mechanisms of drug resistance in *Mycobacterium tuberculosis*: Classical and new drugs. *J. Antimicrob. Chemother.*, 66, 1417–1430.

Angeby, K.A., Klintz, L., and Hoffner, S.E. 2002. Rapid and inexpensive drug susceptibility testing of *Mycobacterium tuberculosis* with a nitrate reductase assay. *J. Clin. Microbiol.*, 40, 553–555.

Angeby, K.A., Hoffner, S.E., and Diwan, V.K. 2004. Should the 'bleach microscopy method' be recommended for improved case detection of tuberculosis? Literature review and key person analysis. *Int. J. Tuberc. Lung Dis.*, 8, 806–815.

Anthony, R.M., Kolk, A.H., Kuijper, S., and Klatser, P.R. 2006. Light emitting diodes for auramine O fluorescence microscopic screening of *Mycobacterium tuberculosis*. *Int. J. Tuberc. Lung Dis.*, 10, 1060–1062.

Ba, F., and Rieder, H.L. 1999. A comparison of fluorescence microscopy with the Ziehl–Neelsen technique in the examination of sputum for acid-fast bacilli. *Int. J. Tuberc. Lung Dis.*, 3, 1101–1105.

Bach, H., Papavinasasundaram, K.G., Wong, D., Hmama, Z., and Av-Gay, Y. 2008. *Mycobacterium tuberculosis* virulence is mediated by PtpA dephosphorylation of human vacuolar protein sorting 33B. *Cell. Host Microbe*, 3, 316–322.

Barczak, A.K., and Hung, D.T. 2009. Productive steps toward an antimicrobial targeting virulence. *Curr. Opin. Microbiol.*, 12, 490–496.

Barnard, M., Albert, H., Coetzee, G., O'Brien, R., and Bosman, M.E. 2008. Rapid molecular screening for multidrug-resistant tuberculosis in a high-volume public health laboratory in South Africa. *Am. J. Respir. Crit. Care Med.*, 177, 787–792.

Bergmann, J.S., and Woods, G.L. 1998. Clinical evaluation of the BDProbeTec strand displacement amplification assay for rapid diagnosis of tuberculosis. *J. Clin. Microbiol.*, 36, 2766–2768.

Birkness, K.A., Guarner, J., Sable, S.B., Tripp, R.A., Kellar, K.L., Bartlett, J., and Quinn, F.D. 2007. An in vitro model of the leukocyte interactions associated with granuloma formation in *Mycobacterium tuberculosis* infection. *Immunol. Cell Biol.*, 85, 160–168.

Biswas, D., Deb, A., Gupta, P., Prasad, R., and Negi, K.S. 2008. Evaluation of the usefulness of phage amplification technology in the diagnosis of patients with paucibacillary tuberculosis. *Indian J. Med. Microbiol.*, 26, 75–78.

Boehme, C.C., Nabeta, P., Henostroza, G., Raqib, R., Rahim, Z., Gerhardt, M., Sanga, E., Hoelscher, M., Notomi, T., Hase, T., and Perkins, M.D. 2007. Operational feasibility of using loop-mediated isothermal amplification for diagnosis of pulmonary tuberculosis in microscopy centers of developing countries. *J. Clin. Microbiol.*, 45, 1936–1940.

Boehme, C.C., Nabeta, P., Hillemann, D., Nicol, M.P., Shenai, S., Krapp, F., Allen, J., Tahirli, R., Blakemore, R., Rustomjee, R., Milovic, A., Jones, M., O'Brien, S.M., Persing, D.H., Ruesch-Gerdes, S., Gotuzzo, E., Rodrigues, C., Alland, D., and Perkins, M.D. 2010. Rapid molecular detection of tuberculosis and rifampin resistance. *N. Engl. J. Med.*, 363, 1005–1015.

Bonnet, M., Gagnidze, L., Guerin, P.J., Bonte, L., Ramsay, A., Githui, W., and Varaine, F. 2011. Evaluation of combined LED-fluorescence microscopy and bleach sedimentation for diagnosis of tuberculosis at peripheral health service level. *PLoS One*, 6, e20175.

Boum, Y., Orikiriza, P., Rojas-Ponce, G., Riera-Montes, M., Atwine, D., Nansumba, M., Bazira, J., Tuyakira, E., De Beaudrap, P., Bonnet, M., and Page, A.L. 2013. Use of colorimetric culture methods for detection of *Mycobacterium tuberculosis* complex isolates from sputum samples in resource-limited settings. *J. Clin. Microbiol.*, 51, 2273–2279.

Brossier, F., Veziris, N., Aubry, A., Jarlier, V., and Sougakoff, W. 2010. Detection by GenoType MTBDRsl test of complex mechanisms of resistance to second-line drugs and ethambutol in multidrug-resistant *Mycobacterium tuberculosis* complex isolates. *J. Clin. Microbiol.*, 48, 1683–1689.

Cattamanchi, A., Davis, J.L., Pai, M., Huang, L., Hopewell, P.C., and Steingart, K.R. 2010. Does bleach processing increase the accuracy of sputum smear microscopy for diagnosing pulmonary tuberculosis? *J. Clin. Microbiol.*, 48, 2433–2439.

Caviedes, L., Lee, T.S., Gilman, R.H., Sheen, P., Spellman, E., Lee, E.H., Berg, D.E., and Montenegro-James, S. 2000. Rapid, efficient detection and drug susceptibility testing of *Mycobacterium tuberculosis* in sputum by microscopic observation of broth cultures. The Tuberculosis Working Group in Peru. *J. Clin. Microbiol.*, 38, 1203–1208.

Chang, K., Lu, W., Wang, J., Zhang, K., Jia, S., Li, F., Deng, S., and Chen, M. 2012. Rapid and effective diagnosis of tuberculosis and rifampicin resistance with Xpert MTB/RIF assay: A meta-analysis. *J. Infect.*, 64, 580–588.

Chao, J., Wong, D., Zheng, X., Poirier, V., Bach, H., Hmama, Z., and Av-Gay, Y. 2010. Protein kinase and phosphatase signaling in *Mycobacterium tuberculosis* physiology and pathogenesis. *Biochim. Biophys. Acta*, 1804, 620–627.

Cobelens, F.G, Egwaga, S.M., van Ginkel, T., Muwinge, H., Matee, M.I., and Borgdorff, M.W. 2006. Tuberculin skin testing in patients with HIV infection: Limited benefit of reduced cutoff values. *Clin. Infect. Dis.*, 43, 634–639.

Cole, S.T., Brosch, R., Parkhill, J., Garnier, T., Churcher, C., Harris, D., Gordon, S.V., Eiglmeier, K., Gas, S., Barry 3rd, C.E., Tekaia, F., Badcock, K., Basham, D., Brown, D., Chillingworth, T., Connor, R., Davies, R., Devlin, K., Feltwell, T., Gentles, S., Hamlin, N., Holroyd, S., Hornsby, T., Jagels, K., Krogh, A., McLean, J., Moule, S., Murphy, L., Oliver, K., Osborne, J., Quail, M.A., Rajandream, M.A., Rogers, J., Rutter, S., Seeger, K., Skelton, J., Squares, R., Squares, S., Sulston, J.E., Taylor, K., Whitehead, S., and Barrell, B.G. 1998. Deciphering the biology of *Mycobacterium tuberculosis* from the complete genome sequence. *Nature*, 393, 537–544.

Coll, P., Garrigó, M., Moreno, C., and Martí, N. 2003. Routine use of Gen-Probe Amplified Mycobacterium Tuberculosis Direct (MTD) test for detection of *Mycobacterium tuberculosis* with smear-positive and smear-negative specimens. *Int. J. Tuberc. Lung Dis.*, 7, 886–891.

Cuevas, L.E., Al-Sonboli, N., Lawson, L., Yassin, M.A., Arbide, I., Al-Aghbari, N., Sherchand, J.B., Al-Absi, A., Emenyonu, E.N., Merid, Y., Okobi, M.I., Onuoha, J.O., Aschalew, M., Aseffa, A., Harper, G., de Cuevas, R.M., Theobald, S.J., Nathanson, C.M., Joly, J., Faragher, B., Squire, S.B., and Ramsay, A. 2011. LED fluorescence microscopy for the diagnosis of pulmonary tuberculosis: A multi-country cross-sectional evaluation. *PLoS Med.*, 8, e1001057.

Dalovisio, J.R., Montenegro-James, S., Kemmerly, S.A., Genre, C.F., Chambers, R., Greer, D., Pankey, G.A., Failla, D.M., Haydel, K.G., Hutchinson, L., Lindley, M.F., Nunez, B.M., Praba,

A., Eisenach, K.D., and Cooper, E.S. 1996. Comparison of the amplified *Mycobacterium tuberculosis* (MTB) direct test, Amplicor MTB PCR, and IS*6110*-PCR for detection of MTB in respiratory specimens. *Clin. Infect. Dis.*, 23, 1099–1106.

Davis, J.M., and Ramakrishnan, L. 2009. The role of the granuloma in expansion and dissemination of early tuberculous infection. *Cell*, 136, 37–49.

De Beenhouwer, H., Lhiang, Z., Jannes, G., Mijs, W., Machtelinckx, L., Rossau, R., Traore, H., and Portaels, F. 1995. Rapid detection of rifampicin resistance in sputum and biopsy specimens from tuberculosis patients by PCR and line probe assay. *Tuber. Lung Dis.*, 76, 425–430.

Diel, R., Loddenkemper, R., and Nienhaus, A. 2010. Evidence-based comparison of commercial interferon-gamma release assays for detecting active TB: A meta-analysis. *Chest*, 137, 952–968.

Dinnes, J., Deeks, J., Kunst, H., Gibson, A., Cummins, E., Waugh, N., Drobniewski, F., and Lalvani, A. 2007. A systematic review of rapid diagnostic tests for the detection of tuberculosis infection. *Health Technol. Assess.*, 11, 1–196.

Driver, E.R., Ryan, G.J., Hoff, D.R., Irwin, S.M., Basaraba, R.J., Kramnik, I., and Lenaerts, A.J. 2012. Evaluation of a mouse model of necrotic granuloma formation using C3HeB/FeJ mice for testing of drugs against *Mycobacterium tuberculosis*. *Antimicrob. Agents Chemother.*, 56, 3181–3195.

ECDC. 2011. *Use of Interferon-gamma Release Assays in Support of TB Diagnosis*. Stockholm: ECDC.

El-Hajj, H.H., Marras, S.A., Tyagi, S., Kramer, F.R., and Alland, D. 2001. Detection of rifampin resistance in *Mycobacterium tuberculosis* in a single tube with molecular beacons. *J. Clin. Microbiol.*, 39, 4131–4137.

Fallahi-Sichani, M., El-Kebir, M., Marino, S., Kirschner, D.E., and Linderman, J.J. 2011. Multiscale computational modeling reveals a critical role for TNF-α receptor 1 dynamics in tuberculosis granuloma formation. *J. Immunol.*, 186, 3472–3483.

Farhat, M., Greenaway, C., Pai, M., and Menzies, D. 2006. False-positive tuberculin skin tests: What is the absolute effect of BCG and non-tuberculous mycobacteria? *Int. J. Tuberc. Lung Dis.*, 10, 1192–1204.

Feng, Y., Liu, S., Wang, Q., Wang, L., Tang, S., Wang, J., and Lu, W. 2013. Rapid diagnosis of drug resistance to fluoroquinolones, amikacin, capreomycin, kanamycin and ethambutol using genotype MTBDRsl assay: A meta-analysis. *PLoS One*, 8, e55292.

Festjens, N., Bogaert, P., Batni, A., Houthuys, E., Plets, E., Vanderschaeghe, D., Laukens, B., Asselbergh, B., Parthoens, E., De Rycke, R., Willart, M.A., Jacques, P., Elewaut, D., Brouckaert, P., Lambrecht, B.N., Huygen, K., and Callewaert, N. 2011. Disruption of the SapM locus in *Mycobacterium bovis* BCG improves its protective efficacy as a vaccine against *M. tuberculosis*. *EMBO Mol. Med.*, 3, 222–234.

Flores, L.L., Pai, M., Colford Jr., J.M., and Riley, L.W. 2005. In-house nucleic acid amplification tests for the detection of *Mycobacterium tuberculosis* in sputum specimens: Meta-analysis and meta-regression. *BMC Microbiol.*, 5, 55.

Gagneux, S., and Small, P.M. 2007. Global phylogeography of *Mycobacterium tuberculosis* and implications for tuberculosis product development. *Lancet Infect. Dis.*, 7, 328–337.

Garcia de Viedma, D. 2003. Rapid detection of resistance in *Mycobacterium tuberculosis*: A review discussing molecular approaches. *Clin. Microbiol. Infect.*, 9, 349–359.

Giannoni, F., Iona, E., Sementilli, F., Brunori, L., Pardini, M., Migliori, G.B., Orefici, G., and Fattorini, L. 2005. Evaluation of a new line probe assay for rapid identification of *gyrA* mutations in *Mycobacterium tuberculosis*. *Antimicrob. Agents Chemother.*, 49, 2928–2933.

Gil, O., Díaz, I., Vilaplana, C., Tapia, G., Díaz, J., Fort, M., Cáceres, N., Pinto, S., Caylà, J., Corner, L., Domingo, M., and Cardona, P.J. 2010. Granuloma encapsulation is a key factor for containing tuberculosis infection in minipigs. *PLoS One*, 5, e10030.

Goto, M., Honda, E., Ogura, A., Nomoto, A., and Hanaki, K. 2009. Colorimetric detection of loop-mediated isothermal amplification reaction by using hydroxy naphthol blue. *Biotechniques*, 46, 167–172.

Greco, S., Rulli, M., Girardi, E., Piersimoni, C., and Saltini, C. 2009. Diagnostic accuracy of in-house PCR for pulmonary tuberculosis in smear-positive patients: Meta-analysis and meta-regression. *J. Clin. Microbiol.*, 47, 569–576.

Grundner, C., Ng, H.L., and Alber, T. 2005. *Mycobacterium tuberculosis* protein tyrosine phosphatase PtpB structure reveals a diverged fold and a buried active site. *Structure*, 13, 1625–1634.

Guirado, E., and Schlesinger, L.S. 2013. Modeling the *Mycobacterium tuberculosis* granuloma – the critical battlefield in host immunity and disease. *Front. Immunol.*, 4, 98.

Habtamu, M., van den Boogaard, J., Ndaro, A., Buretta, R., Irongo, C.F., Lega, D.A., Nyombi, B.M., and Kibiki, G.S. 2012. Light-emitting diode with various sputum smear preparation techniques to diagnose tuberculosis. *Int. J. Tuberc. Lung Dis.*, 16, 402–407.

Harper, J., Skerry, C., Davis, S.L., Tasneen, R., Weir, M., Kramnik, I., Bishai, W.R., Pomper, M.G., Nuermberger, E.L., and Jain, S.K. 2012. Mouse model of necrotic tuberculosis granulomas develops hypoxic lesions. *J. Infect Dis.*, 205, 595–602.

Heifets, L. 1997. Mycobacteriology laboratory. *Clin. Chest Med.*, 18, 35–53.

Helb, D., Jones, M., Story, E., Boehme, C., Wallace, E., Ho, K., Kop, J., Owens, M.R., Rodgers, R., Banada, P., Safi, H., Blakemore, R., Lan, N.T., Jones-López, E.C., Levi, M., Burday, M., Ayakaka, I., Mugerwa, R.D., McMillan, B., Winn-Deen, E., Christel, L., Dailey, P., Perkins, M.D., Persing, D.H., and Alland, D. 2010. Rapid detection of *Mycobacterium tuberculosis* and rifampin resistance by use of on-demand, near-patient technology. *J. Clin. Microbiol.*, 48, 229–237.

Heneberg, P. 2012. Finding the smoking gun: Protein tyrosine phosphatases as tools and targets of unicellular microorganisms and viruses. *Curr. Med. Chem.*, 19, 1530–1566.

Hernández-Neuta, I., Varela, A., Martin, A., von Groll, A., Jureen, P., López, B., Imperiale, B., Skenders, G., Ritacco, V., Hoffner, S., Morcillo, N., Palomino, J.C., and Del Portillo, P. 2010. Rifampin–isoniazid oligonucleotide typing: An alternative format for rapid detection of multidrug-resistant *Mycobacterium tuberculosis*. *J. Clin. Microbiol.*, 48, 4386–4391.

Hillemann, D., Rusch-Gerdes, S., and Richter, E. 2007. Evaluation of the GenoType MTBDRplus assay for rifampin and isoniazid susceptibility testing of *Mycobacterium tuberculosis* strains and clinical specimens. *J. Clin. Microbiol.*, 45, 2635–2640.

Hillemann, D., Rüsch-Gerdes, S., and Richter, E. 2009. Feasibility of the GenoTypeMTBDRsl assay for fluoroquinolone, amikacin-capreomycin, and ethambutol resistance testing of *Mycobacterium tuberculosis* strains and clinical specimens. *J. Clin. Microbiol.*, 47, 1767–1772.

Imperiale, B.R., Zumárraga, M.J., Weltman, G., Zudiker, R., Cataldi, A.A., and Morcillo, N.S. 2012. First evaluation in Argentina of the GenoType® MTBDRplus assay for multidrug-resistant *Mycobacterium tuberculosis* detection from clinical isolates and specimens. *Rev. Argent. Microbiol.*, 44(4), 283–289.

Kaewphinit, T., Arunrut, N., Kiatpathomchai, W., Santiwatanakul, S., Jaratsing, P., and Chansiri, K. 2013. Detection of *Mycobacterium tuberculosis* by using loop-mediated isothermal amplification combined with a lateral flow dipstick in clinical samples. *Biomed. Res. Int.*, 2013, 926230. doi: 10.1155/2013/926230.

Lalvani, A., and Pareek, M. 2004. Interferon gamma release assays: Principles and practice. *Enferm. Infecc. Microbiol. Clin.*, 28, 245–252.

Lawn, S.D., and Nicol, M.P. 2011. Xpert® MTB/RIF assay: Development, evaluation and implementation of a new rapid molecular diagnostic for tuberculosis and rifampicin resistance. *Future Microbiol.*, 6, 1067–1082.

Lee, M.F., Chen, Y.H., Hsu, H.J., and Peng, C.F. 2010. One-tube loop-mediated isothermal amplification combined with restriction endonuclease digestion and ELISA for colorimetric detection of resistance to isoniazid, ethambutol and streptomycin in *Mycobacterium tuberculosis* isolates. *J. Microbiol Methods*, 83, 53–58.

Liu, D., Liang, G., Zhang, Q., and Chen, B. 2013. Detection of *mycobacterium tuberculosis* using a capillary-array microsystem with integrated DNA extraction, loop-mediated isothermal amplification, and fluorescence detection. *Anal. Chem.*, 85, 4698–4704.

Marais, B.J., Brittle, W., Painczyk, K., Hesseling, A.C., Beyers, N., Wasserman, E., van Soolingen, D., and Warren, R.M. 2008. Use of light-emitting diode fluorescence microscopy to detect acid-fast bacilli in sputum. *Clin. Infect. Dis.*, 47, 203–207.

Marino, S., Linderman, J.J., and Kirschner, D.E. 2011. A multifaceted approach to modeling the immune response in tuberculosis. *Wiley Interdisc. Rev. Syst. Biol. Med.*, 3, 479–489.

Martin, A., Montoro, E., Lemus, D., Simboli, N., Morcillo, N., Velasco, M., Chauca, J., Barrera, L., Ritacco, V., Portaels, F., and Palomino, J.C. 2005a. Multicenter evaluation of the nitrate reductase assay for drug resistance detection of *Mycobacterium tuberculosis*. *J. Microbiol. Methods*, 63, 145–150.

Martin, A., Morcillo, N., Lemus, D., Montoro, E., Telles, M.A., Simboli, N., Pontino, M., Porras, T., León, C., Velasco, M., Chacon, L., Barrera, L., Ritacco, V., Portaels, F., and Palomino, J.C. 2005b. Multicenter study of MTT and resazurin assays for testing susceptibility to first-line anti-tuberculosis drugs. *Int. J. Tuberc. Lung Dis.*, 9, 901–906.

Martin, A., Portaels, F., and Palomino, J.C. 2007. Colorimetric redox-indicator methods for the rapid detection of multidrug resistance in *Mycobacterium tuberculosis*: A systematic review and meta-analysis. *J. Antimicrob. Chemother.*, 59, 175–183.

Martin, A., Panaiotov, S., Portaels, F., Hoffner, S., Palomino, J.C., and Angeby, K. 2008. The nitrate reductase assay for the rapid detection of isoniazid and rifampicin resistance in *Mycobacterium tuberculosis*: A systematic review and meta-analysis. *J. Antimicrob. Chemother.*, 62, 56–64.

Martin, A., Munga Waweru, P., Babu Okatch, F., Amondi Ouma, N., Bonte, L., Varaine, F., and Portaels, F. 2009. Implementation of the thin layer agar method for diagnosis of smear-negative pulmonary tuberculosis in a setting with a high prevalence of human immunodeficiency virus infection in Homa Bay, Kenya. *J. Clin. Microbiol.*, 47, 2632–2634.

Maschmann, Rde A., Sá Spies, F., Nunes Lde, S., Ribeiro, A.W., Machado, T.R., Zaha, A., and Rossetti, M.L. 2013. Performance of the GenoType MTBDRplus assay directly on sputum specimens from Brazilian patients with tuberculosis treatment failure or relapse. *J. Clin. Microbiol.*, 51, 1606–1608.

Mejia, G.I., Castrillon, L., Trujillo, H., and Robledo, J.A. 1999. Microcolony detection in 7H11 thin layer culture is an alternative for rapid diagnosis of *Mycobacterium tuberculosis* infection. *Int. J. Tuberc. Lung Dis.*, 3, 138–142.

Metchock, B.G., Nolte, F.S., and Wallace Jr., R.J. 1999. Mycobacterium. In: Murray, P.R., editor. *Manual of Clinical Microbiology*. Washington, DC: American Society for Microbiology. pp. 399–437;

Michos, A.G., Daikos, G.L., Tzanetou, K., Theodoridou, M., Moschovi, M., Nicolaidou, P., Petrikkos, G., Syriopoulos, T., Kanavaki, S., and Syriopoulou, V.P. 2006. Detection of *Mycobacterium tuberculosis* DNA in respiratory and nonrespiratory specimens by the Amplicor MTB PCR. *Diagn. Microbiol. Infect. Dis.*, 54(2), 121–126.

Migliori, G.B., Centis, R., D'Ambrosio, L., Spanevello, A., Borroni, E., Cirillo, D.M., and Sotgiu, G. 2012. Totally drug-resistant and extremely drug-resistant tuberculosis: The same disease? *Clin. Infect. Dis.*, 54, 1379–1380.

Minion, J., Leung, E., Menzies, D., and Pai, M. 2010. Microscopic-observation drug susceptibility and thin layer agar assays for the detection of drug resistant tuberculosis: A systematic review and meta-analysis. *Lancet Infect. Dis.*, 10, 688–698.

Minion, J., Pai, M., Ramsay, A., Menzies, D., and Greenaway, C. 2011. Comparison of LED and conventional fluorescence microscopy for detection of acid fast bacilli in a low-incidence setting. *PLoS One*, 6, e22495.

Mole, R., Trollip, A., Abrahams, C., Bosman, M., and Albert, H. 2007. Improved contamination control for a rapid phage-based rifampicin resistance test for *Mycobacterium tuberculosis*. *J. Med. Microbiol.*, 56(Pt 10), 1334–1339.

Montoro, E., Lemus, D., Echemendia, M., Martin, A., Portaels, F., and Palomino, J.C. 2005. Comparative evaluation of the nitrate reduction assay, the MTT test, and the resazurin microtitre assay for drug susceptibility testing of clinical isolates of *Mycobacterium tuberculosis*. *J. Antimicrob. Chemother.*, 55, 500–505.

Moore, D.A., Evans, C.A., Gilman, R.H., Caviedes, L., Coronel, J., Vivar, A., Sanchez, E., Piñedo, Y., Saravia, J.C., Salazar, C., Oberhelman, R., Hollm-Delgado, M.G., LaChira, D., Escombe, A.R., and Friedland, J.S. 2006. Microscopic-observation drug-susceptibility assay for the diagnosis of TB. *N. Engl. J. Med.*, 355, 1539–1350.

Mori, Y., Nagamine, K., Tomita, N., and Notomi, T. 2001. Detection of loop-mediated isothermal amplification reaction by turbidity derived from magnesium pyrophosphate formation. *Biochem. Biophys. Res. Commun.*, 289, 150–154.

Morris, K. 2011. WHO recommends against inaccurate tuberculosis tests. *Lancet*, 377, 113–114.

Mshana, R.N., Tadesse, G., Abate, G., and Miörner, H. 1998. Use of 3-(4,5-dimethylthiazol-2-yl)-2,5-diphenyl tetrazolium bromide for rapid detection of rifampin-resistant *Mycobacterium tuberculosis*. *J. Clin. Microbiol.*, 36, 1214–1219.

Musa, H.R., Ambroggi, M., Souto, A., and Angeby, K.A. 2005. Drug susceptibility testing of *Mycobacterium tuberculosis* by a nitrate reductase assay applied directly on microscopy-positive sputum samples. *J. Clin. Microbiol.*, 43, 3159–3161.

Notomi, T., Okayama, H., Masubuchi, H., Yonekawa, T., Watanabe, K., Amino, N., and Hase, T. 2000. Loop-mediated isothermal amplification of DNA. *Nucleic Acids Res.*, 28, E63.

Ozkutuk, A., Kirdar, S., Ozden, S., and Esen, N. 2006. Evaluation of Cobas Amplicor MTB test to detect *Mycobacterium tuberculosis* in pulmonary and extrapulmonary specimens. *New Microbiol.*, 29, 269–273.

Pai, M., Riley, L.W., and Colford Jr., J.M. 2004. Interferon-gamma assays in the immunodiagnosis of tuberculosis: A systematic review. *Lancet Infect. Dis.*, 4, 761–776.

Pai, M., Kalantri, S., and Dheda, K. 2006. New tools and emerging technologies for the diagnosis of tuberculosis: Part II. Active tuberculosis and drug resistance. *Exp. Rev. Mol. Diagn.*, 6, 423–432.

Pai, M., Zwerling, A., and Menzies, D. 2008. Systematic review: T-cell based assays for the diagnosis of latent tuberculosis infection – an update. *Ann. Intern. Med.*, 149, 177–184.

Palomino J.C. 2005. Nonconventional and new methods in the diagnosis of tuberculosis: Feasibility and applicability in the field. *Eur. Respir. J.*, 26, 339–350.

Palomino, J.C. 2009. Molecular detection, identification and drug resistance detection in *Mycobacterium tuberculosis*. *FEMS Immunol. Med. Microbiol.*, 56, 103–111.

Palomino, J.C. 2012. Current developments and future perspectives for TB diagnostics. *Future Microbiol.*, 7, 59–71.

Palomino, J.C., Martin, A., Camacho, M., Guerra, H., Swings, J., and Portaels, F. 2002. Resazurin microtiter assay plate: Simple and inexpensive method for detection of drug resistance in *Mycobacterium tuberculosis*. *Antimicrob. Agents Chemother.*, 46, 2720–2722.

Palomino, J.C., Martin, A., Von Groll, A., and Portaels, F. 2008. Rapid culture-based methods for drug-resistance detection in *Mycobacterium tuberculosis*. *J. Microbiol. Methods*, 75, 161–166.

Pandey, B.D., Poudel, A., Yoda, T., Tamaru, A., Oda, N., Fukushima, Y., Lekhak, B., Risal, B., Acharya, B., Sapkota, B.,
Nakajima, C., Taniguchi, T., Phetsuksiri, B., and Suzuki, Y. 2008. Development of an in-house loop-mediated isothermal amplification (LAMP) assay for detection of *Mycobacterium tuberculosis* and evaluation in sputum samples of Nepalese patients. *J. Med. Microbiol.*, 57(Pt 4), 439–443.

Pfyffer, G.E., Funke-Kissling, P., Rundler, E., and Weber, R. 1999. Performance characteristics of the BDProbeTec system for direct detection of *Mycobacterium tuberculosis* complex in respiratory specimens. *J. Clin. Microbiol.*, 37, 137–140.

Piersimoni, C., and Scarparo, C. 2003. Relevance of commercial amplification methods for direct detection of *Mycobacterium tuberculosis* complex in clinical samples. *J. Clin. Microbiol.*, 41, 5355–5365.

Piuri, M., Jacobs Jr., W.R., and Hatfull, G.F. 2009. Fluoromycobacteriophages for rapid, specific, and sensitive antibiotic susceptibility testing of *Mycobacterium tuberculosis*. *PLoS One*, 4, e4870.

Raja, S., Ching, J., Xi, L., Hughes, S.J., Chang, R., Wong, W., McMillan, W., Gooding, W.E., McCarty Jr., K.S., Chestney, M., Luketich, J.D., and Godfrey, T.E. 2005. Technology for automated, rapid, and quantitative PCR or reverse transcription-PCR clinical testing. *Clin. Chem.*, 51, 882–890.

Reiling, N., Homolka, S., Walter, K., Brandenburg, J., Niwinski, L., Ernst, M., Herzmann, C., Lange, C., Diel, R., Ehlers, S., and Niemann, S. 2013. Clade-specific virulence patterns of *Mycobacterium tuberculosis* complex strains in human primary macrophages and aerogenically infected mice. *MBio*, 4, e00250-13.

Roberts, G.D., Goodman, N.L., Heifets, L., Larsh, H.W., Lindner, T.H., McClatchy, J.K., McGinnis, M.R., Siddiqi, S.H., and Wright, P. 1983. Evaluation of the BACTEC radiometric method for recovery of mycobacteria and drug susceptibility testing of *Mycobacterium tuberculosis* from acid-fast smear-positive specimens. *J. Clin. Microbiol.*, 18, 689–696.

Robledo, J.A., Mejía, G.I., Morcillo, N., Chacón, L., Camacho, M., Luna, J., Zurita, J., Bodon, A., Velasco, M., Palomino, J.C., Martin, A., and Portaels, F. 2006. Evaluation of a rapid culture method for tuberculosis diagnosis: A Latin American multicenter study. *Int. J. Tuberc. Lung Dis.*, 10, 613–619.

Rodrigues, C., Shenai, S., Sadani, M., Sukhadia, N., Jani, M., Ajbani, K., Sodha, A., and Mehta, A. 2009. Evaluation of the Bactec MGIT 960 TB system for recovery and identification of *Mycobacterium tuberculosis* complex in a high through put tertiary care centre. *Indian J. Med. Microbiol.*, 27, 217–221.

Rondón, L., Piuri, M., Jacobs Jr., W.R., de Waard, J., Hatfull, G.F., and Takiff, H.E. 2011. Evaluation of fluoromycobacteriophages for detecting drug resistance in *Mycobacterium tuberculosis*. *J. Clin. Microbiol.*, 49, 1838–1842.

Rusch-Gerdes, S., and Richter, E. 2004. Clinical evaluation of the semi-automated BDProbeTec ET System for the detection of *Mycobacterium tuberculosis* in respiratory and non-respiratory specimens. *Diagn. Microbiol. Infect. Dis.*, 48, 265–270.

Saunders, B.M., and Cooper, A.M. 2000. Restraining mycobacteria: Role of granulomas in mycobacterial infections. *Immunol. Cell Biol.*, 78, 334–341.

Seitzer, U., and Gerdes, J. 2003. Generation and characterization of multicellular heterospheroids formed by human peripheral blood mononuclear cells. *Cells Tissues Organs*, 174, 110–116.

Senna, S.G., Gomes, H.M., Ribeiro, M.O., Kristki, A.L., Rossetti, M.L., and Suffys, P.N. 2006. In house reverse line hybridization assay for rapid detection of susceptibility to rifampicin in isolates of *Mycobacterium tuberculosis*. *J. Microbiol. Methods*, 67, 385–389.

Shikama, M.L., Ferro e Silva, R., Villela, G., Sato, D.N., Martins, M.C., Giampaglia, C.M., da Silva, R.F., Ferro e Silva, P., da Silva Telles, M.A., Martin, A., and Palomino, J.C. 2009. Multicentre study of nitrate reductase assay for rapid detection of rifampicin-resistant *M. tuberculosis*. *Int. J. Tuberc. Lung Dis.*, 13, 377–380.

Singh, R., Rao, V., Shakila, H., Gupta, R., Khera, A., Dhar, N., Singh, A., Koul, A., Singh, Y., Naseema, M., Narayanan, P.R., Paramasivan, C.N., Ramanathan, V.D., and Tyagi, A.K. 2003. Disruption of mptpB impairs the ability of *Mycobacterium tuberculosis* to survive in guinea pigs. *Mol. Microbiol.*, 50, 751–762.

Stehle, T., Sreeramulu, S., Löhr, F., Richter, C., Saxena, K., Jonker, H.R., and Schwalbe, H. 2012. The apo-structure of the low molecular weight protein-tyrosine phosphatase A (MptpA) from *Mycobacterium tuberculosis* allows for better target-specific drug development. *J. Biol. Chem.*, 287, 34569–34582.

Steingart, K.R., Henry, M., Laal, S., Hopewell, P.C., Ramsay, A., Menzies, D., Cunningham, J., Weldingh, K., and Pai, M. 2007. A systematic review of commercial serological antibody detection tests for the diagnosis of extrapulmonary tuberculosis. *Thorax*, 62, 911–918.

Syre, H., Phyu, S., Sandven, P., Bjorvatn, B., and Grewal, H.M. 2003. Rapid colorimetric method for testing susceptibility of *Mycobacterium tuberculosis* to isoniazid and rifampin in liquid cultures. *J. Clin. Microbiol.*, 41, 5173–5177.

Tiruviluamala, P., and Reichman, L.B. 2002. Tuberculosis. *Ann. Rev. Pub. Health*, 23, 403–426.

Torrado, E., and Cooper, A.M. 2010. IL-17 and Th17 cells in tuberculosis. *Cytokine Growth Factor Rev.*, 21, 455–462.

Tortoli, E., and Marcelli, F. 2007. Use of the INNO LiPARif.TB for detection of *Mycobacterium tuberculosis* DNA directly in clinical specimens and for simultaneous determination of rifampin susceptibility. *Eur. J. Clin. Microbiol. Infect. Dis.*, 26, 51–55.

Tortoli, E., Urbano, P., Marcelli, F., Simonetti, T.M., and Cirillo, D.M. 2012. Is real-time PCR better than conventional PCR for *Mycobacterium tuberculosis* complex detection in clinical samples? *J. Clin. Microbiol.*, 50, 2810–2813.

Traore, H., van Deun, A., Shamputa, I.C., Rigouts, L., and Portaels, F. 2006. Direct detection of *Mycobacterium tuberculosis* complex DNA and rifampin resistance in clinical specimens from tuberculosis patients by line probe assay. *J. Clin. Microbiol.*, 44, 4384–4388.

Tyagi, S., and Kramer, F.R. 1996. Molecular beacons: Probes that fluoresce upon hybridization. *Nat. Biotechnol.*, 14, 303–308.

Udwadia, Z.F., Amale, R.A., Ajbani, K.K., and Rodrigues, C. 2012. Totally drug-resistant tuberculosis in India. *Clin. Infect. Dis.*, 54, 579–581.

Varma-Basil, M., El-Hajj, H., Colangeli, R., Hazbón, M.H., Kumar, S., Bose, M., Bobadilla-del-Valle, M., García, L.G., Hernández, A., Kramer, F.R., Osornio, J.S., Ponce-de-León, A., and Alland, D. 2004. Rapid detection of rifampin resistance in *Mycobacterium tuberculosis* isolates from India and Mexico by a molecular beacon assay. *J. Clin. Microbiol.*, 42, 5512–5516.

Velayati, A.A., Masjedi, M.R., Farnia, P., Tabarsi, P., Ghanavi, J., Ziazarifi, A.H., and Hoffner, S.E. 2012. Emergence of new forms of totally drug-resistant tuberculosis bacilli: Super extensively drug-resistant tuberculosis or totally drug-resistant strains in Iran. *Chest*, 136, 420–425.

Vergne, I., Chua, J., Lee, H.H., Lucas, M., Belisle, J., and Deretic, V. 2005. Mechanism of phagolysosome biogenesis block by viable *Mycobacterium tuberculosis*. *Proc. Natl. Acad. Sci. USA*, 102, 4033–4038.

Wong, D., Chao, J.D., and Av-Gay, Y. 2013. *Mycobacterium tuberculosis*-secreted phosphatases: From pathogenesis to targets for TB drug development. *Trends Microbiol.*, 21, 100–109.

World Health Organization. 2007. WHO Strategic and Technical Advisory Group for Tuberculosis. Report on Conclusions and Recommendations. http://www.who.int/tb/events/stag_report_2007.pdf

World Health Organization. 2008. *Molecular Line Probe Assays for Rapid Screening of Patients at Risk of Multi-drug Resistant Tuberculosis (MDRTB)*. Geneva, Switzerland: World Health Organization. http://www.who.int/tb/features_archive/policy_statement.pdf. Accessed June 20, 2013.

World Health Organization. 2009. Report of the Ninth Meeting of the Strategic and Technical Advisory Group for Tuberculosis (STAG-TB). http://www.who.int/tb/advisory_bodies/stag_tb_report_2009.pdf. Accessed June 26, 2013.

World Health Organization. 2011. *Fluorescent Light-emitting Diode (LED) Microscopy for Diagnosis of Tuberculosis*. WHO/HTM/TB/2011.8. Geneva, Switzerland: WHO.

World Health Organization. 2012. *Global Tuberculosis Report 2012*. WHO/HTM/TB/2012.6. Geneva, Switzerland: WHO.

Yeager Jr., H., Lacy, J., Smith, L.R., and LeMaistre, C.A. 1967. Quantitative studies of mycobacterial populations in sputum and saliva. *Am. Rev. Respir. Dis.*, 95, 998–1004.

Young, D.B., Perkins, M.D., Duncan, K., and Barry 3rd, C.E. 2008. Confronting the scientific obstacles to global control of tuberculosis. *J. Clin. Invest.*, 118, 1255–1265.

Zhu, R.Y., Zhang, K.X., Zhao, M.Q., Liu, Y.H., Xu, Y.Y., Ju, C.M., Li, B., and Chen, J.D. 2009. Use of visual loop-mediated isotheral amplification of *rimM* sequence for rapid detection of *Mycobacterium tuberculosis* and *Mycobacterium bovis*. *J. Microbiol. Methods*, 78, 339–343.

Chapter 46

Pathogenesis of *Orientia tsutsugamushi* Infection in Humans

Wiwit Tantibhedhyabgkul[1,2] and Jean-Louis Mege[1]

[1]Unité de Recherche sur les Maladies Infectieuses Tropicales et Emergentes, Aix-Marseille Université, Centre National de la Recherche Scientifique Unité Mixte de Recherche 7278, Institut National de la Santé et de la Recherche Scientifique Unité 1095, Marseille, France

[2]Department of Immunology, Faculty of Medicine Siriraj Hospital, Mahidol University, Bangkok, Thailand

46.1 Introduction

Orientia tsutsugamushi is an obligate intracellular Gram-negative bacterium, which replicates in the cytosol of host cells, particularly endothelial cells and mononuclear phagocytes. It belongs to the class of alpha-Proteobacteria, the order of Rickettsiales and the family of Rickettsiaceae. *O. tsutsugamushi* had formerly been classified in the genus *Rickettsia* but was re-classified in a new genus, due to numerous genotypic and phenotypic differences with *Rickettsia*. Unlike *Rickettsia* species, the cell wall of *O. tsutsugamushi* consists of several unique outer membrane proteins, but lacks lipopolysaccharide (LPS) and peptidoglycan. Therefore, it is very fragile, unstable in extracellular environment and highly sensitive to sonication or osmotic shock. In addition, *O. tsutsugamushi* is highly resistant to beta-lactam antibiotics due to the absence of peptidoglycan. The size of bacteria is around 0.5–0.8 × 1.2–3.0 µm. The organisms appear as blue purple coccobacilli after Giemsa or Diff-Quick staining, but are not visualized by Gimenez staining (Eremeeva and Dasch, 2001; Seong et al., 2001; Tamura et al., 1995). *O. tsutsugamushi* grows in several cell lines such as L929, Vero, and ECV304 cells incubated at 34–37°C. The doubling time in L929 cells is about 8–12 hours.

The complete genome sequencing of *O. tsutsugamushi* strains Boryong and Ikeda was completed a few years ago. Their genome exhibits several striking characteristics. It is about 2 Mbp long, contains almost 2000 potential protein-coding genes and is thus larger than that of *Rickettsia* spp. It also contains a large number of repetitive genetic elements, which account for approximately 40% of the genome. Thus, the genome of *O. tsutsugamushi* is recognized as the most highly repeated bacterial genome ever sequenced up to now. Remarkably, the genome sequence also reveals several amplified repetitive sequences of the genes encoding conjugative type IV secretion systems (T4SS) (*tra* genes), which mediate the horizontal gene transfer and are classified as type IVA systems in other bacteria. T4SS genes may be involved in the evolutionary process and intracellular lifestyle of *O. tsutsugamushi* (Cho et al., 2007; Nakayama et al., 2008). In *Rickettsia* spp., it is hypothesized that these genes may function as type IVB systems or effector translocators that secrete effector proteins into the cytosol of host cells, thus interfering with cellular defense and promoting bacterial survival (Balraj et al., 2009).

O. tsutsugamushi has been the sole species in this genus for a long time. Recently, *Orientia chuto* has been proposed as a second species (Izzard et al., 2010). Based on serotypic methods, *O. tsutsugamushi* is classified into different antigenic strains, among which Karp, Gilliam, and Kato are the main prototypes. Other antigenic strains such as Kawasaki, Kuroki, Shimokoshi, Boryong, TA678, TA686, TA716, and TA763 have been reported in several different countries. Outer membrane proteins of *O. tsutsugamushi* are composed of several proteins with different molecular weights, namely 22-, 47-, 56-, and 110-kDa antigens. Among these proteins, the 56-kDa antigen is the most abundantly expressed membrane

Human Emerging and Re-emerging Infections: Bacterial & Mycotic Infections, Volume II, First Edition. Edited by Sunit K. Singh.
© 2016 John Wiley & Sons, Inc. Published 2016 by John Wiley & Sons, Inc.

protein and the immunodominant antigen that strongly induces both T cell and B cell responses (Seong et al., 2001). Due to the high degree of nucleotide polymorphisms, it is also commonly used for genotyping. In addition, the 56-kDa protein can bind to host extracellular matrices such as fibronectin, and thereby mediating bacterial adhesion and invasion into the host cells (Lee et al., 2008).

46.2 Epidemiology of infection

O. tsutsugamushi organisms are maintained in the nature by their life cycles in both mite vectors and vertebrate hosts. Trombiculid mites (*Leptotrombidium* spp.) acquire the organisms by transovarian transmission or after feeding on blood of infected animals during the larval stage. The so-called "chigger," a six-legged larva of mite, is an ectoparasite of vertebrate hosts such as small mammals and ground-feeding birds and feeds only once on blood. After a blood meal, it develops into a nymph and an adult stage, both of which are eight-legged free-living arthropods in the soil. The adult mite transmits *O. tsutsugamushi* organisms to eggs and offspring by transovarian transmission, thereby maintaining the cycle of disease transmission. Small mammals, particularly rodents, are considered reservoir hosts, whereas humans are an accidental host. As a chigger feeds on blood only once, the disease is transmitted to humans only by the chigger that gets infected with *O. tsutsugamushi* via transovarian route (Jeong et al., 2007; Seong et al., 2001).

Scrub typhus is confined to the certain geographic areas called "tsutsugamushi triangle." This triangle includes several countries in East, Southeast, South Asia, and Northern Australia. The disease also poses a risk to foreign tourists who travel to the endemic areas (Jensenius et al., 2004). In tropical countries in Southeast Asia, the incidence of scrub typhus is highest during rainy season (Suputtamongkol et al., 2009), probably because rice production in this season leads to an increase in rodent populations. In addition, humans come into close contact with rodents and mite vectors. In countries in temperate zone such as Korea, the incidence is highest in autumn, from October to November. This period correlates with an increase in the population of mite vectors (Kweon et al., 2009). The disease transmission occurs in agricultural fields, small forests (or scrub), or any location where the surrounding environment is favorable to rodents and mite vectors. Humans acquire the infection by occupational exposure or during their activities of daily living.

46.3 Clinical features

The clinical manifestations of scrub typhus are usually mild and self-limited. In few cases, the disease may be severe or even fatal. The most common clinical manifestations are fever, headache, and myalgia.

Respiratory and gastrointestinal symptoms such as cough, nausea, vomiting, and abdominal pain are quite common as well. Rash, lymphadenopathy, and hepatosplenomegaly are variably observed among patients in different studies. Eschar at the infected site, which is a specific sign of arachnid-borne rickettsioses, is variably present, ranging from less than 1/4 to approximately 2/3 of patients. Complications including mild hepatic dysfunction and interstitial pneumonitis are frequently reported. In severe cases, the patients may develop serious complications such as meningoencephalitis, cardiac complications (myocarditis or congestive heart failure), acute renal failure, respiratory failure, multi-organ dysfunction, and death. In mild cases, patients spontaneously recover within 2–3 weeks (Lee et al., 2013; Silpapojakul et al., 2004; Sirisanthana et al., 2003; Suputtamongkol et al., 2009). The differential diagnoses for patients with acute undifferentiated fever in endemic areas of scrub typhus should include viral infection (e.g., dengue infection), malaria, leptospirosis, primary bacteremia, and other rickettsioses.

46.4 Pathogenesis of *O. tsutsugamushi* infection

O. tsutsugamushi has been reported to infect many types of cells belonging or not to immune system (Fukuhara et al., 2005; Ge and Rikihisa, 2011; Rikihisa and Ito, 1982). A consensus exists that endothelial cells are the most frequently observed cell targets in patients with scrub typhus and in animals infected with *O. tsutsugamushi* (Ge and Rikihisa, 2011). It is likely that the infection and activation of endothelial cells contribute to the pathophysiology of scrub typhus as reported in rickettsioses (Bechah et al., 2011). A recent study showed a tropism for dendritic cells and monocytes in skin biopsies from patients with scrub typhus (Paris et al., 2012). In addition, *O. tsutsugamushi* is found in tissues rich in endothelial cells and macrophages (Moron et al., 2001) and in circulating monocytes from patients with scrub typhus (Walsh et al., 2001). The immunohistochemical examination of eschar lesions showed a prominent infiltrate at the dermoepidermal junction with a preferential bacterial location in Langerhans cells, dermal dendritic cells, and activated macrophages (Paris et al., 2012). These findings suggest that *O. tsutsugamushi* may use dendritic cells and macrophages as a route for bacterial dissemination and/or interfere with mounting a protective immune response.

46.4.1 Invasion and activation of endothelial cells by *O. tsutsugamushi*

The tropism of *O. tsutsugamushi* (and the *Rickettsia* species) for endothelial cells is shared with other bacteria such as *Listeria monocytogenes* (Vazquez-Boland et al., 2001), *Bartonella* species (Minnick and Battisti, 2009), *Chlamydia pneumoniae* (Gaydos, 2000), *Neisseria meningitidis* (Carbonnelle et al., 2009), and invasive strains of *Streptococcus pyogenes* (Amelung et al., 2011). As the literature on invasion of endothelial cells is relatively scarce, the analysis of data should integrate those from endothelial cells and other non-phagocytic cell lines. The lack of LPS in the cell wall of *O. tsutsugamushi* begs the question of how the bacterium is recognized by endothelial cells. Two bacterial ligands have been described so far. The 56-kDa outer membrane protein of *O. tsutsugamushi* interacts with extracellular matrix components, such as fibronectin, which requires engagement of integrins for entry into endothelial cells (Lee et al., 2008). The ScaC autotransporter membrane protein promotes the adherence of *O. tsutsugamushi* to non-phagocytic cell lines, including the human endothelial cell line ECV304; when ScaC is expressed in *Escherichia coli*, bacterial adherence is increased but bacterial invasion is not promoted (Ha et al., 2011). The eukaryotic receptors for *O. tsutsugamushi* include the heparin sulfate moiety of the heparan sulfate proteoglycan (HSPG) (Ge and Rikihisa, 2011) and integrins such as the α5β1 integrin in HeLa cells (Cho et al., 2010b). *O. tsutsugamushi* invades ECV304 and L929 cells by clathrin-mediated endocytosis. Once inside the cells, the bacterium rapidly escapes from the phagosome into the cytosol (Chu et al., 2006) via a hemolysin gene, tlyC, and a gene that potentially encodes phospholipase D (Cho et al., 2007, 2010a). Within the cytosol, *O. tsutsugamushi* moves to the microtubule-organizing center in the perinuclear region (Kim et al., 2001). The intracellular mobility of *O. tsutsugamushi* is distinct from that of other bacteria with an intracytosolic location, as *O. tsutsugamushi* does not mobilize filamentous actin (Stevens et al., 2006). Similarly to other intracellular bacteria, *O. tsutsugamushi* infection induces autophagic response in infected cells, but the bacteria develop some strategies that enable them to escape from autophagy and replicate in cytosol. Treatment with tetracycline prevents this escape and promotes the bacterial entrapment in autophagosomes (Ko et al., 2013), suggesting that *O. tsutsugamushi* may secrete effector proteins that interfere with cellular autophagy.

The interaction of *O. tsutsugamushi* with endothelial cells leads to their activation including cytoskeletal reorganization (Cho et al., 2010b) and production of cytokines and chemokines. In humans, dermal microvascular endothelial cells (HMEC-1), CCL2 (monocyte chemotactic protein-1, MCP-1), CCL5 (regulated upon activation, normal T-cell expressed and secreted, RANTES), and CXCL8 [interleukin (IL)-8] are produced in response to *O. tsutsugamushi* infection (Cho et al., 2001). In ECV304 cells, a large panel of cytokines [IL-1, IL-6, IL-10, IL-15, tumor necrosis factor (TNF), and lymphotoxin alpha] and chemokines [CXCL1-3 (growth-related oncogene, Gro), CCL5, CCL17 (thymus and activation-regulated chemokine, TARC), and CXCL8] are also induced (Cho et al., 2001). In addition, ECV304 cells infected

with *O. tsutsugamushi* secrete IL-32 in a nucleotide oligomerization domain (NOD)-1-dependent manner although the absence of proteoglycans questions the nature of NOD-1 ligand (Cho et al., 2010c). The production of IL-1β and IL-32 (Nold-Petry et al., 2009) and the cascade that they initiate may be critical for endothelial injury and inflammation, which are involved in the clinical complications of patients with scrub typhus.

46.4.2 Interaction of *O. tsutsugamushi* with innate immune cells

46.4.2.1 Infection of monocytes and macrophages by *O. tsutsugamushi*

In vivo, the monocytes are likely targeted by *O. tsutsugamushi* since bacteria are detected in the peripheral blood mononuclear cells (Walsh et al., 2001) and buffy coats (Kim et al., 2006a) of patients with scrub typhus. *Ex vivo*, *O. tsutsugamushi* infects and replicates in human monocytes (Tantibhedhyabgkul et al., 2011). The invasion of monocytes may be beneficial for bacterial survival and bacterial dissemination because *O. tsutsugamushi* is very unstable and quickly loses viability in an extracellular environment (Kim et al., 2002).

O. tsutsugamushi is a potent activator of monocytes. Transcriptional profiling studies revealed that *O. tsutsugamushi*-infected monocytes are polarized toward a sustained M1 profile and exhibit a type I interferon (IFN) response. The M1 status of infected monocytes is associated with the expression of a large panel of inflammatory cytokines and chemokines (Tantibhedhyabgkul et al., 2011). The M1 polarization of monocytes is relatively unusual because the monocyte polarization status is thought to be transient, in contrast to macrophages (Benoit et al., 2008; Mehraj et al., 2013). Thus, this status may contribute to the pathogenicity of *O. tsutsugamushi* rather than to the protection.

The type I IFN response that is induced by *O. tsutsugamushi* involves type I IFN genes (IFN-β and some subtypes of IFN-α) and IFN-stimulated genes (ISGs), such as 2'–5'-oligoadenylate synthetase (OAS) and myxovirus resistance genes (MX1 and MX2) (Tantibhedhyabgkul et al., 2011). The type I IFN response requires live bacteria and likely bacterial DNA but it remains unclear how bacterial DNA is released into the cytosol of host cells. The fragility of *O. tsutsugamushi* due to the absence of peptidoglycan and LPS may account for cell wall rupture after bacterial escape from the phagosome.

We also found that monocytes release high levels of IL-1β in response to live *O. tsutsugamushi* (Tantibhedhyabgkul et al., 2011). These findings suggest an inflammasome activation followed by caspase-1 activation and IL-1β release; both responses have also been observed in *O. tsutsugamushi*-infected ECV304 cells (Cho et al., 2010a). The molecule absent in melanoma (AIM)-2 is present in the inflammasome, recognizes cytosolic DNA, and is critical for the innate immune response to cytosolic bacteria, including *Francisella tularensis* (Rathinam et al., 2010) and *L. monocytogenes* (Fernandes-Alnemri et al., 2010). AIM2 is also an IFN-inducible protein (Choubey and Panchanathan, 2008) and is highly upregulated following live *O. tsutsugamushi* infection (Tantibhedhyabgkul et al., 2011). Because the local and systemic inflammation can be induced by IL-1β (Dinarello, 2011), the IL-1β secreted by monocytes may be critical in the pathogenesis of scrub typhus. The activation of inflammasome accounts for cell death in *O. tsutsugamushi* infection. We have shown that a small proportion of monocytes and macrophages exhibit features of cell death (Tantibhedhyangkul et al., 2011, 2013). Other mechanisms are possible: ISGs, including those for TRAIL/Apo2L, XAF-1, FAS, and PML, that are associated to cell death are upregulated following *O. tsutsugamushi* infection and may promote apoptosis (Chawla-Sarkar et al., 2003). Fas/FasL interaction and perforin/granzyme secretion by cytotoxic lymphocytes may also contribute to apoptosis *in vivo* (Tantibhedhyangkul et al., 2012).

Because inflammatory monocytes are the precursors of M1/inflammatory macrophages, we questioned whether the response of macrophages to *O. tsutsugamushi* could be specific. Reports from human autopsy specimens have demonstrated the presence of *O. tsutsugamushi* in Kupffer cells and macrophages located in the spleen, lymph nodes, and liver capsule (Tseng et al., 2008). However, it is unclear whether these infected macrophages were derived from infected monocytes or were directly infected by *O. tsutsugamushi*. The answer to this question was provided by the use of murine macrophage cell lines, peritoneal macrophages, and human macrophages. *O. tsutsugamushi* induces the expression of chemokine genes, including those encoding CCL2, CCL3 [macrophage inflammatory protein (MIP)-1α], CCL4 (MIP-1β), CCL5, and CXCL2 (MIP-2), in J774A.1 macrophages in a nuclear factor (NF)-κB-dependent manner (Cho et al., 2000; Koo et al., 2009). *O. tsutsugamushi* also activates mitogen-activated protein kinase (MAPK) pathways, leading to the expression of TNF (Yun et al., 2009) and IFN-β (Koo et al., 2009). As observed in monocytes, the expression of TNF does not require live bacteria, in contrast to the expression of IFN-β (Koo et al., 2009). The secretion of TNF by peritoneal macrophages stimulated with *O. tsutsugamushi* does not depend on toll-like receptors (TLRs), a major family of pattern recognition receptors (PRRs), such as TLR-4, which is reminiscent of findings obtained with monocytes infected by *Rikettsia akari* (Quevedo-Diaz et al., 2010; Yun et al., 2009). *O. tsutsugamushi* also induces the production of anti-inflammatory mediators such as IL-10 (Kim et al., 2006b). *O. tsutsugamushi* may use this suppressive mechanism to promote its own survival within macrophages. Indeed, the depletion of macrophages enhance lymphocyte proliferation,

suggesting that macrophages mediate the inhibition of the lymphocyte response (Jerrells, 1985). We showed that the response of human monocyte-derived macrophage to *O. tsutsugamushi* (Tantibhedhyangkul et al., 2013) is similar to what we had previously observed in circulating monocytes (Tantibhedhyabgkul et al., 2011). *O. tsutsugamushi* induces M1 polarization, the type I IFN response, and IL-1β secretion in human monocyte-derived macrophages and modulates the chemokinome (Bromley et al., 2008; Viola et al., 2008). Thus, it is likely that the type I IFN response combined with the chemokine network plays a critical role in mononuclear cell infiltration.

46.4.2.2 Infection of dendritic cells by *O. tsutsugamushi*

The histological examination of eschars from patients with scrub typhus revealed the presence of *O. tsutsugamushi* and dermal infiltrates of inflammatory cells. The infiltrates exhibit a strong predominance of antigen-presenting cells with more than 80% consisting of monocytes/macrophages and dendritic cells. The presence of dermal dendritic cells is more intense at the dermal–epidermal junction and the main cells associated with *O. tsutsugamushi* infection are HLA-DR+ antigen-presenting cells, including dermal dendritic cells and Langerhans cells (Paris et al., 2012). *O. tsutsugamushi* induces the expression of costimulatory molecules in murine bone-marrow-derived dendritic cells but their relative levels remain lower than those induced by LPS. In addition, *O. tsutsugamushi* actively evades the cellular autophagic system and impairs the migration of dendritic cells (Choi et al., 2013). It was recently shown that the coculture of *O. tsutsugamushi*-infected dendritic cells with autologous CD4+ T cells enhances the production of IFN-γ (Chu et al., 2013), suggesting that dendritic cells become at least partially mature in response to *O. tsutsugamushi*.

46.5 Immune response to *O. tsutsugamushi*

The role of immunity in the resolution of *O. tsutsugamushi* infection is based on findings obtained with animal models and from patients with scrub typhus. The mouse models of *O. tsutsugamushi* infection have shown that the susceptibility or the resistance to *O. tsutsugamushi* is genetically controlled. After challenge with the Gilliam *O. tsutsugamushi* strain, susceptible C3H/HeN mice possess a greater number of macrophages and neutrophils but a lower number of lymphocytes in the peritoneal cavity than resistant BALB/c mice (Yun et al., 2005). The susceptible mice produce higher levels of cytokines and chemokines than do resistant mice after challenge with the Gilliam and Karp *O. tsutsugamushi* strains (Koh et al., 2004; Yun et al., 2005). The type I immune response induced by *O. tsutsugamushi* plays likely a role in the control of the infection. Indeed, the bacterial burden was reduced in cytokine-treated macrophages during the initial phase of the infection (Nacy and Osterman, 1979). The pretreatment of murine macrophages with TNF, IFN-γ, or their combination decreases the percentage of infected cells as compared with control macrophages (Geng and Jerrells, 1994). IFN-γ has been shown to partially inhibit the growth of the Gilliam *O. tsutsugamushi* strain in mouse fibroblasts (Hanson, 1991), whereas TNF, but not IFN-γ, was shown to inhibit the growth of the Karp strain in mouse embryonic cell lines (Geng and Jerrells, 1994). IFN-γ is produced by several cell types, including the antigen-responsive T cells found in the draining lymph nodes and spleens of infected mice (Palmer et al., 1984a, 1984b). The IFN-γ-producing T cells during *O. tsutsugamushi* infection are efficient to provide protection when adoptively transferred (Kodama et al., 1987). It is generally believed that cytosolic bacteria engage CD8$^+$ T cells known to produce IFN-γ and to be cytotoxic (CTLs) (Pamer, 2004). The splenic T lymphocytes from mice infected with *O. tsutsugamushi* were able to lyse target cells in a major histocompatibility complex (MHC)-restricted manner (Rollwagen et al., 1986). These results are similar to those from mouse models of rickettsial infections where resistance depends on the expression of markers of CD8$^+$-mediated anti-infectious immunity, including MHC class I molecules, perforin, and IFN-γ (Walker et al., 2001).

Human studies have also revealed the importance of the inflammatory response during scrub typhus. In patients with scrub typhus, the serum levels of TNF, granulocyte colony-stimulating factor (G-CSF), macrophage CSF (M-CSF), IL-12p40, IL-15, IL-18, IFN-γ, and IL-10 were elevated (Chierakul et al., 2004; Chung et al., 2008; Iwasaki et al., 1997; Kramme et al., 2009), and the mRNA expression of IL-1β, TNF, IL-6, IFN-γ, and IL-10 was increased (Chung et al., 2008). Another study also reported elevated levels of CXCL9 (also known as monokine induced by gamma interferon, Mig), CXCL10 (interferon gamma-induced protein 10, IP-10), and granzymes A and B (de Fost et al., 2005). CXCL9 and CXCL10 preferentially attract CTLs and natural killer (NK) cells, which express CXCR3, whereas granzymes A and B are degranulation products of CTLs. The levels of TNF, IL-1β, IL-6, IFN-γ, and IL-10 rapidly decrease following treatment with doxycycline, which correspond to early defervescence and the rapid improvement of symptoms (Chung et al., 2008). An investigation of the transcriptomic profiles of mononuclear cells from patients with scrub typhus revealed the modulation of more than 1000 genes (Tantibhedhyabgkul et al., 2011). Most of the highly expressed genes were found to correspond to biological processes, including DNA metabolism, cell cycle and immune function. For the genes related to immune function, IFN-γ, AIM2, guanylate-binding protein 1 (GBP1), IFN-inducible factor 16 (IFI16), and indoleamine 2,3-dioxygenase (INDO) were identified as upregulated genes. The transcriptome profiles of mononuclear cells from patients with scrub typhus were compared to that of monocytes infected *in vitro* with *O. tsutsugamushi*. This comparison revealed that the differential expression of genes in scrub typhus, such as those for type I and II IFNs and M1-associated genes, was directly related to *O. tsutsugamushi* infection. Finally, a comparison of the transcriptome profiles between scrub typhus patients and patients with murine typhus, malaria, or dengue highlighted the specificity of the scrub typhus signature relating to upregulation of CD8-responsive genes (Tantibhedhyabgkul et al., 2011). This specific signature of scrub typhus supports the role for CD8$^+$ T cells in the protection against *O. tsutsugamushi* in mice (Rollwagen et al., 1986) and the activation of CTLs in patients with scrub typhus (de Fost et al., 2005).

46.6 Diagnosis of *O. tsutsugamushi* infection

The most commonly used methods for diagnosis of scrub typhus are indirect immunofluorescence assay (IFA) and polymerase chain reaction (PCR), although they require special laboratory facilities and needs some expertise to interpret the results. PCR, especially real-time PCR, is quite sensitive for scrub typhus. For serologic diagnosis by IFA, acute and convalescent sera should be collected 2 weeks apart. The criteria for diagnosis by IFA are a fourfold increase in IgG titer and/or IgM titer in paired sera and/or the IgM titer in single serum equal to or higher than 1:400 ratio. However, each country in endemic areas may use the different cut-off value for seropositivity (Blacksell et al., 2007). Antigens for IFA can be prepared from either cells infected with *O. tsutsugamushi* or purified organisms and should comprise at least three prototypic strains: Karp, Gilliam, and Kato. The local antigenic strains in each country may be additionally included. In a sero-survey study in Korea, the IgG titer increases rapidly during the first 2 weeks, reaches a peak at 4 weeks, and is maintained at low level after 6–18 months. The IgM titer gradually increases during the period of 2–3 weeks, reaches the peak level at 4 weeks, and decreases after 5 weeks (Kim et al., 2008). However, in other countries, the dynamics of antibody titers may be different. For example, some patients in Thailand have high and persistent titers of both IgM and IgG, whereas other patients exhibit an IgM titer higher than of IgG (WT, unpublished observation).

IFA test can be used to distinguish between primary and secondary infection. Most patients have serologic profiles suggestive of primary infection that is characterized by a typical fourfold rise in both IgG and IgM in paired sera. In secondary infection, very high IgG titer appears within the first 1–2 weeks of illness, whereas the IgM titer appears at low level or is below the cut-off value. Secondary infection is observed in a minority of patients (Chen et al., 2011). However, there are few patients whose IFA results are equivocal, such as ones with high and sustained IgG and IgM titers since the early stage of illness (TW, an unpublished observation in Thailand). This finding is probably due to past or recent infection, frequent re-infection, or inappropriate timing of blood collection. Therefore, PCR is helpful to diagnosis in patients with equivocal IFA results or at the early stage of infection before the sero-conversion. In contrast to *Rickettsia* spp., PCR for *O. tsutsugamushi* detection is quite sensitive, probably due to the tropism of the organisms for circulating monocytes. Specimens for PCR can be either buffy coats or whole blood with ethylenediaminetetraacetic acid (EDTA) or sodium citrate. Real-time PCR of the genes encoding the 47-kDa protein (Jiang et al., 2004), GroEL operon (Paris et al., 2009), or 16s rRNA or nested PCR targeting the gene encoding the 56-kDa antigen have been reported with good sensitivity, ranging from 60 to 85% (Kim et al., 2006a, 2011; Singhsilarak et al., 2005). When possible, PCR of eschar biopsy is more sensitive; however, as mentioned earlier, eschar is quite rare in some countries. Other commercial serologic tests, such as ELISA or rapid immunochromatographic test (ICT) have been developed. Rapid ICT is beneficial to bed-side diagnosis in hospitals where facilities for IFA or PCR are not available. However, the result should be interpreted carefully, because ICT is qualitative, not a semi-quantitative test. In addition, ICT tests from different companies have been demonstrated to have varying sensitivity (Blacksell et al., 2010a, 2010b). IgG and IgM ELISA kits for scrub typhus also exhibit a good sensitivity and specificity and therefore may be an alternative to IFA tests (Coleman et al., 2002; Land et al., 2000). However, due to the antigenic variation in each endemic area, these tests should be evaluated before routine use.

46.7 Treatment of *O. tsutsugamushi* infection

Because scrub typhus is a very common systemic infection in rural endemic areas and that the precise diagnosis is difficult to make, especially at the early stage of infection, the empirical treatment should be initiated in patients with acute undifferentiated febrile illness when viral infection or malaria is excluded. Doxycycline remains the drug of choice for the treatment of scrub typhus because it has good bioavailability, long half-life, and low MICs, ranging from 0.06 to 0.25 µg/mL (Eremeeva and Dasch, 2001; WT, unpublished data). The efficacy of doxycycline against rickettsiae is well supported by *in vitro*, animal models and clinical studies. The dose of doxycycline is 100 mg twice daily for 7 days. Azithromycin and chloramphenicol are alternative drugs in very young children or pregnant women. In countries where intravenous doxycycline is not available, intravenous azithromycin or chloramphenicol is often initially given to patients with severe disease who cannot take or tolerate oral doxycycline. The patients usually recover within 24–48 hours after initiation of doxycycline treatment. The rapid defervescence of fever is the hallmark of scrub typhus and rickettsioses and is thus considered a diagnostic clue. A third-generation cephalosporin (cefotaxime or ceftriaxone) is usually combined with doxycycline when primary bacteremia or leptospirosis cannot be excluded (Suputtamongkol et al., 2009).

46.8 Prevention and control of *O. tsutsugamushi* infection

Since *O. tsutsugamushi* has an intracellular lifestyle in cytoplasm of host cells, the protective immunity requires both Th1 and cytotoxic T cells. It is still problematic to develop any non-live vaccine that elicits strong cell-mediated response. The only prevention strategy is to avoid arthropod bite. Topical insect repellent DEET (*N,N*-diethyl-*meta*-toluamide) can be applied on skin to prevent arthropod bite. Clothing can be treated with permethrin to kill the attached arthropods (Parola and Raoult, 2006).

References

Amelung, S., Nerlich, A., Rohde, M., Spellerberg, B., Cole, J.N., Nizet, V., Chhatwal, G.S., and Talay, S.R. 2011. The FbaB-type fibronectin-binding protein of *Streptococcus pyogenes* promotes specific invasion into endothelial cells. *Cell Microbiol.*, 13, 1200–1211.

Balraj, P., Renesto, P., and Raoult, D. 2009. Advances in rickettsia pathogenicity. *Ann. N. Y. Acad. Sci.*, 1166, 94–105.

Bechah, Y., Capo, C., and Mege, J.L. 2011. Vasculitis: Endothelial dysfunction during rickettsial infection. In: Amezcua-Guerra, L.M., editor. *Advances in the Etiology, Pathogenesis and Pathology of Vasculitis*. Croatia: InTech. pp. 57–70.

Benoit, M., Desnues, B., and Mege, J.L. 2008. Macrophage polarization in bacterial infections. *J. Immunol.*, 181, 3733–3739.

Blacksell, S.D., Bryant, N.J., Paris, D.H., Doust, J.A., Sakoda, Y., and Day, N.P. 2007. Scrub typhus serologic testing with the indirect immunofluorescence method as a diagnostic gold standard: A lack of consensus leads to a lot of confusion. *Clin. Infect. Dis.*, 44, 391–401.

Blacksell, S.D., Jenjaroen, K., Phetsouvanh, R., Tanganuchitcharnchai, A., Phouminh, P., Phongmany, S., Day, N.P., and Newton, P.N. 2010a. Accuracy of rapid IgM-based immunochromatographic and immunoblot assays for diagnosis of acute scrub typhus and murine typhus infections in Laos. *Am. J. Trop. Med. Hyg.*, 83, 365–369.

Blacksell, S.D., Jenjaroen, K., Phetsouvanh, R., Wuthiekanun, V., Day, N.P., Newton, P.N., and Ching, W.M. 2010b. Accuracy of AccessBio immunoglobulin M and total antibody rapid immunochromatographic assays for the diagnosis of acute scrub typhus infection. *Clin. Vaccine Immunol.*, 17, 263–266.

Bromley, S.K., Mempel, T.R., and Luster, A.D. 2008. Orchestrating the orchestrators: Chemokines in control of T cell traffic. *Nat. Immunol.*, 9, 970–980.

Carbonnelle, E., Hill, D.J., Morand, P., Griffiths, N.J., Bourdoulous, S., Murillo, I., Nassif, X., and Virji, M. 2009. Meningococcal interactions with the host. *Vaccine*, 27(suppl. 2), B78–B89.

Chawla-Sarkar, M., Lindner, D.J., Liu, Y.F., Williams, B.R., Sen, G.C., Silverman, R.H., and Borden, E.C. 2003. Apoptosis and interferons: Role of interferon-stimulated genes as mediators of apoptosis. *Apoptosis*, 8, 237–249.

Chen, H.W., Zhang, Z., Huber, E., Mutumanje, E., Chao, C.C., and Ching, W.M. 2011. Kinetics and magnitude of antibody responses against the conserved 47-kilodalton antigen and the variable 56-kilodalton antigen in scrub typhus patients. *Clin. Vaccine Immunol.*, 18, 1021–1027.

Chierakul, W., de Fost, M., Suputtamongkol, Y., Limpaiboon, R., Dondorp, A., White, N.J., and van der Poll, T. 2004. Differential expression of interferon-gamma and interferon-gamma-inducing cytokines in Thai patients with scrub typhus or leptospirosis. *Clin. Immunol.*, 113, 140–144.

Cho, N.H., Seong, S.Y., Huh, M.S., Han, T.H., Koh, Y.S., Choi, M.S., and Kim, I.S. 2000. Expression of chemokine genes in murine macrophages infected with *Orientia tsutsugamushi*. *Infect. Immun.*, 68, 594–602.

Cho, N.H., Seong, S.Y., Choi, M.S., and Kim, I.S. 2001. Expression of chemokine genes in human dermal microvascular endothelial cell lines infected with *Orientia tsutsugamushi*. *Infect. Immun.*, 69, 1265–1272.

Cho, N.H., Kim, H.R., Lee, J.H., Kim, S.Y., Kim, J., Cha, S., Darby, A.C., Fuxelius, H.H., Yin, J., Kim, J.H., Lee, S.J., Koh, Y.S., Jang, W.J., Park, K.H., Andersson, S.G., Choi, M.S., and Kim, I.S. 2007. The *Orientia tsutsugamushi* genome reveals massive proliferation of conjugative type IV secretion system and host–cell interaction genes. *Proc. Natl. Acad. Sci. USA*, 104, 7981–7986.

Cho, B.A., Cho, N.H., Min, C.K., Kim, S.Y., Yang, J.S., Lee, J.R., Jung, J.W., Lee, W.C., Kim, K., Lee, M.K., Kim, S., Kim, K.P., Seong, S.Y., Choi, M.S., and Kim, I.S. 2010a. Global gene expression profile of *Orientia tsutsugamushi*. *Proteomics*, 10, 1699–1715.

Cho, B.A., Cho, N.H., Seong, S.Y., Choi, M.S., and Kim, I.S. 2010b. Intracellular invasion by *Orientia tsutsugamushi* is mediated by integrin signaling and actin cytoskeleton rearrangements. *Infect. Immun.*, 78, 1915–1923.

Cho, K.A., Jun, Y.H., Suh, J.W., Kang, J.S., Choi, H.J., and Woo, S.Y. 2010c. *Orientia tsutsugamushi* induced endothelial cell activation via the NOD1-IL-32 pathway. *Microb. Pathog.*, 49, 95–104.

Choi, J.H., Cheong, T.C., Ha, N.Y., Ko, Y., Cho, C.H., Jeon, J.H., So, I., Kim, I.K., Choi, M.S., Kim, I.S., and Cho, N.H. 2013. *Orientia tsutsugamushi* subverts dendritic cell functions by escaping from autophagy and impairing their migration. *PLoS Negl. Trop. Dis.*, 7, e1981.

Choubey, D., and Panchanathan, R. 2008. Interferon-inducible Ifi200-family genes in systemic lupus erythematosus. *Immunol. Lett.*, 119, 32–41.

Chu, H., Lee, J.H., Han, S.H., Kim, S.Y., Cho, N.H., Kim, I.S., and Choi, M.S. 2006. Exploitation of the endocytic pathway by *Orientia tsutsugamushi* in nonprofessional phagocytes. *Infect. Immun.*, 74, 4246–4253.

Chu, H., Park, S.M., Cheon, I.S., Park, M.Y., Shim, B.S., Gil, B.C., Jeung, W.H., Hwang, K.J., Song, K.D., Hong, K.J., Song, M., Jeong, H.J., Han, S.H., and Yun, C.H. 2013. *Orientia tsutsugamushi* infection induces CD4+ T cell activation via human dendritic cell activity. *J. Microbiol. Biotechnol.*, 23, 1159–1166.

Chung, D.R., Lee, Y.S., and Lee, S.S. 2008. Kinetics of inflammatory cytokines in patients with scrub typhus receiving doxycycline treatment. *J. Infect.*, 56, 44–50.

Coleman, R.E., Sangkasuwan, V., Suwanabun, N., Eamsila, C., Mungviriya, S., Devine, P., Richards, A.L., Rowland, D., Ching, W.M., Sattabongkot, J., and Lerdthusnee, K. 2002. Comparative evaluation of selected diagnostic assays for the detection of IgG and IgM antibody to *Orientia tsutsugamushi* in Thailand. *Am. J. Trop. Med. Hyg.*, 67, 497–503.

de Fost, M., Chierakul, W., Pimda, K., Dondorp, A.M., White, N.J., and Van der Poll, T. 2005. Activation of cytotoxic lymphocytes in patients with scrub typhus. *Am. J. Trop. Med. Hyg.*, 72, 465–467.

Dinarello, C.A. 2011. A clinical perspective of IL-1beta as the gatekeeper of inflammation. *Eur. J. Immunol.*, 41, 1203–1217.

Eremeeva, M.E., and Dasch, G.A. 2001. *Rickettsia* and *Orientia*. In: Sussman, M., editor. *Molecular Medical Microbiology*. London: Academic Press. pp. 2177–2217.

Fernandes-Alnemri, T., Yu, J.W., Juliana, C., Solorzano, L., Kang, S., Wu, J., Datta, P., McCormick, M., Huang, L., McDermott, E., Eisenlohr, L., Landel, C.P., and Alnemri, E.S. 2010. The AIM2 inflammasome is critical for innate immunity to *Francisella tularensis*. *Nat. Immunol.*, 11, 385–393.

Fukuhara, M., Fukazawa, M., Tamura, A., Nakamura, T., and Urakami, H. 2005. Survival of two *Orientia tsutsugamushi* bacterial strains that infect mouse macrophages with varying degrees of virulence. *Microb. Pathog.*, 39, 177–187.

Gaydos, C.A. 2000. Growth in vascular cells and cytokine production by *Chlamydia pneumoniae*. *J. Infect. Dis.*, 181(suppl. 3), S473–S478.

Ge, Y., and Rikihisa, Y. 2011. Subversion of host cell signaling by *Orientia tsutsugamushi*. *Microbes Infect.*, 13, 638–648.

Geng, P., and Jerrells, T.R. 1994. The role of tumor necrosis factor in host defense against scrub typhus rickettsiae. I. Inhibition of growth of *Rickettsia tsutsugamushi*, Karp strain in cultured murine embryonic cells and macrophages by recombinant tumor necrosis factor-alpha. *Microbiol. Immunol.*, 38, 703–711.

Ha, N.Y., Cho, N.H., Kim, Y.S., Choi, M.S., and Kim, I.S. 2011. An autotransporter protein from *Orientia tsutsugamushi* mediates adherence to nonphagocytic host cells. *Infect. Immun.*, 79, 1718–1727.

Hanson, B. 1991. Comparative susceptibility to mouse interferons of *Rickettsia tsutsugamushi* strains with different virulence in mice and of *Rickettsia rickettsii*. *Infect. Immun.*, 59, 4134–4141.

Iwasaki, H., Takada, N., Nakamura, T., and Ueda, T. 1997. Increased levels of macrophage colony-stimulating factor, gamma interferon, and tumor necrosis factor alpha in sera of patients with *Orientia tsutsugamushi* infection. *J. Clin. Microbiol.*, 35, 3320–3322.

Izzard, L., Fuller, A., Blacksell, S.D., Paris, D.H., Richards, A.L., Aukkanit, N., Nguyen, C., Jiang, J., Fenwick, S., Day, N.P., Graves, S., and Stenos, J. 2010. Isolation of a novel *Orientia* species (*O. chuto* sp. nov.) from a patient infected in Dubai. *J. Clin. Microbiol.*, 48, 4404–4409.

Jensenius, M., Fournier, P.E., and Raoult, D. 2004. Rickettsioses and the international traveler. *Clin. Infect. Dis.*, 39, 1493–1499.

Jeong, Y.J., Kim, S., Wook, Y.D., Lee, J.W., Kim, K.I., and Lee, S.H. 2007. Scrub typhus: Clinical, pathologic, and imaging findings. *Radiographics*, 27, 161–172.

Jerrells, T.R. 1985. Immunosuppression associated with the development of chronic infections with *Rickettsia tsutsugamushi*: Adherent suppressor cell activity and macrophage activation. *Infect. Immun.*, 50, 175–182.

Jiang, J., Chan, T.C., Temenak, J.J., Dasch, G.A., Ching, W.M., and Richards, A.L. 2004. Development of a quantitative real-time polymerase chain reaction assay specific for *Orientia tsutsugamushi*. *Am. J. Trop. Med. Hyg.*, 70, 351–356.

Kim, S.W., Ihn, K.S., Han, S.H., Seong, S.Y., Kim, I.S., and Choi, M.S. 2001. Microtubule- and dynein-mediated movement of *Orientia tsutsugamushi* to the microtubule organizing center. *Infect. Immun.*, 69, 494–500.

Kim, M.K., Odgerel, Z., Chung, M.H., Lim, B.U., and Kang, J.S. 2002. Characterization of monoclonal antibody reacting exclusively against intracellular *Orientia tsutsugamushi*. *Microbiol. Immunol.*, 46, 733–740.

Kim, D.M., Yun, N.R., Yang, T.Y., Lee, J.H., Yang, J.T. Shim, S.K., Choi, E.N., Park, M.Y., and Lee, S.H. 2006a. Usefulness of nested PCR for the diagnosis of scrub typhus in clinical practice: A prospective study. *Am. J. Trop. Med. Hyg.*, 75, 542–545.

Kim, M.J., Kim, M.K., and Kang, J.S. 2006b. *Orientia tsutsugamushi* inhibits tumor necrosis factor alpha production by inducing interleukin 10 secretion in murine macrophages. *Microb. Pathog.*, 40, 1–7.

Kim, D.M., Lee, Y.M., Back, J.H., Yang, T.Y., Lee, J.H., Song, H.J., Shim, S.K., Hwang, K.J., and Park, M.Y. 2008. A serosurvey of *Orientia tsutsugamushi* from patients with scrub typhus. *Clin. Microbiol. Infect.*, 16, 447–451.

Kim, D.M., Park, G., Kim, H.S., Lee, J.Y., Neupane, G.P., Graves, S., and Stenos, J. 2011. Comparison of conventional, nested, and real-time quantitative PCR for diagnosis of scrub typhus. *J. Clin. Microbiol.*, 49, 607–612.

Ko, Y., Choi, J.H., Ha, N.Y., Kim, I.S., Cho, N.H., and Choi, M.S. 2013. Active escape of *Orientia tsutsugamushi* from cellular autophagy. *Infect. Immun.*, 81, 552–559.

Kodama, K., Kawamura, S., Yasukawa, M., and Kobayashi, Y. 1987. Establishment and characterization of a T-cell line specific for *Rickettsia tsutsugamushi*. *Infect. Immun.*, 55, 2490–2495.

Koh, Y.S., Yun, J.H., Seong, S.Y., Choi, M.S., and Kim, I.S. 2004. Chemokine and cytokine production during *Orientia tsutsugamushi* infection in mice. *Microb. Pathog.*, 36, 51–57.

Koo, J.E., Yun, J.H., Lee, K.H., Hyun, J.W., Kang, H.K., Jang, W.J., Park, K.H., and Koh, Y.S. 2009. Activation of mitogen-activated protein kinases is involved in the induction of interferon beta gene in macrophages infected with *Orientia tsutsugamushi*. *Microbiol. Immunol.*, 53, 123–129.

Kramme, S., An le, V., Khoa, N.D., Trin le, V., Tannich, E., Rybniker, J., Fleischer, B., Drosten, C., and Panning, M. 2009. *Orientia tsutsugamushi* bacteremia and cytokine levels in Vietnamese scrub typhus patients. *J. Clin. Microbiol.*, 47, 586–589.

Kweon, S.S., Choi, J.S., Lim, H.S., Kim, J.R., Kim, K.Y., Ryu, S.Y., Yoo, H.S., and Park, O. 2009. Rapid increase of scrub typhus, South Korea, 2001–2006. *Emerg. Infect. Dis.*, 15, 1127–1129.

Land, M.V., Ching, W.M., Dasch, G.A., Zhang, Z., Kelly, D.J., Graves, S.R., and Devine, P.L. 2000. Evaluation of a commercially available recombinant-protein enzyme-linked immunosorbent assay for detection of antibodies produced in scrub typhus rickettsial infections. *J. Clin. Microbiol.*, 38, 2701–2705.

Lee, J.H., Cho, N.H., Kim, S.Y., Bang, S.Y., Chu, H., Choi, M.S., and Kim, I.S. 2008. Fibronectin facilitates the invasion of *Orientia tsutsugamushi* into host cells through interaction with a 56-kDa type-specific antigen. *J. Infect. Dis.*, 198, 250–257.

Lee, J.H., Lee, J.H., Chung, K.M., Kim, E.S., Kwak, Y.G., Moon, C., and Lee, C.S. 2013. Dynamics of clinical symptoms in patients with scrub typhus. *Japan J. Infect. Dis.*, 66, 155–157.

Mehraj, V., Textoris, J., Ben Amara, A., Ghigo, E., Raoult, D., Capo, C., and Mege, J.L. 2013. Monocyte responses in the context of Q fever: From a static polarized model to a kinetic model of activation. *J. Infect. Dis.*, 208, 942–951.

Minnick, M.F., and Battisti, J.M. 2009. Pestilence, persistence and pathogenicity: Infection strategies of *Bartonella*. *Future Microbiol.*, 4, 743–758.

Moron, C.G., Popov, V.L., Feng, H.M., Wear, D., and Walker, D.H. 2001. Identification of the target cells of *Orientia tsutsugamushi* in human cases of scrub typhus. *Mod. Pathol.*, 14, 752–759.

Nacy, C.A., and Osterman, J.V. 1979. Host defenses in experimental scrub typhus: Role of normal and activated macrophages. *Infect. Immun.*, 26, 744–750.

Nakayama, K., Yamashita, A., Kurokawa, K., Morimoto, T., Ogawa, M., Fukuhara, M., Urakami, H., Ohnishi, M., Uchiyama, I., Ogura, Y., Ooka, T., Oshima, K., Tamura, A., Hattori, M., and Hayashi, T. 2008. The whole-genome sequencing of the obligate intracellular bacterium *Orientia tsutsugamushi* revealed massive gene amplification during reductive genome evolution. *DNA Res.*, 15, 185–199.

Nold-Petry, C.A., Nold, M.F., Zepp, J.A., Kim, S.H., Voelkel, N.F., and Dinarello, C.A. 2009. IL-32-dependent effects of IL-1beta on endothelial cell functions. *Proc. Natl. Acad. Sci. USA*, 106, 3883–3888.

Palmer, B.A., Hetrick, F.M., and Jerrells, T.J. 1984a. Production of gamma interferon in mice immune to *Rickettsia tsutsugamushi*. *Infect. Immun.*, 43, 59–65.

Palmer, B.A., Hetrick, F.M., and Jerrells, T.R. 1984b. Gamma interferon production in response to homologous and heterologous strain antigens in mice chronically infected with *Rickettsia tsutsugamushi*. *Infect. Immun.*, 46, 237–244.

Pamer, E.G. 2004. Immune responses to *Listeria monocytogenes*. *Nat. Rev. Immunol.*, 4, 812–823.

Paris, D.H., Aukkanit, N., Jenjaroen, K., Blacksell, S.D., and Day, N.P. 2009. A highly sensitive quantitative real-time PCR assay based on the groEL gene of contemporary Thai strains of *Orientia tsutsugamushi*. *Clin. Microbiol. Infect.*, 15, 488–495.

Paris, D.H., Phetsouvanh, R., Tanganuchitcharnchai, A., Jones, M., Jenjaroen, K., Vongsouvath, M., Ferguson, D.P., Blacksell, S.D., Newton, P.N., Day, N.P., and Turner, G.D. 2012. *Orientia tsutsugamushi* in human scrub typhus eschars shows tropism for dendritic cells and monocytes rather than endothelium. *PLoS Negl. Trop. Dis.*, 6, e1466.

Parola, P., and Raoult, D. 2006. Tropical rickettsioses. *Clin. Dermatol.*, 24, 191–200.

Quevedo-Diaz, M.A., Song, C., Xiong, Y., Chen, H., Wahl, L.M., Radulovic, S., and Medvedev, A.E. 2010. Involvement of TLR2 and TLR4 in cell responses to *Rickettsia akari*. *J. Leukoc. Biol.*, 88, 675–685.

Rathinam, V.A., Jiang, Z., Waggoner, S.N., Sharma, S., Cole, L.E., Waggoner, L., Vanaja, S.K., Monks, B.G., Ganesan, S., Latz, E., Hornung, V., Vogel, S.N., Szomolanyi-Tsuda, E., and Fitzgerald, K.A. 2010. The AIM2 inflammasome is essential for host defense against cytosolic bacteria and DNA viruses. *Nat. Immunol.*, 11, 395–402.

Rikihisa, Y., and Ito, S. 1982. Entry of *Rickettsia tsutsugamushi* into polymorphonuclear leukocytes. *Infect. Immun.*, 38, 343–350.

Rollwagen, F.M., Dasch, G.A., and Jerrells, T.R. 1986. Mechanisms of immunity to rickettsial infection: Characterization of a cytotoxic effector cell. *J. Immunol.*, 136, 1418–1421.

Seong, S.Y., Choi, M.S., and Kim, I.S. 2001. *Orientia tsutsugamushi* infection: Overview and immune responses. *Microbes Infect.*, 3, 11–21.

Silpapojakul, K., Varachit, B., and Silpapojakul, K. 2004. Paediatric scrub typhus in Thailand: A study of 73 confirmed cases. *Trans. R. Soc. Trop. Med. Hyg.*, 98, 354–359.

Singhsilarak, T., Leowattana, W., Looareesuwan, S., Wongchotigul, V., Jiang, J., Richards, A.L., and Watt, G. 2005. Short report: Detection of *Orientia tsutsugamushi* in clinical samples by quantitative real-time polymerase chain reaction. *Am. J. Trop. Med. Hyg.*, 72, 640–641.

Sirisanthana, V., Puthanakit, T., and Sirisanthana, T. 2003. Epidemiologic, clinical and laboratory features of scrub typhus in thirty Thai children. *Pediatr. Infect. Dis. J.*, 22, 341–345.

Stevens, J.M., Galyov, E.E., and Stevens, M.P. 2006. Actin-dependent movement of bacterial pathogens. *Nat. Rev. Microbiol.*, 4, 91–101.

Suputtamongkol, Y., Suttinont, C., Niwatayakul, K., Hoontrakul, S., Limpaiboon, R., Chierakul, W., Losuwanaluk, K., and Saisongkork, W. 2009. Epidemiology and clinical aspects of rickettsioses in Thailand. *Ann. N. Y. Acad. Sci.*, 1166, 172–179.

Tamura, A., Ohashi, N., Urakami, H., and Miyamura, S. 1995. Classification of *Rickettsia tsutsugamushi* in a new genus, *Orientia* gen. nov., as *Orientia tsutsugamushi* comb. nov. *Int. J. Syst. Bacteriol.*, 45, 589–591.

Tantibhedhyabgkul, W., Prachason, T., Waywa, D., El Filali, A., Ghigo, E., Thongnoppakhun, W., Raoult, D., Suputtamongkol, Y., Capo, C., Limwongse, C., and Mege, J.L. 2011. *Orientia tsutsugamushi* stimulates an original gene expression program in monocytes: Relationship with gene expression in patients with scrub typhus. *PLoS Neglect. Trop. Dis.*, 5, e1028.

Tantibhedhyangkul, W., Capo, C., Ghigo, E., and Mege, J.L. 2012. Pathogenesis of *Orientia tsutsugamushi* infection. In: Ghigo, E., Mottola, G., and Mege, J.L., editors. *Pathogen Interaction: At the Frontier of the Cellular Microbiology*, 1st ed. India: Research SignPost/Transworld Research Network. pp. 203–216.

Tantibhedhyangkul, W., Ben Amara, A., Textoris, J., Gorvel, L., Ghigo, E., Capo, C., and Mege, J.L. 2013. *Orientia tsutsugamushi*, the causative agent of scrub typhus, induces an inflammatory program in human macrophages. *Microb. Pathog.*, 55, 55–63.

Tseng, B.Y., Yang, H.H., Liou, J.H., Chen, L.K., and Hsu, Y.H. 2008. Immunohistochemical study of scrub typhus: A report of two cases. *Kaohsiung J. Med. Sci.*, 24, 92–98.

Vazquez-Boland, J.A., Kuhn, M., Berche, P., Chakraborty, T., Dominguez-Bernal, G., Goebel, W., Gonzalez-Zorn, B., Wehland, J., and Kreft, J. 2001. *Listeria* pathogenesis and molecular virulence determinants. *Clin. Microbiol. Rev.*, 14, 584–640.

Viola, A., Molon, B., and Contento, R.L. 2008. Chemokines: Coded messages for T-cell missions. *Front. Biosci.*, 13, 6341–6353.

Walker, D.H., Olano, J.P., and Feng, H.M. 2001. Critical role of cytotoxic T lymphocytes in immune clearance of rickettsial infection. *Infect. Immun.*, 69, 1841–1846.

Walsh, D.S., Myint, K.S., Kantipong, P., Jongsakul, K., and Watt, G. 2001. *Orientia tsutsugamushi* in peripheral white blood cells of patients with acute scrub typhus. *Am. J. Trop. Med. Hyg.*, 65, 899–901.

Yun, J.H., Koh, Y.S., Lee, K.H., Hyun, J.W., Choi, Y.J., Jang, W.J., Park, K.H., Cho, N.H., Seong, S.Y., Choi, M.S., and Kim, I.S. 2005. Chemokine and cytokine production in susceptible C3H/HeN mice and resistant BALB/c mice during *Orientia tsutsugamushi* infection. *Microbiol. Immunol.*, 49, 551–557.

Yun, J.H., Koo, J.E., and Koh, Y.S. 2009. Mitogen-activated protein kinases are involved in tumor necrosis factor alpha production in macrophages infected with *Orientia tsutsugamushi*. *Microbiol. Immunol.*, 53, 349–355.

Chapter 47

Lyme Borreliosis

John J. Halperin[1,2]

[1]Department of Neurosciences, Overlook Medical Center, Summit, NJ, USA
[2]Sidney Kimmel Medical College of Thomas Jefferson University, Philadelphia, PA, USA

47.1 Introduction

Their corkscrew shape seems a metaphor for the role spirochetal infections have played in western history. In counterpoint to the disastrous consequences of the introduction of measles and smallpox on the indigenous populations of the Americas, the return gift of syphilis is thought to have had major effects on the evolution of European civilization – both in terms of its initial lethality and later in terms of its neurobehavioral consequences. Although by historical standards infection with *Borrelia burgdorferi* – commonly referred to as Lyme disease – is but a minor footnote, it has played a remarkable role in the ongoing tension between evidence-based medicine and the more traditional approach based on anecdotal observation, bolstered both by the American tradition of grassroots populism and by the newer notion of supposed "crowd sourced intelligence." As such, this disease makes a fascinating study both in its own right, and as an example of the challenges remaining in promoting scientific literacy.

47.2 History

Manifestations of the disease that we now refer to as Lyme disease were first described in the European dermatologic literature over a century ago – first acrodermatitis chronica atrophicans, then erythema chronicum migrans (Afzelius, 1921). In 1922, Garin and Bujadoux described a patient with lymphocytic meningitis and painful polyradiculitis following a tick bite and rash (Garin and Bujadoux, 1922), presumed it to be a spirochetal infection and treated him – apparently successfully – with neoarsphenamine. The clinical syndrome was then well recognized by European neurologists; in the 1950s, decades before the identification of the etiologic agent, penicillin was shown to provide effective treatment (Hollstrom, 1951).

The history in the United States was similar. A 1970 case report (Scrimenti, 1970) described a patient with erythema migrans (EM). In the mid-1970s, in what initially appeared to be an unrelated observation, a surprising number of children in Lyme and Old Lyme Connecticut were diagnosed as having juvenile rheumatoid arthritis. Unsatisfied with their physicians' explanation of this apparent coincidence, the mothers of several affected children communicated with both the Centers for Disease Control and the Division of Rheumatology at Yale, leading to the series of investigations that ultimately identified the causative agent (Steere et al., 1977).

This historical sequence, though hardly unique in medicine, is critical to an understanding of the current debate over Lyme disease, its diagnosis, and its treatment. For over a century the illness was recognized as a clinical syndrome, with neither a clear understanding of its pathophysiology, nor a definitive diagnostic test. Depending on a clinician's inclinations, this could result in either a very restrictive definition of disease (tick bite +/− EM +/− meningitis with radiculoneuritis +/− large joint relapsing arthritis) or a very loose and expansive one. Identification of the causative organism (Benach et al., 1983; Steere et al., 1983a) and development of ultimately reliable diagnostic tests enabled more precise biology-based diagnosis. By then, however, many were heavily invested in the notion of this being a primarily clinical diagnosis – defining "clinical diagnosis" as anything diagnosed by a confident clinician, regardless of supporting epidemiologic or laboratory considerations.

The identification of *B. burgdorferi* coincided with three important events in medical sociology. First, public activism was playing an increasing role in the approach to HIV infection, strongly reinforcing the notion of grassroots pressure increasing attention to a disease. Second, medicine in general was in the midst of a transition from a longstanding tradition of basing conclusions on small case series and extrapolations from anecdotal observations to the use of large statistically validated scientific studies. Third, electronic communication was facilitating the spread of information and misinformation, allowing the perpetuation of conclusions without any benefit of even skepticism, let alone validation. This set the stage for the current "Lyme disease controversy."

47.3 Pathogen (*B. burgdorferi*)

Lyme disease, or Lyme borreliosis, is caused by infection with the tick-borne spirochete, *B. burgdorferi*. Initially identified in patients with Lyme disease (Benach et al., 1983; Burgdorfer et al., 1982; Steere et al., 1983a), within a year virtually identical organisms were isolated from Scandinavian patients with acrodermatitis (Asbrink et al., 1984). Originally thought to be a single species, at least 17 genospecies have now been identified (Toledo and Benach, 2011). Human disease is primarily caused by three species – *B. burgdorferi sensu stricto*, the only species that causes human disease in the United States, also responsible for a minority of European cases, and *B. garinii* and *B. afzelii*, responsible for most infections in Europe. Of the latter two, the former disproportionately causes nervous system involvement, the latter cutaneous. Different species, subspecies, and strains appear to have different organotropisms (Wormser and Halperin, 2013), accounting for some of the variations in symptomatology. All are transmitted by the bites of hard-shelled *Ixodes* ticks.

47.4 Epidemiology

Lyme borreliosis is a zoonosis – an infection transmitted among species. As such it is focally endemic – requiring co-localization of the causative organism, vectors, reservoir hosts, and humans. Incidence of the disease has grown as humans have spent more time in less developed areas and as the range of infected reservoir hosts has expanded (Bacon et al., 2008). Only some species serve as effective reservoirs – a key requirement is that they have a relatively asymptomatic spirochetemia, allowing ticks feeding on them to become infected.

B. burgdorferi infection is transmitted to humans virtually exclusively by bites of hard-shelled *Ixodes* ticks – *I. scapularis* in most of the United States, *I. pacificus* in California, *I. ricinus* in Europe, and *I. persulcatus* in Asia. The ticks, which occur predominantly in moist temperate ecosystems, undergo a four-stage, usually 2-year life cycle. The first, the egg, is uninfected as there is no transovarial transmission of infection. Newly hatched and uninfected larvae will have a single meal, typically on a small mammal, most often a field mouse. If the mouse is spirochetemic, the tick can become infected. Spirochetes then remain in the tick gut as the tick matures into a nymph. The nymph will then have its one meal, again typically on a small

mammal but potentially on many species including humans. Again, if feeding on a spirochetemic host, the tick can acquire infection. However, if the tick is already infected, it can transmit infection to this host. Warm ingested blood triggers spirochete proliferation in the tick gut. Spirochetes then migrate through the tick ultimately reaching its salivary glands. The tick continuously injects saliva containing anticoagulants, local anesthetic, and other substances during feeding; once spirochetes have reached the salivary gland, they can be injected as well. Since spirochete proliferation and migration takes time, ticks usually must be attached for 24–48 hours before there is significant risk of infection. This prolonged attachment provides an opportunity to identify ticks early and remove them, preventing infection.

Following this second meal, the tick will mature into an adult, when it will have one final meal, lay its eggs, and die. The adult may feed on a human or other large species. Common preferred hosts include deer, sheep, or bears, giving rise to the common names applied to these ticks. Elimination of these large hosts (Perkins et al., 2006) or treating them with acaricide (Fish and Childs, 2009) can significantly deplete the tick population, lowering the incidence of infection. However, since other hosts can be used as well, infection does not typically disappear entirely.

In the United States 96% of cases occur along the eastern seaboard (2011) – in Connecticut, Delaware, Maine, Maryland, Massachusetts, New Hampshire, New Jersey, New York, Pennsylvania, Vermont, and Virginia, as well as Minnesota and Wisconsin (CDC, 2013). In Europe cases occur throughout temperate regions from Scandinavia through France, Germany, Austria, Slovenia, among others. Typically about 30,000 cases per year are confirmed in the United States; European estimates are typically about double that but since Lyme borreliosis is not consistently reportable across Europe numbers are less complete.

Ticks are poikilotherms and only feed when the temperature is above about 10°C (mid-50s °F). Humans are most likely to present with acute infections in spring and summer. In California and other locales where there is less temperature variation during the year, less temporal variation occurs.

Incidence is typically inversely related to human population density, as the preferred reservoir, the field mouse, is much less prevalent in cities. For example, although several hundred cases per year are reported from New York City's five boroughs, many of these are likely imported from nearby more rural areas.

47.5 Clinical features
47.5.1 Cutaneous

As with any illness, the clinical features of Lyme disease range from the quite specific to the totally nonspecific. The greater the specificity of a given finding, the higher its positive predictive value for the diagnosis, and the less the need for confirmatory testing. Unfortunately, few clinical findings have sufficient specificity that the diagnosis can be made purely on the basis of phenomenology. For any feature, a key element is plausible exposure to infected ticks. In individuals with acute symptoms in the middle of winter in Fargo, North Dakota, the likelihood of Lyme disease is zero, regardless of clinical phenomenology, serology, or anything else.

The one nearly – but not completely – pathognomonic finding is the cutaneous lesion known as EM (initially erythema chronicum migrans, ECM; Afzelius, 1921). This rash typically develops at the site of the tick bite, slowly expanding centrifugally from the site of spirochete inoculation, its leading edge reflecting the reaction to outwardly migrating spirochetes. It is generally easily differentiated from allergic reactions to tick saliva which, much like mosquito bites, occur rapidly (within minutes to hours of the bite) at the site of the bite, are quite pruritic, do not enlarge beyond an inch or 2, and subside in a few days. In contrast, EM usually begins a few days (up to 30) after the bite, the delay representing the time it takes the initial inoculum of spirochetes to multiply sufficiently to start migrating outward. The characteristic feature is the daily enlargement, expanding, and persisting over days to weeks, attaining a quite large size – a minimum of 5 cm diameter for epidemiologic reporting purposes but often many inches to a foot or more in diameter. If the reaction to the migrating spirochetes subsides centrally as more peripheral areas become actively inflamed, the erythroderm can develop rings, resembling a target, although it can also be quite homogeneous in appearance (Figure 47-1). Despite its erythematous appearance it is often asymptomatic – neither pruritic nor painful. The latter probably accounts for the fact that many adults never recall seeing the rash – if it involves the back or other not readily visualized body part, patients may never notice it. Despite this, nearly 70% of US cases confirmed by the CDC report an EM (Bacon et al., 2008). In small children with Lyme disease, who are more often inspected by watchful parents, EM is noted in up to 90% (Pediatric Lyme Disease Study Group et al., 1996).

In some patients – approximately 25% in the United States and a smaller proportion in Europe – spirochetes disseminate hematogenously from the primary EM. Each nidus of newly disseminated spirochetes can then establish a secondary EM, indistinguishable from a primary one except for the absence of the central puncta where the tick bites.

EM is virtually unique. In the central United States a similar rash – referred to as STARI (southern tick associated rash illness) – can follow tick bites but not be accompanied by *B. burgdorferi* infection or other sequelae (Feder et al., 2011). Occasionally a drug eruption can appear similar although these are not typically slowly expansile.

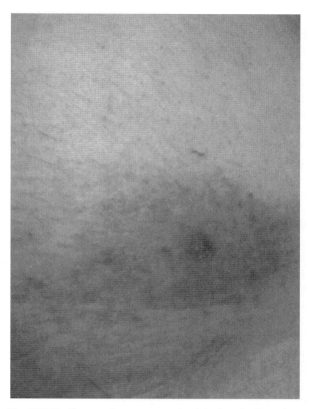

Fig. 47-1 Erythema migrans, approximately 8 cm in diameter. For a color version of this figure, see the color plate section.

EM develops very early in infection. Since serologic diagnosis requires the presence of measurable peripheral blood antibody – which often requires 3–6 weeks to develop – over 50% of patients with EM will be seronegative (Aguero-Rosenfeld et al., 1996). Given the uniqueness of EM, patients with this rash should be immediately treated for Lyme disease without even obtaining laboratory tests.

Two other cutaneous abnormalities occur in European patients, but rarely if ever in the United States. Acrodermatitis chronica atrophicans, from which European *B. afzelii* was first isolated (Asbrink et al., 1984), typically occurs late in infection and involves skin of the distal extremity. Skin becomes tissue paper thin and wrinkly; microscopically innumerable spirochetes are present and patients are strongly seropositive. Borrelia lymphocytoma occurs quite early and consists of dense lymphocytic infiltration of the skin of the areola of the breast or the earlobe. Skin is usually red to purplish, swollen, and tender. Again, patients are usually seropositive.

47.5.2 Nervous system

About 10%–15% of infected individuals develop nervous system involvement (Bacon et al., 2007) very similar to that described by Garin and Bujadoux, and subsequently by the German neurologist Bannwarth (Bannwarth, 1941) (who also described "rheumatism" in affected patients). Patients develop one or more elements of a triad – lymphocytic meningitis, cranial neuritis, and painful radiculoneuritis (Reik et al., 1979; Steere et al., 1983b). Meningitis is typical of "aseptic meningitis" although onset may be a little less abrupt (Garro et al., 2009). Cranial neuritis – inflammation of one or more of the cranial nerves – is probably the most common neurologic manifestation, affecting 7%–10% of patients. Of these as many as 80% involve the facial nerve, causing a typical facial nerve palsy (FNP; Halperin and Golightly, 1992). Up to one-fourth of these will be bilateral. Bilateral FNP in small children is so unusual that in epidemiologically appropriate settings this is virtually pathognomonic. Although in endemic areas summertime unilateral FNP can be attributable to Lyme disease in about a quarter of affected adults, it is important to remember that this means that three-fourths are not related to Lyme disease. Thus, supportive diagnostic testing – that is, positive serology – is required to attribute unilateral FNP (or bilateral FNP in an adult) to Lyme disease.

The other entity, originally described by Garin and Bujadoux, but probably considerably underrecognized today, is painful radiculitis. About 5% of patients develop dysesthetic burning radicular pain, often indistinguishable from a mechanical radiculopathy – or the pain of shingles. This is often accompanied by segmental weakness and reflex changes or, less commonly, sensory loss. The European literature suggests this is most likely to affect the limb that was the site of the tick bite. There has been no systematic analysis in US patients. Pain can be agonizing and intractable, persisting for weeks or even months. When involving truncal dermatomes, it can be mistaken for visceral disease, leading to multiple unproductive assessments for intra-abdominal or thoracic pathology.

Both cranial neuritis and radiculitis appear to be manifestations of the same pathophysiologic process – a mononeuropathy multiplex – a disorder in which multiple individual nerves are separately involved and that can have as many clinical presentations as there are peripheral nerves (Halperin et al., 1990b). Consequently, patients, particularly those with untreated infection of longer duration, can present with other isolated mononeuropathies, lumbosacral, or brachial plexopathies, or even a picture mimicking a diffuse polyneuropathy, caused by what is known as a confluent mononeuropathy multiplex in which so many small nerves are damaged that patients develop typical stocking glove distal sensorimotor loss. The latter process, similar to a diabetic neuropathy, probably only occurs in patients with longstanding untreated infection.

In early studies rare patients – generally estimated at about one patient per thousand untreated individuals per year in endemic areas – developed inflammation within the substance of the brain or spinal cord (Ackermann et al., 1988; Halperin et al., 1989). Best described in European patients, this is probably least rare in patients with Garin–Bujadoux–Bannwarth syndrome, who may develop spinal cord inflammation at the same segmental level as the involved nerve roots. Other patients – also primarily reported before early antimicrobial therapy was widely used – develop focal inflammation in the brain. Collectively these disorders represent forms of encephalomyelitis. As such patients have focal findings on neurologic examination and on MRI imaging reflecting the site of involvement. As with any infectious brain process, cerebrospinal fluid is almost invariably inflammatory with both non-specific and more specific findings reflecting the infection (Halperin et al., 1989). In more recent years such cases have become vanishingly rare.

Finally, the entity known as Lyme encephalopathy, again seen infrequently now that Lyme disease is typically diagnosed and treated quite early, is often misattributed to nervous system infection (Halperin et al., 1990a; Krupp et al., 1991). The disorder was initially described in patients with longstanding symptomatic infection. Like patients with a myriad of other active inflammatory states – ranging from urinary tract infections and pneumonia to active rheumatoid arthritis to flu – such individuals were often aware of cognitive slowing and difficulties with concentration and memory. Initially a focus of interest as a model of other inflammation-associated encephalopathies, it rapidly became apparent that there was nothing unique about this state, and that most such patients had nothing to suggest central nervous system infection. Unfortunately by then the notion had become widespread that this was both characteristic of Lyme disease and evidence of a progressive neurodegenerative process and became the basis for many to diagnose Lyme disease "clinically". Since similar cognitive and related symptoms occur to a disruptive extent in 1–2% of the general population (Luo et al., 2005) at any given time, it should be apparent that the positive predictive value of these symptoms is extremely low for Lyme disease. Consequently, patients treated presumptively for Lyme disease with these symptoms as the only diagnostic criterion can be expected to have limited responses to conventional treatment. This in turn has created the irrational notion that Lyme disease is difficult to treat and requires months if not years of antibiotics.

47.5.3 Post Lyme disease treatment symptoms

Although, as with antibiotic treatment of any infection, there may be occasional treatment failures, many have focused on concerns about persisting fatigue, cognitive slowing, and other non-specific symptoms following treatment for Lyme disease. Some consider this evidence of a persistent infection and have treated this with ever longer, more complex and potentially dangerous regimens. Some have argued that this is a post treatment state, perhaps mediated by persisting immune dysfunction. An alternative question might be whether there is evidence that such a state bears any causal relationship at all to *B. burgdorferi* infection – directly or indirectly.

The symptom complex in question occurs with a remarkable frequency in the general population (Luo et al., 2005), affecting about 2% of individuals to a severe extent, and moderately severe in 7.7% at any given time. Uncontrolled studies suggest such symptoms persist in 20% (Eikeland et al., 2013) to 31% (Ljostad and Mygland, 2009) of patients treated for *neuroborreliosis*, a highly biased subset of all patients with Lyme disease. Interestingly in controlled studies, persisting subjective symptoms were reported in 42% of treated children but 38% of matched controls (Skogman et al., 2012). In controlled studies in adults (Seltzer et al., 2000) similarly the frequency in patients (19%) was not statistically significantly different from that in controls (15%). If one ignores the controlled studies, and assumes, as many proponents suggest, that this state represents a persistent nervous system infection, then one would infer that it occurs in at most 31% of 12% (the proportion with nervous system infection) of patients with *B. burgdorferi* infection, or about 3–4% of Lyme disease patients. Whether this is significantly different from the 2% (severe, 7.7% moderately severe) incidence in the general population is clearly debatable.

47.5.4 Rheumatologic

Unlike the European focus on nervous system involvement in extracutaneous infection, the earliest US reports described rheumatologic abnormalities, initially recognized as what appeared to be juvenile rheumatoid arthritis. It appears that *B. burgdorferi sensu stricto* has a tropism for joints. Specifically Lyme arthritis (Bacon et al., 2008) is reported to affect almost a third of patients with CDC-confirmed Lyme disease. In Europe, this number is substantially lower.

Lyme arthritis has fairly characteristic features (Steere et al., 1987). Unlike many other arthritides, it preferentially affects large joints, usually involving just one joint at a time. It tends to wax and wane spontaneously, perhaps with unprovoked redness and swelling of a knee, subsiding over weeks, followed later by similar involvement of an elbow, shoulder, or hip. Importantly it is an arthritis – involved joints are painful, red, and swollen. Although patients can develop arthralgias – pain without accompanying inflammatory signs, this is so non-specific as to be uninformative (perhaps analogous to the differentiation between focal encephalomyelitis with inflammatory changes on MRI, neurologic exam, and CSF exam and encephalopathy with none of these objective correlates). As a later manifestation of *B. burgdorferi* infection, patients with Lyme arthritis are invariably seropositive – usually quite strongly so.

47.5.5 Cardiac

Early US series suggested up to 5% of infected patients developed cardiac abnormalities (Steere et al., 1977) – primarily an otherwise unexplained heart block, including potentially complete heart block requiring a temporary pacemaker. This has almost always been completely reversible. Although in theory this could be one mechanism by which Lyme disease could be lethal, in fact death from Lyme disease is so rare as to be questionable whether it occurs (Kugeler et al., 2011). More recent data from the CDC suggests that as early treatment has become more widespread, the incidence of cardiac involvement has declined (as have other later manifestations), now affecting fewer than 1% of CDC-confirmed patients (Bacon et al., 2008).

47.6 Diagnosis

Diagnosis of Lyme disease requires three essential elements – epidemiologic likelihood of exposure, clinical phenomenology, and laboratory support for the diagnosis. The oft quoted aphorism that the diagnosis of Lyme disease is a clinical one refers to the need for an appropriate synthesis of all clinical, epidemiologic, and laboratory data, basing conclusions and recommendations on scientifically valid information.

Laboratory support for the diagnosis is critically important. Similar to the other well-known spirochetosis, syphilis, spirochetes can be demonstrated microscopically in skin lesions. However, EM is so characteristic that this is rarely necessary. Unlike in syphilis, the organism can be grown *in vitro*; however, this is quite challenging. The required special medium is not available in most laboratories, and incubation must be both prolonged and at lower temperatures than those generally used in clinical laboratories. More importantly, even in the best of laboratories, sensitivity of culture of biologically accessible fluids and tissues – other than the acute skin lesion – is quite low, probably because there are very few spirochetes in blood or CSF. Even using the tremendous technical sensitivity of polymerase chain reaction has not substantially improved diagnostic sensitivity. In fact given common contamination issues in laboratories, PCR has essentially no clinical applicability in the diagnosis of *B. burgdorferi* infection (Halperin et al., 1996).

As a result, as in syphilis and numerous other infections, laboratory confirmation of the diagnosis rests almost entirely on demonstrating a specific antibody response. This approach in general has a number of inherent limitations. First, it takes time for the antibody response to mature sufficiently to be diagnostically useful. In the case of Lyme borreliosis, antibodies are often undetectable for the first 3–6 weeks of infection. As EM typically occurs in the first 30 days, over 50% of patients with EM are seronegative (Aguero-Rosenfeld et al., 1993). As a result, patients with EM should be treated without even checking a serologic test. At the other end of the disease spectrum, once any infection has resolved, measurable antibodies typically persist for an extended period of time – nominally to protect against reinfection. As a result a positive serology can only be interpreted as evidence of infection, past or present, not a measure of disease activity, or for that matter treatment response.

The preceding are inherent biologic limitations of all serologic testing. Developing tests for antibodies to any given organism always involves technical considerations that balance test sensitivity and specificity. Issues with early assays in the 1980s raised questions about possible persistent seronegativity despite ongoing infection (Dattwyler et al., 1988). As techniques have improved, it is now widely believed these observations – which were not subsequently confirmed – were an artifact of early test methodology (Halperin et al., 2013).

Issues concerning testing are somewhat different in the United States and Europe. All US-acquired infections are caused by *B. burgdorferi sensu stricto*. Although there are numerous substrains, there is sufficient antigenic homogeneity that a two-tier approach – starting with a screening ELISA, followed by a more specific Western blot of positive or borderline ELISA results, has proven highly sensitive and specific (Dressler et al., 1993). Refining serologic criteria for European borreliosis, caused by this strain as well as by *B. garinii* and *B. afzelii*, has been considerably more challenging.

In the United States, using a Western blot for added specificity has allowed the ELISA to evolve to provide increased sensitivity. Criteria for interpreting Western blots were developed based on statistical analyses of results in large numbers of patients (CDC, 1995; Dressler et al., 1993). Some immunoreactivities, such as those to outer surface proteins A and B, while relatively specific occur so rarely that they are not helpful in diagnosis and therefore are excluded from the 3 IgM and 10 IgG bands that form the basis for diagnosis (Table 47-1). Importantly, criteria for interpreting Western blots were developed in patients with

positive or borderline ELISAs. Interpretation is undefined in seronegative patients and the test should not be performed either without an ELISA or if the ELISA is negative.

Different laboratories use different antigens in their assays. Many use whole borrelia; some use recombinant antigens for ELISAs, Western blots or both. One antigen, known as the Vmp-like sequence, expressed (VlsE) protein (Zhang et al., 1997), or a portion of it, the C6 peptide, is proving useful as it seems quite similar if not identical in all Lyme borrelia strains, making it particularly useful in European testing. Recent work has suggested it could replace Western blots as the second-tier test to confer added specificity (Branda et al., 2013; Schoen, 2013).

Because Lyme disease testing is performed so frequently, even in patients in whom the diagnosis is inherently unlikely, it is important to consider these tests' positive predictive value. In highly endemic US counties, Lyme disease incidence is typically up to about 300 cases/100,000 population per year. A three standard deviation cutoff for Lyme serologic testing would lead to a false positive rate of about 0.1%. If testing were performed indiscriminately (which it often is) in the entire population, there would be about 100 false positives/100,000 population, indicating a positive serology has about a 75% chance of diagnosing infection, but still a 25% false positive rate. In areas with substantially lower endemic rates, this problem grows dramatically. In patients with nonspecific symptoms such as headaches – a symptom that affects in excess of 10% of the general population – false-positive serologies would occur in at least 10/100,000 (10% of 100 false positives). If all 12% of actual Lyme disease patients with nervous system involvement had headaches, this would result in 36/100,000 cases – that is, the presence of this nonspecific symptom decreases the proportion that are false positives from 25% to 22% – a marginal improvement at best, meaning that this non-specific symptom adds virtually nothing to the test's positive predictive value.

Since some argue that non-specific symptoms can be diagnostic of Lyme disease even with negative serologic tests, it is worth exploring this possibility as well. Some consider persistent cognitive complaints a diagnostic criterion for Lyme disease. Population studies (Luo et al., 2005) suggest that at any given time approximately 2% of the population suffer from this symptom to a disabling extent. Although there is no evidence that any patients with *B. burgdorferi* infection lack demonstrable serum antibodies after the first 4–6 weeks of infection, suppose that false seronegatives occur in 10% of infections and assume that symptoms persist in 20% of treated patients. In highly endemic areas this would result in 30 seronegative patients/100,000 population per year, of whom 6/100,000 might have persistent symptoms. In contrast, in the general population there would be 2000/100,000 population with the same symptom

Table 47-1 Western blot criteria

IgM (any 2 of 3) (kD)	IgG (any 5 of 10) (kD)
23	18
39	23
41	28
	31
	39
	41
	45
	58
	66
	93
Only in the first 4–6 weeks of infection	Required if infection persists for more than 4–6 weeks

complex, giving the symptoms a 0.3% chance of being diagnostic of Lyme disease.

47.7 Treatment

Borrelia are highly sensitive *in vitro* to penicillins, third-generation cephalosporins, tetracyclines, and macrolides, although the last are less effective clinically. Numerous clinical trials have confirmed the efficacy of oral regimens for most cases of Lyme borreliosis – including arthritis, meningitis, cranial neuritis, and radiculoneuritis (Halperin et al., 2007; Wormser et al., 2006) (Table 47-2). Although oral treatment of these nervous system manifestations has only been tested in European studies, antimicrobial sensitivities of European and North American strains are so similar that it is reasonable to use oral agents in US patients as well. Parenteral treatment is usually reserved for patients who fail to respond to oral medication, who have parenchymal central nervous system disease, who have a complete heart block, or who appear severely ill. Treatment in pregnancy is a special case – since studies of oral therapy have usually used tetracyclines, which are best avoided in pregnancy because of their potential effects on developing bones and teeth, parenteral high dose penicillin is sometime recommended.

Treatment duration is much debated. The existing evidence strongly supports 2–3 weeks for most regimens (Halperin et al., 2007; Wormser et al., 2006). The common but not evidence-based practice of using ever longer courses has driven many to prescribe 4-week courses to combat the anxiety induced in many patients by Internet misinformation. Use of longer courses has been studied in four randomized, controlled, blinded studies (Fallon et al., 2008; Klempner et al., 2001; Krupp et al., 2003). The evidence clearly shows that the risk to benefit ratio of prolonged treatment is not in patients' best interest – that is, there is no evidence of any lasting benefit and substantial risk of harm (Halperin, 2008).

Finally, antibiotic treatment of patients who have had tick bites may be of some potential benefit. As discussed above, *Ixodes* ticks must be attached for 24–48 hours for there to be significant risk of *borrelia* transmission (although *ehrlichia* and *babesia* may be more rapidly transmitted). A meta-analysis of treatment trials (Warshafsky et al., 2010) (all from areas endemic for Lyme disease, with treatment started within 72 hours of the tick bite using oral regimens ranging from two 100 mg doxycycline capsules to 10 days of treatment) demonstrated a statistically significant but clinically small effect. Of patients receiving placebo, 2.2% developed Lyme disease versus 0.2% of antibiotic-treated individuals – 49 patients would need to be treated to prevent a single case of Lyme disease.

47.8 Prevention and control

As with any infectious disease, prevention is generally simpler than treatment. Since *B. burgdorferi*-infected ticks are geographically localized, avoidance of their habitats is highly effective. For those for whom this is impossible, wearing light-colored clothing makes ticks easier to see. The recommendation that in summer children wear long pants tucked into their socks and long sleeve shirts tucked into their pants would be hypothetically helpful but, as any parent will attest, is impractical at best. On the other hand, given the requirement for prolonged tick attachment and feeding before the risk of infection is substantial, a thorough tick check at the end of the day can be all that is needed. Attached ticks are best removed by placing a thin tweezer between the tick body and the skin, then slowly pulling. Efforts to smother the tick in petrolatum or volatile solvents, or to burn them off with cigarettes or other heat sources, make for dramatic theater but are at best unhelpful and at worst dangerous. Following tick removal, mouthparts are often left behind. Efforts to extract them are unnecessary and can lead to super-infection. Watching the site of the bite for several weeks to assure no EM develops is usually all that is needed.

Since ticks require different hosts at different life stages, control of hosts and vectors is a plausible

Table 47-2 Treatment recommendations (Wormser et al., 2006)

	Adult	Pediatric (should never exceed adult dose)
Oral (3, possibly 4 weeks) (skin, neuroborreliosis without brain or spinal cord involvement)		
Doxycycline[a]	100 mg PO b.i.d. to q.i.d.	4 mg/kg/day in two divided doses
Limited data in disseminated disease but probably effective		
Amoxicillin	500 mg PO t.i.d.	50 mg/kg/day in three divided doses
Cefuroxime axetil	500 mg PO b.i.d	30 mg/kg/day in two divided doses
Parenteral (2–4 weeks): oral treatment failure, parenchymal CNS disease, severe disease		
Ceftriaxone[b]	2 g/day IV	50–75 mg/kg/day
Cefotaxime	2 g q8/IV	150–200 mg/kg/day in three to four divided doses
Penicillin	20–24 million units IV/day	300,000 units/kg/day divided, every 4 hours

[a]Doxycycline: not in pregnant women or children <8 years old.
[b]Some advise that ceftriaxone not be used late in pregnancy.

strategy. Decreasing deer populations is often politically problematic and needs to be quite substantial to impact tick density, as the ticks will find alternative hosts (Perkins et al., 2006). As discussed above, allowing the deer to treat themselves with acaricide can be somewhat helpful (Fish and Childs, 2009). Eliminating field mice is impractical; efforts to provide them with acaricide-impregnated nesting material showed some promise in ridding them of infected ticks (Daniels et al., 1991). Unfortunately unless an entire neighborhood adopts the same strategy, the effect of this is short-lived. Perhaps the most unusual strategy was the use of the no-longer-available Lyme disease vaccine (Steere et al., 1998). Although, as in syphilis, the presence of antibodies to *B. burgdorferi* does not appear to be terribly effective in protecting humans against reinfection, the concept underlying these vaccines was different. Patients were immunized against a protein not expressed by *B burgdorferi* in early human infection, but strongly expressed in the tick gut. When the tick ingested blood containing a high titer of antibody, this cured the tick, before the patient could become infected! Unfortunately, since this required persisting high antibody titers, annual revaccination was necessary for ongoing efficacy. For this and other reasons the vaccine never gained broad acceptance and was withdrawn, although several new variants are currently being studied.

47.9 Conclusions

Infection with the tick-borne spirochete *B. burgdorferi* causes Lyme disease. Infection can often be prevented with simple measures. Infection, which can be localized in a single EM skin lesion, or disseminated involving joints, heart, or the nervous system, is highly responsive to straightforward antibiotic regimens – in most cases oral. Serologic testing is quite reliable – after the first 3–6 weeks of infection – but is not helpful in assessing treatment efficacy, as seropositivity can remain for years following microbiologic cure.

References

Ackermann, R., Rehse, K.B., Gollmer, E., and Schmidt, R. 1988. Chronic neurologic manifestations of erythema migrans borreliosis. *Ann. N. Y. Acad. Sci.*, 539, 16–23.

Afzelius, A. 1921. Erythema chronicum migrans. *Acta Derm. Venereol. (Stockh.)*, 2, 120–125.

Aguero-Rosenfeld, M.E., Nowakowski, J., McKenna, D.F., Carbonaro, C.A., and Wormser, G.P. 1993. Serodiagnosis in early Lyme disease. *J. Clin. Microbiol.*, 31, 3090–3095.

Aguero-Rosenfeld, M.E., Nowakowski, J., Bittker, S., Cooper, D., Nadelman, R.B., and Wormser, G.P. 1996. Evolution of the serologic response to *Borrelia burgdorferi* in treated patients with culture-confirmed erythema migrans. *J. Clin. Microbiol.*, 34, 1–9.

Asbrink, E., Hederstedt, B., and Hovmark, A. 1984. The spirochetal etiology of acrodermatitis chronica atrophicans Herxheimer. *Acta Derm. Venereol.*, 64, 506–512.

Bacon, R.M., Kugeler, K.J., Griffith, K.S., and Mead, P.S. 2007. Lyme disease — United States, 2003–2005. *MMWR Morb. Mortal. Wkly. Rep.*, 56, 573–576.

Bacon, R.M., Kugeler, K.J., and Mead, P.S. 2008. Surveillance for Lyme disease — United States, 1992–2006. *MMWR Morb. Mortal. Wkly. Rep.*, 57, 1–9.

Bannwarth, A. 1941. Chronische lymphocytare meningitis, entzundliche polyneuritis und "rheumatismus". *Arch. Psychiatr. Nervenkr.*, 113, 284–376.

Benach, J.L., Bosler, E.M., Hanrahan, J.P., Coleman, J.L., Habicht, G.S., Bast, T.F., Cameron, D.J., Ziegler, J.L., Barbour, A.G., Burgdorfer, W., Edelman, R., and Kaslow, R.A. 1983. Spirochetes isolated from the blood of two patients with Lyme disease. *N. Engl. J. Med.*, 308, 740–742.

Branda, J.A., Strle, F., Strle, K., Sikand, N., Ferraro, M.J., and Steere, A.C. 2013. Performance of United States serologic assays in the diagnosis of Lyme borreliosis acquired in Europe. *Clin. Infect. Dis.*, 57, 333–340.

Burgdorfer, W., Barbour, A.G., Hayes, S.F., Benach, J.L., Grunwaldt, E., and Davis, J.P. 1982. Lyme disease: A tick borne spirochetosis? *Science*, 216, 1317–1319.

CDC. 1995. Recommendations for test performance and interpretation from the Second National Conference on Serologic Diagnosis of Lyme Disease. *MMWR Morb. Mortal. Wkly. Rep.*, 44, 590–591.

CDC. 2013. Lyme disease data, p. overview of US incidence.

Daniels, T.J., Fish, D., and Falco, R.C. 1991. Evaluation of host-targeted acaricide for reducing risk of Lyme disease in southern New York state. *J. Med. Entomol.*, 28, 537–543.

Dattwyler, R.J., Volkman, D.J., Luft, B.J., Halperin, J.J., Thomas, J., and Golightly, M.G. 1988. Seronegative Lyme disease. Dissociation of specific T- and B-lymphocyte responses to *Borrelia burgdorferi*. *N. Engl. J. Med.*, 319, 1441–1446.

Dressler, F., Whalen, J.A., Reinhardt, B.N., and Steere, A.C. 1993. Western blotting in the serodiagnosis of Lyme disease. *J. Infect. Dis.*, 167, 392–400.

Eikeland, R., Mygland, A., Herlofson, K., and Ljostad, U. 2013. Risk factors for a non-favorable outcome after treated European neuroborreliosis. *Acta Neurol. Scand.*, 127, 154–160.

Fallon, B.A., Keilp, J.G., Corbera, K.M., Petkova, E., Britton, C.B., Dwyer, E., Slavov, I., Cheng, J., Dobkin, J., Nelson, D.R., and Sackeim, H.A. 2008. A randomized, placebo-controlled trial of repeated IV antibiotic therapy for Lyme encephalopathy. *Neurology*, 70, 992–1003.

Feder Jr., H.M., Hoss, D.M., Zemel, L., Telford 3rd, S.R., Dias, F., and Wormser, G.P. 2011. Southern Tick-Associated Rash Illness (STARI) in the North: STARI following a tick bite in Long Island, New York. *Clin. Infect. Dis.*, 53, e142–e146.

Fish, D., and Childs, J.E. 2009. Community-based prevention of Lyme disease and other tick-borne diseases through topical application of acaricide to white-tailed deer: Background and rationale. *Vector Borne Zoonotic Dis.*, 9, 357–364.

Garin, C., and Bujadoux, A. 1922. Paralysie par les tiques. *J. Med. Lyon*, 71, 765–767.

Garro, A.C., Rutman, M., Simonsen, K., Jaeger, J.L., Chapin, K., and Lockhart, G. 2009. Prospective validation of a clinical prediction model for Lyme meningitis in children. *Pediatrics*, 123, e829–e834.

Halperin, J.J. 2008. Prolonged Lyme disease treatment. *Neurology*, 70, 986–987.

Halperin, J.J., and Golightly, M. 1992. Lyme borreliosis in Bell's palsy. Long Island Neuroborreliosis Collaborative Study Group. *Neurology*, 42, 1268–1270.

Halperin, J.J., Luft, B.J., Anand, A.K., Roque, C.T., Alvarez, O., Volkman, D.J., and Dattwyler, R.J. 1989. Lyme neuroborreliosis: Central nervous system manifestations. *Neurology*, 39, 753–759.

Halperin, J.J., Krupp, L.B., Golightly, M.G., and Volkman, D.J. 1990a. Lyme borreliosis-associated encephalopathy. *Neurology*, 40, 1340–1343.

Halperin, J.J., Luft, B.J., Volkman, D.J., and Dattwyler, R.J. 1990b. Lyme neuroborreliosis – Peripheral nervous system manifestations. *Brain*, 113, 1207–1221.

Halperin, J., Logigian, E., Finkel, M., and Pearl, R. 1996. Practice parameter for the diagnosis of patients with nervous system Lyme borreliosis (Lyme disease). *Neurology*, 46, 619–627.

Halperin, J.J., Shapiro, E.D., Logigian, E.L., Belman, A.L., Dotevall, L., Wormser, G.P., Krupp, L.B., Gronseth, G., and Bever, C. 2007. Practice parameter: Treatment of nervous system Lyme disease. *Neurology*, 69, 91–102.

Halperin, J.J., Baker, P., and Wormser, G.P. 2013. Common misconceptions about Lyme disease. *Am. J. Med.*, 126, 264.

Hollstrom, E. 1951. Successful treatment of erythema migrans Afzelius. *Acta Derm. Venereol.*, 31, 235–243.

Klempner, M.S., Hu, L.T., Evans, J., Schmid, C.H., Johnson, G.M., Trevino, R.P., Norton, D., Levy, L., Wall, D., McCall, J., Kosinski, M., and Weinstein, A. 2001. Two controlled trials of antibiotic treatment in patients with persistent symptoms and a history of Lyme disease. *N. Engl. J. Med.*, 345, 85–92.

Krupp, L.B., Masur, D., Schwartz, J., Coyle, P.K., Langenbach, L.J., Fernquist, S., Jandorf, L., and Halperin, J.J. 1991. Cognitive functioning in late Lyme borreliosis. *Arch. Neurol.*, 48, 1125–1129.

Krupp, L.B., Hyman, L.G., Grimson, R., Coyle, P.K., Melville, P., Ahnn, S., Dattwyler, R., and Chandler, B. 2003. Study and treatment of post Lyme disease (STOP-LD): A randomized double masked clinical trial. *Neurology*, 60, 1923–1930.

Kugeler, K.J., Griffith, K.S., Gould, L.H., Kochanek, K., Delorey, M.J., Biggerstaff, B.J., and Mead, P.S. 2011. A review of death certificates listing Lyme disease as a cause of death in the United States. *Clin. Infect. Dis.*, 52, 364–367.

Ljostad, U., and Mygland, A. 2009. Remaining complaints 1 year after treatment for acute Lyme neuroborreliosis: Frequency, pattern and risk factors. *Eur. J. Neurol.*, 17, 118–123.

Luo, N., Johnson, J., Shaw, J., Feeny, D., and Coons, S. 2005. Self-reported health status of the general adult U.S. population as assessed by the EQ-5D and Health Utilities Index. *Med. Care*, 43, 1078–1086.

Pediatric_Lyme_Disease_Study_Group, Gerber, M.A., Shapiro, E.D., Burke, G.S., Parcells, V.J., and Bell, G.L. 1996. Lyme disease in children in southeastern Connecticut. *N. Engl. J. Med.*, 335, 1270–1274.

Perkins, S.E., Cattadori, I.M., Tagliapietra, V., Rizzoli, A.P., and Hudson, P.J. 2006. Localized deer absence leads to tick amplification. *Ecology*, 87, 1981–1986.

Reik, L., Steere, A.C., Bartenhagen, N.H., Shope, R.E., and Malawista, S.E. 1979. Neurologic abnormalities of Lyme disease. *Medicine*, 58, 281–294.

Schoen, R.T. 2013. Better laboratory testing for Lyme disease: No more Western blot. *Clin. Infect. Dis.*, 57, 341–343.

Scrimenti, R.J. 1970. Erythema chronicum migrans. *Arch. Dermatol.*, 102, 104–105.

Seltzer, E.G., Gerber, M.A., Cartter, M.L., Freudigman, K., and Shapiro, E.D. 2000. Long-term outcomes of persons with Lyme disease. *JAMA*, 283, 609–616.

Skogman, B.H., Glimaker, K., Nordwall, M., Vrethem, M., Odkvist, L., and Forsberg, P. 2012. Long-term clinical outcome after Lyme neuroborreliosis in childhood. *Pediatrics*, 130, 262–269.

Steere, A.C., Malawista, S.E., Hardin, J.A., Ruddy, S., Askenase, W., and Andiman, W.A. 1977. Erythema chronicum migrans and Lyme arthritis. The enlarging clinical spectrum. *Ann. Intern. Med.*, 86, 685–698.

Steere, A.C., Grodzicki, R.L., Kornblatt, A.N., Craft, J.E., Barbour, A.G., Burgdorfer, W., Schmid, G.P., Johnson, E., and Malawista, S.E. 1983a. The spirochetal etiology of Lyme disease. *N. Engl. J. Med.*, 308, 733–740.

Steere, A.C., Pachner, A.R., and Malawista, S.E. 1983b. Neurologic abnormalities of Lyme disease: Successful treatment with high-dose intravenous penicillin. *Ann. Intern. Med.*, 99, 767–772.

Steere, A.C., Schoen, R.T., and Taylor, E. 1987. The clinical evolution of Lyme arthritis. *Ann. Intern. Med.*, 107, 725–731.

Steere, A.C., Sikand, V.K., Meurice, F., Parenti, D.L., Fikrig, E., Schoen, R.T., Nowakowski, J., Schmid, C.H., Laukamp, S., Buscarino, C., and Krause, D.S. 1998. Vaccination against Lyme disease with recombinant Borrelia burgdorferi. *N. Engl. J. Med.*, 339, 209–215.

Toledo, A., and Benach, J.L. 2011. Borrelia: Biology of the organism. In: Halperin, J.J., editor. *Lyme Disease—An Evidence Based Approach*. Wallingford: CABI.

Warshafsky, S., Lee, D.H., Francois, L.K., Nowakowski, J., Nadelman, R.B., and Wormser, G.P. 2010. Efficacy of antibiotic prophylaxis for the prevention of Lyme disease: An updated systematic review and meta-analysis. *J. Antimicrob. Chemother.*, 65, 1137–1144.

Wormser, G.P., Dattwyler, R.J., Shapiro, E.D., Halperin, J.J., Steere, A.C., Klempner, M.S., Krause, P.J., Bakken, J.S., Strle, F., Stanek, G., Bockenstedt, L., Fish, D., Dumler, J.S., and Nadelman, R.B. 2006. The clinical assessment, treatment, and prevention of Lyme disease, human granulocytic anaplasmosis, and babesiosis: Clinical practice guidelines by the Infectious Diseases Society of America. *Clin. Infect. Dis.*, 43, 1089–1134.

Wormser, G.P., and Halperin, J.J. 2013. Toward a better understanding of European Lyme neuroborreliosis. *Clin. Infect. Dis.*, 57, 510–512.

Zhang, J.R., Hardham, J.M., Barbour, A.G., and Norris, S.J. 1997. Antigenic variation in Lyme disease borreliae by promiscuous recombination of VMP like sequence cassettes. *Cell*, 89, 275–285.

Chapter 48

NDM-Type Carbapenemases in Gram-Negative Rods

Laurent Dortet[1], Laurent Poirel[1,2], and Patrice Nordmann[1,2]

[1]INSERM U914 "Emerging Resistance to Antibiotics", Le Kremlin-Bicêtre, Paris, France
[2]Medical and Molecular Microbiology Unit, Department of Medicine, Faculty of Science, University of Fribourg, Fribourg, Switzerland

48.1 Worldwide dissemination of NDM-producing bacteria

One of the most recent and most clinically significant carbapenemases is NDM-1 (New Delhi metallo-β-lactamase). This carbapenemase belongs to the class B of the Ambler classification of β-lactamases, which includes the metallo-β-lactamases (MBLs). As observed for the other MBLs, NDM enzymes exhibit a broad-spectrum hydrolytic activity, including penicillins, cephalosporins, and carbapenems, sparing monobactams such as aztreonam. The hydrolysis activity of class B β-lactamases depends on the interaction of the β-lactam molecule with Zn^{2+} ion(s) in their active site. As a consequence, their activity is inhibited by chelators of divalent cations, such as EDTA (ethylenediaminetetraacetic acid). The currently available β-lactamase inhibitors (clavulanic acid, tazobactam, and sulbactam) do not antagonize the activity of MBLs.

NDM-1 was first identified in 2008 in a *Klebsiella pneumoniae* isolate recovered from a Swedish patient who had been previously hospitalized in New Delhi, India (Yong et al., 2009). Since then, NDM carbapenemases are the focus of worldwide attention due to their rapid dissemination among Enterobacteriaceae, *Acinetobacter* spp., and *Pseudomonas aeruginosa*. Indeed, after the first description of NDM producers, an extended survey was shortly performed in India, Pakistan, Bangladesh, and the United Kingdom during 2008–2009. The authors identified 180 NDM-1-producing enterobacterial isolates (Kumarasamy et al., 2010), most of them (*n* = 143) being recovered from patients located at different locations over the Indian subcontinent. The 37 remaining strains were recovered from 29 UK patients and half of them (17 out of 29) had a history of travel to India or Pakistan, and 14 of them had been hospitalized in those countries. This obvious link with the Indian subcontinent was then confirmed in most of the following NDM-1 reports. Accordingly, from 2008 to 2010, a total of 77 cases were reported from 13 countries (Struelens et al., 2010). Among 55 cases with a recorded travel history, 31 had previously traveled or had been admitted to a hospital in India or Pakistan. In addition, among the 235 enterobacterial isolates collected from intra-abdominal infections as part of the SMART study and selected for their reduced susceptibility to ertapenem, the most common carbapenemase was NDM-1 (50%), and all the bla_{NDM-1}-positive isolates were from India (Lascols et al., 2011). Since then, the high local prevalence of NDM-1 producers was pointed out by several reports from Indian studies on their own (Krishna, 2010; Mochon et al., 2011; Muir and Weinbren, 2010; Raghunath, 2010; Roy et al., 2011a, 2011b). Data from a main hospital in Mumbai reported a 5–8% prevalence rate of NDM producers among Enterobacteriaceae (Deshpande et al., 2010a, 2010b), then confirmed by the 6.9% prevalence rate observed among 780 consecutive non-duplicate enterobacterial isolates recovered from hospitalized patients and outpatients from February to July 2010 in Varanasi (North India) (Seema et al., 2011), and finally by a 7% prevalence rate observed among 885 Gram-negative bacteria isolated from clinical samples (pus, blood, sputum, fluids) received at the Microbiology Department of Sassoon Hospital (Pune, India) from August to December 2010 (Bharadwaj et al., 2012). The screening of 200 hospitalized patients and outpatients attending two military hospitals in Rawalpindi, Pakistan, highlighted a higher prevalence of fecal carriage of 18.5% (Perry et al., 2011). Since all studies focused on the widespread prevalence of NDM-1 producers in the Indian subcontinent, a retrospective survey performed on Indian collections by the SENTRY Antimicrobial Surveillance System traced this early dissemination of NDM-1-producing *Enterobacteriaceae* in Indian hospitals as early as in 2006 (Castanheira et al., 2011). In addition, the bla_{NDM-1} gene was not only detected in patient samples, but also in drinking-water samples and seepage samples in New Delhi (Walsh et al., 2011). The presence of NDM-1-producing bacteria in environmental samples in New Delhi is significant for people living in the city, who often rely on public water and poor sanitation facilities.

In the SENTRY study reporting the dissemination of NDM-1-producing Enterobacteriaceae in Europe, 5 out of the 55 patients with recorded travel history had been hospitalized in the Balkan region (Struelens et al., 2010), indicating that this area might be a secondary reservoir of NDM-1 producers (Gecaj-Gashi et al., 2011; Livermore et al., 2011b, Nordmann et al., 2011). Accordingly, several studies report that patients colonized or infected with NDM-1 producers actually originate from the Balkan states (Halaby et al., 2012; Mazzariol et al., 2012; Poirel et al., 2010; Zarfel et al., 2011). Recent reports also suggested that the Middle East might be another reservoir of NDM producers (Dortet et al., 2012b; Jamal et al., 2012; Poirel et al., 2011a; Shibl et al., 2013; Sonnevend et al., 2013; Zowawi et al., 2013). This dissemination of NDM producers in the Middle East could mostly be linked to the population exchange between the Middle East and the Indian subcontinent. However, since the mid-2010s, NDM-1-producing bacteria have been reported worldwide (Al-Agamy et al., 2013; Arpin et al., 2012; Arya and Agarwal, 2011; Barantsevich et al., 2013; Barguigua et al., 2013; Barrios et al., 2013; Ben Nasr et al., 2013; Birgy et al., 2011; Bogaerts et al., 2011; Brink et al., 2012; Cabanes et al., 2012; Chen et al., 2012b; Darley et al., 2012; Denis et al., 2012; Dortet et al., 2012b, 2013; El-Herte et al., 2012; Escobar Perez et al., 2013; Fischer et al., 2013; Gaibani et al., 2011; Green et al., 2013; Halaby et al., 2012; Ho et al., 2011, 2012; Huo, 2010; Islam et al., 2012; Isozumi et al., 2012; Jamal et al., 2012; Kim et al., 2012; Leski et al., 2012; Liu et al., 2013; Lowman et al., 2011; Mazzariol et al., 2012; McDermott et al., 2012; Mirovic et al., 2012; Mochon et al., 2011; Nielsen et al., 2012; Nordmann et al., 2012b; Oteo et al., 2012; Pasteran et al., 2012; Peirano et al., 2011a; Pfeifer et al., 2011; Poirel et al., 2010, 2011a, 2011b, 2011d, 2011e, 2011f, 2011g, 2012b, 2012c; Rimrang et al., 2012; Samuelsen et al., 2011, 2013; Savard et al., 2011; Shibl et al., 2013; Sole et al., 2011; Sonnevend et al., 2013; Tada et al., 2013; Tsang et al., 2012; Wang et al., 2013; Williamson et al., 2012; Yamamoto et al., 2011; Yong et al., 2009; Yoo et al., 2013) with a rapid dissemination from the two previously described reservoirs, namely the Indian subcontinent and the Balkan states (Figure 48-1).

Until now, there is no evidence that NDM-producing bacteria are more virulent than other strains (Fuursted et al., 2012; Peirano et al., 2011b, 2013). Indeed, no specific virulence factor was described onto bla_{NDM-1}-carrying plasmids (Bonnin et al., 2012b; Dolejska et al., 2013; Hishinuma et al., 2013; McGann et al., 2012; Sekizuka et al., 2011; Yamamoto et al., 2013). NDM producers were mainly described in Enterobacteriaceae that have been recovered from many clinical settings, reflecting the vast disease panel of these pathogens, including urinary tract infections, pulmonary infections, peritonitis, septicemia, soft tissue infections, and device-associated infections. However, NDM-1 was rarely produced by virulent bacteria such as *Salmonella* (Cabanes et al., 2012; Fischer et al., 2013; Savard et al., 2011) and *Vibrio cholerae* (Darley et al., 2012; Walsh et al., 2011).

Both hospital- and community-acquired infections have been reported. It is highly probable that colonization of the gut flora might precede the infection by NDM producers and oro-fecal transmission in the community might occur mostly through hand contamination, food, and water. This hypothesis is reinforced by the environmental contamination observed in New Delhi and by the frequent recovering of NDM-1-positive bacteria from the gut flora of travelers returning from the Indian subcontinent who were investigated microbiologically for diarrheal symptoms (Leverstein-Van Hall et al., 2010).

Among the NDM-1-producing Enterobacteriaceae, *Klebsiella pneumoniae* and *Escherichia coli* were the most often described species. However, this carbapenemase is also frequently described in other enterobacterial species, including *Klebsiella oxytoca*, *Enterobacter cloacae*, *Citrobacter freundii*, *Proteus mirabilis*, *Salmonella* spp., and *Providencia* spp. Although most of NDM-producing bacteria were Enterobacteriaceae, this carbapenemase was also described in *Acinetobacter* spp. (Bogaerts et al., 2012; Bonnin et al., 2012a; Boulanger et al., 2012; Chen et al., 2011, 2012a; Espinal et al., 2011, 2013; Ghazawi et al., 2012; Glasner et al., 2013; Hammerum et al., 2012; Hrabak et al., 2012a; Hu et al., 2012; Kaase et al., 2011; Nakazawa et al., 2013; Nemec and Krizova, 2012; Sun et al., 2013; Yang et al., 2012; Zhang et al., 2013; Zhou et al., 2012) and in rare cases in *Pseudomonas aeruginosa* (Flateau et al., 2012; Jovcic et al., 2011). Notably, two reservoirs of NDM-producing *Acinetobacter* spp. have been identified in China (Chen et al., 2011, 2012a; Hu et al., 2012; Sun et al., 2013; Yang et al., 2012; Zhang et al., 2013; Zhou et al., 2012) and in the Middle East (Boulanger et al., 2012; Espinal et al., 2011, 2013; Ghazawi et al., 2012; Hammerum et al., 2012; Hrabak et al., 2012a; Kaase et al., 2011). As observed with the dissemination of NDM-1-producing Enterobacteriaceae in the environment of New Delhi, NDM-producing *Acinetobacter* might be soon recovered in the environment in China since several NDM-1-producing *A. baumannii* have already been isolated from the sewage of the

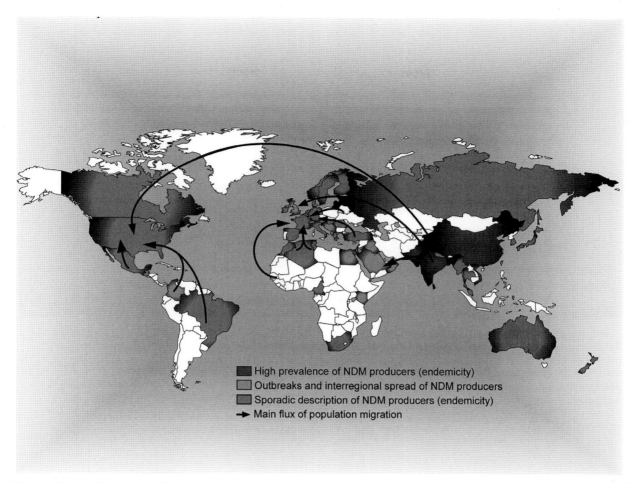

Fig. 48-1 Geographical spread of NDM producers. The worldwide dissemination of the bla_{NDM} genes and the main flux of population migration are represented. For a color version of this figure, see the color plate section.

hospitals in Beijing (Zhang et al., 2013). This large distribution of NDM-1 among Enterobacteriaceae, *Acinetobacter*, and *Pseudomonas* mirrors the association of bla_{NDM-1} with mobile genetic structures, namely plasmids, which is discussed below.

48.2 Antibiotic resistance patterns and clinical impact for the treatment

As stated above, NDM-1, like all MBLs, confers resistance to a broad range of β-lactams, including penicillins, cephalosporins, and carbapenems. However, a peculiar resistance trait observed in almost all NDM producers (Enterobacteriaceae, *Acinetobacter*, and *Pseudomonas*) is their quite systematic association with other antibiotic resistance determinants, such as those encoding other types of carbapenemases (OXA-48-, VIM-, KPC-types), AmpC cephalosporinases, clavulanic-acid-inhibited expanded-spectrum β-lactamases, resistance to aminoglycosides (16S RNA methylases), macrolides (esterases), rifampicin (rifampicin-modifying enzymes), quinolones (Qnr), chloramphenicol, and sulfamethoxazole (Barguigua et al., 2013; Castanheira et al., 2011; Dolejska et al., 2013; Poirel et al., 2011c, 2011f; Samuelsen et al., 2013). Consequently, most of the NDM-1 producers are resis-

tant to all β-lactams, including carbapenems, to all aminoglycosides and fluoroquinolones, to nitrofurantoin, and to sulphonamides, and remain susceptible only to two bactericidal antibiotics (fosfomycin and colistin) and a single bacteriostatic antibiotic (tigecycline) (Falagas et al., 2011; Rogers et al., 2013) (Figure 48-2).

Therapies using colistin plus fosfomycin, colistin plus tigecycline, or fosfomycin plus tigecycline are not a common medical practice. *In-vitro* synergy combination assays performed with NDM-1 producers with those three antibiotic molecules showed a synergistic activity of colistin and fosfomycin, of colistin and tigecycline in rare cases, while most of the associations remained neutral for most of the tested isolates (Berçot et al., 2011). The efficacy of EDTA (Ca-EDTA), as an inhibitor of NDM-1 activity, has been evaluated in a mouse model of sepsis caused by an NDM-1-producing *E. coli*. It showed that the bacterial inoculum was reduced by a combination therapy using imipenem/cilastatin sodium (IPM/CS) and Ca-EDTA, as compared to IPM/CS alone (Yoshizumi et al., 2013). These results suggest that there might be a possibility to use Ca-EDTA in clinical therapeutics. Since NDM-1 is not able to hydrolyze aztreonam, a combination therapy including aztreonam and

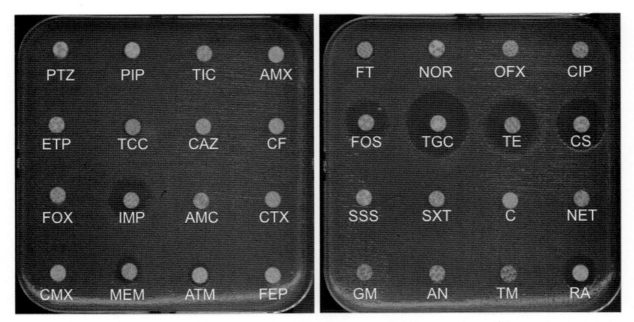

Fig. 48-2 Antibiogram of an NDM-1-producing *K. pneumoniae* isolate. The *bla*$_{NDM-1}$ gene was located onto an IncHIIB plasmid of ~200 kb in that strain, which also harbored two additional β-lactamase genes (*bla*$_{CTX-M-15}$, *bla*$_{OXA-1}$), and an aminoglycoside methylase (*armA*) responsible for high-level resistance to all aminoglycosides. PTZ, piperacillin + tazobactam; PIP, piperacillin; TIC, ticarcillin; AMX, amoxicillin; ETP, ertapenem; TCC, ticarcillin + clavulanic acid; CAZ, ceftazidime; CF, cefalotin; FOX, cefoxitin; IMP, imipenem; AMC, amoxicillin + clavulanic acid; CTX, cefotaxime; CMX, cefuroxime; MEM, meropenem; ATM, aztreonam; FEP, cefepime; FT, nitrofurantoin; NOR, norfloxacin; OFX, ofloxacin; CIP, ciprofloxacin; FOS, fosfomycin; TGC, tigecycline; TE, tetracycline; CS, colistin; SSS, sulfonamide; SXT, sulfamethoxazole + trimethoprim; C, chloramphenicol; NET, netilmicin; GM, gentamicin; AN, amikacin; TM, tobramycin; RA, rifampicin.

β-lactamase inhibitors such as NXL 104 may inhibit the activity of extended spectrum β-lactamases that are mostly present in NDM-1 producers (Livermore et al., 2011a; Shakil et al., 2011).

48.3 Genetics of NDM carbapenemases

The *bla*$_{NDM-1}$ gene is mainly located onto conjugative plasmid in Enterobacteriaceae. However, an investigation of a worldwide collection of NDM-1-producing enterobacterial isolates showed that the current spread of the *bla*$_{NDM-1}$ gene is related to the spread of specific clones, specific plasmids, or sometimes to the dissemination of a single genetic structure (Poirel et al., 2011c). Accordingly, *E. coli* and *K. pneumoniae* producing NDM-1 belong to a diversity of Sequence Types (ST). Although several NDM-1-producing ST101 *E. coli* have been described mainly in the United Kingdom, Pakistan, Canada, and South Korea, this association might be due to the endemicity of this ST in India. However, NDM-1 was also found in *E. coli* of the ST131 lineage, which is known to play a critical role for the worldwide dissemination of the ESBL CTX-M-15 (Coque et al., 2008; Peirano et al., 2011b). Plasmids bearing the *bla*$_{NDM-1}$ gene and belonging to diverse incompatibility groups have been described (Bonnin et al., 2012b; Dolejska et al., 2013; Ho et al., 2011; Hu et al., 2012; McGann et al., 2012; Poirel et al., 2011c; Sekizuka et al., 2011; Sonnevend et al., 2013).

Those plasmids co-harbor multiple resistance determinants, including the *bla*$_{TEM}$, *bla*$_{OXA}$, *bla*$_{CMY}$, *bla*$_{SHV}$, and *bla*$_{CTX-M}$ genes (β-lactamase genes); *qnr* genes (encoding quinolone resistance); *armA*, *rmtA*, *rmtC*, *rmtD*, and *rmtF* (encoding 16S RNA methylases conferring resistance to almost all aminoglycosides); *ereC* (encoding macrolide resistance); *cmlA* (encoding chloramphenicol resistance), *sul* genes (encoding sulfonamide resistance dihydropteroate synthetase), and *arr-2* (encoding rifampicin ribosyltransferase). The evaluation of the conjugation properties and host specificity of five *bla*$_{NDM-1}$-positive plasmids of different incompatibility groups (IncL/M, FII, A/C, and two untypeable plasmids) from clinical Enterobacteriaceae underlined how efficient the spread of the *bla*$_{NDM-1}$ carbapenemase gene could be among Enterobacteriaceae (Potron et al., 2011). Although the *bla*$_{NDM}$-type genes were almost always located on plasmids in Enterobacteriaceae, those genes are either plasmid- or chromosome-encoded in *Acinetobacter* spp. or chromosome-encoded in the rare NDM producers identified in *P. aeruginosa* (Flateau et al., 2012; Jovcic et al., 2011). In several studies related to NDM-producing *A. baumannii*, the *bla*$_{NDM}$ gene was located between two copies of the IS*Aba125* element, forming a composite transposon named Tn*125* (Bogaerts et al., 2012; Bonnin et al., 2012a; Boulanger et al., 2012; Espinal et al., 2011; Hrabak et al., 2012a; Kaase et al., 2011; Pfeifer et al., 2011; Poirel et al., 2012a) (Figure 48-3). However, investigations on the immediate genetic environment of *bla*$_{NDM}$ genes among

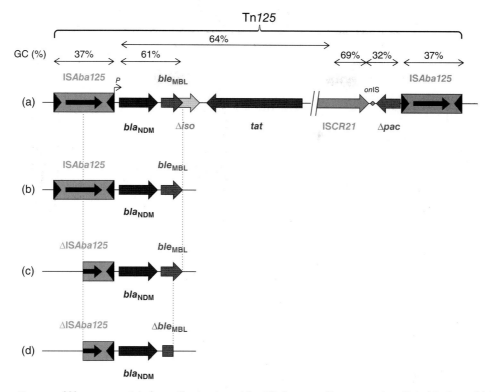

Fig. 48-3 Schematic map of bla_{NDM}-associated genetic structures identified among Gram-negative clinical isolates. (a) Structure found in *A. baumannii*, where the bla_{NDM} gene is part of the composite transposon Tn*125*. (b) Structure in which IS*Aba125* is presented as a full element with the ble_{MBL} gene (bleomycin resistance encoding gene) also being present. (c) Structure in which IS*Aba125* is present as a truncated element with an entire ble_{MBL} gene. (d) Structure in which IS*Aba125* is presented as a truncated element with a truncated ble_{MBL} gene. Genes and their corresponding transcription orientations are represented by horizontal arrows. *ori*IS of ISCR21 is indicated by a circle. The bla_{NDM} promoter is indicated (P). IS, insertion sequence; gene names are abbreviated according to their corresponding proteins: ble_{MBL}, bleomycin resistance gene; Δ*iso*, truncated phosphoribosylanthranilate isomerase; *tat*, the twin-arginine translocation pathway signal sequence protein; Δ*pac*, truncated phospholipid acetyltransferase.

enterobacterial isolates revealed that they were associated always at their 5′-end to the complete or truncated insertion sequence IS*Aba125*, and at their 3′-end to the ble_{MBL} gene (Dortet et al., 2012a) (Figure 48-3). The ble_{MBL} and bla_{NDM} genes form an operon which expression is under the control of a unique promoter located upstream of bla_{NDM} in the right inverted repeat of IS*Aba125*. The ble_{MBL} gene encodes a functional bleomycin resistance protein responsible for the resistance to bleomycin and to bleomycin related-molecules. Since bleomycin refers to a family of structurally related glycopeptides naturally produced by *Streptomyces* spp., it might be found in the environment and thus acts as a selective agent for NDM-producing isolates. In addition, bleomycin has not only a slight antibacterial activity, but is also commonly used in anti-cancer chemotherapy. Consequently, it might be postulated that the selective pressure responsible for the spread of NDM producers might be the result of a β-lactam-related selective pressure, but also of the use of anti-cancer drugs that could be found in hospital sewages. This hypothesis was recently reinforced by the description of a higher isolation rate of NDM-1-producing *A. baumannii* from the hospital sewages in Beijing, whereas not a single NDM-1 producer was recovered from samples obtained from river, drinking, or fishpond waters in the same city (Zhang et al., 2013).

48.4 From the progenitor of bla_{NDM} to the NDM-producing clinical isolates

While the source of the bla_{NDM-1} gene remains unknown, it has been suggested that this gene has been captured from the chromosome of its original environmental progenitor. Considering the high guanine–cytosine (GC) content of the bla_{NDM-1} gene (62%) compared to those of *Acinetobacter* species (38% for *Acinetobacter calcoaceticus* and up to 42% for *Acinetobacter lwoffii*), the possible progenitor of the bla_{NDM} gene is considered to be phylogenetically distant from *A. baumannii*. Likewise, the GC content of IS*Aba125* is of 37% and the fact that IS*Aba125* has been identified separately (regardless of any association with bla_{NDM-1}) seems to indicate an independent acquisition of the bla_{NDM-1} and IS*Aba125* genes (Figure 48-3). Transfer of the bla_{NDM-1} gene to *A. baumannii* is likely the result of a transfer from an unknown bacterial species to *A. baumannii*, both the donor and the recipient being likely present concomitantly in the same environment. In *A. baumannii*, the bla_{NDM-1} gene is part of the Tn*125* composite transposon made of two copies of IS*Aba125* (Poirel et al., 2012a) (Figure 48-3). Downstream of *bla*~NDM-1~, eight open reading frames have been identified (Figure 48-3). The first corresponds to the *ble*~MBL~ gene, encoding a 121-amino-acid-long protein conferring resistance to bleomycin, previously found associated with the *bla*~NDM-1~ gene in enterobacterial isolates (Dortet et al., 2012a). This *ble*~MBL~ gene possesses a GC content of 61% close to the GC content of the *bla*~NDM-1~ gene, suggesting a common progenitor. According to the similarities observed between the *ble*~MBL~ gene and several genes encoding putative bleomycin resistance, the progenitor might be close to the *Brevundimonas* and *Xanthomonas* genera (Poirel et al., 2012a). In the 3′-end of Tn*125*, a gene encoding the putative transposase of an ISCR-like element (ISCR21) was identified, sharing 93% with that of ISCR1 (Naas et al., 2008). ISCR elements are insertion sequences able to mobilize DNA fragments located at their left-hand extremity by a rolling-circle transposition process. At the right-hand extremity of ISCR21 and before the second copy of IS*Aba125* of Tn*125*, a truncated gene encoding a putative phospholipid acetyltransferase was identified, with the corresponding protein sequence sharing 91% amino acid identity with that of *Acinetobacter junii* (Poirel et al., 2012a) (Figure 48-3). This finding likely indicates that an intermediate state may have occurred in an *Acinetobacter* species different from *A. baumannii*. Our working hypothesis is that this unknown *Acinetobacter* species may have acquired concomitantly the *bla*~NDM~ and *ble*~MBL~ genes from an environmental species, likely through a natural transformation process. This transformation

process is known in a wide range of bacterial species that may uptake exogenous DNA through the binding of double-strand DNA on specific membrane receptors, followed by the penetration of a single-strand DNA, and an homologous recombination step necessary to incorporate the exogenous DNA. In that *Acinetobacter* species, successive insertions of two copies of IS*Aba125* might have been at the origin of the building of the transposon Tn*125*. After its dissemination among *Acinetobacter* species, including in *A. baumannii*, an interspecies transfer via a broad-host range plasmid allowed the acquisition of Tn*125* by Enterobacteriaceae and *P. aeruginosa*. This step may be preliminary to the large dissemination of the *bla*~NDM~ genes in Enterobacteriaceae.

Interestingly, another hypothesis has been recently suggested concerning the *de novo* construction of the *bla*~NDM-1~ gene in *Acinetobacter* (Toleman et al., 2012). Precise genetic analysis of the *bla*~NDM-1~ gene itself revealed that it possesses the 5′-end sequence of the aminoglycoside resistance gene *aphA6* (Toleman et al., 2012). Consequently, the *bla*~NDM-1~ gene might be considered as a chimera gene made of the *aphA6* gene (5′-end) and the 3′-end of a β-lactamase gene naturally present in an unknown environmental bacterial species.

48.5 Biochemistry of the NDM carbapenemases

NDM-1 has a molecular mass of 28 kDa, is monomeric, and hydrolyzes all β-lactams except

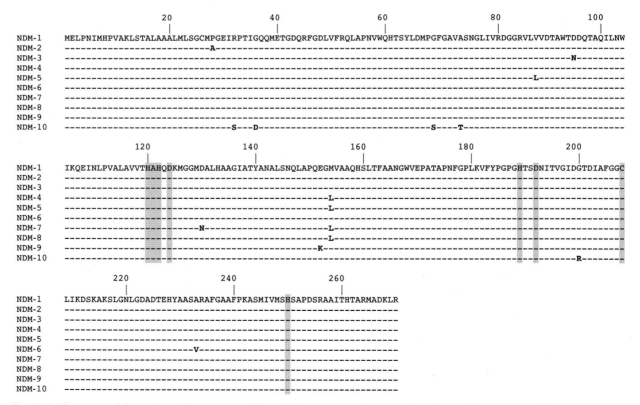

Fig. 48-4 Alignment of the amino acid sequences of the 10 NDM variants. Conserved residues of the active site of MBLs are highlighted in grey.

aztreonam. NDM-1 shares very little identity with other MBLs, the most similar MBLs being VIM-1/VIM-2, with which it shares only 32.4% identity. Compared to the other MBLs, NDM-1 possesses a unique HXHXD motif among the mobile MBLs, as it contains an alanine residue between the two histidine residues. NDM-1 also possesses a tyrosine residue at position 222 instead of the tryptophan residues in MBLs. NDM-1 shares the additional loop of the VIM enzymes at positions 34–47, but also possesses a unique additional sequence at positions 162–166. Compared to VIM-2, NDM-1 displays a tighter binding to most cephalosporins, in particular to cefuroxime, cefotaxime, and cephalothin, but also to penicillins. NDM-1 does not bind to carbapenems as tightly as IMP-1 or VIM-2 does, and the turnover rate of carbapenem hydrolysis is similar to that of VIM-2. As previously described for other MBLs, the loop 1 of NDM-1, a conserved loop adjacent to the active site, was found to play a crucial role in substrate recognition and binding (Kim et al., 2011; Zhu et al., 2013). Similar to the other MBLs, the active site of NDM-1 contains two metal ion binding sites: the His and Cys sites. Accordingly, a 3D-structure modeling of the NDM-1 enzyme showed that two zinc ions were present at both the His and Cys sites with a distance of 4.20 Å (Wang and Chou, 2013). Comparison of IMP-1, VIM-2, and NDM-1 by an *in silico* approach revealed that NDM-1 might have a greater drug profile and catalytic efficiency than IMP-1 and VIM-2 due to a larger pocket opening and a lower distance between the Zn-I ion and β-lactam oxygen of the carbapenem (Pal and Tripathi, 2013).

48.6 NDM variants

Since the first description of NDM-1, 10 variants of this enzyme have been assigned (NDM-1 to NDM-10). Among those 10 variants, only 8 are currently published (NDM-1 to NDM-8). The first variant of NDM-1, named NDM-2, was produced by an *A. baumannii* isolate recovered from a patient transferred to Germany from an Egyptian hospital (Kaase et al., 2011). This point mutation variant had a C to G substitution at position 82 resulting in an amino acid substitution of a proline to an alanine residue at position 28 (Pro → Ala). This point mutation was located at the last amino acid of the peptide leader of the enzyme. Minimum inhibitory concentration (MIC) values of β-lactams, including carbapenems, showed no significant difference between NDM-1 and NDM-2 producers. NDM-2 has been recovered from several *A. baumannii* strains (Espinal et al., 2011, 2013; Ghazawi et al., 2012; Kaase et al., 2011) and not from Enterobacteriaceae so far. NDM-3 was described from an *E. coli* isolate responsible for a community-acquired urinary tract infection in an Australian patient who had traveled in the Indian subcontinent. This enzyme variant differs from NDM-1 by a single nucleotide change conferring a putative peptide sequence change at position 95 (Asp → Asn) (Rogers et al., 2013). The NDM-4 variant was first described in one *E. coli* isolate recovered from a urinary culture of a patient hospitalized in India in January 2010. NDM-4 differs from NDM-1 by a single amino acid substitution at position 154 (Met → Leu). Although this amino acid substitution was not located in the known active sites of NDM-1, kinetic data showed that NDM-4 hydrolyzed imipenem and meropenem at a higher level than NDM-1 does. Higher catalytic efficiencies were also observed for cefalotin, ceftazidime, and cefotaxime for NDM-4, whereas cefepime was less hydrolyzed (Nordmann et al., 2012a). NDM-4 was further identified in one *E. coli* isolated from a rectal swab specimen collected from a patient transferred from Cameroon to France (Dortet et al., 2012c), and in one *Enterobacter cloacae* isolate from a Czech patient previously hospitalized in Sri Lanka (Papagiannitsis et al., 2013). The patient transferred from Cameroon to France was colonized with another *E. coli* isolate producing NDM-1, which was recovered during a rectal screening, indicating that this patient was colonized by two different NDM-producing strains (Dortet et al., 2012c). Interestingly, the patient had a history of Hodgkin lymphoma treated by eight sessions of chemotherapy based on bleomycin 1 year prior to his hospitalization. That anti-cancer drug is widely distributed throughout the body following intravenous administration and the serum concentration increases in proportion with increase of its dose. As the patient was successively treated with 30 mg of bleomycin, and according to our hypotheses, the bleomycin serum concentrations (~2 to 5 µg/mL) might have contributed to the selection of the ble_{MBL} gene, which is always associated with the bla_{NDM} gene, and conferring a high level resistance to that anti-cancer molecule. The NDM-5 variant was identified in a multi-drug-resistant *E. coli* ST648 isolate recovered from the perineum and throat of a patient in the United Kingdom with a recent history of hospitalization in India. NDM-5 differs from NDM-1 by two substitutions at positions 88 (Val → Leu) and 154 (Met → Leu). As previously observed for NDM-4, the substitution of a methionine by an alanine residue at position 154 enhanced the hydrolytic activity of the enzyme toward expanded-spectrum cephalosporins and carbapenems (Hornsey et al., 2011). The NDM-6 variant differs from NDM-1 by a single amino acid substitution at position 233 (Ala → Val). NDM-6 produced by an *E. coli* isolate was recovered from the rectal swab of a New Zealand patient who had been previously hospitalized in New Delhi (India) (Williamson et al., 2012). As observed for NDM-4, the patient was colonized with two different NDM-producing isolates, namely one *P. mirabilis* (bla_{NDM-1}) and the other *E. coli* (bla_{NDM-6}). The NDM-7 variant was concomitantly described from an *E. coli* isolate recovered from the urine of a French patient who had traveled to Burma (Cuzon et al., 2013), and an *E. coli* isolate recovered from the wounds, throat, and rectum of a Yemeni patient who was hospitalized at the Frankfurt University Hospital in Germany (Gottig et al., 2013). The bla_{NDM-7} gene differs from bla_{NDM-1} by two point mutations at positions 388 (G → A) and 460 (A → C) corresponding to amino acid substitutions at positions 130 (Asp → Asn) and 154 (Met → Leu), respectively. As previously observed for NDM-4 and NDM-5, both those studies highlight that the amino acid substitutions at position 154 (Met → Leu) increase the hydrolysis activity of the enzyme (Cuzon et al., 2013; Gottig et al., 2013). The last published NDM variant, namely NDM-8, was identified from a multi-drug-resistant *E. coli* isolate recovered from the respiratory tract of a patient in Nepal. The amino acid sequence of NDM-8 has substitutions at positions 130 (Asp → Gly) and 154 (Met → Leu) compared with NDM-1 (Tada et al., 2013). In this study, the critical impact of the amino acid substitution at position 154 (Met → Leu) was not analyzed by the authors.

48.7 Identification of NDM producers

The identification of NDM producers relies primarily on the identification of carbapenemase production. Detection of carbapenemase producers in clinical specimens is currently based on a preliminary analysis of susceptibility testing. The US guidelines (CLSI) (updated in 2013) retained as breakpoints for imipenem and meropenem, susceptibility (S) ≤ 1 and resistance (R) ≥ 4 mg/L, and for ertapenem, S ≤ 0.5 and R ≥ 2 mg/L, whereas the European guidelines (EUCAST) (updated in 2013) proposed breakpoints for imipenem and meropenem as follows: susceptibility (S) ≤ 2 and resistance (R) ≥ 8 mg/L, and for ertapenem, S ≤ 0.5 and R ≥ 1 mg/L. MIC values of ertapenem are often higher than those of other carbapenems with NDM producers. Consequently, ertapenem seems to be the best molecule for suspecting most of the carbapenemase producers. However, detection of carbapenemase producers based only on MIC values actually lacks sensitivity. Indeed, susceptibility to carbapenems is observed for several carbapenemase producers, and there are only few clinical successes of carbapenem-containing regimens for treating infections due to carbapenemase producers that are susceptible to carbapenems *in vitro*. Accordingly, additional tests for carbapenemase detection are needed to accurately detect carbapenemase producers, including NDM producers. Non-molecular techniques have been proposed for *in-vitro* identification of carbapenemase production. One of the commonly used techniques is the modified Hodge test (MHT), which has been used for years. MHT has an excellent sensitivity for detecting enterobacterial isolates producing Ambler class A (KPC) and class D (OXA-48) carbapenemases, but it has been proved to lack sensitivity for detecting NDM-1 producers (50%). Noticeably, this sensitivity might be increased

to 85.7% by adding $ZnSO_4$ (100 µg/mL) in the culture medium (Girlich et al., 2012). However, this test has a low specificity with *Enterobacter* spp. overexpressing a chromosomal cephalosporinase (Carvalhaes et al., 2010). In addition, it is time-consuming (at least 24 hours). Other detection methods based on the inhibitory properties of several divalent ions chelators (e.g., EDTA and dipicolinic acid) do exist to identify MBL producers, including NDM-producing strains. Among those methods, the Etest MBL strip is a two-sided strip containing gradients of imipenem alone on one side, and imipenem and EDTA on the other side. This test is considered to be positive when the MIC of imipenem is reduced by at least three doubling dilutions in the presence of EDTA (Walsh et al., 2002). A disk-diffusion test based on the detection of a synergy between a substrate-containing disk (imipenem, ceftazidime, or meropenem) and a disk containing an MBL inhibitor (EDTA or mercaptopropionic acid, or dipicolinic acid) might also be performed (Arakawa et al., 2000). The MBL is detected with the EDTA disk-diffusion test when EDTA (10 µL of a 0.1 M solution at pH 8) is added to the imipenem disk (Nordmann et al., 2011). Production of an MBL is suspected when a 5 mm increase of the inhibition diameter around the disk containing imipenem plus EDTA compared to imipenem alone is observed. However, those two phenotypic methods are time-consuming and false-negative results can often arise, especially when a low level of resistance is observed (Nordmann et al., 2011). Several alternative and reliable methods have been developed such as UV spectrophotometry analysis of carbapenem hydrolysis (Bernabeu et al., 2011). Also, analysis of carbapenem hydrolysis by using the MALDI-TOF technology has been shown to infer carbapenemase production in a few hours. This technique is based on the detection of the imipenem spectrum and of its main derivative resulting from carbapenem hydrolysis. After 3 hours of incubation, the candidate culture is put in the presence of imipenem. Hydrolysis of imipenem corresponds to a disappearance of the peak corresponding to the native carbapenem and appearance of the peak corresponding to the metabolite (Burckhardt and Zimmermann, 2011; Hrabak et al., 2011, 2012b, 2013; Kempf et al., 2012). This test has excellent sensitivity and specificity. However, it requires trained microbiologists and expensive equipment. The most promising technique is the rapid Carba NP test. It is based on the detection of the hydrolysis of imipenem by a color change of a pH indicator. This test is 100% sensitive and 100% specific for the detection of any type of carbapenemase produced by Enterobacteriaceae (Nordmann et al., 2012d). The Carba NP test allows the detection of NDM producers in 15 minutes to 1 hour. The Carba NP test has also been validated for the detection of most carbapenemase-producing *Pseudomonas* spp., including all NDM producers (Dortet et al., 2012d). A second version of the

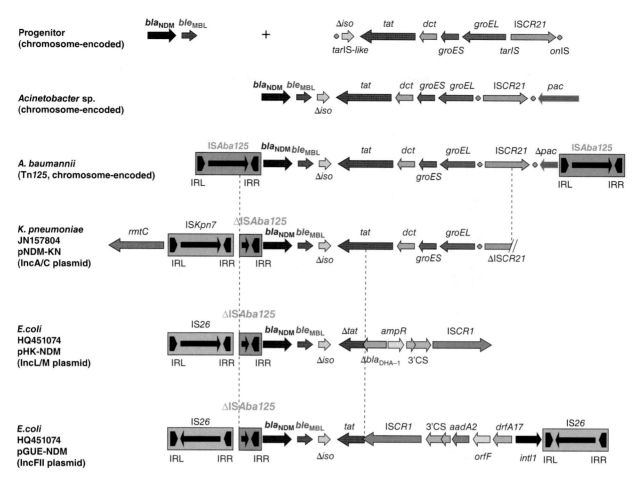

Fig. 48-5 Schematic representation of bla_{NDM} acquisition from the progenitor to Enterobacteriaceae. Genes and their corresponding transcription orientations are represented by horizontal arrows. The lengths of the target genes and the exact location of the target sites are not to scale. *ori*IS of IS*CR21* is indicated by a circle. Gene names are abbreviated according to their corresponding proteins: Δ*iso* for truncated phosphoribosylanthranilate isomerase; *tat* for the twin-arginine translocation pathway signal sequence protein; *dct* for the divalent cation tolerance protein; Δ*pac* for truncated phospholipid acetyltransferase; *orf* for an unknown open reading frame. The *groES* and *groEL* genes encode chaperonin proteins. The *rmtC* and *aad*A2 genes encode a 16S RNA methylase and an aminoglycoside adenyltransferase, respectively, responsible for aminoglycosides resistance. 3'CS corresponds to the 3' conserved sequence of an integron. The *dfr*A17 gene encodes a dihydrofolate reductase responsible for trimethoprim resistance. IRL and IRR indicate inverted repeat left and right of the insertion sequences, respectively. Tn*125* corresponds to a composite transposon bracked by two copies of the IS*Aba125* insertion sequence and found to be chromosome-encoded in *A. baumannii* isolates.

Carba NP test (the Carba NP test II) has been developed to rapidly identify the different carbapenemase types found in Enterobacteriaceae and *P. aeruginosa*, and further ongoing developments of the test may lead to its use for carbapenemase producers in *A. baumannii*. The inhibition properties of EDTA coupled with the high efficiency of the Carba NP test permits the identification of any type of class B MBL producers, including all NDM producers (Dortet et al., 2012e). Recently, the Carba NP test has been evaluated to detect carbapenemase-producing Enterobacteriaceae (*n* = 193) directly from spiked blood cultures. The proposed strategy allowed detection of all NDM producers (*n* = 33) in less than 5 hours, with sensitivity and specificity of 100%, respectively (Dortet et al., 2014).

All the previous techniques can detect the carbapenemase production and in some cases more precisely the production of an MBL, but none of them is able to specifically identify an NDM enzyme or its corresponding gene. Therefore, a number of genotypic approaches have been reported based on PCR techniques to detect NDM producers. Those methods, however, have the disadvantage of being unable to identify totally novel carbapenemase genes, and are quite expensive. Commercial DNA microarray methods are marketed and increase the convenience of those tests (Cuzon et al., 2012), although they cannot overcome general limitations of genotypic techniques. Finally, molecular amplification of the bla_{NDM} gene followed by sequencing is the unique way to identify an NDM variant at the molecular level.

48.8 Detection of infected and colonized patients

The prevention of the spread of carbapenemase producers relies on an early and accurate

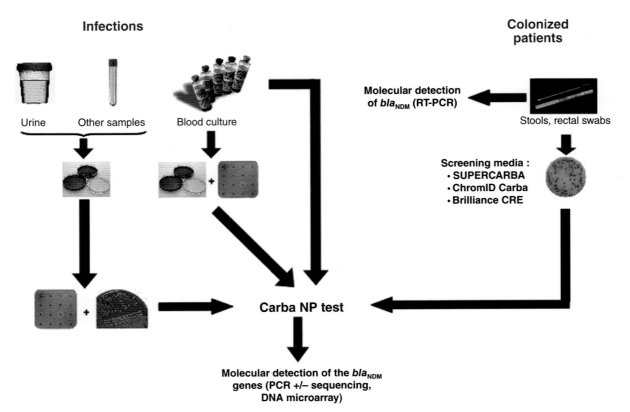

Fig. 48-6 Strategies for the detection of NDM producers in infected and colonized patients.

detection of carriers in hospital units or on admission/discharge, either in a hospital or in a specific unit. Accordingly, recommendations for screening colonized patients have been introduced in several countries. Although some details may differ between countries (ref): "at-risk" patients are at least those hospitalized in intensive-care units, in transplantation units, and immunocompromised patients. In addition, patients transferred from a foreign hospital have to be screened as soon as possible on the day of their hospitalization, and be considered as potential carriers until the results of the screening have been obtained.

Since the intestinal flora is the main reservoir of Enterobacteriaceae, stools and rectal swabs are the most suitable specimens for performing this screening. These specimens may be plated on a screening medium, either directly or after an 18-hour enrichment procedure in a broth containing imipenem 0.5–1 mg/L or ertapenem 0.5 mg/L (Adler et al., 2011; Landman et al., 2005). Although it has been shown to improve the detection of KPC producers, the usefulness of this enrichment step has not been evaluated for NDM producers. The main disadvantage of this enrichment step is that it delays the time needed to confirm or reject carbapenemase detection by 18–24 hours. On the opposite, during outbreak situations, it might increase the sensitivity of the screening, and consequently reduce the number of potential false-negative results.

Regardless of the enrichment step, the specimens have to be plated on selective media. The level of carbapenem resistance displayed by carbapenemase producers may significantly vary, making their detection difficult (Nordmann et al., 2011). Although NDM producers have often a high level of resistance to carbapenems, several isolates exhibited MICs comprised between 0.5 and 1 mg/L, making their detection difficult on screening media containing a high concentration of carbapenems. Several screening media have been evaluated and compared to the screening of carriers of NDM producers. One of the first tested media was the ChromID ESBL culture medium (bioMérieux, La Balmes-les-Grottes, France) containing cefpodoxime as a selector and which is routinely used to screen ESBL producers. Since NDM enzymes have a broad-spectrum activity, they hydrolyze not only carbapenems, but also expanded-spectrum cephalosporins very efficiently. Therefore, detection of NDM-producing isolates using ChromID ESBL (aimed to detect ESBL producers) is possible with a low limit detection being between 8×10^0 and 5×10^2 CFU/mL (Nordmann et al., 2011). Although ChromID ESBL is highly sensitive for the detection of NDM producers, it possesses a low specificity since the selective agent is a cephalosporin and not a specific carbapenemase substrate (e.g., a carbapenem). Several media supplemented with a carbapenem have been evaluated for the screening of carbapenemase producers, including NDM producers. The first marketed screening medium for the detection of KPC, another carbapenemase, was the CHROMagar KPC medium that contains meropenem (CHROMagar, Paris, France) (Moran Gilad et al.,

2011). It detects carbapenem-resistant bacteria if they exhibit a relatively high level of resistance to carbapenems. Its main disadvantage remains in its lack of sensitivity, since it does not detect carbapenemase producers with low-level carbapenem resistance, as observed for several NDM producers (Girlich et al., 2012, 2013; Nordmann et al., 2011). Another screening medium for carbapenemase producers, Colorex KPC, also contains meropenem (E&O Laboratories, Bonnybridge, UK). Since Colorex KPC is reported to have a content identical to that of CHROMagar KPC, only NDM producers with high-level resistance to carbapenems are detected. Consequently, it has been reported that only 57–64% of NDM-producing Enterobacteriaceae were detected using this medium (Perry et al., 2011; Wilkinson et al., 2012). A third screening medium also contains a carbapenem (CRE Brilliance, Thermo Fisher Scientific, Glasgow, UK). Depending on the study, the CRE Brilliance medium was found to be 63–85% sensitive for the detection of patients colonized with NDM producers (Day et al., 2013; Wilkinson et al., 2012). Another screening medium also containing a carbapenem is ChromID CARBA (bioMérieux, La Balmes-les-Grottes, France). This commercially available medium has been found to be more sensitive (87.5–94%) than the others for the detection of NDM-producing Enterobacteriaceae (Day et al., 2013; Perry et al., 2011; Wilkinson et al., 2012). Finally, the SUPERCARBA medium, which is a homemade screening medium containing ertapenem, cloxacillin, and zinc, showed excellent sensitivity and specificity for the detection of carbapenemase producers, including NDM producers. The low ertapenem concentration and the zinc supplementation allow detection of all NDM producers, regardless of their level of resistance to carbapenems (high or low) (Girlich et al., 2013; Nordmann et al., 2012c). Consequently, using the SUPERCARBA medium and then performing the Carba NP test on isolated colonies might be proposed as the recommended strategy for screening of carbapenemase producers, including NDM producers (Nordmann et Poirel, JAC 2012).

Even though those screening media are easy to use, it requires an additional 24–48 hours before the carriage status of the patient can be established when using them. Thus, an in-house quantitative real-time PCR assay using the TaqMan chemistry has been developed to detect the NDM-encoding genes directly from spiked stool samples. The bacterial extraction from stool samples was performed manually or adapted to a fully automated extraction system. This assay was found to be 100% specific and sensitive with detection limits reproducible below 1×10^1 CFU/100 mf of feces (Naas et al., 2011). However, this technology remains expensive and is thus considered to be a valuable tool in the follow-up of an outbreak and cohorting of colonized patients.

48.9 Conclusion

Several interesting features can be drawn from the analysis of the spread of NDM producers. This spread has been occurring at a high rate at least since 2006 when NDM producers were likely present only in the Indian subcontinent. Within less than 8 years, NDM producers have been identified worldwide. Therefore, this is not an old outbreak but a recent phenomenon that will not stop spontaneously, as opposed, for example, to the seasonal flu outbreaks. The reservoir at least in the Indian subcontinent is expanding continuously. Therefore, multi-drug resistance will increase irreversibly, on a daily basis. Whatever the adequate antibiotic policy is followed in any country, it may just delay the spread of those NDM producers. The analysis of the antibiotic resistance phenotype and genotype of NDM producers shows that those strains have accumulated many unrelated resistance mechanisms to distinct families of antibiotics. Those multi-drug-resistant strains are identified now not only among nosocomial pathogens (mostly *K. pneumoniae*), but also among community-acquired pathogens (*E. coli*). It should be kept in mind that *E. coli* is the main source of infections for humans with a high rate of mortality, still in 2013. Natural dissemination of *E. coli* in the environment and in the normal human flora will make the control of NDM producers particularly difficult. This explains why the outbreak of NDM producers is much more serious than an outbreak with other carbapenemase producers (KPC) that remain mostly in nosocomial-acquired *K. pneumoniae* pathogens.

From a genetic point of view, an NDM outbreak does not correspond to the spread of a single gene on a single genetic background. The diversity of genetic support and variants of *bla* NDM genes in non-related Gram-negative species will make their control very difficult (devious dissemination). Co-location of the NDM genes with a bleomycin resistance gene raises the issue of an anti-cancer drug as a source of selection of antibiotic resistance. In addition, detailed genetic analysis of the *bla*NDM gene in *A. baumannii* indicates that this species may act as an important reservoir for transferring this resistance gene to Enterobacteriaceae. Therefore, this suggests that *A. baumanni* may be more important than expected, although being considered as a minor pathogen in humans for years

Finally, the rapid development of the outbreak of NDM producers has pushed the development of novel rapid techniques for identification of resistance genes, such as the Carba NP test. Such techniques shall now be implemented worldwide to prevent as much as possible the spread of those multi-drug resistant strains that may become pandrug resistant on a short term. It shall also be an incentive for the discovery and development of novel antibiotics we urgently need.

References

Adler, A., Navon-Venezia, S., Moran-Gilad, J., Marcos, E., Schwartz, D., and Carmeli, Y. 2011. Laboratory and clinical evaluation of screening agar plates for detection of carbapenem-resistant Enterobacteriaceae from surveillance rectal swabs. *J. Clin. Microbiol.*, 49, 2239–2242.

Al-Agamy, M.H., Shibl, A.M., Elkhizzi, N.A., Meunier, D., Turton, J.F., and Livermore, D.M. 2013. Persistence of *Klebsiella pneumoniae* clones with OXA-48 or NDM carbapenemases causing bacteraemias in a Riyadh hospital. *Diagn. Microbiol. Infect. Dis.*, 76, 214–216.

Arakawa, Y., Shibata, N., Shibayama, K., Kurokawa, H., Yagi, T., Fujiwara, H., and Goto, M. 2000. Convenient test for screening metallo-β-lactamase-producing gram-negative bacteria by using thiol compounds. *J. Clin. Microbiol.*, 38, 40–43.

Arpin, C., Noury, P., Boraud, D., Coulange, L., Manetti, A., Andre, C., M'Zali, F., and Quentin, C. 2012. NDM-1-producing *Klebsiella pneumoniae* resistant to colistin in a French community patient without history of foreign travel. *Antimicrob. Agents Chemother.*, 56, 3432–3434.

Arya, S.C., and Agarwal, N. 2011. International travel with acquisition of multi-drug resistant Gram negative bacteria containing the New Delhi metallo-β-lactamase gene, bla_{NDM-1}. *Travel Med. Infect. Dis.*, 9, 47–48.

Barantsevich, E.P., Churkina, I.V., Barantsevich, N.E., Pelkonen, J., Schlyakhto, E.V., and Woodford, N. 2013. Emergence of *Klebsiella pneumoniae* producing NDM-1 carbapenemase in Saint Petersburg, Russia. *J. Antimicrob. Chemother.*, 68, 1204–1206.

Barguigua, A., El Otmani, F., Lakbakbi El Yaagoubi, F., Talmi, M., Zerouali, K., and Timinouni, M. 2013. First report of a *Klebsiella pneumoniae* strain coproducing NDM-1, VIM-1 and OXA-48 carbapenemases isolated in Morocco. *APMIS*, 121, 675–677.

Barrios, H., Garza-Ramos, U., Reyna-Flores, F., Sanchez-Perez, A., Rojas-Moreno, T., Garza-Gonzalez, E., Llaca-Diaz, J.M., Camacho-Ortiz, A., Guzman-Lopez, S., and Silva-Sanchez, J. 2013. Isolation of carbapenem-resistant NDM-1-positive *Providencia rettgeri* in Mexico. *J. Antimicrob. Chemother.*, 68, 1934–1936.

Ben Nasr, A., Decre, D., Compain, F., Genel, N., Barguellil, F., and Arlet, G. 2013. Emergence of NDM-1 in association with OXA-48 in a *Klebsiella pneumoniae* from Tunisia. *Antimicrob. Agents Chemother.*, 57, 4089–4090.

Berçot, B., Poirel, L., Dortet, L., and Nordmann, P. 2011. In vitro evaluation of antibiotic synergy for NDM-1-producing Enterobacteriaceae. *J. Antimicrob. Chemother.*, 66, 2295–2297.

Bernabeu, S., Poirel, L., and Nordmann, P. 2012. Spectrophotometry-based detection of carbapenemase producers among Enterobacteriaceae. *Diagn. Microbiol. Infect. Dis.*, 74, 88–90.

Bharadwaj, R., Joshi, S., Dohe, V., Gaikwad, V., Kulkarni, G., and Shouche, Y. 2012. Prevalence of New Delhi metallo-β-lactamase (NDM-1)-positive bacteria in a tertiary care centre in Pune, India. *Int. J. Antimicrob. Agents*, 39, 265–266.

Birgy, A., Doit, C., Mariani-Kurkdjian, P., Genel, N., Faye, A., Arlet, G., and Bingen, E. 2011. Early detection of colonization by VIM-1-producing *Klebsiella pneumoniae* and NDM-1-producing *Escherichia coli* in two children returning to France. *J. Clin. Microbiol.*, 49, 3085–3087.

Bogaerts, P., Bouchahrouf, W., de Castro, R.R., Deplano, A., Berhin, C., Pierard, D., Denis, O., and Glupczynski, Y. 2011. Emergence of NDM-1-producing Enterobacteriaceae in Belgium. *Antimicrob. Agents Chemother.*, 55, 3036–3038.

Bogaerts, P., Rezende de Castro, R., Roisin, S., Deplano, A., Huang, T.D., Hallin, M., Denis, O., and Glupczynski, Y. 2012. Emergence of NDM-1-producing *Acinetobacter baumannii* in Belgium. *J. Antimicrob. Chemother.*, 67, 1552–1553.

Bonnin, R.A., Naas, T., Poirel, L., and Nordmann, P. 2012a. Phenotypic, biochemical, and molecular techniques for detection of metallo-∫-lactamase NDM in *Acinetobacter baumannii*. *J. Clin. Microbiol.*, 50, 1419–1421.

Bonnin, R.A., Poirel, L., Carattoli, A., and Nordmann, P. 2012b. Characterization of an IncFII plasmid encoding NDM-1 from *Escherichia coli* ST131. *PloS One*, 7, e34752.

Boulanger, A., Naas, T., Fortineau, N., Figueiredo, S., and Nordmann, P. 2012. NDM-1-producing *Acinetobacter baumannii* from Algeria. *Antimicrob. Agents Chemother.*, 56, 2214–2215.

Brink, A.J., Coetzee, J., Clay, C.G., Sithole, S., Richards, G.A., Poirel, L., and Nordmann, P. 2012. Emergence of New Delhi metallo-β-lactamase (NDM-1) and *Klebsiella pneumoniae* carbapenemase (KPC-2) in South Africa. *J. Clin. Microbiol.*, 50, 525–527.

Burckhardt, I., and Zimmermann, S. 2011. Using matrix-assisted laser desorption ionization-time of flight mass spectrometry to detect carbapenem resistance within 1 to 2.5 hours. *J. Clin. Microbiol.*, 49, 3321–3324.

Cabanes, F., Lemant, J., Picot, S., Simac, C., Cousty, J., Jalin, L., Naze, F., Boisson, V., Cresta, M.P., Andre, H., Thibault, L., Tixier, F., Winer, A., Antok, E., and Michault, A. 2012. Emergence of *Klebsiella pneumoniae* and *Salmonella* metallo-β-lactamase (NDM-1) producers on reunion island. *J. Clin. Microbiol.*, 50, 3812.

Carvalhaes, C.G., Picao, R.C., Nicoletti, A.G., Xavier, D.E., and Gales, A.C. 2010. Cloverleaf test (modified Hodge test) for detecting carbapenemase production in *Klebsiella pneumoniae*: Be aware of false positive results. *J. Antimicrob. Chemother.*, 65, 249–251.

Castanheira, M., Deshpande, L.M., Mathai, D., Bell, J.M., Jones, R.N., and Mendes, R.E. 2011. Early dissemination of NDM-1- and OXA-181-producing Enterobacteriaceae in Indian hospitals: Report from the SENTRY Antimicrobial Surveillance Program, 2006–2007. *Antimicrob. Agents Chemother.*, 55, 1274–1278.

Chen, Y., Zhou, Z., Jiang, Y., and Yu, Y. 2011. Emergence of NDM-1-producing *Acinetobacter baumannii* in China. *J. Antimicrob. Chemother.*, 66, 1255–1259.

Chen, Y., Cui, Y., Pu, F., Jiang, G., Zhao, X., Yuan, Y., Zhao, W., Li, D., Liu, H., Li, Y., Liang, T., Xu, L., Wang, Y., Song, Q., Yang, J., Liang, L., Yang, R., Han, L., and Song, Y. 2012a. Draft genome sequence of an *Acinetobacter* genomic species 3 strain harboring a bla_{NDM-1} gene. *J. Bacteriol.*, 194, 204–205.

Chen, Y.T., Lin, A.C., Siu, L.K., and Koh, T.H. 2012b. Sequence of closely related plasmids encoding bla_{NDM-1} in two unrelated *Klebsiella pneumoniae* isolates in Singapore. *PloS One*, 7, e48737.

Coque, T.M., Novais, A., Carattoli, A., Poirel, L., Pitout, J., Peixe L., Baquero, F., Canton, R., and Nordmann, P. 2008. Dissemination of clonally related *Escherichia coli* strains expressing extended-spectrum β-lactamase CTX-M-15. *Emerg. Infect. Dis.*, 14, 195–200

Cuzon, G., Bonnin, R.A., and Nordmann, P. 2013. First identification of novel NDM carbapenemase, NDM-7, in *Escherichia coli* in France. *PloS One*, 8, e61322.

Cuzon, G., Naas, T., Bogaerts, P., Glupczynski, Y., and Nordmann, P. 2012. Evaluation of a DNA microarray for the rapid detection of extended-spectrum β-lactamases (TEM, SHV and CTX-M), plasmid-mediated cephalosporinases (CMY-2-like, DHA, FOX, ACC-1, ACT/MIR and CMY-1-like/MOX) and carbapenemases (KPC, OXA-48, VIM, IMP and NDM). *J. Antimicrob. Chemother.*, 67, 1865–1869.

Darley, E., Weeks, J., Jones, L., Daniels, V., Wootton, M., MacGowan, A., and Walsh, T. 2012. NDM-1 polymicrobial infections including *Vibrio cholerae*. *Lancet*, 380, 1358.

Day, K.M., Salman, M., Kazi, B., Sidjabat, H.E., Silvey, A., Lanyon, C.V., Cummings, S.P., Ali, M.N., Raza, M.W., Paterson, D.L., and Perry, J.D. 2013. Prevalence of NDM-1 carbapenemase in patients with diarrhoea in Pakistan and evaluation of two chromogenic culture media. *J. Appl. Microbiol.*, 114, 1810–1816.

Denis, C., Poirel, L., Carricajo, A., Grattard, F., Fascia, P., Verhoeven, P., Gay, P., Nuti, C., Nordmann, P., Pozzetto, B., and Berthelot, P. 2012. Nosocomial transmission of NDM-1-producing *Escherichia coli* within a non-endemic area in France. *Clin. Microbiol. Infec.*, 18, E128–E130.

Deshpande, P., Rodrigues, C., Shetty, A., Kapadia, F., Hedge, A., and Soman, R. 2010a. New Delhi metallo-β-lactamase (NDM-1) in Enterobacteriaceae: Treatment options with carbapenems compromised. *J. Assoc. Physicians India*, 58, 147–149.

Deshpande, P., Shetty, A., Kapadia, F., Hedge, A., Soman, R., and Rodrigues, C. 2010b. New Delhi metallo 1: Have carbapenems met their doom? *Clin. Infect. Dis.*, 51, 1222.

Dolejska, M., Villa, L., Poirel, L., Nordmann, P., and Carattoli, A. 2013. Complete sequencing of an IncHI1 plasmid encoding the carbapenemase NDM-1, the ArmA 16S RNA methylase and a resistance-nodulation-cell division/multidrug efflux pump. *J. Antimicrob. Chemother.*, 68, 34–39.

Dortet, L., Nordmann, P., and Poirel, L. 2012a. Association of the emerging carbapenemase NDM-1 with a bleomycin resistance protein in Enterobacteriaceae and *Acinetobacter baumannii. Antimicrob. Agents Chemother.*, 56, 1693–1697.

Dortet, L., Poirel, L., Al Yaqoubi, F., and Nordmann, P. 2012b. NDM-1, OXA-48 and OXA-181 carbapenemase-producing Enterobacteriaceae in Sultanate of Oman. *Clin. Microbiol. Infec.*, 18, E144–E148.

Dortet, L., Poirel, L., Anguel, N., and Nordmann, P. 2012c. New Delhi metallo-beta-lactamase 4-producing *Escherichia coli* in Cameroon. *Emerg. Infect. Dis.*, 18, 1540–1542.

Dortet, L., Poirel, L., and Nordmann, P. 2012d. Rapid detection of carbapenemase-producing Pseudomonas spp. *J. Clin. Microbiol.*, 50, 3773–3776.

Dortet, L., Poirel, L., and Nordmann, P. 2012e. Rapid identification of carbapenemase types in Enterobacteriaceae and *Pseudomonas* spp. by using a biochemical test. *Antimicrob. Agents Chemother.*, 56, 6437–6440.

Dortet, L., Brechard, L., Grenet, K., Nguessan, M.S., and Nordmann, P. 2013. Sri Lanka, another country from the Indian subcontinent with NDM-1-producing *Enterobacteriaceae. J. Antimicrob. Chemother.*, 68, 2172–2173.

Dortet, L., Brechard, L., Poirel, L., and Nordmann, P. 2014. Rapid detection of carbapenemase-producing Enterobacteriaceae from blood cultures. *Clin. Microbiol. Infect.*, 20, 340–344.

El-Herte, R.I., Araj, G.F., Matar, G.M., Baroud, M., Kanafani, Z.A., and Kanj, S.S. 2012. Detection of carbapenem-resistant *Escherichia coli* and *Klebsiella pneumoniae* producing NDM-1 in Lebanon. *J. Infect. Dev. Countr.*, 6, 457–461.

Escobar Perez, J.A., Olarte Escobar, N.M., Castro-Cardozo, B., Valderrama Marquez, I.A., Garzon Aguilar, M.I., Martinez de la Barrera, L., Barrero Barreto, E.R., Marquez-Ortiz, R.A., Moncada Guayazan, M.V., and Vanegas Gomez, N. 2013. Outbreak of NDM-1-producing *Klebsiella pneumoniae* in a neonatal unit in Colombia. *Antimicrob. Agents Chemother.*, 57, 1957–1960.

Espinal, P., Fugazza, G., Lopez, Y., Kasma, M., Lerman, Y., Malhotra-Kumar, S., Goossens, H., Carmeli, Y., and Vila, J. 2011. Dissemination of an NDM-2-producing *Acinetobacter baumannii* clone in an Israeli rehabilitation center. *Antimicrob. Agents Chemother.*, 55, 5396–5398.

Espinal, P., Poirel, L., Carmeli, Y., Kaase, M., Pal, T., Nordmann, P., and Vila, J. 2013. Spread of NDM-2-producing *Acinetobacter baumannii* in the Middle East. *J. Antimicrob. Chemother.*, 68, 1928–1930.

Falagas, M.E., Karageorgopoulos, D.E., and Nordmann, P. 2011. Therapeutic options for infections with Enterobacteriaceae producing carbapenem-hydrolyzing enzymes. *Future Microbiol.*, 6, 653–666.

Fischer, J., Schmoger, S., Jahn, S., Helmuth, R., and Guerra, B. 2013. NDM-1 carbapenemase-producing *Salmonella enterica* subsp. enterica serovar Corvallis isolated from a wild bird in Germany. *J. Antimicrob. Chemother.*, 68, 2954–2956.

Flateau, C., Janvier, F., Delacour, H., Males, S., Ficko, C., Andriamanantena, D., Jeannot, K., Merens, A., and Rapp, C. 2012. Recurrent pyelonephritis due to NDM-1 metallo-β-lactamase producing *Pseudomonas aeruginosa* in a patient returning from Serbia, France, 2012. *Euro Surveill.*, 17, pii: 20311.

Fuursted, K., Scholer, L., Hansen, F., Dam, K., Bojer, M.S., Hammerum, A.M., Dagnaes-Hansen, F., Olsen, A., Jasemian, Y., and Struve, C. 2012. Virulence of a *Klebsiella pneumoniae* strain carrying the New Delhi metallo-β-lactamase-1 (NDM-1). *Microbes Infect.*, 14, 155–158.

Gaibani, P., Ambretti, S., Berlingeri, A., Cordovana, M., Farruggia, P., Panico, M., Landini, M.P., and Sambri, V. 2011. Outbreak of NDM-1-producing Enterobacteriaceae in northern Italy, July to August 2011. *Euro. Surveill.*, 16, 20027.

Gecaj-Gashi, A., Hasani, A., Bruqi, B., and Mulliqi-Osmani, G. 2011. Balkan NDM-1: Escape or transplant? *Lancet Infect. Dis.*, 11, 586.

Ghazawi, A., Sonnevend, A., Bonnin, R.A., Poirel, L., Nordmann, P., Hashmey, R., Rizvi, T.A., B Hamadeh, M., and Pal, T. 2012. NDM-2 carbapenemase-producing *Acinetobacter baumannii* in the United Arab Emirates. *Clin. Microbiol. Infec.*, 18, E34–E36.

Girlich, D., Poirel, L., and Nordmann, P. 2012. Value of the modified Hodge test for detection of emerging carbapenemases in Enterobacteriaceae. *J. Clin. Microbiol.*, 50, 477–479.

Girlich, D., Poirel, L., and Nordmann, P. 2013. Comparison of the SUPERCARBA, CHROMagar KPC, and Brilliance CRE screening media for detection of Enterobacteriaceae with reduced susceptibility to carbapenems. *Diagn. Microbiol. Infect. Dis.*, 75, 214–217.

Glasner, C., Albiger, B., Buist, G., Tambić Andrasević, A., Canton, R., Carmeli, Y., Friedrich, A.W., Giske, C.G., Glupczynski, Y., Gniadkowski, M., Livermore, D.M., Nordmann, P., Poirel, L., Rossolini, G.M., Seifert, H., Vatopoulos, A., Walsh, T., Woodford, N., Donker, T., Monnet, D.L., and Grundmann, H., European Survey on Carbapenemase-Producing Enterobacteriaceae (EuSCAPE) Working Group. 2013. Carbapenemase-producing Enterobacteriaceae in Europe: A survey among national experts from 39 countries, February 2013. *Euro. Surveill.*, 18, pii: 20525.

Gottig, S., Hamprecht, A.G., Christ, S., Kempf, V.A., and Wichelhaus, T.A. 2013. Detection of NDM-7 in Germany, a new variant of the New Delhi metallo-β-lactamase with increased carbapenemase activity. *J. Antimicrob. Chemother.*, 68, 1737–1740.

Green, D.A., Srinivas, N., Watz, N., Tenover, F.C., Amieva, M., and Banaei, N. 2013. A pediatric case of New Delhi metallo-β-lactamase (NDM-1)- producing Enterobacteriaceae in the United States. *Pediatr. Infect. Dis. J.*, 32, 1291–1294.

Halaby, T., Reuland, A.E., Al Naiemi, N., Potron, A., Savelkoul, P.H., Vandenbroucke-Grauls, C.M., and Nordmann, P. 2012. A case of New Delhi metallo-β-lactamase 1 (NDM-1)-producing *Klebsiella pneumoniae* with putative secondary transmission from the Balkan region in the Netherlands. *Antimicrob. Agents Chemother.*, 56, 2790–2791.

Hammerum, A.M., Larsen, A.R., Hansen, F., Justesen, U.S., Friis-Moller, A., Lemming, L.E., Fuursted, K., Littauer, P., Schonning, K., Gahrn-Hansen, B., Ellermann-Eriksen, S., and Kristensen, B. 2012. Patients transferred from Libya to Denmark carried OXA-48-producing *Klebsiella pneumoniae*, NDM-1-producing *Acinetobacter baumannii* and meticillin-resistant *Staphylococcus aureus. Int. J. Antimicrob. Agents*, 40, 191–192.

Hishinuma, A., Yoshida, A., Suzuki, H., Okuzumi, K., and Ishida, T. 2013. Complete sequencing of an IncFII NDM-1 plasmid in *Klebsiella pneumoniae* shows structural features shared with other multidrug resistance plasmids. *J. Antimicrob. Chemother.*, 68, 2415–2417.

Ho, P.L., Lo, W.U., Yeung, M.K., Lin, C.H., Chow, K.H., Ang, I., Tong, A.H., Bao, J.Y., Lok, S., and Lo, J.Y. 2011. Complete sequencing of pNDM-HK encoding NDM-1 carbapenemase from a multidrug-resistant *Escherichia coli* strain isolated in Hong Kong. *PloS One*, 6, e17989.

Ho, P.L., Li, Z., Lai, E.L., Chiu, S.S., and Cheng, V.C. 2012. Emergence of NDM-1-producing Enterobacteriaceae in China. *J. Antimicrob. Chemother.*, 67, 1553–1555.

Hornsey, M., Phee, L., and Wareham, D.W. 2011. A novel variant, NDM-5, of the New Delhi metallo-beta-lactamase in a multidrug-resistant *Escherichia coli* ST648 isolate recovered from a patient in the United Kingdom. *Antimicrob. Agents Chemother.*, 55, 5952–5954.

Hrabak, J., Walkova, R., Studentova, V., Chudackova, E., and Bergerova, T. 2011. Carbapenemase activity detection by matrix-assisted laser desorption ionization-time of flight mass spectrometry. *J. Clin. Microbiol.*, 49, 3222–3227.

Hrabak, J., Stolbova, M., Studentova, V., Fridrichova, M., Chudackova, E., and Zemlickova, H. 2012a. NDM-1 producing *Acinetobacter baumannii* isolated from a patient repatriated to the Czech Republic from Egypt, July 2011. *Euro. Surveill.*, 17(7).

Hrabak, J., Studentova, V., Walkova, R., Zemlickova, H., Jakubu, V., Chudackova, E., Gniadkowski, M., Pfeifer, Y., Perry, J.D., Wilkinson, K., and Bergerova, T. 2012b. Detection of NDM-1, VIM-1, KPC, OXA-48, and OXA-162 carbapenemases by matrix-assisted laser desorption ionization-time of flight mass spectrometry. *J. Clin. Microbiol.*, 50, 2441–2443.

Hrabak, J., Chudackova, E., and Walkova, R. 2013. Matrix-assisted laser desorption ionization-time of flight (MALDI-TOF) mass spectrometry for detection of antibiotic resistance mechanisms: From research to routine diagnosis. *Clin. Microbiol. Rev.*, 26, 103–114.

Hu, H., Hu, Y., Pan, Y., Liang, H., Wang, H., Wang, X., Hao, Q., Yang, X., Yang, X., Xiao, X., Luan, C., Yang, Y., Cui, Y., Yang, R., Gao, G.F., Song, Y., and Zhu, B. 2012. Novel plasmid and its variant harboring both a bla_{NDM-1} gene and type IV secretion system in clinical isolates of *Acinetobacter lwoffii*. *Antimicrob. Agents Chemother.*, 56, 1698–1702.

Huo, T.I. 2010. The first case of multidrug-resistant NDM-1-harboring Enterobacteriaceae in Taiwan: Here comes the superbacteria! *J. Chin. Med. Assoc.*, 73, 557–558.

Islam, M.A., Talukdar, P.K., Hoque, A., Huq, M., Nabi, A., Ahmed, D., Talukder, K.A., Pietroni, M.A., Hays, J.P., Cravioto, A., and Endtz, H.P. 2012. Emergence of multidrug-resistant NDM-1-producing Gram-negative bacteria in Bangladesh. *Eur. J. Clin. Microbiol. Infect. Dis.*, 31, 2593–2600.

Isozumi, R., Yoshimatsu, K., Yamashiro, T., Hasebe, F., Nguyen, B.M., Ngo, T.C., Yasuda, S.P., Koma, T., Shimizu, K., and Arikawa, J. 2012. bla_{NDM-1}-positive *Klebsiella pneumoniae* from environment, Vietnam. *Emerg. Infect. Dis.*, 18, 1383–1385.

Jamal, W., Rotimi, V.O., Albert, M.J., Khodakhast, F., Udo, E.E., and Poirel, L. 2012. Emergence of nosocomial New Delhi metallo-β-lactamase-1 (NDM-1)-producing *Klebsiella pneumoniae* in patients admitted to a tertiary care hospital in Kuwait. *Int. J. Antimicrob. Agents*, 39, 183–184.

Jovcic, B., Lepsanovic, Z., Suljagic, V., Rackov, G., Begovic, J., Topisirovic, L., and Kojic, M. 2011. Emergence of NDM-1 metallo-β-lactamase in *Pseudomonas aeruginosa* clinical isolates from Serbia. *Antimicrob. Agents Chemother.*, 55, 3929–3931.

Kaase, M., Nordmann, P., Wichelhaus, T.A., Gatermann, S.G., Bonnin, R.A., and Poirel, L. 2011. NDM-2 carbapenemase in *Acinetobacter baumannii* from Egypt. *J. Antimicrob. Chemother.*, 66, 1260–1262.

Kempf, M., Bakour, S., Flaudrops, C., Berrazeg, M., Brunel, J.M., Drissi, M., Mesli, E., Touati, A., and Rolain, J.M. 2012. Rapid detection of carbapenem resistance in *Acinetobacter baumannii* using matrix-assisted laser desorption ionization-time of flight mass spectrometry. *PloS One*, 7, e31676.

Kim, Y., Tesar, C., Mire, J., Jedrzejczak, R., Binkowski, A., Babnigg, G., Sacchettini, J., and Joachimiak, A. 2011. Structure of apo- and monometalated forms of NDM-1–a highly potent carbapenem-hydrolyzing metallo-β-lactamase. *PloS One*, 6, e24621.

Kim, M.N., Yong, D., An, D., Chung, H.S., Woo, J.H., Lee, K., and Chong, Y. 2012. Nosocomial clustering of NDM-1-producing *Klebsiella pneumoniae* sequence type 340 strains in four patients at a South Korean tertiary care hospital. *J. Clin. Microbiol.*, 50, 1433–1436.

Krishna, B.V. 2010. New Delhi metallo-β-lactamases: A wake-up call for microbiologists. *Indian J. Med. Microbiol.*, 28, 265–266.

Kumarasamy, K.K., Toleman, M.A., Walsh, T.R., Bagaria, J., Butt, F., Balakrishnan, R., Chaudhary, U., Doumith, M., Giske, C.G., Irfan, S., Krishnan, P., Kumar, A.V., Maharjan, S., Mushtaq, S., Noorie, T., Paterson, D.L., Pearson, A., Perry, C., Pike, R., Rao, B., Ray, U., Sarma, J.B., Sharma, M., Sheridan, E., Thirunarayan, M.A., Turton, J., Upadhyay, S., Warner, M., Welfare, W., Livermore, D.M., and Woodford, N. 2010. Emergence of a new antibiotic resistance mechanism in India, Pakistan, and the UK: A molecular, biological, and epidemiological study. *Lancet Infect. Dis.*, 10, 597–602.

Landman, D., Salvani, J.K., Bratu, S., and Quale, J. 2005. Evaluation of techniques for detection of carbapenem-resistant *Klebsiella pneumoniae* in stool surveillance cultures. *J. Clin. Microbiol.*, 43, 5639–5641.

Lascols, C., Hackel, M., Marshall, S.H., Hujer, A.M., Bouchillon, S., Badal, R., Hoban, D., and Bonomo, R.A. 2011. Increasing prevalence and dissemination of NDM-1 metallo-β-lactamase in India: Data from the SMART study (2009). *J. Antimicrob. Chemother.*, 66, 1992–1997.

Leski, T., Vora, G.J., and Taitt, C.R. 2012. Multidrug resistance determinants from NDM-1-producing *Klebsiella pneumoniae* in the USA. *Int. J. Antimicrob. Agents*, 40, 282–284.

Leverstein-Van Hall, M.A., Stuart, J.C., Voets, G.M., Versteeg, D., Tersmette, T., and Fluit, A.C. 2010. Global spread of New Delhi metallo-β-lactamase 1. *Lancet Infect. Dis.*, 10, 830–831.

Liu, Z., Li, W., Wang, J., Pan, J., Sun, S., Yu, Y., Zhao, B., Ma, Y., Zhang, T., Qi, J., Liu, G., and Lu, F. 2013. Identification and characterization of the first *Escherichia coli* strain carrying NDM-1 gene in China. *PloS One*, 8, e66666.

Livermore, D.M., Mushtaq, S., Warner, M., Zhang, J., Maharjan, S., Doumith, M., and Woodford, N. 2011a. Activities of NXL104 combinations with ceftazidime and aztreonam against carbapenemase-producing Enterobacteriaceae. *Antimicrob. Agents Chemother.*, 55, 390–394.

Livermore, D.M., Walsh, T.R., Toleman, M., and Woodford, N. 2011b. Balkan NDM-1: Escape or transplant? *Lancet Infect. Dis.*, 11, 164.

Lowman, W., Sriruttan, C., Nana, T., Bosman, N., Duse, A., Venturas, J., Clay, C., and Coetzee, J. 2011. NDM-1 has arrived: First report of a carbapenem resistance mechanism in South Africa. *S. Afr. Med. J.*, 101, 873–875.

Mazzariol, A., Bosnjak, Z., Ballarini, P., Budimir, A., Bedenic, B., Kalenic, S., and Cornaglia, G. 2012. NDM-1-producing *Klebsiella pneumoniae*, Croatia. *Emerg. Infect. Dis.*, 18, 532–534.

McDermott, H., Morris, D., McArdle, E., O'Mahony, G., Kelly, S., Cormican, M., and Cunney, R. 2012. Isolation of NDM-1-producing *Klebsiella pnemoniae* in Ireland, July 2011. *Euro. Surveill.*, 17.

McGann, P., Hang, J., Clifford, R.J., Yang, Y., Kwak, Y.I., Kuschner, R.A., Lesho, E.P., and Waterman, P.E. 2012. Complete sequence of a novel 178-kilobase plasmid carrying bla_{NDM-1} in a *Providencia stuartii* strain isolated in Afghanistan. *Antimicrob. Agents Chemother.*, 56, 1673–1679.

Milillo, M., Kwak, Y.I., Snesrud, E., Waterman, P.E., Lesho, E., and McGann, P. 2013. Rapid and simultaneous detection of bla_{KPC} and bla_{NDM} by use of multiplex real-time PCR. *J. Clin. Microbiol.*, 51, 1247–1249.

Mirovic, V., Tomanovic, B., Lepsanovic, Z., Jovcic, B., and Kojic, M. 2012. Isolation of *Klebsiella pneumoniae* producing NDM-1 metallo-β-lactamase from the urine of an outpatient baby boy receiving antibiotic prophylaxis. *Antimicrob. Agents Chemother.*, 56, 6062–6063.

Mochon, A.B., Garner, O.B., Hindler, J.A., Krogstad, P., Ward, K.W., Lewinski, M.A., Rasheed, J.K., Anderson, K.F.,

Limbago, B.M., and Humphries, R.M. 2011. New Delhi metallo-β-lactamase (NDM-1)-producing *Klebsiella pneumoniae*: Case report and laboratory detection strategies. *J. Clin. Microbiol.*, 49, 1667–1670.

Moran Gilad, J., Carmeli, Y., Schwartz, D., and Navon-Venezia, S. 2011. Laboratory evaluation of the CHROMagar KPC medium for identification of carbapenem-nonsusceptible *Enterobacteriaceae*. *Diagn. Microbiol. Infect. Dis.*, 70, 565–567.

Muir, A., and Weinbren, M.J. 2010. New Delhi metallo-β-lactamase: A cautionary tale. *J. Hosp. Infect.*, 75, 239–240.

Naas, T., Namdari, F., Bogaerts, P., Huang, T.D., Glupczynski, Y., and Nordmann, P. 2008. Genetic structure associated with bla_{OXA-18}, encoding a clavulanic acid-inhibited extended-spectrum oxacillinase. *Antimicrob. Agents Chemother.*, 52, 3898–3904.

Naas, T., Ergani, A., Carrer, A., and Nordmann, P. 2011. Real-time PCR for detection of NDM-1 carbapenemase genes from spiked stool samples. *Antimicrob. Agents Chemother.*, 55, 4038–4043.

Nakazawa, Y., Ii, R., Tamura, T., Hoshina, T., Tamura, K., Kawano, S., Kato, T., Sato, F., Horino, T., Yoshida, M., Hori, S., Sanui, M., Ishii, Y., and Tateda, K. 2013. A case of NDM-1-producing *Acinetobacter baumannii* transferred from India to Japan. *J. Infect. Chemother.*, 19, 330–332.

Nemec, A., and Krizova, L. 2012. Carbapenem-resistant *Acinetobacter baumannii* carrying the NDM-1 gene, Czech Republic 2011. *Euro. Surveill.*, 17.

Nielsen, J.B., Hansen, F., Littauer, P., Schonning, K., and Hammerum, A.M. 2012. An NDM-1-producing *Escherichia coli* obtained in Denmark has a genetic profile similar to an NDM-1-producing *E. coli* isolate from the UK. *J. Antimicrob. Chemother.*, 67, 2049–2051.

Nordmann, P., and Poirel, L. 2012. Strategy for identification of carbapenemase-producing *Enterobacteriaceae*. *J. Antimicrob. Chemother.*, 68, 487–489.

Nordmann, P., Poirel, L., Carrer, A., Toleman, M.A., and Walsh, T.R. 2011. How to detect NDM-1 producers. *J. Clin. Microbiol.*, 49, 718–721.

Nordmann, P., Boulanger, A.E., and Poirel, L. 2012a. NDM-4 metallo-β-lactamase with increased carbapenemase activity from *Escherichia coli*. *Antimicrob. Agents Chemother.*, 56, 2184–2186.

Nordmann, P., Couard, J.P., Sansot, D., and Poirel, L. 2012b. Emergence of an autochthonous and community-acquired NDM-1-producing *Klebsiella pneumoniae* in Europe. *Clin. Infect. Dis.*, 54, 150–151.

Nordmann, P., Girlich, D., and Poirel, L. 2012c. Detection of carbapenemase producers in Enterobacteriaceae by use of a novel screening medium. *J. Clin. Microbiol.*, 50, 2761–2766.

Nordmann, P., Poirel, L., and Dortet, L. 2012d. Rapid detection of carbapenemase-producing Enterobacteriaceae. *Emerg. Infect. Dis.*, 18, 1503–1507.

Oteo, J., Domingo-Garcia, D., Fernandez-Romero, S., Saez, D., Guiu, A., Cuevas, O., Lopez-Brea, M., and Campos, J. 2012. Abdominal abscess due to NDM-1-producing *Klebsiella pneumoniae* in Spain. *J. Med. Microbiol.*, 61, 864–867.

Pal, A., and Tripathi, A. 2013. An in silico approach for understanding the molecular evolution of clinically important metallo-β-lactamases. *Infect. Genet. Evol.*, 20C, 39–47.

Papagiannitsis, C.C., Studentova, V., Chudackova, E., Bergerova, T., Hrabak, J., Radej, J., and Novak, I. 2013. Identification of a New Delhi metallo-β-lactamase-4 (NDM-4)-producing Enterobacter cloacae from a Czech patient previously hospitalized in Sri Lanka. *Folia Microbiol. (Praha)*, 58, 547–549.

Pasteran, F., Albornoz, E., Faccone, D., Gomez, S., Valenzuela, C., Morales, M., Estrada, P., Valenzuela, L., Matheu, J., Guerriero, L., Arbizu, E., Calderon, Y., Ramon-Pardo, P., and Corso, A. 2012. Emergence of NDM-1-producing *Klebsiella*

pneumoniae in Guatemala. *J. Antimicrob. Chemother.*, 67, 1795–1797.

Peirano, G., Pillai, D.R., Pitondo-Silva, A., Richardson, D., and Pitout, J.D. 2011a. The characteristics of NDM-producing *Klebsiella pneumoniae* from Canada. *Diagn. Microbiol. Infect. Dis.*, 71, 106–109.

Peirano, G., Schreckenberger, P.C., and Pitout, J.D. 2011b. Characteristics of NDM-1-producing *Escherichia coli* isolates that belong to the successful and virulent clone ST131. *Antimicrob. Agents Chemother.*, 55, 2986–2988.

Peirano, G., Mulvey, G.L., Armstrong, G.D., and Pitout, J.D. 2013. Virulence potential and adherence properties of *Escherichia coli* that produce CTX-M and NDM β-lactamases. *J. Med. Microbiol.*, 62, 525–530.

Perry, J.D., Naqvi, S.H., Mirza, I.A., Alizai, S.A., Hussain, A., Ghirardi, S., Orenga, S., Wilkinson, K., Woodford, N., Zhang, J., Livermore, D.M., Abbasi, S.A., and Raza, M.W. 2011. Prevalence of faecal carriage of Enterobacteriaceae with NDM-1 carbapenemase at military hospitals in Pakistan, and evaluation of two chromogenic media. *J. Antimicrob. Chemother.*, 66, 2288–2294.

Pfeifer, Y., Witte, W., Holfelder, M., Busch, J., Nordmann, P., and Poirel, L. 2011. NDM-1-producing *Escherichia coli* in Germany. *Antimicrob. Agents Chemother.*, 55, 1318–1319.

Poirel, L., Lagrutta, E., Taylor, P., Pham, J., and Nordmann, P. 2010. Emergence of metallo-β-lactamase NDM-1-producing multidrug-resistant *Escherichia coli* in Australia. *Antimicrob. Agents Chemother.*, 54, 4914–4916.

Poirel, L., Al Maskari, Z., Al Rashdi, F., Bernabeu, S., and Nordmann, P. 2011a. NDM-1-producing *Klebsiella pneumoniae* isolated in the Sultanate of Oman. *J. Antimicrob. Chemother.*, 66, 304–306.

Poirel, L., Benouda, A., Hays, C., and Nordmann, P. 2011b. Emergence of NDM-1-producing *Klebsiella pneumoniae* in Morocco. *J. Antimicrob. Chemother.* 66, 2781–2783.

Poirel, L., Dortet, L., Bernabeu, S., and Nordmann, P. 2011c. Genetic features of bla_{NDM-1}-positive Enterobacteriaceae. *Antimicrob. Agents Chemother.*, 55, 5403–5407.

Poirel, L., Fortineau, N., and Nordmann, P. 2011d. International transfer of NDM-1-producing *Klebsiella pneumoniae* from Iraq to France. *Antimicrob. Agents Chemother.*, 55, 1821–1822.

Poirel, L., Revathi, G., Bernabeu, S., and Nordmann, P. 2011e. Detection of NDM-1-producing *Klebsiella pneumoniae* in Kenya. *Antimicrob. Agents Chemother.*, 55, 934–936.

Poirel, L., Ros, A., Carricajo, A., Berthelot, P., Pozzetto, B., Bernabeu, S., and Nordmann, P. 2011f. Extremely drug-resistant *Citrobacter freundii* isolate producing NDM-1 and other carbapenemases identified in a patient returning from India. *Antimicrob. Agents Chemother.*, 55, 447–448.

Poirel, L., Schrenzel, J., Cherkaoui, A., Bernabeu, S., Renzi, G., and Nordmann, P. 2011g. Molecular analysis of NDM-1-producing enterobacterial isolates from Geneva, Switzerland. *J. Antimicrob. Chemother.*, 66, 1730–1733.

Poirel, L., Bonnin, R.A., Boulanger, A., Schrenzel, J., Kaase, M., and Nordmann, P. 2012a. Tn125-related acquisition of bla_{NDM}-like genes in *Acinetobacter baumannii*. *Antimicrob. Agents Chemother.*, 56, 1087–1089.

Poirel, L., Lascols, C., Bernabeu, S., and Nordmann, P. 2012b. NDM-1-producing *Klebsiella pneumoniae* in Mauritius. *Antimicrob. Agents Chemother.*, 56, 598–599.

Poirel, L., Ozdamar, M., Ocampo-Sosa, A.A., Turkoglu, S., Ozer, U.G., and Nordmann, P. 2012c. NDM-1-producing *Klebsiella pneumoniae* now in Turkey. *Antimicrob. Agents Chemother.*, 56, 2784–2785.

Potron, A., Poirel, L., and Nordmann, P. 2011. Plasmid-mediated transfer of the bla_{NDM-1} gene in Gram-negative rods. *FEMS Microbiol. Lett.*, 324, 111–116.

Raghunath, D. 2010. New metallo β-lactamase NDM-1. *Indian J. Med. Res.*, 132, 478–481.

Rimrang, B., Chanawong, A., Lulitanond, A., Wilailuckana, C., Charoensri, N., Sribenjalux, P., Phumsrikaew, W.,

Wonglakorn, L., Kerdsin, A., and Chetchotisakd, P. 2012. Emergence of NDM-1- and IMP-14a-producing Enterobacteriaceae in Thailand. *J. Antimicrob. Chemother.*, 67, 2626–2630.

Rogers, B.A., Sidjabat, H.E., Silvey, A., Anderson, T.L., Perera, S., Li, J., and Paterson, D.L. 2013. Treatment options for New Delhi metallo-β-lactamase-harboring Enterobacteriaceae. *Microb. Drug Resist.*, 19, 100–103.

Roy, S., Singh, A.K., Viswanathan, R., Nandy, R.K., and Basu, S. 2011a. Transmission of imipenem resistance determinants during the course of an outbreak of NDM-1 *Escherichia coli* in a sick newborn care unit. *J. Antimicrob. Chemother.*, 66, 2773–2780.

Roy, S., Viswanathan, R., Singh, A.K., Das, P., and Basu, S. 2011b. Sepsis in neonates due to imipenem-resistant *Klebsiella pneumoniae* producing NDM-1 in India. *J. Antimicrob. Chemother.*, 66, 1411–1413.

Samuelsen, O., Thilesen, C.M., Heggelund, L., Vada, A.N., Kummel, A., and Sundsfjord, A. 2011. Identification of NDM-1-producing Enterobacteriaceae in Norway. *J. Antimicrob. Chemother.*, 66, 670–672.

Samuelsen, O., Naseer, U., Karah, N., Lindemann, P.C., Kanestrom, A., Leegaard, T.M., and Sundsfjord, A. 2013. Identification of Enterobacteriaceae isolates with OXA-48 and coproduction of OXA-181 and NDM-1 in Norway. *J. Antimicrob. Chemother.*, 68, 1682–1685.

Savard, P., Gopinath, R., Zhu, W., Kitchel, B., Rasheed, J.K., Tekle, T., Roberts, A., Ross, T., Razeq, J., Landrum, B.M., Wilson, L.E., Limbago, B., Perl, T.M., and Carroll, K.C. 2011. First NDM-positive *Salmonella* sp. strain identified in the United States. *Antimicrob. Agents Chemother.*, 55, 5957–5958.

Seema, K., Ranjan Sen, M., Upadhyay, S., and Bhattacharjee, A. 2011. Dissemination of the New Delhi metallo-β-lactamase-1 (NDM-1) among Enterobacteriaceae in a tertiary referral hospital in north India. *J. Antimicrob. Chemother.*, 66, 1646–1647.

Sekizuka, T., Matsui, M., Yamane, K., Takeuchi, F., Ohnishi, M., Hishinuma, A., Arakawa, Y., and Kuroda, M. 2011. Complete sequencing of the *bla*NDM-1-positive IncA/C plasmid from *Escherichia coli* ST38 isolate suggests a possible origin from plant pathogens. *PloS One*, 6, e25334.

Shakil, S., Azhar, E.I., Tabrez, S., Kamal, M.A., Jabir, N.R., Abuzenadah, A.M., Damanhouri, G.A., and Alam, Q. 2011. New Delhi metallo-β-lactamase (NDM-1): An update. *J. Chemother.*, 23, 263–265.

Shibl, A., Al-Agamy, M., Memish, Z., Senok, A., Khader, S.A., and Assiri, A. 2013. The emergence of OXA-48- and NDM-1-positive *Klebsiella pneumoniae* in Riyadh, Saudi Arabia. *Int. J. Infect. Dis.*, 17, e1130–e1133.

Sole, M., Pitart, C., Roca, I., Fabrega, A., Salvador, P., Munoz, L., Oliveira, I., Gascon, J., Marco, F., and Vila, J. 2011. First description of an *Escherichia coli* strain producing NDM-1 carbapenemase in Spain. *Antimicrob. Agents Chemother.*, 55, 4402–4404.

Sonnevend, A., Al Baloushi, A., Ghazawi, A., Hashmey, R., Girgis, S., Hamadeh, M.B., Al Haj, M., and Pal, T. 2013. Emergence and spread of NDM-1 producer Enterobacteriaceae with contribution of IncX3 plasmids in the United Arab Emirates. *J. Med. Microbiol.*, 62, 1044–1050.

Struelens, M.J., Monnet, D.L., Magiorakos, A.P., Santos O'Connor, F., Giesecke, J., and European, N.D.M.S.P. 2010. New Delhi metallo-β-lactamase 1-producing Enterobacteriaceae: Emergence and response in Europe. *Euro. Surveill.*, 15.

Sun, Y., Song, Y., Song, H., Liu, J., Wang, P., Qiu, S., Chen, S., Zhu, L., Ji, X., Wang, Z., Liu, N., Xia, L., Chen, W., and Feng, S. 2013. Complete genome sequence of an *Acinetobacter* strain harboring the NDM-1 gene. *Genome Announc.* 1, e0002312.

Tada, T., Miyoshi-Akiyama, T., Dahal, R.K., Sah, M.K., Ohara, H., Kirikae, T., and Pokhrel, B.M. 2013. NDM-8 metallo-β-lactamase in a multidrug-resistant *Escherichia coli* strain isolated in Nepal. *Antimicrob. Agents Chemother.*, 57, 2394–2396.

Toleman, M.A., Spencer, J., Jones, L., and Walsh, T.R. 2012. *bla*NDM-1 is a chimera likely constructed in *Acinetobacter baumannii*. *Antimicrob. Agents Chemother.*, 56, 2773–2776.

Tsang, K.Y., Luk, S., Lo, J.Y., Tsang, T.Y., Lai, S.T., and Ng, T.K. 2012. Hong Kong experiences the 'Ultimate superbug': NDM-1 Enterobacteriaceae. *Hong Kong Med. J.*, 18, 439–441.

Walsh, T.R., Bolmstrom, A., Qwarnstrom, A., and Gales, A. 2002. Evaluation of a new Etest for detecting metallo-β-lactamases in routine clinical testing. *J. Clin. Microbiol.*, 40, 2755–2759.

Walsh, T.R., Weeks, J., Livermore, D.M., and Toleman, M.A. 2011. Dissemination of NDM-1 positive bacteria in the New Delhi environment and its implications for human health: An environmental point prevalence study. *Lancet Infect. Dis.*, 11, 355–362.

Wang, J.F., and Chou, K.C. 2013. Metallo-β-lactamases: Structural features, antibiotic recognition, inhibition, and inhibitor design. *Curr. Top. Med. Chem.*, 13, 1242–1253.

Wang, S.J., Chiu, S.H., Lin, Y.C., Tsai, Y.C., and Mu, J.J. 2013. Carbapenem resistant *Enterobacteriaceae* carrying New Delhi metallo-β-lactamase gene (NDM-1) in Taiwan. *Diagn. Microbiol. Infect. Dis.*, 76, 248–249.

Wilkinson, K.M., Winstanley, T.G., Lanyon, C., Cummings, S.P., Raza, M.W., and Perry, J.D. 2012. Comparison of four chromogenic culture media for carbapenemase-producing Enterobacteriaceae. *J. Clin. Microbiol.*, 50, 3102–3104.

Williamson, D.A., Sidjabat, H.E., Freeman, J.T., Roberts, S.A., Silvey, A., Woodhouse, R., Mowat, E., Dyet, K., Paterson, D.L., Blackmore, T., Burns, A., and Heffernan, H. 2012. Identification and molecular characterisation of New Delhi metallo-β-lactamase-1 (NDM-1)- and NDM-6-producing Enterobacteriaceae from New Zealand hospitals. *Int. J. Antimicrob. Agents*, 39, 529–533.

Yamamoto, T., Takano, T., Iwao, Y., and Hishinuma, A. 2011. Emergence of NDM-1-positive capsulated *Escherichia coli* with high resistance to serum killing in Japan. *J. Infect. Chemother.*, 17, 435–439.

Yamamoto, T., Takano, T., Fusegawa, T., Shibuya, T., Hung, W.C., Higuchi, W., Iwao, Y., Khokhlova, O., and Reva, I. 2013. Electron microscopic structures, serum resistance, and plasmid restructuring of New Delhi metallo-β-lactamase-1 (NDM-1)-producing ST42 *Klebsiella pneumoniae* emerging in Japan. *J. Infect. Chemother.*, 19, 118–127.

Yang, J., Chen, Y., Jia, X., Luo, Y., Song, Q., Zhao, W., Wang, Y., Liu, H., Zheng, D., Xia, Y., Yu, R., Han, X., Jiang, G., Zhou, Y., Zhou, W., Hu, X., Liang, L., and Han, L. 2012. Dissemination and characterization of NDM-1-producing *Acinetobacter pittii* in an intensive care unit in China. *Clin. Microbiol. Infec.*, 18, E506–E513.

Yong, D., Toleman, M.A., Giske, C.G., Cho, H.S., Sundman, K., Lee, K., and Walsh, T.R. 2009. Characterization of a new metallo-β-lactamase gene, *bla*NDM-1, and a novel erythromycin esterase gene carried on a unique genetic structure in *Klebsiella pneumoniae* sequence type 14 from India. *Antimicrob. Agents Chemother.*, 53, 5046–5054.

Yoo, J.S., Kim, H.M., Koo, H.S., Yang, J.W., Yoo, J.I., Kim, H.S., Park, H.K., and Lee, Y.S. 2013. Nosocomial transmission of NDM-1-producing *Escherichia coli* ST101 in a Korean hospital. *J. Antimicrob. Chemother.*, 68, 2170–2172.

Yoshizumi, A., Ishii, Y., Livermore, D.M., Woodford, N., Kimura, S., Saga, T., Harada, S., Yamaguchi, K., and Tateda, K. 2013. Efficacies of calcium-EDTA in combination with imipenem in a murine model of sepsis caused by *Escherichia coli* with NDM-1 β-lactamase. *J. Infect. Chemother.*, 19, 992–995.

Zarfel, G., Hoenigl, M., Wurstl, B., Leitner, E., Salzer, H.J., Valentin, T., Posch, J., Krause, R., and Grisold, A.J. 2011. Emergence of carbapenem-resistant Enterobacteriaceae in Austria, 2001–2010. *Clin. Microbiol. Infec.*, 17, E5–E8.

Zhang, C., Qiu, S., Wang, Y., Qi, L., Hao, R., Liu, X., Shi, Y., Hu, X., An, D., Li, Z., Li, P., Wang, L., Cui, J., Wang, P.,

Huang, L., Klena, J.D., and Song, H. 2013. Higher isolation of NDM-1 producing *Acinetobacter baumannii* from the sewage of the hospitals in Beijing. *PloS One*, 8, e64857.

Zhou, Z., Guan, R., Yang, Y., Chen, L., Fu, J., Deng, Q., Xie, Y., Huang, Y., Wang, J., Wang, D., Liao, C., Gong, S., and Xia, H. 2012. Identification of New Delhi metallo-β-lactamase gene (NDM-1) from a clinical isolate of *Acinetobacter junii* in China. *Can. J. Microbiol.*, 58, 112–115.

Zhu, K., Lu, J., Ye, F., Jin, L., Kong, X., Liang, Z., Chen, Y., Yu, K., Jiang, H., Li, J.Q., and Luo, C. 2013. Structure-based computational study of the hydrolysis of New Delhi metallo-β-lactamase-1. *Biochem. Biophys. Res. Commun.*, 431, 2–7.

Zowawi, H.M., Balkhy, H.H., Walsh, T.R., and Paterson, D.L. 2013. β-Lactamase production in key Gram-negative pathogen isolates from the Arabian peninsula. *Clin. Microbiol. Rev.*, 26, 361–380.

Chapter 49

Pathogenesis of Cryptococcosis in Humans

Ricardo Negroni and Alicia I. Arechavala

Mycology Unit, Francisco J. Muniz Hospital, Buenos Aires City, Argentine Republic

49.1 Introduction

The genus *Cryptococcus* includes about 100 species, of which only *Cryptococcus neoformans* and *C. gattii* can be considered pathogenic for human beings (Baddley and Dismukes, 2003; Viviani and Tortorano, 2009). However, occasional infections due to different species of the genus have been reported, such as *C. albidus*, *C. laurentii*, *C. uniguttulatus*, *C. humicola*, *C. curvatus*, *C. luteolus*, *C. adeliensis*, *C. magnus*, and *C. chernovi* (Fonseca et al., 2011; Heilman et al., 2010; Khan et al., 2011; Rimek et al., 2004). *C. neoformans* is found in two varieties: *C. neoformans* var. *grubii*, which corresponds to serotype A; *C. neoformans* var. *neoformans*, which is serotype D; and a hybrid between both varieties, which is serotype AD (Fonseca et al., 2011; Heilman et al., 2010). *C. gattii* includes serotypes B and C and can also have interspecies hybrids with *C. neoformans* var. *neoformans* (serotype BD) and *C. neoformans* var. *grubii* (serotypes BA), both recently described (Aminnejad et al., 2012; Bovers et al., 2008).

They are encapsulated yeasts belonging to the division Basidiomycota. Teleomorphic (sexual) forms correspond to *Filobasidiella neoformans* for *Cryptococcus neoformans* and to *Filobasidiella bacillispora* for *Cryptococcus gattii* (Chayakulkeeree and Perfect, 2006; Fonseca et al., 2011). They are environmental yeasts which sexual activity is heterothallic; strains of a mating type conjugate with those of a mating type, with both strains occurring naturally. Nevertheless, more than 95% of human infections are produced by α mating type (Meyer et al., 2003).

Since the 1980s a significant increase in the incidence of cryptococcosis has taken place. That was due to the outbreak of the HIV pandemic. More than 90% of the infections associated with HIV/AIDS are caused by *C. neoformans* var. *grubii* (Negroni, 2012).

Now, the incidence of AIDS-associated cryptococcosis has decreased in industrialized countries owing to anti-retroviral treatment, but it continues to be a major problem in developing countries (Negroni, 2008). About one million new cases of this mycosis occur yearly, and more than 650,000 of them die (Park et al., 2009). *C. neoformans* var. *neoformans* causes infections in the South of Europe and *C. gattii* can cause invasive mycosis in both immunocompetent and immunocompromised individuals. Initially it was thought to be restricted to tropical or subtropical areas only, but the Vancouver epidemic, which broke out in the 1990s, and its further extension to the north-west of the United States showed a dramatic change in the habitat of this species (Mac Dougall et al., 2011; Perfect and Casadevall, 2002).

The sources of infection for *C. neoformans* are pigeons' and other birds' droppings, as well as barks of several species of trees (Bartlett et al., 2012; Grover et al., 2007; Kwon-Chung and Bennett, 1992; Viviani and Tortorano, 2009). *C. gattii* has been isolated from the barks, leaves, and fruits of more than 30 tree species (Grover et al., 2007; Negroni, 2012).

The infection is produced from the spores that these fungi produce in the environment. Two types can be found: those produced as teleomorphic forms, called basidiospores, and those generated by anamorphic yeasts (asexual state), known as blastoconidiae. Studies in experimental mice models have shown that both types of spores are able to produce infection by inhalation (Casadevall and Perfect, 1998). Basidiospores are more resistant to oxidative stress, high temperatures, chemical aggressions, and desiccation. They produce sprays in the environment. These spores have a thick cover, which is different from the cell wall of the yeasts. When inoculated intranasally to mice, basidiospores reach the pulmonary alveoli

Human Emerging and Re-emerging Infections: Bacterial & Mycotic Infections, Volume II, First Edition. Edited by Sunit K. Singh.
© 2016 John Wiley & Sons, Inc. Published 2016 by John Wiley & Sons, Inc.

and cause an infection, which can later spread out to other organs, especially the central nervous system (CNS) through the bloodstream (hematogenous dissemination). Basidiospores are more effective as infective propagules, as only 500 spores can generate a fatal infection in the mouse, which amounts to 10 times less than the number of yeasts needed to produce the same effect (Botts and Hull, 2010; Casadevall and Perfect, 1998; Voelz and May, 2010).

The main portal of entry for the infection is the respiratory tract by inhalation of the spores, as pointed out above. The spores are lodged in the lung, where they produce a focal pneumonitis, which progresses by the lymphatic route to the pulmonary hilum and mediastinum lymph nodes (Casadevall and Perfect, 1998; Littman and Zimmerman, 1956). Most of the infections are self-limited and, as a result, pulmonary or nodular granuloma remains, which contains scarce viable yeasts inside the macrophages (MØ; Casadevall and Perfect, 1998; Chayakulkeeree and Perfect, 2006; Negroni, 2012). This status is known as latent infection. It is asymptomatic and can only be diagnosed at necropsy or upon radiology screenings. Indeed, necropsies of individuals who died from other causes have shown granulomas in small subpleural nodules, with encapsulated yeasts inside. In most of the granulomas the diameter was about 1.5 cm. Many of them are associated with hilar adenopathies, similarly to what is observed in the tuberculosis primary complex (Kwon-Chung and Bennett, 1992; Littman and Zimmerman, 1956). These latent lesions can become the starting point for the reactivation and subsequent spreading of the infection to other organs (Perfect and Casadevall, 2002; Voelz and May, 2010).

There are other signs indicative of asymptomatic infection by *Cryptococcus* spp., namely the presence of nodules of a bigger diameter, called cryptococcomas, which are also asymptomatic; the confirmation of positive delayed hypersensitivity skin tests with *C. neoformans* antigens in people without a previous or current history of cryptococcosis; and the detection of antibodies anti-*C. neoformans* capsular antigens in these subjects (Casadevall and Perfect, 1998; Negroni, 2012).

The skin tests were performed with an antigen obtained by filtration of *C. neoformans* cultures in a liquid medium; this antigen has a high concentration of capsular polysaccharides. Also, the supernatant of broken hypocapsulated yeasts had been used. These yeast cells were broken by ultrasound or other mechanical procedures. This reactive is rich in mannoproteins (MPs). Both have shown to have low sensitivity to delayed hypersensitivity skin reactions and besides they are barely specific as they show crossed reactions with other fungal antigens (Muchmore et al., 1968, 1970). However, studies performed using these tests have made it possible to prove that these reactions turn out to be positive more often in pigeon fanciers and in laboratory workers who handle cultures of *Cryptococcus* spp. Yet, these

people do not show a higher incidence of progressive cryptococcosis when compared with the remaining population. Antigens rich in MPs yielded the best skin tests results. Some patients with disseminated cryptococcosis who showed a negative skin test to cryptococcin before treatment became positive after a successful antifungal medication (Casadevall and Perfect, 1998). Antibodies anti-glucuronoxylomannan (main polysaccharide in the capsule) and the MPs have also been detected in subjects without a previous history of cryptococcosis (Champa Saha et al., 2008; Voelz and May, 2010).

Transcutaneous cryptococcal infection is also possible in human beings, but it is considered uncommon. To be recognized as such there must be a single lesion or a chancriform lymphocutaneous syndrome located in exposed areas of the body alongside a history of trauma or injuries exposed to an environmental source of infection (Negroni, 2012; Wilson and Plunkett, 1965). Patients are most likely elderly immunocompetent adults. Most of the skin lesions observed in cryptococcosis are produced by hematogenous dissemination from a pulmonary focus, in the course of a disseminated cryptococcosis. Lesions are multiple and vary in clinical aspect, ranging from ulcerated papules or tubercles to molluscoid papules, subcutaneous nodules, gummas, verrucous skin plaques, etc. (Chayakulkeeree and Perfect, 2006).

The digestive primary infection is not common, although it has not been totally ruled out. It could occur by food intake, especially contaminated fruits (Baddley and Dismukes, 2003; Casadevall and Perfect, 1998).

The course of the infection depends on the infective load, on the virulence of the strain, and on the performance of the defensive host barriers, as well as its innate and adaptive immunity.

49.2 Growth temperature

C. neoformans and *C. gattii* are the only species of the genus capable of developing at 37°C and even at higher temperatures, with the essential participation of the mitochondrial enzyme superoxide-dismutase (SOD_2p) (Chaturvedi and Chaturvedi, 2011). *C. neoformans* is more tolerant to temperature than *C. gattii*, the latter not growing at temperatures higher than 37°C. This property is controlled by genes and it is regulated by two signals in the transduction pathways, namely the RAS cascade and the one dependent on calcineurin, both being crucial to make possible the growth of these yeasts at high temperatures (Romani, 2011). As the development of these fungi is stopped at temperatures higher than 38°C, fever could be an important defense mechanism.

49.3 Virulence factors

The virulence of a microorganism is its ability to resist the natural defensive barriers of the host and to cause infection (Casadevall and Perfect, 1998). Mycoses are an accidental fact in the life cycle of the environmental pathogenic fungi (Sabiiti and May, 2012). Virulence is controlled genetically and different strains of the same fungal species differ in terms of its infective power. The *C. neoformans–C. gattii* complex has the ability to cause infections both in immunocompetent and immunosuppressed individuals and it is pathogenic to various animal species (Castellá et al., 2008; Voelz and May, 2010). Experimental cryptococcosis in animal models has been developed in mice, rats, and rabbits to allow for the assessment of the virulence factors, pathogenesis, immunity, and histopathology of this mycosis.

49.3.1 Capsule

The capsule is a dynamic structure, rich in polysaccharides, and it is a virulence factor of great importance to *C. gattii* and *C. neoformans*. It is made up of polymers of $1 \rightarrow 3$ αD-mannopyranan with ramifications of xylose and glucuronic acid monosaccharides. The formation of the capsule is genetically controlled; several genes promoting the production of the capsule have been identified: CAP10 CAP59, CAP60, and CAP64; they are located among the genes of the ribosomes and the mitochondria (Obayashi et al., 2006).

The main component of the capsule is glucuronoxylomannan (GXM) and it interferes with the immune response in many different ways. Acapsulated or hypocapsulated mutants have low virulence and are more easily phagocytozed by neutrophils and MØ (Casadevall and Perfect, 1998; Chayakulkeeree and Perfect, 2006).

The distribution of GXM results in changes in the density of the capsule, which is not dependent on the composition of the sugars. In studies with monoclonal antibodies, fluctuations in the affinity for binding by antibodies were observed, but not in their binding site, which suggests that the spatial epitope organization (high and low affinity) changes according to its radial position (Maxson et al., 2007).

GXM prevents the appropriate opsonization of the yeasts by means of a masking mechanism, acting as a barrier between opsonins and its receptors, particularly complement receptors (CRs). Furthermore, it contributes to the activation of the complement by an alternate pathway, leading to the deposit of C3b in the capsule. In addition, the capsule delivers GXM, which forms immunocomplexes with the antibodies and the complement before they opsonize the yeasts (Voelz and May, 2010). The capsule protects the yeasts from the oxidative and nitric oxide bursts, thus enabling yeasts to survive in the MØ phagosomes. During the infection the capsule undergoes a phenotypic switching, colonies turn mucoid and thus survive longer inside the MØ than the original smooth strains (Heilman et al., 2010; Jain and Fries, 2008). This phenomenon turns *C. gattii* and *C. neoformans* into intracellular microorganisms with the ability to

survive and reproduce inside the MØ and the giant cells, similarly to *Mycobacterium tuberculosis*. As we will see below, this characteristic allows them to leave the lung without causing any damage hereto and to penetrate other tissues through the bloodstream, protected by the MØ (Trojan horse model) (Casadevall and Perfect, 1998; Voelz and May, 2010).

GXM depresses the activity of CD4+ (helper) T cells and enhances the function of CD8+ (cytotoxic) T cells. An excess of capsular antigen prevents the activation of the MØ by pro-inflammatory cytokines such as IL-2, IL-12, and INF-γ and the formation of compact epithelioid granuloma, which can control the infection efficaciously (Perfect and Casadevall, 2002; Uicker et al., 2005).

The ability of GXM to induce the apoptosis of lymphocytes and MØ has been shown in normal rats; there is an increase in nitric oxide, which induces apoptosis of various cellular types, and this would be one of the mechanisms of immunosuppression occurring during infection by *Cryptococcus* (Chiapello et al., 2003, 2008).

Another capsular component is MP, which is the immunodominant antigen inducing responses of cell-mediated immunity and the proliferation of mononuclear cells (Casadevall and Perfect, 1998; Voelz and May, 2010).

49.3.2 Phenotypic switching

Phenotypic switching is defined as a reversible change undergone by a colony that shows morphological alterations to a greater extent than would occur in a somatic mutation. Microorganisms living in a fluctuating environment must change in order to survive in the host and to manage to be a successful pathogen causing permanent disease.

Cryptococcus causes changes in the capsule and thus triggers various inflammatory responses according to the phenotypic variant, produces an altered immune response in the production of antibodies, and influences the response to different antifungal agents, according to studies performed in mice. This switching is one of the processes contributing to the microevolutionary changes. Smooth variants are predominant in serotypes A and D and account for 96% of environmental *C. neoformans* var. *grubii* isolates, while 100% of *C. gattii* occurs as a mucoid variant. The yeasts of a smooth *C. neoformans* variant undergo a one-way transformation to the mucoid variant as they get into the host, while cells of *C. gattii* will access the body as a mucoid variant and remain that way as they go past the lung, but must become smooth to be able to go through the blood–brain barrier and then reverse back to the mucoid variant (Jain and Fries, 2008).

The strain RC2 (*C. neoformans* serotype D) and the strain NP1 (*C. gattii* VGI) have been used to study the phenotypic changes in both species. Switching produces changes in the composition of GXM and the biophysical and biochemical properties are modified, so that the mucoid variants have a longer survival inside the MØ and also avoid the antibodies or complement-mediated phagocytosis (Jain and Fries, 2008). The strain NP1-SM (smooth variant) resulted in an appropriate inflammatory response in the lung with the presence of MØ and lymphocytes, while mice infected with the mucoid variant NP1-MC were not able to cause inflammation and the accumulation of yeasts immersed in a great amount of polysaccharide (cryptococcomas) was observed (Casadevall and Perfect, 1998; Jain and Fries, 2008).

Cryptococcus is also able to form biofilms consisting of a complex array of yeasts embedded in a polysaccharide matrix. These biofilms are stronger when the parental cells belong to the smooth variant and adhere better to polystyrene than the biofilms produced by the mucoid or rough variants. The cells of the biofilm have a higher switching rate. In chronic infection, the formation of biofilms and the phenotypic switching increase, so that the smooth variants have a stronger adherence and are more efficient upon the invasion and they will then change their phenotype to the mucoid variant, which allows for their survival (Martinez et al., 2008).

In studies on the virulence factors, it was observed that the capsule of *C. gattii* is of a greater size, and also that the phenotypic switching occurs more often (Sánchez et al., 2008).

49.3.3 Melanin

This pigment is related to the virulence of yeasts. Only the complex *C. neoformans*–*C. gattii* has an enzymatic system that is able to metabolize precursors as catechol to turn it into melanin from different substances such as dopa, dopamine, norepinephrine, and epinephrine (Casadevall and Perfect, 1998).

The complex *C. neoformans*–*C. gattii* does not have tyrosinase, so it is not able to produce endogenous dihydroxyphenols, but it can oxidize the hydroxyphenol by means of the phenoloxidase following the Mason–Roper model (Casadevall and Perfect, 1998; Chayakulkeeree and Perfect, 2006). Melanin is concentrated in the internal side of the yeasts cell wall. The enzymes leading to its production are called laccases. These enzymes belong to the Cu-containing oxidases that transform catecholamines into quinones, which will then polymerize by self-oxidation to form melanin and similar pigments. Melanin protects the fungal cell from oxidative stress, the action of nitric oxide, and the antimicrobial peptides. It also participates in the maintenance of the cell wall integrity and together with the capsule, is able to alter the electric charge of the wall, so that the yeast ends up having a negative charge. It alters the immune mechanism because it interferes with the response of the T cells, it precludes antibody-mediated phagocytosis, and it protects the cell from extreme temperatures; furthermore, the presence of melanin would be accountable

for decreased sensitivity to some antifungal agents such as amphotericin B and caspofungin (VanDuin et al., 2002; Voelz and May, 2010).

The production of melanin is controlled by the genes $CnLAC_1$ and $CnLAC_2$; there are also environmental signals regulating the activity of the laccase such as nutrient deprivation, multivalent cations, and the stress due to temperature; in addition, it is mediated by multiple signals of transduction patterns. In its environmental niche, melanin protects *Cryptococcus* from ultraviolet radiation. The protection provided by melanin to *Cryptococcus* inside the MØ is also supplied by glutathione, mannitol, and sphingolipids (Casadevall and Perfect, 1998; Heilman et al., 2010; Ikeda et al., 2002).

One of the endogenous substrates of the laccase is 3-hydroxyantranilate, which allows for iron capture and promotes the development of the yeast in culture media lacking Fe^{3+} and protects *Cryptococcus* from the strong oxidants (Casadevall and Perfect, 1998; Romani, 2011).

The protein kinase A G-α-protein-c-AMP (PKA_1 in *C. neoformans* and PKA_2 in *C. gattii*) has an important role in virulence, as it regulates the nutrients during mating, the capsule and melanin production (Chaturvedi and Chaturvedi, 2011).

49.3.4 Urease

The urease is an important promoter of brain invasion by *C. neoformans*/*C. gattii* in murine cryptococcosis models. The strains producing urease cause an infection with a fungal load 100 times higher than the strains lacking urease; they are associated with a type Th_2 cytokines response producing lung eosinophilia, elevated IgG levels, higher concentrations of IL-4 and IL-13, with decreased concentrations of IL-2 and INF-γ and deficient activation of MØ with low levels of IL-6 in these cells. Urease also promotes the generation of immature dendritic cells (DCs), both in the lungs and in lymph nodes of the pulmonary hilum. This results in an inappropriate presentation of the *Cryptococcus* antigens.

This enzyme plays a very important role in the damage generated in the vascular endothelium at the beginning of the brain invasion. In this phase, urease acts together with the VEGF (vascular endothelial growth factor), in damaging blood capillaries in the brain, which leads to an increase in filtration in the blood–brain barrier; it promotes the formation of a compact group of yeasts both inside and outside the MØ in the lumen of the blood vessels and the increase of the yeast adherence to the vascular endothelium, as well as the translocation of *Cryptococcus* to the brain parenchyma. In the cryptococcal meningoencephalitis capillaritis occurs, then encephalitis is the first step in the CNS infection and the second step is the invasion of the subarachnoid space.

In the brain, the strains with urease give rise to the formation of pseudo-cystic lesions in the cerebral cortex with a high amount of encapsulated yeasts and scarce inflammatory response. Histopathological studies have shown microemboli inside blood capillaries, endothelial cells proliferation and capillary wall damage (Olszewski et al., 2004; Osterholzer et al., 2009). This alteration is probably generated by the ammonia produced by this enzyme, which could promote the toxicity of endothelial cells and astrocytes, leading to opening of the union among those cells and impairing the integrity of the blood–brain barrier, and then affects the adherence of *Cryptococcus* to the endothelial cells by means of a mechanical effect or by induction of adhesion molecules (Osterholzer et al., 2009).

Urease and laccases are essential to the survival of *C. neoformans/C. gattii* in the brain and act synergistically (Liu et al., 2013). Urease increases the survival of *Cryptococcus* inside the phagosomes by inhibiting the action of nitric oxide and altering the response of the adaptive cell-mediated immunity (Olszewski et al., 2004).

C. neoformans strains have more urease activity than *C. gattii* strains, according to studies in which that activity was measured (Torres Rodríguez et al., 2008).

49.3.5 Other metabolites that increase virulence

Iron is required by many pathogenic microorganisms; the *C. neoformans–C. gattii* complex uses it to form the capsule and to produce melanin. These fungi obtain iron (Fe) through several mechanisms such as permeases, ferroxidases, siderophore transporters, and a system of low-affinity uptake-agents. They can acquire iron from the transferrin by means of high-affinity uptake-agents and also use the iron of heme groups.

Copper is essential for laccase production and its transport is encoded by the CTR2 gene. A high copper concentration was observed in the MØ with yeasts inside, especially in the CNS.

Besides urease, other hydrolytic enzymes such as proteases and phospholipases damage tissues and vascular endothelia, thus contributing to the dissemination of these microorganisms (Sabiiti and May, 2012; Vidotto et al., 2006).

Phospholipases hydrolyze esters of glycerolphospholipids. They produce destabilization of membranes, cellular lysis, and the release of secondary lipid messengers that promote the invasion of pulmonary interstice, its dissemination through the lymphatic vessels, and finally, through the bloodstream (Sabiiti and May, 2012). Phospholipase B produces dipalmytoil phosphatidylcholine, which is an important component of the pulmonary surfactant substance, which increases the adhesion of *Cryptococcus* to the lung MØ and to the alveolar epithelial cells, thereby promoting dissemination.

Phospholipase B1 takes arachidonic acid from MØ and generates compounds that suppress the immune response and promote survival of *Cryptococcus* inside the MØ (Sabiiti and May, 2012).

Mannitol is produced by *Cryptococcus* yeasts; it acts as an antioxidant contributing to the survival of fungal cells, and it increases the osmotic pressure in the subarachnoid space, which triggers intracranial hypertension.

Inositol participates in the formation of lipid membranes and in the mechanisms of signals transduction; it is an important intracellular osmotic substance and plays a prominent role in the development and growth of eukaryotic cells. It regulates the biosynthesis of phospholipids. The CNS is rich in inositol (Liu et al., 2013). *Cryptococcus* is capable of assimilating inositol as the only source of carbon; although it is not a preferred nutrient, it also produces inositol from cell glucose. Inositol phosphorylceramide synthase, an enzyme of the fungal sphingolipid cycle, is a *Cryptococcus* virulence factor and regulates the integrity of its cell wall (Ishibashi et al., 2012). These fungi have developed a sophisticated mechanism to obtain inositol including a chain of over 10 genes for the transport of this substance. As outlined above, several factors are involved in the passage of *Cryptococcus* through the blood–brain barrier, including urease, phospholipase B1, and plasmin. Recently this microorganism has shown to interact with the lipids of the endothelial cells in the brain microvessels, thus promoting the invasion by a glycoprotein CD44-dependent procedure. It has been shown that hyaluronic acid produced by *Cryptococcus* acts as a ligand with the glycoprotein CD44 during the interactions of this fungus with its host (Khan, 2009). Inositol, one of the most important metabolites of the CNS, is located mainly in the glia, particularly in the astrocytes, and acts as an osmotic substance. Astrocytes have a close relationship with brain microvessels and inositol acts as a facilitator of the passage of *Cryptococcus* yeasts through the cytoplasm of the vascular endothelium cells, which allows for its exocytosis and thus promotes the invasion of the brain parenchyma. The crossing through the blood–brain barrier is controlled by the genes of the transporters Itr1a and Itr3c. In addition, inositol is important in the regulation of the production of phospholipids by *Cryptococcus*, which play a key role in the transmigration of this microorganism through the blood–brain barrier (Sabiiti and May, 2012).

C. gattii has other virulence factors in addition to the ones mentioned above. It produces metabolites such as acetoin and dihydroxiacetoin, which promote pro-inflammatory responses to a lesser extent and contribute to its survival and reproduction in the host. The superoxide dismutase (SOD_1p) and the trehalose have an antioxidant action. Genes TPS1p y TPS2p that control the biosynthesis of trehalose-6-phosphate synthetase are essential to maintain the integrity of the capsule and the melanin of *C. gattii*, as well as thermotolerance of both species (Chaturvedi and Chaturvedi, 2011).

49.4 Ways of signaling

The signaling cascade of the mitogen-activated protein kinase (MAPK) controls the pheromones during mating and regulates haploid fructification and virulence. The cascade Pbs2-Hog1-MAPK controls the morphological differentiation and the virulence factors in *C. neoformans* serotype A strains but not in serotype D strains, which indicates that the role of the signaling cascades is serotype-specific.

The sphingolipid pathway represents a reservoir with a high content of signaling molecules. The reactions regulated by inositol phosphorylceramide impact on several virulence levels. Glucosylceramide (GlCer) is not produced by mammalians and is fungus-specific. It plays a crucial role in the survival of *Cryptococcus* in the extracellular medium, such as the case of the alveolar space, since it is necessary for the yeast to cause the disease once the fungus has entered intranasally. Mice infected with mutant strains (which do not produce GlCer) form granulomas and do not die because of the infection. Furthermore, GlCer is necessary for *Cryptococcus* to leave the lung and disseminate through the blood; it takes part in budding and cellular division, and plays an important role in the tolerance to alkali, in the elongation of hyphae, and in the susceptibility to defensins of plants. The host produces antibodies against GlCer. The dysfunction in glycol cerebrosidase, which participates in its catabolism, triggers the accumulation of GlCer and some precursors, and leads to an altered capsule formation (Ishibashi et al., 2012; Qureshi et al., 2011; Rodrigues et al., 2000).

49.5 Defenses of the host against *Cryptococcus* spp.

49.5.1 Physical barriers and innate immunity

The skin is an efficient protective barrier; transcutaneous infection by *Cryptococcus* is uncommon and gives rise to localized lesions or a lymphangitic-nodular syndrome (Casadevall and Perfect, 1998; Wilson and Plunkett, 1965).

Rarely is the digestive mucosa the entrance door for the human infection. Although it has not been totally ruled out as a possibility, it would only occur by the ingestion of contaminated fruits. The saliva stops the development of these encapsulated yeasts. *C. neoformans* has been cultured from oral mucosa swabs in 6% of the patients with cancer without a current or previous history of cryptococcosis. Therefore, the digestive mucosa also seems to be an efficient barrier against this mycosis (Casadevall and Perfect, 1998).

The upper airways, especially the nasal mucosa, can be an important portal of entry for the infection in felines. In human beings this seems to be uncommon; yet, *C. neoformans/C. gattii* can adhere to the nasal epithelium and invade it, which fact, added to the closeness of the nose to the base of the brain, suggests that it can be a way of penetration into the CNS.

The most common portal of entry of cryptococcosis is the lung. Bronchi retain the particles greater than 5 μm in diameter that are suspended in the air, thanks to the air flows in their lumen and the action of the cilia. Particles 0.2–5 μm in diameter easily reach the pulmonary alveoli, as it happens with *Cryptococcus* basidiospores and less frequently with desiccated yeasts. The first defensive barrier is the alveolar MØ which spores adhere to, possibly because of the glycosphingolipid lactosylceramide, which acts as a receptor (Chayakulkeeree and Perfect, 2006; Sabiiti and May, 2012).

49.5.2 Cellular defenses

Cellular defense mechanisms participate both in innate immunity and in adaptive cell-mediated immunity and they are the most important elements in controlling this mycosis.

Phagocytosis is triggered directly by *Cryptococcus* antigens or by their recognition by the cellular receptors, the complement, or the antibodies (Voelz and May, 2010). The capsule is recognized by the "toll-like" receptors TLR2 and TLR4, while mannose receptors of DCs recognize the MP of the *Cryptococcus* cell wall. The cells of these microorganisms opsonized by the serum are recognized by complement receptors CR_1 (CD38) or by the heterodimer β-interleukin CR_3 (CD11/CD18) or by CR_4 (CD11/CD35). The yeasts opsonized by the antibodies are recognized by Fcγ receptors (FcγRs) that are expressed in the surface of MØ, DCs, and neutrophils (García Rodas and Zaragoza, 2011).

49.5.3 Dendritic cells

Antigens are mainly presented by DCs. They reduce the concentration of antigens in the medium and modulate adaptive immunity. DCs present the cryptococcal antigens to CD4+ T cells, together with the major histocompatibility complex type II (MHC II), which thus stimulates the beginning of the adaptive immune response (Voelz and May, 2010).

49.5.4 Polymorphonuclear neutrophils (PMNs)

They are the first cells to arrive to the infection focus and they have an important microbicidal activity based on two mechanisms, mainly the oxidative burst and antimicrobial peptides and proteins called defensins. In cryptococcosis they are partially efficacious; they reduce the fungal burden in the initial phases of the infection and then they are replaced by the MØ (Casadevall and Perfect, 1998).

49.5.5 Natural killer cells (NK cells)

NK cells lyse microorganisms without phagocytozing them by means of the proteins called perforins and granulosins, which, when released from NK cells, increase the permeability of the yeasts and lead to their lysis. They are stimulated by INF-γ (Casadevall and Perfect, 1998; Voelz and May, 2010).

49.5.6 Macrophages

They are the most important effector cells of the immunity against *C. neoformans/C. gattii.* They produce the phagocytosis of the yeasts; they generate cytokines and chemokines, and are able to present antigens. Phagocytosis depends on the opsonins. The capsule of *Cryptococcus* has antiphagocytic properties. A great proportion of yeasts is inside the MØ during cryptococcosis (García Rodas and Zaragoza, 2011). Although the MØ and the PMN are able to kill *Cryptococcus*, these microorganisms have developed several strategies to evade the lethal action of these cells. However, in experimental murine cryptococcosis, the mice that survive the infection are those that do not increase a lot the amount of MØ, as if an excessive replication of these cells has an effect deemed adverse to the host defenses (McQuiston and Williamson, 2010).

The phagocytosis of *C. neoformans/C. gattii* by MØ results in three phenomena of great importance in the evolution of cryptococcosis: (1) it decreases the fungal load in the host; (2) it has an alternative dissemination mechanism, as the pathogen travels inside the MØ and thus protects itself from other defensive mechanisms in the host (Trojan horse model); and (3) it allows for the existence of a latent infection, which can be reactivated and causes severe and disseminated disease (García Rodas and Zaragoza, 2011).

C. neoformans/C. gattii can avoid the internalization in the MØ through mechanisms either dependent on or independent of the capsule. Upon binding to the mannose receptor, CRs, and others, the epitopes of the *Cryptococcus* cell wall induce phagocytosis, whereas the capsule of this fungus makes it difficult and acts as a barrier that prevents any contact and the recognition of the cell wall antigens by the above-mentioned receptors (Voelz and May, 2010).

In addition, the capsular polysaccharides bind to TLR2, TLR4, and CD14. This union inhibits the production of the tumoral necrosis factor alpha (TNF-α) and produces the deficient activation of the MØ, while also increasing the expression of the Fas ligands that induce the apoptosis of T cells (Heilman et al., 2010; Monari et al., 2008).

There are anti-phagocytic mechanisms that are independent of the capsule, such as the APP_1 (anti-phagocytic protein$_1$) which expression is stimulated by the diacylglycerol signaling pathway and which inhibits the CR-mediated phagocytosis. This protein has been found in the serum of patients with active cryptococcosis. Another anti-phagocytic component is the protein Gat 201, which acts as a regulator of the virulence of murine cryptococcosis models (García Rodas and Zaragoza, 2011).

The serum opsonins are complements and antibodies. The latter bind their Fc fraction to the Fc receptor (FcR) of the MØ.

The complement can be activated through the alternate pathway by the *Cryptococcus* capsule; C_3b is deposited on the capsule and the activation of the phagocytosis mediated by the CRs, especially CR_3, occurs. This form of opsonization is not efficacious for the yeasts with a big capsule, because the union with the complement is produced in the innermost part of the capsule, which ends up being very distant (Casadevall and Perfect, 1998).

The epitopes of *Cryptococcus* capsular polysaccharides can eventually be exposed during the infection and are recognized by CR_3 and specifically by CD18. The yeasts cell wall glucans can also bind to these receptors in the MØ, PMN, and NK. This union increases the cytotoxicity of the targets containing ligands iC_3b on their surface. The Fab fragments of the antibodies induce phagocytosis through the union of the capsule with CR_3.

Enlargement of the capsule leads to a status defined as non-immune response. Its main component, GXM, interacts with TLR2, TLR4, and CD14 by depressing the activation of the MØ and DC. Mannose receptors play a very important role in the uptake of *Cryptococcus* yeasts by the MØ (Charlier et al., 2005; Romani, 2011).

Once inside the MØ the yeasts migrate to the phagosomes; the latter join the lysosomes and give rise to the phagolysosomes, which decrease the pH and produce hydrolytic enzymes that are capable of killing many microorganisms and of digesting the materials of these phagocyted pathogens. However, *C. neoformans* and *C. gattii*, as well as other intracellular pathogenic agents, have mechanisms that attenuate the damage done by the hostile medium of the phagolysosomes and they are able to multiply at acidic pH. The yeasts of these fungi reproduce more rapidly in the interior of the phagolysosome than in the extracellular medium. If the pH of the phagolysosome increases, the enzyme inositol phosphoryl-ceramidase stops acting and the reproduction speed of the yeasts decreases. The phagolysosomes of the MØ that have phagocytized yeasts increase their size and the permeability of their membranes due to the appearance of holes; there is also a reduction of the microbicidal capacity and the action of the nitric oxide. These phagolysosomal membranes turn filtrating and allow the exit of the capsular polysaccharide and its deposit in the MØ cytoplasm, where it binds to glycolytic substances, forms the phosphofructokinase, and interferes in the normal metabolism of the MØ (Charlier et al., 2009; García Rodas and Zaragoza, 2011).

The increase in the size of the capsule, which occurs when the yeasts access the MØ, facilitates their survival inside the cells because of the antioxidant properties of the capsule, and its greater resistance to antimicrobial peptides and to free radicals. On the other hand, when the size and the permeability of the membrane increase, the phagolysosomes dilute their contents and a physical separation between fungal cells and the phagolysosome membranes occurs (García Rodas and Zaragoza, 2011; Voelz and May, 2010).

As pointed out before, melanin protects the yeasts from the action of the oxidative burst, nitric oxide, and antimicrobial peptides.

Other substances that facilitate the survival of the *Cryptococcus* inside the MØ are glutathione, mannitol, and sphingolipids.

Cryptococcus sp. are not only able to survive and reproduce themselves inside the MØ, but they can also exit from the MØ without causing any damage to them, the transfer between MØ, and the fusion of MØ to result in giant cells (García Rodas and Zaragoza, 2011; Sabiiti and May, 2012).

Through phospholipase B the yeasts can leave the MØ, by means of a process called "extrusion", which is a non-lytic exocytosis and which does not cause any damage, neither to the MØ nor to the yeast. This process is regulated by the action of the actin that is polymerized around the phagolysosomes that contain the yeasts and delay its expulsion. Their expulsion ultimately occurs in groups when they have been opsonized by antibodies or otherwise individually. The fungal cells can also be transferred from an MØ to another; this can only be carried out by living yeasts and this action, called lateral transference, is mediated by actin (McQuiston and Williamson, 2010; Romani, 2011).

The ability to survive inside the MØ means a protection to *Cryptococcus*, because it isolates them from the anti-fungal substances of the serum, the antibodies, and the circulating phagocytes while it favors the dissemination to other organs.

49.6 Humoral immunity

Blood serum contains an antifungal non-dialyzable substance, digestible by the trypsin, which is linked neither to specific antibodies, nor to the complement; it takes part in iron metabolism and it is linked to α 2-globulines and γ 2-globulines, probably similar to that acting as fungistatic against *Candida* spp. and dermatophytes (Casadevall and Perfect, 1998).

The complement is a set of proteins acting as a non-specific defense system against microorganisms by promoting phagocytosis through opsonins and production of chemotactic substances that attract inflammatory cells to the infection focus. The *C. neoformans*–*C. gattii* complex activates the complement both by the alternate and the classic pathways. The former occurs by the adherence of C_3 to the peripheral zones of the capsule and the latter is mediated by anti-glucan antibodies and is generated by the action of cell wall antigens. The complement is detected early in the pulmonary alveoli in the nasal infection in mice and in the yeasts capsule. Within a few hours it

produces a ring of neutrophils around the yeasts, which are later replaced by MØ. The absence of a complement reduces the survival of mice experimentally infected by *C. neoformans* and complement levels are decreased in disseminated human cryptococcosis due to an excessive intake (Voelz and May, 2010). The complement levels are low in the CNS, which partially accounts for the tropism of this infection for the brain and meninges.

The activation of the complement by the alternate pathway is the most important one, because it facilitates the opsonization in an antibodies-free medium and allows the penetration of yeasts inside the phagocytes (Casadevall and Perfect, 1998; Voelz and May, 2010). The fractions C_5a and C_3a are chemotactic for inflammatory cells and C_5b initiates the formation of attack membrane complexes with the activation of the remaining fractions (C_6, C_7, C_8, and C_9) (Romani, 2011).

49.6.1 Antibodies

The role of the antibodies in cryptococcosis is not fully understood. There is evidence that the anti-capsule antibodies are protective in murine models of experimental cryptococcosis. However, there are protective and non-protective antibodies, and antibodies that increase the pathogenicity of the yeasts. The protection mediated by antibodies depends on T cells, which makes it difficult to evaluate its function in patients with cell-mediated immunity compromise. The importance of the antibodies-mediated defense is greater against strongly encapsulated strains (Casadevall and Perfect, 1998; Heilman et al., 2010). When they bind to the capsule, antibodies produce structural changes in it that increase fungistatic activity of the phagocytes such as neutrophils, MØ, eosinophils, microglia, and NK cells by facilitating phagocytosis and the production of nitric oxide and oxidative stress.

The defensive capacity of the antibodies depends on the isotype of the immunoglobulin; IgG_1 has a high affinity for capsular polysaccharides of *C. neoformans* var. *grubii*, whereas IgG_3 is not protective and increases the pathogenicity of this microorganism (Casadevall and Perfect, 1998; Heilman et al., 2010).

In experimental murine cryptococcosis, it has been shown that animals treated with antibodies exhibit a greater number of pulmonary granulomas as well as granulomas in other organs, due to the activation of Fc receptors (Voelz and May, 2010). In murine experimental cryptococcosis, the defenses produced by antibodies IgG_1 are especially important in mice with a deficiency of CD8/NK cells and C_3 complement fraction, whereas they are non-protective in animals deficient in CD4/INF-γ cells and cytokines Th_1/Th_2.

A serologic test that enables the recognition of protective antibodies is yet to be developed. In human beings, IgG_1 antibodies increase the production of co-stimulating CD80 molecules by the MØ and they are a protective factor (Casadevall and Perfect, 1998). IgM antibodies against glucosylceramide, detected by ELISA techniques, yield positive results only in patients with severe progressive forms of cryptococcosis; this serology test is likely to be used in the future as a diagnostic tool (Qureshi et al., 2011; Rodrigues et al., 2000).

49.6.2 Adaptive cell-mediated immunity

The importance of cell-mediated immunity in the evolution of cryptococcosis is clearly shown by the increase of the incidence and severity of this mycosis in patients with AIDS, sarcoidosis, lymphoproliferative processes, and prolonged treatments with high doses of corticosteroids. Advanced infection by HIV is the main underlying cause of disseminated cryptococcosis. In these patients, MØ show a decrease in their fungicidal ability, both for oxygen-dependent mechanisms and for those that are not dependent on oxygen. In addition, HIV-glycoprotein gp120 inhibits the fungicidal ability of the normal MØ (Casadevall and Perfect, 1998; Negroni, 2012; Viviani and Tortorano, 2009; Voelz and May, 2010).

In the experimental field, athymic mice (nu/nu) show a greater fungal load, more disseminated lesions, and less survival than animals with thymus. The effects of the absence of a thymus can be avoided with a thymic transplant or by the transference of splenic T cells from an animal with thymus (Casadevall and Perfect, 1998). However, in human cryptococcosis other factors are deemed to be influential because only between 3% and 30% of the HIV-positive patients with less than 100 CD4+ T cells per microliter show disseminated cryptococcosis (Negroni, 2008).

The activation of CD4+ T cells, which stimulate the expression of pro-inflammatory Th_1 type cytokines such as IL-2, IL-12, INF-γ, and TNF-α, produce the following effects: (1) an increase in the fungistatic activity "in vitro" of CD4+, CD8+, and NK cells, the most important aspect being the classic activation of the MØ, which leads to the destruction of the yeasts due to their increased exposure to oxidative stress and to nitric oxide action; (2) the inducement of the formation of compact epithelioid granulomas with giant cells, which turn out to be very efficient in controlling the infection; and (3) an increase in the production of protective IgG_1 antibodies (Heilman et al., 2010).

In contrast, the expression of Th_2 anti-inflammatory cytokines such as IL-4, IL-10, and IL-13 reduces the ability of the host to control the infection (Voelz and May, 2010).

IL-17 plays an important role in the regulation of the cell-mediated immune response. This cytokine is produced by CD17+ cells and acts by regulating the expression of Th_1 and Th_2 cytokines (Uicker et al., 2005; Voelz and May, 2010).

The exaggerated production of pro-inflammatory cytokines has undesired effects on the host in infections with high fungal burden, as it happens in the inflammatory restoration immune syndrome in patients with AIDS who have undergone antiretroviral therapy (Negroni, 2008, 2012).

The polarization of the production of cytokines towards those of the Th_2 type allows a greater reproduction of the intracellular yeasts, which results in increasing their survival and reducing expulsion outside the MØ. These phenomena as a whole are called alternative activation of the MØ and this facilitates the dissemination of cryptococcosis.

The survival of *Cryptococcus* inside the MØ has importance not only in terms of the fungal load, but also regarding the ability to disseminate itself and to invade other organs in addition to the lung (Charlier et al., 2009). These fungi can go through the vascular endothelia of the brain, because they internalize themselves in endothelial cells, enter the cytoplasm, and emerge out of the endothelial cells in the opposite surface without suffering any damage. This process is mediated by the membrane lipids of the vascular endothelium and is facilitated by the urease. The arrival to the brain capillary vessels is aided by the MØ, which function as carriers and protectors under the "Trojan horse dissemination model" (Casadevall and Perfect, 1998; García Rodas and Zaragoza, 2011; Romani, 2011).

There are two phenomena related to cell-mediated immunity: delayed hypersensitivity and lymphoblastic transformation. Delayed hypersensitivity occurs 2 weeks after the infection and can be confirmed by means of the intradermic injection of cryptococcal antigens, and its reaction is read 48 hours later. A positive reaction produces erythematous papules greater than 5 mm in diameter. Two types of antigens have been used: filtrates from the development of *Cryptococcus* in liquid media rich in capsular polysaccharides and cytoplasmatic supernatant of broken yeasts with a high concentration of MPs. The latter is the best reactive for skin tests. However, none of them has shown to be sensitive and specific enough to allow its use in epidemiological studies of subclinical infections, as in other mycoses. As we have already said, despite this problem a greater percentage of positive tests has been shown in individuals who are keen on breeding pigeons and laboratory workers who handle *Cryptococcus* cultures (Casadevall and Perfect, 1998; Muchmore et al., 1968, 1970). Delayed hypersensitivity is associated with a greater production of Th_1 cytokines such as IL-2 and INF-γ. The skin test is not currently used in clinical practice.

The lymphoblastic transformation could be performed with *Cryptococcus* antigens as stimulants. The most effective is a 105 kDa MP. Although this antigen is different from the one determining delayed hypersensitivity, both tests show parallel results and both are strongly positive in the infected people without symptomatic disease (Casadevall and Perfect, 1998). The cells that proliferate are both CD4+ and CD8+, but while the former proliferate by simple contact with the antigens, CD8+ require IL-2 mediation (Voelz and May, 2010).

There are two suppressive circuits of the immune response, one of them mediated by the *Cryptococcus* capsule, which has already been commented upon, and the other dependent on the CD8+ cells that produce the TsF (T-cell suppressive factor), which is a protein weighing 70 kDa that binds to the capsular polysaccharide. Only a low proportion of MØ respond to the TsF that inhibits the phagocytosis dependent upon the FcR and the mannose receptors, but it does not affect the one mediated by the CRs (Casadevall and Perfect, 1998). The TsF originates in the lymph nodes and is stimulated by the *Cryptococcus* capsule. The intensity of immunosuppression produced is consistent with the amount of capsular polysaccharide and decreases the delayed hypersensitivity and the granuloma production speed.

The fungal burden plays a very important role in the inflammatory reaction: when it is high, inflammation observed is little and without defined granulomas, whereas when it decreases with the antifungal treatment, the same patient produces epithelioid granulomas with giant cells (Negroni, 2012; Perfect and Casadevall, 2002).

49.7 Inflammatory response in the tissues and formation of granulomas

The inflammatory response depends on the immune status of the host, on the fungal load, and on the affected tissues. In the same host, with infections produced by the same amount of yeasts, there are organs that generate granulomas more frequently, such as the lung and the liver, and others that produce pseudocystic lesions, with little inflammation and high fungal burden, such as the brain and the kidneys (Salfelder et al., 1990).

The inflammatory response produced by *C. neoformans*/*C. gattii* is very similar to that brought about by *M. tuberculosis*. Both can cause chronic infections; adaptive cell-mediated immunity is the main mechanism that controls and resolves the infection, through Th$_1$-type cytokines produced by CD4+ T cells, which activate the MØ and form the compact epithelioid granulomas, with giant cells and a peripheral fibrotic ring (Casadevall and Perfect, 1998; Voelz and May, 2010). The encapsulated yeasts can be found inside the MØ or in the giant cells or free outside the cells. In immunodeficient patients and in infections produced by highly encapsulated strains, the inflammatory response is scarce and the fungal load is high, with the high quantity of capsular material conferring an aspect of mucoid cysts to these lesions.

The activation of the complement by the alternate pathway, produced by the capsular antigens, causes the release of chemotactic substances that attract PMN to the site of the infection. These cells form a ring around the encapsulated yeasts and 2–3 days later they are replaced by MØ. The latter become epithelioid cells and initiate the formation of the granuloma. The granuloma will have been completed with the production of a peripheral halo of mononuclear cells, including CD4+ and CD8+ T lymphocytes, NK, DC, and plasma cells (Heilman et al., 2010).

C. gattii produces more frequently these compact granulomas and causes pseudotumoral lesions (cryptococcomas), both in the lungs and in the brain.

During the infection *Cryptococcus* changes the capsular aspect; it usually increases in size and these big capsules are associated with a low inflammatory response and a greater amount of fungi, as it was pointed out before. Lesions can have yeasts both alive and dead; the vitality of the yeasts can be estimated by two indexes: the budding index, percentage of yeasts with blastoconidiae, and the carminophile index, percentage of yeasts stained with mucicarmine in their capsules (Casadevall and Perfect, 1998).

In the lungs diverse types of lesions can be observed: the granulomatous pneumonia, which corresponds to the histopathological expression of primary symptomatic infections; the peripheral granuloma, which includes little subpleural pulmonary nodules that give rise to the pulmonary latent infection, sequelae of the primo-infection; the capillary and interstitial lesions and the mass infection, which are linked to the hematogenous dissemination of *Cryptococcus* and are frequently observed in patients with AIDS; the cryptococcoma, a pseudotumoral granulomatous lesion with a fibrous ring and encapsulated yeasts that can be viable or not; it is an histopathological expression of the latent infection or sequelae of the primary infection and the excavated pneumonia, which present a poorly organized inflammatory reaction, with very lax granulomas, a great amount of encapsulated yeasts, and central necrosis areas; it is observed in HIV-positive patients with CD4+ T cell counts below 100 per microliter (Corti et al., 2008; Negroni, 2008, 2012). The pleurisy is often serofibrinous, unilateral, and close to subpleural pulmonary lesions. Sometimes the pleural cavity is occupied by masses of encapsulated yeasts with scarce inflammatory response.

In the lung NK, CD4+, and CD8+ T cells play a very important role to restrict the infection to this organ and to prevent hematogenous dissemination (Charlier et al., 2009; Voelz and May, 2010). On the other hand, *Cryptococcus* urease promotes the accumulation of immature DC and the response of anti-inflammatory Th$_2$-type cytokines that contribute to progressive infections and the dissemination from the lungs to other organs.

In the CNS *Cryptococcus* produces meningitis, meningoencephalitis, and brain granulomas. Meningitis is always a meningoencephalitis, since the invasion to the CNS starts by the capillary vessels of the cerebral cortex microvasculature. After causing damage to the capillary vessels the yeasts get to the brain parenchyma and from there to the subarachnoid space. They occupy mainly the base of the encephalon and the meninges that surround the spinal cord. In the brain parenchyma perivascular lesions are produced, with scarce inflammatory response, even in immunocompetent patients. In the meninges *Cryptococcus* causes congestion, edema, lymphocytic pleocytosis, and, sometimes, fibrin precipitation. The CSF in immunocompetent patients shows hyperproteinrrhachia, hypoglucorrhachia, and mild lymphocytic pleocytosis, up to 200 cells per microliter. In HIV+ patients with less than 100 CD4+ T cells per microliter, the CSF does not show marked chemical or cytological alterations, abundant encapsulated yeasts are observed, and the opening pressure during lumbar puncture tends to exceed 25 cm of water. Cell counts below 20 cells per microliter and an opening pressure >25 cm of water are associated with an unfavorable prognosis. The intracranial hypertension is one of the main causes of death in patients with AIDS-related cryptococcosis, as well as in patients suffering from meningitis due to *C. gattii*. The cause is the increase of osmotic pressure exerted by the yeast capsules, the presence of substances such as mannitol and inositol, which increase osmolarity, and the obstruction of ventricular holes by *Cryptococcus* yeasts. Therefore, the pathophysiology of endocranial hypertension in cryptococcosis is very different from that of tuberculous meningitis, where the main cause is the inflammatory process (Graybill et al., 2000; Perfect et al., 2010).

The immune response of the CNS in the experimental murine model cryptococcosis has shown to be different from that observed in other organs due to the existence of only one effector, the astrocytes and microglia, as well as the presence of the blood–brain barrier. In immunocompetent mice infected or immunized with cryptococcal antigens, there can be observed a greater expression of Th_1-type cytokines, especially INF-γ, IL-1β, IL-2, as well as chemokines β MCP_1 (macrophages chemo-attractant $protein_1$) and MIF_1 (macrophage inflammatory $protein_1$). These substances attract MØ and monocytes to the infection foci, activate the MØ by the classical pathway, and produce protective granulomas that lower the fungal load and tend to control the infection. However, the inflammatory response is less intense and effective than in other organs (Salfelder et al., 1990; Uicker et al., 2005).

In the skin, most of the lesions are produced by the hematogenous dissemination of *Cryptococcus* in immunocompromised hosts. In patients with AIDS, skin lesions are detected in 6% of the cases (Negroni, 2012) and appear as ulcerated or molluscoid papules, tubercles, infiltrated skin plaques, and nodules that evolve to gummas. These latter lesions, as well as the skin plaques looking like gangrenous pyoderma, are more frequently observed in solid organ transplant recipients (Negroni, 2012). In all these cases, a scarce inflammatory response, loose granulomas, and high fungal load are observed. In skin primoinfections the inflammatory reaction is intense, with a small amount of encapsulated yeasts and formation of compact granulomas. As we have already mentioned, these lesions are located in exposed parts of the body.

References

Aminnejad, M., Díaz, M., Arabatzis, M., Castañeda, E., Lazera, M., Velegraki, A., Marriott, D., Sorrell, T.C., and Meyer, W. 2012. Identification of novel hybrids between *Cryptococcus neoformans* var. *grubii* VNI and *Cryptococcus gattii* VGII. *Mycopathologia*, 173, 337–346.

Baddley, J.W., and Dismukes, W.F. 2003. Cryptococcosis. In: Dismukes, W.E., Pappas, P.G., and Sobel, J.D., editors. *Medical Mycology*. Oxford: Oxford University Press. pp. 188–205.

Bartlett, K.H., Cheng, P.Y., and Duncan, C. 2012. Decade experience: *Cryptococcus gattii* in British Columbia. *Mycopathologia*, 173, 311–319.

Botts, M., and Hull, Ch. 2010. Dueling in the lung: How *Cryptococcus* spores race the host for survival. *Curr. Op. Microbiol.*, 13, 437–442.

Bovers, M., Hagen, F., Kuramae, E.E., Hoogveld, H.L., Dromer, F., St-Germain, G., and Boekhout, T. 2008. AIDS patient death caused by novel *Cryptococcus neoformans* × *C. gattii* hybrid. *Emerg. Infect. Dis.*, 14, 1105–1108.

Casadevall, A., and Perfect, J. 1998. *Cryptococcus neoformans*. Washington, DC: ASM Press.

Castellá, G., Abarca, M.L., and Cabañes, F.J. 2008. Criptococosis y animales de compañía. *Rev. Iberoam. Micol.*, 25, S19–S24.

Champa Saha, D.C., Xess, I., Yoang Zeng, W., and Goldman, D.L. 2008. Antibody responses to *Cryptococcus neoformans* in Indian patients with cryptococcosis. *Med. Mycol.*, 46, 457–463.

Charlier, C., Chretien, F., Baudrimont, M., Mordelet, C., Lortholary, O., and Dromer, F. 2005. Capsule structure changes associated with *Cryptococcus neoformans* crossing the blood-brain barrier. *Am. J. Pathol.*, 166, 429–432.

Charlier, C., Nielsen, K., Daou, S., Brigitte, M., Chretien, F., and Dromer, F. 2009. Evidence of a role for monocytes in dissemination and brain invasion by *Cryptococcus neoformans*. *Infect. Immun.*, 77, 120–127.

Chaturvedi, V., and Chaturvedi, S. 2011. *Cryptococcus gattii*: A resurgent fungal pathogen. *Trends Microbiol.*, 19, 564–571.

Chayakulkeeree, M., and Perfect, J.R. 2006. Cryptococcosis. *Infect. Dis. Clin. N. Am.*, 20, 507–544.

Chiapello, L.S., Aoki, M.P., Rubinstein, H.R., and Masih, D.T. 2003. Apoptosis induction by glucuronoxylomannan of *Cryptococcus neoformans*. *Med. Mycol.*, 41, 347–353.

Chiapello, L.B., Baronetti, J.L., Garro, A.P., Spesso, M.F., and Masih, D.T. 2008. *Cryptococcus neoformans* glucuronoxilomannan induces macrophage apoptosis mediated by nitric oxide in a caspase-independent pathway. *Int. Immunol.*, 20, 1527–1541.

Corti, M., Trione, N., Semorile, K., Palmieri, O., Negroni, R., and Arechavala, A. 2008. Neumonía cavitaria por *Cryptococcus neoformans* en un paciente con SIDA. *Rev. Iberoam. Micol.*, 25, 41–44.

Fonseca, A., Boekhout, T., and Fell, J.W. 2011. *Cryptococcus vuillemin* (1901). In: Kurtzman, C.P., Fell, J.W., and Boekhout, T. editors. *The Yeasts, a Taxonomic Study*, 5th ed. Elsevier. pp. 1661–1737.

García Rodas, R., and Zaragoza, O. 2011. Catch me if you can: Phagocytosis and killing avoidance by

Cryptococcus neoformans. FEMS Immunol. Med. Microbiol., 64(2), 147–161. doi: 10/1111/j1574-695 x 2011.00871

Graybill, J.R., Sobel, J., Saag, M. van Der Horst, C., Powderly, W., Cloud, G., Riser, L., Hamil, R., and Dismukes, W. 2000. Diagnosis and management of increased intracranial pressure in patients with AIDS and cryptococcal meningitis. The NIAID Mycoses Study Group and AIDS Cooperative Treatment Groups. *Clin. Infect. Dis.,* 30, 47–54.

Grover, N., Nawage, S.R., Naidu, J., Singh, S.M., and Sharma, A. 2007. Ecological niche of *Cryptococcus neoformans* var. *grubii* and *Cryptococcus gattii* in decaying wood of trunk hollows of living trees in Yabalpur City of Central India. *Mycopathologia,* 164, 159–170.

Heilman, J., Kozel, T., Kwon-Chung, K.J., Perfect, J., and Casadevall, A. 2010. *Cryptococcus: From human pathogen to model yeast.* Washington, DC: ASM Press.

Ikeda, R., Sugita, T., Jacobson, E.S., and Shinoda, T. 2002. Laccase and melanization in clinically important *Cryptococcus* species other than *Cryptococcus neoformans. J. Clin. Microbiol.,* 40, 1214–1218.

Ishibashi, Y., Ikeda, K., Sabaguchi, K., Okino, N., Taguchi, R., and Ito, M. 2012. Quality control of fungi-specific glucosylceramide in *Cryptococcus neoformans* by a novel glucocerebrosidase EGCrP1. *J. Biol. Chem.,* 287(1), 368–381.

Jain, N., and Fries, B. 2008. Phenotypic switching of *Cryptococcus neoformans* and *Cryptococcus gattii. Mycopathologia,* 166, 181–188.

Khan, S. 2009. Invasive *Cryptococcus* infection in immunocompetent individual may suggest defects in CD40/CD40 L signaling pathway. *Indian J. Pathol. Microbiol.,* 52, 458–459.

Khan, Z., Mokaddas, E., Ahmad, S., and Burhamah, M.H.A. 2011. Isolation of *Cryptococcus magnus* and *Cryptococcus chernovii* from nasal cavities of pediatric patients with acute lymphoblastic leukemia. *Med. Mycol.,* 49, 439–443.

Kwon-Chung, K.J., and Bennett, J.E. 1992. Cryptococcosis. In: Kwon-Chung, K.J., and Bennett, J.E., editors. *Medical Mycology.* Philadelphia, PA: Lea & Febiger. pp. 397–446.

Littman, M., and Zimmerman, L.E. 1956. *Cryptococcosis.* New York: Grune & Strattron.

Liu, T-B., Kim, J-Ch., Wang, Y., Toffaletti, D.L., Eugenin, E., Perfect, J., Kim, K.Y., and Xue, Ch. 2013. Brain inositol in a novel stimulator for promoting *Cryptococcus* penetration of the blood-brain. *PLoS Pathog.,* 9(4), e1003247. doi: 10.1371/journal.ppat.1003247,Epub 2013 Apr 4.

Mac Dougall, L., Fyfe, M., Rommey, M., Starr, M., and Galanis, E. 2011 Risk factors for *Cryptococcus gattii* infection. British Columbia. *Canada Emerg. Infect. Dis.,* 17, 193–199.

Martinez, L.R., Iborn, D.C., Casadevall, A., and Fries, B.C. 2008. Characterization of phenotypic switching in *Cryptococcus neoformans* biofilms. *Mycopathologia,* 166, 175–180.

Maxson, M.E., Dadachova, E., Casadevall, A., and Zaragoza, O. 2007. Radial mass density, charge, and epitope distribution in the *Cryptococcus neoformans* capsule. *Eukariotic Cell,* 6, 95–109.

McQuiston, T.J., and Williamson, P.R. 2010. Paradoxical roles of alveolar macrophages in the host response to *Cryptococcus neoformans. Future Microbiol.,* 5, 1269–1288.

Meyer, W., Castañeda, A., Jackson, S., Huynh, M., and Castañeda, E., Ibero-American Cryptococcal Study Group. 2003. Molecular typing of Iberoamerican *Cryptococcus neoformans* isolates. *Emerg. Infect. Dis.,* 9, 189–195.

Monari, C., Paganelli, F., Bistoni, F., Kozel, T.R., and Vecchiarelli, A. 2008. Capsular polysaccharide induction of apoptosis by intrinsic and extrinsic mechanisms. *Cell. Microbiol.,* 10, 2129–2137.

Muchmore, H.G., Felton, F.G., Salvin, S.B., and Rhoades, E.R. 1968. Delayed hypersensitivity to cryptococcin in man. *Sabouraudia,* 6, 285–288.

Muchmore, H.G., Felton, F.G., Salvin, S.B., and Rhoades, E.R. 1970. Ecology and epidemiology of cryptococcosis. Proceedings of the International Symposium on Mycosis. P.A.H.O. Washington, DC. pp. 202–206.

Negroni, R. 2008. *Criptococosis.* In: Benetucci, J., editor. *Sida y Enfermedades Asociadas. Diagnóstico, Clínica y Tratamiento.* 3rd ed. Buenos Aires: FUNDAI. pp. 332–336.

Negroni, R. 2012. Cryptococcosis. *Clin. Dermatol.,* 30, 599–609.

Obayashi, K., Kano, R., Nakamura, Y., Watanabe, S., and Hasegawa, A. 2006. Capsule-associated genes of serotypes of *Cryptococcus neoformans* especially serotype AD. *Med. Mycol.,* 44, 127–132.

Olszewski, M.A., Noverr, M.C., Chen, G.H., Toews, G.B., Cox, G.M., Perfect, J.R., and Huffnagle, G.B. 2004. Urease expression by *Cryptococcus neoformans* promote microvascular sequestration, thereby enhancing central nervous system invasion. *Am. J. Pathol.,* 264, 1761–1771.

Osterholzer, J.J., Surana, R., Milam, J.E., Montano, G.T., Chen, G.H., Sonstein, J., Curtis, J.L., Huffnagle, G.B., Toews, G.B., and Olszewski, M.A. 2009. Cryptococcal urease promotes accumulation of immature dentritic cells and non-protective T2 immune response within the lung. *Am. J. Pathol.,* 174, 932–943.

Park, B.J., Wannemuehlker, K.A., Govender, N., Pappas, P.G., and Chiller, T.M. 2009. Estimation of the current global burden of cryptococcal meningitis among persons living with HIV/AIDS. *AIDS,* 23, 525–530.

Perfect, J., and Casadevall, A. 2002. Cryptococcosis. *Infect. Dis. N. Am.,* 16, 837–874.

Perfect, J.R., Dismukes, W.E., Dromer, F., Goldman, D.L., Graybill, J.R., Hamill, R.J., Harrison, T.S., Larsen, R.A., Lortholary, O., Nguyen, M.H., Pappas, P.G., Powderly, W.G., Singh, N., Sobel, J.D., and Sorrell, T.C. 2010. Clinical practice guidelines for the management of cryptococcal disease. 2010 update by the Infectious Diseases Society of America. *Clin. Infect. Dis.,* 50, 291–322.

Qureshi, A., Wray, D., Rhome, R., Barry, W., and Del Poeta, M. 2011. Detection of antibody against glucosylceramide in immunocompromised patients: A potential new diagnostic approach for cryptococcosis. *Mycopathologia,* 173, 419–425.

Rimek, D., Haase, G., Luck, A., Casper, J., and Podbielski, A. 2004. First report of a case of meningitis caused by *Cryptococcus adeliensis* in a patient with acute myeloid leukemia. *J. Clin. Microbiol.,* 42, 481–483.

Rodrigues, M.L., Travassos, L.R., Miranda, K.R., Franzen, A.J., Rozental, S., de Souza, W., Alviano, C.S., and Barreto-Bergter, E. 2000. Human antibodies against a purified glucosylceramide from Cryptococcus neoformans inhibit cell budding and fungal growth. *Infect. Immu.,* 68(12), 7049–7060.

Romani, L. 2011. Immunity to fungal infections. *Nat. Rev.,* 11, 275–288.

Sabiiti, W., and May, R. 2012. Mechanisms of infection by the human pathogen *Cryptococcus neoformans. Fut. Microbiol.,* 7, 1–17.

Salfelder, K., de Liscano, T.R., and Sauerteig, E. 1990. *Cryptococosis. Atlas of Fungal Pathology.* Boston, MA: Kluwer Academic Publishers.

Sánchez, A., Escandón, P., and Castañeda, E. 2008. Determinación in vitro de la actividad de los factores asociados con la virulencia de aislamientos clínicos del complejo Cryptococcus neoformans. *Rev. Iberoam. Micol.,* 25, 145–149.

Torres Rodríguez, J.M., Alvarado Ramírez, E., and Gutiérrez Gallego, R. 2008. Diferencias en la actividad de la enzima ureasa entre *Cryptococcus neoformans* y *Cryptococcus gattii. Rev. Iberoam. Micol.,* 25, 27–31.

Uicker, W., Doyle, H.A., Mc Cracken, J.P., Langlois, M., and Buchanan, K.L. 2005. Cytokine and chemokine expression in the central nervous system associated with protective cell-mediated immunity against *Cryptococcus neoformans. Med. Mycol.,* 43, 27–38.

VanDuin, D., Casadevall, A., and Nosanduk, J.D. 2002. Melanization of *Cryptococcus neoformans* and *Histoplasma capsulatum* reduces their susceptibilities to AMB and Caspofungin. *Antimicrob. Agents Chemother.,* 156, 171–176.

Vidotto, V., Ito-Kuwa, S., Nakamura, K., Aoki, S., Melhem, M., Fukushima, K., and Bollo, E. 2006. Extracellular enzymatic activities in *Cryptococcus neoformans* strains isolated from AIDS patients in different countries. *Rev. Iberoam. Micol.*, 23, 216–220.

Viviani, M.A., and Tortorano, A.M. 2009. Cryptococcosis. In: Anaisse, E.J., Mc Ginnis, M.R., and Pfaller, M.A., editors. *Clinical Mycology*. 2nd ed. New York: Churchill Livingstone-Elsevier. pp. 231–249.

Voelz, K., and May, R.C. 2010. Cryptococcal interactions with the host immune system. *Eukaryotic Cell*, 9, 835–846.

Wilson, J.W., and Plunkett, O.A. 1965. Cryptococcosis. In: Wilson, J.W., and Plunkett, O.A., editors. *The Fungous Diseases of Man*. University of California Press. pp. 111–129.

Chapter 50

Candida albicans: Clinical Relevance, Pathogenesis, and Host Immunity

Alexandra Correia[1,2], Paula Sampaio[3], Manuel Vilanova[1,2], and Célia Pais[3]

[1]Instituto de Biologia Molecular e Celular, Universidade do Porto, Porto, Portugal

[2]Instituto de Ciências Biomédicas de Abel Salazar, Universidade do Porto, Porto, Portugal

[3]Department of Biology, Centre of Molecular and Environmental Biology (CBMA), University of Minho, Braga, Portugal

50.1 Introduction

The genus *Candida* is composed of a very heterogeneous group of more than 200 yeast species, yet only a few are recognized human pathogens. The species most commonly encountered in medical practices are *C. albicans*, *C. glabrata*, *C. tropicalis*, *C. parapsilosis*, and *C. krusei*. Besides the yeast form, the majority of the members of the genus can display a filamentous type of growth as pseudohyphae, and *C. albicans* is also able to produce true hyphae.

Even though the prevalence of infections caused by non-*albicans Candida* species is rising, the majority of candidiasis cases, regardless of clinical setting and geographical location, are still caused by *C. albicans*. This yeast commonly colonizes the human gastrointestinal (GI), respiratory, and reproductive tracts, and the skin, generating no obvious pathology. However, under certain conditions, *C. albicans* is capable of causing a wide spectrum of infections, from superficial thrush to life-threatening systemic candidiasis, making it the most prevalent fungal pathogen in humans. A shift from colonization to mucosal invasion and/or systemic dissemination depends upon host and fungal factors. Infection of a host starts with the adherence of fungi to the epithelial surface layers and further dissemination to different host sites. Although the immune status of the host is important for the transition from commensal to pathogen, the adaptation of *C. albicans* to new niches within the host is also critical for this transition. This opportunistic pathogen is able to respond to environmental alterations by inducing transcriptional and translational changes that promote survival and fitness under the new conditions.

50.2 Types of *Candida* infections

Clinical disease can be divided into two broad categories, mucocutaneous and systemic infection, each with its own risk factors.

50.2.1 Mucocutaneous candidiasis

Infections of the skin and mucous membranes due to *Candida* spp. are growing in incidence and can occur either in immunocompromised or in immunocompetent patients (Soll, 2002).

50.2.2 Oropharyngeal candidiasis and oesophagitis

Oropharyngeal candidiasis (OPC) is very common in HIV-positive individuals affecting nearly 90% of subjects at some stage during the course of HIV disease progression. *Candida albicans* is always the most prevalent species in this type of infection, followed by *C. glabrata*, *C. dubliniensis*, and *C. tropicalis* (Patel et al., 2012). The epidemiology of OPC is less intensively studied in non-HIV-infected individuals. Predisposing factors in these patients include concomitant treatments with corticosteroids, antibiotics, or immunosuppressive and anticancer drugs, diabetes, elderly, and infancy (Vazquez, 2010). The most important predisposing factor for OPC, other than T-cell immunodeficiency, is the wearing of removable dentures (Scully et al., 1994).

Candida oesophagitis occurs in patients with chronic diseases, most of who have been previously treated with antibiotics or steroids. The disease most frequently occurs in those with advanced HIV infection and affects 10% of patients with AIDS (Chocarro Martinez et al., 2000; Vazquez, 2010). Some of these patients have concomitant OPC (Kliemann et al., 2008).

50.2.3 Vulvovaginal candidiasis

Vulvovaginal candidiasis (VVC) is a mucosal infection caused by several Candida spp. with C. albicans as the most common yeast obtained from vaginal cultures. Approximately 70%–75% of women experience at least one episode of VVC in their lifetime, and 20% suffer from recurrence (Sobel, 2007). Usually recurrences are caused by the same single strain through sequential infections (Lockhart et al., 1996; Sampaio et al., 2003). Subjects at higher risk for VVC are pregnant HIV-infected women and, to a lesser extent, HIV-negative pregnant women (Burns et al., 1997), yet the influence of CD4 T-cell numbers in the occurrence of VVC is not consensual. Other risk factors are diabetes, antibacterial vaginal or systemic therapy, and immunogenetic alterations that lead to an increased susceptibility to VVC (Jaeger et al., 2013).

50.2.4 Chronic mucocutaneous candidiasis

Chronic mucocutaneous candidiasis (CMC) consists of a heterogeneous group of disorders characterized by an unexplained susceptibility to debilitating chronic and/or recurrent infections of the skin, nails, and mucous membranes due to Candida spp., usually C. albicans. The unifying feature of these disorders appears to be impaired cell-mediated immunity (CMI) against Candida species (Notarangelo et al., 2006).

50.2.5 Disseminated candidiasis

Candida species have emerged as an important cause of bloodstream and deep tissue infections. Bloodstream infections caused by these organisms are also designated as candidemia, which often lead to Candida spread to internal organs, a condition known as disseminated or systemic candidiasis. Most cases occur in hospitalized patients and up to half are associated with intensive care units (ICUs; Bouza and Munoz, 2008; Prowle et al., 2011; Vardakas et al., 2009). As a whole, the incidence of nosocomial fungal infections (defined as invasive fungal infections acquired in a health care-associated setting) has dramatically increased in the last decade. As the population of immunosuppressed individuals has augmented (secondarily to the increased prevalence of cancer, chemotherapy, organ transplantation, and autoimmune diseases), so has the incidence of Candida invasive infections (Pappas, 2011).

50.3 Epidemiology and clinical relevance

Candida species are by far the most common fungi causing invasive disease in humans (Alangaden, 2011; Perlroth et al., 2007). Data from recent studies in the United States indicate that Candida spp. are the third or fourth most frequent nosocomial bloodstream isolates (Wisplinghoff et al., 2003a, 2003b, 2004). However, in Europe, Candida species are considered to be between the fifth and tenth most common causative pathogen of bloodstream infections (Bouza and Munoz, 2008). The true incidence of disseminated candidiasis may nevertheless be markedly underrepresented in studies focusing on blood cultures, because of the difficult diagnostic in these samples (Kami et al., 2002) and thus, studies which reflect clinical diagnosis of disseminated candidiasis, rather than relying upon blood cultures, have indicated a predicted value of 24 cases per 100,000 population in the United States (Wilson et al., 2002). Hence, an estimated number of 60,000–70,000 cases of disseminated candidiasis occur per year in the United States alone, with a health care cost associated of $2–$4 billion per year (Perlroth et al., 2007; Wilson et al., 2002).

Through the late 1980s, the predominant species causing invasive Candida infections was C. albicans. However, since the 1990s there has been a steady increase in the relative frequencies of non-albicans species of Candida causing disseminated candidiasis and this epidemiological trend has profound consequences for selection of empiric antifungal therapy. Candida albicans has been responsible for approximately 50% of invasive yeast infections, with C. glabrata generally as the second most common cause of infection in the United States and much of Europe, causing 15%–29% of cases (Cugno and Cesaro, 2012; Diekema et al., 2012; Hajjeh et al., 2004; Leroy et al., 2009; Perlroth et al., 2007; Pfaller et al., 2012; Ylipalosaari et al., 2012). In contrast, in Latin America, Portugal, and Spain, C. parapsilosis is the second most common cause of invasive candidiasis (Colombo and Guimaraes, 2003; Costa-de-Oliveira et al., 2008; Cuenca-Estrella et al., 2005; Fortun et al., 2012; Sabino et al., 2010). Candida tropicalis was acknowledged as causative of 10–20% of cases in most studies. The frequency of other species remains low, except in major cancer centres where widespread azole prophylaxis is used. In such centres, C. krusei may cause 10% of cases of invasive Candida infections (Antoniadou et al., 2003; Sabino et al., 2010; Safdar et al., 2001).

50.4 Origin of the infections and main risk factors

The origin of the infecting strain in the establishment of systemic infections is still controversial. Although the GI tract of most people is colonized by *Candida*, it is not clear whether the strains that colonize healthy hosts are responsible for causing subsequent invasive disease when those hosts acquire the appropriate risk factors, or if infections are caused by acquisition of more virulent strains from environmental sources in the nosocomial setting. Available data suggest that in most cases the source of an infecting *C. albicans* strain is indeed endogenous flora (Marco et al., 1999; Stephan et al., 2002), but that in certain circumstances exogenous transmission may occur in the nosocomial setting (Fanello et al., 2001; Pertowski et al., 1995).

The majority of patients who develop disseminated candidiasis are not immunosuppressed in the classical sense (such as neutropenic, corticosteroid-treated, and infected with HIV) (Bouza and Munoz, 2008; Cugno and Cesaro, 2012; Hajjeh et al., 2004; Ylipalosaari et al., 2012). Rather, the predominant risk factors for disseminated candidiasis are common iatrogenic and/or nosocomial conditions. In particular, 65–90% of patients with disseminated candidiasis harboured a central venous catheter (Almirante et al., 2005; Hajjeh et al., 2004; Tortorano et al., 2006). This and other medical devices are easily colonized by candidal cells from mucosal surfaces and provide the opportunity for these cells to form biofilms, which are more resistant to drugs and capable of greater tissue invasion, enabling fungal spread from one tissue site to another (Douglas, 2003). Hospitalization in the ICU provides the opportunity for transmission of *Candida* among patients and has been shown to be an additional independent risk factor. Another important independent risk factor for development of disseminated candidiasis is colonization by *Candida* spp. (Marco et al., 1999; Martino et al., 1989; Pittet et al., 1994). Patients with higher colonization burdens and more sites colonized have a proportional higher risk of developing hematogenously disseminated disease (Martino et al., 1989; Tran et al., 1997), and treatments that lower colonization burden simultaneously decrease the risk of candidemia (Eggimann et al., 1999). Broad-spectrum antibiotic therapy may alter the growth of normal bacterial flora, resulting in increased *Candida* colonization burden (Maraki et al., 2001; Mavromanolakis et al., 2001) and, consequently, increasing the risk of disseminated candidiasis (Jarvis, 1995; Saiman et al., 2000). Additionally, disruption of normal skin barriers, for example, by burn injury or percutaneous catheter placement, and disruption of gut mucosal barriers by abdominal surgery, instrumentation, induction of mucositis, or parenteral nutrition are major risk factors for invasive *Candida* infections (Kralovicova et al., 1997; Pappo et al., 1994; Vardakas et al., 2009; Ylipalosaari et al., 2012). Furthermore, cardiac surgery has also been described as a major risk for disseminated candidiasis (Giamarellou, 2002; Verghese et al., 1998).

Neutropenia or abnormalities in neutrophil function greatly increases the risk of developing disseminated candidiasis and the expected mortality rate (Cugno and Cesaro, 2012; Nucci et al., 1995; Ylipalosaari et al., 2012). Concordant with their well-characterized suppression of phagocyte function, glucocorticoids also increase the risk of disseminated candidiasis (Heidenreich et al., 1994; Nucci et al., 1995). Similarly, diabetes markedly increases the incidence of both mucocutaneous and disseminated candidiasis (Duerr et al., 2003; Sobel et al., 2011).

Candidemia in cancer patients is also thought to develop from initial GI colonization with subsequent translocation into the bloodstream after administration of chemotherapy. In a murine model of GI candidiasis, systemic chemotherapy with cyclophosphamide, which causes simultaneously neutropenia and GI mucosal damage, led to disseminated fungal infection and 100% mortality ensued in mice previously colonized with *C. albicans* (Koh et al., 2008).

HIV infection is not an independent risk factor for disseminated candidiasis. The occurrence of disseminated candidiasis in patients infected with HIV is attributable to the increased incidence of the usual risk factors for candidemia, including indwelling catheters, broad-spectrum antibiotics, hospitalization in an ICU, parenteral nutrition, and neutropenia. Patients infected with HIV who do not have additional risk factors for disseminated candidiasis are not at increased risk of developing the disease (Launay et al., 1998; Tumbarello et al., 1999).

50.5 Diagnosis of candidemia

It has been demonstrated that patients receiving very early antifungal therapy show a significant survival benefit (Garey et al., 2006; Morrell et al., 2005). However, it is difficult to diagnose fungal infections at an initial stage, and performing a correct diagnosis on the onset of the infection remains a challenge.

Examination of tissue sections is a reliable method of establishing a diagnosis of systemic fungal infection. *Candida* species usually appear as yeasts and hyphae or pseudohyphae or as yeast cells alone in infected tissues. The presence of *Candida* spp. in stained tissue sections can be used as a diagnostic parameter for invasive candidiasis. However, despite the accuracy of these diagnostic tools their impact on clinical decisions to treat patients is limited.

Detection of *Candida* species in the bloodstream is one of the laboratory keys for the diagnosis of invasive *Candida* infection. Blood culture systems range from in-house or commercial manual systems to fully manual systems that use rich culture media, allowing the recovery of very small numbers of yeasts (Richardson and Carlisle, 2002). However, in many cases these blood cultures fail to detect yeasts, either because the culture medium used is not optimal for fungal growth, or due to the low number of organisms present in the blood. Thus, although positive blood cultures are extremely important, they are positive only in less than 50% of the cases with autopsy-proven candidiasis (Berenguer et al., 1993; Fridkin, 2005;

Giamarellou and Antoniadou, 1996). A number of clinical and technical factors may affect the isolation of yeasts from blood samples, regardless of the system employed, namely the method of collection, number and timing of samples, previous antimicrobial therapy, volume of sample used, culture media, incubation time and temperature, among others (Richardson and Carlisle, 2002). Furthermore, blood cultures seem to be more effective in detecting candidemia in intensive care patients than in patients with hematological malignancies, suggesting that the condition of the patients may also be a limiting factor. All positive specimens should be sub-cultured and plated onto a solid medium, such as chocolate agar or blood agar (for rapid recovery of most yeasts species) or in Sabouraud agar, and also in chromogenic media. The incorporation of chromogenic substrates directly into the growth agar media to reveal species-specific enzyme activity allows for easier discrimination of *Candida* species colonies in mixed yeast populations than does Sabouraud agar, and also allows rapid identification of *Candida* species directly on the primary isolation medium based on the colony colour.

When cultures are obtained in non-chromogenic media, it becomes important to identify the species present by using other methodologies. The germ tube test is a quick and inexpensive method used to differentiate *C. albicans* from non-*albicans Candida* species.

In clinical settings the most frequently used methods for *Candida* species identification consists of strips or plates for carbohydrate assimilation and/or enzyme detection, which are commercially available in an assortment of different formats. Additional methods for *Candida* spp. identification include automated biochemical systems such as the ID32C strip system (bioMerieux), the Vitek Yeast Biochemical Card system (bioMerieux Vitek, Inc., Hazelwood, MO, USA), and the Vitek 2 ID-YST card system (bioMerieux Vitek), among others (Ellepola and Morrison, 2005; Freydiere et al., 2001). Differences among all these systems are especially based on the number of yeast species identified and time needed for a correct identification.

Besides the described culture dependent methods presently non-culture based laboratory methods rely on detection of antibodies against *Candida* antigens, detection of *Candida* antigens, and detection of non-antigenic *Candida* components, such as the metabolites D-arabinitol, and mannose, the cell wall component $(1 \rightarrow 3)$-β-D-glucan, or *Candida* nucleic acids (Pontón et al., 2002).

50.5.1 Molecular diagnosis

New diagnostic tools based on polymerase chain reaction (PCR) methods have been developed in the last two decades, and are becoming increasingly important in the clinical mycology laboratory to overcome most of the limitations found when using the classic approaches.

Recognition of fungal DNA directly from clinical materials offers several advantages compared to conventional methods. Nevertheless, *Candida* PCR-diagnosis directly from blood has some technical problems. First, blood and other samples have large amounts of host nucleic acid that has the potential to interfere with primer hybridization and amplification. Also, since many constituents of blood are inhibitors of *Taq* DNA polymerase, these must first be removed by lysing the erythrocytes and white blood cells (Sullivan and Coleman, 2002). Furthermore, before PCR amplification, *Candida* cells must be disrupted to release the template nucleic acid, which is not always simple, due to the difficulty to lyse the complex fungal cell wall. *Candida*-specific sequences need to be used as PCR targets and the most commonly used target for diagnostic PCR primers is the ribosomal RNA gene operon, encoding the 18S, 5.8S, and 28S RNA subunits, of which there are hundreds of copies in the *Candida* genome, improving the sensitivity of the diagnosis. Moreover, these genes have conserved and divergent regions that allowed the design of panfungal and species-specific primers (Sullivan and Coleman, 2002). Although rDNA genes are highly conserved, the ITS regions are distinctive. The universal fungal oligonucleotide primer pair, ITS3 and ITS4, amplifies portions of the 5.8S and 28S rDNA subunits, and the ITS2 region, allowing species identification.

Improvement of diagnosis has been further achieved by the development of molecular techniques such as real-time PCR for the diagnosis of *Candida* invasive infections. Real-Time PCR (RT-PCR) assays such as TaqMan (PerkinElmer) and the Light Cycler (Roche Molecular Systems) require no post-amplification manipulations and can be automated for all steps from DNA extraction to final PCR amplicon detection and quantitation (Ellepola and Morrison, 2005). These techniques are able to quantify the amounts of fungal nucleic acids present in target samples, thus potentially allowing the microbial load to be measured. Several studies have been performed using the RT-PCR technologies for *Candida* diagnosis (Klingspor and Jalal, 2006; Maaroufi et al., 2003, 2004a, 2004b; Metwally et al., 2007). Using the RT-PCR LightCycler assay Klingspor and Jalal (2006) analysed 1650 consecutive samples from patients with suspected invasive fungal infections, concluding that an RT-PCR assay allows sensitive and specific detection. A recent study evaluated prospectively, in non-neutropenic ICU patients, three TaqMan-based PCR assays and the results showed 90.9% sensitivity and 100% specificity, suggesting a potential usefulness of this method for the diagnosis of candidemia (McMullan et al., 2008).

50.6 Therapeutic strategies

Not only are invasive *Candida* infections extremely difficult to diagnose, they are also difficult to treat. Even with first-line antifungal therapy, disseminated candidiasis has an attributable mortality of up to 40% (Gudlaugsson et al., 2003; Ostrosky-Zeichner and Pappas, 2006), or even over 50% attributable mortality in patients that undergone myeloablative chemotherapy (Martino et al., 1989; Nucci et al., 1995). Unfortunately, newer therapies, such as voriconazole, lipid-based amphotericin formulations, and echinocandins, have not considerably improved survival of patients with candidemia, compared to the classical amphotericin B deoxycholate (Blau and Fauser, 2000; Kullberg et al., 2005; Mora-Duarte et al., 2002).

Because of the difficulties in confirming the diagnosis with laboratory studies, empiric therapy administration must often be based on a clinical diagnosis of disseminated candidiasis. This type of diagnosis is typically made in a patient with signs, symptoms, and laboratory features consistent with infection, who does not respond to broad-spectrum antibiotics, and who has risk factors for disseminated candidiasis. In such patients, early empiric therapy is appropriately administered. If a clinical response is seen, a clinical diagnosis of disseminated candidiasis can be made retrospectively. Consensus guidelines on the empiric treatment of disseminated candidiasis are available (Pagano and Lumb, 2011; Pappas et al., 2009). In the guidelines of the European Society of Clinical Microbiology and Infectious Diseases (ESCMID), it is clear that all three echinocandins (anidulafungin, caspofungin, and micafungin) are likely equivalently efficacious for the treatment of disseminated candidiasis in ICU patients. The use of fluconazole has only marginal support and voriconazole and liposomal amphotericin B have moderate support. Combination therapy is not recommended, either for these patients or for hemato-oncological patients. For the last group of patients, caspofungin and micafungin are the most appropriate antifungal drugs for treatment of invasive candidiasis and candidemia in neutropenic patients or hematopoietic stem cell transplant recipients. For empirical treatment in neutropenic patients, the use of caspofungin and liposomal amphotericin B have a strong recommendation, while there is moderate support for voriconazole and micafungin (Pagano and Lumb, 2011). From these guidelines is also clear that catheter removal is not mandatory if echinocandin is being used (Nucci et al., 2010; Pagano and Lumb, 2011).

50.7 *Candida albicans* genome features

Candida albicans is a diploid organism, with eight chromosome pairs and genome size around 32 Mb. The diploid genome sequence of the *C. albicans* strain SC5314, a clinical isolate that is the parent of mutants widely used for molecular analysis, was published in 2004 (Jones et al., 2004). This strain was chosen for large-scale sequencing because of its widespread use, virulence in animal models, and apparent standard diploid electrophoretic karyotype (Fonzi and Irwin, 1993). As in *Saccharomyces cerevisiae*, a relatively small fraction of *C. albicans* genes contain introns and the intron structure is generally similar to that of *Saccharomyces*. Unlike some other fungal species, *C. albicans* does not appear to have extensively spliced genes. Additionally, *C. albicans* translates the codon CUG as serine rather than the usual leucine in nuclear genes (Santos et al., 1993). Approximately two-thirds of the open reading frames (ORFs) make use of this unusual codon (Jones et al., 2004).

The polymorphisms in the *C. albicans* genome are distributed quite unevenly across its genome. Eleven highly polymorphic regions were identified, being the largest of these the mating-type-like locus, *MTL*. Unlike other pathogens *C. albicans* does not possess large gene families obviously connected to antigenic variation, but it does have a number of gene families related to pathogenesis. These include the agglutinin-like sequence (*ALS*) (Hoyer, 2001), iron transport (Ramanan and Wang, 2000), secreted aspartyl protease (*SAP*) (Naglik et al., 2003a), and secreted lipase (*LIP*) (Hube et al., 2000) genes. Several additional gene families may play a role in infection, including oligopeptide transport (eight or nine genes), eight genes related to an estrogen-binding protein, and four acid sphingomyelinases. Also present are 12 cytochrome P450s, many more than in *S. cerevisiae* or *Schizosaccharomyces pombe* (Jones et al., 2004).

A great variation in genome size and composition was found among six *Candida* species whose genome was sequenced more recently. Comparing the genome sequences of related pathogens and non-pathogenic species, significant expansions of cell wall, secreted and transporter gene families were found in pathogenic species, suggesting adaptations associated with virulence (Butler et al., 2009).

50.8 *Candida albicans* pathogenicity

Candida pathogenicity is a complex and highly regulated multifactorial process. A combination of properties, rather than a single putative virulence factor, are usually required for a fungus to successfully adapt to the different host niches met during the infectious process. These virulence attributes must help the pathogen to grow in different host niches, facilitate adherence, penetration and dissemination, or assist in resistance against or evasion from innate immune defences (Brown et al., 2007; Mayer et al., 2013).

50.9 Morphogenesis

One of the most unique characteristics of *C. albicans* is its ability to reversibly change between the unicellular yeast and filamentous growth modes (Whiteway and Bachewich, 2007). Dimorphism of *C. albicans* is known to be one of the most important virulence factors of the fungus and has been associated with tissue destruction and invasion, although both *C. albicans* yeast and hyphal cells are found in infected host tissues (Jacobsen et al., 2012). The switch from yeast to hyphal growth is controlled by the environment and by numerous other stimuli. The standard trigger of hyphal growth are nutrient poor media, a rise in temperature or CO_2, neutral to alkaline environmental pH, and presence of N-acetyl-glucosamine (GlcNAc) or serum (Whiteway and Bachewich, 2007). In the host, C. albicans switches to the hyphal form to adhere to and penetrate through tissues, and this fungal plasticity has been demonstrated in several studies to be strongly required for C. albicans pathogenesis (Jacobsen et al., 2012; Saville et al., 2008).

The yeast–hypha transition process is regulated by different interconnected signal transduction pathways (Brown and Gow, 1999; Carlisle and Kadosh, 2013). Efg1 and Cph1 have been identified as key transcriptional regulators for the cAMP-protein kinase (PKA) and the mitogen-activated protein kinase (MAPK) pathways, respectively (Bockmuhl et al., 2001; Stoldt et al., 1997). The role of these factors in morphogenesis and their relevance to Candida infections was demonstrated by using strains which lacked either functional Efg1, Cph1 or both factors. Mutants lacking EFG1 failed to produce hyphal cells under most conditions investigated and had significantly attenuated virulence in several infection models, reduced ability to invade or damage host cells (Jacobsen et al., 2012), and to colonize successfully polyurethane central venous catheters (Lewis et al., 2002) and the GI tract (Pierce et al., 2013).

The transcription factor Rim101 regulates pH-dependent transition and mutant strains lacking RIM101 also have a defective virulence phenotype (Davis et al., 2000; Villar et al., 2007). The TEA/ATTS transcription factor Tec1 is essential for serum-induced filamentation (Schweizer et al., 2000) and the Δtup1 null mutant was found to be locked in the filamentous form, with negative consequences in its ability to cause disease (Murad et al., 2001). Other transcription factors (Czf1, Flo8, Hap5 Efh1, Ace2 Mcm1, Ash1, and Cph2) are involved in filamentation under specific growth conditions (Whiteway and Bachewich, 2007). The attenuated virulence of the mutant strains lacking any of these genes seemed to be caused not only by the interference in the filamentation process, but also by a reduction or loss of expression of virulence-associated genes under their regulation (Jacobsen et al., 2012; Lo et al., 1997; Staib et al., 2002).

50.10 Phenotypic switching

Colonies of C. albicans show reversible morphological variation that occurs spontaneously in stress, resulting in changes in cell surface, colony appearance, and metabolic, biochemical, and molecular attributes, with impact on fungal virulence (Kvaal et al., 1997; Odds, 1997; Soll, 2002). Interestingly, strains isolated from vaginitis or systemically infected patients showed higher frequencies of switching, indicating a

strong role for the switching phenomenon in establishing disease (Kvaal et al., 1999). Many switching phenotypes have been described in C. albicans but the white-opaque system in the strain WO-1 is the most studied (Miller and Johnson, 2002; Soll, 1997). Although transition occurs at very low levels, switching can be induced by environmental signals, as the anaerobic conditions found in the mammalian intestine (Ramirez-Zavala et al., 2008).

Macrophages and polymorphonuclear leukocytes (PMN) seem to preferentially phagocytose white cells (Lohse and Johnson, 2008; Sasse et al., 2012) and dendritic cells phagocytosed white and opaque phase cells equally (Sasse et al., 2012). However, white cells are known to be more resistant than opaque cells under many conditions (Kolotila and Diamond, 1990; Kvaal et al., 1997; Slutsky et al., 1987; Vargas et al., 2000). Nevertheless, even though opaque cells are generally considered as less virulent than white cells in several murine models, they were found to be more able than white cells to infect skin (Kvaal et al., 1999).

50.11 Adhesins and biofilm formation

Adherence of C. albicans cells to host tissues is a complex multifactorial phenomenon employing several types of adhesins expressed on morphogenetically changing cell surfaces.

The C. albicans ALS gene family is composed of at least eight genes encoding proteins bound to the fungal cell surface by glycosylphosphatidylinositol (GPI) anchor-motives, and have long been associated with adhesion of C. albicans cells to host tissues and to abiotic surfaces (Hoyer et al., 2008; Zhao et al., 2005). High allelic variability has been shown in ALS genes, leading to strikingly different adherence profiles (Hoyer et al., 2008; Sheppard et al., 2004). Of the ALS family members, the ALS1 and ALS3 genes encode adhesins with the broadest array of substrate affinity (Sheppard et al., 2004). The expression of some ALS genes has also been involved in biofilm formation (Nobile and Mitchell, 2006; Nobile et al., 2008) and in other growth-related functions (Hoyer et al., 2008). ALS gene family members were differentially expressed in response to specific stimuli in vitro and have been associated with hyphal morphogenesis (Argimon et al., 2007; Hoyer et al., 1998; Hoyer, 2001). Interestingly, ALS expression patterns determined by RT-PCR in clinical samples were similar to those observed in the corresponding animal models of oral, vaginal and systemic candidiasis, and in reconstituted human epithelial (RHE) models (Green et al., 2004, 2005, 2006). Als3, and to a lesser extent Als1 and Als5, was proved important for the epithelial and endothelial invasion (Almeida et al., 2008; Phan et al., 2007; Sheppard et al., 2004).

HWP1 (hyphal wall protein) encodes a cell-surface adhesin expressed during hyphal development (Kim et al., 2007; Staab and Sundstrom, 1998) that promotes strong interactions between C. albicans and host cells

(Staab et al., 1999). Studies with a Δ*hwp1* null *C. albicans* mutant have reported its reduced adherence and mortality in murine models (Chaffin et al., 1998; Staab et al., 1999). The *HWP1* expression has been analysed in clinical samples and *HWP1* appears to be important not only for invasion, but also for benign interactions of *C. albicans* with the host (Cheng et al., 2003; Naglik et al., 2006).

Biofilms are surface-associated microbial communities with significant environmental and medical impact. *Candida albicans* is able to form biofilms on catheters, endotracheal tubes, pacemakers, and other prosthetic devices and has been involved in the establishment of device-associated nosocomial infections. Moreover, biofilms have reduced susceptibility to several components of the host immune response and to antifungal drug therapy (Bonhomme and d'Enfert, 2013; Srikantha et al., 2013). *Candida albicans* biofilms are heterogeneous three-dimensional structures containing cells with altered phenotype, growth rate, and gene expression compared to planktonic cells, enclosed in an extracellular polymer matrix consisting of polysaccharides and proteins (Bonhomme and d'Enfert, 2013; Dominic et al., 2007; Ramage et al., 2005). Biofilm formation was found to be linked to dimorphism, phenotypic switching, and secreted aspartyl protease activity, well-known virulence traits for candidal cells (Bonhomme and d'Enfert, 2013; Puri et al., 2012; Ramage et al., 2005; Srikantha et al., 2013). A variety of genes, such as *ALS1* and *ALS3*, *HWP1*, and genes involved in iron homeostasis, are regulated by the transcription factor Bcr1 (biofilm and cell wall regulator) and play a role in biofilm impermeability and drug resistance (Nobile and Mitchell, 2005; Nobile et al., 2006; Srikantha et al., 2013). Both *HWP1* and *ALS* proteins might function as complementary adhesins in biofilm formation (Nobile et al., 2008) and the hypha-associated *EFG1* gene is also required for normal biofilm growth (Lewis et al., 2002; Watamoto et al., 2009). Biofilms are under tight regulation of gene expression that is controlled through quorum sensing molecules (Ramage et al., 2002).

50.12 Hydrolytic enzymes

Three types of secreted hydrolytic enzymes have been described in *C. albicans*: secreted aspartyl proteases, phospholipases, and lipases.

50.12.1 Secreted aspartyl proteases

The *C. albicans SAP* family is composed of 10 different isoenzymes, which are responsible for the fungus extracellular proteolytic activity (Hube et al., 1994). *SAP* production is involved in *C. albicans* pathogenicity and has been associated with nutrient acquisition, adherence, invasion and even interference with and evasion of the host immune responses (Gropp et al., 2009; Meiller et al., 2009; Naglik et al., 2003a; Pietrella et al., 2013; Puri et al., 2012; Wu et al., 2013).

Unlike *SAP1–SAP8*, which are secreted, *SAP9* and *SAP10* are cell-surface GPI-anchored (Albrecht et al., 2006). The different *SAP* genes are differentially regulated and expressed under a variety of laboratory growth conditions, during experimental *C. albicans* infections, and also *in vivo* (Hube et al., 1994; Naglik et al., 1999, 2003b; Schaller et al., 2001; Schaller et al., 1998). While *SAP1–SAP3* are mainly expressed in the yeast form, *SAP4–SAP6* were shown to be almost exclusively expressed during hyphal formation at near neutral pH values *in vitro* (Hube et al., 1994; White and Agabian, 1995).

Although the specific host targets of *SAPs in vivo* are not accurately known, some conclusions can be inferred from *in vitro* studies and from several models of infection using protease inhibitors or *SAP* null mutants. Most studies have focused on *SAP2*, showing that it is able to degrade human mucosal surface proteins (Naglik et al., 2004), as well as proteins of the host immune response (Gropp et al., 2009; Meiller et al., 2009; Naglik et al., 2004).

The overall contribution of *SAPs* to the *C. albicans* infectious process has been estimated by using aspartyl inhibitors, pepstatin A in particular, and studies involving *SAP* null mutants have highlighted the relevance of distinct *SAPs* in different models of *Candida* infection (Naglik et al., 2003a; Schaller et al., 2005a). Digestion of several human proteins by *SAPs* (Naglik et al., 2003a; Villar et al., 2007), adherence of *C. albicans* to human mucosa and epidermis, and also tissue invasion (Naglik et al., 2003a) could be reduced by pepstatin A. The evidence supporting a protective role for pepstatin in mucosal animal models is quite convincing (Naglik et al., 2008; Schaller et al., 2005a, 2005b), but no real protective effect was achieved after an intravenous challenge with *C. albicans* (Fallon et al., 1997; Ruchel et al., 1990; Zotter et al., 1990), indicating that pepstatin can prevent disseminated infections by inhibiting *Candida* penetration through the mucosal route.

In the last two decades, the treatment of HIV-positive patients with highly active antiretroviral therapy (HAART), which includes a cocktail of HIV reverse transcriptase and HIV protease inhibitors, has proved successful in delaying the onset of AIDS. *Candida* proteases and the HIV protease are members of the same aspartyl protease family (Cauda et al., 1999) and several HIV protease inhibitors are able to inhibit *SAP* activity (Cassone et al., 1999; Gruber et al., 1999; Korting et al., 1999), and to reduce the adherence of *C. albicans* to epithelial cells. These observations further support that protease activity is involved in the attachment of *Candida* cells to host surfaces (Borg-von Zepelin et al., 1999; Cassone et al., 2002) and could explain the dramatically reduced incidence of OPC in HIV patients treated with HAART (Cassone et al., 2002; Munro and Hube, 2002). Regardless the strong inhibition of *SAP* expression in the oral cavity, these HIV protease inhibitors had a limited effect on *C. albicans* viability, thus having a limited potential as therapeutic agents in the treatment of *C. albicans* infections (De Bernardis et al., 2004).

The role played by distinct *C. albicans* proteases in mucosal and systemic infections is controversial. The results from the first studies have indicated that *SAP1–SAP3*, but not *SAP4–SAP6*, were important for *C. albicans* virulence in several mucosal infection models (De Bernardis et al., 1999; Jayatilake et al., 2006; Schaller et al., 1998, 1999, 2000, 2003). However, more recently, two different groups, using independently constructed *SAP* null mutants, have shown that even mutants lacking all of the *SAP1–SAP3* or the *SAP4–SAP6* genes invaded and damaged both oral and vaginal reconstituted human epithelium (RHE) to the same extent as their wild-type parental strain (Lermann and Morschhauser, 2008; Naglik et al., 2008). *SAP9* and *SAP10* were proved essential for maximal pathogenicity during interaction with oral epithelial tissue (Albrecht et al., 2006). In contrast, in murine models of peritonitis (Felk et al., 2002; Kretschmar et al., 1999), and in guinea pig and murine models of disseminated candidiasis, upon intravenous infection, *SAP4–SAP6* appeared to contribute more to infection than did *SAP1–SAP3* (Hube et al., 1997; Sanglard et al., 1997). However, Correia et al. have re-evaluated the role of *SAP1–SAP6*, in a similar murine model, using new sets of *SAP* null mutants, and reported that disruption of *SAP1–SAP6* had a limited influence in the infection outcome (Correia et al., 2010). In these models of disseminated candidiasis, a Δ*sap7* null mutant was partially attenuated in virulence (Taylor et al., 2005) and disruption of *SAP9* and *SAP10* did not affect *C. albicans* virulence (Albrecht et al., 2006).

It has become clear that the importance of *SAP*s for the virulence of *C. albicans* varies strongly, depending on the infection model. Even minor differences in the experimental setup may significantly impact on the dependence on protease activity for successful invasion of the different host niches.

50.12.2 Phospholipases B

Phospholipases are enzymes that hydrolyze ester linkages of glycophospholipids and also possess lysophospholipase transacylase activity. Phospholipase B (PLB) activity was the major phospholipase activity found in *C. albicans* culture supernatants (McLain and Dolan, 1997). They play a central role in cellular processes such as signal transduction and inflammation through their effect on the metabolism of phospholipids and lysophospholipids. *Candida albicans* extracellular PLBs are encoded by five genes. *PLB1* (Hoover et al., 1998), *PLB2* (Sugiyama et al., 1999), and *PLB5* (Theiss et al., 2006) have already been cloned and described. Differential PLB expression has been detected *in vitro*, in clinical samples and in mucosal and systemic infection models (Samaranayake et al., 2006; Schaller et al., 2005a) and phospholipase activity highly correlated with virulence

(Schaller et al., 2005a; Theiss et al., 2006). Most likely, PLBs contribute to the pathogenicity of *C. albicans* by damaging host cell membranes and by aiding the fungus to invade host tissues, since adherence of the yeast cells to human endothelial or epithelial cells was not affected in Δ*plb* null mutants (Ghannoum, 2000; Prakobphol et al., 1997). To this hypothesis certainly contribute the findings that PLB secretion is mainly concentrated at the growing tips of mature and developing hyphae (Mukherjee et al., 2001; Pugh and Cawson, 1975).

50.12.3 Lipases

At least 10 members constitute the *C. albicans* LIP gene family, and there is a plenty of evidence that this gene family is differentially expressed both *in vitro* and *in vivo*. Expression of *LIP5*, *LIP6*, *LIP8*, and *LIP9* was detected in a mouse model of *C. albicans* peritonitis (Hube et al., 2000), whereas *LIP1*, *LIP3*, and *LIP9* were found in infected gastric tissues, but undetected in oral mucosa (Schofield et al., 2005). Another study indicated that *LIP4* preferentially played a role in superficial infections (Stehr et al., 2004). It is likely that many or even all of these enzymes are involved in *C. albicans* virulence, but involvement of *LIP*s in the infectious process has only been evaluated for *LIP8* (Gacser et al., 2007).

50.13 Metabolic and environmental stress adaptation

The new genome-wide expression profiling studies examine the global transcriptional response of *C. albicans* to the host using *ex vivo* and *in vivo* infection models, rather than focusing on putative virulence factors. These studies have revealed that most infection-related changes in *C. albicans* gene expression reflect environmental adaptation and that the initial contacts with the host, and also disease progression, are highly associated with metabolic and stress adaptation (Mayer et al., 2013). *Candida albicans* nitrogen metabolism changes following exposure to host immune defences or growth in biofilms (Fradin et al., 2005; Garcia-Sanchez et al., 2004; Lorenz et al., 2004; Rubin-Bejerano et al., 2003). Also, amino acid biosynthetic genes are induced following PMN attack, but not during tissue invasion (Thewes et al., 2007). This is consistent with the observation that the inactivation of the amino acid starvation response or specific amino acid biosynthetic pathways does not attenuate the virulence of *C. albicans* (Brand et al., 2004; Noble and Johnson, 2005). Other genes differentially expressed during *C. albicans* tissue invasion include those involved in iron and phosphate assimilation and in pH sensing (Fradin et al., 2003, 2005; Thewes et al., 2007). These transcript profiling data are in agreement with previous mutant studies indicating that iron assimilation and pH sensing are required for the overall virulence of *C. albicans* (De Bernardis et al., 1998; Ramanan and Wang, 2000).

Candida albicans appears to activate adaptive stress responses in a niche-specific fashion during disease establishment and progression. The yeast must adapt and respond to thermal, osmotic, oxidative, and nitrosative stresses, through activation of McK1-, Hog1-, and Cek1-MAPK pathways. Stress adaptation is known to be essential, since mutants in any of the three referred pathways were all attenuated in virulence in murine models of infection (Alonso-Monge et al., 1999; Csank et al., 1998; Diez-Orejas et al., 1997). Overall, genomic studies show that although virulence factors are important, environmental adaptation appears to be the main key to pathogenicity.

50.14 *Candida albicans* cell wall

The yeast cell wall is a crucial extracellular structure that maintains the viability of cells by protecting them from lysis during environmental stress and morphogenesis. It corresponds to the primary way in which the organism interacts with its host and contributes to pathogenesis by mediating interactions with host cells and eliciting host immune responses (Netea et al., 2008). Fungal cell walls combine skeletal and matrix components. The skeletal component of the *C. albicans* cell wall is based on a core structure of β-(1,3)-glucan covalently linked to β-(1,6)-glucan and chitin (poly-β-(1,4)-*N*-acetylglucosamine (GlcNAc)), which are located towards the inside of the cell wall. The outer layer is enriched with cell wall proteins that are attached to this skeleton. Despite being buried beneath the mannoprotein outer layer, β-glucans and chitin can nevertheless become exposed at the cell surface in the bud scars (Gantner et al., 2005). The *Candida albicans* cell wall contains a matrix mainly composed of glycosylated proteins that represent 30%–40% of the cell wall dry weight (Klis et al., 2001). Lipids are minor components of the cell wall, but still have important functions (Calderone and Braun, 1991). In fact, phospholipomannan (PLM) is an important glycolipid of *C. albicans*, with linear β-1,2-oligomannose chains as the major component. β-1,2-Oligomannosides of PLM have a role in virulence and immunomodulation (Trinel et al., 1999) and are thought to be strong immunogens (Faille et al., 1990). Curiously, β-1,2-oligomannosides are expressed only in *C. albicans* and *C. tropicalis*, which are the most pathogenic *Candida* species.

50.15 Immunity to *C. albicans* infections

50.15.1 Innate immunity

Candida albicans yeasts or hyphae display a diverse set of surface or intracytoplasmic pathogen-associated molecular patterns (PAMPs) that might be recognized by specific pattern recognition receptors (PRRs). The later include soluble molecules, such as collectins and ficolins (Brummer and Stevens, 2010; Runza et al., 2008) and cell-surface molecules expressed on host epithelial cells or phagocytes (Cheng et al., 2012). Cellular PRRs comprise Toll-like receptors (TLR2, TLR4, and TLR9), C-type lectin receptors (Dectin-1, Dectin-2, mannose receptor (MR), dendritic cell-specific intercellular adhesion molecule 3-grabbing nonintegrin (DC-SIGN), Mincle, galectin-3), and NOD-like receptors (NLRs) (Wuthrich et al., 2012). Detailed reviews on host PRRs and respective *C. albicans* ligands can be found elsewhere (Cheng et al., 2012; Gauglitz et al., 2012). The panoply of PRRs able to recognize *C. albicans* components already exposes the complexity of the fungal–host interaction. Moreover, reported synergies between these receptors add to that complexity (Gantner et al., 2003).

Epithelial cells provide a first line of defence against *C. albicans* not only by providing a physical barrier preventing its dissemination into the host, but also by producing immune effectors, as detected in human oral candidiasis (Abiko et al., 2002). Human bucal and vaginal epithelial cells, which can directly inhibit *C. albicans* growth (Barousse et al., 2001; Steele et al., 2000), express PRRs through which they can interact with this fungus (Naglik and Moyes, 2011). Studies in which a model of oral candidosis based on RHE was used showed that keratinocytes respond to *C. albicans* with a strong expression of proinflammatory cytokines, associated with chemoattraction of PMNs to the site of infection (Schaller et al., 2004). The recruited PMNs could protect the RHE from the cell injury caused by *C. albicans*, independently of phagocytosis or PMN transmigration. A subsequent study using a similar model showed that *C. albicans*-infected oral epithelium and PMNs caused PMN-mediated upregulation of epithelial TLR4, which in turn mediated a protective response of the mucosal surface involving the production of the proinflammatory cytokine tumor necrosis-α (TNF-α) (Weindl et al., 2007). The fungal-induced epithelial upregulation of TLR4 expression was not found in the absence of PMNs (Pivarcsi et al., 2003), showing that the protective role of these myeloid cells in oral candidiasis goes beyond their phagocytic and microbicidal capacity. The response of human oral epithelial cells triggered by the recognition of *C. albicans* cell wall components encompasses the activation of both the nuclear factor-kappaB (NF-κB) and mitogen-activated protein kinases (MAPK) pathways, leading to the production of proinflammatory mediators. Interestingly, the MAPK pathway activation was found to depend on fungal burdens and in such a way that allows discrimination between the yeast and hyphal forms of *C. albicans* (Moyes et al., 2010). The epithelial barrier may also benefit from cytokines produced by T cells, such as interleukin-22 (IL-22), which in combination with TNF-α increases epidermal integrity upon *C. albicans* infection (Eyerich et al., 2011).

Once epithelial barriers are overcome, the innate immune response mediated by PMNs and macrophages has a preponderant role and may be sufficient for controlling fungal infections in the immunosufficient host (Koh et al., 2008; van 't Wout et al., 1988). Phagocytosed *C. albicans* cells are killed within the PMN phagolysosomes by hydrolytic enzymes, antimicrobial peptides, or oxidants such as reactive oxygen species (ROS) (Cheng et al., 2012). Extracellularly, killing of *Candida* species may also occur by releasing of neutrophil extracellular traps (NETs) containing the antifungal peptide calprotectin (Urban et al., 2009). *Candida albicans* opsonization with mannose binding lectin (MBL) and subsequent opsonophagocytosis has been shown to stimulate the production of ROS by human PMNs, mediated by intracellular Dectin-1 signalling (Li et al., 2012). The host protective role of MBL was highlighted in mice deficient in that collectin, which presented a higher mortality than wild-type counterparts upon infection with *C. albicans* (Hogaboam et al., 2004), and in patients with MBL variant genotypes that present a higher risk for *Candida* infections (Liu et al., 2006; van Till et al., 2008). The importance of oxidative mechanisms in the immune defence against *C. albicans* is stressed by the higher incidence of infections caused by this fungus observed in chronic granulomatous disease patients, which have a defective generation of ROS in PMNs, macrophages, and eosinophils (Song et al., 2011). As could be expected, neutropenic patients are among the ones presenting a higher risk for disseminated *C. albicans* infections (Perlroth et al., 2007). Contrastingly, the role of monocytes and macrophages in eliminating *C. albicans* infections is apparently limited (Cheng et al., 2012). Although these leukocyte cells are capable of phagocytosing *C. albicans*, they do not seem to be as effective as PMNs in eliminating this pathogen (van't Wout et al., 1988). Moreover, PMNs have the unique capacity among blood phagocytes to inhibit *C. albicans* germ-tube formation (Fradin et al., 2005).

Despite the crucial role of phagocytic cells, full protection against *C. albicans* infections also requires the adaptive immune response, which is usually initiated and shaped by dendritic cells (DCs). DCs play a pivotal role in the early steps of the immune response by sensing pathogens through PRRs and subsequently shaping the adaptive T-cell-mediated immune response. The observation that murine DCs could discriminate between *C. albicans* yeast and hyphal forms and the subsequent implications of this differential recognition for the polarization of the specific T-cell response (d'Ostiani et al., 2000) was a landmark in the understanding of the fungus interaction with the host immune system. Distinct intracellular pathways are activated in the different DC subsets upon recognition of different *C. albicans* morphotypes. Yeast cells may activate inflammatory DCs via adaptor myeloid differentiation factor-88 (MyD88), initiating highly proinflammatory T helper (Th) 17 responses, while hyphae-stimulated DCs may promote the differentiation of Th1 or counterinflammatory T regulatory cells (Treg) through the Toll/IL-1 receptor domain-containing adaptor, inducing

IFN- β (TRIF) signalling. Additionally, signal transducer and activator of transcription 3 (STAT3) by affecting the balance between canonical and non-canonical activation of the transcription factor NF-κB and indoleamine 2,3-dioxygenase (IDO) may also contribute to DC plasticity and differential effector function (Bonifazi et al., 2009). IDO, an enzyme of the tryptophan metabolic pathway, contributes to the differentiation of Treg driven by DCs, which may tame excessive inflammation in response to fungal stimuli (Romani, 2011). Interestingly, *C. albicans*-induced tolerogenic DCs ameliorated experimental colitis, indicating a putative immunomodulatory role for this fungus achieved by distinct receptor-signalling pathways in DCs (Bonifazi et al., 2009).

DCs, as well as monocytes/macrophages and PMNs, express Dectin-1, which is the main receptor for the *C. albicans* cell wall β-glucan (Brown and Gordon, 2001). Signalling from Dectin-1 is mediated by the cytoplasmic immunoreceptor tyrosine-based activation-like motif and involves several downstream pathways, including those mediated by Syk/CARD9 and Raf-1 kinase (Reid et al., 2009). Patients with a complete deficiency of either Dectin-1 or its adaptor molecule CARD9 display persistent mucosal infections with *C. albicans* (Ferwerda et al., 2009; Glocker et al., 2009), stressing the crucial role played by this receptor in host defence against this fungal pathogen. Despite the non-redundant role played by the Dectin-1/CARD9 signalling pathway in mucosal immunity to *C. albicans*, patients harbouring the single-nucleotide polymorphisms Dectin-1 Y238X and CARD9 S12N did not show a higher prevalence of candidemia (Rosentul et al., 2011). Dectin-2, which recognizes fungal mannans, has also been shown to be essential for the immune response to *C. albicans* (Saijo et al., 2010). Other C-type lectin receptors such as MR recognizing mannan and N-linked mannosyl residues, Mincle (glycolipids), or DC-SIGN (high mannose structures) have been shown to be important for *C. albicans* recognition and/or host resistance to this pathogen (Vautier et al., 2010). The mechanisms through which these receptors participate in the host immune response to *C. albicans* may encompass antigen uptake/phagocytosis (Cambi et al., 2003; Vijayan et al., 2012) or T helper cell polarization (van de Veerdonk et al., 2009). Nevertheless, the specific role of these receptors for antifungal resistance in the human host just started to be uncovered.

50.15.2 T-cell-mediated immunity

T cells play a pivotal role in the immune response to *C. albicans*. Glanzman and Riniker evidenced this by describing the cases of two infants with severe infections and diarrhoea that eventually died with disseminated candidiasis. Postmortem examination evidenced a marked depletion of lymphoid tissue in these patients (Glanzmann and Riniker, 1950). This likely represented the first description of severe combined immune deficiency in humans, a heterogeneous group of genetic conditions characterized by lack of T cells. Another link between defects in the T-cell compartment and an increased susceptibility to *C. albicans* infections is provided by HIV-infected patients, which present a high incidence of oesophageal and OPC (Cassone and Cauda, 2012). OPC is actually one of the first clinical signs of underlying HIV infection that eventually occurred in up to 90%–95% of HIV-infected individuals prior to the availability of antiretroviral therapy (Rabeneck et al., 1993). OPC is also a common manifestation of CMC and it may also occur in patients with other underlying immunosuppressive conditions (Horgan and Powderly, 2003).

Of the diverse T-cell subsets, Th1, Th17, and Treg have been shown to be important in avoiding *C. albicans* infection or pathology resulting from infection. A schematic view of DC-driven T-cell subset differentiation and their respective effector functions is presented in Figure 50-1.

Pioneer studies in mice associated an IL-12-driven Th1-type immune response, characterized by production of IFN-γ, with host protection to systemic or mucosal *C. albicans* infection, while a Th2-type immune response with IL-4 and IL-10 production was associated with host susceptibility to this pathogen (Romani, 1999). In line with these early findings, an impaired Th1-type responsiveness was observed in peripheral blood mononuclear cells from CMC patients stimulated *in vitro* with *C. albicans* antigens (Eyerich et al., 2007; Kobrynski et al., 1996; Lilic et al., 1996). Moreover, a dominant Th2-type cytokine profile was detected in the saliva of HIV-positive individuals presenting OPC (Leigh et al., 1998) or a Th0/Th2 response to the *C. albicans* antigen in the blood and lesioned mucosa of HIV-infected patients with pseudomembranous OPC (Vultaggio et al., 2008). Later, in addition to IL-12, IL-18 was shown to also play a major role in promoting IFN-γ production, the hallmark of the Th1-type immune response (Netea et al., 2003).

Th17 cells produce IL-17 and/or IL-22, which stimulates lung and gut epithelial cells, inducing the expression of several antimicrobial effectors, chemokines, including neutrophil chemoattractant CXC chemokines, and G-CSF (Liu et al., 2009a). In this way, IL-17 and IL-22 promote the neutrophilic influx into the mucosa that, altogether with intact epithelial layers, is an important mechanism in controlling the dissemination of *C. albicans* (Koh et al., 2008). While IL-17 has been shown to play a more determinant role in protecting mice from OPC (Conti et al., 2009),

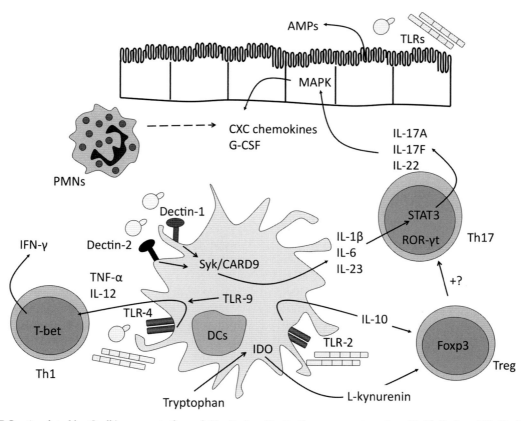

Fig. 50-1 DCs stimulated by *C. albicans* yeasts through Dectin-1 or Dectin-2 engagement produce IL-1β, IL-6, and IL-23 via Syk/CARD9-dependent pathways, which may promote Th17 (ROR-γt+) cell differentiation. Th17 cells, by producing Il-17A, IL-17F, and IL-22, may stimulate distinct epithelial cells to produce antimicrobial peptides (AMPs), such as β-defensins, or PMNs-chemoattractant CXC chemokines, and Granulocyte-colony-stimulating factor (G-CSF). PMNs recruited into infection sites can then use diverse mechanisms to help contain host invasion by fungal cells. TLR-4 and TLR-9 engagement promotes the differentiation of Th1 (T-bet+ cells), which produces host protective proinflammatory cytokine interferon-γ (IFN- γ). Treg differentiated through L-kynurenin, produced by IDO+ DCs, or by IL-10 produced upon TLR-2 engagement by *C. albicans* hyphae may counterbalance excessive inflammation or putatively contribute to differentiation of Th17 cells.

another study highlighted the role of IL-22 in the early stages of the mucosal immune response to *C. albicans* yeast forms (De Luca et al., 2010). Human CD4 T cells with a memory cell phenotype have been shown to produce IL-22 upon stimulation with heat-inactivated *C. albicans* yeasts or hyphae, highlighting its role as a defence mechanism against this pathogen (Liu et al., 2009b).

The importance of IL-17 in the immune response to *C. albicans* infections was first shown in murine studies (Conti et al., 2009; Huang et al., 2004; LeibundGut-Landmann et al., 2007). However, the involvement of IL-17-producing cells in the human immune response to *C. albicans* also became evident (Acosta-Rodriguez et al., 2007; Eyerich et al., 2008) and particularly highlighted in patients with autosomal dominant IL-17F or autosomal recessive IL-17RA deficiencies, which are prone to CMC (Puel et al., 2011). The CMC phenotype may also result from mutations that do not directly affect IL-17 or IL-17R expression but nevertheless cause an impaired development of IL-17-producing T cells. Patients with autosomic dominant CMC have been shown to carry heterozygous missense mutations affecting STAT1 (Liu et al., 2011) and similar mutations in other patients with CMC have been identified in an independent study (van de Veerdonk et al., 2011b). These mutations account for the gain-of-function of STAT1, a key factor in mediating cellular responses to IFN-α, IFN-β, IFN-γ (Stark et al., 1998), and IL-27 as well (Seita et al., 2008). As murine studies have shown that these cytokines could inhibit IL-17 expression by CD4 T cells, it was hypothesized that gain-of-function mutations in STAT1 may promote the impairment of host protective responses to *C. albicans* by inhibiting the development of IL-17-producing T cells (Casanova et al., 2012). Recently, a uniallelic STAT1 mutation was also found in children with polyendocrinopathy, enteropathy, and dermatitis reminiscent of X-linked syndrome that presented a variety of mucosal and disseminated fungal infections (Uzel et al., 2013). The implication of a type I interferon pathway as a central component in the human host defence against *C. albicans* has been shown in a recent study using a combination of transcriptional analysis and functional genomics (Smeekens et al., 2013). Interestingly, the increased IFN-α and IFN-β responsiveness may also account for autoimmune-like features presented by some patients with CMC (Crow, 2011).

Dominant-negative STAT3 mutations have been identified as the cause of autosomic dominant hyper-IgE syndrome (HIES). Patients affected by HIES present an impaired development of IL-17-producing CD4 T cells (Milner et al., 2008) and a higher susceptibility to *C. albicans* (Grimbacher et al., 1999). The development of CMC in autosomic dominant HIES patients may thus result from impaired IL-17-dependent immune mechanisms (Casanova et al., 2012; Liu et al., 2011; Puel et al., 2011). The lack of IL-17, which is a crucial cytokine in inducing the

expression of antimicrobial effectors by respiratory epithelial cells and keratinocytes, may account for the higher incidence of CMC in these patients (Minegishi et al., 2009). Nevertheless, as the IL-12/IL-18/IFNγ axis is impaired in patients with HIES, this may also account for their higher susceptibility to *C. albicans* infection (Netea et al., 2005).

The protective role of T-cell-derived IL-17 in the host defence against *C. albicans* is now supported by ample evidence. The immune mechanisms leading to its production in the context of *C. albicans* infection are being progressively uncovered, as exemplified by recent studies on the role played in this regard by the nucleotide-binding domain leucine-rich repeat-containing receptors such as Nlrp3 and Nlrp10 (Joly et al., 2012; van de Veerdonk et al., 2011a). However, the cell sources of this cytokine may differ in various types of infections caused by this pathogen. Murine studies indicate that besides Th17 cells, other T-cell populations such as γδ T cells or innate lymphoid cells may have a role in host defence as IL-17 producers. Interestingly, IL-17/IFN-γ double-producer Vδ1 T lymphocytes found expanded in the blood of HIV-1-infected patients were shown to respond *in vitro* to *C. albicans* (Fenoglio et al., 2009). However, their putative role in the defence against *C. albicans* opportunistic infections still needs to be demonstrated.

Th1 and Th17 cells are undoubtedly important in host resistance to the different forms of human candidiasis. However, their action must be counterbalanced in order to avoid excessive inflammation deleterious to the host. Studies in the murine model have shown that Treg are an important player in limiting the inflammatory immune response to *C. albicans* in systemic and mucosal candidiasis (De Luca et al., 2007; Netea et al., 2004). However, as discussed below, their role might exceed a counterinflammatory function.

Autoimmune polyendocrine syndrome type 1 (APS-1), also called autoimmune polyendocrinopathy-candidiasis-ectodermal dystrophy (APECED), is a multiorgan autoimmune disease caused by mutations in the autoimmune regulator (AIRE) gene that impairs clonal deletion of autoreactive thymocytes. Interestingly, it has been reported that patients with AIRE mutations also present a deficit in Treg (Kekalainen et al., 2007). In addition to the autoimmune manifestations, CMC is a common feature in patients with AIRE mutations, which may present an increased production of IL-17A in response to *C. albicans* (Ahlgren et al., 2011). Despite the exacerbated production of this proinflammatory cytokine, neutralizing antibodies specific for IL-17 cytokines have also been found in these patients, which may account for the impairment of a Th17-type immune response and susceptibility to CMC (Kisand et al., 2010). Nevertheless, the key mechanisms responsible for the appearance of CMC in APS-1 patients have yet to be identified. Interestingly, *C. albicans* infections may also occur in patients with other immunodeficiencies linked to a deficit in Treg such as immune dysregulation, polyendocrinopathy, enteropathy, X-linked syndrome (Coutinho and Carneiro-Sampaio, 2008), or IL-2 receptor α-chain deficiency (Roifman, 2000). The recent finding that in addition to their counterinflammatory function, Treg might promote rather than suppress Th17 cell differentiation could provide an explanation for why patients with Treg defects are prone to *C. albicans* infections (Pandiyan et al., 2011). It would be interesting to clarify whether impaired differentiation of Treg, which was associated with lethal susceptibility of B-cell-deficient mice to a secondary intravenous challenge with *C. albicans,* could also be associated with impaired Th17 differentiation (Montagnoli et al., 2003).

It is now well established that a successful host immune response to *C. albicans* infection relies on the concerted and balanced action of Th1, Th17, and Treg. Thus, genetic factors that could affect these populations may pose an additional risk for the occurrence of *C. albicans* infections. Conversely, as a recent study has shown, human genetic variants for proinflammatory IL-22 or counterinflammatory IDO1 were found to be associated with heightened resistance to recurrent VVC (De Luca et al., 2013).

The specific role of the different T-cell populations or their produced effector molecules in different types of candidiasis may be time and location dependent and this has been more extensively demonstrated in the murine model than in human patients. As an illustrative example, a systemically induced Th1-type immune response was not important in the protective immunity to vaginal candidiasis (Fidel et al., 1994). However, in humans, primary immunodeficiencies have been and will certainly be important to help elucidate the mechanisms underlying successful immunity and immune regulation in the context of *C. albicans* infections.

50.15.3 B cell and antibodies

Studies in the murine model indicated that B cells were not determinant in the immune response to *C. albicans* as mice depleted of B cells by means of anti-IgM antibody treatment did not show an increased susceptibility to candidiais (Sinha et al., 1987) and congenitally B-cell-deficient mice survived a primary intravenous challenge with this pathogen (Montagnoli et al., 2003). In agreement, primary B cell immunodeficiencies such as X-linked agammaglobulinemia do not have an obvious association with invasive fungal infections, including those caused by *C. albicans* (Lionakis, 2012). Nevertheless, a study showed that B-cell-deficient mice were more susceptible than wild-type counterparts to an infection with filamentous *C. albicans* (Saville et al., 2008). Moreover, passive administration of opsonizing antibodies increased the fungal clearance in these mice and

modified the outcome of DC stimulation by the infecting pathogen. Besides mediating the interaction with phagocytic cells, the host protective role of *C. albicans*-specific antibodies may encompass an impairment of fungal adhesion to mucosal surfaces such as in the human bucal mucosa (Epstein et al., 1982). In addition, diverse immunization studies have shown that monoclonal antibodies or polyclonal antibodies raised upon immunization with *C. albicans* structural or secreted antigens may confer protection to infectious challenges with this pathogen (De Bernardis et al., 1997; Vilanova et al., 2004; Xin and Cutler, 2011). Therefore, although the murine and human evidence points towards a secondary role for B cells in the host immune defence against candidiasis, the effector function of these cells may complement cell-mediated mechanisms of immune resistance to candidiasis.

50.16 Immunoprotective strategies against candidiasis

For many years the development of fungal vaccines did not receive much attention from the pharmaceutical companies. One reason was the general conviction that most patients who develop life-threatening fungal infections have profound defects in immunity. One major concern about vaccinating those patients against invasive fungal infections was the belief that the immune systems of such patients were unlikely to respond protectively to vaccination and might suffer from aggravation of the immunological disorder following the immunostimulation by vaccine antigens and adjuvants. However, only 10%–20% of patients who develop *Candida* bloodstream infection are seriously immunocompromised. Vaccination of high-risk groups of patients is a particularly promising strategy to prevent disseminated fungal infections because risk factors are clearly identified and correspond to common iatrogenic and/or nosocomial conditions that result in a substantial increase in the colonization burden of *Candida* spp. or in the disruption of protective anatomical barriers, rather than to severe immunosuppression (Cutler et al., 2007; Spellberg, 2011). Furthermore, development of these risk factors precedes infection, affording a window of opportunity to vaccinate highly-at-risk patients before the establishment of infection. Such patients – including those candidates to transplants, patients who will undergo gastric or cardiac surgery, and those affected by tumours – could benefit of advance active immunization against *Candida*, since the therapy or medical proceedings predispose to fungal infection (Sheppard et al., 2004). Given that many of these risk factors are of relatively short duration, normally 4–6 weeks, an immunization approach would need to protect patients just for the short period of time during their increased susceptibility. In theory, antibodies can be induced by vaccination in at-risk subjects before predisposing therapy or immunosuppression (Spellberg, 2011). Because of the relative longevity of IgG (from weeks to months, depending on the IgG isotype), its persistence at a good protective level in the circulation, even during a relatively prolonged immunosuppression or hospitalization period, is to be expected (Cassone, 2008). Recently, a renewed interest in the development of vaccination strategies against both mucosal and disseminated *Candida* infections has arisen, including induction of CMI by active immunization and also passive vaccination with antibodies (Edwards, 2012; Spellberg, 2011; Vecchiarelli et al., 2012).

50.17 Proposed *Candida* vaccines

Protection against disseminated candidiasis has been reported following both active vaccination and passive transfer of antibodies (Table 50-1) and it is clear that antigens targeted for vaccination need not be restricted to virulence factors, which markedly increases the antigen repertoire available for testing. The nature of the immunizing antigen and its immunodominant epitopes, interaction with APCs (usually DCs), antigen processing through MHC class II or MHC class I pathways, and the type of adjuvant, all determine the nature of elicited immunity and its outcome in terms of protection.

Table 50-1 Vaccines proposed against *Candida albicans* infections*

Antigen	Protection[a] mucosal; disseminated	Nature of protection[b]
Mannan, mannoproteins	+; ND	Abs, B cells, CMI – Th1
1,3-β-Glucan (Laminarin)	+; +	Abs
Secreted aspartyl protease 2	+; +[c]	Abs
Agglutin-like sequences 1 and 3	+; +	CMI – Th1, Th17
Phosphoglicerate kinase	+; ND	Abs
Enolase	ND; +	Abs, CMI – Th1
Low virulent strains	–; +	CMI
Candida membrane antigen	ND; +[d]	Abs, CMI
Heat shock proteins	+; +, −	Abs
Heat killed cells	ND; +	Unknown
Hyr-1	ND; +	Abs
β-1,2-Mannotriose and mannobiose	ND; +	Abs
C. albicans ds DNA	+[e]; ND	CMI – Th1
DCs transfected with fungal RNA	ND; +	CMI – Th1

[a]+: protective; −: not protective, may enhance virulence; ND: not determined.

[b]Abs: antibodies; CMI: cell mediated immunity.

[c]*C. albicans* peritonitis.

[d]Disseminated of endogenous origin in newborns.

[e]Gastrointestinal.

*More detailed information on this subject can be found in the following reviews: Edwards (2012); Spellberg (2011); and Vecchiarelli et al. (2012).

The protective mechanism for most active vaccines against disseminated or mucosal candidiasis studied to date relies on the induction of cell-mediated, proinflammatory, Th1 or Th17 responses, which improve phagocytic killing of the fungus (Cassone, 2008; Cutler et al., 2007; Spellberg, 2011). However, antibody participation in antifungal protection cannot be ruled out when CMI responses are elicited by vaccination. In fact, a type-1 cytokine response could be necessary for the formation of some protective antibodies against *C. albicans* proteins and most polysaccharide antigens, which, in murine models, are of the IgG2a isotype (Casadevall et al., 2002). The best protective effects observed to date have been obtained by immunization with viable cells from virulent or avirulent *C. albicans* strains (Spellberg, 2011). Whole cell or DNA vaccines maintain a persistent activation of CMI effectors and, while positively controlling the infectious agent, they may nonetheless be also inducing strong inflammation with potential undesirable effects (Cassone, 2008). Moreover, *C. albicans* is a member of the normal microbiota and vaccines against this opportunistic pathogen might result in unwanted inflammation (Cutler et al., 2007; Mochon and Cutler, 2005). Vaccination approaches currently under Phase I clinical trials (using Sap2p or rAls3p as immunogens) may be helpful in understanding putative advantages and drawbacks of *C. albicans* vaccination in the human host (Edwards, 2012).

Antibodies have long been considered irrelevant in host defence against disseminated candidiasis. The existence of inhibitory antibodies, rather than the absence of protective ones, has mostly contributed to this assumption (Casadevall and Pirofski, 2007). The clinical evidence that antibodies are protective against fungal infections is limited to a few cases (Burnie and Matthews, 2004; Matthews and Burnie, 2001, 2004), but over the last two decades, a number of antibodies directed against *C. albicans* cell wall polysaccharides and glycopeptides, proteins and peptide epitopes, have been shown to confer protection in experimental models (Table 50-1). The demonstration of protective anti-*Candida* antibodies makes possible their use in immunoprophylaxis or immunotherapy against disseminated *Candida* infections through passive vaccination, an intervention that would have some advantages. There are several examples of recombinant antibodies against fungal infection (Matthews and Burnie, 2005; Zhang et al., 2006). Some protective antibodies can be generated in a human format devoid of Fc component, suggesting that they can work without the cooperation of the immune system, and thus could be appropriate for use in severely immunocompromised patients. Other monoclonal antibodies against *Candida* need the Fc component and complement activation and deposition on the cell surface for protection (Casadevall et al., 2002; Mochon and Cutler, 2005).

Great efforts have been made in the characterization of the protective immune mechanisms against

Candida infections and in the vaccination field. Still, no therapeutic vaccine successfully used to fight *Candida* infections has been so far provided. Nevertheless, the increasing evidence, from clinical observations and animal models of candidiasis, that some *Candida*-specific antibodies can be immunoprotective during infection points to the viability of an immunotherapeutic approach for the treatment and management of candidiasis, particularly in severely immunosuppressed patients, in combination with antifungal therapy (Casadevall et al., 2002; Cassone, 2008).

References

Abiko, Y., Jinbu, Y., Noguchi, T., Nishimura, M., Kusano, K., Amaratunga, P., Shibata, T., and Kaku, T. 2002. Upregulation of human beta-defensin 2 peptide expression in oral lichen planus, leukoplakia and candidiasis. An immunohistochemical study. *Pathol. Res. Pract.*, 198, 537–542.

Acosta-Rodriguez, E.V., Rivino, L., Geginat, J., Jarrossay, D., Gattorno, M., Lanzavecchia, A., Sallusto, F., and Napolitani, G. 2007. Surface phenotype and antigenic specificity of human interleukin 17-producing T helper memory cells. *Nat. Immunol.*, 8, 639–646.

Ahlgren, K.M., Moretti, S., Lundgren, B.A., Karlsson, I., Ahlin, E., Norling, A., Hallgren, A., Perheentupa, J., Gustafsson, J., Rorsman, F., Crewther, P.E., Ronnelid, J., Bensing, S., Scott, H.S., Kampe, O., Romani, L., and Lobell, A. 2011. Increased IL-17A secretion in response to *Candida albicans* in autoimmune polyendocrine syndrome type 1 and its animal model. *Eur. J. Immunol.*, 41, 235–245.

Alangaden, G.J. 2011. Nosocomial fungal infections: Epidemiology, infection control, and prevention. *Infect. Dis. Clin. North Am.*, 25, 201–225.

Albrecht, A., Felk, A., Pichova, I., Naglik, J.R., Schaller, M., de Groot, P., Maccallum, D., Odds, F.C., Schafer, W., Klis, F., Monod, M., and Hube, B. 2006. Glycosylphosphatidylinositol-anchored proteases of *Candida albicans* target proteins necessary for both cellular processes and host-pathogen interactions. *J. Biol. Chem.*, 281, 688–694.

Almeida, R.S., Brunke, S., Albrecht, A., Thewes, S., Laue, M., Edwards, J.E., Filler, S.G., and Hube, B. 2008. The hyphal-associated adhesin and invasin Als3 of *Candida albicans* mediates iron acquisition from host ferritin. *PLoS Pathog.*, 4, e1000217.

Almirante, B., Rodriguez, D., Park, B.J., Cuenca-Estrella, M., Planes, A.M., Almela, M., Mensa, J., Sanchez, F., Ayats, J., Gimenez, M., Saballs, P., Fridkin, S.K., Morgan, J., Rodriguez-Tudela, J.L., Warnock, D.W., and Pahissa, A. 2005. Epidemiology and predictors of mortality in cases of *Candida* bloodstream infection: Results from population-based surveillance, Barcelona, Spain, from 2002 to 2003. *J. Clin. Microbiol.*, 43, 1829–1835.

Alonso-Monge, R., Navarro-Garcia, F., Molero, G., Diez-Orejas, R., Gustin, M., Pla, J., Sanchez, M., and Nombela, C. 1999. Role of the mitogen-activated protein kinase Hog1p in morphogenesis and virulence of *Candida albicans*. *J. Bacteriol.*, 181, 3058–3068.

Antoniadou, A., Torres, H.A., Lewis, R.E., Thornby, J., Bodey, G.P., Tarrand, J.P., Han, X.Y., Rolston, K.V., Safdar, A., Raad, I.I., and Kontoyiannis, D.P. 2003. Candidemia in a tertiary care cancer center: In vitro susceptibility and its association with outcome of initial antifungal therapy. *Medicine*, 82, 309–321.

Argimon, S., Wishart, J.A., Leng, R., Macaskill, S., Mavor, A., Alexandris, T., Nicholls, S., Knight, A.W., Enjalbert, B., Walmsley, R., Odds, F.C., Gow, N.A., and Brown, A.J. 2007. Developmental regulation of an adhesin gene during cellular morphogenesis in the fungal pathogen *Candida albicans*. *Eukaryot. Cell*, 6, 682–692.

Barousse, M.M., Steele, C., Dunlap, K., Espinosa, T., Boikov, D., Sobel, J.D., and Fidel Jr., P.L. 2001. Growth inhibition of *Candida albicans* by human vaginal epithelial cells. *J. Infect. Dis.*, 184, 1489–1493.

Berenguer, J., Buck, M., Witebsky, F., Stock, F., Pizzo, P.A., and Walsh, T.J. 1993. Lysis-centrifugation blood cultures in the detection of tissue-proven invasive candidiasis. Disseminated versus single-organ infection. *Diagn. Microbiol. Infect. Dis.*, 17, 103–109.

Blau, I.W., and Fauser, A.A. 2000. Review of comparative studies between conventional and liposomal amphotericin B (Ambisome) in neutropenic patients with fever of unknown origin and patients with systemic mycosis. *Mycoses*, 43, 325–332.

Bockmuhl, D.P., Krishnamurthy, S., Gerads, M., Sonneborn, A., and Ernst, J.F. 2001. Distinct and redundant roles of the two protein kinase A isoforms Tpk1p and Tpk2p in morphogenesis and growth of *Candida albicans*. *Mol. Microbiol.*, 42, 1243–1257.

Bonhomme, J., and d'Enfert, C. 2013. *Candida albicans* biofilms: Building a heterogeneous, drug-tolerant environment. *Curr. Opin. Microbiol.*, 16(4), 398–403.

Bonifazi, P., Zelante, T., D'Angelo, C., De Luca, A., Moretti, S., Bozza, S., Perruccio, K., Iannitti, R.G., Giovannini, G., Volpi, C., Fallarino, F., Puccetti, P., and Romani, L. 2009. Balancing inflammation and tolerance in vivo through dendritic cells by the commensal *Candida albicans*. *Mucosal Immunol.*, 2, 362–374.

Borg-von Zepelin, M., Meyer, I., Thomssen, R., Wurzner, R., Sanglard, D., Telenti, A., and Monod, M. 1999. HIV-protease inhibitors reduce cell adherence of *Candida albicans* strains by inhibition of yeast secreted aspartic proteases. *J. Invest. Dermatol.*, 113, 747–751.

Bouza, E., and Munoz, P. 2008. Epidemiology of candidemia in intensive care units. *Int. J. Antimicrob. Agents*, 32(Suppl 2), S87–S91.

Brand, A., MacCallum, D.M., Brown, A.J., Gow, N.A., and Odds, F.C. 2004. Ectopic expression of URA3 can influence the virulence phenotypes and proteome of *Candida albicans* but can be overcome by targeted reintegration of URA3 at the RPS10 locus. *Eukaryot. Cell*, 3, 900–909.

Brown, G.D., and Gordon, S. 2001. Immune recognition. A new receptor for beta-glucans. *Nature*, 413, 36–37.

Brown, A.J., and Gow, N.A. 1999. Regulatory networks controlling *Candida albicans* morphogenesis. *Trends Microbiol.*, 7, 333–338.

Brown, A.J., Odds, F.C., and Gow, N.A. 2007. Infection-related gene expression in *Candida albicans*. *Curr. Opin. Microbiol.*, 10, 307–313.

Brummer, E., and Stevens, D.A. 2010. Collectins and fungal pathogens: Roles of surfactant proteins and mannose binding lectin in host resistance. *Med. Mycol.*, 48, 16–28.

Burnie, J., and Matthews, R. 2004. Genetically recombinant antibodies: New therapeutics against candidiasis. *Expert Opin. Biol. Ther.*, 4, 233–241.

Burns, D.N., Tuomala, R., Chang, B.H., Hershow, R., Minkoff, H., Rodriguez, E., Zorrilla, C., Hammill, H., and Regan, J. 1997. Vaginal colonization or infection with *Candida albicans* in human immunodeficiency virus-infected women during pregnancy and during the postpartum period. Women and Infants Transmission Study Group. *Clin. Infect. Dis.*, 24, 201–210.

Butler, G., Rasmussen, M.D., Lin, M.F., Santos, M.A., Sakthikumar, S., Munro, C.A., Rheinbay, E., Grabherr, M., Forche, A., Reedy, J.L., Agrafioti, I., Arnaud, M.B., Bates, S., Brown, A.J., Brunke, S., Costanzo, M.C., Fitzpatrick, D.A., de Groot, P.W., Harris, D., Hoyer, L.L., Hube, B., Klis, F.M., Kodira, C., Lennard, N., Logue, M.E., Martin, R., Neiman, A.M., Nikolaou, E., Quail, M.A., Quinn, J., Santos, M.C., Schmitzberger,

F.F., Sherlock, G., Shah, P., Silverstein, K.A., Skrzypek, M.S., Soll, D., Staggs, R., Stansfield, I., Stumpf, M.P., Sudbery, P.E., Srikantha, T., Zeng, Q., Berman, J., Berriman, M., Heitman, J., Gow, N.A., Lorenz, M.C., Birren, B.W., Kellis, M., and Cuomo, C.A. 2009. Evolution of pathogenicity and sexual reproduction in eight *Candida* genomes. *Nature*, 459, 657–662.

Calderone, R.A., and Braun, P.C. 1991. Adherence and receptor relationships of *Candida albicans*. *Microbiol. Rev.*, 55, 1–20.

Cambi, A., Gijzen, K., de Vries, J.M., Torensma, R., Joosten, B., Adema, G.J., Netea, M.G., Kullberg, B.J., Romani, L., and Figdor, C.G. 2003. The C-type lectin DC-SIGN (CD209) is an antigen-uptake receptor for *Candida albicans* on dendritic cells. *Eur. J. Immunol.*, 33, 532–538.

Carlisle, P.L., and Kadosh, D. 2013. A genome-wide transcriptional analysis of morphology determination in *Candida albicans*. *Mol. Biol. Cell*, 24, 246–260.

Casadevall, A., and Pirofski, L.A. 2007. Antibody-mediated protection through cross-reactivity introduces a fungal heresy into immunological dogma. *Infect. Immun.*, 75, 5074–5078.

Casadevall, A., Feldmesser, M., and Pirofski, L.A. 2002. Induced humoral immunity and vaccination against major human fungal pathogens. *Curr. Opin. Microbiol.*, 5, 386–391.

Casanova, J.L., Holland, S.M., and Notarangelo, L.D. 2012. Inborn errors of human JAKs and STATs. *Immunity*, 36, 515–528.

Cassone, A. 2008. Fungal vaccines: Real progress from real challenges. *Lancet Infect. Dis.*, 8, 114–124.

Cassone, A., and Cauda, R. 2012. *Candida* and candidiasis in HIV-infected patients: Where commensalism, opportunistic behavior and frank pathogenicity lose their borders. *AIDS*, 26, 1457–1472.

Cassone, A., De Bernardis, F., Torosantucci, A., Tacconelli, E., Tumbarello, M., and Cauda, R. 1999. In vitro and in vivo anticandidal activity of human immunodeficiency virus protease inhibitors. *J. Infect. Dis.*, 180, 448–453.

Cassone, A., Tacconelli, E., De Bernardis, F., Tumbarello, M., Torosantucci, A., Chiani, P., and Cauda, R. 2002. Antiretroviral therapy with protease inhibitors has an early, immune reconstitution-independent beneficial effect on *Candida* virulence and oral candidiasis in human immunodeficiency virus-infected subjects. *J. Infect. Dis.*, 185, 188–195.

Cauda, R., Tacconelli, E., Tumbarello, M., Morace, G., De Bernardis, F., Torosantucci, A., and Cassone, A. 1999. Role of protease inhibitors in preventing recurrent oral candidosis in patients with HIV infection: A prospective case-control study. *J. Acquir. Immune. Defic. Syndr.*, 21, 20–25.

Chaffin, W.L., Lopez-Ribot, J.L., Casanova, M., Gozalbo, D., and Martinez, J.P. 1998. Cell wall and secreted proteins of *Candida albicans*: Identification, function, and expression. *Microbiol. Mol. Biol. Rev.*, 62, 130–180.

Cheng, S., Clancy, C.J., Checkley, M.A., Handfield, M., Hillman, J.D., Progulske-Fox, A., Lewin, A.S., Fidel, P.L., and Nguyen, M.H. 2003. Identification of *Candida albicans* genes induced during thrush offers insight into pathogenesis. *Mol. Microbiol.*, 48, 1275–1288.

Cheng, S.C., Joosten, L.A., Kullberg, B.J., and Netea, M.G. 2012. Interplay between *Candida albicans* and the mammalian innate host defense. *Infect. Immun.*, 80, 1304–1313.

Chocarro Martinez, A., Galindo Tobal, F., Ruiz-Irastorza, G., Gonzalez Lopez, A., Alvarez Navia, F., Ochoa Sangrador, C., and Martin Arribas, M.I. 2000. Risk factors for esophageal candidiasis. *Eur. J. Clin. Microbiol. Infect. Dis.*, 19, 96–100.

Colombo, A.L., and Guimaraes, T. 2003. Epidemiology of hematogenous infections due to *Candida* spp. *Rev. Soc. Bras. Med. Trop.*, 36, 599–607.

Conti, H.R., Shen, F., Nayyar, N., Stocum, E., Sun, J.N., Lindemann, M.J., Ho, A.W., Hai, J.H., Yu, J.J., Jung, J.W., Filler, S.G., Masso-Welch, P., Edgerton, M., and Gaffen, S.L. 2009. Th17 cells and IL-17 receptor signaling are essential for mucosal host defense against oral candidiasis. *J. Exp. Med.*, 206, 299–311.

Correia, A., Lermann, U., Teixeira, L., Cerca, F., Botelho, S., da Costa, R.M., Sampaio, P., Gartner, F., Morschhäuser, J., Vilanova, M., and Pais, C. 2010. Limited role of secreted aspartyl proteinases Sap1 to Sap6 in *Candida albicans* virulence and host immune response in murine hematogenously disseminated candidiasis. *Infect. Immun.*, 78, 4839–4849.

Costa-de-Oliveira, S., Pina-Vaz, C., Mendonca, D., and Goncalves Rodrigues, A. 2008. A first Portuguese epidemiological survey of fungaemia in a university hospital. *Eur. J. Clin. Microbiol. Infect. Dis.*, 27, 365–374.

Coutinho, A., and Carneiro-Sampaio, M. 2008. Primary immunodeficiencies unravel critical aspects of the pathophysiology of autoimmunity and of the genetics of autoimmune disease. *J. Clin. Immunol.*, 28 (Suppl 1), S4–S10.

Crow, Y.J. 2011. Type I interferonopathies: A novel set of inborn errors of immunity. *Ann. N. Y. Acad. Sci.*, 1238, 91–98.

Csank, C., Schroppel, K., Leberer, E., Harcus, D., Mohamed, O., Meloche, S., Thomas, D.Y., and Whiteway, M. 1998. Roles of the *Candida albicans* mitogen-activated protein kinase homolog, Cek1p, in hyphal development and systemic candidiasis. *Infect. Immun.*, 66, 2713–2721.

Cuenca-Estrella, M., Rodriguez, D., Almirante, B., Morgan, J., Planes, A.M., Almela, M., Mensa, J., Sanchez, F., Ayats, J., Gimenez, M., Salvado, M., Warnock, D.W., Pahissa, A., and Rodriguez-Tudela, J.L. 2005. In vitro susceptibilities of bloodstream isolates of *Candida* species to six antifungal agents: Results from a population-based active surveillance programme, Barcelona, Spain, 2002–2003. *J. Antimicrob. Chemother.*, 55, 194–199.

Cugno, C., and Cesaro, S. 2012. Epidemiology, risk factors and therapy of candidemia in pediatric hematological patients. *Pediatr. Rep.*, 4, e9.

Cutler, J.E., Deepe Jr., G.S., and Klein, B.S. 2007. Advances in combating fungal diseases: Vaccines on the threshold. *Nat. Rev. Microbiol.*, 5, 13–28.

Davis, D., Edwards Jr., J.E., Mitchell, A.P., and Ibrahim, A.S. 2000. *Candida albicans* RIM101 pH response pathway is required for host–pathogen interactions. *Infect. Immun.*, 68, 5953–5959.

De Bernardis, F., Boccanera, M., Adriani, D., Spreghini, E., Santoni, G., and Cassone, A. 1997. Protective role of antimannan and anti-aspartyl proteinase antibodies in an experimental model of *Candida albicans* vaginitis in rats. *Infect. Immun.*, 65, 3399–3405.

De Bernardis, F., Muhlschlegel, F.A., Cassone, A., and Fonzi, W.A. 1998. The pH of the host niche controls gene expression in and virulence of *Candida albicans*. *Infect. Immun.*, 66, 3317–3325.

De Bernardis, F., Arancia, S., Morelli, L., Hube, B., Sanglard, D., Schafer, W., and Cassone, A. 1999. Evidence that members of the secretory aspartyl proteinase gene family, in particular SAP2, are virulence factors for *Candida* vaginitis. *J. Infect. Dis.*, 179, 201–208.

De Bernardis, F., Tacconelli, E., Mondello, F., Cataldo, A., Arancia, S., Cauda, R., and Cassone, A. 2004. Anti-retroviral therapy with protease inhibitors decreases virulence enzyme expression in vivo by *Candida albicans* without selection of avirulent fungus strains or decreasing their antimycotic susceptibility. *FEMS Immunol. Med. Microbiol.*, 41, 27–34.

De Luca, A., Montagnoli, C., Zelante, T., Bonifazi, P., Bozza, S., Moretti, S., D'Angelo, C., Vacca, C., Boon, L., Bistoni, F., Puccetti, P., Fallarino, F., and Romani, L. 2007. Functional yet balanced reactivity to *Candida albicans* requires TRIF, MyD88, and IDO-dependent inhibition of Rorc. *J. Immunol.*, 179, 5999–6008.

De Luca, A., Zelante, T., D'Angelo, C., Zagarella, S., Fallarino, F., Spreca, A., Iannitti, R.G., Bonifazi, P., Renauld, J.C., Bistoni, F., Puccetti, P., and Romani, L. 2010. IL-22 defines a novel immune pathway of antifungal resistance. *Mucosal Immunol.*, 3, 361–373.

De Luca, A., Carvalho, A., Cunha, C., Iannitti, R.G., Pitzurra, L., Giovannini, G., Mencacci, A., Bartolommei, L., Moretti, S., Massi-Benedetti, C., Fuchs, D., De Bernardis, F., Puccetti, P., and Romani, L. 2013. IL-22 and IDO1 affect immunity and tolerance to murine and human vaginal candidiasis. *PLoS Pathog.*, 9, e1003486.

d'Ostiani, C.F., Del Sero, G., Bacci, A., Montagnoli, C., Spreca, A., Mencacci, A., Ricciardi-Castagnoli, P., and Romani, L. 2000. Dendritic cells discriminate between yeasts and hyphae of the fungus *Candida albicans*. Implications for initiation of T helper cell immunity in vitro and in vivo. *J. Exp. Med.*, 191, 1661–1674.

Diekema, D., Arbefeville, S., Boyken, L., Kroeger, J., and Pfaller, M. 2012. The changing epidemiology of healthcare-associated candidemia over three decades. *Diagn. Microbiol. Infect. Dis.*, 73, 45–48.

Diez-Orejas, R., Molero, G., Navarro-Garcia, F., Pla, J., Nombela, C., and Sanchez-Perez, M. 1997. Reduced virulence of *Candida albicans* MKC1 mutants: A role for mitogen-activated protein kinase in pathogenesis. *Infect. Immun.*, 65, 833–837.

Dominic, R.M., Shenoy, S., and Baliga, S. 2007. *Candida* biofilms in medical devices: Evolving trends. *Kathmandu Univ. Med. J. (KUMJ)*, 5, 431–436.

Douglas, L.J. 2003. *Candida* biofilms and their role in infection. *Trends Microbiol.*, 11, 30–36.

Duerr, A., Heilig, C.M., Meikle, S.F., Cu-Uvin, S., Klein, R.S., Rompalo, A., and Sobel, J.D. 2003. Incident and persistent vulvovaginal candidiasis among human immunodeficiency virus-infected women: Risk factors and severity. *Obstet. Gynecol.*, 101, 548–556.

Edwards Jr., J.E. 2012. Fungal cell wall vaccines: An update. *J. Med. Microbiol.*, 61, 895–903.

Eggimann, P., Francioli, P., Bille, J., Schneider, R., Wu, M.M., Chapuis, G., Chiolero, R., Pannatier, A., Schilling, J., Geroulanos, S., Glauser, M.P., and Calandra, T. 1999. Fluconazole prophylaxis prevents intra-abdominal candidiasis in high-risk surgical patients. *Crit. Care Med.*, 27, 1066–1072.

Ellepola, A.N., and Morrison, C.J. 2005. Laboratory diagnosis of invasive candidiasis. *J. Microbiol.*, 43, 65–84.

Epstein, J.B., Kimura, L.H., Menard, T.W., Truelove, E.L., and Pearsall, N.N. 1982. Effects of specific antibodies on the interaction between the fungus *Candida albicans* and human oral mucosa. *Arch. Oral. Biol.*, 27, 469–474.

Eyerich, K., Rombold, S., Foerster, S., Behrendt, H., Hofmann, H., Ring, J., and Traidl-Hoffmann, C. 2007. Altered, but not diminished specific T cell response in chronic mucocutaneous candidiasis patients. *Arch. Dermatol. Res.*, 299, 475–481.

Eyerich, K., Foerster, S., Rombold, S., Seidl, H.P., Behrendt, H., Hofmann, H., Ring, J., and Traidl-Hoffmann, C. 2008. Patients with chronic mucocutaneous candidiasis exhibit reduced production of Th17-associated cytokines IL-17 and IL-22. *J. Invest. Dermatol.*, 128, 2640–2645.

Eyerich, S., Wagener, J., Wenzel, V., Scarponi, C., Pennino, D., Albanesi, C., Schaller, M., Behrendt, H., Ring, J., Schmidt-Weber, C.B., Cavani, A., Mempel, M., Traidl-Hoffmann, C., and Eyerich, K. 2011. IL-22 and TNF-alpha represent a key cytokine combination for epidermal integrity during infection with *Candida albicans*. *Eur. J. Immunol.*, 41, 1894–1901.

Faille, C., Michalski, J.C., Strecker, G., Mackenzie, D.W., Camus, D., and Poulain, D. 1990. Immunoreactivity of neoglycolipids constructed from oligomannosidic residues of the *Candida albicans* cell wall. *Infect. Immun.*, 58, 3537–3544.

Fallon, K., Bausch, K., Noonan, J., Huguenel, E., and Tamburini, P. 1997. Role of aspartic proteases in disseminated *Candida albicans* infection in mice. *Infect. Immun.*, 65, 551–556.

Fanello, S., Bouchara, J.P., Jousset, N., Delbos, V., and LeFlohic, A.M. 2001. Nosocomial *Candida albicans* acquisition in a geriatric unit: Epidemiology and evidence for person-to-person transmission. *J. Hosp. Infect.*, 47, 46–52.

Felk, A., Kretschmar, M., Albrecht, A., Schaller, M., Beinhauer, S., Nichterlein, T., Sanglard, D., Korting, H.C., Schafer, W.,

and Hube, B. 2002. *Candida albicans* hyphal formation and the expression of the Efg1-regulated proteinases Sap4 to Sap6 are required for the invasion of parenchymal organs. *Infect. Immun.*, 70, 3689–3700.

Fenoglio, D., Poggi, A., Catellani, S., Battaglia, F., Ferrera, A., Setti, M., Murdaca, G., and Zocchi, M.R. 2009. Vdelta1 T lymphocytes producing IFN-gamma and IL-17 are expanded in HIV-1-infected patients and respond to *Candida albicans*. *Blood*, 113, 6611–6618.

Ferwerda, B., Ferwerda, G., Plantinga, T.S., Willment, J.A., van Spriel, A.B., Venselaar, H., Elbers, C.C., Johnson, M.D., Cambi, A., Huysamen, C., Jacobs, L., Jansen, T., Verheijen, K., Masthoff, L., Morre, S.A., Vriend, G., Williams, D.L., Perfect, J.R., Joosten, L.A., Wijmenga, C., van der Meer, J.W., Adema, G.J., Kullberg, B.J., Brown, G.D., and Netea, M.G. 2009. Human dectin-1 deficiency and mucocutaneous fungal infections. *N. Engl. J. Med.*, 361, 1760–1767.

Fidel Jr., P.L., Lynch, M.E., and Sobel, J.D. 1994. Effects of preinduced *Candida*-specific systemic cell-mediated immunity on experimental vaginal candidiasis. *Infect. Immun.*, 62, 1032–1038.

Fonzi, W.A., and Irwin, M.Y. 1993. Isogenic strain construction and gene mapping in *Candida albicans*. *Genetics*, 134, 717–728.

Fortun, J., Martin-Davila, P., Gomez-Garcia de la Pedrosa, E., Pintado, V., Cobo, J., Fresco, G., Meije, Y., Ros, L., Alvarez, M.E., Luengo, J., Agundez, M., Belso, A., Sanchez-Sousa, A., Loza, E., and Moreno, S. 2012. Emerging trends in candidemia: A higher incidence but a similar outcome. *J. Infec.*, 65, 64–70.

Fradin, C., Kretschmar, M., Nichterlein, T., Gaillardin, C., d'Enfert, C., and Hube, B. 2003. Stage-specific gene expression of *Candida albicans* in human blood. *Mol. Microbiol.*, 47, 1523–1543.

Fradin, C., De Groot, P., MacCallum, D., Schaller, M., Klis, F., Odds, F.C., and Hube, B. 2005. Granulocytes govern the transcriptional response, morphology and proliferation of *Candida albicans* in human blood. *Mol. Microbiol.*, 56, 397–415.

Freydiere, A.M., Guinet, R., and Boiron, P. 2001. Yeast identification in the clinical microbiology laboratory: Phenotypical methods. *Med. Mycol.*, 39, 9–33.

Fridkin, S.K. 2005. Candidemia is costly – Plain and simple. *Clin. Infect. Dis.*, 41, 1240–1241.

Gacser, A., Stehr, F., Kroger, C., Kredics, L., Schafer, W., and Nosanchuk, J.D. 2007. Lipase 8 affects the pathogenesis of *Candida albicans*. *Infect. Immun.*, 75, 4710–4718.

Gantner, B.N., Simmons, R.M., Canavera, S.J., Akira, S., and Underhill, D.M. 2003. Collaborative induction of inflammatory responses by dectin-1 and Toll-like receptor 2. *J. Exp. Med.*, 197, 1107–1117.

Gantner, B.N., Simmons, R.M., and Underhill, D.M. 2005. Dectin-1 mediates macrophage recognition of *Candida albicans* yeast but not filaments. *EMBO J.*, 24, 1277–1286.

Garcia-Sanchez, S., Aubert, S., Iraqui, I., Janbon, G., Ghigo, J.M., and d'Enfert, C. 2004. *Candida albicans* biofilms: A developmental state associated with specific and stable gene expression patterns. *Eukaryot. Cell*, 3, 536–545.

Garey, K.W., Rege, M., Pai, M.P., Mingo, D.E., Suda, K.J., Turpin, R.S., and Bearden, D.T. 2006. Time to initiation of fluconazole therapy impacts mortality in patients with candidemia: A multi-institutional study. *Clin. Infect. Dis.*, 43, 25–31.

Gauglitz, G.G., Callenberg, H., Weindl, G., and Korting, H.C. 2012. Host defence against *Candida albicans* and the role of pattern-recognition receptors. *Acta Derm. Venereol.*, 92, 291–298.

Ghannoum, M.A. 2000. Potential role of phospholipases in virulence and fungal pathogenesis. *Clin. Microbiol. Rev.*, 13, 122–143.

Giamarellou, H. 2002. Nosocomial cardiac infections. *J. Hosp. Infect.*, 50, 91–105.

Giamarellou, H., and Antoniadou, A. 1996. Epidemiology, diagnosis, and therapy of fungal infections in surgery. *Infect. Control Hosp. Epidemiol.*, 17, 558–564.

Glanzmann, E., and Riniker, P. 1950. Essential lymphocytophthisis: New clinical aspect of infant pathology. *Ann. Paediatr.*, 175, 1–32.

Glocker, E.O., Hennigs, A., Nabavi, M., Schaffer, A.A., Woellner, C., Salzer, U., Pfeifer, D., Veelken, H., Warnatz, K., Tahami, F., Jamal, S., Manguiat, A., Rezaei, N., Amirzargar, A.A., Plebani, A., Hannesschlager, N., Gross, O., Ruland, J., and Grimbacher, B. 2009. A homozygous CARD9 mutation in a family with susceptibility to fungal infections. *N. Engl. J. Med.*, 361, 1727–1735.

Green, C.B., Cheng, G., Chandra, J., Mukherjee, P., Ghannoum, M.A., and Hoyer, L.L. 2004. RT-PCR detection of *Candida albicans* ALS gene expression in the reconstituted human epithelium (RHE) model of oral candidiasis and in model biofilms. *Microbiology*, 150, 267–275.

Green, C.B., Zhao, X., and Hoyer, L.L. 2005. Use of green fluorescent protein and reverse transcription-PCR to monitor *Candida albicans* agglutinin-like sequence gene expression in a murine model of disseminated candidiasis. *Infect. Immun.*, 73, 1852–1855.

Green, C.B., Marretta, S.M., Cheng, G., Faddoul, F.F., Ehrhart, E.J., and Hoyer, L.L. 2006. RT-PCR analysis of *Candida albicans* ALS gene expression in a hyposalivatory rat model of oral candidiasis and in HIV-positive human patients. *Med. Mycol.*, 44, 103–111.

Grimbacher, B., Holland, S.M., Gallin, J.I., Greenberg, F., Hill, S.C., Malech, H.L., Miller, J.A., O'Connell, A.C., and Puck, J.M. 1999. Hyper-IgE syndrome with recurrent infections – An autosomal dominant multisystem disorder. *N. Engl. J. Med.*, 340, 692–702.

Gropp, K., Schild, L., Schindler, S., Hube, B., Zipfel, P.F., and Skerka, C. 2009. The yeast *Candida albicans* evades human complement attack by secretion of aspartic proteases. *Mol. Immunol.*, 47, 465–475.

Gruber, A., Berlit, J., Speth, C., Lass-Florl, C., Kofler, G., Nagl, M., Borg-von Zepelin, M., Dierich, M.P., and Wurzner, R. 1999. Dissimilar attenuation of *Candida albicans* virulence properties by human immunodeficiency virus type 1 protease inhibitors. *Immunobiology*, 201, 133–144.

Gudlaugsson, O., Gillespie, S., Lee, K., Vande Berg, J., Hu, J., Messer, S., Herwaldt, L., Pfaller, M., and Diekema, D. 2003. Attributable mortality of nosocomial candidemia, revisited. *Clin. Infect. Dis.*, 37, 1172–1177.

Hajjeh, R.A., Sofair, A.N., Harrison, L.H., Lyon, G.M., Arthington-Skaggs, B.A., Mirza, S.A., Phelan, M., Morgan, J., Lee-Yang, W., Ciblak, M.A., Benjamin, L.E., Sanza, L.T., Huie, S., Yeo, S.F., Brandt, M.E., and Warnock, D.W. 2004. Incidence of bloodstream infections due to *Candida* species and in vitro susceptibilities of isolates collected from 1998 to 2000 in a population-based active surveillance program. *J. Clin. Microbiol.*, 42, 1519–1527.

Heidenreich, S., Kubis, T., Schmidt, M., and Fegeler, W. 1994. Glucocorticoid-induced alterations of monocyte defense mechanisms against *Candida albicans*. *Cell. Immunol.*, 157, 320–327.

Hogaboam, C.M., Takahashi, K., Ezekowitz, R.A., Kunkel, S.L., and Schuh, J.M. 2004. Mannose-binding lectin deficiency alters the development of fungal asthma: Effects on airway response, inflammation, and cytokine profile. *J. Leukoc. Biol.*, 75, 805–814.

Hoover, C.I., Jantapour, M.J., Newport, G., Agabian, N., and Fisher, S.J. 1998. Cloning and regulated expression of the *Candida albicans* phospholipase B (PLB1) gene. *FEMS Microbiol. Lett.*, 167, 163–169.

Horgan, M.M., and Powderly, W.G. 2003. Oral fungal infections. In: Anaissie, E.J., McGinnis, M.R., and Pfaller, M.A., editors. *Clinical Mycology*. Philadelphia, PA: Churchill Livingstone. pp. 443–455.

Hoyer, L.L. 2001. The ALS gene family of *Candida albicans*. *Trends Microbiol.*, 9, 176–180.

Hoyer, L.L., Payne, T.L., Bell, M., Myers, A.M., and Scherer, S. 1998. *Candida albicans* ALS3 and insights into the nature of the ALS gene family. *Curr. Genet.*, 33, 451–459.

Hoyer, L.L., Green, C.B., Oh, S.H., and Zhao, X. 2008. Discovering the secrets of the *Candida albicans* agglutinin-like sequence (ALS) gene family – A sticky pursuit. *Med. Mycol.*, 46, 1–15.

Huang, W., Na, L., Fidel, P.L., and Schwarzenberger, P. 2004. Requirement of interleukin-17A for systemic anti-*Candida albicans* host defense in mice. *J. Infect. Dis.*, 190, 624–631.

Hube, B., Monod, M., Schofield, D.A., Brown, A.J., and Gow, N.A. 1994. Expression of seven members of the gene family encoding secretory aspartyl proteinases in *Candida albicans*. *Mol. Microbiol.*, 14, 87–99.

Hube, B., Sanglard, D., Odds, F.C., Hess, D., Monod, M., Schafer, W., Brown, A.J., and Gow, N.A. 1997. Disruption of each of the secreted aspartyl proteinase genes SAP1, SAP2, and SAP3 of *Candida albicans* attenuates virulence. *Infect. Immun.*, 65, 3529–3538.

Hube, B., Stehr, F., Bossenz, M., Mazur, A., Kretschmar, M., and Schafer, W. 2000. Secreted lipases of *Candida albicans*: Cloning, characterisation and expression analysis of a new gene family with at least ten members. *Arch. Microbiol.*, 174, 362–374.

Jacobsen, I.D., Wilson, D., Wachtler, B., Brunke, S., Naglik, J.R., and Hube, B. 2012. *Candida albicans* dimorphism as a therapeutic target. *Expert Rev. Anti. Infect. Ther.*, 10, 85–93.

Jaeger, M., Plantinga, T.S., Joosten, L.A., Kullberg, B.J., and Netea, M.G. 2013. Genetic basis for recurrent vulvo-vaginal candidiasis. *Curr. Infect. Dis. Rep.*, 15, 136–142.

Jarvis, W.R. 1995. Epidemiology of nosocomial fungal infections, with emphasis on *Candida* species. *Clin. Infect. Dis.*, 20, 1526–1530.

Jayatilake, J.A., Samaranayake, Y.H., Cheung, L.K., and Samaranayake, L.P. 2006. Quantitative evaluation of tissue invasion by wild type, hyphal and SAP mutants of *Candida albicans*, and non-albicans *Candida* species in reconstituted human oral epithelium. *J. Oral Pathol. Med.*, 35, 484–491.

Joly, S., Eisenbarth, S.C., Olivier, A.K., Williams, A., Kaplan, D.H., Cassel, S.L., Flavell, R.A., and Sutterwala, F.S. 2012. Cutting edge: Nlrp10 is essential for protective antifungal adaptive immunity against *Candida albicans*. *J. Immunol.*, 189, 4713–4717.

Jones, T., Federspiel, N.A., Chibana, H., Dungan, J., Kalman, S., Magee, B.B., Newport, G., Thorstenson, Y.R., Agabian, N., Magee, P.T., Davis, R.W., and Scherer, S. 2004. The diploid genome sequence of *Candida albicans*. *Proc. Natl. Acad. Sci. USA*, 101, 7329–7334.

Kami, M., Machida, U., Okuzumi, K., Matsumura, T., Mori Si, S., Hori, A., Kashima, T., Kanda, Y., Takaue, Y., Sakamaki, H., Hirai, H., Yoneyama, A., and Mutou, Y. 2002. Effect of fluconazole prophylaxis on fungal blood cultures: An autopsy-based study involving 720 patients with haematological malignancy. *Br. J. Haematol.*, 117, 40–46.

Kekalainen, E., Tuovinen, H., Joensuu, J., Gylling, M., Franssila, R., Pontynen, N., Talvensaari, K., Perheentupa, J., Miettinen, A., and Arstila, T.P. 2007. A defect of regulatory T cells in patients with autoimmune polyendocrinopathy-candidiasis-ectodermal dystrophy. *J. Immunol.*, 178, 1208–1215.

Kim, S., Wolyniak, M.J., Staab, J.F., and Sundstrom, P. 2007. A 368-base-pair cis-acting HWP1 promoter region, HCR, of *Candida albicans* confers hypha-specific gene regulation and binds architectural transcription factors Nhp6 and Gcf1p. *Eukaryot. Cell*, 6, 693–709.

Kisand, K., Boe Wolff, A.S., Podkrajsek, K.T., Tserel, L., Link, M., Kisand, K.V., Ersvaer, E., Perheentupa, J., Erichsen, M.M., Bratanic, N., Meloni, A., Cetani, F., Perniola, R., Ergun-Longmire, B., Maclaren, N., Krohn, K.J., Pura, M., Schalke, B., Strobel, P., Leite, M.I., Battelino, T., Husebye, E.S., Peterson, P., Willcox, N., and Meager, A. 2010. Chronic mucocutaneous candidiasis in APECED or thymoma patients correlates with autoimmunity to Th17-associated cytokines. *J. Exp. Med.*, 207, 299–308.

Kliemann, D.A., Pasqualotto, A.C., Falavigna, M., Giaretta, T., and Severo, L.C. 2008. *Candida* esophagitis: Species distribution and risk factors for infection. *Rev. Inst. Med. Trop. Sao. Paulo*, 50, 261–263.

Klingspor, L., and Jalal, S. 2006. Molecular detection and identification of *Candida* and *Aspergillus* spp. from clinical samples using real-time PCR. *Clin. Microbiol. Infec.*, 12, 745–753.

Klis, F.M., de Groot, P., and Hellingwerf, K. 2001. Molecular organization of the cell wall of *Candida albicans*. *Med. Mycol.*, 39(Suppl 1), 1–8.

Kobrynski, L.J., Tanimune, L., Kilpatrick, L., Campbell, D.E., and Douglas, S.D. 1996. Production of T-helper cell subsets and cytokines by lymphocytes from patients with chronic mucocutaneous candidiasis. *Clin. Diagn. Lab. Immunol.*, 3, 740–745.

Koh, A.Y., Kohler, J.R., Coggshall, K.T., Van Rooijen, N., and Pier, G.B. 2008. Mucosal damage and neutropenia are required for *Candida albicans* dissemination. *PLoS Pathog.*, 4, e35.

Kolotila, M.P., and Diamond, R.D. 1990. Effects of neutrophils and in vitro oxidants on survival and phenotypic switching of *Candida albicans* WO-1. *Infect. Immun.*, 58, 1174–1179.

Korting, H.C., Schaller, M., Eder, G., Hamm, G., Bohmer, U., and Hube, B. 1999. Effects of the human immunodeficiency virus (HIV) proteinase inhibitors saquinavir and indinavir on in vitro activities of secreted aspartyl proteinases of *Candida albicans* isolates from HIV-infected patients. *Antimicrob. Agents Chemother.*, 43, 2038–2042.

Kralovicova, K., Spanik, S., Oravcova, E., Mrazova, M., Morova, E., Gulikova, V., Kukuckova, E., Koren, P., Pichna, P., Nogova, J., Kunova, A., Trupl, J., and Krcmery Jr., V. 1997. Fungemia in cancer patients undergoing chemotherapy versus surgery: Risk factors, etiology and outcome. *Scand. J. Infect. Dis.*, 29, 301–304.

Kretschmar, M., Hube, B., Bertsch, T., Sanglard, D., Merker, R., Schroder, M., Hof, H., and Nichterlein, T. 1999. Germ tubes and proteinase activity contribute to virulence of *Candida albicans* in murine peritonitis. *Infect. Immun.*, 67, 6637–6642.

Kullberg, B.J., Sobel, J.D., Ruhnke, M., Pappas, P.G., Viscoli, C., Rex, J.H., Cleary, J.D., Rubinstein, E., Church, L.W., Brown, J.M., Schlamm, H.T., Oborska, I.T., Hilton, F., and Hodges, M.R. 2005. Voriconazole versus a regimen of amphotericin B followed by fluconazole for candidaemia in non-neutropenic patients: A randomised non-inferiority trial. *Lancet*, 366, 1435–1442.

Kvaal, C., Lachke, S.A., Srikantha, T., Daniels, K., McCoy, J., and Soll, D.R. 1999. Misexpression of the opaque-phase-specific gene PEP1 (SAP1) in the white phase of *Candida albicans* confers increased virulence in a mouse model of cutaneous infection. *Infect. Immun.*, 67, 6652–6662.

Kvaal, C.A., Srikantha, T., and Soll, D.R. 1997. Misexpression of the white-phase-specific gene WH11 in the opaque phase of *Candida albicans* affects switching and virulence. *Infect. Immun.*, 65, 4468–4475.

Launay, O., Lortholary, O., Bouges-Michel, C., Jarrousse, B., Bentata, M., and Guillevin, L. 1998. Candidemia: A nosocomial complication in adults with late-stage AIDS. *Clin. Infect. Dis.*, 26, 1134–1141.

LeibundGut-Landmann, S., Gross, O., Robinson, M.J., Osorio, F., Slack, E.C., Tsoni, S.V., Schweighoffer, E., Tybulewicz, V., Brown, G.D., Ruland, J., and Reis e Sousa, C. 2007. Syk- and CARD9-dependent coupling of innate immunity to the induction of T helper cells that produce interleukin 17. *Nat. Immunol.*, 8, 630–638.

Leigh, J.E., Steele, C., Wormley Jr., F.L., Luo, W., Clark, R.A., Gallaher, W., and Fidel Jr., P.L. 1998. Th1/Th2 cytokine expression in saliva of HIV-positive and HIV-negative individuals: A pilot study in HIV-positive individuals with oropharyngeal candidiasis. *J. Acquir. Immune Defic. Syndr. Hum. Retrovirol.*, 19, 373–380.

Lermann, U., and Morschhauser, J. 2008. Secreted aspartic proteases are not required for invasion of reconstituted human epithelia by *Candida albicans*. *Microbiology*, 154, 3281–3295.

Leroy, O., Gangneux, J.P., Montravers, P., Mira, J.P., Gouin, F., Sollet, J.P., Carlet, J., Reynes, J., Rosenheim, M., Regnier, B., and Lortholary, O. 2009. Epidemiology, management, and risk factors for death of invasive *Candida* infections in critical care: A multicenter, prospective, observational study in France (2005–2006). *Crit. Care Med.*, 37, 1612–1618.

Lewis, R.E., Lo, H.J., Raad, II, and Kontoyiannis, D.P. 2002. Lack of catheter infection by the efg1/efg1 cph1/cph1 double-null mutant, a *Candida albicans* strain that is defective in filamentous growth. *Antimicrob. Agents Chemother.*, 46, 1153–1155.

Li, D., Dong, B., Tong, Z., Wang, Q., Liu, W., Wang, Y., Chen, J., Xu, L., Chen, L., and Duan, Y. 2012. MBL-mediated opsonophagocytosis of *Candida albicans* by human neutrophils is coupled with intracellular Dectin-1-triggered ROS production. *PloS One*, 7, e50589.

Lilic, D., Cant, A.J., Abinun, M., Calvert, J.E., and Spickett, G.P. 1996. Chronic mucocutaneous candidiasis. I. Altered antigen-stimulated IL-2, IL-4, IL-6 and interferon-gamma (IFN-gamma) production. *Clin. Exp. Immunol.*, 105, 205–212.

Lionakis, M.S. 2012. Genetic susceptibility to fungal infections in humans. *Curr. Fungal Infect. Rep.*, 6, 11–22.

Liu, F., Liao, Q., and Liu, Z. 2006. Mannose-binding lectin and vulvovaginal candidiasis. *Int. J. Gynaecol. Obstet.*, 92, 43–47.

Liu, J.Z., Pezeshki, M., and Raffatellu, M. 2009a. Th17 cytokines and host-pathogen interactions at the mucosa: Dichotomies of help and harm. *Cytokine*, 48, 156–160.

Liu, Y., Yang, B., Zhou, M., Li, L., Zhou, H., Zhang, J., Chen, H., and Wu, C. 2009b. Memory IL-22-producing CD4+ T cells specific for *Candida albicans* are present in humans. *Eur. J. Immunol.*, 39, 1472–1479.

Liu, L., Okada, S., Kong, X.F., Kreins, A.Y., Cypowyj, S., Abhyankar, A., Toubiana, J., Itan, Y., Audry, M., Nitschke, P., Masson, C., Toth, B., Flatot, J., Migaud, M., Chrabieh, M., Kochetkov, T., Bolze, A., Borghesi, A., Toulon, A., Hiller, J., Eyerich, S., Eyerich, K., Gulacsy, V., Chernyshova, L., Chernyshov, V., Bondarenko, A., Grimaldo, R.M., Blancas-Galicia, L., Beas, I.M., Roesler, J., Magdorf, K., Engelhard, D., Thumerelle, C., Burgel, P.R., Hoernes, M., Drexel, B., Seger, R., Kusuma, T., Jansson, A.F., Sawalle-Belohradsky, J., Belohradsky, B., Jouanguy, E., Bustamante, J., Bue, M., Karin, N., Wildbaum, G., Bodemer, C., Lortholary, O., Fischer, A., Blanche, S., Al-Muhsen, S., Reichenbach, J., Kobayashi, M., Rosales, F.E., Lozano, C.T., Kilic, S.S., Oleastro, M., Etzioni, A., Traidl-Hoffmann, C., Renner, E.D., Abel, L., Picard, C., Marodi, L., Boisson-Dupuis, S., Puel, A., and Casanova, J.L. 2011. Gain-of-function human STAT1 mutations impair IL-17 immunity and underlie chronic mucocutaneous candidiasis. *J. Exp. Med.*, 208, 1635–1648.

Lo, H.J., Kohler, J.R., DiDomenico, B., Loebenberg, D., Cacciapuoti, A., and Fink, G.R. 1997. Nonfilamentous *C. albicans* mutants are avirulent. *Cell*, 90, 939–949.

Lockhart, S.R., Reed, B.D., Pierson, C.L., and Soll, D.R. 1996. Most frequent scenario for recurrent *Candida* vaginitis is strain maintenance with "substrain shuffling": Demonstration by sequential DNA fingerprinting with probes Ca3, C1, and CARE2. *J. Clin. Microbiol.*, 34, 767–777.

Lohse, M.B., and Johnson, A.D. 2008. Differential phagocytosis of white versus opaque *Candida albicans* by Drosophila and mouse phagocytes. *PloS One*, 3, e1473.

Lorenz, M.C., Bender, J.A., and Fink, G.R. 2004. Transcriptional response of *Candida albicans* upon internalization by macrophages. *Eukaryot. Cell*, 3, 1076–1087.

Maaroufi, Y., Heymans, C., De Bruyne, J.M., Duchateau, V., Rodriguez-Villalobos, H., Aoun, M., and Crokaert, F. 2003. Rapid detection of *Candida albicans* in clinical blood samples by using a TaqMan-based PCR assay. *J. Clin. Microbiol.*, 41, 3293–3298.

Maaroufi, Y., Ahariz, N., Husson, M., and Crokaert, F. 2004a. Comparison of different methods of isolation of DNA of commonly encountered *Candida* species and its quantitation by using a real-time PCR-based assay. *J. Clin. Microbiol.*, 42, 3159–3163.

Maaroufi, Y., De Bruyne, J.M., Duchateau, V., Georgala, A., and Crokaert, F. 2004b. Early detection and identification of commonly encountered *Candida* species from simulated blood cultures by using a real-time PCR-based assay. *J. Mol. Diagn.*, 6, 108–114.

Maraki, S., Mouzas, I.A., Kontoyiannis, D.P., Chatzinikolaou, I., Tselentis, Y., and Samonis, G. 2001. Prospective evaluation of the impact of amoxicillin, clarithromycin and their combination on human gastrointestinal colonization by *Candida* species. *Chemotherapy*, 47, 215–218.

Marco, F., Lockhart, S.R., Pfaller, M.A., Pujol, C., Rangel-Frausto, M.S., Wiblin, T., Blumberg, H.M., Edwards, J.E., Jarvis, W., Saiman, L., Patterson, J.E., Rinaldi, M.G., Wenzel, R.P., and Soll, D.R. 1999. Elucidating the origins of nosocomial infections with *Candida albicans* by DNA fingerprinting with the complex probe Ca3. *J. Clin. Microbiol.*, 37, 2817–2828.

Martino, P., Girmenia, C., Venditti, M., Micozzi, A., Santilli, S., Burgio, V.L., and Mandelli, F. 1989. *Candida* colonization and systemic infection in neutropenic patients. A retrospective study. *Cancer*, 64, 2030–2034.

Matthews, R., and Burnie, J. 2001. Antifungal antibodies: A new approach to the treatment of systemic candidiasis. *Curr. Opin. Investig. Drugs*, 2, 472–476.

Matthews, R.C., and Burnie, J.P. 2004. Recombinant antibodies: A natural partner in combinatorial antifungal therapy. *Vaccine*, 22, 865–871.

Matthews, R.C., and Burnie, J.P. 2005. Human recombinant antibody to HSP90: A natural partner in combination therapy. *Curr. Mol. Med.*, 5, 403–411.

Mavromanolakis, E., Maraki, S., Cranidis, A., Tselentis, Y., Kontoyiannis, D.P., and Samonis, G. 2001. The impact of norfloxacin, ciprofloxacin and ofloxacin on human gut colonization by *Candida albicans*. *Scand. J. Infect. Dis.*, 33, 477–478.

Mayer, F.L., Wilson, D., and Hube, B. 2013. *Candida albicans* pathogenicity mechanisms. *Virulence*, 4, 119–128.

McLain, N., and Dolan, J.W. 1997. Phospholipase D activity is required for dimorphic transition in *Candida albicans*. *Microbiology*, 143 (Pt 11), 3521–3526.

McMullan, R., Metwally, L., Coyle, P.V., Hedderwick, S., McCloskey, B., O'Neill, H.J., Patterson, C.C., Thompson, G., Webb, C.H., and Hay, R.J. 2008. A prospective clinical trial of a real-time polymerase chain reaction assay for the diagnosis of candidemia in nonneutropenic, critically ill adults. *Clin. Infect. Dis.*, 46, 890–896.

Meiller, T.F., Hube, B., Schild, L., Shirtliff, M.E., Scheper, M.A., Winkler, R., Ton, A., and Jabra-Rizk, M.A. 2009. A novel immune evasion strategy of *Candida albicans*: Proteolytic cleavage of a salivary antimicrobial peptide. *PloS One*, 4, e5039.

Metwally, L., Hogg, G., Coyle, P.V., Hay, R.J., Hedderwick, S., McCloskey, B., O'Neill, H.J., Ong, G.M., Thompson, G., Webb, C.H., and McMullan, R. 2007. Rapid differentiation between fluconazole-sensitive and -resistant species of *Candida* directly from positive blood-culture bottles by real-time PCR. *J. Med. Microbiol.*, 56, 964–970.

Miller, M.G., and Johnson, A.D. 2002. White-opaque switching in *Candida albicans* is controlled by mating-type locus homeodomain proteins and allows efficient mating. *Cell*, 110, 293–302.

Milner, J.D., Brenchley, J.M., Laurence, A., Freeman, A.F., Hill, B.J., Elias, K.M., Kanno, Y., Spalding, C., Elloumi, H.Z., Paulson, M.L., Davis, J., Hsu, A., Asher, A.I., O'Shea, J., Holland, S.M., Paul, W.E., and Douek, D.C. 2008. Impaired T(H)17 cell differentiation in subjects with autosomal dominant hyper-IgE syndrome. *Nature*, 452, 773–776.

Minegishi, Y., Saito, M., Nagasawa, M., Takada, H., Hara, T., Tsuchiya, S., Agematsu, K., Yamada, M., Kawamura, N., Ariga, T., Tsuge, I., and Karasuyama, H. 2009. Molecular explanation for the contradiction between systemic Th17 defect and localized bacterial infection in hyper-IgE syndrome. *J. Exp. Med.*, 206, 1291–1301.

Mochon, A.B., and Cutler, J.E. 2005. Is a vaccine needed against *Candida albicans*? *Med. Mycol.*, 43, 97–115.

Montagnoli, C., Bozza, S., Bacci, A., Gaziano, R., Mosci, P., Morschhauser, J., Pitzurra, L., Kopf, M., Cutler, J., and Romani, L. 2003. A role for antibodies in the generation of memory antifungal immunity. *Eur. J. Immunol.*, 33, 1193–1204.

Mora-Duarte, J., Betts, R., Rotstein, C., Colombo, A.L., Thompson-Moya, L., Smietana, J., Lupinacci, R., Sable, C., Kartsonis, N., and Perfect, J. 2002. Comparison of caspofungin and amphotericin B for invasive candidiasis. *N. Engl. J. Med.*, 347, 2020–2029.

Morrell, M., Fraser, V.J., and Kollef, M.H. 2005. Delaying the empiric treatment of *Candida* bloodstream infection until positive blood culture results are obtained: A potential risk factor for hospital mortality. *Antimicrob. Agents Chemother.*, 49, 3640–3645.

Moyes, D.L., Runglall, M., Murciano, C., Shen, C., Nayar, D., Thavaraj, S., Kohli, A., Islam, A., Mora-Montes, H., Challacombe, S.J., and Naglik, J.R. 2010. A biphasic innate immune MAPK response discriminates between the yeast and hyphal forms of *Candida albicans* in epithelial cells. *Cell Host Microbe*, 8, 225–235.

Mukherjee, P.K., Seshan, K.R., Leidich, S.D., Chandra, J., Cole, G.T., and Ghannoum, M.A. 2001. Reintroduction of the PLB1 gene into *Candida albicans* restores virulence in vivo. *Microbiology*, 147, 2585–2597.

Munro, C.A., and Hube, B. 2002. Anti-fungal therapy at the HAART of viral therapy. *Trends Microbiol.*, 10, 173–177.

Murad, A.M., Leng, P., Straffon, M., Wishart, J., Macaskill, S., MacCallum, D., Schnell, N., Talibi, D., Marechal, D., Tekaia, F., d'Enfert, C., Gaillardin, C., Odds, F.C., and Brown, A.J. 2001. NRG1 represses yeast-hypha morphogenesis and hypha-specific gene expression in *Candida albicans*. *EMBO J.*, 20, 4742–4752.

Naglik, J.R., and Moyes, D. 2011. Epithelial cell innate response to *Candida albicans*. *Adv. Dent. Res.*, 23, 50–55.

Naglik, J.R., Newport, G., White, T.C., Fernandes-Naglik, L.L., Greenspan, J.S., Greenspan, D., Sweet, S.P., Challacombe, S.J., and Agabian, N. 1999. In vivo analysis of secreted aspartyl proteinase expression in human oral candidiasis. *Infect. Immun.*, 67, 2482–2490.

Naglik, J.R., Challacombe, S.J., and Hube, B. 2003a. *Candida albicans* secreted aspartyl proteinases in virulence and pathogenesis. *Microbiol. Mol. Biol. Rev.*, 67, 400–428.

Naglik, J.R., Rodgers, C.A., Shirlaw, P.J., Dobbie, J.L., Fernandes-Naglik, L.L., Greenspan, D., Agabian, N., and Challacombe, S.J. 2003b. Differential expression of *Candida albicans* secreted aspartyl proteinase and phospholipase B genes in humans correlates with active oral and vaginal infections. *J. Infect. Dis.*, 188, 469–479.

Naglik, J., Albrecht, A., Bader, O., and Hube, B. 2004. *Candida albicans* proteinases and host/pathogen interactions. *Cell Microbiol.*, 6, 915–926.

Naglik, J.R., Fostira, F., Ruprai, J., Staab, J.F., Challacombe, S.J., and Sundstrom, P. 2006. *Candida albicans* HWP1 gene expression and host antibody responses in colonization and disease. *J. Med. Microbiol.*, 55, 1323–1327.

Naglik, J.R., Moyes, D., Makwana, J., Kanzaria, P., Tsichlaki, E., Weindl, G., Tappuni, A.R., Rodgers, C.A., Woodman, A.J., Challacombe, S.J., Schaller, M., and Hube, B. 2008. Quantitative expression of the *Candida albicans* secreted aspartyl proteinase gene family in human oral and vaginal candidiasis. *Microbiology*, 154, 3266–3280.

Netea, M.G., Vonk, A.G., van den Hoven, M., Verschueren, I., Joosten, L.A., van Krieken, J.H., van den Berg, W.B., Van der

Meer, J.W., and Kullberg, B.J. 2003. Differential role of IL-18 and IL-12 in the host defense against disseminated *Candida albicans* infection. *Eur. J. Immunol.*, 33, 3409–3417.

Netea, M.G., Sutmuller, R., Hermann, C., Van der Graaf, C.A., Van der Meer, J.W., van Krieken, J.H., Hartung, T., Adema, G., and Kullberg, B.J. 2004. Toll-like receptor 2 suppresses immunity against *Candida albicans* through induction of IL-10 and regulatory T cells. *J. Immunol.*, 172, 3712–3718.

Netea, M.G., Kullberg, B.J., and van der Meer, J.W. 2005. Severely impaired IL-12/IL-18/IFNgamma axis in patients with hyper IgE syndrome. *Eur. J. Clin. Invest.*, 35, 718–721.

Netea, M.G., Brown, G.D., Kullberg, B.J., and Gow, N.A. 2008. An integrated model of the recognition of *Candida albicans* by the innate immune system. *Nat. Rev. Microbiol.*, 6, 67–78.

Nobile, C.J., and Mitchell, A.P. 2005. Regulation of cell-surface genes and biofilm formation by the *C. albicans* transcription factor Bcr1p. *Curr. Biol.*, 15, 1150–1155.

Nobile, C.J., and Mitchell, A.P. 2006. Genetics and genomics of *Candida albicans* biofilm formation. *Cell Microbiol.*, 8, 1382–1391.

Nobile, C.J., Nett, J.E., Andes, D.R., and Mitchell, A.P. 2006. Function of *Candida albicans* adhesin Hwp1 in biofilm formation. *Eukaryot. Cell*, 5, 1604–1610.

Nobile, C.J., Schneider, H.A., Nett, J.E., Sheppard, D.C., Filler, S.G., Andes, D.R., and Mitchell, A.P. 2008. Complementary adhesin function in *C. albicans* biofilm formation. *Curr. Biol.*, 18, 1017–1024.

Noble, S.M., and Johnson, A.D. 2005. Strains and strategies for large-scale gene deletion studies of the diploid human fungal pathogen *Candida albicans*. *Eukaryot. Cell*, 4, 298–309.

Notarangelo, L.D., Gambineri, E., and Badolato, R. 2006. Immunodeficiencies with autoimmune consequences. *Adv. Immunol.*, 89, 321–370.

Nucci, M., Pulcheri, W., Spector, N., Bueno, A.P., Bacha, P.C., Caiuby, M.J., Derossi, A., Costa, R., Morais, J.C., and de Oliveira, H.P. 1995. Fungal infections in neutropenic patients. A 8-year prospective study. *Rev. Inst. Med. Trop. Sao. Paulo*, 37, 397–406.

Nucci, M., Anaissie, E., Betts, R.F., Dupont, B.F., Wu, C., Buell, D.N., Kovanda, L., and Lortholary, O. 2010. Early removal of central venous catheter in patients with candidemia does not improve outcome: Analysis of 842 patients from 2 randomized clinical trials. *Clin. Infect. Dis.*, 51, 295–303.

Odds, E.C. 1997. Switch of phenotype as an escape mechanism of the intruder. *Mycoses*, 40(Suppl 2), 9–12.

Ostrosky-Zeichner, L., and Pappas, P.G. 2006. Invasive candidiasis in the intensive care unit. *Crit. Care Med.*, 34, 857–863.

Pagano, L., and Lumb, J. 2011. Update on invasive fungal disease. *Future Microbiol.*, 6, 985–989.

Pandiyan, P., Conti, H.R., Zheng, L., Peterson, A.C., Mathern, D.R., Hernandez-Santos, N., Edgerton, M., Gaffen, S.L., and Lenardo, M.J. 2011. CD4(+)CD25(+)Foxp3(+) regulatory T cells promote Th17 cells in vitro and enhance host resistance in mouse *Candida albicans* Th17 cell infection model. *Immunity*, 34, 422–434.

Pappas, P.G. 2011. Candidemia in the intensive care unit: Miles to go before we sleep. *Crit. Care Med.*, 39, 884–885.

Pappas, P.G., Kauffman, C.A., Andes, D., Benjamin Jr., D.K., Calandra, T.F., Edwards Jr., J.E., Filler, S.G., Fisher, J.F., Kullberg, B.J., Ostrosky-Zeichner, L., Reboli, A.C., Rex, J.H., Walsh, T.J., and Sobel, J.D. 2009. Clinical practice guidelines for the management of candidiasis: 2009 update by the Infectious Diseases Society of America. *Clin. Infect. Dis.*, 48, 503–535.

Pappo, I., Polacheck, I., Zmora, O., Feigin, E., and Freund, H.R. 1994. Altered gut barrier function to *Candida* during parenteral nutrition. *Nutrition*, 10, 151–154.

Patel, P.K., Erlandsen, J.E., Kirkpatrick, W.R., Berg, D.K., Westbrook, S.D., Louden, C., Cornell, J.E., Thompson, G.R., Vallor, A.C., Wickes, B.L., Wiederhold, N.P., Redding, S.W., and Patterson, T.F. 2012. The changing epidemiology of oropharyngeal candidiasis in patients with HIV/AIDS in the era of antiretroviral therapy. *AIDS Res. Treat.*, 2012, 262471.

Perlroth, J., Choi, B., and Spellberg, B. 2007. Nosocomial fungal infections: Epidemiology, diagnosis, and treatment. *Med. Mycol.*, 45, 321–346.

Pertowski, C.A., Baron, R.C., Lasker, B.A., Werner, S.B., and Jarvis, W.R. 1995. Nosocomial outbreak of *Candida albicans* sternal wound infections following cardiac surgery traced to a scrub nurse. *J. Infect. Dis.*, 172, 817–822.

Pfaller, M., Neofytos, D., Diekema, D., Azie, N., Meier-Kriesche, H.U., Quan, S.P., and Horn, D. 2012. Epidemiology and outcomes of candidemia in 3648 patients: Data from the Prospective Antifungal Therapy (PATH Alliance(R)) registry, 2004–2008. *Diagn. Microbiol. Infect. Dis.*, 74, 323–331.

Phan, Q.T., Myers, C.L., Fu, Y., Sheppard, D.C., Yeaman, M.R., Welch, W.H., Ibrahim, A.S., Edwards Jr., J.E., and Filler, S.G. 2007. Als3 is a *Candida albicans* invasin that binds to cadherins and induces endocytosis by host cells. *PLoS Biol.*, 5, e64.

Pierce, J.V., Dignard, D., Whiteway, M., and Kumamoto, C.A. 2013. Normal adaptation of *Candida albicans* to the murine gastrointestinal tract requires Efg1p-dependent regulation of metabolic and host defense genes. *Eukaryot. Cell*, 12, 37–49.

Pietrella, D., Pandey, N., Gabrielli, E., Pericolini, E., Perito, S., Kasper, L., Bistoni, F., Cassone, A., Hube, B., and Vecchiarelli, A. 2013. Secreted aspartic proteases of *Candida albicans* activate the NLRP3 inflammasome. *Eur. J. Immunol.*, 43, 679–692.

Pittet, D., Monod, M., Suter, P.M., Frenk, E., and Auckenthaler, R. 1994. *Candida* colonization and subsequent infections in critically ill surgical patients. *Ann. Surg.*, 220, 751–758.

Pivarcsi, A., Bodai, L., Rethi, B., Kenderessy-Szabo, A., Koreck, A., Szell, M., Beer, Z., Bata-Csorgoo, Z., Magocsi, M., Rajnavolgyi, E., Dobozy, A., and Kemeny, L. 2003. Expression and function of Toll-like receptors 2 and 4 in human keratinocytes. *Int. Immunol.*, 15, 721–730.

Pontón, J., Moragues, M.D., and Quindós, G. 2002. Non-culture based diagnostics. In: Calderone, R., editor. *Candida and Candidiasis*, 1st ed. Washington, DC: ASM Press. pp. 395–425.

Prakobphol, A., Leffler, H., Hoover, C.I., and Fisher, S.J. 1997. Palmitoyl carnitine, a lysophospholipase-transacylase inhibitor, prevents *Candida* adherence in vitro. *FEMS Microbiol. Lett.*, 151, 89–94.

Prowle, J.R., Echeverri, J.E., Ligabo, E.V., Sherry, N., Taori, G.C., Crozier, T.M., Hart, G.K., Korman, T.M., Mayall, B.C., Johnson, P.D., and Bellomo, R. 2011. Acquired bloodstream infection in the intensive care unit: Incidence and attributable mortality. *Crit. Care*, 15, R100.

Puel, A., Cypowyj, S., Bustamante, J., Wright, J.F., Liu, L., Lim, H.K., Migaud, M., Israel, L., Chrabieh, M., Audry, M., Gumbleton, M., Toulon, A., Bodemer, C., El-Baghdadi, J., Whitters, M., Paradis, T., Brooks, J., Collins, M., Wolfman, N.M., Al-Muhsen, S., Galicchio, M., Abel, L., Picard, C., and Casanova, J.L. 2011. Chronic mucocutaneous candidiasis in humans with inborn errors of interleukin-17 immunity. *Science*, 332, 65–68.

Pugh, D., and Cawson, R.A. 1975. The cytochemical localization of phospholipase A and lysophospholipase in *Candida albicans*. *Sabouraudia*, 13(Pt 1), 110–115.

Puri, S., Kumar, R., Chadha, S., Tati, S., Conti, H.R., Hube, B., Cullen, P.J., and Edgerton, M. 2012. Secreted aspartic protease cleavage of *Candida albicans* Msb2 activates Cek1 MAPK signaling affecting biofilm formation and oropharyngeal candidiasis. *PLoS One*, 7, e46020.

Rabeneck, L., Crane, M.M., Risser, J.M., Lacke, C.E., and Wray, N.P. 1993. A simple clinical staging system that predicts progression to AIDS using CD4 count, oral thrush, and night sweats. *J. Gen. Intern. Med.*, 8, 5–9.

Ramage, G., Bachmann, S., Patterson, T.F., Wickes, B.L., and Lopez-Ribot, J.L. 2002. Investigation of multidrug efflux pumps in relation to fluconazole resistance in *Candida albicans* biofilms. *J. Antimicrob. Chemother.*, 49, 973–980.

Ramage, G., Saville, S.P., Thomas, D.P., and Lopez-Ribot, J.L. 2005. *Candida* biofilms: An update. *Eukaryot. Cell*, 4, 633–638.

Ramanan, N., and Wang, Y. 2000. A high-affinity iron permease essential for *Candida albicans* virulence. *Science*, 288, 1062–1064.

Ramirez-Zavala, B., Reuss, O., Park, Y.N., Ohlsen, K., and Morschhauser, J. 2008. Environmental induction of white-opaque switching in *Candida albicans*. *PLoS Pathog.*, 4, e1000089.

Reid, D.M., Gow, N.A., and Brown, G.D. 2009. Pattern recognition: Recent insights from Dectin-1. *Curr. Opin. Immunol.*, 21, 30–37.

Richardson, M.D., and Carlisle, P.L. 2002. Culture and non-culture based diagnostics for *Candida* species. In: Calderone, R.A., editor. *Candida and Candidiasis*, 1st ed. Washington, DC: ASM Press. pp. 387–394.

Roifman, C.M. 2000. Human IL-2 receptor alpha chain deficiency. *Pediatr. Res.*, 48, 6–11.

Romani, L. 1999. Immunity to *Candida albicans*: Th1, Th2 cells and beyond. *Curr. Opin. Microbiol.*, 2, 363–367.

Romani, L. 2011. Immunity to fungal infections. *Nat. Rev. Immunol.*, 11, 275–288.

Rosentul, D.C., Plantinga, T.S., Oosting, M., Scott, W.K., Velez Edwards, D.R., Smith, P.B., Alexander, B.D., Yang, J.C., Laird, G.M., Joosten, L.A., van der Meer, J.W., Perfect, J.R., Kullberg, B.J., Netea, M.G., and Johnson, M.D. 2011. Genetic variation in the dectin-1/CARD9 recognition pathway and susceptibility to candidemia. *J. Infect. Dis.*, 204, 1138–1145.

Rubin-Bejerano, I., Fraser, I., Grisafi, P., and Fink, G.R. 2003. Phagocytosis by neutrophils induces an amino acid deprivation response in *Saccharomyces cerevisiae* and *Candida albicans*. *Proc. Natl. Acad. Sci. USA*, 100, 11007–11012.

Ruchel, R., Ritter, B., and Schaffrinski, M. 1990. Modulation of experimental systemic murine candidosis by intravenous pepstatin. *Zentralbl. Bakteriol.*, 273, 391–403.

Runza, V.L., Schwaeble, W., and Mannel, D.N. 2008. Ficolins: Novel pattern recognition molecules of the innate immune response. *Immunobiology*, 213, 297–306.

Sabino, R., Verissimo, C., Brandao, J., Alves, C., Parada, H., Rosado, L., Paixao, E., Videira, Z., Tendeiro, T., Sampaio, P., and Pais, C. 2010. Epidemiology of candidemia in oncology patients: A 6-year survey in a Portuguese central hospital. *Med. Mycol.*, 48, 346–354.

Safdar, A., Chaturvedi, V., Cross, E.W., Park, S., Bernard, E.M., Armstrong, D., and Perlin, D.S. 2001. Prospective study of *Candida* species in patients at a comprehensive cancer center. *Antimicrob. Agents Chemother.*, 45, 2129–2133.

Saijo, S., Ikeda, S., Yamabe, K., Kakuta, S., Ishigame, H., Akitsu, A., Fujikado, N., Kusaka, T., Kubo, S., Chung, S.H., Komatsu, R., Miura, N., Adachi, Y., Ohno, N., Shibuya, K., Yamamoto, N., Kawakami, K., Yamasaki, S., Saito, T., Akira, S., and Iwakura, Y. 2010. Dectin-2 recognition of alpha-mannans and induction of Th17 cell differentiation is essential for host defense against *Candida albicans*. *Immunity*, 32, 681–691.

Saiman, L., Ludington, E., Pfaller, M., Rangel-Frausto, S., Wiblin, R.T., Dawson, J., Blumberg, H.M., Patterson, J.E., Rinaldi, M., Edwards, J.E., Wenzel, R.P., and Jarvis, W. 2000. Risk factors for candidemia in neonatal intensive care unit patients. The National Epidemiology of Mycosis Survey Study Group. *Pediatr. Infect. Dis. J.*, 19, 319–324.

Samaranayake, Y.H., Dassanayake, R.S., Cheung, B.P., Jayatilake, J.A., Yeung, K.W., Yau, J.Y., and Samaranayake, L.P. 2006. Differential phospholipase gene expression by *Candida albicans* in artificial media and cultured human oral epithelium. *APMIS*, 114, 857–866.

Sampaio, P., Gusmao, L., Alves, C., Pina-Vaz, C., Amorim, A., and Pais, C. 2003. Highly polymorphic microsatellite for identification of *Candida albicans* strains. *J. Clin. Microbiol.*, 41, 552–557.

Sanglard, D., Hube, B., Monod, M., Odds, F.C., and Gow, N.A. 1997. A triple deletion of the secreted aspartyl proteinase genes SAP4, SAP5, and SAP6 of *Candida albicans* causes attenuated virulence. *Infect. Immun.*, 65, 3539–3546.

Santos, M.A., Keith, G., and Tuite, M.F. 1993. Non-standard translational events in *Candida albicans* mediated by an unusual seryl-tRNA with a 5'-CAG-3' (leucine) anticodon. *EMBO J.*, 12, 607–616.

Sasse, C., Hasenberg, M., Weyler, M., Gunzer, M., and Morschhauser, J. 2012. White-opaque switching of *Candida albicans* allows immune evasion in an environment-dependent fashion. *Eukaryot. Cell*, 12(1), 50–58.

Saville, S.P., Lazzell, A.L., Chaturvedi, A.K., Monteagudo, C., and Lopez-Ribot, J.L. 2008. Use of a genetically engineered strain to evaluate the pathogenic potential of yeast cell and filamentous forms during *Candida albicans* systemic infection in immunodeficient mice. *Infect. Immun.*, 76, 97–102.

Schaller, M., Schafer, W., Korting, H.C., and Hube, B. 1998. Differential expression of secreted aspartyl proteinases in a model of human oral candidosis and in patient samples from the oral cavity. *Mol. Microbiol.*, 29, 605–615.

Schaller, M., Hube, B., Ollert, M.W., Schafer, W., Borg-von Zepelin, M., Thoma-Greber, E., and Korting, H.C. 1999. In vivo expression and localization of *Candida albicans* secreted aspartyl proteinases during oral candidiasis in HIV-infected patients. *J. Invest. Dermatol.*, 112, 383–386.

Schaller, M., Schackert, C., Korting, H.C., Januschke, E., and Hube, B. 2000. Invasion of *Candida albicans* correlates with expression of secreted aspartic proteinases during experimental infection of human epidermis. *J. Invest. Dermatol.*, 114, 712–717.

Schaller, M., Januschke, E., Schackert, C., Woerle, B., and Korting, H.C. 2001. Different isoforms of secreted aspartyl proteinases (Sap) are expressed by *Candida albicans* during oral and cutaneous candidosis in vivo. *J. Med. Microbiol.*, 50, 743–747.

Schaller, M., Bein, M., Korting, H.C., Baur, S., Hamm, G., Monod, M., Beinhauer, S., and Hube, B. 2003. The secreted aspartyl proteinases Sap1 and Sap2 cause tissue damage in an in vitro model of vaginal candidiasis based on reconstituted human vaginal epithelium. *Infect. Immun.*, 71, 3227–3234.

Schaller, M., Boeld, U., Oberbauer, S., Hamm, G., Hube, B., and Korting, H.C. 2004. Polymorphonuclear leukocytes (PMNs) induce protective Th1-type cytokine epithelial responses in an in vitro model of oral candidosis. *Microbiology*, 150, 2807–2813.

Schaller, M., Borelli, C., Korting, H.C., and Hube, B. 2005a. Hydrolytic enzymes as virulence factors of *Candida albicans*. *Mycoses*, 48, 365–377.

Schaller, M., Korting, H.C., Borelli, C., Hamm, G., and Hube, B. 2005b. *Candida albicans* – Secreted aspartic proteinases modify the epithelial cytokine response in an in vitro model of vaginal candidiasis. *Infect. Immun.*, 73, 2758–2765.

Schofield, D.A., Westwater, C., Warner, T., and Balish, E. 2005. Differential *Candida albicans* lipase gene expression during alimentary tract colonization and infection. *FEMS Microbiol. Lett.*, 244, 359–365.

Schweizer, A., Rupp, S., Taylor, B.N., Rollinghoff, M., and Schroppel, K. 2000. The TEA/ATTS transcription factor CaTec1p regulates hyphal development and virulence in *Candida albicans*. *Mol. Microbiol.*, 38, 435–445.

Scully, C., el-Kabir, M., and Samaranayake, L.P. 1994. *Candida* and oral candidosis: A review. *Crit. Rev. Oral Biol. Med.*, 5, 125–157.

Seita, J., Asakawa, M., Ooehara, J., Takayanagi, S., Morita, Y., Watanabe, N., Fujita, K., Kudo, M., Mizuguchi, J., Ema, H., Nakauchi, H., and Yoshimoto, T. 2008. Interleukin-27 directly induces differentiation in hematopoietic stem cells. *Blood*, 111, 1903–1912.

Sheppard, D.C., Yeaman, M.R., Welch, W.H., Phan, Q.T., Fu, Y., Ibrahim, A.S., Filler, S.G., Zhang, M., Waring, A.J., and Edwards Jr., J.E. 2004. Functional and structural diversity in the Als protein family of *Candida albicans. J. Biol. Chem.*, 279, 30480–30489.

Sinha, B.K., Prasad, S., and Monga, D.P. 1987. Studies of the role of B-cells in the resistance of mice to experimental candidiasis. *Zentralblatt Bakteriol., Mikrobiol. Hygiene. Ser. A*, 266, 316–322.

Slutsky, B., Staebell, M., Anderson, J., Risen, L., Pfaller, M., and Soll, D.R. 1987. "White-opaque transition": A second high-frequency switching system in *Candida albicans. J. Bacteriol.*, 169, 189–197.

Smeekens, S.P., Ng, A., Kumar, V., Johnson, M.D., Plantinga, T.S., van Diemen, C., Arts, P., Verwiel, E.T., Gresnigt, M.S., Fransen, K., van Sommeren, S., Oosting, M., Cheng, S.C., Joosten, L.A., Hoischen, A., Kullberg, B.J., Scott, W.K., Perfect, J.R., van der Meer, J.W., Wijmenga, C., Netea, M.G., and Xavier, R.J. 2013. Functional genomics identifies type I interferon pathway as central for host defense against *Candida albicans. Nat. Commun.*, 4, 1342.

Sobel, J.D. 2007. Vulvovaginal candidosis. *Lancet*, 369, 1961–1971.

Sobel, J.D., Fisher, J.F., Kauffman, C.A., and Newman, C.A. 2011. *Candida* urinary tract infections – epidemiology. *Clin. Infect. Dis.*, 52(Suppl 6), S433–S436.

Soll, D.R. 1997. Gene regulation during high-frequency switching in *Candida albicans. Microbiology*, 143(Pt 2), 279–288.

Soll, D.R. 2002. *Candida* commensalism and virulence: The evolution of phenotypic plasticity. *Acta Trop.*, 81, 101–110.

Song, E., Jaishankar, G.B., Saleh, H., Jithpratuck, W., Sahni, R., and Krishnaswamy, G. 2011. Chronic granulomatous disease: A review of the infectious and inflammatory complications. *Clin. Mol. Allergy*, 9, 10.

Spellberg, B. 2011. Vaccines for invasive fungal infections. *F1000 Med. Rep.*, 3, 13.

Srikantha, T., Daniels, K.J., Pujol, C., Kim, E., and Soll, D.R. 2013. Identification of genes upregulated by the transcription factor Bcr1 that are involved in impermeability, impenetrability, and drug resistance of *Candida albicans* A/alpha biofilms. *Eukaryot. Cell*, 12, 875–888.

Staab, J.F., and Sundstrom, P. 1998. Genetic organization and sequence analysis of the hypha-specific cell wall protein gene HWP1 of *Candida albicans. Yeast*, 14, 681–686.

Staab, J.F., Bradway, S.D., Fidel, P.L., and Sundstrom, P. 1999. Adhesive and mammalian transglutaminase substrate properties of *Candida albicans* Hwp1. *Science*, 283, 1535–1538.

Staib, P., Kretschmar, M., Nichterlein, T., Hof, H., and Morschhauser, J. 2002. Transcriptional regulators Cph1p and Efg1p mediate activation of the *Candida albicans* virulence gene SAP5 during infection. *Infect. Immun.*, 70, 921–927.

Stark, G.R., Kerr, I.M., Williams, B.R., Silverman, R.H., and Schreiber, R.D. 1998. How cells respond to interferons. *Annu. Rev. Biochem.*, 67, 227–264.

Steele, C., Leigh, J., Swoboda, R., and Fidel Jr., P.L. 2000. Growth inhibition of *Candida* by human oral epithelial cells. *J. Infect. Dis.*, 182, 1479–1485.

Stehr, F., Felk, A., Gacser, A., Kretschmar, M., Mahnss, B., Neuber, K., Hube, B., and Schafer, W. 2004. Expression analysis of the *Candida albicans* lipase gene family during experimental infections and in patient samples. *FEMS Yeast Res.*, 4, 401–408.

Stephan, F., Bah, M.S., Desterke, C., Rezaiguia-Delclaux, S., Foulet, F., Duvaldestin, P., and Bretagne, S. 2002. Molecular diversity and routes of colonization of *Candida albicans* in a surgical intensive care unit, as studied using microsatellite markers. *Clin. Infect. Dis.*, 35, 1477–1483.

Stoldt, V.R., Sonneborn, A., Leuker, C.E., and Ernst, J.F. 1997. Efg1p, an essential regulator of morphogenesis of the human pathogen *Candida albicans*, is a member of a conserved class of bHLH proteins regulating morphogenetic processes in fungi. *EMBO J.*, 16, 1982–1991.

Sugiyama, Y., Nakashima, S., Mirbod, F., Kanoh, H., Kitajima, Y., Ghannoum, M.A., and Nozawa, Y. 1999. Molecular cloning of

a second phospholipase B gene, caPLB2 from *Candida albicans. Med. Mycol.*, 37, 61–67.

Sullivan, D.J., and Coleman, D.C. 2002. Molecular approaches to identification and typing of *Candida* species. In: Calderone, R.A., editor. *Candida and Candidiasis*, 1st ed. Washington, DC: ASM Press. pp. 427–442.

Taylor, B.N., Hannemann, H., Sehnal, M., Biesemeier, A., Schweizer, A., Rollinghoff, M., and Schroppel, K. 2005. Induction of SAP7 correlates with virulence in an intravenous infection model of candidiasis but not in a vaginal infection model in mice. *Infect. Immun.*, 73, 7061–7063.

Theiss, S., Ishdorj, G., Brenot, A., Kretschmar, M., Lan, C.Y., Nichterlein, T., Hacker, J., Nigam, S., Agabian, N., and Kohler, G.A. 2006. Inactivation of the phospholipase B gene PLB5 in wild-type *Candida albicans* reduces cell-associated phospholipase A2 activity and attenuates virulence. *Int. J. Med. Microbiol.*, 296, 405–420.

Thewes, S., Kretschmar, M., Park, H., Schaller, M., Filler, S.G., and Hube, B. 2007. In vivo and ex vivo comparative transcriptional profiling of invasive and non-invasive *Candida albicans* isolates identifies genes associated with tissue invasion. *Mol. Microbiol.*, 63, 1606–1628.

Tortorano, A.M., Kibbler, C., Peman, J., Bernhardt, H., Klingspor, L., and Grillot, R. 2006. Candidaemia in Europe: Epidemiology and resistance. *Int. J. Antimicrob. Agents*, 27, 359–366.

Tran, L.T., Auger, P., Marchand, R., Carrier, M., and Pelletier, C. 1997. Epidemiological study of *Candida* spp. colonization in cardiovascular surgical patients. *Mycoses*, 40, 169–173.

Trinel, P.A., Plancke, Y., Gerold, P., Jouault, T., Delplace, F., Schwarz, R.T., Strecker, G., and Poulain, D. 1999. The *Candida albicans* phospholipomannan is a family of glycolipids presenting phosphoinositolmannosides with long linear chains of beta-1,2-linked mannose residues. *J. Biol. Chem.*, 274, 30520–30526.

Tumbarello, M., Tacconelli, E., de Gaetano Donati, K., Morace, G., Fadda, G., and Cauda, R. 1999. Candidemia in HIV-infected subjects. *Eur. J. Clin. Microbiol. Infect. Dis.*, 18, 478–483.

Urban, C.F., Ermert, D., Schmid, M., Abu-Abed, U., Goosmann, C., Nacken, W., Brinkmann, V., Jungblut, P.R., and Zychlinsky, A. 2009. Neutrophil extracellular traps contain calprotectin, a cytosolic protein complex involved in host defense against *Candida albicans. PLoS Pathog.*, 5, e1000639.

Uzel, G., Sampaio, E.P., Lawrence, M.G., Hsu, A.P., Hackett, M., Dorsey, M.J., Noel, R.J., Verbsky, J.W., Freeman, A.F., Janssen, E., Bonilla, F.A., Pechacek, J., Chandrasekaran, P., Browne, S.K., Agharahimi, A., Gharib, A.M., Mannurita, S.C., Yim, J.J., Gambineri, E., Torgerson, T., Tran, D.Q., Milner, J.D., and Holland, S.M. 2013. Dominant gain-of-function STAT1 mutations in FOXP3 wild-type immune dysregulation-polyendocrinopathy-enteropathy-X-linked-like syndrome. *J. Allergy Clin. Immunol.*, 131, 1611–1623.

van de Veerdonk, F.L., Marijnissen, R.J., Kullberg, B.J., Koenen, H.J., Cheng, S.C., Joosten, I., van den Berg, W.B., Williams, D.L., van der Meer, J.W., Joosten, L.A., and Netea, M.G. 2009. The macrophage mannose receptor induces IL-17 in response to *Candida albicans. Cell Host Microbe*, 5, 329–340.

van de Veerdonk, F.L., Joosten, L.A., Shaw, P.J., Smeekens, S.P., Malireddi, R.K., van der Meer, J.W., Kullberg, B.J., Netea, M.G., and Kanneganti, T.D. 2011a. The inflammasome drives protective Th1 and Th17 cellular responses in disseminated candidiasis. *Eur. J. Immunol.*, 41, 2260–2268.

van de Veerdonk, F.L., Plantinga, T.S., Hoischen, A., Smeekens, S.P., Joosten, L.A., Gilissen, C., Arts, P., Rosentul, D.C., Carmichael, A.J., Smits-van der Graaf, C.A., Kullberg, B.J., van der Meer, J.W., Lilic, D., Veltman, J.A., and Netea, M.G. 2011b. STAT1 mutations in autosomal dominant chronic mucocutaneous candidiasis. *N. Engl. J. Med.*, 365, 54–61.

van 't Wout, J.W., Linde, I., Leijh, P.C., and van Furth, R. 1988. Contribution of granulocytes and monocytes to resistance against experimental disseminated *Candida albicans* infection. *Eur. J. Clin. Microbiol. Infect. Dis.*, 7, 736–741.

van Till, J.W., Modderman, P.W., de Boer, M., Hart, M.H., Beld, M.G., and Boermeester, M.A. 2008. Mannose-binding lectin deficiency facilitates abdominal *Candida* infections in patients with secondary peritonitis. *Clin. Vaccine Immunol.*, 15, 65–70.

Vardakas, K.Z., Michalopoulos, A., Kiriakidou, K.G., Siampli, E.P., Samonis, G., and Falagas, M.E. 2009. Candidaemia: Incidence, risk factors, characteristics and outcomes in immunocompetent critically ill patients. *Clin. Microbiol. Infec.*, 15, 289–292.

Vargas, K., Messer, S.A., Pfaller, M., Lockhart, S.R., Stapleton, J.T., Hellstein, J., and Soll, D.R. 2000. Elevated phenotypic switching and drug resistance of *Candida albicans* from human immunodeficiency virus-positive individuals prior to first thrush episode. *J. Clin. Microbiol.*, 38, 3595–3607.

Vautier, S., Sousa Mda, G., and Brown, G.D. 2010. C-type lectins, fungi and Th17 responses. *Cytokine Growth Factor Rev.*, 21, 405–412.

Vazquez, J.A. 2010. Optimal management of oropharyngeal and esophageal candidiasis in patients living with HIV infection. *HIV AIDS (Auckl)*, 2, 89–101.

Vecchiarelli, A., Pericolini, E., Gabrielli, E., and Pietrella, D. 2012. New approaches in the development of a vaccine for mucosal candidiasis: Progress and challenges. *Front. Microbiol.*, 3, 294.

Verghese, S., Mullasari, A., Padmaja, P., Sudha, P., Sapna, M.C., and Cherian, K.M. 1998. Fungal endocarditis following cardiac surgery. *Indian Heart J.*, 50, 418–422.

Vijayan, D., Radford, K.J., Beckhouse, A.G., Ashman, R.B., and Wells, C.A. 2012. Mincle polarizes human monocyte and neutrophil responses to *Candida albicans*. *Immunol. Cell Biol.*, 90, 889–895.

Vilanova, M., Teixeira, L., Caramalho, I., Torrado, E., Marques, A., Madureira, P., Ribeiro, A., Ferreira, P., Gama, M., and Demengeot, J. 2004. Protection against systemic candidiasis in mice immunized with secreted aspartic proteinase 2. *Immunology*, 111, 334–342.

Villar, C.C., Kashleva, H., Nobile, C.J., Mitchell, A.P., and Dongari-Bagtzoglou, A. 2007. Mucosal tissue invasion by *Candida albicans* is associated with E-cadherin degradation, mediated by transcription factor Rim101p and protease Sap5p. *Infect. Immun.*, 75, 2126–2135.

Vultaggio, A., Lombardelli, L., Giudizi, M.G., Biagiotti, R., Mazzinghi, B., Scaletti, C., Mazzetti, M., Livi, C., Leoncini, F., Romagnani, S., Maggi, E., and Piccinni, M.P. 2008. T cells specific for *Candida albicans* antigens and producing type 2 cytokines in lesional mucosa of untreated HIV-infected patients with pseudomembranous oropharyngeal candidiasis. *Microbes Infect.*, 10, 166–174.

Watamoto, T., Samaranayake, L.P., Jayatilake, J.A., Egusa, H., Yatani, H., and Seneviratne, C.J. 2009. Effect of filamentation and mode of growth on antifungal susceptibility of *Candida albicans*. *Int. J. Antimicrob. Agents*, 34, 333–339.

Weindl, G., Naglik, J.R., Kaesler, S., Biedermann, T., Hube, B., Korting, H.C., and Schaller, M. 2007. Human epithelial cells establish direct antifungal defense through TLR4-mediated signaling. *J. Clin. Invest.*, 117, 3664–3672.

White, T.C., and Agabian, N. 1995. *Candida albicans* secreted aspartyl proteinases: Isoenzyme pattern is determined by cell type, and levels are determined by environmental factors. *J. Bacteriol.*, 177, 5215–5221.

Whiteway, M., and Bachewich, C. 2007. Morphogenesis in *Candida albicans*. *Annu. Rev. Microbiol.*, 61, 529–553.

Wilson, L.S., Reyes, C.M., Stolpman, M., Speckman, J., Allen, K., and Beney, J. 2002. The direct cost and incidence of systemic fungal infections. *Value Health*, 5, 26–34.

Wisplinghoff, H., Seifert, H., Tallent, S.M., Bischoff, T., Wenzel, R.P., and Edmond, M.B. 2003a. Nosocomial bloodstream infections in pediatric patients in United States hospitals: Epidemiology, clinical features and susceptibilities. *Pediatr. Infect. Dis. J.*, 22, 686–691.

Wisplinghoff, H., Seifert, H., Wenzel, R.P., and Edmond, M.B. 2003b. Current trends in the epidemiology of nosocomial bloodstream infections in patients with hematological malignancies and solid neoplasms in hospitals in the United States. *Clin. Infect. Dis.*, 36, 1103–1110.

Wisplinghoff, H., Bischoff, T., Tallent, S.M., Seifert, H., Wenzel, R.P., and Edmond, M.B. 2004. Nosocomial bloodstream infections in US hospitals: Analysis of 24,179 cases from a prospective nationwide surveillance study. *Clin. Infect. Dis.*, 39, 309–317.

Wu, H., Downs, D., Ghosh, K., Ghosh, A.K., Staib, P., Monod, M., and Tang, J. 2013. *Candida albicans* secreted aspartic proteases 4–6 induce apoptosis of epithelial cells by a novel Trojan horse mechanism. *FASEB J.*, 27, 2132–2144.

Wuthrich, M., Deepe Jr., G.S., and Klein, B. 2012. Adaptive immunity to fungi. *Annu. Rev. Immunol.*, 30, 115–148.

Xin, H., and Cutler, J.E. 2011. Vaccine and monoclonal antibody that enhance mouse resistance to candidiasis. *Clin. Vaccine Immunol.*, 18, 1656–1667.

Ylipalosaari, P., Ala-Kokko, T.I., Karhu, J., Koskela, M., Laurila, J., Ohtonen, P., and Syrjala, H. 2012. Comparison of the epidemiology, risk factors, outcome and degree of organ failures of patients with candidemia acquired before or during ICU treatment. *Crit. Care*, 16, R62.

Zhang, M.X., Bohlman, M.C., Itatani, C., Burton, D.R., Parren, P.W., St Jeor, S.C., and Kozel, T.R. 2006. Human recombinant antimannan immunoglobulin G1 antibody confers resistance to hematogenously disseminated candidiasis in mice. *Infect. Immun.*, 74, 362–369.

Zhao, X., Oh, S.H., Yeater, K.M., and Hoyer, L.L. 2005. Analysis of the *Candida albicans* Als2p and Als4p adhesins suggests the potential for compensatory function within the Als family. *Microbiology*, 151, 1619–1630.

Zotter, C., Haustein, U.F., Schonborn, C., Grimmecke, H.D., and Wand, H. 1990. Effect of pepstatin A on *Candida albicans* infection in the mouse. *Dermatol. Monatsschr.*, 176, 189–198.

Chapter 51

Pathogenesis of *Pneumocystis jirovecii* Pneumonia

Taylor Eddens, Aniket Kaloti, and Jay K. Kolls

Richard King Mellon Foundation Institute for Pediatric Research, Children's Hospital of Pittsburgh of UPMC, Pittsburgh, PA, USA

51.1 Introduction

Pneumocystis is a fungus that causes a diffuse, interstitial pneumonia in immunocompromised individuals. Over the past few decades, *Pneumocystis* has garnered a lot of attention due to its relationship with HIV-infected individuals. In fact, despite the advent of highly active antiretroviral therapy (HAART), *Pneumocystis* remains one of the most common opportunistic infections in individuals diagnosed with HIV. More recently, however, *Pneumocystis* infection has been associated with both the immunosuppression used for organ transplantation as well as newer immunosuppressive medications used to treat autoimmune diseases and malignancies. As the immunosuppressed population continues to grow as a result of these new disease-modifying therapeutic options, *Pneumocystis* may prove to be a re-emerging infectious disease that clinicians of various subspecialties must be able to recognize. In this chapter, we will discuss the epidemiology and pathogenesis of *Pneumocystis* infection, along with the immunologic defenses that are typically protective in an immunocompetent host. We will also examine the clinical features of *Pneumocystis* infection in the immunosuppressed patient, the current therapy against *Pneumocystis* infection, and the future of *Pneumocystis* research.

51.2 *Pneumocystis*: The pathogen

Pneumocystis was first described histologically in 1909 by Dr. Carlos Chagas, who originally classified *Pneumocystis* in the genus *Trypanosoma*. In 1912, Dr. Antonio Carini found similar cysts in the lungs of rats and at that time, *Pneumocystis carinii* became its own biological entity. Using the crude microscopy available in the early 1900s, *Pneumocystis* was classified as a protozoan given its unique biphasic life cycle, which had both cystic and trophic forms. The first documented cases of *Pneumocystis* infection in humans came at the end of World War II, when malnourished infants in orphanages began presenting with diffuse interstitial pneumonia (as reviewed in Chabe et al., 2011).

In the 1980s, following the emergence of the AIDS epidemic, molecular evidence began to arise that supported the notion that *Pneumocystis* belonged in the group fungi as opposed to being classified as a protozoan. Edman et al. (1988) were the first to molecularly confirm the homology of *Pneumocystis* to fungi, such as *Saccharomyces cerevisiae*, by using alignments of the 16S rRNA. Several other groups corroborated the homology of the 16S rRNA to fungi and established homology between *Pneumocystis* and fungal mitochondrial gene sequences as well (Pixley et al., 1991; Stringer et al., 1989; Wakefield et al., 1992). Furthermore, one group demonstrated the presence of elongation factor-3 (EF-3), a protein specific to fungi, in *Pneumocystis* (Ypma-Wong et al., 1992). Although more molecular evidence was uncovered, perhaps the most substantial evidence came in the context of the *Pneumocystis* genome project, when Cushion et al. (2007) made cDNA libraries of *P. carinii* and subsequently sequenced the contigs; the vast majority of these sequences aligned to fungal species such as *S. cerevisiae* and *Schizosaccharomyces pombe* as opposed to protozoa.

Along with reclassifying the genus *Pneumocystis* into the group fungi, researchers also discovered that genetic diversity existed between species of *Pneumocystis* isolated from different hosts. Originally, *Pneumocystis* was believed to be transmitted in a zoonotic fashion, until 1976 when Frenkel (1976) proposed that each *Pneumocystis* species was host-specific. After several other studies validated this theory, the genus *Pneumocystis* grew to contain several species. The genus *Pneumocystis* encompasses five named species: *P. carinii, P. wakefieldiae, P. murina, P. oyrctolagi,* and *Pneumocystis jirovecii* (Aliouat-Denis et al., 2008). *P. carinii*, which is isolated from rats, was the first to be discovered and is the most extensively studied organism within the genus. *P. wakefieldiae* is also isolated from rats and is most genetically similar to *P. carinii. P. murina, P. oyctolagi,* and *P. jirovecii* appear to be the sole species found in mice, rabbits, and humans, respectively (as reviewed in Aliouat-Denis et al., 2008). Interestingly, the amount of genetic variation between *P. carinii* and *P. jirovecii* suggests that these organisms diverged over 90–100 million years ago, which may explain why host-specificity appears to be so paramount to these organisms (Keely et al., 2003). In addition, several other unnamed species appear to have been isolated from other hosts such as nonhuman primates, ferrets, and horses (Peters et al., 1994; Wright et al., 1995a, 1995b).

Although researchers now recognize several species of *Pneumocystis*, one unifying aspect of *Pneumocystis* also happens to be one of the more unique features of the organism: the life cycle (Figure 51-1). The two most prominent stages in the life cycle of *Pneumocystis* are the cyst and troph forms. Cysts, or asci, are ovoid structures 4–7 μm in diameter with a thick wall (Aliouat-Denis et al., 2009; Chabe et al., 2011). Within the cyst, four nuclear divisions results in the formation of eight ascospores, which are ultimately thought to leave the cyst through a small pore and become trophs (Aliouat-Denis et al., 2009; Itatani, 1994). The new trophs are irregular in shape with a diameter of 2–8 μm and have numerous filopodia that allow for firm attachment to type

I pneumocytes in the lung (Aliouat-Denis et al., 2009). Trophic forms that leave the cyst are initially haploid and are thought to mate both sexually and asexually. The asexual phase of the life cycle is most likely binary fission, as budding, the typical asexual pathway for fungi, has not been shown to occur in *Pneumocystis* (Cushion, 2004). Binary fission results in the division of one troph into two genetically identical clones. Trophs can also undergo sexual reproduction within the host organism. This process begins with the firm attachment of two trophs through the use of pheromone receptors, similar to the process utilized by other fungi. Following attachment, the two trophs fuse to form a diploid early sporocyte. The early sporocytes have been shown to have nuclear synaptonemal complexes, indicating that this life form likely utilizes meiosis to divide (Chabe et al., 2011). Meiosis is then followed by mitosis, leading to the development of intermediate and late sporocytes. After mitosis results in eight nuclei, the wall of the late sporocyte thickens and develops into the cyst, thus completing the life cycle (Aliouat-Denis et al., 2009).

Pneumocystis cysts contain eight ascospores, which leave the cyst to become trophs. Trophs can reproduce asexually through binary fission, a process that results in two identical troph clones. Sexual reproduction can also occur, as two different trophs can fuse to become an early sporocyte. The early sporocyte will then undergo meiosis, forming four nuclei within the late sporocyte. The late sporocyte then thickens its wall and undergoes mitosis, resulting in a cyst.

All of the stages of the *Pneumocystis* life cycle occur in the alveolar space of the host lung. Within the alveolus, *Pneumocystis* uses filopodia to attach to the type I pneumocyte; the thickness of the filopodia is specific to each *Pneumocystis* species and may be one

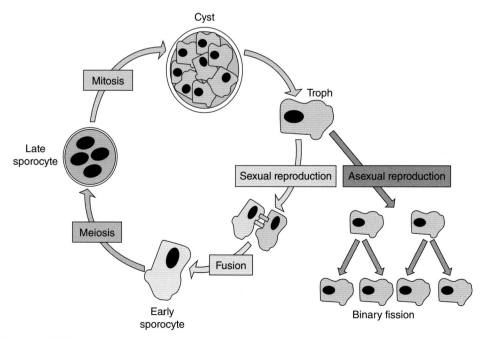

Fig. 51-1 The *Pneumocystis* life cycle.

mechanism of host specificity in the lung (Aliouat-Denis et al., 2008). These filopodia then penetrate, but do not fuse with, the membrane of the type I pneumocyte, allowing for a firm attachment ultimately resulting in a nidus of infection (Dei-Cas et al., 1991). *Pneumocystis* propagation appears to occur in both immunocompetent and immunosuppressed individuals, although the latter has greater clinical ramifications (see sections 51.3 and 51.9). Rarely, *Pneumocystis* can have extrapulmonary spread to sites like the ear, eye, thyroid, and bone marrow, which can complicate the course of the disease and can lead to a worse prognosis (Ng et al., 1997). Fortunately, the advent of HAART and the use of prophylaxis typically limit the incidence of extrapulmonary spread, and as such, we will primarily focus on pulmonary infection. It should also be noted that while the human pathogen is *P. jirovecii*, pneumonia caused by *Pneumocystis* in humans is still abbreviated as PCP (*Pneumocystis* pneumonia), although PJP may be used interchangeably (Stringer et al., 2002). We will refer to *Pneumocystis* pneumonia in humans as PCP for the remainder of the chapter.

51.3 Epidemiology

51.3.1 Current risk factors

Throughout 1960s and 1970s, *Pneumocystis* pneumonia (PCP) was most commonly described in patients with primary immune deficiencies and in patients undergoing immunosuppressive therapy for malignancies, organ transplant, and collagen vascular disorders. At that time, *Pneumocystis* was a relatively uncommon condition causing at most 100 cases per year in the United States (Schliep and Yarrish, 1999). More recently, rates of PCP in patients not receiving prophylactic therapy are 5–25% in transplant patients,

2–6% in patients with collagen vascular disease, and 1–25% in patients with cancer (Morris et al., 2004).

The Centers for Disease Control and Prevention (CDC) in the United States published the first four noted cases of PCP among homosexual men in Los Angeles in June 1981 (Centers for Disease Control and Prevention, 1981). During the AIDS epidemic of the early to mid-1980s, PCP was a diagnosed as a complication of 80% of AIDS cases and as the AIDS-defining illness in 60% cases. PCP remained the most common opportunistic infection among AIDS patients in the United States throughout the 1980s and early 1990s (Schliep and Yarrish, 1999). In these early days of the HIV/AIDS epidemic, PCP rates were as high as 20 per 100 person-years for those with CD4+ cell counts <200 cells/µL (Morris et al., 2004).

The introduction of anti-PCP prophylaxis in 1989 (Morris et al., 2004) led to a reduction in the incidence of PCP among HIV/AIDS patients in the United States and Western Europe. Following the advent of HAART in 1996, the incidence has continued to decline. Despite this progress, PCP still remains the most common serious opportunistic infection among HIV/AIDS patients in the United States and is the second most common AIDS-defining illness (Antiretroviral Therapy Cohort et al., 2009; Huang et al., 2011).

With the reduction in HIV/AIDS-associated PCP, other immune-compromised conditions have become important risk factors for PCP. The conditions that are well-known risk factors for PCP are listed in Table 51-1.

51.3.2 Current incidence

51.3.2.1 HIV/AIDS-associated PCP

As noted in Section 51.3.1, PCP is still one of the most serious opportunistic infections in the HIV/AIDS population in the developed world. Several reports in

Table 51-1 The most common risk factors associated with the development of *Pneumocystis* pneumonia (PCP)

Risk factor	Incidence of PCP due to risk factor in the absence of TMP-SMX prophylaxis	Incidence of PCP due to risk factor with TMP-SMX prophylaxis
HIV+ with CD4 count <200	>50%	1–3%
Hematological malignancy with immunosuppressive medication	16%	0.17–0.34%
Solid tumors	1–25%	1.3–1.7% in lung or breast cancer with CNS metastases and glucocorticoid injection;1–2% in patients with primary CNS tumors
Hematopoietic stem cell transplantation (including bone marrow)	5–15%	<1%
Solid organ transplantation	5–15%	≤1%
Rheumatoid arthritis	1–2%	2–4 cases/100,000 rheumatoid arthritis patients
Severe combined immunodeficiency	27–42%	Exact incidence rates are not available
Collagen vascular diseases	<2%	Exact incidence rates are not available
Dermatomyositis and polymyositis	6%	Exact incidence rates are not available
Wegener's granulomatosis	3.5–12%	Exact incidence rates are not available
Inflammatory bowel disease	4–4.6%	Exact incidence rates are not available

the 1980s and 1990s indicated that it was less common in the developing world, but that seems to be changing more recently (de Armas Rodriguez et al., 2011). Here, we review the current epidemiology of HIV-associated PCP.

51.3.2.1.1 Developed world

Morris et al. (2004) reviewed data from PCP epidemiology studies in the pre- and post-HAART era and demonstrated that the incidence of PCP was drastically reduced by the implementation of HAART. According to the data from the Adult and Adolescent Spectrum of HIV Disease (ASD) Project, PCP cases decreased 3.4% per year from 1992 through 1995 (pre-HAART); the rate of decline of PCP increased to 21.5% per year from 1996 through 1998 (post-HAART). Similarly in Europe, the incidence of PCP fell from 4.9 cases per 100 person-years before March 1995 to 0.3 cases per 100 person-years after March 1998 (Morris et al., 2004).

51.3.2.1.2 Developing world

de Armas Rodriguez et al. (2011) have recently reviewed the epidemiology of HIV-associated PCP in the developing world.

Africa: Some estimates have suggested very low incidence of PCP among black HIV positive adults in South Africa (Karstaedt, 1992). Several new studies have indicated a much higher incidence. Malin et al. (1995) detected PCP in 33% patients from bronchoalveolar lavage (BAL) fluid of 64 HIV-infected patients with acute diffuse pneumonia in southern Africa. Furthermore, Ruffini and Madhi (2002) identified *P. jirovecii* cysts in induced sputum or nasopharyngeal aspirates from 48.6% of 105 children with known or suspected HIV hospitalized with severe pneumonia in South Africa. In Kenya, Chakaya et al. (2003) tested 51 HIV-positive patients presenting with sub-acute onset cough and dyspnea and found that 37.2% patients were positive for *P. jirovecii* cysts as detected by toluidine blue staining on BAL samples. Aderaye et al. (2008) tested induced sputum and BAL samples of 131 HIV-infected patients in Ethiopia and found a total of 56 (42.7%) patients tested positive for *P. jirovecii* by polymerase chain reaction (PCR), 39 (29.4%) by immunofluorescence, and 28 (21.4%) by toluidine blue stain. Several reports indicate high PCP incidence rates in African children aged 3–6 months on autopsy specimens (de Armas Rodriguez et al., 2011).

Asia: Early estimates in the HIV era indicated low incidence of PCP in several parts of Asia (de Armas Rodriguez et al., 2011; Morris et al., 2004). In a prospective study in Northern Thailand including a total of 59 HIV/AIDS patients with interstitial infiltrates on chest radiographs, tuberculosis was the most common diagnosis (44%), followed by PCP (25.4%) (Tansuphasawadikul et al., 2005). Induced sputum samples studied by PCR detected PCP in 21% of 52 HIV/AIDS patients suspected of

PCP (de Armas Rodriguez et al., 2011). In a recent study in Iran, nested PCR and PCR-RFLP assays were used to amplify mtLSU rRNA and DHPS genes from 126 patients, leading to a diagnosis of PCP in 11.9% cases despite PCP prophylaxis and HAART (Sheikholeslami et al., 2013). In India, many studies have indicated a low number of PCP cases, with rates of 5–6.1% described in some reports (de Armas Rodriguez et al., 2011; Kumarasamy et al., 2003; Lanjewar and Duggal, 2001; Rajagopalan et al., 2009). Other reports have indicated higher PCP incidence (13%) among HIV-positive admissions in India, while an additional study noted a higher mortality (22%) rate among Indian HIV patients with PCP (Kumarasamy et al., 2010; Udwadia et al., 2005). In a Malaysian cohort, PCP was found in 12.2% of patients with HIV (Nissapatorn et al., 2004).

51.3.2.1.3 Mexico, Central and South America

de Armas Rodriguez et al. (2011) note that only a few reports of PCP are available in Mexico and Central America, including the Caribbean Islands. In Venezuela, PCP was found in 36.6% of HIV-infected patients with respiratory symptoms treated between 2001 and 2006 (Panizo et al., 2008). Argentina had the sixth largest number of cumulative pediatric cases of AIDS in the Americas in 2002 and PCP was the AIDS-defining illness in 35% of cases observed in a population of 132 pediatric HIV patients (Fallo et al., 2002). In Chile, PCP was the most common lung disease in HIV-infected patients, causing 37.7% of respiratory episodes in 236 assessed patients (Chernilo et al., 2005). In a Brazilian cohort, Soeiro et al. (2008) observed *Pneumocystis* among 27% of lung autopsies of HIV/AIDS patients.

51.3.2.2 Non-HIV PCP

As noted in Section 51.3.2, many PCP predisposing conditions exist other than HIV/AIDS. In the HAART and PCP prophylaxis era, non-HIV PCP is becoming an increasingly important problem in the developed world, with prophylaxis still being uncommon in many other vulnerable populations (Saltzman et al., 2012). Table 51-1 addresses this case for the United States, where most data are available. Similar rates are expected in the rest of the developed world, although few epidemiological studies have been conducted in other countries.

51.4 Morbidity and mortality

PCP infection in either HIV or non-HIV immunocompromised patients is associated with significant morbidity and mortality and can often result in intensive care admission and mechanical ventilation. The overall mortality from PCP in HIV-negative patients is significantly higher (~30–60%), than the 10–20% mortality rate seen in HIV-positive patients, a trend common across the United States, Europe, and Japan (Ewig et al., 1995; Reid et al., 2011; Sepkowitz, 2002; Stamp and Hurst, 2010). Several recent reports have indicated an even lower mortality rate (<10%) among HIV patients presenting with PCP, with some studies finding rates as low as 0% (Ainoda et al., 2012; Enomoto et al., 2010). Stamp and Hurst (2010) note the estimated mortality from PCP in patients with underlying rheumatic disease to be 62.5% in Wegener's granulomatosis, 57.7% in inflammatory myopathy, 47.6% in polyarteritis nodosa, 30.8% in RA, and 16.7% in systemic sclerosis. There have also been reports stating that a discontinuation of PCP prophylaxis in transplant patients can lead to the development of PCP and is associated with higher mortality rate (De Castro et al., 2005). Similarly, failure to prescribe prophylaxis in HIV-positive patients is associated with increased mortality even in the HAART era (Lim et al., 2012). Furthermore, prophylaxis failure due to trimethoprim–sulfamethoxazole (TMP-SMX) resistance via dihydropteroate synthase (DHPS) or dihydrofolate reductase (DHFR) gene mutations has been well documented (Huang et al., 2011).

Morbidity and mortality in PCP is commonly associated with several clinical findings such as pneumothorax, ICU admission, tracheal intubation, mechanical ventilation, and acute respiratory distress syndrome (ARDS) (Enomoto et al., 2010; Fei et al., 2009; Forrest et al., 1999; Radhi et al., 2008). These factors seem to have remained unchanged from the pre-HAART to the HAART era (Radhi et al., 2008). Several biomarkers have also been correlated with increased mortality, including high serum lactate dehydrogenase (LDH) and high alveolar–arterial oxygen gradient (Arozullah et al., 2000; Fei et al., 2009; Matsumura et al., 2011). Presence of underlying interstitial lung disease and non-HIV PCP status are highly correlated with development of respiratory failure and ICU admission, and in turn, results in a higher mortality rate (Enomoto et al., 2010). Some investigators have correlated increased BAL fluid neutrophil count with poorer oxygenation and worse survival rates (Limper et al., 1989).

51.5 Emerging risk factors

In the era of HAART and PCP prophylaxis as well as HIV awareness, non-HIV-related risk factors are becoming more prominent. Furthermore, introduction of new classes of immune-suppressive or immune-modulatory regimens have led to new risk factors for non-HIV-associated PCP. Rituximab, a chimeric monoclonal antibody against the CD20 protein, is used in several hematological and solid organ malignancies as well as collagen vascular diseases. Reid et al. (2011) note an increased risk of PCP in patients receiving Rituximab alone or in combination with glucocorticoids. Among organ transplant recipients, sirolimus and tacrolimus are used to prevent transplant rejection and PCP has been reported in conjunction with the administration of these two agents

(Reid et al., 2011). The authors also note an increased risk for PCP in patients receiving anti-TNF monoclonal antibodies (infliximab, adalimumab) or TNF inhibitors (etanercept) as additional risk factors for PCP in patients with chronic inflammatory diseases such as rheumatoid arthritis or systemic lupus erythematosus. Furthermore, patients with transplants, particularly lung transplant, are at long-term risk for developing PCP. Unlike in the case of HIV/AIDS, there is no well-defined biomarker (such as the CD4 lymphocyte count) that quantifies the risk, thus making it harder to develop criteria to start or stop prophylaxis. Given the adverse events profiles of the first and second line prophylaxis agents, these emerging risk factors and groups constitute a major shift in our thinking about prophylaxis and treatment of PCP.

51.6 PCP in pediatric population

Since the first noticed cases of human PCP among malnourished children and preterm infants in Germany, there has been remarkable progression in our understanding of the pathogenesis of pediatric PCP. Maternal to fetal transmission of HIV led to an upshot in PCP diagnoses among infants and pediatric population in the pre-HAART and PCP prophylaxis era in the United States. Currently, organ transplant and childhood hematological malignancies constitute the major risk factors for the development of PCP. Saltzman et al. (2012) conducted a retrospective review of 80 pediatric patients with PCP at Children's Hospital of Philadelphia admitted from 1996 to 2006. In this cohort, HIV was the single most common associated underlying condition, accounting for 39% of the cases overall, but only in 15% of the cases since 1998. Transplant recipients and oncology patients together comprised another 39% of the cases, while another 9% of cases were attributed to primary immune deficiency. The remaining 9% of cases were associated with less well-recognized causes of susceptibility. High mortality was noted in the HIV and organ transplant groups. The probability of ICU admission and tracheal intubation was greater than 50% in all cases, although there was no increase in mortality associated with ICU admission or tracheal intubation in patients with underlying cancer or primary immune deficiency when compared with HIV-positive PCP. The cause of this discrepancy is unclear, although HIV patients could have had more complicating factors (Saltzman et al., 2012). Saltzman et al. (2012) also noted that prophylaxis provided incomplete protection in the pediatric population, irrespective of the underlying condition, although they particularly noted the failure of prophylaxis among HIV patients. Pediatric HIV-associated PCP is also a common problem in Argentina, which had the sixth largest pediatric AIDS population and a very high PCP attack rate (de Armas Rodriguez et al., 2011; Fallo et al., 2002).

51.7 Pneumocystis transmission and colonization

The most widely accepted theory of *Pneumocystis* transmission is that of aerosolized particles from host to host (Carmona and Limper, 2011). However, various models of infection are still debated. The *reactivation of latent infection theory* postulates that initial infection occurs in infancy or early childhood, with reactivation later in life in the context of immune suppression. *Pneumocystis* DNA has been identified in the air surrounding apple orchards and the surface of pond water, indicating that it may be a ubiquitous organism in nature (Catherinot et al., 2010). Furthermore, seroprevalence of antibodies to *Pneumocystis* is very common among children and has been observed in many parts of the world, thus supporting the theory that children are exposed to *Pneumocystis* early (Carmona and Limper, 2011; Norris and Morris, 2011). It is thought that in health, the infection is self-contained, but in the case of immune suppression, the infection may reactivate.

The *re-acquisition or de novo infection theory* suggests *Pneumocystis* infection is actually re-acquired later in life after early childhood infection clearance, in the context of immune suppression. There is considerable evidence in support of this theory. PCP cases are known to have geographic clustering in the urban setting and genotypes found in patients with PCP are similar to those found in the current patient's environment but not those from his or her place of birth (Catherinot et al., 2010). It is believed now that the mode of transmission is airborne infection through person to person transmission (Thomas and Limper, 2004). Many studies have shown asymptomatic carrier state, particularly resulting by transmission of *Pneumocystis* from PCP patients to healthy hospital workers (Miller et al., 2001; Vargas et al., 2000). Additional evidence comes from animal studies that have demonstrated the transmission of *Pneumocystis* between immunocompetent and immunocompromised animals (Morris and Norris, 2012). Furthermore, asymptomatic carriage state has recently been observed in patients with non-HIV immunosuppression and pulmonary disease (Fily et al., 2011). Although asymptomatic carriage is common to both theories, the biggest difference between the *de novo* infection and the latent infection models is the length of time of colonization; in the latent infection model, colonization occurs early and may be stable at a low level until immune suppression, while the *de novo* infection model suggests that infection is acquired after immune suppression.

51.8 PCP and development of chronic lung disease

Gingo et al. (2011) first documented respiratory abnormalities such as diffusion impairment and irreversible airway obstruction in patients with HIV in the era of combination antiretroviral therapy. The development of chronic obstructive pulmonary disease (COPD) in nonhuman primates with persistent *Pneumocystis* colonization has also been demonstrated (Shipley et al., 2010). Further, infection with *Pneumocystis* has been shown to be associated with the development of COPD in human subjects with and without HIV infection (Norris and Morris, 2011).

51.9 Clinical features

PCP typically presents with subtle onset of progressive dyspnea, nonproductive cough, and low-grade fever. Physical examination typically reveals tachypnea, tachycardia, and normal findings on lung auscultation. Acute dyspnea with pleuritic chest pain may indicate the development of a pneumothorax (Thomas and Limper, 2004). Most authors have noted differences in the PCP presentation between HIV and non-HIV patients (Catherinot et al., 2010; Carmona and Limper, 2011; Thomas and Limper, 2004, 2007). HIV patients with PCP can present with sub-acute onset of progressive dyspnea, nonproductive to minimally productive cough, and low-grade fever and malaise. However, non-HIV patients with PCP can have a more acute presentation such as the acute onset of progressive dyspnea, fever and malaise, and a greater degree of hypoxemia and widened A–a gradient.

51.10 Pathogenesis and immunity

The inability to reliably culture *Pneumocystis* organisms *in vitro* has limited the field to experimental work with the pathogen using both clinical specimens and animal models of infection (Cushion et al., 1991). Human infection with *Pneumocystis* is associated more with defects in cell-mediated immunity than neutrophil dysfunction. As a result, *Pneumocystis* infections have become a particular clinical problem in association with HIV infection, where progressive loss of CD4+ helper T lymphocytes results in profound suppression of cell-mediated immunity. In fact, as noted in Section 51.3, individual risk of an HIV-infected adult for PCP shows a linear and inverse correlation with the number of circulating CD4+ lymphocytes (Stansell et al., 1997). This relationship has also been demonstrated in pediatric PCP infection, although the relative CD4 count may be higher in children (Simonds et al., 1995). The importance of the CD4+ T lymphocyte in host defense against *P. murina* is further supported by work with animal models. Immune-competent mice inoculated with *P. murina* are able to resolve the infection without treatment, while mice depleted of CD4+ T lymphocytes with an anti-CD4 monoclonal antibody develop progressive PCP (Beck et al., 1991b; Shellito et al., 1990). When administration of the monoclonal antibody is stopped and CD4+ lymphocytes are restored in lymphoid tissue, *P. murina* organisms are cleared from lung tissue and infection resolves (Shellito et al., 1990). Collectively, this work along with the work of others (Harmsen and Stankiewicz, 1990; Roths and Sidman, 1992) supports a key role for the CD4+ helper T lymphocyte in host defense against *Pneumocystis*.

51.10.1 Role of Th1 immunity

One mechanism through which CD4+ lymphocytes may mediate host defense against *P. carinii* is through elaboration of cytokines such as interferon (IFN). Lymphocytes exposed to *Pneumocystis* organisms (Theus et al., 1993) or the major surface glycoprotein of *Pneumocystis* (Theus et al., 1997) *in vitro* elaborate IFN. Moreover, lymphocytes from patients with AIDS demonstrate a reduced capacity to produce this cytokine after antigenic or mitogenic stimulation (Murray et al., 1984; Rudy et al., 1988). Although, IFN is not directly lethal to *P. carinii*, it can activate macrophages *in vitro* to kill the organism (Pesanti, 1991). However, evidence supporting an *in vivo* role for IFN in host defense is conflicting. Chen et al. (1992) showed that *in vivo* neutralization of IFN with an antibody does not alter clearance of *P. murina* in reconstituted scid mice. In confirmation of these results, this group has also investigated reconstituting mice with splenocytes from mice with a homozygous deletion of the IFN gene (Garvy et al., 1997a). Using this system, Garvy et al. (1997a) reported that scid mice reconstituted with IFN −/− splenocytes cleared the infection similarly as control mice, but had a prolonged lung inflammatory response. In contrast, administration of IFN by osmotic pump to steroid-treated rats (Shear et al., 1990) or by aerosolization to CD4-depleted mice infected with *P. murina* (Beck et al., 1991a) reduced the intensity of infection. Beck and colleagues have postulated that one potential target cell for exogenous IFN is the alveolar macrophage (AM), since aerosolized IFN can augment class Ia expression of these cells when given by this route (Beck et al., 1991a). Possible mechanisms for IFN to bolster host defense include upregulation of TNF production (Burchett et al., 1988), increased generation of superoxide (Drath, 1986), or increased release of reactive nitrogen intermediates (Sherman et al., 1991). In further support of the role of the AM, it has been recently demonstrated that depletion of AMs leads to delayed clearance of *P. carinii* in the rat lung (Limper et al., 1997). Additionally, overexpression of IFN with gene delivery results in augmented clearance which is in part dependent on the augmented recruitment of CXCR3+ CD8+ T-cells (Kolls et al., 1999; McAllister et al., 2004, 2005). Interestingly, despite these data showing that pharmacological administration of IFN is therapeutic, endogenous IFN is not required as IFN-γ knockout (KO) mice clear *Pneumocystis* infection (Garvy et al., 1997b).

CXCR3+ CD8+ T-cells have a phenotype consistent with Tc1 cells in that they produce IFNγ and GM-CSF (McAllister et al., 2004). GM-CSF deficient mice are susceptible to *Pneumocystis* infection (Paine et al., 2000) and recombinant GM-CSF can augment macrophage killing of the organism *in vitro* (McAllister et al., 2004). Exogenous GM-CSF also augments *Pneumocystis* clearance in mice (Mandujano et al., 1995). Humans that have auto-antibodies against GM-CSF develop pulmonary alveolar proteinosis (Borie et al., 2011) and they have increased susceptibility to *Pneumocystis* infection.

51.10.2 Role of Th2 immunity

It is also of interest that mice deficient in Th2 immunity, such as the IL-4 KO mice, clear *Pneumocystis* (Garvy et al., 1997b). We have recently constructed IL-4Rα (the receptor for IL-4 and IL-13), IL-12p40 double KO mice which are defective in IL-12 and IL-23 production as well as IL-4/IL-13 signaling. These IL-12p40 double KO mice show delayed, but complete, clearance of *Pneumocystis* infection (unpublished observations). Furthermore, it has been shown recently that IL-13 can also augment AM clearance of *Pneumocystis in vitro* (Nelson et al., 2011).

51.10.3 Role of Th17 immunity

Th17 cells are a newly described subset of CD4+ T-cells which express IL-17A, IL-17F, and IL-22 (Chen and Kolls, 2013) and can be regulated upstream by IL-23. *Pneumocystis* inoculation in mice results in the expression of both IL-23 and IL-17 in the lung (Rudner et al., 2007). IL-23-deficient mice had increased organism burden at 2 and 3 weeks but ultimately cleared the infection; these results demonstrate that IL-23 and IL-17 play a role in host immunity to the infection but are ultimately dispensable for organism clearance.

51.10.4 Role of B-cells

Although *Pneumocystis* is not a common complication of inherited defects in immunoglobulin production, such as common variable hypogammaglobulinemia, case reports suggest that it can rarely occur (Esolen et al., 1992; Raj et al., 2013). Moreover, B-cell-deficient mice are permissive for chronic *Pneumocystis* infection and appear to play a role in antigen presentation and T-cell priming in the lung (Marcotte et al., 1996). Patients with mutations in CD40L, a protein required for class switch recombination in B-cells, are also susceptible to *Pneumocystis* infection (Johnson et al., 1993). Treatment with anti-CD20 antibodies which deplete B-cells has also been associated with the development PCP (Martin-Garrido et al., 2013). Further evidence that antibodies play a role in the infection is demonstrated by the fact that the passive transfer of polyclonal sera from *Pneumocystis* vaccinated mice protects scid mice against infection (Zheng et al., 2001). Specific monoclonal antibodies raised against Kex1 (an antigen expressed in *Pneumocystis*) administered intranasally can also prevent infection in rodent models (Gigliotti et al., 2002). Additional studies in primates support the notion that antibody-mediated responses are protective, as anti-Kex1 antibody titers are associated with reduced rates of *Pneumocystis* colonization (Kling et al., 2010).

51.11 Diagnosis

51.11.1 Radiological features

On chest radiograph, PCP usually presents with bilateral, diffuse, reticular, or granular interstitial and alveolar infiltrates, predominantly in the perihilar region (Figure 51-2). CT typically shows ground-glass opacities with patchy distribution, again mainly in the perihilar regions. Multiple and bilateral cysts are observed in 10–34% of cases. Some atypical manifestations include multiple thick-walled cavitary nodules and noncavitary nodules. Lymphadenopathy and effusions are less frequent. In the setting of *Pneumocystis* prophylaxis with aerosolized pentamidine, patients are more likely to present with predominantly upper-lobe disease, pneumothorax, or cyst formation (Catherinot et al., 2010; Carmona and Limper, 2011; Thomas and Limper, 2004).

In line with the differences in clinical features between HIV and non-HIV PCP, there are several differences in the radiological presentation between those groups (Catherinot et al., 2010; Carmona and Limper, 2011; Thomas and Limper, 2004). HIV-positive patients usually have significantly more

Pneumocystis cysts and trophic forms in BAL fluid than non-HIV patients. Non-HIV patients present with more inflammation with higher BAL neutrophils than HIV patients. On chest x-ray, diffuse lung infiltrates consistent with severe lung injury are common among non-HIV patients, but are ultimately rare in HIV patients.

51.11.2 Microbiological diagnosis

Pneumocystis cannot be currently cultured in the laboratory and as such microscopic demonstration of *Pneumocystis* organisms in respiratory specimens such as BAL fluid or induced sputum remains the definitive diagnostic test to date. Trophic forms can be detected with modified Papanicolaou, Wright-Giemsa, or Gram-Weigert stains, while cysts can be stained with Gomori-methenamine silver (GMS) (Figure 51-2), cresyl echt violet, toluidine blue O, or Calcofluor white (Thomas and Limper, 2004). Fluorescein-labeled monoclonal anti-*Pneumocystis* antibodies are available for detecting *Pneumocystis* and possess higher sensitivity and specificity in induced-sputum samples than conventional stains. They are also able to stain both the cysts and trophic forms, unlike most dyes (Thomas and Limper, 2004). Giemsa staining, GMS staining, and fluorescein-conjugated monoclonal antibodies are the most common techniques used currently in laboratories for PCP diagnosis (Carmona and Limper, 2011).

Sputum induction combined with conventional staining has a sensitivity range of 55–66% in HIV-infected patients, whereas fluorescein-conjugated monoclonal antibodies increase the sensitivity of induced sputum to more than 90% in HIV-infected patients. PCP patients without AIDS generally have

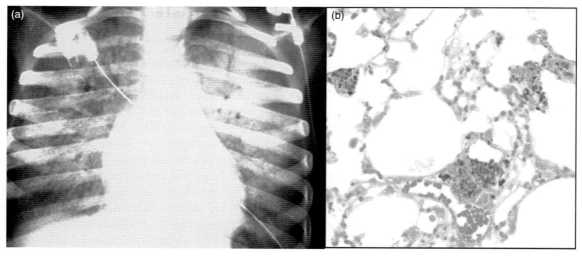

Fig. 51-2 (a) Chest x-ray of a patient with hyper IgM syndrome due to a mutation in CD40L with severe *Pneumocystis* pneumonia (PCP). The chest x-ray shows diffuse bilateral ground-glass infiltrates with distinct air bronchograms consistent with the alveolar nature of the infection. (b) Grocott's methenamine staining of a mouse lung infected with *Pneumocystis*. *Pneumocystis* asci can be seen residing within normal lung architecture.

lower burdens of *P. jirovecii* than those with AIDS, which makes detection more difficult. In these patients, bronchoscopy with BAL is the preferred diagnostic procedure, with reported sensitivity from 89% to 98% using tinctorial or immunofluorescent monoclonal antibody staining (Catherinot et al., 2010).

51.11.3 Molecular diagnosis

P. jirovecii DNA detection in clinical samples using a single-round PCR of the *P. jirovecii* mitochondrial small subunit rRNA (mtLSU rRNA) target gene has been used since 1990 (Reid et al., 2011; Wakefield et al., 1990). Since then, numerous gene targets and at least two types of PCR techniques have been used.

The first method is *conventional PCR* (single-step, nested or semi-nested), which has been used to detect *P. jirovecii* DNA in BAL or sputum samples. Several genes have been targeted for conventional PCR amplification, including mtLSU rRNA (single step and nested), ITS rRNA (nested), DHPS (single step and nested), MSG (hemi-nested), and 18S rRNA (nested) (Reid et al., 2011). In one large study, mtLSU rRNA nested PCR was found to be the most sensitive, with low false-negative rate, and had the highest correlation to histology (Robberts et al., 2007). Nested PCR methods offer superior sensitivities (approaching 100%) over single-step PCR, especially when tested on sputum (Reid et al., 2011; Robberts et al., 2007). In general, PCR diagnosis of PCP is significantly more sensitive (100%) than microscopy (60%) including immunofluorescence, however it is less specific than microscopy (87% for PCR vs. 100% for microscopy) (Flori et al., 2004). Other samples used for *P. jirovecii* DNA PCR include oropharyngeal washes and nasopharyngeal aspirates. One limitation of standard PCR methods is that it is unable to differentiate PCP from asymptomatic *Pneumocystis* colonization (Flori et al., 2004).

On the other hand, *quantitative real-time PCR* may allow differentiation between PCP infection and chronic carrier state by providing information regarding the burden of infection (Flori et al., 2004). These assays also have rapid turnaround time (<3 h) and reduced possibility of run-to-run contamination (Reid et al., 2011). Several gene targets have been used for real-time PCR: mtLSU rRNA, MSG, DHPS, ITS2, β-tubulin, Kex-1, and 5.8S rRNA (Reid et al., 2011). Quantitative real-time PCR with mtLSU rRNA, DHPS, and MSG have shown high sensitivity (90–100%) and high specificity (83–100%). On the other hand, ITS2 has shown lesser sensitivity and specificity (Reid et al., 2011). The remaining genes have been evaluated to a much lesser extent. Although PCR assays have clear advantages over traditional methods, they have typically been custom-designed in laboratories and not widely available as standardized commercial kits, making it difficult to benchmark and use them more generally.

51.11.4 Serum markers

β-(1,3)-D-glucan (BDG): β-glucans are structural components of the cell walls of most fungi (thus lacking specificity for *Pneumocystis*) including *P. jirovecii*, where it is particularly abundant in the cysts (Damiani et al., 2011). Substantially increased BDG levels have been found in patients with *Pneumocystis* infection and therefore could potentially aid in the diagnosis in patients unable to undergo invasive diagnostic procedures (Desmet et al., 2009; Onishi et al., 2012; Reid et al., 2011; Tasaka et al., 2007). A wide range of sensitivity and specificity values have been recorded, although these are dependent on the threshold set by the investigators. For example, Desmet and colleagues (2009) found a sensitivity of 100% and a specificity of 96.4% using a cut-off value of 100 pg/mL.

As noted above, BDG assays are nonspecific and are present in several fungal species. Some investigators have attempted to study the relative value of BDG as a diagnostic test. Onishi and colleagues (2012) performed a systematic analysis of several previous trials and found that BDG had a 96% sensitivity and 84% specificity for PCP, while registering 80% sensitivity and 82% specificity for invasive fungal infections such as candidiasis or aspergillosis. The comparable specificity for PCP versus invasive fungal infection means that BDG assays are best used as screening tools as opposed to a definitive diagnostic test. Damiani and colleagues (2011) found evidence for BDG level increase in primary PCP in immunocompetent infants, but were unable to find evidence to detect elevation in the case of colonization. HIV-positive PCP patients may have higher BDG detection rates than non-HIV PCP patients (Nakamura et al., 2009); however, some investigators have observed a reverse trend in the case of PCP in the settings of hematological malignancies (Desmet et al., 2009). Thus, elevated serum BDG level may be a relevant biomarker for PCP screening, as well as a diagnostic aid in patients with severe respiratory failure where obtaining BAL or induced sputum samples is not possible.

KL-6 is a mucin-like glycoprotein that is expressed on type II pneumocytes. Nakamura and colleagues performed a rigorous analysis of the role of KL-6 in PCP diagnosis. In their cohort, serum KL-6 levels were significantly higher in patients with HIV-related PCP than in those with non-HIV PCP and other pulmonary infections. While the serum KL-6 levels were significantly higher in non-HIV PCP patients than in the controls, these levels were not significantly different from those of patients with other infectious lung diseases. They also found lower sensitivity and specificity of KL-6 compared with BDG for both HIV and non-HIV PCP (Nakamura et al., 2009).

Table 51-2 Current treatment regimens for *Pneumocystis* pneumonia (PCP).

Regimen	Dosage	Treatment failure rate	Common adverse effects	Mortality
Trimethoprim–sulfamethoxazole (first choice)	15–20 mg/kg, IV or orally	21%	Cytopenia, skin reactions, hepatitis, pancreatitis, renal insufficiency, anaphylaxis, GI disturbance	Up to 15%
Dapsone + trimethoprim	5 mg/kg three times daily orally	39%	Skin rash, fever, methemoglobinemia, GI disturbance	~20%
Atovaquone	750 mg 2–3 times orally	36%	Skin rash, fever, GI disturbance, hepatitis	~20%
Clindamycin and primaquine	600 mg four times daily IV or 350–400 mg four times a day orally	35%	Skin rash, fever, neutropenia, GI disturbance, hepatitis	19%
Pentamidine (alternate choice for moderate to severe PCP)	4 mg/kg daily IV	40% or higher	Hypotension, cardiac arrhythmias, calcium, magnesium and potassium level disturbance, renal insufficiency, pancreatitis, hypoglycemia, neutropenia, hepatitis	≥24%

51.12 Treatment

Current treatment is listed in Table 51-2. For hospitalized PCP, first-line therapy remains TMP-SMX at 20 mg/kg/day (of the trimethoprim component) IV for 21 days. Sulfa allergy is typically managed with a second-line drug such as Pentamidine 4 mg/kg IV once daily for 21 days but the patients need to be monitored for nephrotoxicity, glucose abnormalities, pancreatitis, and cardiac arrhythmias. Other second-line therapies include Dapsone 100 mg PO once daily plus trimethoprim 15 mg/kg PO per day in three divided doses for 21 days or Clindamycin 600–900 mg IV every 6–8 h plus primaquine 15–30 mg PO once daily for 21 days. An additional alternative is Atovaquone 750 mg PO BID for 21 days, but as this is only available in oral form, it is indicated in mild-to-moderate disease.

Each of these treatments has its limitations. As alluded to above, the greatest experience in terms of treatment and prophylaxis is with TMP-SMX, but up to 8% of non-HIV-infected patients will be unable to tolerate the drug due to a sulfa allergy (Jick, 1982). Infection with HIV seems to increase the likelihood of having an adverse reaction to sulfa drugs, which can make adequate prophylaxis and treatment of PCP difficult (Phillips and Mallal, 2007). Adding to the challenge, second-line treatment like Pentamidine is typically associated with greater toxicity than TMP-SMX. Moreover, both first- and second-line therapies have a high clinical failure rate (Table 51-2).

51.13 Areas of further research and innovation

Since the discovery of *Pneumocystis* a little more than 100 years ago, the fungus has emerged as an incredible threat to the immunocompromised patient. The emergence of *Pneumocystis* corresponded with the HIV epidemic, although we have since learned of several other risk factors associated with PCP. This emergence led to the adoption of an anti-*Pneumocystis* prophylactic regimen, which helped limit the number of cases and deaths from PCP in the HIV population.

One of the most important future clinical challenges associated with *Pneumocystis* could be the emergence of TMP-SMX-resistant strains (Rabodonirina et al., 2013). Like other drug-resistant microbes, the prevalence of TMP-SMX-resistant strains will need to be monitored and drug susceptibility may have to be evaluated before prescribing the appropriate antimicrobial agent. Furthermore, evaluating the incidence of PCP in non-HIV-infected individuals will become even more important as new immunosuppressive medications and regimens continue to be developed. Identifying the most common risk factors, and considering prophylaxis in those populations, will be crucial to controlling and limiting the clinical sequelae of *Pneumocystis* infection in the future.

While the clinical advances outlined above are certainly of the utmost importance, it is also imperative to build on the foundation outlined by the *Pneumocystis* genome project and increase our knowledge of the pathogen. One hurdle that limits the current study of *Pneumocystis* is the lack of an *in vitro* culture system. However, understanding the molecular mechanisms underlying the virulence and transmission of *Pneumocystis* will be crucial to designing more targeted therapies in the future. For example, although the general process of the *Pneumocystis* life cycle is accepted and understood, the molecular processes driving the life cycle remains unclear. Elucidating this mechanism could lead to the development of newer, more targeted pharmaceuticals designed to block *Pneumocystis* replication in the lung. While this is certainly a lofty example, it is merely meant to illustrate that furthering our understanding of the organism itself could feed forward into the clinical realm and impact treatment and patient care.

The incidences of PCP with regards to these risk factors are reported both in presence and absence of trimethoprim-sulfamethoxazole (TMP-SMX) prophylaxis, when available (Antiretroviral Therapy Cohort et al., 2009; Carmona and Limper, 2011; Catherinot et al., 2010; De Castro et al., 2005; Falagas et al., 2007; Goto and Oka, 2011; Louie et al., 2010; Mussini

et al., 2008; Neff et al., 2009; Poppers and Scherl, 2008; Rodriguez and Fishman, 2004; Sepkowitz, 2002; Stamp and Hurst, 2010; Wolfe et al., 2010).

Common side effects of each medication are listed, as well as the percentage of patients requiring a change in treatment plan as a result of treatment failure or toxicity (Bellamy, 2008; Helweg-Larsen et al., 2009). Three-month mortality of PCP associated with each treatment is also listed (Helweg-Larsen et al., 2009).

References

Aderaye, G., Woldeamanuel, Y., Asrat, D., Lebbad, M., Beser, J., Worku, A., Fernandez, V., and Lindquist, L. 2008. Evaluation of Toluidine Blue O staining for the diagnosis of *Pneumocystis jiroveci* in expectorated sputum sample and bronchoalveolar lavage from HIV-infected patients in a tertiary care referral center in Ethiopia. *Infection*, 36, 237–243.

Ainoda, Y., Hirai, Y., Fujita, T., Isoda, N., and Totsuka, K. 2012. Analysis of clinical features of non-HIV *Pneumocystis jirovecii* pneumonia. *J. Infect. Chemother.*, 18, 722–728.

Aliouat-Denis, C.M., Chabe, M., Demanche, C., Aliouat el, M., Viscogliosi, E., Guillot, J., Delhaes, L., and Dei-Cas, E. 2008. Pneumocystis species, co-evolution and pathogenic power. *Infect. Genet. Evol.*, 8, 708–726.

Aliouat-Denis, C.M., Martinez, A., Aliouat el, M., Pottier, M., Gantois, N., and Dei-Cas, E. 2009. The Pneumocystis life cycle. *Mem. Inst. Oswaldo Cruz*, 104, 419–426.

Antiretroviral Therapy Cohort, C., Mocroft, A., Sterne, J.A., Egger, M., May, M., Grabar, S., Furrer, H., Sabin, C., Fatkenheuer, G., Justice, A., Reiss, P., d'Arminio Monforte, A., Gill, J., Hogg, R., Bonnet, F., Kitahata, M., Staszewski, S., Casabona, J., Harris, R., and Saag, M. 2009. Variable impact on mortality of AIDS-defining events diagnosed during combination antiretroviral therapy: Not all AIDS-defining conditions are created equal. *Clin. Infect. Dis.*, 48, 1138–1151.

Arozullah, A.M., Yarnold, P.R., Weinstein, R.A., Nwadiaro, N., McIlraith, T.B., Chmiel, J.S., Sipler, A.M., Chan, C., Goetz, M.B., Schwartz, D.N., and Bennett, C.L. 2000. A new preadmission staging system for predicting inpatient mortality from HIV-associated *Pneumocystis carinii* pneumonia in the early highly active antiretroviral therapy (HAART) era. *Am. J. Respir. Crit. Care Med.*, 161, 1081–1086.

Beck, J.M., Liggitt, H.D., Brunette, E.N., Fuchs, H.J., Shellito, J.E., and Debs, R.J. 1991a. Reduction in intensity of *Pneumocystis carinii* pneumonia in mice by aerosol administration of gamma interferon. *Infect. Immun.*, 59, 3859–3862.

Beck, J.M., Warnock, M.L., Curtis, J.L., Sniezek, M.J., Arraj-Peffer, S.M., Kaltreider, H.B., and Shellito, J.E. 1991b. Inflammatory responses to *Pneumocystis carinii* in mice selectively depleted of helper T lymphocytes. *Am. J. Respir. Cell. Mol. Biol.*, 5, 186–197.

Bellamy, R.J. 2008. HIV: Treating *Pneumocystis* pneumonia (PCP). *BMJ Clin. Evid.*, 2008, 2501.

Borie, R., Danel, C., Debray, M.P., Taille, C., Dombret, M.C., Aubier, M., Epaud, R., and Crestani, B. 2011. Pulmonary alveolar proteinosis. *Eur. Respir. Rev.*, 20, 98–107.

Burchett, S.K., Weaver, W.M., Westall, J.A., Larsen, A., Kronheim, S., and Wilson, C.B. 1988. Regulation of tumor necrosis factor/cachectin and IL-1 secretion in human mononuclear phagocytes. *J. Immunol.*, 140, 3473–3481.

Carmona, E.M., and Limper, A.H. 2011. Update on the diagnosis and treatment of *Pneumocystis* pneumonia. *Ther. Adv. Respir. Dis.*, 5, 41–59.

Catherinot, E., Lanternier, F., Bougnoux, M.E., Lecuit, M., Couderc, L.J., and Lortholary, O. 2010. *Pneumocystis jirovecii* pneumonia. *Infect. Dis. Clin. North Am.*, 24, 107–138.

Centers for Disease Control and Prevention. 1981. *Pneumocystis* pneumonia–Los Angeles. *MMWR Morb. Mortal. Wkly. Rep.*, 30, 250–252.

Chabe, M., Aliouat-Denis, C.M., Delhaes, L., Aliouat el, M., Viscogliosi, E., and Dei-Cas, E. 2011. Pneumocystis: From a doubtful unique entity to a group of highly diversified fungal species. *FEMS Yeast Res.*, 11, 2–17.

Chakaya, J.M., Bii, C., Ng'ang'a, L., Amukoye, E., Ouko, T., Muita, L., Gathua, S., Gitau, J., Odongo, I., Kabanga, J.M., Nagai, K., Suzumura, S., and Sugiura, Y. 2003. *Pneumocystis carinii* pneumonia in HIV/AIDS patients at an urban district hospital in Kenya. *East Afr. Med. J.*, 80, 30–35.

Chen, K., and Kolls, J.K. 2013. T cell-mediated host immune defenses in the lung. *Annu. Rev. Immunol.*, 31, 605–633.

Chen, W., Havell, E.A., and Harmsen, A.G. 1992. Importance of endogenous tumor necrosis factor alpha and gamma interferon in host resistance against *Pneumocystis carinii* infection. *Infect. Immun.*, 60, 1279–1284.

Chernilo, S., Trujillo, S., Kahn, M., Paredes, M., Echevarria, G., and Sepulveda, C. 2005. [Lung diseases among HIV infected patients admitted to the "Instituto Nacional del Torax" in Santiago, Chile]. *Rev. Med. Chil.*, 133, 517–524.

Cushion, M.T. 2004. Pneumocystis: Unraveling the cloak of obscurity. *Trends Microbiol.*, 12, 243–249.

Cushion, M.T., Smulian, A.G., Slaven, B.E., Sesterhenn, T., Arnold, J., Staben, C., Porollo, A., Adamczak, R., and Meller, J. 2007. Transcriptome of *Pneumocystis carinii* during fulminate infection: Carbohydrate metabolism and the concept of a compatible parasite. *PLoS One*, 2, e423.

Cushion, M.T., Stringer, J.R., and Walzer, P.D. 1991. Cellular and molecular biology of *Pneumocystis carinii*. *Int. Rev. Cytol.*, 131, 59–107.

Damiani, C., Le Gal, S., Lejeune, D., Brahimi, N., Virmaux, M., Nevez, G., and Totet, A. 2011. Serum (1->3)-beta-D-glucan levels in primary infection and pulmonary colonization with *Pneumocystis jirovecii*. *J. Clin. Microbiol.*, 49, 2000–2002.

de Armas Rodriguez, Y., Wissmann, G., Muller, A.L., Pederiva, M.A., Brum, M.C., Brackmann, R.L., Capo de Paz, V., and Calderon, E.J. 2011. *Pneumocystis jirovecii* pneumonia in developing countries. *Parasite*, 18, 219–228.

De Castro, N., Neuville, S., Sarfati, C., Ribaud, P., Derouin, F., Gluckman, E., Socie, G., and Molina, J.M. 2005. Occurrence of *Pneumocystis jiroveci* pneumonia after allogeneic stem cell transplantation: A 6-year retrospective study. *Bone Marrow Transplant.*, 36, 879–883.

Debs, R.J., Fuchs, H.J., Philip, R., Montgomery, A.B., Brunette, E.N., Liggitt, D., Patton, J.S., and Shellito, J.E. 1988. Lung-specific delivery of cytokines induces sustained pulmonary and systemic immunomodulation in rats. *J. Immunol.*, 140, 3482–3488.

Dei-Cas, E., Jackson, H., Palluault, F., Aliouat, E.M., Hancock, V., Soulez, B., and Camus, D. 1991. Ultrastructural observations on the attachment of *Pneumocystis carinii* in vitro. *J. Protozool.*, 38, 205S–207S.

Desmet, S., Van Wijngaerden, E., Maertens, J., Verhaegen, J., Verbeken, E., De Munter, P., Meersseman, W., Van Meensel, B., Van Eldere, J., and Lagrou, K. 2009. Serum (1–3)-beta-D-glucan as a tool for diagnosis of *Pneumocystis jirovecii* pneumonia in patients with human immunodeficiency virus infection or hematological malignancy. *J. Clin. Microbiol.*, 47, 3871–3874.

Drath, D.B. 1986. Modulation of pulmonary macrophage superoxide release and tumoricidal activity following activation by biological response modifiers. *Immunopharmacology*, 12, 117–126.

Edman, J.C., Kovacs, J.A., Masur, H., Santi, D.V., Elwood, H.J., and Sogin, M.L. 1988. Ribosomal RNA sequence shows *Pneumocystis carinii* to be a member of the fungi. *Nature*, 334, 519–522.

Enomoto, T., Azuma, A., Kohno, S., Kaneko, K., Saito, H., Kametaka, M., Usuki, J., Gemma, A., Kudoh, S., and Nakamura, S. 2010. Differences in the clinical characteristics of *Pneumocystis jirovecii* pneumonia in immunocompromised patients with and without HIV infection. *Respirology*, 15, 126–131.

Esolen, L.M., Fasano, M.B., Flynn, J., Burton, A., and Lederman, H.M. 1992. *Pneumocystis carinii* osteomyelitis in a patient with common variable immunodeficiency. *N. Engl. J. Med.*, 326, 999–1001.

Ewig, S., Bauer, T., Schneider, C., Pickenhain, A., Pizzulli, L., Loos, U., and Luderitz, B. 1995. Clinical characteristics and outcome of *Pneumocystis carinii* pneumonia in HIV-infected and otherwise immunosuppressed patients. *Eur. Respir. J.*, 8, 1548–1553.

Falagas, M.E., Manta, K.G., Betsi, G.I., and Pappas, G. 2007. Infection-related morbidity and mortality in patients with connective tissue diseases: A systematic review. *Clin. Rheumatol.*, 26, 663–670.

Fallo, A.A., DobrzanskiNisiewicz, W., Sordelli, N., Cattaneo, M.A., Scott, G., and Lopez, E.L. 2002. Clinical and epidemiologic aspects of human immunodeficiency virus-1-infected children in Buenos Aires, Argentina. *Int. J. Infect. Dis.*, 6, 9–16.

Fei, M.W., Kim, E.J., Sant, C.A., Jarlsberg, L.G., Davis, J.L., Swartzman, A., and Huang, L. 2009. Predicting mortality from HIV-associated *Pneumocystis* pneumonia at illness presentation: An observational cohort study. *Thorax*, 64, 1070–1076.

Fily, F., Lachkar, S., Thiberville, L., Favennec, L., and Caron, F. 2011. *Pneumocystis jirovecii* colonization and infection among non HIV-infected patients. *Med. Maladies Infect.*, 41, 526–531.

Flori, P., Bellete, B., Durand, F., Raberin, H., Cazorla, C., Hafid, J., Lucht, F., and Sung, R.T. 2004. Comparison between real-time PCR, conventional PCR and different staining techniques for diagnosing *Pneumocystis jiroveci* pneumonia from bronchoalveolar lavage specimens. *J. Med. Microbiol.*, 53, 603–607.

Forrest, D.M., Zala, C., Djurdjev, O., Singer, J., Craib, K.J., Lawson, L., Russell, J.A., and Montaner, J.S. 1999. Determinants of short- and long-term outcome in patients with respiratory failure caused by AIDS-related *Pneumocystis carinii* pneumonia. *Arch. Intern. Med.*, 159, 741–747.

Frenkel, J.K. 1976. *Pneumocystis jiroveci* n. sp. from man: Morphology, physiology, and immunology in relation to pathology. *Natl. Cancer Inst. Monogr.*, 43, 13–30.

Garvy, B.A., Ezekowitz, R.A., and Harmsen, A.G. 1997a. Role of gamma interferon in the host immune and inflammatory responses to *Pneumocystis carinii* infection. *Infect. Immun.*, 65, 373–379.

Garvy, B.A., Wiley, J.A., Gigliotti, F., and Harmsen, A.G. 1997b. Protection against *Pneumocystis carinii* pneumonia by antibodies generated from either T helper 1 or T helper 2 responses. *Infect. Immun.*, 65, 5052–5056.

Gigliotti, F., Haidaris, C.G., Wright, T.W., and Harmsen, A.G. 2002. Passive intranasal monoclonal antibody prophylaxis against murine *Pneumocystis carinii* pneumonia. *Infect. Immun.*, 70, 1069–1074.

Gingo, M.R., Lucht, L., Daly, K.R., Djawe, K., Palella, F.J., Abraham, A.G., Bream, J.H., Witt, M.D., Kingsley, L.A., Norris, K.A., Walzer, P.D., and Morris, A. 2011. Serologic responses to pneumocystis proteins in HIV patients with and without *Pneumocystis jirovecii* pneumonia. *J. Acquir. Immune. Defic. Syndr.*, 57, 190–196.

Goto, N., and Oka, S. 2011. *Pneumocystis jirovecii* pneumonia in kidney transplantation. *Transpl. Infect. Dis.*, 13, 551–558.

Harmsen, A.G., and Stankiewicz, M. 1990. Requirement for CD4+ cells in resistance to *Pneumocystis carinii* pneumonia in mice. *J. Exp. Med.*, 172, 937–945.

Helweg-Larsen, J., Benfield, T., Atzori, C., and Miller, R.F. 2009. Clinical efficacy of first- and second-line treatments for HIV-associated *Pneumocystis jirovecii* pneumonia: A tri-centre cohort study. *J. Antimicrob. Chemother.*, 64, 1282–1290.

Huang, L., Cattamanchi, A., Davis, J.L., den Boon, S., Kovacs, J., Meshnick, S., Miller, R.F., Walzer, P.D., Worodria, W., Masur, H., International HIV-Associated Opportunistic Pneumonias Study, and Lung HIV Study. 2011. HIV-associated *Pneumocystis* pneumonia. *Proc. Am. Thorac. Soc.*, 8, 294–300.

Itatani, C.A. 1994. Ultrastructural demonstration of a pore in the cyst wall of *Pneumocystis carinii*. *J. Parasitol.*, 80, 644–648.

Jick, H. 1982. Adverse reactions to trimethoprim-sulfamethoxazole in hospitalized patients. *Rev. Infect. Dis.*, 4, 426–428.

Johnson, J., Filipovich, A.H., and Zhang, K. 1993. X-linked hyper IgM syndrome. In: Pagon, R.A., Adam, M.P., Bird, T.D., Dolan, C.R., Fong, C.T., and Stephens, K., editors. *Gene Reviews*. Seattle, WA: University of Washington.

Karstaedt, A.S. 1992. AIDS–The Baragwanath experience. Part III. HIV infection in adults at Baragwanath Hospital. *S. Afr. Med. J.*, 82, 95–97.

Keely, S.P., Fischer, J.M., and Stringer, J.R. 2003. Evolution and speciation of *Pneumocystis*. *J. Eukaryot. Microbiol.*, 50(Suppl), 624–626.

Kling, H.M., Shipley, T.W., Patil, S.P., Kristoff, J., Bryan, M., Montelaro, R.C., Morris, A., and Norris, K.A. 2010. Relationship of *Pneumocystis jiroveci* humoral immunity to prevention of colonization and chronic obstructive pulmonary disease in a primate model of HIV infection. *Infect. Immun.*, 78, 4320–4330.

Kolls, J.K., Habetz, S., Shean, M.K., Vazquez, C., Brown, J.A., Lei, D., Schwarzenberger, P., Ye, P., Nelson, S., Summer, W.R., and Shellito, J.E. 1999. IFN-gamma and CD8+ T cells restore host defenses against *Pneumocystis carinii* in mice depleted of CD4+ T cells. *J. Immunol.*, 162, 2890–2894.

Kumarasamy, N., Solomon, S., Flanigan, T.P., Hemalatha, R., Thyagarajan, S.P., and Mayer, K.H. 2003. Natural history of human immunodeficiency virus disease in southern India. *Clin. Infect. Dis.*, 36, 79–85.

Kumarasamy, N., Venkatesh, K.K., Devaleenol, B., Poongulali, S., Yephthomi, T., Pradeep, A., Saghayam, S., Flanigan, T., Mayer, K.H., and Solomon, S. 2010. Factors associated with mortality among HIV-infected patients in the era of highly active antiretroviral therapy in southern India. *Int. J. Infect. Dis.*, 14, e127–e131.

Lanjewar, D.N., and Duggal, R. 2001. Pulmonary pathology in patients with AIDS: An autopsy study from Mumbai. *HIV Med.*, 2, 266–271.

Lim, P.L., Zhou, J., Ditangco, R.A., Law, M.G., Sirisanthana, T., Kumarasamy, N., Chen, Y.M., Phanuphak, P., Lee, C.K., Saphonn, V., Oka, S., Zhang, F., Choi, J.Y., Pujari, S., Kamarulzaman, A., Li, P.C., Merati, T.P., Yunihastuti, E., Messerschmidt, L., Sungkanuparph, S., and Treat Asia HIV Observational Database 2012. Failure to prescribe pneumocystis prophylaxis is associated with increased mortality, even in the cART era: Results from the Treat Asia HIV observational database. *J. Int. AIDS Soc.*, 15, 1.

Limper, A.H., Hoyte, J.S., and Standing, J.E. 1997. The role of alveolar macrophages in *Pneumocystis carinii* degradation and clearance from the lung. *J. Clin. Invest.*, 99, 2110–2117.

Limper, A.H., Offord, K.P., Smith, T.F., and Martin 2nd, W.J. 1989. *Pneumocystis carinii* pneumonia. Differences in lung parasite number and inflammation in patients with and without AIDS. *Am. Rev. Respir. Dis.*, 140, 1204–1209.

Louie, G.H., Wang, Z., and Ward, M.M. 2010. Trends in hospitalizations for *Pneumocystis jiroveci* pneumonia among patients with rheumatoid arthritis in the US: 1996–2007. *Arthritis Rheum.*, 62, 3826–3827.

Malin, A.S., Gwanzura, L.K., Klein, S., Robertson, V.J., Musvaire, P., and Mason, P.R. 1995. *Pneumocystis carinii* pneumonia in Zimbabwe. *Lancet*, 346, 1258–1261.

Mandujano, J.F., D'Souza, N.B., Nelson, S., Summer, W.R., Beckerman, R.C., and Shellito, J.E. 1995. Granulocyte-macrophage colony stimulating factor and *Pneumocystis carinii* pneumonia in mice. *Am. J. Respir. Crit. Care Med.*, 151, 1233–1238.

Marcotte, H., Levesque, D., Delanay, K., Bourgeault, A., de la Durantaye, R., Brochu, S., and Lavoie, M.C. 1996. *Pneumocystis carinii* infection in transgenic B cell-deficient mice. *J. Infect. Dis.*, 173, 1034–1037.

Martin-Garrido, I., Carmona, E.M., Specks, U., and Limper, A.H. 2013. *Pneumocystis* pneumonia in patients treated with rituximab. *Chest*, 144, 258–265.

Matsumura, Y., Shindo, Y., Iinuma, Y., Yamamoto, M., Shirano, M., Matsushima, A., Nagao, M., Ito, Y., Takakura, S., Hasegawa, Y., and Ichiyama, S. 2011. Clinical characteristics of *Pneumocystis* pneumonia in non-HIV patients and prognostic factors including microbiological genotypes. *BMC Infect. Dis.*, 11, 76.

McAllister, F., Steele, C., Zheng, M., Shellito, J.E., and Kolls, J.K. 2005. In vitro effector activity of *Pneumocystis murina*-specific T-cytotoxic-1 CD8+ T cells: Role of granulocyte-macrophage colony-stimulating factor. *Infect. Immun.*, 73, 7450–7457.

McAllister, F., Steele, C., Zheng, M., Young, E., Shellito, J.E., Marrero, L., and Kolls, J.K. 2004. T cytotoxic-1 CD8+ T cells are effector cells against pneumocystis in mice. *J. Immunol.*, 172, 1132–1138.

Miller, R.F., Ambrose, H.E., and Wakefield, A.E. 2001. *Pneumocystis carinii* f. sp. hominis DNA in immunocompetent health care workers in contact with patients with *P. carinii* pneumonia. *J. Clin. Microbiol.*, 39, 3877–3882.

Morris, A., Lundgren, J.D., Masur, H., Walzer, P.D., Hanson, D.L., Frederick, T., Huang, L., Beard, C.B., and Kaplan, J.E. 2004. Current epidemiology of *Pneumocystis* pneumonia. *Emerg. Infect. Dis.*, 10, 1713–1720.

Morris, A., and Norris, K.A. 2012. Colonization by *Pneumocystis jirovecii* and its role in disease. *Clin. Microbiol. Rev.*, 25, 297–317.

Murray, H.W., Rubin, B.Y., Masur, H., and Roberts, R.B. 1984. Impaired production of lymphokines and immune (gamma) interferon in the acquired immunodeficiency syndrome. *N. Engl. J. Med.*, 310, 883–889.

Mussini, C., Manzardo, C., Johnson, M., Monforte, A., Uberti-Foppa, C., Antinori, A., Gill, M.J., Sighinolfi, L., Borghi, V., Lazzarin, A., Miro, J.M., Sabin, C., and Late Presenter, I. 2008. Patients presenting with AIDS in the HAART era: A collaborative cohort analysis. *Aids*, 22, 2461–2469.

Nakamura, H., Tateyama, M., Tasato, D., Haranaga, S., Yara, S., Higa, F., Ohtsuki, Y., and Fujita, J. 2009. Clinical utility of serum beta-D-glucan and KL-6 levels in *Pneumocystis jirovecii* pneumonia. *Intern. Med.*, 48, 195–202.

Neff, R.T., Jindal, R.M., Yoo, D.Y., Hurst, F.P., Agodoa, L.Y., and Abbott, K.C. 2009. Analysis of USRDS: Incidence and risk factors for *Pneumocystis jiroveci* pneumonia. *Transplantation*, 88, 135–141.

Nelson, M.P., Christmann, B.S., Werner, J.L., Metz, A.E., Trevor, J.L., Lowell, C.A., and Steele, C. 2011. IL-33 and M2a alveolar macrophages promote lung defense against the atypical fungal pathogen *Pneumocystis murina*. *J. Immunol.*, 186, 2372–2381.

Ng, V.L., Yajko, D.M., and Hadley, W.K. 1997. Extrapulmonary pneumocystosis. *Clin. Microbiol. Rev.*, 10, 401–418.

Nissapatorn, V., Lee, C.K., Rohela, M., and Anuar, A.K. 2004. Spectrum of opportunistic infections among HIV-infected patients in Malaysia. *Southeast Asian J. Trop. Med. Public Health*, 35(Suppl 2), 26–32.

Norris, K.A., and Morris, A. 2011. Pneumocystis infection and the pathogenesis of chronic obstructive pulmonary disease. *Immunol. Res.*, 50, 175–180.

Onishi, A., Sugiyama, D., Kogata, Y., Saegusa, J., Sugimoto, T., Kawano, S., Morinobu, A., Nishimura, K., and Kumagai, S. 2012. Diagnostic accuracy of serum 1,3-beta-D-glucan for *Pneumocystis jiroveci* pneumonia, invasive candidiasis, and invasive aspergillosis: Systematic review and meta-analysis. *J. Clin. Microbiol.*, 50, 7–15.

Paine 3rd, R., Preston, A.M., Wilcoxen, S., Jin, H., Siu, B.B., Morris, S.B., Reed, J.A., Ross, G., Whitsett, J.A., and Beck, J.M. 2000. Granulocyte-macrophage colony-stimulating factor in the innate immune response to *Pneumocystis carinii* pneumonia in mice. *J. Immunol.*, 164, 2602–2609.

Panizo, M.M., Reviakina, V., Navas, T., Casanova, K., Saez, A., Guevara, R.N., Caceres, A.M., Vera, R., Sucre, C., and Arbona, E. 2008. Pneumocystosis in Venezuelan patients: Epidemiology and diagnosis (2001–2006). *Rev. Iberoam. Micol.*, 25, 226–231.

Pesanti, E.L. 1991. Interaction of cytokines and alveolar cells with *Pneumocystis carinii* in vitro. *J. Infect. Dis.*, 163, 611–616.

Peters, S.E., Wakefield, A.E., Whitwell, K.E., and Hopkin, J.M. 1994. *Pneumocystis carinii* pneumonia in thoroughbred foals: Identification of a genetically distinct organism by DNA amplification. *J. Clin. Microbiol.*, 32, 213–216.

Phillips, E., and Mallal, S. 2007. Drug hypersensitivity in HIV. *Curr. Opin. Allergy Clin. Immuno.*, 7, 324–330.

Pixley, F.J., Wakefield, A.E., Banerji, S., and Hopkin, J.M. 1991. Mitochondrial gene sequences show fungal homology for *Pneumocystis carinii*. *Mol. Microbiol.*, 5, 1347–1351.

Poppers, D.M., and Scherl, E.J. 2008. Prophylaxis against *Pneumocystis* pneumonia in patients with inflammatory bowel disease: Toward a standard of care. *Inflamm. Bowel. Dis.*, 14, 106–113.

Rabodonirina, M., Vaillant, L., Taffe, P., Nahimana, A., Gillibert, R.P., Vanhems, P., and Hauser, P.M. 2013. *Pneumocystis jirovecii* genotype associated with increased death rate of HIV-infected patients with pneumonia. *Emerg. Infect. Dis.*, 19, 21–28; quiz 186.

Radhi, S., Alexander, T., Ukwu, M., Saleh, S., and Morris, A. 2008. Outcome of HIV-associated *Pneumocystis* pneumonia in hospitalized patients from 2000 through 2003. *BMC Infect. Dis.*, 8, 118.

Raj, D., Bhutia, T.D., Mathur, S., Kabra, S.K., and Lodha, R. 2013. Pulmonary alveolar proteinosis secondary to *Pneumocystis jiroveci* infection in an infant with common variable immunodeficiency. *Indian J. Pediatr.*, 81, 929–931.

Rajagopalan, N., Suchitra, J.B., Shet, A., Khan, Z.K., Martin-Garcia, J., Nonnemacher, M.R., Jacobson, J.M., and Wigdahl, B. 2009. Mortality among HIV-infected patients in resource limited settings: A case controlled analysis of inpatients at a community care center. *Am. J. Infect. Dis.*, 5, 219–224.

Reid, A.B., Chen, S.C., and Worth, L.J. 2011. *Pneumocystis jirovecii* pneumonia in non-HIV-infected patients: New risks and diagnostic tools. *Curr. Opin. Infect. Dis.*, 24, 534–544.

Robberts, F.J., Liebowitz, L.D., and Chalkley, L.J. 2007. Polymerase chain reaction detection of *Pneumocystis jiroveci*: Evaluation of 9 assays. *Diagn. Microbiol. Infect. Dis.*, 58, 385–392.

Rodriguez, M., and Fishman, J.A. 2004. Prevention of infection due to *Pneumocystis* spp. in human immunodeficiency virus-negative immunocompromised patients. *Clin. Microbiol. Rev.*, 17, 770–782, table of contents.

Roths, J.B., and Sidman, C.L. 1992. Both immunity and hyperresponsiveness to *Pneumocystis carinii* result from transfer of CD4+ but not CD8+ T cells into severe combined immunodeficiency mice. *J. Clin. Invest.*, 90, 673–678.

Rudner, X.L., Happel, K.I., Young, E.A., and Shellito, J.E. 2007. Interleukin-23 (IL-23)-IL-17 cytokine axis in murine *Pneumocystis carinii* infection. *Infect. Immun.*, 75, 3055–3061.

Rudy, T., Opelz, G., Gerlach, R., Daniel, V., and Schimpf, K. 1988. Correlation of in vitro immune defects with impaired gamma interferon response in human-immunodeficiency-virus-infected individuals. *Vox Sang.*, 54, 92–95.

Ruffini, D.D., and Madhi, S.A. 2002. The high burden of *Pneumocystis carinii* pneumonia in African HIV-1-infected children hospitalized for severe pneumonia. *Aids*, 16, 105–112.

Saltzman, R.W., Albin, S., Russo, P., and Sullivan, K.E. 2012. Clinical conditions associated with PCP in children. *Pediatr. Pulmonol.*, 47, 510–516.

Schliep, T.C., and Yarrish, R.L. 1999. *Pneumocystis carinii* pneumonia. *Semin. Respir. Infect.*, 14, 333–343.

Sepkowitz, K.A. 2002. Opportunistic infections in patients with and patients without acquired immunodeficiency syndrome. *Clin. Infect. Dis.*, 34, 1098–1107.

Shear, H.L., Valladares, G., and Narachi, M.A. 1990. Enhanced treatment of *Pneumocystis carinii* pneumonia in rats with interferon-gamma and reduced doses of trimethoprim/sulfamethoxazole. *J. Acquir. Immune. Defic. Syndr.*, 3, 943–948.

Sheikholeslami, M.F., Sadraei, J., Farnia, P., Forozandeh Moghadam, M., and Emadi Kochak, H. 2013. Typing of *Pneumocystis jirovecii* isolates from Iranian immunosuppressed patients based on the internal transcribed spacer (ITS) region of the rRNA gene. *Med. Mycol.*, 51, 843–850.

Shellito, J., Suzara, V.V., Blumenfeld, W., Beck, J.M., Steger, H.J., and Ermak, T.H. 1990. A new model of *Pneumocystis carinii* infection in mice selectively depleted of helper T lymphocytes. *J. Clin. Invest.*, 85, 1686–1693.

Sherman, M.P., Loro, M.L., Wong, V.Z., and Tashkin, D.P. 1991. Cytokine- and *Pneumocystis carinii*- induced L-arginine oxidation by murine and human pulmonary alveolar macrophages. *J. Protozool.*, 38, 234S–236S.

Shipley, T.W., Kling, H.M., Morris, A., Patil, S., Kristoff, J., Guyach, S.E., Murphy, J.E., Shao, X., Sciurba, F.C., Rogers, R.M., Richards, T., Thompson, P., Montelaro, R.C., Coxson, H.O., Hogg, J.C., and Norris, K.A. 2010. Persistent pneumocystis colonization leads to the development of chronic obstructive pulmonary disease in a nonhuman primate model of AIDS. *J. Infect. Dis.*, 202, 302–312.

Simonds, R.J., Hughes, W.T., Feinberg, J., and Navin, T.R. 1995. Preventing *Pneumocystis carinii* pneumonia in persons infected with human immunodeficiency virus. *Clin. Infect. Dis.*, 21(Suppl 1), S44–S48.

Soeiro Ade, M., Hovnanian, A.L., Parra, E.R., Canzian, M., and Capelozzi, V.L. 2008. Post-mortem histological pulmonary analysis in patients with HIV/AIDS. *Clinics*, 63, 497–502.

Stamp, L.K., and Hurst, M. 2010. Is there a role for consensus guidelines for *P. jiroveci* pneumonia prophylaxis in immunosuppressed patients with rheumatic diseases? *J. Rheumatol.*, 37, 686–688.

Stansell, J.D., Osmond, D.H., Charlebois, E., LaVange, L., Wallace, J.M., Alexander, B.V., Glassroth, J., Kvale, P.A., Rosen, M.J., Reichman, L.B., Turner, J.R., and Hopewell, P.C. 1997. Predictors of *Pneumocystis carinii* pneumonia in HIV-infected persons. Pulmonary complications of HIV Infection Study Group. *Am. J. Respir. Crit. Care Med.*, 155, 60–66.

Stringer, J.R., Beard, C.B., Miller, R.F., and Wakefield, A.E. 2002. A new name (*Pneumocystis jiroveci*) for Pneumocystis from humans. *Emerg. Infect. Dis.*, 8, 891–896.

Stringer, S.L., Hudson, K., Blase, M.A., Walzer, P.D., Cushion, M.T., and Stringer, J.R. 1989. Sequence from ribosomal RNA of *Pneumocystis carinii* compared to those of four fungi suggests an ascomycetous affinity. *J. Protozool.*, 36, 14S–16S.

Tansuphasawadikul, S., Pitisuttithum, P., Knauer, A.D., Supanaranond, W., Kaewkungwal, J., Karmacharya, B.M., and Chovavanich, A. 2005. Clinical features, etiology and short term outcomes of interstitial pneumonitis in HIV/AIDS patients. *Southeast Asian J. Trop. Med. Public Health*, 36, 1469–1478.

Tasaka, S., Hasegawa, N., Kobayashi, S., Yamada, W., Nishimura, T., Takeuchi, T., and Ishizaka, A. 2007. Serum indicators for the diagnosis of *Pneumocystis* pneumonia. *Chest*, 131, 1173–1180.

Theus, S.A., Linke, M.J., Andrews, R.P., and Walzer, P.D. 1993. Proliferative and cytokine responses to a major surface glycoprotein of *Pneumocystis carinii*. *Infect. Immun.*, 61, 4703–4709.

Theus, S.A., Smulian, A.G., Sullivan, D.W., and Walzer, P.D. 1997. Cytokine responses to the native and recombinant forms of the major surface glycoprotein of *Pneumocystis carinii*. *Clin. Exp. Immunol.*, 109, 255–260.

Thomas Jr., C.F., and Limper, A.H. 2004. *Pneumocystis* pneumonia. *N. Engl. J. Med.*, 350, 2487–2498.

Thomas Jr., C.F., and Limper, A.H. 2007. Current insights into the biology and pathogenesis of *Pneumocystis* pneumonia. *Nat. Rev. Microbiol.*, 5, 298–308.

Udwadia, Z.F., Doshi, A.V., and Bhaduri, A.S. 2005. *Pneumocystis carinii* pneumonia in HIV infected patients from Mumbai. *J. Assoc. Physicians India*, 53, 437–440.

Vargas, S.L., Ponce, C.A., Gigliotti, F., Ulloa, A.V., Prieto, S., Munoz, M.P., and Hughes, W.T. 2000. Transmission of *Pneumocystis carinii* DNA from a patient with *P. carinii* pneumonia to immunocompetent contact health care workers. *J. Clin. Microbiol.*, 38, 1536–1538.

Wakefield, A.E., Peters, S.E., Banerji, S., Bridge, P.D., Hall, G.S., Hawksworth, D.L., Guiver, L.A., Allen, A.G., and Hopkin, J.M. 1992. *Pneumocystis carinii* shows DNA homology with the ustomycetous red yeast fungi. *Mol. Microbiol.*, 6, 1903–1911.

Wakefield, A.E., Pixley, F.J., Banerji, S., Sinclair, K., Miller, R.F., Moxon, E.R., and Hopkin, J.M. 1990. Detection of *Pneumocystis carinii* with DNA amplification. *Lancet*, 336, 451–453.

Wolfe, R.A., Roys, E.C., and Merion, R.M. 2010. Trends in organ donation and transplantation in the United States, 1999–2008. *Am. J. Transplant. (AJT)*, 10, 961–972.

Wright, T.W., Bissoondial, T.Y., Haidaris, C.G., Gigliotti, F., and Haidaris, P.J. 1995a. Isoform diversity and tandem duplication of the glycoprotein A gene in ferret *Pneumocystis carinii*. *DNA Res.*, 2, 77–88.

Wright, T.W., Gigliotti, F., Haidaris, C.G., and Simpson-Haidaris, P.J. 1995b. Cloning and characterization of a conserved region of human and rhesus macaque *Pneumocystis carinii* gpA. *Gene*, 167, 185–189.

Ypma-Wong, M.F., Fonzi, W.A., and Sypherd, P.S. 1992. Fungus-specific translation elongation factor 3 gene present in *Pneumocystis carinii*. *Infect. Immun.*, 60, 4140–4145.

Zheng, M., Shellito, J.E., Marrero, L., Zhong, Q., Julian, S., Ye, P., Wallace, V., Schwarzenberger, P., and Kolls, J.K. 2001. CD4+ T cell-independent vaccination against *Pneumocystis carinii* in mice. *J. Clin. Invest.*, 108, 1469–1474.

Chapter 52

Pathogenesis of Aspergillosis in Humans

George Dimopoulos[1], Dimitrios K. Matthaiou[1], Nikolaos Moussas[1], Olympia Apostolopoulou[1], George Arabatzis[2], Aristea Velegraki[2], and Garyphalia Poulakou[3]

[1]Department of Critical Care Medicine, Medical School, University of Athens, Chaidari-Athens, Greece

[2]Mycology Research Laboratory, Department of Microbiology, Medical School, National and Kapodistrian University, Athens, Greece

[3]Department of Internal Medicine, Medical School, University of Athens, Chaidari-Athens, Greece

52.1 Introduction

Aspergillus is a hyaline mold ubiquitously found in soil, water, air, and food worldwide (Walsh and Dixon, 1989). It is the causative agent of infections with a great variety of clinical severity. Mild forms range from colonization to the formation of aspergillomas and the development of allergic bronchopulmonary aspergillosis (ABPA). Severe forms of disease present as invasive pulmonary or other organ aspergillosis with underlying tissue necrosis. *Aspergillus* is considered an opportunistic pathogen and the entity of invasive aspergillosis (IA) affects mostly — if not exclusively — immunocompromised hosts (Dutkiewicz and Hage 2010; Kousha et al., 2011). Biomedical advancement led to the recognition of more than 250 species in 7 subgenera of this mold that was first recognized in 1729 in Florence by Micheli (Balajee et al., 2007). IA is a destructive infection of affected tissue and organs and is most commonly caused by the species of *Aspergillus fumigatus*. Less commonly found species are *A. flavus*, *A. terreus*, and *A. niger* (Patterson et al., 2000). Despite the advances in modern mycology, establishing the diagnosis of aspergillosis particularly in the immunocompromised host remains very difficult. Furthermore, the diagnosis of IA in current clinical practice represents a significant challenge for physicians and high-risk patients from a therapeutic and prognostic standpoint, being associated with poor outcomes. The rapidly changing epidemiology of this pathogen, as new species emerge among diverse clinical settings, poses also additional challenges (Cuenc-Estrella et al., 2011; Kousha et al., 2011; Ostrosky-Zeichner, 2012).

52.2 Definitions of invasive fungal disease

The European Organization for Research and Treatment of Cancer/Invasive Fungal Infections Cooperative Group (EORTC) and the National Institute of Allergy and Infectious Diseases Mycoses Study Group (MSG) published in 2002 definitions of invasive fungal infections (IFI). These definitions are employed as a tool to assist to the homogenization of patient groups for epidemiologic and clinical research purposes; additionally, they guide the design of clinical trials evaluating new drugs and management strategies. The definitions apply to immunocompromised cancer patients and hematopoietic stem cell transplants (HSCT) recipients and include three levels of probability for the diagnosis of IFI: proven, probable, and possible IFI. The identification of a fungus by histology or culture from a tissue taken from the site of the disease is required for a "proven IFI"; culture should be drawn from a normally sterile site. Probable or possible IFI categories are based upon the presence of three elements: (1) a host factor that identified the patients at risk; (2) clinical signs and symptoms consistent with the disease; and (3) mycological evidence: culture, microscopic analysis, or indirect tests (e.g., antigen detection) (Asioglu et al., 2002). In the revised definitions of 2008, the panelists from EORTC/MSG reaffirmed that the utility of definitions is only to assist in research. Host factors were further clarified as those characteristics by which patients predisposed to acquire IFI can be recognized; host factors are not identical to risk factors. Fever was removed from host factors since it is actually a

Table 52-1 Criteria for probable invasive fungal infection (endemic mycoses are excluded)

Host factors	Clinical criteria
Recent history of neutropenia ($<0.5 \times 10^9$ NEU/L or <500 neutrophils/mm^3 for >10 days) temporally related to the onset of fungal disease Receipt of an allogeneic stem cell transplant Prolonged use of corticosteroids (excluding among patients with ABPA) at a mean minimum dose of 0.3 mg/kg/day of prednisone equivalent for >3 weeks Treatment with other recognized T-cell immunosuppressants, such as cyclosporine, TNF-α blockers, specific monoclonal antibodies (such as alemtuzumab), or nucleoside analogs during the past 90 days Inherited severe immunodeficiency (such as chronic granulomatous disease or severe combined immunodeficiency)	*Lower respiratory tract fungal disease* The presence of one of the following three signs on CT of the lungs: – dense, well-circumscribed lesion(s) with or without a halo sign – air-crescent sign – cavity *Tracheobronchitis* Tracheobronchial ulceration, nodule, pseudomembrane, plaque, or eschar seen on bronchoscopy *Sinonasal infection* Imaging showing sinusitis plus at least one of the following three signs: – acute localized pain (including pain radiating to the eye) – nasal ulcer with black eschar – extension from the paranasal sinus across bony barriers, intruding into the orbit
Mycological criteria	*CNS infection* One of the following signs: – focal lesions on imaging – meningeal enhancement on MRI or CT
Direct test (cytology, direct microscopy, or culture) Mold in sputum, BAL fluid, bronchial brush, or sinus aspirate samples, indicated by one of the following: presence of fungal elements indicating a mold Recovery by culture of a mold (e.g., *Aspergillus*, *Fusarium*, *Zygomycetes*, or *Scedosporium* spp.) Indirect tests (detection of antigen or cell-wall constituents) Aspergillosis – Galactomannan antigen detected in plasma, serum, BAL fluid, or CSF Invasive fungal disease other than cryptococcosis and zygomycoses β-D-glucan detected in serum	*Disseminated candidiasis* At least one of the following two entities after an episode of Candidemia within the previous 2 weeks: – small, target-like abscesses (bull's eye lesions) in liver or spleen – Progressive retinal exudates on ophthalmologic examination

Note: Probable IFI requires a host factor, a clinical criterion, and a mycological criterion. Cases that meet host factor criteria plus a clinical criterion but lack mycological criteria are considered possible IFI.
Adapted from De Pauw et al. (2008).

clinical feature and also nonspecific for IFI. Host factors were extended with the addition of solid organ transplant recipients, hereditary immunodeficiencies, connective tissue disorders, and therapy with immunosuppressive agents such as corticosteroids or T-cell immunosuppressants (Table 52-1).

Proven IFI requires proof of IFI by demonstration of fungal elements in affected tissue for most conditions (culture and identification). By definition, this category can be valid irrespective of host factors or clinical features (Table 52-2).

Probable IFI needs the presence of three elements in order to be defined: (1) a host factor; (2) clinical features; and (3) mycological evidence (as defined in Table 52-1).

Possible IFI as such are classified as the cases that have appropriate host factors with sufficient clinical evidence consistent with IFI, but without mycological support (Table 52-1).

The utility of the definitions is to only assist in clinical practice but more importantly, if applied consistently, to foster communication, enhance the understanding of the epidemiology of IFI, and facilitate the

assessment of therapeutic regimens and strategies (De Pauw et al., 2008).

52.3 Epidemiology

52.3.1 Distribution of species and emerging species

The commonest species of *Aspergillus* recovered from invasive disease is *A. fumigatus,* to which the burden of the disease is attributed, followed by *A. flavus, A. terreus,* and less commonly *A. niger* (Patterson et al., 2000). The emergence of less common species such as *A. terreus* and *A. udagawae* is observed worldwide. *Neosartorya (Aspergillus) udagawae,* first identified in Brazilian soil in the 1990s, was not identified as a disease etiologic factor since the late 2000 (Balajee et al., 2006; Horie and Coelho 1995). However, the data emerged in 2009 from a retrospective study highlighted the importance of emerging *Aspergillus* spp. in clinical practice, where 11% of total cases of IA identified in the National Institutes of Health clinical microbiology database from 2000 through 2008 were

Table 52-2 Criteria for proven invasive fungal infection excluding endemic mycoses

	Specimens and methods			
MOLDS[a]	Microscopic analysis (normally sterile material)	Culture (normally sterile material)	Blood	Serology: CSF
	Histopathologic, cytopathologic, or direct microscopic examination[b] of a specimen obtained by needle aspiration or biopsy on which hyphae or melanized yeast-like forms are observed, accompanied by evidence of associated tissue damage	Recovery of a mold or "black yeast" by culture of a specimen obtained by a sterile procedure from a normally sterile and clinically or radiologically abnormal site consistent with an infectious disease process, excluding bronchoalveolar lavage fluid, a cranial sinus cavity specimen, and urine	Blood culture that yields a mold[c] (e.g., *Fusarium* spp.) in the context of a compatible infectious disease process	Not applicable
YEASTS[a]	Histopathologic, cytopathologic, or direct microscopic examination[b] of a specimen obtained by needle aspiration or biopsy from a normally sterile site (other than mucous membranes) showing yeast cells – for example, *Cryptococcus* spp. presenting as encapsulated budding yeasts or *Candida* spp. showing pseudohyphae or true hyphae[d]	Recovery of a yeast by culture of a sample obtained by a sterile procedure (including a freshly placed – less than 24 h – drain) from a normally sterile site showing a clinical or radiological abnormality consistent with an infectious disease process	Blood culture that yields yeast (e.g., *Cryptococcus* or *Candida* spp.) or yeast-like fungi (e.g., *Trichosporon* spp.)	Cryptococcal antigen in CSF indicates disseminated cryptococcosis

Adapted from De Pauw, et al. (2008).

[a]Identification at genus or species level should be appended on culture results if culture is available.

[b]Tissues and cells submitted for histopathologic or cytopathologic studies should be stained by Grocott-Gomori methenamine silver stain or by periodic acid Schiff stain to facilitate recognition of fungal structures. Whenever possible, wet mounts of specimens from foci related to invasive fungal disease should be stained with a fluorescent dye (e.g., calcofluor or blankophor).

[c]Recovery of *Aspergillus* sp. from blood cultures represents always contamination.

[d]*Candida*, *Trichosporon*, and yeast-like *Geotrichum* species and *Blastoschizomyces capitatus* may also form pseudohyphae or true hyphae.

found to be actually infections with *N. udagawae* (Vinh et al., 2009). Among these patients, one had a diagnosis of myelodysplastic syndrome and the remainders had chronic granulomatous disease (CGD). Interestingly, disease by *N. udagawae* needed a sevenfold median of time to respond compared with *A. fumigatus*, whereas the minimal inhibitory concentrations (MICs) of *N. udagawae* to voriconazole and Amphotericin B (AmB) are elevated indicating a profile of resistance to first-line antifungals (Vinh et al., 2009; Wingard et al., 2008). Therefore, disease by *N. udagawae* could represent refractory IA.

IA in case of patients with CGD is primarily caused by *A. fumigatus* followed in terms of frequency by *A. nidulans*, a species not uncommonly resistant to AmB (Segal et al., 1998). Third most common species is *N. udagawae*, for which a reduced pathogenicity was demonstrated compared with *A. fumigatus*. *N. udagawae* was more efficiently cleared by neutrophils than *A. fumigatus* and therefore a higher survival rate of infected hosts was observed. Furthermore, while *A. fumigatus* appears only slightly susceptible to reactive oxygen species (ROS) generated from healthy phagocytes, *N. udagawae* strains are highly susceptible (Sugui et al., 2010).

52.4 Risk factors

IA has been historically described among patients with low levels of neutrophils, already in the 1970s (Meyer et al., 1973). The observation that low numbers of neutrophils were effective against *Aspergillus* hyphae *in vitro* led to the conclusion that neutrophil deficit is the basis of IA pathophysiology (Diamond et al., 1978). This neutrophil deficit can be different in each case concerning IA, but mainly it has been documented in hematologic malignancies and as a consequence of myeloablative chemotherapy in the context of stem cell transplantation (Lortholary et al., 2011). Profound and prolonged neutropenia has been demonstrated as an important risk factor for the development of IA. The risk of these patients is reported to vary across studies and underlying conditions. In patients with acute myelogenous leukemia, incidence varies from 2% to more than 28% with a mean incidence from 4% to 7% (Cornely et al., 2007a; De Pauw and Donnell 2007; Pagano et al., 2006). The use of growth factors and the current treatment patterns have limited the percentage of patients with prolonged profound neutropenia; its impact on the incidence of IA in high-risk patients remains to be examined in future studies (Wald et al., 1997).

Table 52-3 Main risk factors associated with the development of aspergillosis

Risk factor	Comments
Prolonged neutropenia	Especially in absolute neutrophil count of 500 cells/mm³; the risk correlates strongly with the duration and severity of neutropenia, as is estimated to increase by 1% per day for the first 3 weeks and then by 4% per day thereafter
Transplantation	The highest risk is with lung transplantation and HSCT; the risk is much higher following allogeneic rather than autologous HSCT (incidences of 2.3–15% and 0.5–4%, respectively); in allogeneic HSCT, the highest risk is in patients with severe GVHD (grade III–IV); the timeline of IPA in these patients follows a bimodal distribution, with a peak in the first month following HSCT, which is associated with neutropenia; the second peak is during the treatment of GVHD (median 78–112 days post-transplantation) and becomes more significant especially with the higher incidence of GVHD associated with unrelated allogeneic transplantation and treatment with intensive immunosuppressive therapy, including corticosteroids, cyclosporine A, anti-TNF agents, and other T-cell-depleting strategies
Corticosteroid therapy	Prolonged (3 weeks) and high dose
Hematological malignancy	Risk is higher with leukemia
Chemotherapy	–
Advanced AIDS disease	IPA is relatively uncommon in patients with HIV infection, especially with the routine use of highly active anti-retroviral therapy (3.5 cases/1000 person-years; risk factors are a low CD4 count (<100 cells/mm³), coexistent neutropenia or corticosteroid therapy, and advanced AIDS; there is increased incidence of tracheobronchial involvement in these patients in addition to the usual clinical picture of IPA
Chronic granulomatous disease	Neutrophil dysfunction that is seen primarily in CGD is a risk factor for IPA, an important cause of mortality in these patients
Critically ill patients	The mortality of the disease is equally high in the group of ICU patients, while the clinical signs and symptoms are nonspecific and the diagnosis remains a challenge

GVHD, graft-versus-host disease; HSCT, hematopoietic stem cell transplantation; IPA, invasive pulmonary aspergillosis.

Solid organ transplantation, especially of liver and lung, is considered a risk factor for IA (Hosseini-Moghaddam and Husain 2012; Pacholczyk et al., 2011). Among transplantation patients, the lung transplant recipients are more vulnerable, because lung is exposed to *Aspergillus* conidia of the environment. The use of corticosteroids, especially in the process of stem cell transplantation is one of the most important risk factors (Rosenhagen et al., 2009). Interestingly, a large number of conditions not related to neutropenia have recently been recognized to predispose for IA. These conditions include chronic obstructive pulmonary disease (COPD), H1N1 influenza virus infection, admission in the intensive care unit (ICU), severe liver disease, major surgery, AIDS, and CGD. Patients with genetic disorders that affect some Toll-like receptors with various mutations and deficiency in mannose-binding lectin most likely are at high risk for IA (Barton, 2013). IA occurs also in ICU patients with an independent pattern related to other risk factors (Table 52-3; Vanderwoude et al., 2006).

52.4.1 Incidence and outcomes

IA portends mortality rates exceeding 80% among bone marrow transplantation (BMT) patients, 90% among solid organ (liver) transplant recipients, and 49% among patients with blood and lymphoid tissue malignancies (Lin et al., 2003; Sing, 2000). According to data published in 2005 by the Transplant Associated Infections Surveillance Network (TransNET), the incidence rates of IA among patients with different types of stem cell transplants vary from 0.5% to 3.8%, being higher than in patients with different solid organ transplants (0.1–2.4%) (Morgan et al., 2005). Among patients with HSCT, incidence rates range from less than 5% for autologous to 5–10% for the allogeneic transplants (Marr et al., 2002a, 2002b). The rates of IA in HIV patients as well as in patients with hematologic malignancy were reported at 0.4% (Denning, 1998). The application of more intense therapeutic protocols even among elderly patients has significantly increased the population at-risk for IA. Despite our better understanding of the pathophysiology and the availability of an expanded antifungal armamentarium, the incidence of IA is increasing. Autopsy data verify that IA has risen from 1% in 1978 to 7% in 1992; the percentage of IA among all mycoses found on autopsy increased also from 17% to 60% (Tong et al., 2009). Data published in 2002 from a large BMT center indicate a rise in the risk of IA from 4% to 12% (Marr et al., 2002b). Total inpatient and outpatient costs were evaluated at $674.1 million, with an overall hospitalization cost of $36,867 per incident in a study of 2002 (Wilson et al., 2002). Based on 38 million hospital discharges in 2003 in the United States, a national incidence of 36 aspergillosis cases per million per year was estimated. The diagnosis of aspergillosis was the primary diagnosis in 29.4% of these cases with a reported mortality rate of 17.1%; patients with pneumonia represented 37.6% of total cases. Hospital mortality is increased when the diagnosis is *Aspergillus* pneumonia (25.1% compared with 12.4% among cases without pneumonia); the same stands true for length of stay (Tong et al., 2009). Hospitalized BMT patients had a diagnosis of aspergillosis 2.5–7 times more often than any other risk categories, namely hematologic malignancy, HIV infection, chemotherapy, BMT, reticuloendothelial, and immunity disorders. Complicated courses increasing the post-transplant hospitalization were also associated with increased rates of aspergillosis in both lung transplant surgery and BMT. The diagnosis of aspergillosis among BMT patients doubled the median estimated costs ($74,945 compared with $33,055 for non-aspergillosis BMT patients). Similarly, diagnosis of aspergillosis among hematologic malignancy and chemotherapy patients carried a 0% and 800% incremental cost, respectively (Tong et al., 2009).

A double peak is reported in the incidence of IA during the period of stem cell transplant and BMT. The first peak is noted early after transplantation (<40 days) and the second in >100 days following transplantation (Cornely et al., 2007b; Morgan et al., 2005). In recent studies, only 31% of the stem cell transplant recipients that presented IA were neutropenic. Non-myeloablative transplantation procedures were considered a key factor contributing to the change of epidemiology in this group of patients. This practice shifted apparently the major risk from neutropenia to the usage of corticosteroids in high doses for the management of graft-versus-host disease (GVHD; Wald et al., 1997).

52.4.2 Environmental exposure

Until recently, infections due to *Aspergillus* were believed to be of nosocomial origin, mostly because of prolonged periods of hospitalization and often associated with constructions in the healthcare settings. Contemporary data indicate that acquisition of *Aspergillus* infection can be done by airborne and waterborne sources, given the fact that *Aspergillus* is a very common mold of the environment (Mullins et al., 1976). It has been recorded that various meteorological conditions and environmental patterns of temperature affect the virulence and also the prevalence of fungal diseases. Weather and environmental conditions also affect human susceptibility and immune mechanism through light–dark cycles and melatonin secretion. *Coccidioides immitis,* a geographically restricted fungus, represents another recognized example of fluctuation according to climate conditions (Comrie, 2005; Dowell, 2001; Roenneberg and Merrow 2001). A cohort study of over 8000 patients who received HSCT in different regions of the United States provided data for seasonal and climatic influence on *Aspergillus* acquisition and infection. Two sub-cohorts were compared including patients of Seattle, Fred Hutchinson Cancer Research Center (FHCRC) and of Houston, MD Anderson Cancer Center (MDACC). In the case of Seattle cohort, IA risk within the first 3 months post-HSCT was higher during the warm season as studied during two 5-year periods (1992–1997 and 1998–2003), whereas in the case of MDACC no significant seasonal impact was

noted. For the period 1992–1998, the person-month IA incidence in Seattle was found to be positively associated with spore count ($P = 0.006$), but negatively with precipitation ($P = 0.03$). The person-month IA rate had no climatic association at the MDACC in Houston (Panackal et al., 2010). The high spore counts in Seattle coincide with high temperatures and low precipitation which leads to high incidence of IA right after these periods of climatic phenomena. Spore counts, in general, correlate with temperature and humidity. Higher counts of spores can be found in outdoor air during the summer and autumn (Hansen et al., 2008). Outbreaks of IA in patients exposed to aerosols of *Aspergillus* conidia and patients who were associated with construction activities or other environmental factors are also documented (Haiduven, 2008). Contaminated water and aerosols can both expose severely immunosuppressed patients to risk for IA.

52.5 The pathogenic aspergilli

The anamorphic fungal genus *Aspergillus* is known to include more than 250 species today (Domsch et al., 2007), and approximately 40 of them have been isolated from human or animal infections. The medically important members of the genus *Aspergillus* are related to seven teleomorph genera: *Emericella, Eurotium, Fennellia, Hemicarpenteles, Neopetromyces, Neosartorya,* and *Petromyces* (Summerbell, 2003). The majority of the *Aspergillus* infections are related to one species, *A. fumigatus* (Figures 52-1a and 52-1b), whereas more than 95% of the total number of aspergilloses are associated with five species: *A. fumigatus, A. flavus, A. niger, A. terreus,* and *Aspergillus* (teleomorph *Emericella*) *nidulans*. Other species implicated in infections are *A. aculeatus, A. alliaceus* (teleomorph *Petromyces alliaceus*), *A. avenaceus, A. caesiellus, A. candidus, A. carneus, A. (Eurotium) chevalieri, A. clavato-nanica, A. clavatus, A. conicus, A. deflectus, A. (Neosartorya) fischeri, A. (Fennellia) flavipes, A. glaucus (Eurotium herbariorum), A. granulosus, A. (Neosartorya) hiratsukae, A. janus, A. japonicus, A. niveus (Fennellia nivea), A. ochraceus, A. oryzae, A. persii, A. (Neosartorya) pseudofischeri, A. reptans (Eurotium repens), A. quadrilineatus (Emericella quadrilineata), A. sclerotiorum, A. sydowii, A. tamarii, A. tubingensis, A. (Emericella) unguis, A. ustus* (the medically important strains of this species have recently been described as *A. calidoustus*), *A. varians, A. versicolor,* and *Aspergillus vitis (Eurotium amstelodami)* (Summerbell, 2003). Recently, sequence-based identification of clinical *Aspergillus* isolates also implicated as infectious causes *A. coreanus (Neosartorya coreana), A. (Neosartorya) fennelliae, A. fumigatiaffinis, A. fumisynnematus, A. lentulus, A. spinosus (Neosartorya spinosa), A. (Neosartorya) udagawae,* and *A. viridinutans* (Balajee et al., 2007).

The genus *Aspergillus* is microscopically characterized by erect, aseptate conidiophores which terminate in an apical vesicle covered with palisading phialides or short branches (metulae) bearing phialides

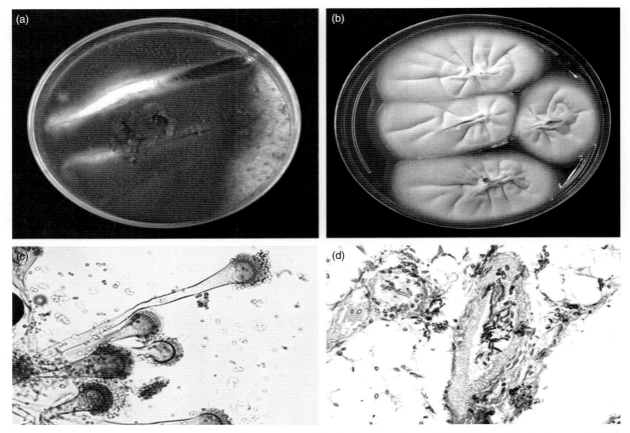

Fig. 52-1 (a) *Aspergillus fumigatus*, (b) *Aspergillus flavus*, (c) *Aspergillus fumigatus* conidiae in BAL (d) *Aspergillus fumigatus* conidiae in bronchial cultures.

(Figure 52-1c) (Domsch et al., 2007; Summerbell, 2003). The phialides are producing chains of hydrophobic conidia. The typical morphology can sometimes be seen during growth in a host, if the fungus can sporulate in the interface between host tissue substrate and air (e.g., inside bronchi). In invasive disease, only hyaline, branching at 45°, hyphae are observed (Figure 52-1d).

52.5.1 Pathogenesis and immunity

In nature, *Aspergillus* species are generally saprobic, living on decaying organic and plant material in niches such as compost piles, decaying vegetation, stored grains, and in warmer geographical areas, soil (Domsch et al., 2007). Members of the genus are additionally capable of being opportunistic pathogens for insects, animals, and humans.

Aspergilli are capable of opportunistically causing three distinctive patterns of disease in humans: invasive disease, allergic disease, and chronic colonization of preformed cavities. The main group of susceptible patients to *Aspergillus* invasive disease are those with hematological malignancies (mainly leukemia), recipients of stem-cell or solid organ transplants, and patients admitted to the ICU (Thompson and Patterson 2008). The main risk factors for invasive disease are prolonged neutropenia, treatment with high doses of cortisone or leukemia, all factors that disrupt the mononuclear and polymorphonuclear

phagocytic systems. Patients with CGD due to mutations in any of the genes encoding the four subunits of NAD(P)H oxidase present increased susceptibility to aspergillosis as they have a deficiency in ROS formation such as superoxide, hydrogen peroxide, hypochlorous acid, resulting in inefficient killing of conidia and hyphae (Morgenstern et al., 1997).

The lungs and the lower respiratory tract are the predominant primary sites of invasive infection by aspergilli (invasive pulmonary aspergillosis, IPA), but paranasal sinuses, skin, or cornea can be also infected. Aspergilli are characteristically angioinvasive, and in IPA their hyphae can invade the vascular wall in the lungs and then enter the lumen, become systemic, and disseminate to the brain or other organs. The angioinvasive process can cause arterial thrombosis, hemorrhagic infarcts, and abscesses. Opportunistic local invasive disease can also appear in immunocompetent as well in immunocompromised patients after accidental trauma (skin, cornea, peritoneal and pleural cavities, etc.), planned trauma (postoperative), or related to a foreign body (prosthetic valve, vascular graft).

Aspergilli are capable of causing a variety of allergic forms of disease such as ABPA, allergic fungal sinusitis, and asthma. Patients with asthma or cystic fibrosis (CF) are at increased risk of ABPA. In these disease entities, *Aspergillus* grows inside plugs of mucus within the bronchial lumen, without invasion. Total serum (polyclonal) IgE concentration is elevated; IgE/IgG serum antibodies to *A. fumigatus* and eosinophilia are present. Aspergilli can also colonize and grow in preformed lung cavities, commonly of tuberculous origin but also of other etiology. The same kind of lesion (aspergilloma) can appear in asthmatic patients when the fungus grows inside a bronchus. In patients with aspergilloma, it is often assumed that there is a subtle immunological defect, such as genetic polymorphisms in the collagen region of surfactant protein SP-A (Vaid et al., 2007).

Aspergilli are coming in contact with the host through propagating asexual spores (conidia) that are continuously discharged into the air. The lower respiratory tract is their main portal of entry into the human host. Appropriately small-sized conidia, usually 2–5 μm, can easily enter and subsequently be trapped in alveoli as the carina excludes many particles larger than 3.5 μm (Schlesinger and Lippmann, 1972). Paranasal sinuses, the skin, and the eyes (cornea) can be secondary entry points. Conidia that are trapped in the mucus are removed from the lungs through the continuous movement of respiratory tract cilia. Alveolar surfactant proteins SP-A and SP-D bind to conidia, facilitating their phagocytosis (Wright, 2005). Adherence of conidia to alveoli is mediated through binding to a variety of host matrix glycoproteins such as fibronectin, fibrinogen, laminin, and type I and type IV collagen, through fungal lectins, hydrophobin (*rodA* gene), and other ill-defined fungal cell wall components.

Normally, humans are highly resistant to aspergilli through a number of innate defense mechanisms. Professional phagocytes such as macrophages/monocytes and neutrophils/eosinophils constitute the main line of defense against aspergilli (Schaffner et al., 1982). These cells are recognizing conidia/hyphae through a variety of receptors such as the C-Type lectins dectin-1 and mannose receptors (MR), pentraxin 3 (PTX3), or Toll-like (TLR-2, TLR-4) receptors. As a first line of defense, resident alveolar macrophages phagocytize and destroy inhaled conidia, thus preventing their germination and outgrowing of potentially invasive hyphae. Killing of the phagocytized conidia is attributed to the lytic enzymes of the phagolysosomes and secondary to ROS. Failing to suppress conidial germination inside the phagolysosome, hyphae grow, penetrate the cell membrane, and destroy the macrophage. Once germinative conidia or hyphae have been formed, neutrophils and eosinophils recruited to alveoli are the only cell types able to destroy germinative conidia or hyphae, by attaching to them and releasing ROS and their cytotoxic granule content (Levitz and Diamond, 1984).

Other innate immune system cells such as natural killer cells (NKs) are able to contribute to hyphae killing producing cytotoxic substances and cytokines activating phagocytic cells, though studies in mice failed to demonstrate an essential role for NK cells in the early host response to fungi (Romani, 2004).

Dendritic cells (DCs) are able to ingest *Aspergillus* conidia and hyphae (through receptors DC-SIGN, MRs, CR3, and FcγR), migrate to draining lymph nodes and present fungal antigens to T-helper cells (Th1), initiating TH cell differentiation and functional commitment (Romani, 2004). A dominant Th1 response is critical in the optimal activation of phagocytes through increased production of inflammatory cytokines IFN-γ, IL-2, and IL-12. On the contrary, a Th2 response with cytokines IL-4 and IL-5 leading to B-cell activation is dominant in ABPA. This disease represents a mixed immune response of the type 1, 3, and 4 but its pathogenesis is not totally clear. Humoral immunity seems to have a limited role in protective immunity against *Aspergillus*.

52.5.2 Virulence determinants of *Aspergillus*

Because aspergilli are opportunistic pathogens and the host immune system is the most important factor in the host–fungus interaction, virulence factors are defined in the broad sense of gene products/cellular aspects that support the pathogens' capacity to cause disease in the immunocompromised host. Such general factors as the small size of conidia (2–3 μm in *A. fumigatus*) that permits their access to alveoli, the primary point of entry in invasive pulmonary disease, the capacity to survive and grow at high temperatures (37°C) inside the host (thermotolerance), and the capacity to grow hyphae (the invasive form) are regarded as important virulence determinants.

The recent decipherment of the *A. fumigatus* genome and of the other important *Aspergillus* pathogens has led to a number of studies that have greatly increased our understanding of how aspergilli interact with the host to produce a wide range of diseases (Gibbons and Rokas 2013). Thus, virulence factors have been identified in animal models in thermotolerance (*thtA* and *cgrA* genes), in primary metabolism (homoaconitase, *lysF* gene; methylcitrate synthase, *mcsA*; PABA synthetase, *pabaA*; OMP decarboxylase, *pyrG*), in iron acquisition (siderophore biosynthesis, *sidA*), and in signal transduction and regulation (calcineurin, *calA/cnaA*; cAMP signaling, *acyA*, *gpaB*, *pkaC1*, *pkaR*; nutrient sensing, Ras/Rheb GTPase, *rhbA*; pH signaling, *pacC*; nitrogen metabolism, *areA*) (Krappmann, 2008). Extracellular proteases are related to virulence as they can cause epithelial cell detachment from respiratory epithelium, help conidia evade phagocytosis and killing, or facilitate tissue invasion (elastase) (Tomee et al., 1997). From the ROS-neutralizing systems, mycelial catalase (*cat1*, *cat2*) and the 1,8-dihydroxynaphthalene (DHN)-like conidial melanin (polyketide synthetase, *pksP/alb1*) have been shown to augment virulence (Krappmann, 2008).

Of the *Aspergillus* mycotoxins, gliotoxin, produced by strains of *A. fumigatus*, *A. niger*, and *A. terreus*, has the ability to interfere with the host immune system and inhibit the movement of respiratory cilia.

Detection of the toxin in the blood of patients and animals with IPA, its production from more clinical than environmental *A. fumigatus* strains, and recent animal experiments with gliotoxin-deficient *A. fumigatus* mutants support the classic concept of gliotoxin as a virulence determinant (Kwon-Chung and Sugui, 2009). Aspergilli are also producing a number of protein allergens, 20 of which have been cloned and studied in *A. fumigatus*. Asp f1 (ribotoxin), Asp f5 (metalloprotease), and Asp f10 (aspartic protease) are regarded as genus-specific allergens (Krappmann, 2008).

52.6 Clinical features

There are five major clinical forms of IA, of which rhinocerebral and pulmonary infections are the most common.

52.6.1 Rhinocerebral aspergillosis

It is commonly seen in neutropenic cancer patients, HSCT recipients, and patients with diabetic ketoacidosis (Saah et al., 1994). Symptoms may include unilateral facial swelling, headaches, nasal or sinus congestion or pain, serosanguinous nasal discharge, and fever. As the infection spreads, ptosis (drooping eyelid), proptosis (eye dislocation), loss of extraocular muscle function, and vision disturbance may occur. Necrotic black lesions on the hard palate or nasal turbinate (nasal cavities) and drainage of black pus from eyes are useful diagnostic signs (Gupta et al., 2012). Chronic invasive sinonasal aspergillosis and

chronic granulomatous *Aspergillus* sinusitis have also been documented in immunocompetent patients living in dry-air climates, such as India, Saudi Arabia, and Sudan predominantly caused by *A. flavus*. This infection may progress to invasion of the orbit and other craniofacial structures and ultimately to intracranial involvement.

52.6.2 Invasive pulmonary aspergillosis

The symptoms include fever, cough, pleuritic chest pain (due to vascular invasion leading to thromboses that cause small pulmonary infarcts), and dyspnea. Angioinvasion results in necrosis of tissue, which may ultimately lead to cavitation and/or hemoptysis usually mild, but can be severe also (Figure 52-5). IPA is one of the most common causes of hemoptysis in neutropenic patients, and may be associated with cavitation that occurs with neutrophil recovery (Figure 52-2). IPA is an emerging infection in patients with COPD because chronic lung disease predisposes to airway colonization by *Aspergillus* spp., and in the development of invasive disease with high mortality (Bulpa et al., 2007; Soubani et al., 2004). Several factors, including structural changes in lung architecture, prolonged use of corticosteroid therapy (Lionakis and Kontoyiannis, 2003), frequent hospitalization, broad-spectrum antibiotic treatment, invasive procedures, mucosal lesions, impaired mucociliary clearance, and comorbid illnesses such as diabetes mellitus, alcoholism, and malnutrition have been identified. It is also possible that abnormalities or deficiencies in surfactant proteins, alveolar macrophages, and Toll-like

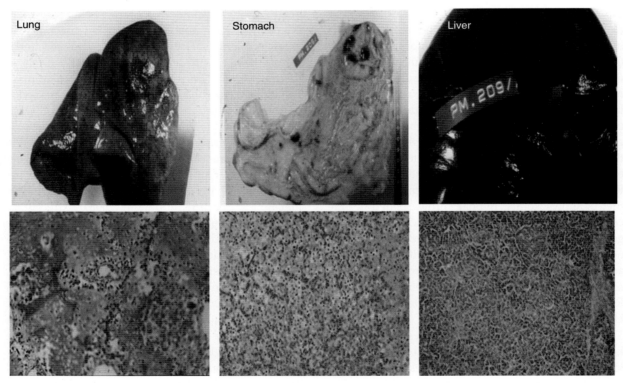

Fig. 52-2 A patient with HIV, admitted in the ICU because of coma and respiratory failure. She received empirically Amphotericin B. She died 48 h later. Autopsy (no whole body) revealed disseminated aspergillosis. For a color version of this figure, see the color plate section.

receptors play a role in the pathogenesis of IPA in some patients with COPD (Hawgood et al., 2002; Sarir et al., 2009). Since the clinical and radiological presentation of IPA in patients with COPD is nonspecific, to avoid delay in the diagnosis and management, a high index of suspicion is warranted. The isolation of *Aspergillus* spp. from a lower respiratory tract specimen should not be routinely dismissed as colonization (Ader et al., 2005; Samarakoon and Soubani, 2008). IPA is also becoming an important disease in ICU patients without meeting however the classical risk factors (neutropenia, leukemia, HSCT) (Figure 52-4). Critically ill patients are prone to developing disturbances in immunoregulation during their stay in the ICU which renders them more vulnerable to fungal infections (Dimopoulos et al., 2012). Risk factors such as COPD, systemic corticosteroid therapy, nonhematological malignancy, chronic renal disease, liver failure, diabetes mellitus, near-drowning, HIV infection, autoimmune diseases, malnutrition, and extensive burns have been described (Dimopoulos et al., 2003; Meersseman et al., 2004; Vandewoude et al., 2004, 2006; Wu et al., 2010). The clinical signs and symptoms of IPA and its radiographic features are often nonspecific in ICU patients. The finding of *Aspergillus* spp. in respiratory tract samples in these patients should not be routinely discarded as colonization, even if these patients are immunocompetent (Blot et al., 2012; Vanderwoude et al., 2006). Therefore, in order not to miss a critical window of therapeutic opportunity, adapted clinical diagnostic criteria should be used for this category and further diagnostic evaluation with early antifungal therapy should be considered once IPA is suspected (Garnacho-Montero and Amaya-Villar 2006; Tasci et al., 2006). IPA in this patient population carries an attributable mortality of 18.9% after adjusting for confounding factors while the isolation of *Aspergillus* from lower respiratory tract samples has been reported to be associated with a worse ICU outcome (Khasawneh et al., 2006; Meersseman et al., 2004). *Aspergillus* tracheobronchitis (ATB) is a unique feature of IPA reflecting an invasion of the tracheobronchial tree by *Aspergillus* spp. Predisposing factors for ATB are similar to those for IPA, but certain patient groups are more likely to develop this entity including lung transplantation recipients, patients with AIDS, and cancer patients with mediastinal involvement (Meersseman et al., 2007). ATB is classified in three forms: (1) the obstructive form, which is characterized by thick mucus plugs of *Aspergillus* spp. without macroscopic bronchial inflammation and poor prognosis; (2) the pseudomembranous form characterized by extensive inflammation of the tracheobronchial tree, development of a membrane overlaying the mucosa and containing *Aspergillus* spp., cough, dyspnea, hemoptysis (though not frequently), nonspecific radiological findings, and segmental or lobar collapse (the prognosis is poor with mortality reaching 78% while the need for mechanical ventilation is a predictor of

high mortality in these patients (Garnacho-Montero et al., 2005)); and (3) the ulcerative form usually found at the suture line in lung transplantation recipients (the outcome is generally favorable with antifungal therapy). The diagnosis of ATB is usually made by the characteristic findings on bronchoscopy combined with microscopic analysis of respiratory specimens obtained during the procedure. High index of suspicion, early diagnosis, and prompt antifungal therapy may be associated with improved outcome (Meersseman et al., 2007).

52.6.3 Gastrointestinal aspergillosis

It is less common, arising from ingestion of the organism and occurring in severely malnourished patients and in transplant recipients (Gonzalez-Vicent et al., 2008; Kami et al., 2002). The stomach, colon, and ileum are the most commonly involved sites. *Aspergillus* peritonitis is considered as a complication of chronic ambulatory peritoneal dialysis. Although *Candida* species are the most common cause, aspergillosis of the esophagus has been found to be relatively common in advanced cases of disseminated IA. Hepatic *aspergillosis* may occur as single or multiple parenchymal lesions, as a process of dissemination from the gastrointestinal tract along the portal venous system, as a component of general systemic dissemination, or as cholangitis. The symptoms depend on the site affected, but nonspecific abdominal pain and distention associated with nausea and vomiting are the most common. Fever, diarrhea, constipation, or masses in the gastrointestinal tract are also described. The diagnosis requires the microscopic examination and culture of biopsy specimens (Eggimann et al., 2006; Segal and Walsh, 2006; Shah et al., 2001).

52.6.4 Cutaneous aspergillosis

It may be primary or secondary. Primary infection is usually caused by direct inoculation of the fungus into disrupted skin and is most often seen in patients with burns or other forms of local skin trauma. The lesions may appear red and indurated and often progress to black eschars. Secondary cutaneous infection is generally seen when the pathogen spreads hematogenously. In hematogenously disseminated infection, the lesions typically begin as an erythematous, indurated, and painful cellulitis and then progress to an ulcer covered with a black eschar (Mohapatra et al., 2009; Ozer et al., 2009; Segal, 2009; Zhang et al., 2005).

52.6.5 Disseminated aspergillosis

It may follow hematogenously any of the four forms of aspergillosis described previously, but is usually seen in neutropenic patients with a pulmonary infection. The most common site of spread is the brain, but metastatic necrotic lesions have also been found in the spleen, heart, and other organs as kidneys, pleura, esophagus, liver, and skin. Clinical manifestations include lethargy, changes in mental status, or obtundation with or without a sudden onset of focal neurological signs, seizures, ring-enhancing lesions, cerebral infarctions, intracranial hemorrhage, meningitis, and epidural abscesses (Kousha et al., 2011; Singh et al., 2006b).

Cardiac invasion may present as pericarditis, endocarditis (valvular vegetations are common either on prosthetic or normal valves, particularly in injection drug users, may be large and pedunculated, with a high risk for embolic complications, particularly CNS-related complications), or myocarditis. *Aspergillus* myocarditis may manifest as myocardial infarction, cardiac arrhythmias, or myoepicarditis. In *Aspergillus* pericarditis, pericardial tamponade may rapidly ensue, leading to hemodynamic deterioration and cardiac arrest.

In *Aspergillus* osteomyelitis, the vertebral bodies and the intervertebral disks are the most common site of infection. *Aspergillus* arthritis arises as an extension from a contiguous focus of *Aspergillus* osteomyelitis.

Aspergillus keratitis is a locally IFI of the cornea that is characterized by ocular pain, potentially rapid loss of vision, and potential development of endophthalmitis if not recognized and treated promptly. *Aspergillus* endophthalmitis is a devastating infection that may result in irreparable loss of vision and rapid destruction of the eye.

Renal aspergillosis usually presents as parenchymal abscesses, as a result of hematogenous dissemination or, less commonly, as fungal balls in the pelvis of the kidney and may cause hematuria, ureteral obstruction, and perinephric abscess with extension into surrounding tissues.

Aspergillus otomycosis is a saprophytic process that usually involves the external auditory canal. Symptoms include pruritus, pain, hypoacusis, and otic discharge. Complications in case of perforated tympanic membrane include the involvement of middle ear, and mastoiditis.

In immunocompetent hosts Aspergillus spp. cause mainly allergic symptoms, such as wheezing, coughing, and allergic rhinitis without invasion and destruction of body tissues (Figure 52-4). The commonest manifestations are

1. ABPA: A common syndrome in asthmatic and CF patients which is considered as the result of hypersensitivity reaction of the tracheobronchial tree to *Aspergillus* colonization (mostly *A. fumigatus*) (Zmeili and Soubani, 2007). The thick mucus found in the airways of these patients may make the clearance of inhaled *Aspergillus* spores difficult. Genetic evidence of susceptibility has been reported. Patients with certain HLA alleles, particularly HLA-DR2, are highly susceptible to ABPA, whereas the presence of HLA-DQ2 appears to be protective (Virnig and Bush, 2007). Fever, pulmonary infiltrates unresponsive to antibacterial therapy, cough, mucous plugs forming bronchial casts, and hemoptysis are the commonest clinical findings. Patients with asthma and ABPA have poor control of the disease and experience difficulty in tapering of oral corticosteroids. ABPA may occur in conjunction with allergic fungal sinusitis, with symptoms including chronic sinusitis with purulent sinus drainage.

2. Chronic necrotizing pulmonary aspergillosis (CNPA) also called semi-invasive or subacute IA, is an indolent, cavitating, infectious process of the lung parenchyma secondary to local invasion by *Aspergillus* species, usually *A. fumigatus* (Binder et al., 1982; Gefter et al., 1981). This syndrome is rare, and the available literature is based on case reports and small case series (Binder et al., 1982; Gefter et al., 1981; Saraceno et al., 1997). CNPA manifests as a subacute pneumonia unresponsive to antibiotic therapy, which progresses and forms cavitations over weeks or months. Patients with CNPA have various comorbidities such as steroid-dependent COPD, alcoholism, previous thoracic surgery, collagen-vascular disease, or CGD. Fever, chronic productive cough, mild-to-severe hemoptysis, night sweats, and weight loss are the commonest clinical findings (Denning, 2001). The patients usually have received prolonged courses of antibiotic therapy and sometimes empiric antituberculous therapy without response prior to diagnosis via biopsy or culture.

3. Aspergilloma is the best-recognized form of pulmonary involvement usually occurring in a preexisting cavity in the lung. *Aspergilloma* is a fungus ball (composed of fungal hyphae, inflammatory cells, fibrin, mucus, and tissue debris) that may manifest as an asymptomatic radiographic abnormality in a patient with preexisting cavitary lung disease due to sarcoidosis, tuberculosis, or other necrotizing pulmonary processes such as bronchiectasis, bronchial cysts and bullae, ankylosing spondylitis, neoplasm, and pulmonary infection (Kawamura et al., 2000; Kousha et al., 2011). The most common species of *Aspergillus* recovered from such lesions is *A. fumigatus* but other fungi, such as Zygomycetes and *Fusarium* spp., may also cause the formation of fungal balls. About 40–60% of patients experience hemoptysis, which may be massive and life-threatening, originating from the bronchi, and caused by local invasion of blood vessels lining the cavity, by endotoxins released from the fungus, by mechanical irritation of the exposed vasculature inside the cavity, or by the mobile fungus ball (Kousha et al., 2011). Less common symptoms include cough and fever. Aspergilloma may also develop in patients with IA, CNPA, or in patients with HIV disease in cystic areas resulting from prior *Pneumocystis jirovecii* pneumonia. The lesion remains stable in the majority of cases, but it may decrease in size or resolve spontaneously

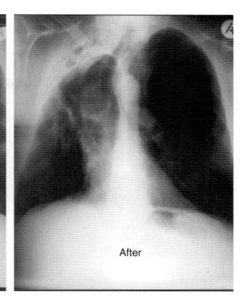

Fig. 52-3 A case of chronic cavitary pulmonary aspergillosis concerning a COPD patient with TBC history under steroid treatment and antibiotics for acute exacerbations of chronic bronchitis. Despite antibiotics fever of unknown origin persisted for >4 days. Bronchial secretions and Ag GM were positive for *A. fumigatus*. Treatment with Voriconazole was added (Before: Prior to the treatment, After: treatment with voriconazole).

without treatment in 10% of cases (Kousha et al., 2011). Rarely, the aspergilloma may increase in size.

52.7 Treatment

Three classes of antifungal agents are available for the treatment of aspergillosis: polyenes, azoles, and echinocandins. Historically, AmB deoxycholate was the major antifungal drug used in patients with IA but currently the drug of choice is voriconazole (Cornely et al., 2007a; Herbrecht et al., 2002). Lipid formulations of AmB are also used (Walsh et al., 2008). Echinocandins have a limited role against *Aspergillus* spp. because they have not been studied adequately for the initial treatment of the disease.

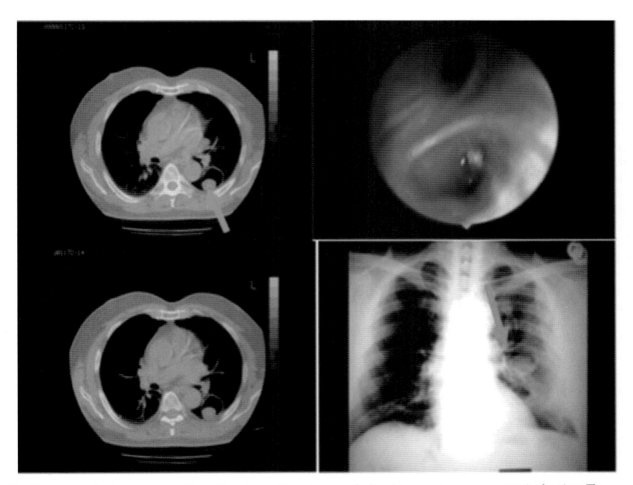

Fig. 52-4 A case of pulmonary aspergilloma (imaging and bronchoscopic findings) in a non-immunocompromised patient. The green arrow indicates to mycetoma.

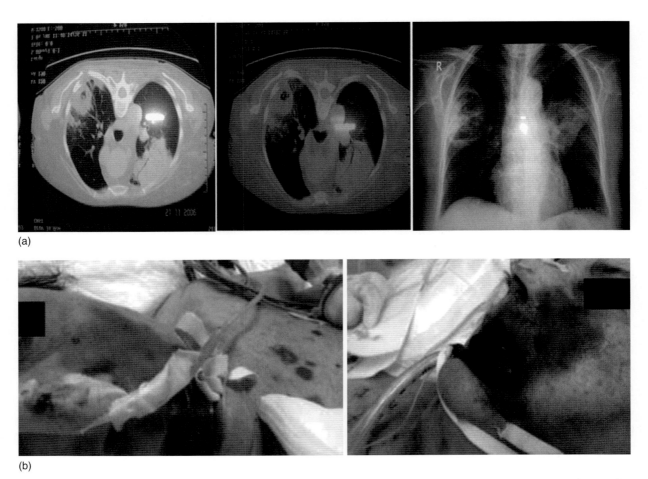

(a)

(b)

Fig. 52-5 (a) A case of pulmonary aspergillosis in a patient with leukemia. The air-crescent sign is apparent. (b) A case of a mixed fungal infection (invasive pulmonary aspergillosis and facial infection due to *Mucor* spp.). For a color version of this figure, see the color plate section.

52.7.1 Triazoles

The triazoles against aspergillosis include voriconazole, posaconazole, and itraconazole. Voriconazole is the drug of choice for IPA and play a major role in the setting of central nervous system disease, for which the mortality rate historically has approached 100% (Pascual et al., 2008; Sipsas and Kontoyiannis, 2006; Spanakis et al., 2006; Upton et al., 2007; Walsh et al., 2008) (Figure 52-3). It is available in both intravenous and oral formulations (Howard et al., 2008; Spanakis et al., 2006; Walsh et al., 2008). The recommended dosing regimen is 6 mg/kg IV twice a day on day 1, followed by 4 mg/kg IV twice daily with the option to switch to oral dosing at 200 mg daily when the patient is able to take oral medications. The target voriconazole concentrations evaluated by trough concentration after 1 week into therapy, with a goal range of >1 µg/mL and <5.5 µg/mL is fully suggested (Roerig, 2008). In patients who do not respond to voriconazole, antifungal resistance should be considered. Adverse events due to voriconazole include visual changes, hallucinations, and photosensitive skin rash. Posaconazole is a broad-spectrum triazole similar to voriconazole in its activity against *Aspergillus* species (Howard et al., 2011; Raad et al.,

2006; Torres et al., 2005; Walsh et al., 2007). The most common side effect of posaconazole is gastrointestinal disturbance. Effective absorption requires oral intake with a high-fat meal (Marr, 2002; Raad et al., 2006). Itraconazole is considered as a second-line agent for the treatment of aspergillosis, used in selected patients with mild immunosuppression and non-life-threatening *Aspergillus* infection or in patients who had already been stabilized with AmB. Oral itraconazole requires an acidic environment for absorption, has poor bioavailability, and severe drug interactions. Antifungal resistance should be considered in patients who do not respond to voriconazole and in patients infected with certain *Aspergillus* species that have reduced antifungal susceptibility profiles (e.g., *A. terreus, A. ustus, A. lentulus, N. udagawae*). *A. terreus* is less susceptible to AmB *in vitro* and in animal models, and clinical reports suggest that outcomes are better with the use of voriconazole (Balajee et al., 2006; Steinbach et al., 2004).

52.7.2 Amphotericin B

AmB deoxycholate is the most extensively used drug in the treatment of aspergillosis, but it is associated with severe nephrotoxicity. The usual dose is 0.5–1 mg/kg/day. There are three marketed lipid formulations of AmB: (a) AmB–lipid complex (ABLC), (b) liposomal AmB, and (c) AmB–cholesteryl sulfate complex (AmB colloidal dispersion, ABCD). They are less toxic when administered in large doses without having definitively shown better outcomes compared with conventional AmB (Bowden et al., 2002). The initial dose of AmB lipid complex is 5 mg/kg IV per day and of liposomal AmB is 3–5 mg/kg IV per day. Using higher doses of lipid formulations of AmB does not result in better response rates (Walsh et al., 2008).

52.7.3 Echinocandins

Echinocandins include caspofungin, micafungin, and anidulafungin. Caspofungin has been approved for the treatment of IA in patients who cannot tolerate or who are refractory to standard therapy, but it is not utilized in primary treatment because of lack of data in this setting. However, it is used for salvage therapy, often in combination with another antifungal agents (Kontoyiannis et al., 2003; Petraitiene et al., 2004). Caspofungin administration requires an intravenous loading dose of 70 mg, followed by 50 mg/day once daily. It is well tolerated and has few drug interactions (Kauffman, 2004). Micafungin and anidulafungin are not FDA-approved for the treatment of IA, although they are active against *Aspergillus* spp.

The treatment of IPA is classified as:

1. Prophylaxis (primary and secondary)

    Antifungal primary prophylaxis includes posaconazole (recommended in HSCT recipients with GVHD at high risk for IA and in patients with acute myelogenous leukemia or myelodysplastic syndrome who are at high risk for IA), itraconazole (effective, but low tolerability limits its use), and *inhaled administration of AmB formulations* which may reduce the incidence of aspergillosis in patients with hematologic malignancies and prolonged neutropenia (Krishna et al., 2007). The use of prophylactic antifungal treatment is balanced with the potential drawbacks including toxicity and drug interactions, pharmaceutical cost, and the potential emergence of resistance. According to observational data, inhaled AmB is often used in lung transplant recipients during the early post-transplant period (Rijnders et al., 2008). Oral drugs often are poorly absorbed in patients with gastrointestinal tract mucositis and/or GVHD (Marr, 2002). It is unclear which specific patients will benefit from prophylactic treatment. It partly depends on patient characteristics and the epidemiology of IFIs at individual institutions (Cordonnier et al., 2004; Winston et al., 2003, 2004). *Secondary prophylaxis* is the re-initiation of antifungal therapy after the completion of primary treatment during periods of increased risk of relapse such as after chemotherapy or hematopoietic cell transplantation. Voriconazole is the most widely used antifungal treatment.

2. Empirical treatment

    Empirical antifungal therapy is recommended for high-risk patients with prolonged neutropenia who remain persistently febrile despite broad-spectrum antibiotic therapy. Conventional AmB, liposomal AmB, itraconazole, voriconazole, or caspofungin may be used. Empirical antifungal therapy is not recommended for patients who are anticipated to have short durations of neutropenia (duration of neutropenia <10 days), unless other findings indicate the presence of an IFI.

3. Preemptive therapy

    It is an early treatment strategy proposed as an alternative to empiric therapy and involves antifungal therapy based upon the results of serial screening for aspergillosis mainly with serial biomarkers (serum galactomannan assay) or polymerase chain reaction (PCR) testing in combination with other clinical indicators. Preemptive therapy has been compared with empiric therapy suggesting a limited benefit of the preemptive strategy (Cordonnier et al., 2004, 2009; Walsh et al., 2008).

4. Combination treatment

    It is used mainly as salvage therapy in patients who do not respond to monotherapy with either voriconazole or liposomal AmB. Such strategies include the use of an echinocandin, such as caspofungin, micafungin, or anidulafungin, in combination with voriconazole or liposomal AmB (Marr et al., 2004). The use of AmB with a triazole for combination therapy is not recommended. The addition of flucytosine or rifampin to AmB has not been shown to be beneficial (Baddley and Pappas, 2005; Caillot et al., 2007; Kontoyiannis et al., 2005; Limper et al., 2011; Marr et al., 2004; Meletiadis et al., 2006; Perkhofer et al., 2007; Singh et al., 2006a; Sionov et al., 2005; Ullmann et al., 2007; Vazquez, 2008).

5. Other adjunctive therapies

    The hypothesis for the administration of immunomodulatory therapies is based on benefits due to the decreased degree of immunosuppression (Karp et al., 1988; Sipsas and Kontoyiannis, 2006). Granulocyte colony-stimulating factor (G-CSF) and interferon-gamma increase the innate neutrophil fungicidal capacity and damage *Aspergillus* hyphae, but their routine use is not recommended. Granulocyte transfusions have been also used in patients with neutropenia and IA.

6. Surgery

    Surgical resection of *Aspergillus*-infected tissue may be useful in patients with lesions that are contiguous with the great vessels or pericardium, lesions causing hemoptysis from a single focus, and lesions causing erosion into the pleural space or ribs.

Surgical excision (primary or adjunctive) is usually required for the treatment of *Aspergillus* rhinosinusitis and in settings in which antifungals cannot be delivered to large necrotic lesions and/or when there is an imminent threat to vessels (as in endocarditis, extensive necrotic cutaneous lesions, pericardial infection, empyema, invasion of the chest wall from a contiguous pulmonary lesion, pulmonary lesions in proximity to great vessels or pericardium, infected prosthetic devices, and osteomyelitis) (Limper et al., 2011).

The duration of treatment is not well defined. The treatment of IPA should be continued for a minimum of 6–12 weeks, but in immunosuppressed patients, therapy should be continued throughout the period of immunosuppression and until lesions have resolved. Radiographic abnormalities should have been stabilized and signs of active infection should have disappeared before treatment is discontinued. For most immunosuppressed patients, antifungal therapy will continue for months or even years in some cases. The therapeutic monitoring of aspergillosis includes clinical evaluation of symptoms and signs, as well as performance of radiographic imaging with CT scans at regular intervals (the frequency cannot be defined and should be individualized). The use of serial serum galactomannan assays for therapeutic monitoring remains debatable but a progressive increase in *Aspergillus* antigen levels over time signifies a poor prognosis. The resolution of galactomannan antigenemia to normal levels is not sufficient as a sole criterion for discontinuation of antifungal therapy. The pathogenesis of relapsing IA is due to reactivation of a latent, subclinical infection that had not been fully eradicated because of the angioinvasive nature of the organism or the lack of sterilization due to poor drug penetration (e.g., in foreign bodies, vegetations, or lung and bone sequestra). Factors that predispose patients to relapsing IA include site of infection (e.g., sinuses), use of systemic glucocorticoids, lack of remission of underlying hematologic malignancy, duration of neutropenia, receipt of an unrelated HSCT, and variations in innate immunity including polymorphisms in the genes encoding TLR-4, dectin-1, and mannose-binding lectin. Re-initiation of antifungal therapy is recommended if chemotherapy is reinitiated and the patient is expected to become neutropenic or if immunosuppression is intensified.

Prophylaxis and Treatment of Pulmonary Aspergillosis

Prophylaxis against aspergillosis
Posaconazole (200 mg every 8 h)

Alternative treatment
Itraconazole (200 mg every 12 h IV for 2 days, then 200 mg every 24 h IV) or
Itraconazole (200 mg PO every 12 h)
Micafungin (50 mg/day)

Treatment of IPA and tracheobronchial aspergillosis

Definite therapy
Voriconazole (6 mg/kg IV every 12 h for 1 day, followed by 4 mg/kg IV every 12 h; oral dosage is 200 mg every 12 h)

Alternative therapy
L-AMB (3–5 mg/kg/day IV), ABLC (5 mg/kg/day IV)
Caspofungin (70 mg day 1 IV and 50 mg/day IV thereafter)
Micafungin (IV 100–150 mg/day; dose not established)
Posaconazole (200 mg QID initially, then 400 mg BID PO after stabilization of disease)
Itraconazole (dosage depends upon formulation)

Empirical and preemptive treatment
L-AMB (3 mg/kg/day IV)
Caspofungin (70 mg day1 IV and 50 mg/day IV thereafter)
Itraconazole (200 mg every day IV or 200 mg BID)
Voriconazole (6 mg/kg IV every 12 h for 1 day, followed by 3 mg/kg IV every 12 h; oral dosage is 200 mg every 12 h)

Chronic Necrotizing Pulmonary Aspergillosis (CNPA)
Because CNPA requires a long-course treatment, an orally administered triazole, such as voriconazole or itraconazole, would be preferred over a parenterally administered agent.

Treatment of other forms of aspergillosis
1. *Aspergillus* peritonitis
   – Removal of peritoneal dialysis catheter and intraperitoneal dialysis with AMB
   – Itraconazole or an extended-spectrum azole (voriconazole or posaconazole)
2. Esophageal and gastrointestinal aspergillosis
   No clear indication of optimal therapy. Combination of medical (antifungals as for disseminated apergillosis) and surgical therapy to prevent the complications of potentially fatal hemorrhage, perforation, obstruction, and infarction.
3. **Hepatic aspergillosis**
   Systemic antifungal treatment plus surgical resection for hepatic abscesses.
   For extrahepatic or perihepatic biliary obstruction, surgical intervention is warranted.
4. **Cutaneous aspergillosis**
   Treatment of cutaneous aspergillosis is similar to IPA. Surgical intervention, particularly for primary cutaneous infection may be useful. Biopsy is very important to distinguish other potential pathogens (e.g., *Fusarium* species and Zygomycetes).

5. **Renal aspergillosis**

All the available antifungal agents with activity against aspergillosis penetrate renal parenchyma but none of them is excreted primarily into the pelvis of the kidney or urine. For this reason, the management of pelvicalyceal and ureteral infection may require nephrostomy with instillation of AMB.

6. **Disseminated aspergillosis**

Similar to that described for invasive pulmonary aspergillosis.

7. *Aspergillus* **endocarditis, pericarditis, and myocarditis**

Antifungal chemotherapy and surgical resection of the infected valve or mural lesion. Conventional Amphotericin–B (D-AMB) is the preferred initial treatment for a minimum of 6 weeks after surgical intervention (B-III). After the replacement of an infected prosthetic valve, an antifungal triazole, such as oral voriconazole or posaconazole should be given lifelong.

8. *Aspergillus* **osteomyelitis and septic arthritis**

Combined medical (with voriconazole) and surgical intervention for a minimum of 6–8 weeks in non-immunocompromised patients. For immunocompromised patients, long-term suppressive therapy or treatment throughout the duration of immunosuppression is needed.

9. *Aspergillus* **endophthalmitis and** *Aspergillus* **keratitis**

IV AMB and intravitreal AMB plus pars plana vitrectomy. Voriconazole intravitreally or systemically is an alternative regiment.

10. **Rhinocelebral and CNS aspergillosis**

Similar to IPA.

11. **Allergic bronchopulmonary aspergillosis (ABPA)**

Corticosteroids plus itraconazole PO. The use of anti-IgE monoclonal antibody omalizumab in patients with ABPA has been reported.

12. **Treatment of allergic** *Aspergillus* **sinusitis (AAS)**

Endoscopic drainage in patients with obstructive symptoms, voriconazole (200 mg PO every 12 h) or posaconazole (400 mg PO BID) or Itraconazole PO and nasal or systemic corticosteroids in some patients.

Treatment of aspergilloma
Chronic cavitary pulmonary aspergilloma (CCPA)

– Medical therapy (itraconazole, voriconazole, or presumably posaconazole) when patients become symptomatic, usually with hemoptysis
– Long-term, perhaps lifelong, antifungal treatment
– Intracavitary treatment, using CT-guided, percutaneously placed catheters to instill amphotericin alone or in combination with other drugs, including acetylcysteine and aminocaproic acid
– Bronchial artery embolization; for life-threatening hemoptysis in patients thought to have insufficient pulmonary reserve to tolerate surgery or in patients with recurrent hemoptysis

Single aspergilloma

– Surgical resection

References

Ader, F., Nseir, S., Le Berre, R., Leroy, S., Tillie-Leblond, I., Marquette, C.H., et al. 2005. Invasive pulmonary aspergillosis in chronic obstructive pulmonary disease: An emerging fungal pathogen. *Clin. Microbiol. Infec.*, 11(6), 427–429.

Asioglu, S., Rex, J., Pauw, B.D., Bennett, J., Bille, J., Crokaert, F., et al. 2002. Defining opportunistic invasive fungal infections in immunocompromised patients with cancer and hematopoietic stem cell transplants: An international consensus. *Clin. Infect. Dis.*, 34, 7–14.

Baddley, J.W., and Pappas, P.G. 2005. Antifungal combination therapy: Clinical potential. *Drugs*, 65(11), 1461–1480.

Balajee, S.A., Houbraken, J., Verweij, P.E., Hong, S.B., Yaghuchi, T., Varga, J., et al. 2007. Aspergillus species identification in the clinical setting. *Stud Mycol.*, 59, 39–46.

Balajee, S.A., Nickle, D., Varga, J., and Marr, K.A. 2006. Molecular studies reveal frequent misidentification of *Aspergillus fumigatus* by morphotyping. *Eukaryot. Cell*, 5, 1705–1712.

Barton, R.C. 2013. Laboratory diagnosis of invasive Aspergillosis: From diagnosis to prediction of outcome. *Scientifica*, 2013, 459405.

Binder, R.E., Faling, L.J., Pugatch, R.D., Mahasaen, C., and Snider, G.L. 1982. Chronic necrotizing pulmonary aspergillosis: A discrete clinical entity. *Medicine*, 61(2), 109–124.

Blot, S.I., Taccone, F.S., Van den Abeele, A.M., Bulpa, P., Meersseman, W., Brusselaers, N., et al. 2012. A clinical algorithm to diagnose invasive pulmonary aspergillosis in critically ill patients. *Am. J. Respir. Crit. Care Med.*, 186(1), 56–64.

Bowden, R., Chandrasekar, P., White, M.H., Li, X., Pietrelli, L., Gurwith, M., et al. 2002. A double-blind, randomized, controlled trial of amphotericin B colloidal dispersion versus amphotericin B for treatment of invasive aspergillosis in immunocompromised patients. *Clin. Infect. Dis.*, 35(4), 359–366.

Bulpa, P., Dive, A., and Sibille, Y. 2007. Invasive pulmonary aspergillosis in patients with chronic obstructive pulmonary disease. *Eur. Respir. J.*, 30(4), 782–800.

Caillot, D., Thiebaut, A., Herbrecht, R., de Botton, S., Pigneux, A., Bernard, F., et al. 2007. Liposomal amphotericin B in combination with caspofungin for invasive aspergillosis in patients with hematologic malignancies: A randomized pilot study (Combistrat trial). *Cancer*, 110(12), 2740–2746.

Comrie, A.C. 2005. Climate factors influencing coccidioidomycosis seasonality and outbreaks. *Envirom. Health Perspect.*, 113, 688–692.

Cordonnier, C., Maury, S., Pautas, C., Bastie, J.N., Chehata, S., Castaigne, S., et al. 2004. Secondary antifungal prophylaxis with voriconazole to adhere to scheduled treatment in leukemic patients and stem cell transplant recipients. *Bone Marrow Transplant.*, 33(9), 943–948.

Cordonnier, C., Pautas, C., Maury, S., Vekhoff, A., Farhat, H., Suarez, F., et al. 2009. Empirical versus preemptive antifungal therapy for high-risk, febrile, neutropenic patients: A randomized, controlled trial. *Clin. Infect. Dis.*, 48(8), 1042–1051.

Cornely, O.A., Maertens, J., Bresnik, M., Ebrahimi, R., Ullmann, A.J., Bouza, E., et al. 2007a. Liposomal amphotericin B as initial therapy for invasive mold infection: A randomized trial comparing a high-loading dose regimen with standard dosing (AmBiLoad trial). *Clin. Infect. Dis.*, 44(10), 1289–1297.

Cornely, O.A., Maertens, J., Winston, D.J., Perfect, J., Ullmann, A.J., Walsh, T.J., et al. 2007b. Possaconazole vs fluconazole or itraconazole prophylaxis in patients with neutropenia. *N. Engl. J. Med.*, 356, 348–359.

Cuenc-Estrella, M., Bassetti, M., Lass-Florl, Z., Racil, Z., Richardson, M., and Rogers, T.R. 2011. Detection and investigation of invasive mould disease. *J. Antimicrob. Chemother.* 66, i15–i24.

De Pauw, B., Walsh, T.J., Donnelly, J.P., Stevens, D.A., Edwards, J.E., Calandra, T., et al. 2008. Revised difinitions of invasive fungal disease from the European Organization for Research and Treatment of Caner/Invasive Fungal Infections Cooperative group and the National Institute of Allergy and Infectious Diseases Mycoses Study Group (EORTC/MSG). *Cin. Infect. Dis.*, 46, 1813–1821.

De Pauw, B.E., and Donnell, J.P. 2007. Prophylaxis and Aspergillosis – Has the principle been proven? *N. Engl. J. Med.*, 356, 409–411.

Denning, D.W. 1998. Invasive Aspergillosis. *Clin. Infect. Dis.*, 26, 781–805.

Denning, D.W. 2001. Chronic forms of pulmonary aspergillosis. *Clin. Microbiol. Infec.*, 7(Suppl 2), 25–31.

Diamond, R.D., Krzesicki, B., Epstein, B., and Jao, W. 1978. Damage to hyphal forms of fungi by human leukocytes in vitro. A possible host defence mechanism in aspergillosis and mucormycosis. *Am. J. Pathol.*, 91, 313–328.

Dimopoulos, G., Frantzeskaki, F., Poulakou, G., and Armaganidis, A. 2012. Invasive aspergillosis in the intensive care unit. *Ann. N. Y. Acad. Sci.*, 1272, 31–39.

Dimopoulos, G., Piagnerelli, M., Berre, J., Eddafali, B., Salmon, I., and Vincent, J.L. 2003. Disseminated aspergillosis in intensive care unit patients: An autopsy study. *J. Chemother.*, 15(1), 71–75.

Domsch, K.H., Gams, W., Anderson, T.H., and Aspergillus, P. 2007. Micheli ex link. In: *Compendium of Soil Fungi*. Eching, Germany: IHW–Verlag. pp. 68–99.

Dowell, S.F. 2001. Seasonal variation in host susceptibility and cycles of certain infectious diseases. *Emerg. Infect. Dis.*, 7, 369–374.

Dutkiewicz, R., and Hage, A. 2010. Aspergillus infections in the critically ill. *Am Thorac Soc.*, 7, 204–209.

Eggimann, P., Chevrolet, J.C., Starobinski, M., Majno, P., Totsch, M., Chapuis, B., et al. 2006. Primary invasive aspergillosis of the digestive tract: Report of two cases and review of the literature. *Infection*, 34(6), 333–388.

Garnacho-Montero, J., and Amaya-Villar, R. 2006. A validated clinical approach for the management of aspergillosis in critically ill patients: Ready, steady, go! *Crit. Care*, 10(2), 132.

Garnacho-Montero, J., Amaya-Villar, R., Ortiz-Leyba, C., Leon, C., Alvarez-Lerma, F., Nolla-Salas, J., et al. 2005. Isolation of Aspergillus spp. from the respiratory tract in critically ill patients: Risk factors, clinical presentation and outcome. *Crit. Care*, 9(3), R191–R199.

Gefter, W.B., Weingrad, T.R., Epstein, D.M., Ochs, R.H., and Miller, W.T. 1981. "Semi-invasive" pulmonary aspergillosis: A new look at the spectrum of aspergillus infections of the lung. *Radiology*, 140(2), 313–321.

Gibbons, J.G., and Rokas, A. 2013. The function and evolution of the Aspergillus genome. *Trends Microbiol.*, 21(1), 14–22.

Gonzalez-Vicent, M., Diaz, M.A., Colmenero, I., Sevilla, J., and Madero, L. 2008. Primary gastrointestinal aspergillosis after autologous peripheral blood progenitor cell transplantation: An unusual presentation of invasive aspergillosis. *Transpl. Infect. Dis.*, 10(3), 193–196.

Gupta, A., Bansal, S., and Rijuneeta, B.G. 2012. Invasive fungal sinusitis. *Clin. Rhinol. An. Int. J.*, 5(2), 63–71.

Haiduven, D. 2008. Nosocomial Aspergillosis and building construction. *Med. Mycol.*, 47, 1–7.

Hansen, D., Blahout, B., Benner, D., and Popp, W. 2008. Environmental sampling of particulate matter and fungal spores during demolition of a building on a hospital area. *J. Hosp. Infect.*, 70, 259–264.

Hawgood, S., Ochs, M., Jung, A., Akiyama, J., Allen, L., Brown, C., et al. 2002. Sequential targeted deficiency of SP-A and -D leads to progressive alveolar lipoproteinosis and emphysema. *Am. J. Physiol. Lung Cell. Mol. Physiol.*, 283(5), L1002–L1010.

Herbrecht, R., Denning, D.W., Patterson, T.F., Bennett, J.E., Greene, R.E., Oestmann, J.W., et al. 2002. Voriconazole versus amphotericin B for primary therapy of invasive aspergillosis. *N. Engl. J. Med.*, 347(6), 408–415.

Horie, Y., and Coelho, K. 1995. New and interesting species of Neosartorya from Brazilian soil. *Mycoscience*, 36, 199–204.

Hosseini-Moghaddam, S.M., and Husain, S. 2012. Fungi and molds following lung transplantation. *Semin. Respir. Crit. Care Med.*, 31, 222–223.

Howard, A., Hoffman, J., and Sheth, A. 2008. Clinical application of voriconazole concentrations in the treatment of invasive aspergillosis. *Ann. Pharmacother.*, 42(12), 1859–1864.

Howard, S.J., Lestner, J.M., Sharp, A., Gregson, L., Goodwin, J., Slater, J., et al. 2011. Pharmacokinetics and pharmacodynamics of posaconazole for invasive pulmonary aspergillosis: Clinical implications for antifungal therapy. *J. Infect. Dis.*, 203(9), 1324–1332.

Kami, M., Hori, A., Takaue, Y., and Mutou, Y. 2002. The gastrointestinal tract is a common target of invasive aspergillosis in patients receiving cytotoxic chemotherapy for hematological malignancy. *Clin. Infect. Dis.*, 35(1), 105–106; author reply 06–7.

Karp, J.E., Burch, P.A., and Merz, W.G. 1988. An approach to intensive antileukemia therapy in patients with previous invasive aspergillosis. *Am. J. Med.*, 85(2), 203–206.

Kauffman, C.A. 2004. New antifungal agents. *Semin. Respir. Crit. Care Med.*, 25(2), 233–239.

Kawamura, S., Maesaki, S., Tomono, K., Tashiro, T., and Kohno, S. 2000. Clinical evaluation of 61 patients with pulmonary aspergilloma. *Intern. Med.*, 39(3), 209–212.

Khasawneh, F., Mohamad, T., Moughrabieh, M.K., Lai, Z., Ager, J., and Soubani, A.O. 2006. Isolation of Aspergillus in critically ill patients: A potential marker of poor outcome. *J. Crit. Care*, 21(4), 322–327.

Kontoyiannis, D.P., Boktour, M., Hanna, H., Torres, H.A., Hachem, R., and Raad, I.I. 2005. Itraconazole added to a lipid formulation of amphotericin B does not improve outcome of primary treatment of invasive aspergillosis. *Cancer*, 103(11), 2334–2337.

Kontoyiannis, D.P., Hachem, R., Lewis, R.E., Rivero, G.A., Torres, H.A., Thornby, J., et al. 2003. Efficacy and toxicity of caspofungin in combination with liposomal amphotericin B as primary or salvage treatment of invasive aspergillosis in patients with hematologic malignancies. *Cancer*, 98(2), 292–299.

Kousha, M., Tadi, R., and Soubani, A.O. 2011. Pulmonary aspergillosis: A clinical review. *Eur Respir Rev*, 20, 156–174.

Krappmann, S. 2008. Pathogenicity determinants and allergens. In: Goldman, G.H., and Osmani, S.A., editors. *The Aspergilli: Genomics, Medical Aspects, Biotechnology, and Research Methods*. Boca Raton, FL: CRC Press. pp. 377–400.

Krishna, G., Martinho, M., Chandrasekar, P., Ullmann, A.J., and Patino, H. 2007. Pharmacokinetics of oral posaconazole in allogeneic hematopoietic stem cell transplant recipients with graft-versus-host disease. *Pharmacotherapy*, 27(12), 1627–1636.

Kwon-Chung, K.J., and Sugui, J.A. 2009. What do we know about the role of gliotoxin in the pathobiology of *Aspergillus fumigatus*? *Med. Mycol.*, 47(Suppl 1), S97–S103.

Levitz, S.M., and Diamond, R.D. 1984. Killing of *Aspergillus fumigatus* spores and *Candida albicans* yeast phase by the iron-hydrogen peroxide-iodide cytotoxic system: Comparison with the myeloperoxidase-hydrogen peroxide-halide system. *Infect. Immun.*, 43(3), 1100–1102.

Limper, A.H., Knox, K.S., Sarosi, G.A., Ampel, N.M., Bennett, J.E., Catanzaro, A., et al. 2011. An official American Thoracic Society statement: Treatment of fungal infections in adult pulmonary and critical care patients. *Am. J. Respir. Crit. Care Med.*, 183(1), 96–128.

Lin, S., Schranz, J., and Teutsch, S.M. 2003. Aspergillosis case-fatality rate: Systematic review of the literature. *Clin. Infect. Dis.*, 32, 358–366.

Lionakis, M.S., and Kontoyiannis, D.P. 2003. Glucocorticoids and invasive fungal infections. *Lancet*, 362(9398), 1828–1838.

Lortholary, O., Gangneux, J.P., Sitbon, K., Lebeau, B., de Monbrison, F., Le Strat, Y., et al. 2011. Epidemiological trends in invasive aspergillosis in France: The SAIF network (2005–2007). *Clin. Microbiol. Infect.*, 17, 1882–1889.

Marr, K.A. 2002. Empirical antifungal therapy–New options, new tradeoffs. *N. Engl. J. Med.*, 346(4), 278–280.

Marr, K.A., Boeckh, M., Carter, R.A., Kim, H.W., and Corey, L. 2004. Combination antifungal therapy for invasive aspergillosis. *Clin. Infect. Dis.*, 39(6), 797–802.

Marr, K.A., Carter, R.A., Boeckh, M., Martin, P., and Corey, L. 2002a. Invasive aspergillosis in allogeneic stem cell transplant recipients: Changes in epidemiology and risk factors. *Blood*, 100, 4358–4366.

Marr, K.A., Carter, R.A., Crippa, F., Wald, A., and Corey, L. 2002b. Epidemiology and outcome of mould infections in hematopoietic stem cell transplant recipients. *Clin. Infect. Dis.*, 34, 909–917.

Meersseman, W., Lagrou, K., Maertens, J., and Van Wijngaerden, E. 2007. Invasive aspergillosis in the intensive care unit. *Clin. Infect. Dis.*, 45(2), 205–216.

Meersseman, W., Vandecasteele, S.J., Wilmer, A., Verbeken, E., Peetermans, W.E., and Van Wijngaerden, E. 2004. Invasive aspergillosis in critically ill patients without malignancy. *Am. J. Respir. Crit. Care Med.*, 170(6), 621–625.

Meletiadis, J., Petraitis, V., Petraitiene, R., Lin, P., Stergiopoulou, T., Kelaher, A.M., et al. 2006. Triazole-polyene antagonism in experimental invasive pulmonary aspergillosis: In vitro and in vivo correlation. *J. Infect. Dis.*, 194(7), 1008–1018.

Meyer, R.D., Young, L.S., Armstrong, D., and Yu, B. 1973. Aspergillosis complicating neoplastic disease. *Am. J. Med.*, 54, 6–15.

Mohapatra, S., Xess, I., Swetha, J.V., Tanveer, N., Asati, D., Ramam, M., et al. 2009. Primary cutaneous aspergillosis due to *Aspergillus niger* in an immunocompetent patient. *Indian J. Med. Microbiol.*, 27(4), 367–370.

Morgan, J., Wammemuehler, K.A., Marr, K.A., Hadley, S., Kontoyiannis, D.P., Walsh, T.J., et al. 2005. Incidence of invasive aspergillosis following hematopoietic stem cell and solid organ transplantation: Interim results of a prospective multicenter surveillance program. *Med. Mycol.*, 43, S49–S58.

Morgenstern, D.E., Gifford, M.A., Li, L.L., Doerschuk, C.M., and Dinauer, M.C. 1997. Absence of respiratory burst in X-linked chronic granulomatous disease mice leads to abnormalities in both host defense and inflammatory response to *Aspergillus fumigatus*. *J. Exp. Med.*, 185(2), 207–218.

Mullins, J., Harvey, R., and Seaton, A. 1976. Sources and incidence of airborne *Aspergillus fumigatus* (FRES). *Clin. Allergy*, 6, 209–217.

Ostrosky-Zeichner, L. 2012. Invasive mycoses: Diagnostic challenges. *Am J Med.*, 125, S14–S24.

Ozer, B., Kalaci, A., Duran, N., Dogramaci, Y., and Yanat, A.N. 2009. Cutaneous infection caused by *Aspergillus terreus*. *J. Med. Microbiol.*, 58(Pt 7), 968–970.

Pacholczyk, M., Lagiewska, B., Lisik, W., Wasiak, D., and Chmura, A. 2011. Invasive fungal infections following liver transplantation- risk factors, incidence and outcome. *Ann. Transplant.*, 16, 14–16.

Pagano, L., Fianchi, L., and Leone, G. 2006. Fungal pneumonia due to molds in patients with hematological malignancies. *J. Chemother.*, 18, 339–352.

Panackal, A.A., Li, H., Kontoyiannis, D.P., Mori, M., Perego, C.A., Boeckh, M., et al. 2010. Geoclimatic influences on invasive aspergillosis after hematopoietic stem cell transplantation. *Clin. Infect. Dis.*, 50, 1588–1597.

Pascual, A., Calandra, T., Bolay, S., Buclin, T., Bille, J., and Marchetti, O. 2008. Voriconazole therapeutic drug monitoring in patients with invasive mycoses improves efficacy and safety outcomes. *Clin. Infect. Dis.*, 46(2), 201–211.

Patterson, T.F., Kirkpatrick, W.R., White, M., Hiemenz, J.W., Wingard, J.R., Dupont, B., et al. 2000. Invasive aspergillosis: Disease spectrum, treatment, practices and outcomes. 13 Aspergillus study group. *Medicine (Baltimore)*, 79, 250–260.

Perkhofer, S., Lugger, H., Dierich, M.P., and Lass-Florl, C. 2007. Posaconazole enhances the activity of amphotericin B against *Aspergillus* hyphae in vitro. *Antimicrob. Agents Chemother.*, 51(2), 791–793.

Petraitiene, R., Petraitis, V., Lyman, C.A., Groll, A.H., Mickiene, D., Peter, J., et al. 2004. Efficacy, safety, and plasma pharmacokinetics of escalating dosages of intravenously administered ravuconazole lysine phosphoester for treatment of experimental pulmonary aspergillosis in persistently neutropenic rabbits. *Antimicrob. Agents Chemother.*, 48(4), 1188–1196.

Raad, II, Graybill, J.R., Bustamante, A.B., Cornely, O.A., Gaona-Flores, V., Afif, C., et al. 2006. Safety of long-term oral posaconazole use in the treatment of refractory invasive fungal infections. *Clin. Infect. Dis.*, 42(12), 1726–1734.

Rijnders, B.J., Cornelissen, J.J., Slobbe, L., Becker, M.J., Doorduijn, J.K., Hop, W.C., et al. 2008. Aerosolized liposomal amphotericin B for the prevention of invasive pulmonary aspergillosis during prolonged neutropenia: A randomized, placebo-controlled trial. *Clin. Infect. Dis.*, 46(9):1401–1408.

Roenneberg, T., and Merrow, M. 2001. Seasonality and photoperiodism in fungi. *J. Biol. Rhythms*, 16, 403–414.

Roerig, a Division of Pfizer, Inc. Vfend® package insert. New York, NY, 2008.

Romani, L. 2004. Immunity to fungal infections. *Nat. Rev. Immunol.*, 4(1), 1–23.

Rosenhagen, M., Feldhues, S., Schmidt, J., Hoppe-Tichy, T., and Geiss, H.K. 2009. A risk profile for invasive aspergillosis in liver transplant recipients. *Infection*, 37, 313–319.

Saah, D., Drakos, P.E., Elidan, J., Braverman, I., Or, R., and Nagler, A. 1994. Rhinocerebral aspergillosis in patients undergoing bone marrow transplantation. *Ann. Otol. Rhinol. Laryngol.*, 103(4 Pt 1), 306–310.

Samarakoon, P., and Soubani, A. 2008. Invasive pulmonary aspergillosis in patients with COPD: A report of five cases and systematic review of the literature. *Chron. Respir. Dis.*, 5(1), 19–27.

Saraceno, J.L., Phelps, D.T., Ferro, T.J., Futerfas, R., and Schwartz, D.B. 1997. Chronic necrotizing pulmonary aspergillosis: Approach to management. *Chest*, 112(2), 541–548.

Sarir, H., Mortaz, E., Karimi, K., Kraneveld, A.D., Rahman, I., Caldenhoven, E., et al. 2009. Cigarette smoke regulates the expression of TLR4 and IL-8 production by human macrophages. *J. Inflamm. (Lond)*, 6, 12.

Schaffner, A., Douglas, H., and Braude, A. 1982. Selective protection against conidia by mononuclear and against mycelia by polymorphonuclear phagocytes in resistance to Aspergillus. Observations on these two lines of defense in vivo and in vitro with human and mouse phagocytes. *J. Clin. Invest.*, 69(3), 617–631.

Schlesinger, R.B., and Lippmann, M. 1972. Particle deposition in casts of the human upper tracheobronchial tree. *Am. Ind. Hyg. Assoc. J.*, 33(4), 237–251.

Segal, B.H. 2009. Aspergillosis. *N. Engl. J. Med.*, 360(18), 1870–1884.

Segal, B.H., Decarlo, E.S., Kwon-Chung, K.J., Malech, H.L., Gallin, J.I., and Holland, S.M. 1998. *Aspergillus nidulans* infections in chronic granulomatous disease. *Medicine (Baltimore)*, 77, 345–354.

Segal, B.H., and Walsh, T.J. 2006. Current approaches to diagnosis and treatment of invasive aspergillosis. *Am. J. Respir. Crit. Care Med.*, 173(7), 707–717.

Shah, S.S., Birnbaum, B.A., and Jacobs, J.E. 2001. Disseminated aspergillosis inciting intestinal ischaemia and obstruction. *Br. J. Radiol.*, 74(888), 1145–1147.

Sing, N. 2000. The current management of infectious diseases in liver transplant recipient. *Clin. Liver Dis.*, Aug, 657–73ix.

Singh, N., Limaye, A.P., Forrest, G., Safdar, N., Munoz, P., Pursell, K., et al. 2006a. Combination of voriconazole and caspofungin as primary therapy for invasive aspergillosis in solid organ transplant recipients: A prospective, multicenter, observational study. *Transplantation*, 81(3), 320–326.

Singh, S., Chowdhury, V., and Dixit, R. 2006b. CNS aspergillosis. *Indian J. Radiol. Imaging*, 16(4), 749–752.

Sionov, E., Mendlovic, S., and Segal, E. 2005. Experimental systemic murine aspergillosis: Treatment with polyene and caspofungin combination and G-CSF. *J. Antimicrob. Chemother.*, 56(3), 594–597.

Sipsas, N.V., and Kontoyiannis, D.P. 2006. Clinical issues regarding relapsing aspergillosis and the efficacy of secondary antifungal prophylaxis in patients with hematological malignancies. *Clin. Infect. Dis.*, 42(11), 1584–1591.

Soubani, A.O., Khanchandani, G., and Ahmed, H.P. 2004. Clinical significance of lower respiratory tract Aspergillus culture in elderly hospitalized patients. *Eur. J. Clin. Microbiol. Infect. Dis.*, 23(6), 491–494.

Spanakis, E.K., Aperis, G., and Mylonakis, E. 2006. New agents for the treatment of fungal infections: Clinical efficacy and gaps in coverage. *Clin. Infect. Dis.*, 43(8), 1060–1068.

Steinbach, W.J., Benjamin Jr., D.K., Kontoyiannis, D.P., Perfect, J.R., Lutsar, I., Marr, K.A., et al. 2004. Infections due to *Aspergillus terreus*: A multicenter retrospective analysis of 83 cases. *Clin. Infect. Dis.*, 39(2), 192–198.

Sugui, J.A., Vind, D.C., Nardone, G., Shea, Y.R., Chang, Y.C., Zelazny, A.M., et al. 2010. *Neosartoya udagawae* (*Aspergillus udagawae*), an emerging agent of aspergillosis: How different is it from *Aspergillus fumigatus*? *J. Clin. Microbiol.*, 48, 220–228.

Summerbell, R. 2003. Ascomycetes. In: Howard, D.H., editor. *Pathogenic Fungi in Humans and Animals*. New York: Marcel Dekker. pp. 237–498.

Tasci, S., Glasmacher, A., Lentini, S., Tschubel, K., Ewig, S., Molitor, E., et al. 2006. Pseudomembranous and obstructive Aspergillus tracheobronchitis – Optimal diagnostic strategy and outcome. *Mycoses*, 49(1), 37–42.

Thompson 3rd, G.R., and Patterson, T.F. 2008. Pulmonary aspergillosis. *Semin. Respir. Crit. Care Med.*, 29(2), 103–110.

Tomee, J.F., Wierenga, A.T., Hiemstra, P.S., and Kauffman, H.K. 1997. Proteases from *Aspergillus fumigatus* induce release of proinflammatory cytokines and cell detachment in airway epithelial cell lines. *J. Infect. Dis.*, 176(1), 300–303.

Tong, K.B., Lau, C.J., Murtagh, K., Layton, A.J., and Seifeldin, R. 2009. The economic impact of aspergillosis: Analysis of hospital expeditures across patient subgroups. *Int. J. Infect. Dis.*, 13, 24–36.

Torres, H.A., Hachem, R.Y., Chemaly, R.F., Kontoyiannis, D.P., and Raad, I.I. 2005. Posaconazole: A broad-spectrum triazole antifungal. *Lancet Infect. Dis.*, 5(12), 775–785.

Ullmann, A.J., Lipton, J.H., Vesole, D.H., Chandrasekar, P., Langston, A., Tarantolo, S.R., et al. 2007. Posaconazole or fluconazole for prophylaxis in severe graft-versus-host disease. *N. Engl. J. Med.*, 356(4), 335–347.

Upton, A., Kirby, K.A., Carpenter, P., Boeckh, M., and Marr, K.A. 2007. Invasive aspergillosis following hematopoietic cell transplantation: Outcomes and prognostic factors associated with mortality. *Clin. Infect. Dis.*, 44(4), 531–540.

Vaid, M., Kaur, S., Sambatakou, H., Madan, T., Denning, D.W., and Sarma, P.U. 2007. Distinct alleles of mannose-binding lectin (MBL) and surfactant proteins A (SP-A) in patients with chronic cavitary pulmonary aspergillosis and allergic bronchopulmonary aspergillosis. *Clin. Chem. Lab. Med.*, 45(2), 183–186.

Vanderwoude, K.H., Blot, S.I., Depuydt, P., Benoit, D., Temmerman, W., Colardyn, F., et al. 2006. Clinical relevance of aspergillus isolation from respiratory tract samples in critically ill patients. *Crit. Care*, 10, R31.

Vandewoude, K.H., Blot, S.I., Benoit, D., Colardyn, F., and Vogelaers, D. 2004. Invasive aspergillosis in critically ill patients: Attributable mortality and excesses in length of ICU stay and ventilator dependence. *J. Hosp. Infect.*, 56(4), 269–276.

Vazquez, J.A. 2008. Clinical practice: Combination antifungal therapy for mold infections: Much ado about nothing? *Clin. Infect. Dis.*, 46(12), 1889–1901.

Vinh, D.C., Shea, Y.R., Sugui, J.A., Parrilla-Castellar, E.R., Freeman, A.F., Campbell, J.W., et al. 2009. Invasive aspergillosis due to *Neosartoya udagawae*. *Clin. Infect. Dis.*, 49, 102–111.

Virnig, C., and Bush, R.K. 2007. Allergic bronchopulmonary aspergillosis: A US perspective. *Curr. Opin. Pulm. Med.*, 13(1), 67–71.

Wald, A., Leisenring, W., van Burik, J.A., and Bowden, R.A. 1997. Epidemiology of aspergillus infections in a large cohort of patients undergoing bone marrow transplantation. *J. Infect. Dis.*, 175, 1459–1461.

Walsh, T.J., Anaissie, E.J., Denning, D.W., Herbrecht, R., Kontoyiannis, D.P., Marr, K.A., et al. 2008. Treatment of aspergillosis: Clinical practice guidelines of the Infectious Diseases Society of America. *Clin. Infect. Dis.*, 46(3), 327–360.

Walsh, T.J., and Dixon, D.M. 1989. Nosocomial aspergillosis: Environmental microbiology, hospital epidemiology, diagnosis and treatment. *Eur. J. Epidemiol.*, 5, 131–142.

Walsh, T.J., Raad, I., Patterson, T.F., Chandrasekar, P., Donowitz, G.R., Graybill, R., et al. 2007. Treatment of invasive aspergillosis with posaconazole in patients who are refractory to or intolerant of conventional therapy: An externally controlled trial. *Clin. Infect. Dis.*, 44(1), 2–12.

Wilson, L.S., Reyes, C.M., Stolpman, M., Speckman, J., Allen, K., and Beney, J. 2002. The direct cost and incidence of systemic fungal infections. *Value Health*, 5, 26–34.

Wingard, J.R., Ribaud, P., Sclamm, H.T., and Herbrecht, R. 2008. Changes in causes of death after treatment for invasive aspergillosis. *Cancer*, 112, 2309–2312.

Winston, D.J., Emmanouilides, C., Bartoni, K., Schiller, G.J., Paquette, R., and Territo, M.C. 2004. Elimination of Aspergillus infection in allogeneic stem cell transplant recipients with long-term itraconazole prophylaxis: Prevention is better than treatment. *Blood*, 104(5), 1581; author reply 82.

Winston, D.J., Maziarz, R.T., Chandrasekar, P.H., Lazarus, H.M., Goldman, M., Blumer, J.L., et al. 2003. Intravenous and oral itraconazole versus intravenous and oral fluconazole for long-term antifungal prophylaxis in allogeneic hematopoietic stem-cell transplant recipients. A multicenter, randomized trial. *Ann. Intern. Med.*, 138(9), 705–713.

Wright, J.R. 2005. Immunoregulatory functions of surfactant proteins. *Nat. Rev. Immunol.*, 5(1), 58–68.

Wu, N., Huang, Y., Li, Q., Bai, C., Huang, H.D., and Yao, X.P. 2010. Isolated invasive Aspergillus tracheobronchitis: A clinical study of 19 cases. *Clin. Microbiol. Infec.*, 16(6), 689–695.

Zhang, Q.Q., Li, L., Zhu, M., Zhang, C.Y., and Wang, J.J. 2005. Primary cutaneous aspergillosis due to *Aspergillus flavus*: A case report. *Chin. Med. J.*, 118(3), 255–257.

Zmeili, O.S., and Soubani, A.O. 2007. Pulmonary aspergillosis: A clinical update. *QJM*, 100(6), 317–334.

Index

Note: Page numbers with suffix "f" refer to figures; those with suffix "t" to tables.